D0073149

Introduction to Mathematical Analysis

Steven A. Douglass

ADDISON-WESLEY PUBLISHING COMPANY

Reading, Massachusetts • Menlo Park, California • New York
Don Mills, Ontario • Wokingham, England • Amsterdam • Bonn • Sydney
Singapore • Tokyo • Madrid • San Juan • Milan • Paris

Sponsoring Editor: Laurie Rosatone
Senior Production Coordinator: Kathleen A. Manley
Senior Marketing Coordinator: Benjamin D. Rivera
Senior Marketing Manager: Andrew Fisher
Cover Design: Peter Blaiwas
Cover Design Supervisor: Eileen R. Hoff
Manufacturing Coordinator: Evelyn Beaton
Senior Manufacturing Manager: Roy E. Logan
Compositor: Eigentype Compositors
Technical Illustrator: Scientific Illustrators

This text is in the Addison-Wesley Higher Mathematics Series.
For more information about Addison-Wesley Mathematics books, access our World Wide Web site at http://www.aw.com/he/.

Library of Congress Cataloging-in-Publication Data

Douglass, Steven A.
Introduction to mathematical analysis / Steven A. Douglass.
p. cm.
Includes bibliographical references and index.
ISBN 0-201-50897-4
1. Mathematical analysis. I. Title.
QA300.D67 1996 95-47899
515–dc20 CIP

Printed in the United States of America.

1 2 3 4 5 6 7 8 9 10–MA–0099989796

To
Dr. Anna Penk
upon her retirement.
An inspiring teacher, steadfast
friend, and a woman of integrity.

Preface

Year after year, when teaching the sequence of courses in undergraduate analysis, I have noticed an ever-widening gulf among the texts available on the market, the expected content of these courses, and the observed capabilities of students in my classes. The classic texts I most admire set a standard rather beyond the reach of many of my students; other texts seem, for various reasons, not quite suitable for my classes. My goal has been to write a book that is mathematically rigorous, that treats those topics generally considered to comprise the central core of analysis at the junior/senior level, and that is accessible to a broad spectrum of these beginning students. I offer this text as my attempt to achieve these goals and to bridge this gap. This text is suitable for use in a one-semester, two-quarter, or year-long course in analysis at the undergraduate level. For students at this stage of development it is rigorous throughout.

Mastery of the material presented here provides a solid initial training for students intending to proceed to graduate study in pure or applied mathematics. Students of engineering and the physical sciences, for whom these courses in analysis are terminal, will receive an essential and pragmatic introduction to classic, nineteenth-century mathematics.

I have attempted to soften the hard edge of the mathematics by adopting a somewhat gentle style of presentation, especially in the early chapters. As the student progresses into the text, that style gradually hardens to more nearly approximate the usual terse mode of modern mathematical discourse.

Having found that most of my students are intrigued by the historical development of mathematics and by the characters in the vast, illustrious cast, I have included brief historical discussions in the text body to serve as motivation. Most students know little of the history of our discipline and often are delighted to realize that mathematics is a purely human endeavor and that all of us have traversed the path from the simple to the complex.

An ample number of examples serve to illustrate the standard techniques and maneuvers used in analysis and to guide the beginner in the composition of careful proofs. I have included a rich supply of exercises ranging in degree of difficulty from the straightforward to the challenging.

Opinions differ, of course, as to what must be included in an introductory analysis text and how that material ought to be presented. Since no one text can adequately treat this vast subject (and still be portable), inevitable compromises are required. Several features of this text deserve mention.

We begin, as we must, with the completeness of $\mathbb{R}$ and promptly obtain the Bolzano–Weierstrass theorem. These essential topological concepts form an unfamiliar and uncharted region for the beginning student of analysis. Moreover, beginners tend to think almost automatically in terms of sequences rather than in terms of the more subtle concept of the continuum. Therefore we immediately discuss sequences in $\mathbb{R}$ and prove that the real numbers are Cauchy complete.

In Chapter 2 we generalize these concepts to $\mathbb{R}^n$ in order to emphasize that the language being developed applies, with only the slightest of adjustments, as well to n-dimensional space as to the real line. This approach prepares the student to learn the basic facts of point set topology in Euclidean spaces and the various equivalent formulations of completeness. Connectedness is treated briefly. Chapter 2 concludes with the crucial concept of compactness.

The decision to treat the topology of Euclidean spaces in this early chapter is the first significant compromise. The topology of $\mathbb{R}$ must be discussed early on; the same theorems must eventually be obtained for $\mathbb{R}^n$. Rather than repeat the identical arguments in a later chapter and thereby increase the bulk of the final product, I have chosen to state and prove these topological theorems for $\mathbb{R}^n$ with the understanding that, by letting $n = 1$, all the proofs remain valid as written.

Instructors who wish to postpone treatment of multivariate analysis can selectively omit much of the material in Chapter 2. In this case, Section 2.1 would be omitted and throughout Section 2.2 $\mathbb{R}^n$ would be read as $\mathbb{R}$; the notation has been designed so the proofs are valid in either case. Section 2.3 would be omitted except for Cantor's nested interval theorem (Theorem 2.3.1 and Corollary 2.3.2), which is proved independently for this purpose. Likewise Sections 2.4 and 2.5 can be used by replacing $\mathbb{R}^n$ by $\mathbb{R}$. The Heine–Borel theorem is proved in $\mathbb{R}$ before its n-dimensional version is discussed. Later, when undertaking analysis in $\mathbb{R}^n$, instructors can return briefly to Chapter 2 to cover the necessary topological facts.

Once the topological preliminaries are established, the way is clear to treat continuity in generality. Chapter 3 deals with continuity in topological terms so that the groundwork is laid for an efficient treatment of functions of a vector variable in Chapters 8 and 9. Again, instructors who wish to postpone the analysis of functions of a vector variable can read $\mathbb{R}$ for $\mathbb{R}^n$ and defer those portions of Chapter 3 applicable only in $\mathbb{R}^n$. The full import of continuity tends to be downplayed in the teaching of introductory calculus courses in the interest of getting on with more "practical" topics; Chapter 3 is intended to rectify this tendency.

The remainder of the text is partitioned into three general areas. Chapters 4 through 7 treat functions of a single real variable. Chapters 8 through 10 discuss functions of a vector variable. Chapters 11 through 14 present the representation of functions by infinite series and integrals. An instructor who so wishes can proceed directly from Chapter 6 to Chapter 11 and return to Chapter 8 at a later time.

The results of paramount importance in Chapter 4 are, of course, the Mean Value Theorem and Taylor's theorem. Chapter 5 treats functions of bounded variation. This material is used in Chapters 6 and 7 and therefore is not optional. These are functions well known to the masters of analysis in the nineteenth century that deserve to be included in the education of every student of mathematics. Functions of bounded variation provide an accessible model of functions more general than the "nice" functions encountered in elementary calculus. As such, they provide a tool to enlarge the student's field of mathematical vision. Moreover, many of the functions that arise in serious applications of mathematics in physics, chemistry, and engineering are of bounded variation.

Many instructors, faced with the inevitable time constraints, will cover the Riemann integral in Chapter 6 but will consider Chapter 7 to be optional. The text has been structured so that this latter chapter can be omitted without disrupting the study of subsequent chapters. It can be used as a resource for self-study by those students who want a deeper understanding of integration. Anyone intending to continue in graduate study in pure or applied mathematics will find Chapter 7 of value. The extension of integration theory to include Lebesgue's integral seems easy and natural to one already familiar with Stieltjes's generalization of Riemann's work.

Chapters 8 and 9 present the differential calculus of real and vector-valued functions of a vector variable. These chapters especially require familiarity with linear algebra. Chapter 10 treats double and triple integrals. Since multiple integrals, although important, are used only in a subordinate way in this text, just the basic facts are proved. A classic and accessible proof of the role played by the Jacobian in the transformation of multiple integrals is included.

Chapters 11 through 14 comprise the heart of the text; they treat infinite series, improper integrals, and the representation of functions by power series, integrals, and Fourier series. These topics introduce the student to one of the central goals of analysis: the development of tools for the careful analysis of the behavior of functions. Here too a compromise was required. This aspect of analysis is vast. To remain within the usual constraints on time and space, I have chosen to present the central core of basic facts and several classic examples but, reluctantly, to omit some favorite topics. For example, rather than attempt to treat both the Laplace and Fourier transforms (and therefore be forced to treat each somewhat superficially), I have chosen to discuss the Laplace transform in detail, including the convolution product, and have left treatment of the Fourier transform for in a course in complex analysis. Someone who has worked through the details presented here will be able to apply these methods to the study of other transforms. Chapter 14 includes Jordan's elegant conditions for the pointwise convergence of Fourier series and Fejér's treatment of the Cesàro summability of these series.

This book would not have seen the light of day without the continuous encouragement of my friends and colleagues to whom I owe a great debt. I especially want to acknowledge the inspiring example of Dr. Anna Penk, who loves mathematics and taught it with passion and who graciously suggested valuable improvements to the various drafts of this text. I must also acknowledge my long-suffering students who struggled, sometimes under duress, with the early versions, whose candid responses helped me reshape my thinking and thereby improve the final product. They reminded me, sometimes quite pointedly, that successful teaching requires close attention to the student's point of view. I appreciate the advice given by those who carefully reviewed the manuscript: Ralph Grimaldi, Rose–Hulman Institute of Technology; Ho Kuen Ng, San Jose State University; Harvey Greenwald, California Polytechnic State University; Ken Johnson, University of Georgia; Jutta Hausen, University of Houston; John D'Angelo, University of Illinois, Urbana-Champaign; Howard Sherwood, University of Central Florida; Mark McConnell, Oklahoma State University; Peter Colwell, Iowa State University; Eric Hayashi, San Francisco State University; John Schiller, Temple University. Typographical errors, like spider mites, seem to thrive and swarm in the process of writing a book; I've tried to snag them all. Any that remain and any mathematical errors that persist are my responsibility alone.

I owe special thanks to Laurie Rosatone, my editor at Addison–Wesley. She nursed this project from its inception to the present, boosting my morale with her enthusiasm, wit, and commitment and guiding me through the rough parts. Many of the improvements from the original draft to the final product are the result of her wise counsel. Her assistant, Ranjani Srinivasan, kept me smiling with her sprightly good cheer and generous assistance. Kathy Manley, responsible for the production phase, helped enormously with the complex task of transforming my work into polished form. My sincere thanks to you all.

SAD

A Note to the Student

This book is designed to be the basic reference for the first year-long sequence of courses in mathematical analysis. These courses, together with studies in abstract algebra, typically provide your first exposure to rigorous mathematical thought. Your subsequent work in mathematics, applied mathematics, or engineering will draw not only on the *content* of this body of knowledge but on the *intellectual skills* you will gain as well. It is fair to say that the most important benefit to you will be these new mental skills and habits. In a sense, our purpose is to transform the way you think. Our goal—your goal—is to attain, at the end of these studies, a plateau marked by a new level of sophistication and mathematical maturity. To achieve this goal, you will need to learn to read and write the language of modern mathematics.

At the outset you face a steep uphill climb. Except for the rare prodigy—and I can't speak for them—we have all found it so. Yet I promise you: If you will persevere, if you will accept the demands of the language without resistance, you will arrive eventually—sooner one hopes rather than later—at a plateau where exquisitely beautiful mathematics is to be found. Once there you can meander as you wish, gathering to yourself the riches of our mathematical heritage. This book offers an introduction to the path; the key to your safe arrival is your own determined effort.

Mathematics, like most human languages, is written linearly. Yet, apparently, human beings do not typically learn best in such a restrictive mode. Usually we grapple with new ideas, only partially grasping their content before moving on to new tasks. Later, new understanding of old ideas emerges. When the new insight occurs abruptly, we enjoy one of those "Aha!" experiences so full of excitement. You can increase the likelihood of your having such experiences by studying this text in a nonlinear way. Read ahead so that you will know where you are headed; double back and review earlier sections, letting your mind search for new connections and interpretations. If, upon attempting an exercise, you can't find a solution, set it aside for a while; later return to it with renewed effort and a new point of view. Avoid plodding through this text page by page; if you adopt a pedestrian style in your study of analysis, you risk missing the larger design and becoming entangled in the small details in which analysis abounds. Most emphatically, learn the definitions accurately and precisely. You can no more learn mathematics without knowing the definitions than you can speak a foreign language without knowing its vocabulary.

Be patient with yourself. Learning analysis is difficult; almost none of it comes easily to beginners. Its arguments are often clever and subtle, framed as they were by some of the giants in the history of mathematics. Invariably those arguments are riddled with delicate maneuvers. And with inequalities. From the novice's point of view, those inequalities come wafting in from the ephemeral blue; to the beginner, they are confusing, unfamiliar, and mysterious. They are not fun.

I don't pretend to be able to make the study of analysis easy or fun; that's a contradiction in terms. But I have provided detailed examples showing you how to maneuver with inequalities, how to construct a sound proof, and how to write your proof in the correct, formal mathematical style. I have also included motivational and historical discussions especially in the earlier chapters, in hopes that, from your point of view, this will be a readable text.

An ample supply of exercises is provided. In the first five chapters they are placed at the end of each section, near the material needed for their solution, a placement that may help you identify a strategy for attacking them. In the remaining chapters the exercises are located at the end of the chapter; they often require material from several of the preceding chapters and sections.

We assume that your knowledge of the mechanics of calculus, learned in your introductory courses, is more or less intact. Our purpose here is to lay the rigorous foundation for that knowledge and to extend your understanding to new levels. In one sense, you are required to set aside your formal knowledge of calculus in order to focus on the underlying theory that justifies it. But you cannot be expected to abandon all you have learned; rather, use your previous exposure to the methods of calculus to guide your intuition as you study the logical framework in which calculus fits. In addition, familiarity with the concepts of elementary linear algebra is assumed. At the very least concurrent study of elementary linear algebra is required. Modern analysis is so firmly grounded in the theory of vector spaces that an efficient, rigorous treatment without reference to linear algebra is inconceivable. We also make free use of elementary set theory; if you are unfamiliar with the relevant concepts and notation, they are provided in the appendices.

Finally, reference is made in the text to concepts from abstract algebra, namely from ring theory. No special preparation is required of you since these references are purely descriptive in nature; your success learning analysis will not depend on any of this material. If you have studied abstract algebra, you will recognize here examples worthy of your attention; if you have not, our examples will help motivate your future study. The point is that abstract algebra and analysis, as is true of most areas of mathematics, are so thoroughly intertwined that the attempt to isolate either from the other seems artificial, catering more to the demands of course catalogs than to the spirit of mathematics. Virtually all the major analytic structures presented in this text prove to be algebraic systems called *rings* and are identified as such.

Most of the students I have known over the decades have experienced an emotional roller coaster during the early stages of their studies of analysis. You probably will too. So you may as well accept that fact beforehand and get on with your climb. Most of those same students, after persistent and admirable effort, have also achieved the plateau of mathematical maturity that is the goal. They became intellectually tough, resilient, and fluent in the demanding language of mathematics. That has been an observable fact. If you persist, you too can emerge from your studies with those same intangible but invaluable traits. Good luck on your climb. I hope this text serves you well, as it is intended, as a resource and as a guide.

Steven A. Douglass
Monmouth, OR

Contents

1

The Structure of the Real Numbers: Sequences 1

1.1 Completeness of the Real Numbers 1
1.2 Neighborhood and Limit Points 11
1.3 The Limit of a Sequence 17
1.4 Cauchy Sequences 35
1.5 The Algebra of Convergent Sequuences 45
1.6 Cardinality 53

2

Euclidean Spaces 61

2.1 Euclidean n-Spaces 61
2.2 Open and Closed Sets 79
2.3 Completeness 93
2.4 Relative Topology and Connectedness 97
2.5 Compactness 105

3

Continuity 115

3.1 Limit and Continuity 115
3.2 The Topological Description of Continuity 137
3.3 The Algebra of Continuous Functions 148
3.4 Uniform Continuity 150
3.5 The Uniform Norm: Uniform Convergence 157
3.6 Vector-Valued Functions on $\mathbb{R}^n$ 174

4

Differentiation 181

4.1 The Derivative 181
4.2 Composition of Functions: The Chain Rule 189
4.3 The Mean Value Theorem 190
4.4 L'Hôpital's Rule 198
4.5 Taylor's Theorem 203

5

Functions of Bounded Variations 213

5.1 Partitions 213
5.2 Monotone Functions on $[a, b]$ 215
5.3 Functions of Bounded Variations 221
5.4 Total Variation as a Function 226
5.5 Continuous Functions of Bounded Variation 230

6

The Riemann Integral 233

6.1 Definition of the Riemann Integral 234
6.2 Existence of the Riemann Integral 239
6.3 The Fundamental Theorem of Calculus 252
6.4 Techniques of Integration 269
6.5 Uniform Convergence and the Integral 277
Exercises 285

7

The Riemann–Stieltjes Integral 299

7.1 Definition of the Riemann–Stieltjes Integral 299
7.2 Techniques of Integration 302
7.3 Existence of the Riemann–Stieltjes Integral 305
7.4 Fundamental Theorems of the Riemann–Stieltjes Integration 320
Exercises 326

8

Differential Calculus in $\mathbb{R}^n$ 331

8.1 Differentiability 331
8.2 The Algebra of Differentiable Functions 348
8.3 Differentiability of Vector-Valued Functions 352
8.4 The Chain Rule 355
8.5 The Mean Value Theorem 366
8.6 Higher-Order Partial Derivatives 368
8.7 Taylor's Theorem 373
8.8 Extreme Values of Differentiable Functions 378
Exercises 390

9

Vector-Valued Functions 401

9.1 The Jacobian 402
9.2 The Inverse Function Theorem 406
9.3 The Implicit Function Theorem 414
9.4 Constrained Optimization 423
Exercises 434

10

Multiple Integrals 439

10.1 The Double Integral 439
10.2 Evaluation of Double Integrals 448
10.3 Transformations: Change of Variables 454
10.4 Multiple Integrals in $\mathbb{R}^3$ 473
Exercises 486

11

Infinite Series 493

11.1 Preliminaries 493
11.2 Convergence Tests (Positive Series) 496
11.3 Absolute Convergence 502
11.4 Conditional Convergence 505
11.5 The Cauchy Product 515
11.6 Cesàro Summability 520
Exercises 523

12

Series of Functions 533

12.1 Preliminaries 533
12.2 Uniform Convergence 535
12.3 Tests for Uniform Convergence 540
12.4 Power Series 545
12.5 The Taylor Series Representation of Functions 563
12.6 Solutions of First-Order Differential Equations 572
Exercises 579

13

Improper Integrals 593

13.1 Preliminaries 593
13.2 Improper Integrals of the First Kind 595
13.3 Improper Integrals of the Second Kind 603
13.4 Uniform Convergence of Improper Integrals 605
13.5 Functions Defined by Improper Integrals 617
13.6 The LaPlace Transform 628
Exercises 653

14

Fourier Series 665

14.1 Convergence of the Mean 667
14.2 Trigonometric Series 678
14.3 Convergence of Trigonometric Series 682

14.4 The Cesàro Summability of Fourier Series 689
14.5 Additional Topics 699
Exercises 708

A

Axioms for the Real Numbers $\mathbb{R}$ 715

B

Set Theory 719

C

Functions 723

D

Polynomials 725

References and Additional Readings 729

Index 731

1

The Structure of the Real Numbers: Sequences

As a branch of mathematics, analysis is an elaboration of the concept of limit. You first encountered this essential idea during your earliest studies of calculus and later discovered that all subsequent concepts and techniques in that subject derived from it. Yet the concept probably remained elusive. Our first goal is to establish a solid footing for the study of limit. To achieve this end, we need to step back to a more primitive level and examine the structure of the real number system. We assume you are familiar with the arithmetic and order properties of the set $\mathbb{R}$ of real numbers, provided in Appendix A for your reference, and will use them without comment. We will concentrate on those *structural* properties of this remarkable system that have been identified and carefully described only during the last one hundred and fifty years.

1.1 Completeness of the Real Numbers

As you are aware, the set $\mathbb{R}$ of real numbers is the union of two disjoint sets: the set $\mathbb{Q}$ of rational numbers and its complement, the set of irrationals. $\mathbb{Q}$ is easily described as the set of all quotients of integers with nonzero denominator. In your early studies you learned that the simplest description of the irrationals utilizes decimal expansions of real numbers: A real number is irrational if and only if its decimal expansion is nonrepeating. The trouble with this decimal description of irrational numbers is that it contributes little to our understanding of the structure of $\mathbb{R}$.

Historically, perhaps the first real number discovered to be irrational was $\sqrt{5}$. That fifth-century B.C. discovery, ascribed to Hippasus of Mesopontos, a member of the Pythagorean School, triggered a remarkable response. The doctrines of that philosophical school held that all measurable phenomena in the universe can be reduced to whole numbers or their ratios and Hippasus produced in $\sqrt{5}$ a number whose existence proved those doctrines false. According to the apocryphal tale, Hippasus's colleagues, distraught at the metaphysical implications, attempted to suppress the discovery by casting him into the Aegean Sea.

The existence of irrational numbers could not long be suppressed. After all, a square, the length of whose side is one unit, has a diagonal of length $\sqrt{2}$; the Pythagorean Theorem guarantees it. But $\sqrt{2}$ is irrational although the sides of the square have not merely rational but *integer* lengths. How can an irrational quantity result from so simple a situation? And if irrationality can be found so easily, perhaps irrational quantities lurk throughout the physical world. That thought rattled the philosophical cage of the ancient world.

Here is the charming little proof, found in older editions of Euclid's *Elements* as Proposition 117 of Book X, that $\sqrt{2}$ is irrational. The proof probably is not original with Euclid but was recorded as part of the mathematical heritage of the time. Suppose, said Euclid, that $\sqrt{2}$ is a rational number. That is, suppose that there were to exist numbers p and q in the set $\mathbb{N} = \{1, 2, 3, \ldots\}$ of natural numbers so that $(p/q)^2 = 2$. Suppose also that p and q have no positive common factors other than 1. The fraction p/q is reduced to lowest terms, just as you learned to reduce fractions in elementary school. The innocent-appearing stipulation that p/q is reduced proves to be essential. Now we play with the algebra. It follows from $(p/q)^2 = 2$ that $2q^2 = p^2$. Hence p^2 must be even. Since the square of any odd integer is odd, this implies that p must also be even. In other words, there exists an integer k such that $p = 2k$. Thus, $p^2 = 4k^2$. It follows that $2q^2 = 4k^2$, implying that $q^2 = 2k^2$. Thus q^2, and hence q, is also even. The assumption that $\sqrt{2} = p/q$ requires that both p and q must be even, having 2 as a common factor. But this contradicts the assumption that p/q is reduced to lowest terms. We conclude that $\sqrt{2}$ cannot be rational and must therefore be irrational.

Perhaps you believe that the rational and irrational numbers alternate as you move along the real line, thereby filling in the line, making it solid. This image of $\mathbb{R}$ is deceptive and misleading. First, the rationals are not discretely scattered along the number line with gaps between them to be filled in by irrationals. If you choose any x in $\mathbb{Q}$, there is no rational number closest or immediately adjacent to it. For if there were such a rational point y, then the midpoint $z = (x + y)/2$ would exist in $\mathbb{Q}$ and would lie halfway between x and y. Since this contradicts the assumed property of y—that it is immediately adjacent to x—we must conclude that y cannot exist. There is no gap between adjacent rationals devoid of other rationals, no gap to be "filled in" by irrationals. In fact, as we will see, between any two real numbers, whether rational or irrational, there exist infinitely many of both types of numbers. These sets of numbers, rather than being interspersed in an alternating fashion, are thoroughly blended together. To understand how the irrationals fill in the real line, we need to establish what we mean by "filled in" or "solid." While these words have intuitive meaning in your everyday life, their meaning in any mathematical context must be logically established before they can be used. The ideas that we require in order to give meaning to the concept of "solid" will be used throughout the text.

DEFINITION 1.1.1 A nonempty set S of real numbers is said to be *bounded above* if there exists some real number M such that $x \leq M$ for every x in S. In this case M is said to be an *upper bound* for S. Likewise, S is said to be *bounded below* if there exists some number m such that $m \leq x$ for all x in S; m is called a *lower bound* for S. We call S *bounded* if it is bounded both above and below. ●

Observe that an upper bound for a set S does not tell us the size of the largest element of S. All that an upper bound for S does, and all we expect it to do, is to place an upper limit on the numbers in S. That upper limit may be crude or it may be fairly accurate; we'll turn to that question in just a moment. For now, to ensure that you have the idea clearly in mind, consider the set

$$S = \{(-1)^k \left(1 - \frac{1}{k}\right) : k \text{ in } \mathbb{N}\}$$

$$= \{0, \frac{1}{2}, -\frac{2}{3}, \frac{3}{4}, -\frac{4}{5}, \ldots\}.$$

It is easy to see that no element of S is larger than 1 or smaller than -1. That is, S is contained in the closed interval $[-1, 1]$. It is also true, however, that no element of S is larger than 5, 50, or even 1000. Each of these numbers is an upper bound for S. Likewise, each of -1, -5, -50, and -1000 is a lower bound for S.

To identify just how large the largest number of S may be, we need to find the smallest, or least, upper bound. To this end, we will use a property of $\mathbb{R}$ that will have frequent application here and elsewhere. (See Fig. 1.1.) Later in this section we will introduce tools to enable you to prove the following assertion as an exercise. For the moment we will demonstrate the assertion to show you how helpful it can be.

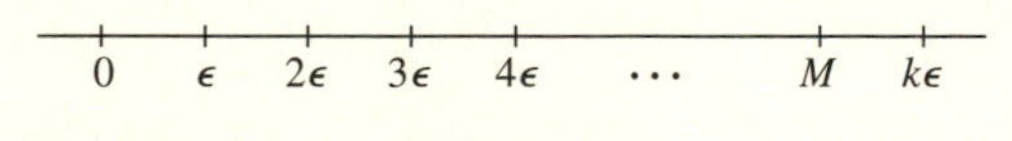

Figure 1.1

Archimedes' Principle[1] Let ϵ and M be any two positive real numbers. There exists a k in $\mathbb{N}$ such that $M < k\epsilon$. ●

In the preceding example $S = \{(-1)^k(1 - 1/k): k \text{ in } \mathbb{N}\}$. It is easy to see that 1 is the least upper bound for S. First, 1 is an upper bound for S because $(-1)^k(1 - 1/k) < 1$ for all k in $\mathbb{N}$. Second, if x is any number less than 1, then we can apply Archimedes' principle to obtain a k in $\mathbb{N}$ such that $1 < 2(1 - x)k$. Consequently, $1/(2k) < 1 - x$ and, therefore,

$$x < 1 - \frac{1}{2k} = (-1)^{2k} \left(1 - \frac{1}{2k}\right).$$

[1] You may have learned Archimedes' principle in a slightly different form: The significance of sufficiently many small mistakes exceeds that of one large one.

Since $(-1)^{2k}[1 - 1/(2k)]$ is a point of S, we conclude that x cannot possibly be an upper bound for S. (See Fig. 1.2.) That is, no number less than 1 can be an upper bound for S, so the smallest upper bound of S is 1.

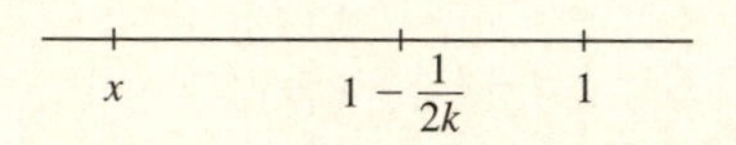

Figure 1.2

In general, suppose S is any nonempty set of real numbers that is bounded above. Among all the upper bounds for S, is there a smallest one and, if so, how should we describe it? To be the smallest of upper bounds of S, an upper bound μ must have the property that, if M is any upper bound for S, then $\mu \le M$. That is,

DEFINITION 1.1.2 Let S be a nonempty set of real numbers that is bounded above. A *supremum* or *least upper bound* of S, denoted $\sup S$, is a real number μ such that

i) $x \le \mu$, for all x in S.

ii) If M is an upper bound for S, then $\mu \le M$. ●

A supremum of S is also sometimes given the acronym for least upper bound and is denoted lub S. It is an easy exercise to prove that $\sup S$, if it exists, is unique. Once you have proved this in Exercise 1.4, $\sup S$ will be referred to as *the* supremum or *the* least upper bound of S.

There is a completely analogous notion of greatest lower bound or infimum of S for any nonempty set S that is bounded below.

DEFINITION 1.1.3 Let S be a nonempty set of real numbers that is bounded below. An *infimum* or *greatest lower bound* of S, denoted $\inf S$, is a real number ν such that

i) $\nu \le x$, for all x in S.

ii) If m is any lower bound for S, then $\nu \ge m$. ●

Sometimes the infimum of S is also denoted by its acronym glb S. Again, $\inf S$, if it exists, is unique and will be called *the* infimum of S.

EXAMPLE 1 If we let $S = \{(-1)^k(1 - 1/k)\colon k \text{ in } \mathbb{N}\}$ as before, then $\sup S = 1$, as we have seen. Imitating our earlier argument, you can prove that $\inf S = -1$. Notice that neither $\sup S$ nor $\inf S$ is in S. ●

EXAMPLE 2 Consider the open interval

$$S = (a, b) = \{x \text{ in } \mathbb{R}\colon a < x < b\}.$$

We claim that $\inf S = a$ and $\sup S = b$. Clearly, $x < b$ for all x in S, so b is an upper bound for S. However, if $a < x < b$, then $y = (x + b)/2$ is in S and is larger than x.

Thus x cannot be an upper bound for S and $b = \sup S$. An analogous argument proves that $a = \inf S$. Notice that neither $\sup S$ nor $\inf S$ is an element of S. ●

EXAMPLE 3 Fix $r > 1$ and let $S = \{r^k\colon k \text{ in } \mathbb{N}\}$. We claim that S is not bounded above and hence $\sup S$ does not exist in $\mathbb{R}$. Write $r = 1 + c$ where $c > 0$. By Bernoulli's inequality (see Exercise 1.16), $r^k = (1+c)^k \geq 1 + kc$ for all k in $\mathbb{N}$. Now let M be any (large) positive number. By Archimedes' principle, choose k in $\mathbb{N}$ such that $M - 1 < kc$. Therefore

$$r^k = (1+c)^k \geq 1 + kc > M.$$

Thus M cannot be an upper bound for S. Since M is an arbitrary positive number, we deduce that S cannot be bounded above. ●

EXAMPLE 4 Let S be the closed interval

$$[-\sqrt{2}, \sqrt{2}] = \{x \text{ in } \mathbb{R}\colon -\sqrt{2} \leq x \leq \sqrt{2}\}.$$

As in Example 2, it is straightforward to show that $\sup S = \sqrt{2}$ and $\inf S = -\sqrt{2}$, both of which are elements of S. ●

The all-important question arises: If S is bounded above, does $\sup S$ exist? If we were to have only the rational numbers available, a variant of Example 4 provides some insight. If, in that example, we were to take $S = \{x \text{ in } \mathbb{Q}\colon -\sqrt{2} \leq x \leq \sqrt{2}\}$, then S is bounded above, say by 1.42. But $\sup S = \sqrt{2}$ does not exist in $\mathbb{Q}$, whereas it does exist in $\mathbb{R}$. The set of real numbers enjoys a subtle property, that of *completeness*, which is lacking in $\mathbb{Q}$ and which will prove to be essential to all our future work.

We will assume as an axiom of the real number system the existence of $\sup S$ for any nonempty set S of real numbers that is bounded above. In a course devoted to the foundations of mathematics, this axiom is often derived from yet more primitive properties of $\mathbb{R}$, but here we will be satisfied to assume it. (See Fig. 1.3.)

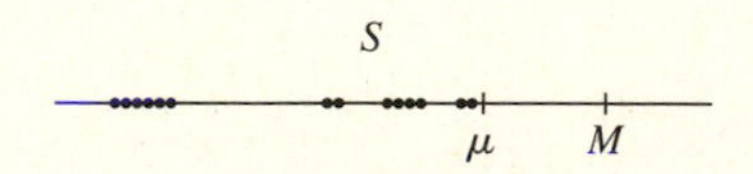

Figure 1.3

AXIOM 1.1.1 The Completeness Axiom for $\mathbb{R}$. If S is a nonempty set of real numbers that is bounded above, then $\sup S$ exists in $\mathbb{R}$. ●

From this statement, deceptive in its simplicity, flows a remarkable stream of facts, indeed all the contents of this text. Observe that this axiom provides us with exactly the tool we need to make firm your intuitive sense that the real line is "solid." There can be no gaps or holes in the real line. To see this, imagine cutting the real line into two disjoint pieces. Let S consist of all those points strictly to the left of the cut. (See Fig. 1.4 on page 6.) Clearly, S is not empty. Furthermore, S is bounded above by any real number to the right of the cut. Hence, by the Completeness Axiom,

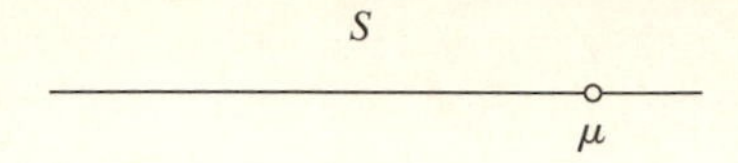

Figure 1.4

S has a least upper bound μ in $\mathbb{R}$. The point μ must be the cut point; that is, the cut could not have occurred at a "hole." In other words, $\mathbb{R}$ has no "holes."

The Completeness Axiom for $\mathbb{R}$ implies that, if S is a nonempty set of real numbers that is bounded below, then $\inf S$ exists in $\mathbb{R}$. To see this, we need merely let

$$T = \{-s : s \text{ in } S\}.$$

The set T is nonempty and is bounded above. We conclude by Axiom 1.1.1 that $\mu = \sup T$ exists. It is straightforward to verify that $\nu = -\mu$ is the greatest lower bound for S. (In Exercise 1.10, you are asked to fill in the details.)

EXAMPLE 5 Fix $c > 0$ and n in $\mathbb{N}$. We prove here that $c^{1/n}$, the positive nth root of c, exists in $\mathbb{R}$. Form the set

$$S = \{x \text{ in } \mathbb{R} : \ 0 < x \text{ and } x^n < c\}.$$

First we show that S is nonempty. Since $c > 0$, Archimedes' principle ensures the existence of a k in $\mathbb{N}$ such that $1 < ck$. Thus $1 < ck^n$ or, equivalently, $1/k^n < c$. Consequently, $1/k$ is in S. This proves that $S \neq \emptyset$.

Next we verify that S is bounded above. Choose any $M > \max\{c, \ 1\}$. Then, for any x in S, $0 < x^n < c < M < M^n$. Since $0 < x^n < M^n$ and since x and M are both positive, we deduce that $x < M$. (We ask you to prove this in Exercise 1.17.) It follows that M is an upper bound for S. By Axiom 1.1.1, $\mu = \sup S$ exists. Because all the numbers in S are positive, μ must also be positive.

To show that μ is the nth root of c, we will eliminate, in turn, the two possibilities $\mu^n < c$ and $\mu^n > c$. That will leave only the possibility $\mu^n = c$ and will prove that μ is the nth root of c.

Suppose that $\mu^n < c$. Let $M = \sum_{j=1}^{n} C(n, j)\mu^{n-j}$, where, for $j = 0, 1, 2, \ldots, n$, the numbers

$$C(n, j) = \frac{n!}{j!(n-j)!}$$

are the binomial coefficients. Since $(c - \mu^n)$ is positive, Archimedes' principle ensures the existence of a k in $\mathbb{N}$ such that $M/k < (c - \mu^n)$. If we expand $(\mu + 1/k)^n$ by the binomial theorem and simplify the result, we obtain

$$\begin{aligned}\left(\mu + \frac{1}{k}\right)^n &= \sum_{j=0}^{n} C(n, j)\mu^{n-j}\left(\frac{1}{k}\right)^j \\ &= \mu^n + \frac{1}{k}\sum_{j=1}^{n} C(n, j)\mu^{n-j}\left(\frac{1}{k}\right)^{j-1}\end{aligned}$$

$$\leq \mu^n + \frac{1}{k}\sum_{j=1}^{n} C(n, j)\mu^{n-j}$$

$$= \mu^n + \frac{M}{k} < \mu^n + (c - \mu^n) = c.$$

Consequently, $\mu + 1/k$ is in S but is larger than $\sup S$. The contradiction proves that μ^n cannot be less than c.

Suppose that $\mu^n > c$. By Archimedes' principle, choose k_1 in $\mathbb{N}$ so that $1 < \mu k_1$. Likewise, choose k_2 in $\mathbb{N}$ so that $n\mu^{n-1} < (\mu^n - c)k_2$. Choose any $k \geq \max\{k_1, k_2\}$. For this choice of k, we have $1 < \mu k$ so that $\mu - 1/k$ is positive. We also have $n\mu^{n-1}/k < \mu^n - c$. In Exercise 1.16 you are asked to show that

$$\left(\mu - \frac{1}{k}\right)^n \geq \mu^n - \frac{n\mu^{n-1}}{k}$$

$$> \mu^n - (\mu^n - c)$$

$$= c > x^n,$$

for all x in S. Consequently $\mu - 1/k$ must be an upper bound for S, contradicting the fact that μ is the smallest upper bound for S. We deduce that μ^n cannot be larger than c. The only remaining option is that $\mu^n = c$. Therefore, $\mu = c^{1/n}$, the positive nth root of c. ●

EXAMPLE 6 Let $S = \{x_k = \sum_{j=0}^{k} 1/j! : \text{k in } \mathbb{N}\}$. We claim that S is bounded above. To prove this assertion, we begin with the inequality $2^{j-1} \leq j!$ for all j in $\mathbb{N}$. (The inductive proof is left for you in Exercise 1.19.) Therefore, $1/j! \leq 1/2^{j-1}$. We use this last inequality and the formula for the sum of a finite geometric series to obtain, for any k in $\mathbb{N}$,

$$x_k = 1 + \frac{1}{1!} + \frac{1}{2!} + \frac{1}{3!} + \cdots + \frac{1}{k!}$$

$$\leq 1 + 1 + \frac{1}{2} + \frac{1}{2^2} + \cdots + \frac{1}{2^{k-1}}$$

$$= 1 + \frac{1 - (1/2)^k}{1 - (1/2)}$$

$$= 1 + 2\left[1 - \left(\frac{1}{2}\right)^k\right] < 3.$$

Consequently, S is bounded above by 3. Clearly S is not empty. By Axiom 1.1.1, $\sup S$ exists with $\sup S \leq 3$. S is bounded below by 1 and is therefore contained in the closed interval $[1, 3]$. It is easy to see that $\inf S = 1$, but the value of $\sup S$ is not at all apparent. In fact, $\sup S$ is the number e, an irrational number whose approximate value is 2.71828, a number of profound importance throughout mathematics. ●

EXAMPLE 7 Define $S = \{(1 + 1/k)^k\colon\ k \text{ in } \mathbb{N}\}$. As you may prove in Exercise 1.20, $(1 + 1/k)^k \leq \sum_{j=0}^{k} 1/j!$ for every k in $\mathbb{N}$. Furthermore, for all k in $\mathbb{N}$,

$\sum_{j=0}^{k} 1/j! < e$, by Example 6. Consequently, e is an upper bound for S. Remarkably, and we will prove this in Section 1.3, $\sup S = e$ also. ●

These several examples show us that $\mu = \sup S$ may or may not belong to S. Observe that whenever μ is in S, it is the maximum of S. The converse is also true: If S has a largest element M, then $M = \sup S$ and is in S. The analogous statements hold for $\inf S$ and the minimum of S. Thus we see that the notions of least upper bound and greatest lower bound are generalizations of maximum and minimum, respectively.

Our first theorem identifies precisely the way in which supremum generalizes maximum and infimum generalizes minimum. We will invoke this important result at several crucial points throughout the text.

THEOREM 1.1.1 Let S be a nonempty set of real numbers and let μ and ν be real numbers.

i) Suppose that S is bounded above. Then $\mu = \sup S$ if and only if μ is an upper bound for S and, for every $\epsilon > 0$, there exists an x in S such that $\mu - \epsilon < x \leq \mu$.

ii) Suppose that S is bounded below. Then $\nu = \inf S$ if and only if ν is a lower bound for S and, for every $\epsilon > 0$, there exists an x in S such that $\nu \leq x < \nu + \epsilon$.

Proof. For (i), note first that, if $\mu = \sup S$, then $x \leq \mu$ for all x in S. If there were no point of S in the interval $(\mu - \epsilon, \mu]$, then $\mu - \epsilon$ would be an upper bound for S, contradicting the fact that μ is the least such. (See Fig. 1.5.) Consequently, there must exist an x in S that is also in the interval $(\mu - \epsilon, \mu]$.

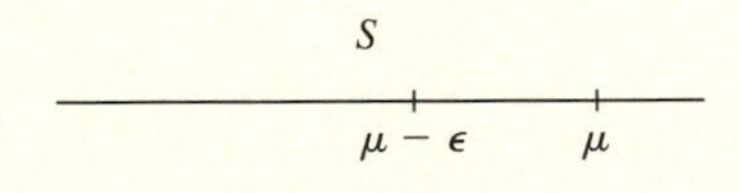

Figure 1.5

Conversely, suppose that μ is an upper bound for S with the property that, for every $\epsilon > 0$, there is a point of S in $(\mu - \epsilon, \mu]$. Let M be any upper bound for S. We want to show that $\mu \leq M$. We assume to the contrary that $M < \mu$ and obtain a contradiction. Let $\epsilon = \mu - M$. By our assumption, ϵ is positive. Consequently, there must be a point x in S such that $\mu - \epsilon < x \leq \mu$. But $\mu - \epsilon = \mu - (\mu - M) = M$. Therefore the point x is in S and $x > M$. This inequality contradicts the assumption that M is an upper bound for S. Thus $\mu \leq M$ for any upper bound M of S. Hence μ must be the least upper bound of S. The analogous proof of (ii) is left as an exercise. ●

In Exercise 1.5 you are asked to prove that Axiom 1.1.1 together with Theorem 1.1.1 implies Archimedes' principle for $\mathbb{R}$. In Exercise 1.6 you are challenged to show that Archimedes' principle is equivalent to the following variant: For any positive c in $\mathbb{R}$, there exists a k in $\mathbb{N}$ such that $k - 1 \leq c < k$. Once you complete this proof, we can prove an extremely important fact, one that will be used at several key points later.

THEOREM 1.1.2 Between any two real numbers there is a rational number.

Proof. Suppose that c and d are real numbers with $0 < c < d$. Since $d - c > 0$, there exists, by Archimedes' principle, a q in $\mathbb{N}$ such that $1 < (d - c)q$. That is,

$$cq + 1 < dq. \tag{1.1}$$

Since cq is positive, the variant of Archimedes' principle stated above implies that there exists a p in $\mathbb{N}$ such that

$$p - 1 \leq cq < p. \tag{1.2}$$

Combining the inequalities (1.1) and (1.2) gives

$$p - 1 \leq cq < p \leq cq + 1 < dq.$$

Extract the inequality $cq < p < dq$ and divide by q to conclude that $c < p/q < d$, proving the theorem in this case. We leave the remaining cases for Exercise 1.7. ●

EXERCISES

1.1. Prove that $\sqrt{3}$ is irrational.

1.2. Prove that a real number is irrational if and only if its decimal representation is nonrepeating. Equivalently, a real number is rational if and only if its decimal representation is repeating. (A decimal representation that terminates is actually repeating with the digit 0 repeated.) *Hint:* To show that a repeating decimal expansion for x is rational, suppose the repeating block of digits has length k, multiply x by 10^k, and show that $10^k x - x$ is a terminating decimal. Then solve for x as a fraction.

1.3. Show that, for any finite set S of real numbers, both $\sup S$ and $\inf S$ exist and are elements of S. *Hint:* Use induction on k, the number of elements in S.

1.4. Suppose S is a nonempty set of real numbers that is bounded above. Let $\mu = \sup S$. Prove that μ is unique. (Assume there exists a second number having the properties of $\mu = \sup S$ and show that it must equal μ.)

1.5. Use Axiom 1.1.1 and Theorem 1.1.1 to derive Archimedes' principle. *Hint:* Given $\epsilon > 0$ and any positive constant M, form the set $S = \{k\epsilon : k \text{ in } \mathbb{N}\}$. Show that M cannot possibly be an upper bound for S.

1.6. Show that Archimedes' principle is equivalent to the following statement: For any $c > 0$, there exists a k in $\mathbb{N}$ such that $k - 1 \leq c < k$.

1.7. Use Theorem 1.1.2 to treat the cases when $c \leq 0 < d$ and $c < d \leq 0$ and thus show that, if $c < d$, there exists a rational number x such that $c < x < d$.

1.8. Suppose that c and d are real numbers with $c < d$. Show that there exists an irrational number x such that $c < x < d$.

1.9. Fix any $a \neq 0$ and let $S = \{ax : x \text{ in } \mathbb{Q}\}$. Show that between any two distinct points c and d in $\mathbb{R}$ there exists a point of S.

1.10. Suppose that c is a real number such that, for all $\epsilon > 0$, it is true that $|c| < \epsilon$. Prove that c must be 0.

1.11. Suppose that S is a nonempty set that is bounded below.

a) Let $T = \{-x\colon x \text{ in } S\}$. Show that T is bounded above.

b) Let $\mu = \sup T$ and $\nu = -\mu$. Prove that $\nu = \inf S$.

c) Prove that $\nu = \inf S$ is unique.

1.12. If S is a nonempty, bounded set in $\mathbb{R}$ and if T is a nonempty subset of S, prove that

$$\inf S \le \inf T \le \sup T \le \sup S.$$

1.13. Let S be any nonempty, bounded set in $\mathbb{R}$.

a) For fixed c in $\mathbb{R}$ define $c + S = \{c + x\colon x \text{ in } S\}$. Prove that $\sup (c + S) = c + \sup S$ and $\inf (c + S) = c + \inf S$.

b) For a fixed $c > 0$, define $cS = \{cx\colon x \text{ in } S\}$. Prove that $\sup (cS) = c \sup S$ and $\inf (cS) = c \inf S$.

c) If, in part (b), $c < 0$, show that $\sup (cS) = c \inf S$ and $\inf (cS) = c \sup S$.

1.14. Suppose that S_1 and S_2 are nonempty, bounded sets in $\mathbb{R}$. Show that $\sup (S_1 \cup S_2) = \max \{\sup S_1, \sup S_2\}$ and that $\inf (S_1 \cup S_2) = \min \{\inf S_1, \inf S_2\}$. Is an analogous theorem for intersections possible? Explain.

1.15. Let S_1 and S_2 be two nonempty, bounded sets in $\mathbb{R}$. Define $S_1 + S_2 = \{x_1 + x_2 : x_i \text{ in } S_i, i = 1, 2\}$. Prove that $S_1 + S_2$ is bounded and that $\sup (S_1 + S_2) = \sup S_1 + \sup S_2$ and $\inf (S_1 + S_2) = \inf S_1 + \inf S_2$.

1.16. **a)** **Bernoulli's Inequality** Prove by induction that $(1 + x)^n \ge 1 + nx$ for all $x > -1$ and all n in $\mathbb{N}$.

b) Use part (a) to show that, as in Example 5, if $1 < \mu k$, then

$$\left(\mu - \frac{1}{k}\right)^n \ge \mu^n - \frac{n\mu^{n-1}}{k}$$

for all n in $\mathbb{N}$.

1.17. Prove that if x and y are positive numbers such that $x^n < y^n$ for some n in $\mathbb{N}$, then $x < y$. (We say that the taking of positive nth roots *preserves order* on $\mathbb{R}^+$.)

1.18. If the positive integer n is even and $c < 0$, then c has no real nth root. However, if n is odd and $c < 0$, then c has a negative nth root. Prove this assertion.

1.19. Prove by induction that, for every j in $\mathbb{N}$, $2^{j-1} \le j!$.

1.20. **a)** Use the Binomial Theorem to show that $(1 + 1/k)^k$ expands to become

$$1 + 1 + \sum_{j=2}^{k} \frac{1}{j!}\left(1 - \frac{1}{k}\right)\left(1 - \frac{2}{k}\right)\cdots\left(1 - \frac{j-1}{k}\right).$$

b) Prove that, for each $j = 2, 3, \ldots, k$,

$$\frac{1}{j!}\left(1 - \frac{1}{k}\right)\left(1 - \frac{2}{k}\right)\cdots\left(1 - \frac{j-1}{k}\right) \le \frac{1}{j!}.$$

c) Show that $(1 + 1/k)^k \le \sum_{j=0}^{k} 1/j!$ for all k in $\mathbb{N}$.

1.21. Prove part (ii) of Theorem 1.1.1.

1.22. Explain how the image of cutting the real line into two disjoint sets as in the discussion following Axiom 1.1.1 affirms your belief that the real line is solid, that is, is a *continuum*.

1.2 NEIGHBORHOODS AND LIMIT POINTS

One of the essential, original ideas of Newton and Leibnitz in creating calculus was the concept of *arbitrarily close.* This concept is central to all of analysis. Understanding it is the key to understanding the concept of *limit* and the more primitive notion of *limit point*; hence it is the key to mastery of this text. Let's examine this concept with some care.

Let x be a point in $\mathbb{R}$ and let S be any nonempty subset of $\mathbb{R}$. What is meant by saying, "x is arbitrarily close to S"? For x to be close to S means that the distance between x and some point of S is small. That is, there exists a y in S such that $|x - y| < \epsilon$, where ϵ is some small, positive number.

Of course, *small* is an ambiguous word. A small error in measuring the shortest distance between New York and Paris has a meaning far different from a small error in adjusting a fine watch. The unadorned, unqualified word *small* has little meaning for us when used out of context. Once you modify it by saying "arbitrarily small," however, the situation changes. Thus modified, it means that regardless of the context, regardless of the frame of reference, the quantity is small. How small? As small as you—or anyone else—may wish.

Thus, if x is arbitrarily close to S, then you can specify a small distance $\epsilon > 0$, no matter how small, and be confident that there will be a point y in S, which will depend on the ϵ prescribed, with the property that the distance between x and y is smaller than ϵ. That is, $|x - y| < \epsilon$.

For much of the history of calculus the question of carefully defining "arbitrarily close" was recognized as being of paramount importance. Newton, discussing his theory of *fluxions*, considered ratios of two quantities (fluxions) each of which tends to 0. He determined what happened to these ratios as both the numerator and the denominator moved simultaneously toward 0. The result, which today we call limit, he called the "ultimate ratio of evanescent quantities" by which, in his words,

> *is to be understood the ratio of the quantities, not before they vanish, not after, but that with which they vanish.*

Mathematicians today recognize the glimmer of brilliance in his words, but would despair of using this sort of description as the foundation on which to build the edifice of analysis. Some contemporaries of Newton mocked the new theory of fluxions. Bishop George Berkeley, in his essay *The Analyst, or a Discourse Addressed to an Infidel Mathematician*, wrote:

> *And what are these fluxions? The velocities of evanescent increments. And what are these same evanescent increments? They are neither finite quantities, not quantities infinitely small, not yet nothing. May we not call them the ghosts of departed quantities?*

Mathematicians of the seventeenth and eighteenth centuries, though embracing the calculus as an unequaled tool for attacking previously intractable problems, certainly felt ill at ease with these ephemeral "ghosts" lurking among the shaky underpinnings of the theory. The list of those who grappled with the issue of adequately defining "limit" reads like a Who's Who of the discipline. Gradually, the mist cleared, the central issue came into focus. In 1821, about one hundred and fifty years after the creation of the methods of calculus, Augustin-Louis Cauchy, whose monumental work *Cours d'analyse algébrique* forms the early logical basis for much of modern analysis, proposed the following definition:

> *When the values successively attributed to a particular variable approach indefinitely a fixed value, so as to end by differing from it by as little as one wishes, this latter is called the limit of all the others.*

A few decades later, in 1859, Karl Weierstrass, at the University of Berlin, quantified the definition of limit and presented it in the form now appearing in every elementary calculus text. Actually, Weierstrass had begun his major work laying a rigorous foundation for analysis much earlier, from 1841 to 1856, while teaching high school mathematics, but his approach became publicly known only in 1859. The following definition provides us with a crucial tool with which to work with the concepts *arbitrarily small* and *limit point* and hence *limit.* It appears that Cauchy was the first to use the word *neighborhood* systematically to capture the idea of *close*.

DEFINITION 1.2.1 Let x be any real number. A *neighborhood* of x with positive radius r is the set

$$N(x; r) = \{y \text{ in } \mathbb{R}: |y - x| < r\}. \quad \bullet$$

Clearly, in $\mathbb{R}$, the neighborhood $N(x; r)$ is merely the open interval $(x - r, x + r)$, but we adopt this notation in order to generalize our discussion later to spaces of higher dimension.

DEFINITION 1.2.2 A *deleted neighborhood* of x in $\mathbb{R}$, denoted $N'(x; r)$, is the neighborhood $N(x; r)$ with the point x itself removed. $\bullet$

Clearly, in $\mathbb{R}$, the deleted neighborhood $N'(x; r)$ is merely the union of the open intervals $(x - r, x)$ and $(x, x + r)$. When the magnitude of the radius r is immaterial, we will omit mention of it and merely write $N(x)$ and $N'(x)$ to denote a neighborhood and a deleted neighborhood, respectively, of the point x.

A second concept, that of limit point, is introduced in the following definition. It emerges from our previous discussion regarding a point x being arbitrarily close to a nonempty set S, and it will play a central role throughout the remainder of the text. It is essential that you master this idea.

DEFINITION 1.2.3 Given a nonempty set S in $\mathbb{R}$, a point x in $\mathbb{R}$ is said to be a *limit point* of S if, for each $\epsilon > 0$, the deleted neighborhood $N'(x; \epsilon)$ contains at least one point of S. $\bullet$

A limit point x of S is also called an *accumulation point* of S (because points of S accumulate near x) or a *cluster point* of S (because points of S cluster around x).

If x is a limit point of S, then x must be arbitrarily close to S: For any $\epsilon > 0$, there must exist a point y in S that is distinct from x and within a distance of ϵ from x. Whether x does or does not belong to S is immaterial; y is in the *deleted* neighborhood $N'(x; \epsilon)$.

If x is a point of S that is not a limit point of S, then there must exist an $\epsilon > 0$ for which $N'(x; \epsilon) \cap S$ is empty. In this case, x is said to be an *isolated point* of S. Evidently, a point in S is either a limit point of S or an isolated point of S. But a set S may well have limit points that do not belong to the set.

Every deleted neighborhood of a limit point x of S contains at least one point of S. In fact, every deleted neighborhood of x must contain *infinitely* many points of S. To prove this claim, suppose, to the contrary, that $N'(x) \cap S$ were to contain only finitely many points. That is, suppose that $N'(x) \cap S = \{x_1, x_2, \ldots, x_k\}$. Then the set

$$S_1 = \{|x_1 - x|, |x_2 - x|, \ldots, |x_k - x|\}$$

is a finite set of positive numbers. By Exercise 1.3, $\inf S_1 = \min S_1$ exists and is an element of S_1. Let x_j be a point of $N'(x) \cap S$ such that $|x_j - x| = \inf S_1$. Then x_j is a point of $N'(x) \cap S$ that is closest to x. Choosing a positive ϵ less than $|x_j - x|$ yields a deleted neighborhood $N'(x; \epsilon)$ that contains no point of S. This contradicts the assumption that x is a limit point of S. Therefore, $N'(x)$ cannot contain only finitely many points of S. We conclude that $N'(x) \cap S$ must contain infinitely many points.

EXAMPLE 8 Let $S = \{(-1)^k(1 + 1/k)\colon k \text{ in } \mathbb{N}\}$. Then 1 and -1 are each limit points of S. To see that 1, for example, is a limit point of S is easy; a simple application of Archimedes' principle does the trick. (See Fig. 1.6.) Fix any $\epsilon > 0$. By that principle, choose k in $\mathbb{N}$ so large that $1/(2k) < \epsilon$. Then, $1 < 1 + 1/(2k) < 1 + \epsilon$. Therefore, $1 + 1/(2k) = (-1)^{2k}[1 + 1/(2k)]$ is in $N'(1; \epsilon) \cap S$. Since ϵ is arbitrary and since the resulting deleted neighborhood $N'(1; \epsilon)$ contains a point of S, the point 1 is a limit point of S. A similar argument applies at the limit point -1. Note that neither 1 nor -1 is a point of S. ●

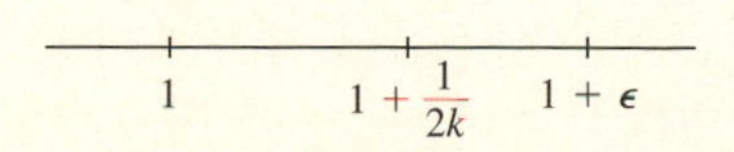

Figure 1.6

To simplify some arguments, it is sometimes helpful to have the following technical theorem available. It emphasizes that, when dealing with the question of a limit point of a set, we are interested only in *small* values of ϵ. It addresses a minor point that should not be allowed to detract from your enjoyment of analysis.

THEOREM 1.2.1 Let S be a nonempty set in $\mathbb{R}$. Then a point x is a limit point of S if and only if there is an $\epsilon_0 > 0$ such that, for all ϵ with $0 < \epsilon < \epsilon_0$, the set $N'(x; \epsilon)$ contains at least one point of S.

Proof. First we deal with the trivial part of the theorem. Suppose that x is a limit point of S. Choose any $\epsilon_0 > 0$. Then, for any ϵ with $0 < \epsilon < \epsilon_0$, the set $N'(x; \epsilon) \cap S \neq \emptyset$ because x is a limit point of S.

Suppose, on the other hand, that there exists an $\epsilon_0 > 0$ such that, for all ϵ satisfying $0 < \epsilon < \epsilon_0$, the set $N'(x; \epsilon)$ contains points of S. Let ϵ_1 be any positive number. We have to show that $N'(x; \epsilon_1) \cap S \neq \emptyset$. To achieve this, simply choose any positive $\epsilon < \min\{\epsilon_0, \epsilon_1\}$. By the hypothesis, $N'(x; \epsilon) \cap S \neq \emptyset$. But $N'(x; \epsilon) \subset N'(x; \epsilon_1)$. Therefore we conclude that $N'(x; \epsilon_1) \cap S \neq \emptyset$ and that x is thus a limit point of S. This proves the theorem. ●

EXAMPLE 9 Let a and b be real numbers with $a < b$ and let $S = (a, b)$. Every point in $[a, b]$ is a limit point of S. Clearly, if x is any point in (a, b) and ϵ is any positive number, then $N'(x; \epsilon)$ contains points of S. In fact, if $\epsilon_0 < \min\{x - a,\ b - x\}$, then $N'(x; \epsilon_0)$ is contained entirely in (a, b). (See Fig. 1.7.) Thus, for every positive ϵ less than ϵ_0, $N'(x; \epsilon) \cap (a, b) \neq \emptyset$. Consequently, by Theorem 1.2.1, x is a limit point of (a, b). That is, each point of (a, b) is a limit point of (a, b). There remains only to check the endpoints a and b. To show that b, for example, is a limit point of S, fix any positive ϵ_0 less than $b - a$ and choose any positive ϵ less than ϵ_0. (See Fig. 1.8.) Observe that any x in the interval $(b - \epsilon, b)$ is in both S and $N'(b; \epsilon)$. By Theorem 1.2.1, b is a limit point of S. Note that b itself does not belong to S. Use a similar argument to treat the endpoint a. ●

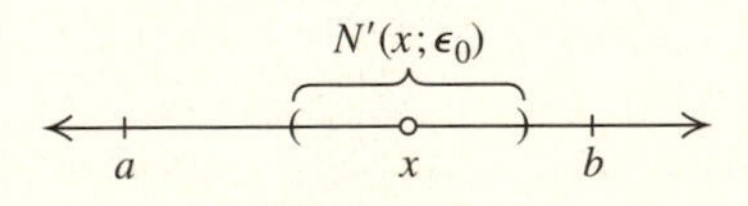

Figure 1.7

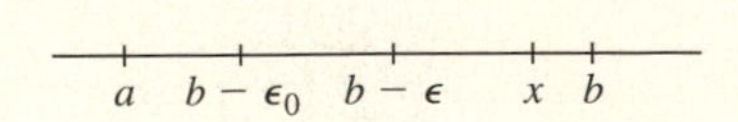

Figure 1.8

EXAMPLE 10 As a variant of Example 9, let a and b be real numbers with $a < b$. Let $S = \{x \text{ in } \mathbb{Q}\colon\ a < x < b\} = \mathbb{Q} \cap (a, b)$ denote the set of all rational numbers in the open interval (a, b). Again, every point in the closed interval $[a, b]$ is a limit point of S. Fix any x in (a, b) and any positive ϵ_0 less than $\min\{x - a, b - x\}$. Choose any positive ϵ less than ϵ_0. By Theorem 1.1.2, there exists a rational number in the interval $(x - \epsilon, x)$ (and, by the way, also another in $(x, x + \epsilon)$). Hence, the deleted neighborhood $N'(x; \epsilon)$ contains a point of S. By Theorem 1.2.1, x is a limit point of S. Thus every point of (a, b), whether rational or irrational, is a limit point of S. The endpoints are treated similarly. ●

EXAMPLE 11 Let $S = \mathbb{Z}$, the set of all integers. Then $\mathbb{Z}$ has no limit points in $\mathbb{R}$. To prove this, note first that, if k is any integer, if $0 < \epsilon_0 < 1$, and if $0 < \epsilon < \epsilon_0$, then $N'(k; \epsilon)$ is contained entirely in $(k - 1, k) \cup (k, k + 1)$, a set devoid of integers. Thus k cannot be a limit point of $\mathbb{Z}$. If x is not an integer, we let $\lfloor x \rfloor$ denote the *integer part* of x, the largest integer less than or equal to x. Then x is in $(\lfloor x \rfloor, \lfloor x \rfloor + 1)$, an open interval containing no integer. (See Fig. 1.9.) If we choose a positive ϵ_0 less than the smaller of $x - \lfloor x \rfloor$ and $\lfloor x \rfloor + 1 - x$ and if we choose any positive ϵ less than ϵ_0, then $N'(x; \epsilon)$ is contained entirely in $(\lfloor x \rfloor, \lfloor x \rfloor + 1)$ and hence contains no integer. Thus x cannot be a limit point of $\mathbb{Z}$ either. That is, no point of $\mathbb{R}$ is a limit point of $\mathbb{Z}$. ●

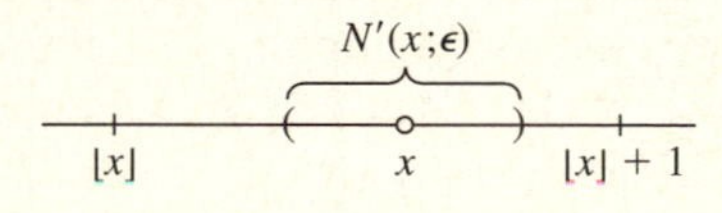

Figure 1.9

DEFINITION 1.2.4 If a nonempty set S in $\mathbb{R}$ has no limit points, then S is said to be *discrete*. ●

We hope our earlier discussion will help you appreciate the clarity of the definition of limit point. It is straightforward, unambiguous, and, as you will see, ideally tailored to establish a sturdy foundation for analysis. For example, here we establish a basic existence theorem that will be indispensable throughout the text.

THEOREM 1.2.2 The Bolzano–Weierstrass Theorem If S is a bounded, infinite subset of $\mathbb{R}$, then S has a limit point in $\mathbb{R}$.

Proof. Since S is bounded, $\nu = \inf S$ and $\mu = \sup S$ exist. Clearly, $S \subseteq [\nu, \mu]$. To establish efficient recursive notation, let $a_0 = \nu$, $b_0 = \mu$, and $I_0 = [a_0, b_0]$. Let c_0 denote the midpoint $(a_0 + b_0)/2$ of I_0. Either $[a_0, c_0]$ or $[c_0, b_0]$ (or both) contains infinitely many points of S, for otherwise S would be a finite set. Let $I_1 = [a_1, b_1]$ denote whichever of the intervals $[a_0, c_0]$ or $[c_0, b_0]$ contains infinitely many points of S. If both subintervals do so, choose the leftmost interval. Note that $I_1 \subset I_0$ and that the length of I_1 is $(b_0 - a_0)/2$.

Bisect the interval $I_1 = [a_1, b_1]$ at the midpoint $c_1 = (a_1 + b_1)/2$. Again, either $[a_1, c_1]$ or $[c_1, b_1]$ contains infinitely many points of S because we chose I_1 so that $I_1 \cap S$ is an infinite set. Let $I_2 = [a_2, b_2]$ denote whichever of these two intervals contains infinitely many points of S. If both satisfy this requirement, choose the leftmost subinterval. Note that $I_2 \subset I_1$. The length of I_2 is $(b_1 - a_1)/2 = (b_0 - a_0)/2^2$.

Continue this procedure, repeatedly bisecting each interval $[a_k, b_k]$ and choosing $I_{k+1} = [a_{k+1}, b_{k+1}]$ to be the half of $[a_k, b_k]$ that contains infinitely many points of S, or, if both halves contain infinitely many points of S, choose the leftmost half. In this way we obtain a sequence $\{I_k\} = \{[a_k, b_k]\}$ of closed intervals with the following properties: For each k in $\mathbb{N}$,

i) $S \cap I_k$ is infinite.

ii) $I_{k+1} \subset I_k$.

iii) The length of I_k is $b_k - a_k = (b_0 - a_0)/2^k$.

The first two properties are self-evident from our construction of the sequence of intervals $\{I_k\}$. The inductive proof of (iii) is left for you as an exercise.

From property (ii), we deduce that, for any $k = 0, 1, 2, \ldots$ and for any $m \geq k$, $I_m \subseteq I_k$. The inductive proof is straightforward and is also left as an exercise. As a consequence, for any k and m with $m \geq k$, we have

$$a_0 \leq a_k \leq a_m \leq b_m \leq b_k \leq b_0. \tag{1.3}$$

Let $A = \{a_k\colon k \text{ in } \mathbb{N}\}$ be the set of all left endpoints of the intervals I_k and let $B = \{b_k\colon k \text{ in } \mathbb{N}\}$ be the set of all right endpoints. Neither A nor B is empty. What's more, A is bounded above by b_0. Likewise, B is bounded below by a_0. By the Completeness Axiom for $\mathbb{R}$, it follows that $\alpha = \sup A$ and $\beta = \inf B$ exist in $\mathbb{R}$. We claim that $\alpha = \beta$ and that this common value is a limit point of S.

It is easy to see that $\alpha \leq \beta$; here's the slick, canonical proof. By (1.3), for every $k = 0, 1, 2, \ldots, a_k$ is a lower bound for B. Therefore, $a_k \leq \beta = \inf B$ for every $k \geq 0$. But this says that β is an upper bound for A. Thus $\alpha = \sup A \leq \beta$.

To show that α and β are actually equal, note first that, for any $k = 0, 1, 2, \ldots,$ we have $a_k \leq \alpha \leq \beta \leq b_k$. Thus $\beta - \alpha \leq b_k - a_k = (b_0 - a_0)/2^k$, for all $k \geq 0$. For arbitrary $\epsilon > 0$, we can use Archimedes' principle and an inductive proof to choose a k in $\mathbb{N}$ such that $(b_0 - a_0)/2^k < \epsilon$. (See Exercise 1.24.) It follows that, for every $\epsilon > 0$, we have $0 < \beta - \alpha < \epsilon$. Since $\beta - \alpha$ is a constant, we conclude by Exercise 1.10 that $\beta - \alpha = 0$. Therefore, $\alpha = \beta$. Let $x_0 = \alpha = \beta$.

We claim that x_0 must be a limit point of S. Assign any positive value to ϵ. By Archimedes' principle, choose a k in $\mathbb{N}$ such that $(b_0 - a_0)/2^k < \epsilon$. The corresponding interval I_k, of length less than ϵ, contains both x_0 and infinitely many points of S. Now, for our choice of k, we have $b_k - a_k < \epsilon$ and $a_k < x_0 < b_k$. Therefore,

$$x_0 + \epsilon > x_0 + (b_k - a_k) = b_k + (x_0 - a_k) \geq b_k.$$

Likewise, $x_0 - \epsilon < a_k$. We deduce that $I_k \subseteq N(x_0; \epsilon)$. Consequently, $N(x_0; \epsilon)$ also contains infinitely many points of S. Removing the one point x_0 still leaves infinitely many points in $N'(x_0; \epsilon) \cap S$. We deduce that, for every $\epsilon > 0$, the deleted neighborhood $N'(x_0; \epsilon)$ contains at least one point of S. Thus x_0 is a limit point of S. This proves the Bolzano–Weierstrass theorem. ●

In accord with the following definition, we say that $\mathbb{R}$ has the Bolzano–Weierstrass property.

DEFINITION 1.2.5 We say that a nonempty set X has the *Bolzano–Weierstrass property* if every bounded, infinite subset S of X has a limit point in X. ●

EXERCISES

1.23. Suppose that S is a bounded, infinite subset of $\mathbb{R}$. Let $\mu = \sup S$. Is μ necessarily a limit point of S? If so, show that, for every k in $\mathbb{N}$, there exists a point x in S such that $\mu - 1/k < x \leq \mu$. If not, exhibit a bounded, infinite set S in $\mathbb{R}$ whose supremum is not a limit point of S.

1.24. **a)** Show by induction that, in the proof of the Bolzano–Weierstrass theorem, the length of the interval I_k is $(b_0 - a_0)/2^k$.

b) Prove that, for k in $\mathbb{N}$, we have $k \leq 2^k$.

c) In the proof of the Bolzano–Weierstrass theorem, use Archimedes' principle and part (b) to prove that, for each $\epsilon > 0$, there is a k in $\mathbb{N}$ such that $(b_0 - a_0)/2^k < \epsilon$.

1.25. If x is any nonzero number and if $0 < \epsilon < |x|$, prove that only finitely many numbers of the form $(-1)^k/2^k$ are in $N(x; \epsilon)$.

1.26. Prove that a nonempty finite set has no limit points.

1.27. Find all limit points of the following sets. In each case prove that your answer is correct.

a) The set $S = \{x\colon x \text{ is irrational and } a < x < b\}$ where $a < b$

b) The set $\mathbb{Q}$ of all rational numbers

c) The set of all irrational numbers

d) The set $S = \{p/2^k\colon\ p \text{ in } \mathbb{Z}, k \text{ in } \mathbb{N}\}$

e) The set $S = \{1/m + 1/n\colon\ m, n \text{ in } \mathbb{N}\}$

1.28. In the proof of the Bolzano–Weierstrass theorem, prove inductively that, for any $k = 0, 1, 2, \ldots$ and for any $m > k$, $I_m \subseteq I_k$.

1.29. Suppose that S is an unbounded, infinite set in $\mathbb{R}$. Does S necessarily have a limit point? If so, prove it. If not, provide an example of an unbounded, infinite set having no limit points.

1.30. Suppose that S is a bounded, infinite subset of $\mathbb{R}$ having exactly one limit point x_0. Prove that, for any $\epsilon > 0$, the neighborhood $N'(x_0; \epsilon)$ contains all but finitely many of the points of S.

1.3 THE LIMIT OF A SEQUENCE

From your studies in elementary calculus you are familiar with what is meant by saying that a sequence $\{x_k\}$ of real numbers converges to a limit x_0, that is, $\lim_{k\to\infty} x_k = x_0$. Intuitively, it means that eventually x_k is arbitrarily near x_0. While we want you to build on your intuitive sense of limit, you surely understand that it alone will not suffice for your present studies of analysis. Let's take the time now to examine this concept more closely.

The English word *eventually* means "beyond some point in time." If you think of the indices 1, 2, 3, ... as a primitive model for time, then *eventually* translates into the following mathematical phrase:

There exists an index k_0 in $\mathbb{N}$ such that, if $k \geq k_0 \ldots$

Some statement $P(k)$ that depends on k is *eventually* valid if there exists a k_0 such that, for $k \geq k_0$, $P(k)$ holds true. Similarly, in this text we will translate the English word *frequently* into the phrase:

For every index k in $\mathbb{N}$, there is at least one $k_1 > k$ such that...

We will say that a statement $P(k)$ is *frequently* valid if, for every k in $\mathbb{N}$, there exists at least one $k_1 > k$ such that $P(k_1)$ holds true. Clearly, if a statement $P(k)$ is eventually valid, then it is frequently valid.

For the sake of simplicity, first suppose that the points x_k of a sequence are distinct and that x_0 differs from all of them. What does it mean to say that the set of numbers $S = \{x_1, x_2, x_3, \ldots\}$ is arbitrarily near x_0? As we saw in Section 1.2, it means that any deleted neighborhood $N'(x_0; \epsilon)$ of x_0 contains at least one and hence infinitely many points of S, that is, $0 < |x_k - x_0| < \epsilon$ for infinitely many of the indices k. Consequently, x_0 is a limit point of S.

Now, it may well happen for a given sequence that not all the x_k are distinct or that x_0 is not distinct from all of them. The sequence with $x_k = c$ for all k is an extreme example of this phenomenon, yet we surely would say that this sequence converges to the limit $x_0 = c$. The essential concept, on which we will base our study of sequences, is identified in the following definition.

DEFINITION 1.3.1 We say that c is a *cluster point* of the sequence $\{x_k\}$ if, for every $\epsilon > 0$ and every k in $\mathbb{N}$, there is a $k_1 > k$ such that x_{k_1} is in $N(c; \epsilon)$. ●

Put briefly, c is a cluster point of $\{x_k\}$ if terms of the sequence are frequently arbitrarily near c. Note that a limit point of the set $S = \{x_1, x_2, x_3, \ldots\}$ is automatically a cluster point of the sequence $\{x_k\}$. The two notions differ only in that for limit points the neighborhood is deleted, whereas for cluster points it is not. The distinction is introduced only to treat the possibility that the terms of a sequence may be repeated frequently. The set S of *numerical values* of the sequence—its range—might even be a finite set and have no limit points, whereas the sequence itself always has infinitely many terms and may have cluster points. For a simple example, let $x_k = (-1)^k$ for k in $\mathbb{N}$. Then the set S is simply $\{-1, 1\}$, consisting of two points and having no limit points. But both -1 and 1 are cluster points of the sequence $\{x_k\}$. The distinction is a simple one of which you want to be aware without putting too fine a point on it.

Consider again your intuitive notion of what it means for a sequence $\{x_k\}$ to converge to x_0: Eventually x_k is arbitrarily near x_0. First, x_0 must be a cluster point of the sequence and, more, beyond some index k_0, all the terms x_k—the *tail of the sequence*—must cluster arbitrarily near x_0. Combining these two ideas, we are led to the following definition.

DEFINITION 1.3.2 The sequence $\{x_k\}$ *converges* to x_0 and we say that x_0 is the *limit* of $\{x_k\}$ if, for each neighborhood $N(x_0; \epsilon)$, there exists an index k_0 such that, whenever $k \geq k_0$, x_k is in $N(x_0; \epsilon)$. We write $\lim_{k\to\infty} x_k = x_0$. If a sequence $\{x_k\}$ fails to converge, for whatever reason, then we say that it *diverges*. ●

Clearly, the limit x_0 of a sequence, if it exists, is a cluster point of the sequence: Given any neighborhood of the limit x_0, eventually the tail of the sequence is contained entirely in it.

It is also worth mentioning that the issue addressed by Theorem 1.2.1 arises here as well. Again, when considering the convergence of a sequence, we are interested only in *small* values of ϵ. If we can find an $\epsilon_0 > 0$ with the property that, for any positive ϵ less than ϵ_0, there exists a k_0 in $\mathbb{N}$ such that, for all $k \geq k_0$, x_k is in $\mathbb{N}(x_0; \epsilon)$, then we can conclude that $\lim_{k\to\infty} x_k = x_0$. The proof is similar to that of Theorem 1.2.1 and is left for you.

We apply the definite article to the limit x_0 because a sequence cannot converge to two different numbers.

THEOREM 1.3.1 The limit of a convergent sequence in $\mathbb{R}$ is unique.

Proof. We have to prove that if $\{x_k\}$ converges to x_0 and also converges to x_0', then $x_0 = x_0'$. If, to the contrary, we assume that $x_0 \neq x_0'$, then, by choosing ϵ to be a positive number less than $|x_0 - x_0'|/2$, we will obtain a contradiction.

Because $\lim_{k\to\infty} x_k = x_0$, we can choose a k_1 in $\mathbb{N}$ such that, when $k \geq k_1$, then x_k is in $N(x_0; \epsilon)$. Similarly, we can choose k_2 in $\mathbb{N}$ such that, when $k \geq k_2$, it follows that x_k is in $N(x_0'; \epsilon)$. Let $k_0 = \max\{k_1, k_2\}$. For any $k \geq k_0$, it follows that x_k is in both $N(x_0; \epsilon)$ and $N(x_0'; \epsilon)$. Thus, for $k \geq k_0$, we obtain the contradictory inequality

$$\begin{aligned}|x_0 - x_0'| &= |x_0 - x_k + x_k - x_0'| \leq |x_0 - x_k| + |x_k - x_0'| \\ &< \epsilon + \epsilon = 2\epsilon < |x_0 - x_0'|.\end{aligned}$$

The contradiction results from our having assumed, falsely, that $x_0 \neq x_0'$. This proves the theorem. ●

By slightly modifying the proof of Theorem 1.3.1, you can easily prove the following theorem as an exercise.

THEOREM 1.3.2 If a sequence has two (or more) cluster points, then it must diverge. ●

Our next result provides a logical fulcrum on which many of our subsequent arguments will hinge.

THEOREM 1.3.3 A convergent sequence is bounded.

Proof. Suppose that $\{x_k\}$ converges to x_0. We must show that there exists some real number M such that $|x_k| \leq M$, for $k = 1, 2, 3, \ldots$. Since we merely want a bound and any sufficiently large M will do, we can afford to be somewhat crude. Choose, for example, $\epsilon = 1$. Since $\{x_k\}$ converges to x_0, there exists an index k_0 such that, if $k \geq k_0$, then x_k belongs to $N(x_0; 1)$. For $k \geq k_0$,

$$|x_k| = |x_k - x_0 + x_0| \leq |x_k - x_0| + |x_0| < 1 + |x_0|.$$

It follows that x_k is in $N(0; |x_0| + 1)$. (See Fig. 1.10 on page 20.)

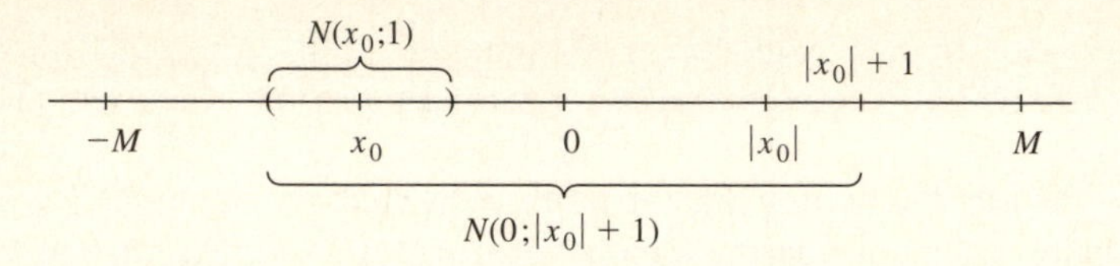

Figure 1.10

Consider the k_0 numbers $|x_1|, |x_2|, \ldots, |x_{k_0-1}|$, and $|x_0| + 1$. We let M be any real number greater than the maximum of these k_0 numbers. It is easy to see that $\{x_k\}$ must be contained in $N(0; M)$ and is therefore bounded. •

The contrapositive of Theorem 1.3.3 provides us with the following useful fact.

THEOREM 1.3.4 If a sequence is unbounded, then it must diverge. •

EXAMPLE 12 For each k in $\mathbb{N}$, let $x_k = 1/k$. Intuitively, it is evident that $\lim_{k\to\infty} x_k = 0$, but you must learn to prove all your assertions. This limit is most easily handled by using Archimedes' principle. Apply that principle to this example by fixing any $\epsilon > 0$ and choosing the smallest positive integer k_0 such that $1 < k_0\epsilon$. Equivalently, $0 < 1/k_0 < \epsilon$. For any $k \geq k_0$,

$$|x_k - 0| = x_k = 1/k \leq 1/k_0 < \epsilon.$$

Thus, for $k \geq k_0$, x_k is in $N(0; \epsilon)$. This proves that $\lim_{k\to\infty} x_k = 0$. •

EXAMPLE 13 For each k in $\mathbb{N}$, let $x_k = 1 - 1/2^k$. We claim that $\lim_{k\to\infty} x_k = 1$. To prove this rigorously, fix $\epsilon > 0$. By Archimedes' principle, choose k_0 in $\mathbb{N}$ such that $1 < \epsilon k_0$. Then, for $k \geq k_0$, we have $\epsilon k_0 < \epsilon k$. By part (b) of Exercise 1.24, we know that $k < 2^k$ for all k in $\mathbb{N}$. Thus, for $k \geq k_0$, we have $1 < \epsilon k < \epsilon 2^k$; it follows that $0 < 1/2^k < \epsilon$. Therefore, for $k \geq k_0$,

$$|x_k - 1| = \left|\left(1 - \frac{1}{2^k}\right) - 1\right| = \frac{1}{2^k} < \epsilon.$$

Thus, x_k is in $N(1; \epsilon)$. This proves that $\lim_{k\to\infty}(1 - 1/2^k) = 1$. •

EXAMPLE 14 The sequence defined by $x_k = (-1)^k$ has two cluster points and therefore diverges. •

EXAMPLE 15 Let $\{x_k\}$ be any sequence that converges to 0 and let c be any positive number. Let $y_k = (-1)^k(c + x_k)$. We claim that both c and $-c$ are cluster points of $\{y_k\}$ and that the sequence must diverge. (See Fig. 1.11.) To show that c is a cluster point of $\{y_k\}$, choose ϵ such that $0 < \epsilon < c$. Since $\lim_{k\to\infty} x_k = 0$, we can choose k_0 in $\mathbb{N}$ such that, for $k \geq k_0$, $|x_k| < \epsilon$. For values of $k \geq k_0$ that

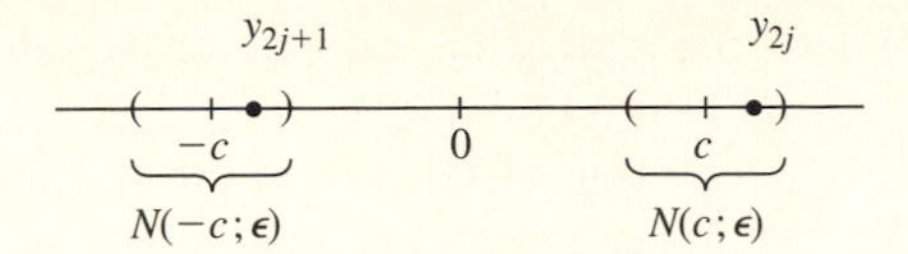

Figure 1.11

are even—that is, with $k = 2j$—the term y_k of the sequence is in $N(c; \epsilon)$ as the following computation shows:

$$|y_{2j} - c| = |(c + x_{2j}) - c| = |x_{2j}| < \epsilon.$$

Therefore c is a cluster point of $\{y_k\}$. Similarly, by considering the terms of $\{y_k\}$ with odd index, you can show that $-c$ is also a cluster point of $\{y_k\}$. Having these two cluster points, $\{y_k\}$ must diverge. ●

EXAMPLE 16 The sequence defined by $x_k = k$ must diverge since it is unbounded. ●

Notice that, when confirming that a sequence $\{x_k\}$ converges to a limit x_0, we first specify an arbitrary $\epsilon > 0$ and subsequently identify the value of k_0 (usually in terms of ϵ) with the property that, for $k \geq k_0$, the distance from x_k to x_0 is certain to be less than the prescribed ϵ. Finding the correct k_0 may involve algebraic manipulations, the application of appropriate inequalities, relevant information about the sequence under consideration, and any of the mathematical tools in your kit. Often, when finding k_0, you will find it useful to work *backward.* Having found k_0 in this way, you must then provide a *forward* proof to confirm that it has the required property. Learning to write correct mathematics in the modern style is tricky, a skill that will require practice. There is some slight room for expressing your individual style, but not much. Our next two examples are intended to show you how to find the key to solving a problem by first working backward. Then, by reversing the steps, we present a formal, forward proof.

EXAMPLE 17 For each k in $\mathbb{N}$, define $x_k = k/(k+1)$. Evidently, $\lim_{k\to\infty} x_k = 1$. We will prove this using our current language. We begin by working backward. Fix $\epsilon > 0$. We want to show that, eventually, $|x_k - 1| = |k/(k+1) - 1| < \epsilon$ or, equivalently, that

$$1 - \epsilon < \frac{k}{k+1} < 1 + \epsilon.$$

We already know that $k/(k+1) < 1$, so the right-hand inequality will be satisfied automatically. Algebraically manipulating the left-hand inequality, we obtain

$$(k+1)(1-\epsilon) < k$$

or, equivalently,

$$1 - \epsilon < k - k(1-\epsilon) = k\epsilon.$$

Thus, working backward, we recognize that the appropriate value of k_0 is the smallest natural number greater than $(1-\epsilon)/\epsilon$.

Now begins the forward version of our proof; our discussion to this point has been preparation for the following, correct presentation of our argument. Fix $\epsilon > 0$ and choose k_0 to be the smallest natural number greater than $(1-\epsilon)/\epsilon$. We want to confirm that, for any $k \geq k_0$, x_k is in $N(1;\epsilon)$. Choose any $k \geq k_0 > (1-\epsilon)/\epsilon$. Multiplying by ϵ yields $k\epsilon > 1-\epsilon$. Therefore

$$k - k(1-\epsilon) = k\epsilon > 1-\epsilon.$$

It follows that $k > k(1-\epsilon) + (1-\epsilon) = (k+1)(1-\epsilon)$. Dividing by $k+1$, we obtain

$$\frac{k}{k+1} > 1-\epsilon.$$

Combine this last inequality with $k/(k+1) < 1$ to deduce that $x_k = k/(k+1)$ is in $N(1;\epsilon)$. We conclude that $\lim_{k\to\infty} k/(k+1) = 1$ as initially asserted. ●

EXAMPLE 18 Fix any r such that $-1 < r < 1$. We claim that $\lim_{k\to\infty} r^k = 0$. First we assume that $0 < r < 1$ and assign an arbitrary positive value to ϵ. We need to exhibit an index k_0 with the property that, if $k \geq k_0$, then r^k is in $N(0;\epsilon)$. Since $r > 0$, we need only show that $0 < r^k < \epsilon$. First note that, since $0 < r < 1$, it follows that $1/r > 1$. Let $c = (1/r) - 1$. Observe that $c > 0$ and that $r = 1/(1+c)$. By Bernoulli's inequality (Exercise 1.16), we know that, for all k in $\mathbb{N}$, $(1+c)^k \geq 1 + kc$. Therefore

$$r^k = \frac{1}{(1+c)^k} \leq \frac{1}{1+kc} < \frac{1}{ck},$$

for all k in $\mathbb{N}$. These last inequalities provide us with the needed clue to the resolution of the question. By Archimedes' principle, choose k_0 in $\mathbb{N}$ such that $1/(\epsilon c) < k_0$.

For the forward version of our proof, we fix $\epsilon > 0$. Let k_0 be the smallest natural number greater than $1/(\epsilon c)$, where $c = (1/r) - 1$. For $k \geq k_0$, we have $k > 1/(\epsilon c)$; equivalently, $1/(ck) < \epsilon$. For such k, we have

$$0 < r^k = \frac{1}{(1+c)^k} \leq \frac{1}{1+kc} < \frac{1}{ck} < \epsilon.$$

Thus, when $k \geq k_0$, r^k is in $N(0;\epsilon)$. This proves that $\lim_{k\to\infty} r^k = 0$.

In case $r = 0$, it is immediate that $\lim_{k\to\infty} r^k = 0$. If $-1 < r < 0$, then the convergence of $\{r^k\}$ to 0 follows from a more general fact contained in the next theorem. Its proof is left as an exercise. ●

THEOREM 1.3.5 A sequence $\{x_k\}$ converges to 0 if and only if $\{|x_k|\}$ converges to 0 also. ●

An elementary fact that we will use frequently is contained in the following theorem. Its proof is straightforward and is left as an exercise.

THEOREM 1.3.6 If $\{x_k\}$ converges to x_0 and if c is any constant, then $\{cx_k\}$ converges to cx_0. ●

Perhaps you have noticed in your first studies of sequences that, in practice, resolving the question of convergence of a specific sequence separates into two distinct problems. One is the issue of convergence itself without regard to the possible value of the limit; the other is the identification of that limit. These two facets will persist in your present studies.

You may also have noticed that relatively few general techniques are available for resolving the questions of convergence and the determination of the value of the limit. Several methods will be discussed in this chapter and in Chapter 11 dealing with infinite series. Generally speaking, each sequence must be dealt with individually, in terms of its particular definition and characteristics. The following principle flows directly from the Axiom 1.1.1 and enables you to decide convergence.

THEOREM 1.3.7 A bounded, monotone sequence of real numbers converges. ●

Recall that $\{x_k\}$ is said to be a *monotone increasing* sequence if $x_k \leq x_{k+1}$ for all k in $\mathbb{N}$. If this latter inequality is reversed, then $\{x_k\}$ is said to be *monotone decreasing.* The sequence is said to be *monotone* if it is either monotone increasing or monotone decreasing. If the inequalities are strict, then the sequence is said to be *strictly monotone.*

The theorem surely feels intuitively plausible to you. Indeed, it is an alternate form of the completeness of $\mathbb{R}$. The strategy for its proof follows from basic principles, for if $\{x_k\}$ is a bounded, monotone increasing sequence, then it must converge to $\sup \{x_k\colon k \text{ in } \mathbb{N}\}$. Likewise, if $\{x_k\}$ is a bounded, monotone decreasing sequence, then it converges to $\inf \{x_k\colon k \text{ in } \mathbb{N}\}$. The proofs of these assertions depend on Theorem 1.1.1 and are left as an exercise. Generally, however, this theorem provides no clue about the value of the limit except to say it is the supremum or the infimum of the sequence as the case may be; its value may or may not be readily apparent. Of course, many sequences, even if bounded, are not monotone; this theorem applies only to monotone sequences.

Perhaps the most useful, basic tool is the so-called pinching theorem or the Squeeze Play.

THEOREM 1.3.8 **The Squeeze Play** If $\{x_k\}$, $\{y_k\}$, and $\{z_k\}$ are three sequences such that $x_k \leq y_k \leq z_k$, for k in $\mathbb{N}$, and if $\lim_{k\to\infty} x_k$ and $\lim_{k\to\infty} z_k$ exist and are equal, then $\lim_{k\to\infty} y_k$ also exists and is equal to this common value.

The proof of this principle relies entirely on the definition of limit. The sequences to which it is applied need not be monotone. Hence this principle is more primitive and more widely useful. The difficulty in using it to determine the limit of the given sequence $\{y_k\}$ is that, from knowledge about $\{y_k\}$, you must create the sequences $\{x_k\}$ and $\{z_k\}$ to satisfy the conditions of the theorem. Clearly, you must understand the behavior of $\{y_k\}$ rather well and be able to maneuver rather skillfully to meet these requirements. Still, several classic limits can be obtained by this method.

Proof of Theorem 1.3.8. To prove the Squeeze Play, let L denote the common limit of $\{x_k\}$ and $\{z_k\}$. We must show that eventually y_k is arbitrarily near L. How do we proceed? First, we assign an arbitrarily small, positive value to ϵ. We need to confirm the existence of an index k_0 such that, for $k \geq k_0$, the number y_k is in $N(L; \epsilon)$.

We use what we know. Because $\lim_{k\to\infty} x_k = L$, there exists an index k_1 such that, for $k \geq k_1$, x_k is in $N(L; \epsilon)$. Likewise, there exists an index k_2 such that, for $k \geq k_2$, z_k is in $N(L; \epsilon)$. If we let $k_0 = \max\{k_1, k_2\}$, then both conditions hold simultaneously. For $k \geq k_0$, both x_k and z_k are in $N(L; \epsilon)$. Since, for all k, the number y_k is trapped between x_k and z_k it must follow that for $k \geq k_0$, y_k is in $N(L; \epsilon)$ also. (See Fig. 1.12.) In other words, $\lim_{k\to\infty} y_k = L$. ●

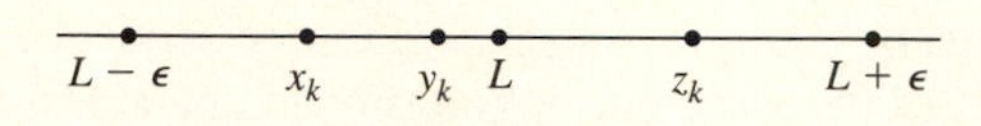

Figure 1.12

In a similar vein we have the following theorem whose proof is left as an exercise.

THEOREM 1.3.9 Limit is order-preserving on convergent sequences. That is, if both $\{x_k\}$ and $\{y_k\}$ converge and if for each k in $\mathbb{N}$ we have $x_k \leq y_k$, then $\lim_{k\to\infty} x_k \leq \lim_{k\to\infty} y_k$. ●

It is worth noting that the hypothesized inequalities in Theorems 1.3.7, 1.3.8, and 1.3.9 need only apply to the *tail* of the sequences involved, that is, for k greater than some index k_1. The proofs of these assertions are assigned in the exercises.

EXAMPLE 19 Fix $r \neq 0$ in $\mathbb{R}$. We claim that $\lim_{k\to\infty} r^k/k! = 0$. To verify our claim, choose k_0 to be the smallest natural number greater than $\lfloor |r| \rfloor$, the integer part of $|r|$. Let $R = |r|/(k_0 + 1)$ and note that $0 < R < 1$. Let c denote the constant $(k_0 + 1)^{k_0}/k_0!$. Fix any $k \geq k_0$. Notice that, for any $j = 1, 2, \ldots, k - k_0$, $1/(k_0 + j) \leq 1/(k_0 + 1)$. Thus we obtain

$$
\begin{aligned}
0 < \frac{|r|^k}{k!} &= |r|^k \frac{1}{k_0!} \cdot \frac{1}{(k_0+1)(k_0+2)\cdots(k-1)k} \\
&\leq |r|^k \frac{1}{k_0!} \cdot \frac{1}{(k_0+1)^{k-k_0}} \\
&= \frac{(k_0+1)^{k_0}}{k_0!} \cdot \left[\frac{|r|}{k_0+1}\right]^k = cR^k.
\end{aligned}
$$

By Example 18 and Theorem 1.3.6, we know that $\lim_{k\to\infty} cR^k = 0$. Therefore we have $0 \leq \lim_{k\to\infty} |r|^k/k! \leq \lim_{k\to\infty} cR^k = 0$. By the Squeeze Play (Theorem 1.3.8), we deduce that $\lim_{k\to\infty} |r|^k/k! = 0$. Invoking Theorem 1.3.5, we conclude that $\lim_{k\to\infty} r^k/k! = 0$ and our initial claim is proved. ●

EXAMPLE 20 For k in $\mathbb{N}$, let

$$x_k = 1 + \frac{1}{1!} + \frac{1}{2!} + \frac{1}{3!} + \cdots + \frac{1}{k!}$$

and

$$y_k = \left(1 + \frac{1}{k}\right)^k.$$

We will show that $\lim_{k\to\infty} x_k = \lim_{k\to\infty} y_k$. The common limit is the irrational number e discussed in Examples 6 and 7 of Section 1.1.

It is easy to see that $\{x_k\}$ is strictly monotone increasing. We saw in Example 6 of Section 1.1 that $\{x_k\}$ is bounded above by 3. Consequently, by Theorem 1.3.7, $\{x_k\}$ converges to $\sup\{x_k\colon k \text{ in } \mathbb{N}\}$. We denote this limit by e.

Consider next the sequence $\{y_k\}$. In Exercise 1.20 we showed that

$$y_k = 1 + 1 + \sum_{j=2}^{k} \frac{1}{j!}\left(1 - \frac{1}{k}\right)\left(1 - \frac{2}{k}\right)\cdots\left(1 - \frac{j-1}{k}\right) \tag{1.4}$$

and that, for every k in $\mathbb{N}$, $y_k \le x_k \le e$. Thus $\{y_k\}$ is bounded above.

Notice that $1 - i/k < 1 - i/(k+1)$, for every $i = 1, 2, \ldots, j-1$. From this it follows that, for every $j = 2, 3, \ldots, k$,

$$\frac{1}{j!}\prod_{i=1}^{j-1}\left(1 - \frac{i}{k}\right) < \frac{1}{j!}\prod_{i=1}^{j-1}\left(1 - \frac{i}{k+1}\right),$$

where $\prod_{i=1}^{p} a_i$ denotes the product $a_1 a_2 \cdots a_p$. Referring to (1.4), observe that

$$y_k = 1 + 1 + \sum_{j=2}^{k} \frac{1}{j!}\prod_{i=1}^{j-1}\left(1 - \frac{i}{k}\right) < 1 + 1 + \sum_{j=2}^{k} \frac{1}{j!}\prod_{i=1}^{j-1}\left(1 - \frac{i}{k+1}\right)$$

$$< 1 + 1 + \sum_{j=2}^{k} \frac{1}{j!}\prod_{i=1}^{j-1}\left(1 - \frac{i}{k+1}\right) + \frac{1}{(k+1)!}\prod_{i=1}^{k}\left(1 - \frac{i}{k+1}\right).$$

This last sum is simply formula (1.4) for y_{k+1}, showing that $y_k < y_{k+1}$ and proving that $\{y_k\}$ is strictly monotone increasing. Since the sequence $\{y_k\}$ is bounded above by e, we are guaranteed by Theorem 1.3.7 that $\{y_k\}$ converges. Denote the limit of $\{y_k\}$ by y_0. Since $y_k \le x_k$ for all k, we can immediately deduce by Theorem 1.3.9 that

$$y_0 \le \lim_{k\to\infty} x_k = e.$$

To obtain the reverse inequality, we create a third sequence from the fragments of information at hand. Fix any k in $\mathbb{N}$. For each $m > k$ define

$$z_m = 1 + 1 + \sum_{j=2}^{k} \frac{1}{j!}\prod_{i=1}^{j-1}\left(1 - \frac{i}{m}\right).$$

This choice of z_m is motivated in part by the observation that $\lim_{m\to\infty} z_m = x_k$.[2]

[2] We verify this observation using the algebraic properties of limit discussed in Section 1.5: As m tends to ∞, each factor in $\prod_{i=1}^{j-1}[1 - i/m]$ tends to 1. There are $j-1$ factors; thus $\lim_{m\to\infty}(1/j!)\prod_{i=1}^{j-1}[1 - i/m] = 1/j!$.

Referring again to (1.4), note that z_m is obtained from y_m by omitting the final $m - k$ positive summands of y_m. Thus, for every $m > k$, we have $z_m < y_m$. Since $\lim_{m\to\infty} z_m = x_k$ and $\lim_{m\to\infty} y_m = y_0$ we deduce by Theorem 1.3.9 that $x_k \le y_0$ for each k in $\mathbb{N}$. Thus y_0 is an upper bound for the sequence $\{x_k\}$. Consequently, we know that

$$e = \lim_{k\to\infty} x_k = \sup\{x_k\colon k \text{ in } \mathbb{N}\} \le y_0.$$

Since we already know that $y_0 \le e$, we conclude that $y_0 = e$. We have proved that $\lim_{k\to\infty}(1 + 1/k)^k = \lim_{k\to\infty} \sum_{j=0}^{k} 1/j! = e$.

To approximate e to seven-place accuracy, take $k = 10$ and use your calculator to compute $x_{10} = \sum_{j=0}^{10} 1/j! \doteq 2.7182818$. By contrast, $y_{10} = (1 + 1/10)^{10} = 2.5937425$. This contrast exhibits the fact that $\{x_k\}$ converges to e more rapidly than does $\{y_k\}$. To achieve seven-place accuracy using the sequence $\{y_k\}$ we must take k much larger than 10. Nevertheless, by taking k sufficiently large, y_k can be made to differ from e by as little as we may wish. ●

To refine our understanding of the distinction between convergence and divergence, it is helpful to explore another concept, that of a subsequence of a sequence.

DEFINITION 1.3.3 Let $\{x_k\}$ be any sequence. Choose any strictly monotone increasing sequence $k_1 < k_2 < k_3 < \cdots$ of natural numbers. For each j in $\mathbb{N}$, let $y_j = x_{k_j}$. The sequence $\{y_j\} = \{x_{k_j}\}$ is called a *subsequence* of $\{x_k\}$. ●

A subsequence is extracted from $\{x_k\}$ simply by choosing some infinite number of the terms of the sequence, omitting others, and retaining the order of the indices inherited from $\mathbb{N}$. For example, if $x_k = (-1)^k(1 + 1/k)$, then one subsequence can be chosen by letting $k_j = 2j$; by these choices we select the subsequence $\{x_2, x_4, x_6, \ldots\}$ with even indices. This subsequence converges to 1. Another subsequence can be chosen by letting $k = 2j - 1$ and by extracting the subsequence with odd indices. This second subsequence converges to -1. Yet a third subsequence can be chosen by letting $k = 3j$; the subsequence $\{x_3, x_6, x_9, \ldots\}$ diverges, since it has both -1 and 1 as cluster points. Clearly, any sequence has an enormous number of subsequences. In practice, we will be interested in extracting only those subsequences that have some desired property.

THEOREM 1.3.10 The point c is a cluster point of $\{x_k\}$ if and only if there exists a subsequence $\{x_{k_j}\}$ that converges to c.

Proof. The proof introduces you to a classic technique that you may find useful elsewhere. It involves recursively selecting the desired subsequence using the fact that c is a cluster point of $\{x_k\}$. Let $\{\epsilon_k\}$ be any sequence of positive numbers that decreases strictly monotonically to 0. Choose k_1 in $\mathbb{N}$ such that x_{k_1} is in $N(c; \epsilon_1)$. By the definition of cluster point, having chosen k_1, we can choose $k_2 > k_1$ such that x_{k_2} is in $N(c; \epsilon_2)$. In general, suppose that we have chosen $k_1 < k_2 < \cdots < k_j$ such that, for $i = 1, 2, \ldots, j$, the point x_{k_i} is in $N(c; \epsilon_i)$. By the definition of a cluster point, choose an index $k_{j+1} > k_j$ such that $x_{k_{j+1}}$ is in $N(c; \epsilon_{j+1})$. By this procedure, we recursively extract a subsequence $\{x_{k_j}\}$ of $\{x_k\}$.

We have to show that $\lim_{j\to\infty} x_{k_j} = c$. To this end, first fix any $\epsilon > 0$. Since $\lim_{j\to\infty} \epsilon_j = 0$, there is a j_0 such that for $j \geq j_0$, we have $0 < \epsilon_j < \epsilon$. Therefore, $N(c; \epsilon_j) \subset N(c; \epsilon)$ for all $j \geq j_0$. Now, x_{k_j} is in $N(c; \epsilon_j)$. Thus, for all $j \geq j_0$, the point x_{k_j} is in $N(c; \epsilon)$. We conclude that $\{x_{k_j}\}$ converges to c as desired.

Conversely, suppose that $\{x_{k_j}\}$ is a convergent subsequence of $\{x_k\}$ with $\lim_{j\to\infty} x_{k_j} = c$. We must show that c is a cluster point of $\{x_k\}$. Fix $\epsilon > 0$ and choose a j_0 such that, if $j \geq j_0$, then x_{k_j} is in $N(c; \epsilon)$. Choose any k in $\mathbb{N}$ and choose any $j \geq j_0$ such that $k_j > k$. Then x_{k_j} is in $N(c; \epsilon)$. Therefore c is a cluster point of $\{x_k\}$. ●

Theorem 1.3.10 enables us to prove the following theorem; it is a key link in our overall program and deserves your close attention.

THEOREM 1.3.11 Any bounded sequence has a cluster point.

Proof. Let $S = \{x_1, x_2, x_3, \ldots\}$ denote the set of numerical values of the sequence. S is a bounded set in $\mathbb{R}$. If S is infinite, then the Bolzano–Weierstrass theorem guarantees the existence of a limit point of S. Any limit point of S is a cluster point of $\{x_k\}$ and the result follows.

If S is finite, then at least one of the values of the sequence must be repeated infinitely often. Choosing the subsequence of all terms having that value produces a constant, convergent subsequence whose limit is a cluster point of $\{x_k\}$. This proves the theorem. ●

By recasting Theorem 1.3.11 as a contrapositive, we have the following theorem.

THEOREM 1.3.12 If a sequence has no cluster points, then it is unbounded. ●

Elegant characterizations of convergence are provided by the next two theorems. The proofs are straightforward and are left as exercises.

THEOREM 1.3.13 A sequence $\{x_k\}$ converges to x_0 if and only if every subsequence of $\{x_k\}$ converges to x_0. ●

THEOREM 1.3.14 A bounded sequence converges if and only if it has exactly one cluster point. ●

In turn we can now formulate the following simple and useful characterization of divergence.

THEOREM 1.3.15 A sequence $\{x_k\}$ diverges if and only if at least one of the following conditions holds:

i) $\{x_k\}$ has at least two cluster points.

ii) $\{x_k\}$ is unbounded.

Proof. Suppose first that $\{x_k\}$ is bounded and divergent. Since it is bounded, $\{x_k\}$ has at least one cluster point by Theorem 1.3.11. If it had exactly one cluster point, then by Theorem 1.3.14 it must converge. The contradiction proves it must have at least two cluster points, so if $\{x_k\}$ is divergent, then either (i) or (ii) must hold.

Conversely, suppose first that $\{x_k\}$ has at least two cluster points. By Theorem 1.3.2, $\{x_k\}$ diverges. Suppose next that $\{x_k\}$ is unbounded. Then the sequence diverges by Theorem 1.3.4. ●

There is yet another way to view the question of convergence of a sequence, one that often provides powerful insight. Given a sequence, let C denote the set of all its cluster points. As we have seen, C might be empty or consist of a single point or several points. If the sequence is bounded, then, by Theorem 1.3.11, C is nonempty. Further, if the sequence is bounded, then the set C of its cluster points is also bounded. You will be challenged to prove this as an exercise. Therefore, if the sequence is bounded, then, by Axiom 1.1.1, $\sup C$ and $\inf C$ exist.

DEFINITION 1.3.4 Let C denote the set of cluster points of a bounded sequence $\{x_k\}$.

i) The *limit superior* of $\{x_k\}$ is defined to be $\sup C$ and is denoted $\limsup x_k$.

ii) The *limit inferior* of $\{x_k\}$ is defined to be $\inf C$ and is denoted $\liminf x_k$. ●

If $\{x_k\}$ is unbounded above, we set $\limsup x_k = \infty$. Likewise, if $\{x_k\}$ is unbounded below, we set $\liminf x_k = -\infty$. If $\{x_k\}$ is bounded above but unbounded below and if C is empty, we set $\limsup x_k = -\infty$. (For example, let $x_k = -k$ for each k in $\mathbb{N}$.) Likewise, if $\{x_k\}$ is bounded below but unbounded above and if C is empty, we set $\liminf x_k = \infty$. (For example, let $x_k = k$ for each k in $\mathbb{N}$.) These definitions cover all possible cases and give meaning to limit superior and limit inferior for all sequences.

We want to explore the properties of limit superior and limit inferior to discover how these concepts might provide us with insight about sequences. First, it is evident from the definitions that, in all cases $\liminf x_k \leq \limsup x_k$.

Suppose that $\mu = \limsup x_k$ is finite. We claim that μ is a cluster point of $\{x_k\}$. To see this, fix any $\epsilon > 0$ and k in $\mathbb{N}$. We need to show that there is some $k_1 > k$ such that x_{k_1} is in $N(\mu; \epsilon)$. The finiteness of μ implies that C is nonempty and bounded above and that $\mu = \sup C$. By Theorem 1.1.1, we can choose a cluster point c in C such that $\mu - \epsilon < c \leq \mu$. If $c = \mu$, then μ is in C and we are finished. Otherwise, $\mu - \epsilon < c < \mu$. Choose any positive ϵ_1 less than $\min\{c - (\mu - \epsilon), \mu - c\}$. (See Fig. 1.13.) Note that $\epsilon_1 < \epsilon$. Because c is a cluster point of $\{x_k\}$, there exists a $k_1 > k$ such that x_{k_1} is in $N(c; \epsilon_1)$. But $N(c; \epsilon_1)$ is contained entirely in $N(\mu; \epsilon)$. Therefore x_{k_1} is in $N(\mu; \epsilon)$. Thus μ is a cluster point of $\{x_k\}$. That is, μ is in C. A completely analogous argument proves the second assertion of the following theorem.

THEOREM 1.3.16 Let $\{x_k\}$ be any sequence.

i) If $\mu = \limsup x_k$ is finite, then μ is in C.

ii) If $\nu = \liminf x_k$ is finite, then ν is in C. ●

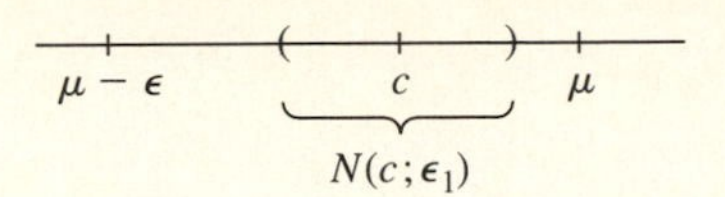

Figure 1.13

Intuitively, $\limsup x_k$ and $\liminf x_k$ are the largest and the smallest cluster points, respectively, of a bounded sequence $\{x_k\}$. To be precise, suppose that $\mu = \limsup x_k$ is finite. Fix any positive ϵ. There can exist only finitely many terms of the sequence to the right of $\mu + \epsilon$; otherwise, there would be a cluster point of $\{x_k\}$—that is, a point of C—greater than $\sup C$. We deduce that there must exist a k_0 such that, for $k \geq k_0$, we have $x_k < \mu + \epsilon$. That is, the terms of the sequence $\{x_k\}$ are eventually less than $\mu + \epsilon$. (See Fig. 1.14.)

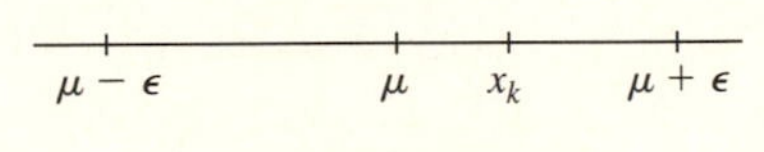

Figure 1.14

On the other hand, since $\mu = \limsup x_k$ is a cluster point of $\{x_k\}$, given any $\epsilon > 0$, there must exist infinitely many terms of the sequence to the right of $\mu - \epsilon$. Thus given any k, there exists a $k_1 > k$ such that $x_{k_1} > \mu - \epsilon$. The terms of the sequence $\{x_k\}$ are frequently greater than $\mu - \epsilon$. (See Fig. 1.15.) Analogous reasoning applied to $\nu = \liminf x_k$ yields analogous conclusions and proves our next theorem.

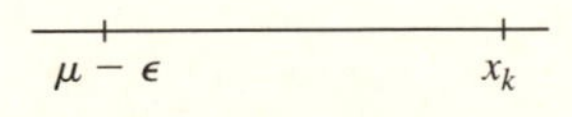

Figure 1.15

THEOREM 1.3.17 Let $\{x_k\}$ be any bounded sequence in $\mathbb{R}$. Fix any $\epsilon > 0$.

i) Let $\mu = \limsup x_k$.

- **a)** There exists a k_0 such that for all $k \geq k_0$, $x_k < \mu + \epsilon$.
- **b)** For any k, there exists a $k_1 > k$ such that $x_{k_1} > \mu - \epsilon$.

ii) Let $\nu = \liminf x_k$.

- **a)** There exists a k_0 such that for all $k \geq k_0$, $x_k > \nu - \epsilon$.
- **b)** For any k, there exists a $k_1 > k$ such that $x_{k_1} < \nu + \epsilon$. ●

Finally, we have the following complete description of convergence and divergence of any sequence expressed in terms of the properties of $\liminf x_k$ and $\limsup x_k$. Its proof is a summary of all we have achieved in this rather lengthy section and will provide you with an exercise to test your understanding.

THEOREM 1.3.18 Let $\{x_k\}$ be any sequence in $\mathbb{R}$.

i) $\{x_k\}$ converges to x_0 if and only if

$$\liminf x_k = \limsup x_k = x_0.$$

ii) $\{x_k\}$ diverges if and only if one of the following conditions holds:

a) Either $\liminf x_k$ or $\limsup x_k$ is infinite.

b) Both $\liminf x_k$ and $\limsup x_k$ are finite and $\liminf x_k < \limsup x_k$.

●

As we know, if $\{x_k\}$ is unbounded, then it must diverge. Two special cases deserve attention. If, for every $M > 0$, there is a k_0 such that $x_k > M$ for all $k \geq k_0$, then we say that $\{x_k\}$ *diverges to* $+\infty$. We write $\lim_{k\to\infty} x_k = \infty$ (Fig. 1.16). In this case, $\liminf x_k = \limsup x_k = \infty$. If $\lim_{k\to\infty} x_k = \infty$, then eventually the terms of the sequence become arbitrarily large and positive.

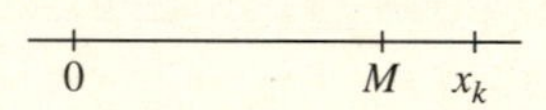

Figure 1.16

Likewise, if, for every $M > 0$, there is a k_0 such that $x_k < -M$ for all $k \geq k_0$, then we say that $\{x_k\}$ *diverges to* $-\infty$. (See Fig. 1.17.) We write $\lim_{k\to\infty} x_k = -\infty$. In this case, $\liminf x_k = \limsup x_k = -\infty$. If $\lim_{k\to\infty} x_k = -\infty$, then eventually the terms of the sequence become arbitrarily large in magnitude and negative.

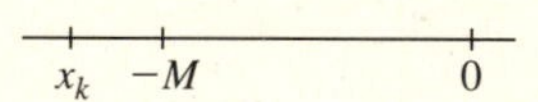

Figure 1.17

EXAMPLE 21 For each k in $\mathbb{N}$ let $x_k = k/(\sqrt{k} + 3)$. We show that $\{x_k\}$ diverges to ∞. Fix $M > 0$. Our task is to find a k_0 such that, for $k \geq k_0$, $x_k > M$. To this end, notice that, for $k > 9$, we have $k/(\sqrt{k} + 3) > k/(2\sqrt{k}) = \sqrt{k}/2$. (See Fig. 1.18.) We need merely choose k_0 in $\mathbb{N}$ to be greater than $\max\{9, 4M^2\}$. For $k > k_0$, we have

$$x_k = \frac{k}{\sqrt{k} + 3} > \frac{\sqrt{k}}{2} > \frac{\sqrt{k_0}}{2} > M.$$

We conclude that $\lim_{k\to\infty} x_k = \infty$. ●

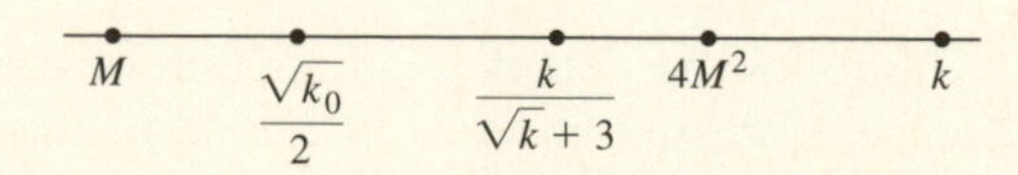

Figure 1.18

EXAMPLE 22 For each k in $\mathbb{N}$ let $x_k = \sum_{j=1}^{k} 1/j$. It is evident that $\{x_k\}$ is strictly monotone increasing. Its convergence hinges entirely on whether it is bounded. We show here that this sequence is unbounded above and hence that $\lim_{k\to\infty} \sum_{j=1}^{k} 1/j = \infty$. Fix $M > 0$. Consider indices of the form $k = 2^m$. We will show that, for sufficiently large m, $x_{2^m} > M$. Working backward, we group the summands as follows:

$$x_{2^m} = 1 + \frac{1}{2} + \left(\frac{1}{3} + \frac{1}{4} + \frac{1}{5} + \frac{1}{6} + \frac{1}{7} + \frac{1}{8}\right) + \cdots$$
$$+ \cdots + \left(\frac{1}{2^{m-1}+1} + \frac{1}{2^{m-1}+2} + \cdots + \frac{1}{2^m}\right)$$
$$= 1 + \sum_{p=1}^{m} \left[\sum_{j=1}^{2^{p-1}} \frac{1}{2^{p-1}+j}\right]. \tag{1.5}$$

For each $j = 1, 2, 3, \ldots, 2^{p-1}$, we have $1/(2^{p-1} + j) > 1/2^p$. Therefore

$$\sum_{j=1}^{2^{p-1}} \frac{1}{2^{p-1}+j} > 2^{p-1}\left(\frac{1}{2^p}\right) = \frac{1}{2}.$$

It follows from (1.5) that $x_{2^m} > 1 + m(1/2) = 1 + m/2$. Now we can identify the required index k_0. The inequality $1 + m/2 > M$ is equivalent to $m > 2(M - 1)$. Choose m_0 to be the smallest natural number greater than $2(M - 1)$ and let $k_0 = 2^{m_0}$. For any $k > k_0 = 2^{m_0}$,

$$x_k = (x_k - x_{k_0}) + x_{k_0} > x_{k_0} > 1 + \frac{m_0}{2} > M.$$

Consequently, $\lim_{k\to\infty} x_k = \infty$. ●

Of course, a divergent sequence need not be of one of these pure types; it can be both unbounded and yet have several cluster points. For example, we could define,

$$x_k = \begin{cases} k, & \text{if } k = 3j \\ 1/k, & \text{if } k = 3j - 1 \\ (-1)^k(1 + 1/k), & \text{if } k = 3j - 2 \end{cases}$$

for j in $\mathbb{N}$. The sequence $\{x_k\}$ thus defined is unbounded above; hence $\limsup x_k = \infty$ and it must diverge. This sequence has three cluster points and $\liminf x_k = -1$. Therefore the sequence does not diverge to ∞.

EXERCISES

1.31. For each of the following sequences show that $\lim_{k\to\infty} x_k = x_0$ where x_0 is the specified limit. Assign an arbitrary positive value to ϵ. Algebraically solve the relevant inequalities to find the smallest index k_0 such that x_{k_0} is in $N(x_0; \epsilon)$. Then reverse your steps to provide a detailed forward proof that, for each $k \geq k_0$, x_k is in $N(x_0; \epsilon)$.

a) Let $x_k = 1/(k^2 + 2)$ and $x_0 = 0$.

b) Let $x_k = (3k + 1)/(2k + 5)$ and $x_0 = 3/2$.

c) Let $x_k = (2k^2 - 1)/(k^2 + 3)$ and $x_0 = 2$.

d) Let $x_k = \sqrt{4k/(k+1)}$ and $x_0 = 2$.

e) Let $x_k = \sqrt{k}/(2\sqrt{k}+1)$ and $x_0 = 1/2$.

1.32. Use the Squeeze Play and Archimedes' principle to prove that $\lim_{k\to\infty} \sin(\sqrt{k})/\sqrt{k} = 0$.

1.33. Fix c such that $0 < c < 1$. Show that $\lim_{k\to\infty} kc^k = 0$ by completing the following steps.

a) Show that there exists a $b > 0$ such that $c = 1/(1+b)$.

b) Use the binomial theorem to show that, for $k \geq 2$,

$$(1+b)^k > \tfrac{1}{2}k(k-1)b^2.$$

c) Show that, for $k \geq 2$, $kc^k < 2/[(k-1)b^2]$.

d) Use Archimedes' principle to show that

$$\lim_{k\to\infty} \frac{2}{(k-1)b^2} = 0.$$

e) Use the Squeeze Play to show that $\lim_{k\to\infty} kc^k = 0$.

1.34. Fix c such that $0 < c < 1$. Modify Exercise 1.33 to show that $\lim_{k\to\infty} k^2c^k = 0$.

1.35. Complete the following steps proving that $\lim_{k\to\infty} k^{1/k} = 1$:

a) For each $k \geq 2$, show that there exists a positive y_k such that $k^{1/k} = 1 + y_k$.

b) Use the binomial theorem to show that, for $k \geq 2$,

$$k > 1 + \tfrac{1}{2}k(k-1)y_k^2.$$

c) Show that $y_k < \sqrt{2/k}$ for all $k \geq 2$.

d) Use Archimedes' principle to show that $\lim_{k\to\infty} \sqrt{2/k} = 0$.

e) Use the Squeeze Play to show that $\lim_{k\to\infty} k^{1/k} = 1$.

1.36. Prove that $\lim_{k\to\infty} k^2/k! = 0$ by completing the following steps:

a) Show that, for $k \geq 3$, $k^2/k! \leq k/(k^2-3k+2)$.

b) Use the Squeeze Play to conclude that $\lim_{k\to\infty} k^2/k! = 0$.

1.37. Fix $c > 0$. Prove that $\lim_{k\to\infty} c^{1/k} = 1$. Treat the cases $0 < c < 1$, $c = 1$, and $c > 1$ separately. Imitate Exercise 1.35.

1.38. Let $x_k = (k+1)^{1/2} - k^{1/2}$. Prove that $\{x_k\}$ and $\{\sqrt{k}x_k\}$ converge. Find the limits of these two sequences. [In each case multiply and divide x_k by $(k+1)^{1/2} + k^{1/2}$.]

1.39. Prove that a sequence $\{x_k\}$ in $\mathbb{R}$ converges to 0 if and only if $\{|x_k|\}$ converges to 0.

1.40. Prove that, if $\lim_{k\to\infty} x_k = x_0$ and if c is a constant, then $\lim_{k\to\infty} cx_k = cx_0$.

1.41. Let x_0 be a limit point of a nonempty subset S in $\mathbb{R}$. Prove that there exists a sequence of distinct points of S that converges to x_0.

1.42. Determine the convergence or divergence of each of the following sequences. If the sequence converges, find its limit x_0 and an index k_0 in terms of ϵ such that, for $k \geq k_0$, we have x_k is in $N(x_0; \epsilon)$. If the sequence diverges, identify the type of divergence and prove your assertion.

a) $x_k = (k^{3/2}+1)/(k^{1/2}+1)$.

b) $x_k = (k^{3/2}-1)/(k^{3/2}+1)$.

c) $x_k = (2k+1)/\sqrt{k+1}$.

d) $x_k = (k^{5/2} + 2k + 1)/(k^2 - 2k - 2)$.

e) $x_k = (-1)^k(k + 1/k)$.

f) $x_k = \sum_{j=1}^{k} 1/[j(j+1)]$. *Hint:* $1/[j(j+1)] = 1/j - 1/(j+1)$.

1.43. Prove that $\{x_k\}$ converges to x_0 if and only if there exists an ϵ_0 with the following property:

> For all ϵ with $0 < \epsilon < \epsilon_0$, there exists a k_0 in $\mathbb{N}$ such that, when $k \geq k_0$, it follows that x_k is in $N(x_0; \epsilon)$.

1.44. **a)** Suppose that $\{x_k\}$ is a bounded sequence that is *eventually monotone increasing,* that is, there exists a k_1 such that, for all $k \geq k_1$, we have $x_k \leq x_{k+1}$. Prove that $\{x_k\}$ converges.

b) State and prove an analogous statement for eventually monotone decreasing sequences.

1.45. Suppose that $\{x_k\}$, $\{y_k\}$, and $\{z_k\}$ are three sequences with the following properties:

i) There exists an index k_1 such that, for $k \geq k_1$, $x_k \leq y_k \leq z_k$.

ii) $\lim_{k\to\infty} x_k = \lim_{k\to\infty} z_k = L$.

Prove that $\lim_{k\to\infty} y_k = L$ also.

1.46. Assume that $\lim_{k\to\infty} x_k = x_0$ and $\lim_{k\to\infty} y_k = y_0$. Prove that, if there is a k_1 such that $x_k \leq y_k$ for all $k \geq k_1$, then $x_0 \leq y_0$.

1.47. Suppose that $\{x_k\}$ is a sequence of positive numbers for which $\lim_{k\to\infty} x_{k+1}/x_k = L$ exists.

a) Prove that if $L < 1$, then $\{x_k\}$ converges. Find $\lim_{k\to\infty} x_k$.

b) Prove that, if $L > 1$, then $\{x_k\}$ diverges to ∞.

c) Exhibit two sequences $\{y_k\}$ and $\{z_k\}$ such that

$$\lim_{k\to\infty} \frac{y_{k+1}}{y_k} = \lim_{k\to\infty} \frac{z_{k+1}}{z_k} = 1$$

and such that $\{y_k\}$ converges while $\{z_k\}$ diverges. Thus, confirm that $L = 1$ gives no information.

d) Apply this test to determine the convergence of $\{k/2^k\}$.

1.48. Suppose that $\{x_k\}$ is a sequence of positive numbers such that $\lim_{k\to\infty}(x_k)^{1/k} = L$ exists.

a) Prove that if $L < 1$, then $\{x_k\}$ converges. Find $\lim_{k\to\infty} x_k$.

b) Prove that if $L > 1$, then $\{x_k\}$ diverges to ∞.

c) Prove that if $L = 1$, then no information is obtained about the convergence of $\{x_k\}$.

1.49. Suppose that $\{x_k\}$ is a sequence no subsequence of which diverges to ∞ or to $-\infty$. Prove that $\{x_k\}$ is bounded.

1.50. **Cantor's Nested Interval Theorem**. For each k in $\mathbb{N}$, let $I_k = [a_k, b_k]$ be a closed interval such that for each k, $I_k \supseteq I_{k+1}$. Apply Theorem 1.3.7 to the sequences $\{a_k\}$ and $\{b_k\}$ to prove that $\bigcap_{k=1}^{\infty} I_k \neq \emptyset$. Furthermore, prove that if $\{a_k\}$ and $\{b_k\}$ both converge to x_0, then $\bigcap_{k=1}^{\infty} I_k = \{x_0\}$.

1.51. Suppose that, for each k in $\mathbb{N}$, I_k is the interval $[k, \infty)$. Find $\bigcap_{k=1}^{\infty} I_k$. Does your answer contradict Cantor's Nested Interval Theorem? Explain. (See Exercise 1.50.)

1.52. Suppose that a sequence $\{x_k\}$ has at least two cluster points. Show that it must diverge. *Hint:* Explain why it suffices to show that the following two assertions are mutually contradictory: (i) $\lim_{k\to\infty} x_k = x_0$ exists. (ii) c is a cluster point of $\{x_k\}$ distinct from x_0. Then prove that these two assertions are, in fact, mutually contradictory.

1.53. Prove that $\{x_k\}$ converges to x_0 if and only if every subsequence of $\{x_k\}$ converges to x_0.

1.54. Prove that a bounded sequence $\{x_k\}$ converges if and only if it has exactly one cluster point.

1.55. Suppose that $\{x_k\}$ is a sequence with exactly one cluster point x_0. Is it necessarily true that $\lim_{k\to\infty} x_k = x_0$? If so, prove it. If not, provide a counterexample.

1.56. Find $\liminf x_k$ and $\limsup x_k$ for each of the following sequences $\{x_k\}$.

a) $x_k = 2^{(-1)^k}(1 + 1/k)$.

b) $x_k = k2^{(-1)^k k}$.

c) $x_k = (1 + 1/k)(1 + (-1)^k)^{1/k}$.

d) $x_k = (-1)^k(1 - 1/k) + (-1)^{k+1}(1 + 1/k)$.

e) $x_k = [k + (-1)^k(2k + 1)]/k$.

1.57. Prove that, if $\{x_k\}$ is a bounded sequence, then the set C of its cluster points is also bounded. Is the converse also true? If so, prove it. If not, provide a counterexample.

1.58. Let $\{x_k\}$ be a sequence such that $\nu = \liminf x_k$ is finite. Prove that ν is a cluster point of $\{x_k\}$.

1.59. Let $\{x_k\}$ be any sequence for which $\nu = \liminf x_k$ is finite. Fix any $\epsilon > 0$.

a) Prove that there exists a k_0 in $\mathbb{N}$ such that, for all $k \geq k_0$, $x_k > \nu - \epsilon$.

b) Prove that, for any k in $\mathbb{N}$, there exists a $k_1 > k$ such that $x_{k_1} < \nu + \epsilon$.

1.60. Let $\{x_k\}$ be any sequence of real numbers. Prove that $\{x_k\}$ converges to x_0 if and only if

$$\liminf x_k = \limsup x_k = x_0.$$

1.61. Let $\{x_k\}$ be any sequence of real numbers. Prove that $\{x_k\}$ diverges if and only if either of the following conditions hold.

i) Either $\liminf x_k$ or $\limsup x_k$ is infinite.

ii) Both $\liminf x_k$ and $\limsup x_k$ are finite and $\liminf x_k < \limsup x_k$.

1.62. Prove that a sequence diverges to ∞ if and only if $\liminf x_k = \limsup x_k = \infty$. State and prove the analogous theorem for sequences that diverge to $-\infty$.

1.63. Prove that a sequence $\{x_k\}$ diverges if and only if at least one of the following conditions hold:

i) Some subsequence of $\{x_k\}$ diverges to ∞ or to $-\infty$.

ii) There are two subsequences of $\{x_k\}$ that converge to different limits.

1.64. Suppose that a sequence $\{x_k\}$ has a cluster point c in $\mathbb{R}$. What conclusion, if any, can be drawn about either $\liminf x_k$ or $\limsup x_k$?

1.65. Let $\{y_k\}$ be any bounded sequence in $\mathbb{R}$. For each k in $\mathbb{N}$ let $x_k = \inf\{y_j : j \geq k\}$ and $z_k = \sup\{y_j : j \geq k\}$.

a) Prove that $x_k \leq y_k \leq z_k$ for all k in $\mathbb{N}$.

b) Prove that $\{x_k\}$ is monotone increasing and bounded above and that $\{z_k\}$ is monotone decreasing and bounded below.

c) Prove that $\{y_k\}$ converges if and only if $\lim_{k\to\infty} x_k = \lim_{k\to\infty} z_k$ in which case $\lim_{k\to\infty} y_k$ is equal to this common limit.

1.66. Let $\{x_k\}$ be a bounded sequence in $\mathbb{R}$. For each k in $\mathbb{N}$, let $y_k = \sup\{x_j : j \geq k\}$. Let $c = \inf\{y_k : k \text{ in } \mathbb{N}\}$. That is, $c = \inf\{\sup\{x_j : j \geq k\} : k \text{ in } \mathbb{N}\}$. Show that there exists a subsequence of $\{x_k\}$ that converges to c.

1.67. Suppose that $\{x_k\}$ is a sequence such that the two subsequences $\{x_{2j}\}$ and $\{x_{2j-1}\}$ converge to x_0. Prove that $\{x_k\}$ converges to x_0 also.

1.68. **a)** Suppose that $\{x_k\}$ is a sequence which has two subsequences $\{x_{k_j}\}$ and $x_{k'_j}$ with the following properties:

i) $\{k_j : j \text{ in } \mathbb{N}\} \cup \{k'_j : j \text{ in } \mathbb{N}\} = \mathbb{N}$.

ii) $\lim_{j\to\infty} x_{k_j} = \lim_{j\to\infty} x_{k'_j} = x_0$.

Prove that $\{x_k\}$ converges to x_0 also.

b) Generalize the result in part (a) to the case where there are a finite number of subsequences all converging to x_0.

1.69. Let $\{x_k\}$ be a bounded sequence of positive numbers. For each k in $\mathbb{N}$ define $y_k = x_{k+1}/x_k$ and $z_k = (x_k)^{1/k}$.

a) Prove that

$$\liminf y_k \leq \liminf z_k \leq \limsup z_k \leq \limsup y_k.$$

b) Give examples to show that the inequalities in part (a) can be strict.

c) Prove that, if $\{y_k\}$ converges, then $\{z_k\}$ also converges.

d) Find a sequence $\{x_k\}$ such that $\{z_k\}$ converges but $\{y_k\}$ diverges.

1.70. Let $\{x_k\}$ be any bounded sequence. For each k in $\mathbb{N}$ define $\sigma_k = (1/k)\sum_{j=1}^{k} x_j$, the arithmetic average of the first k terms of the sequence. If $\{\sigma_k\}$ converges to x_0, then $\{x_k\}$ is said to converge $(C, 1)$ to x_0.

a) Prove that

$$\liminf x_k \leq \liminf \sigma_k \leq \limsup \sigma_k \leq \limsup x_k.$$

b) Show by examples that the inequalities in part (a) can be strict.

c) Prove that if $\{x_k\}$ converges, then $\{x_k\}$ converges $(C, 1)$.

d) Find a divergent sequence that converges $(C, 1)$.

1.71. Let $\{x_k\}$ be a bounded sequence of positive numbers. For each k in $\mathbb{N}$, define $\tau_k = (x_1 x_2 x_3 \ldots x_k)^{1/k}$, the *geometric average* of the first k terms. Show that, if $\{x_k\}$ converges, then so also does $\{\tau_k\}$. Find a divergent sequence $\{x_k\}$ for which $\{\tau_k\}$ converges.

1.4 Cauchy Sequences

Suppose that $\{x_k\}$ is a convergent sequence with limit x_0. Intuitively it must be clear that eventually any two terms of the sequence, each being near x_0, are near each other. Explicitly, fix any positive ϵ. Suppose that k_0 is chosen so that, for $k \geq k_0$, x_k

is in $N(x_0; \epsilon/2)$. Choose any two indices k and m each greater than k_0. It follows that x_k is in $N(x_0; \epsilon/2)$ and x_m is in $N(x_0; \epsilon/2)$. Consequently,

$$\begin{aligned}|x_k - x_m| &= |x_k - x_0 + x_0 - x_m| \le |x_k - x_0| + |x_0 - x_m| \\ &< \frac{\epsilon}{2} + \frac{\epsilon}{2} = \epsilon.\end{aligned}$$

In other words, if k and m are each greater than k_0, then $|x_k - x_m| < \epsilon$. This observation, credited to Cauchy, leads to the following definition.

DEFINITION 1.4.1 The Cauchy Condition A sequence $\{x_k\}$ in $\mathbb{R}$ is called a *Cauchy sequence* if, for any $\epsilon > 0$, there exists a k_0 such that, for k and m both greater than or equal to k_0, it follows that $|x_k - x_m| < \epsilon$. ●

Our preceding discussion proves that a convergent sequence in $\mathbb{R}$ is also a Cauchy sequence. *Convergent* always implies *Cauchy*.

THEOREM 1.4.1 If $\{x_k\}$ is a convergent sequence of real numbers, then $\{x_k\}$ is a Cauchy sequence. ●

You will notice that, in this definition, no mention is made of a value for the limit of the sequence. While this omission initially may appear to be a drawback, in fact it is this very feature of Cauchy sequences that makes them most useful. In many applications it is easier to prove that a sequence is Cauchy than to prove that it is convergent, especially if the value of the limit is unclear. But, as we will prove shortly, if a sequence of real numbers is Cauchy, then it must be convergent. Hence, if you can prove that $\{x_k\}$ is a Cauchy sequence, then you may conclude that the sequence is convergent without knowing or even referring to its limit. Often that limit is an idealized quantity whose value may not be significant; often merely its existence is needed. That a sequence satisfies the Cauchy condition suffices to prove this existence.

The key to achieving the proof of the central theorem of this section (Theorem 1.4.4) is the following fact; all else follows from it.

THEOREM 1.4.2 If a sequence is Cauchy, then it is bounded.

Proof. The proof closely imitates that of Theorem 1.3.3; hence we will be brief. Assume that $\{x_k\}$ is a Cauchy sequence. Let $\epsilon = 1$. By the Cauchy condition choose a k_0 such that, whenever $k, m \ge k_0$, we have $|x_k - x_m| < 1$. Let $m = k_0$ and observe that, for any $k \ge k_0$,

$$\begin{aligned}|x_k| &= |x_k - x_{k_0} + x_{k_0}| \\ &\le |x_k - x_{k_0}| + |x_{k_0}| < 1 + |x_{k_0}|.\end{aligned}$$

Let $M = \max\{|x_1|, |x_2|, \ldots, |x_{k_0-1}|, |x_{k_0}| + 1\}$. Then it is immediate that $|x_k| \le M$ for all k in $\mathbb{N}$. Therefore $\{x_k\}$ is bounded. ●

Since any Cauchy sequence in $\mathbb{R}$ is bounded, it must have at least one cluster point. (Theorem 1.3.11.) In fact, it can have only one. For suppose that c_1 and

c_2 were two distinct cluster points of the Cauchy sequence $\{x_k\}$. (See Fig. 1.19.) Choose ϵ with $0 < \epsilon < |c_1 - c_2|/3$. Choose a k_0 in $\mathbb{N}$ such that, if $k, m \geq k_0$, then $|x_k - x_m| < \epsilon$. Since c_1 and c_2 are cluster points of $\{x_k\}$, there exist $k_1, k_2 > k_0$ such that x_{k_1} is in $N(c_1; \epsilon)$ and x_{k_2} is in $N(c_2; \epsilon)$. Use the inequality $|a + b| \geq |a| - |b|$ twice to obtain the following contradiction:

$$\begin{aligned} \epsilon > |x_{k_1} - x_{k_2}| &= |c_1 - c_2 + x_{k_1} - c_1 + c_2 - x_{k_2}| \\ &\geq |c_1 - c_2| - |x_{k_1} - c_1| - |c_2 - x_{k_2}| \\ &> 3\epsilon - \epsilon - \epsilon = \epsilon. \end{aligned}$$

The contradiction proves that the sequence cannot have two distinct cluster points and proves the following theorem.

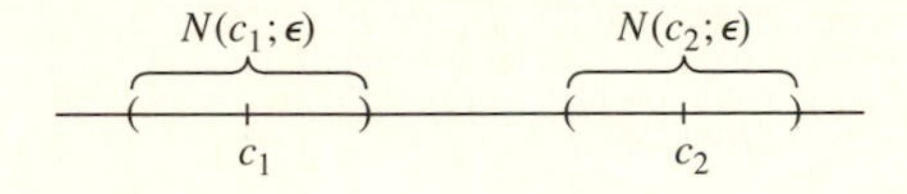

Figure 1.19

THEOREM 1.4.3 A Cauchy sequence in $\mathbb{R}$ has exactly one cluster point. ●

As an immediate consequence, we obtain the central theorem of this section.

THEOREM 1.4.4 A sequence in $\mathbb{R}$ is convergent if and only if it is a Cauchy sequence.

Proof. A convergent sequence is a Cauchy sequence by Theorem 1.4.1. For the converse, suppose that $\{x_k\}$ is a Cauchy sequence in $\mathbb{R}$. By Theorem 1.4.2, $\{x_k\}$ is bounded. By Theorem 1.4.3, $\{x_k\}$ has exactly one cluster point. By Theorem 1.3.14 the sequence $\{x_k\}$ converges. Consequently, a sequence in $\mathbb{R}$ converges if and only if it is Cauchy. ●

DEFINITION 1.4.2 A set X is said to be *Cauchy complete* if every Cauchy sequence in X converges to a point of X. ●

We have proved that $\mathbb{R}$ is Cauchy complete. We will make repeated use of this fact throughout the text.

EXAMPLE 23 Consider the decimal expansion of the irrational number $\sqrt{2} = 1.4142136\ldots$. Of course, this is a nonrepeating decimal representation, so we cannot ever know the digits exactly. But we can write down the first few terms of a Cauchy sequence that converges to $\sqrt{2}$ as follows:

$$\begin{aligned} x_1 &= 1.4000000 & x_4 &= 1.4142000 \\ x_2 &= 1.4100000 & x_5 &= 1.4142100 \\ x_3 &= 1.4140000 & x_6 &= 1.4142130 \\ & & &\vdots \end{aligned}$$

Then $\{x_k\}$ is a Cauchy sequence and converges to $\sqrt{2}$. The proof of this assertion is provided in the more general setting of the next example. ●

EXAMPLE 24 Let $x_0 = n_0 \cdot d_1 d_2 d_3 \cdots d_k \cdots$ be the decimal expansion of any positive real number, where n_0 is the integer part of x_0 and, for $k = 1, 2, 3, \ldots$, the digits d_k are chosen from the set $\{0, 1, 2, 3, 4, 5, 6, 7, 8, 9\}$. Then x_0 can be written as an infinite series as follows:

$$x_0 = n_0 + \sum_{j=1}^{\infty} \frac{d_j}{10^j}.$$

For each $k \geq 1$, let $x_k = n_0 + \sum_{j=1}^{k} d_j/10^j$. We claim that $\{x_k\}$ is a Cauchy sequence that converges to x_0. To see this, we work backward. Choose any indices k and m with $m > k$. Then

$$|x_m - x_k| = \sum_{j=k+1}^{m} \frac{d_j}{10^j} \leq 9\left(\frac{1}{10}\right)^{k+1} \sum_{j=0}^{m-k-1} \left(\frac{1}{10}\right)^j.$$

This last summation is a finite geometric series with ratio $r = 1/10$ and sum

$$\frac{1 - \left(\frac{1}{10}\right)^{m-k}}{1 - \frac{1}{10}} = \frac{10}{9}\left[1 - \left(\frac{1}{10}\right)^{m-k}\right].$$

This sum, in turn, is less than 10/9. Therefore,

$$|x_m - x_k| < 9\left(\frac{1}{10}\right)^{k+1}\left(\frac{10}{9}\right) = \left(\frac{1}{10}\right)^k.$$

This last inequality provides the clue for our forward proof. Given any $\epsilon > 0$, use Archimedes' principle to choose k_0 in $\mathbb{N}$ so large that $(1/10)^{k_0} < \epsilon$. Then, for $m > k \geq k_0$,

$$|x_m - x_k| < \left(\frac{1}{10}\right)^k \leq \left(\frac{1}{10}\right)^{k_0} < \epsilon.$$

Thus $\{x_k\}$ is a Cauchy sequence. It converges to x_0. In fact, this example is the justification that you implicitly use whenever you approximate the irrational number x_0 by a terminating decimal. To confirm that $\lim_{k\to\infty} x_k = x_0$, let ϵ and k_0 be as above and choose any $k \geq k_0$. Then a slight variation of our previous argument gives

$$\begin{aligned} |x_0 - x_k| &= \sum_{j=k+1}^{\infty} \frac{d_j}{10^j} \leq 9\left(\frac{1}{10}\right)^{k+1} \sum_{j=0}^{\infty} \left(\frac{1}{10}\right)^j \\ &= 9\left(\frac{1}{10}\right)^{k+1}\left(\frac{10}{9}\right) = \left(\frac{1}{10}\right)^k < \epsilon. \end{aligned}$$

This proves that $\lim_{k\to\infty} x_k = x_0$. ●

EXAMPLE 25 Recursively construct a sequence of real numbers as follows. Let $x_1 = 0$, $x_2 = 1$, $x_3 = 1/2$, $x_4 = (1 + 1/2)/2 = 3/4$, and in general, for $k =$

2, 3, 4, ..., let the next term be the arithmetic average of the previous two: $x_{k+1} = (x_{k-1} + x_k)/2$.

If you compute and plot a few more values of the sequence $\{x_k\}$ thus defined, you will certainly be convinced intuitively that it converges, yet you may not immediately be able to identify the limiting value x_0. We show that $\{x_k\}$ is a Cauchy sequence, thereby proving that it converges. You are challenged to find $\lim_{k\to\infty} x_k$ in Exercise 1.78.

Choose any two (large) indices k and m. Without loss of generality, $m > k$. Using the telescoping nature of the sum, compute:

$$\begin{aligned}|x_m - x_k| &= |(x_m - x_{m-1}) + (x_{m-1} - x_{m-2}) + \cdots + (x_{k+1} - x_k)| \\ &= |\sum_{j=k}^{m-1}(x_{j+1} - x_j)| \le \sum_{j=k}^{m-1} |x_{j+1} - x_j|. \end{aligned} \tag{1.6}$$

Notice that, for each $j = 2, 3, 4, \ldots,$

$$|x_{j+1} - x_j| = \tfrac{1}{2}|x_j - x_{j-1}|.$$

Therefore it follows by induction that

$$|x_{j+1} - x_j| = \left(\frac{1}{2}\right)^{j-1}. \tag{1.7}$$

(See Exercise 1.72.) Applying (1.6), (1.7), and the formula for the sum of a finite geometric series, we have

$$\begin{aligned}|x_m - x_k| &\le \sum_{j=k}^{m-1} |x_{j+1} - x_j| = \sum_{j=k}^{m-1} \left(\frac{1}{2}\right)^{j-1} \\ &= \left(\frac{1}{2}\right)^{k-1} \left[1 + \frac{1}{2} + \frac{1}{2^2} + \cdots + \frac{1}{2^{m-k-1}}\right] \\ &= \left(\frac{1}{2}\right)^{k-1} \frac{1 - \left(\frac{1}{2}\right)^{m-k}}{1 - \frac{1}{2}} < \left(\frac{1}{2}\right)^{k-2}. \end{aligned}$$

Now assign any positive value to ϵ. Choose k_0 such that $(1/2)^{k_0-2} < \epsilon$. If k and m are both greater than k_0 with $m > k$, then

$$|x_m - x_k| < \left(\frac{1}{2}\right)^{k-2} \le \left(\frac{1}{2}\right)^{k_0-2} < \epsilon.$$

Consequently, $\{x_k\}$ is a Cauchy sequence. Theorem 1.4.4 guarantees that $\{x_k\}$ converges. ●

EXAMPLE 26 For each k in $\mathbb{N}$, let $x_k = \sum_{j=0}^{k} (-1)^j / j!$. We claim that $\{x_k\}$ is Cauchy. We begin by working backward. Choose any two indices m and k with $m > k$. Then

$$|x_m - x_k| = \left| \sum_{j=k+1}^{m} \frac{(-1)^j}{j!} \right| \leq \sum_{j=k+1}^{m} \frac{1}{j!}.$$

In Exercise 1.19 you proved that, for j in $\mathbb{N}$, we have $2^{j-1} \leq j!$. Applying these inequalities and the sum for a finite geometric series, we have

$$\begin{aligned} |x_m - x_k| &\leq \sum_{j=k+1}^{m} \frac{1}{2^{j-1}} = \frac{1}{2^k} \sum_{j=0}^{m-k-1} \left(\frac{1}{2}\right)^j \\ &= \frac{1}{2^k} \frac{1 - \left(\frac{1}{2}\right)^{m-k}}{1 - \frac{1}{2}} \\ &= \frac{1}{2^{k-1}} \left[1 - \left(\frac{1}{2}\right)^{m-k} \right] < \frac{1}{2^{k-1}}. \end{aligned}$$

For any $\epsilon > 0$, use Archimedes' principle to choose the least natural number k_0 such that $1/2^{k_0 - 1} < \epsilon$. For $k, m \geq k_0$ with $m > k$, we have, by the foregoing calculation

$$|x_m - x_k| < \frac{1}{2^{k-1}} \leq \frac{1}{2^{k_0 - 1}} < \epsilon.$$

Consequently, the sequence $\{x_k\}$ is Cauchy. It is therefore convergent. You can compute an approximation to the limit of this sequence, which is accurate to seven decimal places, by taking $k = 10$. We obtain $x_{10} = \sum_{j=0}^{10} (-1)^j / j! \doteq 0.3678794$. Compare this number with the value given by your calculator for the number $1/e$. ●

EXAMPLE 27 Define a sequence $\{x_k\}$ by letting $x_k = \sum_{j=1}^{k} 1/j$ for each k in $\mathbb{N}$. As we saw in Example 22 of Section 1.3, the sequence $\{x_k\}$ is not convergent; therefore it is not Cauchy. ●

A special type of sequence, those that are contractive, is identified in the following definition. Such sequences arise in a remarkable variety of settings. As we will prove, they always converge.

DEFINITION 1.4.3 A sequence $\{x_k\}$ is said to be *contractive* if there exists a constant C, with $0 < C < 1$, such that

$$|x_{k+1} - x_k| \leq C |x_k - x_{k-1}|,$$

for all $k = 2, 3, 4, \ldots$. ●

Remark. The requirement that $C < 1$ is essential. If $C = 1$, then none of our results that follow need be true. For example, if, for each k in $\mathbb{N}$, we define $x_k = k$, then $|x_{k+1} - x_k| = |x_k - x_{k-1}|$, but this sequence is *not* contractive. Exercises at the end of this section deal with this issue.

EXAMPLE 28 Define a sequence recursively as follows. Let x_1 be any positive number. For $k \geq 1$, let $x_{k+1} = 1/(4 + x_k)$. The resulting sequence is contractive. To prove this, we must exhibit a constant C strictly between 0 and 1 such that, for $k \geq 2$, $|x_{k+1} - x_k| \leq C|x_k - x_{k-1}|$. Notice that $x_k > 0$ for all k and that $1 = x_k(4 + x_{k-1})$ for all $k \geq 2$. Fix any $k \geq 2$. Begin with the left-hand side and compute:

$$
\begin{aligned}
|x_{k+1} - x_k| &= \left| \frac{1}{4 + x_k} - x_k \right| \\
&= \frac{|1 - x_k(4 + x_k)|}{4 + x_k} \\
&= \frac{|x_k(4 + x_{k-1}) - x_k(4 + x_k)|}{4 + x_k} \\
&= \frac{x_k}{4 + x_k} |x_{k-1} - x_k| \\
&= \left(1 - \frac{4}{4 + x_k}\right) |x_{k-1} - x_k|.
\end{aligned}
$$

Now, for $k \geq 2$, $x_k = 1/(4 + x_{k-1}) < 1/4$. Therefore $4 + x_k < 17/4$ and $4/(4 + x_k) > 16/17$. We conclude that

$$
0 < 1 - \frac{4}{4 + x_k} < \frac{1}{17}.
$$

We take $C = 1/17$. We have proved that, for $k \geq 2$,

$$
|x_{k+1} - x_k| \leq \frac{1}{17} |x_k - x_{k-1}|.
$$

Consequently, $\{x_k\}$ is contractive. ●

We have singled out the class of contractive sequences because, as we now show, every such sequence is Cauchy.

THEOREM 1.4.5 If $\{x_k\}$ is a contractive sequence in $\mathbb{R}$, then it is Cauchy.

Proof. Suppose that C is a constant with $0 < C < 1$ such that, for $k \geq 2$, $|x_{k+1} - x_k| \leq C|x_k - x_{k-1}|$. Without loss of generality, $|x_2 - x_1| > 0$. It is straightforward to prove by induction that $|x_{k+1} - x_k| \leq C^{k-1}|x_2 - x_1|$ for $k \geq 2$. Now choose any two indices k and m with $m > k$. Using the formula for the sum of a finite geometric series, compute

$$
\begin{aligned}
|x_m - x_k| &= |x_m - x_{m-1} + x_{m-1} - x_{m-2} + \cdots + x_{k+1} - x_k| \\
&\leq \sum_{j=k}^{m-1} |x_{j+1} - x_j| \leq \sum_{j=k}^{m-1} C^{j-1}|x_2 - x_1| \\
&= C^{k-1}|x_2 - x_1| \sum_{j=0}^{m-k-1} C^j
\end{aligned}
$$

$$= C^{k-1}|x_2 - x_1|\frac{1 - C^{m-k}}{1 - C}$$
$$< \frac{C^{k-1}}{1 - C}|x_2 - x_1|.$$

Having worked backward to this point, we now recognize how to proceed. Since $0 < C < 1$, $\lim_{k\to\infty} C^k = 0$ by Example 18. Thus, given any $\epsilon > 0$, we can choose k_0 such that, if $k \geq k_0$, then

$$C^{k-1} < \frac{(1 - C)\epsilon}{|x_2 - x_1|}.$$

Choose any $k, m \geq k_0$ with $m > k$. The preceding computation yields

$$|x_m - x_k| < \frac{C^{k-1}}{1 - C}|x_2 - x_1| < \epsilon. \quad \textbf{(1.8)}$$

This proves that $\{x_k\}$ is Cauchy. ●

Let $\{x_k\}$ be a contractive sequence that, because it is Cauchy, is also convergent. Let $x_0 = \lim_{k\to\infty} x_k$. The inequality (1.8) leads to an estimate for the error $|x_0 - x_k|$ between the limit and the approximation x_k at the kth stage of iteration. In (1.8), let m tend to ∞ and invoke Theorem 1.3.9. We will show in the next section that $\lim_{m\to\infty} |x_m - x_k| = |x_0 - x_k|$. Temporarily granting this result, we obtain

$$|x_0 - x_k| \leq \frac{C^{k-1}}{1 - C}|x_2 - x_1| \quad \textbf{(1.9)}$$

for all $k \geq 2$. This provides us with an upper bound for the error $|x_0 - x_k|$.

Often sequences are designed to solve practical problems. Implementation of the mathematics on a computer provides efficient and highly accurate numerical approximation to some desired quantity. In the branch of mathematics called numerical analysis, elaborate and useful procedures have been developed. Here we will discuss one simple example, intended only to give you a glimpse of a vast area of application.

EXAMPLE 29 The cubic polynomial $x^3 - 4x + 2$ has a root between 0 and 1. Suppose that we want to find it. This root happens to be an irrational number, so we can never find it exactly. By using a Cauchy sequence, however, we can approximate it as closely as we may wish. Here's how. First, we rearrange the equation $x^3 - 4x + 2 = 0$ and write $x = (x^3 + 2)/4$. Then we choose x_1 to be *any* first approximation to the root in the interval (0, 1). We define a recursive relationship derived from the original equation by setting $x_{k+1} = (x_k^3 + 2)/4$, for $k = 1, 2, 3, \ldots$. Thus we have designed a sequence that, we claim, is contractive, hence Cauchy, hence convergent. What is more, it converges to the desired root.

To see that $\{x_k\}$ is contractive is easy. First note that, since x_1 was initially chosen in the interval (0, 1), it follows that x_k is in (0, 1) for all k; the inductive proof is straightforward. Next, observe that

$$|x_{k+1} - x_k| = \left| \frac{(x_k^3 + 2)}{4} - \frac{(x_{k-1}^3 + 2)}{4} \right| = \frac{|x_k^3 - x_{k-1}^3|}{4}$$

$$= \frac{|(x_k - x_{k-1})(x_k^2 + x_k x_{k-1} + x_{k-1}^2)|}{4} < \frac{3}{4}|x_k - x_{k-1}|.$$

We conclude that $\{x_k\}$ is contractive with $C = 3/4$. Therefore $\{x_k\}$ is convergent. Let x_0 be its limit. Of course, we claim that x_0 is the desired root. In Section 1.5 we will derive the algebraic properties of limit that will justify our taking the limit of both sides of the recursive relation $x_{k+1} = (x_k^3 + 2)/4$ to obtain the equation $x_0 = (x_0^3 + 2)/4$. Consequently, $x_0^3 - 4x_0 + 2 = 0$ and x_0 is the root we sought. For example, we set $x_1 = .5$ and we compute the following values for $x_{k+1} = (x_k^3 + 2)/4$:

$$\begin{array}{lll} x_1 = .5000000 & x_5 = .5391081 & x_9 = .5391887 \\ x_2 = .5312500 & x_6 = .5391713 & x_{10} = .5391888 \\ x_3 = .5374832 & x_7 = .5391850 & x_{11} = .5391889 \\ x_4 = .5388181 & x_8 = .5391880 & x_{12} = .5391889 \end{array}$$

Clearly, $x_0 \doteq .5391889$, accurate to seven decimal places.

Since $C = 3/4$, the error between x_0 and the kth approximation x_k is, by the inequality (1.9), no more than $4(3/4)^{k-1}|x_2 - x_1|$. Our computed value for $|x_2 - x_1|$ is .03125. Thus, for any $k \geq 2$, an upper bound for the error $|x_0 - x_k|$ is $(3/4)^{k-1}(.125)$. This is a fairly crude upper bound as you can see by taking, say, $k = 8$. The actual error at the eighth stage is approximately 9×10^{-7} while the upper bound for this error is $(3/4)^7(.125) = .0166855$. We will ask you to derive an alternate approximation of the error in the exercises. ●

EXERCISES

1.72. Prove by induction that, in Example 25,

$$|x_{j+1} - x_j| = \left(\frac{1}{2}\right)^{j-1},$$

for all j in $\mathbb{N}$.

1.73. Let $\{x_k\}$ be a sequence of real numbers. For each k in $\mathbb{N}$, define $y_k = \sum_{j=1}^{k} x_j$ and $z_k = \sum_{j=1}^{k} |x_j|$. Prove that, if $\{z_k\}$ is a Cauchy sequence, then $\{y_k\}$ is also. Is the converse true? If so, prove it; if not, provide a counterexample.

1.74. Fix any integer $p \geq 2$. Let $\{c_k\}$ be any sequence chosen from the set $\{0, 1, 2, \ldots, p - 1\}$. For each k in $\mathbb{N}$ define $x_k = \sum_{j=1}^{k} c_j/p^j$. Prove that $\{x_k\}$ is a Cauchy sequence that converges to some number in [0, 1].

1.75. Suppose that $\{x_k\}$ is a sequence with the property that $\lim_{k\to\infty}(x_{k+1} - x_k) = 0$. Does it follow that $\{x_k\}$ is Cauchy? If so, prove it; if not, provide a counterexample.

1.76. Suppose that $\{x_k\}$ is a sequence with the property that, for all k in $\mathbb{N}$, $|x_{k+1} - x_k| < (2/3)^k$. Does it follow that $\{x_k\}$ is Cauchy? If so, prove it; if not, provide a counterexample.

1.77. For k in $\mathbb{N}$, define $x_k = \sum_{j=0}^{k}(-1)^j 1/(2j)!$. Prove that $\{x_k\}$ is Cauchy and compute an approximation, accurate to five decimal places, of the value of its limit. Compare with the value given by your calculator for the cosine of 1 radian.

1.78. Fix any number w in (1, 0). Fix any two numbers x_1 and x_2. For $k \geq 2$, define $x_{k+1} = wx_{k-1} + (1 - w)x_k$. (The numbers w and $1 - w$ are called *weights* and x_{k+1} is said to be a *weighted average* of x_{k-1} and x_k.) Compute several terms of this sequence and identify the general form of x_k in terms of x_1, x_2, and w. Prove that $\{x_k\}$ is Cauchy. Find $\lim_{k\to\infty} x_k$.

1.79. Fix any $c > 0$. Define a sequence $\{x_k\}$ recursively as follows: Let x_1 be any positive number and, for k in $\mathbb{N}$, let $x_{k+1} = 1/(c + x_k)$. Prove that $\{x_k\}$ is a contractive sequence.

1.80. Following Theorem 1.4.5, we obtained an upper bound for the error $|x_0 - x_k|$ at the kth recursive step of a contractive sequence, namely,

$$|x_0 - x_k| \leq \frac{C^{k-1}}{1 - C}|x_2 - x_1|. \tag{1.10}$$

Its usefulness depends on the fact that $\lim_{k\to\infty} C^{k-1}/(1 - C) = 0$.

a) Show that, for all $k \geq 2$,

$$|x_0 - x_k| \leq \frac{C}{1 - C}|x_k - x_{k-1}|.$$

b) The usefulness of the inequality in part (a) depends on the fact that $\lim_{k\to\infty} |x_k - x_{k-1}| = 0$. Return to the table of computed values in Example 29 and compute the estimated error for each $k \leq 12$ using the estimate in part (a). Compare these with the corresponding approximate error given by the upper bound in (1.10).

1.81. The cubic polynomial $p(x) = x^3 - 2x^2 - 9x + 4$ has a root in the interval (0, 1).

a) Design a contractive sequence $\{x_k\}$ that will converge to this root. Prove that $\{x_k\}$ is contractive and find its contractive constant C.

b) Let $x_0 = \lim_{k\to\infty} x_k$. Prove that x_0 is a root of $p(x)$.

c) Use your calculator to compute successive values of x_k until you are assured that $|x_0 - x_k| < 5 \times 10^{-5}$.

d) At each stage in your calculations in part (c), compute each of the two forms of the estimate for $|x_0 - x_k|$ discussed in Exercise 1.80. Which estimate is more accurate?

1.82. Suppose that $\{x_k\}$ is a sequence in $\mathbb{R}$. Suppose also that there exists a strictly monotone increasing sequence $\{C_k\}$ such that $\lim_{k\to\infty} C_k = 1$ and with the property that, for all $k \geq 2$,

$$|x_{k+1} - x_k| \leq C_k|x_k - x_{k-1}|.$$

Determine whether $\{x_k\}$ is necessarily Cauchy. If so, prove it; if not, provide a counterexample.

1.83. Suppose that $\{x_k\}$ is a sequence in $\mathbb{R}$. Suppose also that there exists a strictly monotone decreasing sequence $\{C_k\}$ that converges to some number in [0, 1) and with the property that, for all $k \geq 2$,

$$|x_{k+1} - x_k| \leq C_k|x_k - x_{k-1}|.$$

Determine whether $\{x_k\}$ is necessarily Cauchy. If so, prove it; if not, provide a counterexample.

1.84. Suppose that S is any nonempty set in $\mathbb{R}$ and that x_0 is any limit point of S. Prove that there exists a Cauchy sequence of *distinct* points in S that converges to x_0.

1.85. Consider the following possible theorem:

> A sequence $\{x_k\}$ in $\mathbb{R}$ is Cauchy if and only
> if each of its proper subsequences is Cauchy.

Determine whether this statement is, in fact, a theorem. If it is, prove it; if not, explain exactly how it fails.

1.86. We used the Bolzano–Weierstrass theorem implicitly in the proof of Theorem 1.4.4. Identify exactly how we used it.

1.87. Experience shows that, while learning analysis, one often gets lost in the details and loses sight of the overall design. To help you clarify your overview, return to Axiom 1.1.1 and review for yourself our proofs, to this point, of the following implications. Write a brief essay summarizing the essential connections. Omit the technical details but otherwise be thorough.

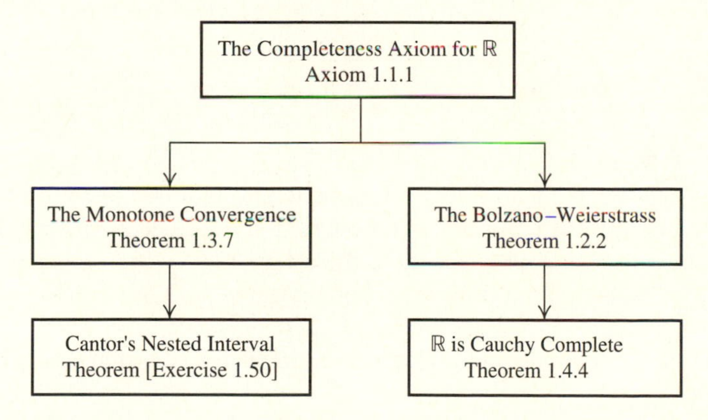

Prove that, if we replace $\mathbb{R}$ by $\mathbb{Q}$ in every statement, then each of these statements becomes false.

1.5 The Algebra of Convergent Sequences

Here we want to establish a secondary theme that will recur throughout this text: the algebraic structure that is interwoven throughout analysis. By this we do not refer merely to the algebraic manipulations you learned in high school but to those unifying algebraic concepts and systems studied in any course in abstract algebra. Here we examine briefly one of several systems in analysis that historically motivated the development of the algebraic branch of mathematics. We will study others later in the text.

If you have already begun your study of abstract algebra, you will recognize here a specific instance of the systems you studied and will benefit from the additional insight those systems provide. If you have not studied abstract algebra, we want to reassure you that the concepts discussed here are simple generalizations of the ideas already familiar to you from arithmetic.

In the present context, let **C** denote the collection of all convergent sequences of real numbers. Several immediate observations can be made that are algebraic in nature. We can add and multiply any two sequences in **C** and obtain sequences in **C**. For sequences $\boldsymbol{x} = \{x_k\}$ and $\boldsymbol{y} = \{y_k\}$ in **C**, define the sum of $\boldsymbol{x}$ and $\boldsymbol{y}$ to be the sequence $\boldsymbol{z} = \boldsymbol{x} + \boldsymbol{y}$ by letting

$$z_k = x_k + y_k,$$

for each k in $\mathbb{N}$. We will prove that the sequence $\boldsymbol{z} = \{z_k\}$ thus defined converges, so $\boldsymbol{z}$ is in **C**. Thus **C** is closed under addition. Furthermore, addition is *commutative:* $\boldsymbol{x} + \boldsymbol{y} = \boldsymbol{y} + \boldsymbol{x}$ for all $\boldsymbol{x}$, $\boldsymbol{y}$ in **C**.

Define the *product* of $\boldsymbol{x} = \{x_k\}$ and $\boldsymbol{y} = \{y_k\}$ to be the sequence $\boldsymbol{z} = \boldsymbol{x}\boldsymbol{y}$ by letting

$$z_k = x_k y_k,$$

for each k in $\mathbb{N}$. We will also prove that the sequence $\boldsymbol{z} = \{z_k\}$ thus defined converges, so $\boldsymbol{z}$ is in **C**. Thus **C** is closed under multiplication. Multiplication is *commutative:* $\boldsymbol{x}\boldsymbol{y} = \boldsymbol{y}\boldsymbol{x}$ for all $\boldsymbol{x}$ and $\boldsymbol{y}$ in **C**.

It is easy to see that both addition and multiplication are *associative*:

$$(\boldsymbol{x} + \boldsymbol{y}) + \boldsymbol{z} = \boldsymbol{x} + (\boldsymbol{y} + \boldsymbol{z}) \qquad \text{and} \qquad (\boldsymbol{x}\boldsymbol{y})\boldsymbol{z} = \boldsymbol{x}(\boldsymbol{y}\boldsymbol{z}),$$

for all $\boldsymbol{x}$, $\boldsymbol{y}$, and $\boldsymbol{z}$ in **C**.

The constant sequence, every term of which is 0, is the *zero* element $\mathbf{0}$ of **C**; it has the property that $\boldsymbol{x} + \mathbf{0} = \boldsymbol{x}$ for every $\boldsymbol{x}$ in **C**. Every sequence $\boldsymbol{x}$ in **C** has an *additive inverse* $-\boldsymbol{x}$ in **C**: If $\boldsymbol{x} = \{x_k\}$, then $-\boldsymbol{x} = \{-x_k\}$. The sequence $-\boldsymbol{x}$ has the property that $\boldsymbol{x} + (-\boldsymbol{x}) = \mathbf{0}$. We can subtract one sequence $\boldsymbol{x}$ from another sequence $\boldsymbol{y}$ in the usual way: $\boldsymbol{y} - \boldsymbol{x}$ is defined to be the sequence $\boldsymbol{y} + (-\boldsymbol{x})$.

Addition and multiplication intertwine in such a way that the *distributive laws* are satisfied:

$$\boldsymbol{x}(\boldsymbol{y} + \boldsymbol{z}) = \boldsymbol{x}\boldsymbol{y} + \boldsymbol{x}\boldsymbol{z} \qquad \text{and} \qquad (\boldsymbol{x} + \boldsymbol{y})\boldsymbol{z} = \boldsymbol{x}\boldsymbol{z} + \boldsymbol{y}\boldsymbol{z},$$

for all $\boldsymbol{x}$, $\boldsymbol{y}$, and $\boldsymbol{z}$ in **C**.

The constant sequence, every term of which is 1, is the *multiplicative identity* $\mathbf{1}$ for **C**; it has the property that $\mathbf{1}\boldsymbol{x} = \boldsymbol{x}$ for every $\boldsymbol{x}$ in **C**. Though the terminology is not important in this text, in the language of abstract algebra, **C** is said to be a *commutative ring with a multiplicative identity.*

To this point, we can add, subtract, and multiply convergent sequences and obtain another as a result. In general, we cannot expect to be able to divide one sequence by another without some restriction on the divisor. Specifically, suppose that $\boldsymbol{x} = \{x_k\}$ is a convergent sequence, no term of which is 0, whose limit x_0 is not 0. We shall prove that the sequence $\boldsymbol{y} = \{1/x_k\}$ is also a convergent sequence with limit $1/x_0$. A simple computation will show you that $\boldsymbol{x}\boldsymbol{y} = \mathbf{1}$. In other words, the *multiplicative inverse* of $\boldsymbol{x}$ exists in **C**. It is typically denoted $\boldsymbol{x}^{-1}$. These are the so-called *units* in the ring **C**. If $\boldsymbol{x}$ has a multiplicative inverse in **C** and if $\boldsymbol{y}$ is in **C**, then we can "divide" $\boldsymbol{y}$ by $\boldsymbol{x}$ and obtain another convergent sequence $\boldsymbol{y}/\boldsymbol{x}$, or more properly, $\boldsymbol{y}\boldsymbol{x}^{-1} = \{y_k/x_k\}$.

It is worth noting here that, if $x_k \neq 0$ for all k in $\mathbb{N}$ but $\lim_{k\to\infty} x_k = 0$, then there will exist sequences $\boldsymbol{y}$ in $\mathbf{C}$ such that $\lim_{k\to\infty} y_k/x_k$ does exist. In other words, it will be possible to divide some sequences by $\boldsymbol{x}$ even though $\boldsymbol{x}^{-1}$ does not exist in $\mathbf{C}$. This possibility depends on the sequence $\boldsymbol{y}$ converging at a sufficiently rapid rate to 0 also. The issue of the rate of convergence of a sequence in $\mathbf{C}$ is a subtle question requiring delicate analysis. It is the issue that lurked beneath all those problems you worked in the section on l'Hôpital's rule in your elementary calculus text.

We have postponed the work long enough. It is time to prove the following assertions.

THEOREM 1.5.1 Suppose that $\boldsymbol{x} = \{x_k\}$ and $\boldsymbol{y} = \{y_k\}$ are in $\mathbf{C}$ with $\lim_{k\to\infty} x_k = x_0$ and $\lim_{k\to\infty} y_k = y_0$. Then

- **i)** The limit of a sum is the sum of the limits: $\boldsymbol{x} + \boldsymbol{y}$ is in $\mathbf{C}$ with $\lim_{k\to\infty}(x_k + y_k) = x_0 + y_0$.
- **ii)** The limit of a product is the product of the limits: $\boldsymbol{xy}$ is in $\mathbf{C}$ with $\lim_{k\to\infty} x_k y_k = x_0 y_0$.
- **iii)** If $x_k \neq 0$ for all k in $\mathbb{N}$ and if $x_0 \neq 0$, then $\boldsymbol{x}^{-1} = \{1/x_k\}$ is in $\mathbf{C}$ with $\lim_{k\to\infty} 1/x_k = 1/x_0$.
- **iv)** The limit of a quotient is the quotient of the limits, provided the denominator is a unit. If $\boldsymbol{x}^{-1}$ exists, then $\boldsymbol{yx}^{-1} = \{y_k/x_k\}$ is in $\mathbf{C}$ with $\lim_{k\to\infty} y_k/x_k = y_0/x_0$.

Proof. To prove (i), first notice that

$$\begin{aligned} |(x_k + y_k) - (x_0 + y_0)| &= |(x_k - x_0) + (y_k - y_0)| \\ &\leq |x_k - x_0| + |y_k - y_0|. \end{aligned} \tag{1.11}$$

Therefore, if we can make each of these last summands eventually arbitrarily small, we will have (i) proved. We can achieve this by making each of the summands, $|x_k - x_0|$ and $|y_k - y_0|$, small independently. This last we can do since we are assuming that $\{x_k\}$ converges to x_0 and that $\{y_k\}$ converges to y_0. This sketches our strategy.

Now assign any positive value to ϵ. Choose k_1 so that, for $k \geq k_1$, x_k is in $N(x_0; \epsilon)$. Choose k_2 so that, for $k \geq k_2$, y_k is in $N(y_0; \epsilon)$. Let $k_0 = \max\{k_1, k_2\}$. For $k \geq k_0$, both conditions hold true, so $|x_k - x_0| < \epsilon$ and $|y_k - y_0| < \epsilon$. Therefore, from (1.11), we conclude that

$$|(x_k + y_k) - (x_0 + y_0)| < 2\epsilon.$$

Thus $\lim_{k\to\infty}(x_k + y_k) = x_0 + y_0$ and (i) is proved.

Before proceeding, let's identify a style we want to be free to adopt throughout the text and one you will want to learn. In the previous argument, we obtained 2ϵ rather than ϵ as an upper bound for the difference

$$|(x_k + y_k) - (x_0 + y_0)|.$$

Formally, Definition 1.3.2 requires greater precision. Had we initially chosen our neighborhoods to be $N(x_0; \epsilon/2)$ and $N(y_0; \epsilon/2)$, then chosen k_1, k_2, and $k_0 =$

$\max\{k_1, k_2\}$ as above, our eventual upper bound would have been ϵ as required by Definition 1.3.2. But this seems to us to represent an unnecessary technicality and one that interferes with the beginner's enjoyment of analysis. The number ϵ is arbitrarily small. Twice an arbitrarily small number is still arbitrarily small. Why quibble? In fact, suppose that K is any positive constant and that ϵ is any arbitrarily small positive number. Then $K\epsilon$ is likewise arbitrarily small. We hasten to emphasize that K must be a positive *constant*. To prove that a sequence $\{x_k\}$ converges to x_0 it suffices to show that for any $\epsilon > 0$ and any positive constant K, there is an index k_0 such that, for $k \geq k_0$, x_k is in $N(x_0; K\epsilon)$. In our proof of (i), $K = 2$ is certainly a constant. You will have the opportunity to confirm your understanding of this point as we now prove (ii).

Suppose that $\{x_k\}$ and $\{y_k\}$ are in **C** with limits x_0 and y_0, respectively. We want to show that $x_k y_k$ is eventually arbitrarily near $x_0 y_0$. By a standard maneuver, we complicate matters to make them simple by introducing and removing the mixed product as follows:

$$
\begin{aligned}
|x_k y_k - x_0 y_0| &= |x_k y_k - x_0 y_k + x_0 y_k - x_0 y_0| \\
&\leq |x_k - x_0||y_k| + |x_0||y_k - y_0|. \qquad \textbf{(1.12)}
\end{aligned}
$$

We can make $|x_k - x_0|$ and $|y_k - y_0|$ as small as we may wish; $|x_0|$ is merely a constant, and the sequence $\{y_k\}$, being convergent, is bounded. These observations lead us to our strategy. Let M be a bound for $\{y_k\}$, so $|y_k| \leq M$ for all k in $\mathbb{N}$. Given $\epsilon > 0$, choose k_1 so that, for $k \geq k_1$, x_k is in $N(x_0; \epsilon)$. Likewise, choose k_2 so that, for $k \geq k_2$, y_k is in $N(y_0; \epsilon)$. Let $k_0 = \max\{k_1, k_2\}$. Applying (1.12), we have, for $k \geq k_0$,

$$
\begin{aligned}
|x_k y_k - x_0 y_0| &\leq |x_k - x_0||y_k| + |x_0||y_k - y_0| \\
&< \epsilon M + |x_0|\epsilon = (M + |x_0|)\epsilon.
\end{aligned}
$$

Since $K = M + |x_0|$ is a positive constant, we conclude that $\lim_{k\to\infty} x_k y_k = x_0 y_0$ and (ii) is proved. Notice that initially we could have taken the neighborhoods to be $N(x_0; \epsilon/2M)$ and $N(y_0; \epsilon/2|x_0|)$ (assuming $x_0 \neq 0$) had we wished. The crucial fact that makes this proof work is that the bound M and the number $|x_0|$ exist *a priori* and therefore independent of the variable index k_0.

To prove (iii), we need to handle a different type of complication. It is this. We want to be able to ensure that

$$
\left|\frac{1}{x_k} - \frac{1}{x_0}\right| = \frac{|x_0 - x_k|}{|x_k||x_0|}
$$

is eventually arbitrarily small. Now, we can make the difference $|x_0 - x_k|$ as small as we may wish and $1/|x_0|$ is just a positive constant that presents no problem. The issue is that we need $1/|x_k|$ to be bounded above. We know that $\{|x_k|\}$ itself is bounded above, but that will not help us since taking reciprocals of positive numbers reverses inequalities. To bound $\{1/|x_k|\}$ above is equivalent to bounding $\{|x_k|\}$ away from 0 and for this we need to use the additional hypothesis in (iii), namely, that $\lim_{k\to\infty} x_k = x_0 \neq 0$ and that none of the terms of the sequence $\{x_k\}$ is 0.

Consider the neighborhood $N(x_0; |x_0|/2)$. It contains all but finitely many of the terms of the sequence. That is, there is an index k_1 such that, for $k \geq k_1$, x_k is in

$N(x_0; |x_0|/2)$. None of the first $k_1 - 1$ terms of the sequence is 0. Consequently, if we let

$$m = \min \left\{ |x_1|, |x_2|, \ldots, |x_{k_1-1}|, \frac{|x_0|}{2} \right\},$$

then, as you can confirm, $0 < m \leq |x_k|$, for all k in $\mathbb{N}$. It follows that $1/|x_k| \leq 1/m$, for all k in $\mathbb{N}$. In other words, $M = 1/m$ is a positive upper bound for $\{1/|x_k|\}$.

Now we can proceed. Let M be any upper bound for $\{1/|x_k|\}$. Given any $\epsilon > 0$, choose k_0 so that, for $k \geq k_0$, x_k is in $N(x_0; \epsilon)$. Then, for such k,

$$\left| \frac{1}{x_k} - \frac{1}{x_0} \right| \leq \left(\frac{M}{|x_0|} \right) |x_k - x_0| < \left(\frac{M}{|x_0|} \right) \epsilon.$$

Since ϵ is arbitrarily small and since $K = M/|x_0|$ is a positive constant, we conclude that $\lim_{k\to\infty}(1/x_k) = 1/x_0$ and (iii) is proved.

Statement (iv) is proved by combining (ii) and (iii) and is left as an exercise. ●

These proofs link the algebraic structure of the ring **C** with the analytic properties of convergence. Similar connections will recur throughout the text.

For future reference, we formalize the insight gained en route to proving part (iii) of Theorem 1.5.1.

THEOREM 1.5.2 If $\boldsymbol{x} = \{x_k\}$ has a reciprocal $\boldsymbol{x}^{-1}$ in **C**; that is, if $x_k \neq 0$ for all $k = 0, 1, 2, \ldots$, then there exists a positive constant m such that $m \leq |x_k|$ for all k in $\mathbb{N}$. ●

We can build on the results obtained in Theorem 1.5.1 and derive additional facts about convergent sequences that are essentially algebraic in nature.

THEOREM 1.5.3 Let $\{x_k\}$ be any convergent sequence in $\mathbb{R}$ with limit x_0.

i) For any m in $\mathbb{N}$, $\lim_{k\to\infty} x_k^m = x_0^m$.

ii) For any polynomial $p(x) = c_0 + c_1 x + c_2 x^2 + \cdots + c_r x^r$,

$$\lim_{k\to\infty} p(x_k) = p(x_0).$$

iii) For any rational function, that is, for a function of the form $p(x)/q(x)$, where p and q are polynomials,

$$\lim_{k\to\infty} \frac{p(x_k)}{q(x_k)} = \frac{p(x_0)}{q(x_0)}$$

provided that $q(x_k) \neq 0$ for all k in $\mathbb{N}$ and that $q(x_0) \neq 0$.

Proof. For (i) use an inductive proof, applying part (ii) of Theorem 1.5.1. For (ii) use Theorem 1.3.6, part (i) of the present theorem, and part (i) of Theorem 1.5.1 in a proof by induction on the degree r of the polynomial. Part (iii) is proved by applying part (iv) of Theorem 1.5.1 and part (ii) of the present theorem. ●

Parts (ii) and (iii) of Theorem 1.5.3 can be expressed by writing $\lim_{k\to\infty} f(x_k) = f(\lim_{k\to\infty} x_k)$ and are often summarized by saying "polynomials and rational functions preserve limit." Such behavior, of course, is not restricted merely to these classes of functions. For many functions it is true that $\lim_{k\to\infty} f(x_k) = f(\lim_{k\to\infty} x_k)$. We will study this phenomenon in a more general setting in Chapter 3. Here we want to establish this property for a few elementary functions.

THEOREM 1.5.4 Let $\{x_k\}$ be a convergent sequence of nonnegative terms in $\mathbb{R}$ with limit x_0.

i) $\lim_{k\to\infty} \sqrt{x_k} = \sqrt{x_0}$.

ii) For any n in $\mathbb{N}$,

$$\lim_{k\to\infty} x_k^{1/n} = x_0^{1/n},$$

where $c^{1/n}$ denotes the positive nth root of c.

iii) Let m/n be any rational number. Then

$$\lim_{k\to\infty} x_k^{m/n} = x_0^{m/n}.$$

If m/n is negative, then we require, in addition, that $x_k \neq 0$ for all k in $\mathbb{N}$ and that $x_0 \neq 0$.

Proof. Both (i) and (ii) follow from the factorization

$$a^n - b^n = (a-b)\sum_{j=1}^{n} a^{n-j}b^{j-1}, \tag{1.13}$$

which you can confirm simply by multiplying out the right-hand side and telescoping the resulting sum. We will prove (ii), noting that (i) is simply the special case when $n = 2$.

We treat the case $x_0 = 0$ separately. If $\lim_{k\to\infty} x_k = 0$, then, given any $\epsilon > 0$, we choose a k_0 in $\mathbb{N}$ such that, for all $k \geq k_0$, we have $0 \leq x_k < \epsilon^n$. Since the taking of positive nth roots preserves order, we deduce that $0 \leq x_k^{1/n} < \epsilon$. Therefore, $\lim_{k\to\infty} x_k^{1/n} = 0$.

There remains the case $\lim_{k\to\infty} x_k = x_0 > 0$. Let $m = x_0/2$. There exists an index k_1 such that, for all $k \geq k_1$, $x_k > m$. Clearly, $x_0 > m$ also. Again, since the taking of positive nth roots is order-preserving, we have $x_0^{1/n} > m^{1/n}$ and, for all $k \geq k_1$, $x_k^{1/n} > m^{1/n}$.

Next, since $\{x_k\}$ converges to x_0, given $\epsilon > 0$, we can choose a k_2 such that, for $k \geq k_2$, we have $|x_k - x_0| < \epsilon$. Let $k_0 = \max\{k_1, k_2\}$ and choose any $k \geq k_0$. Let $a = x_k^{1/n}$ and $b = x_0^{1/n}$. By substituting these values for a and b in (1.13) we obtain

$$|x_k - x_0| = |x_k^{1/n} - x_0^{1/n}| \sum_{j=1}^{n} x_k^{(n-j)/n} x_0^{(j-1)/n}. \tag{1.14}$$

Each summand of the sum in (1.14) is bounded below by $m^{(n-1)/n}$. There are n summands. Therefore

$$\sum_{j=1}^{n} x_k^{(n-j)/n} x_0^{(j-1)/n} \geq nm^{(n-1)/n} > 0.$$

It follows from (1.14) that

$$|x_k - x_0| \geq |x_k^{1/n} - x_k^{1/n}|\, nm^{(n-1)/n} > 0.$$

Let K denote the positive constant $1/(nm^{(n-1)/n})$. We deduce that, for each $k \geq k_0$,

$$|x_k^{1/n} - x_0^{1/n}| \leq K|x_k - x_0| < K\epsilon.$$

Therefore $\lim_{k\to\infty} x_k^{1/n} = x_0^{1/n}$. This completes the proof of part (ii).

For part (iii) apply part (i) of Theorem 1.5.3 and part (ii) of the present theorem. If m/n is negative, use part (iii) of Theorem 1.5.1 as well. ●

COROLLARY 1.5.5 Suppose that $\{x_k\}$ and $\{y_k\}$ are convergent sequences with limits x_0 and y_0, respectively. Then

$$\lim_{k\to\infty} |x_k + y_k| = |x_0 + y_0|.$$

Remark. This corollary justifies an assertion made following the proof of Theorem 1.4.5, which led to our estimate of the error at the kth stage of a contractive sequence.

Proof. For all a and b in $\mathbb{R}$, we have $|a+b| = \sqrt{(a+b)^2}$. Now $\{x_k + y_k\}$ converges to $x_0 + y_0$. Thus $\{(x_k + y_k)^2\}$ converges to $(x_0 + y_0)^2$ by parts (i) and (ii) of Theorem 1.5.1. Therefore $\{\sqrt{(x_k + y_k)^2}\}$ converges to $\sqrt{(x_0 + y_0)^2}$ by part (i) of Theorem 1.5.4. That is, $\lim_{k\to\infty} |x_k + y_k| = |x_0 + y_0|$. This proves the corollary. ●

EXAMPLE 30 Suppose that $\{x_k\}$ is any convergent sequence with limit x_0. Let $y_k = 1/\sqrt{x_k^2 + 1}$. Then $\{y_k\}$ converges to $1/\sqrt{x_0^2 + 1}$. The proof is easy, consisting merely of applying pertinent facts that we have already proved. First, $\{x_k^2 + 1\}$ converges to $x_0^2 + 1$ by Theorem 1.5.3. By part (iii) of Theorem 1.5.4, with $m/n = -1/2$, we deduce that

$$\lim_{k\to\infty} \frac{1}{\sqrt{x_k^2 + 1}} = \frac{1}{\sqrt{x_0^2 + 1}}.$$

Although each individual piece of this puzzle is simple, by combining them creatively you can prove with relative ease that many complicated sequences converge, thereby possibly avoiding laborious proofs which rely on first principles. ●

Earlier we asked you to take some results of this sort on faith. Let's revisit those examples.

EXAMPLE 31 In Example 20 of Section 1.3, we used the fact that for any fixed j, the product $\prod_{i=1}^{j-1}[1 - i/m]$ has limit 1 as m tends to ∞ . For each factor, $\lim_{m\to\infty}[1 - i/m] = 1$. There are only $j - 1$ factors and, by inductively using part (ii) of Theorem 1.5.1, we know that the product of them all has limit 1 also. ●

EXAMPLE 32 In Example 29 of Section 1.4, we passed to the limit in the recursive relation $x_{k+1} = (x_k^3 + 2)/4$, knowing that $\lim_{k\to\infty} x_k = x_0$. We concluded that $x_0 =$

$(x_0^3+2)/4$. Technically, we needed to invoke part (ii) of Theorem 1.5.3 to justify $\lim_{k\to\infty}(x_k^3+2)/4 = (x_0^3+2)/4$. ●

EXAMPLE 33 In Exercise 1.79 at the end of Section 1.4, we defined a sequence recursively by $x_{k+1} = 1/(c+x_k)$ where c is a positive constant and $x_1 > 0$. You proved that $\{x_k\}$ is contractive, hence convergent to some limit x_0. But you were not asked to find x_0. Since rational functions preserve limit,

$$\lim_{k\to\infty} \frac{1}{c+x_k} = \frac{1}{c+x_0}.$$

Therefore, taking the limit in the recursive relation, we see that x_0 satisfies the equation $x_0 = 1/(c+x_0)$ and hence the quadratic equation $x_0^2 + cx_0 - 1 = 0$. Since x_0 is positive, we conclude that $\{x_k\}$ converges to $x_0 = [-c+\sqrt{c^2+4}]/2$. ●

EXAMPLE 34 We know that $\lim_{k\to\infty}(1+1/k)^k = e$. If we want to find $\lim_{k\to\infty}[1+1/(2k)]^k$, we simply write

$$\left[1+\frac{1}{2k}\right]^k = \left(\left[1+\frac{1}{2k}\right]^{2k}\right)^{1/2}.$$

Note that the sequence $\{[1+1/(2k)]^{2k}\}$ is the subsequence of $\{(1+1/k)^k\}$ having even indices. This subsequence converges to e also. By part (i) of Theorem 1.5.4,

$$\lim_{k\to\infty}\left[1+\frac{1}{2k}\right]^k = e^{1/2}. \quad ●$$

EXERCISES

1.88. Prove part (iv) of Theorem 1.5.1.

1.89. To confirm that $\lim_{k\to\infty} x_k = x_0$, prove that it suffices to show that for $\epsilon > 0$ and any positive constant K, there is an index k_0 such that, for $k \geq k_0$, x_k is in $N(x_0; K\epsilon)$.

1.90. Prove directly that the sum and product of two Cauchy sequences is Cauchy.

1.91. Prove that, if $\{x_k\}$ converges to x_0, then $\{|x_k|\}$ converges to $|x_0|$. Is the converse true? If so, prove it; otherwise, provide a counterexample.

1.92. Let $x_1 = 1$ and define $x_{k+1} = \sqrt{2x_k}$ for k in $\mathbb{N}$. Prove that $\{x_k\}$ converges and find its limit.

1.93. Let $x_1 = 1$ and define $x_{k+1} = (2x_k+3)/4$ for k in $\mathbb{N}$. Determine whether $\{x_k\}$ converges. If it does, find its limit; otherwise, prove that it diverges.

1.94. Choose any $x_1 > 0$ and define $x_{k+1} = x_k + 1/x_k$ for k in $\mathbb{N}$. Determine whether $\{x_k\}$ converges. If it does, find its limit; otherwise prove that it diverges.

1.95. Fix any $c > 0$. Let x_1 be any positive number and define $x_{k+1} = (x_k + c/x_k)/2$.

a) Prove that $\{x_k\}$ converges and find its limit.

b) Use this sequence to calculate $\sqrt{5}$, accurate to six decimal places.

1.96. Let x_1 be any positive number. For k in $\mathbb{N}$, define $x_{k+1} = 3 + 4/x_k$. Determine whether $\{x_k\}$ converges. If so, find its limit; otherwise, show that it diverges.

1.97. Find $\lim_{k\to\infty}(1 + 1/(k+1))^k$.

1.98. Find $\lim_{k\to\infty}(1 + 3/k)^k$.

1.99. Let a and b be constants such that $0 < a < b$. Prove that $\lim_{k\to\infty}(a^k + b^k)^{1/k} = b$.

1.100. Let **B** denote the collection of all bounded sequences of real numbers. Define addition and multiplication as in C in the text and show that **B** is a commutative ring with multiplicative identity.

1.6 Cardinality

We began as children to understand the process of counting things. We touch or pick up one item after another while simultaneously voicing the next integer: one, two, three, ... In childhood, at least, every set of things that we counted was finally exhausted and we could declare how many objects were in the set. What we were doing, although it would be the rare child who explained it thus, was constructing a one-to-one correspondence between the elements in the set S of objects being counted and a subset $\{1, 2, \ldots, n\}$ of the natural numbers $\mathbb{N}$. Having identified n, we would say that there were n elements (jellybeans, toys, popsicles, ...) in the set S. Provided n is not too large, this is easy for you now, but at the time, you passed through a most astounding intellectual transformation, one apparently beyond the mental capacities of most species and even some humans. Of course, if the set S is "large," determining the number of elements in it might, even now, prove daunting. You were presented always with sets of objects that were "countable" and you bravely set about counting, confident that you would arrive at an end to the process. Actually, to be more precise, you were always presented with a set that was *finite*—a special case of "countable"—and that is why you always could finish the task of counting it. It will come as no surprise that we intend to complicate these matters by plunging into the depths.

Historically, there has lurked in the shadows a concept called *infinite*, meaning merely "not finite," that has fascinated humankind for millennia. The early Greeks toyed with the concept, crafting a number of paradoxes that, while amusing, reveal a flaw in the understanding of infinite processes. One of the more famous of these paradoxes, that of Achilles and the Tortoise, is attributed to Zeno who proposed the following scenario. Suppose Achilles, who can run twice as fast as his friend the Tortoise, agrees to allow him to start a race a certain distance d ahead. Before Achilles can overtake the Tortoise, he must first arrive at the point where the Tortoise began the race. Meanwhile, the Tortoise will have advanced half the distance d and thus will still be leading. Achilles must now make up this difference before overtaking the Tortoise, but by the time he has done so, the Tortoise will again have advanced $d/4$ and thus will still be winning the race. Repeating this argument *ad infinitum*, says Zeno, proves that Achilles can never overtake the Tortoise and thus cannot win the race. Contrasting this argument with observable reality gives the paradox. The ancients had no explanation to resolve it.

During the Middle Ages, with its sharp distinction between sensible, earthly experience and the heavenly ideal, the "infinite" became entangled with the "eternal," a topic considered inappropriate for study by mere mathematicians. Indeed, theologians and philosophers of medieval and Renaissance Europe responded with outright hostility to the persistent appearance of the infinite in mathematical discourse. For example, the Neoplatonist Nicholas of Cusa (1401–1464), rejecting the possibility of mathematical truth regarding the infinite, asserted:

> *A finite intellect cannot by means of comparison reach the absolute truth of things. Being by nature indivisible, truth excludes "more" or "less," so that nothing but truth itself can be the exact measure of truth. . . . In consequence, our intellect, which is not the truth, never grasps the truth with such precision that it could not be comprehended with infinitely greater precision.*

Earlier, in a similar if more mundane vein, Duns Scotus (1266–1308) had rejected the idea that lines and curves are composed of infinitely many points. He reached this conclusion by considering two concentric circles and observing that the points of either can be put into one-to-one correspondence with the points of the other by rays emanating from their common center. He concluded that, were these curves composed of infinitely many points, their circumferences, being composed of the same number of points, must be equal, an obvious falsehood. Galileo countered that the two circles did, in fact, consist of the same (infinite) number of points but that the larger circle also contained infinitely many "infinitely small" gaps that did not occur in the smaller circle.

Hobbes (1588–1679) asserted that, "When it is asked if the world is finite or infinite, there is nothing in the mind corresponding to the vocable world; . . . whatever we imagine is *ipso facto* finite." And the empiricist Locke (1632–1704) declared, "If a man had a positive idea of infinite, he could add two infinites together; nay, make one infinite infinitely bigger than another, absurdities too gross to be confuted."

Ridicule and hostility persisted over the centuries until, in the late nineteenth century, Georg Cantor came to grips with the elusive properties of the "nonfinite." Cantor began his career as an applied mathematician, concerned with the convergence properties of Fourier series as they apply to the central issues of mathematical physics in his day, the conduction of heat and the propagation of light. The technical difficulties he encountered in his researches forced him to examine the nature of "infinity." Bertrand Russell, in recognition of Cantor's genius, wrote:

> *For over two thousand years the human intellect was baffled by the problem of [infinity]. A long line of philosophers, from Zeno to Bergson, have based much of their metaphysics upon the supposed impossibility of infinite collections. The definitive solution of the difficulties is due to Georg Cantor.*

As is true of most genuine mathematical advances, Cantor's approach was both subtle and simple. The trick was to find the correct underlying principle. He did that for us.

DEFINITION 1.6.1 A set S is called *finite* if it is in one-to-one correspondence with some subset $\{1, 2, 3, \ldots, n\}$ of the natural numbers $\mathbb{N}$. S is called *infinite* if it is not finite. A set is called *countably infinite* if it is in one-to-one correspondence with the set of natural numbers $\mathbb{N}$. A set is called *countable* either if it is finite or if it is countably infinite. An infinite set that is not countably infinite is said to be *uncountable.* ●

This definition creates three categories of sets determined by the "number of elements," the *cardinality*, of the set: finite, countably infinite, and uncountable. Actually, the category "uncountable" is an amalgam of infinitely many degrees of infinitude just as Locke, two centuries before, had scoffingly predicted, but for our purposes it will suffice to consider just these three categories.

At this stage, it may not be clear to you that any uncountable sets exist. As we have seen, over the centuries many denied the existence even of countably infinite sets, let alone yet larger ones. Even today it is commonplace to balk at the notion that such huge sets exist. If you are having trouble grasping the idea, know that you are in good company.

The initial response to Cantor's work was mixed. The German mathematician Kronecker, who asserted that "God made the integers; all else is the work of man," reacted with bitter acrimony, even attempting to block publication of Cantor's work. By contrast, Hilbert (1862–1943), who in a 1900 speech to the International Congress of Mathematicians established much of the program for twentieth century mathematical research, declared, "No one shall be able to drive us from the paradise that Cantor has created for us."

To give you a first glimpse of Cantor's original work, let us begin by considering countably infinite sets. As we proceed, do not be surprised if we obtain results that seem counterintuitive. For example, the function $f(k) = 2k$ maps the set of natural numbers one-to-one onto the set of all even natural numbers. It follows immediately that these two sets have the same cardinality; there are the "same number" of even natural numbers as there are natural numbers despite the fact that the former set is a proper subset of $\mathbb{N}$. Likewise, the function f defined by

$$f(k) = \begin{cases} 1 & \text{if } k = 0 \\ 2|k| & \text{if } k < 0 \\ 2k + 1 & \text{if } k > 0, \end{cases}$$

for k in $\mathbb{Z}$ maps the set $\mathbb{Z}$ of all integers one-to-one onto $\mathbb{N}$. Therefore $\mathbb{Z}$ is countably infinite.

It will be convenient to have an alternative way to describe countably infinite sets: Observe that a set S is countably infinite if and only if it can be written as a sequence of distinct points $S = \{s_1, s_2, s_3, \ldots, s_k, \ldots\}$ by virtue of the one-to-one correspondence between S and $\mathbb{N}$. As an immediate application, suppose that S and T are each countably infinite and are disjoint. We want to show that $S \cup T$ is also countably infinite. By the previous observation, write $S = \{s_1, s_2, s_3, \ldots\}$ and $T = \{t_1, t_2, t_3, \ldots\}$. Define a function f with domain $S \cup T$ and range $\mathbb{N}$ as follows:

$$f(x) = \begin{cases} 2k & \text{if } x = s_k \\ 2k - 1 & \text{if } x = t_k. \end{cases}$$

It is straightforward to see that f maps $S \cup T$ one-to-one onto $\mathbb{N}$ and therefore that $S \cup T$ is countably infinite. By induction on n, if $S_1, S_2, \ldots, S_n$ are countably infinite and disjoint, then $S_1 \cup S_2 \cup S_3 \cup \cdots \cup S_n$ is also countably infinite.

It is easy to see that if S is a subset of a countable set, then S itself must also be countable. (See Exercise 1.102.)

You are familiar with the Cartesian product of two sets, $S \times T = \{(s, t) : s \text{ in } S, t \text{ in } T\}$. Cantor observed that, if each of S and T is countably infinite, then $S \times T$ is also countably infinite. His trick is to display the elements of $S \times T$ as follows

$$\begin{array}{ccccccccc}
(s_1, t_1) & \rightarrow & (s_1, t_2) & & (s_1, t_3) & \rightarrow & (s_1, t_4) \ldots & & \\
& \swarrow & & \nearrow & & \swarrow & & \nearrow & \\
(s_2, t_1) & & (s_2, t_2) & & (s_2, t_3) & & (s_2, t_4) \ldots & & \\
\downarrow & \nearrow & & \swarrow & & \nearrow & & & \\
(s_3, t_1) & & (s_3, t_2) & & (s_3, t_3) & & (s_3, t_4) \ldots & & \\
& \swarrow & & \nearrow & & \swarrow & & & \\
(s_4, t_1) & & (s_4, t_2) & & (s_4, t_3) & & (s_4, t_4) \ldots & & \\
\vdots & \nearrow & \vdots & \swarrow & \vdots & & \vdots & &
\end{array}$$

and to display a one-to-one function from $\mathbb{N}$ onto $S \times T$ starting in the upper left-hand corner and zig-zagging diagonally as indicated. Without writing down the function explicitly, we conclude that $S \times T$ is countably infinite. By induction on n, if each of $S_1, S_2, \ldots,$ and S_n is countably infinite, then $S_1 \times S_2 \times \cdots \times S_n$ is also countably infinite.

Now, any positive rational number $x = p/q$, with p and q relatively prime (that is, p/q is reduced to lowest terms), can be identified with the ordered pair (p, q) in $\mathbb{N} \times \mathbb{N}$. Since $\mathbb{N} \times \mathbb{N}$ is countably infinite and since

$$\{(p, q) : p, q \text{ in } \mathbb{N}, p \text{ and } q \text{ relatively prime}\}$$

is a subset of $\mathbb{N} \times \mathbb{N}$, we conclude that the set of positive rationals, $\mathbb{Q}^+$, is countably infinite. Likewise, the set, $\mathbb{Q}^-$ of all negative rationals is countably infinite and therefore $\mathbb{Q}$—the union of $\mathbb{Q}^+$, $\mathbb{Q}^-$, and $\{0\}$—is also countably infinite.

To this point, we have discovered that each of $\mathbb{N}$, $\mathbb{Z}$, and $\mathbb{Q}$ has the same cardinality; each is countably infinite. When we turn our attention to $\mathbb{R}$, obtained from $\mathbb{Q}$ by adjoining the set of irrational numbers, we discover a new phenomenon, one of the first of Cantor's many triumphs. As we will soon demonstrate, $\mathbb{R}$ is uncountable. We begin, following Cantor, with the closed interval $S = [0, 1]$ and show that S is uncountable. Cantor's technique in this proof is especially worthy of your attention because variants of it are used throughout analysis; in honor of Cantor's cleverness, it is called *Cantor's diagonalization technique.* Since S is clearly not finite, it must be either countably infinite or uncountable. If we assume S is countably infinite and obtain a logically untenable conclusion, then we will be forced to conclude that S is uncountable.

Suppose then that $S = [0, 1]$ is countably infinite. We will represent numbers in S by infinite decimals of the form $0.d_1 d_2 d_3 \ldots$, where d_k is any of the digits $\{0, 1, 2, \ldots, 9\}$. Since this representation is ambiguous for repeating decimals of the form $x = 0.d_1 d_2 d_3 \ldots d_k \overline{9}$ with $d_k \neq 9$, we exclude this form of representation of x and use instead

$$x = 0.d_1 d_2 d_3 \ldots (d_k + 1)\overline{0}$$

obtained by "rounding up" the kth digit. The number 1 is to be represented by $1 = .9999\ldots$. With this stipulation, there is no ambiguity in the representation of the numbers in S as infinite decimals.

Since we are assuming that S is countably infinite, we are assuming that there exists a function f mapping $\mathbb{N}$ one-to-one onto S. For k in $\mathbb{N}$, denote

$$f(k) = x_k = 0.d_{k1}d_{k2}d_{k3}\ldots d_{kk}\ldots.$$

When we display these decimal representations of the elements of S, we obtain the following array:

$$\begin{array}{l}
x_1 = 0.d_{11}d_{12}d_{13}d_{14}\ldots d_{1k}\ldots \\
x_2 = 0.d_{21}d_{22}d_{23}d_{24}\ldots d_{2k}\ldots \\
x_3 = 0.d_{31}d_{32}d_{33}d_{34}\ldots d_{3k}\ldots \\
x_4 = 0.d_{41}d_{42}d_{43}d_{44}\ldots d_{4k}\ldots \\
\vdots \\
x_k = 0.d_{k1}d_{k2}d_{k3}d_{k4}\ldots d_{kk}\ldots \\
\vdots
\end{array}$$

We are assuming that every number in $[0, 1]$ is listed in this array; that is, we are assuming f maps $\mathbb{N}$ onto S. Cantor focused on the digits d_{kk} on the main diagonal of this array (hence the name *diagonal technique*). Fix any two digits in $\{1, 2, 3, 4, 5, 6, 7, 8\}$, say, 2 and 7. We will construct a real number $y = 0.c_1c_2c_3\cdots c_k\cdots$ that is in S but that is not, indeed cannot be, listed in the above array. We will construct y by selecting the digits c_k for each $k = 1, 2, 3, \ldots$ according to the following rule:

$$\text{If } d_{kk} = 2, \quad \text{let } c_k = 7; \qquad \text{if } d_{kk} \neq 2, \quad \text{let } c_k = 2.$$

Let $y = 0.c_1c_2c_3\ldots c_k\ldots$. The digits c_k in the decimal representation of y are either 2 or 7, so y is a number in S. But Cantor's construction guarantees that y cannot occur in the preceding array; $y \neq x_1$ because these two numbers differ in the first digit. Likewise, $y \neq x_2$ because they differ in the second digit. In general, $y \neq x_k$ because these numbers differ in the kth digit. We are forced to conclude that the function f cannot map $\mathbb{N}$ onto S, although we initially assumed it does. The contradiction forces us to conclude that there can exist no function mapping $\mathbb{N}$ onto $S = [0, 1]$. Therefore $S = [0, 1]$ cannot be countable; it must be uncountable.

Removing either or both endpoints has no significant effect. Each of the intervals $[0, 1)$, $(0, 1]$, and $(0, 1)$ is also uncountably infinite. Moreover, there is nothing special about $[0, 1]$ in the preceding argument; simplicity of notation led to our choosing it. Given any interval $[a, b]$, the function $f(x) = (b - a)x + a$ maps the interval $[0, 1]$ one-to-one onto the interval $[a, b]$. (See Fig. 1.20 on page 58.) Thus any interval $[a, b]$ is uncountable, each with the same cardinality. By Theorem 1.1.2, there exists a rational number in $[a, b]$. Likewise each of $[a, b)$, $(a, b]$, and (a, b) is uncountable and contains rational numbers. In particular, every neighborhood $N(c; r)$ is uncountable and contains rational numbers. Finally, the function

$$f(x) = \frac{2x - 1}{x(x - 1)}$$

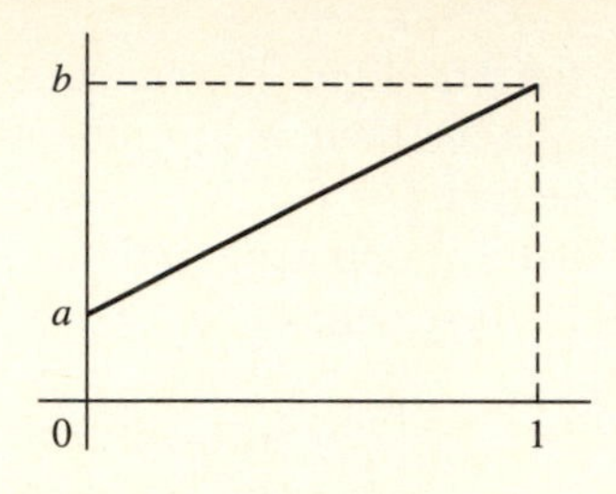

Figure 1.20

maps the interval (0, 1) one-to-one onto all of $\mathbb{R}$. We ask you to prove this in Exercise 1.103. (See Fig. 1.21.) Thus $\mathbb{R}$ is uncountable with the same cardinality as any interval (a, b). While the set of rationals $\mathbb{Q}$ is merely countably infinite, the set $\mathbb{R}$ is uncountably infinite and thus is vastly more huge than $\mathbb{Q}$. Nevertheless, as we have observed, given any neighborhood $N(x)$ of any point x in $\mathbb{R}$, there exists a rational number in $N(x)$. We condense this observation, in accord with the following definition, into the phrase "$\mathbb{Q}$ is dense in $\mathbb{R}$."

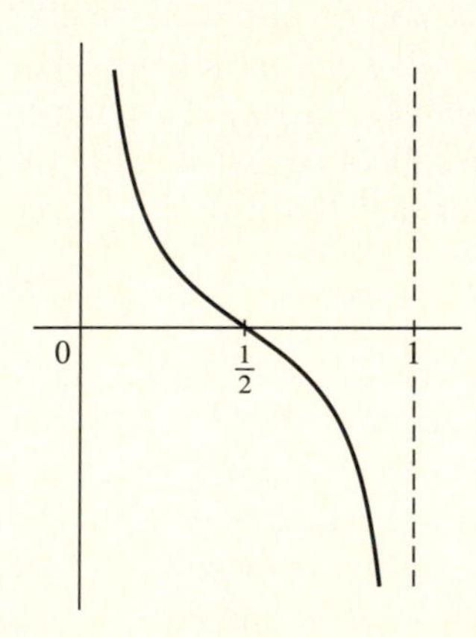

Figure 1.21

DEFINITION 1.6.2 We say a subset S of T is *dense* in T if every neighborhood of any point x of T contains points of S. ●

We summarize the essential facts of this section in the following theorem.

THEOREM 1.6.1

i) $\mathbb{N}$, $\mathbb{Z}$, and $\mathbb{Q}$ are countably infinite.

ii) $\mathbb{R}$ is uncountable.

iii) $\mathbb{Q}$ is dense in $\mathbb{R}$. ●

EXERCISES

1.101. Resolve Zeno's Paradox of Achilles and the Tortoise.

1.102. Prove that a subset of a countable set is countable.

1.103. Prove that $f(x) = (2x - 1)/[x(x - 1)]$ maps $(0, 1)$ one-to-one onto $\mathbb{R}$.

1.104. Prove that the set of all irrational numbers is uncountable.

1.105. Prove that the union of finitely many countably infinite sets is countably infinite.

1.106. Prove that the Cartesian product of finitely many countably infinite sets is countably infinite.

1.107. Prove that the union of a countable collection of countable sets is itself countable. [Let $S_k = \{s_{kj} : j \text{ in } \mathbb{N}\}$ and consider $f(s_{kj}) = 2^k 3^j$ in $\mathbb{N}$.]

1.108. **a)** Suppose that $S = \{s_1, s_2, s_3, \ldots, s_n\}$ is any set with n elements. The power set of S, denoted here $P(S)$, is the collection of all subsets of S, including the empty set and S itself. Let $\mathbf{B}_n = \{(b_1, b_2, b_3, \ldots, b_n) : b_j = 0 \text{ or } 1\}$ denote the set of all binary strings of length n. Show that the cardinality of $P(S)$ is the same as that of $\mathbf{B}_n$.

b) Prove by induction on n that $P(S)$ has cardinality 2^n. (For this reason, $P(S)$ is often denoted 2^S.)

1.109. **a)** Suppose that $S = \{s_1, s_2, s_3, \ldots.\}$ is any countably infinite set and let $P(S)$ denote the collection of all subsets of S. Let $\mathbf{B} = \{(b_1, b_2, b_3, \ldots) : b_j = 0 \text{ or } 1\}$ denote the set of all infinite binary sequences. Show that the cardinality of $P(S)$ is the same as that of $\mathbf{B}$. (Again, $P(S)$ is sometimes denoted 2^S for this reason.)

b) Use Cantor's diagonalization technique to show that $\mathbf{B}$ is uncountable.

c) Hence show that a countably infinite set S has uncountably many subsets.

1.110. Prove that an uncountable set of real numbers must have a limit point.

2

Euclidean Spaces

We will garner significant benefits by freeing ourselves from the narrow confines of the real line and by expanding our horizon to include n-dimensional Euclidean space $\mathbb{R}^n$. As we will see in this chapter, virtually all the concepts introduced in Chapter 1, except the linear order of $\mathbb{R}$ and the resultant concepts of supremum and infimum, lift naturally and easily to $\mathbb{R}^n$. Moreover, we will gain significant power from this more general approach.

2.1 EUCLIDEAN n-SPACE

Euclidean n-space $\mathbb{R}^n$ is the Cartesian product of n copies of $\mathbb{R}$. It consists of all n-tuples $\boldsymbol{x} = (x_1, x_2, \ldots, x_n)$, where, for each coordinate index j, x_j is in $\mathbb{R}$. Because $\mathbb{R}^n$ is a real vector space when endowed with the following operations, points of $\mathbb{R}^n$ will also be called *vectors*.

We define the *sum* of two vectors $\boldsymbol{x} = (x_1, x_2, \ldots, x_n)$ and $\boldsymbol{y} = (y_1, y_2, \ldots, y_n)$ to be the vector

$$\boldsymbol{x} + \boldsymbol{y} = (x_1 + y_1, x_2 + y_2, \ldots, x_n + y_n).$$

This definition produces an operation, *addition*, that is commutative and associative: For all $\boldsymbol{x}$, $\boldsymbol{y}, \boldsymbol{z}$ in $\mathbb{R}^n$,

$$\boldsymbol{x} + \boldsymbol{y} = \boldsymbol{y} + \boldsymbol{x}$$

and

$$\boldsymbol{x} + (\boldsymbol{y} + \boldsymbol{z}) = (\boldsymbol{x} + \boldsymbol{y}) + \boldsymbol{z}.$$

The *zero vector* is $\mathbf{0} = (0, 0, \ldots, 0)$. It has the property that $\boldsymbol{x} + \mathbf{0} = \boldsymbol{x}$ for all $\boldsymbol{x}$ in $\mathbb{R}^n$. The negative of the vector $\boldsymbol{x} = (x_1, x_2, \ldots, x_n)$ is $-\boldsymbol{x} = (-x_1, -x_2, \ldots, -x_n)$. It has the property that $\boldsymbol{x} + (-\boldsymbol{x}) = \mathbf{0}$.

Furthermore, $\mathbb{R}^n$ has a natural *scalar multiplication* of a real number c times a vector $\boldsymbol{x} = (x_1, x_2, \ldots, x_n)$, denoted $c\boldsymbol{x}$ and defined by

$$c\boldsymbol{x} = (cx_1, cx_2, \ldots, cx_n).$$

Scalar multiplication has the following properties: For all $\boldsymbol{x}$ and $\boldsymbol{y}$ in $\mathbb{R}^n$ and all c and d in $\mathbb{R}$,

i) $c(\boldsymbol{x} + \boldsymbol{y}) = c\boldsymbol{x} + c\boldsymbol{y}$.

ii) $c(d\boldsymbol{x}) = (cd)\boldsymbol{x}$.

iii) $(c + d)\boldsymbol{x} = c\boldsymbol{x} + d\boldsymbol{x}$.

iv) $0\boldsymbol{x} = \mathbf{0}$, $1\boldsymbol{x} = \boldsymbol{x}$, and $(-1)\boldsymbol{x} = -\boldsymbol{x}$.

All these properties follow easily from the definitions. In fact, we assume that you are already familiar with these ideas from your first courses in calculus and, preferably, from your first course in linear algebra. The set $\mathbb{R}^n$, together with addition and scalar multiplication, becomes a vector space.

The attempt by Hamilton in the mid-1800s to discover a way to multiply two vectors led, first, to the creation of two forms of vector multiplication, both of which we will discuss in this section and in the exercises, and, second, to the foundation of that branch of mathematics now known as linear algebra. In general, however, for an arbitrary n, there is no obvious way to multiply two vectors and obtain another vector as a product. One of Hamilton's discoveries was the *outer product*, now most commonly called the *cross product* of two vectors in $\mathbb{R}^3$. The other was the *inner product*, which is often also called the *dot product* because of the notation $\boldsymbol{x} \cdot \boldsymbol{y}$ commonly adopted by mathematical physicists.

DEFINITION 2.1.1 The *inner product* of two vectors $\boldsymbol{x} = (x_1, x_2, \ldots, x_n)$ and $\boldsymbol{y} = (y_1, y_2, \ldots, y_n)$ in $\mathbb{R}^n$ is

$$\langle \boldsymbol{x}, \boldsymbol{y} \rangle = \sum_{j=1}^{n} x_j y_j. \quad \bullet$$

Notice that $\langle \boldsymbol{x}, \boldsymbol{y} \rangle$ is a function from $\mathbb{R}^n \times \mathbb{R}^n$ to $\mathbb{R}$. In passing, note also that, if either $\boldsymbol{x}$ or $\boldsymbol{y}$ is the $\mathbf{0}$ vector, then $\langle \boldsymbol{x}, \boldsymbol{y} \rangle = 0$. The converse is false. Further, note that $\langle \boldsymbol{x}, \boldsymbol{x} \rangle$ is a nonnegative real number. The essential properties of the inner product are given in the following theorem.

THEOREM 2.1.1 If $\boldsymbol{x}$, $\boldsymbol{y}$, and $\boldsymbol{z}$ are arbitrary vectors in $\mathbb{R}^n$ and if a and b are real numbers, then the following hold:

i) The inner product is *additive* in both its variables:

$$\langle \boldsymbol{x} + \boldsymbol{y}, \boldsymbol{z} \rangle = \langle \boldsymbol{x}, \boldsymbol{z} \rangle + \langle \boldsymbol{y}, \boldsymbol{z} \rangle$$

$$\langle \boldsymbol{x}, \boldsymbol{y} + \boldsymbol{z} \rangle = \langle \boldsymbol{x}, \boldsymbol{y} \rangle + \langle \boldsymbol{x}, \boldsymbol{z} \rangle.$$

ii) The inner product is *symmetric*: $\langle \boldsymbol{x}, \boldsymbol{y} \rangle = \langle \boldsymbol{y}, \boldsymbol{x} \rangle$.

iii) The inner product is *homogeneous* in both its variables: $\langle a\boldsymbol{x}, b\boldsymbol{y} \rangle = ab\langle \boldsymbol{x}, \boldsymbol{y} \rangle$.

Proof. The first and third properties follow directly from simple computations using the distributive and associative laws for $\mathbb{R}$, the second from the commutative law for multiplication in $\mathbb{R}$. ●

DEFINITION 2.1.2 The *Euclidean norm* of a vector $\boldsymbol{x}$ in $\mathbb{R}^n$ is

$$||\boldsymbol{x}|| = \langle \boldsymbol{x}, \boldsymbol{x} \rangle^{1/2}. \quad \bullet$$

To derive the essential properties of the Euclidean norm, we need the following fact; its proof is a little gem that relies only on elementary algebra and the simplest properties of the inner product.

THEOREM 2.1.2 **The Cauchy–Schwarz Inequality** If $\boldsymbol{x}$ and $\boldsymbol{y}$ are any two vectors in $\mathbb{R}^n$, then

$$|\langle \boldsymbol{x}, \boldsymbol{y} \rangle| \leq ||\boldsymbol{x}|| \cdot ||\boldsymbol{y}||.$$

Proof. For t in $\mathbb{R}$, form the vector $z = t\boldsymbol{x} + \boldsymbol{y}$ and compute the inner product of z with itself. Using the additivity, symmetry, and homogeneity of the inner product, we have

$$\begin{aligned} 0 \leq ||z||^2 = \langle z, z \rangle &= \langle t\boldsymbol{x} + \boldsymbol{y}, t\boldsymbol{x} + \boldsymbol{y} \rangle \\ &= \langle t\boldsymbol{x}, t\boldsymbol{x} \rangle + \langle t\boldsymbol{x}, \boldsymbol{y} \rangle + \langle \boldsymbol{y}, t\boldsymbol{x} \rangle + \langle \boldsymbol{y}, \boldsymbol{y} \rangle \\ &= t^2\langle \boldsymbol{x}, \boldsymbol{x} \rangle + 2t\langle \boldsymbol{x}, \boldsymbol{y} \rangle + \langle \boldsymbol{y}, \boldsymbol{y} \rangle \\ &= ||\boldsymbol{x}||^2 t^2 + 2\langle \boldsymbol{x}, \boldsymbol{y} \rangle t + ||\boldsymbol{y}||^2. \end{aligned}$$

Let $a = ||\boldsymbol{x}||^2$, $b = \langle \boldsymbol{x}, \boldsymbol{y} \rangle$, and $c = ||\boldsymbol{y}||^2$. Then the preceding inequality becomes $at^2 + 2bt + c \geq 0$, for all t in $\mathbb{R}$. As a consequence, the discriminant $\Delta = (2b)^2 - 4ac$ of the quadratic must be nonpositive. Thus $b^2 \leq ac$. Substituting the values for a, b, and c, we obtain

$$0 \leq \langle \boldsymbol{x}, \boldsymbol{y} \rangle^2 \leq ||\boldsymbol{x}||^2 ||\boldsymbol{y}||^2.$$

Taking the square root of each side of this inequality proves the theorem. ●

With the Cauchy–Schwarz inequality in hand, we can now obtain the essential properties of the Euclidean norm.

THEOREM 2.1.3 For vectors $\boldsymbol{x}$ and $\boldsymbol{y}$ in $\mathbb{R}^n$ and any c in $\mathbb{R}$, the Euclidean norm has the following properties:

i) *Positive Definiteness* $||\boldsymbol{x}|| \geq 0$; $||\boldsymbol{x}|| = 0$ if and only if $\boldsymbol{x} = \boldsymbol{0}$.

ii) *Absolute Homogeneity* $||c\boldsymbol{x}|| = |c| \cdot ||\boldsymbol{x}||$.

iii) *Subadditivity* $||\boldsymbol{x} + \boldsymbol{y}|| \leq ||\boldsymbol{x}|| + ||\boldsymbol{y}||$.

Proof. For part (i), we have already seen that $||\boldsymbol{x}|| \geq 0$ for all $\boldsymbol{x}$ in $\mathbb{R}^n$. Clearly, $||\boldsymbol{x}||^2 = \sum_{j=1}^{n} x_j^2 = 0$ if and only if $x_j^2 = 0$ for each $j = 1, 2, \ldots, n$ because the only way for a sum of nonnegative real numbers to vanish is for each summand to be 0. Equivalently, $x_j = 0$ for each j. That is, $\boldsymbol{x} = \boldsymbol{0}$.

For part (ii), merely compute

$$||c\boldsymbol{x}||^2 = \langle c\boldsymbol{x}, c\boldsymbol{x}\rangle = c^2\langle \boldsymbol{x}, \boldsymbol{x}\rangle = c^2||\boldsymbol{x}||^2.$$

Taking the square root gives (ii).

The third assertion of the theorem follows from the Cauchy–Schwarz inequality:

$$\begin{aligned} 0 \leq ||\boldsymbol{x}+\boldsymbol{y}||^2 &= \langle \boldsymbol{x}+\boldsymbol{y}, \boldsymbol{x}+\boldsymbol{y}\rangle \\ &= ||\boldsymbol{x}||^2 + 2\langle \boldsymbol{x}, \boldsymbol{y}\rangle + ||\boldsymbol{y}||^2 \\ &\leq ||\boldsymbol{x}||^2 + 2||\boldsymbol{x}||\,||\boldsymbol{y}|| + ||\boldsymbol{y}||^2 \\ &= (||\boldsymbol{x}|| + ||\boldsymbol{y}||)^2. \end{aligned}$$

Taking the square root gives $||\boldsymbol{x}+\boldsymbol{y}|| \leq ||\boldsymbol{x}|| + ||\boldsymbol{y}||$. This completes the proof of the theorem. ●

DEFINITION 2.1.3 A *norm* on $\mathbb{R}^n$ is any function $n(x)$ from $\mathbb{R}^n$ to $\mathbb{R}$ that is positive definite, absolutely homogeneous, and subadditive. ●

There are many norms on $\mathbb{R}^n$. All are interesting; several are also useful. We will ask you to examine two especially interesting norms on $\mathbb{R}^n$ in the exercises. The Euclidean norm, $||\boldsymbol{x}|| = \langle \boldsymbol{x}, \boldsymbol{x}\rangle^{1/2}$, is the one special norm we will use in $\mathbb{R}^n$ throughout this text. It is the norm generated by the inner product.

DEFINITION 2.1.4 A *metric* $d(\boldsymbol{x}, \boldsymbol{y})$ on $\mathbb{R}^n$ is a function from $\mathbb{R}^n \times \mathbb{R}^n$ to $\mathbb{R}$ having the following properties:

i) *Positive Definiteness* $d(\boldsymbol{x}, \boldsymbol{y}) \geq 0$ for all $\boldsymbol{x}$, $\boldsymbol{y}$ in $\mathbb{R}^n$; $d(\boldsymbol{x}, \boldsymbol{y}) = 0$ if and only if $\boldsymbol{x} = \boldsymbol{y}$.

ii) *Symmetry* $d(\boldsymbol{x}, \boldsymbol{y}) = d(\boldsymbol{y}, \boldsymbol{x})$ for all $\boldsymbol{x}$, $\boldsymbol{y}$ in $\mathbb{R}^n$.

iii) *The Triangle Inequality* $d(\boldsymbol{x}, \boldsymbol{z}) \leq d(\boldsymbol{x}, \boldsymbol{y}) + d(\boldsymbol{y}, \boldsymbol{z})$ for all $\boldsymbol{x}$, $\boldsymbol{y}$, and $\boldsymbol{z}$ in $\mathbb{R}^n$. ●

A metric systematically gives a numerical measure of the "distance" between two vectors. If you will translate the three properties of any metric in terms of your intuitive concept of "distance," you will see that these three properties capture exactly the fundamental characteristics of that concept: Distances are nonnegative and the distance from here to there is 0 if and only if there is here. The distance from here to there is the same in either direction. And, finally, distance may only be increased by traveling first to some intermediate location.

Later we will have occasion to consider a variety of metrics on various vector spaces. Here our interest centers on the Euclidean metric for $\mathbb{R}^n$.

DEFINITION 2.1.5 The *Euclidean metric* on $\mathbb{R}^n$ is defined by

$$d(\boldsymbol{x}, \boldsymbol{y}) = ||\boldsymbol{x} - \boldsymbol{y}|| = \left[\sum_{j=1}^{n}(x_j - y_j)^2\right]^{1/2},$$

for $\boldsymbol{x}$ and $\boldsymbol{y}$ in $\mathbb{R}^n$. ●

THEOREM 2.1.4 The function $d(\boldsymbol{x}, \boldsymbol{y}) = ||\boldsymbol{x} - \boldsymbol{y}||$, for $\boldsymbol{x}$ and $\boldsymbol{y}$ in $\mathbb{R}^n$, is a metric on $\mathbb{R}^n$.

Proof. The theorem follows directly from the properties of the norm. For (i), $||\boldsymbol{x} - \boldsymbol{y}|| \geq 0$ for all $\boldsymbol{x}$ and $\boldsymbol{y}$ in $\mathbb{R}^n$. Further, $||\boldsymbol{x} - \boldsymbol{y}|| = 0$ if and only if $\boldsymbol{x} - \boldsymbol{y} = \boldsymbol{0}$ or $\boldsymbol{x} = \boldsymbol{y}$. Symmetry follows from the simple observation that

$$||\boldsymbol{x} - \boldsymbol{y}|| = ||(-1)(\boldsymbol{y} - \boldsymbol{x})|| = ||\boldsymbol{y} - \boldsymbol{x}||.$$

The triangle inequality results from the subadditivity of the norm:

$$\begin{aligned} d(\boldsymbol{x}, \boldsymbol{z}) &= ||\boldsymbol{x} - \boldsymbol{z}|| = ||(\boldsymbol{x} - \boldsymbol{y}) + (\boldsymbol{y} - \boldsymbol{z})|| \\ &\leq ||\boldsymbol{x} - \boldsymbol{y}|| + ||\boldsymbol{y} - \boldsymbol{z}|| = d(\boldsymbol{x}, \boldsymbol{y}) + d(\boldsymbol{y}, \boldsymbol{z}). \end{aligned}$$

This proves the theorem. ●

Notice that the proof of Theorem 2.1.4 used only the three abstracted properties of the norm rather than any specific characteristic of the Euclidean norm. Consequently, any norm $n(x)$ generates a metric defined by $n(x - y)$; the preceding proof applies verbatim to verify the required properties of a metric. In all cases, the norm of a vector is its distance from $\boldsymbol{0}$.

The Cauchy–Schwarz inequality can be interpreted geometrically to enable us to introduce the notion of the *angle between two vectors*. Suppose that $\boldsymbol{x}$ and $\boldsymbol{y}$ are two distinct, nonzero vectors. The three points—$\boldsymbol{x}$, $\boldsymbol{y}$, and $\boldsymbol{0}$—form a triangle T in space. (See Fig. 2.1.) The sides of the triangle have lengths $||\boldsymbol{x}||$, $||\boldsymbol{y}||$, and $||\boldsymbol{x} - \boldsymbol{y}||$. From the Cauchy–Schwarz inequality,

$$-||\boldsymbol{x}||\,||\boldsymbol{y}|| \leq \langle \boldsymbol{x}, \boldsymbol{y}\rangle \leq ||\boldsymbol{x}||\,||\boldsymbol{y}||.$$

Therefore

$$-1 \leq \frac{\langle \boldsymbol{x}, \boldsymbol{y}\rangle}{||\boldsymbol{x}|| \cdot ||\boldsymbol{y}||} \leq 1.$$

From this inequality we deduce that there exists a unique angle ϕ in $[0, \pi]$ such that $\cos\phi = \langle \boldsymbol{x}, \boldsymbol{y}\rangle / ||\boldsymbol{x}|| \cdot ||\boldsymbol{y}||$. Thus

$$\langle \boldsymbol{x}, \boldsymbol{y}\rangle = ||\boldsymbol{x}||\,||\boldsymbol{y}|| \cos\phi. \tag{2.1}$$

The angle ϕ is the same as the angle θ formed by the line segments from $\boldsymbol{0}$ to $\boldsymbol{x}$ and from $\boldsymbol{0}$ to $\boldsymbol{y}$, as indicated in Fig. 2.1. We can see this by applying the Law of Cosines to the triangle T. The Law of Cosines states that

$$||\boldsymbol{x} - \boldsymbol{y}||^2 = ||\boldsymbol{x}||^2 + ||\boldsymbol{y}||^2 - 2||\boldsymbol{x}||\,||\boldsymbol{y}|| \cos\theta.$$

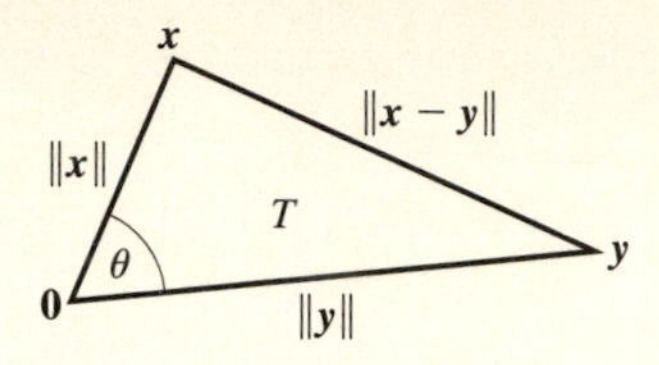

Figure 2.1

Expressing this identity in terms of the coordinates of x and y, we have

$$\sum_{j=1}^{n}(x_j - y_j)^2 = \sum_{j=1}^{n} x_j^2 + \sum_{j=1}^{n} y_j^2 - 2||x|| \cdot ||y|| \cos\theta.$$

Expanding and simplifying, we obtain

$$\langle x, y\rangle = \sum_{j=1}^{n} x_j y_j = ||x|| \cdot ||y|| \cos\theta. \tag{2.2}$$

Comparing (2.1) and (2.2), we see that $\cos\phi = \cos\theta$. Since both ϕ and θ are in $[0, \pi]$, we conclude that $\phi = \theta$. Therefore the angle θ between the line segments joining $\mathbf{0}$ to x and $\mathbf{0}$ to y is the unique angle in $[0, \pi]$ for which

$$\langle x, y\rangle = ||x||\ ||y|| \cos\theta.$$

We refer to this angle as the angle between x and y. Notice that these two line segments are perpendicular if and only if $\theta = \pi/2$. This observation leads to our next definition.

DEFINITION 2.1.6 Two vectors x and y in $\mathbb{R}^n$ are said to be *orthogonal* if $\langle x, y\rangle = 0$. ●

With the Euclidean norm and metric, the inner product, and this notion of orthogonality, $\mathbb{R}^n$ is a vector space in which all of Euclidean geometry applies. We will freely draw upon that geometry in the sequel. For example, the Euclidean metric enables us to lift the concept of neighborhood to $\mathbb{R}^n$.

DEFINITION 2.1.7 A *neighborhood* $N(x; r)$ of x in $\mathbb{R}^n$ with radius r is the set

$$N(x; r) = \{y \text{ in } \mathbb{R}^n : ||x - y|| < r\}.$$

A *deleted neighborhood* $N'(x; r)$ of x is obtained by removing the center point x. ●

In $\mathbb{R}^2$ the neighborhood $N(x; r)$ is the circular disk centered at $x = (x_1, x_2)$ with radius r, omitting the circular boundary. In $\mathbb{R}^3$ the neighborhood $N(x; r)$ is the ball centered at the point $x = (x_1, x_2, x_3)$ with radius r, omitting the spherical boundary. (See Fig. 2.2.) It will come as no surprise that in $\mathbb{R}^n$ there is no everyday language to describe $N(x; r)$; we refer to the neighborhood $N(x; r)$ as a "hypersphere centered at $x = (x_1, x_2, \ldots, x_n)$ with radius r, omitting the hyperspherical boundary."

To develop your intuition, we suggest that you read Abbott's amusing and satirical little book, *Flatland*. In this charming tale, Abbott, an early advocate of

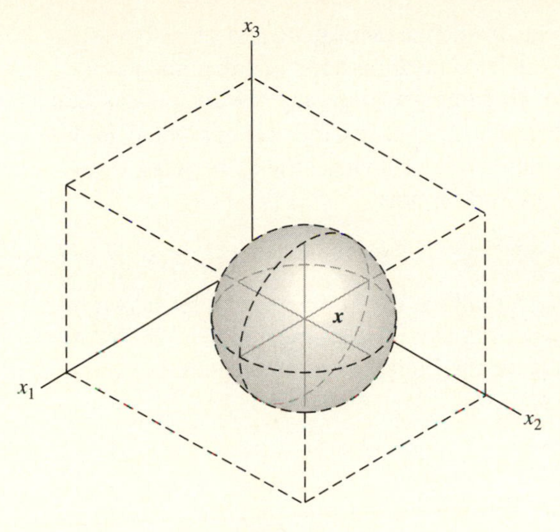

Figure 2.2

universal suffrage and women's rights, tweaked his readers' imagination regarding the structure of space while poking fun at Victorian attitudes on "women's *proper place* in society."

To define $N(\boldsymbol{x}; r)$ in $\mathbb{R}^n$, we have used the Euclidean metric but each metric d on $\mathbb{R}^n$ can be used to define a neighborhood:

$$N_d(\boldsymbol{x}; r) = \{\boldsymbol{y} \text{ in } \mathbb{R}^n : d(\boldsymbol{x}, \boldsymbol{y}) < r\}.$$

While the formal definition remains unchanged, the geometric description of the physical shape of the neighborhood changes to fit what is meant by "distance." (See Exercises 2.2 to 2.5.)

With this definition of a neighborhood in $\mathbb{R}^n$, we are prepared to define the concept of limit point in $\mathbb{R}^n$.

DEFINITION 2.1.8 Let S be a nonempty subset of $\mathbb{R}^n$. We say that $\boldsymbol{x}$ is a *limit point* of S if, for every $\epsilon > 0$, there exists a point of S in $N'(\boldsymbol{x}; \epsilon)$. ●

You will notice that the definition merely repeats verbatim that of a limit point of a set in $\mathbb{R}$. This is a phenomenon we wish to emphasize: The essential concepts of analysis of one real variable, when properly articulated, lift verbatim to $\mathbb{R}^n$.

By imitating the proof of the corresponding fact in $\mathbb{R}$, it is easy to show that, if $\boldsymbol{x}$ is a limit point of S, then every deleted neighborhood of $\boldsymbol{x}$ contains infinitely many points of S. (See Exercise 2.8.) It follows, of course, that for S to have a limit point, it must be not merely nonempty but must have infinitely many points. Again, it is immaterial whether the limit point $\boldsymbol{x}$ belongs to S; every *deleted* neighborhood of $\boldsymbol{x}$ must contain points of S.

EXAMPLE 1 Let $S = \{(x_1, x_2) \text{ in } \mathbb{R}^2\colon x_1 \text{ and } x_2 \text{ in } \mathbb{Q} \cap (0, 1)\}$ denote the points inside the unit square in $\mathbb{R}^2$ with rational coordinates. The set of limit points of S

consists of all points inside the unit square and on its boundary. It is straightforward to show that every deleted neighborhood of any point $\boldsymbol{y} = (y_1, y_2)$ inside the square contains points of S. Merely choose any positive ϵ less than $\min\{y_1, y_2, 1 - y_1, 1 - y_2\}$. For each $j = 1, 2$, choose a rational number x_j in $N'(y_j; \epsilon/\sqrt{2})$. The point $\boldsymbol{x} = (x_1, x_2)$ has rational coordinates strictly between 0 and 1 and thus is a point of S. (See Fig. 2.3.) Furthermore,

$$||\boldsymbol{x} - \boldsymbol{y}||^2 = (x_1 - y_1)^2 + (x_2 - y_2)^2 < \epsilon^2/2 + \epsilon^2/2 = \epsilon^2.$$

Therefore, $\boldsymbol{x}$ is in $S \cap N'(\boldsymbol{y}; \epsilon)$. Thus every deleted neighborhood of $\boldsymbol{y}$ contains points of S. This proves that $\boldsymbol{y}$ is a limit point of S. We leave as an exercise the proof that every $\boldsymbol{y}$ on the boundary of the square is also a limit point of S. (See Exercise 2.9.) ●

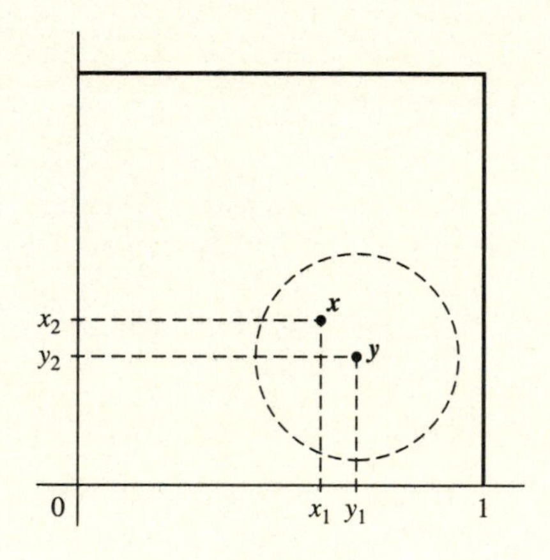

Figure 2.3

The technique used in Example 1 suggests a strategy for a proof of the following theorem.

THEOREM 2.1.5 $\mathbb{Q}^n$ is dense in $\mathbb{R}^n$.

Proof. $\mathbb{Q}^n$ denotes the set of all points in $\mathbb{R}^n$ with rational coordinates. Given any $\boldsymbol{y} = (y_1, y_2, \ldots, y_n)$ in $\mathbb{R}^n$ and any $\epsilon > 0$, we must exhibit an $\boldsymbol{x}$ in $\mathbb{Q}^n$ such that $\boldsymbol{x}$ is also in $N'(\boldsymbol{y}; \epsilon)$. But this is easy; we need merely invoke Theorem 1.1.2 n times as follows: For each $j = 1, 2, \ldots, n$, choose a rational x_j in $N'(y_j; \epsilon/\sqrt{n})$ and form $\boldsymbol{x} = (x_1, x_2, \ldots, x_n)$, a point in $\mathbb{Q}^n$. To show that $\boldsymbol{x}$ is in $N(\boldsymbol{y}; \epsilon)$, compute

$$||\boldsymbol{x} - \boldsymbol{y}||^2 = \sum_{j=1}^{n}(x_j - y_j)^2 < n(\epsilon/\sqrt{n})^2 = \epsilon^2.$$

Taking square roots shows that $\boldsymbol{x}$ is in $N(\boldsymbol{y}; \epsilon)$. Therefore $\boldsymbol{y}$ is a limit point of $\mathbb{Q}^n$. This proves that $\mathbb{Q}^n$ is dense in $\mathbb{R}^n$. ●

For $n \geq 2$, there is no linear order on $\mathbb{R}^n$, so there can be no concept of upper or lower bound for a set. Consequently, we have no notion of supremum or of infimum. But it is easy to lift the concept of boundedness to $\mathbb{R}^n$; again, we merely imitate the approach we took in $\mathbb{R}$.

DEFINITION 2.1.9 A subset S of $\mathbb{R}^n$ is said to be *bounded* if there exists an M in $\mathbb{R}^+$ such that $||\boldsymbol{x}|| \leq M$ for all $\boldsymbol{x}$ in S. ●

Equivalently, S is bounded if it is contained in some hypersphere $N(\mathbf{0}; M_1)$ of radius M_1 and centered at $\mathbf{0}$.

EXAMPLE 2 The union of finitely many bounded sets $S_1, S_2, S_3, \ldots, S_k$ is bounded. Suppose that, for each $j = 1, 2, \ldots, k$, M_j is a bound for S_j. That is, $S_j \subseteq N(\mathbf{0}; M_j)$. Let M be the largest of $M_1, M_2, \ldots, M_k$. Then

$$\bigcup_{j=1}^{k} S_j \subseteq \bigcup_{j=1}^{k} N(\mathbf{0}; M_j) = N(\mathbf{0}; M).$$

Therefore, $\cup_{j=1}^{k} S_j$ is also bounded. ●

EXAMPLE 3 The solid n-dimensional hyper-rectangle

$$S = I_1 \times I_2 \times \cdots \times I_n$$

is obtained by taking the Cartesian product of intervals $I_1, I_2, \ldots, I_n$, one in each of the coordinate axes. If, for each j, I_j has length s_j, then the dimensions of S, as a geometric solid, are $s_1, s_2, \ldots, s_n$. If $s_j = s$ for all j, then the hyper-rectangle is called an n-dimensional cube, or more simply, an n-cube, $C(s)$, with side s. The sides of the n-cube are parallel to the coordinate axes. (See Fig. 2.4.) You can prove inductively, with repeated use of the Pythagorean Theorem in Exercise 2.10, that,

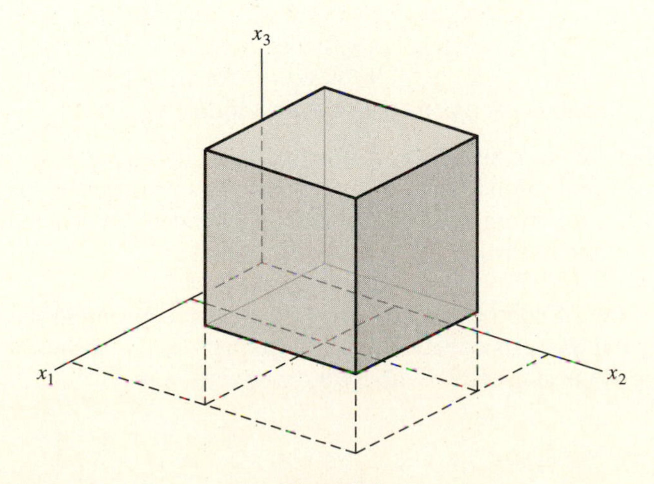

Figure 2.4

if $C(s)$ is any n-cube with side s, then the length of its main diagonal[1] is $s\sqrt{n}$.

If, as a special case, we let $\boldsymbol{x} = (x_1, x_2, x_3, \ldots, x_n)$, and if, in each coordinate axis, we take the interval I_j to be the neighborhood $N(x_j; r) = (x_j - r, x_j + r)$, then the resulting n-cube—$C(2r) = N(x_1; r) \times N(x_2; r) \times \ldots \times N(x_n; r)$,—is centered at $\boldsymbol{x}$, and has main diagonal of length $2r\sqrt{n}$. Consequently, if for any $\epsilon > 0$ we let $r = \epsilon/\sqrt{n}$, then the resulting n-cube $C(2\epsilon/\sqrt{n})$ is *inscribed* in the hypersphere $N(\boldsymbol{x}; \epsilon)$. If we let $r = \epsilon$, then $C(2\epsilon)$ is the n-cube *circumscribed* about the hypersphere $N(\boldsymbol{x}; \epsilon)$.

If each of the intervals I_j is a bounded interval with endpoints a_j and b_j, then the resulting hyper-rectangle S is also bounded. Take $M = \max\{|a_j|, |b_j| : j = 1, 2, \ldots, n\}$. If $\boldsymbol{y} = (y_1, y_2, \ldots, y_n)$ is any point of S, then $|y_j| \leq M$ for every $j = 1, 2, \ldots, n$. Therefore $||\boldsymbol{y}||^2 = \sum_{j=1}^{n} y_j^2 \leq nM^2$. Taking square roots yields $||\boldsymbol{y}|| \leq M\sqrt{n}$. Thus S is bounded. ●

2.1.1 Sequences in $\mathbb{R}^n$

Consider now a sequence $\{\boldsymbol{x}_k\}$ of points in $\mathbb{R}^n$. Take pencil and paper and randomly form a sequence of dots, any sequence of dots, the first one, the second one, the third one .... Immediately you will notice that you have enormous freedom in the choice of each successive dot; you can choose the next dot anywhere in the plane of the paper. You will also notice that *there is no notion of* "*monotone.*" But there are three concepts pertaining to sequences that retain their meaning in higher-dimensional spaces. That a sequence is *bounded* still makes sense. Merely invoke Definition 2.1.9. The idea of a *cluster point* of a sequence also still makes sense.

DEFINITION 2.1.10 We say that $\boldsymbol{c}$ is a *cluster point* of the sequence $\{\boldsymbol{x}_k\}$ in $\mathbb{R}^n$ if, for every $\epsilon > 0$ and every k in $\mathbb{N}$, there is a $k_1 > k$ such that $\boldsymbol{x}_{k_1}$ is in $N(\boldsymbol{c}; \epsilon)$. ●

Again, we say that terms of the sequence are *frequently* arbitrarily near $\boldsymbol{c}$. And our all-important concept of *limit* still makes sense.

DEFINITION 2.1.11 The sequence $\{\boldsymbol{x}_k\}$ in $\mathbb{R}^n$ *converges* to $\boldsymbol{x}_0$ and we say that the *limit* of $\{\boldsymbol{x}_k\}$ is $\boldsymbol{x}_0$ if, for each neighborhood $N(\boldsymbol{x}_0; \epsilon)$ of $\boldsymbol{x}_0$, there exists an index k_0 such that, whenever $k \geq k_0$, $\boldsymbol{x}_k$ is in $N(\boldsymbol{x}_0; \epsilon)$. We write $\lim_{k\to\infty} \boldsymbol{x}_k = \boldsymbol{x}_0$. If the sequence $\{\boldsymbol{x}_k\}$ fails to converge, then we say that it *diverges*. ●

Clearly, the limit $\boldsymbol{x}_0$ of a sequence, if it exists, is a cluster point of the sequence. As in the case of real-valued sequences, the limit of a convergent sequence in $\mathbb{R}^n$ is unique. Likewise, a convergent sequence in $\mathbb{R}^n$ is bounded. We leave proof of these assertions and of the following theorem for the exercises.

THEOREM 2.1.6 Suppose that $\{\boldsymbol{x}_k\}$ and $\{\boldsymbol{y}_k\}$ are two convergent sequences in $\mathbb{R}^n$ with limits $\boldsymbol{x}_0$ and $\boldsymbol{y}_0$, respectively. Suppose also that c and d are two real numbers. Then $\{c\boldsymbol{x}_k + d\boldsymbol{y}_k\}$ is convergent with limit $c\boldsymbol{x}_0 + d\boldsymbol{y}_0$. ●

---

[1] The *main diagonal* is the line segment from $(a_1, a_2, \ldots, a_n)$ to $(b_1, b_2, \ldots, b_n)$ where $I_j = (a_j, b_j)$.

Of the elementary methods to confirm the convergence of a sequence in $\mathbb{R}$, which can be lifted to $\mathbb{R}^n$? Since there is no concept of "monotone," we cannot generalize Theorem 1.3.7 to $\mathbb{R}^n$. More, because there is no linear order, we cannot generalize the Squeeze Play to $\mathbb{R}^n$. We have eliminated our most basic tools to determine whether a sequence converges.

Fortunately, not all is lost. As we now prove, a sequence of points $\boldsymbol{x}_k = (x_1^{(k)}, x_2^{(k)}, \ldots, x_n^{(k)})$ converges to $\boldsymbol{x}_0 = (x_1^{(0)}, x_2^{(0)}, \ldots, x_n^{(0)})$ if and only if each coordinate sequence $\{x_j^{(k)}\}$ of real numbers converges to $x_j^{(0)}$.

First, suppose that $\lim_{k\to\infty} \boldsymbol{x}_k = \boldsymbol{x}_0$. Fix $\epsilon > 0$ and choose an index k_0 such that, for $k \geq k_0$, it follows that $\boldsymbol{x}_k$ is in $N(\boldsymbol{x}_0; \epsilon)$. Choose $k \geq k_0$. For any coordinate index j,

$$[x_j^{(k)} - x_j^{(0)}]^2 \leq \sum_{i=1}^{n} (x_i^{(k)} - x_i^{(0)})^2 = ||\boldsymbol{x}_k - \boldsymbol{x}_0||^2 < \epsilon^2.$$

Thus, for $k \geq k_0$, it follows that $x_j^{(k)}$ is in $N(x_j^{(0)}; \epsilon)$. This proves that $\lim_{k\to\infty} x_j^{(k)} = x_j^{(0)}$ for each coordinate index $j = 1, 2, \ldots, n$.

Conversely, suppose that, for each coordinate index $j = 1, 2, \ldots, n$, the sequence $\{x_j^{(k)}\}$ converges to $x_j^{(0)}$. Fix $\epsilon > 0$. For each j, choose k_j in $\mathbb{N}$ such that, for $k \geq k_j$ we have $x_j^{(k)}$ in $N(x_j^{(0)}; \epsilon/\sqrt{n})$. That is, $|x_j^{(k)} - x_j^{(0)}| < \epsilon/\sqrt{n}$. Let $k_0 = \max\{k_1, k_2, \ldots, k_n\}$. For $k \geq k_0$, $\boldsymbol{x}_k = (x_1^{(k)}, x_2^{(k)}, \ldots, x_n^{(k)})$ is in the n-cube $C(2\epsilon/\sqrt{n})$ centered at $\boldsymbol{x}_0$. From Example 3, we know that this n-cube is inscribed in $N(\boldsymbol{x}_0; \epsilon)$. (See Fig. 2.5 for the case $n = 2$.) Therefore, $\boldsymbol{x}_k$ is in $N(\boldsymbol{x}_0; \epsilon)$ for all $k \geq k_0$. This proves that $\lim_{k\to\infty} \boldsymbol{x}_k = \boldsymbol{x}_0$. We have proved the following theorem.

THEOREM 2.1.7 Let $\boldsymbol{x}_k = (x_1^{(k)}, x_2^{(k)}, \ldots, x_n^{(k)})$ for each k in $\mathbb{N}$. Let $\boldsymbol{x}_0 = (x_1^{(0)}, x_2^{(0)}, \ldots, x_n^{(0)})$. The sequence $\{\boldsymbol{x}_k\}$ converges to $\boldsymbol{x}_0$ if and only if, for each $j = 1, 2, \ldots, n$, the sequence $\{x_j^{(k)}\}$ converges to $x_j^{(0)}$. ●

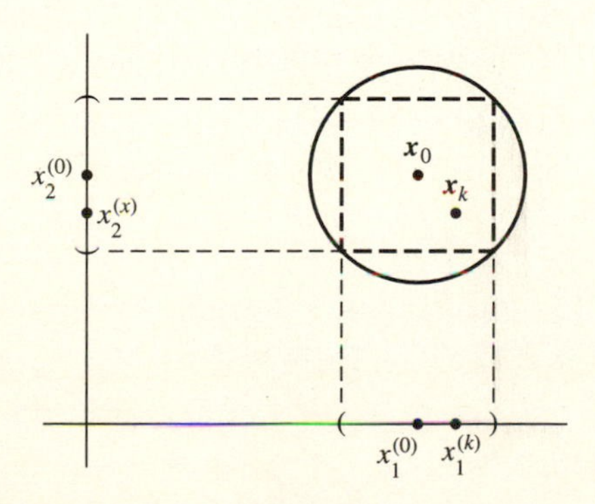

Figure 2.5

Consequently, we can determine the convergence of a sequence $\{\boldsymbol{x}_k\}$ by examining each coordinate sequence $\{x_j^{(k)}\}$. If each of these n latter sequences has a limit, then the original sequence $\{\boldsymbol{x}_k\}$ converges. This fact will prove to be the key link in our program to lift major theorems from $\mathbb{R}$ to $\mathbb{R}^n$. See, for example, our proof of Theorem 2.1.8.

The Cauchy condition, which proved so useful in $\mathbb{R}$, lifts naturally to $\mathbb{R}^n$.

DEFINITION 2.1.12 The Cauchy Condition A sequence $\{\boldsymbol{x}_k\}$ in $\mathbb{R}^n$ is a *Cauchy sequence* if, for every $\epsilon > 0$, there exists a k_0 such that, if k and m are greater than or equal to k_0, then $||\boldsymbol{x}_k - \boldsymbol{x}_m|| < \epsilon$. ●

With the concept of Cauchy sequence in hand, we are ready for the first profoundly important theorem of this section.

THEOREM 2.1.8 Cauchy's Completeness Theorem A sequence $\{\boldsymbol{x}_k\}$ in $\mathbb{R}^n$ is Cauchy if and only if it converges. $\mathbb{R}^n$ is Cauchy complete.

Proof. One half of the proof consists of proving that, if $\{\boldsymbol{x}_k\}$ converges in $\mathbb{R}^n$, then it is a Cauchy sequence. We leave this part as an exercise; merely imitate the proof of Theorem 1.4.1.

For the other half, let $\{\boldsymbol{x}_k\}$ be any Cauchy sequence in $\mathbb{R}^n$. We must show that there exists a point $\boldsymbol{x}_0$ in $\mathbb{R}^n$ such that $\lim_{k\to\infty} \boldsymbol{x}_k = \boldsymbol{x}_0$. Let the kth term of the Cauchy sequence have coordinates $\boldsymbol{x}_k = (x_1^{(k)}, x_2^{(k)}, \ldots, x_n^{(k)})$.

For each coordinate index j and any two indices k and m, it is evident that

$$|x_j^{(k)} - x_j^{(m)}| \le \left[\sum_{i=1}^{n}(x_i^{(k)} - x_i^{(m)})^2\right]^{1/2} = ||\boldsymbol{x}_k - \boldsymbol{x}_m||. \tag{2.3}$$

Since $\{\boldsymbol{x}_k\}$ is assumed to be a Cauchy sequence, given $\epsilon > 0$, we can choose k_0 such that, for $k, m \ge k_0$, $||\boldsymbol{x}_k - \boldsymbol{x}_m|| < \epsilon$. From (2.3) it follows that, for each coordinate index j and for $k, m \ge k_0$, $|x_j^{(k)} - x_j^{(m)}| < \epsilon$ also. Therefore each coordinate sequence $\{x_j^{(k)} : k \text{ in } \mathbb{N}\}$ is a Cauchy sequence of real numbers. The Cauchy completeness of $\mathbb{R}$ (Theorem 1.4.4) guarantees that, for each $j = 1, 2, \ldots, n$, the limit $x_j^{(0)}$ of $\{x_j^{(k)}\}$ exists in $\mathbb{R}$. Let $\boldsymbol{x}_0 = (x_1^{(0)}, x_2^{(0)}, \ldots, x_n^{(0)})$. Theorem 2.1.7 ensures that $\{\boldsymbol{x}_k\}$ converges to $\boldsymbol{x}_0$. Therefore, $\mathbb{R}^n$ is Cauchy complete. ●

The Cauchy completeness of $\mathbb{R}^n$ enables us to establish several essential facts about n-dimensional space. For example, it immediately enables us to prove the next important theorem.

THEOREM 2.1.9 The Generalized Bolzano–Weierstrass Theorem Every bounded infinite set in $\mathbb{R}^n$ has a limit point in $\mathbb{R}^n$.

Proof. Suppose that S is any bounded, infinite set in $\mathbb{R}^n$. Being bounded, S is contained in some n-cube

$$C(2M) = [-M, M] \times [-M, M] \times \cdots \times [-M, M]$$

centered at $\mathbf{0}$. The strategy for our proof consists of bisecting repeatedly, as in the proof of Theorem 1.2.2, the intervals in each of the coordinate axes. Thereby we will construct a sequence of nested n-cubes. The main diagonals of these n-cubes will have lengths that tend to 0. By recursively choosing a point from each n-cube, we will obtain a sequence in $\mathbb{R}^n$. Since the diagonals of the n-cubes will shrink to zero, the resulting sequence will be Cauchy, hence convergent, in $\mathbb{R}^n$ and will yield the desired limit point. (See Fig. 2.6 for the case when $n = 2$.)

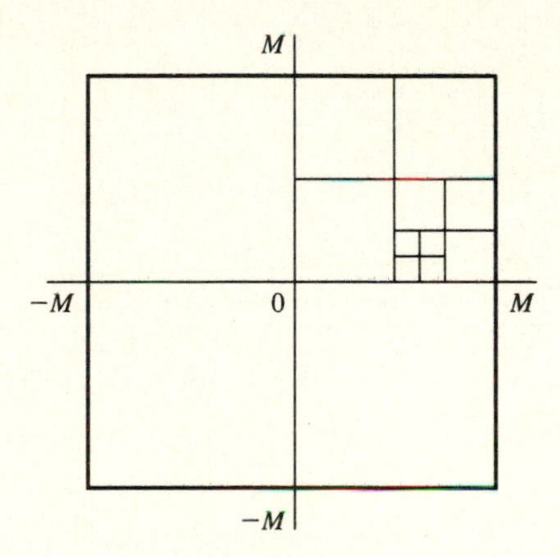

Figure 2.6

Let $C_0 = C(2M)$, the original n-cube containing the infinite set S. Its main diagonal has length $d_0 = 2Mn^{1/2}$. In each coordinate axis, bisect the interval $[-M, M]$ into two closed intervals $I_j(1, 1)$ and $I_j(1, 2)$. Form the 2^n n-cubes obtained by taking the Cartesian products

$$I_1(1, i) \times I_2(1, i) \times \cdots \times I_n(1, i),$$

where i is either 1 or 2.

The main diagonal of each of these 2^n n-cubes has length $d_1 = Mn^{1/2}$. The union of all of them is C_0 and at least one of them contains infinitely many points of S. Otherwise S would be finite. Let

$$C_1 = J_1(1) \times J_2(1) \times \cdots \times J_n(1)$$

denote any one of these n-cubes containing infinitely many points of S. [$J_j(1)$ is simply that interval in the jth coordinate axis, either $I_j(1, 1)$ or $I_j(1, 2)$, which was used to form this particular n-cube.] Choose any $\boldsymbol{x}_1$ in $C_1 \cap S$.

In each coordinate axis, bisect the interval $J_j(1)$ into two closed intervals, $I_j(2, 1)$ and $I_j(2, 2)$. Form the 2^n n-cubes

$$I_1(2, i) \times I_2(2, i) \times \cdots \times I_n(2, i),$$

where i is either 1 or 2. The main diagonal of each of these n-cubes has length $d_2 = Mn^{1/2}/2$, the union of all of them is C_1, and at least one of them contains infinitely many points of S. This last assertion follows from our having chosen C_1 to contain infinitely many points of S. Let

$$C_2 = J_1(2) \times J_2(2) \times \cdots \times J_n(2)$$

denote any one of these n-cubes that contains infinitely many points of S. Choose $\boldsymbol{x}_2$ in $C_2 \cap S$ distinct from $\boldsymbol{x}_1$.

Continue this process. Suppose we have chosen the first k n-cubes $C_1, C_2, \ldots, C_k$ such that

i) $C_1 \supset C_2 \supset C_3 \supset \cdots \supset C_k$ with

$$C_m = J_1(m) \times J_2(m) \times \cdots \times J_n(m)$$

for $m = 1, 2, \ldots, k$.

ii) The main diagonal of C_m has length $d_m = Mn^{1/2}/2^{m-1}$ for $m = 1, 2, \ldots, k$.

iii) $C_m \cap S$ is infinite for $m = 1, 2, \ldots, k$.

Suppose also that we have chosen distinct points $\boldsymbol{x}_1, \boldsymbol{x}_2, \ldots, \boldsymbol{x}_k$ such that, for $m = 1, 2, \ldots, k$, $\boldsymbol{x}_m$ is in $C_m \cap S$.

For each coordinate index j, bisect $J_j(k)$ into two closed intervals $I_j(k+1, 1)$ and $I_j(k+1, 2)$. Form the 2^n n-cubes

$$I_1(k+1, i) \times I_2(k+1, i) \times \cdots \times I_n(k+1, i),$$

where i is either 1 or 2. Again, the main diagonal of each of these n-cubes has length $d_{k+1} = Mn^{1/2}/2^k$. The union of all of them is C_k and at least one of them contains infinitely many points of S because we chose C_k so that $C_k \cap S$ is infinite. Let

$$C_{k+1} = J_1(k+1) \times J_2(k+1) \times \cdots \times J_n(k+1)$$

denote any one of these n-cubes containing infinitely many points of S. Choose $\boldsymbol{x}_{k+1}$ in $C_{k+1} \cap S$ distinct from all the preceding $\boldsymbol{x}_1, \boldsymbol{x}_2, \ldots, \boldsymbol{x}_k$. With this construction we have recursively created a sequence $\{\boldsymbol{x}_k\}$ in S and sufficient machinery to prove that it is Cauchy.

Given $\epsilon > 0$, by Archimedes' principle, we can choose k_0 in $\mathbb{N}$ such that $0 < d_k = M\sqrt{n}/2^{k_0-1} < \epsilon$. Choose any two indices k and m both greater than k_0. Since C_k and C_m are contained in C_{k_0}, it follows that $\boldsymbol{x}_k$ and $\boldsymbol{x}_m$ are in C_{k_0}. Thus $||\boldsymbol{x}_k - \boldsymbol{x}_m||$, being no larger than the length d_{k_0} of the main diagonal of C_{k_0}, must be less than ϵ. Therefore, $\{\boldsymbol{x}_k\}$ is Cauchy. By Theorem 2.1.8, $\{\boldsymbol{x}_k\}$ must converge to some point $\boldsymbol{x}_0$ in $\mathbb{R}^n$. Clearly $\boldsymbol{x}_0$ is a limit point of S since every deleted neighborhood $N'(\boldsymbol{x}_0)$ contains at least one of the distinct points of the sequence $\{\boldsymbol{x}_k\}$, all of which belong to S. We have proved that $\mathbb{R}^n$ has the Bolzano–Weierstrass property. ●

Another concept introduced in Chapter 1 that lifts easily to $\mathbb{R}^n$ is that of a subsequence of a sequence. It plays the same essential role here as it did in our earlier discussion.

DEFINITION 2.1.13 Let $\{\boldsymbol{x}_k\}$ be any sequence in $\mathbb{R}^n$. Choose a strictly increasing, infinite sequence $k_1 < k_2 < k_3 < \cdots$ of natural numbers. For each j in $\mathbb{N}$, let $\boldsymbol{y}_j = \boldsymbol{x}_{k_j}$. The sequence $\{\boldsymbol{y}_j\} = \{\boldsymbol{x}_{k_j}\}$ is called a *subsequence* of $\{\boldsymbol{x}_k\}$. ●

Knowing that $\mathbb{R}^n$ has the Bolzano–Weierstrass property enables us to prove a string of theorems with ease.

THEOREM 2.1.10 A point $\boldsymbol{c}$ in $\mathbb{R}^n$ is a cluster point of $\{\boldsymbol{x}_k\}$ in $\mathbb{R}^n$ if and only if there exists a subsequence $\{\boldsymbol{x}_{k_j}\}$ that converges to $\boldsymbol{c}$.

Proof. Repeat the proof of Theorem 1.3.10 verbatim. ●

THEOREM 2.1.11 Any bounded sequence in $\mathbb{R}^n$ has a cluster point.

Proof. Repeat the proof of Theorem 1.3.11 verbatim. ●

THEOREM 2.1.12 If a sequence in $\mathbb{R}^n$ has no cluster points, then it is unbounded. ●

THEOREM 2.1.13 A sequence $\{\boldsymbol{x}_k\}$ in $\mathbb{R}^n$ converges to $\boldsymbol{x}_0$ if and only if every subsequence of $\{\boldsymbol{x}_k\}$ converges to $\boldsymbol{x}_0$.

Proof. Repeat your proof of Exercise 1.53 verbatim. ●

THEOREM 2.1.14 A bounded sequence in $\mathbb{R}^n$ converges if and only if it has exactly one cluster point.

Proof. Repeat your proof of Exercise 1.54 verbatim. ●

THEOREM 2.1.15 A sequence $\{\boldsymbol{x}_k\}$ in $\mathbb{R}^n$ diverges if and only if at least one of the following two conditions holds:

i) $\{\boldsymbol{x}_k\}$ has at least two cluster points.

ii) $\{\boldsymbol{x}_k\}$ is unbounded.

Proof. Repeat the proof of Theorem 1.3.15 verbatim. ●

This flurry of theorems and their proofs must convince you that, as we said earlier, the major theorems of analysis in $\mathbb{R}$, when properly articulated, lift easily and naturally to $\mathbb{R}^n$. We will see more examples of this phenomenon in the rest of the text.

We close this section with a summary of the implications we have achieved to this point connecting various theorems relating to completeness.

Clearly, by taking $n = 1$, Theorem 2.1.9 ($\mathbb{R}^n$ has the Bolzano–Weierstrass property) implies Theorem 1.2.2 ($\mathbb{R}$ has the Bolzano–Weierstrass property). We have already proved directly the following chain of implications: Theorem 1.2.2 implies Theorem 1.4.4 implies Theorem 2.1.8 which, in turn, implies Theorem 2.1.9. Consequently these four theorems say exactly the same thing, though in different words. They merely describe different facets of the same central mathematical idea. Therefore we have established the following flow of implications.

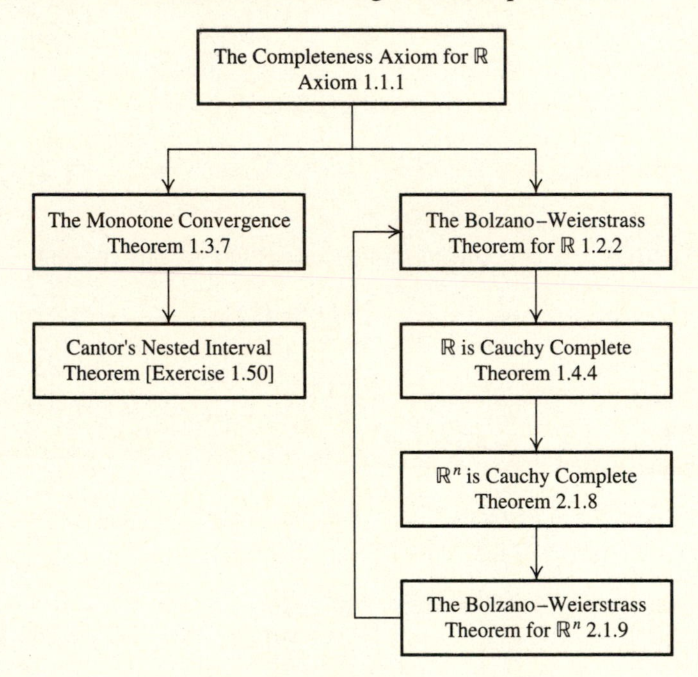

By now you might suspect that any one of these statements implies each of the others. If so, you are correct. In Section 2.3 we will complete the circle of proofs so that, later, whenever we appeal to the completeness of $\mathbb{R}$ or $\mathbb{R}^n$, we mean each of these statements.

Any one of these statements could have been taken as the one axiom for all of this book; all others would then have been derived as theorems. We chose to begin with Axiom 1.1.1 because it requires the least explanatory preparation and seems to be the version of completeness most intuitively accessible to beginners just learning analysis.

EXERCISES

2.1. Let $\boldsymbol{x}$ and $\boldsymbol{y}$ be two distinct points in $\mathbb{R}^n$.

a) Show that the directed line segment from $\boldsymbol{x}$ to $\boldsymbol{y}$ is

$$[\boldsymbol{x}, \boldsymbol{y}] = \{\boldsymbol{z} = (1-t)\boldsymbol{x} + t\,\boldsymbol{y} :\ 0 \le t \le 1\}.$$

b) Let $\boldsymbol{u} = (\boldsymbol{y} - \boldsymbol{x})/||\boldsymbol{y} - \boldsymbol{x}||$. Show that $||\boldsymbol{u}|| = 1$ and that the directed line segment from $\boldsymbol{x}$ to $\boldsymbol{y}$ is given by the equation $\boldsymbol{z} = \boldsymbol{x} + t\boldsymbol{u}$ with $0 \le t \le ||\boldsymbol{y} - \boldsymbol{x}||$.

c) Let $\boldsymbol{x}$ and $\boldsymbol{y}$ be distinct points in a neighborhood $N(\boldsymbol{x}_0; r)$. Prove that the line segment $[\boldsymbol{x}, \boldsymbol{y}]$ is contained in $N(\boldsymbol{x}_0; r)$.

2.2. For $\boldsymbol{x} = (x_1, x_2)$ in $\mathbb{R}^2$, define $||\boldsymbol{x}||_1 = |x_1| + |x_2|$. Prove that $||\cdot||_1$ is a norm on $\mathbb{R}^2$.

2.3. For $\boldsymbol{x} = (x_1, x_2)$ and $\boldsymbol{y} = (y_1, y_2)$ in $\mathbb{R}^2$, define

$$d_1(\boldsymbol{x}, \boldsymbol{y}) = ||\boldsymbol{x} - \boldsymbol{y}||_1.$$

a) Show that d_1 is a metric on $\mathbb{R}^2$.

b) Using this metric, define a 1-neighborhood $N_1(\boldsymbol{x}; r)$ of $\boldsymbol{x} = (x_1, x_2)$ to be $N_1(\boldsymbol{x}; r) = \{\boldsymbol{y} \text{ in } \mathbb{R}^2 : d_1(\boldsymbol{x}, \boldsymbol{y}) < r\}$. Sketch the neighborhood $N_1(\boldsymbol{x}; r)$.

c) Let $N(\boldsymbol{x}; r)$ be any (Euclidean) neighborhood of $\boldsymbol{x}$. Show that there exist positive r_1 and r_2 such that

$$N_1(\boldsymbol{x}; r_1) \subset N(\boldsymbol{x}; r) \subset N_1(\boldsymbol{x}; r_2).$$

d) Let $N_1(\boldsymbol{x}; r)$ be any 1-neighborhood of $\boldsymbol{x}$. Show that there are Euclidean neighborhoods with radii r_1 and r_2 such that

$$N(\boldsymbol{x}; r_1) \subset N_1(\boldsymbol{x}; r) \subset N(\boldsymbol{x}; r_2).$$

2.4. For $\boldsymbol{x} = (x_1, x_2)$ in $\mathbb{R}^2$, define

$$||\boldsymbol{x}||_\infty = \max\{|x_1|, |x_2|\}.$$

Prove that $||\cdot||_\infty$ is a norm on $\mathbb{R}^2$.

2.5. For $\boldsymbol{x} = (x_1, x_2)$ and $\boldsymbol{y} = (y_1, y_2)$ in $\mathbb{R}^2$, define

$$d_\infty(\boldsymbol{x}, \boldsymbol{y}) = ||\boldsymbol{x} - \boldsymbol{y}||_\infty.$$

a) Show that d_∞ is a metric on $\mathbb{R}^2$.

b) Using this metric, define an ∞-neighborhood $N_\infty(\boldsymbol{x}; r)$ of $\boldsymbol{x}$ to be the set $N_\infty(\boldsymbol{x}; r) = \{\boldsymbol{y} \text{ in } \mathbb{R}^2 : d_\infty(\boldsymbol{x}, \boldsymbol{y}) < r\}$. Sketch the neighborhood $N_\infty(\boldsymbol{x}; r)$.

c) Let $N(\boldsymbol{x}; r)$ be any (Euclidean) neighborhood of $\boldsymbol{x}$. Show that there exist positive r_1 and r_2 such that

$$N_\infty(\boldsymbol{x}; r_1) \subset N(\boldsymbol{x}; r) \subset N_\infty(\boldsymbol{x}; r_2).$$

d) Let $N_\infty(\boldsymbol{x}; r)$ be any ∞-neighborhood of $\boldsymbol{x}$. Show there are Euclidean neighborhoods with radii r_1 and r_2 so that

$$N(\boldsymbol{x}; r_1) \subset N_\infty(\boldsymbol{x}; r) \subset N(\boldsymbol{x}; r_2).$$

2.6. For any point $\boldsymbol{x}$ and any subset S of $\mathbb{R}^n$, define $\boldsymbol{x} + S = \{\boldsymbol{x} + \boldsymbol{y} : \boldsymbol{y} \text{ in } S\}$. Prove that $\boldsymbol{x} + N(\boldsymbol{0}; r) = N(\boldsymbol{x}; r)$ for every $\boldsymbol{x}$ in $\mathbb{R}^n$ and every $r > 0$.

2.7. Prove that, for any two vectors $\boldsymbol{x}$ and $\boldsymbol{y}$ in $\mathbb{R}^n$, $||\boldsymbol{x} + \boldsymbol{y}|| \geq ||\boldsymbol{x}|| - ||\boldsymbol{y}||$.

2.8. Let $\boldsymbol{x}$ be a limit point of a nonempty subset S of $\mathbb{R}^n$. Prove that any deleted neighborhood of $\boldsymbol{x}$ contains infinitely many points of S.

2.9. Complete the proof of the assertion in Example 1 that each of the points on the boundary of the unit square is a limit point of $S = \{(x_1, x_2) : x_1 \text{ and } x_2 \text{ in } \mathbb{Q} \cap (0, 1)\}$.

2.10. Prove that the main diagonal of an n-cube $C(s)$ with side s in $\mathbb{R}^n$ has length $sn^{1/2}$.

2.11. Let S be any neighborhood $N(\boldsymbol{x}; r)$ in $\mathbb{R}^2$.

a) Fix any $\boldsymbol{y}$ in S. Prove that $\boldsymbol{y}$ is a limit point of S.

b) Fix any $\boldsymbol{z}$ in $\mathbb{R}^2$ such that $||\boldsymbol{z} - \boldsymbol{x}|| = r$. Prove that $\boldsymbol{z}$ is a limit point of S.

c) Fix any $\boldsymbol{u}$ in $\mathbb{R}^2$ such that $||\boldsymbol{u} - \boldsymbol{x}|| > r$. Prove that $\boldsymbol{u}$ is not a limit point of S.

d) Generalize parts (a), (b), and (c) to a neighborhood of $\boldsymbol{x}$ in $\mathbb{R}^n$.

2.12. **a)** Prove that the limit of a convergent sequence in $\mathbb{R}^n$ is unique.

b) Prove that a convergent sequence in $\mathbb{R}^n$ is bounded.

c) Prove that a convergent sequence in $\mathbb{R}^n$ is Cauchy.

2.13. Suppose that $\{\boldsymbol{x}_k\}$ converges to $\boldsymbol{x}_0$ and $\{\boldsymbol{y}_k\}$ converges to $\boldsymbol{y}_0$ in $\mathbb{R}^n$. Suppose also that c and d are real numbers. Prove that $\{c\boldsymbol{x}_k + d\,\boldsymbol{y}_k\}$ converges to $c\boldsymbol{x}_0 + d\,\boldsymbol{y}_0$.

2.14. **a)** Prove that a sequence $\{\boldsymbol{x}_k\}$ in $\mathbb{R}^n$ converges to $\boldsymbol{0}$ if and only if $\{||\boldsymbol{x}_k||\}$ converges to 0.

b) Prove that if $\{\boldsymbol{x}_k\}$ converges to $\boldsymbol{x}_0$, then $\{||\boldsymbol{x}_k||\}$ converges to $||\boldsymbol{x}_0||$. Is the converse true? Prove your answer.

2.15. Prove that every Cauchy sequence in $\mathbb{R}^n$ is bounded.

2.16. Prove that a bounded sequence in $\mathbb{R}^n$ converges if and only if it has exactly one cluster point.

2.17. Suppose that S is any nonempty set in $\mathbb{R}^n$ and that $\boldsymbol{x}_0$ is any limit point of S. Prove that there exists a Cauchy sequence of distinct points in S that converges to $\boldsymbol{x}_0$.

2.18. Hamilton considered generalized "numbers," called *quaternions*, of the form $\boldsymbol{x} = x_0 + x_1\boldsymbol{i} + x_2\boldsymbol{j} + x_3\boldsymbol{k}$, where x_0, x_1, x_2, and x_3 are in $\mathbb{R}$. The symbols $\boldsymbol{i}, \boldsymbol{j}, \boldsymbol{k}$ are idealized quantities with the following multiplicative properties:

$$\begin{aligned} &\boldsymbol{i}^2 = \boldsymbol{j}^2 = \boldsymbol{k}^2 = -1, \\ &\boldsymbol{ij} = \boldsymbol{k},\ \ \boldsymbol{jk} = \boldsymbol{i},\ \ \boldsymbol{ki} = \boldsymbol{j}, \\ &\boldsymbol{ji} = -\boldsymbol{k},\ \ \boldsymbol{kj} = -\boldsymbol{i},\ \ \boldsymbol{ik} = -\boldsymbol{j}. \end{aligned}$$

Two quaternions are added in accord with vector addition. Two quaternions are multiplied by expanding the product as in elementary algebra, invoking the preceding relations among $\boldsymbol{i}$, $\boldsymbol{j}$, and $\boldsymbol{k}$, and collecting like terms:

$$\begin{aligned} \boldsymbol{x}\,\boldsymbol{y} &= (x_0 + x_1\boldsymbol{i} + x_2\boldsymbol{j} + x_3\boldsymbol{k})(y_0 + y_1\boldsymbol{i} + y_2\boldsymbol{j} + y_3\boldsymbol{k}) \\ &= x_0y_0 - x_1y_1 - x_2y_2 - x_3y_3 \\ &\qquad + [x_0y_1 + x_1y_0 + x_2y_3 - x_3y_2]\boldsymbol{i} \\ &\qquad\qquad + [x_0y_2 - x_1y_3 + x_2y_0 + x_3y_1]\,\boldsymbol{j} \\ &\qquad\qquad\qquad + [x_0y_3 + x_1y_2 - x_2y_1 + x_3y_0]\boldsymbol{k}. \end{aligned}$$

a) Confirm this product and show that, if $x_0 = y_0 = 0$, this product becomes $\boldsymbol{x}\,\boldsymbol{y} = \boldsymbol{x}' \times \boldsymbol{y}' - \langle \boldsymbol{x}', \boldsymbol{y}' \rangle$, where $\langle \boldsymbol{x}', \boldsymbol{y}' \rangle$ is the inner product of $\boldsymbol{x}' = (x_1, x_2, x_3)$ and $\boldsymbol{y}' = (y_1, y_2, y_3)$ and $\boldsymbol{x}' \times \boldsymbol{y}'$ is the cross product of $\boldsymbol{x}'$ and $\boldsymbol{y}'$ defined by

$$\boldsymbol{x}' \times \boldsymbol{y}' = \det \begin{vmatrix} \boldsymbol{i} & \boldsymbol{j} & \boldsymbol{k} \\ x_1 & x_2 & x_3 \\ y_1 & y_2 & y_3 \end{vmatrix}.$$

b) Prove that, in general, this multiplication is not commutative $[\boldsymbol{x}\,\boldsymbol{y} \neq \boldsymbol{y}\boldsymbol{x}]$, but that the distributive laws hold.

2.2 Open and Closed Sets

Open and *closed* are two commonplace words used in various ways in everyday speech. Because they also have technical meanings in mathematics, you will need to detach the colloquial from the technical. You are already familiar with their simple use in describing *open* intervals (a, b) and *closed* intervals $[a, b]$, and you will already have observed that the closed interval $[a, b]$ is obtained from (a, b) by adjoining the two endpoints, the *boundary* of (a, b). Since we will wish to apply these words to sets more general than mere intervals and will want our concepts to lift to $\mathbb{R}^n$, we examine (a, b) and $[a, b]$ more closely.

Choose any point x in $S = (a, b)$ and notice that x is buffered away from the complement of (a, b). It is inside or *interior* to the interval in the following sense: If you choose any positive $r < \min\{x - a, b - x\}$, then the neighborhood $N(x; r)$ lies entirely in (a, b). Points sufficiently near x are also in S. The same cannot be said of either endpoint a or b. In fact, for our purposes, this observation is what distinguishes the boundary points of S from the interior points: Every neighborhood of either a or b contains both points of S and points of the complement S^c of S. With this preliminary discussion to reinforce your intuitive sense of the meanings of these words, you will appreciate the precision of the following definition, applied to any set in $\mathbb{R}^n$.

DEFINITION 2.2.1 Let S be any subset of $\mathbb{R}^n$ and let $\boldsymbol{x}$ be any point in $\mathbb{R}^n$.

i) We say that $\boldsymbol{x}$ is an *interior point* of S if there exists an $r > 0$ such that $N(\boldsymbol{x}; r) \subseteq S$.

ii) If every point of S is an interior point of S, then S is said to be *open*. Thus, for every $\boldsymbol{x}$ in S, there exists a neighborhood $N(\boldsymbol{x})$ contained in S. (See Fig. 2.7.)

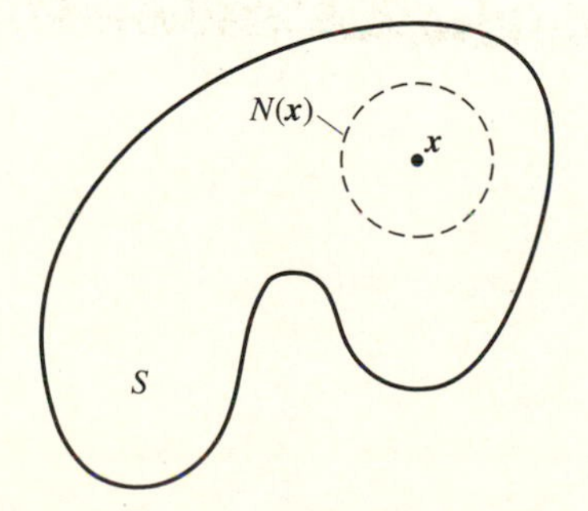

Figure 2.7 x is an interior point of S

iii) We call $\boldsymbol{x}$ a *boundary point* of S if every neighborhood $N(\boldsymbol{x})$ contains points in S and also points not in S. (See Fig. 2.8 on page 80.)

iv) If S contains all its boundary points, then S is said to be *closed*. ●

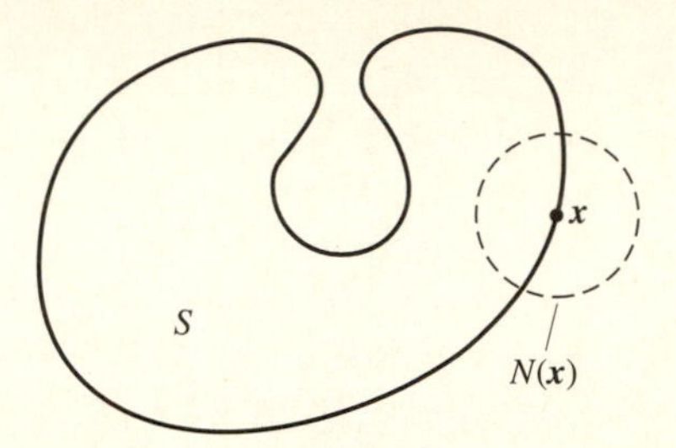

Figure 2.8 x is a boundary point of S

First examine a few simple examples of open sets in $\mathbb{R}$. Every open interval (a, b), with $a < b$, is an open set in $\mathbb{R}$. The empty set $\emptyset$ is trivially open—every point in $\emptyset$ is an interior point because there are no points in $\emptyset$. The entire real line $\mathbb{R}$ is open in $\mathbb{R}$ because every neighborhood of every point in $\mathbb{R}$ is contained entirely in $\mathbb{R}$. A union of open intervals is also open in $\mathbb{R}$. In fact, as you can prove as an exercise, every open set in $\mathbb{R}$ is a union of open intervals. After we have obtained general theorems concerning open and closed sets, we will be in a position to describe the open sets in $\mathbb{R}^n$ as well. Our first theorem provides a first step toward this goal.

THEOREM 2.2.1 The union of any collection of open sets in $\mathbb{R}^n$ is open. The intersection of any finite collection of open sets in $\mathbb{R}^n$ is also open.

Proof. Suppose that $\{U_\alpha : \alpha \text{ in } A\}$ is any collection of open sets in $\mathbb{R}^n$. Let $U = \cup_{\alpha \in A} U_\alpha$. To show that U is open, we must show that any point $\boldsymbol{x}$ in U is an interior point of U. Being in U, $\boldsymbol{x}$ belongs to at least one of the open sets U_α. Since U_α is open, $\boldsymbol{x}$ must be an interior point of U_α. Consequently, there must exist some neighborhood $N(\boldsymbol{x})$ contained in U_α. It follows that $N(\boldsymbol{x})$ must be contained entirely in U, and thus $\boldsymbol{x}$ is an interior point of the set U. Consequently, U must be open as claimed.

To prove the second assertion of the theorem, let U be the intersection of any finite collection $\{U_1, U_2, \ldots, U_k\}$ of open sets and let $\boldsymbol{x}$ be any point of U. To prove that $\boldsymbol{x}$ is an interior point of U, note first that, for each index $j = 1, 2, \ldots, k$, the point $\boldsymbol{x}$ is in the open set U_j. Hence, for each j, there exists a neighborhood $N(\boldsymbol{x}; r_j)$ contained entirely in U_j. Let $r = \min\{r_1, r_2, \ldots, r_k\}$. It follows that $N(\boldsymbol{x}; r) \subseteq N(\boldsymbol{x}; r_j)$ for every index j. Consequently,

$$N(\boldsymbol{x}; r) \subseteq \bigcap_{j=1}^{k} N(\boldsymbol{x}; r_j) \subseteq \bigcap_{j=1}^{k} U_j = U.$$

This proves that $\boldsymbol{x}$ is an interior point of U. Since $\boldsymbol{x}$ is an arbitrary point of U, we have proved that every point of U is an interior point of U. Hence U is open. ●

Be forewarned that the intersection of *infinitely* many open sets may fail to be open. In fact, consider the intersection of the infinitely many neighborhoods $N(\boldsymbol{x}; 1/k)$, $k = 1, 2, 3, \ldots$. This intersection is the set S consisting of the single point $\{\boldsymbol{x}\}$. But S is *not* open. The reason? The set S does not contain any neighborhood of the point $\boldsymbol{x}$.

You are also emphatically forewarned against confusing colloquial usage of the words *open* and *closed* with mathematical usage. In everyday speech these words are often taken to be antonyms, *not open* taken to mean *closed* and *not closed* to mean *open*. In mathematical context, *open* and *closed* are *not* antonyms; to say a set is not open does not mean it is closed. The interval $[a, b)$, for example, is neither open nor closed. Confusion on this point is commonplace among beginners and you will be wise to pay attention from the outset and to reprogram your thinking, especially in view of the following theorem.

THEOREM 2.2.2 A set C in $\mathbb{R}^n$ is closed if and only if its complement C^c is open.

Proof. Suppose that C is closed, so it contains all its boundary points. Suppose that $\boldsymbol{x}$ is any point in C^c. Since $\boldsymbol{x}$ is not in C, it cannot be a boundary point of C. The point $\boldsymbol{x}$ fails to be a boundary point of C only if there exists a neighborhood $N(\boldsymbol{x})$ for which either $N(\boldsymbol{x}) \cap C = \emptyset$ or $N(\boldsymbol{x}) \cap C^c = \emptyset$.

The second possibility cannot occur because $\boldsymbol{x}$ belongs to $N(\boldsymbol{x}) \cap C^c$, so this set is nonempty. Therefore $N(\boldsymbol{x}) \cap C$ is empty. Equivalently, $N(\boldsymbol{x}) \subseteq C^c$. In other words, $\boldsymbol{x}$ is an interior point of C^c. Since $\boldsymbol{x}$ is an arbitrary point in C^c, we see that C^c is open.

Conversely, suppose that C^c is open and that $\boldsymbol{x}$ is a boundary point of C. To conclude that C is closed, we must show that $\boldsymbol{x}$ is in C. Suppose, to the contrary, that $\boldsymbol{x}$ is in C^c. We have assumed that C^c is open and that $\boldsymbol{x}$ must therefore be an interior point of C^c. That is, there exists a neighborhood $N(\boldsymbol{x})$ contained entirely in C^c. But $\boldsymbol{x}$ is a boundary point of C. Thus $N(\boldsymbol{x}) \cap C$ is not empty. This conclusion cannot be reconciled with the containment $N(\boldsymbol{x}) \subseteq C^c$. The contradiction forces us to conclude that $\boldsymbol{x}$ cannot be in C^c. Therefore, an arbitrary boundary point of C is already in C. Consequently, C must be closed. ●

Clearly, in $\mathbb{R}$, the interval $[a, b]$ is closed since its complement $(-\infty, a) \cup (b, \infty)$ is open. Likewise the set consisting of the single point $\{\boldsymbol{x}\}$ is closed; its complement is the open set $(-\infty, \boldsymbol{x}) \cup (\boldsymbol{x}, \infty)$. The empty set $\emptyset$ is closed because its complement $\mathbb{R}$ is open. The set $\mathbb{R}$ is closed since its complement $\emptyset$ is open. You will notice that in $\mathbb{R}$ the empty set and all of $\mathbb{R}$ are both open and closed—in the jargon of the trade these sets are said to be *clopen*—but in $\mathbb{R}$, at least, these are the only sets having both properties. They serve to remind you, however, that the properties of being open and of being closed are not mutually exclusive. Nor are they exhaustive; there exist many sets that are neither open nor closed.

Applying DeMorgan's laws of set theory and combining the content of Theorem 2.2.1 and Theorem 2.2.2, you can prove the following theorem.

THEOREM 2.2.3 The intersection of any collection of closed sets is closed. The union of any finite collection of closed sets is closed. ●

The union of *infinitely* many closed sets may fail to be closed. For example, the union of all $C_k = [1/k, 3 - 1/k]$ for k in $\mathbb{N}$, is the open interval $(0, 3)$, an assertion whose proof is left as an exercise.

Before proceeding further, another warning is essential. The property of being open or of being closed depends on the universal space—the context—of the discussion. For example, in the universal space $\mathbb{R}$ the sets (a, b) and $\mathbb{R}$ *are both open* (in $\mathbb{R}$). However, if the universal space is $\mathbb{R}^n$, with $n \geq 2$, then neither the line segment (a, b) nor the real line $\mathbb{R}$, viewed as subsets of $\mathbb{R}^n$, is *open in* $\mathbb{R}^n$. No point of either set is an interior point since every n-dimensional neighborhood extends beyond the set. Moreover, the line segment (a, b) *is not closed in* $\mathbb{R}^n$ either, whereas the real line $\mathbb{R}$ *is closed in* $\mathbb{R}^n$. In the first instance, the endpoints a and b are boundary points absent from (a, b); in the second, $\mathbb{R}$ contains all its boundary points in $\mathbb{R}^n$. You will want to heed this warning by establishing the universal context of any discussion firmly in your mind from the outset. The following theorem, giving an alternate characterization of closed sets, often helps resolve this question.

THEOREM 2.2.4 A subset C of $\mathbb{R}^n$ is closed if and only if C contains all its limit points.

Proof. Suppose that C is closed and that $\boldsymbol{x}$ is any limit point of C. We want to show that $\boldsymbol{x}$ is in C. Suppose, to the contrary, that $\boldsymbol{x}$ is in C^c. Since C is closed, its complement C^c is open and $\boldsymbol{x}$ must be an interior point of C^c. Thus there is a neighborhood $N(\boldsymbol{x})$ contained entirely in C^c. $N(\boldsymbol{x})$ contains no points of C, contradicting the assumption that $\boldsymbol{x}$ is a limit point of C. The assumption that $\boldsymbol{x}$ is in C^c is untenable. Therefore $\boldsymbol{x}$ must be in C. Thus C must contain all its limit points.

Suppose, conversely, that C contains all its limit points. We want to prove that C must be closed. Equivalently, we want to show that C^c is open. To prove that C^c is open, we need only show that all its points are interior points. This sketches our strategy.

Choose any $\boldsymbol{x}$ in C^c. Since C contains all its limit points and since $\boldsymbol{x}$ does not belong to C, it follows that $\boldsymbol{x}$ is not a limit point of C. Hence there exists a neighborhood $N(\boldsymbol{x})$ that contains no points of C. It follows that $N(\boldsymbol{x}) \subseteq C^c$. But this says that $\boldsymbol{x}$ must be an interior point of C^c. Since $\boldsymbol{x}$ is an arbitrary point of C^c, we have proved that C^c is open. Therefore $C = (C^c)^c$ must be closed. ●

EXAMPLE 4 Let $\{\boldsymbol{x}_k\}$ be any convergent sequence in $\mathbb{R}^n$ with limit $\boldsymbol{x}_0$. The set $S = \{\boldsymbol{x}_k : k \text{ in } \mathbb{N}\} \cup \{\boldsymbol{x}_0\}$ is closed. The reason? S contains all its limit points. (See Exercise 2.25.) ●

EXAMPLE 5 Let $\{\boldsymbol{x}_k\}$ be any sequence in $\mathbb{R}^n$ and let C be the set of cluster points of $\{\boldsymbol{x}_k\}$. We claim that C is closed. To prove our claim, suppose that $\boldsymbol{x}_0$ is a limit point of C. We want to show that $\boldsymbol{x}_0$ itself is a cluster point of $\{\boldsymbol{x}_k\}$ and therefore is in C. Once we have done this, we can conclude that C, containing all its limit points, is closed. This sets our strategy. The point $\boldsymbol{x}_0$ is a limit point of C if, for any $\epsilon > 0$, $N'(\boldsymbol{x}_0; \epsilon)$ contains points of C. Choose any $\boldsymbol{c}$ in $C \cap N'(\boldsymbol{x}_0; \epsilon)$. Choose any positive ϵ_1 less than $\epsilon - ||\boldsymbol{c} - \boldsymbol{x}_0||$. We claim that $N(\boldsymbol{c}; \epsilon_1) \subseteq N(\boldsymbol{x}_0; \epsilon)$. (See Fig. 2.9.) To prove this claim, choose any $\boldsymbol{y}$ in $N(\boldsymbol{c}; \epsilon_1)$. Compute

$$\begin{aligned} ||\boldsymbol{y} - \boldsymbol{x}_0|| &\leq ||\boldsymbol{y} - \boldsymbol{c}|| + ||\boldsymbol{c} - \boldsymbol{x}_0|| \\ &< \epsilon_1 + ||\boldsymbol{c} - \boldsymbol{x}_0|| < \epsilon, \end{aligned}$$

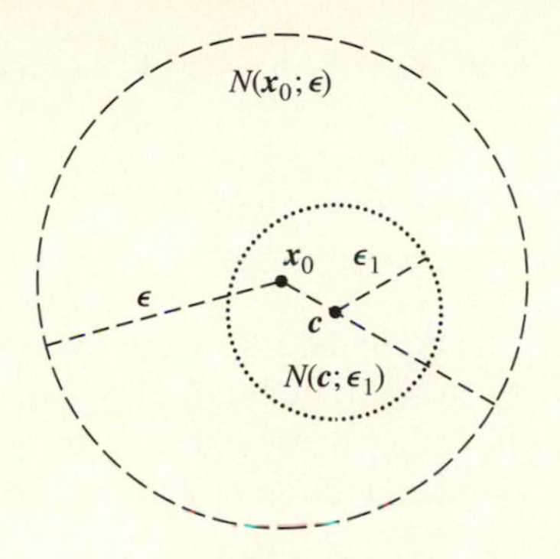

Figure 2.9

since $\epsilon_1 < \epsilon - ||\boldsymbol{c} - \boldsymbol{x}_0||$. Because $\boldsymbol{y}$ is an arbitrary point in $N(\boldsymbol{c}; \epsilon_1)$, we conclude that $N(\boldsymbol{c}; \epsilon_1) \subseteq N(\boldsymbol{x}_0; \epsilon)$.

Since $\boldsymbol{c}$ is a cluster point of $\{\boldsymbol{x}_k\}$, given this ϵ_1 and any k in $\mathbb{N}$, there is a $k_1 > k$ such that $\boldsymbol{x}_{k_1}$ is in $N(\boldsymbol{c}; \epsilon_1)$. Thus $\boldsymbol{x}_{k_1}$ is in $N(\boldsymbol{x}_0; \epsilon)$. We conclude that $\boldsymbol{x}_0$ is a cluster point of sequence $\{\boldsymbol{x}_k\}$ and therefore is in C. We have proved that C is closed. ●

EXAMPLE 6 The boundary of the set $N(\boldsymbol{x}_0; r)$ in $\mathbb{R}^n$ is the hyperspherical surface $B = \{\boldsymbol{x} \text{ in } \mathbb{R}^n : ||\boldsymbol{x} - \boldsymbol{x}_0|| = r\}$. The set $C = N(\boldsymbol{x}_0; r) \cup B = \{\boldsymbol{x} \text{ in } \mathbb{R}^n : ||\boldsymbol{x} - \boldsymbol{x}_0|| \leq r\}$ is closed in $\mathbb{R}^n$. To prove these claims, first show that any point $\boldsymbol{x}$ in B is a boundary point of $N(\boldsymbol{x}_0; r)$. Fix $\boldsymbol{x}$ in B and choose any neighborhood $N(\boldsymbol{x}; \epsilon)$. Form the vector $\boldsymbol{u} = (\boldsymbol{x} - \boldsymbol{x}_0)/r$ and note that $||\boldsymbol{u}|| = 1$. As the real number t varies over $[0, \infty)$, the vector $\boldsymbol{y}(t) = \boldsymbol{x}_0 + t\boldsymbol{u}$ describes the straight line originating at $\boldsymbol{x}_0$ and passing through $\boldsymbol{x}$. (See Fig. 2.10.) For $0 \leq t < r$, $\boldsymbol{y}(t)$ is in $N(\boldsymbol{x}_0; r)$ because

$$||\boldsymbol{y}(t) - \boldsymbol{x}_0|| = ||t\boldsymbol{u}|| = t < r.$$

If $t \geq r$, then a similar computation shows that $\boldsymbol{y}(t)$ is not in $N(\boldsymbol{x}_0; r)$. Thus $\boldsymbol{y}(t)$ is in $N(\boldsymbol{x}_0; r)^c$.

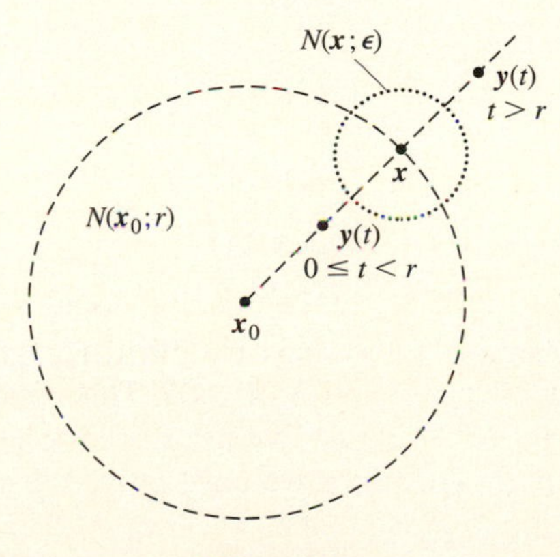

Figure 2.10

If $r - \epsilon < t < r + \epsilon$, then $\boldsymbol{y}(t)$ is in $N(\boldsymbol{x}; \epsilon)$, as the following computation shows:

$$\begin{aligned}||\boldsymbol{y}(t) - \boldsymbol{x}|| &= ||\boldsymbol{x}_0 + t\boldsymbol{u} - \boldsymbol{x}|| = ||\boldsymbol{x}_0 + t\boldsymbol{u} - (\boldsymbol{x}_0 + r\boldsymbol{u})|| \\ &= ||(t - r)\boldsymbol{u}|| = |t - r| < \epsilon.\end{aligned}$$

These computations lay the groundwork for our proof that $\boldsymbol{x}$ is a boundary point of $N(\boldsymbol{x}_0; r)$. Merely choose, on the one hand, any t_1 in $(r - \epsilon, r)$. Then $\boldsymbol{y}(t_1)$ is in $N(\boldsymbol{x}_0; r) \cap N(\boldsymbol{x}; \epsilon)$. On the other hand, choosing t_2 in $(r, r + \epsilon)$ yields a point $\boldsymbol{y}(t_2)$ in $N(\boldsymbol{x}_0; r)^c \cap N(\boldsymbol{x}; \epsilon)$. Hence if $||\boldsymbol{x} - \boldsymbol{x}_0|| = r$, then $\boldsymbol{x}$ is a boundary point of $N(\boldsymbol{x}_0; r)$.

Suppose next that $\boldsymbol{x}$ is any boundary point of $N(\boldsymbol{x}_0; r)$. That is, every neighborhood of $\boldsymbol{x}$ contains points of $N(\boldsymbol{x}_0; r)$ and of $N(\boldsymbol{x}_0; r)^c$. We want to show that $\boldsymbol{x}$ is in B. That is, we want to prove that $||\boldsymbol{x} - \boldsymbol{x}_0|| = r$. Suppose that $||\boldsymbol{x} - \boldsymbol{x}_0|| = s$. By the trichotomy law, either $s < r$, $s > r$, or $s = r$. We want to eliminate the first two possibilities. If $s < r$, then $\boldsymbol{x}$ is in $N(\boldsymbol{x}_0; r)$. Choose any positive $\epsilon < r - s$. For this choice of ϵ, the neighborhood $N(\boldsymbol{x}; \epsilon)$ is contained in $N(\boldsymbol{x}_0; r)$. For, if $\boldsymbol{y}$ is in $N(\boldsymbol{x}; \epsilon)$, then

$$||\boldsymbol{y} - \boldsymbol{x}_0|| \le ||\boldsymbol{y} - \boldsymbol{x}|| + ||\boldsymbol{x} - \boldsymbol{x}_0|| < \epsilon + s < r.$$

Therefore, $\boldsymbol{y}$ is in $N(\boldsymbol{x}_0; r)$. Thus $N(\boldsymbol{x}; \epsilon) \subseteq N(\boldsymbol{x}_0; r)$. (See Fig. 2.11.) If $s > r$, then $\boldsymbol{x}$ is in $N(\boldsymbol{x}_0; r)^c$. Choose a positive $\epsilon < s - r$. For this choice of ϵ, $N(\boldsymbol{x}; \epsilon)$ is contained in $N(\boldsymbol{x}_0; r)^c$. For if $\boldsymbol{y}$ is in $N(\boldsymbol{x}; \epsilon)$, then

$$||\boldsymbol{x}_0 - \boldsymbol{y}|| \ge ||\boldsymbol{x}_0 - \boldsymbol{x}|| - ||\boldsymbol{x} - \boldsymbol{y}|| > s - \epsilon > r.$$

Therefore $\boldsymbol{y}$ is in $N(\boldsymbol{x}_0; r)^c$ and $N(\boldsymbol{x}; \epsilon) \subseteq N(\boldsymbol{x}_0; r)^c$. (See Fig. 2.12.)

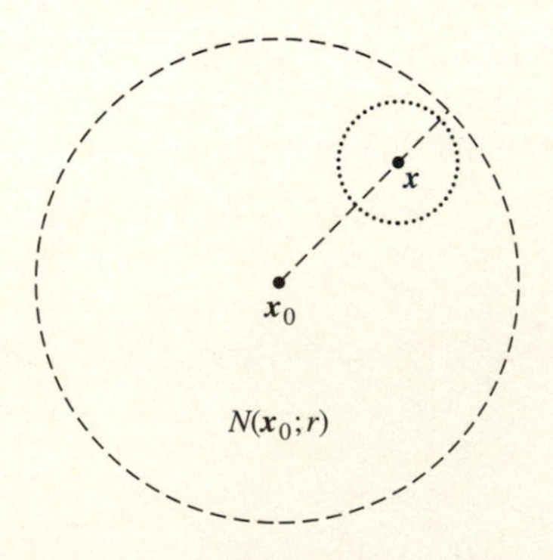

Figure 2.11

In either case ($s < r$ or $s > r$), we have exhibited a neighborhood of $\boldsymbol{x}$ that fails to contain points of both $N(\boldsymbol{x}_0; r)$ and $N(\boldsymbol{x}_0; r)^c$. This contradicts the assumption that $\boldsymbol{x}$ is a boundary point of $N(\boldsymbol{x}_0; r)$. We are forced to conclude that $s = ||\boldsymbol{x} - \boldsymbol{x}_0|| = r$ and that $\boldsymbol{x}$ is in B. Therefore we have proved that B is the boundary of $N(\boldsymbol{x}_0; r)$. We leave for you to prove that $C = N(\boldsymbol{x}_0; r) \cup B$ is closed. ●

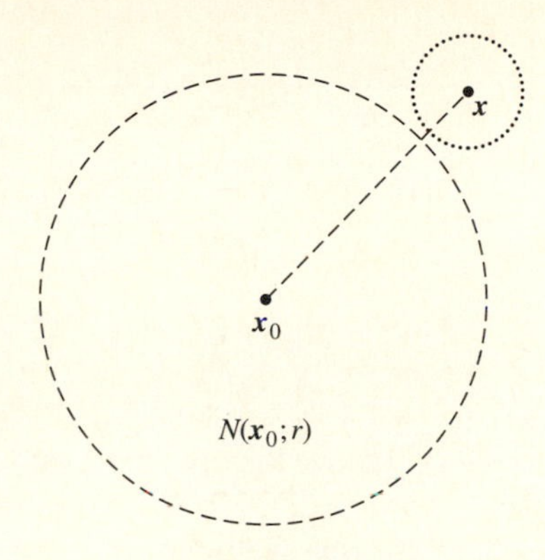

Figure 2.12

EXAMPLE 7 In Section 2.1 we discussed n-dimensional hyper-rectangles $S = I_1 \times I_2 \times \cdots \times I_n$, where each I_j is an interval in the jth coordinate axis. If each of the I_j is an open interval in $\mathbb{R}$, then the resulting hyper-rectangle S is open in $\mathbb{R}^n$. If each of the I_j is a closed interval in $\mathbb{R}$, then the resulting hyper-rectangle is closed in $\mathbb{R}^n$. (See Exercise 2.26.) The open n-cube $C(2r)$ centered at $\boldsymbol{x} = (x_1, x_2, \ldots, x_n)$ in $\mathbb{R}^n$ is

$$C(2r) = N(x_1; r) \times N(x_2; r) \times \cdots \times N(x_n; r). \quad \bullet$$

EXAMPLE 8 Experience shows that, at this early stage of your studies, you will tend to imagine only relatively simple sets. Our last example in this section, Cantor's set, will help remedy that shortcoming. This set is yet another brilliant discovery that Cantor used to display the power of his methods. It is used in more advanced studies to construct astonishing examples of mathematical objects that boggle the imagination.

Begin with the closed unit interval $[0, 1]$ and methodically remove, in stages, certain subintervals. At the completion of the process, Cantor's set will consist of those points remaining.

At the first step, remove the open interval $I(1, 1) = (1/3, 2/3)$. Partition the remaining two subintervals into equal thirds and remove the open middle thirds. That is, remove the intervals

$$I(2, 1) = \left(\frac{1}{9}, \frac{2}{9}\right) \quad \text{and} \quad I(2, 2) = \left(\frac{7}{9}, \frac{8}{9}\right),$$

each of length $1/3^2$. There remain four subintervals, as indicated in Fig. 2.13 on page 86. Partition each of these four remaining subintervals into equal thirds and, again,

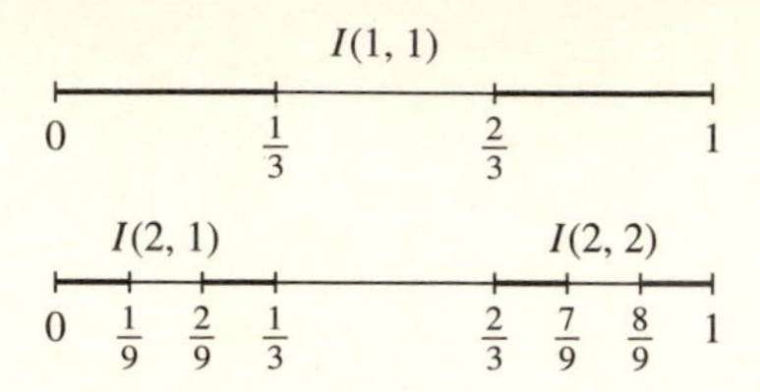

Figure 2.13

remove the middle thirds. That is, remove the open intervals $I(3, 1) = (1/27, 2/27)$, $I(3, 2) = (7/27, 8/27)$, $I(3, 3) = (19/27, 20/27)$, and $I(3, 4) = (25/27, 26/27)$, each of length $1/3^3$. Continue this process methodically. At the kth stage, partition each of the 2^{k-1} intervals, each of length $1/3^{k-1}$, into three subintervals of equal length and remove the open middle third of each. That is, remove the open intervals $I(k, 1), I(k, 2), \ldots,$ and $I(k, 2^{k-1})$, each of length $1/3^k$. Let

$$U_k = \bigcup_{p=1}^{k} \bigcup_{j=1}^{2^{p-1}} I(p, j)$$

denote the disjoint union of all the open intervals removed by the kth stage. Denote the length of $I(p, j)$ by $|I(p, j)|$. For each $p = 1, 2, \ldots, k$ and $j = 1, 2, \ldots, 2^{p-1}$, we have $|I(p, j)| = 1/3^p$. When we compute the total length of the intervals in the union defining U_k, we obtain

$$d_k = \sum_{p=1}^{k} \sum_{j=1}^{2^{p-1}} |I(p, j)| = \sum_{p=1}^{k} \frac{2^{p-1}}{3^p} = \left(\frac{1}{3}\right) \sum_{p=0}^{k-1} \left(\frac{2}{3}\right)^p$$

$$= \left(\frac{1}{3}\right) \frac{1 - (2/3)^k}{1 - (2/3)} = 1 - \left(\frac{2}{3}\right)^k.$$

Continue this process indefinitely. Let $U = \cup_{p=1}^{\infty} \cup_{j=1}^{2^{p-1}} I(p, j)$ denote the set of all points removed from $[0, 1]$. Since U is the union of open intervals, U is open. Hence U^c is closed. The set

$$C = [0, 1] \cap U^c,$$

being the intersection of two closed sets, is closed. The set C is called *Cantor's set* and consists of those points in $[0, 1]$ that were *not removed* by the process just described. Certainly C contains the endpoints of the open intervals $I(p, j)$, p in $\mathbb{N}$, $j = 1, 2, \ldots, 2^{p-1}$, that were removed. Thus the cardinality of C is at least countably infinite.

Cantor's set has several remarkable properties. To discover the first noteworthy property, consider the total d_k of the lengths of the intervals removed from $[0, 1]$ by the kth stage of the construction of C: $d_k = 1 - (2/3)^k$. Since $\lim_{k\to\infty} (2/3)^k = 0$, it follows that $\lim_{k\to\infty} d_k = 1$. In the process of constructing Cantor's set, we removed from $[0, 1]$, itself of length 1, disjoint intervals the sum of whose lengths approaches 1. For this reason you might be tempted to assume that C must be a small set. While C has no "length," as we typically think of it, C is not small. In fact, C is uncountably infinite. We will outline a proof of this remarkable fact in the exercises.

Being infinite and bounded, C has at least one limit point by the Bolzano–Weierstrass theorem. Since C is also closed, it contains all its limit points. But no point of C is an interior point. Assume, to the contrary, that x were to be an interior point of C; there exists a neighborhood $N(x; r)$ contained in C. It follows that $N(x; r) \cap U = \emptyset$. Therefore $N(x; r) \cap U_k = \emptyset$ for all k in $\mathbb{N}$. Since $N(x; r)$ and U_k are disjoint and are contained in $[0, 1]$, we have $d_k + 2r \leq 1$ for each k in $\mathbb{N}$. We will obtain a contradiction. Since $\lim_{k\to\infty}(2/3)^k = 0$, we can choose k in $\mathbb{N}$ such that $(2/3)^k < 2r$. Equivalently, $2r - (2/3)^k > 0$. We deduce that

$$1 \geq d_k + 2r = 1 - \left(\frac{2}{3}\right)^k + 2r > 1,$$

an evident impossibility. As a consequence, we conclude that no point of C is an interior point; every point of C is a boundary point and is, in fact, a limit point of C. ●

By specifying a set S in $\mathbb{R}^n$, we implicitly create several additional sets related to S.

DEFINITION 2.2.2 Let S be any subset of $\mathbb{R}^n$.

i) The *interior* of S, denoted S^0, is the set of all interior points of S.

ii) The *boundary* of S, denoted bd(S), is the set of all boundary points of S.

iii) The *derived set* of S, denoted S', is the set of all limit points of S.

iv) The *closure* of S, denoted $\overline{S}$, is the union of S and S'.

v) The *complement* of S, denoted S^c, is the set of all points in $\mathbb{R}^n$ that are not in S. ●

We want to examine these various sets and identify some of the relationships among them.

THEOREM 2.2.5 Let S be any subset of $\mathbb{R}^n$. The interior of S is the union of all open sets contained in S.

Proof. Let $\mathcal{U} = \{U_\alpha : U_\alpha \text{ open}, U_\alpha \subseteq S\}$ denote the collection of all open sets contained in S. Let $U = \cup_\alpha U_\alpha$. Being the union of open sets, U is also open by Theorem 2.2.1. Since each U_α is contained in S, the same is true of U. We are to show that U consists exactly of all interior points of S.

Suppose that $\boldsymbol{x}$ is an interior point of S. Then there exists a neighborhood $N(\boldsymbol{x})$ that is contained in S. Thus $N(\boldsymbol{x})$ is a set in $\mathcal{U}$. Therefore, $N(\boldsymbol{x}) \subseteq U$, the union of all the sets in $\mathcal{U}$, and, consequently, $\boldsymbol{x}$ is in U.

Suppose, on the other hand, that $\boldsymbol{x}$ is any point of U. Since U is open, $\boldsymbol{x}$ is an interior point of U. Hence there is a neighborhood $N(\boldsymbol{x})$ contained entirely in U. But U is a subset of S. Therefore $N(\boldsymbol{x}) \subseteq S$, proving that $\boldsymbol{x}$ is an interior point of S. These two arguments, taken together, prove that $U = S^0$. ●

COROLLARY 2.2.6 For any set S in $\mathbb{R}^n$, the set S^0 is open. ●

We often say, because of this theorem, that S^0 is the *largest open set* contained in S.

We also have the following analogous description of the closure of any set S in $\mathbb{R}^n$.

THEOREM 2.2.7 The closure of S is the intersection of all closed sets that contain S.

Proof. Let $\mathcal{C} = \{C_\alpha : C_\alpha \text{ closed}, S \subseteq C_\alpha\}$ denote the collection of all closed sets that contain S. Let $C = \cap_\alpha C_\alpha$. By Theorem 2.2.3, C is closed. Clearly $S \subseteq C$. We are to prove that $C = \overline{S}$. Recall that $\overline{S} = S \cup S'$.

We know already that $S \subseteq C$. Therefore, to show first that $\overline{S} \subseteq C$, we need only show that $S' \subseteq C$. To this end, suppose that $\boldsymbol{x}$ is in S'; that is, $\boldsymbol{x}$ is a limit point of S. It is evident that, if T is any set that contains S, then $\boldsymbol{x}$ is a limit point of T also. (You can prove this as an exercise.) We apply this observation by noting that, since $S \subseteq C_\alpha$ for each α, the point $\boldsymbol{x}$ is a limit point of each C_α in $\mathcal{C}$. But each C_α is closed and therefore, by Theorem 2.2.4, contains all its limit points. Consequently, $\boldsymbol{x}$ is in C_α for all α. Therefore, $\boldsymbol{x}$ is in $C = \cap_\alpha C_\alpha$. This shows that $S' \subseteq C$. We conclude that $\overline{S} = S \cup S' \subseteq C$.

To obtain the reverse containment, we show directly that $\overline{S}$ is closed. It suffices, by Theorem 2.2.4, to show that $\overline{S}$ contains all its limit points. To this end, suppose that $\boldsymbol{y}$ is any limit point of $\overline{S}$. Every deleted neighborhood $N'(\boldsymbol{y}; \epsilon)$ of $\boldsymbol{y}$ contains a point $\boldsymbol{z}$ of $\overline{S} = S \cup S'$. Either $\boldsymbol{z}$ is in S or it is in S'.

If $\boldsymbol{z}$ is in S, then $S \cap N'(\boldsymbol{y}; \epsilon) \neq \emptyset$. If $\boldsymbol{z}$ is in S', then $\boldsymbol{z}$ is a limit point of S. In this latter case, choose a positive $\epsilon_1 < \min\{\|\boldsymbol{z} - \boldsymbol{y}\|, \epsilon - \|\boldsymbol{z} - \boldsymbol{y}\|\}$. (See Fig. 2.14.) By imitating our earlier, similar work, it is straightforward to prove that $N(\boldsymbol{z}; \epsilon_1) \subseteq N'(\boldsymbol{y}; \epsilon)$. Since $\boldsymbol{z}$ is a limit point of S, there exists an $\boldsymbol{x}$ in $S \cap N'(\boldsymbol{z}; \epsilon_1)$. Because $S \cap N'(\boldsymbol{z}; \epsilon_1) \subseteq S \cap N'(\boldsymbol{y}; \epsilon)$, it follows that the point $\boldsymbol{x}$ is in $S \cap N'(\boldsymbol{y}; \epsilon)$.

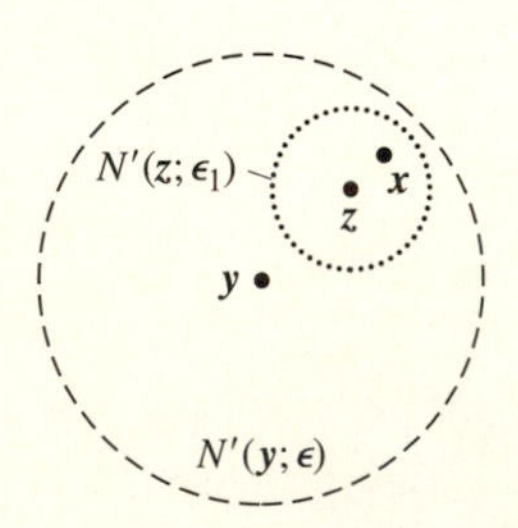

Figure 2.14

In either case, we have shown that $S \cap N'(\boldsymbol{y}; \epsilon) \neq \emptyset$ and that $\boldsymbol{y}$ is therefore a limit point of S itself. That is, $\boldsymbol{y}$ is in S'. Thus $\boldsymbol{y}$ is in $\overline{S}$. We have proved that $\overline{S}$ contains all its limit points. By Theorem 2.2.4, $\overline{S}$ is closed. Since $\overline{S}$ is closed and contains S, we deduce that $\overline{S}$ is one of the sets in $\mathcal{C}$. Therefore $\overline{S} \supseteq C$. Combining this conclusion with the earlier part of the proof, we conclude that $C = \overline{S}$. ●

COROLLARY 2.2.8 For any set S in $\mathbb{R}^n$, $\overline{S}$ is closed. ●

We often say, because of this theorem, that $\overline{S}$ is the *smallest closed set* that contains S.

There are several connections between interiors, closures, boundaries, and complements. In the following theorem we list several; the proof of each assertion is left as an exercise. Once you have worked through the details on your own, you will have a much more solid grasp of these concepts.

THEOREM 2.2.9 Let S be any set in $\mathbb{R}^n$.

i) $(S^0)^0 = S^0$.

ii) $\overline{(\overline{S})} = \overline{S}$.

iii) $S^0 \cap bd(S) = \emptyset$.

iv) $S^0 \cup bd(S) = \overline{S}$.

v) $\overline{S} \cap \overline{(S^c)} = bd(S)$. ●

The notion of the diameter of a bounded set is a concept that will be useful to us at various junctures in the text.

DEFINITION 2.2.3 Let S be any bounded, nonempty set in $\mathbb{R}^n$. The *diameter* of S is defined to be

$$d(S) = \sup\{||\boldsymbol{x} - \boldsymbol{y}|| : \boldsymbol{x}, \boldsymbol{y} \text{ in } S\}. \quad \bullet$$

Since S is assumed to be bounded, there is an M such that $||\boldsymbol{x}|| \leq M$ for all $\boldsymbol{x}$ in S. Then, for any $\boldsymbol{x}$, $\boldsymbol{y}$ in S,

$$||\boldsymbol{x} - \boldsymbol{y}|| \leq ||\boldsymbol{x}|| + ||\boldsymbol{y}|| \leq 2M.$$

Consequently, the set $\{||\boldsymbol{x} - \boldsymbol{y}|| : \boldsymbol{x}, \boldsymbol{y} \text{ in } S\}$ of real numbers is bounded above by $2M$. Therefore its supremum exists in $\mathbb{R}$. This observation proves that every bounded set in $\mathbb{R}^n$ has a diameter and that $d(S) \leq 2M$, where M is a bound for S.

Finally, we come full circle and consider from our new vantage point the concept of the distance from a point $\boldsymbol{x}$ to a set S.

DEFINITION 2.2.4 Let S be any nonempty set in $\mathbb{R}^n$ and let $\boldsymbol{x}$ be any point in $\mathbb{R}^n$. The *distance* from $\boldsymbol{x}$ to S is defined to be

$$d(\boldsymbol{x}, S) = \inf\{||\boldsymbol{x} - \boldsymbol{y}|| : \boldsymbol{y} \text{ in } S\}. \quad \bullet$$

This infimum certainly exists in $\mathbb{R}$ and is nonnegative. If $\boldsymbol{x}$ is already in S, then, of course, $d(\boldsymbol{x}, S) = 0$. More generally, we have the following theorem.

THEOREM 2.2.10 Let S be a nonempty set in $\mathbb{R}^n$ and let $\boldsymbol{x}$ be a point of $\mathbb{R}^n$.

i) Then $d(\boldsymbol{x}, S) = 0$ if and only if $\boldsymbol{x}$ is in $\overline{S}$.

ii) S is closed if and only if $d(\boldsymbol{x}, S) > 0$ for every $\boldsymbol{x}$ in S^c.

iii) If S is closed, then there exists a $\boldsymbol{y}_0$ in S such that $d(\boldsymbol{x}, S) = ||\boldsymbol{x} - \boldsymbol{y}_0||$.

iv) If S is open and if $\boldsymbol{x}$ is in S^c, then there exists no $\boldsymbol{y}$ in S such that $d(\boldsymbol{x}, S) = ||\boldsymbol{x} - \boldsymbol{y}||$.

Proof. We will prove part (i) and leave the remaining parts as exercises. Suppose that $d(\boldsymbol{x}, S) = 0$ and that $\boldsymbol{x}$ is not in S. We want to show that $\boldsymbol{x}$ must be a limit point of S. From the definition of $d(\boldsymbol{x}, S)$, we know, by Theorem 1.1.1, that, for every $\epsilon > 0$, there is a $\boldsymbol{y}$ in S such that $0 \leq ||\boldsymbol{x} - \boldsymbol{y}|| < \epsilon$. Since $\boldsymbol{x}$ is not in S, we know that $\boldsymbol{y} \neq \boldsymbol{x}$. Therefore $\boldsymbol{y}$ is in $N'(\boldsymbol{x}; \epsilon)$. That is, $\boldsymbol{x}$ is a limit point of S. Thus $\boldsymbol{x}$ is in S'. We conclude that $\boldsymbol{x}$ is in $\overline{S}$.

Conversely, if $\boldsymbol{x}$ is a limit point of S, then for all $\epsilon > 0$, there exists a $\boldsymbol{z}$ in S such that $0 < ||\boldsymbol{x} - \boldsymbol{z}|| < \epsilon$. Therefore $d(\boldsymbol{x}, S) = \inf \{||\boldsymbol{x} - \boldsymbol{y}|| : \boldsymbol{y} \text{ in } S\} < \epsilon$. But $d(\boldsymbol{x}, S)$ is a nonnegative constant. By Exercise 1.10, $d(\boldsymbol{x}, S) = 0$. ●

EXERCISES

2.19. Let U be any open set in $\mathbb{R}$. Show that U is the union of open neighborhoods of its points. Likewise, show that any open set in $\mathbb{R}^n$ is the union of neighborhoods of its points.

2.20. Prove that the union of the sets $C_k = [1/k, 3 - 1/k]$, for $k = 3, 4, 5, \ldots$, is the open interval $(0, 3)$.

2.21. Is the set $S = \{(x_1, x_2, 0) : x_1^2 + x_2^2 < r^2\}$ open in $\mathbb{R}^3$? If so, prove it; otherwise, explain why not.

2.22. Prove that the intersection of any collection of closed sets in $\mathbb{R}^n$ is closed.

2.23. Prove that the union of any finite collection of closed sets in $\mathbb{R}^n$ is closed.

2.24. Let $\boldsymbol{x}_1$ and $\boldsymbol{x}_2$ be two distinct points in $\mathbb{R}^n$. Prove that there exist disjoint open sets U_1 and U_2 such that $\boldsymbol{x}_1$ is in U_1 and $\boldsymbol{x}_2$ is in U_2.

2.25. Prove that, if $\{\boldsymbol{x}_k\}$ is a convergent sequence in $\mathbb{R}^n$ with limit $\boldsymbol{x}_0$, then $S = \{\boldsymbol{x}_k : k \text{ in } \mathbb{N}\} \cup \{\boldsymbol{x}_0\}$ is closed in $\mathbb{R}^n$. (Evidently, $\boldsymbol{x}_0$ is a limit point of S. How do you know it is the *only* limit point, not just of $\{\boldsymbol{x}_k : k \text{ in } \mathbb{N}\}$, but of S? Exercise 1.27 and/or Exercise 2.17 are relevant.)

2.26. Prove that the set $C = N(\boldsymbol{x}_0; r) \cup B$ in Example 6 is closed in $\mathbb{R}^n$.

2.27. **a)** Prove that the n-dimensional hyper-rectangle

$$S = (a_1, b_1) \times (a_2, b_2) \times \cdots \times (a_n, b_n)$$

is an open set in $\mathbb{R}^n$.

b) Prove that the n-dimensional hyper-rectangle

$$T = [a_1, b_1] \times [a_2, b_2] \times \cdots \times [a_n, b_n]$$

is a closed set in $\mathbb{R}^n$.

c) Find the relationship between $\overline{S}$ and T. Also find the relationship between S and T^0. Prove your answers.

2.28. Prove that the n-cube

$$C(2r) = N(x_1; r) \times N(x_2; r) \times \cdots \times N(x_n; r)$$

centered at $\boldsymbol{x} = (x_1, x_2, \ldots, x_n)$ is the set

$$N_\infty(\boldsymbol{x}; r) = \{\boldsymbol{y} \text{ in } \mathbb{R}^n : d_\infty(\boldsymbol{x}, \boldsymbol{y}) < r\},$$

where $d_\infty(\boldsymbol{x}, \boldsymbol{y}) = \max \{|x_1 - y_1|, |x_2 - y_2|, \ldots, |x_n - y_n|\}$.

2.29. For each set, find S^0, $\overline{S}$, S', and bd(S).

a) $S = (a, b] = \{x \text{ in } \mathbb{R} : a < x \leq b\}$ in $\mathbb{R}$ where $a < b$.

b) $S = \{(x_1, 0) \text{ in } \mathbb{R}^2 : a < x_1 \leq b\}$ in $\mathbb{R}^2$, where $a < b$.

c) $S = \mathbb{Q}$ in $\mathbb{R}$.

d) $S = \mathbb{Q}^n$ in $\mathbb{R}^n$.

e) $S =$ Cantor's set C in $\mathbb{R}$.

2.30. For each set in Exercise 2.29, show explicitly that $(S^0)^0 = S^0$, $\overline{(\overline{S})} = \overline{S}$, $S^0 \cap \text{bd}(S) = \emptyset$, $S^0 \cup \text{bd}(S) = \overline{S}$, and $\overline{S} \cap \overline{(S^c)} = \text{bd}(S)$.

2.31. Prove that, for any open set U in $\mathbb{R}^n$, $U^0 = U$. Explain why, for any set S in $\mathbb{R}^n$, it follows that $(S^0)^0 = S^0$.

2.32. Prove that, for any closed set C in $\mathbb{R}^n$, $\overline{C} = C$. Explain why, for any set S in $\mathbb{R}^n$, it follows that $\overline{(\overline{S})} = \overline{S}$.

2.33. Prove that, for any set S in $\mathbb{R}^n$, $S^0 \cap \text{bd}(S) = \emptyset$.

2.34. Prove that, for any set S in $\mathbb{R}^n$, $S^0 \cup \text{bd}(S) = \overline{S}$.

2.35. Prove that, for any set S in $\mathbb{R}^n$, $\overline{S} \cap \overline{(S^c)} = \text{bd}(S)$.

2.36. Prove or disprove: For every set S in $\mathbb{R}^n$, $\overline{(S^0)} = \overline{S}$.

2.37. Prove or disprove: For every set S in $\mathbb{R}^n$, bd$(S) =$bd(S^c).

2.38. Prove or disprove: For every set S in $\mathbb{R}^n$, bd$(\overline{S}) =$bd(S).

2.39. Prove that, for any two sets S and T in $\mathbb{R}^n$,

a) $(S \cap T)^0 = S^0 \cap T^0$

b) $\overline{(S \cup T)} = \overline{S} \cup \overline{T}$.

2.40. Prove that, for any two sets S and T in $\mathbb{R}^n$,

a) $(S \cup T)^0 \supseteq S^0 \cup T^0$

b) $\overline{(S \cap T)} \subseteq \overline{S} \cap \overline{T}$.

Find examples of sets S and T for which each of these containments is proper.

2.41. **a)** Let S be a bounded set in $\mathbb{R}^n$. Prove that $\overline{S}$ is also bounded.

b) If M is a bound for S, is it true that M is necessarily a bound for $\overline{S}$ also? Prove your answer.

2.42. Let S be any bounded set in $\mathbb{R}^n$. Prove that $d(\overline{S}) = d(S)$, where $d(A)$ denotes the diameter of the set A.

2.43. Prove that a nonempty subset S of $\mathbb{R}^n$ is closed if and only if $d(\boldsymbol{x}, S) > 0$ for every $\boldsymbol{x}$ in S^c.

2.44. Suppose that S is a closed subset of $\mathbb{R}^n$ and that $\boldsymbol{x}$ is a point of S^c. Prove that there exists a $\boldsymbol{y}_0$ in S such that $d(\boldsymbol{x}, S) = ||\boldsymbol{x} - \boldsymbol{y}_0||$. (*Hint:* Consider separately the two cases when S is a finite set and when S is an infinite set. If S is an infinite set, assume that no such $\boldsymbol{y}_0$ exists. Let $\{\epsilon_k\}$ be a sequence of positive numbers that converges monotonically to 0. Explain how to choose distinct points $\boldsymbol{y}_k$ in S such that $d(\boldsymbol{x}, S) < ||\boldsymbol{x} - \boldsymbol{y}_k|| < d(\boldsymbol{x}, S) + \epsilon_k$ for all k in $\mathbb{N}$. Apply the Bolzano–Weierstrass theorem, then extract a convergent subsequence $\{\boldsymbol{y}_{k_j}\}$. What properties must the limit of the subsequence $\{\boldsymbol{y}_{k_j}\}$ have?)

2.45. Suppose that S is an open subset of $\mathbb{R}^n$ and that $\boldsymbol{x}$ is a point of S^c. Prove that there exists no $\boldsymbol{y}$ in S such that $d(\boldsymbol{x}, S) = ||\boldsymbol{x} - \boldsymbol{y}||$.

2.46. Prove that no point of Cantor's set C is an isolated point of C.

2.47. Here is one way of proving that Cantor's set C is uncountable.

Points in $[0, 1]$ can be expressed as tricimals of the form $x = .t_1t_2t_3\cdots t_k\cdots. = \sum_{j=1}^{\infty} t_j/3^j$, where $t_j = 0, 1$, or 2. As with decimal expansions, for most of the numbers in $[0, 1]$, there is no ambiguity in such expansions. However, as with decimals, expansions of the two forms

$$.t_1t_2t_3\cdots t_k\overline{2} \quad \text{and} \quad .t_1t_2t_3\cdots(t_k+1)\overline{0},$$

where $t_k \neq 2$, represent the same number in $[0, 1]$.

a) Show that

$$.t_1t_2t_3\cdots t_k\overline{2} = \sum_{j=1}^{k}\frac{t_j}{3^j} + \sum_{j=k+1}^{\infty}\frac{2}{3^j}$$

and

$$.t_1t_2t_3\cdots(t_k+1)\overline{0} = \sum_{j=1}^{k-1}\frac{t_j}{3^j} + \frac{(t_k+1)}{3^k}$$

where $t_k \neq 2$, represent the same x in $[0, 1]$.

For the purposes of this problem, if x in $[0, 1]$ has two tricimal expansions, we always choose, if possible, the one not using the digit 1. For example, 1/3 has the two expansions .1000 ... and .0222.... We use the latter. On the other hand, 2/3 can be expanded as .2000 ... and as .1222 ...; for 2/3 we use the former. Some expansions, such as .10222 ... and .11000 ..., describe the same x; choose the form using the fewest number of 1s. *Fix this convention for the remainder of this problem.*

Now revive the notation used in the construction of Cantor's set C. Note that any point in $I(1, 1) = (1/3, 2/3)$ has a tricimal expansion of the form $.1t_2t_3t_4\ldots$. Any point in $I(2, 1) = (1/9, 2/9)$ has tricimal expansion of the form $.01t_3t_4\ldots$. Likewise, any point in $I(2, 2) = (7/9, 8/9)$ has a tricimal expansion of the form $.21t_3t_4t_5\ldots$. As you can check, at the next stage in the construction of Cantor's set, we have

$$\begin{aligned}
&x \text{ in } I(3,1) \text{ has form } .001t_4t_5t_6\cdots\\
&x \text{ in } I(3,2) \text{ has form } .021t_4t_5t_6\cdots\\
&x \text{ in } I(3,3) \text{ has form } .201t_4t_5t_6\cdots\\
&x \text{ in } I(3,4) \text{ has form } .221t_4t_5t_6\cdots.
\end{aligned}$$

b) Show, in general, that an x in $[0, 1]$ is in one of the $I(k, j)$ and is excluded from Cantor's set if and only if the tricimal expansion of x has the form

$$.t_1t_2\cdots t_{k-1}1t_{k+1}t_{k+2}\cdots,$$

where *none* of the $t_1, t_2, \ldots, t_{k-1}$ is 1. (For $j > k$, the t_j are any of 0, 1, 2, using as few 1s as possible in accord with our convention.) Thus show that Cantor's set consists of exactly those x for which the tricimal expansion contains no 1s.

c) Show that Cantor's set C can be put in one-to-one correspondence with $\mathbf{B}$, the set of infinite binary sequences. (Refer to Exercise 1.109.)

d) Hence prove that C is uncountable.

2.3 COMPLETENESS

While the concept of supremum does not generalize to $\mathbb{R}^n$, we have already derived powerful theorems that bypass that difficulty, which apply as easily to $\mathbb{R}^n$ as to $\mathbb{R}$, and whose content guarantees, under mild hypotheses, the existence of a limit point. We refer to the Cauchy completeness of $\mathbb{R}^n$ and to the fact that $\mathbb{R}^n$ has the Bolzano–Weierstrass property. In this section we complete our program of establishing equivalent formulations of completeness. We've seen the ideas introduced by Cauchy, Bolzano, and Weierstrass. Here we examine Cantor's approach to the question.

In an 1884 paper, Cantor began with a sequence $\{I_k\}$ of nonempty, bounded, closed intervals such that $I_k \supseteq I_{k+1}$ for each k in $\mathbb{N}$. He reasoned that $\cap_{k=1}^{\infty} I_k$ cannot be empty. Furthermore, if $I_k = [a_k, b_k]$ and if $\lim_{k\to\infty}(b_k - a_k) = 0$, then, said Cantor, this intersection must reduce to a single point. Here we prove that Cantor's intuition was, once again, correct.

DEFINITION 2.3.1 A sequence $\{S_k\}$ of sets in $\mathbb{R}^n$ such that $S_k \supseteq S_{k+1}$ for each k in $\mathbb{N}$ is said to be *nested.* ●

THEOREM 2.3.1 Cantor's Nested Interval Theorem For each k in $\mathbb{N}$, let $I_k = [a_k, b_k]$ with $a_k < b_k$. Suppose that $\{I_k\}$ is a nested sequence in $\mathbb{R}$. Then

$$\bigcap_{k=1}^{\infty} I_k = [\alpha, \beta],$$

where $\alpha = \sup\{a_k : k \text{ in } \mathbb{N}\}$ and $\beta = \inf\{b_k : k \text{ in } \mathbb{N}\}$.

Proof. The proof is strongly reminiscent of the second half of the proof of Theorem 1.2.2 (The Bolzano–Weierstrass theorem in $\mathbb{R}$), so we will be brief, letting you fill in the details. Let A denote the set of all left endpoints and let B denote the set of all right endpoints of the intervals I_k. The set A is nonempty and is bounded above by any b_k in B. By Axiom 1.1.1, $\alpha = \sup A$ exists in $\mathbb{R}$. Likewise, B is nonempty and is bounded below by any a_k in A. Therefore $\beta = \inf B$ exists in $\mathbb{R}$. It is easy to see that $\alpha \le \beta$; merely repeat the argument presented for the corresponding assertion in the proof of Theorem 1.2.2.

To prove that $\cap_{k=1} I_k = [\alpha, \beta]$, choose any x in $[\alpha, \beta]$. Then, for each k in $\mathbb{N}$, we have $a_k \le \alpha \le x \le \beta \le b_k$. That is, x is in I_k for each k in $\mathbb{N}$. Consequently, x is in the intersection $\cap_{k=1}^{\infty} I_k$. We deduce that $[\alpha, \beta] \subseteq \cap_{k=1}^{\infty} I_k$.

To obtain the reverse containment, note that if x is a point common to all the intervals I_k, then $a_k \le x \le b_k$ for every k in $\mathbb{N}$. We deduce that x is simultaneously an upper bound for A and a lower bound for B. Therefore $\alpha \le x \le \beta$. Consequently, x is in $[\alpha, \beta]$. Thus $\cap_{k=1}^{\infty} I_k \subseteq [\alpha, \beta]$. This proves that $\cap_{k=1}^{\infty} I_k = [\alpha, \beta]$. ●

COROLLARY 2.3.2 If, in the notation of the theorem, we have in addition that $\lim_{k\to\infty}(b_k - a_k) = 0$, then $\cap_{k=1}^{\infty} I_k$ is a single point.

Proof. We need merely show that $\alpha = \beta$. But this is easy. Since $\lim_{k\to\infty}(b_k - a_k) = 0$, given $\epsilon > 0$, we can choose an index k_0 such that, for $k \geq k_0$, $b_k - a_k < \epsilon$. It follows that the constant $\beta - \alpha$ satisfies the inequality

$$0 \leq \beta - \alpha \leq b_k - a_k < \epsilon$$

for all $k \geq k_0$. Since ϵ is arbitrary, we conclude by Exercise 1.10 that $\beta - \alpha = 0$ or $\alpha = \beta$. Therefore, $\bigcap_{k=1} I_k = \{\alpha\} = \{\beta\}$, a set consisting of a single point. ●

Applications of Corollary 2.3.2 often yield a single point x having some special property determined by the context. Thus the corollary provides a powerful *existence theorem*, one that can be used to devise elegant proofs of some of the fundamental theorems of analysis. Keep it in mind when trying to prove the existence of a special point having some particular property.

As usual, we want, if possible, to extract the content of this theorem, free of dimension, and to lift it to $\mathbb{R}^n$.

THEOREM 2.3.3 Cantor's Criterion If $\{C_k\}$ is a nested sequence of closed, bounded, nonempty subsets of $\mathbb{R}^n$, then

$$\bigcap_{k=1}^{\infty} C_k \neq \emptyset.$$

Furthermore, if $\lim_{k\to\infty} d(C_k) = 0$, where $d(C_k)$ is the diameter of C_k, then $\bigcap_{k=1}^{\infty} C_k = \{\boldsymbol{x}_0\}$ for some $\boldsymbol{x}_0$ in $\mathbb{R}^n$.

Proof. In the event that any of the C_k is a finite set, Cantor's criterion is trivial. Hence we assume that each C_k is an infinite set. Choose $\boldsymbol{x}_1$ in C_1. Choose $\boldsymbol{x}_2$ in C_2 distinct from $\boldsymbol{x}_1$. Having chosen distinct points $\boldsymbol{x}_1, \boldsymbol{x}_2, \ldots, \boldsymbol{x}_k$ with $\boldsymbol{x}_j$ in C_j, choose $\boldsymbol{x}_{k+1}$ in C_{k+1} distinct from $\boldsymbol{x}_1, \boldsymbol{x}_2, \ldots, \boldsymbol{x}_k$. By this simple procedure we construct recursively an infinite set

$$S = \{\boldsymbol{x}_k : k \text{ in } \mathbb{N}\}$$

of distinct points with $\boldsymbol{x}_k$ in C_k for each k in $\mathbb{N}$. (See Fig. 2.15.) The set S is contained in C_1. Therefore S is bounded. By the Bolzano–Weierstrass theorem for $\mathbb{R}^n$, S must have a limit point $\boldsymbol{x}_0$ in $\mathbb{R}^n$. We claim that $\boldsymbol{x}_0$ is in each of the sets C_k and thus is in $\bigcap_{k=1}^{\infty} C_k$.

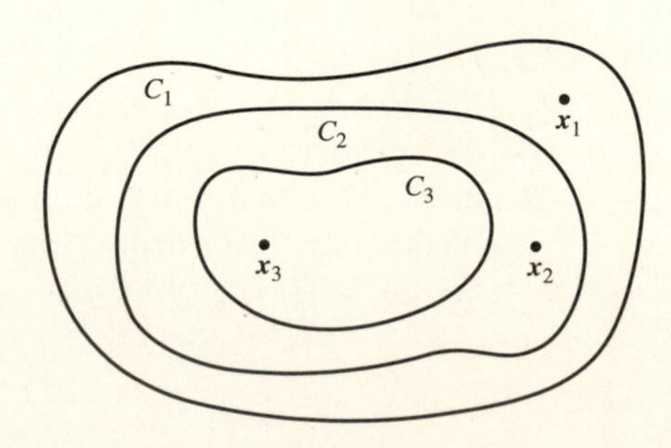

Figure 2.15

Fix any k in $\mathbb{N}$. Choose any deleted neighborhood $N'(\boldsymbol{x}_0)$ of $\boldsymbol{x}_0$. Because $N'(\boldsymbol{x}_0)$ contains infinitely many points of S, there exists a $k_1 > k$ such that $\boldsymbol{x}_{k_1}$ is in $N'(\boldsymbol{x}_0)$. Now, we chose $\boldsymbol{x}_{k_1}$ to be in C_k and $C_{k_1} \subseteq C_k$. Therefore $\boldsymbol{x}_{k_1}$ is in C_k. This proves that $\boldsymbol{x}_{k_1}$ is in $C_k \cap N'(\boldsymbol{x}_0)$. In other words, each deleted neighborhood of $\boldsymbol{x}_0$ contains a point of C_k. We deduce that $\boldsymbol{x}_0$ is a limit point of C_k. But C_k is closed and contains all its limit points. Consequently, $\boldsymbol{x}_0$ is in C_k. This is true for every k in $\mathbb{N}$. We deduce that $\boldsymbol{x}_0$ is in the intersection $\cap_{k=1}^{\infty} C_k$. Therefore this intersection is not empty.

Suppose, in addition, that the diameters $d(C_k)$ of the closed sets C_k have the property that $\lim_{k\to\infty} d(C_k) = 0$. We claim that the intersection of all the C_k must reduce to a single point. Suppose, to the contrary, that this intersection were to contain two points $\boldsymbol{x}$ and $\boldsymbol{y}$. Choose a positive ϵ less than $||\boldsymbol{x} - \boldsymbol{y}||$ and choose k such that $d(C_k) < \epsilon$. (See Fig. 2.16.) Both $\boldsymbol{x}$ and $\boldsymbol{y}$ are C_k so $||\boldsymbol{x} - \boldsymbol{y}|| \leq d(C_k) < \epsilon$, a contradiction. Therefore $\cap_{k=1}^{\infty} C_k$ cannot contain two distinct points. We deduce that this intersection must reduce to a single point. This proves the theorem. ●

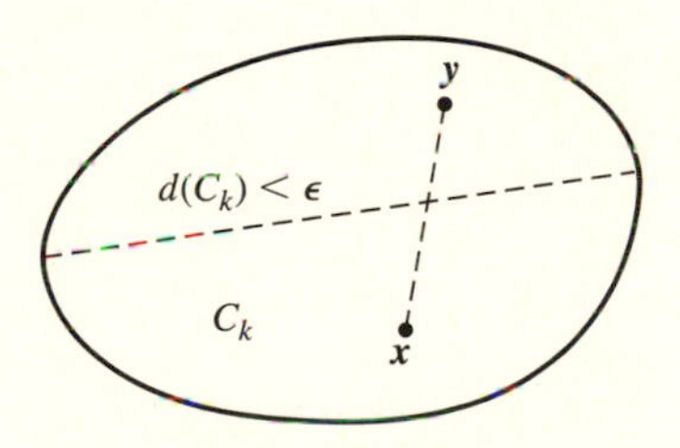

Figure 2.16

With Cantor's criterion proved valid in $\mathbb{R}^n$, we are ready to come full circle. We suspend our earlier proofs but *assume* Cantor's criterion in $\mathbb{R}^n$ and Archimedes' principle in $\mathbb{R}$. Since $\mathbb{R}$ is a closed subset of $\mathbb{R}^n$, it follows that Cantor's criterion holds in $\mathbb{R}$ as well. Cantor's criterion in $\mathbb{R}$, together with Archimedes' principle, implies that any nonempty set in $\mathbb{R}$ that is bounded above has a supremum in $\mathbb{R}$. In other words, if we were to choose as our starting point Cantor's criterion in $\mathbb{R}$ together with Archimedes' principle, then our original starting point, the Completeness Axiom for $\mathbb{R}$ [Axiom 1.1.1], follows as a theorem.

THEOREM 2.3.4 Suppose that Cantor's criterion and Archimedes' principle hold in $\mathbb{R}$. Suppose also that S is a nonempty set in $\mathbb{R}$ that is bounded above. Then sup S exists in $\mathbb{R}$. ●

Since the proof of Theorem 2.3.4 closely resembles that of the Bolzano–Weierstrass theorem, we will outline a proof in the exercises and leave its construction for you.

To summarize our cycle of theorems, we present the flow diagram in Fig. 2.17. Examine our proofs to confirm that, at each step, we have used only Archimedes' principle and the preceding theorem to prove the next in the cycle. Assume that Archimedes' principle is valid in $\mathbb{R}$. Consequently, all of these formulations are equivalent.

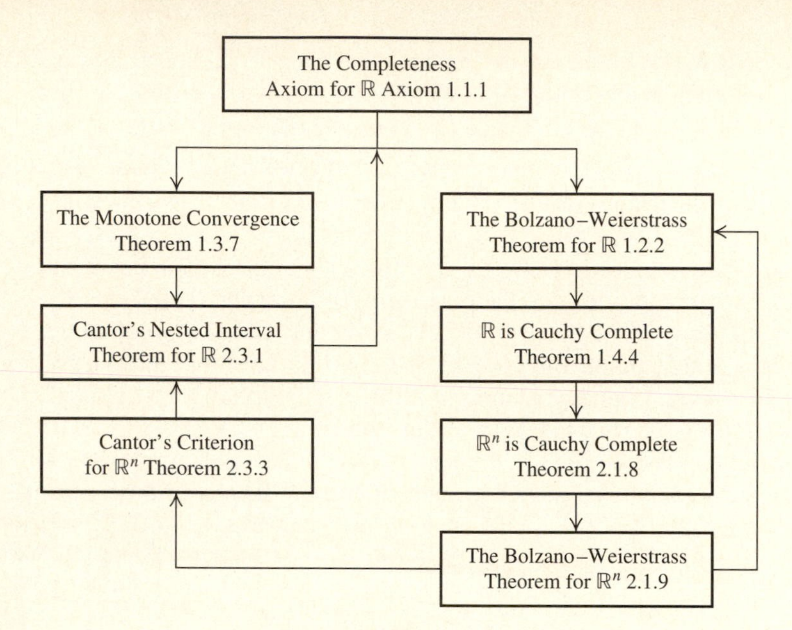

Figure 2.17

Now that we have completed the cycle of proofs, we will dispense with the distinctions and, when we apply the word *complete* to $\mathbb{R}^n$ we will mean any of these statements: the existence of least upper bounds (in $\mathbb{R}$), the Monotone Convergence theorem (in $\mathbb{R}$), Cantor's criterion, the Bolzano–Weierstrass property, or Cauchy completeness—whichever best applies in a given context.

THEOREM 2.3.5 Assuming that Archimedes' principle holds in $\mathbb{R}$, the following are equivalent:

i) Axiom 1.1.1: Every nonempty set in $\mathbb{R}$ that is bounded above has a supremum in $\mathbb{R}$.

ii) Every bounded monotone sequence in $\mathbb{R}$ converges.

iii) $\mathbb{R}$ has the Bolzano–Weierstrass property.

iv) $\mathbb{R}$ is Cauchy complete.

v) $\mathbb{R}^n$ is Cauchy complete.

vi) $\mathbb{R}^n$ has the Bolzano–Weierstrass property.

vii) Cantor's criterion is valid in $\mathbb{R}^n$.

viii) Cantor's criterion is valid in $\mathbb{R}$.

Proof. We have proved that (i) implies (ii) (Theorem 1.3.7); (ii) implies (viii) (Exercise 1.50); and (viii) implies (i) (Theorem 2.3.4). We have also proved that (i) implies (iii) (Theorem 1.2.2); (iii) implies (iv) (Theorem 1.4.4); (iv) implies (v) (Theorem 2.1.8); (v) implies (vi) (Theorem 2.1.9); (vi) implies (vii) (Theorem

2.3.3); (vii) implies (viii) [as a special case of (vii)]; and (viii) implies (i) (Theorem 2.3.4). Consequently, each of the eight statements implies all of the others. ●

EXERCISES

2.48. Find a nested sequence of nonempty, bounded, open sets in $\mathbb{R}$ whose intersection is empty.

2.49. Find a nested sequence of nonempty, closed sets in $\mathbb{R}$ whose intersection is empty.

2.50. Let $\boldsymbol{x}_0$ be an interior point of a set S in $\mathbb{R}^n$. Exhibit a nested sequence $\{C_k\}$ of nonempty, closed, bounded sets, each contained in S such that $\bigcap_{k=1}^{\infty} C_k = \{\boldsymbol{x}_0\}$.

2.51. Let $\boldsymbol{x}_0$ be a limit point of a set S in $\mathbb{R}^n$. Exhibit a nested sequence $\{C_k\}$ of nonempty, closed, bounded sets, each of which contains points of S, whose intersection is $\{\boldsymbol{x}_0\}$.

2.52. Fix $n \geq 2$. Assume that Cantor's criterion holds in $\mathbb{R}^n$. Show in detail that Cantor's criterion holds in $\mathbb{R}$.

2.53. Assume Theorem 2.3.1 and its corollary as an axiom. Also assume Archimedes' principle. Let S be a nonempty set in $\mathbb{R}$ that is bounded above. We prove here that $\sup S$ exists in $\mathbb{R}$.

a) Let B denote the set of all upper bounds for S and let $A = B^c$. Show that $A \neq \emptyset$. Also show that if a is in A and if b is in B, then $a < b$ and $S \cap [a, b] \neq \emptyset$.

b) Use the bisection method to construct a nested sequence $\{[a_k, b_k]\}$ of closed, bounded intervals such that a_k is in A, b_k is in B, and $b_k - a_k = (b_1 - a_1)/2^{k-1}$ for all k in $\mathbb{N}$.

c) Show that $\bigcap_{k=1}^{\infty} [a_k, b_k]$ consists of a single point x_0.

d) Show that x_0 is an upper bound for S.

e) Show that $x_0 \leq b$, for all b in B. Thus $x_0 = \sup S$.

2.54. Let X be a nonempty, closed set in $\mathbb{R}$. Form

$$X^n = X \times X \times \cdots \times X$$

in $\mathbb{R}^n$. Throughout Theorem 2.3.5, replace $\mathbb{R}$ by X and $\mathbb{R}^n$ by X^n. Determine whether that theorem remains valid; that is, are the eight statements still equivalent? Write a brief essay justifying your assertions. If any statement fails to remain equivalent to the others, explain why.

2.4 Relative Topology and Connectedness

We need to take a moment to follow up on comments made in Section 2.2 regarding *context.* Suppose that you are working in $\mathbb{R}^n$ for a particular value of n. You now know what it means for a set S in $\mathbb{R}^n$ to be open or to be closed. We want to emphasize again that, lacking any modifiers, the words *open* and *closed* mean "open relative to $\mathbb{R}^n$" and "closed relative to $\mathbb{R}^n$," respectively. Suppose, however, that the universal set or "space" in which some problem is posed happens to be, not all of $\mathbb{R}^n$, but some proper subset X of $\mathbb{R}^n$. Suppose also that S is a subset of X. Remembering that the universal context is X, consider what is meant by the statement "S is open relative to X" or, more briefly, "S is relatively open." Consider, likewise, the meaning of the statement "S is relatively closed."

DEFINITION 2.4.1 A set S is said to be *relatively open* (in X) if there exists an open set U in $\mathbb{R}^n$ such that $S = U \cap X$. Likewise, a set S is said to be *relatively closed* if there exists a closed set C in $\mathbb{R}^n$ such that $S = C \cap X$. ●

EXAMPLE 9 Let $X = [0, \infty)$ be the universal space under consideration. Let $S = [0, 1)$. Although S is not open in $\mathbb{R}$, it *is relatively open* in X, for $S = (-1, 1) \cap X$ and $U = (-1, 1)$ is open in $\mathbb{R}$. ●

EXAMPLE 10 Let $X = (0, \infty)$ and let $S = (0, 1]$. While S is *not* closed in $\mathbb{R}$, it *is relatively closed* in X because $S = [0, 1] \cap X$ and $C = [0, 1]$ is closed in $\mathbb{R}$. ●

EXAMPLE 11 Let X be any nonempty subset of $\mathbb{R}^n$. If S is any subset of X that is already closed in $\mathbb{R}^n$, then S is automatically relatively closed in X: $S = S \cap X$. Likewise, if S is a subset of X that is open in $\mathbb{R}^n$, then S is relatively open in X. However, there may well be many subsets of X that are relatively open or are relatively closed in X without being open or closed (respectively) in $\mathbb{R}^n$ itself. Restricting our attention to a subspace X may only increase the number of sets that are open or closed (in X). ●

EXAMPLE 12 If X itself is a proper, closed subset of $\mathbb{R}^n$ and if a subset S of X is relatively closed, then $S = C \cap X$ for some closed C in $\mathbb{R}^n$. Therefore S, being the intersection of two closed sets in $\mathbb{R}^n$, is also closed in $\mathbb{R}^n$. However, there may exist relatively open sets in X that are not open in $\mathbb{R}^n$. (See Example 9.) ●

EXAMPLE 13 If X itself is a proper, open subset of $\mathbb{R}^n$ and if a subset S of X is relatively open, then $S = U \cap X$ for some open U in $\mathbb{R}^n$. Therefore S, being the intersection of two open sets in $\mathbb{R}^n$, is also open in $\mathbb{R}^n$. In this case, there may exist relatively closed sets in X that are not closed in $\mathbb{R}^n$. (See Example 10.) ●

All of our earlier concepts are defined in X as in $\mathbb{R}^n$ with the stipulation that all these definitions are taken strictly within X.

DEFINITION 2.4.2 Let X be a nonempty subset of $\mathbb{R}^n$.

i) A *relative neighborhood* of $\boldsymbol{x}$ in X is $N(\boldsymbol{x}; r) \cap X$. A *deleted relative neighborhood* is obtained by omitting the point $\boldsymbol{x}$ from $N(\boldsymbol{x}; r) \cap X$.

ii) A sequence $\{\boldsymbol{x}_k\}$ in X *converges in* X if there exists a point $\boldsymbol{x}_0$ in X such that $\lim_{k\to\infty} \boldsymbol{x}_k = \boldsymbol{x}_0$.

iii) The *relative closure* of S is $\overline{S} \cap X$. ●

That a point $\boldsymbol{x}_0$ in X is a (relative) limit point of S needs no modified definition, but we emphasize that $\boldsymbol{x}_0$ is *required to be in* X. Likewise, we emphasize that, for a sequence $\{\boldsymbol{x}_k\}$ to converge (in X), the limit $\boldsymbol{x}_0$ *must be in* X. The definition of a Cauchy sequence, lacking any mention of its limit, needs no modification.

EXAMPLE 14 Let $X = (0, \infty)$. To ensure that these distinctions are clear in your mind, prove each of the following assertions.

i) If $S = (0, 1)$, then the set of (relative) limit points of S is $(0, 1]$. The relative closure of S is $(0, 1]$.

ii) The sequence $\{1/k\}$ is a Cauchy sequence in X that does not converge (in X).

iii) If, for each k in $\mathbb{N}$, we define $C_k = \overline{N(0; 1/k)} \cap X$, then $\{C_k\}$ is a nested sequence of nonempty, bounded, relatively closed subsets of X whose intersection $\cap_{k=1}^{\infty} C_k$ is empty. ●

EXAMPLE 15 Let $X = \mathbb{Q}$. Again, to strengthen your intuition, prove each of the following assertions:

i) If $S = (a, b) \cap X$ denotes the set of all rational points in the interval (a, b), then the set of all limit points of S (in X) is $[a, b] \cap X$. If an endpoint a or b is rational, it is a limit point of S; otherwise it is not.

ii) Let x_0 be any irrational number and let $\{x_k\}$ be a Cauchy sequence of distinct rational numbers that converges in $\mathbb{R}$ to x_0. (For example, if $x_0 = d_0.d_1d_2d_3\ldots$ is a nonrepeating decimal expansion, let $x_k = \sum_{j=0}^{k} d_j/10^j$.) Then $\{x_k\}$ is a Cauchy sequence in X that does not converge (in X). Viewed as a set, $S = \{x_k : k \text{ in } \mathbb{N}\}$ is a bounded, infinite set in X that has no limit point in X.

iii) Let x_0 be any irrational number and let $\{\epsilon_k\}$ be any sequence of positive numbers that converges monotonically to 0 in $\mathbb{R}$. For each k in $\mathbb{N}$, define $C_k = \overline{N(x_0; \epsilon_k)} \cap X$. Then $\{C_k\}$ is a nested sequence of nonempty, bounded, relatively closed subsets of X whose intersection $\cap_{k=1}^{\infty} C_k$ is empty. (Note that $\cap_{k=1}^{\infty} \overline{N(x_0; \epsilon_k)} = \{x_0\}$ in $\mathbb{R}$.) ●

These examples warn us that, when we restrict to a subset X in $\mathbb{R}^n$, we may lose the desirable properties we have labored to confirm for $\mathbb{R}^n$: Cauchy completeness, the Bolzano–Weierstrass property, and Cantor's criterion. We now know that each of these properties describes *completeness* and the natural question arises: What characteristic of X guarantees that we retain these completeness properties? Our next theorem tells us that we either retain all three properties or we lose them all when we restrict to X.

THEOREM 2.4.1 Let X be any nonempty subset of $\mathbb{R}^n$. The following statements are equivalent:

i) Every Cauchy sequence $\{\boldsymbol{x}_k\}$ in X converges to a point of X. Thus X inherits Cauchy completeness from $\mathbb{R}^n$.

ii) If S is a bounded, infinite subset of X, then S has a limit point in X. Thus X inherits the Bolzano–Weierstrass property from $\mathbb{R}^n$.

iii) If $\{C_k\}$ is any nested sequence of nonempty, bounded, relatively closed subsets of X, then $\cap_{k=1}^{\infty} C_k \neq \emptyset$. Further, if the diameters $d(C_k)$ tend to 0, then $\cap_{k-1}^{\infty} C_k$ is a single point. Thus X inherits Cantor's criterion from $\mathbb{R}^n$.

Proof. We prove first that (i) implies (ii). Suppose that each Cauchy sequence in X converges to a point of X. Let S be any bounded infinite subset of X. By the

Bolzano–Weierstrass theorem in $\mathbb{R}^n$, S has a limit point $\boldsymbol{x}_0$ in $\mathbb{R}^n$. Now imitate our proof of Theorem 1.3.10 (or appeal to your solution of Exercise 2.17) to extract from S, viewed temporarily as a subset of $\mathbb{R}^n$, a Cauchy sequence of points in S whose limit is $\boldsymbol{x}_0$ in $\mathbb{R}^n$. Since we are assuming (i) (that X inherits Cauchy completeness), $\boldsymbol{x}_0$ must be in X because it is the limit of a Cauchy sequence in X. Therefore S has a limit point $\boldsymbol{x}_0$ in X. Consequently, X has the Bolzano–Weierstrass property and statement (i) implies (ii).

We next prove that (ii) implies (iii). Suppose that X has the Bolzano–Weierstrass property: Every infinite, bounded subset S of X has a limit point in X. We want to show that Cantor's criterion is valid in X. To this end, suppose that $\{C_k\}$ is any nested sequence of bounded, nonempty, *relatively closed* subsets of X. We must show that $\bigcap_{k=1}^{\infty} C_k$ is nonempty and that, if $\lim_{k\to\infty} d(C_k) = 0$, then $\bigcap_{k=1}^{\infty} C_k$ consists of a single point. We can assume that each C_k is an infinite set; otherwise the result is trivial.

For each k in $\mathbb{N}$, let $B_k = \overline{C}_k$, the closure of C_k in $\mathbb{R}^n$. The sets B_k have several properties that are pertinent to our problem. First, B_k is closed (Corollary 2.2.8) and nonempty. Next, since C_k is relatively closed, we have $C_k = B_k \cap X$. (We ask you to prove this as an exercise; the delicate point of the proof is to show that, by taking the closure of C_k, we do not obtain any limit points of C_k in X that were not already in C_k.) Because C_k is bounded, we know from Exercise 2.41 that B_k is also bounded.

Furthermore, the sequence $\{B_k\}$ is nested. To confirm this assertion, note that, because the sequence $\{C_k\}$ is assumed to be nested, we have $C_{k+1} \subseteq C_k$ for each k in $\mathbb{N}$. If $\boldsymbol{x}$ is any limit point of C_{k+1} in $\mathbb{R}^n$, then $\boldsymbol{x}$ is also a limit point of C_k. That is, $\overline{C}_{k+1} \subseteq \overline{C}_k$. In other words, $B_{k+1} \subseteq B_k$. Therefore $\{B_k\}$ is nested.

Finally, from Exercise 2.42 we know that $d(B_k) = d(C_k)$. Therefore, if $\lim_{k\to\infty} d(C_k) = 0$, then $\lim_{k\to\infty} d(B_k) \neq 0$ also.

By Cantor's criterion in $\mathbb{R}^n$, $\bigcap_{k=1}^{\infty} B_k \neq \emptyset$. Furthermore, if $\lim_{k\to\infty} d(C_k) = 0$, then $\bigcap_{k=1}^{\infty} B_k$ consists of a single point. Now we are ready to prove (iii). As in the proof of Theorem 2.3.3, recursively choose an infinite set $S = \{\boldsymbol{x}_k : k \text{ in } \mathbb{N}\}$ of distinct points such that, for each k in $\mathbb{N}$, $\boldsymbol{x}_k$ is in C_k. Then S is a bounded, infinite subset of X. Since we have assumed that X inherits the Bolzano–Weierstrass property, we know that S has a limit point $\boldsymbol{x}_0$ in X. But $\boldsymbol{x}_k$ is in $B_k = \overline{C}_k$ for each k in $\mathbb{N}$ and $\{B_k\}$ is a nested sequence of nonempty, closed, bounded sets in $\mathbb{R}^n$. Therefore, exactly as in the proof of Theorem 2.3.3, $\boldsymbol{x}_0$ is in each B_k also. Therefore $\boldsymbol{x}_0$ is in each $C_k = B_k \cap X$. We deduce that $\boldsymbol{x}_0$ is in $\bigcap_{k=1}^{\infty} C_k$. Therefore this intersection is not empty.

If, in addition, $\lim_{k\to\infty} d(C_k) = 0$, then $\lim_{k\to\infty} d(B_k) = 0$. By Cantor's criterion in $\mathbb{R}^n$, we know that $\bigcap_{k=1}^{\infty} B_k = \{\boldsymbol{x}_0\}$. But, for every k in $\mathbb{N}$, $C_k \subseteq B_k$ and therefore $\bigcap_{k=1}^{\infty} C_k \subseteq \bigcap_{k=1}^{\infty} B_k = \{\boldsymbol{x}_0\}$. Since we already know that $\bigcap_{k=1}^{\infty} C_k \neq \emptyset$, we conclude that $\bigcap_{k=1}^{\infty} C_k = \{\boldsymbol{x}_0\}$ also. Thus Cantor's criterion holds in X. We have proved that (ii) implies (iii).

Finally, to show that (iii) implies (i), we assume that Cantor's criterion holds in the subset X of $\mathbb{R}^n$. Let $\{\boldsymbol{x}_k\}$ be a Cauchy sequence in X. We must show that $\{\boldsymbol{x}_k\}$ converges to a point of X. We know already that $\{\boldsymbol{x}_k\}$ converges to some point $\boldsymbol{x}_0$ in

$\mathbb{R}^n$. We have no choice about the limit of this Cauchy sequence; it is whatever it is in $\mathbb{R}^n$. Our task is to show that $\boldsymbol{x}_0$ must already be in X.

Let $\{\epsilon_k\}$ be any sequence of positive numbers that converges monotonically to 0. For each k, let $C_k = \overline{N(\boldsymbol{x}_0; \epsilon_k)} \cap X$. Then $\{C_k\}$ is a nested sequence of bounded and relatively closed sets in X. We need to show that, for each k in $\mathbb{N}$, $C_k \neq \emptyset$. Choose and fix any k_1 in $\mathbb{N}$. Since $\{\boldsymbol{x}_k\}$ converges to $\boldsymbol{x}_0$ in $\mathbb{R}^n$, there exists an index k_0 such that, whenever $k \geq k_0$, it follows that $\boldsymbol{x}_k$ is in $N(\boldsymbol{x}_0; \epsilon_{k_1})$. (Of course, k_0 depends on k_1.) Therefore, for all $k \geq k_0$, $\boldsymbol{x}_k$ is in $C_{k_1} = \overline{N(\boldsymbol{x}_0; \epsilon_{k_1})} \cap X$. This proves that C_{k_1} is not empty. Since k_1 is arbitrary in $\mathbb{N}$, we have proved that $C_k \neq \emptyset$ for each k in $\mathbb{N}$.

Since we are assuming that Cantor's criterion is valid in X, we deduce that $\bigcap_{k=1}^{\infty} C_k$ is not empty. Furthermore, since $d(C_k) \leq 2\epsilon_k$ and since $\lim_{k\to\infty} \epsilon_k = 0$, we know that $\lim_{k\to\infty} d(C_k) = 0$. Consequently, $\bigcap_{k=1}^{\infty} C_k$ is a single point in X.

We offer as one solution to Exercise 2.51, the nested sequence $\{\overline{N(\boldsymbol{x}_0; \epsilon_k)}\}$. That is, $\bigcap_{k=1} \overline{N(\boldsymbol{x}_0; \epsilon_k)} = \{\boldsymbol{x}_0\}$. We leave the proof for you. Finally observe that

$$\bigcap_{k=1}^{\infty} C_k = \bigcap_{k=1}^{\infty} \left(\overline{N(\boldsymbol{x}_0; \epsilon_k)} \cap X \right) \subseteq \bigcap_{k=1}^{\infty} \overline{N(\boldsymbol{x}_0; \epsilon_k)} = \{\boldsymbol{x}_0\}.$$

Therefore, $\bigcap_{k=1}^{\infty} C_k = \{\boldsymbol{x}_0\}$. It follows that $\boldsymbol{x}_0$ is in X. Thus the Cauchy sequence $\{\boldsymbol{x}_k\}$ converges to a point $\boldsymbol{x}_0$ in X. We have proved that X inherits Cauchy completeness. Therefore, (iii) implies (i). This completes the proof of the theorem. ●

In light of Theorem 2.4.1, we are led to the following definition.

DEFINITION 2.4.3 Let X be a nonempty subset of $\mathbb{R}^n$. If any of the equivalent properties of Theorem 2.4.1 hold in the set X, then X is said to be *complete*. ●

Finally, we can identify exactly when, upon restricting to a subset X of $\mathbb{R}^n$, we retain Cauchy completeness, the Bolzano–Weierstrass property, and Cantor's criterion.

THEOREM 2.4.2 A nonempty subset X of $\mathbb{R}^n$ is complete if and only if X is closed in $\mathbb{R}^n$.

Proof. Suppose first that X is closed. Let S be any bounded, infinite set in X. Since S is also a bounded, infinite set in $\mathbb{R}^n$, S must have a limit point $\boldsymbol{x}_0$ in $\mathbb{R}^n$. The point $\boldsymbol{x}_0$ must also be a limit point of X in $\mathbb{R}^n$, but X, being closed in $\mathbb{R}^n$, contains all its limit points. Therefore, $\boldsymbol{x}_0$ is in X. We have shown that any bounded, infinite set S in X has a limit point in X. Therefore X is complete by Definition 2.4.3.

To prove the converse, assume that X is a complete subset of $\mathbb{R}^n$. According to Definition 2.4.3, X inherits Cauchy completeness from $\mathbb{R}^n$. Suppose that $\boldsymbol{x}_0$ is any limit point of X in $\mathbb{R}^n$. We must show that $\boldsymbol{x}_0$ is in X in order to conclude that X is closed. By Exercise 2.17, we know that there exists a Cauchy sequence $\{\boldsymbol{x}_k\}$ of distinct points of X that converges to $\boldsymbol{x}_0$. By the Cauchy completeness of X, this Cauchy sequence converges to a point in X. But this Cauchy sequence converges to $\boldsymbol{x}_0$, so $\boldsymbol{x}_0$ belongs to X. Since X contains all its limit points, X is closed. This completes the proof of the theorem. ●

EXAMPLE 16 The set $\mathbb{Q}$ is not complete. The reason? $\mathbb{Q}$ is not closed. For this reason, to practice analysis, we first "complete" $\mathbb{Q}$ by forming $\mathbb{R}$ and work in complete subsets of $\mathbb{R}$ or $\mathbb{R}^n$. If the underlying space in which we work is not complete, all the issues addressed by analysis become problematic. ●

Remark. We would not bother with these fine points if they did not matter.

Warning. Be aware of your working environment.

There remains one more concept to be discussed in this section, that of *connectedness*. What does it mean to say that a set is connected? Mathematicians, who are often childlike in their curiosity and mental playfulness, have agreed that it means exactly what it meant in childhood: A set is *connected* if it does not separate into two (or more) parts, in other words, if it is not *disconnected.* There is one subtle distinction, however, that would not occur to a normal child.

DEFINITION 2.4.4 A set S is *disconnected* if there exist two nonempty, disjoint open sets U and V such that

i) $S \subseteq U \cup V$ and

ii) $S \cap U \neq \emptyset$ and $S \cap V \neq \emptyset$.

A set S is *connected* if it is not disconnected. ●

EXAMPLE 17 The set $S = [0, 1) \cup (1, 2]$ is disconnected. To see this, let $U = (-\infty, 1)$ and $V = (1, \infty)$. U and V are disjoint open sets with $S \subseteq U \cup V$, $S \cap U \neq \emptyset$, and $S \cap V \neq \emptyset$. ●

Any interval—say $[a, b] = \{x \text{ in } \mathbb{R}: a \leq x \leq b\}$—in $\mathbb{R}$ is connected, an assertion that probably seems self-evident to you, needing no proof. However, if we let $S = [a, b] \cap \mathbb{Q}$, the set of all rational numbers in $[a, b]$, we obtain a set that is disconnected. To see this, choose any irrational x_0 in (a, b); define $U = (-\infty, x_0)$ and $V = (x_0, \infty)$. Then U and V are disjoint open sets, each of which contains points of S. Furthermore, $S \subseteq U \cup V$. This proves that S is disconnected. Consequently, you might suspect that the connectedness of the interval $[a, b]$ flows from the completeness of $\mathbb{R}$ and that the proof might require some sophistication.

THEOREM 2.4.3 An interval I in $\mathbb{R}$ is connected.

Proof. We will prove the theorem for an interval of the form $[a, b]$ and will leave the other cases for you to treat in the exercises.

Let U and V be two disjoint open sets in $\mathbb{R}$ such that $[a, b] \subseteq U \cup V$. Without loss of generality, the endpoint a is in the open set U and is therefore an interior point of U. Hence, there exists a neighborhood $N(a; \epsilon)$ that is contained in U. Without loss of generality, $\epsilon < b - a$. Thus $[a, a + \epsilon) \subseteq U \cap [a, b]$. (See Fig. 2.18.) This containment shows that the set

$$S = \{x \text{ in } [a, b] : [a, x) \subseteq U\}$$

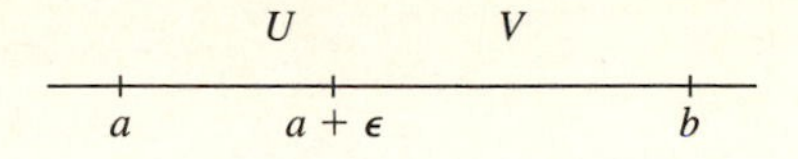

Figure 2.18

is not empty. The set S is bounded above by b. By Axiom 1.1.1, $\mu = \sup S$ exists with

$$a < \mu \leq b.$$

Thus μ belongs to $[a, b]$ and therefore μ belongs either to U or to V. (See Fig. 2.19.) We will show that μ cannot be in V and that, if μ is in U, then μ must equal b. These two statements, taken together, imply that $[a, b] \subseteq U$ and therefore that $[a, b] \cap V = \emptyset$. Our method of proof is akin to squeezing toothpaste.

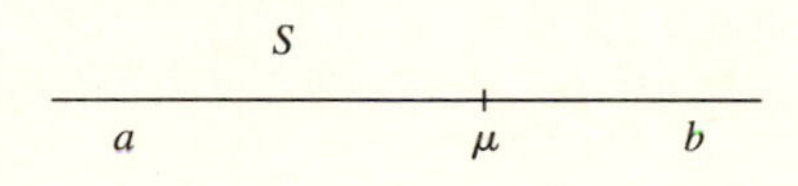

Figure 2.19

LEMMA If y is any point in (a, μ), then y is in S.

Proof. Since $\mu = \sup S$, there exists, by Theorem 1.1.1, an x in S such that $y < x \leq \mu$. Because x is in S, we know that $[a, x) \subseteq U$. Thus $[a, y) \subset [a, x) \subseteq U$. This containment implies that y is in S and proves the lemma.

If we assume that μ belongs to the open set V, then there exists some neighborhood $N(\mu; \epsilon)$ contained in V. Without loss of generality, we can assume that $\epsilon < \mu - a$. Thus $(\mu - \epsilon, \mu] \subseteq [a, b] \cap V$.

Since $\mu = \sup S$, Theorem 1.1.1 guarantees the existence of some point x in S such that $\mu - \epsilon < x \leq \mu$. (See Fig. 2.20.) By the definition of S, we have $[a, x) \subseteq U$. Therefore, $(\mu - \epsilon, x) \subseteq U$. But $(\mu - \epsilon, x) \subseteq V$. Consequently, $(\mu - \epsilon, x)$ is contained in $U \cap V$. This says that $U \cap V \neq \emptyset$, contrary to our initial assumption that U and V are disjoint. Therefore μ does not belong to V.

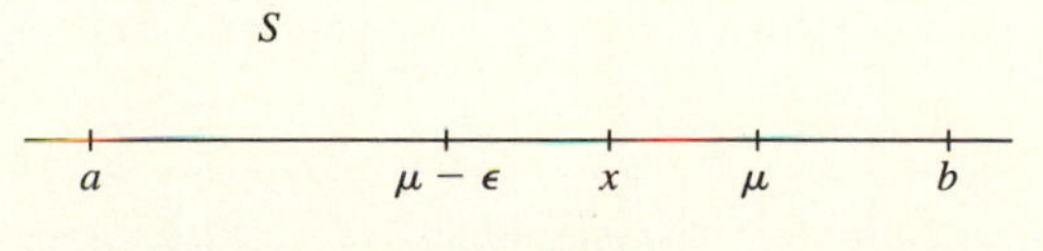

Figure 2.20

Consequently, μ is in U. We know already that $\mu \leq b$. Assume that $\mu < b$. Because U is open, there exists some $\epsilon > 0$ such that $N(\mu; \epsilon) \subseteq U$. Without loss of generality, we can also assume that $\epsilon < b - \mu$. Therefore, $[\mu, \mu + \epsilon)$ is contained in $U \cap [a, b]$. (See Fig. 2.21.)

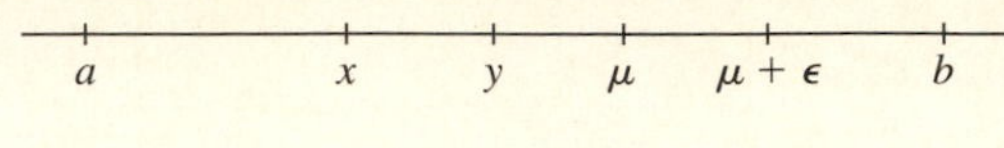

Figure 2.21

But $[a, \mu)$ is also contained in U. To verify this claim, let x be any point in $[a, \mu)$ and choose any y in (x, μ). Thus, y is in $[a, \mu)$; by the lemma, y is in S. Therefore $[a, y) \subseteq U$. Thus the point x, which belongs to $[a, y)$, is in U. It follows that $[a, \mu) \subseteq U$. Consequently,

$$[a, \mu + \epsilon) = [a, \mu) \cup [\mu, \mu + \epsilon)$$

is contained in U. But this implies that $\mu + \epsilon$ is in S, contradicting the fact that $\mu = \sup S$. We are forced to conclude that, if μ is in U, then $\mu = b$.

Finally, to summarize, μ is in $[a, b] \subset U \cup V$, μ cannot be in V, and thus μ must be in U. Therefore $\mu = b$. We conclude that $[a, b] \subseteq U$ and that $[a, b] \cap V = \emptyset$. Thus $[a, b]$ cannot be disconnected and is therefore, connected. ●

EXAMPLE 18 In $\mathbb{R}^2$, the unit square

$$S = \{(x_1, x_2) : 0 \leq x_1 \leq 1,\ 0 \leq x_2 \leq 1\}$$

is connected. More generally, any n-dimensional hyper-rectangle is connected. Proof of these assertions is left for the exercises. ●

In many situations, the space X under consideration, defined by the problem being studied, is connected. When X is disconnected, you will want to take special care to attend to the separate parts of X.

EXERCISES

2.55. Let $\boldsymbol{x}_0$ be any point in $\mathbb{R}^n$. Let $\{\epsilon_k\}$ be any sequence of positive numbers that converge monotonically to 0. For each k in $\mathbb{N}$, let $C_k = \overline{N(\boldsymbol{x}_0; \epsilon_k)}$.

- **a)** Prove that $\{C_k\}$ is a nested sequence of nonempty, bounded, closed subsets of $\mathbb{R}^n$.
- **b)** Prove that $\cap_{k=1}^{\infty} C_k = \{\boldsymbol{x}_0\}$.

2.56. Let $X = (0, \infty)$ in $\mathbb{R}$.

- **a)** Prove that the set of relative limit points of $S = (0, 1)$ is $(0, 1]$.
- **b)** Prove that the sets $C_k = \overline{N(0; 1/k)} \cap X$ form a nested sequence of nonempty, bounded, relatively closed sets in X whose intersection is empty.

2.57. Let $X = \mathbb{Q}$.

- **a)** For a, b in $\mathbb{R}$ with $a < b$, the set $(a, b) \cap X$ is relatively open. Find a relatively open set of the form $(a, b) \cap X$ that is also relatively closed, hence relatively clopen.
- **b)** Exhibit two Cauchy sequences in X whose limit in $\mathbb{R}$ is e but that fail to converge in X.
- **c)** Let x_0 be any irrational number and let $\{\epsilon_k\}$ be any sequence of positive numbers that converges monotonically to 0. Prove that the sets $C_k = \overline{N(x_0; \epsilon_k)} \cap X$ form a nested sequence of nonempty, bounded, relatively closed sets in X whose intersection is empty.

2.58. Let $X = [-1, 0) \cup (0, 1)$.

a) Show that both $[-1, 0)$ and $(0, 1)$ are relatively open in X. Show that both are also relatively closed in X.

b) Let $S = (-1/2, 1/2) \cap X$. Is S relatively open in X? Is S relatively closed in X? Prove your answers.

c) For each k in $\mathbb{N}$, let $x_k = (-1)^k[1 - 1/(k+1)]$. Let $S = \{x_k : k \text{ in } \mathbb{N}\}$. Show that S has exactly one relative limit point x_0 in X. What is it?

d) Thus the sequence $\{x_k\}$ defined in part (c) is a bounded sequence having exactly one cluster point x_0 in X. Does this sequence converge to x_0? If so, prove it; otherwise, explain why not.

e) Exhibit a Cauchy sequence in X that fails to converge in X.

f) Show that the sets $C_k = [-1/k, 1/k] \cap X$ form a nested sequence of bounded relatively closed sets with an empty intersection.

2.59. Let X be a nonempty subset of $\mathbb{R}^n$ and let C be a relatively closed subset of X. Prove that $\overline{C} \cap X = C$.

2.60. Prove that each of the intervals (a, b), $(a, b]$, and $[a, b)$ is connected in $\mathbb{R}$.

2.61. Prove that $\mathbb{R}$ itself is connected.

2.62. Use the result in Exercise 2.61 to prove that the only clopen subsets of $\mathbb{R}$ are $\emptyset$ and $\mathbb{R}$.

2.63. Prove that the closed unit square

$$S = \{\boldsymbol{x} \text{ in } \mathbb{R}^2 : 0 \leq x_1 \leq 1, 0 \leq x_2 \leq 1\}$$

is connected.

2.64. Prove that any n-dimensional closed hyper-rectangle in $\mathbb{R}^n$ is connected.

2.65. Let X be any nonempty subset of $\mathbb{R}^n$. Prove that X is relatively clopen (in X).

2.66. Let X be any nonempty, disconnected subset of $\mathbb{R}^n$. Let U_1 and U_2 be disjoint open sets such that (i) $X \subseteq U_1 \cup U_2$ and (ii) $X_1 = U_1 \cap X \neq \emptyset$ and $X_2 = U_2 \cap X \neq \emptyset$. If X_1 and X_2 are themselves connected, they are called *components* of X. Suppose that X_1 and X_2 are components of X. Prove that both X_1 and X_2 are relatively clopen.

2.5 Compactness

We come at last to one of the most important concepts in analysis, that of *compactness*. It will come into play at several key points in all of the remainder of this text. Consequently, you will want to absorb the content of this section thoroughly.

If S is a bounded subset of $\mathbb{R}$, then every subset of S is also bounded and hence every nonempty subset of S has both a supremum and an infimum in $\mathbb{R}$. If S is also closed, then all those suprema and infima are already in S. What's more, each Cauchy sequence of points in S converges to a point of S. In other words, analytic processes within S produce limits that are also in S. Here we want to examine more closely sets that are both closed and bounded.

We start simply. Historically, the first important class of such sets to be considered were the closed, bounded intervals $[a, b]$. Sets of this type remain among the most important for our purposes. What can we say about $[a, b]$? Well, it is a particularly simple set with which, by now, you are thoroughly familiar; it is closed, bounded, and connected. And it has another property, first identified and used in 1872 by Heinrich Heine and explicitly stated and proved in 1895 by Emil Borel.

Let $S = [a, b]$ be any closed, bounded interval in $\mathbb{R}$. Let $\mathcal{C}$ be any sequence of open intervals, $\{I_k : k \text{ in } \mathbb{N}\}$, with the property that every x in $[a, b]$ is contained in at least one of the sets I_k. The collection $\mathcal{C}$ is called an *open cover* of $[a, b]$. Notice how well chosen this term is: The sets I_k are *open* and the collection of all of them covers $[a, b]$. In mathematical shorthand, $[a, b] \subseteq \cup_{k=1}^{\infty} I_k$.

We want to show that only a finite number of the sets I_k are required to cover $S = [a, b]$. That is, we want to show that a *finite subcover exists*. For each k in $\mathbb{N}$, let U_k be the union of $I_1, I_2, \ldots$, and I_k. Notice that U_k, being the union of open sets, is open. Therefore $C_k = U_k^c$ is closed. Notice also that $U_1 \subseteq U_2 \subseteq U_3 \subseteq \cdots U_k \subseteq \cdots$. Taking complements reverses these containment relationships:

$$C_1 \supseteq C_2 \supseteq C_3 \supseteq \cdots \supseteq C_k \supseteq \cdots.$$

Next, for each k in $\mathbb{N}$, let $D_k = C_k \cap S$. The set D_k consists of those points of $[a, b]$ which are *not covered* by any of $I_1, I_2, \ldots, I_k$. Notice that, for each k, D_k is the intersection of two closed sets and is therefore closed. Also, D_k is bounded, because it is a subset of $[a, b]$. Furthermore, the sequence $\{D_k\}$ is nested, since it inherits this property from the sequence $\{C_k\}$.

If we assume, and here is the crux of our strategy, that none of these closed sets D_k is empty—that is, that none of the sets U_k contain all of $[a, b]$—then we must conclude by Cantor's criterion that the nested sequence $\{D_k\}$ has a nonempty intersection. Choose any point x in $\cap_{k=1}^{\infty} D_k$. This x is a point in the interval $[a, b]$ and it does not belong to *any* of the sets $U_k = \cup_{j=1}^{k} I_j$. Therefore x does not belong to $\cup_{j=1}^{\infty} I_j$. But this says that x is not covered by any of the sets in $\mathcal{C}$, contradicting the fact that $\mathcal{C}$ is an open cover of $[a, b]$. The contradiction results from our having made one glaring assumption: We assumed, falsely, that none of the D_k is empty.

Consequently, there must exist a k in $\mathbb{N}$ such that $D_k = C_k \cap [a, b] = \emptyset$. For this k, the set C_k lies entirely outside $[a, b]$, and its complement U_k must contain all of $[a, b]$. Therefore the collection $\{I_1, I_2, \ldots, I_k\}$ is a finite, open cover of $[a, b]$ that we have extracted from the original open cover $\mathcal{C}$. In summary, we have proved the following theorem.

THEOREM 2.5.1 Heine–Borel Let $S = [a, b]$ be any closed and bounded interval in $\mathbb{R}$ and let $\mathcal{C} = \{I_k : k \text{ in } \mathbb{N}\}$ be any open cover of $[a, b]$ consisting of open intervals. There exists a finite collection $\{I_{k_1}, I_{k_2}, \ldots, I_{k_p}\}$ consisting of sets in $\mathcal{C}$ that also covers S. ●

Throughout this discussion, we have assumed that the open sets I_k, which are used to cover the set S, are *open intervals*. Although in many applications this will be the case, there is no compelling reason to restrict ourselves in this way. Nor is there any compelling reason to restrict ourselves to subsets of the real line $\mathbb{R}$.

DEFINITION 2.5.1 Let S be any nonempty subset of $\mathbb{R}^n$. An *open cover* of S is any collection $\mathcal{C} = \{U_\alpha : \alpha \text{ in } A\}$ of open sets such that $S \subseteq \cup_{\alpha \in A} U_\alpha$. ●

Actually, in $\mathbb{R}^n$ this definition is unnecessarily general, as the following theorem confirms. We need not index an open cover in $\mathbb{R}^n$ with an arbitrary index set A that may be uncountable; we need only index with the natural numbers. Every open cover of any set in $\mathbb{R}^n$ can automatically be reduced to a countable subcover. In more general contexts this may not be true.

THEOREM 2.5.2 Lindelöf's Theorem Let S be any subset of $\mathbb{R}^n$ and let $\mathcal{C} = \{U_\alpha : \alpha \text{ in } A\}$ be any open cover of S. Then some countable subcollection of $\mathcal{C}$ also covers S.

Proof. Recall from Section 1.6 that $\mathbb{Q}^{n+1}$ is countably infinite. Notice that $\mathbb{Q}^{n+1}$ is in one-to-one correspondence with the collection $\{N(y; r) : \boldsymbol{y} \text{ in } \mathbb{Q}^n, r \text{ in } \mathbb{Q}\}$ of neighborhoods in $\mathbb{R}^n$ where the center point $\boldsymbol{y}$ has rational coordinates and the radius r is rational. Hence the collection of all such neighborhoods is countably infinite and can be listed sequentially as $\{N_1, N_2, N_3, \ldots\}$.

Fix any $\boldsymbol{x}$ in S. Because $\mathcal{C}$ is an open cover of S, $\boldsymbol{x}$ is in at least one of the U_α. Choose any one of the U_α containing $\boldsymbol{x}$. We claim that at least one of the neighborhoods N_k contains $\boldsymbol{x}$ and is contained in U_α. To see this, note that U_α is open and that $\boldsymbol{x}$ is an interior point of U_α. Thus there exists an $\epsilon > 0$ such that $N(\boldsymbol{x}; \epsilon) \subseteq U_\alpha$. Because $\mathbb{Q}^n$ is dense in $\mathbb{R}^n$, there exists a $\boldsymbol{y}$ with rational coordinates such that $\boldsymbol{y}$ is in $N(\boldsymbol{x}; \epsilon/2)$. Since $\mathbb{Q}$ is dense in $\mathbb{R}$, there exists a positive, rational number r such that $||\boldsymbol{x} - \boldsymbol{y}|| < r < \epsilon - ||\boldsymbol{x} - \boldsymbol{y}||$. It follows that $N(\boldsymbol{y}; r) \subseteq N(\boldsymbol{x}; \epsilon) \subseteq U_\alpha$. The neighborhood $N(\boldsymbol{y}; r)$ is one of the countable collection of neighborhoods $\{N_k : k \text{ in } \mathbb{N}\}$, contains $\boldsymbol{x}$, and is contained in U_α as claimed.

Let $K = \{k \text{ in } \mathbb{N}: \boldsymbol{x} \text{ is in } N_k \text{ and } N_k \subseteq U_\alpha\}$. K is a nonempty set of natural numbers and, by the Well-Ordering Property of $\mathbb{N}$, the set K contains a least element $k(\boldsymbol{x})$. (The index $k(\boldsymbol{x})$, of course, depends on $\boldsymbol{x}$.) Thus, for our initially chosen $\boldsymbol{x}$ in S, we identify a well-defined neighborhood, $N_{k(\boldsymbol{x})}$ that contains $\boldsymbol{x}$ and is contained in U_α. Repeat this process at each point $\boldsymbol{x}$ as $\boldsymbol{x}$ varies over S.

The collection $\{N_{k(\boldsymbol{x})} : \boldsymbol{x} \text{ in } S\}$ is clearly an open cover of S and, being a subcollection of a countably infinite collection, it is countable. (In general, there will be repetitions in this collection.) For each of the $N_{k(\boldsymbol{x})}$ choose a U_α that contains it and reindex U_α as $U_{k(\boldsymbol{x})}$. The resulting collection $\{U_{k(\boldsymbol{x})} : \boldsymbol{x} \text{ in } S\}$ of open sets is clearly countable and covers S. This proves Lindelöf's theorem. ●

EXAMPLE 19 Let $S = [0, 1)$. Suppose that for each x in S, a neighborhood $N(x; \epsilon_x)$ is specified having some particular property. (In Chapter 3, treating continuity, we shall see explicit instances of this example.) The collection of neighborhoods $\mathcal{C} = \{N(x; \epsilon_x) : x \text{ in } [0, 1)\}$ is an open cover of S that is indexed by an uncountable set. Thus $\mathcal{C}$ is an uncountable open cover of S. Lindelöf's theorem ensures that we can dispense with most of the sets in this open cover and can extract a countably infinite, open subcover $\{N(x_k; \epsilon_{x_k}) : k \text{ in } \mathbb{N}\}$ that will better serve our purposes than will the original cover. ●

In light of Definition 2.5.1 and Lindelöf's theorem, review the proof of the Heine–Borel theorem. Several features of the proof deserve attention. Nowhere did we use the fact that S is an interval. We did use the fact that S is closed in order to guarantee that each of the sets D_k is closed. We also used the fact that S is bounded to guarantee that each D_k is bounded. But no special properties of intervals were invoked. Therefore our proof is still valid if we replace $[a, b]$ by any *closed, bounded set* in $\mathbb{R}$. Note also that we never used the assumption that the sets I_k are intervals. We only required that I_k be open to ensure that the complement C_k of U_k and hence D_k would be closed and bounded. This enabled us to invoke Cantor's Theorem. Consequently, our proof is valid when $\mathcal{C}$ is any collection of *open sets* that cover the closed and bounded set S. Finally, notice that we made no use of the one-dimensionality of $\mathbb{R}$. Every step applies equally well to any closed, bounded subset of $\mathbb{R}^n$ and any open cover of it. After we first extract a countable open subcover by Lindelöf's theorem, our original proof lifts verbatim to prove the following theorem.

THEOREM 2.5.3 The Generalized Heine–Borel Theorem Let S be any closed, bounded set in $\mathbb{R}^n$ and let $\mathcal{C}$ be any open cover of S. Then there exists a finite subcover of S. ●

This little proof is a gem that epitomizes a mental attitude and a style that pervades mathematics. Start simply, develop a proof using only those properties demanded by the situation, then double back to examine the original hypotheses and relax those that are not required. We suggest that you learn to adopt this style for your own.

DEFINITION 2.5.2 A set S in $\mathbb{R}^n$ is said to be *compact* if, for every open cover $\mathcal{C}$ of S, there exists a finite subcover. ●

The Heine–Borel theorem is the following statement: Every closed, bounded set S in $\mathbb{R}^n$ is compact. The converse is also true. As we will prove every compact set S in $\mathbb{R}^n$ is both bounded and closed. To develop your intuition, we look at four examples before launching into the proof.

EXAMPLE 20 Let $S = [0, 1]$. Since S is closed and bounded, it is compact and every open cover $\mathcal{C}$ of S can be reduced to a finite subcover. Let's construct a particular example of an open cover of S. Each x in the open interval $(0, 1)$ is an interior point, so for each such x, we can choose a positive r_x such that $N(x; r_x) \subset S$. Let $N(0)$ and $N(1)$ be any neighborhoods of 0 and of 1, respectively, each with radius less than, say, 1/2. Then the collection

$$\mathcal{C} = \{N(0), N(1)\} \cup \{N(x; r_x) : x \text{ in } (0, 1)\}$$

is an uncountable open cover of $S = [0, 1]$. By Lindelöf's theorem, it can be reduced first to a countable cover. By the Heine–Borel theorem, that cover can be further reduced to a finite subcover. Note that the finite subcover must include the two special neighborhoods $N(0)$ and $N(1)$ because these are the only sets in $\mathcal{C}$ that cover the endpoints of $[0, 1]$. ●

EXAMPLE 21 Cantor's set C, being closed and bounded, is compact. ●

EXAMPLE 22 Let $S = (0, 3)$. For each k in $\mathbb{N}$, let $I_k = (1/k, 3 - 1/k)$. Then $\mathcal{C} = \{I_k : k \text{ in } \mathbb{N}\}$ is an *open cover* of S. To see this, let x be any point in $(0, 3)$. By Archimedes' principle, there exists a k in $\mathbb{N}$ such that

$$0 < 1/k < \min\{x, 3 - x\}.$$

We urge you to confirm that, for this choice of k, x is in I_k. We conclude that S is covered by the sets in $\mathcal{C}$. It is worth noting—in fact, it is the point of this example—that no finite subcollection of $\mathcal{C}$, no matter how carefully chosen, can possibly cover all of $(0, 3)$. To prove this claim, let $\{I_{k_1}, I_{k_2}, \ldots, I_{k_p}\}$ be any finite subcollection of $\mathcal{C}$. Let $k_0 = \max\{k_1, k_2, \ldots, k_p\}$. Then any x in the interval $(0, 1/k_0]$ or in the interval $[3 - 1/k_0, 3)$ is in S but fails to belong to any of the sets $I_{k_1}, I_{k_2}, \ldots, I_{k_p}$. Thus, this finite subcollection cannot possibly be a cover of S. Since we have considered *any finite subcollection* of $\mathcal{C}$, we must conclude that no finite subcover exists and that $S = (0, 3)$ is not compact. In this example, while S is bounded, it is not closed. ●

The Heine–Borel theorem guarantees that such a situation never occurs when covering a closed, bounded set S with open sets; *any* open cover of such a set always contains a finite subcover.

EXAMPLE 23 Here we take any unbounded set S in $\mathbb{R}$. The collection $\mathcal{C}$ of all intervals of the form $I_k = (-k, k)$, with k in $\mathbb{N}$, certainly forms an open cover of S, but again, no finite subcover can possibly exist. For suppose that $\{I_{k_1}, I_{k_2}, \ldots, I_{k_p}\}$ is any finite subcollection of $\mathcal{C}$. Let $k_0 = \max\{k_1, k_2, \ldots, k_p\}$. Observe that I_{k_0} contains each of the I_{k_j}, for $j = 1, 2, \ldots, p$ and is equal to one of them. Therefore the union of all the I_{k_j} is just $I_{k_0} = (-k_0, k_0)$. This interval cannot possibly contain S, for if it did, S would be bounded. Therefore no finite subcover of S can possibly exist. No unbounded set in $\mathbb{R}$ is compact. ●

You may argue, as beginners often do, that some other open cover of S, carefully chosen, can be reduced to a finite subcover. This is certainly true. After all, the open set $(-1, 4)$ is itself a finite open cover of $(0, 3)$. Likewise, the single open set $\mathbb{R}$ is an open cover of any unbounded set S. But this line of argument misses the point entirely. The existence of some *particular finite open cover* will not suffice. Compactness requires that, *regardless of the open cover initially chosen, there must exist a finite subcover.* In each of Examples 22 and 23, we have exhibited one particular open cover that fails this test. Study these last two examples; by them we hope to impress upon you the subtlety of the concept of compactness and the remarkable nature of the Heine–Borel theorem. Moreover, Examples 22 and 23 contain the essential ideas used to prove the following converse of Theorem 2.5.3.

THEOREM 2.5.4 If S is a compact subset of $\mathbb{R}^n$, then S is closed and bounded.

Proof. Suppose that S is a subset of $\mathbb{R}^n$ that is not closed. Then S must have a limit point $\boldsymbol{x}_0$ that is not in S. Let $\{\epsilon_k\}$ be any sequence of positive numbers that converges monotonically to 0. Define $U_k = [\overline{N(\boldsymbol{x}_0; \epsilon_k)}]^c$. Then $\mathcal{C} = \{U_k : k \text{ in } \mathbb{N}\}$ is a collection of open sets in $\mathbb{R}^n$. We claim that $\mathcal{C}$ is an open cover of S. To prove this claim, let $\boldsymbol{x}$ be any point of S. Note that $\boldsymbol{x}_0$ is not in S, so $\boldsymbol{x} \neq \boldsymbol{x}_0$. Therefore, $||\boldsymbol{x} - \boldsymbol{x}_0|| > 0$. Since

$\lim_{k\to\infty} \epsilon_k = 0$, we can choose a k_0 such that, for all $k \geq k_0$, $0 < \epsilon_k < ||\boldsymbol{x} - \boldsymbol{x}_0||$. Thus, for $k \geq k_0$, $\boldsymbol{x}$ does not belong to $\overline{N(\boldsymbol{x}_0; \epsilon_k)}$. Equivalently, for $k \geq k_0$, $\boldsymbol{x}$ *does* belong to $U_k = [\overline{N(\boldsymbol{x}_0; \epsilon_k)}]^c$. This proves that $\{U_k : k \text{ in } \mathbb{N}\}$ is an open cover of S. (See Fig. 2.22.)

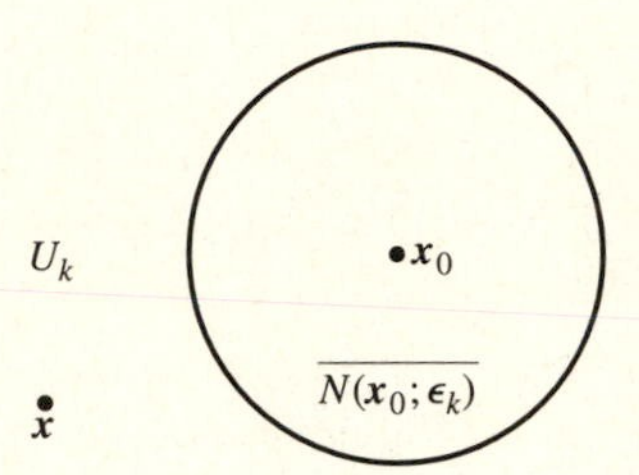

Figure 2.22

We claim that no finite subcollection of $\mathcal{C}$ can cover S. To prove this assertion, begin with any finite collection $\{U_{k_1}, U_{k_2}, \ldots, U_{k_p}\}$ of sets in $\mathcal{C}$. Let k_0 denote the largest of $k_1, k_2, \ldots$, and k_p. Because the sequence $\{\epsilon_k\}$ is monotone decreasing, it follows that $\epsilon_{k_0} \leq \epsilon_{k_j}$ for $j = 1, 2, 3, \ldots, p$. Therefore $\overline{N(\boldsymbol{x}_0; \epsilon_{k_0})} \subseteq \overline{N(\boldsymbol{x}_0; \epsilon_{k_j})}$ for $j = 1, 2, 3, \ldots, p$. It follows, by taking complements, that $U_{k_0} \supseteq U_{k_j}$ for $j = 1, 2, 3, \ldots, p$. Furthermore, $\cup_{j=1}^{p} U_{k_j} = U_{k_0}$. Observe that the distance from any point in U_{k_0} to x_0 is at least ϵ_{k_0}. Choose a positive ϵ less than ϵ_{k_0}. (See Fig. 2.23.) Because $\boldsymbol{x}_0$ is a limit point of S, there exists an $\boldsymbol{x}$ in $S \cap N'(\boldsymbol{x}_0; \epsilon)$. The point $\boldsymbol{x}$ is contained in S but is not contained in U_{k_0}. Therefore the point $\boldsymbol{x}$ is not covered by the finite collection $\{U_{k_j} : j = 1, 2, \ldots, p\}$. Since $\{U_{k_j} : j = 1, 2, \ldots, p\}$ is any finite collection of sets in $\mathcal{C}$, we conclude that no finite subcover can exist. We have proved that, if S is not closed, then there is no finite subcover in the collection $\mathcal{C}$. That is, if S is compact, then S must be closed.

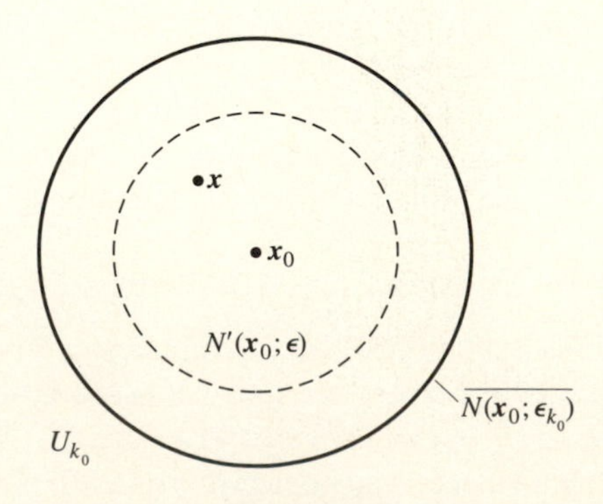

Figure 2.23

Next suppose that S is an unbounded subset of $\mathbb{R}^n$. For each k in $\mathbb{N}$, define $U_k = N(\mathbf{0}; k)$. The union of all the U_k is all of $\mathbb{R}^n$ and certainly contains the set S. Choose any finite subcollection $\{U_{k_1}, U_{k_2}, \ldots, U_{k_p}\}$ of sets in the open cover $\mathcal{C} = \{U_k : k \text{ in } \mathbb{N}\}$. Let $k_0 = \max\{k_1, k_2, \ldots, k_p\}$. Because S is unbounded, there exists an $\boldsymbol{x}$ in S with $||\boldsymbol{x}|| > k_0$. Thus $\boldsymbol{x}$ is not in $U_{k_0} = \cup_{j=1}^{p} U_{k_j}$. We conclude again that no finite subcollection of $\mathcal{C}$ can possibly cover all of S. We have proved that, if S is unbounded, then there is no finite subcover in the collection $\mathcal{C}$. That is, if S is compact, then S must be bounded.

These two arguments, taken together, prove that, if S is compact, then S is closed and bounded. ●

Compactness can also be characterized, in the spirit of the Bolzano–Weierstrass theorem, in terms of the existence of limit points, an alternative formulation that can be extremely powerful in an appropriate context. The following theorem identifies equivalent formulations of compactness in $\mathbb{R}^n$.

THEOREM 2.5.5 Let S be a nonempty subset of $\mathbb{R}^n$. The following statements are equivalent.

i) S is closed and bounded.

ii) S is compact.

iii) Every infinite subset of S has a limit point in S.

Remark. Examine the third statement of this theorem more closely; it looks a great deal like the Bolzano–Weierstrass theorem, a statement equivalent to the *completeness of* $\mathbb{R}^n$. Compactness is conceptually akin to completeness, but *implies boundedness,* whereas in the Bolzano–Weierstrass theorem we had to *assume* the boundedness of the infinite subsets in question. It is also extremely important to emphasize that in (iii), *every* infinite subset of S must have a limit point in S.

Proof. We know, by Theorem 2.5.3 and Theorem 2.5.4, that (i) and (ii) are equivalent. It remains only to prove the equivalence of (i) and (iii) in order to establish the equivalence of all three of the statements of the theorem. We outline such a proof in the exercises. ●

Often someone in your situation will wonder, somewhat dazed, "What is this all about? Why make life so difficult? Who cares about these abstractions anyway?" Here is a hint of what is just around the corner. In your studies of elementary calculus, you learned what it means for a function f to be continuous and you saw a theorem that asserted that, if f is continuous on $[a, b]$, then f assumes a maximum and a minimum value on $[a, b]$. You used this fact to solve a variety of problems, but you never proved it; you were not prepared to do so. Now you are.

You also learned that if f is continuous on $[a, b]$, is negative at one endpoint and positive at the other, then f must have value 0 somewhere between a and b. This theorem, known properly as the Intermediate Value Theorem, seemed self-evident to you; your instructor wanted it to seem so, for you were by no means prepared to deal with its proof either. Now you are.

You may even have noticed that, if f is continuous on $[a, b]$, if

$$m = \min\{f(x) : x \text{ in } [a, b]\}$$

and if

$$M = \max\{f(x) : x \text{ in } [a, b]\},$$

then the range of f is the interval $[m, M]$. It may also occur to you now that the continuous image of the compact set $[a, b]$ is the compact set $[m, M]$, so it may occur to you that we have been constructing exactly the concepts and language you will need in order to deal rigorously with the concept of *continuity.* If so, it will come as no surprise that continuity is the topic of discussion for Chapter 3.

EXERCISES

2.67. Determine whether the following sets are compact in $\mathbb{R}^2$:

a) $S = \{(x_1, x_2) : 0 \leq x_1 \leq 1, 0 \leq x_2 < 1\}$

b) $S = \{\boldsymbol{x} : ||\boldsymbol{x}|| = 1\}$

c) $S = \{(x_1, x_2) : x_2 \geq x_1\}$.

2.68. If possible, give an example of each of the following. If no example exists, explain why.

a) A closed subset S of $\mathbb{R}^n$ that is not compact.

b) A compact subset S of $\mathbb{R}^n$ that is not closed.

c) An open subset S of $\mathbb{R}^n$ that is not compact.

d) A compact subset S of $\mathbb{R}^n$ that is not open.

e) Two compact subsets C_1 and C_2 of $\mathbb{R}^n$ such that $C_1 \cup C_2$ is not compact.

f) Two compact subsets C_1 and C_2 of $\mathbb{R}^n$ such that $C_1 \cap C_2$ is not compact.

2.69. Prove that every closed subset of a compact set in $\mathbb{R}^n$ is compact.

2.70. Let $C_1, C_2, \ldots, C_k$ be compact subsets of $\mathbb{R}^n$.

a) Is $\cup_{j=1}^k C_j$ necessarily compact? If so, prove it; otherwise, provide a counterexample.

b) Is $\cap_{j=1}^k C_j$ necessarily compact? If so, prove it; otherwise, provide a counterexample.

2.71. For $j = 1, 2, \ldots, n$, let $I_j = [a_j, b_j]$. Form the n-dimensional hyper-rectangle $S = I_1 \times I_2 \times \cdots \times I_n$. Prove that S is compact.

2.72. Let C_1 and C_2 be two disjoint compact subsets of $\mathbb{R}^n$. Prove that there exist disjoint open sets U_1 and U_2 such that $C_1 \subseteq U_1$ and $C_2 \subseteq U_2$.

2.73. Complete the proof of Theorem 2.5.5 (that is, show that (i) and (iii) of that theorem are equivalent) by completing the following steps.

a) Let S be a nonempty, closed, bounded set in $\mathbb{R}^n$ and let T be any infinite subset of S. Use the Bolzano–Weierstrass theorem to show that T must have a limit point $\boldsymbol{x}_0$ and that $\boldsymbol{x}_0$ must be a point in S.

b) Let S be a nonempty subset of $\mathbb{R}^n$ with the property that every infinite subset T of S has a limit point in S.

i) Assume that S is unbounded and construct an infinite subset T of S that cannot have a limit point. Thus show that S must be bounded.

ii) Suppose that $\boldsymbol{x}_0$ is a limit point of S. Use Exercise 2.17 to find an infinite subset T of S having $\boldsymbol{x}_0$ as its only limit point. Thus show that $\boldsymbol{x}_0$ must already be in S and that S must be closed.

2.74. Definition. A set S is said to be *locally compact* if, for every $\boldsymbol{x}$ in S, there is an open set U containing x such that $\overline{U}$ is compact. Prove that $\mathbb{R}^n$ is locally compact.

2.75. In 1908 Frigyes Riesz, one of the great mathematicians of the first half of the twentieth century, proposed the following approach to compactness.

Definition. A collection $\mathcal{C} = \{S_\alpha : \alpha \text{ in } A\}$ has the *finite intersection property* if every finite subcollection of $\mathcal{C}$ has a nonempty intersection.

Theorem. F. Riesz. A set S in $\mathbb{R}^n$ is compact if and only if every collection $\mathcal{C} = \{S_\alpha : S_\alpha \text{ closed}, S_\alpha \subseteq S\}$ of closed subsets of S with the finite intersection property has a nonempty intersection: $\bigcap_\alpha S_\alpha \neq \emptyset$.

Prove Riesz's theorem.

2.76. In 1923, P. S. Alexandroff, a leading Soviet mathematician of this century, addressed the following question: How can a space that is not compact be embedded in a space that is compact? He discovered the following elegant and universal construction.

Let X be a space and let the symbol ∞ denote the *point at infinity,* an idealized object that is not in X. Form the set $Y = X \cup \{\infty\}$. A set U in Y is said to be *open in* Y if any one of the following three conditions holds:

i) U is an open set in X.

ii) U is a subset of Y whose complement is a nonempty compact subset of X.

iii) $U = Y$.

a) Prove that $\emptyset$ and Y are open in Y. Prove that any union of sets open in Y is open in Y. Prove that the intersection of finitely many sets open in Y is open in Y.

Alexandroff proved that the space Y, together with this definition of open set, is always compact. The construction is called *Alexandroff's one-point compactification of* X.

b) Suppose that $X = \mathbb{R}$ and that both $+\infty$ and $-\infty$ are identified with the single point at infinity. (Conceptually, the real line is made into a circle, joining $+\infty$ and $-\infty$ at one point, called ∞.) Prove that $Y = \mathbb{R} \cup \{\infty\}$, together with Alexandroff's extended definition of what it means for a set to be open, is compact. That is, prove that every open cover has a finite subcover. (*Hint:* What kind of open set is required to cover the point at infinity?)

3

Continuity

Here we embark on the central issue of analysis, the study of the behavior of functions. We will be primarily interested in functions having as domain some set S in $\mathbb{R}$ or in $\mathbb{R}^n$ and range in $\mathbb{R}$ or $\mathbb{R}^m$. However, our approach is designed to prepare you to study functions having more general domains and ranges as well. Initially, we will concentrate on *local behavior*—behavior in some neighborhood of a point. Later we turn our attention to *global behavior* and draw conclusions about certain classes of functions on the entire shared domain.

3.1 Limit and Continuity

Let f be a real-valued function defined on a subset S of $\mathbb{R}$ and let c be a point in $\overline{S}$. As you learned in your elementary coursework, we say that $\lim_{x \to c} f(x) = L$ provided that $f(x)$ is arbitrarily near L whenever x is sufficiently near but not equal to c. While this essentially intuitive description of limit is one we certainly want you to retain, we will need a more technical formulation of the concept of limit for the careful development of our theory.

DEFINITION 3.1.1 Let f be a real-valued function with domain S in $\mathbb{R}$. Let c be a point in $\overline{S}$. We say that f has *limit* L as x approaches c if, given $\epsilon > 0$, there exists a $\delta > 0$ such that, if $0 < |x - c| < \delta$, then $|f(x) - L| < \epsilon$. We write $\lim_{x \to c} f(x) = L$. ●

Notice that we must require x to be restricted to the domain S of the function f; otherwise, the symbol $f(x)$ makes no sense. The point c, however, may not be in S; it is chosen in $\overline{S}$. Note also that ϵ is specified first; the value of δ usually is to be determined in terms of ϵ.

Recast this definition in the notation we have established in Chapter 1. (See Fig. 3.1.) The inequality $|f(x) - L| < \epsilon$ translates into the statement that $f(x)$ belongs to the the neighborhood $N(L; \epsilon)$. The inequality

$$0 < |x - c| < \delta$$

means that x belongs to the deleted neighborhood $N'(c; \delta)$. For x also to be in S simply means that x is in the intersection of S and $N'(c; \delta)$, a relatively open, deleted neighborhood of c. In other words, $\lim_{x \to c} f(x) = L$ means that, given any neighborhood $N(L; \epsilon)$ of L, there exists a neighborhood $N(c; \delta)$ of c such that, if x is in $S \cap N'(c; \delta)$, then $f(x)$ is in $N(L; \epsilon)$.

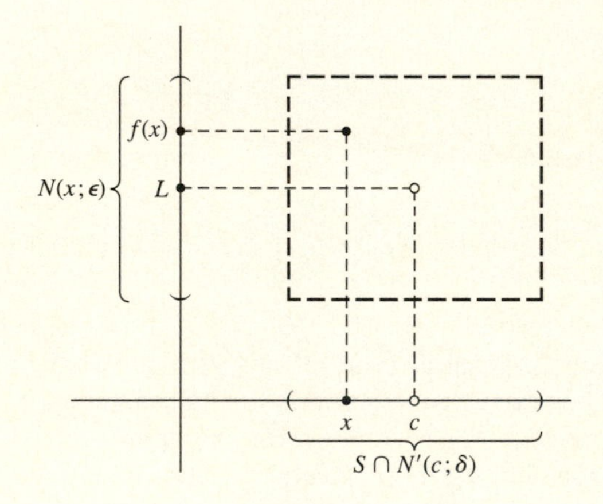

Figure 3.1

Even more briefly, $f(S \cap N'(c; \delta)) \subseteq N(L; \epsilon)$ where the notation $f(A)$ denotes the set $\{f(x) : x \text{ in } A\}$. The function f maps every point in the set $S \cap N'(c; \delta)$ into the neighborhood $N(L; \epsilon)$. For x restricted to $S \cap N'(c; \delta)$, the graph of f remains within the rectangle $N(c; \delta) \times N(L; \epsilon)$.

Here is one final notational variation we will use. If f is a function from S to T and if B is any subset of T, then $f^{-1}(B)$ is the set of x in the domain of f mapped by f into the set B. That is, $f^{-1}(B) = \{x \text{ in } S : f(x) \text{ is in } B\}$. The set $f^{-1}(B)$ is called the *inverse image* or the *preimage* of the set B under the function f. The statement $f(A) \subseteq B$ means the same thing as $A \subseteq f^{-1}(B)$. Using this notation, $\lim_{x \to c} f(x) = L$ means that, for any neighborhood $N(L; \epsilon)$ of L, there exists a neighborhood $N(c; \delta)$ of c such that

$$S \cap N'(c; \delta) \subseteq f^{-1}(N(L; \epsilon)). \tag{3.1}$$

One advantage of this notation is that it enables us to avoid explicit mention of ϵ and δ . They are understood simply to be the radii of the neighborhoods $N(L)$ and $N(c)$ in the range and the domain respectively of f. Thus $\lim_{x \to c} f(x) = L$ means that, for any neighborhood $N(L)$, there exists a neighborhood $N(c)$ such that $S \cap N'(c) \subseteq f^{-1}(N(L))$. This formulation gets immediately to the heart of the matter, expressing the concept of limit in terms of sets, and avoids symbols of secondary importance. In any argument in which ϵ and δ play explicit roles, they can be easily reinstated as in (3.1).

To this point, we have considered only the limit of f at a point c in a subset $\overline{S}$ of $\mathbb{R}$. Suppose now that S is a subset in $\mathbb{R}^n$, that f is a real-valued function defined on S, and that $\boldsymbol{c}$ is a point of $\overline{S}$. Which details of the foregoing must be modified in order to define the limit of f at $\boldsymbol{c}$? Our question is rhetorical; no changes whatever need be made. True, the neighborhood of $\boldsymbol{c}$ is now defined using the Euclidean norm in $\mathbb{R}^n$, rather than absolute value in $\mathbb{R}$, but conceptually nothing changes.

DEFINITION 3.1.2 Let f be a real-valued function with domain S in $\mathbb{R}^n$. Let $\boldsymbol{c}$ be a point in $\overline{S}$. We say that f has *limit* L as $\boldsymbol{x}$ approaches $\boldsymbol{c}$ provided that, for every neighborhood $N(L)$, there exists a neighborhood $N(\boldsymbol{c})$ such that

$$S \cap N'(\boldsymbol{c}) \subseteq f^{-1}(N(L)).$$

We write $\lim_{\boldsymbol{x} \to \boldsymbol{c}} f(\boldsymbol{x}) = L$. ●

A point of $\overline{S}$ is either a limit point or an isolated point of S. Notice that, if $\boldsymbol{c}$ is an isolated point of S, then, for δ sufficiently small, $S \cap N'(\boldsymbol{c}; \delta) = \emptyset$. Since the empty set is always a subset of $f^{-1}(N(f(\boldsymbol{c})))$, whatever may be the function f, it follows that any real-valued function f has limit $f(\boldsymbol{c})$ at any isolated point $\boldsymbol{c}$ of its domain S. This case is trivial and our interest centers on the case when $\boldsymbol{c}$ is a limit point of S. (See Fig. 3.2.) Again, we emphasize that, in this definition, the neighborhood $N(L)$ of the limiting value of f is specified first and that the neighborhood $N(\boldsymbol{c})$ is to be determined second.

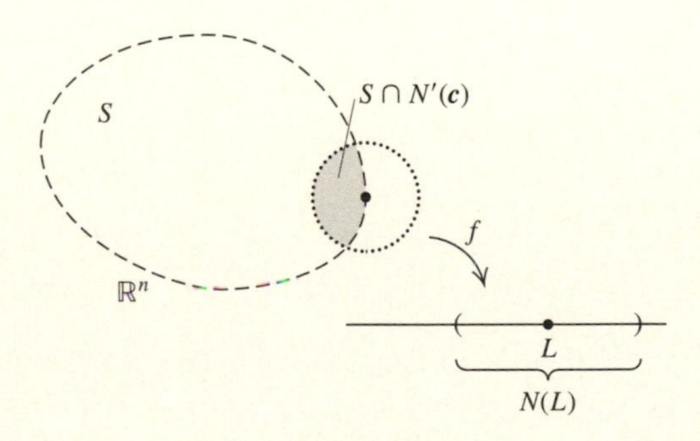

Figure 3.2

We have discussed the meaning of limit at such length because it is one of the concepts on which all our subsequent arguments will be based. It is essential that you master it. We examine first some simple examples, but we caution you that, in practice, the limits we will confront throughout the text are usually more difficult to analyze.

EXAMPLE 1 Let $S = [0, 3]$. For $x \neq 2$, let $f(x) = x^2$. Define $f(2) = 1$. We claim that $\lim_{x\to 2} f(x) = 4$. To prove this, fix $\epsilon > 0$. The point $(4-\epsilon)^{1/2}$ is mapped by f to $4-\epsilon$; the point $(4+\epsilon)^{1/2}$ is mapped by f to $4+\epsilon$. (See Fig. 3.3.) Therefore we choose δ such that

$$0 < \delta < \min\{2 - (4-\epsilon)^{1/2}, (4+\epsilon)^{1/2} - 2\}.$$

In fact, as you can confirm for yourself, this minimum is $(4+\epsilon)^{1/2} - 2$. Therefore, if x is in $N'(2; \delta)$, then $x \neq 2$ and

$$4 - (4+\epsilon)^{1/2} < 2 - \delta < x < 2 + \delta < (4+\epsilon)^{1/2}.$$

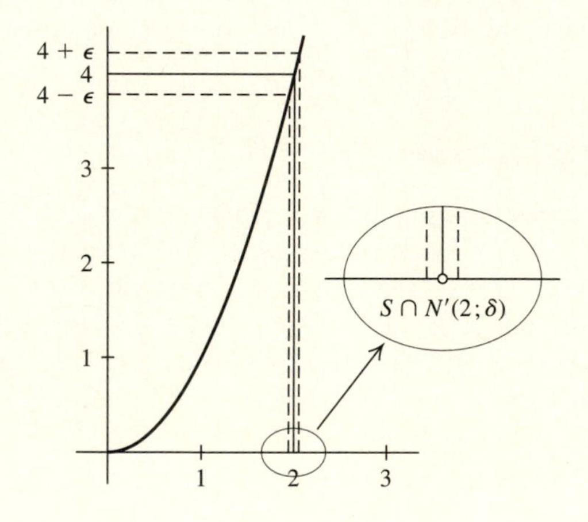

Figure 3.3

Consequently, for any x in $N'(2; \delta)$, we have

$$\begin{aligned} |x^2 - 4| = |x+2|\,|x-2| &< [(4+\epsilon)^{1/2} + 2]\delta \\ &< [(4+\epsilon)^{1/2} + 2][(4+\epsilon)^{1/2} - 2] = \epsilon. \end{aligned}$$

Thus if x is in $N'(2; \delta)$, then $f(x)$ is in $N(4; \epsilon)$. This proves that $\lim_{x\to 2} x^2 = 4$. This simple example shows you that the value of f at $c = 2$ is immaterial; what matters in discussing $\lim_{x\to c} f(x)$, is the *limiting behavior* of f near c, not the *value* of f at c. ●

EXAMPLE 2 For $\boldsymbol{x} = (x_1, x_2) \neq \boldsymbol{0}$ in $\mathbb{R}^2$, define

$$f(\boldsymbol{x}) = \frac{x_1^2 x_2^2}{x_1^2 + x_2^2}.$$

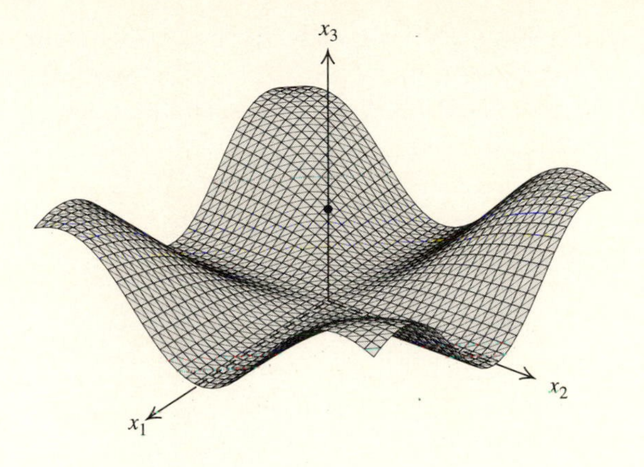

Figure 3.4

Define $f(\mathbf{0}) = 1$. The graph of f is sketched in Fig. 3.4. We claim that $\lim_{\boldsymbol{x}\to\mathbf{0}} f(\boldsymbol{x}) = 0$. To prove this claim, fix any $\epsilon > 0$. Let $\delta = \epsilon^{1/2}$. Since $x_j^2 \geq 0$ for $j = 1, 2$, it follows that, whenever $\boldsymbol{x}$ is in $N'(\mathbf{0}; \delta)$, we have

$$0 \leq \frac{x_1^2 x_2^2}{x_1^2 + x_2^2} \leq x_1^2 \leq x_1^2 + x_2^2 < \delta^2 = \epsilon.$$

Thus if $\boldsymbol{x}$ is in $N'(\mathbf{0}; \delta)$, then $f(\boldsymbol{x})$ is in $N(0; \epsilon)$. This proves that $\lim_{\boldsymbol{x}\to\mathbf{0}} f(\boldsymbol{x}) = 0$. ●

EXAMPLE 3 For $\boldsymbol{x} = (x_1, x_2) \neq \mathbf{0}$, define

$$f(\boldsymbol{x}) = \frac{2x_1 x_2}{(x_1^2 + x_2^2)}.$$

Define $f(\mathbf{0}) = 1$. The graph of f is sketched in Fig. 3.5. Again, we ask for

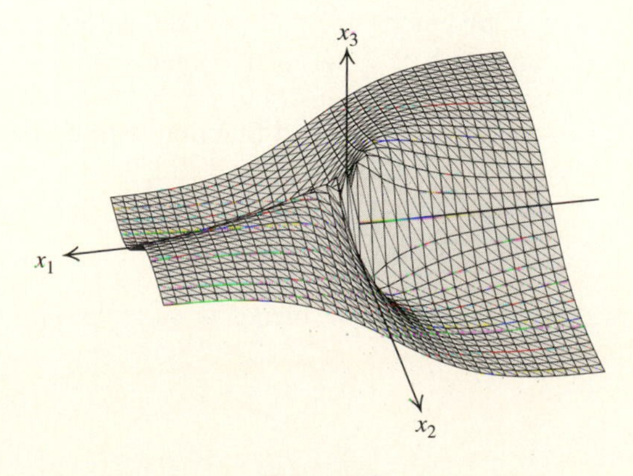

Figure 3.5

$\lim_{\boldsymbol{x}\to\mathbf{0}} f(\boldsymbol{x})$. But this limit does not exist. To prove this claim, choose any L in $[-1, 1]$. Choose c in $\mathbb{R}$ such that $L = 2c/(1+c^2)$.[1] Now choose any deleted neighborhood $N'(\mathbf{0})$ and choose any $\boldsymbol{x}$ in $N'(\mathbf{0})$ such that $x_2 = cx_1$. (See Fig. 3.6.) Evaluate f at this point $\boldsymbol{x}$ to obtain

$$f(\boldsymbol{x}) = \frac{2cx_1^2}{(1+c^2)x_1^2} = L.$$

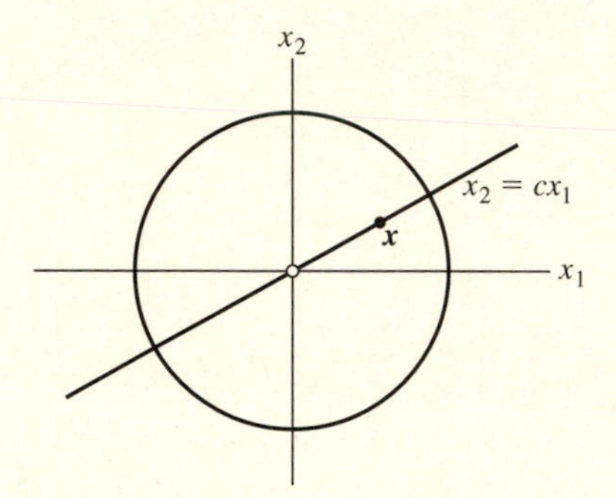

Figure 3.6

In other words, f assumes every value between -1 and 1 on any deleted neighborhood of $\mathbf{0}$. If $\lim_{\boldsymbol{x}\to\mathbf{0}} f(\boldsymbol{x})$ were to exist, with value, say, L_0, and if we were to choose any L in $[-1, 1]$ not equal to L_0, then we need merely choose a positive ϵ less than $|L - L_0|$. By our preceding observation, whatever the choice of $\delta > 0$, there is an $\boldsymbol{x}$ in $N'(\mathbf{0}; \delta)$ such that $f(\boldsymbol{x}) = L$. For such an $\boldsymbol{x}$, $|f(\boldsymbol{x}) - L_0| = |L - L_0| > \epsilon$. Thus $f(\boldsymbol{x})$ is not in $N(L_0; \epsilon)$. This proves that L_0 cannot be $\lim_{\boldsymbol{x}\to\mathbf{0}} f(x)$. Any value assigned to $\lim_{\boldsymbol{x}\to\mathbf{0}} f(\boldsymbol{x})$ gives this same contradiction. Consequently, this limit cannot exist. ●

During your initial studies of calculus, you also learned that a function f is *continuous* at a point $\boldsymbol{c}$ in its domain S if $\lim_{\boldsymbol{x}\to\boldsymbol{c}} f(\boldsymbol{x}) = f(\boldsymbol{c})$. In accord with our definition of limit, this means that $f(\boldsymbol{x})$ is arbitrarily near $f(\boldsymbol{c})$ provided that $\boldsymbol{x}$ is in S and is sufficiently near $\boldsymbol{c}$. We emphasize that, for the symbol $f(\boldsymbol{c})$ to make sense, $\boldsymbol{c}$ must be a point of the domain S of f, not just a point of $\overline{S}$.

DEFINITION 3.1.3 Let f be a real-valued function with domain S in $\mathbb{R}^n$. Let $\boldsymbol{c}$ be a point of S. We say that f is *continuous* at $\boldsymbol{c}$ if $\lim_{\boldsymbol{x}\to\boldsymbol{c}} f(\boldsymbol{x}) = f(\boldsymbol{c})$. Thus, for every neighborhood $N(f(\boldsymbol{c}))$, there exists a neighborhood $N(\boldsymbol{c})$ such that

$$S \cap N(\boldsymbol{c}) \subseteq f^{-1}(N(f(\boldsymbol{c}))).$$

If f is continuous at every point of a set S, then f is said to be *continuous on* S. ●

---

[1] The number c is a solution of $Lc^2 - 2c + L = 0$. Thus if $L \neq 0$, then $c = [1 + (1 - L^2)^{1/2}]/L$. If $L = 0$, then $c = 0$.

In the special case when S is a subset of $\mathbb{R}$, we have some special notation that has evolved over the years; its simplicity and usefulness warrant your learning and using it. We emphasize, however, that, conceptually, nothing new is being introduced at this point; this is merely notational convention. The *right-hand limit* of f at c, denoted

$$f(c^+) = \lim_{x \to c^+} f(x),$$

is obtained, providing that this limit exists, by taking x in S and greater than c. Analogously, the *left-hand limit* of f at c, denoted

$$f(c^-) = \lim_{x \to c^-} f(x),$$

is obtained, again providing that this limit exists, by requiring that x remain in S and be less than c.

Clearly, if c is an interior point of S, then $\lim_{x \to c} f(x)$ exists if and only if both $f(c^+)$ and $f(c^-)$ exist and are equal; $\lim_{x \to c} f(x)$ is the common value of $f(c^+)$ and $f(c^-)$. You may prove this as an exercise. In the special case when S is an interval of the form $[a, b]$, $[a, b)$, $(a, b]$, or (a, b) and c is either a or b, then only $f(a^+)$ and $f(b^-)$ can make any sense. If $f(a^+)$ exists, it is the limit of f as x approaches a. An analogous comment holds at the endpoint b. For the function f to be continuous at a, it is necessary and sufficient, first, that a be in S and, second, that $f(a^+)$ exist and equal $f(a)$. Likewise, f is continuous at b if and only if b is in S, $f(b^-)$ exists and $f(b^-) = f(b)$.

EXAMPLE 1 (revisited) If in Example 1 we redefine f at $c = 2$ to have value 4, then f becomes continuous at $c = 2$. As you can prove, $f(x) = x^2$ is continuous on all of $\mathbb{R}$. ●

EXAMPLE 2 (revisited) If, in Example 2, we redefine f at $\mathbf{c} = \mathbf{0}$ to have value 0, then f becomes continuous at $\mathbf{c} = \mathbf{0}$ because $\lim_{\mathbf{x} \to \mathbf{0}} f(\mathbf{x}) = 0$. As we will see, f is continuous at any point other than $\mathbf{0}$. Consequently, f is continuous on all of $\mathbb{R}^2$. ●

EXAMPLE 3 (revisited) In Example 3, no matter how we may wish to define $f(\mathbf{0})$, the function f cannot be made continuous there. The reason? The limiting behavior of f near $\mathbf{c} = \mathbf{0}$ is such that $\lim_{\mathbf{x} \to \mathbf{0}} f(\mathbf{x})$ does not exist. However, as we will see in Example 7 below the function $f(\mathbf{x}) = 2x_1x_2/(x_1^2 + x_2^2)$ is continuous at all other points in $\mathbb{R}^2$. ●

EXAMPLE 4 Let $f(x) = x^{1/2}$ on $S = [0, \infty)$. We claim that f is continuous at every point of S. Recall that f is strictly monotone increasing on S and, consequently, the function f^{-1} exists on the range of f with $(f^{-1} \circ f)(x) = x$ for all x in S. Clearly, $f^{-1}(x) = x^2$.

First, consider c in $(0, \infty)$. Choose any ϵ such that $0 < \epsilon < c^{1/2}$ and form $f^{-1}(N(c^{1/2}; \epsilon))$. As you can compute

$$f^{-1}(c^{1/2} - \epsilon) = (c^{1/2} - \epsilon)^2 = c - (2\epsilon c^{1/2} - \epsilon^2).$$

Likewise

$$f^{-1}(c^{1/2}+\epsilon) = c + (2\epsilon c^{1/2} + \epsilon^2).$$

(See Fig. 3.7.) Let $\delta = 2\epsilon c^{1/2} - e^2$. We claim that, if x is in $N'(c; \delta)$, then $f(x)$ is in $N(c^{1/2}; \epsilon)$. That x is in $N(c; \delta)$ entails the inequalities

$$\begin{aligned} 0 < (c^{1/2}-\epsilon)^2 &= c - 2\epsilon c^{1/2} + \epsilon^2 \\ &< x < c + 2\epsilon c^{1/2} - \epsilon^2 < c + 2\epsilon c^{1/2} + \epsilon^2 \\ &= (c^{1/2}+\epsilon)^2. \end{aligned}$$

By taking square roots we obtain

$$c^{1/2} - \epsilon < x^{1/2} < c^{1/2} + \epsilon.$$

Therefore if x is in $[0, \infty) \cap N(c; \delta)$, then $f(x)$ is in $N(c^{1/2}; \epsilon)$. This proves that f is continuous at c.

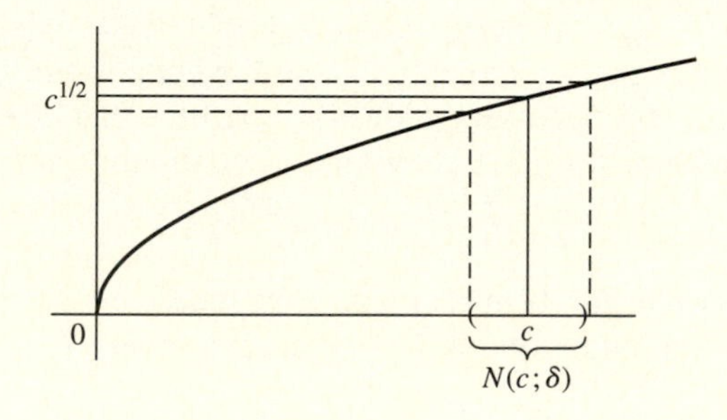

Figure 3.7

To show that $f(x) = x^{1/2}$ is continuous at $c = 0$, we must show that $f(0^+)$ exists and equals $f(0) = 0$. Fix an arbitrary $\epsilon > 0$. Choose $\delta = \epsilon^2$. For $0 < x < \delta = \epsilon^2$, it follows that $0 < f(x) = x^{1/2} < \epsilon$. Consequently, $f(0^+) = 0 = f(0)$ and f is continuous from the right at $c = 0$. ●

Before examining more complicated functions, we establish basic properties of the limiting operation to ease our way. You will note that the theorems resemble those we have already proved in Section 1.5 for sequences. We first prove two auxiliary theorems, of merit in their own right, before establishing the major theorem for limits.

THEOREM 3.1.1 Suppose that f is a real-valued function defined on a set S in $\mathbb{R}^n$. Suppose also that $\boldsymbol{c}$ is a point in $\overline{S}$ where $\lim_{\boldsymbol{x}\to\boldsymbol{c}} f(\boldsymbol{x}) = L$ exists. Then f is locally bounded on some deleted neighborhood of $\boldsymbol{c}$. That is, there exists a $\delta > 0$ and a constant M such that $|f(\boldsymbol{x})| \le M$ for all $\boldsymbol{x}$ in $S \cap N'(\boldsymbol{c}; \delta)$.

Proof. Fix $\epsilon > 0$. Choose $\delta > 0$ such that, if $\boldsymbol{x}$ is in $S \cap N'(c; \delta)$, then $f(\boldsymbol{x})$ is in $N(L; \epsilon)$. It follows that, for such $\boldsymbol{x}$,

$$|f(\boldsymbol{x})| \leq |f(\boldsymbol{x}) - L| + |L| < \epsilon + |L|.$$

Let $M = |L| + \epsilon$. Then, as promised by the theorem, M is a local upper bound for $|f|$ on $S \cap N'(\boldsymbol{c}; \delta)$. ●

THEOREM 3.1.2 Suppose that f is a real-valued function defined on a set S in $\mathbb{R}^n$. Suppose also that $\boldsymbol{c}$ is a point of $\overline{S}$ where $\lim_{\boldsymbol{x}\to\boldsymbol{c}} f(\boldsymbol{x}) = L$ exists and is not 0. Then f is locally bounded away from 0 on some deleted neighborhood of $\boldsymbol{c}$. That is, there exists an $m > 0$ and a $\delta > 0$ such that $|f(\boldsymbol{x})| \geq m$ for all $\boldsymbol{x}$ in $S \cap N'(\boldsymbol{c}; \delta)$.

Proof. Let $\epsilon = |L|/2$. Since L is not zero, $\epsilon > 0$. Choose $\delta > 0$ such that $S \cap N'(\boldsymbol{c}; \delta) \subseteq f^{-1}N(L; |L|/2)$. For $\boldsymbol{x}$ in $S \cap N'(\boldsymbol{c}; \delta)$, $|f(\boldsymbol{x}) - L| < |L|/2$. Thus $L - |L|/2 < f(\boldsymbol{x}) < L + |L|/2$. It follows that $|f(\boldsymbol{x})| > m = |L|/2$ for all $\boldsymbol{x}$ in $S \cap N'(\boldsymbol{c}; \delta)$. This proves the theorem. ●

THEOREM 3.1.3 Suppose that f_1 and f_2 are two real-valued functions with common domain S in $\mathbb{R}^n$, that $\boldsymbol{c}$ is a point of $\overline{S}$, and that $\lim_{\boldsymbol{x}\to\boldsymbol{c}} f_1(\boldsymbol{x}) = L_1$ and $\lim_{\boldsymbol{x}\to\boldsymbol{c}} f_2(\boldsymbol{x}) = L_2$ exist. Then

i) $\lim_{\boldsymbol{x}\to\boldsymbol{c}}[f_1(\boldsymbol{x}) + f_2(\boldsymbol{x})] = L_1 + L_2$.

ii) For any constant a in $\mathbb{R}$, $\lim_{\boldsymbol{x}\to\boldsymbol{c}} af_1(\boldsymbol{x}) = aL_1$.

iii) $\lim_{\boldsymbol{x}\to\boldsymbol{c}} f_1(\boldsymbol{x})f_2(\boldsymbol{x}) = L_1L_2$.

iv) $\lim_{\boldsymbol{x}\to\boldsymbol{c}} 1/f_2(\boldsymbol{x}) = 1/L_2$, provided that $L_2 \neq 0$.

v) $\lim_{\boldsymbol{x}\to\boldsymbol{c}} f_1(\boldsymbol{x})/f_2(\boldsymbol{x}) = L_1/L_2$ provided that $L_2 \neq 0$.

Proof. For (i) fix $\epsilon > 0$. Since $\lim_{\boldsymbol{x}\to\boldsymbol{c}} f_1(\boldsymbol{x}) = L_1$, we can choose $\delta_1 > 0$ for which $S \cap N'(\boldsymbol{c}; \delta_1) \subseteq f_1^{-1}(N(L_1; \epsilon))$. Likewise, choose $\delta_2 > 0$ such that $S \cap N'(\boldsymbol{c}; \delta_2) \subseteq f_2^{-1}(N(L_2; \epsilon))$. Now choose $\delta = \min\{\delta_1, \delta_2\}$. For $\boldsymbol{x}$ in $S \cap N'(\boldsymbol{c}; \delta)$, both conditions hold. That is, $f_1(\boldsymbol{x})$ is in $N(L_1 : \epsilon)$ and $f_2(\boldsymbol{x})$ is in $N(L_2; \epsilon)$. Thus, if $\boldsymbol{x}$ is in $S \cap N'(\boldsymbol{c}; \delta)$, we have

$$\begin{aligned} |f_1(\boldsymbol{x}) + f_2(\boldsymbol{x}) - (L_1 + L_2)| &\leq |f_1(\boldsymbol{x}) - L_1| + |f_2(\boldsymbol{x}) - L_2| \\ &< \epsilon + \epsilon = 2\epsilon. \end{aligned}$$

Therefore, $f_1(\boldsymbol{x}) + f_2(\boldsymbol{x})$ is in $N(L_1 + L_2; 2\epsilon)$. We conclude that $\lim_{\boldsymbol{x}\to\boldsymbol{c}}[f_1(\boldsymbol{x}) + f_2(\boldsymbol{x})] = L_1 + L_2$.

For part (ii), choose $\epsilon > 0$, then choose $\delta > 0$ such that $S \cap N'(\boldsymbol{c}; \delta) \subseteq f_1^{-1}(N(L_1; \epsilon))$. Then for $\boldsymbol{x}$ in $S \cap N'(\boldsymbol{c}; \delta)$, we have

$$|af_1(\boldsymbol{x}) - aL_1| = |a| \cdot |f_1(\boldsymbol{x}) - L_1| < |a|\epsilon.$$

It follows that $\lim_{\boldsymbol{x}\to\boldsymbol{c}} af_1(\boldsymbol{x}) = aL_1$.

To prove (iii) first note that, by Theorem 3.1.1, we can choose a $\delta_0 > 0$ and an $M_1 > 0$ such that, if $\boldsymbol{x}$ is a point in $S \cap N'(\boldsymbol{c}; \delta_0)$, then $|f_1(\boldsymbol{x})| \leq M_1$. Fix $\epsilon > 0$.

Choose $\delta_1 > 0$ such that, if $\boldsymbol{x}$ is in $S \cap N'(\boldsymbol{c}; \delta_1)$, then $f_1(\boldsymbol{x})$ is in $N(L_1; \epsilon)$. Likewise, choose $\delta_2 > 0$ such that, if $\boldsymbol{x}$ is in $S \cap N'(\boldsymbol{c}; \delta_2)$, then $f_2(\boldsymbol{x})$ is in $N(L_2; \epsilon)$. Let $\delta = \min\{\delta_0, \delta_1, \delta_2\}$. For $\boldsymbol{x}$ in $S \cap N'(\boldsymbol{c}; \delta)$, all three of the preceding conditions hold. Thus for $\boldsymbol{x}$ in $S \cap N'(\boldsymbol{c}; \delta)$, we obtain

$$\begin{aligned} |f_1\boldsymbol{x})f_2(\boldsymbol{x}) - L_1L_2| &\leq |f_1(\boldsymbol{x})|\,|f_2(\boldsymbol{x}) - L_2| + |f_1(\boldsymbol{x}) - L_1|\,|L_2| \\ &< M_1\epsilon + \epsilon|L_2| = (M_1 + |L_2|)\epsilon. \end{aligned}$$

Consequently, for $\boldsymbol{x}$ in $S \cap N'(\boldsymbol{c}; \delta)$, $f_1(\boldsymbol{x})f_2(\boldsymbol{x})$ is in the neighborhood $N(L_1L_2; (M_1 + |L_2|)\epsilon)$. Therefore, as desired, $\lim_{\boldsymbol{x}\to\boldsymbol{c}} f_1(\boldsymbol{x})f_2(\boldsymbol{x}) = L_1L_2$.

Part (iv) of the theorem is proved using Theorem 3.1.2 and imitating the argument in the proof of part (iii) of Theorem 1.5.1. Prove part (v) by combining parts (iii) and (iv) of the present theorem. We leave the details for you to complete. ●

EXAMPLE 5 Evidently, $f(\boldsymbol{x}) = ||\boldsymbol{x}||$ is continuous at every $\boldsymbol{c}$ in $\mathbb{R}^n$. The function $g(\boldsymbol{x}) = ||\boldsymbol{x}||^2 = f^2(\boldsymbol{x})$ is also continuous at each $\boldsymbol{c}$ in $\mathbb{R}^n$. A simple application of part (iii) of Theorem 3.1.3 proves this claim as follows:

$$\lim_{\boldsymbol{x}\to\boldsymbol{c}} g(\boldsymbol{x}) = \lim_{\boldsymbol{x}\to\boldsymbol{c}} f^2(\boldsymbol{x}) = [\lim_{\boldsymbol{x}\to\boldsymbol{c}} f(\boldsymbol{x})]^2 = f^2(\boldsymbol{c}) = g(\boldsymbol{c}).$$

It is similarly easy to see that, for a fixed $\boldsymbol{x_0}$ in $\mathbb{R}^n$, the function $h(\boldsymbol{x}) = ||\boldsymbol{x} - \boldsymbol{x_0}||^2$ is also continuous on $\mathbb{R}^n$. By part (iv) of Theorem 3.1.3, it follows that the functions $1/||\boldsymbol{x}||$ and $1/||\boldsymbol{x}||^2$ are continuous on all of $\mathbb{R}^n$ except at $\boldsymbol{c} = \boldsymbol{0}$. Likewise, for any fixed $\boldsymbol{x_0}$ in $\mathbb{R}^n$, $1/||\boldsymbol{x} - \boldsymbol{x_0}||$ and $1/||\boldsymbol{x} - \boldsymbol{x_0}||^2$ are continuous except at $\boldsymbol{c} = \boldsymbol{x_0}$. ●

EXAMPLE 6 By Theorem 3.1.3, the function

$$f(\boldsymbol{x}) = \frac{x_1^2x_2^2}{(x_1^2 + x_2^2)}$$

of Example 2 has limiting value $c_1^2c_2^2/(c_1^2 + c_2^2)$ at any point $\boldsymbol{c} \neq \boldsymbol{0}$ in $\mathbb{R}^2$. Therefore, f is continuous at all $\boldsymbol{c} \neq \boldsymbol{0}$. As we have already seen, by defining $f(\boldsymbol{0}) = 0$, we can make f continuous at $\boldsymbol{0}$ also. With this stipulation, f is continuous on all of $\mathbb{R}^2$. ●

EXAMPLE 7 Similarly, by Theorem 3.1.3, the function $f(\boldsymbol{x}) = 2x_1x_2/(x_1^2 + x_2^2)$ is continuous at all $\boldsymbol{c} \neq \boldsymbol{0}$ in $\mathbb{R}^2$. Recall from Example 3 that no assignment of a value to $f(\boldsymbol{0})$ can cause f to be continuous at $\boldsymbol{c} = \boldsymbol{0}$. ●

The foregoing examples are elementary, but as we proceed we will be able to examine the limiting behavior and the continuity of ever more complicated functions. An extremely useful tool in this regard, as was true in our study of sequences, is the Squeeze Play. By bracketing a given function g by two simple functions f and h whose limiting behavior is known, we can gather information about the limiting behavior of g.

THEOREM 3.1.4 The Squeeze Play Let f, g, and h be three real-valued functions sharing a common domain S in $\mathbb{R}^n$. Let $\boldsymbol{c}$ be a point of $\overline{S}$ where $\lim_{\boldsymbol{x}\to\boldsymbol{c}} f(\boldsymbol{x}) = \lim_{\boldsymbol{x}\to\boldsymbol{c}} h(\boldsymbol{x}) = L$. Suppose also that, for some $\delta_0 > 0$ and for all $\boldsymbol{x}$ in $S \cap N'(\boldsymbol{c}; \delta_0)$, $f(\boldsymbol{x}) \leq g(\boldsymbol{x}) \leq h(\boldsymbol{x})$. Then $\lim_{\boldsymbol{x}\to\boldsymbol{c}} g(\boldsymbol{x}) = L$.

Proof. Given $\epsilon > 0$, first choose $\delta_1 > 0$ such that, for $\boldsymbol{x}$ in $S \cap N'(\boldsymbol{c}; \delta_1)$, $f(\boldsymbol{x})$ is in $N(L; \epsilon)$. Next, choose $\delta_2 > 0$ such that, for $\boldsymbol{x}$ in $S \cap N'(\boldsymbol{c}; \delta_2)$, $h(\boldsymbol{x})$ is in $N(L; \epsilon)$. Let $\delta = \min\{\delta_0, \delta_1, \delta_2\}$. Then, for $\boldsymbol{x}$ in $S \cap N'(\boldsymbol{c}; \delta)$, we have

$$L - \epsilon < f(\boldsymbol{x}) \leq g(\boldsymbol{x}) \leq h(\boldsymbol{x}) < L + \epsilon.$$

Therefore, for such $\boldsymbol{x}$, $g(\boldsymbol{x})$ is in $N(L; \epsilon)$. This proves that $\lim_{\boldsymbol{x}\to\boldsymbol{c}} g(\boldsymbol{x}) = L$, as promised by the theorem. ●

The following theorem, related to the Squeeze Play, confirms that the limiting operation preserves order. Its proof is straightforward and is left as an exercise.

THEOREM 3.1.5 Suppose that f_1 and f_2 are two real-valued functions having common domain S in $\mathbb{R}^n$. Suppose also that $\boldsymbol{c}$ is a point in $\overline{S}$ such that $\lim_{\boldsymbol{x}\to\boldsymbol{c}} f_1(\boldsymbol{x}) = L_1$ and $\lim_{\boldsymbol{x}\to\boldsymbol{c}} f_2(\boldsymbol{x}) = L_2$ exist. Finally, assume that there exists a $\delta_0 > 0$ such that, for all $\boldsymbol{x}$ in $S \cap N'(\boldsymbol{c}; \delta_0)$, we have $f_1(\boldsymbol{x}) \leq f_2(\boldsymbol{x})$. Then $L_1 \leq L_2$. ●

As a typical application of the Squeeze Play, we prove in our next example that the six trigonometric functions are continuous at any point of $\mathbb{R}$ where they are defined. First we prove a simple fact.

LEMMA If $\boldsymbol{x}$ and $\boldsymbol{y}$ are vectors in $\mathbb{R}^n$ and if $\boldsymbol{x}$ and $\boldsymbol{y}$ are orthogonal, then $||\boldsymbol{x}|| \leq ||\boldsymbol{x} - \boldsymbol{y}||$.

Proof. That $\boldsymbol{x}$ and $\boldsymbol{y}$ are orthogonal entails $\langle \boldsymbol{x}, \boldsymbol{y} \rangle = 0$. Therefore,

$$\begin{aligned} ||\boldsymbol{x} - \boldsymbol{y}||^2 &= \langle \boldsymbol{x} - \boldsymbol{y}, \boldsymbol{x} - \boldsymbol{y} \rangle = ||\boldsymbol{x}||^2 - 2\langle \boldsymbol{x}, \boldsymbol{y} \rangle + ||\boldsymbol{y}||^2 \\ &= ||\boldsymbol{x}||^2 + ||\boldsymbol{y}||^2 \geq ||\boldsymbol{x}||^2 \geq 0. \end{aligned}$$

Taking square roots proves the lemma. ●

EXAMPLE 8 The sine and cosine functions are continuous on all of $\mathbb{R}$. The key to the proof is first to show that each of these functions is continuous at 0. We assume first that $0 < \theta < \pi/4$ and show that $\lim_{\theta\to 0^+} \sin\theta = 0 = \sin 0$ and that $\lim_{\theta\to 0^+} \cos\theta = 1 = \cos 0$. That is, we show that sine and cosine are continuous from the right at 0. We leave for you the analogous proof that $\lim_{\theta\to 0^-} \sin\theta = 0$ and that $\lim_{\theta\to 0^-} \cos\theta = 1$. Referring to Fig. 3.8, let $\boldsymbol{x} = (1 - \cos\theta, 0)$ and $\boldsymbol{y} = (0, \sin\theta)$. Clearly, $\boldsymbol{x}$ and $\boldsymbol{y}$ are orthogonal. By the lemma, we have $||\boldsymbol{x}|| \leq ||\boldsymbol{x} - \boldsymbol{y}||$ and $||\boldsymbol{y}|| \leq ||\boldsymbol{x} - \boldsymbol{y}||$. In terms of Fig. 3.8, the legs of the right triangle ABC are each shorter than the hypotenuse. As you can compute, in terms of the coordinates, we have

$$\begin{gathered} 0 < ||\boldsymbol{y}|| = \sin\theta \leq (2 - 2\cos\theta)^{1/2} = ||\boldsymbol{x} - \boldsymbol{y}|| \\ \text{and} \\ 0 < ||\boldsymbol{x}|| = 1 - \cos\theta \leq (2 - 2\cos\theta)^{1/2} = ||\boldsymbol{x} - \boldsymbol{y}||. \end{gathered} \tag{3.2}$$

In Euclidean geometry, the shortest distance between two points is given by the straight-line path. Therefore $||\boldsymbol{x} - \boldsymbol{y}|| = (2 - 2\cos\theta)^{1/2}$ is less than the length of

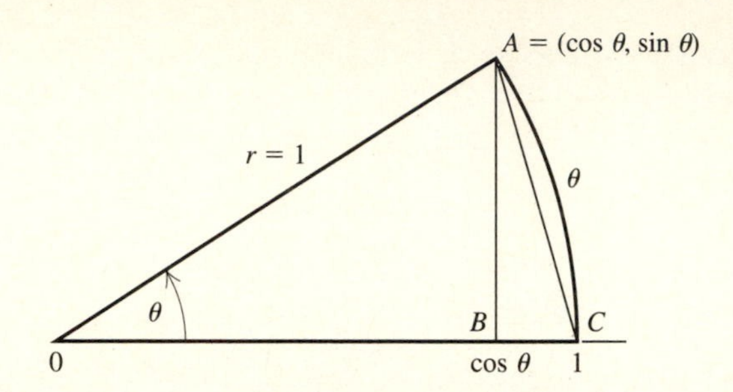

Figure 3.8

the arc $\overset{\frown}{AC}$. Since we measure angles in radians and since the central angle is θ, the length of the arc $\overset{\frown}{AC}$ is θ. Apply these observations to the inequalities (3.2) to obtain

$$0 < \sin\theta \leq \theta \qquad \text{and} \qquad 0 < 1 - \cos\theta \leq \theta.$$

By the Squeeze Play, $0 \leq \lim_{\theta\to 0^+} \sin\theta \leq \lim_{\theta\to 0^+} \theta = 0$, proving that $\lim_{\theta\to 0^+} \sin\theta = 0 = \sin 0$. Therefore the sine function is continuous from the right at 0. Similarly, $0 \leq \lim_{\theta\to 0^+}(1 - \cos\theta) \leq \lim_{\theta\to 0^+} \theta = 0$, proving that $\lim_{\theta\to 0^+} \cos\theta = 1 = \cos 0$. Thus the cosine function is also continuous from the right at 0. As an exercise you may prove that $\lim_{\theta\to 0^-} \sin\theta = 0$ and $\lim_{\theta\to 0^-} \cos\theta = 1$. Thus sine and cosine are continuous at 0.

To prove the continuity of sine and cosine at an arbitrary θ_0 in $\mathbb{R}$, invoke relevant trigonometric identities and Theorem 3.1.3 as in the following computations:

$$\begin{aligned}\lim_{\theta\to\theta_0} \sin\theta &= \lim_{\phi\to 0} \sin(\theta_0 + \phi)\\ &= \lim_{\phi\to 0}[\sin\theta_0 \cos\phi + \cos\theta_0 \sin\phi]\\ &= (\sin\theta_0)\cdot 1 + (\cos\theta_0)\cdot 0 = \sin\theta_0.\end{aligned}$$

This proves that the sine function is continuous on $\mathbb{R}$. Likewise,

$$\begin{aligned}\lim_{\theta\to\theta_0} \cos\theta &= \lim_{\phi\to 0} \cos(\theta_0 + 0)\\ &= \lim_{\phi\to 0}[\cos\theta_0 \cos\phi - \sin\theta_0 \sin\phi]\\ &= (\cos\theta_0)\cdot 1 - (\sin\theta_0)\cdot 0 = \cos\theta_0.\end{aligned}$$

It follows that the cosine function is also continuous on $\mathbb{R}$.

By invoking part (v) of Theorem 3.1.3, we deduce that $\lim_{\theta\to\theta_0} \tan\theta = \lim_{\theta\to\theta_0}[\sin\theta/\cos\theta] = \sin\theta_0/\cos\theta_0 = \tan\theta_0$ and that $\lim_{\theta\to\theta_0} \sec\theta = \lim_{\theta\to\theta_0} 1/\cos\theta = 1/\cos\theta_0 = \sec\theta_0$ for any value of θ_0 where $\cos\theta_0 \neq 0$, that is, provided that θ_0 is not an odd multiple of $\pi/2$. Therefore the tangent and secant functions are continuous except at such points. A completely analogous argument proves that the cotangent and cosecant functions are continuous at all points in $\mathbb{R}$ except integer multiples of π. ●

EXAMPLE 9 A similar application of the Squeeze Play can be used to derive two essential and basic limits,

$$\lim_{\theta \to 0} \frac{\sin \theta}{\theta} = 1 \qquad \text{and} \qquad \lim_{\theta \to 0} \frac{(1 - \cos \theta)}{\theta} = 0.$$

Again, we will assume that $0 < \theta < \pi/4$ and obtain the right-hand limits, leaving for you the analogous derivations of those from the left. In this example, rather than compare lengths, we compare areas. Refer to Fig. 3.9. The triangle AOC, with area $(\sin \theta)/2$, is contained in the sector of the unit circle AOC with area $\theta/2$. The sector, in turn, is contained in the triangle DOC with area $(\tan \theta)/2$. Consequently,

$$\frac{1}{2} \sin \theta \le \frac{1}{2} \theta \le \frac{1}{2} \tan \theta.$$

Multiply these inequalities by the positive number $2/\sin \theta$ to obtain

$$1 \le \frac{\theta}{\sin \theta} \le \frac{1}{\cos \theta}.$$

Taking reciprocals gives

$$\cos \theta \le \frac{\sin \theta}{\theta} \le 1.$$

Finally, passing to the limit, we have

$$1 = \lim_{\theta \to 0^+} \cos \theta \le \lim_{\theta \to 0^+} \frac{\sin \theta}{\theta} \le 1.$$

Therefore, by the Squeeze Play, $\lim_{\theta \to 0^+} (\sin \theta)/\theta = 1$.

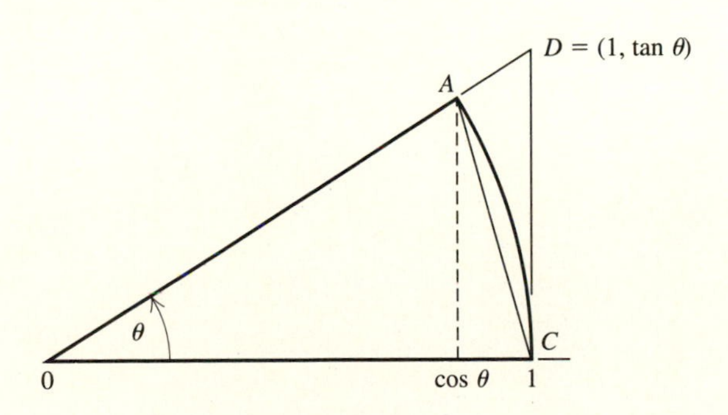

Figure 3.9

The second limit in this example is obtained by using algebraic and trigonometric manipulations and Theorem 3.1.3:

$$\begin{aligned} \lim_{\theta \to 0^+} \frac{1 - \cos \theta}{\theta} &= \lim_{\theta \to 0^+} \frac{(1 - \cos \theta)(1 + \cos \theta)}{\theta(1 + \cos \theta)} \\ &= \lim_{\theta \to 0^+} \frac{\sin^2 \theta}{\theta(1 + \cos \theta)} \end{aligned}$$

$$= \lim_{\theta\to 0^+} \sin\theta \frac{\sin\theta}{\theta}\frac{1}{1+\cos\theta}$$

$$= \lim_{\theta\to 0^+} \sin\theta \lim_{\theta\to 0^+} \frac{\sin\theta}{\theta} \lim_{\theta\to 0^+} \frac{1}{1+\cos\theta}$$

$$= 0 \cdot 1 \cdot \frac{1}{2} = 0.$$

That is, $\lim_{\theta\to 0^+}(1-\cos\theta)/\theta = 0$. ●

3.1.1 Characterizations of Discontinuities in $\mathbb{R}$

Although it is awkward to describe formally every possible way that a function can fail to be continuous at a point $\boldsymbol{c}$ in $\mathbb{R}^n$, it is worthwhile to examine the four types of isolated discontinuity that can occur when f is a function of a single real variable. In the following discussion, f is to be discontinuous at a point c in $\mathbb{R}$ but is to be continuous at all points of some deleted neighborhood $N'(c)$. How can a function fail to be continuous at a point c in $\mathbb{R}$? There are two possibilities. Either (i) $\lim_{x\to c} f(x)$ fails to exist, or (ii) $\lim_{x\to c} f(x)$ exists but fails to equal $f(c)$.

The first possibility can happen in any of three ways. First, it is possible that $\lim_{x\to c} f(x)$ fails to exist because f is unbounded on any deleted neighborhood of c. That is, $\lim_{x\to c} |f(x)| = \infty$. Within this case there are three subcases. It can happen that $\lim_{x\to c} f(x) = \infty$ or that $\lim_{x\to c} f(x) = -\infty$. By the first statement we mean that, for every $M > 0$, there exists a $\delta > 0$ such that, for all x in $S \cap N'(c; \delta)$, we have $f(x) > M$. By the second, we mean that for every $M > 0$, there exists a $\delta > 0$ such that, for all x in $S \cap N'(c; \delta)$, we have $f(x) < -M$. There is also the possibility that, for any $M > 0$, there exists a $\delta > 0$ such that, for all x in $S \cap N'(c; \delta)$, either $f(x) > M$ or $f(x) < -M$. In brief, f may take on both arbitrarily large positive and arbitrarily large negative values in $S \cap N'(c; \delta)$. In all three subcases, we say that f has a *pole* at c. A pole is an *essential discontinuity* in the sense that it describes essential, discontinuous behavior inherent in the function.

EXAMPLE 10 The function $f(x) = 1/x$ has a pole at $c = 0$. More generally, for any $p > 0$, $f(x) = 1/x^p$ has a pole at $c = 0$. We leave the proof of this assertion as an exercise. (See Fig. 3.10.) ●

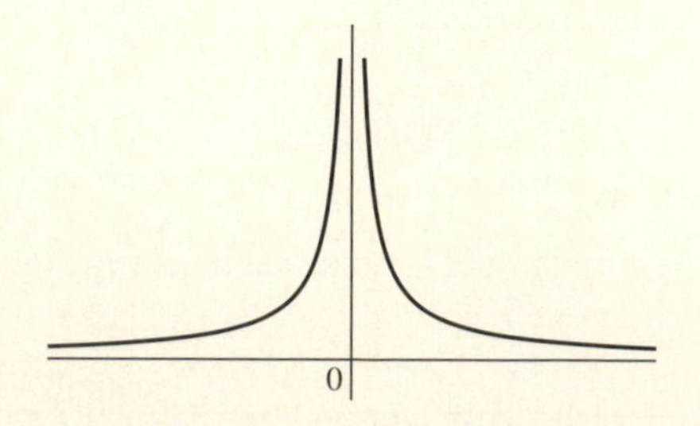

Figure 3.10 $f(x) = x^{-2}$ has a pole at $c = 0$

It is also possible for $\lim_{x \to c} f(x)$ to fail to exist in two additional ways, although f remains bounded. We will consider the simplest, the jump discontinuity, first. If $f(c^-)$ and $f(c^+)$ exist but are not equal, then $\lim_{x \to c} f(x)$ fails to exist; f is said to have a *jump discontinuity* at c. (See Fig. 3.11.)

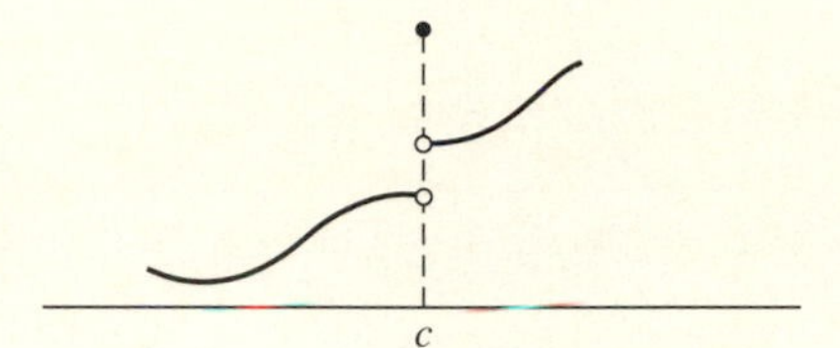

Figure 3.11 f has a jump discontinuity at $x = c$

EXAMPLE 11 For x in $[0, 1)$, let $f(x) = 0$; for x in $(1, 2]$, let $f(x) = 2$. The value of f at $c = 1$ is immaterial. Whatever its value there, f must be discontinuous at c; the left and right-hand limits of f at c differ; therefore, f has a jump discontinuity at $c = 1$. ●

Another type of discontinuity, distinct from the previous two, occurs because the limit fails to exist, although f remains bounded. To see this, consider the following example.

EXAMPLE 12 For $x \neq 0$, let $f(x) = \sin(1/x)$. We can define $f(0)$ any way we may wish, since its value will not affect the discontinuity of f at $c = 0$. For example, let $f(0) = 0$. First, f is bounded; in fact, $|f(x)| \leq 1$ for all x in $\mathbb{R}$ because the sine function has 1 for a bound. But the limit of $\sin(1/x)$ does not exist at 0. We reason as follows. Choose any number b in $[-1, 1]$, and choose y such that $\sin y = b$. Then for any integer k, $\sin(y + 2k\pi) = b$ as well. If we let $x = 1/(y + 2k\pi)$, then $\sin(1/x) = b$. By choosing k sufficiently large, x can be made arbitrarily near 0. Thus in any neighborhood of 0, $\sin(1/x)$ takes on every value between -1 and 1. We conclude that $\lim_{x \to 0} \sin(1/x)$ cannot exist and $\sin(1/x)$ cannot be continuous at $c = 0$. Such a discontinuity is also called an essential discontinuity and results from the oscillatory behavior of f. The function f oscillates dramatically near 0. (See Fig. 3.12 on page 130.) ●

On the other hand, by modifying Example 12 slightly, we can produce a function that oscillates near 0 but is continuous there.

EXAMPLE 13 For $x \neq 0$, define $f(x) = x \sin(1/x)$; define $f(0) = 0$. For all x in $\mathbb{R}$, $|x \sin(1/x)| \leq |x|$, and, therefore, by the Squeeze Play, $\lim_{x \to 0} x \sin(1/x) = 0 = f(0)$. Thus f is continuous at $c = 0$. ●

Our final type of discontinuity is referred to as a *removable discontinuity*. It occurs at a point c in the domain of f where $\lim_{x \to c} f(x)$ exists but has a value different from $f(c)$. The word *removable* is used because, by redefining f at the

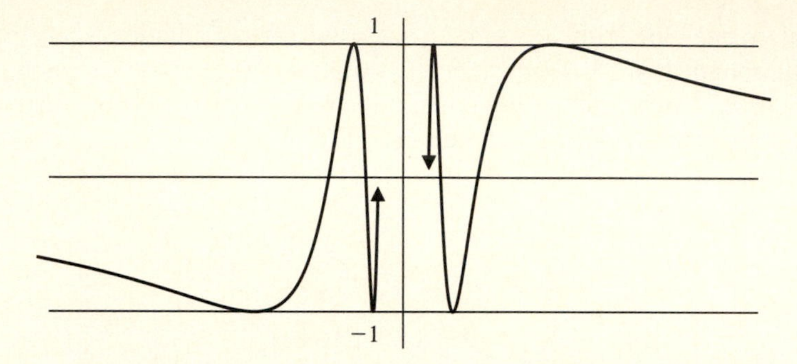

Figure 3.12 $f(x) = \sin \frac{1}{x}$ has an oscillatory discontinuity at $c = 0$

single point c to have the new value $\lim_{x \to c} f(x)$, you can "remove the discontinuity" and produce a new function that is continuous at c and agrees with the original function everywhere else. (See Fig. 3.13.)

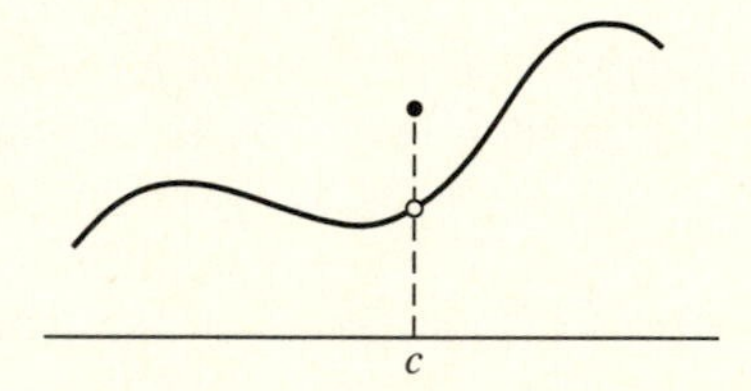

Figure 3.13 f has a removable discontinuity at $x = c$

EXAMPLE 14 For $x \neq 2$, let $f(x) = (x^2 - 4)/(x - 2)$; let $f(2) = 5$. As you can easily see, f is discontinuous at $c = 2$, but the discontinuity is removable. In fact, $\lim_{x \to 2} f(x) = \lim_{x \to 2}(x + 2) = 4$. By redefining f at the single point $c = 2$ so that $f(2) = 4$, we can force the new function to be continuous at $c = 2$. ●

As we have seen, the limiting behavior of a function f at a point $\boldsymbol{c}$ warrants careful attention only when $\boldsymbol{c}$ is a limit point of the domain S of f. When $\boldsymbol{c}$ is such a point, there will exist a Cauchy sequence $\{\boldsymbol{x}_k\}$ in S that converges to $\boldsymbol{c}$. Immediately there arises a second sequence $\{f(\boldsymbol{x}_k)\}$. We ask the natural question, Does this latter sequence necessarily converge and, if so, to what limiting value? The question is resolved completely by the following theorem.

THEOREM 3.1.6 Let S be a nonempty subset of $\mathbb{R}^n$, $\boldsymbol{c}$ a limit point of S, and f a real-valued function with domain S.

i) $\lim_{\boldsymbol{x} \to \boldsymbol{c}} f(\boldsymbol{x}) = L$ if and only if, for every Cauchy sequence $\{\boldsymbol{x}_k\}$ in S such that $\lim_{k \to \infty} \boldsymbol{x}_k = \boldsymbol{c}$, it follows that $\lim_{k \to \infty} f(\boldsymbol{x}_k) = L$.

ii) Consequently, if $\boldsymbol{c}$ is in S, then f is continuous at $\boldsymbol{c}$ if and only if, for every Cauchy sequence $\{\boldsymbol{x}_k\}$ of points of S such that $\lim_{k \to \infty} \boldsymbol{x}_k = \boldsymbol{c}$, it follows that $\lim_{k \to \infty} f(\boldsymbol{x}_k) = f(\boldsymbol{c})$. ●

Developing a proof of this theorem as an exercise will provide you with a good test of your understanding.

Let f be a function with domain S in $\mathbb{R}^n$. At each interior point $\boldsymbol{c} = (c_1, c_2, \ldots, c_j, \ldots, c_n)$ of S, there are n implicitly-defined functions, one for each coordinate, that are related to f. For each $j = 1, 2, \ldots, n$, let

$$g_j(t) = f(c_1, c_2, \ldots, c_{j-1}, t, c_{j+1}, \ldots, c_n).$$

The function g_j is obtained from f by holding all but the jth coordinate fixed and allowing that jth coordinate to vary. Our interest centers on this question: What limiting properties does g_j inherit from f? The entire answer is contained in the following theorem.

THEOREM 3.1.7 Let f be a real-valued function defined on a set S in $\mathbb{R}^n$ and let $\boldsymbol{c} = (c_1, c_2, \ldots, c_n)$ be a point in S^0. For $j = 1, 2, \ldots, n$, let

$$g_j(t) = f(c_1, c_2, \ldots, c_{j-1}, t, c_{j+1}, \ldots, c_n).$$

i) If $\lim_{\boldsymbol{x}\to\boldsymbol{c}} f(\boldsymbol{x}) = L$, then, for each j, $\lim_{t\to c_j} g_j(t) = L$.

ii) If f is continuous at $\boldsymbol{c}$, then, for each j, the function g_j is also continuous at c_j with $\lim_{t\to c_j} g_j(t) = f(\boldsymbol{c})$.

Proof. For (i), since $\boldsymbol{c}$ is an interior point of S, we can choose a $\delta_0 > 0$ such that $N(\boldsymbol{c}; \delta_0) \subseteq S$. Fix $\epsilon > 0$. Choose any $\delta_1 > 0$ such that, if $\boldsymbol{x}$ is in $S \cap N'(\boldsymbol{c}; \delta_1)$, then $f(\boldsymbol{x})$ is in $N(L; \epsilon)$. Let $\delta = \min\{\delta_0, \delta_1\}$. Then $N(\boldsymbol{c}; \delta)$ is contained entirely in S and any $\boldsymbol{x}$ in $N'(\boldsymbol{c}; \delta)$ is mapped by f into $N(L; \epsilon)$. (See Fig. 3.14 on page 132.) Choose any t in $N'(c_j; \delta)$ and form the point $\boldsymbol{x} = (c_1, c_2, \ldots, c_{j-1}, t, c_{j+1}, \ldots, c_n)$. It is easy to see that $\boldsymbol{x}$ is in $N'(\boldsymbol{c}; \delta)$; merely compute

$$0 < ||\boldsymbol{x} - \boldsymbol{c}||^2 = (t - c_j)^2 < \delta^2.$$

Consequently,

$$|f(\boldsymbol{x}) - L| < \epsilon.$$

But for this choice of $\boldsymbol{x}$, we have $f(\boldsymbol{x}) = g_j(t)$. Therefore whenever t is in $N'(c_j; \delta)$, it follows that $g_j(t)$ is in $N(L; \epsilon)$. This proves that $\lim_{t\to c_j} g_j(t) = L$. We leave part (ii) for you to prove as an exercise. ●

It is important for you to understand that the converse of Theorem 3.1.7 is false. To elaborate, suppose that a function f on S in $\mathbb{R}^n$ gives rise to the coordinate functions $g_1, g_2, \ldots, g_n$ at a point $\boldsymbol{c} = (c_1, c_2, \ldots, c_n)$ as defined in the theorem. Suppose also that, for each $j = 1, 2, \ldots, n$, $\lim_{t\to c_j} g_j(t)$ exists and equals L. Can we necessarily conclude that $\lim_{\boldsymbol{x}\to\boldsymbol{c}} f(x) = L$? While you may wish to believe that this is true, the answer is an emphatic no. The reason is that the limiting behavior of the functions g_j merely reflects the limiting behavior of f along each of the n straight lines passing through the point $\boldsymbol{c}$ and parallel to the n coordinate axes. Consider the geometric implications of Theorem 3.1.6: The limiting behavior of f itself near $\boldsymbol{c}$ can be described only by analyzing the behavior of f along all possible paths of approach to $\boldsymbol{c}$, any of which can be followed by a Cauchy sequence $\{\boldsymbol{x}_k\}$

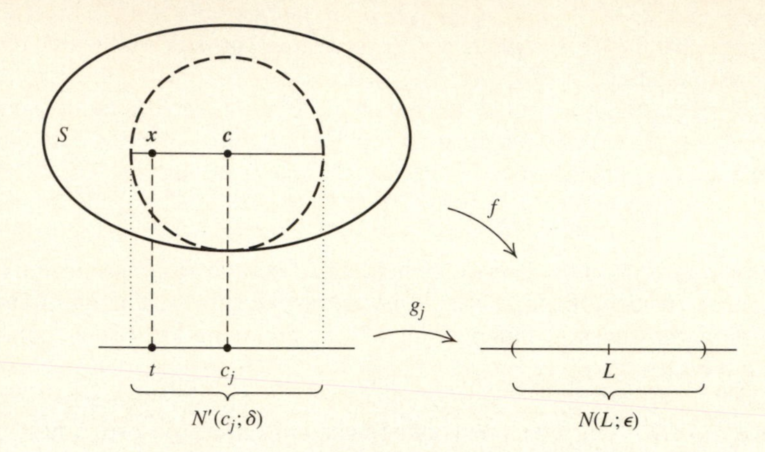

Figure 3.14

with limit $\mathbf{c}$. Infinitely many paths of approach exist other than those parallel to the coordinate axes and the limiting behavior of f may differ on any of them. We ask you to explore this question further in the exercises.

EXERCISES

3.1. Suppose that f is a real-valued function on S in $\mathbb{R}^n$ and that $\mathbf{c}$ is an isolated point of S. Prove that f must be continuous at $\mathbf{c}$.

3.2. Suppose that f is a real-valued function with domain S in $\mathbb{R}$ and that c is an interior point of S. Prove that $\lim_{x\to c} f(x) = L$ if and only if $f(c^-) = f(c^+) = L$. Prove also that f is continuous at c if and only if $f(c^-) = f(c^+) = f(c)$.

3.3. Suppose $f(x) = mx + b$ is a linear function with $m \neq 0$. Prove that f is continuous on all of $\mathbb{R}$. Moreover, show that for any $\epsilon > 0$, the same value of $\delta = \epsilon/|m|$ applies at every point c to ensure the continuity of f.

3.4. Prove parts (iv) and (v) of Theorem 3.1.3.

3.5. **a)** Prove by induction that, for any m in $\mathbb{N}$, $f(x) = x^m$ is continuous on $\mathbb{R}$.

b) Prove by induction that every polynomial $p(x) = \sum_{j=0}^{r} c_j x^j$ is continuous on $\mathbb{R}$.

c) Prove that every rational function $f(x) = p(x)/q(x)$ is continuous at all points where $q(x)$ does not vanish.

3.6. Prove that $f(x) = 1/x^2$ is continuous on $S = (0, \infty)$. For a fixed $\epsilon > 0$ and each c in S, find the largest $\delta_c > 0$ that will work to ensure the continuity of f at c. Holding ϵ fixed, show that $\lim_{c\to 0^+} \delta_c = 0$.

3.7. Prove that, for any rational $p > 0$, the function $f(x) = 1/x^p$ on $(0, \infty)$ has a pole at $c = 0$.

3.8. Let m/n be any rational number. Define $f(x) = x^{m/n}$.

a) Prove that f is continuous at every $c > 0$.

b) If m/n is positive, prove that f is continuous from the right at $c = 0$.

3.9. **a)** Prove that $\lim_{\theta\to 0^-} \sin\theta = 0$.

b) Prove that $\lim_{\theta \to 0^-} \cos\theta = 1$.

3.10. Prove that the cotangent and cosecant functions are continuous at every point in $\mathbb{R}$ not in $\{k\pi : k \text{ in } \mathbb{Z}\}$.

3.11. **a)** Prove that $\lim_{\theta \to 0^-} \sin\theta/\theta = 1$.

b) Prove that $\lim_{\theta \to 0^-} (1 - \cos\theta)/\theta = 0$.

3.12. Find each limit, if it exists; if a limit fails to exist, prove it.

a) $\lim_{x \to c} \dfrac{\sin(x - c)}{(x^2 - c^2)}$

b) $\lim_{x \to 0} \dfrac{(1 - \cos x)}{x^2}$

c) $\lim_{x \to 1} \dfrac{[(1 + x)^{1/2} - (1 - x)^{1/2}]}{x}$

d) $\lim_{x \to 0}[x + \operatorname{sgn}(x)]$, where

$$\operatorname{sgn}(x) = \begin{cases} -1, & \text{if } x < 0 \\ 0, & \text{if } x = 0 \\ 1, & \text{if } x > 0. \end{cases}$$

3.13. For x in $\mathbb{R}$, define $f(x) = 0$ if x is irrational and $f(x) = 1/q$ if $x = p/q$ in lowest terms. Prove that f is continuous at every irrational point and is discontinuous at every rational point.

3.14. For each function f, identify the domain S, identify any points in $\overline{S}$ where f has a limit, find all points where f is continuous, and find all points where f is discontinuous.

a) $f(x) = 1/(x^2 - 1)$.

b) $f(x) = (\sin x)/x$ if $x \neq 0$, $f(0) = 1$.

c) $f(x) = \sin x/\sqrt{x}$, if $x \neq 0$; $f(0) = 1$.

d) $f(x) = (1 - \cos^2 x)/x$, if $x \neq 0$; $f(0) = 0$.

e) $f(x) = x - \lfloor x \rfloor$, where $\lfloor x \rfloor$ is the largest integer $k \leq x$.

f)

$$f(x) = \operatorname{sgn}(x) = \begin{cases} -1, & x < 0 \\ 0, & x = 0 \\ 1, & x > 0. \end{cases}$$

3.15. **a)** Show that, for each k in $\mathbb{N}$, $f_k(x) = 1/(1 + x^{2k})$ is continuous on all of $\mathbb{R}$.

b) Define

$$f_0(x) = \begin{cases} 0, & \text{for } |x| < 1 \\ 1/2, & \text{for } |x| = 1 \\ 1, & \text{for } |x| > 1. \end{cases}$$

Sketch the graph of f_0. Identify where it is continuous, and identify the type and location of discontinuities of f_0. Prove your assertions.

c) Superimpose on the graph of f_0, the graphs of f_k for $k = 1, 2, 3$, and 4.

d) Fix any x in $\mathbb{R}$ and consider the sequence $\{y_k\}$ where, for each k in $\mathbb{N}$, $y_k = f_k(x)$. Show that, whatever the choice of x, the sequence $\{y_k\}$ converges. (Consider the three cases: $|x| > 1$, $|x| = 1$, and $|x| < 1$ separately.)

e) Show that for each x in $\mathbb{R}$, $\lim_{k \to \infty} f_k(x) = f_0(x)$.

3.16. For each $k = 0, 1, 2, \ldots$ and for x in $[-1, 1]$, define $f_k(x) = \cos[k \cos^{-1} x]$.

a) Show that $|f_k(x)| \leq 1$ for all x in $[-1, 1]$, all $k \geq 0$.

b) Let $\theta = \cos^{-1} x$ and use trigonometric identities for $\cos(k+1)\theta$ and $\cos(k-1)\theta$ to show that, for $k \geq 1$,

$$f_{k+1}(x) = 2xf_k(x) - f_{k-1}(x).$$

c) Calculate $f_0(x)$, $f_1(x)$, $f_2(x)$, and $f_3(x)$ as polynomials in x.

d) By induction show that f_k is a polynomial of degree k, with leading coefficient 2^{k-1}. Hence, each f_k is continuous on $[-1, 1]$.

e) Prove that $f_k(1) = 1$ and $f_k(-1) = (-1)^k$ for all $k \geq 0$.

f) Prove that $f_k(x)$ has k roots in the interval $(-1, 1)$. Find them.

g) Sketch the graphs of f_k for $k = 0, 1, 2, \ldots, 6$. The polynomials f_k are variants of *Chebychev's polynomials*.

3.17. Fix any $m > 0$. Define f on $[0, 1]$ to be $f(x) = mx$ if x is rational and $f(x) = m(1-x)$ otherwise. Prove that f is continuous only at $x = 1/2$.

3.18. Let f be continuous on $[a, b]$. Define a function g as follows: $g(a) = f(a)$ and, for x in $(a, b]$,

$$g(x) = \sup\{f(y) : y \text{ in } [a, x]\}.$$

Prove that g is monotone increasing and continuous on $[a, b]$.

3.19. Fix $a > 1$ and $c > 0$. The purpose of this exercise is to define the exponential a^c. If $c = p/q$ is rational, then a^c is already defined to be $(a^p)^{1/q}$, the positive qth root of a^p. For positive, rational c, $a^c > 1$ because $a > 1$.

a) Prove that, if c_1 and c_2 are rational numbers such that $0 < c_1 < c_2$, then $1 < a^{c_1} < a^{c_2}$.

Our interest centers on defining a^c when c is irrational and positive. Let $\{c_k\}$ be any Cauchy sequence of positive, rational numbers such that $\lim_{k\to\infty} c_k = c$. For each k in $\mathbb{N}$, $d_k = a^{c_k}$ is defined. The key to our construction is the fact that $\{d_k\}$ is a Cauchy sequence of real numbers. The following is a programmed proof of this fact.

b) Show that there exists an $M > 0$ such that, for any indices k and m,

$$|d_m - d_k| \leq M(a^{|c_m - c_k|} - 1).$$

(In $|d_m - d_k|$, factor out a^{c_0} where $c_0 = \min\{c_m, c_k\}$.)

c) Explain why, given $\epsilon > 0$, there exists an r_0 in $\mathbb{N}$ such that, for $r \geq r_0$ with r in $\mathbb{N}$, it follows that

$$0 < a^{1/r} - 1 < \frac{\epsilon}{M}.$$

(See Exercise 1.37.)

d) Fix $r \geq r_0$. Explain why there exists a k_0 such that, for $k, m \geq k_0$, it follows that $0 \leq |c_m - c_k| < 1/r$.

e) Show how to conclude that $\{d_k\}$ is a Cauchy sequence.

f) Define $a^c = \lim_{k\to\infty} d_k = \lim_{k\to+\infty} a^{c_k}$. Prove that a^c is *well defined*. (In mathematics the phrase *well defined* has a technical meaning. In this instance it means that regardless of the Cauchy sequence $\{c_k\}$ of positive, rational numbers initially chosen with $\lim_{k\to\infty} c_k = c$, the resulting numerical value for a^c is the same.)

g) Show how to extend the definition of a^c to all values of c in $\mathbb{R}$, consistent with the usual "laws of exponents."

h) Prove that the resulting function $f(x) = a^x$ is continuous on all of $\mathbb{R}$.

i) Prove that, if $a > 1$, then the function a^x is strictly monotone increasing on $\mathbb{R}$. That is, if $x < y$, then $a^x < a^y$.

3.20. Suppose that f is a real-valued function defined on all of $\mathbb{R}$ and satisfying the identity

$$f(x + y) = f(x)f(y),$$

for all x, y in $\mathbb{R}$.

a) Prove that $f(x) \geq 0$ for all x in $\mathbb{R}$.

b) Prove that, if f has value 0 at even one point, then f is identically 0 on $\mathbb{R}$.

c) Prove that, if f has a nonzero value at even one point, then $f(0) = 1$.

d) Prove that, if f is continuous at $x = 0$, then f is continuous on all of $\mathbb{R}$.

e) Prove that, if f is continuous and nonzero, then $f(x) = a^x$, where $a = f(1)$.

3.21. A function f on a set S in $\mathbb{R}^n$ is said to *satisfy a Lipschitz condition of order k* at a point $\boldsymbol{c}$ in S if there exists a constant K and neighborhood $N(\boldsymbol{c})$ such that, for all $\boldsymbol{x}$ in $N(\boldsymbol{c})$,

$$|f(\boldsymbol{x}) - f(\boldsymbol{c})| \leq K||\boldsymbol{x} - \boldsymbol{c}||^k.$$

a) Prove that if f satisfies a Lipschitz condition of order $k > 0$ at some $\boldsymbol{c}$ in its domain, then f is continuous at $\boldsymbol{c}$.

b) Show that a function f on $\mathbb{R}$ satisfies a Lipschitz condition of order $k = 1$ at a point c if and only if, for x in a neighborhood of c, the graph of f lies between the lines $y = K(x - c) + f(c)$ and $y = -K(x - c) + f(c)$.

c) Find an example of a function f on $\mathbb{R}$ that satisfies a Lipschitz condition of order $k = 2$ at $c = 0$. Find a function on $\mathbb{R}$ that satisfies a Lipschitz condition of order $k = 1/2$ at $c = 1$.

3.22. Let $f(x) = (x - 1)/(x^2 - 1)^{1/2}$. Determine the domain S of f. Find $\lim_{x \to c} f(x)$ for each c in S. Where is f continuous? Can you define $f(1)$ to make f continuous at $c = 1$? Can you define $f(-1)$ to make f continuous at $c = -1$? Explain.

3.23. (See Example 5 of Section 2.2 and Exercise 2.42.) Let C denote Cantor's set. Any x in C has a tricimal expansion of the form $x = .t_1t_2t_3 \cdots$ with $t_k = 0$ or 2. Define $f(x) = \sum_{j=1}^{\infty} t_j/2^j$.

a) Show that, if x and y are in C with $x < y$, then

$$f(x) < f(y).$$

b) Show that f maps C onto [0, 1]. (*Hint:* Every number in [0, 1] has a binary expansion $.b_1b_2b_3 \cdots$ where $b_k = 0$ or 1.)

c) Show that, if $a(k, j)$ and $b(k, j)$ are the endpoints of any of the intervals $I(k, j)$ removed from [0, 1] in the construction of Cantor's set, then $f(a(k, j)) = f(b(k, j))$. (Thus, for example, $f(1/3) = f(2/3) = 1/2$, $f(1/9) = f(2/9) = 1/4$, $f(7/9) = f(8/9) = 1/2 + 1/4 = 3/4, \ldots$.)

d) Extend f to be defined on $C^c \cap [0, 1]$ as follows: For x in $I(k, j)$, define $f(x) = f(a(k, j)) = f(b(k, j))$. Thus f is constant on each of the open intervals $I(k, j)$. With this definition, f maps all of [0, 1] onto [0, 1]. Prove that f is continuous on [0, 1]. Sketch the graph of f. The function f is called *Cantor's function* or *the devil's staircase*.

3.24. For $\boldsymbol{x} \neq \mathbf{0}$ in $\mathbb{R}^n$, let $f(\boldsymbol{x}) = \cos(1/||\boldsymbol{x}||)$. Prove that f is continuous at all points where it is defined. Does $\lim_{\boldsymbol{x}\to\mathbf{0}} f(\boldsymbol{x})$ exist? Can you extend the definition of f so that f is continuous at $\mathbf{0}$ as well? Explain.

3.25. For $\boldsymbol{x} = (x_1, x_2) \neq \mathbf{0}$ in $\mathbb{R}^2$, let

$$f(\boldsymbol{x}) = \frac{x_1^2 x_2^2}{x_1^2 + x_2^2}.$$

Define $f(\mathbf{0}) = 0$. Let $\boldsymbol{c} = \mathbf{0}$ and define $g_1(t) = f(t, 0)$ and $g_2(t) = f(0, t)$. Show explicitly, without appealing to Theorem 3.1.7, that for each $j = 1, 2$, $\lim_{t\to 0} g_j(t) = f(\mathbf{0}) = 0$.

3.26. For $\boldsymbol{x} = (x_1, x_2) \neq \mathbf{0}$ in $\mathbb{R}^2$, let

$$f(\boldsymbol{x}) = \frac{2x_1 x_2}{x_1^2 + x_1^2}.$$

Let $f(\mathbf{0}) = 0$. Let $\boldsymbol{c} = \mathbf{0}$ and define $g_1(t) = f(t, 0)$ and $g_2(t) = f(0, t)$. Prove that, for $j = 1, 2$, $\lim_{t\to 0} g_j(t) = f(\mathbf{0})$. Does it follow that $\lim_{\boldsymbol{x}\to\mathbf{0}} f(\boldsymbol{x}) = 0$? (See Example 3.)

3.27. For $\boldsymbol{x} = (x_1, x_2)$ in $\mathbb{R}^2$, define

$$f(\boldsymbol{x}) = \begin{cases} \dfrac{x_1^2 - x_2^2}{x_1^2 + x_2^2}, & \text{for } \boldsymbol{x} \neq \mathbf{0} \\ 0, & \text{for } \boldsymbol{x} = \mathbf{0}. \end{cases}$$

a) Determine whether $\lim_{\boldsymbol{x}\to\mathbf{0}} f(\boldsymbol{x})$ exists. If so, find it; otherwise, explain why not. In either case, prove your assertion.

b) Determine where in $\mathbb{R}^2$ the function f is continuous.

c) Let $\boldsymbol{c} = \mathbf{0}$. Define $g_1(t) = f(t, 0)$ and $g_2(t) = f(0, t)$. For $j = 1, 2$, find $\lim_{t\to 0} g_j(t)$. Are these two limits equal? Does your answer say anything about $\lim_{\boldsymbol{x}\to\mathbf{0}} f(\boldsymbol{x})$?

3.28. Define f on the open first quadrant $(0, \infty) \times (0, \infty)$ by $f(\boldsymbol{x}) = x_1/x_2$, where $\boldsymbol{x} = (x_1, x_2)$. Examine the limiting behavior of f as $\boldsymbol{x}$ approaches $\mathbf{0}$ along various paths. Prove that $\lim_{\boldsymbol{x}\to\mathbf{0}} f(\boldsymbol{x})$ does not exist. Sketch the graph of f.

3.29. Find a function f, other than examples in the text, for which, at some point $\boldsymbol{c} = (c_1, c_2, \ldots, c_n)$ in $\mathbb{R}^n$, the limit of f does not exist but for which the coordinate functions $g_1, g_2, \ldots, g_n$ have the property that $\lim_{t\to c_j} g_j(t)$ are all equal for $j = 1, 2, \ldots, n$.

3.30. Prove that, if f is a real-valued function that is continuous at a point $\boldsymbol{c} = (c_1, c_2, \ldots, c_n)$ in $\mathbb{R}^n$ and if g_j denotes the jth coordinate function

$$g_j(t) = f(c_1, c_2, \ldots, c_{j-1}, t, c_{j+1}, \ldots, c_n),$$

then g_j is continuous at c_j.

3.31. Let f be a real-valued function defined on a rectangle $S = I_1 \times I_2$ in $\mathbb{R}^2$, where $I_j = [a_j, b_j]$. Assume that f factors into a product $f(x_1, x_2) = f_1(x_1) f_2(x_2)$ of functions f_1 on I_1 and f_2 on I_2. Assume also that f_1 is continuous at c_1 in I_1 and that f_2 is continuous at c_2 in I_2. Is f necessarily continuous at $\boldsymbol{c} = (c_1, c_2)$? If so, prove it; otherwise, provide a counterexample.

3.32. Suppose that f is a real-valued function defined on a set S in $\mathbb{R}^n$ and that $\boldsymbol{c}$ is a limit point of S. Prove that $\lim_{\boldsymbol{x}\to\boldsymbol{c}} f(\boldsymbol{x}) = L$ if and only if, for every Cauchy sequence $\{\boldsymbol{x}_k\}$ in S that converges to $\boldsymbol{c}$, $\lim_{k\to\infty} f(\boldsymbol{x}_k) = L$.

3.33. Use Exercise 3.32 to prove that f is continuous at a point $\boldsymbol{c}$ in S if and only if, for every Cauchy sequence $\{\boldsymbol{x}_k\}$ in S which converges to $\boldsymbol{c}$, $\lim_{k\to\infty} f(\boldsymbol{x}_k) = f(\boldsymbol{c})$.

3.34. For each function f identify the domain S in $\mathbb{R}^2$, identify all points in S where f has a limit, find all points in S where f is continuous, and find all points where f is discontinuous. In all cases, $\boldsymbol{x} = (x_1, x_2)$.

a) $f(\boldsymbol{x}) = 1/(x_1 - x_2)$

b) $f(\boldsymbol{x}) = x_1[1 - 1/(x_1x_2)]$

c) $f(\boldsymbol{x}) = [1 - x_1^2 - x_2^2]^{1/2}$

d) $f(\boldsymbol{x}) = |x_1 - x_2|/(x_1^2 - 2x_1x_2 + x_2^2)$

e) $f(\boldsymbol{x}) = x_1/[x_1^2 + x_2^2]^{1/2}$

f) $f(\boldsymbol{x}) = (x_1 - x_2)/[x_1^2 + x_2^2]^{1/2}$

g) $f(\boldsymbol{x}) = \sin[1/(x_1^2 + x_2^2)^{1/2}]$

h) $f(\boldsymbol{x}) = [\sin(x_1x_2)]/[x_1^2 + x_2^2]^{1/2}$

i) $f(\boldsymbol{x}) = \sin(x_1/x_2)$

3.35. Prove that the limit operation is order-preserving. Suppose that f_1 and f_2 are real-valued functions defined on a common domain S in $\mathbb{R}^n$ and that $\boldsymbol{c}$ is a limit point of S. Assume that f_1 and f_2 have limits L_1 and L_2, respectively, at $\boldsymbol{c}$. Assume also that there exists a neighborhood $N(\boldsymbol{c}; \delta_0)$ such that $f_1(\boldsymbol{x}) \le f_2(\boldsymbol{x})$ for all $\boldsymbol{x}$ in $S \cap N'(c; \delta_0)$. Prove that $L_1 \le L_2$.

3.2 The Topological Description of Continuity

Here we explore a powerful way to describe the continuity of a real-valued function f on a subset S of $\mathbb{R}$ or, more generally, of $\mathbb{R}^n$. Suppose that f maps S continuously onto $T = f(S)$. Let U be any relatively open set in T. We claim that $f^{-1}(U)$ is relatively open in S. We will prove this by showing that every point $\boldsymbol{c}$ in $f^{-1}(U)$ is a relative interior point of S—that is, that there exists some neighborhood $N(\boldsymbol{c})$ of $\boldsymbol{c}$ such that $S \cap N(\boldsymbol{c}) \subseteq f^{-1}(U)$.

Let's get oriented. The point $\boldsymbol{c}$ is in S and f is continuous at $\boldsymbol{c}$. Also, $f(\boldsymbol{c})$ is in U, a subset of T. Since U is relatively open, there is a neighborhood $N(f(\boldsymbol{c}))$ such that $T \cap N(f(\boldsymbol{c})) \subseteq U$. Since f is continuous at $\boldsymbol{c}$, there must exist a neighborhood $N(\boldsymbol{c})$ such that $S \cap N(\boldsymbol{c})$ is contained entirely in $f^{-1}(N(f(\boldsymbol{c}))$ that, in turn, is contained entirely in $f^{-1}(U)$. That is, $S \cap N(\boldsymbol{c}) \subseteq f^{-1}(U)$. Thus $S \cap N(\boldsymbol{c})$ is a relatively open neighborhood of $\boldsymbol{c}$ contained in the inverse image of U. Therefore $\boldsymbol{c}$ is a relative interior point of $f^{-1}(U)$. Since $\boldsymbol{c}$ is any point of $f^{-1}(U)$, it follows that $f^{-1}(U)$ is relatively open in S. This proves that, if f is continuous on S, then the inverse image of every relatively open set in $T = f(S)$ is relatively open in S.

The converse is also true. If the inverse image under f of every relatively open set U in T is relatively open in S, then f is continuous on S. For suppose that $\boldsymbol{c}$ is any point of S. Consider the relatively open neighborhood $T \cap N(f(\boldsymbol{c}))$ in $T = f(S)$. Its inverse image under f is assumed to be relatively open in S. The point $\boldsymbol{c}$ belongs to $f^{-1}(T \cap N(f(\boldsymbol{c})))$, a relatively open set in S of which, therefore, $\boldsymbol{c}$ must be a relative interior point. Thus there exists a neighborhood $N(\boldsymbol{c})$ such that $S \cap N(\boldsymbol{c}) \subseteq f^{-1}(N(f(c)))$. But this is precisely the statement that f is continuous at $\boldsymbol{c}$. Since $\boldsymbol{c}$ is an arbitrary point of S, we conclude that f must be continuous on S. We have proved the following theorem.

THEOREM 3.2.1 A function f, mapping the subset S of $\mathbb{R}^n$ onto T, is continuous on S if and only if the inverse image of every relatively open set in T is relatively open in S. ●

With this theorem in hand we are at last in a position to prove several of the most important facts about continuous functions. Here is the first.

THEOREM 3.2.2 If S is a connected subset of $\mathbb{R}^n$ and if f is continuous on S, then $T = f(S)$ is also connected.

Proof. Suppose, to the contrary, that T is disconnected. Then there exist two nonempty, disjoint open sets U and V in $\mathbb{R}$ such that (i) $U \cap T \neq \emptyset$ and $V \cap T \neq \emptyset$, and (ii) $T \subseteq U \cup V$. Since f is continuous on S, $U_1 = f^{-1}(U)$ and $V_1 = f^{-1}(V)$ are nonempty and relatively open in S. As you may show in the exercises, (i) $U_1 \cap V_1 = \emptyset$ and (ii) $S = U_1 \cup V_1$. But these conditions imply that, contrary to our hypothesis, S is not connected. The contradiction implies that T is connected. This proves the theorem. ●

Intuitively you know, or at least believe, that if f is continuous on the interval $S = [a, b]$, if

$$m = \min\{f(x) : x \text{ in } S\}$$

and

$$M = \max\{f(x) : x \text{ in } S\},$$

then $f([a, b]) = [m, M]$. Expressed in terms of the concepts we have developed, the compact set $S = [a, b]$ is mapped by the continuous function f onto the compact set $[m, M]$. Here we prove that the continuous image of a compact set is compact.

Suppose that S is any compact set in $\mathbb{R}^n$ and that f is a continuous, real-valued function on S. Let $T = f(S)$. We want to show that T is compact. To this end, let $\{U_\alpha : \alpha \text{ in } A\}$ be any open cover of T. We want to extract a finite subcover of T from this arbitrary open cover. To do so, we will have to use the linkage provided by the continuous function f to shift back to S and then use the compactness of S. This observation determines our entire strategy.

Since f is assumed to be continuous and, for each α in A, U_α is assumed to be open, $f^{-1}(U_\alpha)$ is a relatively open set in S. Hence the collection $\{f^{-1}(U_\alpha) : \alpha \text{ in } A\}$ forms a (relatively) open cover of the compact set S. We extract a finite subcover $\{f^{-1}(U_{\alpha_1}), f^{-1}(U_{\alpha_2}), \ldots, f^{-1}(U_{\alpha_p})\}$ of S. We claim that $\{U_{\alpha_1}, U_{\alpha_2}, \ldots, U_{\alpha_p}\}$ must form a finite subcover of T. To prove this, all we need show is that each y in $T = f(S)$ is in $\cup_{j=1}^{p} U_{\alpha_j}$. Suppose, then, that y is any point of T. There exists an $\boldsymbol{x}$ in S such that $f(\boldsymbol{x}) = y$. The point $\boldsymbol{x}$ is in at least one of the sets $f^{-1}(U_{\alpha_j})$. Thus

$y = f(\mathbf{x})$ is in U_{α_j}. Since y is arbitrary, we deduce that $T \subseteq \cup_{j=1}^{p} U_{\alpha_j}$. Therefore, $\{U_{\alpha_1}, U_{\alpha_2}, \ldots, U_{\alpha_p}\}$ is a finite subcover of T chosen from the arbitrary open cover $\{U_\alpha : \alpha \text{ in } A\}$ of T. This proves that T must be compact and proves the following theorem.

THEOREM 3.2.3 If f is a continuous, real-valued function on a compact set S in $\mathbb{R}^n$, then $f(S)$ is compact. ●

Notice that our proof is entirely independent of the dimensionality of the domain S. Our theorem is a deep and useful fact of analysis, one we will use repeatedly in the sequel. As immediate consequences we have the following several theorems.

THEOREM 3.2.4 If S is a compact subset of $\mathbb{R}^n$ and if f is a continuous, real-valued function on S, then f has a minimum value and a maximum value on S and f is bounded on S.

Proof. Since S is compact, by Theorem 3.2.3, $T = f(S)$ is also compact and is therefore a closed and bounded subset of $\mathbb{R}$. Since T is bounded, $m = \inf T$ and $M = \sup T$ both exist in $\mathbb{R}$. Since T is closed, m and M both belong to T. Therefore points x_1 and x_2 exist in S so that $f(x_1) = m$ and $f(x_2) = M$. That is, f assumes its minimum and its maximum values on S. ●

THEOREM 3.2.5 The Intermediate Value Theorem

i) If f is a continuous, real-valued function on $[a, b]$ and if $f(a)f(b) < 0$, then f must have value 0 at some point in $[a, b]$.

ii) If f is a continuous, real-valued function on $[a, b]$ and if c is any number between $f(a)$ and $f(b)$, then there exists an x in $[a, b]$ such that $f(x) = c$.

Proof. The assumption $f(a)f(b) < 0$ is simply a quick way to say that f has values of opposite sign, one negative, one positive, at the endpoints a and b. As a consequence,

$$m = \min\{f(x) : x \text{ in } [a, b]\} < 0$$

and

$$M = \max\{f(x) : x \text{ in } [a, b]\} > 0.$$

Since $[a, b]$ is connected and f is continuous, it follows from Theorem 3.2.1 that $f([a, b])$ is also connected. By Theorem 3.2.2, m and M are, respectively, the smallest and the largest numbers in $f([a, b])$. Since $m < 0$ and $M > 0$, the interval $[m, M]$ contains 0. Consequently, there exists an x in $[a, b]$ such that $f(x) = 0$. (See Fig. 3.15.)

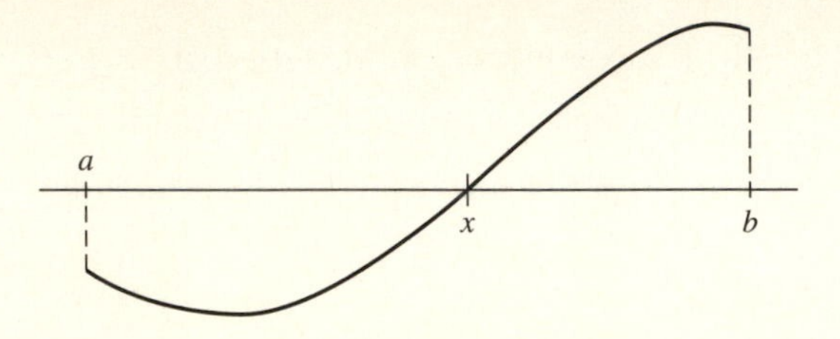

Figure 3.15

For the second assertion of the theorem, replace f by the continuous, real-valued function $g = f - c$ on $[a, b]$. Since c lies between $f(a)$ and $f(b)$, it follows that $g(a)g(b) < 0$. Therefore, by the first part of the theorem, there exists an x in $[a, b]$ such that $g(x) = 0$. Thus $f(x) = c$. ●

THEOREM 3.2.6 The Generalized Intermediate Value Theorem

i) If S is any compact, connected set in $\mathbb{R}^n$, if f is any continuous, real-valued function on S, and if f assumes both positive and negative values on S, then f has value 0 at some point of S.

ii) If S is any compact, connected set in $\mathbb{R}^n$, if f is any continuous, real-valued function on S, if $f(\boldsymbol{x}_1)$ and $f(\boldsymbol{x}_2)$ are any two values of f on S, and if c is any number between $f(\boldsymbol{x}_1)$ and $f(\boldsymbol{x}_2)$, then there exists a point $\boldsymbol{x}$ in S such that $f(\boldsymbol{x}) = c$.

Proof. Since S is connected and compact in $\mathbb{R}^n$ and f is continuous on S, $f(S)$ is also a connected, compact subset of $\mathbb{R}$. Any such subset must be a closed and bounded interval; $f(S)$ can only be $[m, M]$ where, as above, $m = \min\{f(\boldsymbol{x}) : \boldsymbol{x}$ in $S\}$ and $M = \max\{f(\boldsymbol{x}) : \boldsymbol{x}$ in $S\}$. Again, our hypothesis implies that $m < 0$ and $M > 0$, so 0 is in $[m, M]$ and f must take on the value 0 at some point of S. The second assertion is left for you to prove. ●

3.2.1 The Composition of Continuous Functions

Suppose that f is a real-valued function defined on a set S in $\mathbb{R}^n$. Its range, $f(S)$, is a subset of $\mathbb{R}$. Suppose further that g is a real-valued function with domain T in $\mathbb{R}$ and that $T \supseteq f(S)$. Starting with any $\boldsymbol{x}$ in S, we first compute $f(\boldsymbol{x})$ in $f(S) \subseteq T$ and then $g(f(\boldsymbol{x}))$. The result is a function $(g \circ f)$ from S to $\mathbb{R}$ defined by $(g \circ f)(\boldsymbol{x}) = g(f(\boldsymbol{x}))$, for all $\boldsymbol{x}$ in S. The function $(g \circ f)$ is called the *composition* of g with f. As usual, we are interested in identifying the properties inherited by $g \circ f$ from f and g. As we might reasonably hope, continuity is inherited.

THEOREM 3.2.7 If f is continuous at $\boldsymbol{c}$ in S and if g is continuous at $f(\boldsymbol{c})$ in T, then $(g \circ f)$ is continuous at $\boldsymbol{c}$. If f is continuous on S and if g is continuous on $f(S)$, then $(g \circ f)$ is continuous on S.

Proof. Let $d = (g \circ f)(\boldsymbol{c})$. Choose any neighborhood $N(d)$ of d. By the continuity of g at $f(\boldsymbol{c})$, there exists a neighborhood $N(f(\boldsymbol{c}))$ such that $T \cap N(f(\boldsymbol{c})) \subseteq g^{-1}(N(d))$. By the continuity of f at $\boldsymbol{c}$, there exists a neighborhood $N(\boldsymbol{c})$ such that $S \cap N(\boldsymbol{c}) \subseteq f^{-1}(N(f(\boldsymbol{c})))$. That is, for any $\boldsymbol{x}$ in $S \cap N(\boldsymbol{c})$, the point $f(\boldsymbol{x})$ is in $T \cap N(f(\boldsymbol{c}))$. Therefore $(g \circ f)(\boldsymbol{x})$ is in $N(d)$. Thus $S \cap N(\boldsymbol{c}) \subseteq (g \circ f)^{-1}(N((g \circ f)(\boldsymbol{c})))$. This proves that $g \circ f$ is continuous at $\boldsymbol{c}$. The second assertion of the theorem follows immediately. (See Fig. 3.16.) ●

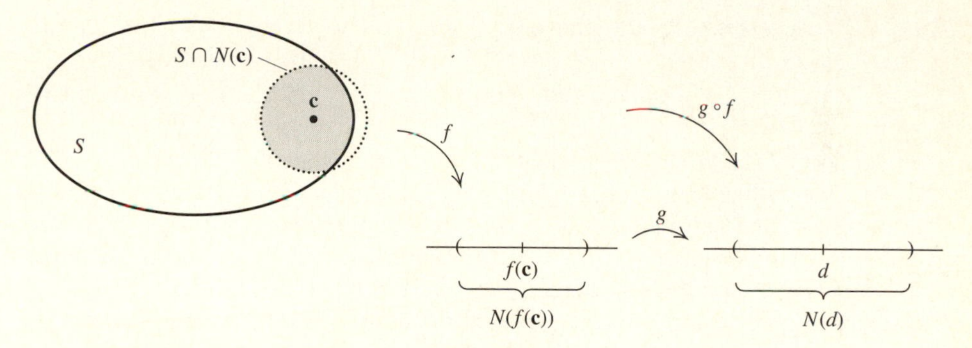

Figure 3.16

EXAMPLE 15 For $\boldsymbol{x} = (x_1, x_2, \ldots, x_n)$ in $\mathbb{R}^n$, let $f(\boldsymbol{x}) = ||\boldsymbol{x}||^2$. For y in $\mathbb{R}$, let $g(y) = e^{-y}$. Then $g \circ f$ is the function $(g \circ f)(\boldsymbol{x}) = \exp[-(x_1^2 + x_2^2 + \cdots + x_n^2)]$, for $\boldsymbol{x}$ in $\mathbb{R}^n$. We know that f is continuous on $\mathbb{R}^n$. In Exercise 3.19 you proved that g is continuous on $\mathbb{R}$. It follows that $g \circ f$ is continuous on $\mathbb{R}^n$. ●

Recall that, if a function f maps a set S one-to-one onto a set T, then the inverse function $g = f^{-1}$ exists and maps T one-to-one onto S. The function f^{-1} has the property that $f^{-1} \circ f$ is the identity function on S and $f \circ f^{-1}$ is the identity function on T. If, in addition, we assume that f is real-valued, that S is the interval $[a, b]$, and that f is continuous on S, then f must be strictly monotone on S and f^{-1} must also be strictly monotone on $T = f([a, b])$. The main point of the following theorem is that f^{-1} must also be continuous on T.

THEOREM 3.2.8 If f is a strictly monotone, continuous real-valued function mapping $S = [a, b]$ onto $[c, d]$, then the inverse function f^{-1} is also strictly monotone, continuous, and maps $[c, d]$ one-to-one onto $[a, b]$.

Proof. We leave for you to prove as an exercise that f^{-1} is strictly monotone, is one-to-one, and maps $[c, d]$ onto $[a, b]$. We will prove that f^{-1} is continuous on the compact set $f(S) = [c, d]$. Denote f^{-1} by g. To prove the continuity of g at a point y in $f(S)$, it suffices, by part (ii) of Theorem 3.1.6, to prove that, whenever a sequence $\{y_k\}$ of distinct points in $f(S)$ converges to y, then $\{g(y_k)\}$ converges to $g(y)$ in S. Choose any such sequence and, for each k in $\mathbb{N}$, let $x_k = g(y_k)$. Since g is one-to-one, the sequence $\{x_k\}$ consists of distinct points of S. Because S is compact, the Bolzano–Weierstrass theorem ensures that the set $\{x_k : k \text{ in } \mathbb{N}\}$ has at least one limit point x. Any limit point of $\{x_k : k \text{ in } \mathbb{N}\}$ is a cluster point of the sequence and must already be in S since S is compact. Thus, x belongs to S. By Theorem 1.3.10, there exists a subsequence $\{x_{k_j}\}$ of $\{x_k\}$ such that $\lim_{j\to\infty} x_{k_j} = x$. The continuity of f on S guarantees that

$$\lim_{j\to\infty} f(x_{k_j}) = f(x). \tag{3.3}$$

But $f(x_{k_j}) = f(g(y_{k_j})) = y_{k_j}$ and, therefore, $\{f(x_{k_j})\}$ is a subsequence of $\{y_k\}$. Since $\{y_k\}$ converges to y, we deduce that

$$\lim_{j\to\infty} f(x_{k_j}) = y. \tag{3.4}$$

Comparing (3.3) and (3.4), we see that $f(x) = y$. Equivalently, $g(y) = x$.

Now we deduce that x must be the only limit point of the sequence $\{x_k\}$ for, if x' were another, then, by repeating the foregoing argument, we have $g(y) = x'$ as well. But g is a function, so $x' = x$. Consequently, $\{x_k\}$ is a bounded sequence with exactly one limit point. Therefore $\{x_k\}$ converges to x. Equivalently, $\{g(y_k)\}$ converges to $g(y)$. This proves that g is continuous at y. Since y is an arbitrary point of $[c, d]$, we conclude that $g = f^{-1}$ is continuous on $[c, d]$. ●

3.2.2 Limiting Behavior at Infinity

If we assume that S is compact, then, by Theorem 3.2.4, any continuous, real-valued function on S is automatically bounded on S. If S is not compact, this is no longer true. The functions $f(x) = x^2$ on $\mathbb{R}$ and $f(x) = 1/[x(1-x)]$ on $(0, 1)$ suffice to confirm this for you. In particular, if S is not bounded, then S is not compact and a continuous function on S need not be bounded there. If we restrict ourselves to considering only those functions in $C(S)$ that are bounded on S, then we obtain the space, $C_\infty(S)$, of all real-valued functions on S that are both continuous and bounded on S. As we will see in Section 3.3, $C_\infty(S)$ is both a vector space and a commutative ring. Further, as we will prove, the reciprocal of a function f in $C_\infty(S)$ is also in $C_\infty(S)$ provided f is also bounded away from 0 on all of S, that is, if there exists an $m > 0$ such that $|f(\boldsymbol{x})| \geq m$ for all $\boldsymbol{x}$ in S. (*Proof:* $1/|f| \leq M$ if and only if $|f| \geq 1/M = m$. Recall, for $1/f$ to be in $C_\infty(S)$, it must be bounded.)

In classical applications of the theory, we often will take S to be an unbounded interval in $\mathbb{R}$, either of the form $[a, \infty)$, (a, ∞), $(-\infty, \infty)$, $(-\infty, b]$, or $(-\infty, b)$. In the first three instances, we say that f approaches L, has limit L, or converges to L, as x tends to ∞ and write $\lim_{x\to\infty} f(x) = L$ provided that, for every $\epsilon > 0$, there exists a positive constant M such that, if x is in S and $x > M$, then $|f(x) - L| < \epsilon$. In other words, if x is in $S \cap (M, \infty)$, then $f(x)$ is in $N(L; \epsilon)$.

It is suggestive to think of the interval (M, ∞) as a "neighborhood $N(\infty; M)$ of ∞," though it is to be understood that such a neighborhood cannot be defined in terms of "distance from ∞." With this understanding, $\lim_{x\to\infty} f(x) = L$ means that, given $\epsilon > 0$, there exists a neighborhood $N(\infty; M)$ such that $S \cap N(\infty; M) \subseteq f^{-1}(N(L; \epsilon))$. This formulation expresses the convergence of f to L as x approaches ∞ in exactly the same form as that developed at the beginning of this chapter. Since it is desirable to develop a unified notation in mathematics, this is an especially attractive formulation.

If S is any of $(-\infty, \infty)$, $(-\infty, b]$, or $(-\infty, b)$, we say, likewise, that $\lim_{x\to-\infty} f(x) = L$ if, for any $\epsilon > 0$, there exists a neighborhood $N(-\infty; -M) = (-\infty, -M)$ of $-\infty$ such that $S \cap N(-\infty; -M) \subseteq f^{-1}(N(L; \epsilon))$. In other words, if x is in S and $x < -M$, then $|f(x) - L| < \epsilon$.

With only minor adjustment of your thinking, these ideas can be lifted to apply to a real-valued function f defined on any unbounded set S of $\mathbb{R}^n$. We write $\lim_{||\boldsymbol{x}||\to\infty} f(\boldsymbol{x}) = L$ if, for every $\epsilon > 0$, there exists a positive constant M such that, if $\boldsymbol{x}$ is in S and $||\boldsymbol{x}|| > M$, then $f(\boldsymbol{x})$ is in $N(L; \epsilon)$. Equivalently, $S \cap N(\infty; M) \subseteq f^{-1}(N(L; \epsilon))$, where $N(\infty; M) = \{\boldsymbol{x} \text{ in } \mathbb{R}^n : ||\boldsymbol{x}|| > M\}$, a set that can be described as a "neighborhood of ∞" in $\mathbb{R}^n$. Thus $|f(\boldsymbol{x}) - L|$ is arbitrarily small off the closure of the hypersphere $N(\mathbf{0}; M)$.

DEFINITION 3.2.1

i) Let S be an unbounded set in $\mathbb{R}$. Let f be a real-valued function defined on S.

- **a)** We say that *f has limit L at* ∞ if, for all $\epsilon > 0$, there exists an $M > 0$ such that $S \cap N(\infty; M) \subseteq f^{-1}(N(L; \epsilon))$ where $N(\infty; M) = \{x \text{ in } \mathbb{R} : M < x < \infty\}$. We write $\lim_{x\to\infty} f(x) = L$.
- **b)** We say that *f has limit L at* $-\infty$ if, for all $\epsilon > 0$, there exists $M > 0$ such that $S \cap N(-\infty; -M) \subseteq f^{-1}(N(L; \epsilon))$, where $N(-\infty; -M) = \{x \text{ in } \mathbb{R} : -\infty < x < -M\}$. We write $\lim_{x\to-\infty} f(x) = L$.

ii) Let S be an unbounded set in $\mathbb{R}^n$. We say that *f has limit L at* ∞, if, for all $\epsilon > 0$, there exists $M > 0$ such that $S \cap N(\infty; M) \subseteq f^{-1}(N(L; \epsilon))$, where $N(\infty; M) = \{\boldsymbol{x} \text{ in } \mathbb{R}^n : ||\boldsymbol{x}|| > M\}$. We write $\lim_{||\boldsymbol{x}||\to\infty} f(\boldsymbol{x}) = L$. ●

With only minor modifications of the proof of Theorem 3.1.3, we can obtain the analog of that theorem in the present context. Thus, if f_1 and f_2 have limits L_1 and L_2 at infinity, then the functions $f_1 + f_2$, $f_1 f_2$, $1/f_2$, and f_1/f_2 have the corresponding limits at infinity. (In the latter two cases, we require that $L_2 \neq 0$.) Likewise, the analog of the Squeeze Play (Theorem 3.1.4) is valid with only a slightly modified proof. We leave the details as exercises.

THEOREM 3.2.9 The Squeeze Play Let f, g, and h be three real-valued functions defined on an unbounded set S in $\mathbb{R}^n$. Suppose that there is an $M > 0$ so that, for $\boldsymbol{x}$ in $S \cap N(\infty; M)$, we have $f(\boldsymbol{x}) \le g(\boldsymbol{x}) \le h(\boldsymbol{x})$. Suppose also that $\lim_{||\boldsymbol{x}||\to\infty} f(\boldsymbol{x}) = \lim_{||\boldsymbol{x}||\to\infty} h(\boldsymbol{x}) = L$. Then $\lim_{||\boldsymbol{x}||\to\infty} g(\boldsymbol{x}) = L$ as well. ●

EXAMPLE 16 Let $f(x) = \exp(-x^2)$. By Example 15, we know that f is continuous on $\mathbb{R}$. We claim that $\lim_{x\to\infty} \exp(-x^2) = 0$ and $\lim_{x\to-\infty} \exp(-x^2) = 0$. To establish these limits, fix $\epsilon > 0$ and let $M = \sqrt{\ln(1/\epsilon)}$. Then, for $x > M$, we have

$$0 < \exp(-x^2) < \exp\left[-\ln\left(\frac{1}{\epsilon}\right)\right] = \epsilon,$$

since the exponential function is monotone increasing. (See part (i) of Exercise 3.19.) This proves that $\lim_{x\to\infty} \exp(-x^2) = 0$. That $\lim_{x\to-\infty} \exp(-x^2) = 0$ follows from the fact that f is an even function, that is, $f(-x) = f(x)$ for all x in $\mathbb{R}$. ●

EXAMPLE 17 For x in $\mathbb{R}$ define two functions $f(x) = e^{-x} \sin x$ and $g(x) = e^{-x} \cos x$. As we will see below, the product of continuous functions is continuous; therefore f and g are continuous on $\mathbb{R}$. Since $|f(x)| \leq e^{-x}$ and $|g(x)| \leq e^{-x}$ and since $f(x) = e^{-x}$ when $\sin x = 1$ and $g(x) = e^{-x}$ when $\cos x = 1$, we deduce that the limiting behavior of f and of g reflects that of e^{-x}. As you can prove, $\lim_{x\to\infty} e^{-x} = 0$ and $\lim_{x\to-\infty} e^{-x} = \infty$. Therefore, by the Squeeze Play, $\lim_{x\to\infty} e^{-x} \sin x = \lim_{x\to\infty} e^{-x} \cos x = 0$. On the other hand, $\lim_{x\to-\infty} e^{-x} \sin x$ and $\lim_{x\to-\infty} e^{-x} \cos x$ do not exist. Each function oscillates dramatically through large positive and negative values as x approaches $-\infty$. ●

EXAMPLE 18 In Exercise 3.49 we ask you to obtain properties of the function $\tan^{-1} x$ on $\mathbb{R}$. After completing that work, you can prove that $\lim_{x\to\infty} \tan^{-1} x = \pi/2$ and $\lim_{x\to-\infty} \tan^{-1} x = -\pi/2$. ●

EXAMPLE 19 For $\boldsymbol{x} = (x_1, x_2)$ in $\mathbb{R}^2$, define a function $f(\boldsymbol{x}) = \tan^{-1}(x_1^2 + x_2^2)$. In Exercise 3.51 we ask you to find all points in $\mathbb{R}^2$ where f is continuous. Since $x_1^2 + x_2^2 \geq 0$, it follows that $0 \leq \tan^{-1}(x_1^2 + x_2^2) < \pi/2$. As in Example 18, you can prove that $\lim_{||\boldsymbol{x}||\to\infty} \tan^{-1}(x_1^2 + x_2^2) = \pi/2$. ●

It is often important to know that a continuous function is bounded on its domain, even when that domain is unbounded. The following theorem provides sufficient conditions that lead to this conclusion.

THEOREM 3.2.10 Let S be a closed, unbounded set in $\mathbb{R}^n$. Suppose that f is a real-valued function that is continuous on S. Suppose also that $\lim_{||\boldsymbol{x}||\to\infty} f(\boldsymbol{x}) = L$ exists. Then f is bounded on S.

Proof. Given $\epsilon > 0$, choose $M > 0$ such that, for $\boldsymbol{x}$ in S with $||\boldsymbol{x}|| > M$, it follows that $|f(\boldsymbol{x}) - L| < \epsilon$. Consequently, for such $\boldsymbol{x}$ we have $|f(\boldsymbol{x})| < |L| + \epsilon$. Since $S \cap \overline{N(\mathbf{0}; M)}$ is a closed, bounded set in $\mathbb{R}^n$, it is compact. Therefore the continuous function f is bounded on $S \cap \overline{N(\mathbf{0}; M)}$, say with $|f(\boldsymbol{x})| \leq K$, for all $\boldsymbol{x}$ in $S \cap \overline{N(\mathbf{0}; M)}$. It follows easily that $|f(\boldsymbol{x})| \leq \max\{K, |L| + \epsilon\}$ for all $\boldsymbol{x}$ in S. This proves that f is bounded on S. ●

EXERCISES

3.36. Prove that, in the notation developed in the proof of Theorem 3.2.2, (i) $U_1 \cap V_1 = \emptyset$ and (ii) $U_1 \cup V_1 = S$.

3.37. Suppose that f is continuous on $[a, b]$ and that $m = \min\{f(x) : x \text{ in } [a, b]\}$ and $M = \max\{f(x) : x \text{ in } [a, b]\}$. Prove that $f([a, b]) = [m, M]$.

3.38. Suppose that f and g are real-valued functions that are each continuous on a common domain S in $\mathbb{R}^n$. Determine whether the set $\{\boldsymbol{x} \text{ in } S : f(\boldsymbol{x}) > g(\boldsymbol{x})\}$ is relatively open in S. If so, prove it; otherwise, provide a counterexample.

3.39. Suppose that f is a real-valued function that is continuous on a nonempty set S in $\mathbb{R}^n$. Suppose that U is a relatively open set in S. Is $f(U) = \{f(\boldsymbol{x}) : \boldsymbol{x} \text{ in } U\}$ necessarily relatively open in $f(S)$? If so, prove it; otherwise, provide a counterexample.

3.40. Let f be a real-valued function that is continuous on a nonempty set S in $\mathbb{R}^n$. Assume that C is a relatively closed subset of $f(S)$. Prove that $f^{-1}(C)$ is relatively closed in S.

3.41. Suppose that f is a real-valued function defined on a nonempty set S in $\mathbb{R}^n$ such that, for every relatively closed subset C of $f(S)$, $f^{-1}(C)$ is relatively closed in S. Does it necessarily follow that f is continuous on S? If so, prove it; otherwise, provide a counterexample.

3.42. Suppose that f is a real-valued function that is continuous on a nonempty set S in $\mathbb{R}^n$ and suppose that $f(S)$ is compact in $\mathbb{R}$. Does it necessarily follow that S must be compact? If so, prove it; otherwise, provide a counterexample.

3.43. Prove part (ii) of Theorem 3.2.6.

3.44. For x in $\mathbb{R}$, define $f(x) = 1/(1 + x^2)$.

a) Prove that f is continuous and bounded on $\mathbb{R}$.

b) Find $\lim_{x\to\infty} f(x)$ and $\lim_{x\to-\infty} f(x)$.

c) Sketch the graph of f.

3.45. For x in $\mathbb{R}$, define $f(x) = x/(1 + x^2)$.

a) Prove that f is continuous and bounded on $\mathbb{R}$.

b) Find $\lim_{x\to\infty} f(x)$ and $\lim_{x\to-\infty} f(x)$.

c) Sketch the graph of f.

3.46. For x in $\mathbb{R}$ and for $k \geq 2$, define $f(x) = x^k/(1 + x^2)$.

a) Prove that, for all values of k, f is continuous on $\mathbb{R}$.

b) Prove that, if $k = 2$, then f is bounded on $\mathbb{R}$ whereas, if $k > 2$, then f is unbounded on $\mathbb{R}$.

c) Find $\lim_{x\to\infty} f(x)$ and $\lim_{x\to-\infty} f(x)$. (Treat the cases $k = 2$ and $k > 2$ separately.)

3.47. Let f be a continuous real-valued function on $[a, b]$ that is one-to-one. Prove that f must be strictly monotone.

3.48. Let f be a continuous, strictly monotone function mapping $[a, b]$ onto $[c, d]$.

a) Prove that f^{-1} maps $[c, d]$ one-to-one onto $[a, b]$.

b) Prove that f^{-1} must also be strictly monotone on $[c, d]$.

3.49. **a)** Prove that the tangent function maps $(-\pi/2, \pi/2)$ one-to-one onto $\mathbb{R}$.

b) Prove that the tangent function is strictly monotone increasing on $(-\pi/2, \pi/2)$.

c) Prove that the inverse tangent function, $\tan^{-1} x$, exists and maps $\mathbb{R}$ one-to-one onto $(-\pi/2, \pi/2)$.

d) Prove that $\tan^{-1} x$ is also a strictly increasing function of x in $\mathbb{R}$.

e) Prove that $\tan^{-1} x$ is continuous on $\mathbb{R}$. (This exercise is not covered directly by Theorem 3.2.8 because $(-\pi/2, \pi/2)$ is not closed and because $(-\infty, \infty)$ is not bounded. But you can apply Theorem 3.2.8 to compact intervals in the domain and the range of the tangent function.)

f) Prove that $\lim_{x\to\infty} \tan^{-1} x = \pi/2$ and $\lim_{x\to-\infty} \tan^{-1} x = -\pi/2$.

3.50. In Exercise 3.19 you proved that, for $a > 1$, the function a^x exists and is continuous on $\mathbb{R}$. When a has the special value $a = e$, we obtain the natural exponential function $f(x) = e^x$, also denoted $\exp(x)$. We know that e^x is continuous on $\mathbb{R}$.

a) Prove that $f(x) = e^x$ is strictly monotone increasing on $\mathbb{R}$ and maps $(-\infty, \infty)$ one-to-one onto $(0, \infty)$.

b) Prove that f^{-1} exists and maps $(0, \infty)$ one-to-one onto $\mathbb{R}$.

c) For simplicity, denote f^{-1} by g. Prove the following properties of g: (i) $g(1) = 0$; (ii) $g(xy) = g(x) + g(y)$ for all x and y in $(0, \infty)$; (iii) $g(x^r) = rg(x)$ for all x in $(0, \infty)$ and all r in $\mathbb{R}$; (iv) $g(x/y) = g(x) - g(y)$ for all x and y in $(0, \infty)$.

The inverse function $g = f^{-1}$ is a version of the logarithm discovered by Napier in about 1594 to assist with the calculations required in the renascent science of astronomy. The properties in part (c) made such calculations, if not easy, at least less difficult. Only when its analytic properties became known did this function assume its present important status. It is now denoted $g(x) = \ln x$ and is called the natural logarithm. Little could Napier have known the profoundly important role this function would come to play in widely divergent areas of mathematics and the sciences.

d) Prove that $\ln x$ is strictly monotone increasing on its domain $(0, \infty)$.

e) Prove that $\ln x$ is continuous on $(0, \infty)$.

f) Find $\lim_{x\to0^+} \ln x$ and $\lim_{x\to\infty} \ln x$.

3.51. Define a function f at points $\boldsymbol{x} = (x_1, x_2)$ in $\mathbb{R}^2$ by $f(\boldsymbol{x}) = \tan^{-1}(x_1^2 + x_2^2)$.

a) Find the domain and the range of f.

b) Find all points $\boldsymbol{x}$ in $\mathbb{R}^2$ where f is continuous.

c) Prove that $\lim_{\|\boldsymbol{x}\|\to\infty} \tan^{-1}(x_1^2 + x_2^2) = \pi/2$.

3.52. Define a function f at points $\boldsymbol{x} = (x_1, x_2)$ in $\mathbb{R}^2$ by $f(\boldsymbol{x}) = \tan^{-1}[1/(x_1^2 + x_2^2)]$.

a) Find the domain and the range of f.

b) Find all points $\boldsymbol{x}$ in $\mathbb{R}^2$ where f is continuous.

c) Does $\lim_{\boldsymbol{x}\to\boldsymbol{0}} f(\boldsymbol{x})$ exist? Is it possible to define $f(\boldsymbol{0})$ in such a way that f is continuous at $\boldsymbol{0}$?

d) Does $\lim_{\|\boldsymbol{x}\|\to\infty} \tan^{-1}[1/(x_1^2 + x_2^2)]$ exist? Prove your answer.

3.53. Define a function f at points $\boldsymbol{x} = (x_1, x_2)$ in $\mathbb{R}^2$ by $f(\boldsymbol{x}) = \tan^{-1}(x_1/x_2)$.

a) Find the domain and the range of f.

b) Find all points $\boldsymbol{x}$ in $\mathbb{R}^2$ where f is continuous.

c) Does $\lim_{\boldsymbol{x}\to\boldsymbol{0}} f(\boldsymbol{x})$ exist? Is it possible to define $f(\boldsymbol{0})$ in such a way that f is continuous at $\boldsymbol{0}$?

d) Does $\lim_{\|\boldsymbol{x}\|\to\infty} \tan^{-1}(x_1/x_2)$ exist? Prove your answer.

3.54. Define a function f at points $\boldsymbol{x} = (x_1, x_2)$ in $\mathbb{R}^2$ by $f(\boldsymbol{x}) = \ln(x_1^2 + x_2^2)$.

a) Find the domain and the range of f.

b) Find all points $\boldsymbol{x}$ in $\mathbb{R}^2$ where f is continuous.

c) Does $\lim_{\boldsymbol{x}\to\boldsymbol{0}} f(\boldsymbol{x})$ exist? Is it possible to define $f(\boldsymbol{0})$ in such a way that f is continuous at $\boldsymbol{0}$?

d) Does $\lim_{\|\boldsymbol{x}\|\to\infty} \ln(x_1^2 + x_2^2)$ exist? Prove your answer.

3.55. Define a function f at points $\boldsymbol{x} = (x_1, x_2)$ in $\mathbb{R}^2$ by $f(\boldsymbol{x}) = \ln(x_1/x_2)$.

a) Find the domain and the range of f.

b) Find all points $\boldsymbol{x}$ in $\mathbb{R}^2$ where f is continuous.

c) Does $\lim_{\boldsymbol{x}\to\boldsymbol{0}} f(\boldsymbol{x})$ exist? Is it possible to define $f(\boldsymbol{0})$ in such a way that f is continuous at $\boldsymbol{0}$?

d) Does $\lim_{\|\boldsymbol{x}\|\to\infty} f(\boldsymbol{x})$ exist? Prove your answer.

3.56. Define a function f at points $\boldsymbol{x} = (x_1, x_2)$ in $\mathbb{R}^2$ by $f(\boldsymbol{x}) = \exp[-(x_1 + x_2)]$.

a) Find the domain and the range of f.

b) Find all points $\boldsymbol{x}$ in $\mathbb{R}^2$ where f is continuous.

c) Find $\lim_{\boldsymbol{x}\to\boldsymbol{0}} f(\boldsymbol{x})$.

d) Does $\lim_{\|\boldsymbol{x}\|\to\infty} f(\boldsymbol{x})$ exist? Prove your answer.

e) Sketch the graph of f.

3.57. Define a function f at points $\boldsymbol{x} = (x_1, x_2)$ in $\mathbb{R}^2$ by $f(\boldsymbol{x}) = \exp[-1/(x_1^2 + x_2^2)]$.

a) Find the domain and the range of f.

b) Find all points $\boldsymbol{x}$ in $\mathbb{R}^2$ where f is continuous.

c) Does $\lim_{\boldsymbol{x}\to\boldsymbol{0}} f(\boldsymbol{x})$ exist? Is it possible to define $f(\boldsymbol{0})$ in such a way that f is continuous at $\boldsymbol{0}$?

d) Does $\lim_{\|\boldsymbol{x}\|\to\infty} f(\boldsymbol{x})$ exist? Prove your answer.

e) Sketch the graph of f.

3.58. Define a function f at points $\boldsymbol{x} = (x_1, x_2)$ in $\mathbb{R}^2$ by $f(\boldsymbol{x}) = \exp[x_1/x_2]$.

a) Find the domain and the range of f.

b) Find all points $\boldsymbol{x}$ in $\mathbb{R}^2$ where f is continuous.

c) Does $\lim_{\boldsymbol{x}\to\boldsymbol{0}} f(\boldsymbol{x})$ exist? Is it possible to define $f(\boldsymbol{0})$ in such a way that f is continuous at $\boldsymbol{0}$?

d) Does $\lim_{\|\boldsymbol{x}\|\to\infty} f(\boldsymbol{x})$ exist? Prove your answer.

3.59. Suppose that f is any continuous function that maps $[a, b]$ into $[a, b]$. Prove that there must exist an x in the interval $[a, b]$ such that $f(x) = x$. (This is your first example of an important class of theorems called *fixed-point theorems*; there is a point x left fixed by f. Such theorems treat questions of *stability* and *equilibrium*.)

3.60. Weaken the hypothesis in Exercise 3.59 as follows: Assume that f is a continuous function on $[a, b]$ and that $f(a) \geq a$ and $f(b) \leq b$. Prove that there must exist an x in (a, b) such that $f(x) = x$. Does the same result follow if $f(a) < a$ and $f(b) > b$? Prove your answer.

3.61. Prove that there exists a unique positive x such that $\cos x = x$.

3.62. Prove that there exists a unique positive x such that $\tan^{-1}(1/x) = x$.

3.63. Prove that there exist exactly three values of x such that $\tan^{-1} x = x$.

3.64. Prove that there exist exactly two values of x such that $\ln(x + 1) = x - 1$.

3.65. Suppose that f maps S in $\mathbb{R}^n$ continuously onto $f(S)$ in $\mathbb{R}$. Next, suppose that g maps $T = f(S)$ continuously onto U in $\mathbb{R}$. Finally, suppose that h maps U continuously onto V in $\mathbb{R}$. Show that $h \circ (g \circ f) = (h \circ g) \circ f = h \circ g \circ f$ is a function from S onto V that is continuous on S.

3.3 THE ALGEBRA OF CONTINUOUS FUNCTIONS

Let S be any nonempty subset of $\mathbb{R}^n$ and let $C(S)$ denote the collection of all real-valued functions that are continuous on all of S. $C(S)$ is a remarkably rich mathematical system. Here we begin to explore its structure. First, we can add or multiply any two functions, f and g, in $C(S)$ and obtain functions in $C(S)$ as follows:

$$(f+g)(\boldsymbol{x}) = f(\boldsymbol{x}) + g(\boldsymbol{x}) \qquad \text{and} \qquad (fg)(\boldsymbol{x}) = f(\boldsymbol{x})g(\boldsymbol{x}),$$

for all $\boldsymbol{x}$ in S. We will prove that the functions $f+g$ and fg, thus defined, are themselves continuous on S and hence are in $C(S)$. Here we identify the algebraic properties that hold true for addition and multiplication in $C(S)$. We will use these properties throughout the text without further comment. It is to be understood that two functions in $C(S)$ are equal if and only if they have the same value at every point of S.

First, because addition and multiplication in $\mathbb{R}$ are both associative and commutative, the same can be said about these operations in $C(S)$. For the associativity of addition, $f + (g + h) = (f + g) + h$, because

$$f(\boldsymbol{x}) + [g(\boldsymbol{x}) + h(\boldsymbol{x})] = [f(\boldsymbol{x}) + g(\boldsymbol{x})] + h(\boldsymbol{x})$$

for all $\boldsymbol{x}$ in S. Likewise, multiplication is associative, that is, $f(gh) = (fg)h$, because, for all $\boldsymbol{x}$ in S,

$$f(\boldsymbol{x})[g(\boldsymbol{x})h(\boldsymbol{x})] = [f(\boldsymbol{x})g(\boldsymbol{x})]h(\boldsymbol{x}).$$

These operations are commutative—that is, $f + g = g + f$ and $fg = gf$—because $f(\boldsymbol{x}) + g(\boldsymbol{x}) = g(\boldsymbol{x}) + f(\boldsymbol{x})$ and $f(\boldsymbol{x})g(\boldsymbol{x}) = g(\boldsymbol{x})f(\boldsymbol{x})$ for all $\boldsymbol{x}$ in S.

The constant function that is identically zero on S is the 0 in $C(S)$; it has the property: $f + 0 = 0 + f = f$ for all f in $C(S)$. The constant function $1(x) = 1$, for all $\boldsymbol{x}$ in S, that is identically one on S is the multiplicative identity 1 in $C(S)$; it has the property that $f1 = 1f = f$ for all f in $C(S)$.

The additive inverse (or negative) of f in $C(S)$ is the function $(-f)(\boldsymbol{x}) = -(f(\boldsymbol{x}))$ for all $\boldsymbol{x}$ in S. Subtraction is defined in $C(S)$ by $f - g = f + (-g)$. The distributive law holds: $f(g+h) = (fg) + (fh)$, for all f, g, and h in $C(S)$. Thus $C(S)$ is a commutative ring with multiplicative identity.

What are the units in $C(S)$? That is, for which f in $C(S)$ is it true that $g = 1/f$ is also in $C(S)$? We claim that, if $f(\boldsymbol{x}) \neq 0$ for all $\boldsymbol{x}$ in S, then $g = 1/f$ also belongs to $C(S)$. Therefore the units in $C(S)$ are those continuous functions that do not vanish on S. We can divide one continuous function, f, by another, g, provided that g is a unit in $C(S)$, that is, provided that g does not vanish on S. The quotient function, (f/g), has value $f(\boldsymbol{x})/g(\boldsymbol{x})$ at $\boldsymbol{x}$ in S.

Notice that every constant function $f(\boldsymbol{x}) = c$ is continuous. Thus any constant multiple of any function in $C(S)$ is also continuous; $C(S)$ is a vector space. Also notice that, since $C(S)$ is closed under multiplication, every positive integer power, f^k, of f in $C(S)$ is also in $C(S)$.

In the special case when S is a subset of $\mathbb{R}$ (rather than of an arbitrary $\mathbb{R}^n$), the function $e(x) = x$ is certainly in $C(S)$. Therefore $p(x) = a_0 + a_1x + a_2x^2 + \cdots + a_rx^r$, a polynomial of degree r, is in $C(S)$. Now, the collection $P(S)$ of all polynomials is itself a commutative ring with identity and, in fact, is a subring of $C(S)$. This will prove to be an extremely important and powerful fact in both theoretical and applied aspects of analysis.

In proving Theorem 3.1.3, we have already done the essential work required to prove all these claims. Return to that proof and, in each part, assume $\boldsymbol{c}$ is a point of S and replace L_1 by $f_1(\boldsymbol{c})$ and L_2 by $f_2(\boldsymbol{c})$. The result is the major theorem of this section.

THEOREM 3.3.1 Let S be any subset of $\mathbb{R}^n$. Let f, f_1 and f_2 belong to $C(S)$. Then the following properties hold:

i) The sum of two continuous functions on S is also continuous on S; $f_1 + f_2$ is in $C(S)$.

ii) For any a in $\mathbb{R}$, af is in $C(S)$.

iii) The product of two continuous functions on S is also continuous on S; $f_1 f_2$ is in $C(S)$.

iv) The reciprocal of a continuous function on S is continuous on S, provided it does not vanish on S; $1/f$ is in $C(S)$ provided that $f(\boldsymbol{x}) \neq 0$ for all $\boldsymbol{x}$ in S.

v) The quotient of two continuous functions on S is continuous on S, provided the denominator does not vanish on S; f_1/f_2 is in $C(S)$ provided that $f_2(\boldsymbol{x}) \neq 0$ for all $\boldsymbol{x}$ in S.

Proof. Fix any $\boldsymbol{c}$ in S. To prove part (i), we have

$$\lim_{\boldsymbol{x}\to\boldsymbol{c}} (f_1 + f_2)(\boldsymbol{x}) = \lim_{\boldsymbol{x}\to\boldsymbol{c}} f_1(\boldsymbol{x}) + \lim_{\boldsymbol{x}\to\boldsymbol{c}} f_2(\boldsymbol{x}) = f_1(\boldsymbol{c}) + f_2(\boldsymbol{c}).$$

For part (ii), $\lim_{\boldsymbol{x}\to\boldsymbol{c}}(af)(\boldsymbol{x}) = a \lim_{\boldsymbol{x}\to\boldsymbol{c}} f(\boldsymbol{x}) = af(\boldsymbol{c})$. Part (iii) follows from the fact that

$$\lim_{\boldsymbol{x}\to\boldsymbol{c}} (f_1 f_2)(\boldsymbol{x}) = [\lim_{\boldsymbol{x}\to\boldsymbol{c}} f_1(\boldsymbol{x})][\lim_{\boldsymbol{x}\to\boldsymbol{c}} f_2(\boldsymbol{x})] = f_1(\boldsymbol{c}) f_2(\boldsymbol{c}).$$

For part (iv), simply write $\lim_{\boldsymbol{x}\to\boldsymbol{c}}(1/f)(\boldsymbol{x}) = 1/\lim_{\boldsymbol{x}\to\boldsymbol{c}} f(\boldsymbol{x}) = 1/f(\boldsymbol{c})$, since $f(\boldsymbol{c}) \neq 0$. Finally, to prove part (v), write

$$\lim_{\boldsymbol{x}\to\boldsymbol{c}} (f_1/f_2)(\boldsymbol{x}) = [\lim_{\boldsymbol{x}\to\boldsymbol{c}} f_1(\boldsymbol{x})]/[\lim_{\boldsymbol{x}\to\boldsymbol{c}} f_2(\boldsymbol{x})] = f_1(\boldsymbol{c})/f_2(\boldsymbol{c}),$$

since $f_2(\boldsymbol{c}) \neq 0$. This completes the proof of the theorem. ●

When applied to a function f that is continuous at a point $\boldsymbol{c}$ in $\mathbb{R}^n$, Theorem 3.1.1 yields the following.

THEOREM 3.3.2 Suppose that f is continuous at a point $\boldsymbol{c}$ in $\mathbb{R}^n$. Then f is locally bounded at $\boldsymbol{c}$. That is, there exists a neighborhood $N(\boldsymbol{c})$ of $\boldsymbol{c}$ and a positive constant M such that $|f(\boldsymbol{x})| \leq M$, for all $\boldsymbol{x}$ in $S \cap N(\boldsymbol{c})$. ●

Likewise, when applied to a function f that is continuous at a point $\boldsymbol{c}$ in $\mathbb{R}^n$, Theorem 3.1.2 yields the following.

THEOREM 3.3.3 Suppose that f is continuous at a point $\boldsymbol{c}$ in $\mathbb{R}^n$ and that $f(\boldsymbol{c}) \neq 0$. Then f is locally bounded away from 0 at $\boldsymbol{c}$. That is, there exists a neighborhood $N(\boldsymbol{c})$ of $\boldsymbol{c}$ and a positive constant m such that $|f(\boldsymbol{x})| \geq m > 0$, for all $\boldsymbol{x}$ in $S \cap N(\boldsymbol{c})$. ●

Finally, if we restrict our attention to the collection of all bounded functions that are continuous on a set S in $\mathbb{R}^n$, then we obtain the space $C_\infty(S)$, which, by Theorem 3.3.1, is closed under addition and multiplication. Further, as we have noted, if a function f in $C_\infty(S)$ is bounded away from 0 on S, then $1/f$ is also in $C_\infty(S)$. Also recall that, if S is compact, then $C_\infty(S) = C(S)$.

EXERCISES

3.66. Let S be a nonempty subset of $\mathbb{R}^n$. Let $B(S)$ denote the collection of all real-valued functions f on S for which

$$\sup\{|f(\boldsymbol{x})| : \boldsymbol{x} \text{ in } S\}$$

is finite. Define addition, multiplication, scalar multiplication, the zero, negatives, the identity, and reciprocals as in $C(S)$.

a) Determine which of the properties (i) through (v) of Theorem 3.3.1 hold true for $B(S)$. Prove those that are true; disprove the others. Find the units of $B(S)$.

b) Do Theorems 3.3.2 and 3.3.3 necessarily remain true for $B(S)$ with the continuity hypothesis removed?

3.4 UNIFORM CONTINUITY

You will notice that the continuity of a function f on a set S in $\mathbb{R}^n$ is defined pointwise, by which we mean that f is continuous at each point $\boldsymbol{c}$ in S. The measure δ of required proximity to $\boldsymbol{c}$ depends not only on ϵ but also on the point $\boldsymbol{c}$ itself. Varying $\boldsymbol{c}$ may require that δ vary also in order that $S \cap N(\boldsymbol{c}; \delta) \subseteq f^{-1}(N(f(\boldsymbol{c}); \epsilon))$. If it should happen that a single $\delta_0 > 0$ can be chosen, independent of $\boldsymbol{c}$ in S, such that $S \cap N(\boldsymbol{c}; \delta_0) \subseteq f^{-1}(N(f(\boldsymbol{c}); \epsilon))$ for all $\boldsymbol{c}$ in S, then f is said to be *uniformly continuous on* S.

DEFINITION 3.4.1 A function f with domain S contained in $\mathbb{R}^n$ is said to be *uniformly continuous on* S if, for every $\epsilon > 0$, there exists a $\delta > 0$ such that

$$S \cap N(\boldsymbol{c}; \delta) \subseteq f^{-1}(N(f(\boldsymbol{c}); \epsilon))$$

for all $\boldsymbol{c}$ in S. ●

It is essential, in this definition, to understand that, since δ is mentioned prior to any mention of the variable $\boldsymbol{c}$ in S, its existence and value is independent of the value of $\boldsymbol{c}$. Therefore δ is taken to be the same regardless of the point $\boldsymbol{c}$ in S. That is, δ depends on ϵ but not on $\boldsymbol{c}$. The word *uniformly* refers to the fact that the uniform continuity of f on S is a *global*, rather than a *local*, property of f; it is a property of f holding simultaneously at each point of the entire set S rather than merely at each point of S separately. In practice the uniform continuity of f on S is often recast in a slightly different form, one that emphasizes its uniform nature.

THEOREM 3.4.1 A real-valued function f on a set S in $\mathbb{R}^n$ is uniformly continuous if and only if, for any $\epsilon > 0$, there exists a $\delta > 0$ such that, whenever $\boldsymbol{x}$ and $\boldsymbol{y}$ are points of S such that $||\boldsymbol{x} - \boldsymbol{y}|| < \delta$, it follows that $|f(\boldsymbol{x}) - f(\boldsymbol{y})| < \epsilon$. ●

It is generally acknowledged that the subtlety inherent in the concept of uniform continuity makes it one of the ideas in analysis most difficult for the beginner to grasp. Yet uniform continuity is one of the most crucial concepts to master. We therefore urge you to concentrate intently on the contents of this section. First, it is immediate, as you can prove as an exercise, that a uniformly continuous function on S is continuous at each point of S. On the other hand, the mere continuity of a function f on a set S does not imply its uniform continuity there. Uniform continuity on a set is a much stronger property than mere continuity.

EXAMPLE 20 The function $f(x) = 1/x$ is continuous on the interval $S = (0, \infty)$, but is not uniformly continuous there. The reason, as we will see, is that, as c in S gets ever closer to 0, the value of δ must become ever smaller to ensure that $S \cap N(c; \delta) \subseteq f^{-1}(N(f(c); \epsilon))$. In this first example, δ depends on both ϵ and on c in S.

If $\epsilon > 0$ is fixed and if we vary c in S, at each stage choosing the largest δ_c that will suffice to ensure that $S \cap N(c; \delta_c) \subseteq f^{-1}(N(f(c); \epsilon))$, we will see this phenomenon explicitly. If we start with $c = 1$, then we can take

$$\delta_1 = 1 - \frac{1}{1+\epsilon} = \frac{\epsilon}{1+\epsilon}.$$

As you can confirm for yourself, if

$$1 - \frac{\epsilon}{1+\epsilon} < x < 1 + \frac{\epsilon}{1+\epsilon},$$

then

$$1 - \epsilon < \frac{1}{x} < 1 + \epsilon.$$

If, however, we take $c = 1/2$, then $\delta_{1/2}$ must be taken no larger than

$$\frac{1}{2} - \frac{1}{2+\epsilon} = \frac{\epsilon}{2(2+\epsilon)}.$$

For this choice of $\delta_{1/2}$, and for no larger δ, it is true that, if x is in $N(1/2; \delta_{1/2})$, then $1/x$ is in $N(2; \epsilon)$. It is easy to see that

$$\delta_1 = \frac{\epsilon}{1+\epsilon} > \frac{\epsilon}{2(2+\epsilon)} = \delta_{1/2}.$$

That is, the largest δ_1 that works for $c = 1$ is bigger than the largest $\delta_{1/2}$ that works for $c = 1/2$. More generally, at any $c > 0$, δ_c can be taken no larger than

$$c - \frac{1}{\epsilon + 1/c} = \frac{c^2\epsilon}{1 + c\epsilon}.$$

For this choice of δ_c, and for no larger δ, it is true that, if x is in $N(c; \delta_c)$, then $1/x$ is in $N(1/c; \epsilon)$. Our point is that

$$0 \le \lim_{c \to 0^+} \delta_c = \lim_{c \to 0^+} \frac{c^2 \epsilon}{1 + c\epsilon} = 0.$$

Remember, ϵ is held constant while computing this limit. Consequently, δ_c will have to be taken ever smaller as c approaches 0; no one δ can possibly suffice for all c in S. See Fig. 3.17.

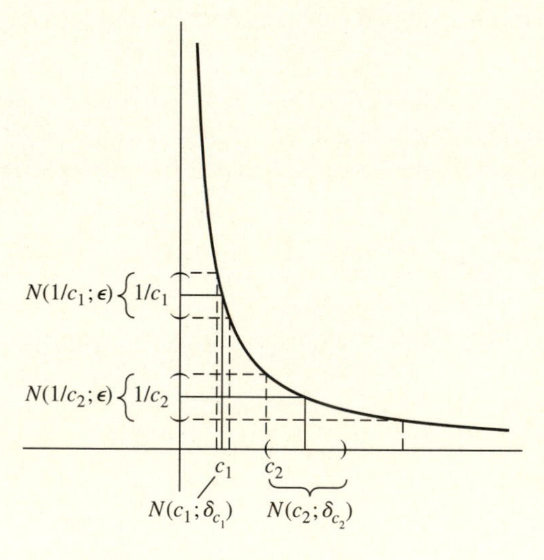

Figure 3.17 $f(x) = 1/x, x > 0$

To prove formally that no single δ_0 can possibly suffice, build on the foregoing discussion as follows. Fix $\epsilon > 0$ and suppose, to the contrary, that there were to exist a single $\delta_0 > 0$ such that, for all c in $(0, \infty)$ and all x in $N(c; \delta_0)$, it follows that $1/x$ is in $N(1/c; \epsilon)$. Since, as we have seen, $\lim_{c \to 0^+} \delta_c = 0$, we can choose $c > 0$ such that $0 < \delta_c < \delta_0$. Now choose any x in $(c - \delta_0, c - \delta_c) \cap (0, \infty)$. On the one hand, our assumption about δ_0 guarantees that $1/x$ is in $N(1/c; \epsilon)$. But, for our chosen x and with $\delta_c = c^2\epsilon/(1 + c\epsilon)$ we have

$$\left| \frac{1}{x} - \frac{1}{c} \right| = \frac{1}{x} - \frac{1}{c} > \frac{1}{c - \delta_c} - \frac{1}{c} = \epsilon.$$

Thus $|1/x - 1/c| > \epsilon$. The contradiction proves that δ_0 cannot exist and that $f(x) = 1/x$ cannot be uniformly continuous on $(0, \infty)$. ●

EXAMPLE 21 If we take the same function $f(x) = 1/x$ and restrict its domain to, say, $S = [1, \infty)$, then the largest δ_c that suffices for the continuity of f at $c = 1$

will also suffice at all the points of $[1, \infty)$. Specifically, take $\delta_0 = \epsilon/(1+\epsilon)$. Then, for any c in $[1, \infty)$ and for any x in $S \cap N(c; \delta_0)$,

$$|f(x) - f(c)| = \left|\frac{1}{x} - \frac{1}{c}\right| = \frac{|c-x|}{cx}$$
$$\leq |c - x| < \delta_0 = \frac{\epsilon}{1+\epsilon} < \epsilon.$$

Consequently, f is uniformly continuous on $[1, \infty)$. ●

In these two examples, the significant difference between $(0, \infty)$ and $[1, \infty)$ is that the latter set is closed and the former is not. Our next two examples will show that the mere closure of the set S will not suffice to ensure uniform continuity of the continuous function f.

EXAMPLE 22 Suppose that $f(x) = x^2$ on the closed set $S = [0, \infty)$. Then f is continuous on S, but is not uniformly continuous there. In confirming the continuity of f at any c in S, we find that, given $\epsilon > 0$, the largest permissible δ_c that will suffice to ensure that $S \cap N(c; \delta_c) \subseteq f^{-1}(N(c^2; \epsilon))$ is $\delta_c = (c^2 + \epsilon)^{1/2} - c$. However, as c approaches ∞, the largest permissible value of δ_c will again tend to 0, as the following computation confirms.

$$\lim_{c\to\infty} \delta_c = \lim_{c\to\infty} [(c^2+\epsilon)^{1/2} - c] = \lim_{c\to\infty} \frac{\epsilon}{(c^2+\epsilon)^{1/2} + c} = 0.$$

Again, remember that ϵ is held constant while computing this limit. (See Fig. 3.18 on page 154.) Therefore a single value of δ that will work for all c in S cannot exist. You can design a formal proof modeled on that in Example 20. In other words, f cannot be uniformly continuous on all of $S = [0, \infty)$, although $[0, \infty)$ is closed. ●

EXAMPLE 23 If we modify the domain of the function $f(x) = x^2$ in Example 22 by restricting f to the set $S = [0, b]$, where b is any positive number, then we claim that f is uniformly continuous on S. In fact, given any $\epsilon > 0$, we claim that the single value $\delta_0 = (b^2 + \epsilon)^{1/2} - b$ will suffice to ensure that $S \cap N(c; \delta_0) \subseteq f^{-1}(N(c^2; \epsilon))$ for all c in S. As you can compute, for any c in S and for any x in $S \cap N(c; \delta_0)$,

$$|x^2 - c^2| < (c + \delta_0)^2 - c^2 = 2c\delta_0 + \delta_0^2$$
$$= 2(b - c)[b - (b^2 + \epsilon)^{1/2}] + \epsilon < \epsilon,$$

since $(b - c) \geq 0$ and $b - (b^2 + \epsilon)^{1/2} < 0$. Thus if x is in $S \cap N(c; \delta_0)$, then x^2 is in $N(c^2; \epsilon)$ for all c in S simultaneously. That is, f is uniformly continuous on $S = [0, b]$. ●

EXAMPLE 24 As you saw in Exercise 3.3, any linear function $f(x) = mx + b$ is uniformly continuous on $\mathbb{R}$. If $m \neq 0$ and if $\epsilon > 0$ is given, then we can take $\delta_0 = \epsilon/|m|$. For this choice of δ_0, if x is in $N(c; \delta_0)$, then $mx + b$ is in $N(mc + b; \epsilon)$ for every c in $\mathbb{R}$ simultaneously. If $m = 0$, then any choice of δ_0 will suffice. ●

As we will see in our next theorem, continuity of f on a compact set S guarantees its uniform continuity. Our examples suffice to warn you that the compactness of

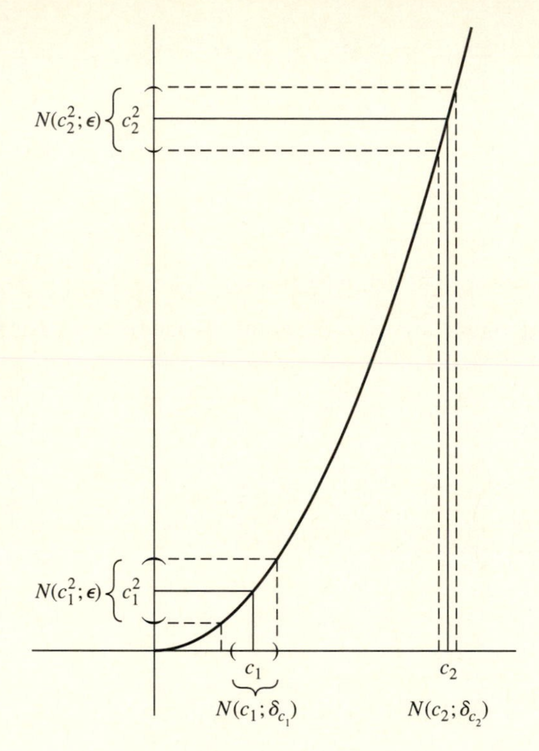

Figure 3.18 $f(x) = x^2, x \geq 0$

S is a sufficient but not a necessary condition for the uniform continuity of the continuous function f on S. Repeatedly, throughout the text, the uniform continuity of a function f will prove to be an especially desirable property; hence we will appeal to the following theorem at several crucial junctures.

THEOREM 3.4.2 Suppose that f is continuous on a compact subset S of $\mathbb{R}^n$. Then f is uniformly continuous on S.

Proof. Given $\epsilon > 0$, our task is to find, in accord with the definition of uniform continuity, a single $\delta_0 > 0$ such that

$$S \cap N(\boldsymbol{c}; \delta_0) \subseteq f^{-1}(N(f(\boldsymbol{c}); \epsilon))$$

for every $\boldsymbol{c}$ in S. The foregoing examples warn us that we must use both the continuity of f at every $\boldsymbol{c}$ in S and the compactness of S. Our strategy will involve several steps. First, we construct an open cover of the compact set S. Then we extract a finite subcover from this cover. Finally, we choose the minimum of a finite number of positive values that will arise. We can be assured that this minimum value is positive, since we take the minimum of only finitely many positive numbers. This minimum value will provide us with the δ_0 that will serve globally to ensure the uniformity of the continuity of f. Thus you will see the power of the Heine–Borel characterization of compactness.

Since f is continuous at each point of S, for each $\boldsymbol{c}$ in S, we can choose $\delta(\boldsymbol{c}) > 0$ such that

$$S \cap N(\boldsymbol{c}; \delta(\boldsymbol{c})) \subseteq f^{-1}\left(N\left(f(\boldsymbol{c}); \frac{\epsilon}{2}\right)\right).$$

(The adjustment, using $\epsilon/2$ as radius here, is necessary only to yield a tidy computation at the culmination of the proof.) The collection $\mathcal{C} = \{N(\boldsymbol{c}; \delta(\boldsymbol{c})/2) : \boldsymbol{c}$ in $S\}$ forms an uncountable, open cover of the compact set S. (The adjustment, using $\delta(\boldsymbol{c})/2$ as radius here, is essential to obtain a necessary containment at the crux of the proof.) By the definition of compactness, there exists a finite subcover

$$\mathcal{C}_1 = \left\{N\left(\boldsymbol{c}_1; \frac{\delta(\boldsymbol{c}_1)}{2}\right), N\left(\boldsymbol{c}_2; \frac{\delta(\boldsymbol{c}_2)}{2}\right), \ldots, N\left(\boldsymbol{c}_k; \frac{\delta(\boldsymbol{c}_k)}{2}\right)\right\}$$

of S. Let $\delta_0 = \min\{\delta(\boldsymbol{c}_1)/2, \delta(\boldsymbol{c}_2)/2, \ldots, \delta(\boldsymbol{c}_k)/2\}$. We want to show that $S \cap N(\boldsymbol{c}; \delta_0) \subseteq f^{-1}(N(f(\boldsymbol{c}); \epsilon))$ for all $\boldsymbol{c}$ in S simultaneously.

Choose any $\boldsymbol{c}$ in S. Since $\mathcal{C}_1$ is a subcover of S, there exists an index j in $\{1, 2, \ldots, k\}$ such that $\boldsymbol{c}$ is in the neighborhood $N(\boldsymbol{c}_j; \delta(\boldsymbol{c}_j)/2)$. For any $\boldsymbol{x}$ in $N(\boldsymbol{c}; \delta_0)$,

$$||\boldsymbol{x} - \boldsymbol{c}_j|| \leq ||\boldsymbol{x} - \boldsymbol{c}|| + ||\boldsymbol{c} - \boldsymbol{c}_j|| < \delta_0 + \frac{\delta(\boldsymbol{c}_j)}{2} \leq \delta(\boldsymbol{c}_j)$$

because $\delta_0 \leq \delta(\boldsymbol{c}_j)/2$. It follows that

$$N(\boldsymbol{c}; \delta_0) \subseteq N(\boldsymbol{c}_j; \delta(\boldsymbol{c}_j)).$$

Finally, choose any $\boldsymbol{x}$ in $S \cap N(\boldsymbol{c}; \delta_0)$ and compute

$$\begin{aligned} |f(\boldsymbol{x}) - f(\boldsymbol{c})| &= |f(\boldsymbol{x}) - f(\boldsymbol{c}_j) + f(\boldsymbol{c}_j) - f(\boldsymbol{c})| \\ &\leq |f(\boldsymbol{x}) - f(\boldsymbol{c}_j)| + |f(\boldsymbol{c}_j) - f(\boldsymbol{c})| \\ &< \frac{\epsilon}{2} + \frac{\epsilon}{2} = \epsilon, \end{aligned}$$

because both $\boldsymbol{x}$ and $\boldsymbol{c}$ are in $N(\boldsymbol{c}_j; \delta(\boldsymbol{c}_j))$. In other words, if $\boldsymbol{x}$ is in $S \cap N(\boldsymbol{c}; \delta_0)$, then $f(\boldsymbol{x})$ is in $N(f(\boldsymbol{c}); \epsilon)$. Since δ_0 was selected independently of the subsequent choice of $\boldsymbol{c}$, f must be uniformly continuous on S. This proves our theorem. ●

EXERCISES

3.67. Prove that, if f is uniformly continuous on a subset S of $\mathbb{R}^n$, then f is continuous at each point $\boldsymbol{c}$ in S.

3.68. Show that the function $f(x) = 1/(1 + x^2)$ is uniformly continuous on $\mathbb{R}$.

3.69. Show that the function $f(x) = x/(1 + x^2)$ is uniformly continuous on $\mathbb{R}$.

3.70. Prove that the function $f(x) = \tan^{-1} x$ is uniformly continuous on $\mathbb{R}$.

3.71. For x in $S = (0, \infty)$, let $f(x) = \sin(1/x)$.

a) Prove that f is in $C_\infty(S)$.

b) Prove that f is not uniformly continuous on S.

c) Fix any $a > 0$. Prove that f is uniformly continuous on $[a, \infty)$.

d) Does $\lim_{x\to\infty} f(x)$ exist? Prove your answer.

3.72. For x in $S = [0, \infty)$, define $f(x) = x \sin(1/x)$ and $f(0) = 0$.

a) Prove that f is in $C(S)$.

b) Prove that f is not bounded on S and that $\lim_{x\to\infty} f(x)$ does not exist.

c) Prove that, for any $b > 0$, f is uniformly continuous on $[0, b]$.

3.73. For $x \neq 0$, define $f(x) = (\sin x)/x$. Define $f(0) = 1$.

a) Show that f is continuous on $\mathbb{R}$.

b) Show that f is bounded on $\mathbb{R}$.

c) Find $\lim_{x\to-\infty} f(x)$ and $\lim_{x\to\infty} f(x)$.

d) Determine whether f is uniformly continuous on $\mathbb{R}$.

3.74. For $x = 0$, define $f(x) = (\sin x)/x^2$.

a) Show that $\lim_{x\to 0} f(x)$ does not exist.

b) Show that f is continuous on $S = (-\infty, 0) \cup (0, \infty)$.

c) Find $\lim_{x\to-\infty} f(x)$ and $\lim_{x\to\infty} f(x)$.

d) Show that, for any $r > 0$, f is uniformly continuous on $N(0; r)^c$.

3.75. For x in $\mathbb{R}$, let $f(x) = e^x$.

a) Determine whether f is uniformly continuous on $\mathbb{R}$.

b) Prove that, for any b in $\mathbb{R}$, f is uniformly continuous on $(-\infty, b]$.

3.76. For x in $S = (0, \infty)$, let $f(x) = \ln x$.

a) Determine whether f is uniformly continuous on S.

b) Fix any $a > 0$. Determine whether f is uniformly continuous on $[a, \infty)$.

3.77. Prove that $f(x) = \exp(-x^2)$ is uniformly continuous on $\mathbb{R}$.

3.78. For x in $\mathbb{R}$, let $f(x) = \exp(-x^2)\sin x$ and $g(x) = \exp(-x^2)\cos x$. Prove that both f and g are uniformly continuous on $\mathbb{R}$.

3.79. Prove that $f(\mathbf{x}) = \exp[-(x_1^2 + x_2^2)]$ is uniformly continuous on $\mathbb{R}^2$.

3.80. Prove that $f(\mathbf{x}) = \tan^{-1}(x_1^2 + x_2^2)$ is uniformly continuous on $\mathbb{R}^2$.

3.81. Define $S = \{\mathbf{x} \text{ in } \mathbb{R}^2 : \mathbf{x} \neq 0\}$. Define

$$f(\mathbf{x}) = \frac{2x_1x_2}{x_1^2 + x_2^2}.$$

a) Show that f is in $C_\infty(S)$.

b) Prove that f is not uniformly continuous on S.

c) Prove that $\lim_{\|x\|\to\infty} f(\mathbf{x})$ does not exist.

d) Prove that, for any $r > 0$, f is uniformly continuous on $N(\mathbf{0}; r)^c$.

3.82. Let $f(x) = p(x)/q(x)$ be a rational function such that deg $p <$ deg q and such that the polynomial q has no real roots.

a) Prove that f is in $C_\infty(\mathbb{R})$.

b) Prove that $\lim_{x\to-\infty} f(x) = \lim_{x\to\infty} f(x) = 0$.

c) Determine whether f is uniformly continuous on $\mathbb{R}$.

3.83. A function f on $\mathbb{R}$ is said to be periodic with period p if $f(x+p) = f(x)$ for all x in $\mathbb{R}$.

a) Prove that a continuous, periodic function on $\mathbb{R}$ is bounded and uniformly continuous on $\mathbb{R}$.

b) Prove that the sine and the cosine functions are uniformly continuous on $\mathbb{R}$.

3.84. Let S be a closed, unbounded set in $\mathbb{R}$ and let f be a function in $C_\infty(S)$. Suppose that $\lim_{x\to-\infty} f(x)$ and $\lim_{x\to\infty} f(x)$ exist. Prove that f is uniformly continuous on S.

3.85. Prove that, if f is uniformly continuous on a bounded set S in $\mathbb{R}^n$, then f is bounded on S.

3.86. Suppose that f is uniformly continuous on a subset S of $\mathbb{R}^n$ and that $\boldsymbol{c}$ is a limit point of S that is not in S. Prove that f can be extended uniquely to be defined at $\boldsymbol{c}$ in such a way that f is continuous at $\boldsymbol{c}$.

3.87. Suppose that f is a real-valued function that is uniformly continuous on a nonempty set S in $\mathbb{R}^n$ and that g is a real-valued function that is uniformly continuous on $f(S)$. Prove that $g \circ f$ is uniformly continuous on S.

3.88. Suppose that f and g are uniformly continuous, real-valued functions on a set S in $\mathbb{R}^n$. Prove that, if f and g are bounded on S, then fg, defined by $(fg)(\boldsymbol{x}) = f(\boldsymbol{x})g(\boldsymbol{x})$, is also uniformly continuous on S. Can we drop the boundedness of either f or g and still draw the same conclusion? Prove your answer.

3.5 The Uniform Norm: Uniform Convergence

Let S be a nonempty subset of $\mathbb{R}^n$. $B(S)$ denotes the vector space and ring of all bounded, real-valued functions on S. For each f in $B(S)$, $\{|f(\boldsymbol{x})| : \boldsymbol{x} \text{ in } S\}$ is a nonempty, bounded set of real numbers. Therefore $\sup\{|f(\boldsymbol{x})| : \boldsymbol{x} \text{ in } S\}$ exists and is nonnegative. We denote this number by $||f||_\infty$ and refer to it as the *uniform norm* of f. As we shall see, this concept provides us with an essential, useful tool in all that follows.

DEFINITION 3.5.1 The *uniform norm* of a function f in $B(S)$ is

$$||f||_\infty = \sup\{|f(\boldsymbol{x})| : \boldsymbol{x} \text{ in } S\}. \quad \bullet$$

In this definition, the word *norm* is properly used. That is, as our next theorem confirms, the uniform norm on $B(S)$ has the three essential properties of any norm.

THEOREM 3.5.1 The uniform norm is a norm. That is, it has the following properties for all f and g in $B(S)$ and all c in $\mathbb{R}$:

i) *Positive Definiteness* $||f||_\infty \geq 0$ and $||f||_\infty = 0$ if and only if $f = 0$

ii) *Absolute Homogeneity* $||cf||_\infty = |c|\,||f||_\infty$

iii) *Subadditivity* $||f+g||_\infty \leq ||f||_\infty + ||g||_\infty$

Proof. Only (iii) requires a few comments by way of proof. The essential observation is that for any point $\boldsymbol{x}$ in S,

$$|(f+g)(\boldsymbol{x})| \leq |f(\boldsymbol{x})| + |g(\boldsymbol{x})| \leq ||f||_\infty + ||g||_\infty.$$

Hence, $||f||_\infty + ||g||_\infty$ is an upper bound for the set

$$\{|(f+g)(\boldsymbol{x})| : \boldsymbol{x} \text{ in } S\}.$$

Taking the supremum, we have $||f+g||_\infty \le ||f||_\infty + ||g||_\infty$. ●

The uniform norm induces a metric d_∞ on $B(S)$ as in the following definition.

DEFINITION 3.5.2 The *uniform metric* on $B(S)$ is

$$d_\infty(f, g) = ||f - g||_\infty,$$

for all f and g in $B(S)$. ●

By imitating the proof of Theorem 2.1.4, we can prove that the function d_∞ defined above is, in fact, a metric. That is, d_∞ has the three essential properties of a metric. We leave the proof for you.

THEOREM 3.5.2 The function d_∞ is a metric on $B(S)$. That is, the following properties hold for all f, g, and h in $B(S)$:

i) *Positive Definiteness* $d_\infty(f, g) \ge 0$; $d_\infty(f, g) = 0$ if and only if $f = g$

ii) *Symmetry* $d_\infty(f, g) = d_\infty(g, f)$

iii) *The Triangle Inequality* $d_\infty(f, g) \le d_\infty(f, h) + d_\infty(h, g)$ ●

With this metric in hand we can lift all the central concepts introduced in Chapters 1 and 2 to $B(S)$. By doing so we will obtain useful tools for the analysis of the behavior of bounded, real-valued functions on S.

DEFINITION 3.5.3 A *[uniform] neighborhood* $N(f; r)$ of f with radius r is the set $N(f; r) = \{g \text{ in } B(S) : d_\infty(f, g) < r\}$. A *deleted [uniform] neighborhood of* f is obtained from $N(f; r)$ by removing the function f. ●

DEFINITION 3.5.4

i) A function f_0 defined on a set S in $\mathbb{R}^n$ is said to be a *[uniform] limit point* of a set F in $B(S)$ if every deleted [uniform] neighborhood $N'(f_0)$ of f_0 contains a function in F.

ii) A sequence $\{f_k : k \text{ in } \mathbb{N}\}$ of functions, each defined on a set S in $\mathbb{R}^n$, is said to *converge uniformly* to f_0 on S if, for every $\epsilon > 0$, there exists a k_0 such that, for $k \ge k_0$, it follows that f_k is in $N(f_0; \epsilon)$; that is, $||f_k - f_0||_\infty < \epsilon$. We write

$$\lim_{k\to\infty} f_k = f_0 \text{ [uniformly]}.$$

iii) A sequence $\{f_k : k \text{ in } \mathbb{N}\}$ of functions, each defined on a set S in $\mathbb{R}^n$, is said to *converge pointwise* to f_0 on S if, for each $\boldsymbol{c}$ in S and for every $\epsilon > 0$, there exists a k_0 such that, when $k \ge k_0$, $|f_k(\boldsymbol{c}) - f_0(\boldsymbol{c})| < \epsilon$. We write

$$\lim_{k\to\infty} f_k = f_0 \text{ [pointwise]}.$$

iv) A sequence $\{f_k : k \text{ in } \mathbb{N}\}$ of functions, each defined on a set S in $\mathbb{R}^n$, is said to be *[uniformly] Cauchy* if, for every $\epsilon > 0$, there exists a k_0 such that whenever $k, m \geq k_0$, it follows that $||f_m - f_k||_\infty < \epsilon$. ●

We can visualize a (uniform) neighborhood $N(f_0; r)$ of a function f_0 by the following images. First visualize the graph $G(f_0)$ of f_0. This graph is a curve, a surface, or a hypersurface, depending on whether $n = 1, 2$, or larger. Enclose $G(f_0)$ in a band of width $2r$ with $G(f_0)$ at its center. The boundaries of the band are curves (or surfaces or hypersurfaces) parallel to $G(f_0)$. The neighborhood $N(f_0; r)$ consists of all functions f in $B(S)$ whose graphs lie completely within the boundaries of this band. (See Fig. 3.19 for the case when $n = 1$, Fig. 3.20 for $n = 2$.) That f lies in $N(f_0; r)$ means that, at each point $\boldsymbol{x}$ in S, $|f(\boldsymbol{x}) - f_0(\boldsymbol{x})| < r$. Clearly this is a global property on S; it holds at every point of S simultaneously. For this reason the word *uniform* is used.

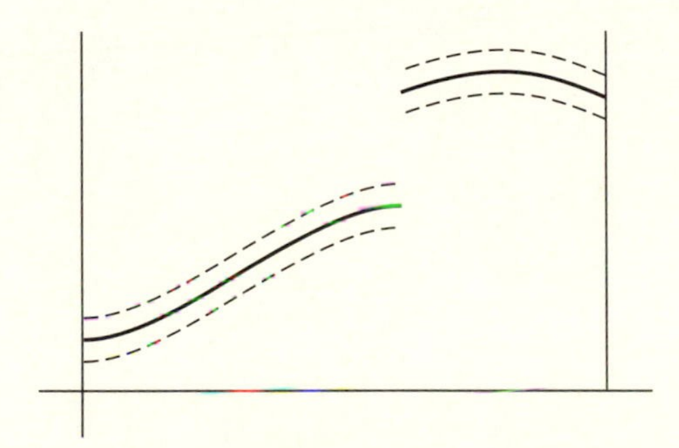

Figure 3.19

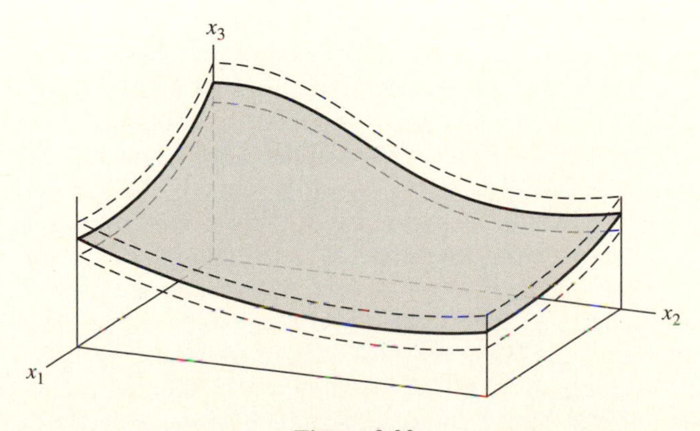

Figure 3.20

That f_0 is a uniform limit point of a collection F of functions means that, for every $\epsilon > 0$, there is a function f in F whose graph lies entirely within this band whose width is 2ϵ. Clearly, this is a global property on S; the value of f differs from the value of f_0 by less than ϵ at all the points of S simultaneously.

That a sequence $\{f_k\}$ of functions converges uniformly to f_0 means that, given $\epsilon > 0$, eventually f_k is in $N(f_0; \epsilon)$. That is, for k sufficiently large, the entire graph of f_k lies in this band of width 2ϵ. Again, uniform convergence is a global property on S. The uniform convergence of $\{f_k\}$ to f_0 strongly restricts the behavior of f_k for large k. The function f_k must closely approximate f_0 on the entire set S simultaneously. (See Fig. 3.21.)

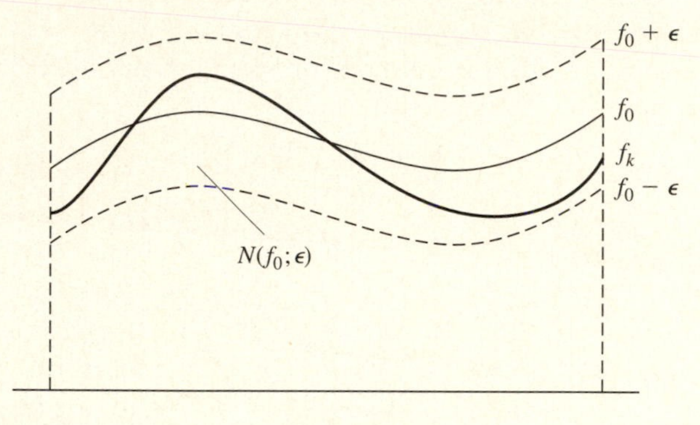

Figure 3.21

That the convergence of $\{f_k\}$ to f_0 is uniform is expressed in the definition by the specification of an index k_0 without mention of a point of S. The index k_0 depends on ϵ but is independent of any point of S that might be considered. Therefore the property is global. This is a subtle point in the language of mathematics over which beginners often stumble; we hope this informal discussion helps to guide you.

The situation becomes dramatically different when we consider the pointwise convergence of a sequence $\{f_k\}$ of functions to f_0. In that definition, first choose a point $\boldsymbol{c}$ in S. Next form the sequence $\{f_k(\boldsymbol{c})\}$ of real numbers. According to the definition, $\{f_k(\boldsymbol{c})\}$ converges to $f_0(\boldsymbol{c})$. The rate of convergence is revealed by the relative size of the index k_0. If, for a given ϵ, the index k_0 must be relatively large, then $\{f_k(\boldsymbol{c})\}$ converges to $f_0(\boldsymbol{c})$ slowly. If, for this same ϵ, the index k_0 can be taken to be relatively small, then the convergence of $\{f_k(\boldsymbol{c})\}$ to $f_0(\boldsymbol{c})$ is rapid. As $\boldsymbol{c}$ varies over S, each sequence $\{f_k(\boldsymbol{c})\}$ converges to $f_0(\boldsymbol{c})$, but the definition of pointwise convergence does not guarantee that the rate of convergence must be the same at each point. For each $\boldsymbol{c}$ in S and for each $\epsilon > 0$, there does exist an index k_0 such that, for $k \geq k_0$, the point $f_k(\boldsymbol{c})$ is in $N(f_0(\boldsymbol{c}); \epsilon)$. But generally, the index k_0 depends not only on ϵ but on $\boldsymbol{c}$ as well. In the definition of pointwise convergence, this dependence is implicit. First $\boldsymbol{c}$ and ϵ are specified; the existence of k_0 is declared subsequently. Accordingly, k_0 is understood to depend on both $\boldsymbol{c}$ and ϵ. While it is true that, for each ϵ and each $\boldsymbol{c}$ in S, the point $(\boldsymbol{c}, f_k(\boldsymbol{c}))$ eventually lies in the band of width 2ϵ centered at f_0, there may exist no function f_k in the sequence whose entire graph lies completely within the band. There may be no term f_k of the sequence in $N(f_0; \epsilon)$. However, and you can prove this as an exercise, if $\{f_k\}$ converges uniformly to f_0 on S, then $\{f_k\}$ converges pointwise to f_0 as well.

EXAMPLE 25 Let S be the interval $[0, 1]$. For each k in $\mathbb{N}$, let $f_k(x) = x^k$. We claim that the sequence $\{f_k\}$ converges pointwise to the function f_0, where $f_0(x) = 0$, for x in $[0, 1)$, and $f_0(1) = 1$. However, the convergence of $\{f_k\}$ to f_0 is not uniform. (See Fig. 3.22.) To confirm that $\lim_{k\to\infty} f_k = f_0$ [pointwise],

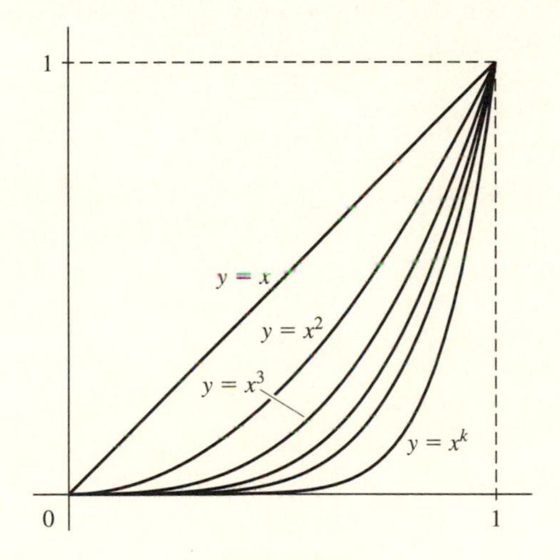

Figure 3.22

choose any c in $[0, 1)$. Then $\lim_{k\to\infty} f_k(c) = \lim_{k\to\infty} c^k = 0 = f_0(c)$. When $c = 1$ it is evident that $f_k(1) = 1 = f_0(1)$ for all k. Therefore, $\lim_{k\to\infty} f_k(1) = f_0(1)$ also. Thus $\{f_k\}$ converges pointwise to f_0. To see that the convergence cannot be uniform, fix any $\epsilon > 0$. Choose any k in $\mathbb{N}$. We want to show that there is a point x in $(0, 1)$ such that $|f_k(x) - f_0(x)| = x^k > \epsilon$. The inequality $x^k > \epsilon$ is equivalent to $x > \epsilon^{1/k}$. We can assume that $\epsilon < 1$; it follows that $\epsilon^{1/k} < 1$. Now choose any x in $(\epsilon^{1/k}, 1)$. Notice that x depends on both ϵ and k. For this choice of x, we have $|f_k(x) - f_0(x)| = x^k > \epsilon$. (See Fig. 3.23 on page 162.) Thus f_k is not in $N(f_0; \epsilon)$. But k was initially chosen arbitrarily in $\mathbb{N}$. No matter how large k may be, f_k is not in $N(f_0; \epsilon)$. We conclude that the convergence of $\{f_k\}$ to f_0 cannot be uniform. This simple example confirms that a sequence of functions can converge pointwise but not uniformly.

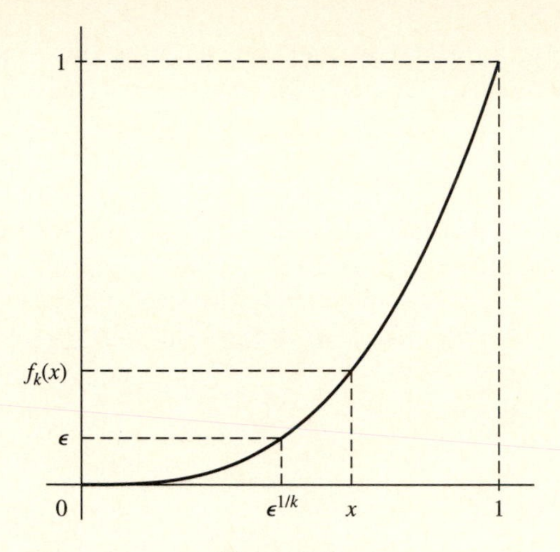

Figure 3.23

Notice that, in this example, the functions f_k are continuous on the set $S = [0, 1]$, but the pointwise limiting function f_0 is not continuous at $c = 1$. Mere pointwise convergence of a sequence $\{f_k\}$ of continuous functions does not ensure the continuity of the limit function f_0. ●

EXAMPLE 26 For each k in $\mathbb{N}$ and each x in $[0, \infty)$, let

$$f_k(x) = \frac{1}{1 + x^{2k}}.$$

You proved in Exercise 3.15(e) that $\{f_k\}$ converges pointwise to the function

$$f_0(x) = \begin{cases} 0, & \text{for } x > 1 \\ \frac{1}{2}, & \text{for } x = 1 \\ 1, & \text{for } 0 \le x < 1. \end{cases}$$

We claim, again, that the convergence cannot be uniform. To prove this claim, fix any ϵ such that $0 < \epsilon < 1/4$ and form the [uniform] neighborhood $N(f_0; \epsilon)$ as sketched in Fig. 3.24. Fix any k in $\mathbb{N}$. The unique point a_k in $[0, \infty)$ such that $f_k(a_k) = 1 - \epsilon$ is

$$a_k = \left[\frac{\epsilon}{1 - \epsilon}\right]^{1/(2k)}.$$

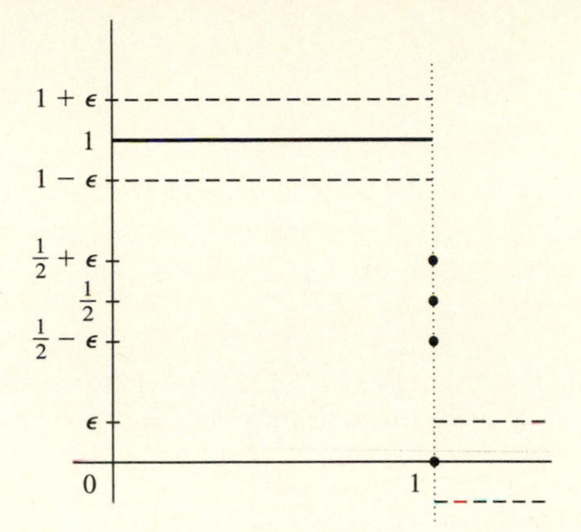

Figure 3.24

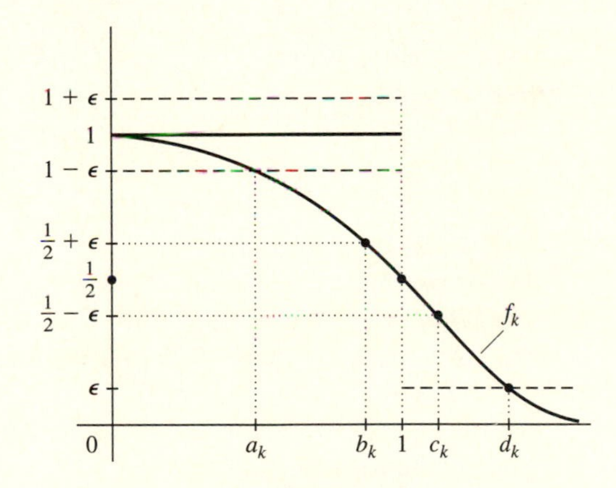

Figure 3.25

Likewise, the unique point b_k in $[0, \infty)$ such that $f_k(b_k) = 1/2 + \epsilon$ is

$$b_k = \left[\frac{1-2\epsilon}{1+2\epsilon}\right]^{1/(2k)}.$$

(See Fig. 3.25.) As you can confirm, $0 < a_k < b_k < 1$. Furthermore, for any x in (a_k, b_k),

$$\frac{1}{2} + \epsilon < f_k(x) < 1 - \epsilon.$$

These observations show that the function f_k fails to lie in the neighborhood $N(f_0; \epsilon)$. Similarly, you can find points c_k and d_k such that $1 < c_k < d_k$ and such that, for any x in (c_k, d_k), we have

$$\epsilon < f_k(x) < \frac{1}{2} - \epsilon.$$

Thus, for values of x in $(a_k, b_k) \cup (c_k, d_k)$, we have

$$|f_k(x) - f_0(x)| > \epsilon.$$

That is, the points $(x, f_k(x))$ lie outside the band of width 2ϵ centered at f_0. In other words, f_k does not belong to $N(f_0; \epsilon)$. Since k was initially chosen to be arbitrary, it follows that, although $\{f_k\}$ converges pointwise to f_0, the convergence cannot be uniform. ●

Note again that, in this example, each function f_k is continuous on all of $\mathbb{R}$, but the limit function f_0 has a jump discontinuity at $c = 1$.

EXAMPLE 27 For each k in $\mathbb{N}$ define $f_k(x) = kx/(1 + kx)$ for x in $S = [0, \infty)$. We claim first that $\{f_k\}$ converges pointwise to the function f_0 where $f_0(0) = 0$ and $f_0(x) = 1$ for $x > 0$. To prove this, observe that $f_k(0) = 0$ for all k. Thus we have $\lim_{k\to\infty} f_k(0) = 0$. Next, fix $x > 0$. Then, $\lim_{k\to\infty} f_k(x) = \lim_{k\to\infty} kx/(1 + kx) = \lim_{k\to\infty} 1 - 1/(1 + kx) = 1^-$. Therefore, $\lim_{k\to\infty} f_k = f_0$ [pointwise] as claimed.

However, the convergence is not uniform on $[0, \infty)$. To prove this, fix any k in $\mathbb{N}$ and any ϵ in $(0, 1/2)$. Then, for x such that

$$\frac{\epsilon}{k(1-\epsilon)} < x < \frac{1-\epsilon}{k\epsilon},$$

we have $\epsilon < f_k(x) < 1 - \epsilon$. For such values of x, $|f_k(x) - f_0(x)| > \epsilon$. Therefore, f_k is not in $N(f_0; \epsilon)$. Since k is arbitrary, we deduce that the convergence cannot be uniform. (See Fig. 3.26.) However, by excluding any small interval containing 0 from the domain S, the convergence will be uniform on the set that remains. Specifically, fix any $a > 0$. We claim that $\lim_{k\to\infty} f_k = f_0$ [uniformly] on $[a, \infty)$. To prove this, simply choose k_0 to be the smallest natural number greater than $(1 - \epsilon)/(a\epsilon)$. Then, as you can confirm for yourself, for $k \geq k_0$ and for any x in $[a, \infty)$,

$$1 > f_k(x) = \frac{kx}{1+kx} \geq \frac{ka}{1+ka} = f_k(a) > 1 - \epsilon.$$

It follows that, for all $k \geq k_0$, we have $||f_k - f_0||_\infty < \epsilon$. Thus, $\lim_{k\to\infty} f_k = f_0$ [uniformly] on $[a, \infty)$. ●

Throughout this text our interest will focus primarily on continuous functions on S. Our examples alert us to the fact that a pointwise convergent sequence in $C_\infty(S)$ may fail to have a limit that is in $C_\infty(S)$. Mere pointwise convergence cannot yield a satisfactory concept of completeness in our present context. As we now prove, however, the uniform convergence of any sequence of continuous functions $\{f_k\}$ ensures the continuity of the limit function.

THEOREM 3.5.3 Suppose that $\{f_k\}$ is a sequence of continuous, real-valued functions on S that converges uniformly to f_0 on S. Then f_0 is continuous on S.

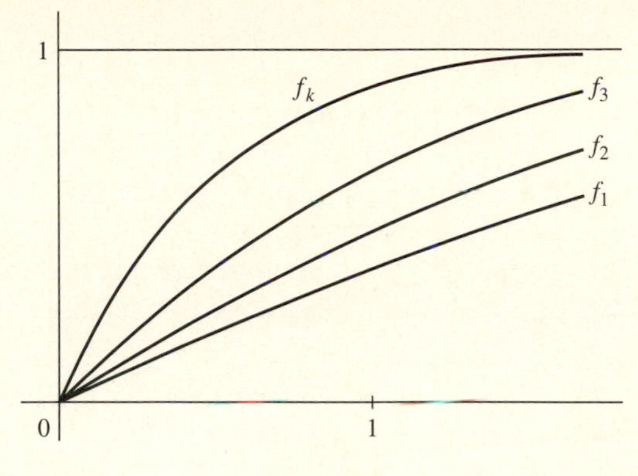

Figure 3.26

Proof. All we need prove is that f_0 is continuous at any point $\boldsymbol{c}$ in S. That is, we need to prove that, for any $\epsilon > 0$ and for any $\boldsymbol{c}$ in S, there exists a $\delta > 0$ such that

$$S \cap N(\boldsymbol{c}; \delta) \subseteq f_0^{-1}(N(f_0(\boldsymbol{c}); \epsilon)).$$

To this end, fix $\boldsymbol{c}$ in S and $\epsilon > 0$. By the uniform convergence of $\{f_k\}$ to f_0, choose a k_0 with the property that, if $k \geq k_0$, then $||f_k - f_0||_\infty < \epsilon/3$. Fix such a k. The function f_k is continuous on all of S, thus, in particular, at $\boldsymbol{c}$. Choose a $\delta > 0$ such that

$$S \cap N(\boldsymbol{c}; \delta) \subseteq f_k^{-1}\left(N\left(f_k(\boldsymbol{c}); \frac{\epsilon}{3}\right)\right).$$

By inserting the usual intermediate summands and using the triangle inequality, we can write, for $\boldsymbol{x}$ in $S \cap N(\boldsymbol{c}; \delta)$,

$$\begin{aligned}&|f_0(\boldsymbol{x}) - f_0(\boldsymbol{c})| \\ &\qquad \leq |f_0(\boldsymbol{x}) - f_k(\boldsymbol{x})| + |f_k(\boldsymbol{x}) - f_k(\boldsymbol{c})| + |f_k(\boldsymbol{c}) - f_0(\boldsymbol{c})|.\end{aligned}$$

The first and third of these summands are bounded above by $||f_k - f_0||_\infty < \epsilon/3$. Since $\boldsymbol{x}$ is in $S \cap N(\boldsymbol{c}; \delta)$, the middle summand is less than $\epsilon/3$. Therefore, if $\boldsymbol{x}$ is in $S \cap N(\boldsymbol{c}; \delta)$, we have

$$|f_0(\boldsymbol{x}) - f_0(\boldsymbol{c})| < \frac{\epsilon}{3} + \frac{\epsilon}{3} + \frac{\epsilon}{3} = \epsilon.$$

We conclude that f_0 is continuous at $\boldsymbol{c}$. Since $\boldsymbol{c}$ is an arbitrary point of S, it follows that f_0 is continuous on all of S. This proves that the uniform limit of continuous functions is continuous. ●

In Definition 3.5.4 we identified what is meant by saying that a sequence $\{f_k\}$ of bounded functions on S is uniformly Cauchy. As we learned in Chapters 1 and 2, Cauchy sequences play a central role in analysis. Here we further develop this central theme, discovering that analogues of our earlier theorems are valid in the present context.

THEOREM 3.5.4 Let S be any subset of $\mathbb{R}^n$. A Cauchy sequence $\{f_k\}$ in $B(S)$ is bounded. That is, there exists an $M > 0$ such that $||f_k||_\infty \leq M$ for all k in $\mathbb{N}$. ●

You can easily prove the theorem as an exercise, merely imitating the proof of Theorem 1.4.1.

As we now prove, every Cauchy sequence $\{f_k\}$ in $C_\infty(S)$ converges uniformly to a function f_0 in $C_\infty(S)$. This will prove that $C_\infty(S)$ is *[uniformly] Cauchy complete* or, more briefly, *complete.* Although we may not be able to know the ideal limit function f_0, we will be able, in applications, to approximate it arbitrarily closely over all of S simultaneously by a single function f_k in the Cauchy sequence. Consequently, this theory is enormously useful in all branches of applied analysis.

THEOREM 3.5.5 $C_\infty(S)$ is complete. That is, given any Cauchy sequence $\{f_k\}$ in $C_\infty(S)$, there exists a function f_0 in $C_\infty(S)$ such that $\lim_{k\to\infty} f_k = f_0$ [uniformly] on S.

Proof. We break the proof into manageable parts. Our first task is to establish the existence of the function f_0. Only thereafter will we prove that $\{f_k\}$ converges uniformly to f_0 and that f_0 is continuous and bounded on S.

First, choose any $\boldsymbol{c}$ in S. The sequence $\{f_k(\boldsymbol{c})\}$ is a Cauchy sequence of real numbers that, by the completeness of $\mathbb{R}$, must converge to a real number. Call its limit $f_0(\boldsymbol{c})$. As $\boldsymbol{c}$ varies over S, we obtain the values of $f_0(\boldsymbol{c})$. This defines the function f_0 mapping S into $\mathbb{R}$. Clearly, by the very construction of f_0, $\lim_{k\to\infty} f_k = f_0$ [pointwise].

We claim that $\{f_k\}$ actually converges uniformly to f_0. To see this, fix any $\epsilon > 0$. Since $\{f_k\}$ is a uniformly Cauchy sequence, we can choose a k_0 such that, for m, $k \geq k_0$ we have $||f_m - f_k||_\infty < \epsilon/3$. Note that k_0 depends only on ϵ. It follows that, for every $\boldsymbol{c}$ in S, $|f_m(\boldsymbol{c}) - f_k(\boldsymbol{c})| < \epsilon/3$.

Temporarily fix a $\boldsymbol{c}$ in S. Now, $\lim_{k\to\infty} f_k(\boldsymbol{c}) = f_0(\boldsymbol{c})$. Thus there exists an m_0 such that, for $m \geq m_0$,

$$|f_m(\boldsymbol{c}) - f_0(\boldsymbol{c})| < \frac{\epsilon}{3}.$$

For $k \geq k_0$ and $m \geq \max\{k_0, m_0\}$,

$$\begin{aligned}|f_k(\boldsymbol{c}) - f_0(\boldsymbol{c})| &\leq |f_k(\boldsymbol{c}) - f_m(\boldsymbol{c})| + |f_m(\boldsymbol{c}) - f_0(\boldsymbol{c})| \\ &< \frac{\epsilon}{3} + \frac{\epsilon}{3} = \frac{2\epsilon}{3}.\end{aligned}$$

Thus we see that $2\epsilon/3$ is an upper bound for $|f_k(\boldsymbol{c}) - f_0(\boldsymbol{c})|$ whenever $k \geq k_0$. Since we can obtain this same upper bound at each $\boldsymbol{c}$ in S, we conclude that $||f_k - f_0||_\infty \leq 2\epsilon/3 < \epsilon$ whenever $k \geq k_0$. Consequently, the sequence $\{f_k\}$ converges uniformly to f_0. That f_0 is continuous on S follows immediately by Theorem 3.5.3.

All that remains is to show that f_0 is also bounded and therefore is in $C_\infty(S)$. But this is easy. By Theorem 3.5.4, there is an $M > 0$ such that $||f_k||_\infty \leq M$ for all k in $\mathbb{N}$. Choose k_0 such that, for $k \geq k_0$, $||f_k - f_0||_\infty < 1$. Then, for $k \geq k_0$, $||f_0||_\infty \leq ||f_0 - f_k||_\infty + ||f_k||_\infty < 1 + M$. Therefore f_0 is in $C_\infty(S)$, proving all our claims. For every subset S in $\mathbb{R}^n$, $C_\infty(S)$ is complete. ●

COROLLARY 3.5.6 If S is a compact subset of $\mathbb{R}^n$, then $C(S)$ is complete.

Proof. If S is compact, then $C(S) = C_\infty(S)$ and $C_\infty(S)$ is complete. ●

In Chapter 4, when we derive Taylor's formula, and in Chapter 12, when we discuss sequences and series of functions in detail, we will have ample opportunity to explore the ramifications of this major theorem. Anticipating the results we will obtain there and building on your knowledge from your introductory calculus courses, we list, without proof, several examples that indicate the importance of Theorem 3.5.5.

EXAMPLE 28 For each k in $\mathbb{N}$, define

$$f_k(x) = 1 + x + \frac{x^2}{2!} + \frac{x^3}{3!} + \cdots + \frac{x^k}{k!}.$$

Then $\{f_k\}$ is a sequence of polynomials that converges uniformly to e^x on any compact subset S of $\mathbb{R}$. ●

EXAMPLE 29 For each k in $\mathbb{N}$, define

$$f_k(x) = 1 - \frac{x^2}{2!} + \frac{x^4}{4!} - \cdots + (-1)^k \frac{x^{2k}}{(2k)!}.$$

Then $\{f_k\}$ is a sequence of polynomials that converges uniformly to $\cos x$ on any compact subset S of $\mathbb{R}$. ●

EXAMPLE 30 For each k in $\mathbb{N}$, define

$$f_k(x) = x - \frac{x^3}{3} + \frac{x^5}{5} - \frac{x^7}{7} + \cdots + (-1)^k \frac{x^{2k+1}}{2k+1}.$$

Then $\{f_k\}$ is a sequence of polynomials that converges uniformly to $\tan^{-1} x$ on any compact subset of $(-1, 1)$. ●

EXAMPLE 31 For each k in $\mathbb{N}$, define

$$f_k(x) = x - \frac{x^2}{2} + \frac{x^3}{3} - \frac{x^4}{4} + \cdots + (-1)^{k-1} \frac{x^k}{k}.$$

Then $\{f_k\}$ is a sequence of polynomials that converges uniformly to $\ln(1 + x)$ on any compact subset of $(-1, 1)$. ●

Perhaps these examples will raise a conjecture in your mind: Suppose that S is a compact subset of $\mathbb{R}$ and that f is any function in $C(S)$. Is it possible that f can be uniformly approximated to any desired degree of accuracy by a polynomial? The affirmative answer, first provided by Weierstrass, is the content of Theorem 3.5.7, one of the most powerful theorems in applied analysis.

DEFINITION 3.5.5 Let S be a nonempty subset of $\mathbb{R}^n$. A collection F of functions in $C_\infty(S)$ is said to be *uniformly dense* in $C_\infty(S)$ if, for every f_0 in $C_\infty(S)$ and every (uniform) neighborhood $N(f_0)$ of f_0, there exists an f in $F \cap N(f_0)$. ●

The definition says formally that F is uniformly dense in $C_\infty(S)$ if every function in $C_\infty(S)$ can be uniformly approximated over all of S, to any desired degree of accuracy, by functions in the collection F. That is, given any f_0 in $C_\infty(S)$ and any $\epsilon > 0$, there exists an f in F such that

$$\|f - f_0\|_\infty < \epsilon.$$

Equivalently, there exists a sequence $\{f_k\}$ of functions in F that converges uniformly to f_0.

If S is a compact set in $\mathbb{R}$ and if $P(S)$ denotes the set of all polynomials in the single variable x, then $P(S)$ is certainly a subset of $C_\infty(S) = C(S)$.

THEOREM 3.5.7 The Weierstrass Approximation Theorem If S is a compact subset of $\mathbb{R}$, then $P(S)$ is uniformly dense in $C(S)$. ●

For our purposes it will suffice to prove Weierstrass's theorem under the assumption that $S = [a, b]$. Although several ingenious proofs of this theorem are now available, most require tools not yet at our disposal. One proof, crafted by Serge Bernštein, is accessible. It is a fairly long proof, involving several subordinate parts, labeled as lemmas. At its culmination, Bernštein draws all the parts together into an abrupt grand finale. (Imagine that you are watching the performance of a skilled juggler and marvel at his virtuosity. The beauty of the theorem is worth the effort.)

Bernštein restricts consideration first to the unit interval $[0, 1]$. Given a continuous function f on $[0, 1]$, construct, for each k in $\mathbb{N}$, the kth Bernštein polynomial for f, defined as follows: For x in $[0, 1]$,

$$B_k(x) = \sum_{j=0}^{k} f\left(\frac{j}{k}\right) C(k, j) x^j (1 - x)^{k-j}.$$

The numbers $C(k, j)$ are the binomial coefficients, $C(k, j) = k!/j!(k - j)!$, and the numbers $f(j/k)$ are the values of f at the equally spaced points $\{0, 1/k, 2/k, 3/k, \ldots, 1\}$ of $[0, 1]$. Note that $B_k(x)$ is a polynomial of degree at most k and that the coefficients of B_k depend on the function f.

LEMMA 1 For any k in $\mathbb{N}$, $\sum_{j=0}^{k} C(k, j) x^j (1 - x)^{k-j} = 1$.

Remark. This gives the kth Bernštein polynomial for the function $f(x) = 1$.

Proof. The identity follows immediately from the binomial theorem:

$$1 = [x + (1 - x)]^k = \sum_{j=0}^{k} C(k, j) x^j (1 - x)^{k-j}. \quad ●$$

LEMMA 2 For any k in $\mathbb{N}$ and any x in $[0, 1]$,

$$x = \sum_{j=0}^{k} \left(\frac{j}{k}\right) C(k, j) x^j (1 - x)^{k-j}.$$

Remark. This gives the kth Bernštein polynomial for the function $f(x) = x$.

Proof. By Lemma 1,

$$x = x1 = x\left\{\sum_{j=0}^{k-1} C(k-1, j)x^j(1-x)^{k-j-1}\right\} \qquad \text{(multiply by } x\text{)}$$

$$= \sum_{j=0}^{k-1} C(k-1, j)x^{j+1}(1-x)^{k-(j+1)} \qquad \text{(change index)}$$

$$= \sum_{j=1}^{k} C(k-1, j-1)x^j(1-x)^{k-j} \qquad \text{(insert } j/k \cdot k/j\text{)}$$

$$= \sum_{j=1}^{k} \left(\frac{j}{k}\right) C(k, j)x^j(1-x)^{k-j} \qquad \text{(insert } j = 0\text{)}$$

$$= \sum_{j=0}^{k} \left(\frac{j}{k}\right) C(k, j)x^j(1-x)^{k-j}. \quad \bullet$$

LEMMA 3 For any k in $\mathbb{N}$ and any x in $[0, 1]$,

$$k(k-1)x^2 = \sum_{j=0}^{k} (j^2 - j)C(k, j)x^j(1-x)^{k-j}.$$

Proof. As you can compute, for any $j = 2, 3, \ldots, k$,

$$(j^2 - j)C(k, j) = k(k-1)C(k-2, j-2). \qquad \mathbf{(3.5)}$$

Therefore,

$$\sum_{j=0}^{k} (j^2 - j)C(k, j)x^j(1-x)^{k-j} = \sum_{j=2}^{k} (j^2 - j)C(k, j)x^j(1-x)^{k-j}$$

$$= x^2 \sum_{j=2}^{k} (j^2 - j)C(k, j)x^{j-2}(1-x)^{k-j} \qquad \text{[apply (3.5)]}$$

$$= x^2 \sum_{j=2}^{k} k(k-1)C(k-2, j-2)x^{j-2}(1-x)^{(k-2)-(j-2)} \qquad \text{(change index)}$$

$$= x^2 k(k-1) \sum_{j=0}^{k-2} C(k-2, j)x^j(1-x)k^{(k-2)-j} \qquad \text{(Lemma 1)}$$

$$= k(k-1)x^2. \quad \bullet$$

LEMMA 4 For any k in $\mathbb{N}$ and any x in $[0, 1]$,

$$\left(1 - \frac{1}{k}\right)x^2 + \frac{x}{k} = \sum_{j=0}^{k} \left(\frac{j}{k}\right)^2 C(k, j)x^j(1-x)^{k-j}.$$

Proof. Play with algebra, apply Lemmas 2 and 3, and telescope

$$\begin{aligned}\left(1-\frac{1}{k}\right)x^2+\frac{x}{k} &= \frac{k(k-1)x^2}{k^2}+\frac{x}{k} \\ &= \frac{1}{k^2}\sum_{j=0}^{k}(j^2-j)C(k,j)x^j(1-x)^{k-j} \\ &\quad +\frac{1}{k}\sum_{j=0}^{k}\left(\frac{j}{k}\right)C(k,j)x^j(1-x)^{k-j} \\ &= \sum_{j=0}^{k}\left(\frac{j}{k}\right)^2 C(k,j)x^j(1-x)^{k-j}. \quad \bullet\end{aligned}$$

LEMMA 5 For any k in $\mathbb{N}$ and any x in $[0, 1]$,

$$\sum_{j=0}^{k}\left(x-\frac{j}{k}\right)^2 C(k,j)x^j(1-x)^{k-j} = \frac{x(1-x)}{k} \le \frac{1}{4k}.$$

Proof. By Lemmas 1, 2, and 4,

$$\begin{aligned}&\sum_{j=0}^{k}\left(x-\frac{j}{k}\right)^2 C(k,j)x^j(1-x)^{k-j} \\ &= \sum_{j=0}^{k}\left[x^2-2x\left(\frac{j}{k}\right)+\left(\frac{j}{k}\right)^2\right]C(k,j)x^j(1-x)^{k-j} \\ &= x^2\sum_{j=0}^{k}C(k,j)x^j(1-x)^{k-j} - 2x\sum_{j=0}^{k}\left(\frac{j}{k}\right)C(k,j)x^j(1-x)^{k-j} \\ &\quad +\sum_{j=0}^{k}\left(\frac{j}{k}\right)^2 C(k,j)x^j(1-x)^{k-j} \\ &= x^2-2x^2+\left(1-\frac{1}{k}\right)x^2+\frac{x}{k} = \frac{x(1-x)}{k}.\end{aligned}$$

Since, for x in $[0, 1]$, the function $x(1-x)$ is nonnegative and has a maximum value of $1/4$ at $x = 1/2$, we have

$$0 \le \sum_{j=0}^{k}\left(x-\frac{j}{k}\right)^2 C(k,j)x^j(1-x)^{k-j} = \frac{x(1-x)}{k} \le \frac{1}{4k}. \quad \bullet$$

THEOREM 3.5.8 Bernštein's Approximation Theorem Let f be any function in $C([0, 1])$. For any $\epsilon > 0$, there exists a k_0 in $\mathbb{N}$ such that, for $k \ge k_0$, $\|f - B_k\|_\infty < \epsilon$.

Proof. The function f is assumed to be continuous on the compact set $[0, 1]$. Theorem 3.3.2 guarantees that f is also uniformly continuous there, a fact that provides the key to our proof. Given $\epsilon > 0$, choose a $\delta > 0$ in accord with the definition of uniform continuity. If x and y are in $[0, 1]$ and if $|x - y| < \delta$, then $|f(x) - f(y)| < \epsilon/2$. Being continuous on the compact set $[0, 1]$, f is bounded there with $|f(x)| \leq ||f||_\infty$ for all x in $[0, 1]$.

Fix any x in $[0, 1]$ and compute the difference between $f(x)$ and its kth Bernštein polynomial, using Lemma 1 at the first step:

$$
\begin{aligned}
&|f(x) - B_k(x)| \\
&= \left| \sum_{j=0}^{k} f(x) C(k, j) x^j (1-x)^{k-j} - \sum_{j=0}^{k} f\left(\frac{j}{k}\right) C(k, j) x^j (1-x)^{k-j} \right| \\
&= \left| \sum_{j=0}^{k} \left[f(x) - f\left(\frac{j}{k}\right) \right] C(k, j) x^j (1-x)^{k-j} \right| \\
&\leq \sum_{j=0}^{k} \left| f(x) - f\left(\frac{j}{k}\right) \right| C(k, j) x^j (1-x)^{k-j} = S_1 + S_2,
\end{aligned}
\tag{3.6}
$$

where S_1 is the sum over the set A_1 of those indices j such that $|x - j/k| < \delta$ and S_2 is the sum over the set A_2 of those indices j such that $|x - j/k| \geq \delta$. We analyze each sum separately and show that, for different reasons, each of S_1 and S_2 must be small provided that k is sufficiently large. To assure the uniform approximation of f by B_k, it will be essential that k not depend on x.

That S_1 is small is easy to see. For j in A_1, we have

$$
\left| f(x) - f\left(\frac{j}{k}\right) \right| C(k, j) x^j (1-x)^{k-j} < \left(\frac{\epsilon}{2}\right) C(k, j) x^j (1-x)^{k-j}.
$$

Therefore, by Lemma 1,

$$
\begin{aligned}
S_1 &= \sum_{j \text{ in } A_1} \left| f(x) - f\left(\frac{j}{k}\right) \right| C(k, j) x^j (1-x)^{k-j} < \left(\frac{\epsilon}{2}\right) \sum_{j \text{ in } A_1} C(k, j) x^j (1-x)^{k-j} \\
&\leq \left(\frac{\epsilon}{2}\right) \sum_{j=0}^{k} C(k, j) x^j (1-x)^{k-j} = \frac{\epsilon}{2}.
\end{aligned}
$$

Notice that this last result required only Lemma 1 and the uniform continuity of f. Therefore, $S_1 < \epsilon/2$ independently of k.

To treat S_2, we note first that, if j is in A_2, then $|x - j/k| \geq \delta$ and thus $1/|x - j/k| \leq 1/\delta$. Consequently,

$$
\begin{aligned}
S_2 &= \sum_{j \text{ in } A_2} \left| f(x) - f\left(\frac{j}{k}\right) \right| C(k, j) x^j (1-x)^{k-j} \\
&\leq 2||f||_\infty \sum_{j \text{ in } A_2} C(k, j) x^j (1-x)^{k-j}
\end{aligned}
$$

$$= 2\|f\|_\infty \sum_{j \text{ in } A_2} \left(x - \frac{j}{k}\right)^2 \frac{1}{\left(x - \frac{j}{k}\right)^2} C(k, j)x^j(1-x)^{k-j}$$

$$\leq \left[\frac{2\|f\|_\infty}{\delta^2}\right] \sum_{j \text{ in } A_2} \left(x - \frac{j}{k}\right)^2 C(k, j)x^j(1-x)^{k-j}$$

$$\leq \left[\frac{2\|f\|_\infty}{\delta^2}\right] \sum_{j=0}^{k} \left(x - \frac{j}{k}\right)^2 C(k, j)x^j(1-x)^{k-j}$$

$$= \left[\frac{2\|f\|_\infty}{\delta^2}\right] \frac{x(1-x)}{k} \leq \frac{\|f\|_\infty}{2k\delta^2},$$

by Lemma 5. Finally, choose k_0 to be the smallest natural number greater than $\|f\|_\infty/(\epsilon\delta^2)$. Notice that k_0 is independent of x in $[0, 1]$. It follows that, for any $k \geq k_0$, we have $S_2 < \epsilon/2$. Thus, by (3.6), for any $k \geq k_0$, we have

$$|f(x) - B_k(x)| \leq S_1 + S_2 < \frac{\epsilon}{2} + \frac{\epsilon}{2} = \epsilon.$$

Since the choice of k_0 was independent of x, it follows that $\|f - B_k\|_\infty < \epsilon$, and Bernštein's approximation theorem is proved. Whew! ●

After all the trickery, Bernštein proves Weierstrass' theorem (Theorem 3.5.7) with breathtaking ease.

Proof of Theorem 3.5.7. Let f be any function in $C([a, b])$. Fix any $\epsilon > 0$. For t in $[0, 1]$, define

$$g(t) = f(a + (b-a)t).$$

This change of variables, consisting of a change of scale and translation, creates a new function g in $C([0, 1])$. Apply Bernštein's approximation theorem to obtain a Bernštein polynomial B_k such that $\|g - B_k\|_\infty < \epsilon$. Now reverse the change of variables: Let $t = (x-a)/(b-a)$. We have $f(x) = g((x-a)/(b-a))$ and

$$p(x) = B_k\left(\frac{x-a}{b-a}\right),$$

a polynomial in $P([a, b])$. Clearly, $\|f - p\|_\infty < \epsilon$. The polynomial p approximates f uniformly on $[a, b]$ with the prescribed accuracy. ●

For over a century only mathematicians appreciated the beauty of this result and its power in theoretical research. With the advent of computers, its importance is now more generally recognized. Essentially, computers can only add and multiply and, consequently, can easily calculate the values of any polynomial. But machines are unable to compute the values of any continuous function f other than polynomials. By first uniformly approximating f with a polynomial as closely as desired, however, one can program a computer to approximate the values of f by calculating the values of the approximating polynomial. The only limitation on the accuracy of the computation results from the limited capabilities of the machine itself, not from the mathematics.

EXERCISES

3.89. Let S be a nonempty subset of $\mathbb{R}^n$. Suppose that $\{f_k\}$ is a uniformly convergent sequence in $B(S)$ with limit f_0. Prove that $\{f_k\}$ converges pointwise to f_0.

3.90. Let $\{c_k\}$ be any sequence of real numbers such that $\lim_{k\to\infty} c_k = 1$. For k in $\mathbb{N}$ and all x in $\mathbb{R}$, let $f_k(x) = c_k x^2$. Determine whether $\{f_k\}$ converges uniformly to $f_0(x) = x^2$ on $\mathbb{R}$. Prove your answer.

3.91. For each k in $\mathbb{N}$ and for x in $[0, \infty)$, define

$$f_k(x) = \frac{kx}{2 + k^2x^2}.$$

a) Determine those values of x for which the sequence $\{f_k\}$ converges pointwise. Identify the limit function f_0.

b) Show that the convergence in part (a) is not uniform on $[0, \infty)$.

c) Fix $a > 0$. Show that $\lim_{k\to\infty} f_k = f_0$ [uniformly] on $[a, \infty)$.

3.92. For each k in $\mathbb{N}$ and for each x in $[0, \infty)$, define

$$f_k(x) = \frac{x^k}{1 + x^k}.$$

a) Find all x where $\lim_{k\to\infty} f_k(x)$ exists. Identify the limit function f_0.

b) Fix any b in $(0, \infty)$. Prove that, if $b < 1$, then the convergence of $\{f_k\}$ to f_0 is uniform on $[0, b]$. Prove also that, if $b \geq 1$, then the convergence fails to be uniform on $[0, b]$.

3.93. For k in $\mathbb{N}$ and x in $\mathbb{R}$, define $f_k(x) = \tan^{-1} kx$.

a) Find the pointwise limit f_0 of $\{f_k\}$ on $\mathbb{R}$.

b) Prove that the convergence in part (a) is not uniform on $\mathbb{R}$.

c) Prove that the convergence is uniform on any closed interval I that does not contain $c = 0$.

3.94. Fix r in $(0, 1]$. For each k in $\mathbb{N}$ and each $x \geq 0$, define $f_k(x) = kx^r e^{-kx}$.

a) Prove that each f_k is continuous on $[0, \infty)$.

b) Determine those values of x for which the sequence $\{f_k\}$ converges pointwise. Identify the limit function f_0. (*Hint:* Refer to Exercise 1.33.)

c) Determine whether the convergence is uniform on $[0, \infty)$.

d) Fix any $a > 0$. Is the convergence in part (b) uniform on $[a, \infty)$? Prove your answer.

3.95. Repeat Exercise 3.94 with $r > 1$, k in $\mathbb{N}$, $x \geq 0$, and $f_k(x) = kx^r e^{-kx}$.

3.96. For each k in $\mathbb{N}$ and each x in $[0, 1]$, let $f_k(x)$ be defined by

$$f_k(x) = \begin{cases} 2k^2x, & \text{for } 0 \leq x \leq \frac{1}{2}k \\ -2k^2x + 2k, & \text{for } \frac{1}{2}k < x < \frac{1}{k} \\ 0, & \text{for } \frac{1}{k} \leq x \leq 1. \end{cases}$$

a) Show that, for each k, f_k is continuous on $[0, 1]$.

b) Determine those values of x in $[0, 1]$ for which the sequence $\{f_k\}$ converges pointwise. Identify the limit function f_0.

c) Show that $\{f_k\}$ does not converge uniformly to f_0.

d) Show that for each k, $\int_0^1 f_k(x)\,dx = 1/2$, but that $\int_0^1 f_0(x)\,dx = 0$. (In Chapter 6, this example will recur to show that $\lim_{k\to\infty} \int_0^1 f_k(x)\,dx$ and $\int_0^1 \lim_{k\to\infty} f_k(x)\,dx$ may not be equal if $\{f_k\}$ converges pointwise but not uniformly.)

3.97. For each k in $\mathbb{N}$ and each $\boldsymbol{x} = (x_1, x_2)$ in $\mathbb{R}^2$, define $f_k(\boldsymbol{x}) = 1/[1 + (x_1^2 + x_2^2)^k]$.

a) Prove that f_k is continuous on $\mathbb{R}^2$.

b) Prove that $\{f_k\}$ converges pointwise to f_0 defined by

$$f_0(x) = \begin{cases} 0, & \text{for } ||\boldsymbol{x}|| > 1 \\ \frac{1}{2}, & \text{for } ||\boldsymbol{x}|| = 1 \\ 1, & \text{for } ||\boldsymbol{x}|| < 1. \end{cases}$$

c) Prove that the convergence is not uniform.

d) Fix a positive $r < 1$ and restrict the functions f_k and f_0 to the compact set $S = \{\boldsymbol{x}$ in $\mathbb{R}^2 : ||\boldsymbol{x}|| \le r\}$. Prove that $\{f_k\}$ converges uniformly to f_0 on S.

3.98. Let S be a nonempty subset of $\mathbb{R}^n$ and let F be a uniformly dense subset of $C_\infty(S)$. Prove that if f_0 is any function in $C_\infty(S)$, then there is a uniformly Cauchy sequence $\{f_k\}$ of functions in F that converges uniformly to f_0.

3.99. Let $\{f_k\}$ be a uniformly convergent sequence in $C_\infty(S)$ with limit f_0.

a) If S is compact, is f_0 necessarily uniformly continuous on S? If so, prove it; otherwise, provide a counterexample.

b) If S is not compact, is f_0 necessarily uniformly continuous on S? If so, prove it; otherwise, provide a counterexample.

3.100. Suppose that f is a continuous, real-valued function on $[a, b]$. Prove that, for any $\epsilon > 0$, there exists a step function s on $[a, b]$ such that $||f - s||_\infty < \epsilon$. (A step function has only finitely many jump discontinuities at $c_1 < c_2 < \cdots < c_p$ in $[a, b]$ and is constant on each interval $[a_1, c_2)$, (c_{j-1}, c_j), and $(c_p, b]$.)

3.101. Suppose that f is a continuous, real-valued function on $[a, b]$. Prove that, for any $\epsilon > 0$, there exists a polygonal function p on $[a, b]$ such that $||f - p||_\infty < \epsilon$. (A polygonal function is a continuous function whose graph consists of straight-line segments joined end to end.)

3.6 VECTOR-VALUED FUNCTIONS ON $\mathbb{R}^n$

The success of the approach adopted in Section 3.2 motivates us to attempt its generalization to a function that maps vectors in $\mathbb{R}^n$ to vectors in $\mathbb{R}^m$. As we will see, being *coordinate free,* many of the results obtained in Section 3.2 lift easily to this more general context. Let $\boldsymbol{f}$ be a function with domain S in $\mathbb{R}^n$ and range T in $\mathbb{R}^m$. For $\boldsymbol{x}$ in S, we write

$$\boldsymbol{f}(\boldsymbol{x}) = \boldsymbol{y} = (y_1, y_2, \ldots, y_m).$$

Then, for each $j = 1, 2, \ldots, m$, there is a real-valued function f_j on S defined by $f_j(\boldsymbol{x}) = y_j$. The functions $f_1, f_2, \ldots, f_m$ are called the *component functions* of $\boldsymbol{f}$. We write $\boldsymbol{f} = (f_1, f_2, \ldots, f_m)$. In this notation, for $\boldsymbol{x}$ in S, we have

$$\boldsymbol{f}(\boldsymbol{x}) = (f_1(\boldsymbol{x}), f_2(\boldsymbol{x}), \ldots, f_m(\boldsymbol{x})) = (y_1, y_2, \ldots, y_m) = \boldsymbol{y}.$$

As we will see, properties of $\boldsymbol{f}$ are reflected in those of the m component functions $f_1, f_2, \ldots, f_m$.

DEFINITION 3.6.1 Let S be a subset of $\mathbb{R}^n$ and let f be a function defined on S with values in $\mathbb{R}^m$.

i) Let c be a point in $\overline{S}$. We say that f has *limit* v as x approaches c, and we write $\lim_{x\to c} f(x) = v$ if, for every neighborhood $N(v)$, there exists a deleted neighborhood $N'(c)$ such that $S \cap N'(c) \subseteq f^{-1}(N(v))$.

ii) Let c be a point of S. We say that f is *continuous* at c if $\lim_{x\to c} f(x) = f(c)$.

iii) We say that f is *continuous on* S if f is continuous at every point of S. ●

Our first theorem links the limiting behavior of f with the simultaneous limiting behavior of all its component functions.

THEOREM 3.6.1 Let $f = (f_1, f_2, \ldots, f_m)$ denote a function defined on a set S in $\mathbb{R}^n$ with values in $\mathbb{R}^m$.

i) Let c be a point of $\overline{S}$. Then $\lim_{x\to c} f(x) = v = (v_1, v_2, \ldots, v_m)$ if and only if, for each $j = 1, 2, \ldots, m$, $\lim_{x\to c} f_j(x) = v_j$.

ii) Let c be a point of S. Then f is continuous at c if and only if, for each $j = 1, 2, \ldots, m$, the component function f_j is continuous at c.

iii) The function f is continuous on S if and only if, for each $j = 1, 2, \ldots, m$, the component function f_j is continuous on S.

Proof. First suppose that $f = (f_1, f_2, \ldots, f_m)$ has limit v as x approaches c. Fix $\epsilon > 0$ and choose $\delta > 0$ such that, if x is in $S \cap N'(c; \delta)$, then $f(x)$ is in $N(v; \epsilon)$. Choose any coordinate index j and any x in $S \cap N'(c; \delta)$. Then we have

$$0 \le [f_j(x) - v_j]^2 \le \sum_{i=1}^{m} [f_i(x) - v_i]^2 = ||f(x) - v||^2 < \epsilon^2.$$

Thus, if x is in $S \cap N'(c; \delta)$, then $f_j(x)$ is in $N(v_j; \epsilon)$. That is, $\lim_{x\to c} f_j(x) = v_j$ for each $j = 1, 2, \ldots, m$.

Conversely, suppose that $\lim_{x\to c} f_j(x) = v_j$ for each $j = 1, 2, \ldots, m$. Fix $\epsilon > 0$. For each coordinate index j choose δ_j so that, when x is in $S \cap N'(c; \delta_j)$, it follows that $f_j(x)$ is in $N(v_j; \epsilon/\sqrt{m})$. Let $\delta = \min\{\delta_1, \delta_2, \ldots, \delta_m\}$. Then, for any x in $S \cap N'(c; \delta)$, we have

$$||f(x) - v||^2 = \sum_{j=1}^{m} [f_j(x) - v_j]^2 < \epsilon^2.$$

Thus $f(x)$ is in $N(v; \epsilon)$. This proves that $\lim_{x\to c} f(x) = v$ and completes the proof of part (i). Part (ii) follows by replacing v by $f(c) = (f_1(c), f_2(c), \ldots, f_m(c))$ throughout the preceding argument. Part (iii) of the theorem follows immediately. ●

Theorem 3.6.1, combined with Theorems 2.1.7 and 3.1.6, yields the following useful result.

COROLLARY 3.6.2 Let $\boldsymbol{f} = (f_1, f_2, \ldots, f_m)$ be a function defined on a set S in $\mathbb{R}^n$ and taking its values in $\mathbb{R}^m$.

i) Let $\boldsymbol{c}$ be a point in $\overline{S}$. Then $\lim_{\boldsymbol{x}\to\boldsymbol{c}} \boldsymbol{f}(\boldsymbol{x}) = \boldsymbol{v}$ if and only if, for every sequence $\{\boldsymbol{x}_k\}$ in S such that $\lim_{k\to\infty} \boldsymbol{x}_k = \boldsymbol{c}$, we have $\lim_{k\to\infty} \boldsymbol{f}(\boldsymbol{x}_k) = \boldsymbol{v}$.

ii) Let $\boldsymbol{c}$ be a point of S. Then $\boldsymbol{f}$ is continuous at $\boldsymbol{c}$ if and only if, for every sequence $\{\boldsymbol{x}_k\}$ in S such that $\lim_{k\to\infty} \boldsymbol{x}_k = \boldsymbol{c}$, we have $\lim_{k\to\infty} \boldsymbol{f}(\boldsymbol{x}_k) = \boldsymbol{f}(\boldsymbol{c})$.

Proof. By Theorem 3.6.1, we know that $\lim_{\boldsymbol{x}\to\boldsymbol{c}} \boldsymbol{f}(\boldsymbol{x}) = \boldsymbol{v}$ if and only if, for each $j = 1, 2, \ldots, m$, we have $\lim_{\boldsymbol{x}\to\boldsymbol{c}} f_j(\boldsymbol{x}) = v_j$ where $\boldsymbol{v} = (v_1, v_2, \ldots, v_m)$. By part (i) of Theorem 3.1.6, for each j the function f_j has limit v_j as $\boldsymbol{x}$ tends to $\boldsymbol{c}$ if and only if, for every sequence $\{\boldsymbol{x}_k\}$ in S such that $\lim_{k\to\infty} \boldsymbol{x}_k = \boldsymbol{c}$, it follows that $\lim_{k\to\infty} f_j(\boldsymbol{x}_k) = v_j$. That is, each component sequence $\{f_j(\boldsymbol{x}_k)\}$ of the sequence $\{\boldsymbol{f}(\boldsymbol{x}_k)\}$ converges to the jth component of $\boldsymbol{v}$. By Theorem 2.1.7, this last condition is equivalent to the convergence of $\{\boldsymbol{f}(\boldsymbol{x}_k)\}$ to $\boldsymbol{v}$.

Part (ii) of the corollary follows by replacing $\boldsymbol{v}$ throughout the foregoing proof by $\boldsymbol{f}(\boldsymbol{c}) = (f_1(\boldsymbol{c}), f_2(\boldsymbol{c}), \ldots, f_m(\boldsymbol{c}))$. ●

The proof of Theorem 3.2.1 lifts verbatim to prove its analog in the present context.

THEOREM 3.6.3 Let $\boldsymbol{f}$ map S in $\mathbb{R}^n$ onto T in $\mathbb{R}^m$. Then $\boldsymbol{f}$ is continuous on S if and only if the inverse image of every relatively open set in T is relatively open in S. ●

Likewise, the proof of Theorem 3.2.2 applies without change to prove

THEOREM 3.6.4 The continuous image of a connected set S in $\mathbb{R}^n$ is connected in $\mathbb{R}^m$. ●

The proof of Theorem 3.2.3 was designed in such a way as to prove, with no change, the following theorem.

THEOREM 3.6.5 If $\boldsymbol{f}$ is a continuous function on the compact set S in $\mathbb{R}^n$ with values in $\mathbb{R}^m$, then $\boldsymbol{f}(S)$ is compact in $\mathbb{R}^m$. ●

COROLLARY 3.6.6 If $\boldsymbol{f}$ is a continuous function mapping a compact set S in $\mathbb{R}^n$ into $\mathbb{R}^m$, then $\boldsymbol{f}(S) = T$ is a bounded set in $\mathbb{R}^m$. That is, there exists an $M > 0$ such that $||\boldsymbol{f}(\boldsymbol{x})|| \leq M$, for all $\boldsymbol{x}$ in S.

Proof. Since S is compact and $\boldsymbol{f}$ is continuous on S, we know by Theorem 3.6.5 that $\boldsymbol{f}(S) = T$ is compact. Therefore, T is bounded. Thus there exists an $M > 0$ such that $||\boldsymbol{y}|| \leq M$ for all $\boldsymbol{y}$ in T. That is, $||\boldsymbol{f}(\boldsymbol{x})|| \leq M$ for all $\boldsymbol{x}$ in S. ●

Since, for $m \geq 2$, there is no linear order on $\mathbb{R}^m$, we cannot expect a generalization of the intermediate value theorem (Theorems 3.2.5 and 3.2.6) to hold in our present context. However, Theorem 3.2.7, regarding the continuity of the composition of continuous functions, and its proof lift verbatim.

THEOREM 3.6.7 Let $\boldsymbol{f}$ map S in $\mathbb{R}^n$ onto $T = \boldsymbol{f}(S)$ in R^m and let $\boldsymbol{g}$ map T onto $\boldsymbol{g}(T)$ in $\mathbb{R}^p$. If $\boldsymbol{f}$ is continuous at a point $\boldsymbol{c}$ in S and if $\boldsymbol{g}$ is continuous at $\boldsymbol{d} = \boldsymbol{f}(\boldsymbol{c})$ in T, then $(g \circ f)$ is continuous at $\boldsymbol{c}$. If $\boldsymbol{f}$ is continuous on S and if $\boldsymbol{g}$ is continuous on $T = \boldsymbol{f}(S)$, then $(g \circ f)$ is continuous on S. ●

Suppose that $\boldsymbol{f}$ is a continuous function that maps a set S in $\mathbb{R}^n$ one-to-one onto a set T in $\mathbb{R}^m$. It follows that the inverse function $\boldsymbol{f}^{-1}$ exists and maps $T = \boldsymbol{f}(S)$ one-to-one onto S. The proof of Theorem 3.2.8 applies verbatim to prove our next result.

THEOREM 3.6.8 Let $\boldsymbol{f}$ be a continuous function mapping a compact set S in $\mathbb{R}^n$ one-to-one into $\mathbb{R}^m$. Then $\boldsymbol{f}^{-1}$ is also continuous on $\boldsymbol{f}(S)$. ●

The important concept of uniformly continuity is also *dimension free* and lifts in a straightforward way to apply to functions defined on $\mathbb{R}^n$ with values in $\mathbb{R}^m$.

DEFINITION 3.6.2 Let $\boldsymbol{f}$ be function mapping a set S in $\mathbb{R}^n$ into $\mathbb{R}^m$. We say that $\boldsymbol{f}$ is *uniformly continuous* on S if, for every $\epsilon > 0$, there exists a $\delta > 0$ such that,

$$S \cap N(\boldsymbol{c}; \delta) \subseteq \boldsymbol{f}^{-1}(N(\boldsymbol{f}(\boldsymbol{c}); \epsilon))$$

for all $\boldsymbol{c}$ in S. ●

Again, since in this definition the value of δ is specified prior to any mention of a point $\boldsymbol{c}$ in S, it is to be understood that δ depends on ϵ but is independent of $\boldsymbol{c}$. The uniform continuity of a vector-valued function $\boldsymbol{f}$ is a global property of $\boldsymbol{f}$. The equivalent reformulation of Definition 3.6.2, analogous to Theorem 3.4.1, is this: Given $\epsilon > 0$, there exists a $\delta > 0$ such that, whenever $\boldsymbol{x}$ and $\boldsymbol{y}$ are points of S with $||\boldsymbol{x} - \boldsymbol{y}|| < \delta$, it follows that $||\boldsymbol{f}(\boldsymbol{x}) - \boldsymbol{f}(\boldsymbol{y})|| < \epsilon$.

We immediately have the following theorem; its proof is straightforward and is left as an exercise.

THEOREM 3.6.9 Let $\boldsymbol{f} = (f_1, f_2, \ldots, f_m)$ be a function defined on a set S in $\mathbb{R}^n$ with values in $\mathbb{R}^m$. Then $\boldsymbol{f}$ is uniformly continuous on S if and only if, for each $j = 1, 2, \ldots, m$, the function f_j is uniformly continuous on S. ●

We conclude this section with the following generalization of Theorem 3.4.2. Its proof imitates that of Theorem 3.4.2 in all details and is left for you to confirm.

THEOREM 3.6.10 Let S be a compact set in $\mathbb{R}^n$ and let $\boldsymbol{f}$ be a continuous function that maps S into $\mathbb{R}^m$. Then $\boldsymbol{f}$ is uniformly continuous on S. ●

EXERCISES

3.102. A curve C in the plane can be described by a continuous function $\boldsymbol{f} = (f_1, f_2)$ mapping an interval I into $\mathbb{R}^2$. Likewise, a curve C in 3-space can be described by a continuous function $\boldsymbol{f} = (f_1, f_2, f_3)$ mapping I into $\mathbb{R}^3$. The component functions of $\boldsymbol{f}$ provide the parametric equations that represent the curve. The variable x in I is called the parameter in the representation. In each of the following, sketch the curve represented by the function $\boldsymbol{f}$.

a) Fix any $r > 0$ and, for x in $[0, 2\pi)$, let $f_1(x) = r\cos x$ and $f_2(x) = r\sin x$. Let $\boldsymbol{f} = (f_1, f_2)$.

b) Fix $a > 0$ and $b > 0$. For x in $[0, 2\pi)$, let $f_1(x) = a\cos x$ and $f_2(x) = b\sin x$. Let $\boldsymbol{f} = (f_1, f_2)$.

c) For x in $[0, 4\pi]$, define $f_1(x) = e^{-x}\cos x$ and $f_2(x) = e^{-x}\sin x$. Let $\boldsymbol{f} = (f_1, f_2)$.

d) Fix $r > 0$. For x in $[0, 2\pi]$, let $f_1(x) = r(x + \sin x)$ and $f_2(x) = r(1 - \cos x)$. Let $\boldsymbol{f} = (f_1, f_2)$.

e) Fix $r > 0$. For x in $[0, 4\pi]$, let $f_1(x) = r\cos x$, $f_2(x) = r\sin x$, and $f_3(x) = x^{1/2}$. Let $\boldsymbol{f} = (f_1, f_2, f_3)$.

f) For x in $[0, 6\pi]$, define $\boldsymbol{f} = (f_1, f_2, f_3)$ by letting $f_1(x) = x + \cos x$, $f_2(x) = x + \sin x$, and $f_3(x) = x$.

3.103. A surface $\mathcal{S}$ in $\mathbb{R}^3$ can be represented parametrically by a continuous function $\boldsymbol{f} = (f_1, f_2, f_3)$ mapping a rectangle R in $\mathbb{R}^2$ into $\mathbb{R}^3$. The variables x_1 and x_2 with $\boldsymbol{x} = (x_1, x_2)$ in $\mathbb{R}$ are called the parameters of the representation. In each of the following, sketch the surface represented by $\boldsymbol{f}$.

a) Fix $r > 0$. For $\boldsymbol{x} = (x_1, x_2)$ in $[0, 2\pi) \times [0, \pi]$, let

$$f_1(\boldsymbol{x}) = r\cos x_1 \sin x_2,$$

$$f_2(\boldsymbol{x}) = r\sin x_1 \sin x_2,$$

$$f_3(\boldsymbol{x}) = r\cos x_2.$$

b) Fix three positive numbers a_1, a_2, and a_3. For $\boldsymbol{x} = (x_1, x_2)$ in $[0, 2\pi) \times [0, \pi]$, define $\boldsymbol{f} = (f_1, f_2, f_3)$ by letting

$$f_1(\boldsymbol{x}) = a_1 \cos x_1 \sin x_2,$$

$$f_2(\boldsymbol{x}) = a_2 \sin x_1 \sin x_2,$$

$$f_3(\boldsymbol{x}) = a_3 \cos x_2.$$

c) For $\boldsymbol{x} = (x_1, x_2)$ in $[0, 2\pi) \times [0, 4]$, define a function $\boldsymbol{f} = (f_1, f_2, f_3)$ by letting

$$f_1(\boldsymbol{x}) = x_2^{1/2}\cos x_1,$$

$$f_2(\boldsymbol{x}) = x_2,$$

$$f_3(\boldsymbol{x}) = x_2^{1/2}\sin x_1.$$

d) Fix r_1 and r_2 with $0 < r_2 < r_1$. For $\boldsymbol{x} = (x_1, x_2)$ in the rectangle $[0, 2\pi) \times [0, 2\pi)$, let

$$f_1(\boldsymbol{x}) = (r_1 + r_2\cos x_2)\cos x_1,$$

$$f_2(\boldsymbol{x}) = (r_1 + r_2\cos x_2)\sin x_1,$$

$$f_3(\boldsymbol{x}) = r_2 \sin x_2.$$

3.104. Let $\boldsymbol{f}$ be a function mapping a set S in $\mathbb{R}^n$ into $\mathbb{R}^m$. Prove that $\boldsymbol{f}$ is continuous on S if and only if, for every relatively open set U in $T = \boldsymbol{f}(S)$, $\boldsymbol{f}^{-1}(U)$ is relatively open in S.

3.105. Prove that, if $\boldsymbol{f}$ is a continuous function that maps a connected set S in $\mathbb{R}^n$ into $\mathbb{R}^m$, then $\boldsymbol{f}(S)$ is connected.

3.106. Let S be a compact set in $\mathbb{R}^n$. Suppose that $\boldsymbol{f}$ is a continuous function mapping S into $\mathbb{R}^m$. Prove that $\boldsymbol{f}(S)$ is compact.

3.107. Let $f = (f_1, f_2, \ldots, f_m)$ be a function defined on a set S in $\mathbb{R}^n$ with values in $\mathbb{R}^m$. Prove that f is uniformly continuous on S if and only if, for each $j = 1, 2, \ldots, m$, the function f_j is uniformly continuous on S.

3.108. Let S be a compact set in $\mathbb{R}^n$. Suppose that f is a continuous function mapping S into $\mathbb{R}^m$. Prove that f is uniformly continuous on S.

4

Differentiation

In your introductory course in calculus, you first studied the operation of differentiation. Here we build on that introduction and provide rigorous proofs of the essential facts you will need throughout the remainder of this text.

4.1 THE DERIVATIVE

Suppose that a real-valued function f is defined in a neighborhood of a point c in $\mathbb{R}$. Consider the line

$$y = m(x - c) + f(c)$$

passing through the point $(c, f(c))$ with slope m. Temporarily consider m to be an undetermined constant. Let x be any point in $N'(c)$ and compare the change $\Delta f = f(x) - f(c)$ in the value of f with the change in the y value on the line:

$$df = y - f(c) = m(x - c) = mh,$$

where $h = x - c$ is the change in x. (See Fig. 4.1 on page 182.) Notice that, as x tends to c, h tends to 0. The error incurred by approximating Δf by df is $\Delta f - df$. We are particularly interested in understanding what condition must hold to ensure that $\Delta f - df$ tends to zero faster than h does. Specifically, we are interested in determining what value of m, if any, can be chosen to ensure such behavior. Mathematically, we express this limiting behavior by writing

$$\lim_{h \to 0} \frac{\Delta f - df}{h} = 0.$$

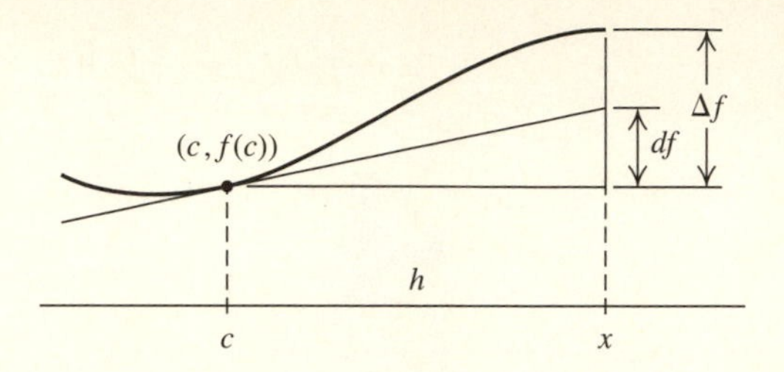

Figure 4.1

We recast this limit in the following equivalent form:

$$\lim_{h\to 0}\left\{\frac{f(c+h)-f(c)}{h}-m\right\}=0.$$

Since m is to be a constant, this says that, if $\lim_{h\to 0}[f(c+h)-f(c)]/h$ exists and if we take m to be this limiting value, then $\Delta f - df$ tends to 0 faster than h. This motivates the following definition.

DEFINITION 4.1.1 Let f be defined on an interval I in $\mathbb{R}$. Let c be a point in I. The *derivative* of f at c is defined to be

$$\lim_{h\to 0}\frac{f(c+h)-f(c)}{h}$$

provided this limit exists. The derivative of f at c is denoted by $f'(c)$. We say that f is *differentiable at* c if $f'(c)$ exists. We say that f is *differentiable on* I if $f'(x)$ exists at each x in I. ●

We can interpret the existence of the derivative $f'(c)$ geometricly as follows. (Refer to Fig. 4.2.) The line L passing through the point $(c, f(c))$ with slope m is represented by the equation

$$y = m(x-c)+f(c).$$

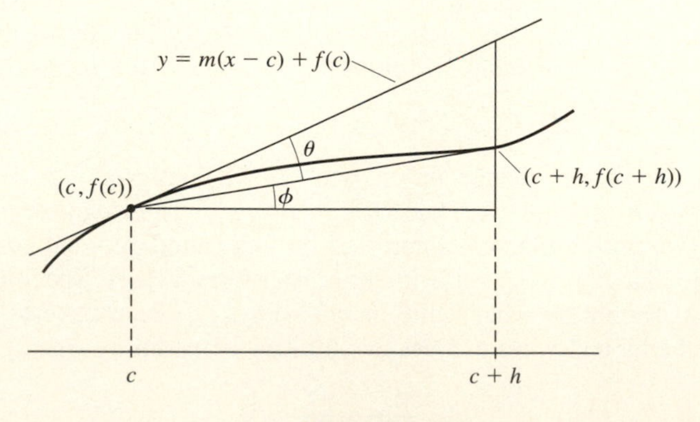

Figure 4.2

Let $\theta = \theta(h)$ denote the angle between this line and the chord joining the points $(c, f(c))$ and $(c+h, f(c+h))$.

DEFINITION 4.1.2 If $\lim_{h\to 0} \theta(h) = 0$, then the line L is said to be *tangent to the graph* of f at the point $(c, f(c))$. ●

The existence of the derivative $f'(c)$ corresponds to the existence of a line tangent to the graph of f at $(c, f(c))$.

THEOREM 4.1.1 The line $y = m(x-c) + f(c)$ is tangent to the graph of f at $(c, f(c))$ if and only if $f'(c)$ exists and $m = f'(c)$.

Proof. Suppose first that f is differentiable at c and that the line L is taken to be the line $y = f'(c)(x-c) + f(c)$. We want to show that $\lim_{h\to 0} \theta(h) = 0$. Equivalently, we want to show that $\lim_{h\to 0} \tan\theta = 0$. To this end, let ϕ denote the angle between the positive x-axis and the chord joining $(c, f(c))$ and $(c+h, f(c+h))$. Note that $\tan\phi = \Delta f/h$ and that $\tan(\theta+\phi) = df/h = m = f'(c)$. Using the trigonometric identity for $\tan(\alpha - \beta)$, we obtain

$$\tan\theta = \tan[(\theta+\phi)-\phi] = \frac{\tan(\theta+\phi) - \tan\phi}{1 + \tan(\theta+\phi)\tan\phi} = \frac{f'(c) - \dfrac{\Delta f}{h}}{1 + f'(c)\dfrac{\Delta f}{h}}.$$

Since we have assumed that $\lim_{h\to 0} \Delta f/h = f'(c)$, we deduce that

$$\lim_{h\to 0} \tan\theta = \lim_{h\to 0} \frac{f'(c) - \dfrac{\Delta f}{h}}{1 + f'(c)\dfrac{\Delta f}{h}} = \frac{f'(c) - f'(c)}{1 + (f'(c))^2} = 0.$$

Consequently, $\lim_{h\to 0} \theta(h) = 0$ and the line L is tangent to the graph of f at $(c, f(c))$.

To prove the converse, suppose that the line L with equation $y = m(x-c) + f(c)$ is tangent to the graph of f. That is, assume that $\lim_{h\to 0} \theta(h) = 0$. Equivalently, $\lim_{h\to 0} \tan\theta = 0$. We want to show that $f'(c) = \lim_{h\to 0} \Delta f/h$ exists and has value m. In the notation already established, $\tan\theta = \Delta f/h$, $\tan(\theta+\phi) = m$, and

$$\tan\theta = \frac{\tan(\theta+\phi) - \tan\phi}{1 + \tan(\theta+\phi)\tan\phi} = \frac{m - \dfrac{\Delta f}{h}}{1 + m\dfrac{\Delta f}{h}}.$$

Solving this last equation for $\Delta f/h$ we obtain

$$\frac{\Delta f}{h} = \frac{m - \tan\theta}{1 + m\tan\theta}.$$

Therefore,

$$\lim_{h\to 0} \frac{\Delta f}{h} = \lim_{h\to 0} \frac{m - \tan\theta}{1 + m\tan\theta} = m.$$

Thus f is differentiable at c and $f'(c) = m$. This completes the proof of the theorem. ●

To lay the groundwork for our discussion in Chapter 8 regarding the differentiability of functions of several variables, let us return to our initial discussion and summarize the insights we gained there.

DEFINITION 4.1.3 Let f be a real-valued function defined on an interval I in $\mathbb{R}$. Suppose that f is differentiable at a point c in I. The function $df(c; t) = f'(c)t$ is called the *differential* of f at c. ●

Notice that, as a function of the real variable t, the differential is defined on all of $\mathbb{R}$. The point c is held fixed. The differential of f is also characterized by the following two simple properties.

THEOREM 4.1.2 Suppose that f is differentiable at a point c in its domain.

i) The differential of f is a linear function of t. That is,

$$df(c; t_1 + t_2) = df(c; t_1) + df(c; t_2),$$

for all t_1 and t_2 in $\mathbb{R}$.

ii) For any $\epsilon > 0$, there exists a deleted neighborhood $N'(0)$ such that,

$$|f(c+t) - f(c) - df(c; t)| < \epsilon |t|,$$

for all t in $N'(0)$. ●

Notice that, if c is the left endpoint of the interval I, then only $f'(c^+)$ makes sense. If this limit exists, then it is taken to be the derivative $f'(c)$. Similarly, if c is the right endpoint of I and if $f'(c^-)$ exists, then $f'(c)$ is taken to be $f'(c^-)$.

On occasion it is more convenient to write the limit defining $f'(c)$ in the equivalent form

$$f'(c) = \lim_{x \to c} \frac{f(x) - f(c)}{x - c}.$$

EXAMPLE 1 For k in $\mathbb{N}$, define $f(x) = x^k$. We compute the derivative of f at any point c in $\mathbb{R}$ by finding

$$\lim_{x \to c} \frac{x^k - c^k}{x - c} = \lim_{x \to c} \frac{(x - c) \sum_{j=0}^{k-1} x^{k-j-1} c^j}{x - c} = kc^{k-1}. \quad ●$$

EXAMPLE 2 Let $f(x) = \sin x$. Then $f'(x) = \cos x$, as the following calculation proves:

$$\begin{aligned}
\lim_{h \to 0} \frac{\sin(x+h) - \sin x}{h} &= \lim_{h \to 0} \frac{\sin x \cos h + \cos x \sin h - \sin x}{h} \\
&= \sin x \lim_{h \to 0} \frac{\cos h - 1}{h} + \cos x \lim_{h \to 0} \frac{\sin h}{h} \\
&= (\sin x) \cdot 0 + (\cos x) \cdot 1 = \cos x.
\end{aligned}$$

A similar calculation shows that, if $f(x) = \cos x$, then $f'(x) = -\sin x$. ●

Notice that, if $f'(x)$ exists at each x in an interval I, then f' is itself a function on I. We will subsequently explore some of the elementary analytic and algebraic properties inherent in Definition 4.1.1. First we need a basic fact.

THEOREM 4.1.3 If f is differentiable at c, then f is continuous at c. If f is differentiable at each point of I, then f is in $C(I)$.

Proof. Write

$$f(x) - f(c) = \frac{f(x) - f(c)}{x - c} \cdot (x - c).$$

Taking the limit as x tends to c, we have

$$\lim_{x \to c} [f(x) - f(c)] = \lim_{x \to c} \frac{f(x) - f(c)}{x - c} \cdot (x - c) = f'(c) \cdot 0 = 0.$$

Therefore, $\lim_{x \to c} f(x) = f(c)$. This proves that f is continuous at c. The second assertion of the theorem follows immediately. ●

We are particularly interested in the effect of this new operation, differentiation, on the algebraic structure already established. The elementary properties are summarized in the following theorem. They are, of course, merely the standard rules of differentiation with which you are already familiar.

THEOREM 4.1.4 Suppose that f and g are two functions each differentiable at c and that a is any real number.

i) The function $f + g$ is also differentiable at c and

$$(f + g)'(c) = f'(c) + g'(c).$$

ii) The function af is also differentiable at c and

$$(af)'(c) = af'(c).$$

iii) The function fg is also differentiable at c and

$$(fg)'(c) = f'(c)g(c) + f(c)g'(c).$$

iv) If $g(c) \neq 0$, then $(1/g)$ is also differentiable at c and

$$\left(\frac{1}{g}\right)'(c) = -\frac{g'(c)}{g^2(c)}.$$

v) If $g(c) \neq 0$, then (f/g) is also differentiable at c and

$$\left(\frac{f}{g}\right)'(c) = \frac{f'(c)g(c) - f(c)g'(c)}{g^2(c)}.$$

Remark. This theorem asserts that the collection of all differentiable functions on an interval I forms a commutative ring with identity.

Proof. The first assertion is simple; it is proved by noting that

$$(f+g)(x)-(f+g)(c)=[f(x)-f(c)]+[g(x)-g(c)],$$

dividing by $x-c$, and passing to the limit as x tends to c.

The proof of (ii) is trivial. Part (iii) requires some care. Using a technique you have seen before, we complicate matters in order to simplify them by writing

$$\begin{aligned}(fg)(x)-(fg)(c)&=f(x)g(x)-f(c)g(x)+f(c)g(x)-f(c)g(c)\\&=[f(x)-f(c)]\,g(x)+f(c)[g(x)-g(c)].\end{aligned}$$

Now, $f(c)$ is a constant and g is differentiable, and hence continuous, at c by Theorem 4.1.3. Thus dividing by $x-c$ and passing to the limit as x tends to c yields

$$\lim_{x\to c}\frac{(fg)(x)-(fg)(c)}{x-c}=f'(c)g(c)+f(c)g'(c).$$

This proves part (iii).

To prove part (iv) we recall that, because g is continuous at c and $g(c)\neq 0$, then by Theorem 3.4.3, g is locally bounded away from 0 at c. That is, there exists a neighborhood $N(c)$ and a positive constant m such that $|g(x)|\geq m>0$ for all x in $N(c)$. Thus, for x in $N(c)$, we can sensibly form the reciprocal $1/g(x)$. Now write

$$\frac{1}{g(x)}-\frac{1}{g(c)}=-\frac{[g(x)-g(c)]}{g(x)g(c)},$$

divide by $x-c$ and pass to the limit as x tends to c. Upon doing so, we obtain, again using the continuity of g at c,

$$\lim_{x\to c}\frac{[1/g(x)-1/g(c)]}{x-c}=-\frac{g'(c)}{g^2(c)}.$$

This argument proves part (iv). The final assertion is proved by combining (iii) and (iv) and is left for you to complete. ●

EXAMPLE 3 The function $f(x)=\tan x=(\sin x)/(\cos x)$ has derivative $f'(x)=\sec^2 x$ at all x where f is defined. This fact is proved by invoking part (v) of Theorem 4.1.4 and using the known derivatives of the sine and cosine functions. Likewise, the derivative of $\cot x$ is $-\csc^2 x$ at all x where $\cot x$ is defined. Using part (iv) of Theorem 4.1.4, we find similarly that the functions $\sec x=1/\cos x$ and $\csc x=1/\sin x$ have derivatives $\sec x\tan x$ and $-\csc x\cot x$, respectively. ●

EXAMPLE 4 Define a function on [0, 2] as follows:

$$f(x)=\begin{cases}x^2, & 0\leq x\leq 1\\(x-2)^2, & 1<x\leq 2.\end{cases}$$

(See Fig. 4.3.) It is easy to see that

$$f'(1^-)=\lim_{x\to 1^-}\frac{f(x)-f(1)}{x-1}=2$$

and that

$$f'(1^+) = \lim_{x\to 1^+} \frac{f(x) - f(1)}{x - 1} = -2.$$

Since $f'(1^-) \neq f'(1^+)$, we deduce that $f'(1)$ fails to exist. Therefore f is not differentiable at $c = 1$. ●

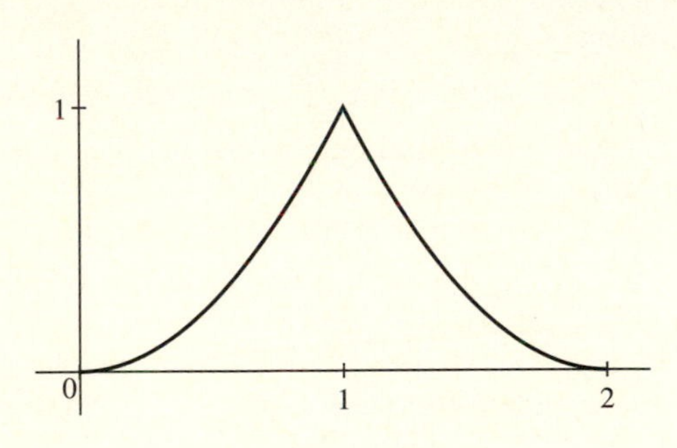

Figure 4.3

DEFINITION 4.1.4 Suppose that f is differentiable at every point of a neighborhood of c. If

$$f''(c) = \lim_{h\to 0} \frac{f'(c+h) - f'(c)}{h}$$

exists, then f'' is called the *second derivative* of f at c and f is said to be *twice differentiable* at c. Recursively, if f', f'', $f^{(3)}, \ldots, f^{(k-1)}$ all exist at every point of a neighborhood of c and if

$$f^{(k)}(c) = \lim_{h\to 0} \frac{f^{(k-1)}(c+h) - f^{(k-1)}(c)}{h}$$

exists, then $f^{(k)}(c)$ is called the kth derivative of f at c and f is said to be *k times differentiable* at c. If $f^{(k)}(c)$ exists for all k in $\mathbb{N}$, then f is said to have *derivatives of all orders* at c. If, for all k in $\mathbb{N}$, $f^{(k)}$ exists at all x in I, then f has *derivatives of all orders* on I. ●

EXERCISES

4.1. Let p be a positive constant and define a function f on $[0, 1]$ by $f(x) = x^p$ if x is rational and $f(x) = 0$ if x is irrational.

a) Prove that f is continuous only at $x = 0$.

b) Prove that if $0 < p \leq 1$, then f is not differentiable at any point of $[0, 1]$.

c) If $p > 1$, prove that f is differentiable only at $x = 0$.

4.2. Define a function f on $[0, 1]$ by setting $f(x) = 0$ if x is irrational and $f(x) = 1/q$ if $x = p/q$ is rational with p and q in lowest terms. Determine where f is differentiable.

4.3. Let C denote Cantor's set. (See Section 2.2, Example 5.) Let f be Cantor's function. (See Exercise 3.23.) Prove that $f'(x) = 0$ for any x in $U = C^c \cap [0, 1]$. Thus Cantor's function is monotone increasing, maps $[0, 1]$ continuously onto $[0, 1]$, and has derivative 0 on a subset of $[0, 1]$ whose total length is 1; f does all its rising on Cantor's set.

4.4. Suppose that a function f is defined on an interval I, that c is a point of I, and that $\{x_k\}$ is any sequence of points in I, no term of which is c, such that $\lim_{k\to\infty} x_k = c$. Define a sequence $\{y_k\}$ by $y_k = [f(x_k) - f(c)]/(x_k - c)$.

a) Prove that $f'(c)$ exists if and only if $\lim_{k\to\infty} y_k$ exists and has the same value for every such sequence $\{x_k\}$.

b) Prove that, if $f'(c)$ exists, then $\lim_{k\to\infty} y_k = f'(c)$ for every such sequence $\{x_k\}$.

4.5. Suppose that a function f is defined on an interval I, that c is a point of I, and that $\{x_k\}$ and $\{y_k\}$ are any two sequences in I such that, for all k, $x_k \neq y_k$ and such that $\lim_{k\to\infty} x_k = \lim_{k\to\infty} y_k = c$. Prove that, if $f'(c)$ exists, then

$$\lim_{k\to\infty} \frac{f(x_k) - f(y_k)}{x_k - y_k} = f'(c).$$

4.6. Suppose that f is a function defined in a neighborhood of a point c and that

$$\lim_{h\to 0} \frac{f(c+h) - f(c-h)}{2h} = L$$

exists. Does it necessarily follow that $f'(c)$ exists? If so, prove it; otherwise, provide a counterexample.

4.7. Suppose that f is differentiable on a neighborhood $N(c)$ of a point c. Prove that

$$\lim_{h\to 0} \frac{f(c+h) - f(c-h)}{2h} = f'(c).$$

4.8. A function f is said to *satisfy a Lipschitz condition of order* k at a point c if there exists a $K > 0$ and a neighborhood $N(c)$ such that, for x in $N(c)$,

$$|f(x) - f(c)| \leq K|x - c|^k.$$

a) Prove that, if f satisfies a Lipschitz condition of order $k > 1$ at c, then f is differentiable at c.

b) Find an example of a function f that satisfies a Lipschitz condition of order $k = 1$ at some point c and that is not differentiable at c.

4.9. Suppose that f and g are two functions that are each differentiable of all orders at a point c. Define $h = fg$. Prove by induction that the kth derivative of h at c is

$$h^{(k)}(c) = \sum_{j=0}^{k} C(k, j) f^{(j)}(c) g^{(k-j)}(c).$$

4.10. Write a brief essay expressing your beliefs about the answer to the following question: If a function f is differentiable at every point of an interval I, does it necessarily follow that f' is continuous on I?

4.11. Let f be a function defined on an interval I. An *antiderivative* of f is a function F defined on I such that $F'(x) = f(x)$ for all x in I. Write a brief essay expressing your beliefs about the answer to the following question: Does every function f on I necessarily have an antiderivative?

4.2 Composition of Functions: The Chain Rule

Now suppose that g is differentiable at an interior point c of the interval I, that $d = g(c)$ is an interior point of $g(I)$, that f is defined on $g(I)$, and that f is differentiable at d. We want to explore the differentiability of the composite function $f \circ g$ at the point c. That is, we want to compute $\lim_{x\to c}[f(g(x)) - f(g(c))]/(x - c)$. We introduce an auxiliary function that will enable us to bypass the possible logical pitfall of dividing by zero. For y in $g(I)$, define

$$h(y) = \begin{cases} \dfrac{f(y) - f(d)}{y - d} - f'(d), & y \neq d \\ 0, & y = d. \end{cases}$$

Since $\lim_{y\to d} h(y) = f'(d) - f'(d) = 0 = h(d)$, the function h is continuous at d. Also, since h is defined on $g(I)$, we can form the composite function $h \circ g$ on I as follows:

$$(h \circ g)(x) = \frac{f(g(x)) - f(g(c))}{g(x) - g(c)} - f'(g(c)),$$

if $g(x) \neq g(c) = d$, and $(h \circ g)(x) = 0$, if $g(x) = g(c) = d$.

Now, g is continuous at c and h is continuous at $d = g(c)$. Therefore, by Theorem 3.2.7, $h \circ g$ is continuous at c with $\lim_{x\to c}(h \circ g)(x) = (h \circ g)(c) = h(g(c)) = h(d) = 0$. From the definition of $h \circ g$ for $g(x) \neq g(c)$, solve for

$$f(g(x)) - f(g(c)) = [(h \circ g)(x) + f'(g(c))][g(x) - g(c)].$$

Dividing by $x - c$ and passing to the limit as x tends to c, we obtain

$$\begin{aligned} \lim_{x\to c} \frac{f(g(x)) - f(g(c))}{x - c} &= \lim_{x\to c}[(h \circ g)(x) + f'(g(c))]\frac{g(x) - g(c)}{x - c} \\ &= [0 + f'(g(c))]g'(c) = f'(g(c))g'(c). \end{aligned}$$

Thus we have proved that $f \circ g$ is differentiable at c with

$$(f \circ g)'(c) = f'(g(c))g'(c).$$

THEOREM 4.2.1 The Chain Rule Under all the preceding conditions, the function $f \circ g$ is differentiable at c with $(f \circ g)'(c) = f'(g(c))g'(c)$. ●

EXERCISES

4.12. **a)** Prove that, for $x > 0$, the derivative of $f(x) = x^{1/2}$ is $x^{-1/2}/2$.

b) Prove that $f'(0^+)$ does not exist.

c) Find the derivative of $h(x) = (1 - x^2)^{1/2}$.

4.13. Let $f(x) = x\sin(1/x)$ for $x \neq 0$ and $f(0) = 0$. Prove that f is differentiable at every x in $\mathbb{R}$ except 0.

4.14. Let $f(x) = x^2\sin(1/x)$ for $x \neq 0$ and $f(0) = 0$. Prove that f is differentiable at all x in $\mathbb{R}$, but that $\lim_{x\to 0} f'(x)$ does not exist.

4.15. Let $f(x) = x^2\sin(1/x^2)$ for $x \neq 0$ and $f(0) = 0$. Prove that f is differentiable at all x in $\mathbb{R}$, but that f' is not bounded on any neighborhood $N(0)$ of 0.

4.16. Suppose that h, g, and f are functions defined on open intervals I_1, I_2, and I_3, respectively, that $h(I_1) \subseteq I_2$, and that $g(I_2) \subseteq I_3$. Suppose also that h is differentiable at a point c in I_1, that g is differentiable at $d = h(c)$ in I_2, and that f is differentiable at $g(d)$ in I_3. Prove that the composite function $(f \circ g \circ h)$ is differentiable at c and that

$$(f \circ g \circ h)'(c) = f'(g(h(c)))g'(h(c))h'(c).$$

4.17. Suppose that a function g maps an interval I into an interval J and that f maps J into $\mathbb{R}$. Suppose also that g is differentiable on a neighborhood $N(c)$ of c and is twice differentiable at c and that f is differentiable on a neighborhood of $d = g(c)$ and is twice differentiable at d. Prove that $f \circ g$ is twice differentiable at c and find a formula for $(f \circ g)''(c)$.

4.3 THE MEAN VALUE THEOREM

Suppose that a function f is continuous on the closed interval $I = [a, b]$. By Theorem 3.2.4, f assumes its maximum value at some point c in $[a, b]$. Suppose that c is an interior point and that f is differentiable at c. What, if anything, can be said about the value of $f'(c)$? More generally, suppose that f has a *local maximum* at an interior point c. By this we mean, there exists a neighborhood $N(c)$ in I such that, if x is in $N(c)$, then $f(x) \leq f(c)$. What, if anything, can we say about the value of $f'(c)$?

To answer the question, first take x in $N'(c)$ with $x < c$. Then $f(x) - f(c) \leq 0$ and $x - c < 0$. It follows that $[f(x) - f(c)]/(x - c) \geq 0$. Therefore $f'(c^-) \geq 0$. On the other hand, if x is in $N'(c)$ and $x > c$, then $f(x) - f(c) \leq 0$ and $x - c > 0$. It follows that $[f(x) - f(c)]/(x - c) \leq 0$. Therefore $f'(c^+) \leq 0$. Since f is assumed to be differentiable at c, we know that $f'(c^-) = f'(c^+) = f'(c)$. Our reasoning leads us to conclude that $f'(c) \leq 0$ and $f'(c) \geq 0$ and, consequently, $f'(c) = 0$. (See Fig. 4.4.) An analogous argument applies in the event that f has a local minimum at a point c where f is differentiable: $f'(c) = 0$. This argument proves the following theorem.

THEOREM 4.3.1 If f has either a local maximum or a local minimum at an interior point c of $I = [a, b]$ and if f is differentiable at c, then $f'(c) = 0$. ●

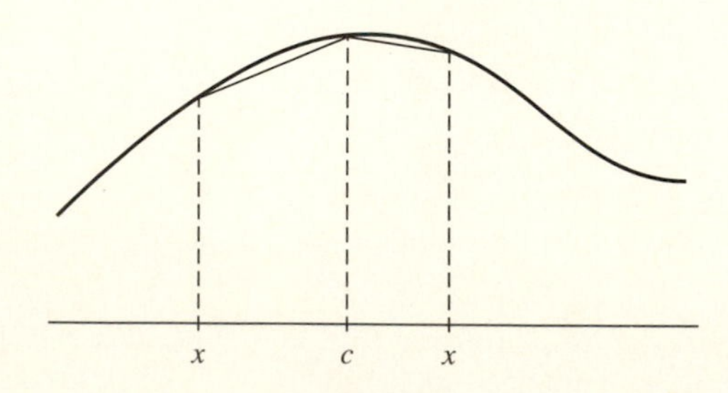

Figure 4.4

DEFINITION 4.3.1 If a function f is differentiable at a point c and if $f'(c) = 0$, then c is called a *critical point* of f. ●

Theorem 4.3.1 tells us that, if f has a local extreme value at a point c and if $f'(c)$ exists, then c is a critical point of f. The converse is false. That $f'(c) = 0$ does not imply that f has a local extremum at c. For example, if $f(x) = x^3$, then $f'(0) = 0$, but f has neither a local maximum nor local minimum at $c = 0$.

The simple observation embedded in Theorem 4.3.1 leads immediately to our next important result.

THEOREM 4.3.2 **Rolle's Theorem** Suppose that f is continuous on $[a, b]$ and is differentiable on (a, b). Suppose further that $f(a) = f(b)$. Then there exists at least one c in (a, b) such that $f'(c) = 0$.

Proof. If $f(x) = f(a)$ for all x in $[a, b]$, then $f' = 0$ at every point of $[a, b]$ and the theorem is proved. Otherwise, f must have either a maximum value or a minimum value at some point c in (a, b). (See Fig. 4.5.) By Theorem 4.3.1, it follows that $f'(c) = 0$. ●

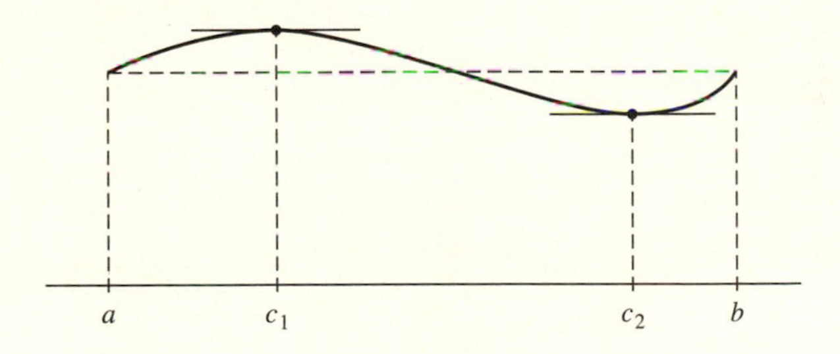

Figure 4.5

Rolle's theorem, in turn, leads immediately to one of the cornerstone theorems of analysis.

THEOREM 4.3.3 **The Mean Value Theorem** Suppose that f is continuous on $[a, b]$ and differentiable on (a, b). There exists a point c in (a, b) such that

$$f'(c) = \frac{f(b) - f(a)}{b - a}.$$

Proof. The line joining the points $(a, f(a))$ and $(b, f(b))$ has equation

$$y = m(x - a) + f(a)$$

where $m = [f(b) - f(a)]/(b - a)$. (See Fig. 4.6 on page 192.) Form the auxiliary function

$$h(x) = f(x) - [m(x - a) + f(a)],$$

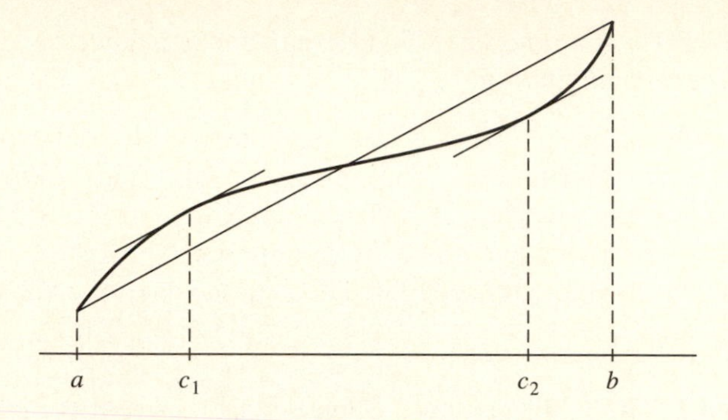

Figure 4.6

the (signed) vertical distance from the line to the graph of f. Note that h is continuous on $[a, b]$, that h is differentiable on (a, b) with $h'(x) = f'(x) - m$, and that $h(a) = h(b) = 0$. Therefore, Rolle's theorem applies to the function h. There exists a point c in (a, b) such that $h'(c) = 0$. That is, $f'(c) - m = 0$. Thus

$$f'(c) = \frac{f(b) - f(a)}{b - a}$$

and the Mean Value Theorem is proved. ●

Essentially, the usefulness of the Mean Value Theorem results from the observation that this is the first major theorem to serve as a bridge between a local property of f (the value of f' at a single point) and a global property of f (the average rate of change of f on an interval $[a, b]$). Henceforth, when you see a quantity of the form $f(x_1) - f(x_2)$, think of the Mean Value Theorem. Many of the subsequent major theorems of this text flow from this one. Later, in retrospect, you will find it illuminating to construct a flow diagram of these results and to confirm for yourself that from this one deceptively simple, even self-evident theorem streams the body of calculus. Here we derive the most simple and immediate corollaries.

COROLLARY 4.3.4 Suppose that f is continuous on $[a, b]$, that f is differentiable on (a, b), and that $f'(x) = 0$ for all x in (a, b). Then f is constant on the interval $[a, b]$.

Proof. Choose any x in $(a, b]$ and apply the Mean Value Theorem to f on the interval $[a, x]$: there exists a c in (a, x) such that $f'(c) = [f(x) - f(a)]/(x - a)$. But, whatever may be the choice of c in (a, b), $f'(c) = 0$. Therefore, $f(x) - f(a) = 0$, or $f(x) = f(a)$ for any x in $(a, b]$. Consequently, f is constant on $[a, b]$. ●

COROLLARY 4.3.5 Suppose that f and g are two functions that are continuous on $[a, b]$ and differentiable on (a, b) with $f'(x) = g'(x)$ for all x in (a, b). Then f and g differ by a constant.

Proof. Let $h = f - g$. (See Fig. 4.7.) The function h is continuous on $[a, b]$, is differentiable on (a, b), and $h'(x) = 0$ for all x in (a, b). By Corollary 4.3.4, h is constant on $[a, b]$. Thus f and g differ by a constant as claimed. ●

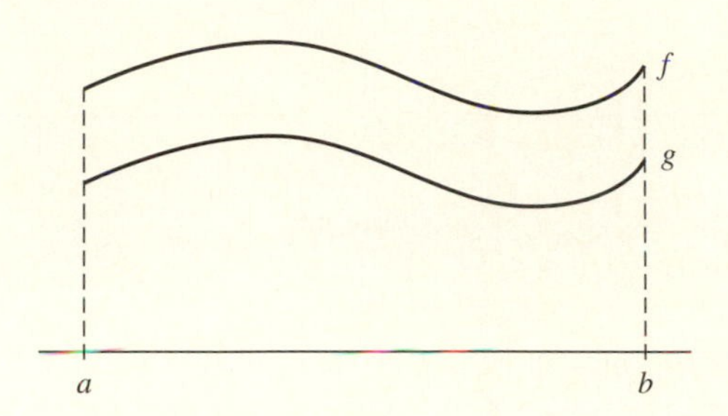

Figure 4.7

COROLLARY 4.3.6 Suppose that f is continuous on $[a, b]$ and is differentiable on (a, b). If $f'(x) > 0$ for all x in (a, b), then f is strictly increasing. If $f'(x) < 0$ for all x in (a, b), then f is strictly decreasing.

Proof. Choose any x_1 and x_2 in $[a, b]$ with $x_1 < x_2$. Apply the Mean Value Theorem to f on $[x_1, x_2]$: There exists a c in (x_1, x_2) such that $f(x_2) - f(x_1) = f'(c)(x_2 - x_1)$. Since $x_2 - x_1 > 0$, the corollary follows immediately. ●

The next consequence of the Mean Value Theorem is of sufficient importance to warrant its identification as a theorem in its own right.

THEOREM 4.3.7 Cauchy's Generalized Mean Value Theorem Suppose that f and g are two functions each of which is continuous on $[a, b]$ and differentiable on (a, b). There exists a point c in (a, b) such that

$$f'(c)[g(b) - g(a)] = g'(c)[f(b) - f(a)].$$

Proof. For x in $[a, b]$, define an auxiliary function

$$h(x) = f(x)[g(b) - g(a)] - g(x)[f(b) - f(a)].$$

Notice that h is continuous on $[a, b]$. Further, note that h is differentiable on (a, b) with

$$h'(x) = f'(x)[g(b) - g(a)] - g'(x)[f(b) - f(a)].$$

Finally, as a brief computation will confirm, $h(a) = h(b)$. Therefore, Rolle's theorem applies and there must exist a point c in (a, b) such that $h'(c) = 0$. That is,

$$f'(c)[g(b) - g(a)] = g'(c)[f(b) - f(a)],$$

as promised by the theorem. ●

Cauchy's Mean Value Theorem can be interpreted geometrically in the following way. Let $y_1 = f_1(x)$ and $y_2 = f_2(x)$ for x in $[a, b]$. Then the function $\boldsymbol{f} = (f_1, f_2)$

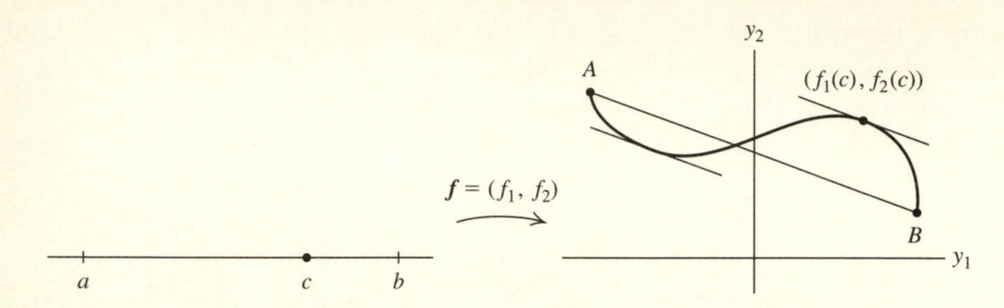

Figure 4.8

maps the interval $[a, b]$ into $\mathbb{R}^2$, resulting in a graph that is a curve. (See Fig. 4.8.) The curve is connected because f is continuous; the curve is smooth because both f_1 and f_2 are differentiable. Let $A = (f_1(a), f_2(a))$ and $B = (f_1(b), f_2(b))$. As the point x moves from a to b, the point $(f_1(x), f_2(x))$ moves along the curve from A to B. The slope of the line joining A to B is

$$\frac{f_2(b) - f_2(a)}{f_1(b) - f_1(a)},$$

assuming that $f_1(a) \neq f_1(b)$. The slope of the line tangent to the curve at the point $(f_1(x), f_2(x))$ is $f_2'(x)/f_1'(x)$ (assuming that $f_1'(x) \neq 0$). According to Cauchy's Mean Value Theorem, there exists a point c in (a, b) such that

$$f_1'(c)[f_2(b) - f_2(a)] = f_2'(c)[f_1(b) - f_1(a)].$$

That is, the point $(f_1(c), f_2(c))$ is a point on the curve where the tangent line is parallel to the chord joining A and B.

EXAMPLE 5 For x in $[0, \pi/2]$, define $f_1(x) = e^x \sin x$ and $f_2(x) = e^x \cos x$. The curve defined by the function $f = (f_1, f_2)$ on $[0, \pi/2]$ is sketched in Fig. 4.9. The initial point is $A = (0, 1)$ and the terminal point is $B = (e^{\pi/2}, 0)$. The slope of the chord joining A and B is $-1/e^{\pi/2}$. According to Cauchy's Mean Value Theorem, there is a point c in $(0, \pi/2)$ such that

$$f_2'(c)/f_1'(c) = \frac{f_2(\pi/2) - f_2(0)}{f_1(\pi/2) - f_1(0)} = -1/e^{\pi/2}.$$

To find a point c in $(0, \pi/2)$ where the line tangent to the curve has slope $-1/e^{\pi/2}$, first compute[1]

$$f_1'(x) = e^x(\cos x + \sin x)$$

and

$$f_2'(x) = e^x(\cos x - \sin x).$$

[1] We assume here that the derivative of the function $f(x) = e^x$ is $f'(x) = e^x$ for all x in $\mathbb{R}$. We will eventually prove this fact in Chapter 6.

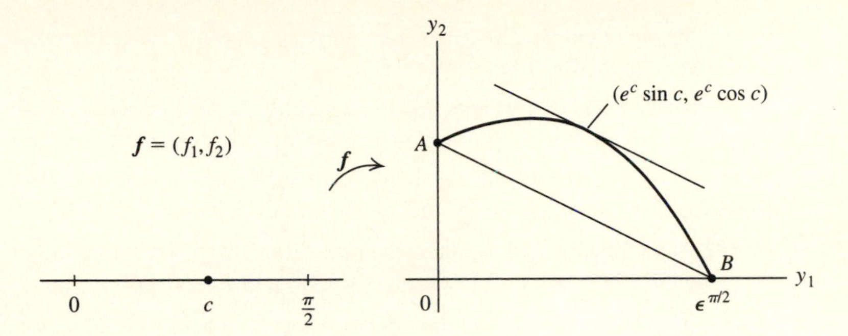

Figure 4.9

Next solve the equation

$$\frac{f_2'(c)}{f_1'(c)} = \frac{e^x(\cos c - \sin c)}{e^x(\cos c + \sin c)}$$
$$= -\frac{1}{e^{\pi/2}}$$

for $\tan c = (e^{\pi/2} + 1)/(e^{\pi/2} - 1)$. Consequently,

$$c = \tan^{-1}\left(\frac{e^{\pi/2} + 1}{e^{\pi/2} - 1}\right) \approx .9903586 \text{ radians.}$$

For this value of c the line tangent to the curve at the point $(f_1(c), f_2(c))$ is parallel to the chord joining A and B in Fig. 4.9.

EXERCISES

4.18. Show that the conclusion of the Mean Value Theorem applied to f on $[x, x+h]$ can be stated as follows: There is a θ in $(0, 1)$ such that $f(x+h) - f(x) = f'(a + \theta h)\, h$.

4.19. Let $f_1(x) = x^3$, $f_2(x) = e^x$, and $f_3(x) = \ln x$.

a) For each of f_1, f_2, and f_3 and any x in the domain of f_j, apply Exercise 4.18 to find $\theta_j = \theta_j(x, h)$ in terms of x and h so that $f_j(x+h) - f_j(x) = f_j'(x + \theta_j h)\, h$. (You may assume that the derivative of e^x is e^x and that the derivative of $\ln x$ is $1/x$.)

b) In each case, holding x fixed, compute $\lim_{h \to 0} \theta_j(x, h)$.

4.20. Suppose that f is continuous on $[0, b]$ and differentiable on $(0, b)$. Suppose also that $f(0) = 0$ and f' is increasing on $(0, b)$. Define $g(x) = f(x)/x$ for x in $(0, b)$. Prove that g is increasing on $(0, b)$.

4.21. Suppose that f and g are differentiable on $[a, \infty)$, that $f(a) = g(a)$ and that $f'(x) < g'(x)$ for all x in (a, ∞). Prove that $f(x) < g(x)$ for all x in (a, ∞).

4.22. Suppose that $r > 0$ and that a function f is differentiable on $(-r, r)$. Prove that f is even on $(-r, r)$ if and only if f' is odd on $(-r, r)$ and that f is odd on $(-r, r)$ if and only if $f(0) = 0$ and f' is even on $(-r, r)$.

4.23. Suppose that f is a function defined on an interval I such that $f''(x) = 0$ for all x in I. Prove that f is a linear function, $f(x) = a_0 + a_1 x$, for all x in I.

4.24. **a)** Suppose that f is monotone increasing on an open interval I and differentiable at a point c in I. Prove that $f'(c) \geq 0$.

b) Suppose that, in part (a), f is assumed to be strictly monotone increasing. Does it follow that $f'(c) > 0$? If so, prove it; otherwise, provide a counterexample.

4.25. Suppose that f is a function defined on an interval I such that f' exists and is bounded on I. Prove that f must be uniformly continuous on I.

4.26. Suppose that f is a function defined on an interval I such that f' exists and is bounded on I. Prove that f must satisfy a Lipschitz condition of order 1 on I.

4.27. Suppose that f' is bounded on $(0, 1]$. Prove that $\lim_{k\to\infty} f(1/k)$ exists.

4.28. Suppose that f, g, and h are continuous on $[a, b]$ and differentiable on (a, b). For x in $[a, b]$, define

$$F(x) = \det \begin{vmatrix} f(x) & f(a) & f(b) \\ g(x) & g(a) & g(b) \\ h(x) & h(a) & h(b) \end{vmatrix}.$$

a) Prove that $F'(c) = 0$ for some c in (a, b).

b) Show that, for an appropriate choice of g and h, the result in part (a) implies the Mean Value Theorem.

c) Show that, for an appropriate choice of h, the result of part (a) implies Cauchy's Generalized Mean Value Theorem.

4.29. Suppose that f is continuous on $[a, \infty)$ and differentiable on (a, ∞).

a) Prove that between adjacent zeros of f, there is a critical point of f. (We say that the zeros of f and f' *intertwine*.)

b) Assume that f'' exists on (a, ∞). Prove that between adjacent critical points of f, there is a point where f'' is 0. (Thus, the zeros of f' and f'' also intertwine.)

4.30. Suppose that a function f is monotone increasing, bounded, and differentiable on (a, ∞). Does it necessarily follow that $\lim_{x\to\infty} f'(x) = 0$? If so, prove it; otherwise, construct a counterexample.

4.31. Suppose that a function f is twice differentiable on $[a, \infty)$. Suppose also that $\lim_{x\to\infty} f(x) = 0$ and that f'' is bounded on $[0, \infty)$. Prove that $\lim_{x\to\infty} f'(x) = 0$.

4.32. Use the Mean Value Theorem to prove Bernoulli's inequality: For every $x > -1$ and every k in $\mathbb{N}$,

$$(1 + x)^k \geq 1 + kx.$$

4.33. Suppose that f is continuous on $[a, b]$, that f is differentiable on (a, b), and that $f(a) = f(b) = 0$. Prove that, for every k in $\mathbb{R}$, there exists a c in (a, b) such that $f'(c) = kf(c)$. (Consider the function $g(x) = e^{-kx} f(x)$ for x in $[a, b]$.)

4.34. Prove *Darboux's Intermediate Value Theorem for derivatives*: If f is differentiable on $[a, b]$ and if d is some number between $f'(a)$ and $f'(b)$, then there exists a c in (a, b) such that $f'(c) = d$. (*Hint*: For x in $[a, b]$, let $g(x) = d(x - a) - f(x)$. Show that g must have a critical point in (a, b).)

4.35. Suppose that f is differentiable on $[a, b]$ and that $f'(x) \neq 0$ for all x in $[a, b]$.

a) Use Exercise 4.34 to prove that either $f' > 0$ for all x in $[a, b]$ or $f'(x) < 0$ for all x in $[a, b]$.

b) Hence prove that f is strictly monotone on $[a, b]$. (Thus f maps $[a, b]$ one-to-one onto $f([a, b])$ and $g = f^{-1}$ exists, mapping $f([a, b])$ one-to-one onto $[a, b]$. Further, $(f \circ g)(y) = y$ for all y in $f([a, b])$.)

c) Prove that g is differentiable on $f([a, b])$ and that

$$g'(y) = \frac{1}{(f' \circ g)(y)}$$

for all y in $f([a, b])$.

4.36. For this exercise, assume that $\lim_{h \to 0} (e^h - 1)/h = 1$.

a) Use this fact to prove that, for all x in $\mathbb{R}$, the derivative of e^x is e^x.

b) Use part (a) and Exercise 4.35 to prove that, for each $x > 0$, the derivative of $\ln x$ is $1/x$.

4.37. For this exercise, assume that, for all $x > 0$, the derivative of $\ln x$ is $1/x$. (We will prove this in Chapter 6.)

a) Use this fact and Exercise 4.35 to prove that, for all x in $\mathbb{R}$, the derivative of e^x is e^x.

b) Use part (a) to prove that $\lim_{h \to 0} (e^h - 1)/h = 1$.

4.38. Let $f(x) = \sin x$ on $[-\pi/2, \pi/2]$.

a) Prove that f maps $[-\pi/2, \pi/2]$ one-to-one onto $[-1, 1]$. Hence prove that $g = f^{-1}$ exists and maps $[-1, 1]$ one-to-one onto $[-\pi/2, \pi/2]$. (The function $g = f^{-1}$ is denoted $\sin^{-1}$ or arcsin.)

b) Use Exercise 4.35 to find the derivative of $\sin^{-1}$ on the interval $(-1, 1)$.

4.39. Let $f(x) = \cos x$ for x in $[0, \pi]$.

a) Prove that f maps $[0, \pi]$ one-to-one onto $[-1, 1]$. Hence prove that $g = f^{-1}$ exists and maps $[-1, 1]$ one-to-one onto $[0, \pi]$. (The function $g = f^{-1}$ is denoted $\cos^{-1}$ or arccos.)

b) Use Exercise 4.35 to find the derivative of $\cos^{-1}$ on the interval $(-1, 1)$.

4.40. Let $f(x) = \tan x$ for x in $(-\pi/2, \pi/2)$.

a) Prove that f maps $(-\pi/2, \pi/2)$ one-to-one onto $\mathbb{R}$. Prove that $g = f^{-1}$ exists and maps $\mathbb{R}$ one-to-one onto $(-\pi/2, \pi/2)$. (The function $g = f^{-1}$ is denoted $\tan^{-1}$ or arctan.)

b) Use Exercise 4.35 to find the derivative of $\tan^{-1}$ on $\mathbb{R}$.

4.41. Let $a_1, a_2, \ldots, a_k$ be real numbers and let f be defined on $\mathbb{R}$ by $f(x) = \sum_{j=1}^{k} (a_j - x)^2$. Find the unique value of x where f has a minimum value.

4.42. Prove that $|\sin x - \sin y| \le |x - y|$ for x and y in $\mathbb{R}$.

4.43. Prove that, for $x > 1$, $(x - 1)/x < \ln x < x - 1$.

4.44. For x in $[0, 2]$, let $f(x) = 4x^2$ and $g(x) = x^3 + 1$. Find all values of c in $(0, 2)$ that satisfy the conclusion of Cauchy's Generalized Mean Value Theorem.

4.45. Let $p(x) = \sum_{j=0}^{k} a_j x^j$ be a polynomial of degree k. Prove that if $\sum_{j=0}^{k} a_j/(j + 1) = 0$, then p has at least one root in the interval $(0, 1)$.

4.46. Suppose that f is three times differentiable on $[a, b]$ and that $f(a) = f(b) = f'(a) = f'(b)$. Prove that there exists a c in (a, b) such that $f^{(3)}(c) = 0$.

4.47. For $0 < a < b$, let f be continuous on $[a, b]$ and differentiable on (a, b). Prove that there exists a c in (a, b) such that $f(b) - f(a) = cf'(c)\ln(b/a)$. (Use Cauchy's mean value theorem.) Hence prove that $\lim_{k\to\infty} k(a^{1/k} - 1) = \ln a$.

4.48. Suppose that f is differentiable on some neighborhood $N(c; r)$ of c and that $f''(c)$ exists. Prove that

$$\lim_{h\to 0} \frac{f(c+h) + f(c-h) - 2f(c)}{h^2}$$

exists and equals $f''(c)$. (For t in $[-r/2, r/2]$, let $F(t) = f(c+t) + f(c-t) - 2f(c)$ and $G(t) = t^2$. Fix h such that $0 < |h| < r/2$. Apply Exercise 4.7 and Cauchy's Generalized Mean Value Theorem on the interval with endpoints 0 and h.)

4.49. Suppose that f is differentiable on (a, b) and that $\lim_{x\to b^-} f(x) = \infty$. Prove that either $\lim_{x\to b^-} f'(x)$ fails to exist or $\lim_{x\to b^-} f'(x) = \infty$.

4.50. Suppose that f is differentiable on (a, b) and that c is a point of (a, b) such that $\lim_{x\to c} f'(x)$ exists. Prove that this limit must be $f'(c)$.

4.51. Suppose that f is continuous on (a, b) and that f' is known to exist everywhere on (a, b) except possibly at the point c. Suppose that $\lim_{x\to c} f'(x) = L$ exists. Prove that f is differentiable at c and that $f'(c) = L$.

4.52. If f is twice differentiable on $[a, a+h]$, prove that there exists a c in $(a, a+h)$ such that

$$f(a+h) = f(a) + hf'(a) + \frac{1}{2}h^2 f''(c).$$

4.53. Suppose that f maps $[a, b]$ continuously into $[a, b]$ and that f is differentiable on (a, b) with

$$||f'||_\infty = \sup\{|f'(x)| : x \text{ in } (a, b)\} < 1.$$

By Exercise 3.56, we know that f has a fixed point x_0 in $[a, b]$, that is, a point such that $f(x_0) = x_0$. Here we show how to compute x_0. First choose any x_1 in $[a, b]$. Define $x_2 = f(x_1)$. In general, for k in $\mathbb{N}$, define $x_{k+1} = f(x_k)$.

a) Prove that the resulting sequence $\{x_k\}$ is contractive, therefore convergent.

b) Prove that $\lim_{k\to\infty} x_k = x_0$ is a fixed point of f.

4.54. Let $f(x) = \cos x$ for x in $I = [0, 3\pi/8]$. Show that f maps I continuously into I and that $||f'||_\infty < 1$. Use your calculator to find the first several terms of a contractive sequence that converges to a fixed point x_0 of f. Compute sufficiently many terms of the sequence to approximate x_0 accurately to six decimal places.

4.4 L'Hôpital's Rule

Suppose that f and g are continuous on some interval $[a, b]$ and are differentiable on (a, b). Suppose further that c is some point of $[a, b]$ where $f(c) = g(c) = 0$. Finally, assume that $g'(x)$ does not vanish in some deleted neighborhood $N'(c)$ of c. The problem addressed here concerns the computation of $\lim_{x\to c} f(x)/g(x)$. Naively, one might assume the limit to be 0/0, an assumption that makes no sense; on the face of it, this problem appears intractable. Cauchy's Generalized Mean Value Theorem, however, saves the day. Notice that $f(x) = f(x) - f(c)$ and $g(x) = g(x) - g(c)$.

Applying Cauchy's Mean Value Theorem to f and g on the interval with endpoints c and x ensures the existence of a point d between c and x such that

$$f'(d)g(x) = f'(d)[g(x) - g(c)] = g'(d)[f(x) - f(c)] = g'(d)f(x).$$

Equivalently, $f(x)/g(x) = f'(d)/g'(d)$. As x tends to c, the point d, trapped between x and c, also tends to c. Therefore, if $\lim_{x\to c} f'(x)/g'(x) = L$ exists, then $\lim_{x\to c} f(x)/g(x)$ also exists and equals L. This argument proves the following theorem.

THEOREM 4.4.1 L'Hôpital's Rule, Form I Suppose that f and g are each continuous on $[a, b]$ and differentiable on (a, b). Suppose that c is a point of $[a, b]$ such that $f(c) = g(c) = 0$. Suppose also that, on some deleted neighborhood $N'(c)$, g' does not vanish. If $\lim_{x\to c} f'(x)/g'(x) = L$ exists, then $\lim_{x\to c} f(x)/g(x)$ also exists and equals L. ●

Two observations are in order. First, if $\lim_{x\to c} f(x)/g(x)$ is *not* of the indeterminate form 0/0 (or some one of the other indeterminate forms that we will discuss), then you *must not apply l'Hôpital's rule*. In this situation, even if $\lim_{x\to c} f'(x)/g'(x)$ does exist, it need not equal $\lim_{x\to c} f(x)/g(x)$. Second, if, upon applying this rule to $\lim_{x\to c} f(x)/g(x)$, you obtain a quotient $f'(x)/g'(x)$, that also tends to an indeterminate form as x tends to c, if f' and g' (as well as f and g) satisfy the conditions of l'Hôpital's rule and if $\lim_{x\to c} f''(x)/g''(x)$ exists, then

$$\lim_{x\to c} \frac{f''(x)}{g''(x)} = \lim_{x\to c} \frac{f'(x)}{g'(x)} = \lim_{x\to c} \frac{f(x)}{g(x)}.$$

Two (or more) applications of l'Hôpital's rule may be necessary to find the value of the original limit.

EXAMPLE 6 To find $\lim_{x\to 0}(\sin x)/x$, notice that $f(x) = \sin x$ and $g(x) = x$ satisfy the hypotheses of l'Hôpital's rule at the point $c = 0$. The ratio $f(x)/g(x)$ gives rise to the indeterminate form 0/0 as x tends to 0. Application of the rule yields $\lim_{x\to 0}(\sin x)/x = \lim_{x\to 0}(\cos x)/1 = 1$. This result is interpreted by saying that $\sin x$ approaches 0 at the same rate as x approaches 0. ●

EXAMPLE 7 The functions $f(x) = \ln(1 + x)$ and $g(x) = x$ satisfy the hypotheses of l'Hôpital's rule at $c = 0$. Therefore

$$\lim_{x\to 0} \frac{\ln(1 + x)}{x} = \lim_{x\to 0} \frac{1}{1 + x} = 1. \tag{4.1}$$

Notice that $[\ln(1 + x)]/x = \ln(1 + x)^{1/x}$ and that the exponential of this function is $\exp(\ln(1 + x)^{1/x}) = (1 + x)^{1/x}$. By the continuity of the exponential function, the limit (4.1), and Theorem 3.2.7, we conclude that

$$\lim_{x\to 0}(1 + x)^{1/x} = \lim_{x\to 0} \exp[\ln(1 + x)^{1/x}] = e.$$

In particular, letting $x = 1/k$, we obtain

$$\lim_{k\to\infty}\left(1+\frac{1}{k}\right)^k = e.$$

We derived this same limit in Section 1.3 with more work and fewer tools available. ●

Limits of the form $\lim_{x\to\infty} f(x)/g(x)$ or $\lim_{x\to-\infty} f(x)/g(x)$ that give rise to an indefinite form can often be treated by use of the following theorem.

THEOREM 4.4.2 L'Hôpital's Rule, Form II Suppose that f and g are differentiable on some neighborhood (M, ∞) of ∞ and suppose that g' does not vanish on (M, ∞). Suppose also that $\lim_{x\to\infty} f(x) = \lim_{x\to\infty} g(x) = \infty$. If

$$\lim_{x\to\infty}\frac{f'(x)}{g'(x)} = L$$

exists, then $\lim_{x\to\infty} f(x)/g(x)$ also exists and equals L.

Proof. Fix $\epsilon > 0$. Choose a neighborhood (M_1, ∞) of ∞ such that, for x in (M_1, ∞), we have $|f'(x)/g'(x) - L| < \epsilon$. Without loss of generality, $M_1 \geq M$, so g' does not vanish in (M_1, ∞). Since $\lim_{x\to\infty} f(x) = \lim_{x\to\infty} g(x) = \infty$, we can assume, as well, that $f(x)$ and $g(x)$ are both positive for all x in $[M_1, \infty)$.

More, there must exist an $M_2 > M_1$ such that, for x in (M_2, ∞), we have $f(x) > f(M_1)$ and $g(x) > g(M_1)$. Choose any $x > M_2$. Apply Cauchy's Mean Value Theorem to f and g on the interval $[M_1, x]$. There exists a c_x in (M_1, x) such that

$$\frac{f'(c_x)}{g'(c_x)} = \frac{f(x) - f(M_1)}{g(x) - g(M_1)} = \frac{f(x)\left[\dfrac{1 - f(M_1)}{f(x)}\right]}{g(x)\left[\dfrac{1 - g(M_1)}{g(x)}\right]} \tag{4.2}$$

Let $h(x) = [1 - g(M_1)/g(x)]/[1 - f(M_1)/f(x)]$. From (4.2) we obtain

$$\frac{f(x)}{g(x)} = \frac{f'(c_x)}{g'(c_x)}h(x).$$

Next, observe that $\lim_{x\to\infty} h(x) = 1$. Thus there exists a neighborhood (M_3, ∞) of ∞ such that, for x in (M_3, ∞), we have $|h(x) - 1| < \epsilon$. Let $M_0 = \max\{M_2, M_3\}$. For any $x > M_0$,

$$\begin{aligned}
\left|\frac{f(x)}{g(x)} - L\right| &= \left|\frac{f'(c_x)}{g'(c_x)}h(x) - L\right| \\
&= \left|\frac{f'(c_x)}{g'(c_x)}h(x) - Lh(x) + Lh(x) - L\right| \\
&\leq \left|\frac{f'(c_x)}{g'(c_x)} - L\right| |h(x)| + |L|\,|h(x) - 1|
\end{aligned}$$

$$< \epsilon(1+\epsilon) + |L|\epsilon = (1+|L|+\epsilon)\,\epsilon.$$

Therefore $\lim_{x\to\infty} f(x)/g(x) = L$. This proves the theorem. ●

EXAMPLE 8 The functions $f(x) = x^k$ and $g(x) = e^x$ satisfy the hypotheses of Theorem 4.4.2. Applying that theorem k times, we have

$$\lim_{x\to\infty} \frac{x^k}{e^x} = \lim_{x\to\infty} \frac{kx^{k-1}}{e^x} = \cdots = \lim_{x\to\infty} \frac{k!}{e^x} = 0.$$

Therefore, the function e^x tends to infinity faster than any positive integer power x^k of x. ●

EXAMPLE 9 Let $f(x) = \ln x$ and $g(x) = x^k$ and consider

$$\lim_{x\to\infty} \frac{\ln x}{x^k},$$

a limit giving rise to the indeterminate form ∞/∞. An application of Theorem 4.4.2 yields

$$\lim_{x\to\infty} \frac{\ln x}{x^k} = \lim_{x\to\infty} \frac{1}{kx^k} = 0.$$

For any k in $\mathbb{N}$, the function x^k grows to infinity faster than does $\ln x$. ●

EXERCISES

4.55. Find $\lim_{x\to 0}(1-\cos x^2)/(x^3 \sin x)$.

4.56. Find $\lim_{x\to 0} \tan 3x/\ln(1+x)$.

4.57. Find $\lim_{x\to 0} [x/(e^x-1)]^{1/x}$.

4.58. Find $\lim_{x\to 0^+} x^x$.

4.59. Find

$$\lim_{x\to\infty} \frac{e^x - 1/x}{e^x + 1/x}.$$

4.60. Suppose that, upon attempting to find $\lim_{x\to c} f(x)g(x)$, you obtain the indeterminate form $0\cdot\infty$. Write

$$\lim_{x\to c} f(x)g(x) = \lim_{x\to c} \frac{f(x)}{\dfrac{1}{g(x)}},$$

a limit giving rise to the indeterminate form 0/0 or ∞/∞. Suppose that l'Hôpital's rule applies to the functions $F = f$ and $G = 1/g$. Show how to compute $\lim_{x\to c} f(x)g(x)$.

4.61. Find $\lim_{x\to 1^+}(x-1)\ln(x-1)$.

4.62. Find $\lim_{x\to 0^+}(e^x-1)\ln x$.

4.63. Find

$$\lim_{x\to 0}\left(1-\frac{\sin x}{x}\right)\csc x.$$

4.64. Suppose that, upon attempting to find

$$\lim_{x \to c} \left[\frac{f_1(x)}{g_1(x)} - \frac{f_2(x)}{g_2(x)} \right],$$

you obtain the indeterminate form $\infty - \infty$. Write

$$\lim_{x \to c} \left[\frac{f_1(x)}{g_1(x)} - \frac{f_2(x)}{g_2(x)} \right] = \lim_{x \to c} \frac{f_1(x)g_2(x) - f_2(x)g_1(x)}{g_1(x)g_2(x)}.$$

Suppose that l'Hôpital's rule applies to the functions $F = f_1 g_2 - f_2 g_1$ and $G = g_1 g_2$. Show how to compute the original limit.

4.65. Find

$$\lim_{x \to 0} \left[\frac{1}{x} - \frac{1}{\ln(x+1)} \right].$$

4.66. Find

$$\lim_{x \to 0} \left[\frac{1}{(e^x - 1)} - \frac{1}{\sin x} \right].$$

4.67. Find

$$\lim_{x \to 1} \left[\frac{x}{(x-1)} - \frac{1}{\ln x} \right].$$

4.68. Suppose that f is differentiable in some neighborhood of c and that $f''(c)$ exists. Use l'Hôpital's rule and Exercise 4.7 to prove that

$$\lim_{h \to 0} \frac{f(c+h) + f(c-h) - 2f(c)}{h^2} = f''(c).$$

4.69. Suppose that f is twice differentiable on some neighborhood of a point c. Prove that

$$\lim_{h \to 0} \frac{f(c+h) - f(c) - f'(c)h}{h^2} = \frac{f''(c)}{2}.$$

4.70. Suppose that f is three times differentiable on some neighborhood of a point c. Prove that

$$\lim_{h \to 0} \frac{f(c+h) - f(c) - f'(c)h - f''(c)h^2/2}{h^3} = \frac{f^{(3)}(c)}{6}.$$

4.71. Define a function f on $\mathbb{R}$ as follows:

$$f(x) = \begin{cases} \exp(-1/x), & \text{for } x > 0 \\ 0, & \text{for } x \le 0. \end{cases}$$

a) If $x \neq 0$, prove that f is differentiable at x and compute $f'(x)$.

b) Use l'Hôpital's rule to compute $\lim_{x \to 0} [f(x) - f(0)]/x$. Hence, show that $f'(0)$ exists.

c) If $x \neq 0$, prove that f' is differentiable at x and compute $f''(x)$.

d) Use l'Hôpital's rule to find $\lim_{x \to 0} [f'(x) - f'(0)]/x$. Hence, show that $f''(0)$ exists.

e) Use induction to prove that $f^{(k)}(0) = 0$ for all k in $\mathbb{N}$.

4.72. For x in $\mathbb{R}$, define $f(x) = \exp(-1/x^2)$ for $x \neq 0$ and $f(0) = 0$.

a) If $x \neq 0$, show that f is differentiable at x and find $f'(x)$.

b) Use l'Hôpital's rule to find $\lim_{x\to 0}[f(x) - f(0)]/x$ and hence show that $f'(0)$ exists.

c) If $x \neq 0$, show that f' is differentiable at x and find $f''(x)$.

d) Use l'Hôpital's rule to find $\lim_{x\to 0}[f'(x) - f'(0)]/x$ and hence show that $f''(0)$ exists.

e) Use induction to prove that $f^{(k)}(0) = 0$ for all k in $\mathbb{N}$.

4.5 TAYLOR'S THEOREM

In light of Weierstrass's approximation theorem (Theorem 3.5.7), we know that a function f, defined and continuous on an interval $[a, b]$, can be uniformly approximated by a polynomial. Bernštein's proof of that theorem provides a method for constructing such a polynomial; in this section we will study another method. As we will see, if f has derivatives of sufficiently high order, then we can develop a systematic process for constructing a polynomial approximation to f. Just as important, we will obtain a formula for the error incurred by this approximation. Taylor's theorem provides us with a powerful method, applicable to a large class of functions.

Assume that f and its first k derivatives exist and are continuous on $[a, b]$ and that $f^{(k+1)}$ exists on (a, b). Fix a point x_0 in $[a, b]$. We seek a kth degree polynomial

$$p_k(x) = a_0 + a_1(x - x_0) + a_2(x - x_0)^2 + \cdots + a_k(x - x_0)^k$$

such that $p_k^{(j)}(x_0) = f^{(j)}(x_0)$ for $j = 0, 1, 2, \ldots, k$. The coefficients of p_k are to be determined. To identify the coefficients $a_0, a_1, a_2, \ldots, a_k$, we use the following procedure.

First set $x = x_0$. It follows that $p_k(x_0) = a_0$. Since we want $p_k(x_0) = f(x_0)$, we let $a_0 = f(x_0)$.

Next we differentiate p_k, shift index, and set $x = x_0$. At the first step,

$$p_k'(x) = \sum_{j=0}^{k} j a_j (x - x_0)^{j-1} = \sum_{i=0}^{k-1} (i+1) a_{i+1} (x - x_0)^i.$$

Setting $x = x_0$, we observe that all terms except the term corresponding to $i = 0$ must vanish. Therefore $p_k'(x_0) = a_1$. Again, since we want $p_k'(x_0) = f'(x_0)$, we let $a_1 = f'(x_0)$.

At the next step, differentiate p_k', again shift index, and set $x = x_0$. Upon doing so, we obtain

$$p_k''(x) = \sum_{i=0}^{k-1} (i+1) i a_{i+1} (x - x_0)^{i-1} = \sum_{i=0}^{k-2} (i+2)(i+1) a_{i+2} (x - x_0)^i$$

$$= \sum_{i=0}^{k-2} \left[\frac{(i+2)!}{i!}\right] a_{i+2} (x - x_0)^i.$$

When $x = x_0$, all terms in this last sum except that corresponding to $i = 0$ must vanish. Therefore $p_k''(x_0) = 2!a_2$. Since we want $p_k''(x_0) = f''(x_0)$, we choose $a_2 = f''(x_0)/2!$. Recursively differentiating, we find that the jth derivative of p_k is

$$p_k^{(j)}(x) = \sum_{i=0}^{k-j} \frac{(i+j)!a_{i+j}}{i!}(x - x_0)^i.$$

When $x = x_0$, all terms of $p_k^{(j)}$ except that corresponding to $i = 0$ must vanish. Therefore $p_k^{(j)}(x_0) = j!a_j$. Since we want $p_k^{(j)}(x_0) = f^{(j)}(x_0)$, we let $a_j = f^{(j)}(x_0)/j!$. Thus, for $j = 0, 1, 2, \ldots, k$, we want to choose

$$a_j = \frac{f^{(j)}(x_0)}{j!}.$$

Consequently, the kth degree polynomial p_k with the property that $p_k^{(j)}(x_0) = f^{(j)}(x_0)$ for $j = 0, 1, 2, \ldots, k$ is

$$p_k(x) = \sum_{j=0}^{k} \frac{f^{(j)}(x_0)}{j!}(x - x_0)^j.$$

DEFINITION 4.5.1 The polynomial p_k is called the kth-*degree Taylor polynomial* of f at $x = x_0$. ●

The error incurred by approximating f by the Taylor polynomial p_k is

$$|f(x) - p_k(x)| = |f(x) - \sum_{j=0}^{k} \frac{f^{(j)}(x_0)}{j!}(x - x_0)^j|.$$

For some functions, this error will be small provided x is sufficiently near x_0; for others it will not. The value of Taylor's theorem is that it provides a formula for the *exact error* resulting from this approximation. This formula will enable us to calculate bounds on that error. Taylor's theorem is a direct consequence of the next generalized version of Cauchy's Mean Value Theorem.

THEOREM 4.5.1 Let f and g be two continuous functions on $[a, b]$ that have continuous kth derivatives on the closed interval $[a, b]$. Suppose that f and g also have $(k+1)$st derivatives on the open interval (a, b). Fix x_0 in $[a, b]$. Then for every x in $[a, b]$ with $x \neq x_0$, there exists a c strictly between x_0 and x such that

$$f^{(k+1)}(c)[g(x) - q_k(x)] = g^{(k+1)}(c)[f(x) - p_k(x)],$$

where p_k and q_k are the kth-degree Taylor polynomials at x_0 of f and g, respectively. That is,

$$p_k(x) = \sum_{j=0}^{k} \frac{f^{(j)}(x_0)}{j!}(x - x_0)^j$$

and

$$q_k(x) = \sum_{j=0}^{k} \frac{g^{(j)}(x_0)}{j!}(x - x_0)^j.$$

Proof. For definiteness, assume that $a \leq x_0 < x \leq b$. For each t in $[x_0, x]$, define

$$F(t) = f(t) + \sum_{j=1}^{k} \frac{f^{(j)}(t)}{j!}(x - t)^j$$

and

$$G(t) = g(t) + \sum_{j=1}^{k} \frac{g^{(j)}(t)}{j!}(x - t)^j.$$

Since all the functions $f^{(j)}$ and $g^{(i)}$, $j = 0, 1, 2, \ldots, k$ are continuous on all of $[a, b]$, both F and G, as functions of t, are continuous on the closed interval $[x_0, x]$. What is more, F and G are differentiable on (x_0, x), since, for $j = 0, 1, 2, \ldots, k - 1$, $f^{(j)}$ and $g^{(j)}$ are continuously differentiable on $[a, b]$ and since $f^{(k)}$ and $g^{(k)}$ are differentiable on (a, b). Therefore Cauchy's Mean Value Theorem (Theorem 4.3.7) applies and ensures the existence of a point c in (x_0, x) such that

$$F'(c)[G(x) - G(x_0)] = G'(c)[F(x) - F(x_0)]. \tag{4.3}$$

Returning to the definitions of $F(t)$ and $G(t)$, notice that $F(x) = f(x)$, $G(x) = g(x)$, $F(x_0) = p_k(x)$ and $G(x_0) = q_k(x)$. When we substitute these quantities in (4.3) we obtain

$$F'(c)[g(x) - q_k(x)] = G'(c)[f(x) - p_k(x)]. \tag{4.4}$$

It remains, then, to compute $F'(t)$ and $G'(t)$. By symmetry, it will suffice to compute $F'(t)$. Using the product rule, shifting the index in the second sum, and telescoping the finite series yields

$$\begin{aligned}
F'(t) &= f'(t) + \sum_{j=1}^{k} \frac{f^{(j+1)}(t)}{j!}(x - t)^j - \sum_{j=1}^{k} \frac{f^{(j)}(t)}{j!} j(x - t)^{j-1} \\
&= f'(t) + \sum_{j=1}^{k} \frac{f^{(j+1)}(t)}{j!}(x - t)^j - \sum_{j=1}^{k} \frac{f^{(j)}(t)}{(j - 1)!}(x - t)^{j-1} \\
&= f'(t) + \sum_{j=1}^{k} \frac{f^{(j+1)}(t)}{j!}(x - t)^j - \sum_{j=0}^{k-1} \frac{f^{(j+1)}(t)}{j!}(x - t)^j \\
&= \frac{f^{(k+1)}(t)}{k!}(x - t)^k.
\end{aligned}$$

Likewise, $G'(t) = [g^{(k+1)}(t)/k!](x-t)^k$. Substituting these quantities in (4.4) yields

$$\frac{f^{(k+1)}(c)}{k!}(x-c)^k[g(x) - q_k(x)] = \frac{g^{(k+1)}(c)}{k!}(x-c)^k[f(x) - p_k(x)].$$

Canceling $k!$ and $(x-c)^k$ we obtain

$$f^{(k+1)}(c)[g(x) - q_k(x)] = g^{(k+1)}(c)[f(x) - p_k(x)],$$

the promised conclusion of the theorem. ●

THEOREM 4.5.2 Taylor's Theorem If f has k continuous derivatives on the closed interval $[a, b]$ and has a $(k+1)$st derivative on (a, b), then, for any x_0 and x in $[a, b]$ with $x \neq x_0$, there exists a c strictly between x_0 and x such that

$$f(x) = \sum_{j=0}^{k} \frac{f^{(j)}(x_0)}{j!}(x - x_0)^j + R_k(x_0; x),$$

where

$$R_k(x_0; x) = \frac{f^{(k+1)}(c)}{(k+1)!}(x - x_0)^{k+1}.$$

Proof. Let $g(x) = (x - x_0)^{k+1}$. We apply Theorem 4.5.1 to the function f and this g. There exists a c strictly between x_0 and x such that

$$f^{(k+1)}(c)[g(x) - q_k(x)] = g^{(k+1)}(c)[f(x) - p_k(x)], \tag{4.5}$$

where p_k and q_k are the kth-degree Taylor polynomials of f and g, respectively. Notice that, for $j = 0, 1, 2, \ldots, k$, $g^{(j)}(x)$ retains $x - x_0$ as a factor, so $g^{(j)}(x_0) = 0$. Consequently, $q_k(x)$ is the zero polynomial. Finally, $g^{(k+1)}(x) = (k+1)!$. Substituting in (4.5) gives

$$f^{(k+1)}(c)[(x - x_0)^{k+1} - 0] = (k+1)![f(x) - p_k(x)].$$

Upon rearranging, we obtain

$$\begin{aligned} f(x) &= p_k(x) + \frac{f^{(k+1)}(c)}{(k+1)!}(x - x_0)^{k+1} \\ &= \sum_{j=0}^{k} \frac{f^{(j)}(x_0)}{j!}(x - x_0)^j + R_k(x_0; x). \end{aligned}$$

The remainder or error caused by approximating $f(x)$ by $p_k(x)$ is

$$R_k(x_0; x) = \frac{f^{(k+1)}(c)}{(k+1)!}(x - x_0)^{k+1},$$

a form of the remainder attributed to Lagrange. This proves the theorem. ●

Having Lagrange's remainder in hand, we can identify a class of functions for which the kth Taylor polynomial gives a good approximation to f when k is sufficiently large. This class includes many of the most important functions you have met in your earlier studies, restricted, of course, to the correct domain. Here we will verify assertions made in the examples of Section 3.5.

To identify some of the functions in this class, suppose that f has derivatives of all orders on an interval $[a, b]$. Automatically, for every k in $\mathbb{N}$, f satisfies the conditions of Taylor's theorem. Further, for each k in $\mathbb{N}$, $||f^{(k)}||_\infty = \sup\{|f^{(k)}(x)| : x \text{ in } [a, b]\}$ is finite. Fix any x_0 in $[a, b]$. Then, for each k in $\mathbb{N}$ and each x in $[a, b]$, we have

$$|R_k(x_0; x)| = \frac{|f^{(k+1)}(c)|}{(k+1)!}|x - x_0|^{k+1}$$
$$\leq \frac{||f^{(k+1)}||_\infty}{(k+1)!}|b - a|^{k+1}.$$

Suppose also that there exists an $M > 0$ such that, for all k in $\mathbb{N}$,

$$||f^{(k)}||_\infty^{1/k} \leq M. \tag{4.6}$$

Since, for any fixed r in $\mathbb{R}$, $\lim_{k\to\infty} r^k/k! = 0$, it follows that

$$0 \leq \lim_{k\to\infty} |R_k(x_0; x)| \leq \lim_{k\to\infty} \frac{||f^{(k+1)}||_\infty}{(k+1)!}|b - a|^{k+1}$$
$$\leq \lim_{k\to\infty} \frac{(M|b - a|)^{k+1}}{(k+1)!} = 0,$$

with $r = M|b - a|$. Thus, if (4.6) holds, then the Squeeze Play ensures that $\lim_{k\to\infty} R_k(x_0; x) = 0$ [uniformly] on $[a, b]$. Consequently, for k sufficiently large, the kth Taylor polynomial p_k approximates f uniformly on $[a, b]$. This proves the following theorem. ●

THEOREM 4.5.3 Let f have derivatives of all orders on $[a, b]$. Suppose that there exists a constant M such that $||f^{(k)}||_\infty^{1/k} \leq M$ for all k in $\mathbb{N}$. For any x_0 in $[a, b]$,

$$\lim_{k\to\infty} \sum_{j=0}^{k} \frac{f^{(j)}(x_0)}{j!}(x - x_0)^j = f(x) \qquad \text{[uniformly]}$$

on $[a, b]$. ●

In particular, if f has derivatives of all orders on $[a, b]$ and if there exists a single M such that $||f^{(k)}||_\infty \leq M$ for all k in $\mathbb{N}$, then the condition (4.6) holds. Thus, given any $\epsilon > 0$, we can choose k_0 so large that, for $k \geq k_0$,

$$\frac{M|b - a|^{k+1}}{(k+1)!} < \epsilon.$$

It follows that, for any x_0 and x in $[a, b]$ and any $k \geq k_0$,

$$|f(x) - p_k(x)| \leq \frac{||f^{(k+1)}||_\infty}{(k+1)!}|x - x_0|^{k+1}$$
$$\leq \frac{M|b - a|^{k+1}}{(k+1)!} < \epsilon.$$

Thus $||f - p_k||_\infty < \epsilon$ for $k \geq k_0$. This proves that the sequence $\{p_k\}$ of Taylor polynomials of f at x_0 converges uniformly to f on $[a, b]$.

COROLLARY 4.5.4 If f has derivatives of all orders on $[a, b]$ and if there exists a constant M such that $||f^{(k)}||_\infty \leq M$ for all k in $\mathbb{N}$, then

$$\lim_{k\to\infty} \sum_{j=0}^{k} \frac{f^{(j)}(x_0)}{j!}(x - x_0)^j = f(x) \qquad \text{[uniformly]}$$

on $[a, b]$. ●

EXAMPLE 10 Let $f(x) = e^x$ on any interval $[a, b]$ containing $x_0 = 0$. For every k, $f^{(k)}(x) = e^x$. It follows that $||f^{(k)}||_\infty = e^b = M$. Let $r = \max\{|a|, |b|\}$. By Taylor's theorem, there is a c between 0 and x such that

$$0 \leq \lim_{k\to\infty} |R_k(0; x)| = \lim_{k\to\infty} \frac{e^c|x|^{k+1}}{(k+1)!} \leq \lim_{k\to\infty} \frac{e^b r^{k+1}}{(k+1)!} = 0.$$

Thus, $\lim_{k\to\infty} R_k(0; x) = 0$ [uniformly] on $[a, b]$. It follows that $\lim_{k\to\infty} p_k = f$ [uniformly] on $[a, b]$. To compute the kth Taylor polynomial of $f(x) = e^x$ at $x_0 = 0$, note that for every $j = 0, 1, 2, \ldots, k$, $f^{(j)}(0) = e^0 = 1$. Therefore,

$$p_k(x) = 1 + x + \frac{x^2}{2!} + \frac{x^3}{3!} + \cdots + \frac{x^k}{k!} = \sum_{j=0}^{k} \frac{x^j}{j!}.$$

With this result in hand, we can prove that the number e is irrational, thereby fulfilling a promise made in Section 1.3. Taking $x = 1$ in the foregoing derivation, we have, for any k in $\mathbb{N}$,

$$e = 1 + 1 + \frac{1}{2!} + \frac{1}{3!} + \cdots + \frac{1}{k!} + \frac{e^c}{(k+1)!}$$

where c is some number between 0 and 1. Suppose that e were rational. That is, assume that $e = p/q$, where p and q are positive integers having no common positive divisors other than 1 and with $q \geq 2$. Take $k > q \geq 2$. We deduce that $k!e = k!(p/q)$ is an integer. Moreover,

$$\begin{aligned} k!e &= 2k! + \frac{k!}{2!} + \frac{k!}{3!} + \cdots + \frac{k!}{k!} + \frac{e^c}{k+1} \\ &= K + \frac{e^c}{k+1}, \end{aligned}$$

where $K = 2k! + k!/2! + k!/3! + \cdots + k!/k!$ is also an integer. Thus $e^c/(k+1) = k!e - K$ is an integer. But here we have a contradiction. Because $0 < c < 1$ and because the exponential function is strictly monotone increasing, it follows that $1 < e^c < e < 3$. Therefore, $e^c/(k+1)$ is an integer in the interval (0, 1). But there is no integer in (0, 1). Therefore e must be irrational. ●

EXAMPLE 11 Let $f(x) = \sin x$ on any interval $[a, b]$ containing $x_0 = 0$. Then f has derivatives of all orders. The derivatives of f of odd order are

$$f^{(2j+1)}(x) = (-1)^j \cos x.$$

Those of even order are

$$f^{(2j)}(x) = (-1)^j \sin x.$$

For each $k \geq 0$, $||f^{(k)}||_\infty \leq 1$. Let $r = \max\{|a|, |b|\}$. Then, for some c between 0 and x, we have

$$\begin{aligned} 0 \leq \lim_{k\to\infty} |R_k(0; x)| &= \lim_{k\to\infty} \frac{|f^{(k+1)}(c)|}{(k+1)!}|x|^{k+1} \\ &\leq \lim_{k\to\infty} \frac{r^{k+1}}{(k+1)!} = 0. \end{aligned}$$

Thus $\lim_{k\to\infty} R_k(0; x) = 0$ [uniformly]. It follows that the sequence of Taylor polynomials of f at 0 converges uniformly to $f(x) = \sin x$. Evaluating the derivatives of f at 0, we have, for $j = 0, 1, 2, \ldots,$

$$f^{(2j+1)}(0) = (-1)^j$$

and

$$f^{(2j)}(0) = 0.$$

Therefore, only the odd powers of x occur with a nonzero coefficient in every Taylor polynomial of f at x_0. It follows that the Taylor polynomial of $f(x) = \sin x$ at $x_0 = 0$ with degree $2k + 1$ is

$$p_{2k+1}(x) = x - \frac{x^3}{3!} + \frac{x^5}{5!} - \cdots \frac{(-1)^j x^{2k+1}}{(2k+1)!} = \sum_{j=0}^{k} \frac{(-1)^j x^{2j+1}}{(2j+1)!}.$$

As we know, $\lim_{k\to\infty} R_k(0; x) = 0$ [uniformly] on $[a, b]$. Therefore,

$$\lim_{k\to\infty} \sum_{j=0}^{k} \frac{(-1)^j x^{2j+1}}{(2j+1)!} = \sin x \quad \text{[uniformly]}$$

on $[a, b]$. ●

EXAMPLE 12 Similarly, the Taylor polynomial of degree $2k$ for $f(x) = \cos x$ at $x_0 = 0$ is

$$p_{2k}(x) = 1 - \frac{x^2}{2!} + \frac{x^4}{4!} - \cdots \frac{(-1)^j x^{2j}}{(2k)!} = \sum_{j=0}^{k} \frac{(-1)^j x^{2j}}{(2j)!}.$$

Further, by Corollary 4.5.4, the sequence of these polynomials converges uniformly to $f(x) = \cos x$ on any interval $[a, b]$ containing $x_0 = 0$. ●

EXAMPLE 13 Let $f(x) = \ln x$ on the closed interval [1, 2] and let $x_0 = 1$. Computing the successive derivatives of f, we have $f'(x) = 1/x$, $f''(x) = -x^{-2}$, $f^{(3)}(x) = 2x^{-3}, \ldots$, and, in general, $f^{(j)}(x) = (-1)^{(j-1)}(j-1)!x^{-j}$. Since

$1 \le x \le 2$, we have $||f^{(j)}||_\infty = (j-1)!$, for $j = 1, 2, \ldots$. By Taylor's theorem, there exists a c in $(1, x)$ such that

$$\ln x = p_k(x) + \frac{f^{(k+1)}(c)(x-1)^{k+1}}{(k+1)!}.$$

Furthermore,

$$\begin{aligned} 0 \le \lim_{k\to\infty} |R_k(1;x)| &= \lim_{k\to\infty} \frac{|f^{(k+1)}(c)||x-1|^{k+1}}{(k+1)!} \\ &\le \lim_{k\to\infty} \frac{||f^{(k+1)}||_\infty}{(k+1)!} = \lim_{k\to\infty} \frac{k!}{(k+1)!} \\ &= \lim_{k\to\infty} \frac{1}{k+1} = 0. \end{aligned}$$

By the Squeeze Play, $\lim_{k\to\infty} R_k(1;x) = 0$ [uniformly] on [1, 2]. Consequently, the sequence of Taylor polynomials of f at $x_0 = 1$ converges uniformly to f on [1, 2]. To compute the kth Taylor polynomial $p_k(x)$ when $x_0 = 1$, evaluate f and its first k derivatives at $x_0 = 1$: $f(1) = 0$, and, for $j = 1, 2, 3, \ldots, k$, $f^{(j)}(1) = (-1)^{j-1}(j-1)!$. Thus the coefficient of $(x-1)^j$ in $p_k(x)$ is $f^{(j)}(1)/j! = (-1)^{j-1}/j$. That is,

$$\begin{aligned} p_k(x) &= (x-1) - \frac{(x-1)^2}{2} + \frac{(x-1)^3}{3} - \cdots + \frac{(-1)^{k-1}(x-1)^k}{k} \\ &= \sum_{j=1}^{k} \frac{(-1)^{j-1}(x-1)^j}{j}. \end{aligned}$$

Since $\lim_{k\to\infty} R_k(1;x) = 0$ [uniformly] on [1, 2], we deduce that

$$\lim_{k\to\infty} \sum_{j=1}^{k} \frac{(-1)^{j-1}(x-1)^j}{j} = \ln x \qquad \text{[uniformly]}$$

on [1, 2]. ●

EXERCISES

4.73. Use the Taylor polynomial for $f(x) = \sin x$ about $x_0 = 0$ to compute an approximation, accurate to within 5×10^{-5}, for the sine of .435 radians. Compare your approximation with that given by your calculator's built-in program for the sine function. How large must the degree of the polynomial be to achieve this degree of accuracy?

4.74. Use the Taylor polynomial for $f(x) = e^x$ about $x_0 = 0$ to compute an approximation, accurate to within 5×10^{-5}, for $e^{1/2}$. Compare your approximation with that given by your calculator's built-in program for the exponential function. How large must the degree of the polynomial be to achieve this degree of accuracy. If, for this k and for any x in $[-1/2, 1/2]$, you compute $p_k(x)$, what can you say about the error $|e^x - p_k(x)|$?

4.75. **a)** Find the kth Taylor polynomial about $x_0 = 0$ for $f(x) = 1/(1+x)$. Find the Lagrange form of the remainder.

b) Use an appropriate substitution in the result of part (a) to find the $2k$th Taylor polynomial of $1/(1+x^2)$ at $x_0 = 0$.

c) Adjust your computations from part (a) to find the kth Taylor polynomial for $1/(1+x)^2$ at $x_0 = 0$.

4.76. Let $f(x) = \tan^{-1} x$ for x in $\mathbb{R}$.

a) Find the Taylor polynomial $p_{2k+1}(x)$ of degree $2k+1$ for f at $x_0 = 0$. (After computing $f'(x)$, use your results from Exercise 4.75 part (b).)

b) Find the smallest k such that $p_{2k+1}(1)$ approximates $\tan^{-1} 1$ to within .0005.

4.77. Let $f(x) = \tan x$ for $-\pi/2 < x < \pi/2$.

a) Compute the Taylor polynomial $p_5(x)$ for $f(x)$ at $x_0 = 0$.

b) Use your calculator to compute $p_5(\pi/8)$ and compare with the value of $\tan(\pi/8)$ given by the built-in program for the tangent function.

4.78. Let $f(x) = \ln(1+x)$ for x in $(-1, 1)$.

a) Find the kth Taylor polynomial p_k of f at $x_0 = 0$.

b) Find Lagrange's form of the remainder $R_k(0; x)$ for f.

c) If $0 < r < 1$ and if x is in $[-r, r]$, then $R_k(0; x)$ converges uniformly to 0. Lagrange's form of the remainder is inadequate to prove this claim however. Lagrange's remainder will suffice to prove that $R_k(0; x)$ converges pointwise to 0 on $(-1/2, 1]$ and that the convergence is uniform on any closed interval $[a, b]$ contained in the open interval $(-1/2, 1)$. Prove these assertions.

4.79. Let $f(x) = (1+x)^{1/2}$ for $-1 < x < 1$. Find the kth Taylor polynomial $p_k(x)$ for f at $x_0 = 0$ and show that it can be written in the form

$$p_k(x) = 1 + \sum_{j=1}^{k} \frac{(-1)^{j-1}(2j-2)!}{2^{2j-1}(j-1)!j!} x^j.$$

Find the corresponding Lagrange form of the remainder.

4.80. Let $f(x) = \sqrt{2}\sin(x + \pi/4)$.

a) Find the Taylor polynomial of degree $2k+1$ for f at $x_0 = 0$ and the corresponding Lagrange form of the remainder $R_{2k+1}(0; x)$.

b) Show that $\lim_{k\to\infty} R_{2k+1}(0; x) = 0$ [uniformly] on $\mathbb{R}$.

c) Compare the Taylor polynomial (of degree $2k+1$) you found in part (a) with the sum of the Taylor polynomials for $\sin x$ (of degree $2k+1$) and $\cos x$ (of degree $2k$). What trigonometric identity applies to explain the relationship you found?

4.81. For x in $\mathbb{R}$, let $f(x) = (x+1)e^x$.

a) Find the kth Taylor polynomial for f at $x_0 = 0$ and the corresponding Lagrange form of the remainder $R_k(0; x)$.

b) Prove that $\lim_{k\to\infty} R_k(0; x) = 0$ [uniformly] on $\mathbb{R}$.

c) Compare the Taylor polynomial of degree $k+1$ you found in part (a) with the product of $(x+1)$ and the kth Taylor polynomial of e^x.

4.82. Find the kth Taylor polynomial $p_k(x)$ and the corresponding Lagrange remainder $R_k(e; x)$ for the function $f(x) = \ln x$ about the point $x_0 = e$.

4.83. Use Taylor's theorem to prove that, for $x > 0$,

$$\ln x + \frac{1}{x} - \frac{1}{2x^2} < \ln(x+1) < \ln x + \frac{1}{x}.$$

4.84. *Newton's generalized binomial theorem* Consider $(a+b)^\alpha$. If $\alpha = n$ is a positive integer, $(a+b)^n$ can be expanded by the binomial theorem. But, if α is not a positive integer, then an expansion of $(a+b)^\alpha$ requires some effort. The discovery of the correct procedure giving an expansion of $(a+b)^\alpha$ was one of four achievements that Newton considered his best. First, without loss of generality, we can assume that $0 < |a| < b$. Let $x = a/b$. Thus, $(a+b)^\alpha = (1+x)^\alpha b^\alpha$, where $-1 < x < 1$. We focus our attention on the task of expanding $(1+x)^\alpha$.

a) Find the kth Taylor polynomial of $f(x) = (1+x)^\alpha$ at $x_0 = 0$.

b) Find Lagrange's form of the remainder $R_k(0; x)$. (In Chapter 12 we will discuss $\lim_{k\to\infty} R_k(0; x)$ in detail.)

4.85. For $x \neq 0$, define $f(x) = \exp(-1/x^2)$. Let $f(0) = 0$. (Refer to Exercise 4.72.)

a) Find the kth Taylor polynomial of f at $x_0 = 0$.

b) Show that the sequence of Taylor polynomials of f at $x_0 = 0$ cannot possibly converge to f except at the one point x_0. Thus no matter how large k may be, the error $|f(x) - p_k(x)|$ cannot be made arbitrarily small for any $x \neq 0$.

5

Functions of Bounded Variation

This chapter introduces material that is strongly algebraic in flavor and lays the foundation for a proper treatment of the integration theory to be developed in Chapters 6 and 7. We first discuss the collection of finite partitions of a closed interval $[a, b]$. Using this information, we next identify and explore yet another ring of functions on $[a, b]$, the functions of *bounded variation.* Intuitively, a function f is of bounded variation if its total amount of wiggling remains bounded.

5.1 PARTITIONS

One everyday meaning of the word *partition* is "to break or divide into pieces." It is this meaning that we will adopt.

DEFINITION 5.1.1 A (finite) *partition* π of the interval $[a, b]$ is a finite collection of points $\{x_0, x_1, \ldots, x_p\}$, called *partition points,* such that

$$a = x_0 < x_1 < x_2 < \cdots < x_p = b.$$

The partition π *partitions* the interval $[a, b]$ into the subintervals $[x_{j-1}, x_j]$, for $j = 1, 2, 3, \ldots, p$. The length of the interval $[x_{j-1}, x_j]$ is denoted $\Delta x_j = x_j - x_{j-1}$. The collection of all (finite) partitions of $[a, b]$ is denoted $\Pi[a, b]$. ●

DEFINITION 5.1.2 A partition $\pi_1 = \{y_0, y_1, y_2, \ldots, y_q\}$ is a *refinement* of $\pi_2 = \{x_0, x_1, x_2, \ldots, x_p\}$ if every partition point x_j in π_2 also belongs to π_1. We write $\pi_2 \preceq \pi_1$. We also say that π_1 is *finer* than π_2 or that π_2 is *coarser* than π_1. ●

Notice that $\pi_2 \preceq \pi_1$ merely means that each partition interval $[y_{i-1}, y_i]$ of π_1 is a subset of one of the partition subintervals $[x_{j-1}, x_j]$ of the partition π_2. If the partition π_2 breaks the interval $[a, b]$ into the subintervals $[x_{j-1}, x_j]$, $j = 1, 2, 3, \ldots, p$, then the more refined partition π_1 merely breaks *some of the subintervals* $[x_{j-1}, x_j]$ into yet smaller subintervals.

DEFINITION 5.1.3 Given any two partitions π_1 and π_2 of $[a, b]$, there exists a third partition, called the *least common refinement* of π_1 and π_2 and denoted $\pi_1 \vee \pi_2$. It is formed by taking the union of the partition points in π_1 and in π_2. ●

It is immediate from the definition that $\pi_1 \preceq \pi_1 \vee \pi_2$ and that $\pi_2 \preceq \pi_1 \vee \pi_2$. You can prove as an exercise that $\pi_1 \vee \pi_2$ is the smallest partition that refines both π_1 and π_2 in the following sense: If π is a refinement of each of π_1 and π_2, then π must also be a refinement of $\pi_1 \vee \pi_2$.

It is worthwhile taking a moment to explore the properties of the relation $\preceq$ and of the operation $\vee$ on $\Pi[a, b]$. First, $\preceq$ is a partial order on $\Pi[a, b]$.

i) It is reflexive: $\pi \preceq \pi$ for all π in $\Pi[a, b]$.

ii) It is antisymmetric: If $\pi_1 \preceq \pi_2$ and $\pi_2 \preceq \pi_1$, then $\pi_1 = \pi_2$.

iii) Finally, $\preceq$ is transitive: If $\pi_1 \preceq \pi_2$ and $\pi_2 \preceq \pi_3$, then $\pi_1 \preceq \pi_3$.

But trichotomy fails, as we see by the following example. If $x_1 = (a + b)/2$ and if $\pi_1 = \{x_0, x_1, x_2\}$, then π_1 is a partition of $[a, b]$. (See Fig. 5.1.) On the other hand, let $y_1 = 2a/3 + b/3$ and $y_2 = a/3 + 2b/3$. Form the partition $\pi_2 = \{y_0, y_1, y_2, y_3\}$. These two partitions are not equal nor is either a refinement of the other. They are simply not comparable. Trichotomy does not hold and $\preceq$ is a partial, but not a linear, order on $\Pi[a, b]$.

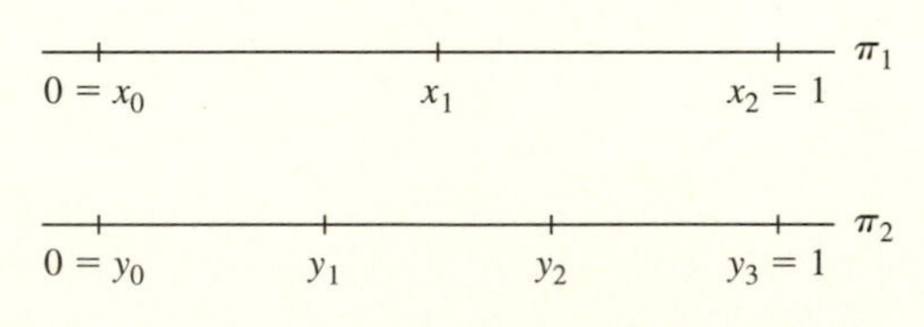

Figure 5.1

The *least refined*, or *coarsest*, partition in $\Pi[a, b]$ is the partition $\pi_0 = \{a = x_0, x_1 = b\}$, which leaves the interval $[a, b]$ intact. We will refer to π_0 as the *trivial partition* of $[a, b]$. Every other partition is a refinement of π_0. The uniform partition of $[a, b]$ with $k + 1$ partition points is constructed as follows. Fix k in $\mathbb{N}$ and let $h = (b - a)/k$. For each $j = 0, 1, 2, 3, \ldots, k$, let $x_j = a + jh$. The partition $\pi = \{x_0, x_1, x_2, \ldots, x_k\}$ is called the *uniform partition* with k subintervals. The uniform partition is the one most frequently used in introductory calculus texts when constructing the Riemann integral; it will not suffice for our purposes.

The operation $\vee$ of taking the least common refinement of two partitions is a commutative, associative, binary operation on $\Pi[a, b]$: $\pi_1 \vee \pi_2 = \pi_2 \vee \pi_1$ and

$(\pi_1 \vee \pi_2) \vee \pi_3 = \pi_1 \vee (\pi_2 \vee \pi_3)$. Furthermore, the trivial partition π_0 already identified is an identity with respect to $\vee$: $\pi_0 \vee \pi = \pi$ for all π in $\Pi[a, b]$. The partial order $\preceq$ and the binary operation $\vee$ are related as follows.

THEOREM 5.1.1 For any two partitions π_1 and π_2 in $\Pi[a, b]$, $\pi_1 \preceq \pi_2$ if and only if $\pi_1 \vee \pi_2 = \pi_2$.

Proof. Suppose that $\pi_1 \preceq \pi_2$ where $\pi_1 = \{x_0, x_1, x_2, \ldots, x_p\}$ and $\pi_2 = \{y_0, y_1, y_2, \ldots, y_q\}$. Then, as sets,

$$\{x_0, x_1, x_2, \ldots, x_p\} \subseteq \{y_0, y_1, y_2, \ldots, y_q\}.$$

The partition $\pi_1 \vee \pi_2$ is formed by taking as partition points the union of these two sets. The union is $\{y_0, y_1, \ldots, y_q\}$. Therefore $\pi_1 \vee \pi_2 = \pi_2$. We leave the proof of the converse as an exercise. ●

DEFINITION 5.1.4 Let $\pi = \{x_0, x_1, x_2, \ldots, x_p\}$ be any partition in $\Pi[a, b]$. For $j = 1, 2, 3, \ldots, p$, let $\Delta x_j = x_j - x_{j-1}$, the length of the jth partition interval. The *gauge* of the partition π, denoted $||\pi||$, is

$$||\pi|| = \max\{\Delta x_j : j = 1, 2, 3, \ldots, p\}. \quad ●$$

The gauge of the trivial partition is $b - a$. The gauge of the uniform partition with $k + 1$ equally spaced partition points is $(b - a)/k$. This observation confirms that there are partitions in $\Pi[a, b]$ with arbitrarily small gauge. In general, notice that, if $\pi_1 \preceq \pi_2$, then $||\pi_2|| \leq ||\pi_1||$. The gauge of a partition can only decrease as it is refined.

EXERCISES

5.1. Prove that, if $\pi_1 \preceq \pi$ and $\pi_2 \preceq \pi$, then $\pi_1 \vee \pi_2 \preceq \pi$.

5.2. Prove that, if $\pi_1 \vee \pi_2 = \pi_2$, then $\pi_1 \preceq \pi_2$.

5.3. Prove that $\preceq$ is a partial order on $\Pi[a, b]$.

5.4. Fix π_1 in $\Pi[a, b]$. The set

$$C(\pi_1) = \{\pi \text{ in } \Pi[a, b] : \pi_1 \preceq \pi\}$$

is called the *cone* determined by π_1.

a) Prove that, if $\pi_1 \preceq \pi_2$, then $C(\pi_1) \subseteq C(\pi_2)$.

b) Prove that, if $||\pi_1|| < \epsilon$, then $||\pi|| < \epsilon$ for all π in $C(\pi_1)$.

c) Prove that, if $||\pi_1|| < \epsilon$ and if π is any partition of $[a, b]$, then $||\pi \vee \pi_1|| < \epsilon$.

5.2 MONOTONE FUNCTIONS ON [a, b]

DEFINITION 5.2.1 Let f be a real-valued function defined on the closed interval $[a, b]$. Then f is said to be *increasing* on $[a, b]$ if, whenever $a \leq x_1 < x_2 \leq b$, it follows that $f(x_1) \leq f(x_2)$. Likewise, f is said to be *strictly increasing* if $f(x_1) < f(x_2)$. Similarly, f is said to be *decreasing* on $[a, b]$ if, whenever $a \leq x_1 < x_2 \leq b$, it follows that $f(x_1) \geq f(x_2)$ and is *strictly decreasing* if $f(x_1) > f(x_2)$. A function

f is said to be (*strictly*) *monotone* on $[a, b]$ if it is either (strictly) increasing or (strictly) decreasing. ●

Our first theorem is a fundamental, though unsurprising, fact about increasing functions. It provides us with an essential tool in the sequel.

THEOREM 5.2.1 Let f be increasing on $[a, b]$ and let $\pi = \{x_0, x_1, x_2, \ldots, x_p\}$ be any partition in $\Pi[a, b]$. Then

$$\sum_{j=0}^{p} [f(x_j^+) - f(x_j^-)] \leq f(b) - f(a).$$

Proof. It is to be understood that, in the statement of the theorem, $f(a^-) = f(a)$ and $f(b^+) = f(b)$. For $j = 0, 1, \ldots, p-1$, choose a point y_j in (x_j, x_{j+1}). (See Fig. 5.2.) Since f is increasing, we have $f(y_{j-1}) \leq f(x_j^-) \leq f(x_j^+) \leq f(y_j)$ for $j = 1, 2, \ldots, p-1$. It follows that, for such j,

$$f(x_j^+) - f(x_j^-) \leq f(y_j) - f(y_{j-1}).$$

Note also that $f(a) \leq f(a^+) \leq f(y_0)$ and that $f(y_{p-1}) \leq f(b^-) \leq f(b)$. By telescoping the summation, we obtain

$$\begin{aligned}
&\sum_{j=0}^{p} [f(x_j^+) - f(x_j^-)] \\
&\qquad = f(a^+) - f(a) + \sum_{j=1}^{p-1} [f(x_j^+) - f(x_j^-)] + f(b) - f(b^-) \\
&\qquad \leq f(y_0) - f(a) + \sum_{j=1}^{p-1} [f(y_j) - f(y_{j-1})] + f(b) - f(y_{p-1}) \\
&\qquad = f(b) - f(a).
\end{aligned}$$

This proves the theorem. ●

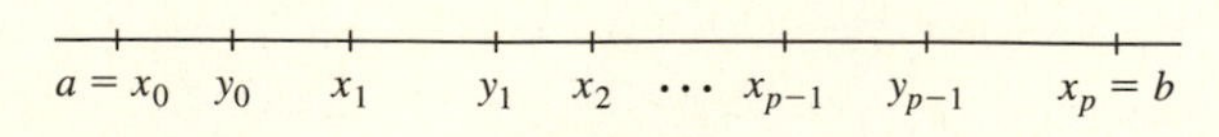

Figure 5.2

The difference $f(x_j^+) - f(x_j^-)$ is the magnitude of the jump discontinuity of f at x_j. If f were to be continuous at x_j, then this magnitude would be 0, of course; otherwise, it would be some positive number (since f is increasing). Theorem 5.2.1 tells us that, for every finite collection of points $\{x_j : j = 0, 1, 2, \ldots, p\}$ in $[a, b]$, the sum of the jumps is always bounded by the one constant $f(b) - f(a)$. This leads directly to a particularly satisfying result.

THEOREM 5.2.2 If f is monotone on $[a, b]$, then the set of discontinuities of f is countable.

Proof. Since f is defined on the entire closed interval $[a, b]$ and since f is monotone, it is clear that the only discontinuities f can have, if any, are jump discontinuities. Assume that f is increasing. For each k in $\mathbb{N}$, let

$$D_k = \{x \text{ in } (a, b) : f(x^+) - f(x^-) > 1/k\}.$$

Thus D_k is the set of all points in (a, b) where the magnitude of the jump discontinuity of f exceeds $1/k$. If $x_1 < x_2 < x_3 < \cdots < x_p$ are p points in D_k, Theorem 5.2.1 implies that $p/k < f(b) - f(a)$. It follows that

$$p < k[f(b) - f(a)].$$

Thus D_k must be a finite set with no more than $k[f(b) - f(a)]$ elements. The set of all points where f has a jump discontinuity is the union of all the D_k. Since the union of a countable number of finite sets is countable, f can have no more than a countable number of discontinuities. If f is decreasing, apply the above argument to the increasing function $g = -f$. This proves the theorem. ●

5.2.1 Saltus Functions (Optional)

Lurking just beneath the surface here is a little gem from classical analysis. Its derivation involves techniques that you may find clever and useful; the result itself will give you additional insight into the class of functions, those of bounded variation, the study of which forms the primary goal of this chapter.

Suppose that f is monotone increasing on $[a, b]$. As we have just seen, the set S of points in $[a, b]$ where f is discontinuous is at most countably infinite. Further, all the discontinuities of f are jumps. It may happen that S is actually finite, but we cannot depend on that, so we use $\mathbb{N}$ to index the points in $S = \{x_1, x_2, x_3, \ldots\}$ with the understanding that the list may terminate. Beware, however: The indexing does not necessarily correspond to the linear order of $\mathbb{N}$. That is, if $j < k$, it need not follow that $x_j < x_k$. If S is actually a finite set, then we could arrange such a linear ordering from the leftmost to the rightmost point of S and the following discussion could be greatly simplified. But if S is countably infinite, as it may well be, we cannot hope for such tidiness. Remember, if S is infinite, then the Bolzano–Weierstrass theorem ensures the existence of a limit point of S in $[a, b]$ and therefore of a Cauchy sequence in S that converges to this limit point. We conclude in this case that some of the points of S are arbitrarily near one another.

Given a monotone increasing function f, we define two functions associated with f on $[a, b]$ as follows:

$$u(x) = \begin{cases} 0, & x = a \\ f(x) - f(x^-), & x \text{ in } (a, b] \end{cases}$$

and

$$v(x) = \begin{cases} f(x^+) - f(x), & x \text{ in } [a, b) \\ 0, & x = b. \end{cases}$$

The function $u(x)$ is the *left-hand jump* of f at x in $(a, b]$. Likewise, $v(x)$ is the *right-hand jump* of f at x in $[a, b)$. Both u and v are nonnegative and their sum at x in (a, b),

$$u(x) + v(x) = f(x^+) - f(x^-),$$

is the total jump of f at x. Furthermore, f is continuous from the left at x in $(a, b]$ if and only if $u(x) = 0$. Likewise, f is continuous from the right at x in $[a, b)$ if and only if $v(x) = 0$. Finally, f is continuous at x in $[a, b]$ if and only if $u(x) = v(x) = 0$. Notice that, because f is increasing on $[a, b]$,

$$\sum_{x_j \in S} u(x_j) + \sum_{x_j \in S} v(x_j) \leq f(b) - f(a),$$

since the summations total the left- and right-hand jumps of f over the set of all the discontinuities of f. Elsewhere f may increase, but it does so continuously. It follows, for any nonempty subset T of $[a, b]$, that

$$\sum_{x_j \in S \cap T} u(x_j) = \sup\left\{ \sum_{x_j \in F} u(x_j) : F \subseteq S \cap T,\ F \text{ finite} \right\}$$

and

$$\sum_{x_j \in S \cap T} v(x_j) = \sup\left\{ \sum_{x_j \in F} v(x_j) : F \subseteq S \cap T,\ F \text{ finite} \right\}.$$

These suprema exist because each of these sums is bounded above by $f(b) - f(a)$.

To simplify notation in our present discussion when the set T above is a subinterval of $[a, b]$, we introduce the following notation:

$$S[x, y] = S \cap [x, y] \qquad S(x, y] = S \cap (x, y]$$

$$S[x, y) = S \cap [x, y) \qquad S(x, y) = S \cap (x, y).$$

In the sequel, pay close attention to the endpoint notation of these sets.

We are ready now to define the *saltus function* associated with the monotone increasing function f. At the endpoint a, define $s_f(a) = f(a)$. For x in $(a, b]$, define

$$s_f(x) = f(a) + \sum_{x_j \in S(a,x]} u(x_j) + \sum_{x_j \in S[a,x)} v(x_j).$$

As you can confirm for yourself, $s_f(x)$ is a generalization of a step function, but one with perhaps countably infinitely many values. For example, if it should happen that for two points x_j and x_k of S with $x_j < x_k$, the set $S(x_j, x_k) = \emptyset$, then the graph of $s_f(x)$ on $[x_j, x_k]$ is as drawn in Fig. 5.3. The jumps of s_f at x_j and x_k equal the corresponding jumps of the function f with which s_f is associated. But you want to be aware that, if S is infinite, then near a limit point of S the horizontal segments of the graph of s_f become arbitrarily short. Also, it may be impossible, given x_j in S, to find an $x_k > x_j$ such that $S(x_j, x_k) = \emptyset$. (Imagine, for example, that $S = Q \cap [a, b]$.) The functional value $s_f(x)$ is $f(a)$ together with the total of the jumps of f in $[a, x)$ and the left-hand jump, $f(x) - f(x^-)$, of f at the point x itself.

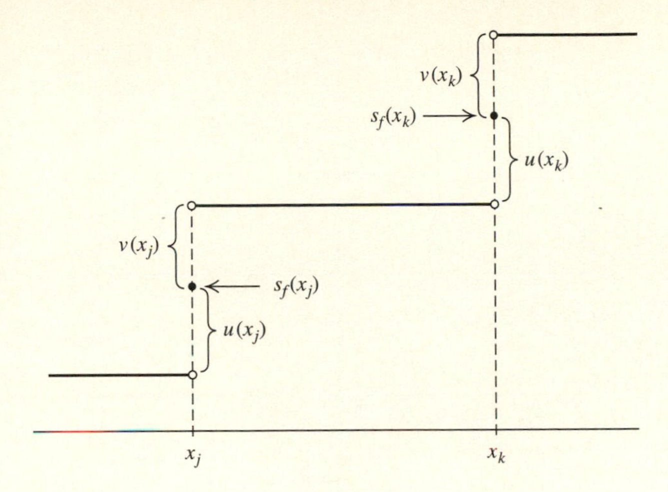

Figure 5.3

We first list several of the properties of the saltus function s_f associated with a function f that is monotone increasing on $[a, b]$.

LEMMA Let f be a monotone increasing function on $[a, b]$.

i) The saltus function s_f is monotone increasing on $[a, b]$.

ii) For any two points x and y such that $a \leq x < y \leq b$, we have

$$0 \leq s_f(y) - s_f(x) \leq f(y) - f(x).$$

iii) Let S denote the set of all discontinuities of f. Then s_f is continuous at every x in $[a, b] \cap S^c$. ●

The proofs of these three properties rely solely on the definition of s_f and are left as exercises. Our purpose in identifying the saltus function of f is this: If we subtract from f its accumulated jump discontinuities, then we ought to obtain a continuous function that behaves, otherwise, much as f does. This, in fact, proves to be the case. Define a function f_c by

$$f_c(x) = f(x) - s_f(x),$$

for all x in $[a, b]$. (See Fig. 5.4 on page 220.) We claim that f_c is monotone increasing and continuous on $[a, b]$. To prove these claims, suppose that $a \leq x < y \leq b$. By part (ii) of the lemma,

$$\begin{aligned} f_c(y) - f_c(x) &= [f(y) - s_f(y)] - [f(x) - s_f(x)] \\ &= f(y) - f(x) - [s_f(y) - s_f(x)] \geq 0. \end{aligned}$$

Therefore, $f_c(x) \leq f_c(y)$. This proves that f_c is monotone increasing.

To see that f_c is continuous, first fix any x in $[a, b)$ and any $\epsilon > 0$. Since $f(x^+) = \lim_{y \to x^+} f(y)$ exists, we can choose a $\delta > 0$ such that, for all y in $(x, x + \delta)$,

$$0 \leq f(y) - f(x^+) < \epsilon.$$

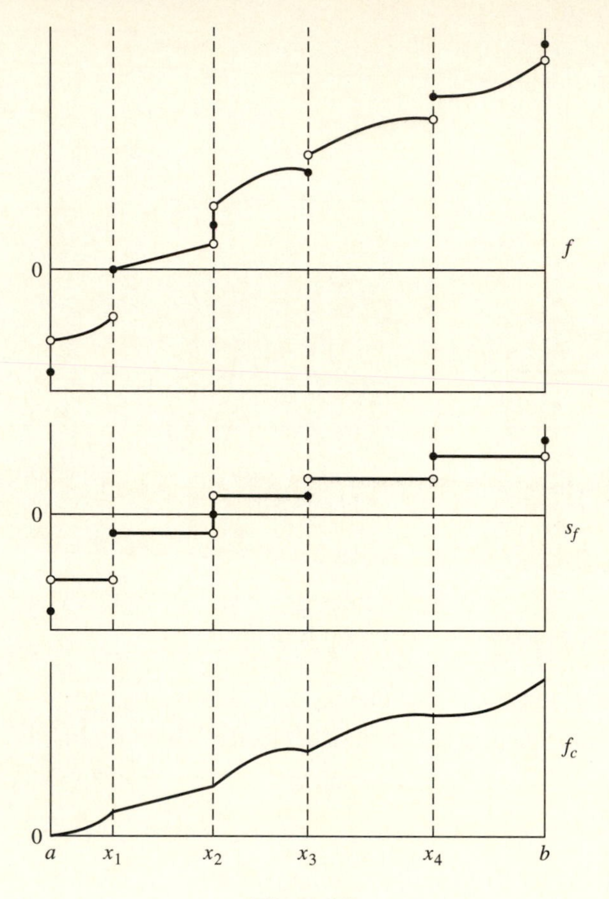

Figure 5.4

Choose any y in $(x, x+\delta)$ and compute

$$
\begin{aligned}
f_c(y) - f_c(x) &= [f(y) - s_f(y)] - [f(x) - s_f(x)] \\
&= f(y) - \left[\sum_{x_j \in S(x,y]} u(x_j) + \sum_{x_j \in S[x,y)} v(x_j) \right] - f(x) \\
&= [f(y) - u(y)] - \sum_{x_j \in S(x,y)} [u(x_j) + v(x_j)] - [v(x) + f(x)] \\
&= f(y^-) - \sum_{x_j \in S(x,y)} [u(x_j) + v(x_j)] - f(x^+) \\
&\le f(y^-) - f(x^+) \le f(y) - f(x^+) < \epsilon.
\end{aligned}
$$

Thus f_c is continuous from the right at x in $[a, b)$. By a completely analogous argument, f_c is continuous from the left at each x in $(a, b]$, because $f(x^-) = \lim_{y \to x^-} f(y)$ exists.

In summary we have proved that every monotone increasing function f can be written $f = f_c + s_f$ where f_c is the continuous part of f and s_f is the saltus function associated with f. Note that, if f is monotone decreasing, then we can apply the preceding reasoning to $-f$ to conclude that every monotone function is the sum of a continuous monotone function of the same type and the associated monotone saltus function of the same type. This classic result has ramifications in integration theory and probability theory.

THEOREM 5.2.3 Let f be a monotone function on $[a, b]$ and let s_f be the saltus function associated with f. Then we can write $f = f_c + s_f$, where f_c is continuous on $[a, b]$. If f is monotone increasing (decreasing), then s_f and f_c are also monotone increasing (decreasing) on $[a, b]$. ●

EXERCISES

5.5. Let f be monotone increasing on $[a, b]$. Prove that the associated saltus function s_f is also monotone increasing on $[a, b]$.

5.6. Let f be monotone increasing on $[a, b]$. Prove that its associated saltus function s_f satisfies the inequality

$$0 \leq s_f(y) - s_f(x) \leq f(y) - f(x)$$

for all x, y in $[a, b]$ with $x < y$. (Begin with any finite number of points such that $x = x_0 < x_1 < x_2 < \cdots < x_p = y$ and show that

$$\sum_{x_j \in F \cap S(x,y]} u(x_j) + \sum_{x_j \in F \cap S[x,y)} v(x_j) \leq f(y) - f(x).$$

Then take the supremum over all such finite sets F to obtain the desired inequality.)

5.7. Let f be monotone increasing on $[a, b]$, let S denote the set of points in $[a, b]$ where f is discontinuous, and let s_f denote the saltus function associated with f. Prove that s_f is continuous at every point in $[a, b] \cap S^c$.

5.8. If f is monotone increasing on $[a, b]$, if s_f is its associated saltus function, and if $f_c = f - s_f$, prove that f_c is continuous from the left at any x in $(a, b]$.

5.9. Define a function f on $[0, 2]$ as follows:

$$f(x) = \begin{cases} -1, & x = 0 \\ x^2, & 0 < x < 1 \\ \frac{7}{4}, & x = 1 \\ (x+3)^{\frac{1}{2}}, & 1 < x < 2 \\ 3, & x = 2. \end{cases}$$

Find the saltus function s_f associated with f. Find the function $f_c = f - s_f$. Sketch the graphs of f, s_f, and f_c.

5.3 FUNCTIONS OF BOUNDED VARIATION

We now have all the tools necessary to describe a new class of functions, those of *bounded variation*. This class forms, as we will see, both a vector space and a commutative ring with identity. In Chapters 6 and 7 we will also see that these functions arise naturally in the theory of integration. As we will demonstrate, most of the elementary functions you know best are included.

DEFINITION 5.3.1 Let f be a real-valued function defined on $[a, b]$ and let $\pi = \{x_0, x_1, x_2, \ldots, x_p\}$ be any partition of $[a, b]$. For $j = 1, 2, \ldots, p$, let $\Delta f_j = f(x_j) - f(x_{j-1})$ denote the change in f on the jth partition interval. If there exists an $M > 0$ such that $\sum_{j=1}^{p} |\Delta f_j| \leq M$ for every π in $\Pi[a, b]$, then f is said to be of *bounded variation* on $[a, b]$. The collection of all functions of bounded variation on $[a, b]$ is denoted $BV(a, b)$. ●

If f is in $BV(a, b)$, then surely f is bounded on $[a, b]$. That is, $||f||_\infty = \sup\{|f(x)| : x \text{ in } [a, b]\}$ is finite. For suppose f is of bounded variation on $[a, b]$. There is an M such that $\sum_{j=1}^{p} |\Delta f_j| \leq M$ for every π in $\Pi[a, b]$. In particular, choose x in (a, b) and form the partition $\pi = \{a, x, b\}$. Then

$$\begin{aligned} |f(x)| - |f(a)| &\leq |f(x) - f(a)| \\ &\leq |f(x) - f(a)| + |f(b) - f(x)| \leq M, \end{aligned}$$

from which it follows that $|f(x)| \leq M + |f(a)|$. If, in the above, we were to choose $x = b$, the same inequality results. It follows then that $||f||_\infty \leq M + |f(a)|$ and f is bounded on $[a, b]$. We have proved the following theorem.

THEOREM 5.3.1 If f is in $BV(a, b)$, then f is bounded on $[a, b]$. ●

Initially, it may not be clear which functions are of bounded variation on $[a, b]$. We begin simply. Assume first that f is increasing on $[a, b]$. Let $\pi = \{x_0, x_1, x_2, \ldots, x_p\}$ be any partition of $[a, b]$. Since f is monotone increasing, we have $\Delta f_j \geq 0$ for $j = 1, 2, \ldots, p$. Thus,

$$\begin{aligned} \sum_{j=1}^{p} |\Delta f_j| = \sum_{j=1}^{p} \Delta f_j &= \sum_{j=1}^{p} [f(x_j) - f(x_{j-1})] \\ &= f(b) - f(a) = M. \end{aligned}$$

It follows immediately that f is of bounded variation. By applying the same argument to $g = -f$ if f is decreasing, we have proved the following theorem.

THEOREM 5.3.2 If f is monotone on $[a, b]$, then f is of bounded variation on $[a, b]$. ●

EXAMPLE 1 The function $f(x) = \cos x$ is monotone decreasing on $[0, \pi]$ and therefore is in $BV(0, \pi)$. ●

What is startling, as we will see subsequently, is that the class $BV(a, b)$ of all functions of bounded variation on $[a, b]$ consists exactly of those functions f that can be written in the form $f = g - h$, where g and h are monotone increasing on $[a, b]$. This is a satisfying and most useful characterization. But before launching on the program that will achieve this result, let's identify another large class of functions in $BV(a, b)$ with which you are well familiar from your early studies.

THEOREM 5.3.3 Suppose that f is continuous on $[a, b]$ and differentiable on (a, b). Suppose also that

$$||f'||_\infty = \sup\{|f'(x)| : x \text{ in } (a, b)\}$$

is finite. Then f is of bounded variation on $[a, b]$.

Proof. Let $\pi = \{x_0, x_1, x_2, \ldots, x_p\}$ be any partition of $[a, b]$. Apply the Mean Value Theorem to f on each subinterval $[x_{j-1}, x_j]$. For $j = 1, 2, 3, \ldots, p$, there exists a c_j in (x_{j-1}, x_j) such that

$$\Delta f_j = f(x_j) - f(x_{j-1}) = f'(c_j)(x_j - x_{j-1}).$$

It follows that

$$\sum_{j=1}^{p} |\Delta f_j| = \sum_{j=1}^{p} |f'(c_j)|(x_j - x_{j-1}) \leq ||f'||_\infty \sum_{j=1}^{p} (x_j - x_{j-1})$$
$$= ||f'||_\infty (b - a) = M.$$

Therefore f is in $BV(a, b)$. ●

EXAMPLE 2 Let $f(x) = \tan^{-1} x$ on any interval $[a, b]$. Since $f'(x) = 1/(1 + x^2)$ and since $||f'||_\infty \leq 1$, we deduce that f is in $BV(a, b)$. ●

That there should exist an M such that $\sum_{j=1}^{p} |\Delta f_j| \leq M$ for every partition π in $\Pi[a, b]$ suffices to guarantee the bounded variation of f on $[a, b]$, but to proceed we need a numerical measure of the exact amount of variation in the values of f. Here the completeness of $\mathbb{R}$ serves us well. If f is in $BV(a, b)$, then the set of real numbers,

$$V = \left\{\sum_{j=1}^{p} |\Delta f_j| : \pi \text{ in } \Pi[a, b]\right\}$$

is bounded above by M. Therefore $\sup V$ exists in $\mathbb{R}$. That is all we need.

DEFINITION 5.3.2 Let f be a function in $BV(a, b)$. The real number,

$$V(f; a, b) = \sup\left\{\sum_{j=1}^{p} |\Delta f_j| : \pi \text{ in } \Pi[a, b]\right\}$$

depending on f, a, and b, is called the *total variation* of f on $[a, b]$. ●

As usual after having defined a new concept, we immediately explore the elementary properties of the quantity $V(f; a, b)$. First, observe that $V(f; a, b) \geq 0$ since every number in V is nonnegative. Furthermore it is clear that $V(f; a, b) = 0$ if and only if f is constant on $[a, b]$. It is also self-evident that the total variation of f is the same as the total variation of $-f$. That is, $V(-f; a, b) = V(f; a, b)$. These properties are simple; the next two are not.

THEOREM 5.3.4 If f and g are in $BV(a, b)$, then $f + g$ and $f - g$ are also in $BV(a, b)$ and

$$V(f \pm g; a, b) \leq V(f; a, b) + V(g; a, b).$$

Proof. Let $\pi = \{x_0, x_1, x_2, \ldots, x_p\}$ be any partition of $[a, b]$. Observe that

$$\begin{aligned}|\Delta(f+g)_j| &= |(f+g)(x_j) - (f+g)(x_{j-1})| \\ &\leq |f(x_j) - f(x_{j-1})| + |g(x_j) - g(x_{j-1})| = |\Delta f_j| + |\Delta g_j|.\end{aligned}$$

Thus

$$\begin{aligned}\sum_{j=1}^{p} |\Delta(f+g)_j| &\leq \sum_{j=1}^{p} |\Delta f_j| + \sum_{j=1}^{p} |\Delta g_j| \\ &\leq V(f; a, b) + V(g; a, b).\end{aligned}$$

As a consequence, taking the supremum of the left-hand side, we have $V(f + g; a, b) \leq V(f; a, b) + V(g; a, b)$. Therefore $f + g$ is in $BV(a, b)$. It follows easily, since $V(-g; a, b) = V(g; a, b)$, that

$$\begin{aligned}V(f - g; a, b) &\leq V(f; a, b) + V(-g; a, b) \\ &= V(f; a, b) + V(g; a, b).\end{aligned}$$

Therefore $f - g$ is also in $BV(a, b)$. This proves that $BV(a, b)$ is closed under addition and subtraction. ●

As we now prove, $BV(a, b)$ is also closed under multiplication.

THEOREM 5.3.5 If f and g are in $BV(a, b)$, then fg is also in $BV(a, b)$ with

$$V(fg; a, b) \leq ||g||_\infty V(f; a, b) + ||f||_\infty V(g; a, b).$$

Proof. Since f and g are in $BV(a, b)$, we know by Theorem 5.3.1 that f and g are bounded on $[a, b]$. Let $h = fg$ and let $\pi = \{x_0, x_1, x_2, \ldots, x_p\}$ be any partition of $[a, b]$. Then for each $j = 1, 2, 3, \ldots, p$,

$$\begin{aligned}|\Delta h_j| &= |f(x_j)g(x_j) - f(x_{j-1})g(x_{j-1})| \\ &\leq |f(x_j) - f(x_{j-1})|\,|g(x_j)| + |f(x_{j-1})|\,|g(x_j) - g(x_{j-1})| \\ &\leq |\Delta f_j|\,||g||_\infty + ||f||_\infty |\Delta g_j|.\end{aligned}$$

Summing on j, we deduce that $\sum_{j=1}^{p} |\Delta h_j|$ is bounded above by

$$||g||_\infty \sum_{j=1}^{p} |\Delta f_j| + ||f||_\infty \sum_{j=1}^{p} |\Delta g_j|,$$

which, in turn, is bounded above by

$$||g||_\infty V(f; a, b) + ||f||_\infty V(g; a, b).$$

Taking the supremum over all partitions in $\Pi[a, b]$, we conclude that

$$V(fg; a, b) \leq ||g||_\infty V(f; a, b) + ||f||_\infty V(g; a, b),$$

as claimed. Therefore $BV(a, b)$ is closed under multiplication. ●

Our results to this point imply that $BV(a, b)$ is a commutative ring with identity $f = 1$ (a function with total variation 0). As we have asked before, we ask now for the units in $BV(a, b)$. That is, for which f in $BV(a, b)$ is it true that $1/f$ is also in $BV(a, b)$? Clearly, if f vanishes at some point of $[a, b]$, then $1/f$ is not even defined on all of $[a, b]$. But it is possible, since f need not be continuous, that f not vanish at any point, yet $1/f$ not be in $BV(a, b)$. For example, if $\lim_{x \to c} f(x) = 0$ when x approaches a point c in $[a, b]$, even if $f(c) \neq 0$, then the function $1/f$ is unbounded. Since every function of bounded variation is already bounded, $1/f$ cannot possibly be in $BV(a, b)$. However, if f is of bounded variation and is bounded away from 0 on all of $[a, b]$, then $1/f$ is also in $BV(a, b)$. We prove this in the following theorem.

THEOREM 5.3.6 Suppose that f is in $BV(a, b)$. Suppose also that there exists an m such that $|f(x)| \geq m > 0$ for all x in $[a, b]$. Then $1/f$ is also in $BV(a, b)$ with

$$V\left(\frac{1}{f}; a, b\right) \leq \frac{V(f; a, b)}{m^2}.$$

Proof. Let $g = 1/f$. Let $\pi = \{x_0, x_1, x_2, \ldots, x_p\}$ be any partition of $[a, b]$. For each $j = 1, 2, 3, \ldots, p$,

$$|\Delta g_j| = \left|\frac{1}{f(x_j)} - \frac{1}{f(x_{j-1})}\right| = \frac{|f(x_{j-1}) - f(x_j)|}{|f(x_j)|\,|f(x_{j-1})|} \leq \frac{|\Delta f_j|}{m^2}.$$

Therefore, $\sum_{j=1}^{p} |\Delta g_j| \leq \sum_{j=1}^{p} |\Delta f_j|/m^2 \leq V(f; a, b)/m^2$. Consequently, taking the supremum of the left-hand side, we obtain $V(g; a, b) \leq V(f; a, b)/m^2$. It follows that $g = 1/f$ is in $BV(a, b)$. ●

EXAMPLE 3 Any polynomial p is in $BV(a, b)$ for any interval $[a, b]$. The polynomial p is a unit in $BV(a, b)$ for any interval $[a, b]$ containing none of its zeros. ●

EXAMPLE 4 Any rational function p/q is in $BV(a, b)$ for any interval $[a, b]$ containing none of the zeros of q. ●

EXAMPLE 5 The exponential function e^x is in $BV(a, b)$ for any interval $[a, b]$. More generally, the function $e^{u(x)}$ is of bounded variation for any interval $[a, b]$ on which $u(x)$ has a bounded derivative or on which $u(x)$ is monotone. ●

EXAMPLE 6 The trigonometric functions $\sin x$ and $\cos x$, having bounded derivatives, are in $BV(a, b)$ for any interval $[a, b]$. Consequently, by Theorem 5.3.6, $\sec x = 1/\cos x$ is in $BV(a, b)$ for any interval $[a, b]$ containing none of the points $\{(2k + 1)\pi/2 : k \text{ in } \mathbb{Z}\}$. It follows, by Theorem 5.3.5, that $\tan x$ is in $BV(a, b)$ for any interval containing none of the points $\{(2k + 1)\pi/2 : k \text{ in } \mathbb{Z}\}$. ●

A review of the stock of standard functions from your introductory courses in calculus will show that almost all of these functions are of bounded variation on appropriately chosen intervals. In fact, you cannot actually draw the graph of any function of unbounded variation; such a function either has a pole or must wiggle too much for you to be able to graph it in a finite amount of time.

EXAMPLE 7 Define $f(x) = \sin(1/x)$ for $x \neq 0$. Let $f(0) = 0$. Then f is not of bounded variation on any interval containing 0. For example, take the interval $[a, b]$ to be $[0, 1]$. Choose any k in $\mathbb{N}$ and form the partition with partition points $x_0 = 0$, $x_j = 2/[(2k-2j+1)\pi]$ for $j = 1, 2, \ldots, k$, and $x_{k+1} = 1$. As you can show as an exercise, for this partition, $\sum_{j=1}^{k+1} |\Delta f_j| \geq 2k-1$. Therefore, $V(f, 0, 1)$ cannot be finite. ●

5.4 TOTAL VARIATION AS A FUNCTION

Heretofore, we have kept the interval $[a, b]$ fixed, but now that we are familiar with the total variation function V, we can derive its more subtle properties. By allowing the interval under consideration to vary, we note first that V is *additive on intervals*.

THEOREM 5.4.1 Suppose that f is in $BV(a, b)$. Let c be any point in (a, b). Then f is in $BV(a, c)$ and f is in $BV(c, b)$. Furthermore

$$V(f; a, b) = V(f; a, c) + V(f; c, b).$$

Proof. Let $\pi_1 = \{a = x_0, x_1, \ldots, x_p = c\}$ be a partition of $[a, c]$ and let $\pi_2 = \{c = y_0, y_1, \ldots, y_q = b\}$ be a partition of $[c, b]$. Then $\pi = \{x_0, x_1, \ldots, x_p = y_0, y_1, \ldots, y_q\}$ is a partition of $[a, b]$. Computing the changes of f over the intervals in this partition, we obtain

$$\sum_{j=1}^{p} |f(x_j) - f(x_{j-1})| + \sum_{j=1}^{q} |f(y_j) - f(y_{j-1})| \leq V(f; a, b).$$

Taking the supremum as π_1 varies in $\Pi[a, c]$, we deduce that

$$V(f; a, c) + \sum_{j=1}^{q} |f(y_j) - f(y_{j-1})| \leq V(f; a, b).$$

Taking the supremum as π_2 varies in $\Pi[c, b]$, we conclude that

$$V(f; a, c) + V(f; c, b) \leq V(f; a, b). \tag{5.1}$$

To obtain the reverse inequality, let $\pi = \{x_0, x_1, \ldots, x_p\}$ be any partition of $[a, b]$. We may assume without loss of generality that c is a partition point of π. Otherwise, adjoin c to π and obtain a refinement of π. Suppose that $c = x_k$. Then π is the union of the partition

$$\pi_1 = \{x_0, x_1, \ldots, x_{k-1}, x_k = c\}$$

of $[a, c]$ and the partition

$$\pi_2 = \{c = x_k, x_{k+1}, x_{k+2}, \ldots, x_p\}$$

of $[c, b]$. It follows that

$$\sum_{j=1}^{p}|\Delta f_j| = \sum_{j=1}^{k}|\Delta f_j| + \sum_{j=k+1}^{p}|\Delta f_j| \le V(f; a, c) + V(f; c, b).$$

Taking the supremum of the left-hand side as π varies over $\Pi[a, b]$ gives

$$V(f; a, b) \le V(f; a, c) + V(f; c, b). \tag{5.2}$$

Combining (5.1) and (5.2), we obtain

$$V(f; a, b) = V(f; a, c) + V(f; c, b),$$

and the theorem is proved. ●

Theorem 5.4.1 enables us to create, for any f in $BV(a, b)$, a new function that measures the variation of f over the interval $[a, x]$.

DEFINITION 5.4.1 Let f be a function in $BV(a, b)$. The *variation* of f is the function V_f on $[a, b]$ defined as follows: $V_f(a) = 0$ and, for x in $(a, b]$,

$$V_f(x) = V(f; a, x) = \sup\left\{\sum_{j=1}^{p}|\Delta f_j| : \pi \text{ in } \Pi[a, x]\right\}. \quad ●$$

The first important property of the function V_f is identified in the next theorem.

THEOREM 5.4.2 If f is in $BV(a, b)$, then $V_f(x)$ is an increasing function of x on $[a, b]$.

Proof. Choose any two points x_1 and x_2 such that $a \le x_1 < x_2 \le b$. We deduce, by Theorem 5.4.1, that f is in both $BV(a, x_1)$ and $BV(a, x_2)$. Another application of Theorem 5.4.1 implies that f is also in $BV(x_1, x_2)$ and that

$$\begin{aligned} V_f(x_2) = V(f; a, x_2) &= V(f; a, x_1) + V(f; x_1, x_2) \\ &= V_f(x_1) + V(f; x_1, x_2). \end{aligned}$$

Therefore, $V_f(x_2) - V_f(x_1) = V(f; x_1, x_2) \ge 0$. It follows that $V_f(x_1) \le V_f(x_2)$, proving the theorem. ●

THEOREM 5.4.3 If f is in $BV(a, b)$, then $V_f - f$ is an increasing function on $[a, b]$.

Proof. Let x_1 and x_2 be any two points with $a \le x_1 < x_2 \le b$. It is straightforward to compute

$$\begin{aligned} (V_f - f)(x_2) - (V_f - f)(x_1) &= V_f(x_2) - V_f(x_1) - [f(x_2) - f(x_1)] \\ &= V(f; a, x_2) - V(f; a, x_1) - [f(x_2) - f(x_1)] \\ &= V(f; x_1, x_2) - [f(x_2) - f(x_1)]. \end{aligned}$$

Now, the total variation of f on $[x_1, x_2]$ is

$$V(f; x_1, x_2) = \sup\left\{\sum_j |\Delta f_j| : \pi \text{ in } \Pi[x_1, x_2]\right\}.$$

Since $\{x_1, x_2\}$ is the trivial partition of $[x_1, x_2]$,

$$f(x_2) - f(x_1) \leq V(f; x_1, x_2).$$

Consequently, $V(f; x_1, x_2) - [f(x_2) - f(x_1)] \geq 0$. Therefore $(V_f - f)(x_1) \leq (V_f - f)(x_2)$, proving that $(V_f - f)$ is an increasing function on $[a, b]$. ●

Theorems 5.4.2 and 5.4.3 combine to enable us to prove the following important and elegant characterization of the functions in $BV(a, b)$.

THEOREM 5.4.4 $BV(a, b)$ consists exactly of those functions f that can be written as the difference of two monotone increasing functions.

Proof. Suppose that $f = g - h$ where g and h are monotone increasing functions on $[a, b]$. By Theorem 5.3.1, both g and h are in $BV(a, b)$. By Theorem 5.3.3, $g - h = f$ is also in $BV(a, b)$.

Conversely, suppose that f is in $BV(a, b)$. For x in $[a, b]$, let $g(x) = V_f(x)$. By Theorem 5.4.2, g is monotone increasing on $[a, b]$ and, by Theorem 5.4.3, the function $h = V_f - f = g - f$ is also monotone increasing on $[a, b]$. We conclude that $f = g - h$ where g and h are monotone increasing on $[a, b]$. ●

EXAMPLE 8 Let $f(x) = \cos x$ on $[0, 2\pi]$. Let

$$g(x) = \begin{cases} 0, & \text{for } 0 \leq x \leq \pi \\ 1 + \cos x, & \text{for } \pi < x \leq 2\pi \end{cases}$$

and

$$h(x) = \begin{cases} -\cos x, & \text{for } 0 \leq x \leq \pi \\ 1, & \text{for } \pi < x \leq 2\pi. \end{cases}$$

Then both g and h are monotone increasing and $f = g - h$. ●

By combining Theorem 5.4.4 with our earlier results on saltus functions, we obtain an especially elegant characterization of every function f of bounded variation. By Theorem 5.4.4, first write $f = g - h$ where both g and h are monotone increasing on $[a, b]$. Let s_g and s_h denote the saltus functions of g and h, respectively. Define

$$g_c = g - s_g \qquad \text{and} \qquad h_c = h - s_h.$$

We know by Theorem 5.2.3 that g_c and h_c are both monotone increasing and continuous on $[a, b]$. More, $g = g_c + s_g$ and $h = h_c + s_h$. Therefore

$$f = g - h = (g_c + s_g) - (h_c + s_h) = (g_c - h_c) + (s_g - s_h).$$

Note that $g_c - h_c$ is a continuous function of bounded variation and $s_g - s_h$ is a saltus function of bounded variation. The function $g_c - h_c$ is the continuous part and $s_g - s_h$ is the saltus part of f.

It is worthwhile noting that the representation of a function f in $BV(a, b)$ as the difference of two monotone increasing functions is not unique, for if $f = g - h$, where g and h are increasing on $[a, b]$, if f_1 is any increasing function on $[a, b]$, and if $g_1 = g + f_1$ and $h_1 = h + f_1$, then $f = g_1 - h_1$ is a second representation of the required form.

EXERCISES

5.10. Define $f(0) = 0$ and $f(x) = \sin(1/x)$ for $x \neq 0$. Confirm the assertion made in Example 7 to show that the function f is not in $BV(0, 1)$.

5.11. Define $f(0) = 0$ and $f(x) = x\sin(1/x)$ for $x \neq 0$. Show that f is of unbounded variation on any interval containing $x_0 = 0$. (See Example 22 in Section 1.3.)

5.12. Define $f(0) = 0$ and $f(x) = x^2\sin(1/x)$ for $x \neq 0$. Show that f is of bounded variation on any interval $[a, b]$, including those that contain $x_0 = 0$.

5.13. Identify all intervals $[a, b]$ on which the functions $\csc x$ and $\cot x$ are of bounded variation.

5.14. The hyperbolic cosine and the hyperbolic sine functions are defined as follows:

$$\cosh x = \frac{1}{2}(e^x + e^{-x})$$

$$\sinh x = \frac{1}{2}(e^x - e^{-x}).$$

a) Identify all intervals $[a, b]$ on which each of these functions is of bounded variation.

b) The remaining hyperbolic trigonometric functions are

$$\tanh x = \frac{\sinh x}{\cosh x} \qquad \operatorname{sech} x = \frac{1}{\cosh x}$$

$$\coth x = \frac{\cosh x}{\sinh x} \qquad \operatorname{csch} x = \frac{1}{\sinh x}.$$

Identify all intervals $[a, b]$ on which each of these functions is of bounded variation.

5.15. Show that every polynomial is of bounded variation on every interval $[a, b]$.

5.16. Let $f(x) = e^{-x}\cos x$. Show that f is of bounded variation on any compact interval. For each k in $\mathbb{N}$, compute $V(f; 0, 2k\pi)$ and show that $\{V(f; 0, 2k\pi) : k \text{ in } \mathbb{N}\}$ is an increasing sequence that is bounded above. Find its limit.

5.17. Dirichlet's dancing function on $[0, 1]$ is defined by

$$f(x) = \begin{cases} 1, & \text{if } x \text{ is rational} \\ 0, & \text{if } x \text{ is irrational.} \end{cases}$$

Prove that f is of unbounded variation on $[0, 1]$.

5.18. Let f be in $BV(a, b)$ and let $\pi = \{x_0, x_1, x_2, \ldots, x_p\}$ be any partition of $[a, b]$. Let $J^+(\pi)$ denote the set of indices j such that $\Delta f_j \geq 0$; let $J^-(\pi)$ denote the set of indices j such that $\Delta f_j < 0$. Define

$$V^+(f; a, b) = \sup\left\{ \sum_{j \in J^+(\pi)} \Delta f_j : \pi \text{ in } \Pi[a, b] \right\}$$

and

$$V^-(f;a,b) = \sup\left\{\sum_{j\in J^-(\pi)} |\Delta f_j| : \pi \text{ in } \Pi[a,b]\right\}.$$

$V^+(f;a,b)$ is called the *positive variation* and $V^-(f;a,b)$ the *negative variation* of f on $[a,b]$. For x in $(a,b]$, define $V_f^+(x) = V^+(f;a,x)$ and $V_f^-(x) = V^-(f;a,x)$. Define $V_f^+(a) = V_f^-(a) = 0$. Prove the following assertions.

a) $V_f(x) = V_f^+(x) + V_f^-(x)$, for all x in $[a,b]$.

b) $0 \le V_f^+(x) \le V_f(x)$ and $0 \le V_f^-(x) \le V_f(x)$, for all x in $[a,b]$.

c) V_f^+ and V_f^- are monotone increasing on $[a,b]$.

d) $f(x) = f(a) + V_f^+(x) - V_f^-(x)$, for all x in $[a,b]$.

Let f be a function defined on $\mathbb{R}$ that is of bounded variation on every compact interval $[a,b]$. Suppose further that the set of real numbers

$$S = \{V(f;a,b) : [a,b] \text{ a compact interval}\}$$

is bounded above. Define $V(f;\mathbb{R})$ to be sup S. Then $V(f;\mathbb{R})$ is called the *total variation* of f on $\mathbb{R}$ and f is said to be of *bounded variation* on $\mathbb{R}$.

5.19. Prove that, if f is of bounded variation on $\mathbb{R}$, then f is bounded on $\mathbb{R}$.

5.20. Prove that, if f and g are of bounded variation on $\mathbb{R}$, then $f+g$, $f-g$, and fg are also of bounded variation on $\mathbb{R}$.

5.21. Prove that if f is of bounded variation on $\mathbb{R}$ and if f is bounded away from 0 on $\mathbb{R}$, then $1/f$ is of bounded variation on $\mathbb{R}$.

5.22. Suppose that f is of bounded variation on $\mathbb{R}$. Let c be any point in $\mathbb{R}$. Prove that f is of bounded variation on $(-\infty, c]$ and on $[c,\infty)$. That is, prove that

$$V(f;-\infty,c) = \sup\{V(f;a,b) : [a,b] \subset (-\infty,c]\}$$

and

$$V(f;c,\infty) = \sup\{V(f;a,b) : [a,b] \subset [c,\infty)\}$$

are finite. Further, prove that

$$V(f;\mathbb{R}) = V(f;-\infty,c) + V(f;c,\infty).$$

5.23. Let f be of bounded variation on $\mathbb{R}$. For all x in $\mathbb{R}$, define $V_f(x) = V(f;-\infty,x)$. Prove that V_f is a monotone increasing function of $\mathbb{R}$ and that $V_f - f$ is also monotone increasing on $\mathbb{R}$.

5.5 Continuous Functions of Bounded Variation

We close this chapter with a brief discussion concerning the continuity of a function of bounded variation. We begin with a pertinent fact that applies to all functions in $BV(a,b)$.

THEOREM 5.5.1 Let f be in $BV(a,b)$. Then $f(x^+)$ exists at every x in $[a,b)$ and $f(x^-)$ exists at every x in $(a,b]$.

Proof. By Theorems 5.4.2 and 5.4.3, the functions V_f and $V_f - f$ are monotone increasing on $[a, b]$. Therefore $V_f(x^+)$ and $(V_f - f)(x^+)$ exist at every point x in $[a, b)$. It follows that $f(x^+) = V_f(x^+) - (V_f - f)(x^+)$ exists at every x in $[a, b)$. The right-hand limits are treated analogously. ●

Observe that, if f is in $BV(a, b)$, then f can have at most countably many discontinuities and that these can only be jump discontinuities. This follows from the fact that f can be written as the difference of two monotone increasing functions. Now we identify in the main theorem of this section necessary and sufficient conditions that f is continuous at a point c in $[a, b]$.

THEOREM 5.5.2 Let f be in $BV(a, b)$. Then f is continuous at a point c in $[a, b]$ if and only if V_f is continuous at c.

Proof. We will treat the case when c is an interior point of $[a, b]$, leaving for you to prove the theorem when $c = a$ or b. Suppose first that f is continuous at c. We will show first that $V_f(c^+) = V_f(c)$ so that V_f is continuous from the right at c. An analogous argument will show that $V_f(c^-) = V_f(c)$. These two arguments, taken together, prove that V_f is continuous at c.

Fix $\epsilon > 0$. By the continuity of f at c, choose $\delta > 0$ such that $[a, b] \cap N(c; \delta) \subseteq f^{-1}(N(f(c); \epsilon))$. By Theorem 1.1.1 choose a partition $\pi = \{x_0, x_1, \ldots, x_p\}$ in $\Pi[c, b]$ such that

$$V(f; c, b) - \epsilon < \sum_{j=1}^{p} |\Delta f_j| \leq V(f; c, b). \tag{5.3}$$

Refining the partition π only increases the sum $\sum_j |\Delta f_j|$. Hence we can assume that π is sufficiently refined so that the gauge $||\pi||$ of π is less than δ. This implies, in particular, that the partition point x_1 of π is in $N(c; \delta)$. Therefore $|f(x_1) - f(c)| < \epsilon$. Isolate the first summand in the inequality (5.3) and note that $\{x_1, x_2, \ldots, x_p\}$ is a partition of $[x_1, b]$. We obtain

$$V(f; c, b) - \epsilon < |f(x_1) - f(c)| + \sum_{j=2}^{p} |\Delta f_j| < \epsilon + V(f; x_1, b).$$

Consequently,

$$V(f; c, b) - V(f; x_1, b) < 2\epsilon.$$

Finally, observe that

$$\begin{aligned} 0 \leq V_f(x_1) - V_f(c) &= V(f; a, x_1) - V(f; a, c) = V(f; c, x_1) \\ &= V(f; c, b) - V(f; x_1, b) < 2\epsilon. \end{aligned}$$

Since $x = x_1$ is entirely arbitrary in $(c, c + \delta)$, we have shown that, if $x > c$ and $x - c < \delta$, then $V_f(x) - V_f(c) < 2\epsilon$. This proves that $V_f(c^+) = V_f(c)$. Thus, V_f is continuous from the right at c.

To prove the continuity of V_f from the left at c, return to the beginning of the above proof and choose a partition $\pi = \{y_0, y_1, \ldots, y_q\}$ of $[a, c]$ with gauge less than δ, having the property that $V(f; a, c) - \epsilon < \sum_{j=1}^{q} |\Delta f_j|$. Again, note that $\{y_0, y_1, \ldots, y_{q-1}\}$ is a partition of $[a, y_{q-1}]$. Therefore,

$$V_f(c) - \epsilon = V(f; a, c) - \epsilon < \sum_{j=1}^{q} |\Delta f_j| = \sum_{j=1}^{q-1} |\Delta f_j| + |f(c) - f(y_{q-1})|$$

$$< V(f; a, y_{q-1}) + \epsilon = V_f(y_{q-1}) + \epsilon.$$

Thus,

$$0 \leq V_f(c) - V_f(y_{q-1}) < 2\epsilon.$$

It follows that $V_f(c^-) = V_f(c)$. That is, V_f is continuous from the left at c. Therefore V_f is continuous at c.

To prove the converse, assume that V_f is continuous at the interior point c. Given $\epsilon > 0$, choose $\delta > 0$ such that, if x is in $[a, b] \cap N(c; \delta)$, then $|V_f(x) - V_f(c)| < \epsilon$. Choose any x in $(c, c + \delta) \cap [a, b]$. Then $\{c, x\}$ is the trivial partition of $[c, x]$. Therefore

$$0 \leq |f(x) - f(c)| \leq V(f; c, x) = V_f(x) - V_f(c) < \epsilon.$$

It follows that $f(c^+) = f(c)$. Similarly, if x is in $(c - \delta, c) \cap [a, b]$, then

$$0 \leq |f(c) - f(x)| \leq V(f; x, c) = V_f(c) - V_f(x) < \epsilon.$$

This proves that f is continuous from the left at c. The two arguments, taken together, prove that, if V_f is continuous at c, then f is also continuous at c. This completes the proof of the theorem. ●

COROLLARY 5.5.3 A function f is in $C([a, b]) \cap BV(a, b)$ if and only if V_f and $V_f - f$ are monotone increasing and continuous on $[a, b]$. ●

EXERCISES

5.24. Let f be in $BV(a, b)$. Let c be either a or b. Prove that f is continuous at c if and only if V_f is continuous at c.

5.25. Let f be of bounded variation on $\mathbb{R}$. Let

$$V_f(x) = V(f; -\infty, x).$$

Prove that f is continuous at c if and only if V_f is continuous at c.

6

The Riemann Integral

During your studies of elementary calculus, you learned the rudiments of integration and discovered several of the known techniques for computing integrals. You also glimpsed some of the applications of the theory to solve practical problems. What you probably did not learn was that, between differentiation and integration, the latter operation is the more important in mathematics, having more far-reaching consequences. Your perception is entirely understandable in that most elementary calculus texts focus first on differentiation as a tool, more accessible, in a sense, to the beginner. A moment's thought, however, might make the distinction between the two processes evident. If a function is differentiable at a point or on an interval, it is automatically continuous there. However, as you may recall and we will discuss subsequently, the integrability of a function over some interval requires neither its continuity nor its differentiability. In other words, integration applies to a larger class of functions than does differentiation. That alone leads one to suspect that integration, being more general, is also more powerful and useful.

In another sense, integration is also more primitive than differentiation. It gives rise to a *smoothing* or an *averaging* operation. As an infant, once you learned to focus your eyes and to visually perceive objects, your brain began integrating. By contrast, it would be the rare infant who began differentiating a few days after birth. But once you began gaining control of your physical movements, your brain began mastering quite complicated differential calculus.

The type of integration you learned is named in honor of Georg Riemann (1826–1866) for his work in 1854 placing the operation on a rigorous foundation. Riemann's theory was completed by Darboux in 1875. In this chapter we will study the Riemann–Darboux integration theory. In Chapter 7 we will consider a brilliant generalization introduced in 1894 by T. J. Stieltjes, the Riemann–Stieltjes integral.

6.1 THE DEFINITION OF THE RIEMANN INTEGRAL

Let f be any bounded function defined on the interval $[a, b]$ and let $\pi = \{x_1, x_2, \ldots, x_p\}$ be any partition of the interval $[a, b]$. Let $\Delta x_j = x_j - x_{j-1}$ denote the length of the jth partition interval. Finally, let s_j be any point in the interval $[x_{j-1}, x_j]$. A sum of the form

$$S(f, \pi) = \sum_{j=1}^{p} f(s_j)\Delta x_j$$

is called a *Riemann sum* for the function f. (See Fig. 6.1.) In general, with no restriction on the function f, the numerical values of such sums may vary widely. However, as we will see, under suitable restrictions on f and with π sufficiently refined, the numbers $S(f, \pi)$ will cluster around some single number I. The number I will be called the *Riemann integral* of f over the interval $[a, b]$ and will be denoted

$$I = \int_a^b f(x)\,dx.$$

DEFINITION 6.1.1 Let f be a bounded, real-valued function on the interval $[a, b]$. Suppose that there exists a number I with the following property:

> *For any $\epsilon > 0$, there exists a partition π_0 in $\Pi[a, b]$ such that, if π is any partition in $\Pi[a, b]$ that is a refinement of π_0, then for any choice of the points s_j in the partition intervals $[x_{j-1}, x_j]$ of π,*
>
> $$|S(f, \pi) - I| < \epsilon.$$

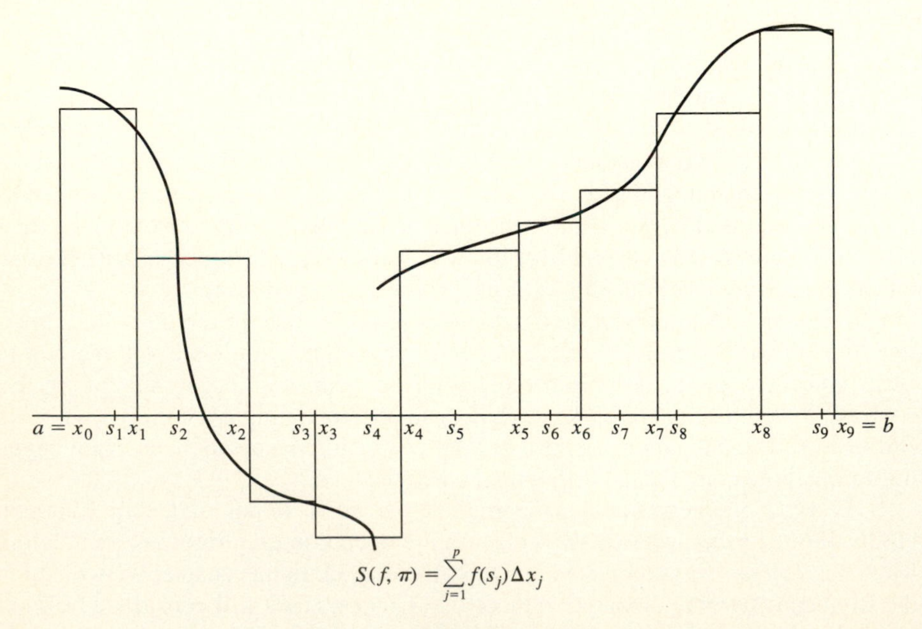

Figure 6.1

Then f is said to be (*Riemann*) *integrable* on $[a, b]$. The number I is called the (*Riemann*) *integral* of f on $[a, b]$ and is denoted $I = \int_a^b f(x)\,dx$. •

We denote by $R[a, b]$ the collection of all bounded functions f that are integrable on $[a, b]$. If f is in $R[a, b]$, then we say that the (Riemann) integral $I = \int_a^b f(x)\,dx$ exists. The function f is called the *integrand*. By convention, we reasonably define

$$\int_b^a f(x)\,dx = -\int_a^b f(x)\,dx \qquad \text{and} \qquad \int_a^a f(x)\,dx = 0.$$

To begin to explore properties of the integral, suppose first that f_1 and f_2 are two functions in $R[a, b]$ and that c_1 and c_2 are any two nonzero real numbers. Consider any Riemann sum for the function $c_1 f_1 + c_2 f_2$:

$$\begin{aligned} S(c_1 f_1 + c_2 f_2, \pi) &= \sum_{j=1}^{p} [c_1 f_1(s_j) + c_2 f_2(s_j)]\Delta x_j \\ &= c_1 \sum_{j=1}^{p} f_1(s_j)\Delta x_j + c_2 \sum_{j=1}^{p} f_2(s_j)\Delta x_j \\ &= c_1 S(f_1, \pi) + c_2 S(f_2, \pi). \end{aligned}$$

Since f_1 is integrable, given any $\epsilon > 0$, there exists a partition π_1 such that, for any refinement π of π_1 and any choice of s_j in $[x_{j-1}, x_j]$,

$$|S(f_1, \pi) - I_1| < \frac{\epsilon}{2|c_1|}, \tag{6.1}$$

where $I_1 = \int_a^b f_1(x)\,dx$. Likewise, since f_2 is integrable, there exists a partition π_2 such that, for any refinement π of π_2,

$$|S(f_2, \pi) - I_2| < \frac{\epsilon}{2|c_2|}, \tag{6.2}$$

where $I_2 = \int_a^b f_2(x)\,dx$.

Let $\pi_0 = \pi_1 \vee \pi_2$, the least common refinement of the two partitions π_1 and π_2. For any refinement π of π_0, it follows that both inequalities (6.1) and (6.2) hold simultaneously and therefore that

$$|S(c_1 f_1 + c_2 f_2, \pi) - (c_1 I_1 + c_2 I_2)| < \frac{\epsilon}{2} + \frac{\epsilon}{2} = \epsilon.$$

According to Definition 6.1.1, this means that $c_1 f_1 + c_2 f_2$ is also in $R[a, b]$ and that

$$\int_a^b [c_1 f_1(x) + c_2 f_2(x)]\,dx = c_1 \int_a^b f_1(x)\,dx + c_2 \int_a^b f_2(x)\,dx.$$

We summarize these observations in the following theorem.

THEOREM 6.1.1 $R[a, b]$ is a vector space and the real-valued function $I(f) = \int_a^b f(x)\,dx$ is a linear function of f. •

The linear properties of the integral extend inductively to any linear combination $\sum_j c_j f_j$ of integrands with each f_j in $R[a, b]$.

It is easy to see that $\int_a^b 1\,dx = (b - a)$ and therefore for any constant c in $\mathbb{R}$, $\int_a^b c\,dx = c(b - a)$.

THEOREM 6.1.2 If the function f is integrable on $[a, b]$, if

$$m = \inf\{f(x) : x \text{ in } [a, b]\},$$

and if

$$M = \sup\{f(x) : x \text{ in } [a, b]\},$$

then

$$m(b - a) \leq \int_a^b f(x)\,dx \leq M(b - a).$$

(See Figs. 6.2 and 6.3.)

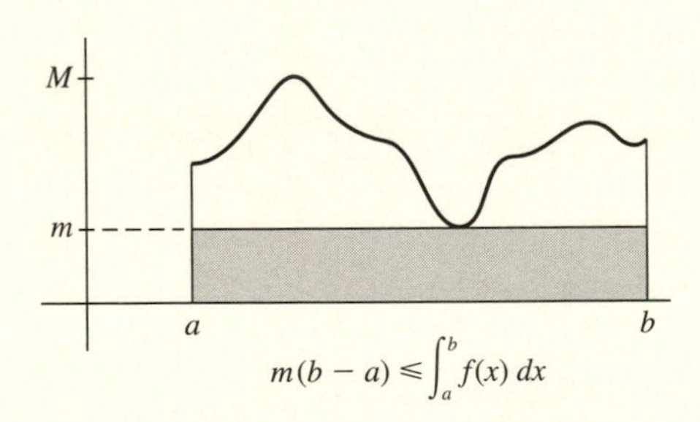

Figure 6.2

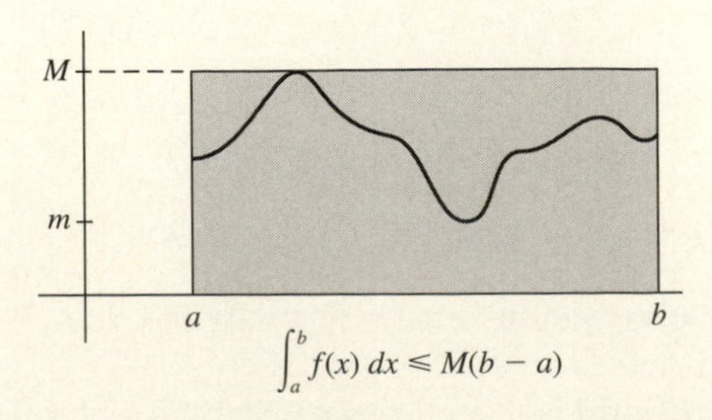

Figure 6.3

Proof. Since $m \leq f(x) \leq M$ for all x in $[a, b]$, it follows that, for any partition π of $[a, b]$ and any choice of s_j in $[x_{j-1}, x_j]$,

$$m(b-a) = \sum_{j=1}^{p} m\Delta x_j \leq \sum_{j=1}^{p} f(s_j)\Delta x_j \leq \sum_{j=1}^{p} M\Delta x_j = M(b-a). \tag{6.3}$$

Given $\epsilon > 0$, choose a partition π_0 in $\Pi[a, b]$ such that, for any refinement π of π_0,

$$\int_a^b f(x)\,dx - \epsilon < S(f, \pi) < \int_a^b f(x)\,dx + \epsilon. \tag{6.4}$$

Combining (6.3) and (6.4) yields

$$m(b-a) < \int_a^b f(x)\,dx + \epsilon \qquad \text{and} \qquad \int_a^b f(x)\,dx - \epsilon < M(b-a).$$

Since these inequalities hold for any $\epsilon > 0$, the theorem follows. ●

THEOREM 6.1.3 The Riemann integral is order-preserving. Let f, f_1, and f_2 be integrable functions on $[a, b]$.

i) If $f \geq 0$ on $[a, b]$, then $\int_a^b f(x)\,dx \geq 0$.

ii) If $f_1(x) \leq f_2(x)$ for all x in $[a, b]$, then

$$\int_a^b f_1(x)\,dx \leq \int_a^b f_2(x)\,dx.$$

Proof. For part (i) note that the hypothesis implies that

$$0 \leq m = \inf\{f(x) : x \text{ in } [a, b]\}$$

and therefore by Theorem 6.1.2, $0 \leq m(b-a) \leq \int_a^b f(x)\,dx$.

For part (ii), $f_2 - f_1 \geq 0$ implies that $\int_a^b [f_2(x) - f_1(x)]\,dx \geq 0$ by part (i). By Theorem 6.1.1,

$$\int_a^b [f_2(x) - f_1(x)]\,dx = \int_a^b f_2(x)\,dx - \int_a^b f_1(x)\,dx$$

and part (ii) follows immediately. ●

There is a second, important linear property of the integral, namely, its *additivity* on nonoverlapping intervals. To be precise, suppose that f is a bounded function on the interval $[a, b]$ and that c is any point in (a, b). We want to show that, if both the integrals $\int_a^c f(x)\,dx$ and $\int_c^b f(x)\,dx$ exist, then $\int_a^b f(x)\,dx$ also exists and

$$\int_a^b f(x)\,dx = \int_a^c f(x)\,dx + \int_c^b f(x)\,dx.$$

(It is this fact that will enable us later to define the indefinite integral $F(x) = \int_a^x f(t)\,dt$, x in $[a, b]$, a concept of utmost importance in the development of the theory and its applications.)

Fix c in (a, b) and let π be any partition of $[a, b]$. Without loss of generality, we may assume c is a partition point of π, for if it were not we could simply insert it. Let π' denote the partition of $[a, c]$ consisting of those partition points of π in the interval $[a, c]$. Likewise, let π'' denote the partition of $[c, b]$ consisting of those partition points of π in the interval $[c, b]$. Thus π is the partition $\{a = x_0, x_1, x_2, \ldots, x_k = c, x_{k+1}, \ldots, x_p = b\}$. The Riemann sum $S(f, \pi)$ over $[a, b]$, with points s_j chosen arbitrarily from $[x_{j-1}, x_j]$, is

$$\sum_{j=1}^{p} f(s_j)\Delta x_j = \sum_{j=1}^{k} f(s_j)\Delta x_j + \sum_{j=k+1}^{p} f(s_j)\Delta x_j = S(f, \pi') + S(f, \pi''), \tag{6.5}$$

where the latter two summands are taken over $[a, c]$ and $[c, b]$, respectively. Since we are assuming that both the integrals $\int_a^c f(x)\,dx$ and $\int_c^b f(x)\,dx$ exist, we know that, given any $\epsilon > 0$, there exist partitions π_0' of $[a, c]$ and π_0'' of $[c, b]$ such that

$$|S(f, \pi') - \int_a^c f(x)\,dx| < \frac{1}{2}\epsilon \tag{6.6}$$

for any partition π' of $[a, c]$ that is a refinement of π_0' and

$$|S(f, \pi'') - \int_c^b f(x)\,dx| < \frac{1}{2}\epsilon \tag{6.7}$$

for any partition π'' of [c, b] that refines π_0''.

Let π_0 denote the partition of $[a, b]$ whose partition points are in either the partition π_0' of $[a, c]$ or π_0'' of $[c, b]$. Clearly, c is a partition point of π_0 and consequently of any refinement of π_0. Choose any partition π of $[a, b]$ that refines π_0. Let π' denote the partition of $[a, c]$ consisting of those points in π that are also in $[a, c]$ and let π'' denote the partition of $[c, b]$ consisting of those points in π that are also in $[c, b]$. It is easy to see that π' is a refinement of π_0' and that π'' is a refinement of π_0'', so the inequalities (6.6) and (6.7) hold on $[a, c]$ and $[c, b]$ separately. Using (6.5), we obtain

$$\begin{aligned}
&\left|S(f, \pi) - \left[\int_a^c f(x)\,dx + \int_c^b f(x)\,dx\right]\right| \\
&\qquad = \left|\left[S(f, \pi') - \int_a^c f(x)\,dx\right] + \left[S(f, \pi'') - \int_c^b f(x)\,dx\right]\right| \\
&\qquad \le \left|S(f, \pi') - \int_a^c f(x)\,dx\right| + \left|S(f, \pi'') - \int_c^b f(x)\,dx\right| \\
&\qquad < \frac{1}{2}\epsilon + \frac{1}{2}\epsilon = \epsilon.
\end{aligned}$$

Thus, by Definition 6.1.1, the Riemann integral $\int_a^b f(x)\,dx$ exists and equals $\int_a^c f(x)\,dx + \int_c^b f(x)\,dx$.

Provided the integrals in question exist—and we will examine existence conditions shortly—the integral, as a function of the sets over which integration is performed, is additive.

THEOREM 6.1.4 Let f be a bounded function on the interval $[a, b]$. Let c be any point in the interval (a, b). If any two of the integrals

$$\int_a^b f(x)\,dx, \int_a^c f(x)\,dx \text{ and } \int_c^b f(x)\,dx$$

exist, then the third integral also exists and

$$\int_a^b f(x)\,dx = \int_a^c f(x)\,dx + \int_c^b f(x)\,dx. \quad \bullet$$

We have proved one case of the theorem. It is left for you as an exercise to prove the remaining cases. We caution you that you may not simply assume the existence of $\int_a^c f(x)\,dx$ or of $\int_c^b f(x)\,dx$ from that of $\int_a^b f(x)\,dx$. You have to prove their existence.

Inductively, you can prove that if, $c_0, c_1, c_2, \ldots, c_k$ are $k+1$ points in $[a, b]$ with $a = c_0 < c_1 < c_2 < \cdots < c_k = b$ and if any k of the $k+1$ integrals

$$\int_a^b f(x)\,dx \quad \text{and} \quad \int_{c_{j-1}}^{c_j} f(x)\,dx, \quad j = 1, 2, 3, \ldots, k$$

exist, then the remaining integral also exists and

$$\int_a^b f(x)\,dx = \sum_{j=1}^{k} \int_{c_{j-1}}^{c_j} f(x)\,dx.$$

6.2 EXISTENCE OF THE RIEMANN INTEGRAL

Now suppose that f is any bounded function on $[a, b]$ and that $\pi = \{x_0, x_1, x_2, \ldots, x_p\}$ is any partition of $[a, b]$. For each $j = 1, 2, 3, \ldots, p$, define

$$M_j = \sup\{f(x) : x \text{ in } [x_{j-1}, x_j]\}$$
$$m_j = \inf\{f(x) : x \text{ in } [x_{j-1}, x_j]\}.$$

The numbers m_j and M_j are finite because f is assumed to be bounded on $[a, b]$. Furthermore, for any s_j in $[x_{j-1}, x_j]$, $m_j \le f(s_j) \le M_j$. Note, in passing, that $M_j - m_j$ is the diameter of the range of f restricted to $[x_{j-1}, x_j]$ so that $M_j - m_j = \sup\{f(x) - f(y) : x, y \text{ in } [x_{j-1}, x_j]\}$. Notice also that

$$\sum_{j=1}^{p} m_j \Delta x_j \le \sum_{j=1}^{p} f(s_j)\Delta x_j \le \sum_{j=1}^{p} M_j \Delta x_j.$$

The numbers $L(f, \pi) = \sum_{j=1}^{p} m_j \Delta x_j$ and $U(f, \pi) = \sum_{j=1}^{p} M_j \Delta x_j$ are called, respectively, the *lower* and the *upper Riemann sums* of f for the partition π. (See Figs. 6.4 and 6.5 on page 240.) It is evident that, for every partition π,

$$L(f, \pi) \le S(f, \pi) \le U(f, \pi).$$

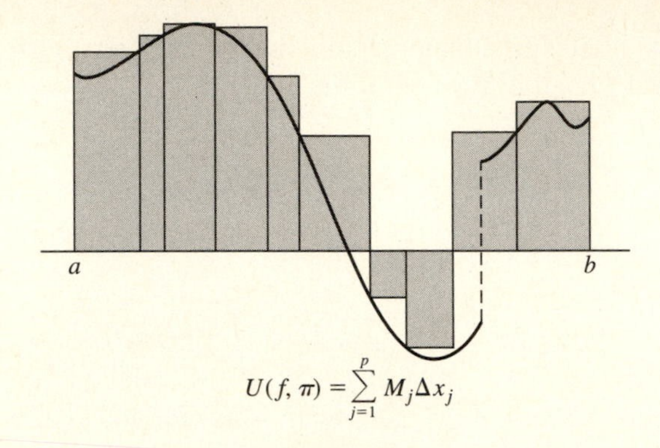

Figure 6.4

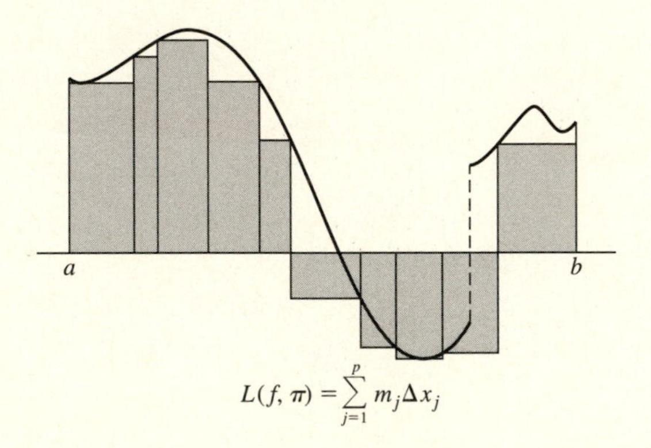

Figure 6.5

Eventually we will be interested in the difference

$$U(f, \pi) - L(f, \pi) = \sum_{j=1}^{p} (M_j - m_j) \Delta x_j.$$

Intuitively, when this difference can be made arbitrarily small by taking π to be sufficiently refined, $S(f, \pi)$, trapped between the two extremes $L(f, \pi)$ and $U(f, \pi)$, must be near some single number I. This will lead to the conclusion that f is integrable on $[a, b]$ and that $\int_a^b f(x)\,dx$ will equal the number I. However, before we can achieve the proof of such an assertion, we must first attend to some easy, technical details.

Suppose that π is any partition of $[a, b]$ and that the refinement π' is obtained from π by inserting one additional partition point x' in, say, the kth partition interval. Thus

$$\pi : a = x_0 < x_1 < \cdots < x_{k-1} < x_k < \cdots < x_p = b$$

and

$$\pi' : a = x_0 < x_1 < \cdots < x_{k-1} < x' < x_k < \cdots < x_p = b.$$

Let

$$m' = \inf\{f(x) : x \text{ in } [x_{k-1}, x']\}$$

and

$$m'' = \inf\{f(x) : x \text{ in } [x', x_k]\}.$$

Notice that $m_k = \min\{m', m''\}$ and that

$$\begin{aligned} m_k \Delta x_k &= m_k(x_k - x_{k-1}) \\ &= m_k(x_k - x') + m_k(x' - x_{k-1}) \\ &\leq m'' \Delta x'' + m' \Delta x', \end{aligned}$$

where $\Delta x' = x' - x_{k-1}$ and $\Delta x'' = x_k - x'$. It follows that

$$\begin{aligned} L(f, \pi') &= \sum_{j \neq k} m_j \Delta x_j + m' \Delta x' + m'' \Delta x'' \\ &\geq \sum_{j \neq k} m_j \Delta x_j + m_k \Delta x_k = L(f, \pi). \end{aligned}$$

Thus $L(f, \pi) \leq L(f, \pi')$. By a completely analogous demonstration, $U(f, \pi) \geq U(f, \pi')$. Induction on the number of inserted partition points will prove the following theorem.

THEOREM 6.2.1 If π' is a refinement of π, then $L(f, \pi) \leq L(f, \pi')$ and $U(f, \pi) \geq U(f, \pi')$. ●

THEOREM 6.2.2 If π_1 and π_2 are any two partitions of $[a, b]$ then

$$L(f, \pi_1) \leq U(f, \pi_2).$$

Proof. Let $\pi = \pi_1 \vee \pi_2$. Since π is a refinement of π_1,

$$L(f, \pi_1) \leq L(f, \pi)$$

by Theorem 6.2.1. Likewise,

$$U(f, \pi) \leq U(f, \pi_2).$$

Consequently,

$$L(f, \pi_1) \leq L(f, \pi) \leq U(f, \pi) \leq U(f, \pi_2)$$

and the theorem is proved. ●

Theorem 6.2.2 provides us with the tool to establish existence criteria for the Riemann integral of a bounded function f on $[a, b]$. Consider first the set $\mathcal{L}$ of real numbers consisting of all lower sums:

$$\mathcal{L} = \{L(f, \pi) : \pi \text{ in } \Pi[a, b]\}.$$

Theorem 6.2.2 says that $\mathcal{L}$ is bounded above by any upper sum $U(f, \pi)$; therefore, by the completeness of R, sup $\mathcal{L}$ exists. Let

$$L(f) = \sup \mathcal{L} = \sup \{L(f, \pi) : \pi \text{ in } \Pi[a, b]\},$$

a number called the *lower Riemann integral* of f on $[a, b]$. It is clear that

$$L(f) \leq U(f, \pi),$$

for every partition π in $\Pi[a, b]$. In an analogous manner, the set

$$\mathcal{U} = \{U(f, \pi) : \pi \text{ in } \Pi[a, b]\}$$

is a set of real numbers bounded below by $L(f)$. Therefore, again by the completeness of $\mathbb{R}$, inf $\mathcal{U}$ exists. Let

$$U(f) = \inf \mathcal{U} = \inf \{U(f, \pi) : \pi \text{in } \Pi[a, b]\},$$

a number called the *upper Riemann integral* of f on $[a, b]$. By the very definitions of $L(f)$ and $U(f)$, it follows that $L(f) \leq U(f)$.

DEFINITION 6.2.1

i) The *lower Riemann integral* of f on $[a, b]$ is

$$L(f) = \sup \{L(f, \pi) : \pi \text{ in } \Pi[a, b]\}.$$

ii) The *upper Riemann integral* of f on $[a, b]$ is

$$U(f) = \inf \{U(f, \pi) : \pi \text{ in } \Pi[a, b]\}. \quad \bullet$$

THEOREM 6.2.3 For every bounded function f on $[a, b]$,

$$L(f) \leq U(f). \quad \bullet$$

Let us recapitulate what we know to this point, given any bounded integrand f.

i) A bounded function f is in $R[a, b]$ if and only if there exists a real number $I = \int_a^b f(x)\,dx$ with the following property: For every $\epsilon > 0$, there exists a partition π_0 such that, for every refinement π of π_0, $|S(f, \pi) - I| < \epsilon$.

ii) For every partition π of $[a, b]$,

$$L(f, \pi) \leq S(f, \pi) \leq U(f, \pi).$$

iii) $L(f) \leq U(f)$.

From these facts, we will derive our fundamental criteria for the existence of the Riemann integral of f.

DEFINITION 6.2.2 Riemann's Condition *Riemann's condition* is said to hold for the bounded function f on $[a, b]$ if, for any $\epsilon > 0$, there exists a partition π_0 such that, for every refinement π of π_0, $U(f, \pi) - L(f, \pi) < \epsilon$. $\bullet$

THEOREM 6.2.4 Let f be any bounded function on $[a, b]$. The following are equivalent.

i) f is in $R[a, b]$.

ii) Riemann's condition holds for f on $[a, b]$.

iii) $L(f) = U(f)$.

Proof. First, suppose that f is in $R[a, b]$ with $I = \int_a^b f(x)\,dx$. Fix $\epsilon > 0$. Choose π_0 so that, for any refinement π of π_0 with $\pi = \{x_0, x_1, x_2, \ldots, x_p\}$ and for any choices of s_j in the partition intervals $[x_{j-1}, x_j]$,

$$|S(f, \pi) - I| < \epsilon/3.$$

Below we will choose two such sets of points s_j and t_j so that

$$\left|\sum_{j=1}^{p} f(s_j)\Delta x_j - I\right| < \frac{\epsilon}{3}$$

and

$$\left|\sum_{j=1}^{p} f(t_j)\Delta x_j - I\right| < \frac{\epsilon}{3}.$$

For any such choices of s_j and t_j, we have

$$\begin{aligned}\left|\sum_{j=1}^{p}[f(s_j) - f(t_j)]\Delta x_j\right| &= \left|\sum_{j=1}^{p} f(s_j)\Delta x_j - \sum_{j=1}^{p} f(t_j)\Delta x_j\right| \\ &\le \left|\sum_{j=1}^{p} f(s_j)\Delta x_j - I\right| + \left|\sum_{j=1}^{p} f(t_j)\Delta x_j - I\right| \\ &< \frac{\epsilon}{3} + \frac{\epsilon}{3} = \frac{2\epsilon}{3}. \end{aligned} \tag{6.8}$$

For $j = 1, 2, 3, \ldots, p$, let $M_j = \sup\{f(x) : x \text{ in } [x_{j-1}, x_j]\}$ and $m_j = \inf\{f(x) : x \text{ in } [x_{j-1}, x_j]\}$. Hence we can choose s_j and t_j in $[x_{j-1}, x_j]$ such that

$$M_j - \frac{\epsilon}{6(b-a)} < f(s_j) \le M_j$$

and

$$m_j \le f(t_j) < m_j + \frac{\epsilon}{6(b-a)}.$$

Therefore

$$M_j - m_j - \frac{\epsilon}{3(b-a)} < f(s_j) - f(t_j). \tag{6.9}$$

Multiplying by $\Delta x_j > 0$ and summing on j give:

$$\sum_{j=1}^{p}[M_j\Delta x_j - m_j\Delta x_j] - \frac{\epsilon}{3(b-a)}\sum_{j=1}^{p}\Delta x_j < \sum_{j=1}^{p}[f(s_j) - f(t_j)]\Delta x_j.$$

By applying (6.8), we simplify this inequality to obtain

$$U(f, \pi) - L(f, \pi) - \frac{\epsilon}{3} < \sum_{j=1}^{p} [f(s_j) - f(t_j)]\Delta x_j < \frac{2\epsilon}{3}.$$

It follows that, if π is a refinement of π_0, then

$$U(f, \pi) - L(f, \pi) < \epsilon.$$

This proves that Riemann's condition holds for f on $[a, b]$. Thus (i) implies (ii).

Next assume that Riemann's condition holds for f on $[a, b]$. We must prove that $L(f) = U(f)$. Fix $\epsilon > 0$ and choose a partition π_0 of $[a, b]$ such that, for any refinement π of π_0, $U(f, \pi) - L(f, \pi) < \epsilon$. That is,

$$U(f, \pi) < L(f, \pi) + \epsilon.$$

From the definitions of $L(f)$ and $U(f)$, we have

$$U(f) \le U(f, \pi) < L(f, \pi) + \epsilon \le L(f) + \epsilon.$$

Thus $0 \le U(f) - L(f) < \epsilon$ for every $\epsilon > 0$. We conclude that $U(f) - L(f) = 0$ and $L(f) = U(f)$: (ii) implies (iii).

Finally, we start with the assumption that the lower and upper Riemann integrals of f on $[a, b]$ are equal and prove that f is in $R[a, b]$ with $I = L(f) = U(f)$. Fix $\epsilon > 0$. Since $I = L(f) = \sup\{L(f, \pi) : \pi \text{ in } \Pi[a, b]\}$, we can choose a partition π_1 such that $I - \epsilon < L(f, \pi_1)$. Likewise, since $I = U(f) = \inf\{U(f, \pi) : \pi$ in $\Pi[a, b]\}$, we can choose a partition π_2 such that $U(f, \pi_2) < I + \epsilon$.

Let $\pi_0 = \pi_1 \vee \pi_2$. If π is a refinement of π_0, then π is a refinement of both π_1 and π_2. Therefore

$$I - \epsilon < L(f, \pi_1) \le L(f, \pi) \le S(f, \pi) \le U(f, \pi) \le U(f, \pi_2) < I + \epsilon.$$

Consequently, for any refinement π of π_0, $|S(f, \pi) - I| < \epsilon$ and f is integrable on $[a, b]$. Thus (iii) implies (i) and the theorem is proved. ●

We will systematically exploit Theorem 6.2.4, first to confirm algebraic properties of $R[a, b]$, then to identify classes of functions integrable on $[a, b]$.

Assume that f is integrable on $[a, b]$ and consider the integrability of the function $|f|$. For any partition $\pi = \{x_0, x_1, x_2, \ldots, x_p\}$ of $[a, b]$, let M_j and m_j denote the supremum and the infimum, respectively, of f on the interval $[x_{j-1}, x_j]$; let M_j' and m_j' denote the supremum and the infimum respectively of $|f|$ on $[x_{j-1}, x_j]$. Then

$$M_j - m_j = \sup\{f(x) - f(y) : x \text{ and } y \text{ in } [x_{j-1}, x_j]\}$$

and

$$M_j' - m_j' = \sup\{|f(x)| - |f(y)| : x \text{ and } y \text{ in } [x_{j-1}, x_j]\}.$$

Since we have, for any real numbers c and d,

$$||c| - |d|| \le |c - d|,$$

it follows that $M_j' - m_j' \le M_j - m_j$. Multiplying by Δx_j and summing on j, we obtain

$$U(|f|, \pi) - L(|f|, \pi) \le U(f, \pi) - L(f, \pi).$$

Since f is assumed to be in $R[a, b]$, Riemann's condition holds for f on $[a, b]$. Therefore, for any $\epsilon > 0$ and for π sufficiently refined,

$$U(|f|, \pi) - L(|f|, \pi) \le U(f, \pi) - L(f, \pi) < \epsilon.$$

Consequently, Riemann's condition holds for $|f|$. Hence, by Theorem 6.2.4, $|f|$ is in $R[a, b]$. Furthermore, since $-|f| \le f \le |f|$, we deduce by Theorem 6.1.3 that

$$\left| \int_a^b f(x)\, dx \right| \le \int_a^b |f(x)|\, dx.$$

(See Figs. 6.6 and 6.7.)

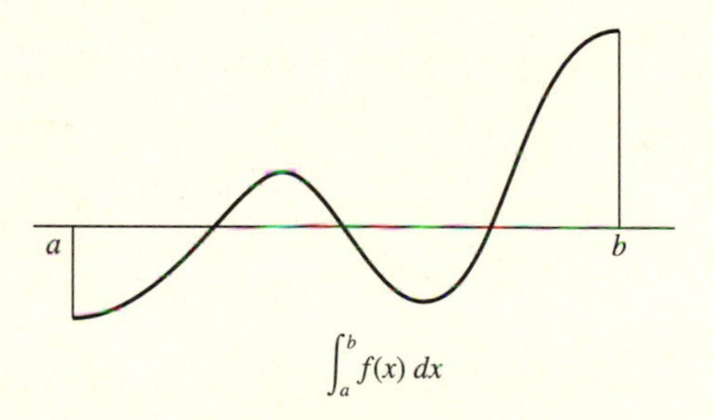

Figure 6.6

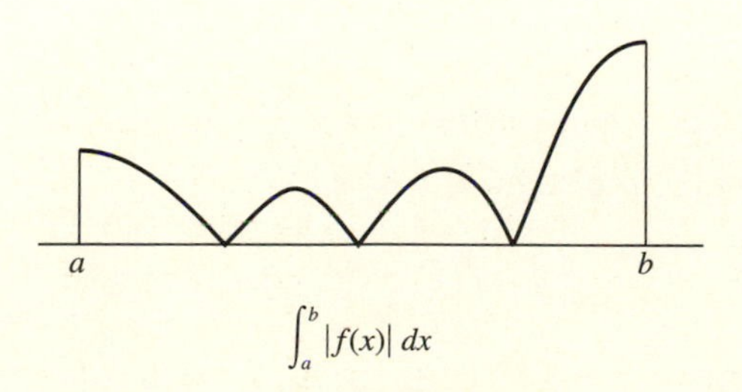

Figure 6.7

A similar argument proves that if f is in $R[a, b]$, then f^2 is also integrable on $[a, b]$. Fix any $\epsilon > 0$. Since f is assumed integrable on $[a, b]$, we can choose a partition π_0 with the property that, for any refinement π of π_0,

$$U(f, \pi) - L(f, \pi) < \frac{\epsilon}{2\|f\|_\infty}.$$

Fix any such partition $\pi = \{x_0, x_1, x_2, \ldots, x_p\}$. Again let m_j and M_j denote the infimum and the supremum of f on the interval $[x_{j-1}, x_j]$; we have $M_j - m_j = \sup\{f(x) - f(y) : x, y \text{ in } [x_{j-1}, x_j]\}$. Let

$$M'_j = \sup\{f^2(x) : x \text{ in } [x_{j-1}, x_j]\}$$

and

$$m'_j = \inf\{f^2(x) : x \text{ in } [x_{j-1}, x_j]\}.$$

In this instance,

$$\begin{aligned} M'_j - m'_j &= \sup\{f^2(x) - f^2(y) : x, y \text{ in } [x_{j-1}, x_j]\} \\ &= \sup\{[f(x) + f(y)][f(x) - f(y)] : x, y \text{ in } [x_{j-1}, x_j]\} \\ &\leq 2\|f\|_\infty \sup\{f(x) - f(y) : x, y \text{ in } [x_{j-1}, x_j]\} \\ &= 2\|f\|_\infty (M_j - m_j). \end{aligned}$$

To confirm that Riemann's condition holds for f^2 on $[a, b]$, consider

$$\begin{aligned} U(f^2, \pi) - L(f^2, \pi) &= \sum_{j=1}^{p} (M'_j - m'_j)\Delta x_j \leq 2\|f\|_\infty \sum_{j=1}^{p} (M_j - m_j)\Delta x_j \\ &= 2\|f\|_\infty [U(f, \pi) - L(f, \pi)] < \epsilon. \end{aligned}$$

Since Riemann's condition holds for f^2 on $[a, b]$, we deduce by Theorem 6.2.4 that f^2 is integrable on $[a, b]$.

This result in turn implies that if f and g are two integrable functions on $[a, b]$, then the product fg is also integrable on $[a, b]$. Simply note that

$$fg = \tfrac{1}{2}[(f + g)^2 - f^2 - g^2].$$

Since f and g are integrable, the same can be said of $f + g$ and the squares and the differences involved in the preceding identity. We summarize these several results in one formal statement for future reference.

THEOREM 6.2.5 Let f and g be two functions in $R[a, b]$.

i) The function $|f|$ is also in $R[a, b]$ and

$$\left| \int_a^b f(x)\,dx \right| \leq \int_a^b |f(x)|\,dx.$$

ii) The functions $f^+ = \max\{f, 0\}$ and $f^- = \max\{-f, 0\}$ are in $R[a, b]$. We have

$$\int_a^b f(x)\,dx = \int_a^b [f^+(x) - f^-(x)]\,dx$$

and

$$\int_a^b |f(x)|\,dx = \int_a^b [f^+(x) + f^-(x)]\,dx.$$

iii) The function f^2 is also in $R[a, b]$.

iv) The function fg is also in $R[a, b]$.

v) If there exist m and M such that $0 < m \le |f| \le M$, then $1/f$ is also in $R[a, b]$.

Remark. This theorem shows that $R[a, b]$ is a (commutative) ring with identity. Part (v) identifies the units in $R[a, b]$.

Proof. To prove part (ii), merely note that

$$f^+ = \tfrac{1}{2}(|f| + f)$$

and

$$f^- = \tfrac{1}{2}(|f| - f).$$

Apply part (i) of the present theorem and Theorem 6.1.1. We leave part (v) as an exercise. ●

Using Riemann's condition, we can strengthen the result in Theorem 6.1.4 to obtain the following theorem.

THEOREM 6.2.6 If f is integrable on $[a, b]$, then f is integrable on any subinterval $[c, d]$ of $[a, b]$.

Proof. Suppose that $a < c < d \le b$. If we can prove that f is integrable on both $[a, c]$ and $[a, d]$, then we can conclude by Theorem 6.1.4 that f is integrable on $[c, d]$. Thus it suffices to prove that f is integrable on any interval $[a, c]$ where c is in $[a, b]$.

Fix $\epsilon > 0$. Choose a partition π_0 in $\Pi[a, b]$ such that, for every refinement π of π_0, $U(f, \pi) - L(f, \pi) < \epsilon$. Without loss of generality, c is a partition point of π_0; if it is missing, simply insert it. The partition points of π_0 that are in the interval $[a, c]$ form a partition π_0' in $\Pi[a, c]$. If π' is any partition in $\Pi[a, c]$ that is a refinement of π_0', then $\pi = \pi_0 \cup \pi'$, consisting of the partition points in π_0 and in π', is a partition in $\Pi[a, b]$ that refines π_0. Let $\pi = \{x_0, x_1, \ldots, x_k = c,\ x_{k+1}, \ldots, x_p\}$. Then

$$\begin{aligned} U(f, \pi') - L(f, \pi') &= \sum_{j=1}^{k}[M_j - m_j]\Delta x_j \le \sum_{j=1}^{p}[M_j - m_j]\Delta x_j \\ &= U(f, \pi) - L(f, \pi) < \epsilon. \end{aligned}$$

It follows that Riemann's condition holds for f on the interval $[a, c]$. Therefore, by Theorem 6.2.4, f is integrable on $[a, c]$ and hence on any subinterval $[c, d]$ of $[a, b]$. ●

We come at last to the two most general existence theorems we will present for the Riemann integral, results that will demonstrate that the integral exists for a large class of functions. Specifically, if f is continuous or is of bounded variation on $[a, b]$, then $\int_a^b f(x)\,dx$ exists. You are cautioned that these results are not exhaustive: There exist Riemann integrable functions that are neither continuous nor of bounded variation. However, the technicalities are better handled in a more advanced treatment of the more general theory of Lebesgue integration and we will be satisfied with the level of generality achieved by our theorems.

THEOREM 6.2.7 If f is continuous on $[a, b]$, then f is integrable on $[a, b]$.

Proof. Suppose that f is continuous on the compact set $[a, b]$. Thus, by Theorem 3.3.1, f is uniformly continuous. That is, given $\epsilon > 0$, there exists a single $\delta > 0$ so that, whenever s and t are points in $[a, b]$ with $|s - t| < \delta$, it follows that $|f(s) - f(t)| < \epsilon/(b - a)$.

Let π_0 be any partition of $[a, b]$ with gauge less than this δ and let π be a refinement of π_0. We denote the partition points in π by $a = x_0 < x_1 < x_2 < \cdots < x_p = b$. Since f is continuous on each partition interval $[x_{j-1}, x_j]$ of π, f actually assumes its maximum and minimum values there. That is, there exist points s_j and t_j in $[x_{j-1}, x_j]$ such that $f(s_j) = M_j$ and $f(t_j) = m_j$. Note that $|s_j - t_j| < \delta$ and therefore by the uniform continuity of f,

$$M_j - m_j = |f(s_j) - f(t_j)| < \frac{\epsilon}{(b - a)},$$

for $j = 1, 2, \ldots, p$. Multiply by Δx_j and sum on j to obtain

$$U(f, \pi) - L(f, \pi) < \frac{\epsilon}{(b - a)} \sum_{j=1}^{p} \Delta x_j = \epsilon.$$

Therefore Riemann's condition holds and, by Theorem 6.2.4, f is integrable on $[a, b]$. ●

Our second existence theorem applies to functions of bounded variation on $[a, b]$.

THEOREM 6.2.8 If f is of bounded variation on $[a, b]$, then f is integrable on $[a, b]$.

Proof. Assume first that f is monotone increasing and that $f(a) < f(b)$. We will confirm that Riemann's condition holds for f on $[a, b]$. To this end, fix any $\epsilon > 0$ and choose a partition π_0 of $[a, b]$ with gauge less than $\epsilon/[f(b) - f(a)]$. Fix any refinement π of π_0 with $\pi = \{x_0, x_1, x_2, \ldots, x_p\}$. Note that the gauge of π is no larger than the gauge of π_0, so $\Delta x_j < \epsilon/[f(b) - f(a)]$ for $j = 1, 2, \ldots, p$.

Note also that $M_j = \sup\{f(x) : x \text{ in } [x_{j-1}, x_j]\} = f(x_j)$ and that $m_j = \inf\{f(x) : x \text{ in } [x_{j-1}, x_j]\} = f(x_{j-1})$ because f is monotone increasing. Consequently,

$$\begin{aligned} U(f, \pi) - L(f, \pi) &= \sum_{j=1}^{p} (M_j - m_j)\Delta x_j \\ &= \sum_{j=1}^{p} [f(x_j) - f(x_{j-1})]\Delta x_j \end{aligned}$$

$$< \left\{ \sum_{j=1}^{p} [f(x_j) - f(x_{j-1})] \right\} \frac{\epsilon}{f(b) - f(a)}$$

$$= [f(b) - f(a)] \frac{\epsilon}{f(b) - f(a)} = \epsilon.$$

Thus Riemann's condition holds for f on $[a, b]$ and therefore, by Theorem 6.2.4, f is integrable on $[a, b]$.

If f is in $BV(a, b)$, Theorem 5.4.4 guarantees that f can be written as the difference of bounded, monotone increasing functions g and h, $f = g - h$. By the foregoing argument, both g and h are integrable on $[a, b]$; therefore the same is true of f. ●

EXAMPLE 1 Let f be "Dirichlet's dancing function" defined on [0, 1] by

$$f(x) = \begin{cases} 0, & \text{if } x \text{ is irrational} \\ 1, & \text{if } x \text{ is rational.} \end{cases}$$

The graph of f is sketched in Fig. 6.8. It is easy to see that f is discontinuous at each x in [0, 1]. Furthermore, f is not Riemann integrable on [0, 1]. To prove this assertion, we let π be any partition of [0, 1]. Then $L(f, \pi) = 0$ and $U(f, \pi) = 1$. Consequently $L(f) = 0$ and $U(f) = 1$ and f is not integrable by Theorem 6.2.4. ●

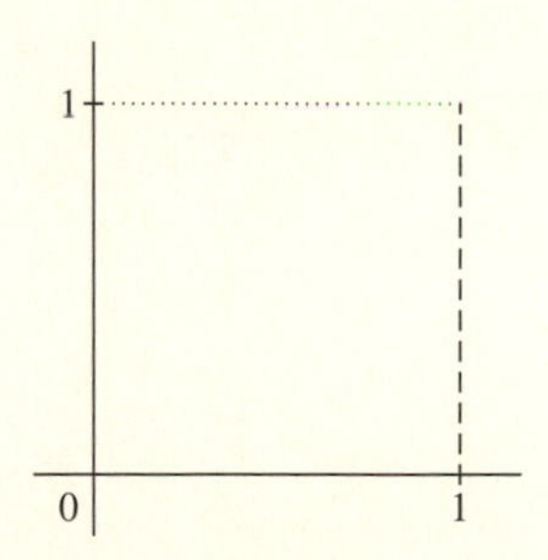

Figure 6.8

EXAMPLE 2 Define f on [0, 1] as follows. If x is irrational, let $f(x) = 0$. For rational $x = p/q$ in lowest terms, let $f(x) = 1/q$. The graph of f is sketched in Fig. 6.9. As you proved in Exercise 3.13, f is continuous at every irrational point and is discontinuous at every rational point in [0, 1]. Although f is discontinuous on a countable, dense subset of [0, 1], f is integrable on [0, 1]. To prove this, first fix $\epsilon > 0$ and choose m in $\mathbb{N}$ such that $1/m < \epsilon/2$. Note that there are only finitely many points in [0, 1] where f has value $> 1/m$, namely at those rational p/q where $1 \leq q < m$ and $1 \leq p \leq q$ with p and q relatively prime. Suppose that there are exactly k points $\{y_1, y_2, \ldots, y_k\}$ where $f(y_j) > 1/m$, indexed so that $0 < y_1 < y_2 < \cdots < y_k = 1$.

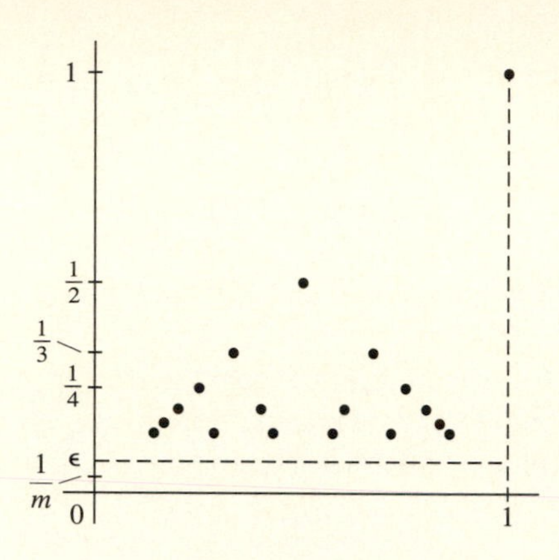

Figure 6.9

We will construct a partition π_0 to enclose these k points in extremely short partition intervals, thus nullifying the effect of f having relatively large values at the points y_j. Choose partition points $x_0, x_1, x_2, \ldots, x_{2k}$ such that

$$0 = x_0 < x_1 < y_1 < x_2 < x_3 < y_2 < x_4 < \cdots < x_{2k-1} < y_k = x_{2k} = 1$$

and such that $\sum_{j=1}^{k}(x_{2j} - x_{2j-1}) < \epsilon/2$. On a partition interval of the form $[x_{2j-1}, x_{2j}]$, $m_{2j} = 0$ (since every interval contains irrational numbers), and $M_{2j} = f(y_j) \le 1$. On a partition interval of the form $[x_{2j}, x_{2j+1}]$, again $m_{2j+1} = 0$ and $M_{2j+1} \le 1/m < \epsilon/2$. Therefore

$$U(f, \pi_0) - L(f, \pi_0) = U(f, \pi_0) = \sum_{j=1}^{k} M_{2j}\Delta x_{2j} + \sum_{j=0}^{k-1} M_{2j+1}\Delta x_{2j+1}$$

$$< \sum_{j=1}^{k} \Delta x_{2j} + \frac{1}{2}\epsilon \sum_{j=0}^{k-1} \Delta x_{2j+1} < \frac{1}{2}\epsilon + \frac{1}{2}\epsilon = \epsilon.$$

For any refinement π of π_0,

$$U(f, \pi) - L(f, \pi) = U(f, \pi) \le U(f, \pi_0) < \epsilon,$$

proving that Riemann's condition holds for f on $[a, b]$ and therefore that f is integrable on $[a, b]$. In fact, by taking $\inf\{U(f, \pi) : \pi \text{ in } \Pi[a, b]\} = 0$, we have proved that $\int_0^1 f(x)\,dx = 0$. ●

This example is meant to provide you with an emphatic warning: Although in Example 2, f is nonnegative and f is strictly positive on an infinite subset of [0, 1], the integral of f has value 0. By contrast, we have the following theorem.

THEOREM 6.2.9 If f is continuous and nonnegative on $[a, b]$ and if $\int_a^b f(x)\,dx = 0$, then $f(x) = 0$ for all x in $[a, b]$.

Proof. Since f is continuous on $[a, b]$, it is integrable on $[a, b]$ by Theorem 6.2.7. Suppose there were to exist a point x_0 in $[a, b]$ where $f(x_0) > 0$. By Theorem 3.4.3, there exist $m > 0$ and a neighborhood $N(x_0)$ such that $0 < m \le f(x)$ for all x in $[a, b] \cap N(x_0)$. Let the interval $[a, b] \cap N(x_0)$ have endpoints c and d with $c < d$. By Theorem 6.2.6, f is integrable on $[c, d]$. Since $f \ge m$ on $[c, d]$,

$$\int_c^d f(x)\,dx \ge m(d - c) > 0.$$

(See Fig. 6.10.) But $f \ge 0$ on all of $[a, b]$. Therefore

$$\int_a^b f(x)\,dx = \int_a^c f(x)\,dx + \int_c^d f(x)\,dx + \int_d^b f(x)\,dx \ge \int_c^d f(x)\,dx > 0,$$

contradicting the hypothesis that $\int_a^b f(x)\,dx = 0$. Therefore there can exist no point in $[a, b]$ where f is not zero. We conclude that f is identically 0 on $[a, b]$. ●

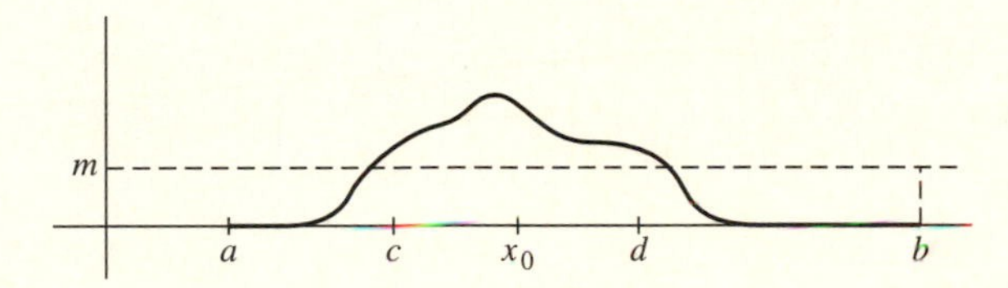

Figure 6.10

Example 2 raises another question that warrants our attention. Suppose that f and g are two bounded functions each of which is integrable on $[a, b]$ and suppose further that $f(x) = g(x)$ for all x in $[a, b]$ except at a finite set of points $\{c_1, c_2, \ldots, c_k\}$. Can we conclude that

$$\int_a^b f(x)\,dx = \int_a^b g(x)\,dx?$$

To resolve the question, consider the following simple, even self-evident, observation.

LEMMA 1 If f is a bounded function on $[a, b]$ such that $f(x) = 0$ for all x in $(a, b]$ and $f(a) \ne 0$, then f is integrable on $[a, b]$ and $\int_a^b f(x)\,dx = 0$.

Proof. Without loss of generality, $f(a) > 0$. Fix $\epsilon > 0$. Choose a positive $\delta < \epsilon/f(a)$ and choose any partition π_0 of $[a, b]$ with gauge less than δ. Fix any partition $\pi = \{x_0, x_1, \ldots, x_p\}$ of $[a, b]$ which refines π_0. It is easy to see that $U(f, \pi) = f(a)\Delta x_1 < f(a)\delta < \epsilon$ and that $L(f, \pi) = 0$. Clearly, Riemann's condition is met and f is integrable on $[a, b]$. Further, $U(f) = L(f) = 0$ so $\int_a^b f(x)\,dx = 0$.

Similarly, if $f(x) = 0$ for x in $[a, b)$ and if $f(b) \ne 0$, then the conclusion of Lemma 1 remains true. ●

LEMMA 2 Suppose that f and g are two bounded, integrable functions on $[a, b]$ such that $f(x) = g(x)$ for all x in the open interval (a, b). Then $\int_a^b f(x)\,dx = \int_a^b g(x)\,dx$.

Proof. Simply fix any c in (a, b) and write

$$\left|\int_a^b f(x)\,dx - \int_a^b g(x)\,dx\right| \le \int_a^b |f(x) - g(x)|\,dx$$

$$= \int_a^c |f(x) - g(x)|\,dx + \int_c^b |f(x) - g(x)|\,dx = 0,$$

by applying Lemma 1 to each of the latter integrals. ●

THEOREM 6.2.10 Suppose that f and g are two bounded functions on $[a, b]$ each of which is integrable on $[a, b]$. If $f(x) = g(x)$ except at finitely many points $c_1, c_2, \ldots, c_k$, then $\int_a^b f(x)\,dx = \int_a^b g(x)\,dx$.

Proof. Index the c_j so that $a = c_0 < c_1 < \cdots < c_k < c_{k+1} = b$. Write

$$\int_a^b f(x)\,dx = \sum_{j=1}^{k+1} \int_{c_{j-1}}^{c_j} f(x)\,dx = \sum_{j=1}^{k+1} \int_{c_{j-1}}^{c_j} g(x)\,dx = \int_a^b g(x)\,dx$$

by applying Lemma 2 to f and g on each interval. ●

In light of Theorem 6.2.10, it is evident that, when computing a Riemann integral, changing the value of the integrand at a finite number of points has no effect on the value of the integral.

6.3 THE FUNDAMENTAL THEOREMS OF CALCULUS

We begin this section by deriving two important results for Riemann integrals, called *mean value theorems*, that link global behavior—the value of an integral of a function over an interval—with the value of a function at a single point. Thus these theorems form a bridge between global and pointwise behavior, comparable in importance to the Mean Value Theorem of differential calculus. It is also worth mentioning that most integrals cannot be evaluated exactly and that these theorems enable us to estimate the values of some integrals.

THEOREM 6.3.1 Suppose that f is continuous on $[a, b]$ and that g is a non-negative and integrable function on $[a, b]$. There exists a point c in $[a, b]$ such that

$$\int_a^b f(x)g(x)\,dx = f(c)\int_a^b g(x)\,dx.$$

Proof. Since f is continuous and g is integrable, the product fg is also integrable on $[a, b]$. Let m and M denote the minimum and maximum values of f respectively on $[a, b]$. Since $m \le f(x) \le M$ for all x in $[a, b]$ and since $g(x) \ge 0$,

$$mg(x) \le f(x)g(x) \le Mg(x)$$

for all x in $[a, b]$. Integrating, we obtain

$$m\int_a^b g(x)\,dx \le \int_a^b f(x)g(x)\,dx \le M\int_a^b g(x)\,dx.$$

If $\int_a^b g(x)\,dx = 0$, then evidently $\int_a^b f(x)g(x)\,dx = 0$ also and any point c in $[a, b]$ satisfies the requirements of the theorem. Otherwise, $\int_a^b g(x)\,dx > 0$. It follows that

$$m \leq \frac{\int_a^b f(x)g(x)\,dx}{\int_a^b g(x)\,dx} \leq M.$$

By the Intermediate Value Theorem applied to f, there exists a point c in $[a, b]$ such that

$$f(c) = \frac{\int_a^b f(x)g(x)\,dx}{\int_a^b g(x)\,dx}$$

from which the conclusion of the theorem follows. Note that the theorem holds also if $g(x) \leq 0$ for all x in $[a, b]$. ●

THEOREM 6.3.2 The Mean Value Theorem for Integrals If f is continuous on $[a, b]$, then there exists a c in $[a, b]$ such that

$$\frac{1}{b-a} \int_a^b f(x)\,dx = f(c).$$

Remark. The quantity $[1/(b-a)] \int_a^b f(x)\,dx$ is called the *average value of* f on $[a, b]$; the value of f at c is the average value of f over the interval $[a, b]$.

Proof. In Theorem 6.3.1, simply set $g(x) = 1$ for all x in $[a, b]$. See Fig. 6.11. ●

We are ready at last for the single most powerful and important theorem in integration theory. There are relatively few "fundamental theorems" in mathematics; this is one of them.

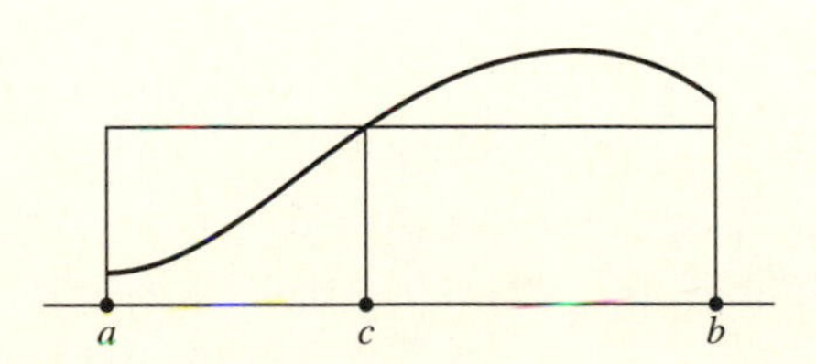

Figure 6.11

THEOREM 6.3.3 The Fundamental Theorem of Calculus Let f be an integrable function on $[a, b]$. Define a function F by setting $F(x) = \int_a^x f(t)\,dt$ for x in $[a, b]$ (see Fig. 6.12).

i) The function F is in $BV(a, b)$.

ii) The function F is in $C([a, b])$.

iii) If f is continuous at c in $[a, b]$, then F is differentiable at c and $F'(c) = f(c)$.

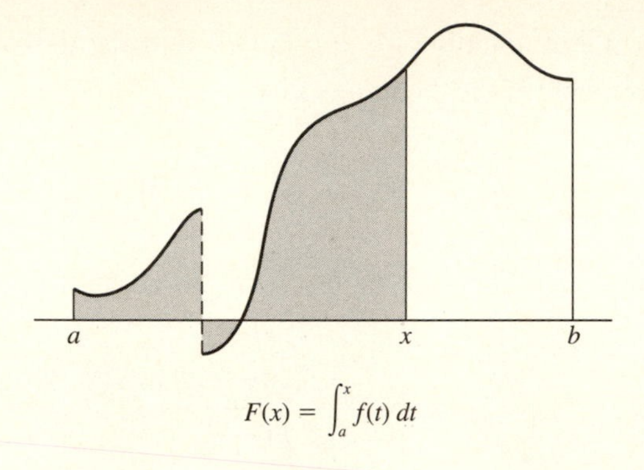

Figure 6.12

Proof. For statement (i), let $\pi = \{x_0, x_1, x_2, \ldots, x_p\}$ be any partition of $[a, b]$. Let m_j and M_j denote the infimum and the supremum of f on the interval $[x_{j-1}, x_j]$. The function f is integrable on $[x_{j-1}, x_j]$ and, by Theorem 6.1.2, for each j,

$$m_j \Delta x_j \leq \int_{x_{j-1}}^{x_j} f(t)\, dt \leq M_j \Delta x_j.$$

Therefore, for each $j = 1, 2, \ldots, p$, $|\int_{x_{j-1}}^{x_j} f(t)\, dt| \leq ||f||_\infty \Delta x_j$. Compute

$$|\Delta F_j| = |F(x_j) - F(x_{j-1})| = \left| \int_a^{x_j} f(t)\, dt - \int_a^{x_{j-1}} f(t)\, dt \right|$$

$$= \left| \int_{x_{j-1}}^{x_j} f(t)\, dt \right| \leq ||f||_\infty \Delta x_j.$$

Summing on j gives $\sum_{j=1}^{p} |\Delta F_j| \leq ||f||_\infty \sum_{j=1}^{p} \Delta x_j = ||f||_\infty (b - a)$. Consequently, $V(F; a, b) = \sup\{\sum_j |\Delta F_j| : \pi \text{ in } \Pi[a, b]\}$, being bounded above by $||f||_\infty (b - a)$, is finite and F is in $BV(a, b)$.

For statement (ii), choose x and y such that $a \leq x < y \leq b$. Let m and M denote the infimum and supremum of f on $[x, y]$. Again note that, by Theorem 6.1.2,

$$m(y - x) \leq \int_x^y f(t)\, dt \leq M(y - x)$$

and therefore $|\int_x^y f(t)\, dt| \leq ||f||_\infty (y - x)$. Consequently,

$$|F(y) - F(x)| = \left| \int_x^y f(t)\, dt \right| \leq ||f||_\infty (y - x).$$

This proves that F satisfies a Lipschitz condition of order 1 at each point of $[a, b]$. Thus F is continuous on $[a, b]$.

To prove statement (iii), suppose that f is continuous at a point c in $[a, b]$. Given $\epsilon > 0$, choose $\delta > 0$ such that, if x is in $[a, b] \cap N(c; \delta)$, then $f(x)$ is in $N(f(c); \epsilon/2)$. Fix any h such that $0 < |h| < \delta$ and such that $c + h$ is in $[a, b]$. Let

m and M denote the infimum and the supremum of f on the closed interval with endpoints c and $c+h$. It follows that

$$f(c) - \frac{1}{2}\epsilon \le m \le f(x) \le M \le f(c) + \frac{1}{2}\epsilon$$

for all x between c and $c+h$. Next, by Theorem 6.1.2, we have $m \le (1/h)\int_c^{c+h} f(x)\,dx \le M$. But

$$\frac{F(c+h) - F(c)}{h} = \frac{1}{h}\int_c^{c+h} f(t)\,dt$$

(see Fig. 6.13).

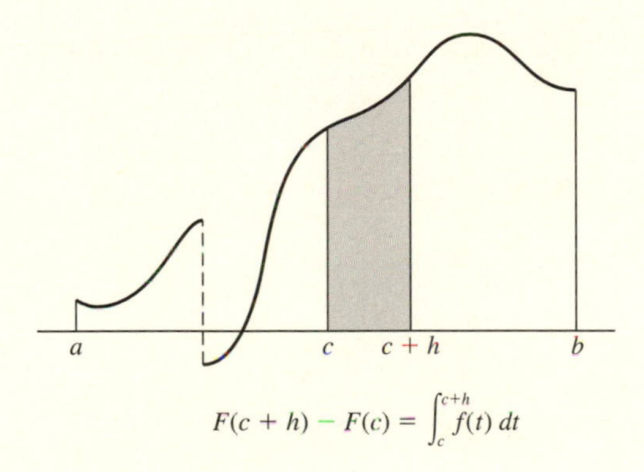

$F(c+h) - F(c) = \int_c^{c+h} f(t)\,dt$

Figure 6.13

Consequently, combining these observations, we have

$$f(c) - \frac{1}{2}\epsilon \le \frac{F(c+h) - F(c)}{h} \le f(c) + \frac{1}{2}\epsilon.$$

We deduce that $|[F(c+h) - F(c)]/h - f(c)| \le \epsilon/2 < \epsilon$. This proves that $F'(c) = \lim_{h\to 0}[F(c+h) - F(c)]/h$ exists and equals $f(c)$. This completes the proof of the theorem. ●

One immediate insight to be gleaned from Theorem 6.3.3 is the observation that the function F, called the *indefinite integral of* f, is always more well behaved than is the integrand f. While f may not be continuous, F always is; while f may be continuous without necessarily being differentiable at a point, the continuity of f ensures the differentiability of F. To carry this one step further, if f is differentiable at a point, then F is twice differentiable there. We summarize this observation by saying that integration is a *smoothing operation*. Theorem 6.3.2 tells us that integration is also an *averaging operation*.

EXAMPLE 3 Define a function f on $[-1, 1]$ by

$$f(x) = \begin{cases} -1, & -1 < x < 0 \\ 0, & x = 0 \\ 1, & 0 < x < 1. \end{cases}$$

The graph of f is sketched in Fig. 6.14. Since f is of bounded variation on $[-1, 1]$, $\int_{-1}^{x} f(t)\,dt$ exists for every x in $[-1, 1]$. It is evident that, for $-1 \le x < 0$,

$$F(x) = (-1)[x - (-1)] = -x - 1.$$

To compute $F(0)$, we simply use Theorem 6.2.10, noting that the function f differs from $g(x) = -1$ on $[-1, 0]$ only at the point $x = 0$. Since $\int_{-1}^{0} g(x)\,dx = -1$, we conclude that $F(0) = \int_{-1}^{0} f(x)\,dx = -1$.

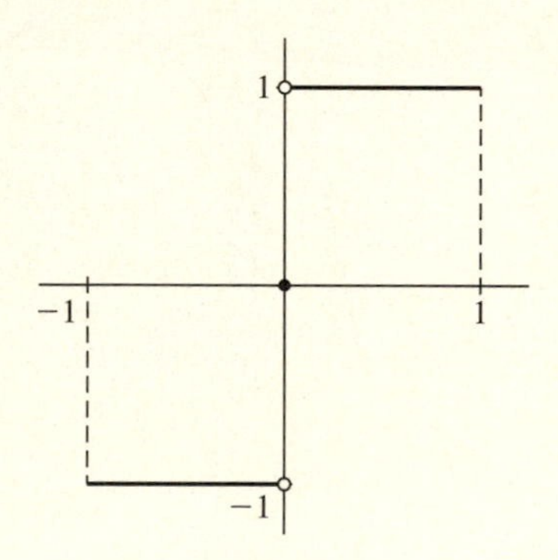

Figure 6.14

For $x > 0$, write $F(x) = F(0) + \int_{0}^{x} f(t)\,dt$. To compute the latter integral, note that the function f differs from the function $g(x) = 1$ on $[0, 1]$ only at the one point $x = 0$. Consequently, since $\int_{0}^{x} g(t)\,dt = x$, it follows, by Theorem 6.2.10, that $\int_{0}^{x} f(t)\,dt = x$. Thus, for $0 < x \le 1$, $F(x) = x - 1$. The graph of F is sketched in Fig. 6.15. You will notice that F has total variation $V(F; -1, 1) = 2$, that F is continuous on $[-1, 1]$, that F is differentiable at every point of $[-1, 1]$ except $x = 0$, and that $F'(x) = f(x)$ for $x \ne 0$, as promised by Theorem 6.3.3. ●

EXAMPLE 4 For x in $(0, 2\pi]$, define $f(x) = (\sin x)/x$ and define $f(0) = 1$. (See Fig. 6.16.) Then f is continuous on $[0, 2\pi]$ and thus is integrable there. Let $F(x) = \int_{0}^{x} f(t)\,dt$. The function F is continuous, of bounded variation, and differentiable on $[0, 2\pi]$. (See Fig. 6.17.) For every x in $(0, 2\pi]$, $F'(x) = (\sin x)/x$ and $F'(0^+) = f(0) = 1$. To confirm this last, note that, for $h > 0$,

$$\frac{F(h) - F(0)}{h} = \frac{F(h)}{h} = \frac{1}{h}\int_{0}^{h} \frac{\sin x}{x}\,dx = \frac{\sin c_h}{c_h},$$

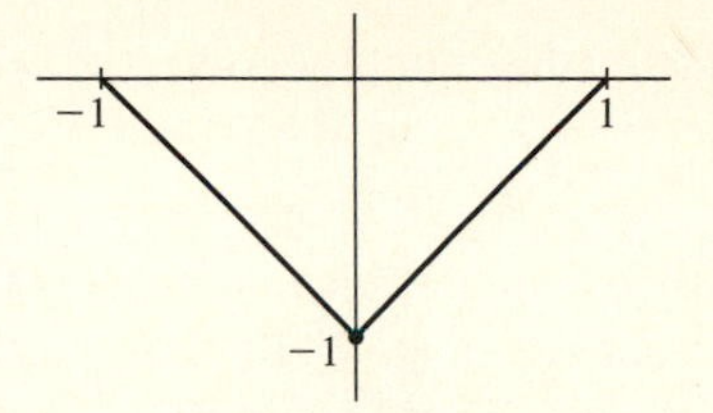

Figure 6.15

for some c_h in $[0, h]$. As h tends to 0^+, so also does c_h and

$$F'(0^+) = \lim_{h \to 0^+} \frac{F(h) - F(0)}{h} = \lim_{c_h \to 0^+} \frac{\sin c_h}{c_h} = 1.$$

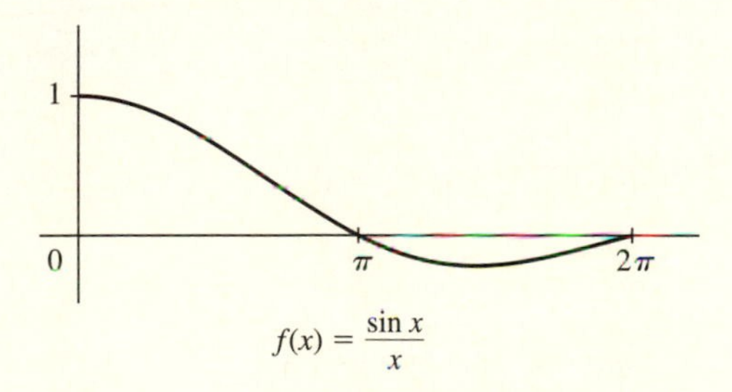

Figure 6.16

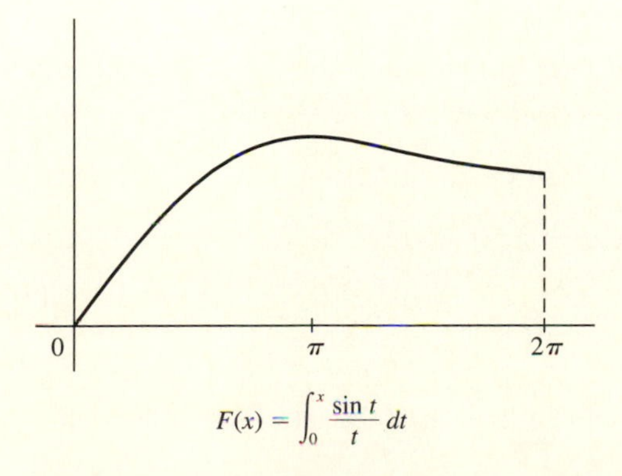

Figure 6.17

Although the integral that defines F is not elementary and cannot be evaluated in terms of elementary functions, we can deduce all these facts by Theorem 6.3.3. To actually compute numerical values for the function F is another matter altogether.

Later we will develop the mathematical theory that will enable us to compute approximate values of F. ●

COROLLARY 6.3.4 Suppose that a function v maps the interval $[c, d]$ into the interval $[a, b]$ and that x_0 is a point of $[c, d]$. Suppose further that f is integrable on $[a, b]$. For x in $[c, d]$ define $G(x) = \int_a^{v(x)} f(t)\,dt$.

i) If v is continuous at x_0, then G is also continuous at x_0.

ii) If v is differentiable at x_0 and if f is continuous at $y_0 = v(x_0)$, then G is differentiable at x_0 and $G'(x_0) = f(v(x_0))v'(x_0)$.

Proof. For y in $[a, b]$, let $F(y) = \int_a^y f(t)\,dt$. Notice that, for x in $[c, d]$, $G(x) = F(v(x))$. To prove (i), note that by Theorem 6.3.3, F is continuous on all of $[a, b]$ and hence at $y_0 = v(x_0)$. If v is continuous at x_0, then G, being the composition of continuous functions, is also continuous at x_0.

For (ii), if f is continuous at y_0, then F is differentiable at y_0 and $F'(y_0) = f(y_0)$. If v is also differentiable at x_0, then, by the chain rule, $G(x) = F(v(x))$ is differentiable at x_0 and

$$G'(x_0) = F'(v(x_0))v'(x_0) = f(v(x_0))v'(x_0)$$

as promised. ●

COROLLARY 6.3.5 Suppose that v_1 and v_2 are two functions that map the interval $[c, d]$ into the interval $[a, b]$ and that f is integrable on $[a, b]$. For x in $[c, d]$, define

$$G(x) = \int_{v_1(x)}^{v_2(x)} f(t)\,dt.$$

i) If v_1 and v_2 are continuous at x_0 in $[c, d]$, then G is also continuous at x_0.

ii) If v_1 and v_2 are differentiable at x_0 and if f is continuous at $y_1 = v_1(x_0)$ and at $y_2 = v_2(x_0)$, then G is differentiable at x_0 and

$$G'(x_0) = f(v_2(x_0))v_2'(x_0) - f(v_1(x_0))v_1'(x_0).$$

Proof. Merely write

$$G(x) = \int_{v_1(x)}^{a} f(t)\,dt + \int_a^{v_2(x)} f(t)\,dt = \int_a^{v_2(x)} f(t)\,dt - \int_a^{v_1(x)} f(t)\,dt$$

and apply Corollary 6.3.4 to each of the latter integrals. ●

EXAMPLE 5 For x in $[0, M]$, define $G(x) = \int_0^{x^2} \sqrt{t}\exp(-t^2/2)\,dt$. Regardless of the limits of integration, this integral is not elementary and cannot be evaluated directly. Nevertheless, we know by Theorem 6.3.3 that G is continuous and of bounded variation on $[0, M]$. More, since the integrand is continuous, G is differentiable at every point of $[0, M]$ with $G'(x) = \sqrt{x^2}\exp(-x^4/2)(2x) = 2x^2\exp(-x^4/2)$. ●

THEOREM 6.3.6 Bonnet's Mean Value Theorem for Integrals Suppose that f is continuous and monotone increasing on $[a, b]$ and that g is nonnegative and integrable on $[a, b]$. Then there exists a c in $[a, b]$ such that

$$\int_a^b f(t)g(t)\,dt = f(a)\int_a^c g(t)\,dt + f(b)\int_c^b g(t)\,dt.$$

Proof. For x in $[a, b]$, let $h(x) = [f(b) - f(a)]\int_x^b g(t)\,dt$. Note that, by Theorem 6.3.3, h is continuous on $[a, b]$. More, since g is nonnegative, h is monotone decreasing on $[a, b]$. Compute $h(a) = [f(b) - f(a)]\int_a^b g(t)\,dt$ and $h(b) = 0$. Next apply Theorem 6.3.1: Choose a c_1 in $[a, b]$ such that

$$\int_a^b f(t)g(t)\,dt = f(c_1)\int_a^b g(t)\,dt.$$

Notice that, because f is monotone increasing, the number $[f(c_1) - f(a)]\int_a^b g(t)\,dt$ lies between the extreme values $h(a)$ and $h(b)$ of the continuous function h. By the Intermediate Value Theorem, there is a c in $[a, b]$ such that

$$\begin{aligned} h(c) &= [f(c_1) - f(a)]\int_a^b g(t)\,dt \\ &= f(c_1)\int_a^b g(t)\,dt - f(a)\int_a^b g(t)\,dt \\ &= \int_a^b f(t)g(t)\,dt - f(a)\int_a^b g(t)\,dt. \end{aligned}$$

From the definition of the function h, we have

$$h(c) = [f(b) - f(a)]\int_c^b g(t)\,dt.$$

Consequently,

$$\int_a^b f(t)g(t)\,dt - f(a)\int_a^b g(t)\,dt = f(b)\int_c^b g(t)\,dt - f(a)\int_c^b g(t)\,dt.$$

Upon rearranging, we conclude that

$$\int_a^b f(t)g(t)\,dt = f(a)\int_a^c g(t)\,dt + f(b)\int_c^b g(t)\,dt$$

and the theorem is proved. ●

COROLLARY 6.3.7 If f is continuous and monotone increasing on $[a, b]$, there exists a c in $[a, b]$ such that

$$\int_a^b f(t)\,dt = f(a)(c - a) + f(b)(b - c).$$

Proof. In Theorem 6.3.6, simply set $g(x) = 1$ for all x in $[a, b]$. (See Fig. 6.18 on page 260.) ●

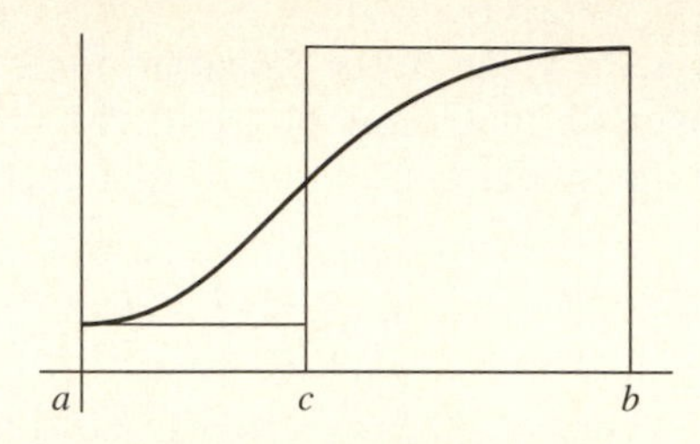

Figure 6.18

Theorem 6.3.3 is appropriately named as a fundamental theorem; it provides the bridge between the operation of differentiation and that of integration. For suppose that the integrand f is continuous on $[a, b]$. By part (iii) of Theorem 6.3.3, the function $F(x) = \int_a^x f(t)\,dt$ is differentiable at every point of $[a, b]$ and $F' = f$. If G is *any* function with $G' = f$ on $[a, b]$—that is, G is an *antiderivative* of f—then F and G have the same derivative and therefore, by Corollary 4.3.5, F and G differ by a constant. Consequently, to compute the integral of a continuous function f, we search first for an antiderivative G of f, then adjust G by adding an appropriate constant to obtain the function F. This observation leads to the second fundamental theorem, credited to Cauchy. In fact, we can weaken the hypothesis and only require the integrand to be integrable and to be the derivative of some function G on $[a, b]$. Whether f has an antiderivative to be found is another matter.

THEOREM 6.3.8 Cauchy's Fundamental Theorem of Calculus Suppose that f is integrable on $[a, b]$ and that G is any antiderivative of f on $[a, b]$. Then

$$\int_a^b f(x)\,dx = G(x)\Big|_a^b = G(b) - G(a).$$

Remark. Given an integrable function f on $[a, b]$, it may not be possible to find an antiderivative G of f on the entire interval; if there is none, then the theorem does not help us. On the other hand, if f is continuous on $[a, b]$, then $F(x) = \int_a^x f(t)\,dt$ is already an antiderivative of f, but this theorem helps us evaluate the integral only if we can identify the function F explicitly in terms of known functions.

Proof. First observe that, since G is differentiable at every point of $[a, b]$, G is continuous on $[a, b]$. Next assign any positive value to ϵ. By the integrability of f on $[a, b]$, choose a partition π_0 such that, for any refinement $\pi = \{x_0, x_1, x_2, \ldots, x_p\}$ of π_0 and for any choice of points s_j in $[x_{j-1}, x_j]$, $|S(f, \pi) - \int_a^b f(x)\,dx| < \epsilon$. Apply the Mean Value Theorem to G on each interval $[x_{j-1}, x_j]$. Choose s_j in (x_{j-1}, x_j) such that

$$G(x_j) - G(x_{j-1}) = G'(s_j)\Delta x_j.$$

Since $G' = f$ on $[a, b]$, $G'(s_j) = f(s_j)$ for $j = 1, 2, \ldots, p$. Summing on j and telescoping, we obtain

$$G(b) - G(a) = \sum_{j=1}^{p} [G(x_j) - G(x_{j-1})] = \sum_{j=1}^{p} G'(s_j)\Delta x_j$$

$$= \sum_{j=1}^{p} f(s_j)\Delta x_j = S(f, \pi).$$

Consequently, $|[G(b) - G(a)] - \int_a^b f(x)\,dx| < \epsilon$. Since ϵ is arbitrary, it follows that $G(b) - G(a) = \int_a^b f(x)\,dx$. ●

COROLLARY 6.3.9 If f is differentiable on $[a, b]$ and if f' is integrable on $[a, b]$, then, for all x in $[a, b]$,

$$\int_a^x f'(t)\,dt = f(x) - f(a). \quad ●$$

EXAMPLE 6 Define $f(x)$ on $[0, 3]$ as follows:

$$f(x) = \begin{cases} x, & 0 \le x \le 1 \\ -(x-2), & 1 < x \le 3. \end{cases}$$

The graph of f is sketched in Fig. 6.19. The function f is continuous on $[0, 3]$, and hence is integrable there. By Theorem 6.3.3, the function $F(x) = \int_0^x f(t)\,dt$ is also

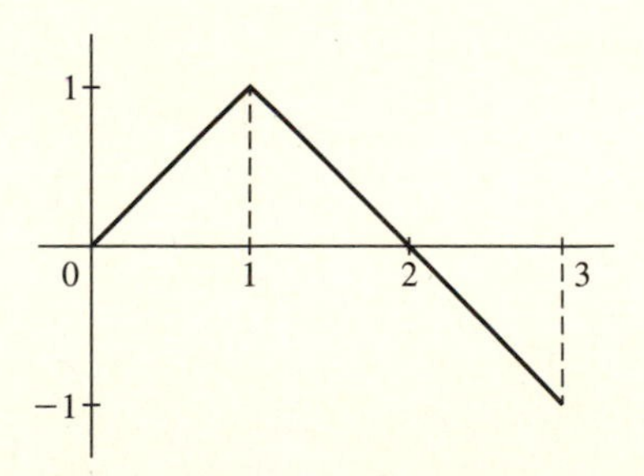

Figure 6.19

continuous and differentiable on $[0, 3]$ with $F'(x) = f(x)$ for all x in $[0, 3]$. We use Theorem 6.3.8 systematically to compute $F(x)$. For $0 \le x \le 1$, let $G(x) = x^2/2$. Then for such x,

$$F(x) = \int_0^x t\,dt = G(x) - G(0) = \frac{1}{2}x^2.$$

For $1 < x \le 3$, integrate over $[0, 1]$ and $[1, x]$ separately to obtain

$$F(x) = \int_0^x f(t)\,dt = \int_0^1 t\,dt - \int_1^x (t-2)\,dt = \frac{1}{2} - \int_1^x (t-2)\,dt.$$

To evaluate this last integral, use Theorem 6.3.8 with $G(t) = (t-2)^2/2$, an antiderivative of f on the interval $[1, x]$. Therefore,

$$\int_1^x (t-2)\,dt = G(x) - G(1) = \frac{(x-2)^2}{2} - \frac{1}{2}$$
$$= \frac{1}{2}x^2 - 2x + \frac{3}{2}.$$

Consequently, for $1 < x \le 3$, $F(x) = -x^2/2 + 2x - 1$. In summary, the function F is defined by the equations

$$F(x) = \begin{cases} \frac{1}{2}x^2, & 0 \le x \le 1 \\ -\frac{1}{2}x^2 + 2x - 1, & 1 < x \le 3. \end{cases}$$

The graph of F is sketched in Fig. 6.20.

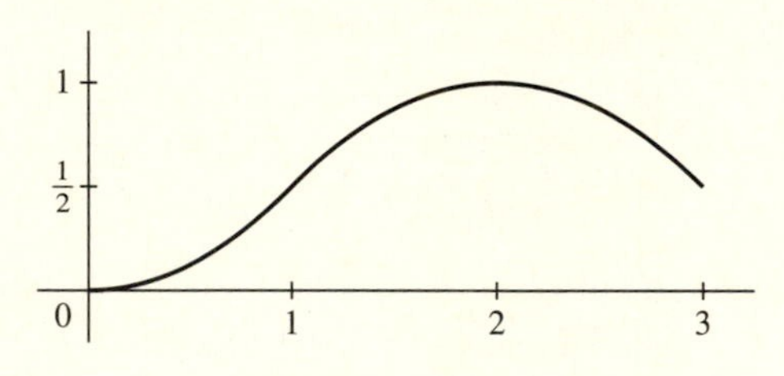

Figure 6.20

It is evident that, for $x \ne 1$, $F'(x) = f(x)$. We confirm that, in accord with Theorem 6.3.3, $F'(1) = f(1) = 1$. If $h < 0$, then

$$\frac{F(1+h) - F(1)}{h} = \frac{(1+h)^2/2 - 1/2}{h} = 1 + \frac{1}{2}h.$$

Therefore, $F'(1^-) = 1$. If $h > 0$, then

$$\frac{F(1+h) - F(1)}{h} = \frac{-(1+h)^2/2 + 2(1+h) - 1 - 1/2}{h} = 1 - \frac{1}{2}h.$$

It follows that $F'(1^+) = 1 = f(1)$. Since $F'(1^-) = F'(1^+) = f(1)$, we conclude that $F'(1) = f(1)$, as claimed. ●

EXAMPLE 7 Define a function f on $[0, 2]$ by

$$f(x) = \begin{cases} x, & 0 \le x < 1 \\ \frac{1}{2}, & x = 1 \\ x - 1, & 1 < x \le 2. \end{cases}$$

The graph of f is sketched in Fig. 6.21. Since f is of bounded variation on $[0, 2]$, f is integrable on $[0, 2]$. Note that f has a jump discontinuity at $x = 1$. For each x

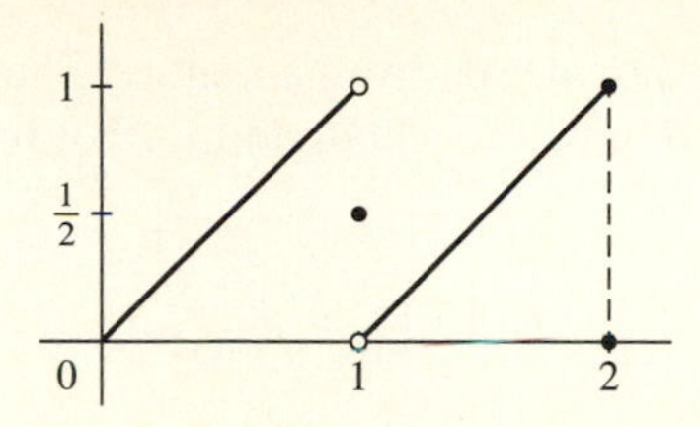

Figure 6.21

in [0, 2], define $F(x) = \int_0^x f(t)\,dt$. By using Theorem 6.3.8 and Theorem 6.2.10, you can show that, for $0 \le x \le 1$,

$$F(x) = \int_0^x t\,dt = \frac{1}{2}x^2.$$

Likewise, for $1 < x \le 2$,

$$F(x) = F(1) + \int_1^x (t-1)\,dt = \frac{1}{2} + \frac{1}{2}(x-1)^2 = \frac{1}{2}(x^2 - 2x + 2).$$

The graph of F is sketched in Fig. 6.22. Again, for $x \ne 1$, $F'(x) = f(x)$. In this example, however, $F'(1)$ fails to exist. To confirm this claim, compute $F'(1^-)$ and $F'(1^+)$. For $h < 0$,

$$\frac{F(1+h) - F(1)}{h} = \frac{(1+h)^2/2 - 1/2}{h} = 1 + \frac{h}{2}.$$

Therefore $F'(1^-) = 1$. However, if $h > 0$,

$$\frac{F(1+h) - F(1)}{h} = \frac{(1+h)^2 - 2(1+h) + 2 - 1}{2h} = \frac{h}{2}.$$

We deduce that $F'(1^+) = 0$. Since $F'(1^-) \ne F'(1^+)$ at the interior point $x = 1$, we conclude that $F'(1)$ fails to exist.

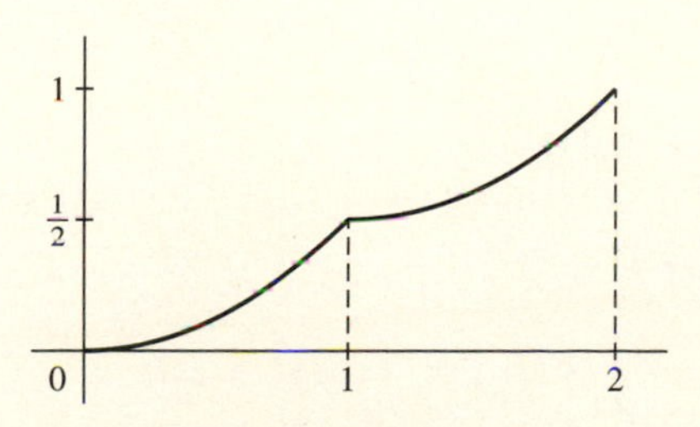

Figure 6.22

Let us briefly consider the integral from an entirely new point of view. Suppose that f is a continuous function on the rectangle $[a, b] \times [c, d]$ in $\mathbb{R}^2$. For each x_2 in $[c, d]$ define

$$F(x_2) = \int_a^b f(t, x_2)\,dt.$$

This integral exists because, as you will recall, for any fixed x_2 in $[c, d]$, $f(x_1, x_2)$ is a continuous function of x_1 in $[a, b]$. (Refer to Theorem 3.1.6.) We claim that F is continuous on $[c, d]$. To prove this, note that, since f is continuous on the compact set $[a, b] \times [c, d]$, f is uniformly continuous there. Given $\epsilon > 0$, choose $\delta > 0$ such that, whenever $\|\boldsymbol{y} - \boldsymbol{x}\| < \delta$, we have $|f(\boldsymbol{y}) - f(\boldsymbol{x})| < \epsilon$. Choose y_2 in $N(x_2; \delta) \cap [c, d]$ and compute

$$\begin{aligned} |F(y_2) - F(x_2)| &= \left| \int_a^b f(t, y_2)\,dt - \int_a^b f(t, x_2)\,dt \right| \\ &\le \int_a^b |f(t, y_2) - f(t, x_2)|\,dt < \epsilon \int_a^b dt = (b - a)\epsilon. \end{aligned}$$

We conclude that F is continuous at an arbitrary point x_2 of $[c, d]$. (Refer to Fig. 6.23.) We have proved the following theorem.

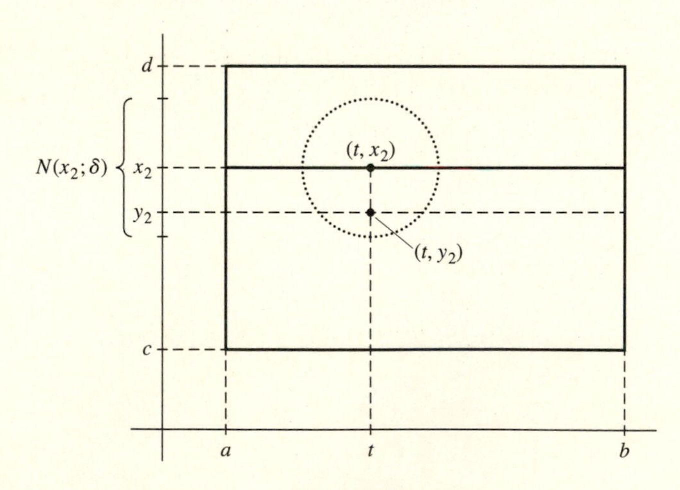

Figure 6.23

THEOREM 6.3.10 If f is continuous on $[a, b] \times [c, d]$, then for x_2 in $[c, d]$, the function F defined by the integral

$$F(x_2) = \int_a^b f(t, x_2)\,dt$$

exists and is continuous on $[c, d]$. ●

We might also ask, in this same vein, for conditions under which the function F defined above is differentiable. Again, suppose that f is a real-valued function defined on $[a, b] \times [c, d]$ in $\mathbb{R}^2$. Suppose that, for each x_2 in $[c, d]$, the function $f(t, x_2)$ is integrable on $[a, b]$; define

$$F(x_2) = \int_a^b f(t, x_2)\, dt.$$

Suppose also that, for each x_1 in $[a, b]$, the function $g(x_2) = f(x_1, x_2)$ is a differentiable function on $[c, d]$. That is, for each x_1 in $[a, b]$,

$$\lim_{h \to 0} \frac{f(x_1, x_2 + h) - f(x_1, x_2)}{h}$$

exists at every x_2 in $[c, d]$. We will denote this derivative by $D_2 f$. (The derivative $D_2 f$ is, of course, simply the *partial derivative of* f with respect to the second variable x_2; it is often denoted $\partial f / \partial x_2$. We will study such derivatives in depth in Chapter 8.) If we assume that $D_2 f$ is a continuous function on the compact set $[a, b] \times [c, d]$, then, as we shall prove in the following theorem, the function F is differentiable at each x_2 in $[c, d]$ and

$$F'(x_2) = \int_a^b D_2 f(t, x_2)\, dt.$$

THEOREM 6.3.11 Leibnitz's Rule Let f be a real-valued function defined on a rectangle $R = [a, b] \times [c, d]$. If, for each x_2 in $[c, d]$, $f(t, x_2)$ is integrable on $[a, b]$ and if $D_2 f$ exists and is continuous on R, then the function F is differentiable on $[c, d]$ and

$$F'(x_2) = \int_a^b D_2 f(t, x_2)\, dt.$$

Remarks. That $D_2 f$ exists on $[a, b] \times [c, d]$ implies that, for each x_1 in $[a, b]$, the function $g(x_2) = f(x_1, x_2)$ is a continuous function on $[c, d]$, but this fact is insufficient to guarantee the existence of $\int_a^b f(t, x_2)\, dt$; we must assume the integrability of $f(t, x_2)$ separately. Also, if we were merely to assume that, for each x_1 in $[a, b]$, $D_2 f(x_1, x_2)$ is a continuous function of x_2, we would be unable to prove our theorem. We need the stronger hypothesis that $D_2 f$ is continuous on the compact rectangle $R = [a, b] \times [c, d]$ in $\mathbb{R}^2$. Thus $D_2 f$ is assumed to be uniformly continuous on R. Given any $\epsilon > 0$, there exists a $\delta > 0$ such that, for any two points $\boldsymbol{x} = (x_1, x_2)$ and $\boldsymbol{y} = (y_1, y_2)$ in R with $\|\boldsymbol{x} - \boldsymbol{y}\| < \delta$, we have $|D_2 f(x_1, x_2) - D_2 f(y_1, y_2)| < \epsilon$. We begin our proof by fixing such a δ.

Proof. Fix any x_2 in $[c, d]$. We want to prove that

$$\lim_{h \to 0} \frac{F(x_2 + h) - F(x_2)}{h}$$

exists and equals $\int_a^b D_2 f(t, x_2)\,dt$. Choose any h such that $0 < |h| < \delta$ and such that $x_2 + h$ is also in $[c, d]$. Note that

$$F(x_2 + h) - F(x_2) = \int_a^b [f(t, x_2 + h) - f(t, x_2)]\,dt.$$

For each t in $[a, b]$, apply the Mean Value Theorem to the continuous and differentiable function $f(t, \cdot)$ on the subinterval of $[c, d]$ with endpoints x_2 and $x_2 + h$; there exists a point c_t between x_2 and $x_2 + h$ such that

$$f(t, x_2 + h) - f(t, x_2) = h D_2 f(t, c_t).$$

It follows that

$$\begin{aligned}\left| \frac{F(x_2 + h) - F(x_2)}{h} - \int_a^b D_2 f(t, x_2)\,dt \right| &= \left| \int_a^b [D_2 f(t, c_t) - D_2 f(t, x_2)]\,dt \right| \\ &\le \int_a^b |D_2 f(t, c_t) - D_2 f(t, x_2)|\,dt.\end{aligned} \tag{6.10}$$

For each t, the distance from (t, x_2) to (t, c_t) is less than δ. Thus, by the uniform continuity of $D_2 f$, the integrand in the last integral above is less than ϵ for every t. By (6.10), if $0 < |h| < \delta$, then

$$\left| \frac{F(x_2 + h) - F(x_2)}{h} - \int_a^b D_2 f(t, x_2)\,dt \right| < (b - a)\epsilon.$$

That is,

$$\lim_{h \to 0} \frac{F(x_2 + h) - F(x_2)}{h} = \int_a^b D_2 f(t, x_2)\,dt.$$

This proves the theorem. ●

In the spirit of Theorems 6.3.10 and 6.3.11, we have the following theorem, which justifies the interchange of two integrations, subject to certain constraints on the integrand f. The hypotheses of the theorem are only sufficient, not necessary.

THEOREM 6.3.12 Let f be a continuous, real-valued function on the rectangle $R = [a, b] \times [c, d]$ in $\mathbb{R}^2$. Define

$$F(x_2) = \int_a^b f(x_1, x_2)\,dx_1, \qquad \text{for each } x_2 \text{ in } [c, d]$$

and

$$G(x_1) = \int_c^d f(x_1, x_2)\,dx_2, \qquad \text{for each } x_1 \text{ in } [a, b].$$

Then F is in $R[c, d]$, G is in $R[a, b]$ and

$$\int_c^d F(x_2)\,dx_2 = \int_a^b G(x_1)\,dx_1.$$

Remark. This theorem assures us that, provided f is continuous on $[a, b] \times [c, d]$,

$$\int_c^d \left[\int_a^b f(x_1, x_2)\,dx_1\right] dx_2 = \int_a^b \left[\int_c^d f(x_1, x_2)\,dx_2\right] dx_1.$$

If f fails to meet this hypothesis, these iterated integrals may not be equal.

Proof. Because f is continuous on R, F is continuous on $[c, d]$ by Theorem 6.3.10. Therefore F is in $R[c, d]$. Likewise, G is in $R[a, b]$.

Since R is compact and since f is continuous, we know that f is uniformly continuous on R. That is, given $\epsilon > 0$, there exists a $\delta > 0$ such that, whenever $\boldsymbol{x} = (x_1, x_2)$ and $\boldsymbol{y} = (y_1, y_2)$ are in R with $||\boldsymbol{x} - \boldsymbol{y}|| < \delta$, it follows that $|f(\boldsymbol{x}) - f(\boldsymbol{y})| < \epsilon$. Fix such a δ. Choose p in $\mathbb{N}$ such that $h_1 = (b - a)/p < \delta/\sqrt{2}$ and such that $h_2 = (d - c)/p < \delta/\sqrt{2}$. Use this p to form the uniform partitions of $[a, b]$ and of $[c, d]$ by setting

$$a_i = a + ih_1$$

and

$$c_j = c + jh_2$$

for $i, j = 0, 1, 2, \ldots, p$. (See Fig. 6.24.) Thus

$$\pi_1 : a = a_0 < a_1 < a_2 < \cdots < a_p = b$$

is the uniform partition of $[a, b]$ with gauge h_1 and

$$\pi_2 : c = c_0 < c_1 < c_2 < \cdots < c_p = d$$

is the uniform partition of $[c, d]$ with gauge h_2. By the additivity of the integral, write

$$\begin{aligned}\int_c^d F(x_2)\,dx_2 &= \int_c^d \left[\int_a^b f(x_1, x_2)\,dx_1\right] dx_2 \\ &= \sum_{j=1}^{p}\sum_{i=1}^{p} \int_{c_{j-1}}^{c_j} \left[\int_{a_{i-1}}^{a_i} f(x_1, x_2)\,dx_1\right] dx_2. \qquad \textbf{(6.11)}\end{aligned}$$

By Theorem 6.3.2 (The Mean Value Theorem for Integrals), for $i = 1, 2, \ldots, p$, we can choose an $x_1^{(i)}$ in $[a_{i-1}, a_i]$ such that $\int_{a_{j-1}}^{a_i} f(x_1, x_2)\,dx_1 = f(x_1^{(i)}, x_2)h_1$. Likewise, for each $j = 1, 2, \ldots, p$, we can choose an $x_2^{(j)}$ in $[c_{j-1}, c_j]$ such that $\int_{c_{j-1}}^{c_j} f(x_1^{(i)}, x_2)\,dx_2 = f(x_1^{(i)}, x_2^{(j)})h_2$. Substituting in (6.11) yields

$$\int_c^d F(x_2)\,dx_2 = \sum_{j=1}^{p}\sum_{i=1}^{p} f(x_1^{(i)}, x_2^{(j)})h_1 h_2. \qquad \textbf{(6.12)}$$

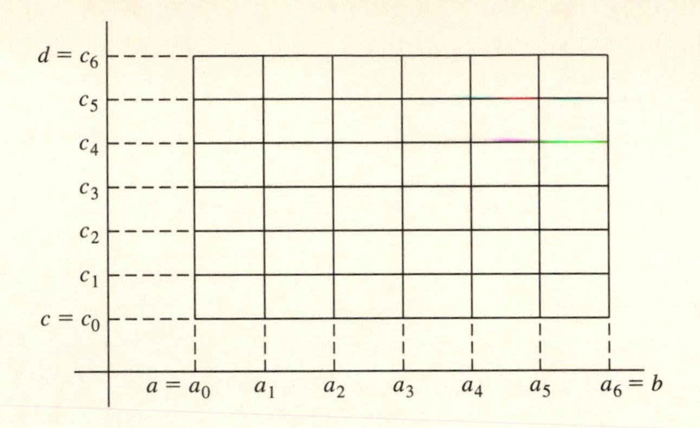

Figure 6.24

Similarly, for $i, j = 1, 2, \ldots, p$, we can choose $y_1^{(i)}$ in $[a_{i-1}, a_i]$ and $y_2^{(j)}$ in $[c_{j-1}, c_j]$ such that

$$\int_a^b G(x_1)\,dx_1 = \sum_{i=1}^{p}\sum_{j=1}^{p} f(y_1^{(i)}, y_2^{(j)})h_1h_2. \tag{6.13}$$

For each $i, j = 1, 2, \ldots, p$, let $\boldsymbol{x}_{ij} = (x_1^{(i)},\ x_2^{(j)})$ and $\boldsymbol{y}_{ij} = (y_1^{(i)},\ y_2^{(j)})$ denote the points in $R = [a, b] \times [c, d]$ determined by this process. (See Fig. 6.25.)

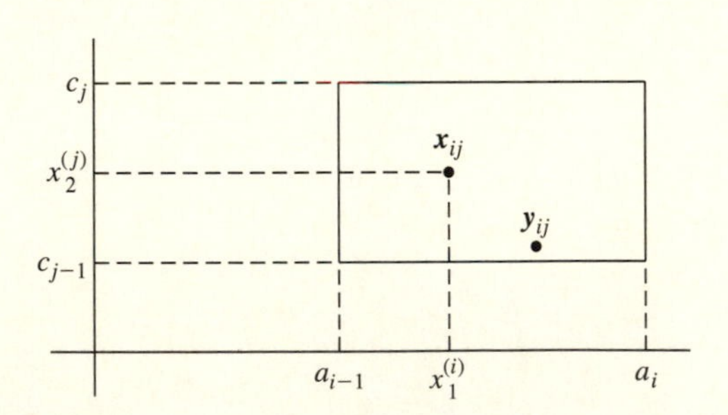

Figure 6.25

Note that, for all $i, j = 1, 2, \ldots, p$,

$$\begin{aligned}\|\boldsymbol{x}_{ij} - \boldsymbol{y}_{ij}\| &= [(x_1^{(i)} - y_1^{(i)})^2 + (x_2^{(j)} - y_2^{(j)})^2]^{1/2}\\ &\le [h_1^2 + h_2^2]^{1/2} < \left[\tfrac{1}{2}\delta^2 + \tfrac{1}{2}\delta^2\right]^{1/2} = \delta.\end{aligned}$$

Consequently,

$$|f(\boldsymbol{x}_{ij}) - f(\boldsymbol{y}_{ij})| = |f(x_1^{(i)}, x_2^{(j)}) - f(y_1^{(i)}, y_2^{(j)})| < \epsilon.$$

From (6.12) and (6.13) we conclude that

$$\left|\int_c^d F(x_2)\,dx_2 - \int_a^b G(x_1)\,dx_1\right| = \left|\sum_{i=1}^{p}\sum_{j=1}^{p}[f(\boldsymbol{x}_{ij}) - f(\boldsymbol{y}_{ij})]h_1h_2\right|$$
$$\leq \sum_{i=1}^{p}\sum_{j=1}^{p}|f(\boldsymbol{x}_{ij}) - f(\boldsymbol{y}_{ij})|h_1h_2$$
$$< \epsilon(b-a)(d-c).$$

Since $\int_c^d F(x_2)\,dx_2 - \int_a^b G(x_1)\,dx_1$ is a constant and since ϵ is arbitrarily small, it follows that $\int_c^d F(x_2)\,dx_2 = \int_a^b G(x_1)\,dx_1$, as promised by the theorem. ●

6.4 TECHNIQUES OF INTEGRATION

During your early work with integral calculus you learned two general techniques for evaluating integrals: Change of variables and integration by parts. Indeed, you probably have integrated scores of functions of varying degrees of complexity using just these two methods. Both techniques are simple; success in using them depends solely on skill and cleverness. Here we prove that, under suitable hypotheses, each of these techniques is valid.

THEOREM 6.4.1 Change of Variables Let v be a continuously differentiable function on $I = [a, b]$. Let $c = v(a)$ and $d = v(b)$. Let f be continuous on the interval $v(I)$. Then for all x in $[a, b]$,

$$\int_a^x f(v(t))v'(t)\,dt = \int_c^{v(x)} f(y)\,dy.$$

In particular,

$$\int_a^b f(v(t))v'(t)\,dt = \int_c^d f(y)\,dy.$$

Proof. Since f, v, and v' are continuous, $(f \circ v)v'$ is also continuous on the interval I. Thus $(f \circ v)v'$ is in $R[a, b]$. Note that v is not assumed to be monotone so that $v(I)$ contains, but may not equal, the interval J with endpoints $c = v(a)$ and $d = v(b)$. However, since f is assumed to be continuous on the interval $v(I)$, it is integrable on every subinterval of $v(I)$. (See Fig. 6.26.) For x in $[a, b]$, define

$$F(x) = \int_a^x f(v(t))v'(t)\,dt.$$

By part (iii) of Theorem 6.3.3, for x in $[a, b]$, we have

$$F'(x) = f(v(x))v'(x).$$

For any y in $v(I)$, define $G(y) = \int_c^y f(t)\,dt$. Since f is continuous, we have $G'(y) = f(y)$. The chain rule implies that the function $G \circ v$ has derivative

$$G'(v(x))v'(x) = f(v(x))v'(x)$$

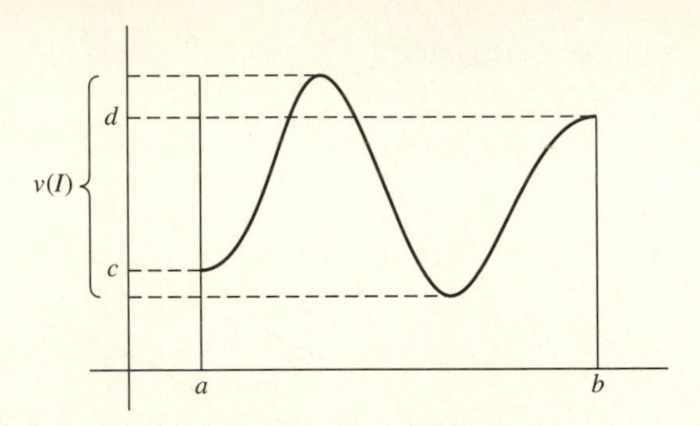

Figure 6.26

at each point x in $[a, b]$. Consequently, $(G \circ v)' = F'$ on $[a, b]$. By Corollary 4.3.5, there exists a constant C such that $G \circ v = F + C$. That is,

$$G(v(x)) = \int_c^{v(x)} f(y)\,dy = F(x) + C = \int_a^x f(v(t))v'(t)\,dt + C.$$

To evaluate C, simply let $x = a$ so that $v(x) = v(a) = c$. We obtain

$$0 = \int_c^c f(t)\,dt = \int_a^a f(v(t))v'(t)\,dt + C = 0 + C$$

and conclude that $C = 0$. To obtain the final assertion of the theorem, simply set $x = b : \int_c^d f(y)\,dy = \int_a^b f(v(t))v'(t)\,dt.$ ●

EXAMPLE 8 Having the Fundamental Theorem of Calculus and the technique of change of variables available, we can finally define the logarithm function and derive its properties. Although we have used the logarithm and exponential functions in earlier examples and exercises, we have done so without having rigorously established all their properties. This example legitimizes our previous computations using $\ln x$ and e^x. For x in $(0, \infty)$, consider the indefinite integral $\int_1^x 1/t\,dt$. Note that the integrand is continuous on $(0, \infty)$; this ensures that the integral exists over the closed interval with endpoints 1 and x. Technically speaking, since we have not yet proved that $d(\ln x)/dx = 1/x$ we cannot appeal to Cauchy's Fundamental Theorem in order to compute the integral $\int_1^x 1/t\,dt$. In fact, we *define* the natural logarithm function to be this integral. (See Figs. 6.27 and 6.28.) For x in $(0, \infty)$, let

$$\ln x = \int_1^x \frac{1}{t}\,dt.$$

All the properties of the logarithm follow from this definition—to show, for example, that the logarithm function transforms products into sums, fix $x > 0$ and $y > 0$ and consider $\ln(xy) = \int_1^{xy} 1/t\,dt$. Write

$$\ln(xy) = \int_1^x \frac{1}{t}\,dt + \int_x^{xy} \frac{1}{t}\,dt.$$

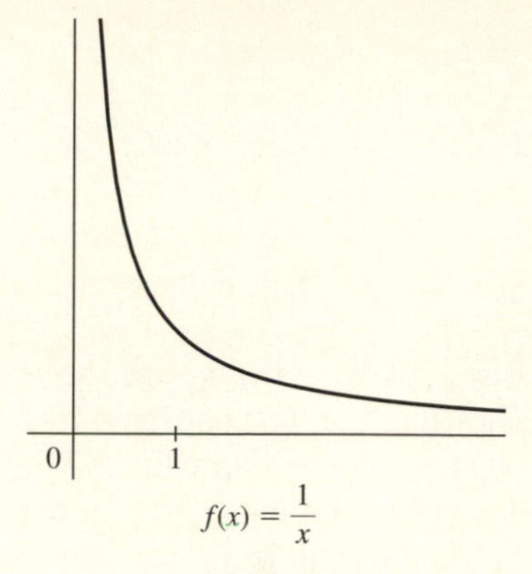

$f(x) = \frac{1}{x}$

Figure 6.27

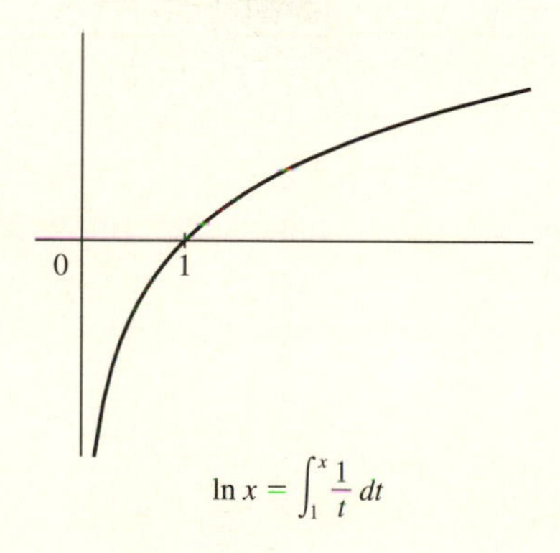

$\ln x = \int_1^x \frac{1}{t}\,dt$

Figure 6.28

The first integral is simply $\ln x$. Change variables in the second by letting $f(t) = 1/t$ and $t = v(u) = xu$. Then v is continuously differentiable on the interval with endpoints 1 and y with $v'(u) = x$, $v(1) = x$, and $v(y) = xy$. Thus

$$\int_x^{xy} f(t)\,dt = \int_x^{xy} \frac{1}{t}\,dt = \int_1^y f(v(u))v'(u)\,du = \int_1^y \frac{1}{xu} x\,du$$
$$= \int_1^y \frac{1}{u}\,du = \ln y.$$

We conclude that $\ln(xy) = \ln x + \ln y$. As an exercise, you can prove similarly that, for $x > 0$ and r in $\mathbb{R}$,

$$\ln x^r = \int_1^{x^r} \frac{1}{t}\,dt = r\int_1^x \frac{1}{u}\,du = r\ln x.$$

By the Fundamental Theorem of Calculus, we know that $\ln x$ is of bounded variation on any compact interval contained in $(0, \infty)$ and that $\ln x$ is continuous at each $x > 0$. More, because the integrand $1/x$ is continuous, we know that $\ln x$ is differentiable at each x in $(0, \infty)$ with derivative $1/x$.

Since the derivative of $\ln x$ is positive on $(0, \infty)$, we know by Corollary 4.3.6 that $\ln x$ is strictly monotone increasing on $(0, \infty)$. Further, the second derivative of $\ln x$ is $-1/x^2$ and is negative on $(0, \infty)$. Therefore the graph of $\ln x$ is concave down. By Theorem 3.2.8 and Exercise 3.48, we know that the logarithm has an inverse function, denoted e^x, which is also monotone increasing and continuous. More, in Exercise 4.37 you showed that, if the derivative of $\ln x$ is $1/x$, then the derivative e^x is e^x.

You have no means to actually *compute* the values of $\ln x$ (except for very special values of x), a fact that surely contributed to your initial unease when dealing with this function, but you now know that, by using Taylor's theorem, you can approximate the values of $\ln x$ as closely as you may choose. Still, you cannot know $\ln x$ in the same way that you can know, say, a polynomial. But, as you surely recognize by now, the natural logarithm is among the most important functions in mathematics. And you will not be surprised if we exploit this technique of using the integral to create new functions from old whenever it suits our purpose. In doing so, we can build functions that you may never have suspected exist. ●

EXAMPLE 9 For k in $\mathbb{N}$, evaluate $\int_1^e (\ln x)^k/x\,dx$ by letting $v(x) = \ln x$ on the interval $[1, e]$. By Example 8, the function $v(x)$ is continuously differentiable on $[1, e]$, with $v(1) = 0$ and $v(e) = 1$. Also, $v'(x) = 1/x$. The function f is the kth power function and the integrand you are to integrate is $f(v(x))v'(x)$. Changing variables, we obtain

$$\int_1^e \frac{(\ln x)^k}{x}\,dx = \int_0^1 y^k\,dy = \frac{y^{k+1}}{k+1}\Big|_0^1 = \frac{1}{k+1},$$

where we have used Theorem 6.3.8 with $G(y) = y^{k+1}/(k+1)$, an antiderivative of y^k. ●

EXAMPLE 10 Evaluate $\int_1^2 x \exp(-x^2)\,dx$ by letting $v(x) = -x^2$ on $[1, 2]$. The function v is continuously differentiable on $[1, 2]$ with $v'(x) = -2x$, $v(1) = -1$, and $v(2) = -4$. We can rewrite the original integral as

$$-\tfrac{1}{2}\int_1^2 e^{-x^2}(-2x)\,dx = -\tfrac{1}{2}\int_1^2 e^{v(x)}v'(x)\,dx,$$

from which it is evident that the function f to be integrated is $f(y) = e^y$. An application of Theorem 6.4.1 yields

$$\begin{aligned}-\tfrac{1}{2}\int_1^2 e^{-x^2}(-2x)\,dx &= -\tfrac{1}{2}\int_{-1}^{-4} e^y\,dy = \tfrac{1}{2}\int_{-4}^{-1} e^y\,dy \\ &= \tfrac{1}{2}e^y\,\Big|_{-4}^{-1} = \tfrac{1}{2}(e^{-1} - e^{-4}).\end{aligned}$$

Again, we have used Theorem 6.3.8 (Cauchy's Fundamental Theorem) with $G(y) = e^y$, an antiderivative of f. ●

6.4.1 Integration by Parts

The technique of integration by parts is an easy consequence of the formula for the derivative of a product (part (iii) of Theorem 4.1.2) combined with Cauchy's Fundamental Theorem of Calculus (Theorem 6.3.8).

THEOREM 6.4.2 Integration by Parts Let f and g be continuously differentiable on $[a, b]$. Then

$$\int_a^b f(x)g'(x)\,dx = f(x)g(x)\Big|_a^b - \int_a^b g(x)f'(x)\,dx.$$

Proof. Since f, f', g, and g' are continuous on $[a, b]$, we are guaranteed that the two integrals in the theorem exist. Furthermore, for all x in $[a, b]$,

$$(fg)'(x) = f'(x)g(x) + f(x)g'(x)$$

or

$$f(x)g'(x) = (fg)'(x) - g(x)f'(x) \tag{6.14}$$

The function fg is an antiderivative for $(fg)'$ and thus by Theorem 6.3.8,

$$\int_a^b (fg)'(x)\,dx = f(x)g(x)\Big|_a^b = f(b)g(b) - f(a)g(a).$$

Integrating (6.14) from a to b and substituting yields

$$\begin{aligned}\int_a^b f(x)g'(x)\,dx &= \int_a^b (fg)'(x)\,dx - \int_a^b g'(x)f(x)\,dx\\ &= f(x)g(x)\Big|_a^b - \int_a^b g'(x)f(x)\,dx,\end{aligned}$$

as claimed in the theorem. ●

EXAMPLE 11 To evaluate $\int_0^\pi e^x \sin x\,dx$, integrate by parts by letting $f(x) = e^x$, $f'(x) = e^x$, $g'(x) = \sin x$, $g(x) = -\cos x$. We obtain

$$\int_0^\pi e^x \sin x\,dx = -e^x \cos x\Big|_0^\pi + \int_0^\pi e^x \cos x\,dx.$$

Integrating this latter integral by parts again with $f(x) = e^x$, $f'(x) = e^x$, $g'(x) = \cos x$, and $g(x) = \sin x$ yields

$$\int_0^\pi e^x \sin x\,dx = [-e^x \cos x + e^x \sin x]\Big|_0^\pi - \int_0^\pi e^x \sin x\,dx.$$

Transposing this last integral to the left-hand side and solving for it gives

$$\int_0^\pi e^x \sin x\,dx = \frac{1}{2}e^x[\sin x - \cos x]\Big|_0^\pi = \frac{1}{2}(e^\pi + 1).$$

This example warns you that more than one application of the technique may be required to evaluate an integral. ●

EXAMPLE 12 To evaluate $\int_0^{\pi/2} x \sin x \, dx$, let $f(x) = x$ and $g'(x) = \sin x$. Then $f'(x) = 1$ and $g(x) = -\cos x$ and the method of integration by parts yields

$$\int_0^{\pi/2} x \sin x \, dx = -x \cos x \Big|_0^{\pi/2} - \int_0^{\pi/2} (-\cos x) \, dx$$
$$= [-x \cos x + \sin x]\Big|_0^{\pi/2} = 1.$$

It is worth noting in this example that, if we had chosen $f(x) = \sin x$ and $g'(x) = x$, then $f'(x) = \cos x$ and $g(x) = x^2/2$. With these choices, integration by parts yields

$$\int_0^{\pi/2} x \sin x \, dx = \tfrac{1}{2}(x^2 \cos x) \Big|_0^{\pi/2} - \tfrac{1}{2} \int_0^{\pi/2} x^2 \cos x \, dx.$$

While this is a valid equation, it does not help us evaluate the integral; the integral on the right is more complicated than the original problem. ●

If our goal is to evaluate the integral, our choices of f and g' in the second alternative solution in Example 12 lead to a dead end. However, if our intent is to use the integral as a tool to discover principles describing the behavior of some function f, such choices may be exactly the right ones.

Suppose, for example, that f is twice continuously differentiable on an interval $[a, b]$ and that x_0 and x are two distinct points in $[a, b]$. We know by Corollary 6.3.9 that $\int_{x_0}^x f'(t) \, dt = f(x) - f(x_0)$, or, equivalently, that

$$f(x) = f(x_0) + \int_{x_0}^x f'(t) \, dt. \tag{6.15}$$

However, if we integrate the same integral by parts, letting $F(t) = f'(t)$ and $G'(t) = 1$, we have $F'(t) = f''(t)$ and we can choose $G(t) = (t - x)$, an antiderivative of the function 1. With these choices,

$$\int_{x_0}^x f'(t) \, dt = f'(t)(t - x)\Big|_{x_0}^x - \int_{x_0}^x (t - x) f''(t) \, dt$$
$$= -f'(x_0)(x_0 - x) + \int_{x_0}^x (x - t) f''(t) \, dt$$
$$= f'(x_0)(x - x_0) + \int_{x_0}^x (x - t) f''(t) \, dt.$$

Substituting in (6.15), we see that

$$f(x) = f(x_0) + f'(x_0)(x - x_0) + \int_{x_0}^x (x - t) f''(t) \, dt.$$

This last equation is reminiscent of Taylor's theorem (which is exactly our point) and was obtained by using the sort of choices while integrating by parts that led to a dead end in Example 12. We generalize this procedure in the following theorem.

THEOREM 6.4.3 Taylor's Theorem with Integral Remainder Let f have $k+1$ continuous derivatives on $[a, b]$. Fix x_0 and x in $[a, b]$. Then

$$f(x) = \sum_{j=0}^{k} \frac{f^{(j)}(x_0)}{j!}(x - x_0)^j + \frac{1}{k!}\int_{x_0}^{x} (x - t)^k f^{(k+1)}(t)\,dt.$$

Remark. The polynomial $p_k(x) = \sum_{j=0}^{k} [f^{(j)}(x_0)/j!](x - x_0)^j$ is, of course, the kth Taylor polynomial of f at x_0. The new feature is that the remainder $R_k(x_0; x)$ is now displayed as an integral: $R_k(x_0; x) = (1/k!) \int_{x_0}^{x} (x - t)^k f^{(k+1)}(t)\,dt$.

Proof. We have already established the formula in case $k = 1$. Suppose the formula holds for some j with $1 \leq j < k$. We prove inductively that it must also hold for $j + 1$. Assume, then, that

$$f(x) = p_j(x) + \frac{1}{j!}\int_{x_0}^{x} (x - t)^j f^{(j+1)}(t)\,dt. \tag{6.16}$$

Integrate the last integral by parts. Let $F(t) = f^{(j+1)}(t)$ and $G'(t) = (x - t)^j$. We have $F'(t) = f^{(j+2)}(t)$ and $G(t) = -(x - t)^{j+1}/(j + 1)$. Integrating by parts, we obtain

$$\int_{x_0}^{x} (x - t)^j f^{(j+1)}(t)\,dt$$

$$= \left.\frac{-f^{(j+1)}(t)(x - t)^{j+1}}{j + 1}\right|_{x_0}^{x} + \frac{1}{j + 1}\int_{x_0}^{x} (x - t)^{j+1} f^{(j+2)}(t)\,dt$$

$$= \frac{f^{(j+1)}(x_0)(x - x_0)^{j+1}}{j + 1} + \frac{1}{j + 1}\int_{x_0}^{x} (x - t)^{j+1} f^{(j+2)}(t)\,dt.$$

Substituting in (6.16) proves that

$$f(x) = p_{j+1}(x) + \frac{1}{(j + 1)!}\int_{x_0}^{x} (x - t)^{j+1} f^{(j+2)}(t)\,dt.$$

To establish the theorem, repeat this argument recursively until $j = k$. ●

EXAMPLE 13 If $f(x) = e^{-x}$ and $x_0 = 0$, then $f^{(j)}(x) = (-1)^j e^{-x}$ for all $j = 0, 1, 2, \ldots$. Consequently, by Theorem 6.4.3, $e^{-x} = p_k(x) + R_k(0; x)$ where $p_k(x) = \sum_{j=0}^{k} (-1)^j x^j / j!$ and

$$R_k(0; x) = \frac{(-1)^{k+1}}{k!}\int_{0}^{x} (x - t)^k e^{-t}\,dt.$$

Fix any $r > 0$. We can show easily that the sequence $\{R_k(0; x)\}$ converges to 0 uniformly on $[-r, r]$. Notice that, for $0 \leq |x| \leq r$, $e^{-x} \leq e^r$. Therefore

$$
\begin{aligned}
|R_k(0; x)| &= \frac{1}{k!}\left|\int_0^x (x-t)^k e^{-t}\,dt\right| \\
&\le \frac{e^r}{k!}\left|\int_0^x (x-t)^k\,dt\right| = \frac{e^r}{k!}\,\frac{|x-t|^{k+1}}{k+1}\bigg|_0^x \\
&= \frac{e^r|x|^{k+1}}{(k+1)!} \le \frac{e^r r^{k+1}}{(k+1)!}.
\end{aligned}
$$

Given $\epsilon > 0$, choose k_0 in $\mathbb{N}$ such that, for $k \ge k_0$, $0 < e^r r^{k+1}/(k+1)! < \epsilon$. Thus, for $k \ge k_0$ and for all x in $[-r, r]$, we have $|R_k(0; x)| < \epsilon$. This proves that $\lim_{k\to\infty} R_k(0; x) = 0$ [uniformly]. We conclude that

$$
\lim_{k\to\infty} \sum_{j=0}^{k} \frac{(-1)^j x^j}{j!} = e^{-x} \quad \text{[uniformly]}
$$

on any interval $[-r, r]$. ●

EXAMPLE 14 With the integral form of the remainder for Taylor's theorem in hand, we can now treat the Taylor polynomial for the logarithm function in detail. Let $f(x) = \ln(1+x)$ and $x_0 = 0$. For each j in $\mathbb{N}$,

$$
f^{(j)}(x) = \frac{(-1)^{j-1}(j-1)!}{(1+x)^j}.
$$

Thus the kth Taylor polynomial of $f(x) = \ln(1+x)$ about $x_0 = 0$ is

$$
p_k(x) = \sum_{j=1}^{k} \frac{(-1)^{j-1} x^j}{j}
$$

and the integral form of the remainder, from Theorem 6.4.3, is

$$
R_k(0; x) = (-1)^k \int_0^x \frac{(x-t)^k}{(1+t)^{k+1}}\,dt.
$$

We want to use this form of the remainder to show that, for any positive $r < 1$, $\lim_{k\to\infty} R_k(0; x) = 0$ [uniformly] on the interval $[-r, 1]$. We treat the intervals $[0, 1]$ and $[-r, 0]$ separately.

First suppose that $0 \le x \le 1$. Since the variable t of integration varies over $[0, x]$, we have $0 \le t \le x \le 1$. For such t we have $1 + t \ge 1$ and

$$
\frac{(x-t)^k}{(1+t)^{k+1}} \le (x-t)^k.
$$

We conclude that, for x in $[0, 1]$,

$$
\begin{aligned}
|R_k(0; x)| &= \int_0^x \frac{(x-t)^k}{(1+t)^{k+1}}\,dt \le \int_0^x (x-t)^k\,dt \\
&= -\frac{(x-t)^{k+1}}{k+1}\bigg|_0^x = \frac{x^{k+1}}{k+1} \le \frac{1}{k+1}.
\end{aligned}
$$

Next, suppose that $-1 < -r \le x \le 0$. For $x \le t \le 0$, observe that $0 \le xt < -t$, so that $x + xt < x - t < 0$. Consequently, $|x - t| < |x + xt| = |x||1 + t|$ or $|(x - t)/(1 + t)| < |x|$. Now for any x in $[-r, 0)$, consider

$$\begin{aligned}|R_k(0; x)| &= \left| \int_0^x \left[\frac{x - t}{1 + t}\right]^k \frac{1}{1 + t}\, dt \right| \\ &< |x|^k \left| \int_0^x \frac{1}{1 + t}\, dt \right| = |x|^k |\ln(1 + x)| \\ &< r^k |\ln(1 - r)|.\end{aligned}$$

To confirm the uniform convergence of $R_k(0; x)$ to 0 on $[-r, 1]$, assign any positive value to ϵ. Choose k_1 in $\mathbb{N}$ such that, for all $k \ge k_1$, $1/(k + 1) < \epsilon$. Choose k_2 in $\mathbb{N}$ such that, for all $k \ge k_2$, $r^k |\ln(1 - r)| < \epsilon$. Let k_0 be the larger of k_1 and k_2. For all $k \ge k_0$ and all x in $[-r, 1]$, $|R_k(0; x)| < \epsilon$, proving that $\lim_{k\to\infty} R_k(0; x) = 0$ [uniformly]. We conclude that

$$\lim_{k\to\infty} \sum_{j=1}^{k} \frac{(-1)^{j-1} x^j}{j} = \ln(1 + x) \quad \text{[uniformly]}$$

on any interval $[-r, 1]$ where $0 < r < 1$. ●

6.5 Uniform Convergence and the Integral

Let $\{f_k\}$ be a sequence of functions in $R[a, b]$ that converges to a function f_0 on $[a, b]$. We want to know whether the sequence $\{I_k\}$ of numbers with terms $I_k = \int_a^b f_k(x)\, dx$ also converges. If it does, we can reasonably ask whether

$$\lim_{k\to\infty} \int_a^b f_k(x)\, dx = \int_a^b f_0(x)\, dx.$$

That is, we are asking whether

$$\lim_{k\to\infty} \int_a^b f_k(x)\, dx = \int_a^b \lim_{k\to\infty} f_k(x)\, dx.$$

The essential question is whether the limit inherent in the construction of the integral can be interchanged with the limit of the sequence $\{f_k\}$. As it stands, the problem is not yet well posed; we need to specify the type of convergence at play when we write $\lim_{k\to\infty} f_k = f_0$. Our first example confirms that, if $\{f_k\}$ merely converges *pointwise* to f_0, then our answer can be in the negative.

EXAMPLE 15 For k in $\mathbb{N}$, define

$$f_k(x) = \begin{cases} 2k^2 x, & 0 \le x \le \frac{1}{2}k \\ -2k^2 x + 2k, & \frac{1}{2}k < x \le \frac{1}{k} \\ 0, & \frac{1}{k} < x \le 1. \end{cases}$$

Each function f_k is continuous on [0, 1] and is therefore integrable. First, we claim that $\lim_{k\to\infty} f_k = 0$ [pointwise]. To see this, fix any point x in (0, 1]. Choose $k_0 = k_0(x)$ in $\mathbb{N}$ such that $1/k_0 < x$. Then, for $k \geq k_0(x)$, $f_k(x) = 0$. If $x = 0$, then $f_k(0) = 0$ for all k in $\mathbb{N}$. In all cases, then, $\lim_{k\to\infty} f_k(x) = 0$ [pointwise]. But the convergence is not uniform on [0, 1]. To prove this claim, fix any $\epsilon > 0$ and note that the maximum value of f_k on [0, 1] is k and is assumed at $x = 1/(2k)$. Thus, for every $k \geq \epsilon$,

$$||f_k - f_0||_\infty = ||f_k||_\infty = k \geq \epsilon,$$

proving that $\{f_k\}$ cannot converge uniformly to $f_0 = 0$. (Refer to Fig. 6.29.)

The point of this example is that the sequence $\{I_k\}$, with $I_k = \int_0^1 f_k(x)\,dx$, does not converge to $\int_0^1 f_0(x)\,dx = 0$. In fact, for every k in $\mathbb{N}$, $I_k = \int_0^1 f_k(x)\,dx = 1/2$, as you can easily confirm. This example alone suffices to show that the mere pointwise convergence of a sequence of integrable functions on $[a, b]$ does not guarantee that $\lim_{k\to\infty} \int_a^b f_k(x)\,dx = \int_a^b \lim_{k\to\infty} f_k(x)\,dx$. ●

EXAMPLE 16 However, if $f_k(x) = x^k$ for x in [0, 1], then, as we already know, $\{f_k\}$ converges pointwise but not uniformly to the function f_0 where $f_0(x) = 0$ for x in [0, 1) and $f_0(1) = 1$. Now, $\int_0^1 f_k(x)\,dx = 1/(k+1)$ and it follows that $\lim_{k\to\infty} \int_0^1 f_k(x)\,dx = \int_0^1 f_0(x)\,dx = 0$. This example serves to warn you that pointwise, nonuniform convergence of $\{f_k\}$ *may* be consistent with $\lim_{k\to\infty} \int_a^b f_k(x)\,dx = \int_a^b \lim_{k\to\infty} f_k(x)\,dx$. ●

The question remains: What reasonable conditions on the convergence of $\{f_k\}$ to f_0 suffice to guarantee that $\lim_{k\to\infty} \int_a^b f_k(x)\,dx = \int_a^b f_0(x)\,dx$? One of several answers, accessible to you at this stage of your studies and of use throughout the text, is provided by our next theorem.

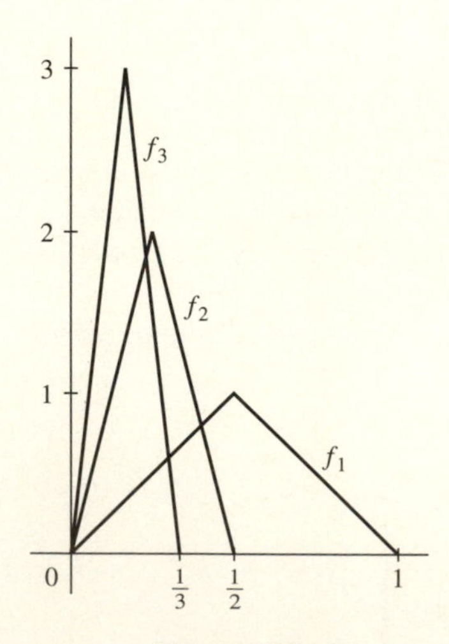

Figure 6.29

THEOREM 6.5.1 If $\{f_k\}$ converges uniformly to f_0 on the compact set $[a, b]$ and if each f_k is integrable on $[a, b]$, then f_0 is also integrable on $[a, b]$. Furthermore,

i) If $F_k(x) = \int_a^x f_k(t)\,dt$, then $\{F_k\}$ converges uniformly to the function $F_0(x) = \int_a^x f_0(t)\,dt$ on $[a, b]$.

ii) In particular, $\lim_{k\to\infty} \int_a^b f_k(x)\,dx = \int_a^b f_0(x)\,dx$.

Proof. Our first task is to prove that f_0 is integrable. To achieve this, we show that Riemann's condition holds for f_0 on $[a, b]$. Fix $\epsilon > 0$. Since $\{f_k\}$ converges uniformly to f_0, we can choose a k_0 in $\mathbb{N}$ such that, when $k \geq k_0$, $||f_k - f_0||_\infty < \epsilon$. Fix any $k \geq k_0$. Because f_k is integrable on $[a, b]$, we know that Riemann's condition holds for f_k on $[a, b]$: There exists a partition π_0 of $[a, b]$ such that, for every partition π that refines π_0, $U(f_k, \pi) - L(f_k, \pi) < \epsilon$. Fix such a partition $\pi = \{x_0, x_1, x_2, \ldots, x_p\}$. For each $j = 1, 2, \ldots, p$, define

$$\begin{aligned} M_j &= \sup\{f_k(x) : x \text{ in } [x_{j-1}, x_j]\} \\ m_j &= \inf\{f_k(x) : x \text{ in } [x_{j-1}, x_j]\} \\ M'_j &= \sup\{f_0(x) : x \text{ in } [x_{j-1}, x_j]\} \\ m'_j &= \inf\{f_0(x) : x \text{ in } [x_{j-1}, x_j]\} \\ M''_j &= \sup\{[f_0 - f_k](x) : x \text{ in } [x_{j-1}, x_j]\} \\ m''_j &= \inf\{[f_0 - f_k](x) : x \text{ in } [x_{j-1}, x_j]\}. \end{aligned}$$

Notice that $-\epsilon < m''_j \leq M''_j < \epsilon$, so that $M''_j - m''_j < 2\epsilon$. Notice also that

$$\begin{aligned} M'_j &= \sup\{[f_0 - f_k](x) + f_k(x) : x \text{ in } [x_{j-1}, x_j]\} \\ &\leq \sup\{[f_0 - f_k](x) : x \text{ in } [x_{j-1}, x_j]\} + \sup\{f_k(x) : x \text{ in } [x_{j-1}, x_j]\} \\ &= M''_j + M_j. \end{aligned}$$

Likewise,

$$\begin{aligned} m'_j &= \inf\{[f_0 - f_k](x) + f_k(x) : x \text{ in } [x_{j-1}, x_j]\} \\ &\geq \inf\{[f_0 - f_k](x) : x \text{ in } [x_{j-1}, x_j]\} + \inf\{f_k(x) : x \text{ in } [x_{j-1}, x_j]\} \\ &= m''_j + m_j. \end{aligned}$$

Therefore $M'_j - m'_j \leq M''_j - m''_j + M_j - m_j < 2\epsilon + M_j - m_j$. Multiply by Δx_j and sum on j to obtain

$$\begin{aligned} U(f_0, \pi) - L(f_0, \pi) &= \sum_{j=1}^{p} (M'_j - m'_j)\Delta x_j < \sum_{j=1}^{p} 2\epsilon\,\Delta x_j + \sum_{j=1}^{p} (M_j - m_j)\Delta x_j \\ &= 2\epsilon(b - a) + U(f_k, \pi) - L(f_k, \pi) < [2(b - a) + 1]\epsilon. \end{aligned}$$

Therefore Riemann's condition holds for f_0 on $[a, b]$. We deduce that f_0 is integrable on $[a, b]$.

The work is behind us; now we can show easily that the sequence $\{F_k\}$ defined by $F_k(x) = \int_a^x f_k(t)\,dt$ converges uniformly to $F_0(x) = \int_a^x f_0(t)\,dt$. For a fixed $\epsilon > 0$, choose k_0 such that, for $k \geq k_0$, $\|f_k - f_0\|_\infty < \epsilon$. For any x in $[a, b]$, compute

$$\begin{aligned} |F_k(x) - F_0(x)| &= \left| \int_a^x f_k(t)\,dt - \int_a^x f_0(t)\,dt \right| \\ &\leq \int_a^x |f_k(t) - f_0(t)|\,dt < \epsilon(x-a) < \epsilon(b-a). \end{aligned}$$

Thus $\lim_{k\to\infty} \|F_k - F_0\|_\infty = 0$ and $\{F_k\}$ converges uniformly to F_0. The final statement of the theorem is obtained by letting $x = b$ and noting that the uniform convergence of $\{F_k\}$ to F_0 implies its pointwise convergence at $x = b$. ●

EXAMPLE 17 Let $f(x) = 1/(1+x^2)$: Let $x_0 = 0$ and let $p_k(x) = \sum_{j=0}^k (-1)^j x^{2j}$ denote the kth Taylor polynomial of f at the point x_0. We know from Exercise 6.85 that $\{p_k\}$ converges uniformly to f on any interval $[-r, r]$ where $0 < r < 1$. For each k in $\mathbb{N}$, and for x in $[-r, r]$, define

$$P_k(x) = \int_0^x p_k(t)\,dt = \sum_{j=0}^k (-1)^j \int_0^x t^{2j}\,dt = \sum_{j=0}^k \frac{(-1)^j x^{2j+1}}{2j+1}.$$

By Theorem 6.5.1, the sequence of polynomials $\{P_k\}$ converges uniformly to the function F_0 defined by $F_0(x) = \int_0^x f(t)\,dt$. That is, the sequence $\{P_k\}$ converges uniformly to the function

$$\int_0^x \frac{1}{1+t^2}\,dt = \tan^{-1} x$$

on $[-r, r]$. Therefore,

$$\lim_{k\to\infty} \sum_{j=0}^k \frac{(-1)^j x^{2j+1}}{2j+1} = \tan^{-1} x \quad \text{[uniformly]}$$

on any interval $[-r, r]$ where $0 < r < 1$. ●

EXAMPLE 18 Let $f(x) = \exp(-x^2/2)$. By an easy modification of Example 13 in Section 6.4, you can show that the sequence of polynomials $p_k(x) = \sum_{j=0}^k (-1)^j x^{2j}/(2^j j!)$ converges uniformly to $\exp(-x^2/2)$ on any interval $[-r, r]$. For x in $[-r, r]$, define

$$P_k(x) = \int_0^x p_k(t)\,dt = \sum_{j=0}^k \frac{(-1)^j}{2^j j!} \int_0^x t^{2j}\,dt = \sum_{j=0}^k \frac{(-1)^j x^{2j+1}}{2^j j!(2j+1)}.$$

By Theorem 6.5.1,

$$\lim_{k\to\infty} \sum_{j=0}^k \frac{(-1)^j x^{2j+1}}{2^j j!(2j+1)} = \int_0^x e^{-t^2/2}\,dt \quad \text{[uniformly]}$$

on any interval $[-r, r]$. By this result we can compute accurate approximations to $(1/\sqrt{2\pi}) \int_0^x \exp(-t^2/2)\,dt$. For a given x, simply evaluate $(1/\sqrt{2\pi}) P_k(x)$ for sufficiently large k. Thus we obtain a numerical approximation to the value of the integral. For example, if $x = 2$ and if we compute successive values of $(1/\sqrt{2\pi})P_k(2)$, we obtain the values in the following table:

k	$(1/\sqrt{2\pi})P_k(2)$	k	$(1/\sqrt{2\pi})P_k(2)$
1	.2659615	8	.4772999
2	.5851153	9	.4772407
3	.4331373	10	.4772514
4	.4922399	11	.4772496
5	.4728972	12	.4772499
6	.4783529	13	.4772498
7	.4770019	14	.4772498

Thus, we can take .4772498 as an approximate value, accurate to seven decimal places, for the integral $(1/\sqrt{2\pi})\int_0^2 \exp(-t^2/2)\,dt$. ●

6.5.1 Another Proof of the Weierstrass Approximation Theorem

Here we present a second proof of the assertion that every continuous function f on $[a, b]$ can be uniformly approximated by a polynomial. We begin with a special case. Assume for the time being that f is a continuous function on $[0, 1]$ and that $f(0) = f(1) = 0$. Our first goal is to find a sequence $\{P_k\}$ of polynomials that converges uniformly to f on $[0, 1]$. Extend f to be defined on all of $\mathbb{R}$ by defining $f(x) = 0$ for x in $[0, 1]^c$. Thus extended, f is uniformly continuous on all of $\mathbb{R}$. Given $\epsilon > 0$, we can choose $\delta > 0$ such that, whenever $|x - y| < \delta$, then $|f(x) - f(y)| < \epsilon/2$. Without loss of generality, $\delta < 1$. Fix such a δ.

We first prove a number of small results, all of which combine eventually to prove Theorem 6.5.2.

LEMMA 1 If $0 < c < 1$, then $\lim_{k\to\infty} \sqrt{k}c^k = 0$.

Proof. By Exercise 1.33, we know that $\lim_{k\to\infty} kc^k = 0$. Therefore, $\lim_{k\to\infty} \sqrt{k}c^k = \lim_{k\to\infty}[kc^k]/\sqrt{k} = 0$. ●

We will have occasion to consider, for each k in $\mathbb{N}$, the integral $\int_{-1}^{1}(1 - x^2)^k\,dx$.

LEMMA 2 $\int_{-1}^{1}(1 - x^2)^k\,dx = \dfrac{2(2^k k!)^2}{(2k+1)!}$.

Proof. We proceed inductively. For $k = 1$, $\int_{-1}^{1}(1 - x^2)\,dx = 4/3 = 2(2^1 1!)^2/3!$. Assume that, for some k in $\mathbb{N}$,

$$\int_{-1}^{1}(1 - x^2)^k\,dx = \frac{2(2^k k!)^2}{(2k+1)!}.$$

Integrate $\int_{-1}^{1}(1-x^2)^{k+1}\,dx$ by parts; let $f(x)=(1-x^2)^{k+1}$ and $g'(x)=1$. We obtain

$$\int_{-1}^{1}(1-x^2)^{k+1}\,dx = x(1-x^2)^{k+1}\Big|_{-1}^{1} + 2(k+1)\int_{-1}^{1} x^2(1-x^2)^k\,dx$$
$$= 2(k+1)\int_{-1}^{1}[(1-x^2)^k-(1-x^2)^{k+1}]\,dx.$$

Consequently,

$$(2k+3)\int_{-1}^{1}(1-x^2)^{k+1}\,dx = 2(k+1)\int_{-1}^{1}(1-x^2)^k\,dx$$
$$= 2(k+1)\left[\frac{2(2^k k!)^2}{(2k+1)!}\right]$$
$$= \frac{2[2^{k+1}(k+1)!]^2}{(2k+2)!}.$$

It follows that $\int_{-1}^{1}(1-x^2)^{k+1}\,dx = 2[2^{k+1}(k+1)!]^2/(2k+3)!$ as required and the lemma is proved by induction. ●

LEMMA 3 For all k in $\mathbb{N}$,

$$\frac{(2k+1)!}{2(2^k k!)^2} < \sqrt{k}.$$

Proof. The inductive proof is straightforward and is left for you. ●

Define a polynomial $p_k(x)$ of degree $2k$ by the formula

$$p_k(x) = \frac{(2k+1)!(1-x^2)^k}{2(2^k k!)^2}.$$

Our next lemma identifies the properties of p_k that are pertinent to our problem.

LEMMA 4

i) $0 \le p_k(x)$ for all x in $[-1, 1]$.

ii) $\int_{-1}^{1} p_k(x)\,dx = 1$.

iii) For all x such that $\delta \le |x| \le 1$, we have

$$p_k(x) < \sqrt{k}(1-\delta^2)^k.$$

iv) $\lim_{k\to\infty} p_k(x) = 0$ [uniformly] on $[-1,-\delta]\cup[\delta,1]$.

Proof. The assertion in (i) is self-evident; (ii) follows from Lemma 2 and the definition of p_k. Part (iii) follows from the fact that, for $\delta \le |x| \le 1$, we have

$$(1 - x^2)^k \le (1 - \delta^2)^k$$

and from Lemma 3.

Only (iv) requires comment. Let $c = 1 - \delta^2$. Then c is in (0, 1) and, by Lemma 1, $\lim_{k\to\infty} \sqrt{k}c^k = 0$. Given $\epsilon > 0$, choose k_0 in $\mathbb{N}$ such that, for $k \ge k_0$, we have $0 < \sqrt{k}c^k < \epsilon$. By part (iii) of this lemma, for any $k \ge k_0$ and for any x in $[-1, -\delta] \cup [\delta, 1]$, we have

$$0 \le p_k(x) < \sqrt{k}(1 - \delta^2)^k = \sqrt{k}c^k < \epsilon.$$

Consequently, $\lim_{k\to\infty} p_k(x) = 0$ [uniformly] for such x. This proves the lemma. ●

We can now define the polynomials P_k that will uniformly approximate f on [0, 1]. For each x in [0, 1], let

$$P_k(x) = \int_{-1}^{1} f(x+t)p_k(t)\,dt.$$

At first glance, it may not be at all apparent that P_k is actually a polynomial, but the following argument reveals this all-important fact.

LEMMA 5 P_k is a polynomial in x of degree no more than $2k$.

Proof. Recall that f has been extended so that $f(x) = 0$ for all x in $[0, 1]^c$. For x in [0, 1], break the integral defining P_k into the sum of three integrals,

$$P_k(x) = \left(\int_{-1}^{-x} + \int_{-x}^{1-x} + \int_{1-x}^{1}\right) f(x+t)p_k(t)\,dt,$$

and consider each of these in turn.

For $-1 \le t \le -x$, we have $x + t \le 0$ and therefore $f(x+t) = 0$. Consequently, $\int_{-1}^{-x} f(x+t)p_k(t)\,dt = 0$, since the integrand vanishes on the interval of integration.

Likewise, for $1 - x \le t \le 1$, we have $x + t \ge 1$ and, again, $f(x+t) = 0$. We conclude that $\int_{1-x}^{1} f(x+t)p_k(t)\,dt = 0$.

These observations allow us to conclude that, for any x in [0, 1],

$$P_k(x) = \int_{-x}^{1-x} f(x+t)p_k(t)\,dt. \tag{6.17}$$

In this last integral, change variables by letting $u = u(t) = x + t$. Then $u(-x) = 0$, $u(1-x) = 1$, and $u'(t) = 1$. By Theorem 6.4.1, the integral (6.17) for P_k is transformed into

$$P_k(x) = \int_0^1 f(u)p_k(u-x)\,du = \frac{(2k+1)!}{2(2^k k!)^2}\int_0^1 f(u)[1-(u-x)^2]^k\,du.$$

Upon expanding the factor $[1-(u-x)^2]^k$ in the integrand and using the linearity of the integral, we deduce [1] that $P_k(x)$ is a polynomial in x of degree no more than $2k$ as claimed in the lemma. ●

THEOREM 6.5.2 The sequence $\{P_k\}$ converges uniformly to f on [0, 1].

Proof. Fix $\epsilon > 0$. By the uniform convergence of $\{p_k\}$ to 0 on $[-1,-\delta] \cup [\delta, 1]$, choose k_0 such that, for $k \ge k_0$,

$$0 \le p_k(t) < \frac{\epsilon}{8\|f\|_\infty}$$

for all t in $[-1,-\delta] \cup [\delta, 1]$. Fix any $k \ge k_0$. For x in [0, 1], write

$$\begin{aligned}
|P_k(x) - f(x)| &= \left| \int_{-1}^{1} f(x+t)p_k(t)\,dt - f(x)\int_{-1}^{1} p_k(t)\,dt \right| \\
&\le \int_{-1}^{1} |f(x+t) - f(x)|p_k(t)\,dt \\
&= \left(\int_{-1}^{-\delta} + \int_{-\delta}^{\delta} + \int_{\delta}^{1} \right) |f(x+t) - f(x)|p_k(t)\,dt.
\end{aligned}$$

We consider each of these three integrals separately. First,

$$\begin{aligned}
\int_{-1}^{-\delta} |f(x+t) - f(x)|p_k(t)\,dt &\le 2\|f\|_\infty \int_{-1}^{-\delta} p_k(t)\,dt \\
&< 2\|f\|_\infty \frac{\epsilon}{8\|f\|_\infty}(1-\delta) < \frac{1}{4}\epsilon. \qquad \textbf{(6.18)}
\end{aligned}$$

Likewise,

$$\int_{\delta}^{1} |f(x+t) - f(x)|p_k(t)\,dt < \frac{1}{4}\epsilon. \qquad \textbf{(6.19)}$$

For $-\delta < t < \delta$, we have $|f(x+t) - f(x)| < \epsilon/2$ and thus

$$\begin{aligned}
\int_{-\delta}^{\delta} |f(x+t) - f(x)|p_k(t)\,dt &< \frac{1}{2}\epsilon \int_{-\delta}^{\delta} p_k(t)\,dt \\
&\le \frac{1}{2}\epsilon \int_{-1}^{1} p_k(t)\,dt = \frac{1}{2}\epsilon. \qquad \textbf{(6.20)}
\end{aligned}$$

---

[1] For example, if $k = 2$,

$$\begin{aligned}
\int_0^1 f(u)[1-(u-x)^2]^2\,du &= \int_0^1 f(u)(1-u^2)^2\,du + 4x\int_0^1 f(u)(u-u^3)\,du \\
&\quad - 2x^2\int_0^1 f(u)(1-3u^2)\,du - 4x^3\int_0^1 f(u)u\,du + x^4\int_0^1 f(u)\,du \\
&= a_0 + a_1x + a_2x^2 + a_3x^3 + a_4x^4.
\end{aligned}$$

Combining (6.18), (6.19), and (6.20), we deduce that, for any x in $[0, 1]$,

$$|P_k(x) - f(x)| < \frac{1}{4}\epsilon + \frac{1}{2}\epsilon + \frac{1}{4}\epsilon = \epsilon.$$

We conclude that $\lim_{k\to\infty} P_k = f$ [uniformly] on $[0, 1]$. ●

THEOREM 6.5.3 The Weierstrass Approximation Theorem The set of all polynomials is uniformly dense in $C([a, b])$.

Proof. First suppose that g is in $C([0, 1])$. For x in $[0, 1]$ define $f(x) = g(x) - g(0) - [g(1) - g(0)]x$. Then f is in $C([0, 1])$ also and $f(0) = f(1) = 0$. Apply Theorem 6.5.2 to f. Given $\epsilon > 0$ we can choose a polynomial P on $[0, 1]$ such that $||f - P||_\infty < \epsilon$. Let

$$Q(x) = P(x) + [g(1) - g(0)]x + g(0).$$

It follows that Q is also a polynomial and $g - Q = f - P$. Thus $||g - Q||_\infty < \epsilon$.

In general, for g in $C([a, b])$, use the transformation $x = a + (b - a)t$, for t in $[0, 1]$, to shift the domain from $[a, b]$ to $[0, 1]$, as in the proof of Theorem 3.5.7, and apply the preceding argument. ●

EXERCISES

6.1. Suppose that f and g are bounded functions on $[a, b]$ such that $f + g$ is in $R[a, b]$. Does it follow that f and g are also in $R[a, b]$? If so, prove it; otherwise provide a counterexample.

6.2. Suppose that f and g are bounded functions on $[a, b]$ such that f and fg are in $R[a, b]$. Does it follow that g is in $R[a, b]$? If so, prove it; otherwise provide a counterexample.

6.3. Let f and g be integrable on $[a, b]$. Prove that $h_1 = \max\{f, g\}$ and $h_2 = \min\{f, g\}$ are also integrable functions on $[a, b]$. (*Hint:* $h_1 = (f - g)^+ + g$.)

6.4. Suppose that f is a bounded function on $[a, b]$ and that c is a point in (a, b) such that $\int_a^b f(x)\,dx$ and $\int_a^c f(x)\,dx$ exist. Prove that $\int_c^b f(x)\,dx$ also exists and that

$$\int_a^b f(x)\,dx = \int_a^c f(x)\,dx + \int_c^b f(x)\,dx.$$

6.5. Suppose that f is in $R[a, b]$ and that there exist m and M such that $0 < m \le |f(x)| \le M$ for all x in $[a, b]$. Prove that $1/f$ is also in $R[a, b]$.

6.6. Suppose that f is a nonnegative function in $R[a, b]$. Prove that $\sqrt{f}$ is also in $R[a, b]$.

6.7. Suppose that g is in $R[a, b]$, that $g([a, b]) \subseteq [c, d]$, and that f is in $R[c, d]$. Does it follow that $f \circ g$ is in $R[a, b]$? *Hint:* Consider the following pair of functions on $[0, 1]$:

$$g(x) = \begin{cases} 1, & x = 0 \\ 1/q, & x = p/q > 0,\ p, q \text{ relatively prime} \\ 0, & x \text{ irrational} \end{cases}$$

and

$$f(y) = \begin{cases} 0, & y = 0 \\ 1, & 0 < y < 1. \end{cases}$$

Identify $f \circ g$ on $[0, 1]$ and determine its integrability.

6.8. Prove that if g is monotone on $[a, b]$ and if f is continuous on $g([a, b])$, then $f \circ g$ is integrable on $[a, b]$.

6.9. Prove that, if g is continuous on $[a, b]$ and if f is integrable on $g([a, b])$, then $f \circ g$ is integrable on $[a, b]$.

6.10. Suppose that f, g, and h are bounded functions on $[a, b]$ such that $f(x) \le g(x) \le h(x)$ for all x in $[a, b]$. Suppose further that f and h are in $R[a, b]$ with

$$\int_a^b f(x)\,dx = \int_a^b h(x)\,dx = I.$$

Prove that g is also in $R[a, b]$ and that $\int_a^b g(x)\,dx = I$.

6.11. Suppose that f is in $R[-a, a]$. Prove that:

a) If f is even, then $\int_{-a}^a f(x)\,dx = 2\int_0^a f(x)\,dx$.

b) If f is odd, then $\int_{-a}^a f(x)\,dx = 0$.

6.12. Suppose that f and g are two continuous functions on $[a, b]$ such that $\int_a^b f(x)\,dx = \int_a^b g(x)\,dx$. Prove that there must exist a c in $[a, b]$ such that $f(c) = g(c)$.

6.13. Suppose that f is a continuous function on $[0, \infty)$ such that, for all $x > 0$, $f(x) > 0$ and $f^2(x) = 2\int_0^x f(t)\,dt$. Prove that $f(x) = x$ for all $x > 0$.

6.14. Suppose that f is continuous on $[a, b]$ and that $\int_a^b f(x)g(x)\,dx = 0$ for all g in $R[a, b]$. Prove that $f(x) = 0$ for all x in $[a, b]$.

6.15. Let f denote the function in Example 2. For x in $[0, 1]$, define $F(x) = \int_0^x f(t)\,dt$. Prove that $F(x) = 0$ for all x in $[0, 1]$.

6.16. Let $f(x)$ denote Heaviside's function

$$f(x) = \begin{cases} 0, & x < 0 \\ 1, & x \ge 0. \end{cases}$$

For each x in $\mathbb{R}$, find $F(x) = \int_0^x f(t)\,dt$. Sketch the graph of F and show that, for $x \ne 0$, $F'(x)$ exists and equals $f(x)$. Show also that $F'(0)$ does not exist.

6.17. Define a function f on $[0, 1]$ by $f(x) = 2x$ for $x \ne 1/2$; define $f(1/2) = 2$. For each x in $[0, 1]$, define $F(x) = \int_0^x f(t)\,dt$. Prove that $F(x) = x^2$ for all x in $[0, 1]$. Show also that $F'(x)$ exists for all x in $[0, 1]$ but that $F'(1/2) \ne f(1/2)$.

6.18. Define a function f on $[0, 1]$ as follows:

$$f(x) = \begin{cases} x, & \text{if } x \text{ is rational} \\ 0, & \text{if } x \text{ is irrational.} \end{cases}$$

Find $L(f)$ and $U(f)$ and determine whether f is in $R[a, b]$.

6.19. Define a function f on $[0, 1]$ as follows:

$$f(x) = \begin{cases} x, & \text{if } x = p/2^q, \text{ for } p, q \text{ in } \mathbb{N} \sim \{0\} \\ 0, & \text{otherwise.} \end{cases}$$

Determine whether f is in $R[0, 1]$. If it is, compute $\int_0^1 f(x)\,dx$; otherwise, prove that f is not in $R[0, 1]$.

6.20. Suppose that f is a bounded function on $[a, b]$ such that $|f|$ is in $R[a, b]$. Does it necessarily follow that f is also in $R[a, b]$? If so, prove it; otherwise provide a counterexample.

6.21. Suppose that f is a bounded function on $[a, b]$ such that either $f^+ = \max\{f, 0\}$ or $f^- = \max\{-f, 0\}$ (but not necessarily both) is in $R[a, b]$. Does it necessarily follow that f is also in $R[a, b]$? If so, prove it; otherwise, provide a counterexample.

6.22. Prove that, if f is in $R[a, b]$, then

$$\int_a^b f(x)\,dx = \int_a^b f^+(x)\,dx - \int_a^b f^-(x)\,dx$$

and

$$\int_a^b |f(x)|\,dx = \int_a^b f^+(x)\,dx + \int_a^b f^-(x)\,dx.$$

6.23. Prove that, for any two bounded functions f and g on $[a, b]$,

$$L(f + g) \le L(f) + L(g) \le U(f) + U(g) \le U(f + g).$$

6.24. Let $f(x) = x$ on $[a, b]$. For each k in $\mathbb{N}$, let π_k denote the uniform partition with gauge $h_k = (b - a)/k$.

a) Compute $L(f, \pi_k)$ and $U(f, \pi_k)$ in closed form by using the identity $\sum_{j=1}^{k} j = k(k + 1)/2$.

b) Use your result in part (a) to show that

$$\lim_{k\to\infty} L(f, \pi_k) = \lim_{k\to\infty} U(f, \pi_k) = (b^2 - a^2)/2.$$

6.25. Let $f(x) = x^2$ on $[a, b]$. For each k in $\mathbb{N}$, let π_k denote the uniform partition with gauge $h_k = (b - a)/k$.

a) Compute $L(f, \pi_k)$ and $U(f, \pi_k)$ in closed form by using the identity $\sum_{j=1}^{k} j^2 = k(k + 1)(2k + 1)/6$.

b) Use your result in part (a) to show that

$$\lim_{k\to\infty} L(f, \pi_k) = \lim_{k\to\infty} U(f, \pi_k) = (b^3 - a^3)/3.$$

6.26. Let $f(x) = x^3$ on $[a, b]$. For each k in $\mathbb{N}$, let π_k denote the uniform partition with gauge $h_k = (b - a)/k$.

a) Compute $L(f, \pi_k)$ and $U(f, \pi_k)$ in closed form by using the identity $\sum_{j=1}^{k} j^3 = [k(k + 1)/2]^2$.

b) Use your result in part (a) to show that

$$\lim_{k\to\infty} L(f, \pi_k) = \lim_{k\to\infty} U(f, \pi_k) = (b^4 - a^4)/4.$$

6.27. Suppose that f is in $R[a, b]$. Prove that there exists a sequence $\{\pi_k\}$ of partitions in $\Pi[a, b]$ such that

i) $\{\|\pi_k\|\}$ converges monotonically to 0.

ii) $\lim_{k\to\infty} L(f, \pi_k) = \lim_{k\to\infty} U(f, \pi_k) = \int_a^b f(x)\,dx$.

6.28. Suppose that f is a bounded function on $[a, b]$ and that there exists a sequence $\{\pi_k\}$ of partitions in $\Pi[a, b]$ such that

i) $\{\|\pi_k\|\}$ converges monotonically to 0.

ii) $\lim_{k\to\infty}[U(f, \pi_k) - L(f, \pi_k)] = 0$.

Prove that f must be integrable on $[a, b]$.

6.29. **a)** Suppose that f is continuous on $[0, 1]$. Prove that

$$\lim_{k\to\infty} \frac{1}{k} \sum_{j=0}^{k-1} f\left(\frac{j}{k}\right) = \lim_{k\to\infty} \frac{1}{k} \sum_{j=1}^{k} f\left(\frac{j}{k}\right) = \int_0^1 f(x)\,dx.$$

b) Suppose that f is continuous on $[a, b]$. For k in $\mathbb{N}$, let $h_k = (b-a)/k$. Prove that

$$\lim_{k\to\infty} \frac{1}{k} \sum_{j=0}^{k-1} f(a + jh_k) = \lim_{k\to\infty} \frac{1}{k} \sum_{j=1}^{k} f(a + jh_k) = \frac{1}{b-a} \int_a^b f(x)\,dx.$$

6.30. Write each of the following sums as Riemann sums for some function f on some interval and compute the indicated limits by computing the corresponding integrals.

a) $\lim_{k\to\infty} \sum_{j=1}^{k} \dfrac{j^2}{k^3}$

b) $\lim_{k\to\infty} \sum_{j=1}^{k} \dfrac{2k}{(k+2j)^2}$

c) $\lim_{k\to\infty} \sum_{j=1}^{k} \dfrac{\pi \sin(j\pi/k)}{k}$

d) $\lim_{k\to\infty} \sum_{j=1}^{k} \dfrac{1}{j+k}$

e) $\lim_{k\to\infty} \sum_{j=1}^{k} \dfrac{k}{(j^2+k^2)}$

f) $\lim_{k\to\infty} \sum_{j=1}^{k} \dfrac{j}{(j^2+k^2)}$

6.31. Suppose that f is monotone increasing on $[a, b]$ and that π_k denotes the uniform partition of $[a, b]$ with gauge $(b-a)/k$. Prove that

$$0 \le U(f, \pi_k) - \int_a^b f(x)\,dx \le [f(b) - f(a)]\frac{b-a}{k}.$$

6.32. Suppose that f is continuously differentiable on $[a, b]$. For any k in $\mathbb{N}$, let $h_k = (b-a)/k$. Prove that

$$\left| \frac{1}{b-a} \int_a^b f(x)\,dx - \frac{1}{k} \sum_{j=1}^{k} f(a + jh_k) \right| \le \|f'\|_\infty h_k.$$

6.33. Suppose that f satisfies a Lipschitz condition of order 1 with Lipschitz constant K at every x in $[0, 1]$. That is, $|f(y) - f(x)| \le K|y - x|$ for all x, y in $[0, 1]$. Prove that f is in $R[0, 1]$ and that, for every k in $\mathbb{N}$,

$$\left| \int_0^1 f(x)\,dx - \frac{1}{k} \sum_{j=1}^{k} f\left(\frac{j}{k}\right) \right| \le \frac{K}{2k}.$$

6.34. Suppose that f is in $R[0, 1]$. Does it necessarily follow that

$$\lim_{k\to\infty} \frac{1}{k} \sum_{j=1}^{k} f\left(\frac{j}{k}\right) = \int_0^1 f(x)\,dx?$$

If so, prove it; otherwise, provide a counterexample.

6.35. Suppose that f is continuous and maps [0, 1] one-to-one onto [0, 1]. Therefore, the inverse function f^{-1} exists. Show that

$$\int_0^1 f(x)\,dx + \int_0^1 f^{-1}(y)\,dy = 1$$

and give a geometric interpretation of this equation.

DEFINITION A step function g on $[a, b]$ is a bounded function on $[a, b]$ that has only finitely many jump discontinuities at $c_0, c_1, c_2, \ldots, c_k$ in $[a, b]$ and is constant on each interval (c_{j-1}, c_j).

6.36. Show that every step function g on $[a, b]$ is in $BV(a, b)$ and hence is integrable. Find a general formula for $\int_a^b g(x)\,dx$ in terms of the points c_j where g has jump discontinuities and the values of g on the intervals (c_{j-1}, c_j).

6.37. Let f be in $R[a, b]$. Given any $\epsilon > 0$, prove that there exist step functions g and h such that

i) $g(x) \le f(x) \le h(x)$, for all x in $[a, b]$.

ii) $0 \le \int_a^b [f(x) - g(x)]\,dx < \epsilon$ and $0 \le \int_a^b [h(x) - f(x)]\,dx < \epsilon$.

6.38. Suppose that f is a bounded function on $[a, b]$ and suppose that, for any $\epsilon > 0$, there exist step functions g and h such that

i) $g(x) \le f(x) \le h(x)$, for all x in $[a, b]$.

ii) $0 \le \int_a^b [h(x) - g(x)]\,dx < \epsilon$.

Prove that f must be in $R[a, b]$.

6.39. Suppose that f is continuous on (a, b) and is bounded on $[a, b]$. Prove that f is in $R[a, b]$.

6.40. Define $f(x) = \sin(1/x)$ for $x \ne 0$ and $f(0) = 0$. Prove that f is in $R[a, b]$ for every interval $[a, b]$.

6.41. Let f be continuously differentiable on $[a, b]$. We know by Theorem 5.3.2 that f is in $BV(a, b)$. Prove that $V(f; a, b) = \int_a^b |f'(x)|\,dx$.

6.42. Let f be a periodic function on $\mathbb{R}$ with period τ that is integrable on $[0, \tau]$. Prove that $\int_x^{x+\tau} f(t)\,dt = \int_0^\tau f(t)\,dt$.

6.43. Compute each of the following integrals, justifying each step.

a) $\int_0^1 \sqrt{x}/[(1 + x)(4 + x)]\,dx$

b) $\int_{-1}^1 (1 + x)/\sqrt{4 - x^2}\,dx$

c) $\int_1^2 (2x + 4)/(x^2 + 3x + 2)\,dx$

d) $\int_{-1}^1 (x + 1)\tan^{-1} x\,dx$

e) $\int_0^1 (x + 1)\ln(x^2 + 1)\,dx$

f) $\int_0^{\pi/2} 1/(a^2 \cos^2 x + b^2 \sin^2 x)\,dx \quad (a, b > 0)$

g) $\int_0^1 x^3 \exp(-x^2/2)\,dx$

h) $\int_0^1 \sqrt{x}/\sqrt{x + 1}\,dx$

6.44. In each case find the derivative $F'(x)$ for suitably restricted values of x.

a) $F(x) = \int_0^x \tan t^2\,dt$

b) $F(x) = \int_0^{x^2} \tan t\,dt$

c) $F(x) = \int_{x^2}^{1} e^{-t}/(1+t)\,dt$

d) $F(x) = \int_0^{\tan x} (1+t)\,dt$

e) $F(x) = \int_1^{\ln x} t^{1/2} e^{-t}\,dt$

f) $F(x) = \int_{-x}^{x} e^{-t} \ln(1+t)\,dt$

6.45. Prove the Cauchy–Schwarz inequality for integrals: If f and g are in $R[a, b]$, then

$$\left[\int_a^b f(x)g(x)\,dx\right]^2 \le \left[\int_a^b f^2(x)\,dx\right]\left[\int_a^b g^2(x)\,dx\right].$$

Hint: Consider, for any t in $\mathbb{R}$, $\int_a^b [tf(x) + g(x)]^2\,dx$ and imitate the proof of Theorem 2.1.2.

6.46. For f in $C([a, b])$, define $||f||_2 = [\int_a^b f^2(x)\,dx]^{1/2}$. Prove that $||\cdot||_2$ is a norm on $C([a, b])$. That is, for all f, g in $C([a, b])$ and all c in $\mathbb{R}$:

i) $||f||_2 \ge 0$; $||f||_2 = 0$ if and only if $f = 0$.

ii) $||cf||_2 = |c|\,||f||_2$.

iii) $||f + g||_2 \le ||f||_2 + ||g||_2$. (Use Exercise 6.45.)

Remark. $||\cdot||_2$ is called the L_2-*norm.* It has properties analogous to those of the Euclidean norm on $\mathbb{R}^n$.

6.47. For f, g in $C([a, b])$, define $d_2(f, g) = ||f - g||_2$. Prove that d_2 is a metric on $C([a, b])$.

Remark. d_2 is called the L_2-*metric* and has properties analogous to those of the Euclidean metric on $\mathbb{R}^n$.

6.48. For f in $C([a, b])$, define $||f||_1 = \int_a^b |f(x)|\,dx$. Prove that $||\cdot||_1$ is a norm on $C([a, b])$.

Remark. $||\cdot||_1$ is called the L_1-*norm.* In general, for $p \ge 1$, $||f||_p = [\int_a^b |f(x)|^p\,dx]^{1/p}$ is also a norm on $C([a, b])$ and is called *the* L_p-*norm.* For our purposes, the norms obtained when $p = 1, 2$ will suffice.

6.49. For f, g in $C([a, b])$, define $d_1(f, g) = ||f - g||_1$. Prove that d_1 is a metric on $C([a, b])$.

Remark. d_1 is called the L_1-*metric* and has properties analogous to those of $||x - y||_1 = \sum_{j=1}^{n} |x_j - y_j|$ on $\mathbb{R}^n$.

6.50. Suppose that f is continuous on $[a, b]$. Prove that

$$\lim_{p\to\infty} \left[\int_a^b |f(x)|^p\,dx\right]^{1/p} = ||f||_\infty.$$

Remark. With this exercise you have proved that the sequence of L_p-norms of f converges to the uniform norm of f.

6.51. Use the trigonometric identities

$$\cos(u + v) = \cos u \cos v - \sin u \sin v$$
$$\cos(u - v) = \cos u \cos v + \sin u \sin v$$
$$\sin(u + v) = \sin u \cos v + \cos u \sin v$$
$$\sin(u - v) = \sin u \cos v - \cos u \sin v$$

to prove the following:

a) For m and n in $\mathbb{N}$,

$$\int_{-\pi}^{\pi} \sin mx \cos nx \, dx = 0.$$

b) For m and n in $\mathbb{N}$ with $m \neq n$,

$$\int_{-\pi}^{\pi} \cos mx \cos nx \, dx = \int_{-\pi}^{\pi} \sin mx \sin nx \, dx = 0.$$

Remark. The functions $\cos mx$, $\sin nx$ are said to be mutually *orthogonal*.

6.52. Use the trigonometric identities

$$\cos^2 u = (1 + \cos 2u)/2$$

and

$$\sin^2 u = (1 - \cos 2u)/2$$

to prove that, for every m in $\mathbb{N}$,

$$\int_{-\pi}^{\pi} \cos^2 mx \, dx = \int_{-\pi}^{\pi} \sin^2 mx \, dx = \pi.$$

6.53. Use the function $x = u(t) = a \cos^2 t + b \sin^2 t$ with a and b nonzero, to change variables and thus evaluate

$$\int_a^b \sqrt{(x-a)(x-b)} \, dx.$$

6.54. **a)** Suppose that f is continuous on $[0, 1]$. Prove that

$$\int_0^{\pi} x f(\sin x) \, dx = \frac{1}{2}\pi \int_0^{\pi} f(\sin x) \, dx.$$

b) Use part (a) to evaluate $\int_0^{\pi} (x \sin x)/(1 + \cos^2 x) \, dx$.

6.55. Let f be continuous on $[0, 1]$. Prove that

$$\int_0^{\pi/2} f(\sin x) \, dx = \int_{\pi/2}^{\pi} f(\sin x) \, dx.$$

Hence prove that $\int_0^{\pi} f(\sin x) \, dx = 2 \int_0^{\pi/2} f(\sin x) \, dx$.

6.56. Prove that, for $|a| \neq |b|$,

$$\lim_{x \to \infty} \frac{1}{x} \int_0^x \sin at \cos bt \, dt = 0.$$

6.57. Let f be twice continuously differentiable on $[a, b]$. Prove that $\int_a^b x f''(x) \, dx = b f'(b) - f(b) + f(a) - a f'(a)$.

6.58. Prove that $\int_0^1 x^m (1-x)^n \, dx = \int_0^1 x^n (1-x)^m \, dx$, for m, n in $\mathbb{N}$.

6.59. Let f be a continuous function on $[a, b]$. Suppose that there exists a positive constant K such that

$$|f(x)| \leq K \int_a^x |f(t)| \, dt,$$

for all x in $[a, b]$. Prove that $f(x) = 0$ for all x in $[a, b]$.

6.60. Let f be a continuous, nonnegative function on $[0, 1]$. Prove that

$$\left[\int_0^1 f(x)\,dx\right]^2 \le \int_0^1 f^2(x)\,dx.$$

6.61. Suppose that f is integrable on $[a, b]$. Define

$$F(\mu) = \int_a^b [f(x) - \mu]^2\,dx.$$

Find the value of μ for which F has a minimum. Prove that your value of μ actually gives a minimum.

6.62. Let f be in $R[a, b]$ and let F be a function with the following property: For every interval $[c, d] \subseteq [a, b]$, there is a ξ in $[c, d]$ such that $F(d) - F(c) = f(\xi)(d - c)$. Prove that $\int_a^b f(x)\,dx = F(b) - F(a)$.

6.63. Define $F(x) = x^2 \sin(1/x)$ for x in $(0, 1]$ and $F(0) = 0$. Prove that $F'(x)$ exists on $[0, 1]$ but F' is not integrable on $[0, 1]$. (*Not every derivative is integrable.*)

6.64. Define a function f on $[0, 2]$ by

$$f(x) = \begin{cases} \dfrac{1}{x+1}, & 0 \le x < 1 \\ \dfrac{1}{x}, & 1 \le x \le 2. \end{cases}$$

a) Sketch the graph of f and explain why f is in $R[a, b]$.

b) For each x in $[0, 2]$, define $F(x) = \int_0^x f(t)\,dt$. Compute $F(x)$ for each x and sketch its graph.

c) Show explicitly that F is continuous at every point in $[0, 2]$ and that $F'(x) = f(x)$ for all $x \ne 1$. Prove that $F'(1)$ fails to exist. (*Not every integrable function is a derivative.*)

6.65. Let f be a bounded, monotone increasing function on $[a, b]$ with a jump discontinuity at a point c in (a, b). Let $F(x) = \int_a^x f(t)\,dt$. Prove that F cannot be differentiable at the point c.

6.66. Prove that, for $x > 0$ and r in $\mathbb{R}$, $\ln x^r = r \ln x$ by showing that $\int_1^{x^r} 1/t\,dt = r \int_1^x 1/y\,dy$.

6.67. Prove that $\lim_{h\to 0} \int_{-a}^{a} h/(h^2 + x^2)\,dx = \pi$.

6.68. Prove that, if f is continuous on $[-1, 1]$, then

$$\lim_{h\to 0} \int_{-1}^{1} \frac{h}{h^2 + x^2} f(x)\,dx = \pi f(0).$$

6.69. Suppose that f is continuous on $[0, 1]$. For each k in $\mathbb{N}$ and for x in $[0, 1]$, define $g_k(x) = f(x^k)$.

a) Prove that each g_k is in $R[0, 1]$.

b) Prove that $\lim_{k\to\infty} \int_0^1 g_k(x)\,dx = f(0)$.

6.70. For x in $[0, 1]$ and k in $\mathbb{N}$, define

$$f_k(x) = kx\,(1 - x)^k.$$

a) Find a function f_0 on $[0, 1]$ such that

$$\lim_{k\to\infty} f_k = f_0 \quad \text{[pointwise]}.$$

b) Determine whether $\{f_k\}$ converges to f_0 uniformly.

c) For each k, compute $\int_0^1 f_k(x)\,dx$ and determine whether

$$\lim_{k\to\infty} \int_0^1 f_k(x)\,dx = \int_0^1 f_0(x)\,dx.$$

6.71. For x in $[-1, 1]$ and k in $\mathbb{N}$, define

$$f_k(x) = \frac{x^{2k}}{1+x^{2k}}.$$

a) Find a function f_0 on $[-1, 1]$ such that $\{f_k\}$ converges pointwise to f_0.

b) Determine whether $\{f_k\}$ converges uniformly to f_0.

c) Use the Squeeze Play to find $\lim_{k\to\infty} \int_{-1}^1 f_k(x)\,dx$.

d) Calculate $\int_{-1}^1 f_0(x)\,dx$ and determine whether

$$\lim_{k\to\infty} \int_{-1}^1 f_k(x)\,dx = \int_{-1}^1 f_0(x)\,dx.$$

6.72. For x in $[0, 1]$ and k in $\mathbb{N}$ define

$$f_k(x) = \frac{kx}{1+kx}.$$

a) Find a function f_0 on $[0, 1]$ such that $\{f_k\}$ converges pointwise to f_0.

b) Determine whether $\{f_k\}$ converges uniformly to f_0.

c) For each k, compute $\int_0^1 f_k(x)\,dx$ and determine whether

$$\lim_{k\to\infty} \int_0^1 f_k(x)\,dx = \int_0^1 f_0(x)\,dx.$$

6.73. For each k in $\mathbb{N}$ and all $x \geq 0$, define $f_k(x) = e^{-kx}/k$.

a) Find a function f_0 on $[0, \infty)$ such that $\{f_k\}$ converges pointwise to f_0.

b) Determine whether the convergence of $\{f_k\}$ to f_0 is uniform.

c) Fix any $r > 0$. For each k in $\mathbb{N}$ and each x in $[0, r]$, compute $\int_0^x f_k(t)\,dt$ and determine whether

$$\lim_{k\to\infty} \int_0^x f_k(t)\,dt = \int_0^x f_0(t)\,dt \quad \text{[uniformly]}$$

for x in the interval $[0, r]$.

6.74. For any fixed $x > 0$, find $\lim_{k\to\infty} \int_0^x \exp(-kt^2/2)\,dt$.

6.75. Inductively evaluate $\lim_{x\to\infty} \int_0^x t^k e^{-t}\,dt$ for each k in $\mathbb{N}$.

6.76. Prove that $\lim_{x\to\infty} \exp(-x^2) \int_0^x \exp(t^2)\,dt = 0$.

6.77. Fix any a in $(0, \pi)$. For each k in $\mathbb{N}$ define $s_k = \int_a^\pi (\sin kx)/kx\,dx$. Prove that $\lim_{k\to\infty} s_k = 0$.

6.78. Find $\lim_{x\to 0} \left[x \int_0^x \exp(-t^2)\,dt\right] / [1 - \exp(-x^2)]$.

6.79. Let f_0 be continuous on $[0, b]$. For k in $\mathbb{N}$ define

$$f_k(x) = \frac{1}{k!} \int_0^x (x-t)^k f_0(t)\,dt.$$

Show that for each k, the kth derivative of f_k exists and equals f_0.

6.80. Let $\{f_k\}$ be a sequence of continuously differentiable functions on $[a, b]$ such that

i) $\lim_{k\to\infty} f_k = f_0$ [pointwise] on $[a, b]$.

ii) $\lim_{k\to\infty} f_k' = g$ [uniformly] on $[a, b]$.

a) Prove that, for x in $[a, b]$,

$$f_0(x) - f_0(a) = \int_a^x g(t)\,dt.$$

b) Prove that $f_0'(x) = g(x)$ for all x in $[a, b]$.

In contrast to the result in Exercise 6.80, we have the following:

6.81. For each k in $\mathbb{N}$ and x in $[0, 1]$, define $f_k(x) = x^k/k$.

a) Show that $\{f_k\}$ converges uniformly to a function f_0 on $[0, 1]$.

b) Show that $\{f_k'\}$ converges pointwise to a function g that differs from the derivative of the function f_0 in part (a).

6.82. For x in $\mathbb{R}$, define two sequences of functions $\{C_k\}$ and $\{S_k\}$ recursively as follows: $C_1(x) = 1$ and

$$S_k(x) = \int_0^x C_k(t)\,dt \qquad \text{and} \qquad C_{k+1}(x) = 1 + \int_0^x S_k(t)\,dt.$$

a) Prove that there exist two functions C and S on $\mathbb{R}$ such that $\{C_k\}$ converges to C and $\{S_k\}$ converges to S.

b) Prove that $C(0) = 1$ and $S(0) = 0$.

c) Prove that $C'(x) = S(x)$ and $S'(x) = C(x)$ for all x in $\mathbb{R}$.

d) Prove that $C''(x) = C(x)$ and $S''(x) = S(x)$ for all x in $\mathbb{R}$.

e) Show that the functions C and S satisfy the identity

$$C^2(x) - S^2(x) = 1$$

for all x in $\mathbb{R}$. [$C(x)$ and $S(x)$ are called the *hyperbolic cosine* and the *hyperbolic sine*, respectively, and are denoted $\cosh x$ and $\sinh x$.]

6.83. Let $\{x_1, x_2, x_3, \ldots\}$ be any sequential listing of the rational numbers in $[0, 1]$. For each k in $\mathbb{N}$, define a function f as follows:

$$f_k(x) = \begin{cases} 1, & \text{if } x = x_j,\ j = 1, 2, 3, \ldots, k \\ 0, & \text{otherwise.} \end{cases}$$

a) Show that each f_k is in $R[0, 1]$ and that $\int_0^1 f_k(x)\,dx = 0$.

b) Show that $f_k \le f_{k+1}$ for all k in $\mathbb{N}$ and that the sequence $\{f_k\}$ converges pointwise to Dirichlet's dancing function of Example 1. Hence conclude that *the pointwise limit of integrable functions may not be integrable.*

6.84. Let $f(x) = 1/(1 + x)$ and $x_0 = 0$.

a) Find the kth Taylor polynomial $p_k(x)$ of f at x_0 and the integral form of the remainder $R_k(0; x)$.

b) Imitate Example 14 in Section 6.4 to show that, for any r in $(0, 1)$, $\lim_{k\to\infty} R_k(0; x) = 0$ [uniformly] on $[-r, r]$.

c) Show that

$$\lim_{k\to\infty} \sum_{j=0}^{k} (-1)^j x^j = \frac{1}{1 + x} \qquad \text{[uniformly]}$$

on any interval $[-r, r]$ where $0 < r < 1$.

6.85. Use Exercise 6.84 to find the kth Taylor polynomial of the function $f(x) = 1/(1+x^2)$ and show that $\lim_{k\to\infty} p_k = f$ [uniformly] on any interval $[-r, r]$ where $0 < r < 1$.

6.86. Use the integral form of the remainder $R_k(0; x)$ for $f(x) = \sin x$ to find a sequence of polynomials that converges uniformly to $\int_0^x [\sin(t)/t]\, dt$ on any compact interval $[-r, r]$. Use this sequence to estimate $\int_0^\pi [\sin(t)/t]\, dt$ with accuracy within 5×10^{-4}.

6.87. Use the integral form of the remainder $R_k(0; x)$ for $f(x) = \cos x$ to find a sequence of polynomials that converges uniformly to $\int_0^x (1 - \cos t)/t\, dt$ on any compact interval $[-r, r]$. Estimate $\int_0^\pi (1 - \cos t)/t\, dt$ with accuracy within 5×10^{-4}.

6.88. Derive *Cauchy's form of the remainder* for Taylor's theorem: If f has $k+1$ continuous derivatives on $[a, b]$ and if x_0 and x are two points in $[a, b]$, then

$$f(x) = p_k(x) + R_k(x_0; x),$$

where $p_k(x)$ is the kth Taylor polynomial of f about x_0 and

$$R_k(x_0; x) = \frac{f^{(k+1)}(c)(x-c)^k(x-x_0)}{k!}$$

for some c between x_0 and x. (*Hint:* Start with the integral form of the remainder. Apply Theorem 6.3.1 by replacing $f(t)$ with $(x-t)^k f^{(k+1)}(t)$ and let $g(t) = 1$.)

6.89. Use the Cauchy form of the remainder for the function $f(x) = \ln(1+x)$ on $(-1, 1]$ to show that $\lim_{k\to\infty} R_k(0; x) = 0$ [uniformly] on $[-r, 1]$, where $0 < r < 1$. (See Exercise 6.88.)

6.90. Let $f(x) = (1+x)^{1/2}$ for x in $[-1, 1]$.

a) Find the kth Taylor polynomial of f about $x_0 = 0$.

b) Find both the integral and the Cauchy form of the remainder $R_k(0; x)$.

c) Show that $\lim_{k\to\infty} R_k(0; x) = 0$ [uniformly] on $[-r, 1]$ for any r in $(0, 1)$.

6.91. Derive the *generalized Cauchy form of the remainder:* Suppose that f has $k+1$ continuous derivatives on $[a, b]$. Let x_0 and x be any two points in $[a, b]$ and let q be any integer between 0 and k. Then $f(x) = p_k(x) + R_k(x_0; x)$, where $p_k(x)$ is the kth Taylor polynomial of f about x_0 and

$$R_k(x_0; x) = \frac{f^{(k+1)}(c)(x-c)^{k-q}(x-x_0)^{q+1}}{(q+1)k!}$$

for some c between x_0 and x. (Note that if $q = 0$, then we obtain the ordinary Cauchy form of the remainder; if $q = k$, we obtain Lagrange's form of the remainder.)

6.92. For any x in $\mathbb{R}$, define

$$F(x) = \frac{1}{\sqrt{2\pi}} \int_0^x e^{-t^2/2}\, dt.$$

(The scalar factor $1/\sqrt{2\pi}$ is simply an adjustment so that the integrand becomes the probability density function for Gauss's standard normal distribution.)

a) Sketch the graph of $f(x) = (1/\sqrt{2\pi}) \exp(-x^2/2)$.

b) Explain why F is continuous and differentiable on $\mathbb{R}$. Find $F'(x)$. Prove that F is strictly monotone increasing on $\mathbb{R}$. Prove that the graph of F is concave down on $(0, \infty)$ and is concave up on $(-\infty, 0)$.

c) Assume that $\lim_{x\to\infty} F(x) = 1/2$. (We will derive this limit in Chapter 13.) Sketch the graph of F.

6.93. For x in $\mathbb{R}$, define $F(x) = \int_0^x t(t-1)\exp(-t^2)\,dt$. Sketch an accurate graph of F, identifying locations of relative maxima and minima, intervals where the graph is monotone, and the limiting behavior at $\pm\infty$. (Assume the validity of the limit given in part (c) of Exercise 6.92.)

6.94. For f in $R[a,b]$, define $F(x) = \int_a^x f(t)\,dt$, for x in $[a,b]$. Define $I(f) = F$. Then I, called the *indefinite integral,* is a linear function mapping $R[a,b]$ into $C([a,b] \cap BV(a,b)$.

a) Prove that I is not one-to-one on $R[a,b]$.

b) Prove that, when restricted to the subset $C([a,b])$ of $R[a,b]$, I is one-to-one. That is, prove that if f_1 and f_2 are in $C([a,b])$ and if $I(f_1) = I(f_2)$, then $f_1 = f_2$.

c) Prove that I is a positive linear function on $R[a,b]$. That is, if f is in $R[a,b]$ and if $f \geq 0$, then $I(f) \geq 0$.

d) The linear function I is said to be *positive definite* if it is positive and if, whenever $I(f) = 0$ for $f \geq 0$, it follows that $f = 0$. Show that I is not positive definite on $R[a,b]$ but that I is positive definite on $C([a,b])$.

e) Prove that, if f is nonnegative and continuous on $[a,b]$, then $F = I(f)$ is nonnegative and monotone increasing on $[a,b]$.

f) Prove that, if f is positive and continuous on $[a,b]$, then $F = I(f)$ is nonnegative and strictly monotone increasing on $[a,b]$.

6.95. **a)** Trace Theorem 6.3.3, the Fundamental Theorem of Calculus, back to Axiom 1.1.1. Identify each intermediate result required along the way to achieve its proof, sketch a flow diagram for the path from the axiom to the theorem, and write an expository essay describing the steps required.

b) Repeat part (a) for Theorem 6.3.8, Cauchy's form of the Fundamental Theorem of Calculus.

c) Repeat part (a) for Theorem 6.4.3, Taylor's theorem with integral form of the remainder.

d) Repeat part (a) for Theorem 6.5.1.

6.96. Consider the following six statements:

p_1 : f is continuous on $[a,b]$.

p_2 : f is uniformly continuous on $[a,b]$.

p_3 : f is differentiable on $[a,b]$.

p_4 : f has an antiderivative on $[a,b]$.

p_5 : f is in $R[a,b]$.

p_6 : f is the indefinite integral of some g in $R[a,b]$.

There are 30 possible sentences of the form $p_i \to p_j$ with $i \neq j$. Which are theorems; which are not? Complete the following table, indicating your answers T or F, for true or false. For example, the statement $p_1 \to p_2$ is a theorem as indicated in the table.

	p_1	p_2	p_3	p_4	p_5	p_6
p_1		T				
p_2						
p_3						
p_4						
p_5						
p_6						

7

The Riemann–Stieltjes Integral

In 1894 T. J. Stieltjes introduced the generalization of the Riemann integral, which now bears his name. This new integral represents a significant advance over Riemann's in that it provides a more flexible tool for modeling physical phenomena. Historically, it also revealed that infinitely many different integrals, not just one, are available for our use. Yet the study of these numerous integrals is unified within one elegant theory. As we will see in the next section, Riemann's integral is subsumed in this theory, representing just one of the choices available to us. The choice of integral for any particular situation is based on the context of the problem.

7.1 Definition of the Riemann–Stieltjes Integral

The material presented here has strong similarities to that in Chapter 6. Many of the techniques and proofs imitate the corresponding manuevers and proofs for the Riemann integral, although the initial starting point for Stieltjes' version of the integral is slightly different. We let f and g be two bounded functions on the interval $[a, b]$ and let $\pi = \{x_1, x_2, x_3, \ldots, x_p\}$ be any partition of $[a, b]$. Let $\Delta g_j = g(x_j) - g(x_{j-1})$ denote the change in g on the interval $[x_{j-1}, x_j]$. Choose any point s_j in each interval $[x_{j-1}, x_j]$ and form the sum

$$S(f, g, \pi) = \sum_{j=1}^{p} f(s_j)\Delta g_j.$$

$S(f, g, \pi)$ is called a *Riemann–Stieltjes sum* for the function f with respect to g. As in the simpler case of Riemann sums, without some restrictions on the functions f and g, the numbers $S(f, g, \pi)$ may vary widely. However, under suitable hypotheses on f and g, the values of $S(f, g, \pi)$ will cluster near some unique number I whenever π is sufficiently refined

As a simple physical example, suppose that a straight, thin wire of *varying density* is laid along the x-axis from a to b. Suppose further that $g(x)$ denotes the total mass of that portion of the wire extending from a to x; Δg_j denotes the mass of the wire in the jth subinterval $[x_{j-1}, x_j]$. (See Fig. 7.1.) To compute, say, the second moment (or the moment of inertia) of the wire about the y-axis, recall that the second moment of a point mass M is $r^2 M$ (where r denotes the radius of rotation of the mass about the axis of rotation). The second moment of the jth segment of the wire is approximately $s_j^2 \Delta g_j$, where s_j is a point in $[x_{j-1}, x_j]$ that measures the approximate distance of the jth segment from the y-axis. The second moment of the entire length of wire is approximated by the sum, $\sum_{j=1}^{p} s_j^2 \Delta g_j$, of the individual second moments. That sum, you will notice, is $S(f, g, \pi)$, with $f(x) = x^2$. As we will see later, the Riemann–Stieltjes theory will enable us to find the second moment exactly as an integral.

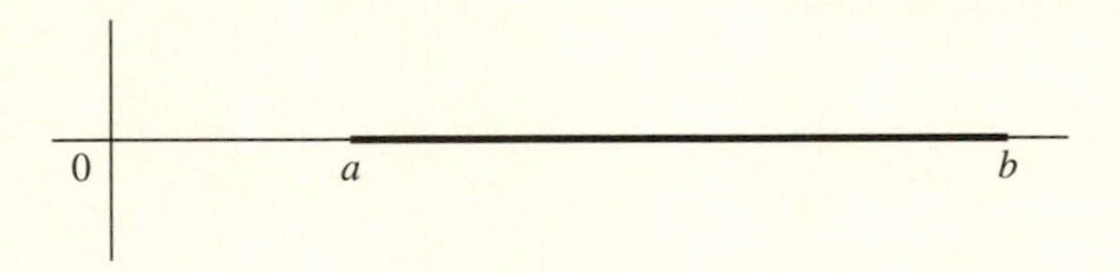

Figure 7.1

DEFINITION 7.1.1 Let f and g be bounded functions on $[a, b]$. For each partition π of $[a, b]$, form $S(f, g, \pi)$. Suppose there exists a number I with the following property:

> *For every $\epsilon > 0$, there exists a partition π_0 of $[a, b]$ such that, for every refinement π of π_0 and for every choice of points s_j in the partition intervals $[x_{j-1}, x_j]$ of π, we have $|S(f, g, \pi) - I| < \epsilon$.*

Then f is said to be (*Riemann–Stieltjes*) *integrable* with respect to g on $[a, b]$; I is called the (*Riemann–Stieltjes*) *integral of f* with respect to g on $[a, b]$ and is denoted

$$I = \int_a^b f(x)\, dg(x). \quad \bullet$$

We call f the *integrand* and g the *integrator*. We denote the collection of all functions f that are integrable with respect to g on $[a, b]$ by $RS[g; a, b]$.

Notice that if $g(x) = x$ on $[a, b]$, then this definition reduces to Definition 6.1.1 and the Riemann–Stieltjes integral of f with respect to g reduces to the Riemann integral of f. For more general integrators g, however, we obtain a substantial generalization of the Riemann integral. Although Riemann–Stieltjes integrals are often more difficult to compute, they enable us to design more realistic models for a variety of physical and mathematical situations.

As usual, we want to identify the elementary properties of this integral. By convention, $\int_a^b f(x)\,dg(x) = -\int_a^b f(x)\,dg(x)$ and $\int_a^a f(x)\,dg(x) = 0$. The following theorem identifies the linear properties of the integral. The proofs consist of straightforward computations as in Theorem 6.1.1 and are left as an exercise.

THEOREM 7.1.1

i) Suppose that f_1 and f_2 are in $RS[g; a, b]$ and that c_1 and c_2 are real numbers. Then $c_1 f_1 + c_2 f_2$ is in $RS[g; a, b]$ and

$$\int_a^b [c_1 f_1(x) + c_2 f_2(x)]\,dg(x) = c_1 \int_a^b f_1(x)\,dg(x) + c_2 \int_a^b f_2(x)\,dg(x).$$

ii) Suppose that f is in both $RS[g_1; a, b]$ and $RS[g_2; a, b]$ and that c_1 and c_2 are real numbers. Then f is also in $RS[c_1 g_1 + c_2 g_2; a, b]$ and

$$\int_a^b f(x)d[c_1 g_1(x) + c_2 g_2(x)] = c_1 \int_a^b f(x)\,dg_1(x) + c_2 \int_a^b f(x)\,dg_2(x).$$

•

As with the Riemann integral, the Riemann–Stieltjes integral is additive over disjoint intervals of integration.

THEOREM 7.1.2 Suppose that f and g are bounded functions on $[a, b]$ and that c is any point in (a, b). If any two of the integrals $\int_a^b f(x)\,dg(x)$, $\int_a^c f(x)\,dg(x)$, and $\int_c^b f(x)\,dg(x)$ exist, then the third integral also exists and

$$\int_a^b f(x)\,dg(x) = \int_a^c f(x)\,dg(x) + \int_c^b f(x)\,dg(x). \quad \bullet$$

The proof is modeled on that of Theorem 6.1.4 and is left as an exercise. By induction on k, you can prove the following generalizations of Theorems 6.1.1 and 6.1.4.

THEOREM 7.1.3

i) If $f_1, f_2, \ldots, f_k$ are bounded functions each of which is in $RS[g; a, b]$ and if $c_1, c_2, \ldots, c_k$ are in $\mathbb{R}$, then $\sum_{j=1}^k c_j f_j$ is also in $RS[g; a, b]$ and

$$\int_a^b \left[\sum_{j=1}^k c_j f_j(x)\right] dg(x) = \sum_{j=1}^k c_j \int_a^b f_j(x)\,dg(x).$$

ii) If $g_1, g_2, \ldots, g_k$ are bounded functions on $[a, b]$, if f is in $\cap_{j=1}^k RS[g_j; a, b]$, and if $c_1, c_2, \ldots, c_k$ are in $\mathbb{R}$, then f is in $RS\left[\sum_{j=1}^k c_j g_j; a, b\right]$ and

$$\int_a^b f(x)d\left[\sum_{j=1}^k c_j g_j\right](x) = \sum_{j=1}^k c_j \int_a^b f(x)\,dg_j(x).$$

iii) If $a = c_0 < c_1 < c_2 < \cdots < c_k = b$ and if any k of the integrals $\int_a^b f(x)\,dg(x)$ and $\int_{c_{j-1}}^{c_j} f(x)\,dg(x)$, $j = 1, 2, \ldots, k$ exist, then the remaining integral also exists and

$$\int_a^b f(x)\,dg(x) = \sum_{j=1}^{k} \int_{c_{j-1}}^{c_j} f(x)\,dg(x). \quad \bullet$$

7.2 Techniques of Integration

We found in Section 6.4 that, in order to evaluate Riemann integrals efficiently, we relied on two techniques—change of variables and integration by parts. Fortunately, both techniques are available to us when faced with a Riemann–Stieltjes integral.

We first consider the method of change of variables. Suppose that f is in $RS[g; a, b]$ and that u is a strictly monotone, continuous function on $[a, b]$. We assume that u is increasing; otherwise, consider $-u$. Let $c = u(a)$, $d = u(b)$. The function u maps the interval $[a, b]$ one-to-one onto the interval $[c, d]$. By Theorem 3.2.8, u has a continuous, strictly increasing inverse $u^{-1} = v$ which maps $[c, d]$ one-to-one onto $[a, b]$. Notice that $v(c) = a$ and $v(d) = b$. It follows that the functions $f \circ v$ and $g \circ v$ are bounded functions on $[c, d]$. Our goal here is to show that $f \circ v$ is in $RS[g \circ v; c, d]$ and that

$$\int_c^d f(v(y))\,dg(v(y)) = \int_a^b f(x)\,dg(x),$$

where $x = v(y)$ for y in $[c, d]$ or, equivalently, $y = u(x)$ for x in $[a, b]$.

Our demonstration will depend on the relationship between partitions in $\Pi[a, b]$ and those in $\Pi[c, d]$. We claim that the functions u and v induce mappings between these two collections of partitions and that those mappings are one-to-one and onto. Let $\pi = \{x_0, x_1, \ldots, x_p\}$ be any partition of $[a, b]$. For $j = 0, 1, 2, \ldots, p$, let $y_j = u(x_j)$. Since we have assumed that u is strictly increasing and maps $[a, b]$ onto $[c, d]$, we have $c = y_0 < y_1 < y_2 < \cdots < y_p = d$. Therefore $\pi' = \{y_0, y_1, y_2, \ldots, y_p\}$ is a partition of $[c, d]$. (See Fig. 7.2.) We will adopt, with no confusion, the convenient notation $\pi' = u(\pi)$. Thus, u maps partitions in $\Pi[a, b]$ to $\Pi[c, d]$. The correspondence between π and $\pi' = u(\pi)$ is one-to-one. Equally important for our purpose, this correspondence is order-preserving: If π_1 is a refinement of π, then $\pi_1' = u(\pi_1)$ is a refinement of $\pi' = u(\pi)$.

Since the inverse function $v = u^{-1}$ is also continuous and strictly monotone increasing on $[c, d]$, it also induces a one-to-one, order-preserving map of $\Pi[c, d]$ onto $\Pi[a, b]$.

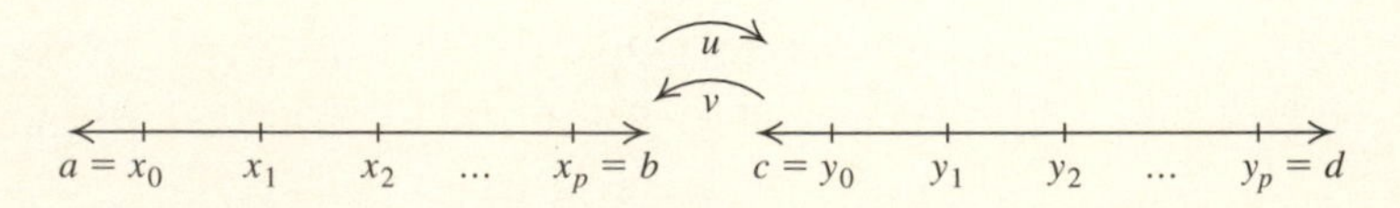

Figure 7.2

THEOREM 7.2.1 If f is in $RS[g; a, b]$ and if u is a strictly monotone, continuous function mapping $[a, b]$ onto $[c, d]$ with inverse $u^{-1} = v$, then $f \circ v$ is in $RS[g \circ v; c, d]$ and

$$\int_c^d (f \circ v)(y) d(g \circ v)(y) = \int_a^b f(x)\, dg(x).$$

Proof. We assume u is strictly monotone increasing. Fix any $\epsilon > 0$. Since f is in $RS[g; a, b]$, we can choose a partition π_0 in $\Pi[a, b]$ such that, for any refinement π of π_0, $|S(f, g, \pi) - \int_a^b f(x)\, dg(x)| < \epsilon$. Let $\pi_0' = u(\pi_0)$. For any refinement $\pi' = \{y_0, y_1, y_2, \ldots, y_p\}$ of π_0', the partition $\pi = v(\pi') = \{x_0, x_1, x_2, \ldots, x_p\}$ is a refinement of π_0. For each $j = 1, 2, \ldots, p$, choose an arbitrary t_j in $[y_{j-1}, y_j]$. Then $s_j = v(t_j)$ is a point in the interval $[x_{j-1}, x_j]$. We compute

$$\begin{aligned} S(f \circ v, g \circ v, \pi') &= \sum_{j=1}^{p} f(v(t_j))[g(v(y_j)) - g(v(y_{j-1}))] \\ &= \sum_{j=1}^{p} f(s_j)[g(x_j) - g(x_{j-1})] = S(f, g, \pi). \end{aligned}$$

Consequently,

$$\left| S(f \circ v, g \circ v, \pi') - \int_a^b f(x)\, dg(x) \right| = \left| S(f, g, \pi) - \int_a^b f(x)\, dg(x) \right| < \epsilon.$$

We have proved that, if $\pi' = \{y_0, y_1, \ldots, y_p\}$ is a refinement of π_0', then, for any choices of t_j in $[y_{j-1}, y_j]$,

$$\left| S(f \circ v, g \circ v, \pi') - \int_a^b f(x)\, dg(x) \right| < \epsilon.$$

That is, we have simultaneously proved that $f \circ v$ is in $RS[g \circ v; c, d]$ and that

$$\int_c^d f(v(y))\, dg(v(y)) = \int_a^b f(x)\, dg(x),$$

as promised by the theorem. ●

As we have seen, integration by parts is an essential tool for our work with integrals. Fortunately, this tool remains available to us when treating Riemann–Stieltjes integrals. In fact, the greater flexibility provided by the Stieltjes theory allows us to relax significantly the hypotheses required. No longer must we rely on the formula for the derivative of a product and the Fundamental Theorem of Calculus as in the proof of Theorem 6.4.2. The relationship implicit in the formula for integration by parts is primitive and does not depend on such sophisticated structure; it reveals a remarkable connection between f and g whenever f is in $RS[g; a, b]$.

THEOREM 7.2.2 Integration by Parts If f is in $RS[g; a, b]$, then g is in $RS[f; a, b]$ and

$$\int_a^b g(x)\,df(x) = f(x)g(x)\Big|_a^b - \int_a^b f(x)\,dg(x),$$

where $f(x)g(x)\Big|_a^b = f(b)g(b) - f(a)g(a)$.

Proof. In part, we must show that there exists a number I such that, if $\epsilon > 0$ is prescribed, there exists a partition π_0 of $[a, b]$ so that, whenever π is a refinement of π_0 and for arbitrary choices of s_j in $[x_{j-1}, x_j]$, $|S(g, f, \pi) - I| < \epsilon$. In fact, the formula given in the theorem alerts us to the required value of I; it will be

$$I = f(b)g(b) - f(a)g(a) - \int_a^b f(x)\,dg(x).$$

Fix any $\epsilon > 0$ and choose a partition π_0 of $[a, b]$ such that, for any refinement π of π_0 and for any choice of s_j in the partition intervals $[x_{j-1}, x_j]$, we have

$$\left| S(f, g, \pi) - \int_a^b f(x)\,dg(x) \right| < \epsilon.$$

Fix any refinement π of π_0. Note first that

$$S(g, f, \pi) = \sum_{j=1}^{p} g(s_j)\Delta f_j = \sum_{j=1}^{p} g(s_j)[f(x_j) - f(x_{j-1})]$$

$$= \sum_{j=1}^{p} g(s_j)f(x_j) - \sum_{j=1}^{p} g(s_j)f(x_{j-1}). \tag{7.1}$$

Note also that, with $a = x_0$ and $b = x_p$, we have, by the telescoping of the right-hand side,

$$f(b)g(b) - f(a)g(a) = \sum_{j=1}^{p} f(x_j)g(x_j) - \sum_{j=1}^{p} f(x_{j-1})g(x_{j-1}). \tag{7.2}$$

Subtracting (7.1) from (7.2) yields

$$f(x)g(x)\Big|_a^b - S(g, f, \pi)$$

$$= \sum_{j=1}^{p} f(x_j)[g(x_j) - g(s_j)] + \sum_{j=1}^{p} f(x_{j-1})[g(s_j) - g(x_{j-1})]. \tag{7.3}$$

Let π_1 denote the partition in $\Pi[a, b]$ having as partition points both the x_j and the $s_j : \pi_1 = \{x_0, s_1, x_1, \ldots, s_p, x_p\}$. Clearly, π_1 is a refinement of π and hence of π_0. Moreover, the two latter summands in (7.3) can be coalesced into a single sum of the form $S(f, g, \pi_1)$, so this sum must be within ϵ of the integral $\int_a^b f(x)\,dg(x)$. It follows from (7.3) that

$$S(f, g, \pi_1) = f(x)g(x)\Big|_a^b - S(g, f, \pi).$$

Therefore

$$\begin{aligned}
\epsilon &> \left| S(f, g, \pi_1) - \int_a^b f(x)\, dg(x) \right| \\
&= \left| f(x)g(x)\Big|_a^b - S(g, f, \pi) - \int_a^b f(x)\, dg(x) \right| \\
&= \left| S(g, f, \pi) - \left\{ f(x)g(x)\Big|_a^b - \int_a^b f(x)\, dg(x) \right\} \right|.
\end{aligned}$$

Consequently,

$$\int_a^b g(x)\, df(x) = f(b)g(b) - f(a)g(a) - \int_a^b f(x)\, dg(x).$$

This proves the theorem. ●

7.3 EXISTENCE OF THE RIEMANN–STIELTJES INTEGRAL

To this point, we have purposely avoided attempting to compute any Riemann–Stieltjes integrals. With only the basic definition and properties available, either the integrals that can be computed are discouragingly simple or the work involved is excessive. But once we develop the tools in this section to address the existence of such integrals, we will also be able to compute increasingly complicated Riemann–Stieltjes integrals. We start simply and work in stages toward greater generality. Note from the outset that the existence and the value of an integral depend on characteristics of both the integrand f and the integrator g.

7.3.1 Step Functions as Integrators

DEFINITION 7.3.1 A bounded function g on $[a, b]$ is a *step function* if g has only finitely many jump discontinuities at $c_0 < c_1 < \cdots < c_k$ and is constant on each interval (c_{j-1}, c_j). ●

Evidently, for a given step function g with jump discontinuities at $c_0, c_1, \ldots, c_k$, $g(c_j^-)$ and $g(c_j^+)$ exist with $g(c_j^-)$ equal to the value of g on (c_{j-1}, c_j) and $g(c_j^+)$ equal to the value of g on (c_j, c_{j+1}). Of course, if $c_0 = a$, then only $g(c_0^+)$ exists; if $c_k = b$, only $g(c_k^-)$ exists. The values of $g(c_j)$ at the points c_j are completely arbitrary. (See Fig. 7.3 on page 306.) We treat the simplest case first.

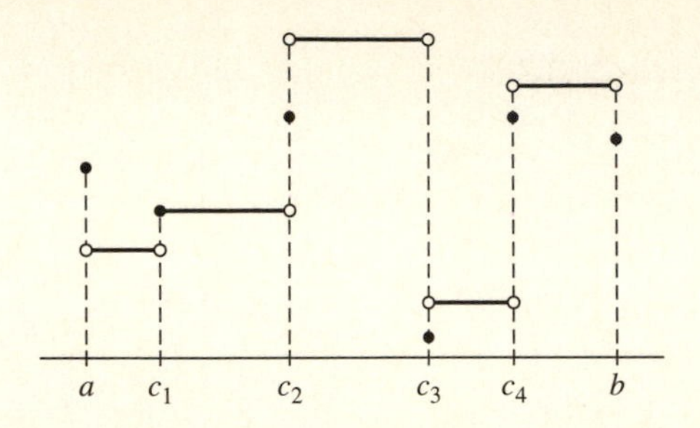

Figure 7.3

LEMMA Suppose that g is a step function on $[a, b]$ with exactly one jump discontinuity at a point c in $[a, b]$. Let f be any bounded function on $[a, b]$.

i) If f is continuous at c, then f is in $RS[g; a, b]$.

ii) If c is in (a, b), if $f(c^-) = f(c)$, and if $g(c^+) = g(c)$, then f is in $RS[g; a, b]$.

iii) If c is in (a, b), if $f(c^+) = f(c)$ and if $g(c^-) = g(c)$, then f is in $RS[g; a, b]$.

For the various cases the value of the integral is:

- If c is in (a, b), then $\int_a^b f(x)\,dg(x) = f(c)[g(c^+) - g(c^-)]$.
- If $c = a$, then $\int_a^b f(x)\,dg(x) = f(a)[g(a^+) - g(a)]$.
- If $c = b$, then $\int_a^b f(x)\,dg(x) = f(b)[g(b) - g(b^-)]$.

Proof. First suppose that c is in (a, b). Let π be any partition of $[a, b]$. If c is not a partition point of π, insert it; thus $c = x_k$ for some k. The only possible contribution to $S(f, g, \pi)$ can come from the intervals $[x_{k-1}, x_k]$ and $[x_k, x_{k+1}]$, since $\Delta g_j = 0$ otherwise. (See Fig. 7.4.) That is,

$$\begin{aligned} S(f, g, \pi) &= f(s_k)\Delta g_k + f(s_{k+1})\Delta g_{k+1} \\ &= f(s_k)[g(c) - g(c^-)] + f(s_{k+1})[g(c^+) - g(c)], \end{aligned}$$

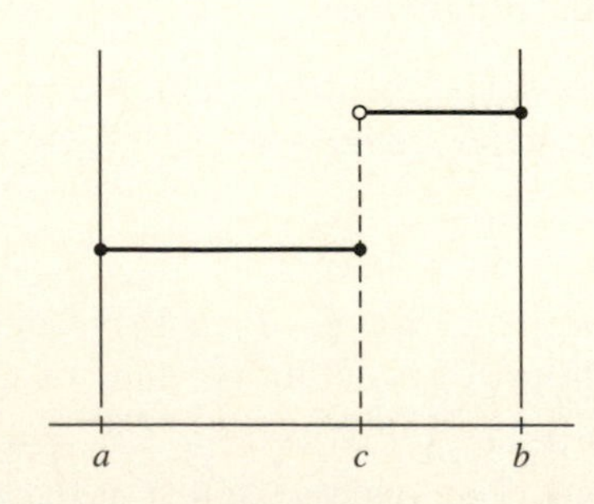

Figure 7.4

where s_k and s_{k+1} are chosen arbitrarily so that

$$x_{k-1} \le s_k \le x_k = c \le s_{k+1} \le x_{k+1}.$$

It follows that

$$\begin{aligned} S(f, g, \pi) &- f(c)[g(c^+) - g(c^-)] \\ &= [f(s_k) - f(c)][g(c) - g(c^-)] \\ &\quad + [f(s_{k+1}) - f(c)][g(c^+) - g(c)]. \end{aligned} \tag{7.4}$$

The assertions of the lemma follow from this identity. To prove part (i), if f is continuous at c, let

$$K = \max\{|g(c) - g(c^-)|, |g(c^+) - g(c)|\} > 0.$$

Given $\epsilon > 0$, choose $\delta > 0$ so that, if x is in $N(c; \delta) \cap [a, b]$, then

$$|f(x) - f(c)| < \frac{\epsilon}{2K}.$$

Choose the gauge of the partition π to be less than this δ. Thus $[x_{k-1}, x_{k+1}] \subset N(c; \delta)$. It follows that s_k and s_{k+1} are in $N(c; \delta)$. Therefore

$$|f(s_k) - f(c)| < \frac{\epsilon}{2K} \qquad \text{and} \qquad |f(s_{k+1}) - f(c)| < \frac{\epsilon}{2K}.$$

From (7.4) we deduce that $|S(f, g, \pi) - f(c)[g(c^+) - g(c^-)]| < \epsilon$. This proves that f is in $RS[g; a, b]$ with

$$\int_a^b f(x)\, dg(x) = f(c)[g(c^+) - g(c^-)].$$

If f is continuous from the left at c and g is continuous from the right at c, then $g(c^+) - g(c) = 0$ and (7.4) reduces to

$$S(f, g, \pi) - f(c)[g(c^+) - g(c^-)] = [f(s_k) - f(c)][g(c) - g(c^-)].$$

The factor $f(s_k) - f(c)$ can be made arbitrarily small by taking the gauge of the partition π to be sufficiently small. Thus

$$|S(f, g, \pi) - f(c)[g(c^+) - g(c^-)]|$$

can be made arbitrarily small by taking the partition π to be sufficiently refined. Again, f is in $RS[g; a, b]$ with

$$\int_a^b f(x)\, dg(x) = f(c)[g(c^+) - g(c^-)].$$

The third assertion of the lemma is proved similarly and is left as an exercise. The case of c as an endpoint is treated by a simple modification of the preceding argument. ●

The proof of the lemma can be extended easily to prove the following theorem.

THEOREM 7.3.1 Suppose that g is a step function on $[a, b]$ with possible jump discontinuities at $a = c_0, c_1, \ldots, c_{k+1} = b$. Suppose also that f is a bounded function on $[a, b]$. If, for each $j = 0, 1, 2, \ldots, k + 1$, either f is continuous at c_j or f is continuous from the left (right) and g is continuous from the right (left) at c_j, then f is in $RS[g; a, b]$ and

$$\int_a^b f(x)\,dg(x) = f(a)[g(a^+) - g(a)]$$

$$+ \sum_{j=1}^{k} f(c_j)[g(c_j^+) - g(c_j^-)] + f(b)[g(b) - g(b^-)]. \quad \bullet$$

One immediate insight to be gained from this theorem is the observation that, in stark contrast to the case of the Riemann integral, changing the value of either the integrand f or the integrator g at even a single point can affect the existence of the integral of f with respect to g. You will have an opportunity to examine this point in the exercises.

7.3.2 Existence Continued: Monotone Integrators

Generalizing the class of integrators one more step, consider now a bounded function g that is monotone on $[a, b]$. We can assume without loss of generality that g is monotone increasing, for if g were decreasing, then $h = -g$ would be increasing and, for every bounded function f and every partition π, $S(f, g, \pi) = -S(f, h, \pi)$. Thus any existence result we obtain for increasing integrators will transfer immediately to the class of decreasing functions. Moreover, working with increasing functions will simplify certain inequalities because, for any partition, $\Delta g_j \geq 0$ for all j.

Many of the proofs of our results for Riemann–Stieltjes integrals with a monotone increasing integrator will duplicate, with only minor modifications, our proofs of the corresponding facts for the Riemann integral; we will leave many of these proofs for you. Often you will need only replace Δx_j with Δg_j and $b - a$ with $g(b) - g(a)$.

Now suppose that f is any bounded function on $[a, b]$ and that $\pi = \{x_0, x_1, x_2, \ldots, x_p\}$ is any partition of $[a, b]$. Again define, for each $j = 1, 2, 3, \ldots, p$,

$$M_j = \sup\{f(x) : x \text{ in } [x_{j-1}, x_j]\}$$

$$m_j = \inf\{f(x) : x \text{ in } [x_{j-1}, x_j]\}.$$

Since g is monotone increasing, $\Delta g_j \geq 0$ for every j and

$$\sum_{j=1}^{p} m_j \Delta g_j \leq \sum_{j=1}^{p} f(s_j) \Delta g_j \leq \sum_{j=1}^{p} M_j \Delta g_j.$$

The numbers $L(f, g, \pi) = \sum_{j=1}^{p} m_j \Delta g_j$ and $U(f, g, \pi) = \sum_{j=1}^{p} M_j \Delta g_j$ are called, respectively, the *lower* and the *upper Riemann–Stieltjes sums of* f with respect to g for the partition π. It is evident that, for every partition π,

$$L(f, g, \pi) \leq S(f, g, \pi) \leq U(f, g, \pi).$$

We will be interested in the difference

$$U(f, g, \pi) - L(f, g, \pi).$$

When this difference can be made arbitrarily small by taking π to be sufficiently refined, $S(f, g, \pi)$, trapped between these two extremes, must be near some single number I. We will conclude that f must be integrable with respect to g on $[a, b]$ and that $\int_a^b f(x)\,dg(x)$ must equal the number I.

The proof of the following theorem closely follows that of Theorem 6.2.1 and is left for you.

THEOREM 7.3.2 Assume g is monotone increasing. If π' is a refinement of π, then $L(f, g, \pi) \leq L(f, g, \pi')$ and $U(f, g, \pi) \geq U(f, g, \pi')$. ●

THEOREM 7.3.3 Assume g is monotone increasing. If π_1 and π_2 are any two partitions, then

$$L(f, g, \pi_1) \leq U(f, g, \pi_2).$$

Proof. The proof is identical with that for Theorem 6.2.2. ●

For bounded functions f and g with g monotone increasing on $[a, b]$, consider the set of all lower sums

$$\mathcal{L} = \{L(f, g, \pi) : \pi \text{ in } \Pi[a, b]\}$$

The set $\mathcal{L}$ is bounded above by any upper sum $U(f, g, \pi)$. Thus by the completeness of $\mathbb{R}$, $\sup \mathcal{L}$ exists. Let

$$L(f, g) = \sup \mathcal{L} = \sup \{L(f, g, \pi) : \pi \text{ in } \Pi[a, b]\}.$$

The number $L(f, g)$ is called the *lower Riemann–Stieltjes integral of* f with respect to g. It is clear that

$$L(f, g) \leq U(f, g, \pi),$$

for every partition π. Likewise, the set

$$\mathcal{U} = \{U(f, g, \pi) : \pi \text{ in } \Pi[a, b]\}$$

is bounded below by $L(f, g)$. Therefore, again by the completeness of $\mathbb{R}$, $\inf \mathcal{U}$ exists. Let

$$U(f, g) = \inf \mathcal{U} = \inf \{U(f, g, \pi) : \pi \text{ in } \Pi[a, b]\}.$$

The number $U(f, g)$ is called the *upper Riemann–Stieltjes integral of* f with respect to g. Clearly, $L(f, g) \leq U(f, g)$.

Let us recapitulate what we know to this point, given any bounded integrand f and any increasing integrator g.

i) The function f is in $RS[g; a, b]$ if there exists a number $I = \int_a^b f(x)\,dg(x)$ with the following property:

For every $\epsilon > 0$, there exists a partition π_0 such that, for every refinement π of π_0, $|S(f, g, \pi) - I| < \epsilon$.

ii) For every partition π of $[a, b]$,

$$L(f, g, \pi) \leq S(f, g, \pi) \leq U(f, g, \pi).$$

iii) $L(f, g) \leq U(f, g)$.

These facts lead to our basic existence theorem for the Riemann–Stieltjes integral with an increasing integrator.

DEFINITION 7.3.2 Riemann's Condition Given bounded functions f and g on $[a, b]$, *with g monotone increasing, Riemann's condition* is said to hold for f with respect to g on $[a, b]$ if, for any $\epsilon > 0$, there exists a partition π_0 of $[a, b]$ such that, for every refinement π of π_0, we have $U(f, g, \pi) - L(f, g, \pi) < \epsilon$. ●

THEOREM 7.3.4 Let f and g be bounded functions on $[a, b]$ with g monotone increasing. The following are equivalent:

i) f is in $RS[g; a, b]$.

ii) Riemann's condition holds for f with respect to g on $[a, b]$.

iii) $L(f, g) = U(f, g)$.

Proof. The proof of Theorem 6.2.4 can be modified in a straightforward way to prove the present theorem. ●

Provided the integrator g is monotone increasing, several useful, order-preserving properties of the integral can be derived. The proofs imitate closely those of the analogous statements of Theorems 6.1.2, 6.1.3, and 6.2.5 and will be left for you. In each case, replace Δx_j by Δg_j and $b - a$ by $g(b) - g(a)$ and proceed as with the earlier proofs.

THEOREM 7.3.5 Suppose that the integrator g is bounded and monotone increasing on $[a, b]$ and that f, f_1, and f_2 are in $RS[g; a, b]$.

i) If $m \leq f(x) \leq M$ for all x in $[a, b]$, then

$$m[g(b) - g(a)] \leq \int_a^b f(x)\, dg(x) \leq M[g(b) - g(a)].$$

ii) If $f_1(x) \leq f_2(x)$ for all x in $[a, b]$, then

$$\int_a^b f_1(x)\, dg(x) \leq \int_a^b f_2(x)\, dg(x).$$

iii) If $f(x) \geq 0$ for all x in $[a, b]$, then $\int_a^b f(x)\, dg(x) \geq 0$.

iv) The function $|f|$ is also in $RS[g; a, b]$ and

$$\left| \int_a^b f(x)\, dg(x) \right| \leq \int_a^b |f(x)|\, dg(x).$$

v) The functions $f^+ = \max\{f, 0\}$ and $f^- = \max\{-f, 0\}$ are also in $RS[g; a, b]$. ●

Again, under the assumption that g is monotone increasing, several useful algebraic properties of the integral can be derived and, again, the proofs imitate with only slight modifications those of the corresponding statements of Theorem 6.2.5. The content of our next theorem is summarized by saying that $RS[g; a, b]$ is a commutative ring with multiplicative identity. Statement (iii) of the theorem identifies the units of $RS[g; a, b]$.

THEOREM 7.3.6 Suppose that the integrator g is bounded and monotone increasing on $[a, b]$ and that f, f_1, and f_2 are in $RS[g; a, b]$.

i) The function f^2 is also in $RS[g; a, b]$.

ii) The function $f_1 f_2$ is also in $RS[g; a, b]$.

iii) If there exist m and M such that $0 < m \leq |f| \leq M$, then $1/f$ is also in $RS[g; a, b]$. ●

Finally, provided the integrator g is monotone increasing, we can integrate any function f in $RS[g; a, b]$ over any subinterval $[c, d]$. As we now know from our work with Riemann integrals, this fact is an essential tool for the study of the indefinite integral $F(x) = \int_a^x f(t)\, dg(t)$.

THEOREM 7.3.7 Suppose that the integrator g is bounded and monotone increasing on $[a, b]$ and that f is in $RS[g; a, b]$. Then, for any subinterval $[c, d]$ of $[a, b]$, f is also in $RS[g; c, d]$.

Proof. In light of Theorem 7.1.2, we can merely imitate the proof of the corresponding fact for Riemann integrals [Theorem 6.2.6], replacing Δx_j with Δg_j. We leave the details for you. ●

7.3.3 Existence Continued: Integrators of Bounded Variation

Suppose now that the integrator g is of bounded variation on $[a, b]$ and that f is in $RS[g; a, b]$. By Theorem 5.4.4 we know that we can decompose g into the difference, $g = h - h_1$, of monotone increasing functions; in fact, as we know, we can achieve such a decomposition of g in infinitely many ways. However, we have no guarantee, and, in fact, it is not generally true, that f is also integrable with respect to either h or h_1. But there is one special decomposition of g given by Theorem 5.4.4 and the discussion preceding it that gives us a bridge between monotone integrators and integrators of bounded variation.

Recall that the decomposition developed in Section 5.4 would have us choose for h the total variation of g on the interval $[a, x]$; that is, define $h(a) = 0$ and, for x in $(a, b]$, define $h(x) = V(g; a, x)$. Then we know that h and $h_1 = h - g$ are both monotone increasing on $[a, b]$ and that $g = h - h_1$. Our objective here is to show that, if f is in $RS[g; a, b]$, then f is in $RS[h; a, b]$ also. Once we have proved this fact, it will follow, since $h_1 = h - g$, that f is in $RS[h_1; a, b]$ also. (Refer to part (ii) of Theorem 7.1.1.) Thus, when using an integrator of bounded variation, we can reduce to a monotone increasing integrator, then piece together the separate integrals.

Let us suppose, then, that g is in $BV(a, b)$, that h is the total variation function of g, and that f is integrable with respect to g on $[a, b]$. We want to prove that f is in $RS[h; a, b]$. Fix any $\epsilon > 0$. Choose first a partition π_1 such that, for any refinement $\pi = \{x_0, x_1, x_2, \ldots, x_p\}$ of π_1 and for any choices of s_j in $[x_{j-1}, x_j]$, we have

$$\left| S(f, g, \pi) - \int_a^b f(x)\, dg(x) \right| < \epsilon.$$

Thus, choosing two sets of points s_j and s_j' in $[x_{j-1}, x_j]$, we have

$$\left| \sum_{j=1}^{p} [f(s_j) - f(s_j')] \Delta g_j \right| < 2\epsilon.$$

Since $h(b) = V(g; a, b) = \sup\{\sum_j |\Delta g_j| : \pi \text{ in } \Pi[a, b]\}$, we know there must exist a partition π_2 of $[a, b]$ such that, for any refinement π of π_2, we have $h(b) - \epsilon < \sum_j |\Delta g_j|$. Let $\pi_0 = \pi_1 \vee \pi_2$ and choose any refinement $\pi = \{x_0, x_1, \ldots, x_p\}$ of π_0. Again, let M_j and m_j denote the suprema and the infima, respectively, of f on the intervals $[x_{j-1}, x_j]$. Invoking the usual inequalities, compute

$$\begin{aligned}
\sum_{j=1}^{p} [M_j - m_j](\Delta h_j - |\Delta g_j|) &\le 2\|f\|_\infty \sum_{j=1}^{p} (\Delta h_j - |\Delta g_j|) \\
&= 2\|f\|_\infty \left(h(b) - \sum_{j=1}^{p} |\Delta g_j| \right) \\
&< 2\|f\|_\infty \epsilon.
\end{aligned}$$

Consider also the quantity $\sum_{j=1}^{p} [M_j - m_j]|\Delta g_j|$; we want to confirm that it is also arbitrarily small. Let $J^+(\pi)$ denote the set of those indices j such that $\Delta g_j \ge 0$ and $J^-(\pi)$ denote those such that $\Delta g_j < 0$. If j is in $J^+(\pi)$, choose s_j and s_j' in $[x_{j-1}, x_j]$ such that

$$f(s_j) - f(s_j') > M_j - m_j - \epsilon.$$

If j is in $J^-(\pi)$, choose s_j and s_j' in $[x_{j-1}, x_j]$ such that

$$f(s_j') - f(s_j) > M_j - m_j - \epsilon.$$

Separating the summands, we have

$$\begin{aligned}
\sum_{j=1}^{p} [M_j - m_j]|\Delta g_j| &< \sum_{j \in J^+(\pi)} [f(s_j) - f(s_j')] \Delta g_j \\
&\quad + \sum_{j \in J^-(\pi)} [f(s_j') - f(s_j)] |\Delta g_j| + \sum_{j=1}^{p} \epsilon |\Delta g_j| \\
&= \sum_{j=1}^{p} [f(s_j) - f(s_j')] \Delta g_j + \epsilon \sum_{j=1}^{p} |\Delta g_j| \\
&< 2\epsilon + \epsilon h(b) = [2 + h(b)]\epsilon.
\end{aligned}$$

Finally, combine all these computations:

$$\sum_{j=1}^{p}[M_j - m_j]\Delta h_j = \sum_{j=1}^{p}[M_j - m_j](\Delta h_j - |\Delta g_j|) + \sum_{j-1}^{p}[M_j - m_j]|\Delta g_j|$$
$$< 2||f||_\infty \epsilon + [2 + h(b)]\epsilon = K\epsilon,$$

where $K = 2||f||_\infty + 2 + h(b)$ is a constant. We conclude that, for any refinement π of π_0,

$$U(f, h, \pi) - L(f, h, \pi) < K\epsilon.$$

Therefore Riemann's condition holds for f with respect to h on $[a, b]$. By Theorem 7.3.4, f is in $RS[h; a, b]$, as claimed in the following theorem.

THEOREM 7.3.8 Let g be in $BV(a, b)$. Let f be in $RS[g; a, b]$. If h denotes the total variation function of g and if $h_1 = h - g$, then f is integrable with respect to both h and h_1 on $[a, b]$. ●

Although we cannot expect order-preserving properties to remain valid in the event g has nontrivial negative variation, the purely algebraic properties of Theorem 7.3.6 do persist when the integrator is of bounded variation.

THEOREM 7.3.9 Suppose that the integrator g is in $BV(a, b)$ and that f, f_1, and f_2 are in $RS[g; a, b]$.

i) The function f^2 is in $RS[g; a, b]$ also.

ii) The function $f_1 f_2$ is in $RS[g; a, b]$ also.

iii) If there exist m and M such that $0 < m \leq |f| \leq M$, then $1/f$ also belongs to $RS[g; a, b]$.

iv) If $[c, d]$ is any subinterval of $[a, b]$, then f is integrable with respect to g on $[c, d]$.

Proof. For (i), since f is in $RS[g; a, b]$, we know that f is integrable with respect to the increasing functions h and h_1 as in Theorem 7.3.8. By Theorem 7.3.6, f^2 is integrable with respect to both h and h_1 on $[a, b]$, so f^2 is integrable with respect to $h - h_1 = g$. This argument gives the flavor of the proofs of the remaining claims of the theorem, the details of which are left as an exercise. To prove (iv), appeal to Theorem 7.3.7. We have proved that, when g is of bounded variation on $[a, b]$, $RS[g; a, b]$ is a commutative ring with a multiplicative identity. ●

We are ready, at last, for the fundamental existence theorem for Riemann–Stieltjes integrals. Our result will give merely *sufficient, not necessary*, conditions for the existence of the integral, deferring the question of finding necessary conditions to the more advanced study of Lebesgue integration theory, where it properly belongs.

THEOREM 7.3.10

i) If f is continuous and if g is of bounded variation on $[a, b]$, then f is integrable with respect to g on $[a, b]$.

ii) If f is of bounded variation and if g is continuous on $[a, b]$, then f is integrable with respect to g on $[a, b]$.

Proof. To prove (i), we assume initially that g is monotone increasing. Without loss of generality, we can also assume that $g(a) < g(b)$. Let $K = g(b) - g(a) > 0$. Since f is assumed to be continuous on the compact set $[a, b]$, f is uniformly continuous. Given $\epsilon > 0$, choose $\delta > 0$ such that, whenever s and t are points of $[a, b]$ with $|s - t| < \delta$, it follows that $|f(s) - f(t)| < \epsilon/K$. Now repeat the remainder of the proof of Theorem 6.2.7 with Δx_j replaced by Δg_j to show that Riemann's condition holds for f with respect to g on $[a, b]$. Thus f is in $RS[g; a, b]$.

To complete the proof of (i), let g be any function in $BV(a, b)$ and write $g = h - h_1$, where $h = V_g$ and $h_1 = V_g - g$ are monotone increasing on $[a, b]$. By Theorem 7.3.8, f is integrable with respect to both h and h_1 on $[a, b]$ and so f is integrable with respect to g on $[a, b]$. Furthermore,

$$\int_a^b f(x)\, dg(x) = \int_a^b f(x)\, dh(x) - \int_a^b f(x)\, dh_1(x).$$

For (ii), observe that, since g is continuous and f is of bounded variation on $[a, b]$, part (i) of the theorem implies that g is in $RS[f; a, b]$. The relationship given by the technique of integration by parts (Theorem 7.2.2) implies that f is in $RS[g; a, b]$, proving the theorem. ●

It is worth noting here that, if g is in $BV(a, b)$ and if f is in $RS[g; a, b]$, then we compute the integral of f with respect to g by decomposing the integral into its "continuous" and its "discrete" parts. By Theorem 5.2.3, write $g = g_c + s_g$, where g_c is continuous on $[a, b]$ and s_g is the saltus function associated with g. Recall that s_g has jump discontinuities of the same magnitude and at the same countably many points $\{c_k\}$ as g has. Assuming that f and s_g satisfy the conditions of Theorem 7.3.1, write

$$\begin{aligned}\int_a^b f(x)\, dg(x) &= \int_a^b f(x)\, dg_c(x) + \int_a^b f(x)\, ds_g(x)\\ &= \int_a^b f(x)\, dg_c(x) + f(a)[s_g(a^+) - s_g(a)]\\ &\quad + \sum_j f(c_j)[s_g(c_j^+) - s_g(c_j^-)] + f(b)[s_g(b) - s_g(b^-)].\end{aligned}$$

To evaluate the original integral, we compute $\int_a^b f(x)\, dg_c(x)$ and the discrete summands separately.

Theorem 7.3.10 ensures that, if g is in $BV(a, b)$, then the vector space $RS[g; a, b]$ of all functions integrable with respect to g contains $C([a, b])$. In fact, it contains many more functions, as Theorem 7.3.1 already implies, even if g is merely a step function. If g is continuous and also of bounded variation on $[a, b]$, then, by both statements in Theorem 7.3.10 taken together, $RS[g; a, b]$ contains both $C([a, b])$ and $BV(a, b)$. Consequently, most functions of interest in classical applications are Riemann–Stieltjes integrable with respect to any continuous integrator g of bounded variation.

There is an elegant connection between the Riemann–Stieltjes integral and Riemann's integral when f is continuous and the integrator g is continuously differentiable on $[a, b]$. Thus g' is bounded on $[a, b]$ and it follows, by Theorem 5.3.2, that g is in $BV(a, b)$. Consequently, by Theorem 7.3.10, f is (Riemann–Stieltjes) integrable with respect to g on $[a, b]$. Since fg' is continuous on $[a, b]$, we know that fg' is (Riemann) integrable on $[a, b]$. We will show that the Riemann–Stieltjes integral $I = \int_a^b f(x)\,dg(x)$ reduces, in this situation, to the ordinary Riemann integral $\int_a^b f(x)g'(x)\,dx$.

THEOREM 7.3.11 If f is continuous and g is continuously differentiable on $[a, b]$, then f is in $RS[g; a, b]$ and fg' is in $R[a, b]$. Furthermore,

$$\int_a^b f(x)\,dg(x) = \int_a^b f(x)g'(x)\,dx.$$

Proof. Fix $\epsilon > 0$. The function g' is continuous, hence uniformly continuous, on $[a, b]$. Given $\epsilon > 0$, choose $\delta > 0$ with the property that, whenever $|s - t| < \delta$, it follows that $|g'(s) - g'(t)| < \epsilon$.

Next, choose any partition π_0 such that, for any refinement $\pi = \{x_0, x_1, \ldots, x_p\}$ of π_0 and for any choice of the points s_j in $[x_{j-1}, x_j]$, it follows that

$$|S(f, g, \pi) - I| < \epsilon.$$

Without loss of generality, we may also suppose that the gauge $||\pi_0||$ of π_0 is less than δ. Thus any refinement of π_0 also has gauge less than δ.

Note that, for each $j = 1, 2, \ldots, p$, the Mean Value Theorem applies to g on $[x_{j-1}, x_j]$. Thus for each j, we can choose t_j in (x_{j-1}, x_j) such that $\Delta g_j = g'(t_j)\Delta x_j$. It follows that

$$S(f, g, \pi) = \sum_{j=1}^{p} f(s_j)\Delta g_j = \sum_{j=1}^{p} f(s_j)g'(t_j)\Delta x_j. \tag{7.5}$$

If we use the same partition π and the same choice of the s_j to form a Riemann sum for fg', we obtain

$$S(fg', \pi) = \sum_{j=1}^{p} f(s_j)g'(s_j)\Delta x_j. \tag{7.6}$$

Subtracting (7.5) from (7.6), we have

$$S(fg', \pi) - S(f, g, \pi) = \sum_{j=1}^{p} f(s_j)[g'(s_j) - g'(t_j)]\Delta x_j. \tag{7.7}$$

Since $||\pi|| < \delta$, we have $|s_j - t_j| < \delta$ for $j = 1, 2, \ldots, p$. Thus $|g'(s_j) - g'(t_j)| < \epsilon$. Consequently, by (7.7), we have

$$|S(fg', \pi) - S(f, g, \pi)| < ||f||_\infty \epsilon \sum_{j=1}^{p} \Delta x_j = ||f||_\infty (b - a)\epsilon.$$

Put all this together: For any refinement π of π_0,

$$|S(fg', \pi) - I| \le |S(fg', \pi) - S(f, g, \pi)| + |S(f, g, \pi) - I|$$
$$< ||f||_\infty (b - a)\epsilon + \epsilon = K\epsilon.$$

where $K = ||f||_\infty (b - a) + 1$. Consequently fg' is Riemann integrable and $I = \int_a^b f(x)g'(x)\,dx$. This argument proves the theorem. ●

COROLLARY 7.3.12 If g is continuously differentiable on $[a, b]$ and if f is in $RS[g; a, b]$, then

$$\int_a^b f(x)\,dg(x) = \int_a^b f(x)g'(x)\,dx.$$

Proof. Nowhere in the proof of Theorem 7.3.11 did we require that f be continuous. All we needed was that f be integrable with respect to g to be able to conclude that $S(fg', \pi)$ is arbitrarily near $I = \int_a^b f(x)\,dg(x)$ provided π is sufficiently refined. ●

EXAMPLE 1 Consider a thin, straight wire of length 1 and total mass 1 located along the line segment [1, 2] in the x-axis. Suppose that the total mass of the wire from 1 to x is given by the function $g(x) = -(x - 1)/(x - 3)$, for x in [1, 2]. Thus the wire's density varies, increasing as the point x moves toward the right endpoint. Suppose that the wire were to be rotated about the vertical axis and suppose we want to compute the second moment of that rotation of the wire. As we saw earlier, the second moment is given by the Riemann–Stieltjes integral $\int_1^2 x^2\,dg(x)$. (See Figs. 7.5 and 7.6.) Our task is to evaluate this integral. To this end, notice that the integrator g is continuously differentiable on [1, 2] with $g'(x) = 2/(x - 3)^2$. By Theorem 7.3.11,

$$\int_1^2 x^2\,dg(x) = \int_1^2 x^2 g'(x)\,dx = \int_1^2 \frac{2x^2}{(x-3)^2}\,dx.$$

This latter integral is an ordinary Riemann integral that we can evaluate using a simple change of variables. Upon completing the details, we obtain

$$\int_1^2 x^2\,dg(x) = 11 - 12\ln 2 \doteq 2.6822338.$$

If the density of the wire were to be uniform with total mass still equal 1, then $g(x) = x - 1$ and the second moment is

$$\int_1^2 x^2\,dg(x) = \int_1^2 x^2\,dx = \frac{x^3}{3}\Big|_1^2 = \frac{7}{3} = 2.\overline{3}.$$

Note that in the nonuniform case the second moment is larger than in the uniform case, reflecting the fact that, in the nonuniform case, more of the mass is distributed farther from the axis of rotation.

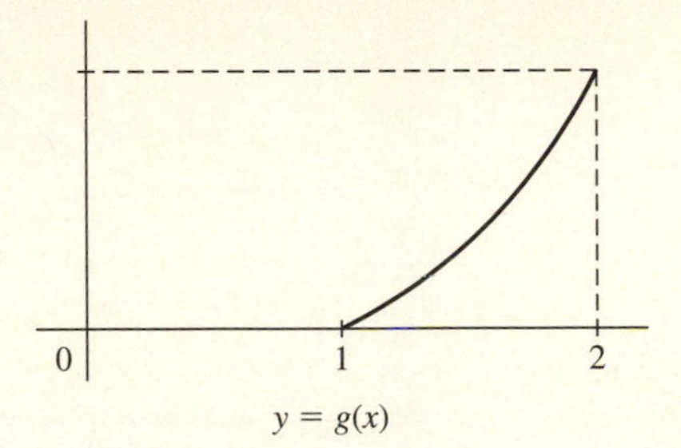

Figure 7.5

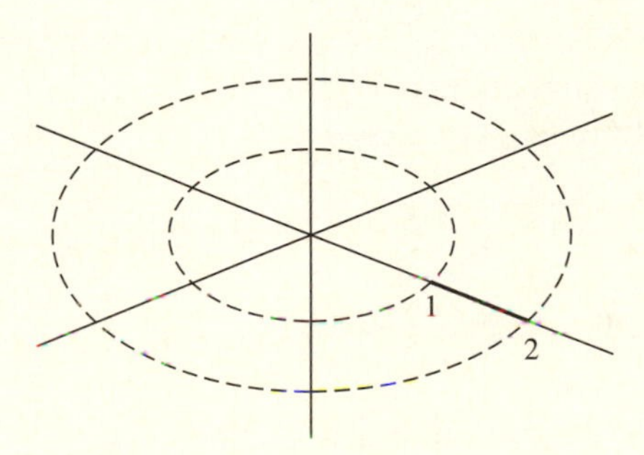

Figure 7.6

EXAMPLE 2 Define a step function g on [2, 12] as follows:

$$g(x) = \begin{cases} \displaystyle\sum_{j=2}^{\lfloor x \rfloor} \frac{(j-1)}{36}, & 2 \le x < 8 \\ \displaystyle\frac{21}{36} + \sum_{j=8}^{\lfloor x \rfloor} \frac{(13-j)}{36}, & 8 \le x \le 12. \end{cases}$$

The graph of g is sketched in Fig. 7.7. If you evaluate g, you might recognize this function as the *cumulative probability distribution function* of a fair pair of dice. The probability of rolling a number $j \le x$ is given by $g(x)$. Notice that g is everywhere continuous from the right.

The integral $\int_a^b x\, dg(x)$ plays a particularly important role in probability theory; it is called the *mean* of the distribution and is denoted μ. To compute μ, we note that $f(x) = x$ is continuous on [2, 12] and that g is a step function with jump discontinuities at $j = 2, 3, \ldots, 12$. By Theorem 7.3.1,

$$\mu = \int_2^{12} x\, dg(x) = \sum_{j=2}^{12} j[g(j^+) - g(j^-)]$$

$$= \frac{2 \cdot 1 + 3 \cdot 2 + 4 \cdot 3 + 5 \cdot 4 + 6 \cdot 5 + 7 \cdot 6 + 8 \cdot 5 + 9 \cdot 4 + 10 \cdot 3 + 11 \cdot 2 + 12 \cdot 1}{36}$$

$$= 7.$$

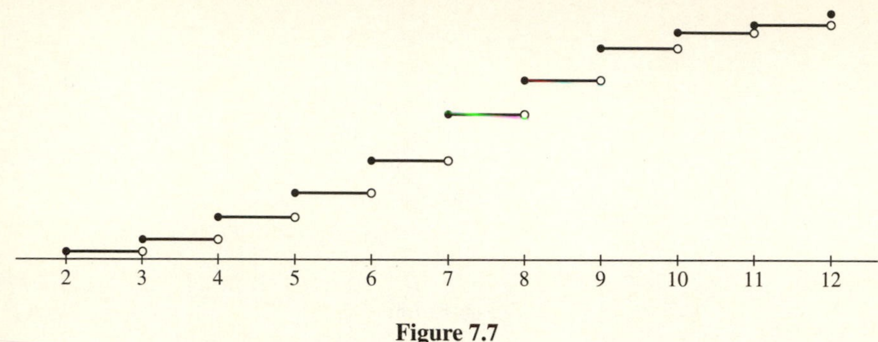

Figure 7.7

Thus the mean, or the *expected value*, of a single roll of a pair of fair dice is $\mu = 7$.

Another significant quantity in probability theory is the *variance*, or the second moment of the distribution about the mean. It is denoted σ^2 and is defined by

$$\sigma^2 = \int_a^b (x-\mu)^2 \, dg(x).$$

In this particular example, we find the variance of the distribution by computing

$$\sigma^2 = \int_2^{12} (x-7)^2 \, dg(x) = \sum_{j=2}^{12} (j-7)^2 [g(j^+) - g(j^-)] = \frac{35}{6}. \quad \bullet$$

EXAMPLE 3 Suppose that the integrator g is defined by

$$g(x) = \begin{cases} x^2, & 0 \le x \le 1 \\ x^2 + 1, & 1 < x \le 2. \end{cases}$$

Then g can be decomposed into the sum $g = g_c + s_g$, where $g_c(x) = x^2$ is the continuous part of g and s_g is the saltus function associated with g defined by

$$s_g(x) = \begin{cases} 0, & 0 \le x \le 1 \\ 1, & 1 < x \le 2. \end{cases}$$

(See Fig. 7.8.)

If we want to evaluate, say, $\int_0^2 e^x \, dg(x)$, we first decompose the integral into the sum of the continuous and the discrete parts:

$$\int_0^2 e^x \, dg(x) = \int_0^2 e^x \, dg_c(x) + \int_0^2 e^x \, ds_g(x).$$

The first of these latter integrals reduces to an ordinary Riemann integral because g_c is continuously differentiable with $g_c'(x) = 2x$. Thus

$$\int_0^2 e^x \, dg_c(x) = \int_0^2 e^x (2x) \, dx = 2(e^2 + 1),$$

where the latter integral is evaluated by integrating by parts.

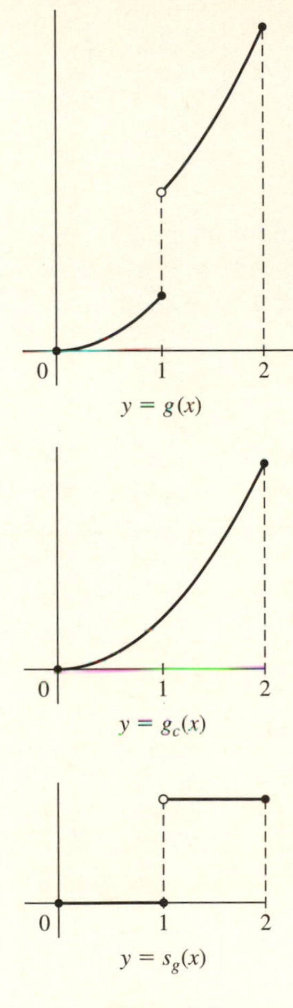

Figure 7.8

The integral $\int_0^2 e^x \, ds_g(x) = e^1[s_g(1^+) - s_g(1^-)] = e$, because the integrator is a step function and the integrand is continuous. Combining these two results, we obtain

$$\int_0^2 e^x \, dg(x) = 2(e^2 + 1) + e.$$

Notice that the original Riemann–Stieltjes integral could have been integrated by parts, using the fact that $f(x) = e^x$ is continuously differentiable on [0, 2] so that Corollary 7.3.12 applies to $\int_0^2 g(x)d(e^x)$. Upon performing the calculations, we obtain

$$\begin{aligned}\int_0^2 e^x \, dg(x) &= e^x g(x)\Big|_0^2 - \int_0^2 g(x)d(e^x)\\ &= [e^2 5 - 1 \cdot 0] - \left[\int_0^1 x^2 e^x \, dx + \int_1^2 (x^2+1)e^x \, dx\right]\end{aligned}$$

$$= 5e^2 - \left[\int_0^2 x^2 e^x dx + \int_1^2 e^x dx\right]$$

$$= 5e^2 - \left[(x^2 - 2x + 2)e^x \Big|_0^2 + e^x \Big|_1^2\right] = 2e^2 + e + 2.$$

EXAMPLE 4 For $x \geq 0$, define an integrator $g(x) = 1 - e^{-x}$. (See Fig 7.9.) Since g is continuously differentiable with derivative $g'(x) = e^{-x}$, the integral $\int_0^x t\, dg(t)$ of $f(t) = t$ with respect to g reduces to an ordinary Riemann integral:

$$F(x) = \int_0^x t\, dg(t) = \int_0^x te^{-t}\, dt.$$

This latter integral can be integrated by parts. For $x \geq 0$, $F(x) = 1 - (x + 1)e^{-x}$. Notice that $\lim_{x\to\infty} F(x) = 1$ and that $F'(x) = xe^{-x} = f(x)g'(x)$.

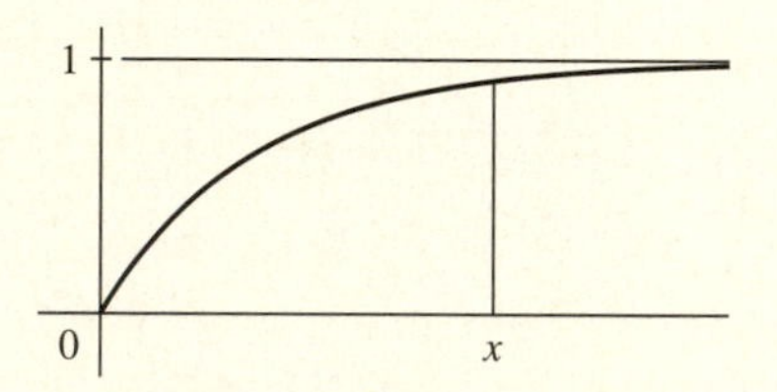

Figure 7.9

7.4 Fundamental Theorems of Riemann–Stieltjes Integration

We begin by deriving *mean value theorems* for Riemann–Stieltjes integrals, theorems that play the same role here as the analogous results in Section 6.3 play in the Riemann theory. These results provide us with a bridge between global and pointwise behavior, comparable to those obtained for Riemann integrals. Riemann–Stieltjes integrals are often difficult to compute exactly; these theorems help us estimate the values of some integrals.

THEOREM 7.4.1 Integral Mean Value Theorem I Let g be monotone increasing on $[a, b]$ and let f be a bounded function in $RS[g; a, b]$. Let $M = \sup\{f(x) : x \text{ in } [a, b]\}$ and $m = \inf\{f(x) : x \text{ in } [a, b]\}$. There exists a K in $[m, M]$ such that

$$\int_a^b f(x)\, dg(x) = K[g(b) - g(a)].$$

If f is continuous on $[a, b]$, then there exists a c in $[a, b]$ such that $f(c) = K$ and

$$\int_a^b f(x)\, dg(x) = f(c)[g(b) - g(a)].$$

Proof. We assume that $g(a) < g(b)$. We have, for any partition π of $[a, b]$,

$$m[g(b) - g(a)] \le L(f, g, \pi) \le U(f, g, \pi) \le M[g(b) - g(a)]$$

and therefore

$$m[g(b) - g(a)] \le \int_a^b f(x)\,dg(x) \le M[g(b) - g(a)].$$

Consequently, the number $K = \int_a^b f(x)\,dg(x)/[g(b) - g(a)]$ is in the interval $[m, M]$. This proves that

$$\int_a^b f(x)\,dg(x) = K[g(b) - g(a)].$$

If f is continuous on $[a, b]$, then the Intermediate Value Theorem guarantees the existence of a c in $[a, b]$ such that $f(c) = K$, proving the second assertion of the theorem. ●

Combining Theorem 7.4.1 with Theorem 7.2.2 (integration by parts) we can easily derive a second form of the Integral Mean Value Theorem.

THEOREM 7.4.2 Integral Mean Value Theorem II Suppose that f is monotone increasing and g is continuous on $[a, b]$. There exists a c in $[a, b]$ such that

$$\int_a^b f(x)\,dg(x) = f(a)[g(c) - g(a)] + f(b)[g(b) - g(c)].$$

Proof. Integrate $\int_a^b f(x)\,dg(x)$ by parts to obtain

$$\int_a^b f(x)\,dg(x) = f(x)g(x)\Big|_a^b - \int_a^b g(x)\,df(x). \tag{7.8}$$

Since f is monotone increasing and g is continuous on $[a, b]$ Theorem 7.4.1 guarantees the existence of a c in $[a, b]$ such that $\int_a^b g(x)\,df(x) = g(c)[f(b) - f(a)]$. Substituting in (7.8), we obtain

$$\begin{aligned}\int_a^b f(x)\,dg(x) &= f(b)g(b) - f(a)g(a) - g(c)[f(b) - f(a)]\\ &= f(a)[g(c) - g(a)] + f(b)[g(b) - g(c)].\end{aligned}$$

This proves the theorem. ●

In Section 6.3 we found that the Fundamental Theorem of Calculus (for Riemann integrals) is an essential analytic tool for the elaboration and application of the theory. A similar theorem plays the same role in the Riemann–Stieltjes theory of integration.

Assume that the integrator g is in $BV(a, b)$ and that f is in $RS[g; a, b]$. By part (iv) of Theorem 7.3.9, we know that f is integrable with respect to g on any subinterval of $[a, b]$. Consequently, for any x in $[a, b]$, we can define the indefinite integral of f with respect to g by

$$F(x) = \int_a^x f(t)\,dg(t).$$

The Fundamental Theorem identifies the critical properties of the function F in terms of those of the integrand f and the integrator g.

THEOREM 7.4.3 The Fundamental Theorem of Riemann–Stieltjes Integration Suppose that g is in $BV(a, b)$ and that f is in $RS[g; a, b]$. For x in $[a, b]$, define

$$F(x) = \int_a^x f(t)\, dg(t).$$

Then

i) F is in $BV(a, b)$.

ii) If g is continuous at c, then F is continuous at c.

iii) If g is monotone and is differentiable at a point c and if f is continuous at c, then F is also differentiable at c and $F'(c) = f(c)g'(c)$.

Proof. We assume first that g is monotone increasing. Suppose that x and y are any two distinct points in $[a, b]$; let I denote the closed interval with x and y as endpoints. By Theorem 7.4.1, applied to f on the interval I, there is a K between $m = \inf\{f(t) : t \text{ in } I\}$ and $M = \sup\{f(t) : t \text{ in } I\}$ such that

$$F(y) - F(x) = \int_x^y f(t)\, dg(t) = K[g(y) - g(x)]. \tag{7.9}$$

With this basic relationship in mind, we first prove part (i). Let $\pi = \{x_0, x_1, x_2, \ldots, x_p\}$ be any partition of $[a, b]$. Computing the variation of F determined by this partition, we have

$$\sum_{j=1}^{p} |\Delta F_j| = \sum_{j=1}^{p} |F(x_j) - F(x_{j-1})| = \sum_{j=1}^{p} |K_j|[g(x_j) - g(x_{j-1})]$$

$$\leq ||f||_\infty \sum_{j=1}^{p} \Delta g_j = ||f||_\infty[g(b) - g(a)].$$

Thus,

$$V(F, a, b) = \sup\left\{\sum_j |\Delta F_j| : \pi \text{ in } \Pi[a, b]\right\}$$

$$\leq ||f||_\infty[g(b) - g(a)].$$

We conclude that F is of bounded variation on $[a, b]$.

Return to (7.9), set $x = c$, and suppose that g is continuous at c. Given $\epsilon > 0$, choose $\delta > 0$ such that, if y is in $N(c; \delta) \cap [a, b]$, then $|g(y) - g(c)| < \epsilon/||f||_\infty$. Consequently, for such y, $|F(y) - F(c)| < \epsilon$. Thus, F is also continuous at c.

Finally, suppose that c is a point in $[a, b]$ where g' exists and where f is continuous. For any $\epsilon > 0$, we can choose a $\delta_1 > 0$ such that, if y is in $N'(c; \delta_1) \cap [a, b]$, then

$$\left|\frac{g(y) - g(c)}{y - c} - g'(c)\right| < \epsilon.$$

Also, by the continuity of f at c, we can choose a $\delta_2 > 0$ such that, if y is in $N(c; \delta_2) \cap [a, b]$, then $|f(y) - f(c)| < \epsilon$.

Let $\delta = \min\{\delta_1, \delta_2\}$ and fix any y in $N'(c; \delta) \cap [a, b]$. Let m and M denote the infimum and the supremum of f on the interval with endpoints c and y. In (7.9), set $x = c$ and note that $f(c) - \epsilon < m \le K \le M < f(c) + \epsilon$ with K as in (7.9). Thus $|K - f(c)| < \epsilon$. Finally, begin with (7.9) and compute:

$$\begin{aligned}\left|\frac{F(y) - F(c)}{y - c} - f(c)g'(c)\right| &= \left|\frac{K[g(y) - g(c)]}{y - c} - f(c)g'(c)\right| \\ &\le |K| \left|\frac{g(y) - g(c)}{y - c} - g'(c)\right| + |K - f(c)|\,|g'(c)| \\ &< ||f||_\infty \epsilon + \epsilon|g'(c)| = (||f||_\infty + |g'(c)|)\epsilon.\end{aligned}$$

Since $||f||_\infty + |g'(c)|$ is a positive constant independent of y, it follows that $\lim_{y \to c}[F(y) - F(c)]/(y - c) = f(c)g'(c)$. This proves that F is differentiable at c and that $F'(c) = f(c)g'(c)$, as promised. (Note Example 4 in Section 7.3.)

To complete the proof, let g be any function of bounded variation. Write $g = g_1 - g_2$ as in Theorem 7.3.8, where $g_1 = V_g$ and $g_2 = V_g - g$ are monotone increasing. Define

$$F_1(x) = \int_a^x f(t)\, dg_1(t)$$

and

$$F_2(x) = \int_a^x f(t)\, dg_2(t).$$

Observe that

$$F(x) = \int_a^x f(t)\, dg(t) = F_1(x) - F_2(x).$$

Both F_1 and F_2 are of bounded variation by our previous discussion. Therefore F is also in $BV(a, b)$.

If g is continuous at c, then, by Theorem 5.5.1, $g_1 = V_g$ is also continuous at c. Further, $g_2 = g_1 - g$ is continuous at c. Since g_1 and g_2 are monotone increasing, our previous argument proves that F_1 and F_2 are also continuous at c, as is $F = F_1 - F_2$. This completes the proof of the theorem. ●

Theorem 7.4.3 enables us to generalize Theorem 7.3.11 in the following way.

THEOREM 7.4.4 Let g be in $BV(a, b)$. For h in $RS[g; a, b]$ and for x in $[a, b]$, define $H(x) = \int_a^x h(t)\, dg(t)$. If f is continuous on $[a, b]$, then f is in $RS[H; a, b]$. Furthermore,

$$\int_a^b f(x)\, dH(x) = \int_a^b f(x)h(x)\, dg(x).$$

Proof. First assume that the integrator g is monotone increasing, that h is in $RS[g; a, b]$, and that f is continuous on $[a, b]$. Construct the function

$$H(x) = \int_a^x h(t)\, dg(t), \qquad x \text{ in } [a, b].$$

Then H is in $BV(a, b)$. Since f is in $C([a, b])$, the integral $\int_a^b f(x)\, dH(x)$ exists. Both f and h are in $RS[g; a, b]$ so the integral $\int_a^b f(x)h(x)\, dg(x)$ also exists.

Fix $\epsilon > 0$. The continuity of f on $[a, b]$ is uniform; we can choose a $\delta > 0$ such that, whenever x and y are in $[a, b]$ with $|x - y| < \delta$, it follows that $|f(x) - f(y)| < \epsilon$. Choose any partition π of $[a, b]$ with gauge less than this δ and choose a point s_j in each of the partition intervals $[x_{j-1}, x_j]$ of π. Note that

$$S(f, H, \pi) = \sum_{j=1}^{p} f(s_j)[H(x_j) - H(x_{j-1})] = \sum_{j=1}^{p} f(s_j) \int_{x_{j-1}}^{x_j} h(x)\, dg(x)$$

and that

$$\int_a^b f(x)h(x)\, dg(x) = \sum_{j=1}^{p} \int_{x_{j-1}}^{x_j} f(x)h(x)\, dg(x).$$

Because g is monotone increasing, we can invoke Theorem 7.3.5 to deduce that

$$\begin{aligned} &\left| S(f, H, \pi) - \int_a^b f(x)h(x)\, dg(x) \right| \\ &\qquad = \left| \sum_{j=1}^{p} \int_{x_{j-1}}^{x_j} f(s_j)h(x)\, dg(x) - \sum_{j=1}^{p} \int_{x_{j-1}}^{x_j} f(x)h(x)\, dg(x) \right| \\ &\qquad \le \sum_{j=1}^{p} \int_{x_{j-1}}^{x_j} |f(s_j) - f(x)|\, |h(x)|\, dg(x). \end{aligned}$$

For each $j = 1, 2, \ldots, p$ and for all x in $[x_{j-1}, x_j]$, we have $|s_j - x| \le ||\pi|| < \delta$. Consequently, $|f(s_j) - f(x)| < \epsilon$ for all x in $[x_{j-1}, x_j]$. Therefore

$$\begin{aligned} \left| S(f, H, \pi) - \int_a^b f(x)h(x)\, dg(x) \right| &< \epsilon \sum_{j=1}^{p} \int_{x_{j-1}}^{x_j} |h(x)|\, dg(x) \\ &= \epsilon \int_a^b |h(x)|\, dg(x) = K\epsilon, \end{aligned}$$

where $K = \int_a^b |h(x)|\, dg(x)$ is a positive constant. It follows that f is in $RS[H; a, b]$ and that

$$\int_a^b f(x)\, dH(x) = \int_a^b f(x)h(x)\, dg(x).$$

To prove the theorem in case g is in $BV(a, b)$, write $g = g_1 - g_2$, where g_1 is the total variation function of g and $g_2 = g_1 - g$. By Theorem 7.3.8, h is integrable with respect to both g_1 and g_2. For x in $[a, b]$, define

$$H_1(x) = \int_a^x h(t)\, dg_1(t)$$

and

$$H_2(x) = \int_a^x h(t)\, dg_2(t).$$

We draw several conclusions. First H_1 and H_2 are in $BV(a, b)$ and $H = H_1 - H_2$. Also f is in $RS[H_1; a, b] \cap RS[H_2; a, b]$ and, hence, f is in $RS[H; a, b]$. Finally, by the foregoing argument,

$$\begin{aligned}\int_a^b f(x)\, dH(x) &= \int_a^b f(x)\, dH_1(x) - \int_a^b f(x)\, dH_2(x)\\ &= \int_a^b f(x)h(x)\, dg_1(x) - \int_a^b f(x)h(x)\, dg_2(x)\\ &= \int_a^b f(x)h(x)\, dg(x).\end{aligned}$$

This completes the proof of the theorem. ●

Theorem 6.3.10 also generalizes to Riemann–Stieltjes integrals. Suppose that f is a continuous function on the rectangle $[a, b] \times [c, d]$ in $\mathbb{R}^2$ and that g is in $BV(a, b)$. For each x_2 in $[c, d]$, $f(x_1, x_2)$ is a continuous function of x_1 in $[a, b]$; therefore we can define

$$F(x_2) = \int_a^b f(t, x_2)\, dg(t).$$

We claim that F is continuous on $[c, d]$. Assume first that g is increasing on $[a, b]$. Since f is continuous on the compact set $[a, b] \times [c, d]$, f is uniformly continuous there. Given $\epsilon > 0$, choose a $\delta > 0$ such that, whenever $\|\boldsymbol{y} - \boldsymbol{x}\| < \delta$, we have $|f(\boldsymbol{y}) - f(\boldsymbol{x})| < \epsilon$. Choose any y_2 in $N(x_2; \delta) \cap [c, d]$ and compute:

$$\begin{aligned}\left|F(y_2) - F(x_2)\right| &= \left|\int_a^b f(t, y_2)\, dg(t) - \int_a^b f(t, x_2)\, dg(t)\right|\\ &\le \int_a^b |f(t, y_2) - f(t, x_2)|\, dg(t) < \epsilon \int_a^b dg(t)\\ &= [g(b) - g(a)]\epsilon.\end{aligned}$$

We conclude that F is continuous at an arbitrary point x_2 of $[c, d]$. If g is of bounded variation, decompose g as the difference $g_1 - g_2$, as usual, and apply the preceding argument to each component of g. We have proved the following theorem.

THEOREM 7.4.5 If f is continuous on $[a, b] \times [c, d]$ and if g is in $BV(a, b)$, then $F(x_2) = \int_a^b f(t, x_2)\, dg(t)$ exists and is continuous on $[c, d]$. ●

Finally, Theorem 6.5.1, concerning the uniform convergence of a sequence of integrals, lifts to the more general setting of Riemann–Stieltjes theory. Let the integrator g be in $BV(a, b)$. Suppose that $\{f_k\}$ is a sequence of functions in $RS[g; a, b]$ that converges uniformly to f_0 on $[a, b]$. Then, as with the Riemann integral, the limit function f_0 is also integrable with respect to g on $[a, b]$ and the sequence of functions $F_k(x) = \int_a^x f_k(t)\, dg(t)$ converges uniformly to $F_0(x) = \int_a^x f_0(t)\, dg(t)$.

THEOREM 7.4.6 Suppose that g is in $BV(a, b)$ and that $\{f_k\}$ is a sequence in $RS[g; a, b]$ that converges uniformly to f_0 on $[a, b]$.

i) The function f_0 is also in $RS[g; a, b]$.

ii) For k in $\mathbb{N}$ and x in $[a, b]$, define

$$F_k(x) = \int_a^x f_k(t)\, dg(t)$$

and

$$F_0(x) = \int_a^x f_0(t)\, dg(t).$$

Then $\lim_{k\to\infty} F_k = F_0$ [uniformly] on $[a, b]$. In particular,

$$\lim_{k\to\infty} \int_a^b f_k(t)\, dg(t) = \int_a^b f_0(t)\, dg(t).$$

Proof. The proof imitates that of Theorem 6.5.1 almost verbatim. Assume first that g is monotone increasing and modify the earlier proof by replacing Δx_j with Δg_j. Then let g be in $BV(a, b)$ and write $g = g_1 - g_2$ where g_1 and g_2 are monotone increasing. We leave the details as an exercise. ●

EXERCISES

7.1. Suppose that $f_1, f_2, \ldots, f_k$ are in $RS[g; a, b]$ and that $c_1, c_2, \ldots, c_k$ are real numbers. Prove that $\sum_{j=1}^k c_j f_j$ is in $RS[g; a, b]$ and that

$$\int_a^b \left[\sum_{j=1}^k c_j f_j(x)\right] dg(x) = \sum_{j=1}^k c_j \int_a^b f_j(x)\, dg(x).$$

7.2. Suppose that $g_1, g_2, \ldots, g_k$ are bounded functions on $[a, b]$, that f is in each $RS[g_j; a, b]$ for $j = 1, 2, \ldots, k$, and that $c_1, c_2, \ldots, c_k$ are real numbers. Prove that f is also in $RS[\sum_{j=1}^k c_j g_j; a, b]$ and that

$$\int_a^b f(x) d\left[\sum_{j=1}^k c_j g_j\right](x) = \sum_{j=1}^k c_j \int_a^b f(x)\, dg_j(x).$$

7.3. Suppose that f and g are bounded functions on $[a, b]$, that c is any point in (a, b), and that $\int_a^b f(x)\, dg(x)$ and $\int_a^c f(x)\, dg(x)$ exist. Prove that $\int_c^b f(x)\, dg(x)$ also exists and that

$$\int_a^b f(x)\, dg(x) = \int_a^c f(x)\, dg(x) + \int_c^b f(x)\, dg(x).$$

7.4. Suppose that u is a strictly increasing, continuous function mapping $[a, b]$ one-to-one onto $[c, d]$. Suppose that for every partition π of $[a, b]$ we let $\pi' = u(\pi)$ be the corresponding partition of $[c, d]$.

a) Prove that u maps $\Pi[a, b]$ one-to-one onto $\Pi[c, d]$.

b) Prove that u is order-preserving: If π_1 and π_2 are partitions of $[a, b]$ and if π_1 is a refinement of π_2, then $u(\pi_1)$ is a refinement of $u(\pi_2)$.

7.5. Suppose that g is defined by

$$g(x) = \begin{cases} 1, & \text{for } 0 \le x < 1 \\ 3, & \text{for } 1 \le x \le 2. \end{cases}$$

a) Let f be the function defined on [0, 2] by

$$f(x) = \begin{cases} x, & x \text{ is rational} \\ 2 - x, & x \text{ is irrational.} \end{cases}$$

Show that f is in $RS[g; a, b]$. Evaluate $\int_0^2 f(x)\, dg(x)$.

b) Suppose that f is any bounded function on [0, 2] that is discontinuous from the left at $c = 1$. Prove that f is not in $RS[g; a, b]$.

7.6. The following are the definitions for three integrators g_1, g_2, and g_3 and three integrands f_1, f_2, and f_3 on [0, 2]. For each pair of indices i and j, determine whether $\int_0^2 f_i(x)\, dg_j\,(x)$ exists. If an integral exists, compute it; otherwise, prove that it fails to exist.

$$g_1(x) = \begin{cases} x, & 0 \le x < 1 \\ x + 1, & 1 \le x \le 2 \end{cases} \qquad f_1(x) = \begin{cases} 1, & 0 \le x < 1 \\ x - 1, & 1 \le x \le 2 \end{cases}$$

$$g_2(x) = \begin{cases} x, & 0 \le x \le 1 \\ x + 1, & 1 < x < 2 \\ 4, & x = 2 \end{cases} \qquad f_2(x) = \begin{cases} 1, & x = 0 \\ x, & 0 < x \le 2 \end{cases}$$

$$g_3(x) = \begin{cases} -1, & x = 0 \\ x, & 0 < x \le 1 \\ x + 1, & 1 < x < 2 \\ 4, & x = 2 \end{cases} \qquad f_3(x) = \begin{cases} 2, & x = 0 \\ 1, & 0 < x < 1 \\ x - 1, & 1 \le x \le 2 \end{cases}$$

7.7. Suppose that f is bounded and that g is increasing on $[a, b]$. Let π' be obtained from the partition π by inserting one point x' in the partition interval (x_{k-1}, x_k). Prove that $L(f, g, \pi) \le L(f, g, \pi')$ and $U(f, g, \pi') \le U(f, g, \pi)$.

7.8. Let f and g be bounded functions on $[a, b]$ with g monotone increasing. Let the partition π' be obtained from the partition π by the insertion of j new partition points. Prove by induction on j that $L(f, g, \pi) \le L(f, g, \pi')$ and that $U(f, g, \pi) \ge U(f, g, \pi')$.

7.9. Let g be monotone increasing on $[a, b]$ and let f be in $RS[g; a, b]$. Prove that $|f|$ is also in $RS[g; a, b]$ and that $|\int_a^b f(x)\, dg(x)| \le \int_a^b |f(x)|\, dg(x)$.

7.10. Let g be in $BV(a, b)$. Assume that f_1 and f_2 are in $RS[g; a, b]$. Prove that $f_1 f_2$ is also in $RS[g; a, b]$.

7.11. Let g be monotone increasing on $[a, b]$ and let f be in $RS[g; a, b]$. Suppose that there exist constants m and M such that $0 < m \le |f| \le M$. Prove that $1/f$ is also in $RS[g; a, b]$. Generalize to the case when g is in $BV(a, b)$.

7.12. Let g be in $BV(a, b)$. Assume that f is in $RS[g; a, b]$. Prove that, for any interval $[c, d] \subseteq [a, b]$, f is also in $RS[g; c, d]$.

7.13. Let g be monotone on $[a, b]$ and let f be a nonnegative function in $RS[g; a, b]$. Prove that $f^{1/2}$ is also in $RS[g; a, b]$. Generalize to the case when g is in $BV(a, b)$.

7.14. Let f be a bounded function on $[a, b]$. Define the *positive part* f^+ and the *negative part* f^- of f to be $f^+(x) = \max\{f(x), 0\}$ and $f^-(x) = \max\{-f(x), 0\}$ for all x in $[a, b]$.

a) Show that if g is monotone increasing on $[a, b]$ and if f is in $RS[g; a, b]$, then f^+ and f^- are also in $RS[g; a, b]$.

b) Suppose that g is in $BV(a, b)$. Does the result in part (a) necessarily remain valid? If so, prove it; if not, provide a counterexample.

7.15. Let $g(x) = x^2$ on $[0, 1]$. Compute $\int_0^1 f(x)\, dg(x)$ when (i) $f(x) = x$; (ii) $f(x) = \sin x$; (iii) $f(x) = \sum_{j=0}^{k} a_j x^j$ is any polynomial; (iv) $f(x) = e^{-x}$.

7.16. Let g denote the *greatest integer function*, $g(x) = \lfloor x \rfloor$, where $\lfloor x \rfloor$ is the largest integer j such that $j \leq x$.

a) Sketch the graph of g on the interval $[0, k]$ for any k in $\mathbb{N}$. Explain why g is of bounded variation on $[0, k]$.

b) Identify exactly those bounded functions on $[0, k]$ that are integrable with respect to g on $[0, k]$ and show that, for any such function, $\int_0^k f(x)\, dg(x) = \sum_{j=1}^{k} f(j)$. (This exercise explicitly demonstrates that integration is a generalization of summation.)

7.17. Integrate $\int_0^3 \lfloor x \rfloor d(e^x)$ by parts.

7.18. Let $g(x) = x - \lfloor x \rfloor$.

a) Sketch the graph of g on any interval $[0, k]$ for any k in $\mathbb{N}$. Show that g is of bounded variation on $[0, k]$ and compute its total variation.

b) Compute $\int_0^k f(x)\, dg(x)$ when (i) $f(x) = x$; (ii) $f(x) = x^2$; (iii) $f(x) = e^{-x}$. Find a general formula for integrating any continuous function with respect to g.

7.19. Let $g(x) = [1 + (-1)^{\lfloor x \rfloor}]/2$ on any interval $[0, k]$ for any k in $\mathbb{N}$.

a) Sketch the graph of g and show that g is of bounded variation on $[0, k]$.

b) Describe the functions that are integrable with respect to g on $[0, k]$ and, for any such function f, compute $\int_0^k f(x)\, dg(x)$.

7.20. Let $g(x) = 2x - x^2$ on $[0, 2]$. For f in $RS[g; 0, 2]$, define $I_g(f) = F$ where $F(x) = \int_0^x f(t)\, dg(t)$.

a) Find a positive function f in $RS[g; 0, 2]$ for which $I_g(f)$ is not a positive function.

b) Find two functions f_1 and f_2 in $RS[g; 0, 2]$ such that $f_1(x) \leq f_2(x)$ for all x in $[0, 2]$ for which it is not true that $I_g(f_1) \leq I_g(f_2)$.

7.21. If f is continuous on $[a, b]$, then f is integrable with respect to every bounded, monotone increasing function g on $[a, b]$. Is the converse also true? If so, prove it; otherwise, provide a counterexample.

7.22. Let g be strictly monotone increasing on $[a, b]$. For f_1 and f_2 in $C([a, b])$, define

$$\langle f_1, f_2 \rangle = \int_a^b f_1(x)\, f_2(x)\, dg(x).$$

a) Prove that $<, >$ is an inner product on $C([a, b])$.

b) For f in $C([a,b])$, define

$$||f||_2 = \sqrt{\langle f, f\rangle} = \left[\int_a^b f^2(x)\,dg(x)\right]^{1/2}.$$

Prove the Cauchy–Schwarz inequality,

$$|\langle f_1, f_2\rangle| \le ||f_1||_2 ||f_2||_2,$$

for all f_1, f_2 in $C([a,b])$.

c) Prove that $||\cdot||_2$ is a norm on $C([a,b])$.

d) Use the norm in part (c) to define a metric d_2 on $C([a,b])$ as follows: $d_2(f_1, f_2) = ||f_1 - f_2||_2$. Prove that d_2 is a metric on $C([a,b])$.

7.23. Suppose that g is strictly monotone and that f is continuous on $[a,b]$. Suppose also that

$$\langle f, h\rangle = \int_a^b f(x)h(x)\,dg(x) = 0$$

for every function h in $RS[g; a, b]$. Prove that f must vanish on $[a,b]$. (This exercise tells us that, for a given strictly monotone integrator g of all continuous functions, only the zero function is orthogonal to every integrable function on $[a,b]$.)

7.24. In Exercise 3.16 we identified variants of the Chebychev polynomials on $[-1,1]$. Namely, $f_k(x) = \cos[k\cos^{-1}x]$ for $k = 0, 1, 2, \ldots$. This exercise is intended to show you that the "natural" integrator (or weight function) for this sequence of polynomials is $g(x) = \sin^{-1}x$.

a) Show that, for $k = 0, 1, 2, \ldots$,

$$\int_{-1}^{1} f_k^2(x)\,dg(x) = \begin{cases} \pi, & \text{for } k = 0 \\ \frac{1}{2}\pi, & \text{for } k \text{ in } \mathbb{N}. \end{cases}$$

b) Show that if $j, k = 0, 1, 2, \ldots$ with $j \ne k$, then

$$\int_{-1}^{1} f_j(x) f_k(x)\,dg(x) = 0.$$

The Chebychev polynomials are said to be *mutually orthogonal relative to the integrator* g.

7.25. Prove the following corollary to Theorem 7.4.5. If f is continuous on $[a,b]\times[c,d]$, if h is Riemann integrable on $[a,b]$, and if, for x_2 in $[c,d]$, we define

$$F(x_2) = \int_a^b f(t, x_2)h(t)\,dt,$$

then F is continuous on $[c,d]$.

7.26. Let g be a function in $BV(a,b)$. Suppose that $\{f_k\}$ is a sequence in $RS[g; a, b]$ that converges uniformly to f_0 on $[a,b]$.

a) Prove that f_0 is in $RS[g; a, b]$.

b) For k in $\mathbb{N}$ and x in $[a,b]$, define $F_k(x) = \int_a^x f_k(t)\,dg(t)$ and $F_0(x) = \int_a^x f_0(t)\,dg(t)$. Prove that $\lim_{k\to\infty} F_k = F_0$ [uniformly] on $[a,b]$.

7.27. Let g be in $BV(a,b)$. For each f in $RS[g; a, b]$ define $I_g(f) = F$, where $F(x) = \int_a^x f(t)\,dg(t)$ for x in $[a,b]$.

a) Show that I_g is a linear function mapping $RS[g; a, b]$ into $BV(a,b)$. Show that, if g is in $BV(a,b)\cap C([a,b])$, then I_g maps $RS(a,b)$ into $BV(a,b)\cap C([a,b])$.

Show that, if g is monotone and differentiable on $[a, b]$, then I_g maps $C([a, b])$ into the space of all differentiable functions of bounded variation.

b) Suppose that g is monotone increasing on $[a, b]$ and that f_1 and f_2 are two functions in $RS[g; a, b]$ such that $I_g(f_1) = I_g(f_2)$. Does it necessarily follow that $f_1 = f_2$? If so, prove it; otherwise, provide a counterexample.

c) Assume that g is strictly monotone on $[a, b]$. Prove that I_g, restricted to $C([a, b])$, is one-to-one. Equivalently, show that, if f is in $C([a, b])$ and if $I_g(f) = 0$, then f is the zero function on $[a, b]$.

d) Prove that, if g is monotone increasing, then I_g is a *positive* linear function. That is, if $f \geq 0$, then

$$F(x) = \int_a^x f(t)\, dg(t) \geq 0$$

for all x in $[a, b]$.

e) We say that I_g is *positive definite* if I_g is positive and $I_g(f) = 0$ if and only if $f = 0$. Prove that I_g is positive definite on $C([a, b])$ if and only if g is strictly monotone increasing on $[a, b]$.

8

Differential Calculus in $\mathbb{R}^n$

The concept of differentiability, discussed in Chapter 4, lifts naturally to functions defined on sets in $\mathbb{R}^n$. We now turn to this question. Our first objective is to find the appropriate definition of differentiability of a function on an open set U in $\mathbb{R}^n$. In this larger context we will want differentiable functions to retain the essential properties of differentiable functions of one variable. In the one-dimensional case the differentiability of f implies its continuity, a property we certainly want to retain. Once we have identified the proper condition for differentiability, we will develop the consequent theory. As you might expect, the chain rule, the Mean Value Theorem, and Taylor's theorem generalize to this larger context. En route, we will consider differentiable functions $\boldsymbol{f}$ mapping an open set U in $\mathbb{R}^m$ to $\mathbb{R}^n$. Such functions have the form $\boldsymbol{f} = (f_1, f_2, \ldots, f_n)$, where each of the component functions f_j is a differentiable function mapping U into $\mathbb{R}$.

8.1 DIFFERENTIABILITY

We assume that f is a real-valued function defined on an open set U contained in $\mathbb{R}^n$ and that $\boldsymbol{c}$ is a point of U. A first reasonable step toward the goal of describing differentiability of f at the point $\boldsymbol{c}$ is to consider the limit of the ratio $[f(\boldsymbol{x}) - f(\boldsymbol{c})]/||\boldsymbol{x} - \boldsymbol{c}||$ as the vector $\boldsymbol{x}$ tends toward the point $\boldsymbol{c}$. Of course, in $\mathbb{R}^n$, as opposed to $\mathbb{R}$, there are infinitely many paths along which $\boldsymbol{x}$ can approach $\boldsymbol{c}$; we must take all of them into account. Moreover, we must ensure that $\boldsymbol{x}$ remains in U. Since $\boldsymbol{c}$ is taken to be an interior point of U, there exists a neighborhood $N(\boldsymbol{c}; r)$ contained entirely in U. We let $\boldsymbol{u} = (u_1, u_2, \ldots, u_n)$ be any unit vector in $\mathbb{R}^n$ and let $\boldsymbol{x} = \boldsymbol{c} + h\boldsymbol{u}$, with $0 < |h| < r$. Then $\boldsymbol{x}$ is in $N'(\boldsymbol{c}; r)$ and thus in U. As h tends to 0, $\boldsymbol{x}$ approaches $\boldsymbol{c}$ along the straight-line path passing through the point $\boldsymbol{c}$ in the direction determined by the vector $\boldsymbol{u}$.

DEFINITION 8.1.1 Let f be a real-valued function defined on an open set U in $\mathbb{R}^n$. The *directional derivative* of f at the point $\boldsymbol{c}$ in U in the direction determined by the unit vector $\boldsymbol{u}$ is

$$D_{\boldsymbol{u}} f(\boldsymbol{c}) = \lim_{h \to 0} \frac{f(\boldsymbol{c} + h\boldsymbol{u}) - f(\boldsymbol{c})}{h},$$

provided this limit exists. ●

EXAMPLE 1 For any $\boldsymbol{x} = (x_1, x_2)$ in $\mathbb{R}^2$, let $f(\boldsymbol{x}) = e^{-(x_1+x_2)}$. The graph of f is sketched in Fig. 8.1. Let $\boldsymbol{u} = (u_1, u_2)$ be any unit vector and let $\boldsymbol{x} = (c_1 + hu_1, c_2 + hu_2)$. The directional derivative of f at $\boldsymbol{c}$ in the direction of $\boldsymbol{u}$ is

$$\begin{aligned} D_{\boldsymbol{u}} f(\boldsymbol{c}) &= \lim_{h \to 0} \frac{e^{-(c_1+hu_1+c_2+hu_2)} - e^{-(c_1+c_2)}}{h} \\ &= \lim_{h \to 0} f(\boldsymbol{c}) \frac{e^{-(u_1+u_2)h} - 1}{h}. \end{aligned}$$

Provided that $u_1 + u_2 \neq 0$, this limit is

$$f(\boldsymbol{c}) \lim_{h \to 0} \frac{e^{-(u_1+u_2)h} - 1}{h} = -(u_1 + u_2) f(\boldsymbol{c}).$$

If $u_1 + u_2 = 0$, then $f(\boldsymbol{x}) = f(\boldsymbol{c})$ for all $h \neq 0$ and therefore $D_{\boldsymbol{u}} f(\boldsymbol{c}) = 0$. In each case, $D_{\boldsymbol{u}} f(\boldsymbol{c}) = -(u_1 + u_2) f(\boldsymbol{c})$. ●

The function we considered in Example 1 is continuous everywhere in $\mathbb{R}^2$. However, as Example 2 shows, the existence of $D_{\boldsymbol{u}} f(\boldsymbol{c})$ for every unit vector $\boldsymbol{u}$ does not guarantee the continuity of f at $\boldsymbol{c}$. This fact warns us that the mere existence of $D_{\boldsymbol{u}} f$ is inadequate as a basis on which to define differentiability.

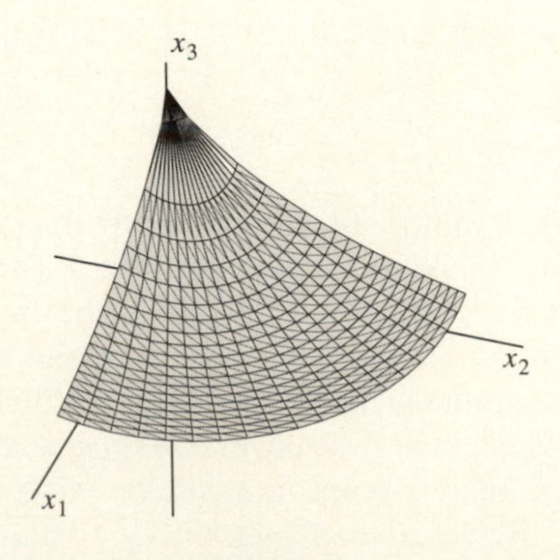

Figure 8.1

EXAMPLE 2 If $\boldsymbol{x} = (x_1, x_2) \neq 0$, define

$$f(\boldsymbol{x}) = \frac{x_1^2 x_2^3}{x_1^4 + x_2^6}$$

and define $f(\mathbf{0}) = 0$. The function f is defined on all of $\mathbb{R}^2$ and is discontinuous at $\mathbf{0}$. To see this, note that, as $\boldsymbol{x}$ approaches $\mathbf{0}$ along the line $\boldsymbol{x} = (x_1, 0)$—that is, along the x_1-axis—we have $\lim_{x_1 \to 0} f(\boldsymbol{x}) = 0$. However, if we let $\boldsymbol{x}$ approach $\mathbf{0}$ along the curve $x_2 = x_1^{2/3}$, then we have

$$\lim_{x_1 \to 0} f(\boldsymbol{x}) = \lim_{x_1 \to 0} \frac{x_1^4}{x_1^4 + x_1^4} = \frac{1}{2}.$$

Since these two limits differ, f cannot be continuous at $\mathbf{0}$. However, for every unit vector $\boldsymbol{u} = (u_1, u_2)$, the directional derivative $D_{\boldsymbol{u}} f(\mathbf{0})$ exists. To see this, suppose first that $u_1 \neq 0$. It follows that

$$\begin{aligned} D_{\boldsymbol{u}} f(\mathbf{0}) &= \lim_{h \to 0} \frac{f(hu_1, hu_2) - f(0, 0)}{h} \\ &= \lim_{h \to 0} \frac{u_1^2 u_2^3 h^5}{h(h^4 u_1^4 + h^6 u_2^6)} \\ &= \lim_{h \to 0} \frac{u_1^2 u_2^3}{u_1^4 + h^2 u_2^6} = \frac{u_2^3}{u_1^2}. \end{aligned}$$

If $u_1 = 0$ and if $\boldsymbol{x} = h\boldsymbol{u} = (0, hu_2)$, then $f(\boldsymbol{x}) = 0$ and $D_{\boldsymbol{u}} f(\mathbf{0}) = 0$. Thus, for any unit vector $\boldsymbol{u}$, $D_{\boldsymbol{u}} f(\mathbf{0})$ exists despite the fact that f is discontinuous at $\mathbf{0}$. ●

The directional derivative $D_{\boldsymbol{u}} f(\boldsymbol{c})$ measures the rate of change at the point $\boldsymbol{c}$ of the functional values of f in the direction determined by $\boldsymbol{u}$. In particular, if we choose for the vector $\boldsymbol{u}$, in turn, the unit vectors $\boldsymbol{e}_1 = (1, 0, \ldots, 0)$, $\boldsymbol{e}_2 = (0, 1, 0, \ldots, 0), \ldots, \boldsymbol{e}_n = (0, \ldots, 0, 1)$ in the directions of each of the positive coordinate axes, we obtain the n special directional derivatives

$$D_j f(\boldsymbol{c}) = \lim_{h \to 0} \frac{f(\boldsymbol{c} + h\boldsymbol{e}_j) - f(\boldsymbol{c})}{h},$$

for $j = 1, 2, \ldots, n$, provided these limits exist. These n derivatives measure the rates of change of f at $\boldsymbol{c}$ in the directions of the coordinate axes.

DEFINITION 8.1.2 Let f be a real-valued function defined on an open set U in $\mathbb{R}^n$. Let $\boldsymbol{c}$ be a point of U. The *partial derivative* of f at $\boldsymbol{c}$ with respect to the jth coordinate variable x_j is

$$D_j f(\boldsymbol{c}) = \lim_{h \to 0} \frac{f(\boldsymbol{c} + h\boldsymbol{e}_j) - f(\boldsymbol{c})}{h},$$

provided this limit exists. ●

Several different notations have been adopted for the partial derivatives of f including $\partial f / \partial x_j$, f_{x_j}, and $D_{x_j} f$. If $n = 2$ or 3 and if the independent variables are denoted x and y or x, y, and z, then $\partial f/\partial x$, $\partial f/\partial y$, and $\partial f/\partial z$ are commonly used to denote $D_1 f$, $D_2 f$, and $D_3 f$, respectively. We will use the notation established in Definition 8.1.2 when developing the general theory and will use alternate notations sparingly.

Since the existence of $D_{\boldsymbol{u}} f(\boldsymbol{c})$ for all unit vectors $\boldsymbol{u}$ fails to ensure the continuity of f at $\boldsymbol{c}$, the mere existence of the n partial derivatives $D_1 f(\boldsymbol{c}), \ldots, D_n f(\boldsymbol{c})$ cannot suffice either. Indeed, as you can easily confirm, the function $f(\boldsymbol{x}) = f(x_1, x_2) = x_1 x_2/(x_1^2 + x_2^2)$ for $\boldsymbol{x} \neq \boldsymbol{0}$ and $f(\boldsymbol{0}) = 0$ has partial derivatives $D_1 f(\boldsymbol{0}) = D_2 f(\boldsymbol{0}) = 0$, but f fails to be continuous at $\boldsymbol{0}$. Therefore, to find an adequate concept of differentiability, we must press on.

DEFINITION 8.1.3 Let f be a real-valued function defined on an open set U in $\mathbb{R}^n$ and let $\boldsymbol{c}$ be a point of U where all the partial derivatives $D_j f(\boldsymbol{c})$, $j = 1, 2, \ldots, n$, exist. The *gradient* of f at $\boldsymbol{c}$, denoted $\nabla \boldsymbol{f}(\boldsymbol{c})$, is the vector

$$\nabla \boldsymbol{f}(\boldsymbol{c}) = (D_1 f(\boldsymbol{c}), D_2 f(\boldsymbol{c}), \ldots, D_n f(\boldsymbol{c})) = \sum_{j=1}^{n} D_j f(\boldsymbol{c}) \boldsymbol{e}_j. \quad \bullet$$

The existence of the gradient of f at $\boldsymbol{c}$ results, of course, simply from the existence of the n partial derivatives of f at $\boldsymbol{c}$. Therefore, as we have seen, the existence of $\nabla \boldsymbol{f}(\boldsymbol{c})$ is an insufficient basis for the definition of differentiability. Instead, we turn to Theorem 4.1.2 in order to identify the properties that characterize differentiability.

DEFINITION 8.1.4 Suppose that f is a real-valued function defined on an open set U in $\mathbb{R}^n$. Fix $\boldsymbol{c}$ in U. Suppose that there exists a function $F(\boldsymbol{c}; \boldsymbol{t})$ defined for all $\boldsymbol{t}$ in $\mathbb{R}^n$ with the following two properties:

i) F is a linear function of $\boldsymbol{t}$. That is, for $\boldsymbol{t}'$ and $\boldsymbol{t}''$ in $\mathbb{R}^n$ and for a_1 and a_2 in $\mathbb{R}$, we have

$$F(\boldsymbol{c}; a_1 \boldsymbol{t}' + a_2 \boldsymbol{t}'') = a_1 F(\boldsymbol{c}; \boldsymbol{t}') + a_2 F(\boldsymbol{c}; \boldsymbol{t}'').$$

ii) For all $\epsilon > 0$, there exists a deleted neighborhood $N'(\boldsymbol{0})$ such that, for $\boldsymbol{t}$ in $N'(\boldsymbol{0})$, $\boldsymbol{c} + \boldsymbol{t}$ is in U and

$$|f(\boldsymbol{c} + \boldsymbol{t}) - f(\boldsymbol{c}) - F(\boldsymbol{c}; \boldsymbol{t})| < \epsilon ||\boldsymbol{t}||.$$

Then $F(\boldsymbol{c}; \boldsymbol{t})$ is called the *differential* of f at $\boldsymbol{c}$. If the differential of f at $\boldsymbol{c}$ exists, then f is said to be *differentiable* at $\boldsymbol{c}$. $\bullet$

The definite article is applied to the differential of f at $\boldsymbol{c}$ because, as we will prove, if such a function $F(\boldsymbol{c}; \boldsymbol{t})$ exists, then it is unique. To be precise, we will prove that, if the differential of f exists at $\boldsymbol{c}$, then all the partial derivatives $D_j f(\boldsymbol{c})$ also exist and

$$F(\boldsymbol{c}; \boldsymbol{t}) = \langle \nabla \boldsymbol{f}(\boldsymbol{c}), \boldsymbol{t} \rangle = \sum_{j=1}^{n} D_j f(\boldsymbol{c}) t_j, \tag{8.1}$$

where $\boldsymbol{t} = (t_1, t_2, \ldots, t_n)$. The differential is often written $df = \sum_{j=1}^{n} \partial f/\partial x_j \, dx_j$, using $\boldsymbol{dx} = (dx_1, dx_2, \ldots, dx_n)$ in place of $\boldsymbol{t} = (t_1, t_2, \ldots, t_n)$. We will use the streamlined notation of (8.1).

THEOREM 8.1.1 Let f be a real-valued function defined on an open set U in $\mathbb{R}^n$ and let $\boldsymbol{c}$ be a point of U. Suppose that the differential $F(\boldsymbol{c}; \boldsymbol{t})$ of f exists at $\boldsymbol{c}$. Then all the partial derivatives of f exist at $\boldsymbol{c}$. Furthermore,

$$F(\boldsymbol{c}; \boldsymbol{t}) = \sum_{j=1}^{n} D_j f(\boldsymbol{c}) t_j = \langle \nabla f(\boldsymbol{c}), \boldsymbol{t} \rangle,$$

for all $\boldsymbol{t} = (t_1, t_2, \ldots, t_n)$ in $\mathbb{R}^n$.

Proof. First we use the linearity of F in the variable $\boldsymbol{t}$. Given any $\boldsymbol{t}$ in $\mathbb{R}^n$, we write $\boldsymbol{t} = \sum_{j=1}^{n} t_j \boldsymbol{e}_j$, where $\boldsymbol{e}_1, \boldsymbol{e}_2, \ldots, \boldsymbol{e}_n$ are the standard unit vectors in the directions of the coordinate axes. By the linearity of F,

$$F(\boldsymbol{c}; \boldsymbol{t}) = F(\boldsymbol{c}; \sum_{j=1}^{n} t_j \boldsymbol{e}_j) = \sum_{j=1}^{n} t_j F(\boldsymbol{c}; \boldsymbol{e}_j).$$

Define $m_j(\boldsymbol{c}) = F(\boldsymbol{c}; \boldsymbol{e}_j)$, a constant that depends on $\boldsymbol{c}$ but not on $\boldsymbol{t}$. With this notation, we have $F(\boldsymbol{c}; \boldsymbol{t}) = \sum_{j=1}^{n} m_j(\boldsymbol{c}) t_j$ for all $\boldsymbol{t} = (t_1, t_2, \ldots, t_n)$ in $\mathbb{R}^n$.

All that remains is to show that, for $j = 1, 2, \ldots, n$, $m_j(\boldsymbol{c}) = D_j f(\boldsymbol{c})$. We use property (ii) of the differential: Given $\epsilon > 0$, choose a deleted neighborhood $N'(\boldsymbol{0}; r)$ such that, for $\boldsymbol{t}$ in $N'(\boldsymbol{0}; r)$ we have $\boldsymbol{c} + \boldsymbol{t}$ in U and

$$|f(\boldsymbol{c} + \boldsymbol{t}) - f(\boldsymbol{c}) - F(\boldsymbol{c}; \boldsymbol{t})| < \epsilon \|\boldsymbol{t}\|. \tag{8.2}$$

Fix any coordinate index j. Choose h such that $0 < |h| < r$. Then $\boldsymbol{t} = h\boldsymbol{e}_j$ is in $N'(\boldsymbol{0}; r)$ and, by (8.2),

$$|f(\boldsymbol{c} + h\boldsymbol{e}_j) - f(\boldsymbol{c}) - F(\boldsymbol{c}; h\boldsymbol{e}_j)| < \epsilon |h|. \tag{8.3}$$

But $F(\boldsymbol{c}; h\boldsymbol{e}_j) = hF(\boldsymbol{c}; \boldsymbol{e}_j) = hm_j(\boldsymbol{c})$. Therefore, dividing by $|h|$ in (8.3), we obtain $|[f(\boldsymbol{c} + h\boldsymbol{e}_j) - f(\boldsymbol{c})]/h - m_j(\boldsymbol{c})| < \epsilon$, for all h in $N'(0; r)$. But this says that

$$D_j f(\boldsymbol{c}) = \lim_{h \to 0} \frac{f(\boldsymbol{c} + h\boldsymbol{e}_j) - f(\boldsymbol{c})}{h}$$

exists and equals $m_j(\boldsymbol{c})$. Thus $m_j(\boldsymbol{c}) = D_j f(\boldsymbol{c})$. We conclude that all the partial derivatives $D_j f(\boldsymbol{c})$ exist and that $F(\boldsymbol{c}; \boldsymbol{t}) = \sum_{j=1}^{n} D_j f(\boldsymbol{c}) t_j$. This proves the theorem. ●

Having established Theorem 8.1.1, we will adopt the convential notation $df(\boldsymbol{c}; \boldsymbol{t})$ for the differential of f at $\boldsymbol{c}$, namely,

$$df(\boldsymbol{c}; \boldsymbol{t}) = \langle \nabla f(\boldsymbol{c}), \boldsymbol{t} \rangle = \sum_{j=1}^{n} D_j f(\boldsymbol{c}) t_j.$$

It will be convenient to reformulate the content of Theorem 8.1.1 in the following form.

COROLLARY 8.1.2 Let f be a real-valued function defined on an open set U in $\mathbb{R}^n$ and let $\boldsymbol{c}$ be a point of U. Suppose also that $\boldsymbol{v}$ is a vector with the following property: Given any $\epsilon > 0$, there is a deleted neighborhood $N'(\mathbf{0})$ such that, for $\boldsymbol{t}$ in $N'(\mathbf{0})$,

$$|f(\boldsymbol{c}+\boldsymbol{t}) - f(\boldsymbol{c}) - \langle \boldsymbol{v}, \boldsymbol{t}\rangle| < \epsilon||\boldsymbol{t}||.$$

Then f is differentiable at $\boldsymbol{c}$, the vector $\boldsymbol{v}$ is $\nabla \boldsymbol{f}(\boldsymbol{c})$, and $df(\boldsymbol{c}; \boldsymbol{t}) = \langle \boldsymbol{v}, \boldsymbol{t}\rangle$ for all $\boldsymbol{t}$ in $\mathbb{R}^n$.

Proof. The function $F(\boldsymbol{c}; \boldsymbol{t})$ defined by $F(\boldsymbol{c}; \boldsymbol{t}) = \langle \boldsymbol{v}, \boldsymbol{t}\rangle$ is linear in $\boldsymbol{t}$. Also, we assume in the hypotheses of the corollary that $F(\boldsymbol{c}; \boldsymbol{t})$ has property (ii) of the differential. Therefore, by Theorem 8.1.1, the partial derivatives $D_j f(\boldsymbol{c})$ exist and

$$F(\boldsymbol{c}; \boldsymbol{t}) = \langle \boldsymbol{v}, \boldsymbol{t}\rangle = \sum_{j=1}^{n} D_j f(\boldsymbol{c}) t_j = df(\boldsymbol{c}; \boldsymbol{t}),$$

for all $\boldsymbol{t} = (t_1, t_2, \ldots, t_n)$ in $\mathbb{R}^n$. Since the differential $df(\boldsymbol{c}; \boldsymbol{t})$ of f at $\boldsymbol{c}$ exists, f is differentiable at $\boldsymbol{c}$. All that remains is to prove that $\boldsymbol{v} = \nabla \boldsymbol{f}(\boldsymbol{c})$. Now,

$$\langle \boldsymbol{v}, \boldsymbol{t}\rangle = \sum_{j=1}^{n} D_j f(\boldsymbol{c}) t_j = \langle \nabla \boldsymbol{f}(\boldsymbol{c}), \boldsymbol{t}\rangle,$$

for all $\boldsymbol{t} = (t_1, t_2, \ldots, t_n)$ in $\mathbb{R}^n$. Equivalently, $\langle \boldsymbol{v} - \nabla \boldsymbol{f}(\boldsymbol{c}), \boldsymbol{t}\rangle = 0$, for all $\boldsymbol{t}$ in $\mathbb{R}^n$. When $t = \boldsymbol{v} - \nabla \boldsymbol{f}(\boldsymbol{x})$, we obtain $0 = \langle \boldsymbol{v} - \nabla \boldsymbol{f}(\boldsymbol{c}), \boldsymbol{v} - \nabla \boldsymbol{f}(\boldsymbol{c})\rangle = ||\boldsymbol{v} - \nabla \boldsymbol{f}(\boldsymbol{c})||^2$. Therefore $\boldsymbol{v} - \nabla \boldsymbol{f}(\boldsymbol{c}) = 0$ or $\boldsymbol{v} = \nabla \boldsymbol{f}(\boldsymbol{c})$. This completes the proof of the corollary. ●

We will apply the corollary in the following context. Suppose that we are trying to prove that a function is differentiable at a point $\boldsymbol{c}$. If we can find a vector $\boldsymbol{v}$ with the property of the corollary, then we can be certain that $\boldsymbol{v}$ must be $\nabla \boldsymbol{f}(\boldsymbol{c})$, that $\langle \boldsymbol{v}, \boldsymbol{t}\rangle$ must be $df(\boldsymbol{c}; \boldsymbol{t})$, and that f must be differentiable at $\boldsymbol{c}$.

As the following theorem confirms, Definition 8.1.4 provides us with exactly the generalization of the concept of differentiability that we want.

THEOREM 8.1.3 Let f be a real-valued function defined on an open set U in $\mathbb{R}^n$ and let $\boldsymbol{c}$ be a point in U. Assume that f is differentiable at $\boldsymbol{c}$.

i) For any unit vector $\boldsymbol{u}$, we have

$$D_{\boldsymbol{u}} f(\boldsymbol{c}) = df(\boldsymbol{c}; \boldsymbol{u}) = \langle \nabla \boldsymbol{f}(\boldsymbol{c}), \boldsymbol{u}\rangle.$$

ii) The function f satisfies a local Lipschitz condition in a neighborhood of $\boldsymbol{c}$. That is, there exists a neighborhood $N(\boldsymbol{c})$ contained in U and a Lipschitz constant $K > 0$ such that $|f(\boldsymbol{x}) - f(\boldsymbol{c})| \le K||\boldsymbol{x} - \boldsymbol{c}||$ for all $\boldsymbol{x}$ in $N(\boldsymbol{c})$.

iii) The function f is continuous at $\boldsymbol{c}$.

Proof. Given $\epsilon > 0$, choose a neighborhood $N(\mathbf{0}; r)$ such that, for all $\boldsymbol{t}$ in $N'(\mathbf{0}; r)$, it follows that $\boldsymbol{c} + \boldsymbol{t}$ is in U and

$$|f(\boldsymbol{c}+\boldsymbol{t}) - f(\boldsymbol{c}) - df(\boldsymbol{c}; \boldsymbol{t})| < \epsilon||\boldsymbol{t}||. \tag{8.4}$$

To prove part (i) of the theorem, fix any unit vector $\boldsymbol{u}$ and choose h so that $0 < |h| < r$. Then $\boldsymbol{t} = h\boldsymbol{u}$ is in $N'(\mathbf{0}; r)$. Using (8.4) and the linearity of the differential, we have

$$|f(\boldsymbol{c} + h\boldsymbol{u}) - f(\boldsymbol{c}) - h\,df(\boldsymbol{c}; \boldsymbol{u})| < \epsilon |h|.$$

Dividing by $|h|$, we deduce that

$$\left| \frac{f(\boldsymbol{c} + h\boldsymbol{u}) - f(\boldsymbol{c})}{h} - df(\boldsymbol{c}; \boldsymbol{u}) \right| < \epsilon$$

for all h in $N'(\mathbf{0}; r)$. Thus $\lim_{h \to 0}[f(\boldsymbol{c} + h\boldsymbol{u}) - f(\boldsymbol{c})]/h$ exists and equals $df(\boldsymbol{c}; \boldsymbol{u})$. Therefore $D_u f(\boldsymbol{c}) = df(\boldsymbol{c}; \boldsymbol{u})$. Furthermore, $D_u f(\boldsymbol{c}) = df(\boldsymbol{c}; \boldsymbol{u}) = \langle \nabla \boldsymbol{f}(\boldsymbol{c}), \boldsymbol{u} \rangle$, completing the proof of (i).

For part (ii) we return to (8.4) and observe that, for all $\boldsymbol{t}$ in $N'(\mathbf{0}; r)$,

$$|f(\boldsymbol{c} + \boldsymbol{t}) - f(\boldsymbol{c})| < |df(\boldsymbol{c}; \boldsymbol{t})| + \epsilon ||\boldsymbol{t}||. \quad \textbf{(8.5)}$$

Since $df(\boldsymbol{c}; \boldsymbol{t}) = \langle \nabla \boldsymbol{f}(\boldsymbol{c}), \boldsymbol{t} \rangle$, the Cauchy–Schwarz inequality implies that $|df(\boldsymbol{c}; \boldsymbol{t})| \leq ||\nabla \boldsymbol{f}(\boldsymbol{c})||\, ||\boldsymbol{t}||$.

Let $N(\boldsymbol{c}; \boldsymbol{r}) = \boldsymbol{c} + N(\mathbf{0}; \boldsymbol{r})$ denote the neighborhood whose existence is promised in part (ii) of the theorem. If $\boldsymbol{x}$ is in $N(\boldsymbol{c}; \boldsymbol{r})$, then $\boldsymbol{t} = \boldsymbol{x} - \boldsymbol{c}$ is a point in $N(\mathbf{0}; r)$. Substituting in (8.5) yields

$$\begin{aligned} |f(\boldsymbol{x}) - f(\boldsymbol{c})| &< |df(\boldsymbol{c}; \boldsymbol{x} - \boldsymbol{c})| + \epsilon ||\boldsymbol{x} - \boldsymbol{c}|| \\ &\leq ||\nabla \boldsymbol{f}(\boldsymbol{c})||\, ||\boldsymbol{x} - \boldsymbol{c}|| + \epsilon ||\boldsymbol{x} - \boldsymbol{c}|| = K ||\boldsymbol{x} - \boldsymbol{c}||, \end{aligned}$$

where $K = ||\nabla \boldsymbol{f}(\boldsymbol{c})|| + \epsilon$. Therefore f satisfies a local Lipschitz condition at $\boldsymbol{c}$.

For part (iii), use the result in (ii): There exists a constant $K > 0$ and a neighborhood $N(\boldsymbol{c}; r)$ contained in U such that

$$|f(\boldsymbol{x}) - f(\boldsymbol{c})| \leq K ||\boldsymbol{x} - \boldsymbol{c}|| \quad \textbf{(8.6)}$$

for all $\boldsymbol{x}$ in $N(\boldsymbol{c}; r)$. Given $\epsilon > 0$, choose a positive δ less than $\min\{r, \epsilon/K\}$. Then, for $\boldsymbol{x}$ in $N(\boldsymbol{c}; \delta)$, we have by (8.6)

$$|f(\boldsymbol{x}) - f(\boldsymbol{c})| < \epsilon.$$

This proves that f is continuous at $\boldsymbol{c}$ and completes the proof of the theorem. ●

COROLLARY 8.1.4 Let f be a real-valued function defined on an open set U in $\mathbb{R}^n$ and let $\boldsymbol{c}$ be a point in U. Assume that f is differentiable at $\boldsymbol{c}$ and that $\nabla f(\boldsymbol{c}) \neq \mathbf{0}$. Then the direction in which $|D_u f(\boldsymbol{c})|$ is a maximum is the direction of the unit vector $\boldsymbol{u}^* = \nabla \boldsymbol{f}(\boldsymbol{c})/||\nabla \boldsymbol{f}(\boldsymbol{c})||$. The maximum value of $|D_u f(\boldsymbol{c})|$ is $||\nabla \boldsymbol{f}(\boldsymbol{c})||$.

Proof. For any unit vector $\boldsymbol{u}$ we have $D_u f(\boldsymbol{c}) = \langle \nabla \boldsymbol{f}(\boldsymbol{c}), \boldsymbol{u} \rangle$ by the theorem. From our discussion of the inner product in Section 2.1, we know that $\langle \nabla \boldsymbol{f}(\boldsymbol{c}), \boldsymbol{u} \rangle = ||\nabla \boldsymbol{f}(\boldsymbol{c})|| \cos\theta$, where θ is the angle between $\nabla f(\boldsymbol{c})$ and the unit vector $\boldsymbol{u}$. We deduce that $|D_u f(\boldsymbol{c})| = ||\nabla \boldsymbol{f}(\boldsymbol{c})||\, |\cos\theta|$ and that $|D_u f(\boldsymbol{c})|$ has its maximum value of $||\nabla \boldsymbol{f}(\boldsymbol{c})||$ when $\theta = 0$, that is, when $\boldsymbol{u}$ is in the direction of $\nabla \boldsymbol{f}(\boldsymbol{c})$. ●

EXAMPLE 3 For $\boldsymbol{x} = (x_1, x_2)$ in $\mathbb{R}^2$, define

$$f(\boldsymbol{x}) = f(x_1, x_2) = \exp(-x_1^2 - x_2^2).$$

Then $D_1 f(\boldsymbol{x}) = -2x_1 f(\boldsymbol{x})$ and $D_2 f(\boldsymbol{x}) = -2x_2 f(\boldsymbol{x})$. Therefore $\nabla \boldsymbol{f}(\boldsymbol{x}) = -2f(\boldsymbol{x})(x_1, x_2) = -2f(\boldsymbol{x})\boldsymbol{x}$. It follows that

$$df(\boldsymbol{x}; \boldsymbol{t}) = \langle \nabla \boldsymbol{f}(\boldsymbol{x}), \boldsymbol{t} \rangle = -2f(\boldsymbol{x})\langle \boldsymbol{x}, \boldsymbol{t} \rangle.$$

To show that f is differentiable at any $\boldsymbol{x}$ in $\mathbb{R}^2$, we let $\boldsymbol{t} = (t_1, t_2)$ and show that

$$\lim_{\boldsymbol{t} \to \boldsymbol{0}} \frac{f(\boldsymbol{x} + \boldsymbol{t}) - f(\boldsymbol{x}) - df(\boldsymbol{x}; \boldsymbol{t})}{||\boldsymbol{t}||} = 0.$$

Now,

$$f(\boldsymbol{x} + \boldsymbol{t}) = \exp[-(x_1 + t_1)^2 - (x_2 + t_2)^2] = f(\boldsymbol{x}) f(\boldsymbol{t}) e^{-2\langle \boldsymbol{x}, \boldsymbol{t} \rangle}$$

and $df(\boldsymbol{x}; \boldsymbol{t}) = -2f(\boldsymbol{x})\langle \boldsymbol{x}, \boldsymbol{t} \rangle$. Consequently

$$|f(\boldsymbol{x} + \boldsymbol{t}) - f(\boldsymbol{x}) - df(\boldsymbol{x}; \boldsymbol{t})| = f(\boldsymbol{x}) |f(\boldsymbol{t}) e^{-2\langle \boldsymbol{x}, \boldsymbol{t} \rangle} - 1 + 2\langle \boldsymbol{x}, \boldsymbol{t} \rangle|. \tag{8.7}$$

We let $s = 2\langle \boldsymbol{x}, \boldsymbol{t} \rangle$ and, before launching into our calculations, we make two observations. First, by the Cauchy–Schwarz inequality, $|s| \leq 2||\boldsymbol{x}|| \, ||\boldsymbol{t}||$. Thus, as the vector $\boldsymbol{t}$ tends to $\boldsymbol{0}$, the number s also tends to 0. Furthermore, $|s|/||\boldsymbol{t}|| \leq 2||\boldsymbol{x}||$. In (8.7) we divide by $||\boldsymbol{t}||$ and write

$$\begin{aligned}
\frac{|f(\boldsymbol{x} + \boldsymbol{t}) - f(x) - df(\boldsymbol{x}; \boldsymbol{t})|}{||\boldsymbol{t}||} &= \frac{f(\boldsymbol{x}) |f(\boldsymbol{t}) e^{-s} - 1 + s|}{||\boldsymbol{t}||} \\
&= \frac{f(\boldsymbol{x}) |f(\boldsymbol{t})(e^{-s} - 1 + s) + [f(\boldsymbol{t}) - 1](1 - s)|}{||\boldsymbol{t}||} \\
&\leq \frac{f(\boldsymbol{x}) f(\boldsymbol{t}) |e^{-s} - 1 + s|}{||\boldsymbol{t}||} + \frac{f(\boldsymbol{x}) |f(\boldsymbol{t}) - 1|(1 + |s|)}{||\boldsymbol{t}||}.
\end{aligned} \tag{8.8}$$

We take up each summand in turn. First, apply Taylor's theorem to write $e^{-s} = 1 - s + e^{-c} s^2/2!$ for some c between 0 and s. Therefore

$$\begin{aligned}
\frac{f(\boldsymbol{x}) f(\boldsymbol{t}) |e^{-s} - 1 + s|}{||\boldsymbol{t}||} &= \frac{f(\boldsymbol{x}) f(\boldsymbol{t}) e^{-c} s^2}{2||\boldsymbol{t}||} \\
&= f(\boldsymbol{x}) f(\boldsymbol{t}) e^{-c} \left(\frac{s}{||\boldsymbol{t}||} \right)^2 \frac{||\boldsymbol{t}||}{2} \\
&\leq f(\boldsymbol{x}) f(\boldsymbol{t}) e^{-c} [2||\boldsymbol{x}||]^2 \frac{||\boldsymbol{t}||}{2} \\
&= 2f(\boldsymbol{x}) f(\boldsymbol{t}) e^{-c} ||\boldsymbol{x}||^2 ||\boldsymbol{t}||.
\end{aligned}$$

Because c lies between 0 and s and because s tends to 0 as $\boldsymbol{t}$ tends to $\boldsymbol{0}$, $\lim_{\boldsymbol{t} \to \boldsymbol{0}} e^{-c} = 1$. Also, $\lim_{\boldsymbol{t} \to \boldsymbol{0}} f(\boldsymbol{t}) = 1$. Therefore

$$\lim_{\boldsymbol{t} \to \boldsymbol{0}} 2f(\boldsymbol{x}) f(\boldsymbol{t}) e^{-c} ||\boldsymbol{x}||^2 ||\boldsymbol{t}|| = 0. \tag{8.9}$$

To treat the second summand in (8.8), we recall that $\lim_{z\to 0}(e^{-z} - 1)/z = 1$. Therefore $\lim_{t\to 0}[f(t) - 1]/||t||^2 = \lim_{t\to 0}[e^{-(t_1^2+t_2^2)} - 1)/(t_1^2 + t_2^2) = 1$. Thus the limit, as t tends to $\mathbf{0}$, of the second summand in (8.8) is

$$\lim_{t\to 0} f(x)\frac{f(t) - 1}{||t||^2}||t||(1 + |s|) = 0. \tag{8.10}$$

Substituting the limits (8.9) and (8.10) in (8.8) and applying the Squeeze Play, we conclude that

$$\lim_{t\to 0} \frac{|f(x+t) - f(x) - df(x; t)|}{||t||} = 0.$$

This proves that f is differentiable at every point x in $\mathbb{R}^2$. ●

The effort required to show that a function is differentiable, even for a function as simple as that in Example 3, compels us to seek useful criteria that enable us to bypass the work. The most useful are provided by the following theorem.

THEOREM 8.1.5 Let f be a real-valued function defined on an open set U in $\mathbb{R}^n$ and let c be a point of U. Suppose that, for every $j = 1, 2, \ldots, n$ and for every x in some neighborhood $N(c; r)$ of c, each partial derivative $D_j f(x)$ exists. Suppose also that each of the partial derivatives $D_j f$ is continuous at c. Then f is differentiable at c.

Proof. Fix $\epsilon > 0$. For each $j = 1, 2, \ldots, n$, there exists a $\delta_j > 0$ such that, for all x in $N(c; \delta_j)$, we have $D_j f(x)$ in $N(D_j f(c); \epsilon/\sqrt{n})$. Let $\delta = \min\{\delta_1, \delta_2, \ldots, \delta_n, r\}$. Fix x in $N'(c; \delta)$. We want to show that

$$|f(x) - f(c) - df(c; x - c)| < \epsilon||x - c||. \tag{8.11}$$

Let $u = (x - c)/||x - c||$. Notice that u is a unit vector in the direction from c to x and therefore that $x = c + hu$ for $0 < h = ||x - c|| < \delta$. We write $u = (u_1, u_2, \ldots, u_n)$ as a unique linear combination of the standard unit vectors $e_1, e_2, \ldots, e_n$ to obtain $u = \sum_{j=1}^n u_j e_j$. Next, we let

$$\begin{aligned} x^{(0)} &= c \\ x^{(1)} &= x^{(0)} + hu_1 e_1 \\ x^{(2)} &= x^{(1)} + hu_2 e_2, \end{aligned}$$

and, recursively,

$$x^{(j)} = x^{(j-1)} + hu_j e_j$$

for $j = 1, 2, \ldots, n$. Notice that $x^{(n)} = x$ and that the points $x^{(0)}, x^{(1)}, x^{(2)}, \ldots, x^{(n)}$ are the successive vertices of a polygonal path originating at c and terminating at x. Notice also that the jth segment of the polygonal path joins $x^{(j-1)}$ to $x^{(j)}$ and is

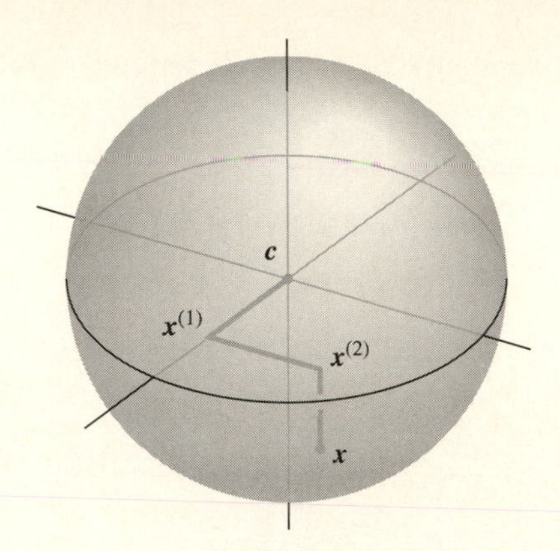

Figure 8.2

parallel to the jth coordinate axis. (See Fig. 8.2, with $n = 3$.) Furthermore, each of these points $\boldsymbol{x}^{(k)}$ is in $N(\boldsymbol{c}; \delta)$ because

$$||\boldsymbol{x}^{(k)} - \boldsymbol{c}||^2 = \left|\left|\sum_{j=1}^{k} hu_j \boldsymbol{e}_j\right|\right|^2 = h^2 \sum_{j=1}^{k} u_j^2 \le h^2 ||\boldsymbol{u}||^2 = h^2 < \delta^2.$$

Finally, observe that the line segment joining $\boldsymbol{x}^{(j-1)}$ to $\boldsymbol{x}^{(j)}$ is contained entirely within $N(\boldsymbol{c}; \delta)$. Briefly put, $N(\boldsymbol{c}; \delta)$ is convex.[1] (See Exercise 2.1.) A point z on the line segment joining $\boldsymbol{x}^{(j-1)}$ to $\boldsymbol{x}^{(j)}$ has the form $z = \boldsymbol{x}^{(j-1)} + t\boldsymbol{e}_j$, with t between 0 and hu_j. Now evaluate f at each of the points $\boldsymbol{c} = \boldsymbol{x}^{(0)}, \boldsymbol{x}^{(1)}, \boldsymbol{x}^{(2)}, \ldots, \boldsymbol{x}^{(n)} = \boldsymbol{x}$. Observe that, by telescoping, we have

$$\begin{aligned} f(\boldsymbol{x}) - f(\boldsymbol{c}) &= \sum_{j=1}^{n} [f(\boldsymbol{x}^{(j)}) - f(\boldsymbol{x}^{(j-1)})] \\ &= \sum_{j=1}^{n} [f(\boldsymbol{x}^{(j-1)} + hu_j \boldsymbol{e}_j) - f(\boldsymbol{x}^{(j-1)})]. \end{aligned} \tag{8.12}$$

We analyze each of these summands.

For each $j = 1, 2, \ldots, n$ and for t in the closed interval I_j with endpoints 0 and hu_j, let

$$g_j(t) = f(\boldsymbol{x}^{(j-1)} + t\boldsymbol{e}_j) - f(\boldsymbol{x}^{(j-1)}).$$

The very form of g_j leads us to think of the Mean Value Theorem. We want to show that $g_j'(t)$ exists at every point of I_j. By the convexity of $N(\boldsymbol{c}; \delta)$, the point

[1]A set C in $\mathbb{R}^n$ is said to be *convex* if, whenever it contains two points, it also contains the line segment joining those two points.

$x^{(j-1)} + t\boldsymbol{e}_j$ is in $N(\boldsymbol{c}; \delta)$ and thus is in $N(\boldsymbol{c}; r)$. Consequently, the partial derivative $D_j f(\boldsymbol{x}^{(j-1)} + t\boldsymbol{e}_j)$ exists. In fact,

$$\begin{aligned} D_j f(\boldsymbol{x}^{(j-1)} + t\boldsymbol{e}_j) &= \lim_{\tau \to 0} \frac{f(\boldsymbol{x}^{(j-1)} + t\boldsymbol{e}_j + \tau\boldsymbol{e}_j) - f(\boldsymbol{x}^{(j-1)} + t\boldsymbol{e}_j)}{\tau} \\ &= \lim_{\tau \to 0} \frac{g_j(t+\tau) - g_j(t)}{\tau} = g_j'(t). \end{aligned}$$

Therefore g_j' exists at every point of I_j and thus is continuous on I_j. By the Mean Value Theorem there exists a point t_j strictly between 0 and hu_j such that

$$g_j(hu_j) - g_j(0) = g_j'(t_j)hu_j.$$

But

$$\begin{aligned} g_j(hu_j) - g_j(0) &= f(\boldsymbol{x}^{(j-1)} + hu_j\boldsymbol{e}_j) - f(\boldsymbol{x}^{(j-1)}) \\ &= f(\boldsymbol{x}^{(j)}) - f(\boldsymbol{x}^{(j-1)}). \end{aligned}$$

By our preceding calculations, we also know that

$$g_j'(t_j)hu_j = D_j f(\boldsymbol{x}^{(j-1)} + t_j\boldsymbol{e}_j)hu_j.$$

We deduce that

$$f(\boldsymbol{x}^{(j)}) - f(\boldsymbol{x}^{(j-1)}) = D_j f(\boldsymbol{x}^{(j-1)} + t_j\boldsymbol{e}_j)hu_j.$$

Substituting in (8.12) we obtain

$$\begin{aligned} f(\boldsymbol{x}) - f(\boldsymbol{c}) &= \sum_{j=1}^{n} [f(\boldsymbol{x}^{(j)}) - f(\boldsymbol{x}^{(j-1)})] \\ &= \sum_{j=1}^{n} D_j f(\boldsymbol{x}^{(j-1)} + t_j\boldsymbol{e}_j)hu_j. \end{aligned} \tag{8.13}$$

Finally, we are ready to prove the differentiability of f at $\boldsymbol{c}$. We substitute (8.13) into (8.11) to obtain

$$\begin{aligned} |f(\boldsymbol{x}) - f(\boldsymbol{c}) - df(\boldsymbol{c}; h\boldsymbol{u})| &= \left| \sum_{j=1}^{n} D_j f(\boldsymbol{x}^{(j-1)} + t_j\boldsymbol{e}_j)hu_j - \sum_{j=1}^{n} D_j f(\boldsymbol{c})hu_j \right| \\ &= \left| \sum_{j=1}^{n} [D_j f(\boldsymbol{x}^{(j-1)} + t_j\boldsymbol{e}_j) - D_j f(\boldsymbol{c})]hu_j \right| \\ &= |\langle \boldsymbol{D}, h\boldsymbol{u} \rangle|, \end{aligned}$$

where $\boldsymbol{D}$ is the vector whose jth coordinate is $D_j f(\boldsymbol{x}^{(j-1)} + t_j\boldsymbol{e}_j) - D_j f(\boldsymbol{c})$. By the Cauchy–Schwarz inequality, $|\langle \boldsymbol{D}, h\boldsymbol{u} \rangle| \le ||\boldsymbol{D}|| \cdot ||h\boldsymbol{u}|| = ||\boldsymbol{D}|| \cdot |h|$. Thus

$$|f(\boldsymbol{x}) - f(\boldsymbol{c}) - df(\boldsymbol{c}; h\boldsymbol{u})| \le \left[\sum_{j=1}^{n} [D_j f(\boldsymbol{x}^{(j-1)} + t_j\boldsymbol{e}_j) - D_j f(\boldsymbol{c})]^2 \right]^{1/2} |h| < \epsilon |h|,$$

since, for each j, $|D_j f(\boldsymbol{x}^{(j-1)} + t_j \boldsymbol{e}_j) - D_j f(\boldsymbol{c})| < \epsilon/\sqrt{n}$. This completes the proof that f is differentiable at $\boldsymbol{c}$. ●

EXAMPLE 3 (Revisited) If $f(\boldsymbol{x}) = \exp(-x_1^2 - x_2^2)$, then the partial derivatives $D_1 f(\boldsymbol{x}) = -2x_1 f(\boldsymbol{x})$ and $D_2 f(\boldsymbol{x}) = -2x_2 f(\boldsymbol{x})$ are continuous on $\mathbb{R}^2$. Therefore, by Theorem 8.1.5, f is differentiable at every point of $\mathbb{R}^2$. ●

An application of Theorem 8.1.5 to determine the differentiability of a function f so dramatically reduces the workload that we identify the special class of functions to which it applies.

DEFINITION 8.1.5 Let f be a real-valued function defined on an open set U in $\mathbb{R}^n$ and let $\boldsymbol{x}$ be a point of U.

i) If all the partial derivatives $D_j f$ of f exist in a neighborhood of $\boldsymbol{x}$ and are continuous at $\boldsymbol{x}$, then f is said to be *continuously differentiable* at $\boldsymbol{x}$.

ii) If f is (continuously) differentiable at every point of U, then f is said to be (*continuously*) *differentiable* on U. The class of all continuously differentiable, real-valued functions on U will be denoted $C^{(1)}(U)$. ●

8.1.1 A Geometric Interpretation of Differentiability

Heretofore we have avoided invoking a geometric interpretation of our results because we have wanted to emphasize that all our proofs must be analytic in nature, relying solely on careful mathematical arguments rather than on geometrically inspired intuition. But our results can be translated into geometric statements. In fact, to feel comfortable with the concept of differentiability, it is essential that you grasp its geometric significance. We shift briefly therefore to a geometric point of view.

In $\mathbb{R}^2$, a straight line L passing through the point $\boldsymbol{c} = (c_1, c_2)$ is represented by a linear equation

$$a_1(x_1 - c_1) + a_2(x_2 - c_2) = 0,$$

with at least one of the coefficients not zero. That is, $a_1^2 + a_2^2 \neq 0$. The vector $\boldsymbol{v} = a_1\boldsymbol{e}_1 + a_2\boldsymbol{e}_2$ is perpendicular to L. Likewise, in $\mathbb{R}^3$, a linear equation of the form

$$a_1(x_1 - c_1) + a_2(x_2 - c_2) + a_3(x_3 - c_3) = 0,$$

where $a_1^2 + a_2^2 + a_3^2 \neq 0$ represents a plane P passing through the point $\boldsymbol{c} = (c_1, c_2, c_3)$. You will recall from your early studies that $\boldsymbol{v} = a_1\boldsymbol{e}_1 + a_2\boldsymbol{e}_2 + a_3\boldsymbol{e}_3$ is a vector perpendicular to the plane P.

In $\mathbb{R}^n$, a hyperplane P passing through the point $\boldsymbol{c} = (c_1, c_2, \ldots, c_n)$ is represented by a linear equation

$$a_1(x_1 - c_1) + a_2(x_2 - c_2) + \cdots + a_n(x_n - c_n) = 0,$$

with $a_1^2 + a_2^2 + \cdots + a_n^2 \neq 0$. Again, we claim that $\boldsymbol{v} = \sum_{j=1}^n a_j \boldsymbol{e}_j$ is a vector perpendicular to the hyperplane P. To prove this claim, simply let $\boldsymbol{x}$ be a point in the hyperplane other than $\boldsymbol{c}$. Form the vector $\boldsymbol{x} - \boldsymbol{c}$. This vector is parallel to the hyperplane because both $\boldsymbol{x}$ and $\boldsymbol{c}$ lie in P. Since

$$\langle \boldsymbol{v}, \boldsymbol{x} - \boldsymbol{c} \rangle = \sum_{j=1}^{n} a_j (x_j - c_j) = 0,$$

we deduce that $\boldsymbol{v}$ is orthogonal to $\boldsymbol{x} - \boldsymbol{c}$. Since $\boldsymbol{x} \neq \boldsymbol{c}$ is an arbitrary point of P, we deduce that $\boldsymbol{v}$ is orthogonal to every vector parallel to $\boldsymbol{P}$. This is what is meant when we say that $\boldsymbol{v}$ is perpendicular to P.

Next, suppose that f is a real-valued function defined on an open set U in $\mathbb{R}^2$. It is often helpful to picture the function f by its graph $G(f)$. We write $z = f(\boldsymbol{x}) = f(x_1, x_2)$ and the point (x_1, x_2, z) as $(\boldsymbol{x}; z)$. The graph of f is the set

$$\begin{aligned} G(f) &= \{(x_1, x_2, z) \text{ in } \mathbb{R}^3 : (x_1, x_2) \text{ in } U, z = f(x_1, x_2)\} \\ &= \{(\boldsymbol{x}; z) \text{ in } \mathbb{R}^3 : \boldsymbol{x} \text{ in } U, z = f(\boldsymbol{x})\}. \end{aligned}$$

It is a common practice to refer to the graph of f as a *surface*,[2] but without continuity restrictions on the function f, its graph need not appear physically like any surface you have ever seen. Nor must the graph of f be *smooth*; it may appear physically to have ridges, crevices, cusps, sharp, needlelike points, a whole range of features other than the smooth surfaces of rolling hills to which you may have been conditioned. A surface is *smooth* if it has a tangent plane at every point on it. This simple statement requires a definition of the words *tangent plane*; the following discussion prepares the ground for that definition.

Fix a point $\boldsymbol{c} = (c_1, c_2)$ in the domain U of f. The point $(\boldsymbol{c}; f(\boldsymbol{c})) = (c_1, c_2, f(c_1, c_2))$ is a point in the graph of f. An arbitrary plane P passing through the point $(\boldsymbol{c}; f(\boldsymbol{c}))$ is represented by an equation of the form

$$a_1(x_1 - c_1) + a_2(x_2 - c_2) + a_3(z - f(c)) = 0, \tag{8.14}$$

where $a_1^2 + a_2^2 + a_3^2 \neq 0$. The plane P is not parallel to the z-axis in $\mathbb{R}^3$ if and only if $a_3 \neq 0$. Multiplying this equation by any nonzero constant yields an equivalent equation that represents the same plane. Thus if P is not parallel to the z-axis, then we can represent P by an equation of the form

$$a_1(x_1 - c_1) + a_2(x_2 - c_2) - (z - f(c)) = 0.$$

(Simply multiply (8.14) by $-1/a_3$ and rename the first two coefficients.) Therefore the vector

$$\boldsymbol{v} = a_1\boldsymbol{e}_1 + a_2\boldsymbol{e}_2 - \boldsymbol{e}_3$$

is perpendicular to the plane P. (See Fig. 8.3.)

[2]A careful definition of the word *surface* and a study of surfaces takes us rather outside the focus of this text. Contrast in your mind the glacially smoothed hillocks of Wisconsin with the jagged crags of the Rocky Mountains. Both have *surfaces*; the description of the former is simple compared to that of the latter, as anyone who reads topographical maps well knows.

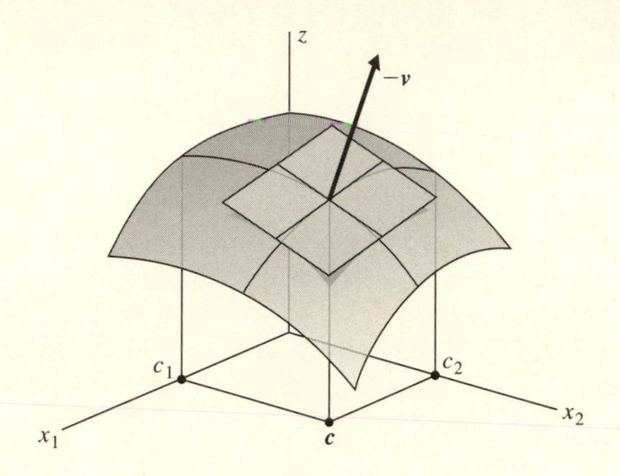

Figure 8.3

Choose a neighborhood $N(\boldsymbol{c}; r)$ contained in U. Let $\boldsymbol{t} = (t_1, t_2)$ be any vector in $N'(\boldsymbol{0}; r)$. Then $\boldsymbol{c} + \boldsymbol{t}$ is in the deleted neighborhood $N'(\boldsymbol{c}; r)$. Form the vector

$$\begin{aligned} \boldsymbol{w}(\boldsymbol{t}) &= (\boldsymbol{c} + \boldsymbol{t}; f(\boldsymbol{c} + \boldsymbol{t})) - (\boldsymbol{c}; f(\boldsymbol{c})) \\ &= t_1\boldsymbol{e}_1 + t_2\boldsymbol{e}_2 + [f(\boldsymbol{c} + \boldsymbol{t}) - f(\boldsymbol{c})]\,\boldsymbol{e}_3. \end{aligned}$$

Form the line segment joining $(\boldsymbol{c}; f(\boldsymbol{c}))$ to $(\boldsymbol{c} + \boldsymbol{t}; f(\boldsymbol{c} + \boldsymbol{t}))$. Note that this line segment is parallel to $\boldsymbol{w}(\boldsymbol{t})$ and has length $||\boldsymbol{w}(\boldsymbol{t})|| = (||\boldsymbol{t}||^2 + [f(\boldsymbol{c} + \boldsymbol{t}) - f(\boldsymbol{c})]^2)^{1/2}$. Let $\phi(\boldsymbol{t})$ denote the angle in $[0, \pi/2]$ between the line segment joining $(\boldsymbol{c}; f(\boldsymbol{c}))$ to $(\boldsymbol{c} + \boldsymbol{t}; f(\boldsymbol{c} + \boldsymbol{t}))$ and the plane P.

DEFINITION 8.1.6 The plane P is said to be *tangent* to the graph of f if $\lim_{\boldsymbol{t}\to\boldsymbol{0}} \phi(\boldsymbol{t}) = 0$. ●

Let $\theta(\boldsymbol{t})$ denote the angle between $\boldsymbol{w}(\boldsymbol{t})$ and the vector $\boldsymbol{v}$. Then $\theta(\boldsymbol{t}) = \pi/2 - \phi(\boldsymbol{t})$. (See Fig. 8.4 for a cross section of the surface and the plane.) From Section 2.1 we know that

$$\langle \boldsymbol{w}(\boldsymbol{t}), \boldsymbol{v}\rangle = ||\boldsymbol{w}(\boldsymbol{t})|| \cdot ||\boldsymbol{v}|| \cos\theta(\boldsymbol{t}).$$

Since $\theta(\boldsymbol{t}) = \pi/2 - \phi(\boldsymbol{t})$, we have $\cos\theta(\boldsymbol{t}) = \cos(\pi/2 - \phi(\boldsymbol{t})) = \sin\phi(\boldsymbol{t})$. Expanding the preceding inner product yields the equation upon which all our reasoning in the proof of Theorem 8.1.6 will hinge:

$$a_1t_1 + a_2t_2 - [f(\boldsymbol{c} + \boldsymbol{t}) - f(\boldsymbol{c})] = ||\boldsymbol{w}(\boldsymbol{t})|| \cdot ||\boldsymbol{v}|| \sin\phi(\boldsymbol{t}). \tag{8.15}$$

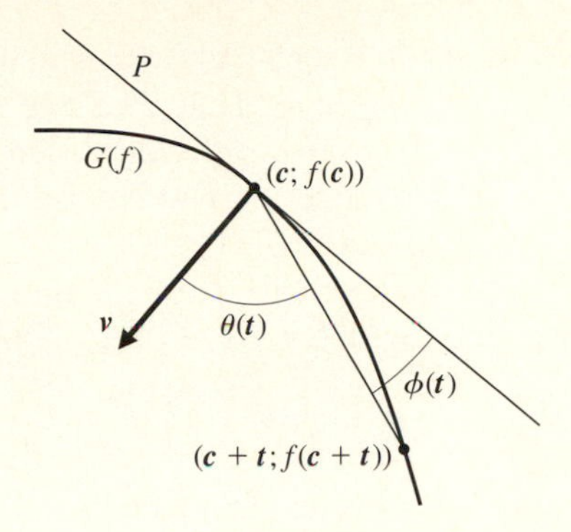

Figure 8.4

We are now ready to prove the following theorem describing the geometric significance of differentiability.

THEOREM 8.1.6 The graph of f has a tangent plane that is not parallel to the z-axis at the point $(\boldsymbol{c};\ f(\boldsymbol{c}))$ if and only if f is differentiable at $\boldsymbol{c}$. If f is differentiable at $\boldsymbol{c}$, then the equation that represents the tangent plane is

$$D_1 f(\boldsymbol{c})(x_1 - c_1) + D_2 f(\boldsymbol{c})(x_2 - c_2) - (z - f(\boldsymbol{c})) = 0.$$

Proof. Suppose first that f is differentiable at $\boldsymbol{c}$. Given any $\epsilon > 0$, choose the neighborhood $N(\mathbf{0}; r)$ so that, for all $\boldsymbol{t}$ in $N'(\mathbf{0}; r)$,

$$|f(\boldsymbol{c}+\boldsymbol{t}) - f(\boldsymbol{c}) - df(\boldsymbol{c};\boldsymbol{t})| < \epsilon ||\boldsymbol{v}|| \cdot ||\boldsymbol{t}||. \tag{8.16}$$

Let P be the plane whose equation is given in the theorem, so $a_j = D_j f(\boldsymbol{c})$, for $j = 1, 2$, and $a_1 t_1 + a_2 t_2 = df(\boldsymbol{c}; \boldsymbol{t})$. We must show that $\lim_{\boldsymbol{t}\to\mathbf{0}} \phi(\boldsymbol{t}) = 0$, where $\phi(\boldsymbol{t})$ is the angle introduced in the preceding discussion. Solving for $|\sin\phi(\boldsymbol{t})|$ in (8.15), we obtain, for any $\boldsymbol{t}$ in $N'(\mathbf{0}; r)$,

$$\begin{aligned}
|\sin\phi(\boldsymbol{t})| &= \frac{|f(\boldsymbol{c}+\boldsymbol{t}) - f(\boldsymbol{c}) - df(\boldsymbol{c};\boldsymbol{t})|}{||v|| \cdot ||\boldsymbol{w}(\boldsymbol{t})||} \\
&= \frac{1}{||\boldsymbol{v}||} \frac{|f(\boldsymbol{c}+\boldsymbol{t}) - f(\boldsymbol{c}) - df(\boldsymbol{c};\boldsymbol{t})|}{||\boldsymbol{t}||} \frac{||\boldsymbol{t}||}{||\boldsymbol{w}(\boldsymbol{t})||} \\
&< \frac{\epsilon ||\boldsymbol{v}||}{||\boldsymbol{v}||} \cdot \frac{||\boldsymbol{t}||}{||\boldsymbol{w}(\boldsymbol{t})||} \le \epsilon,
\end{aligned}$$

since $||\boldsymbol{t}||/||\boldsymbol{w}(\boldsymbol{t})|| \le 1$. Therefore $\lim_{\boldsymbol{t}\to\mathbf{0}} \sin\phi(\boldsymbol{t}) = 0$. But $\phi(\boldsymbol{t})$ is in $[0, \pi/2]$. Consequently, $\lim_{\boldsymbol{t}\to\mathbf{0}} \phi(\boldsymbol{t}) = 0$. We conclude that the plane P is tangent to the graph of f. Because in the equation representing P, the coefficient of $z - f(\boldsymbol{c})$ is not 0, the tangent plane is not parallel to the z-axis.

Conversely, suppose that the graph of f has a tangent plane P at the point $(\boldsymbol{c};\ f(\boldsymbol{c}))$ that is not parallel to the z-axis. Let the equation of P be given by

$$a_1(x_1 - c_1) + a_2(x_2 - c_2) - (z - f(\boldsymbol{c})) = 0.$$

We choose only those $\boldsymbol{t} = (t_1, t_2) \neq \boldsymbol{0}$ such that $\boldsymbol{c} + \boldsymbol{t}$ is in the open domain U of f. Again let $\phi(\boldsymbol{t})$ denote the angle between $\boldsymbol{w}(\boldsymbol{t})$ and P. Since P is tangent to the graph of f at $(\boldsymbol{c};\ f(\boldsymbol{c}))$, we know that $\lim_{\boldsymbol{t}\to\boldsymbol{0}} \phi(\boldsymbol{t}) = 0$. Consequently, $\lim_{\boldsymbol{t}\to\boldsymbol{0}} \sin\phi(\boldsymbol{t}) = 0$. We must show that the partial derivatives of f at $\boldsymbol{c}$ exist, that $a_1 = D_1 f(\boldsymbol{c}),\ a_2 = D_2 f(\boldsymbol{c})$, and that f is differentiable at $\boldsymbol{c}$.

To streamline the following algebraic computations, we let $\alpha(\boldsymbol{t}) = a_1 t_1 + a_2 t_2,\ \beta(t) = ||\boldsymbol{v}|| \sin\phi(\boldsymbol{t})$, and $\Delta f = f(\boldsymbol{c}+\boldsymbol{t}) - f(\boldsymbol{c})$. Since

$$\lim_{\boldsymbol{t}\to\boldsymbol{0}} \beta(\boldsymbol{t}) = \lim_{\boldsymbol{t}\to\boldsymbol{0}} ||\boldsymbol{v}|| \sin\phi(\boldsymbol{t}) = 0,$$

we can also assume, without loss of generality, that the vector $\boldsymbol{t}$ is so restricted that $0 \le \beta(\boldsymbol{t}) < 1$. Also notice that, with this notation, $||\boldsymbol{w}(\boldsymbol{t})|| = (||\boldsymbol{t}||^2 + (\Delta f)^2)^{1/2}$. Substituting in (8.15), we obtain

$$\alpha(\boldsymbol{t}) - \Delta f = (||\boldsymbol{t}||^2 + (\Delta f)^2)^{1/2} \beta(\boldsymbol{t}).$$

Isolate Δf. Square both sides and collect terms to obtain

$$[1 - \beta^2(\boldsymbol{t})](\Delta f)^2 - 2\alpha(\boldsymbol{t})(\Delta f) + \alpha^2(\boldsymbol{t}) - ||\boldsymbol{t}||^2 \beta^2(\boldsymbol{t}) = 0.$$

This is a quadratic in the quantity Δf. The quadratic formula yields, upon simplification,

$$\Delta f = \frac{\alpha(\boldsymbol{t}) \pm \beta(\boldsymbol{t})\sqrt{\alpha^2(\boldsymbol{t}) + ||\boldsymbol{t}||^2[1 - \beta^2(\boldsymbol{t})]}}{1 - \beta^2(\boldsymbol{t})}. \tag{8.17}$$

From (8.17) will follow all the conclusions we seek.

First, let $\boldsymbol{t} = (t_1, 0) = t_1 \boldsymbol{e}_1$. Equation (8.17) becomes

$$f(\boldsymbol{c} + t_1\boldsymbol{e}_1) - f(\boldsymbol{c}) = \frac{a_1 t_1 \pm \beta(\boldsymbol{t})\sqrt{a_1^2 t_1^2 + t_1^2[1 - \beta^2(\boldsymbol{t})]}}{1 - \beta^2(\boldsymbol{t})}.$$

Dividing by t_1 yields

$$\frac{f(\boldsymbol{c} + t_1\boldsymbol{e}_1) - f(\boldsymbol{c})}{t_1} = \frac{a_1 \pm \beta(\boldsymbol{t})\sqrt{a_1^2 + 1 - \beta^2(\boldsymbol{t})}}{1 - \beta^2(\boldsymbol{t})}.$$

Since $\lim_{\boldsymbol{t}\to\boldsymbol{0}} \beta(\boldsymbol{t}) = 0$, we conclude that

$$\lim_{t_1\to 0} \frac{f(\boldsymbol{c} + t_1\boldsymbol{e}_1) - f(\boldsymbol{c})}{t_1} = a_1.$$

This proves that $D_1 f(\boldsymbol{c})$ exists and equals a_1. Likewise, if we let $\boldsymbol{t} = (0, t_2) = t_2\boldsymbol{e}_2$, then

$$\lim_{t_2\to 0} \frac{f(\boldsymbol{c} + t_2\boldsymbol{e}_2) - f(\boldsymbol{c})}{t_2} = a_2.$$

Thus $D_2 f(\boldsymbol{c})$ exists and equals a_2.

Finally, we prove that f is differentiable at $\boldsymbol{c}$. That is, by using the Squeeze Play, we will show that

$$\lim_{\boldsymbol{t}\to\boldsymbol{0}} \frac{|f(\boldsymbol{c}+\boldsymbol{t}) - f(\boldsymbol{c}) - df(\boldsymbol{c};\boldsymbol{t})|}{||\boldsymbol{t}||} = 0.$$

According to what we have just proved,

$$\alpha(\boldsymbol{t}) = D_1 f(\boldsymbol{c})t_1 + D_2 f(\boldsymbol{c})t_2 = df(\boldsymbol{c};\boldsymbol{t}).$$

To abbreviate, we let df denote the differential $df(\boldsymbol{c};\boldsymbol{t})$. Then (8.17) becomes

$$\Delta f = \frac{df \pm \beta(\boldsymbol{t})\sqrt{(df)^2 + ||\boldsymbol{t}||^2[1-\beta^2(\boldsymbol{t})]}}{1-\beta^2(\boldsymbol{t})}.$$

Algebraically manipulating this equation and dividing by $||\boldsymbol{t}||$ yields

$$\frac{|\Delta f - df|}{||\boldsymbol{t}||} \le \frac{(|df|/||\boldsymbol{t}||)\beta^2(\boldsymbol{t}) + \beta(\boldsymbol{t})\sqrt{(df/||\boldsymbol{t}||)^2 + 1 - \beta^2(\boldsymbol{t})}}{1-\beta^2(\boldsymbol{t})}.$$

To obtain an upper bound for $|df|/||\boldsymbol{t}||$, let

$$M = \max\{|D_1 f(\boldsymbol{c})|,\ |D_2 f(\boldsymbol{c})|\}.$$

Then $|df| \le |D_1 f(\boldsymbol{c})||t_1| + |D_2 f(\boldsymbol{c})||t_2| \le M(|t_1| + |t_2|)$. Since, for any two non-negative numbers a and b, we have the inequality $a + b \le [2(a^2+b^2)]^{1/2}$, it follows that $|df| \le M\sqrt{2}||t||$. Therefore,

$$\frac{|df|}{||t||} \le M\sqrt{2}.$$

Consequently,

$$\frac{|\Delta f - df|}{||\boldsymbol{t}||} \le \frac{M\sqrt{2}\beta^2(\boldsymbol{t}) + \beta(\boldsymbol{t})\sqrt{2M^2 + 1 - \beta^2(\boldsymbol{t})}}{1-\beta^2(\boldsymbol{t})}.$$

Again, since $\lim_{\boldsymbol{t}\to\boldsymbol{0}} \beta(\boldsymbol{t}) = 0$, we deduce by the Squeeze Play that

$$\lim_{\boldsymbol{t}\to\boldsymbol{0}} \frac{f(\boldsymbol{c}+\boldsymbol{t}) - f(\boldsymbol{c}) - df(\boldsymbol{c};\boldsymbol{t})}{||\boldsymbol{t}||} = 0.$$

Therefore, f is differentiable at $\boldsymbol{c}$. This completes the proof of the theorem. ●

The preceding proof, including the notation, was designed to apply with only slight modifications to a real-valued function f defined on an open set U in $\mathbb{R}^n$. Let $z = f(\boldsymbol{x})$. The graph of f is the set $\{(\boldsymbol{x}; z) : \boldsymbol{x} \text{ in } U, z = f(\boldsymbol{x})\}$. We have the following theorem.

THEOREM 8.1.7 Let f be a real-valued function defined on an open set U in $\mathbb{R}^n$. The graph of f has a tangent hyperplane that is not parallel to the z-axis at the point $(\boldsymbol{c}; f(\boldsymbol{c}))$ if and only if f is differentiable at $\boldsymbol{c}$. If f is differentiable at $\boldsymbol{c}$, then the tangent hyperplane is represented by the equation $\sum_{j=1}^{n} D_j f(\boldsymbol{c})(x_j - c_j) - (z - f(\boldsymbol{c})) = 0$. ●

8.2 THE ALGEBRA OF DIFFERENTIABLE FUNCTIONS

As we have seen, the gradient ∇f can reasonably be chosen as one appropriate generalization of the derivative to real-valued, differentiable functions f defined on an open set in $\mathbb{R}^n$. For this reason the gradient is also called the *total derivative* of f. We know already that $\langle \nabla \boldsymbol{f}(\boldsymbol{x}), \boldsymbol{u} \rangle = D_u f(\boldsymbol{x})$ and $\langle \nabla \boldsymbol{f}(\boldsymbol{x}), \boldsymbol{t} \rangle = df(\boldsymbol{x}; \boldsymbol{t})$, but these are by no means the only useful and illuminating of its properties. Here we explore further the properties of the gradient and the differential. We will assume that all functions considered are defined and differentiable on an open set U in $\mathbb{R}^n$. To streamline the notation, we will emphasize the functional relationships; when the functions are evaluated at a point $\boldsymbol{x}$ of U, a numerical equation results.

THEOREM 8.2.1 The gradient has the following properties.

i) For any differentiable function f and any a in $\mathbb{R}$,

$$\nabla(\boldsymbol{af}) = a(\nabla \boldsymbol{f}).$$

ii) For two differentiable functions f and g,

$$\nabla(\boldsymbol{f} + \boldsymbol{g}) = \nabla \boldsymbol{f} + \nabla \boldsymbol{g}.$$

iii) For two differentiable functions f and g,

$$\nabla(\boldsymbol{f}\boldsymbol{g}) = (\nabla \boldsymbol{f})g + f(\nabla \boldsymbol{g}).$$

iv) For a differentiable function g,

$$\nabla(1/\boldsymbol{g}) = -(\nabla \boldsymbol{g})/g^2$$

at any point $\boldsymbol{x}$ in U where $g(\boldsymbol{x}) \neq 0$.

v) For two differentiable functions f and g

$$\nabla(\boldsymbol{f}/\boldsymbol{g}) = [(\nabla \boldsymbol{f})g - f(\nabla \boldsymbol{g})]/g^2,$$

at any point $\boldsymbol{x}$ in U where $g(\boldsymbol{x}) \neq 0$.

Proof. The proofs of properties (i) and (ii) are easy and are left for you. For (iii), note that for each $j = 1, 2, \ldots, n$, $D_j(fg) = (D_j f)g + f(D_j g)$, by part (iii) of Theorem 4.1.4. Thus

$$\begin{aligned}
\nabla(\boldsymbol{f}\boldsymbol{g}) &= ((D_1 f)g + f(D_1 g), (D_2 f)g + f(D_2 g), \ldots, (D_n f)g + f(D_n g)) \\
&= (D_1 f, D_2 f, \ldots, D_n f)g + f(D_1 g, D_2 g, \ldots, D_n g) \\
&= (\nabla \boldsymbol{f})g + f(\nabla \boldsymbol{g}).
\end{aligned}$$

For part (iv) we simply observe that, for $j = 1, 2, \ldots, n$, the partial derivative of $1/g$ with respect to the jth variable x_j is $D_j(1/g) = -(D_j g)/g^2$, by part (iv) of Theorem 4.1.4. Consequently,

$$\begin{aligned}
\nabla(1/\boldsymbol{g}) &= (-(D_1 g)/g^2, -(D_2 g)/g^2, \ldots, -(D_n g)/g^2) \\
&= -(D_1 g, D_2 g, \ldots, D_n g)/g^2 = -(\nabla \boldsymbol{g})/g^2.
\end{aligned}$$

Part (v) is proved by combining parts (iii) and (iv) and is left for you to complete. ●

These properties of ∇f generalize the properties of the derivative and, surely, further strengthen your conviction that the gradient is an appropriate generalization to $\mathbb{R}^n$. We can apply the gradient's properties to obtain the corresponding properties of the directional derivative.

THEOREM 8.2.2 The directional derivative has the following properties. Let $\boldsymbol{u}$ be any unit vector in $\mathbb{R}^n$.

i) For any differentiable function f and any a in $\mathbb{R}$,

$$D_u(af) = aD_u f.$$

ii) For any two differentiable functions f and g,

$$D_u(f + g) = D_u f + D_u g.$$

iii) For any two differentiable functions f and g,

$$D_u(fg) = (D_u f)g + f(D_u g).$$

iv) For any differentiable function g,

$$D_u(1/g) = -(D_u g)/g^2$$

at any point $\boldsymbol{x}$ in U where $g(\boldsymbol{x}) \neq 0$.

v) For any two differentiable functions f and g,

$$D_u(f/g) = [(D_u f)g - f(D_u g)]/g^2$$

at any point $\boldsymbol{x}$ in U where $g(\boldsymbol{x}) \neq 0$.

Proof. All these properties follow from the basic equation $D_u f(\boldsymbol{x}) = \langle \nabla f(\boldsymbol{x}), \boldsymbol{u} \rangle$ and the properties of the inner product. The straightforward proofs are left for you. ●

THEOREM 8.2.3 The differential has the following properties.

i) For any differentiable function f and any a in $\mathbb{R}$,

$$d(af) = a(df).$$

ii) For any two differentiable functions f and g,

$$d(f + g) = df + dg.$$

iii) For any two differentiable functions f and g,

$$d(fg) = (df)g + f(dg).$$

iv) For any differentiable function g,

$$d(1/g) = -(dg)/g^2$$

at any point $\boldsymbol{x}$ in U where $g(\boldsymbol{x}) \neq 0$.

v) For any two differentiable functions f and g,

$$d(f/g) = [(df)g - f(dg)]/g^2$$

at any point $\boldsymbol{x}$ in U where $g(\boldsymbol{x}) \neq 0$.

Proof. Again, these properties follow from the properties of the gradient described in Theorem 8.2.1 and the basic equation $df(\boldsymbol{x}; \boldsymbol{t}) = \langle \nabla f(\boldsymbol{x}), \boldsymbol{t} \rangle$. We will prove (iii) and leave the remaining assertions for you to prove as exercises. Fix $\boldsymbol{x}$ in U and let $\boldsymbol{t}$ be any vector in $\mathbb{R}^n$. We have

$$\begin{aligned} d(fg)(\boldsymbol{x}; \boldsymbol{t}) &= \langle \nabla(f g)(\boldsymbol{x}), \boldsymbol{t} \rangle \\ &= \langle [\nabla f(\boldsymbol{x})] g(\boldsymbol{x}) + f(\boldsymbol{x}) [\nabla g(\boldsymbol{x})], \boldsymbol{t} \rangle \\ &= \langle \nabla f(\boldsymbol{x}), \boldsymbol{t} \rangle g(\boldsymbol{x}) + f(\boldsymbol{x}) \langle \nabla g(\boldsymbol{x}), \boldsymbol{t} \rangle \\ &= df(\boldsymbol{x}; \boldsymbol{t}) g(\boldsymbol{x}) + f(\boldsymbol{x}) dg(\boldsymbol{x}; \boldsymbol{t}). \end{aligned}$$

Thus, $d(fg) = (df)g + f(dg)$. This gives you the flavor of the remaining proofs. ●

Theorem 8.2.3 enables us to confirm that the collection of all differentiable, real-valued functions on an open set U in $\mathbb{R}^n$ forms a commutative ring with identity. That is the content of the following theorem.

THEOREM 8.2.4 The sum and the product of any two differentiable functions on U are also differentiable on U. The quotient f/g of differentiable functions is also differentiable at any point $\boldsymbol{x}$ in U where $g(\boldsymbol{x}) \neq 0$.

Proof. Suppose that f and g are differentiable on U. Let $\boldsymbol{x}$ be any point in U. Fix any $\epsilon > 0$. We can choose a neighborhood $N(\boldsymbol{x})$ such that, for every $\boldsymbol{y}$ in $N'(\boldsymbol{x})$,

$$|f(\boldsymbol{y}) - f(\boldsymbol{x}) - df(\boldsymbol{x};\ \boldsymbol{y} - \boldsymbol{x})| < \frac{1}{2}\epsilon ||\boldsymbol{y} - \boldsymbol{x}||$$

and

$$|g(\boldsymbol{y}) - g(\boldsymbol{x}) - dg(\boldsymbol{x};\ \boldsymbol{y} - \boldsymbol{x})| < \frac{1}{2}\epsilon ||\boldsymbol{y} - \boldsymbol{x}||.$$

It follows that

$$\begin{aligned} &|(f+g)(\boldsymbol{y}) - (f+g)(\boldsymbol{x}) - d(f+g)(\boldsymbol{x};\ \boldsymbol{y} - \boldsymbol{x})| \\ &\qquad \le |f(\boldsymbol{y}) - f(\boldsymbol{x}) - df(\boldsymbol{x};\ \boldsymbol{y} - \boldsymbol{x})| + |g(\boldsymbol{y}) - g(\boldsymbol{x}) - dg(\boldsymbol{x},\ \boldsymbol{y} - \boldsymbol{x})| \\ &\qquad < \frac{1}{2}\epsilon ||\boldsymbol{y} - \boldsymbol{x}|| + \frac{1}{2}\epsilon ||\boldsymbol{y} - \boldsymbol{x}|| = \epsilon ||\boldsymbol{y} - \boldsymbol{x}||. \end{aligned}$$

Thus $f + g$ is differentiable at $\boldsymbol{x}$. Since $\boldsymbol{x}$ is arbitrary in U, it follows that $f + g$ is differentiable on U.

To prove that the product fg of two differentiable functions is also differentiable, fix $\boldsymbol{x}$ in U and $\epsilon > 0$. Now, g is differentiable and therefore continuous at $\boldsymbol{x}$. Thus, assuming that $\nabla \boldsymbol{f}(\boldsymbol{x}) \neq 0$, we can choose a neighborhood $N(\boldsymbol{x}; \delta_1)$ such that, for all $\boldsymbol{y}$ in $N(\boldsymbol{x}; \delta_1)$,

$$|g(\boldsymbol{y}) - g(\boldsymbol{x})| < \frac{\epsilon}{3||\nabla \boldsymbol{f}(\boldsymbol{x})||}.$$

We can also choose a neighborhood $N(\boldsymbol{x}; \delta_2)$ such that $|g(\boldsymbol{y})| < |g(\boldsymbol{x})| + 1$ for all $\boldsymbol{y}$ in $N(\boldsymbol{x}; \delta_2)$. Assuming that $f(\boldsymbol{x}) \neq 0$, next choose a neighborhood $N(\boldsymbol{x}; \delta_3)$ such that

$$|g(\boldsymbol{y}) - g(\boldsymbol{x}) - dg(\boldsymbol{x}; \boldsymbol{y} - \boldsymbol{x})| < \frac{\epsilon||\boldsymbol{y} - \boldsymbol{x}||}{3|f(\boldsymbol{x})|},$$

for any $\boldsymbol{y}$ in $N'(\boldsymbol{x}; \delta_3)$. Finally, choose a neighborhood $N(\boldsymbol{x}; \delta_4)$ such that

$$|f(\boldsymbol{y}) - f(\boldsymbol{x}) - df(\boldsymbol{x}; \boldsymbol{y} - \boldsymbol{x})| < \frac{\epsilon||\boldsymbol{y} - \boldsymbol{x}||}{3(|g(\boldsymbol{x})| + 1)}$$

for all $\boldsymbol{y}$ in $N'(\boldsymbol{x}; \delta_4)$. Let $\delta = \min\{\delta_1, \delta_2, \delta_3, \delta_4\}$ and choose any $\boldsymbol{y}$ in $N'(\boldsymbol{x}; \delta)$. Using (iii) of Theorem 8.2.3, we write

$$\begin{aligned}
&|(fg)(\boldsymbol{y}) - (fg)(\boldsymbol{x}) - d(fg)(\boldsymbol{x}; \boldsymbol{y} - \boldsymbol{x})| \\
&= |f(\boldsymbol{y})g(\boldsymbol{y}) - f(\boldsymbol{x})g(\boldsymbol{y}) + f(\boldsymbol{x})g(\boldsymbol{y}) - f(\boldsymbol{x})g(\boldsymbol{x}) \\
&\quad -df(\boldsymbol{x}; \boldsymbol{y} - \boldsymbol{x})g(\boldsymbol{y}) + df(\boldsymbol{x}; \boldsymbol{y} - \boldsymbol{x})g(\boldsymbol{y}) \\
&\quad -df(\boldsymbol{x}; \boldsymbol{y} - \boldsymbol{x})g(\boldsymbol{x}) - f(\boldsymbol{x})dg(\boldsymbol{x}; \boldsymbol{y} - \boldsymbol{x})| \\
&\leq |f(\boldsymbol{y}) - f(\boldsymbol{x}) - df(\boldsymbol{x}; \boldsymbol{y} - \boldsymbol{x})||g(\boldsymbol{y})| \\
&\quad +|f(\boldsymbol{x})||g(\boldsymbol{y}) - g(\boldsymbol{x}) - dg(\boldsymbol{x}; \boldsymbol{y} - \boldsymbol{x})| \\
&\quad +|df(\boldsymbol{x}; \boldsymbol{y} - \boldsymbol{x})||g(\boldsymbol{y}) - g(\boldsymbol{x})|.
\end{aligned}$$

We take up these three summands in turn. First,

$$|f(\boldsymbol{y}) - f(\boldsymbol{x}) - df(\boldsymbol{x}; \boldsymbol{y} - \boldsymbol{x})||g(\boldsymbol{y})| < \epsilon||\boldsymbol{y} - \boldsymbol{x}||/3 \tag{8.18}$$

because $\boldsymbol{y}$ is in both $N(\boldsymbol{x}; \delta_2)$ and $N'(\boldsymbol{x}; \delta_4)$. Next,

$$|f(\boldsymbol{x})||g(\boldsymbol{y}) - g(\boldsymbol{x}) - dg(\boldsymbol{x}; \boldsymbol{y} - \boldsymbol{x})| < \epsilon||\boldsymbol{y} - \boldsymbol{x}||/3 \tag{8.19}$$

because $\boldsymbol{y}$ is in $N'(\boldsymbol{x}; \delta_3)$. Finally, by the Cauchy–Schwarz inequality,

$$|df(\boldsymbol{x}; \boldsymbol{y} - \boldsymbol{x})| = |\langle \nabla \boldsymbol{f}(\boldsymbol{x}), \boldsymbol{y} - \boldsymbol{x} \rangle| \leq ||\nabla \boldsymbol{f}(\boldsymbol{x})|| \cdot ||\boldsymbol{y} - \boldsymbol{x}||.$$

Therefore, since $\boldsymbol{y}$ is in $N(\boldsymbol{x}; \delta_1)$,

$$|df(\boldsymbol{x}; \boldsymbol{y} - \boldsymbol{x})||g(\boldsymbol{y}) - g(\boldsymbol{x})| < \epsilon||\boldsymbol{y} - \boldsymbol{x}||/3. \tag{8.20}$$

Combining (8.18), (8.19), and (8.20), we conclude that

$$|(fg)(\boldsymbol{y}) - (fg)(\boldsymbol{x}) - d(fg)(\boldsymbol{x}; \boldsymbol{y} - \boldsymbol{x})| < \epsilon||\boldsymbol{y} - \boldsymbol{x}||$$

for all $\boldsymbol{y}$ in $N'(\boldsymbol{x}; \delta)$. Therefore fg is differentiable at $\boldsymbol{x}$. Since $\boldsymbol{x}$ is arbitrary in U, fg is differentiable on U.

With the preceding proof as a guide, you can prove that the quotient of differentiable functions is also differentiable wherever the denominator does not vanish. ●

The multiplicative identity of the ring of all differentiable functions on U is, of course, the function that has value 1 at every $\boldsymbol{x}$ in U. The units of the ring are those differentiable functions that do not vanish on U.

8.3 The Differentiability of Vector-Valued Functions

Here we lift the results of Section 8.1 to apply to functions $\boldsymbol{f} = (f_1, f_2, \ldots, f_n)$ defined on an open set U in $\mathbb{R}^m$ with values in $\mathbb{R}^n$. Each component function f_j of $\boldsymbol{f}$ is taken to be a real-valued function defined on U. If, for $j = 1, 2, \ldots, n$ and $\boldsymbol{x}$ in U, we have $f_j(\boldsymbol{x}) = y_j$, then

$$\boldsymbol{f}(\boldsymbol{x}) = (f_1(\boldsymbol{x}), f_2(\boldsymbol{x}), \ldots, f_n(\boldsymbol{x})) = (y_1, y_2, \ldots, y_n) = \boldsymbol{y}.$$

We first establish what is meant by saying that $\boldsymbol{f}$ is differentiable at a point $\boldsymbol{c}$ in U. We then prove the main theorem of this section asserting that the differentiability of $\boldsymbol{f}$ at $\boldsymbol{c}$ is equivalent to the differentiability of all the component functions $f_1, f_2, \ldots,$ and f_n at $\boldsymbol{c}$.

DEFINITION 8.3.1 Let $\boldsymbol{f}$ be a vector-valued function defined on an open set U in $\mathbb{R}^m$ with values in $\mathbb{R}^n$. Let $\boldsymbol{c}$ be a point of U. We say that $\boldsymbol{f}$ is *differentiable* at $\boldsymbol{c}$ if there exists a function $\boldsymbol{F}(\boldsymbol{c}; \boldsymbol{t})$ defined for all $\boldsymbol{t}$ in $\mathbb{R}^m$ with values in $\mathbb{R}^n$ that has the following two properties.

i) $\boldsymbol{F}(\boldsymbol{c}; \boldsymbol{t})$ is a linear function of $\boldsymbol{t}$ on $\mathbb{R}^m$.

ii) For any $\epsilon > 0$, there exists a neighborhood $N(\mathbf{0})$ in $\mathbb{R}^m$ such that, for all t in $N'(\mathbf{0})$, $\boldsymbol{c} + \boldsymbol{t}$ is in U and

$$||\boldsymbol{f}(\boldsymbol{c} + \boldsymbol{t}) - \boldsymbol{f}(\boldsymbol{c}) - \boldsymbol{F}(\boldsymbol{c}; \boldsymbol{t})|| < \epsilon||\boldsymbol{t}||.$$

The function $\boldsymbol{F}(\boldsymbol{c}; \boldsymbol{t})$ is called the *differential* of $\boldsymbol{f}$. ●

One immediate observation to be made is this: Since $\boldsymbol{F}(\boldsymbol{c}; \boldsymbol{t})$ is a linear function from $\mathbb{R}^m$ to $\mathbb{R}^n$, it can be represented by an $n \times m$ matrix $A_F = (a_{jk})$; the value of $\boldsymbol{F}$ at $\boldsymbol{t}$ is computed by the action of A_F on $\boldsymbol{t}$. That is,

$$\boldsymbol{F}(\boldsymbol{c}; \boldsymbol{t}) = \begin{pmatrix} a_{11} & a_{12} & a_{13} & \cdots & a_{1m} \\ a_{21} & a_{22} & a_{23} & \cdots & a_{2m} \\ a_{31} & a_{32} & a_{33} & \cdots & a_{3m} \\ \vdots & \vdots & \vdots & & \vdots \\ a_{n1} & a_{n2} & a_{n3} & \cdots & a_{nm} \end{pmatrix} \begin{pmatrix} t_1 \\ t_2 \\ t_3 \\ \vdots \\ t_m \end{pmatrix} = \begin{pmatrix} \sum_{k=1}^m a_{1k}t_k \\ \sum_{k=1}^m a_{2k}t_k \\ \sum_{k=1}^m a_{3k}t_k \\ \vdots \\ \sum_{k=1}^m a_{nk}t_k \end{pmatrix} = \begin{pmatrix} \langle \boldsymbol{a}_1, \boldsymbol{t} \rangle \\ \langle \boldsymbol{a}_2, \boldsymbol{t} \rangle \\ \langle \boldsymbol{a}_3, \boldsymbol{t} \rangle \\ \vdots \\ \langle \boldsymbol{a}_n, \boldsymbol{t} \rangle \end{pmatrix},$$

where $\boldsymbol{a}_j = (a_{j1}, a_{j2}, \ldots, a_{jn})$ is the vector in the jth row of the matrix A_F. The first task is to identify the entries a_{jk} of the matrix A_F.

THEOREM 8.3.1 Let $\boldsymbol{f} = (f_1, f_2, \ldots, f_n)$ be a function defined on an open set U in $\mathbb{R}^m$ with values in $\mathbb{R}^n$. Let $\boldsymbol{c}$ be a point of U. Suppose that $\boldsymbol{f}$ is differentiable at $\boldsymbol{c}$.

Then, for each $j = 1, 2, \ldots, n$ and $k = 1, 2, \ldots, m$, the partial derivative $D_k f_j(\mathbf{c})$ exists and $a_{jk} = D_k f_j(\mathbf{c})$.

Remark. With this theorem we will have proved that the vector $\mathbf{a}_j$ in the jth row of the matrix A_F is the gradient $\nabla \mathbf{f}_j(\mathbf{c})$ of f_j at $\mathbf{c}$. Thus we will have proved that

$$\langle \mathbf{a}_j, \mathbf{t} \rangle = \langle \nabla \mathbf{f}_j(\mathbf{c}), \mathbf{t} \rangle$$

and that

$$\mathbf{F}(\mathbf{c}; \mathbf{t}) = (\langle \nabla \mathbf{f}_1(\mathbf{c}), \mathbf{t} \rangle, \langle \nabla \mathbf{f}_2(\mathbf{c}), \mathbf{t} \rangle, \ldots, \langle \nabla \mathbf{f}_n(\mathbf{c}), \mathbf{t} \rangle).$$

Proof. Fix $\epsilon > 0$. Choose a neighborhood $N(\mathbf{0}; r)$ in $\mathbb{R}^m$ such that, for all $\mathbf{t}$ in $N'(\mathbf{0}; r)$, $\mathbf{c} + \mathbf{t}$ is in U and we have

$$||\mathbf{f}(\mathbf{c} + \mathbf{t}) - \mathbf{f}(\mathbf{c}) - \mathbf{F}(\mathbf{c}; \mathbf{t})|| < \epsilon ||\mathbf{t}||. \tag{8.21}$$

Fix any $k = 1, 2, \ldots, m$ and choose h in $N'(0; r)$. Let $\mathbf{t} = h\mathbf{e}_k$. Thus $\mathbf{t}$ is in $N'(\mathbf{0}; r)$ and $\mathbf{c} + \mathbf{t}$ is in U. Note that, for this value of $\mathbf{t}$,

$$\begin{aligned}
\mathbf{F}(\mathbf{c}; h\mathbf{e}_k) &= h\mathbf{F}(\mathbf{c}; \mathbf{e}_k) \\
&= h(\langle \mathbf{a}_1, \mathbf{e}_k \rangle, \langle \mathbf{a}_2, \mathbf{e}_k \rangle, \ldots, \langle \mathbf{a}_n, \mathbf{e}_k \rangle) \\
&= h(a_{1k}, a_{2k}, \ldots, a_{nk}) = h \sum_{j=1}^{n} a_{jk} \mathbf{e}_j.
\end{aligned}$$

The inequality (8.21) becomes

$$||\mathbf{f}(\mathbf{c} + h\mathbf{e}_k) - \mathbf{f}(\mathbf{c}) - h \sum_{j=1}^{n} a_{jk} \mathbf{e}_j|| < \epsilon |h|.$$

Since, for any vector $\mathbf{v} = (v_1, v_2, \ldots, v_n)$ and for each $j = 1, 2, \ldots, n$, we have $|v_j| \leq ||\mathbf{v}||$, it follows that

$$|f_j(\mathbf{c} + h\mathbf{e}_k) - f_j(\mathbf{c}) - h a_{jk}|\epsilon < \epsilon |h|, \tag{8.22}$$

for each $j = 1, 2, 3, \ldots, n$. In (8.22) divide by $|h|$. Thus for h in $N'(0; r)$, we have

$$\left| \frac{f_j(\mathbf{c} + h\mathbf{e}_k) - f_j(\mathbf{c})}{h} - a_{jk} \right| < \epsilon.$$

This proves that $D_k f_j(\mathbf{c})$ exists and equals a_{jk} and completes the proof of the theorem. ●

With this theorem in hand we are ready to prove the main theorem of this section.

THEOREM 8.3.2 Let $\mathbf{f} = (f_1, f_2, \ldots, f_n)$ be defined on an open set U in $\mathbb{R}^m$ with values in $\mathbb{R}^n$. Let $\mathbf{c}$ be a point of U. Then $\mathbf{f}$ is differentiable at $\mathbf{c}$ if and only if, for each $j = 1, 2, \ldots, n$, the function f_j is differentiable at $\mathbf{c}$.

Proof. Suppose first that, for each $j = 1, 2, \ldots, n$, the function f_j is differentiable at $\boldsymbol{c}$. Fix $\epsilon > 0$. For each j, choose $N(\boldsymbol{0}; r_j)$ such that, for $\boldsymbol{t}$ in $N'(\boldsymbol{0}; r_j)$, we have $\boldsymbol{c} + \boldsymbol{t}$ is in U and

$$|f_j(\boldsymbol{c}+\boldsymbol{t}) - f_j(\boldsymbol{c}) - df_j(\boldsymbol{c}; \boldsymbol{t})| < \epsilon||\boldsymbol{t}||/\sqrt{n}. \tag{8.23}$$

Let $r = \min\{r_1, r_2, \ldots, r_n\}$ and choose any $\boldsymbol{t}$ in $N'(\boldsymbol{0}; r)$. Then $\boldsymbol{c} + \boldsymbol{t}$ is in U and, for each j, the inequality (8.23) holds. Let $\boldsymbol{F}(\boldsymbol{c}; \boldsymbol{t}) = (df_1(\boldsymbol{c}; \boldsymbol{t}), df_2(\boldsymbol{c}; \boldsymbol{t}), \ldots, df_n(\boldsymbol{c}; \boldsymbol{t}))$. Then

$$\begin{aligned}&||\boldsymbol{f}(\boldsymbol{c}+\boldsymbol{t}) - \boldsymbol{f}(\boldsymbol{c}) - \boldsymbol{F}(\boldsymbol{c}; \boldsymbol{t})||^2 \\ &\quad = \sum_{j=1}^{n} [f_j(\boldsymbol{c}+\boldsymbol{t}) - f_j(\boldsymbol{c}) - df_j(\boldsymbol{c}; \boldsymbol{t})]^2 < \epsilon^2||\boldsymbol{t}||^2.\end{aligned}$$

It follows that $\boldsymbol{f}$ is differentiable at $\boldsymbol{c}$.

For the converse, suppose that $\boldsymbol{f}$ is differentiable at $\boldsymbol{c}$. By Theorem 8.3.1, we know that, for $j = 1, 2, \ldots, n$ and for $k = 1, 2, \ldots, m$, each of the partial derivatives $D_k f_j(\boldsymbol{c})$ exists. Fix $\epsilon > 0$ and choose a neighborhood $N(\boldsymbol{0})$ in $\mathbb{R}^m$ such that, if $\boldsymbol{t}$ is in $N'(\boldsymbol{0})$, then $\boldsymbol{c} + \boldsymbol{t}$ is in U and

$$||\boldsymbol{f}(\boldsymbol{c}+\boldsymbol{t}) - \boldsymbol{f}(\boldsymbol{c}) - \boldsymbol{F}(\boldsymbol{c}; \boldsymbol{t})|| < \epsilon||\boldsymbol{t}||,$$

where $\boldsymbol{F}(\boldsymbol{c}; \boldsymbol{t}) = (\langle\nabla f_1(\boldsymbol{c}), \boldsymbol{t}\rangle, \langle\nabla f_2(\boldsymbol{c}), \boldsymbol{t}\rangle, \ldots, \langle\nabla f_n(\boldsymbol{c}), \boldsymbol{t}\rangle)$. It follows that, for each $j = 1, 2, \ldots, n$ and for any $\boldsymbol{t}$ in $N'(\boldsymbol{0})$, we have $\boldsymbol{c} + \boldsymbol{t}$ in U and

$$|f_j(\boldsymbol{c}+\boldsymbol{t}) - f_j(\boldsymbol{c}) - \langle\nabla f_j(\boldsymbol{c}), \boldsymbol{t}\rangle| < \epsilon||\boldsymbol{t}||.$$

Thus, for each $j = 1, 2, \ldots, n$, f_j is differentiable at $\boldsymbol{c}$. This completes the proof of the theorem. ●

Theorem 8.3.2 enables us to derive a general property of differentiable, vector-valued functions that is pertinent here and elsewhere.

THEOREM 8.3.3 Let $\boldsymbol{f} = (f_1, f_2, \ldots, f_n)$ be defined on an open set U in $\mathbb{R}^m$ and let $\boldsymbol{c}$ be a point of U. Supppose that $\boldsymbol{f}$ is differentiable at $\boldsymbol{c}$.

i) The function $\boldsymbol{f}$ satisfies a local Lipschitz condition at $\boldsymbol{c}$. That is, there exists a neighborhood $N(\boldsymbol{c}; r)$ contained in U and a constant $K > 0$ such that, if $\boldsymbol{x}$ is in $N(\boldsymbol{c}; r)$, then

$$||\boldsymbol{f}(\boldsymbol{x}) - \boldsymbol{f}(\boldsymbol{c})|| \le K||\boldsymbol{x} - \boldsymbol{c}||.$$

ii) The function $\boldsymbol{f}$ is continuous at $\boldsymbol{c}$.

Proof. We know by Theorem 8.3.2 that each function f_j is differentiable at $\boldsymbol{c}$. Therefore, we know by Theorem 8.1.3 that each f_j satisfies a local Lipschitz condition at $\boldsymbol{c}$: There exists a neighborhood $N(\boldsymbol{c}; r_j)$ contained in U and a Lipschitz constant $K_j > 0$ such that, for all $\boldsymbol{x}$ in $N(\boldsymbol{c}; r_j)$,

$$|f_j(\boldsymbol{x}) - f_j(\boldsymbol{c})| \le K_j||\boldsymbol{x} - \boldsymbol{c}||.$$

Let $r = \min\{r_1, r_2, \ldots, r_n\}$ and choose any $\boldsymbol{x}$ in $N(\boldsymbol{c}; r)$. Then

$$\begin{aligned} ||\boldsymbol{f}(\boldsymbol{x}) - \boldsymbol{f}(\boldsymbol{c})|| &= \left[\sum_{j=1}^{n} [f_j(\boldsymbol{x}) - f_j(\boldsymbol{c})]^2\right]^{1/2} \\ &\leq \left[\sum_{j=1}^{n} K_j^2 ||\boldsymbol{x} - \boldsymbol{c}||^2\right]^{1/2} = K||\boldsymbol{x} - \boldsymbol{c}||, \end{aligned}$$

where $K = [\sum_{j=1}^{n} K_j^2]^{1/2}$. Thus $\boldsymbol{f}$ satisfies a local Lipschitz condition at $\boldsymbol{c}$. The continuity of $\boldsymbol{f}$ at $\boldsymbol{c}$ follows easily. ●

8.4 THE CHAIN RULE

By now you are thoroughly familiar with the method of computing the derivative of the composition $g \circ f$ of two real-valued functions of a single variable, where f is differentiable at c in $\mathbb{R}$ and g is differentiable at $y = f(c)$ in $\mathbb{R}$. That method, the chain rule, tells us that $g \circ f$ is also differentiable at c and that $(g \circ f)'(c) = g'(f(c))f'(c)$. (See Fig. 8.5.)

We want to find the appropriate generalization of this chain rule to functions of several variables. Upon considering the problem for a moment, you may realize that, by varying the dimensions of the domains of f and g, the generalized chain rule takes on the *appearance* of three different rules, all of the same general form, all subsumed in one single rule. As we will see, in all cases the composition of a differentiable function f with a differentiable function g produces a differentiable composite $g \circ f$. Before launching into the technical details and proofs, we examine briefly the various forms of the chain rule.

Suppose first that f is a real-valued function defined on an open set U in $\mathbb{R}^n$ that is differentiable at a point $\boldsymbol{c}$ in U. Suppose also that g is defined on an open interval I in $\mathbb{R}$, that $f(\boldsymbol{c})$ is in I, and that g is differentiable at $y = f(\boldsymbol{c})$. (See Fig. 8.6 on page 356.) We can compose f and g to obtain the real-valued function $g \circ f$ with value $g(f(\boldsymbol{c}))$ at $\boldsymbol{c}$. We will prove that $g \circ f$ is also differentiable at $\boldsymbol{c}$. Since $g \circ f$ is a function of the n variables $x_1, x_2, \ldots, x_n$, we are looking first for the n partial derivatives of $g \circ f$ and for its gradient $\nabla(g \circ f)$. In fact, the elementary chain rule suffices to enable us to compute the partial derivatives of $g \circ f$ in this case. For any $j = 1, 2, \ldots, n$, hold all but the jth variable fixed and treat $(g \circ f)(\boldsymbol{x})$ as a function of that single jth variable. By the chain rule (Theorem 4.2.1),

$$D_j(g \circ f)(\boldsymbol{c}) = g'(f(\boldsymbol{c}))D_j f(\boldsymbol{c}).$$

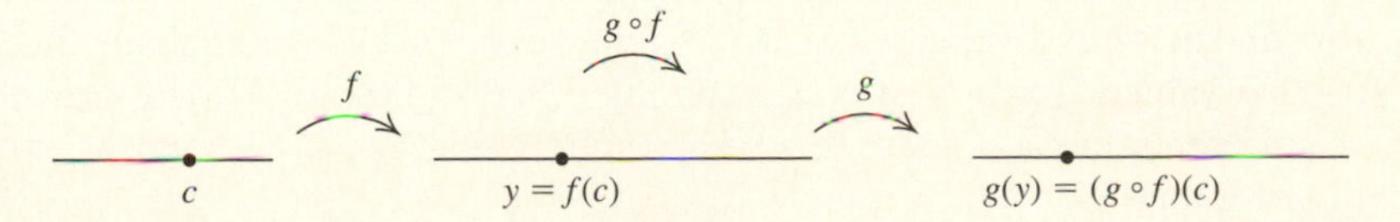

Figure 8.5

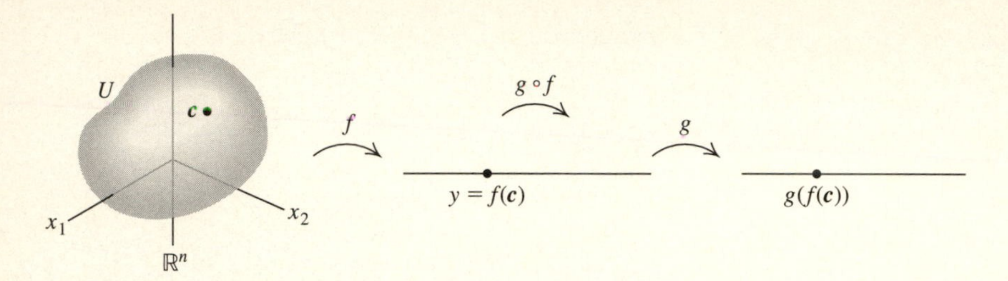

Figure 8.6

Consequently,

$$\begin{aligned}
&\nabla(\boldsymbol{g} \circ \boldsymbol{f})(\boldsymbol{c}) \\
&= (g'(f(\boldsymbol{c}))D_1 f(\boldsymbol{c}), g'(f(\boldsymbol{c}))D_2 f(\boldsymbol{c}), \ldots, g'(f(\boldsymbol{c}))D_n f(\boldsymbol{c})) \\
&= g'(f(\boldsymbol{c}))(D_1 f(\boldsymbol{c}), D_2 f(\boldsymbol{c}), \ldots, D_n f(\boldsymbol{c})) \\
&= g'(f(\boldsymbol{c}))\nabla \boldsymbol{f}(\boldsymbol{c}).
\end{aligned}$$

Notice the strong resemblance between this result and the original chain rule: The gradient of the composite function $g \circ f$ is the product of a derivative and a generalized derivative, the gradient of f. Notice also that, if n were 1, our result would reduce to the original chain rule, since in this case $\nabla \boldsymbol{f}(\boldsymbol{c}) = f'(\boldsymbol{c})$. We emphasize, however, that by these calculations we have not shown that $g \circ f$ is differentiable, but we have discovered the form that its gradient, and therefore its differential must take:

$$\nabla(\boldsymbol{g} \circ \boldsymbol{f})(\boldsymbol{c}) = g'(f(\boldsymbol{c}))\nabla \boldsymbol{f}(\boldsymbol{c})$$

and

$$d(g \circ f)(\boldsymbol{c}; \boldsymbol{t}) = \langle \nabla(\boldsymbol{g} \circ \boldsymbol{f})(\boldsymbol{c}), t \rangle = g'(f(\boldsymbol{c}))\langle \nabla \boldsymbol{f}(\boldsymbol{c}), \boldsymbol{t} \rangle.$$

We could use this formula to prove the differentiability of $f \circ g$ in accord with Definition 8.1.4.

Next, suppose that c is a point in an open interval I on which are defined n real-valued functions $f_1, f_2, \ldots, f_n$, each of which is differentiable at c. We form a function $\boldsymbol{f} = (f_1, f_2, \ldots, f_n)$ from I into $\mathbb{R}^n$. The value of $\boldsymbol{f}$ at x is

$$\boldsymbol{y} = \boldsymbol{f}(x) = (f_1(x), f_2(x), \ldots, f_n(x)).$$

Suppose that g is a real-valued function defined on an open set in $\mathbb{R}^n$ that contains $\boldsymbol{y} = \boldsymbol{f}(c)$ and that g is differentiable at $\boldsymbol{y}$. Again we can form the composite function $g \circ \boldsymbol{f}$, a real-valued function of a real variable. (See Fig. 8.7.) We will prove that $g \circ \boldsymbol{f}$ is differentiable at c. Here this means simply that $(g \circ \boldsymbol{f})'(c)$ exists. In fact, we will prove that

$$(g \circ \boldsymbol{f})'(c) = \langle (\nabla \boldsymbol{g})(\boldsymbol{f}(c)), \boldsymbol{f}'(c) \rangle,$$

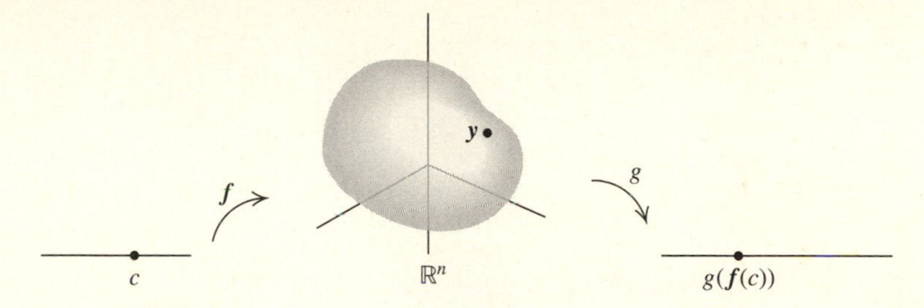

Figure 8.7

where $f'(c) = (f_1'(c), f_2'(c), \ldots, f_n'(c))$, a vector in $\mathbb{R}^n$. Again, notice the resemblance to the elementary chain rule. The gradient of g plays the role of the derivative and the inner product generalizes ordinary multiplication in $\mathbb{R}$.

Finally suppose that $\boldsymbol{f} = (f_1, f_2, \ldots, f_n)$ is a function defined on an open set U in $\mathbb{R}^m$ with values in $\mathbb{R}^n$, that $\boldsymbol{c}$ is a point of U, and that $\boldsymbol{f}$ is differentiable at $\boldsymbol{c}$. By Theorem 8.3.2, we know that each of the component functions f_j is differentiable at $\boldsymbol{c}$.

Next, let g be a real-valued function defined on an open set V in $\mathbb{R}^n$: Assume that $\boldsymbol{f}(U) \subseteq V$, that $\boldsymbol{y} = \boldsymbol{f}(\boldsymbol{c})$ is an interior point of $\boldsymbol{f}(U)$, and that g is differentiable at $\boldsymbol{y}$. The composite function $g \circ \boldsymbol{f}$ maps points of U in $\mathbb{R}^m$ into $\mathbb{R}$. (See Fig. 8.8.)

Again, we will prove that the composite function $g \circ \boldsymbol{f}$ is differentiable at $\boldsymbol{c}$. Since $g \circ \boldsymbol{f}$ is a function of m variables, we want to find its m partial derivatives. We list the formulas for these partial derivatives and extract the unifying pattern. Later we will prove that these formulas are valid. We use $\boldsymbol{y}$ to represent $\boldsymbol{f}(\boldsymbol{c})$:

$$\begin{aligned}
D_1(g \circ \boldsymbol{f})(\boldsymbol{c}) &= D_1 g(\boldsymbol{y}) D_1 f_1(\boldsymbol{c}) + D_2 g(\boldsymbol{y}) D_1 f_2(\boldsymbol{c}) + \cdots + D_n g(\boldsymbol{y}) D_1 f_n(\boldsymbol{c}) \\
D_2(g \circ \boldsymbol{f})(\boldsymbol{c}) &= D_1 g(\boldsymbol{y}) D_2 f_1(\boldsymbol{c}) + D_2 g(\boldsymbol{y}) D_2 f_2(\boldsymbol{c}) + \cdots + D_n g(\boldsymbol{y}) D_2 f_n(\boldsymbol{c}) \\
D_3(g \circ \boldsymbol{f})(\boldsymbol{c}) &= D_1 g(\boldsymbol{y}) D_3 f_1(\boldsymbol{c}) + D_2 g(\boldsymbol{y}) D_3 f_2(\boldsymbol{c}) + \cdots + D_n g(\boldsymbol{y}) D_3 f_n(\boldsymbol{c}) \\
&\vdots \\
D_m(g \circ \boldsymbol{f})(\boldsymbol{c}) &= D_1 g(\boldsymbol{y}) D_m f_1(\boldsymbol{c}) + D_2 g(\boldsymbol{y}) D_m f_2(\boldsymbol{c}) + \cdots + D_n g(\boldsymbol{y}) D_m f_n(\boldsymbol{c}).
\end{aligned}$$

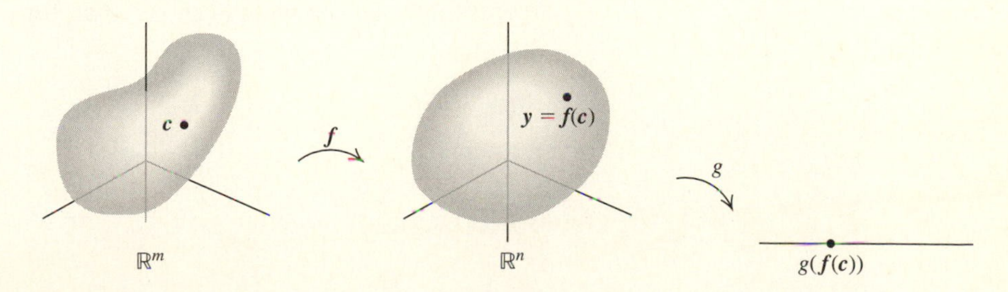

Figure 8.8

We coalesce these m equations into one: for $j = 1, 2, \ldots, m$,

$$D_j(g \circ \boldsymbol{f})(\boldsymbol{c}) = \sum_{k=1}^{n} D_k g(\boldsymbol{y}) D_j f_k(\boldsymbol{c}). \tag{8.24}$$

For each $j = 1, 2, \ldots, m$, we denote the vector

$$(D_j f_1(\boldsymbol{c}),\ D_j f_2(\boldsymbol{c}), \ldots, D_j f_n(\boldsymbol{c})),$$

by $D_j \boldsymbol{f}(\boldsymbol{c})$, the partial derivative, with respect to the jth variable x_j, of the vector-valued function $\boldsymbol{f}$, evaluated at $\boldsymbol{c}$. With this notation, (8.24) becomes

$$D_j(g \circ \boldsymbol{f})(\boldsymbol{c}) = \langle(\nabla \boldsymbol{g})(\boldsymbol{f}(\boldsymbol{c})),\ D_j \boldsymbol{f}(\boldsymbol{c})\rangle, \tag{8.25}$$

for $j = 1, 2, \ldots, m$.

The form of $D_j(g \circ \boldsymbol{f})(\boldsymbol{c})$ obtained in (8.25) again exhibits the same essential pattern of the elementary chain rule. Again the gradient of g plays the role of a derivative that multiplies a derivative of $\boldsymbol{f}$, the product being the inner product in $\mathbb{R}^n$.

Notice that all our previous chain rules are subsumed in this one. For if we take $m = n = 1$, we obtain the elementary chain rule of functions of one variable. If we take $m > 1$ and $n = 1$, we obtain our first generalization above. And if we take $m = 1$ and $n > 1$, we obtain the second new chain rule discussed above. Therefore it will suffice to prove the chain rule for this last case only.

From (8.24) we also can extract the formula for the gradient of $g \circ \boldsymbol{f}$ at $\boldsymbol{c}$. Specifically, we have

$$\begin{aligned}
\nabla(\boldsymbol{g} \circ \boldsymbol{f})(\boldsymbol{c}) &= (D_1(g \circ \boldsymbol{f})(\boldsymbol{c}), D_2(g \circ \boldsymbol{f})(\boldsymbol{c}), \ldots, D_m(g \circ \boldsymbol{f})(\boldsymbol{c})) \\
&= \left(\sum_{k=1}^{n} D_k g(\boldsymbol{y}) D_1 f_k(\boldsymbol{c}), \sum_{k=1}^{n} D_k g(\boldsymbol{y}) D_2 f_k(\boldsymbol{c}), \ldots, \sum_{k=1}^{n} D_k g(\boldsymbol{y}) D_m f_k(\boldsymbol{c})\right) \\
&= \sum_{k=1}^{n} D_k g(\boldsymbol{y})(D_1 f_k(\boldsymbol{c}), D_2 f_k(\boldsymbol{c}), \ldots, D_m f_k(\boldsymbol{c})) \\
&= \sum_{k=1}^{n} D_k g(\boldsymbol{y}) \nabla f_k(\boldsymbol{c}).
\end{aligned} \tag{8.26}$$

With this formula for the gradient of $g \circ \boldsymbol{f}$ in hand, we can finally propose a formula for the differential of $g \circ \boldsymbol{f}$. We propose here, and we will prove, that, for all $\boldsymbol{t}$ in $\mathbb{R}^m$,

$$\begin{aligned}
d(g \circ \boldsymbol{f})(\boldsymbol{c}; \boldsymbol{t}) = \langle\nabla(\boldsymbol{g} \circ \boldsymbol{f})(\boldsymbol{c}), \boldsymbol{t}\rangle &= \left\langle \sum_{k=1}^{n} D_k g(\boldsymbol{y}) \nabla f_k(\boldsymbol{c}), \boldsymbol{t} \right\rangle \\
&= \sum_{k=1}^{n} D_k\, g(\boldsymbol{y}) \langle \nabla \boldsymbol{f}_k(\boldsymbol{c}), \boldsymbol{t}\rangle = \sum_{k=1}^{n} D_k g(\boldsymbol{y}) d f_k(\boldsymbol{c}; \boldsymbol{t}).
\end{aligned} \tag{8.27}$$

Keep in mind, we have proved nothing yet; we have merely manipulated the symbols in accordance with the properties of the inner product obtained in Section

2.1 and of the differential obtained in Section 8.1. But, upon proving Theorem 8.4.1, we will have confirmed that (8.27) is a valid formula for the differential of $g \circ \boldsymbol{f}$ and therefore that (8.26) is a valid formula for the gradient of $g \circ \boldsymbol{f}$. This done, the formulas (8.24) for the m partial derivatives of $g \circ \boldsymbol{f}$ and hence all the forms of the chain rule will follow immediately.

THEOREM 8.4.1 Let U be an open set in $\mathbb{R}^m$ and let $\boldsymbol{c}$ be a point of U. Let $\boldsymbol{f} = (f_1, f_2, \ldots, f_n)$ be a function mapping U into an open set V in $\mathbb{R}^n$ that is differentiable at $\boldsymbol{c}$. Let $\boldsymbol{y} = \boldsymbol{f}(\boldsymbol{c})$. Let g be a real-valued function defined on V that is differentiable at $\boldsymbol{y}$. Then the composite function $g \circ \boldsymbol{f}$ is a real-valued function on U that is differentiable at $\boldsymbol{c}$. The differential of $g \circ \boldsymbol{f}$ at $\boldsymbol{c}$ is

$$d(g \circ \boldsymbol{f})(\boldsymbol{c}; \boldsymbol{t}) = \sum_{k=1}^{n} D_k g(\boldsymbol{f}(\boldsymbol{c}))\, df_k(\boldsymbol{c}; \boldsymbol{t}).$$

Proof. Fix $\epsilon > 0$. We must show that there exists a $\delta > 0$ such that, if $\boldsymbol{h}$ is in $N'(\boldsymbol{0}; \delta)$, then $\boldsymbol{c} + \boldsymbol{h}$ is in U and

$$|(g \circ \boldsymbol{f})(\boldsymbol{c} + \boldsymbol{h}) - (g \circ \boldsymbol{f})(\boldsymbol{c}) - d(g \circ \boldsymbol{f})(\boldsymbol{c}; \boldsymbol{h})| < \epsilon ||\boldsymbol{h}||.$$

Of course, we propose as the differential, $d(g \circ \boldsymbol{f})(\boldsymbol{c}; \boldsymbol{t})$, the quantity

$$\sum_{k=1}^{n} D_k g(\boldsymbol{f}(\boldsymbol{c}))\, df_k(\boldsymbol{c}; \boldsymbol{t}) = \sum_{k=1}^{n} D_k g(\boldsymbol{f}(\boldsymbol{c})) \langle \nabla f_k(\boldsymbol{c}), \boldsymbol{t} \rangle.$$

Any point in U different from $\boldsymbol{c}$ can be written in the form $\boldsymbol{c} + \boldsymbol{h}$ with $\boldsymbol{h} \neq \boldsymbol{0}$. For such a point, let $z = \boldsymbol{f}(\boldsymbol{c} + \boldsymbol{h})$. Eventually, we will choose $\boldsymbol{h}$ with sufficiently small norm so that $\boldsymbol{h}$ is in the required neighborhood $N'(\boldsymbol{0}; \delta)$ or, equivalently, so that $\boldsymbol{c} + \boldsymbol{h}$ is in $N'(\boldsymbol{c}; \delta)$. We fix this notation throughout the proof. (Refer to Fig. 8.9.)

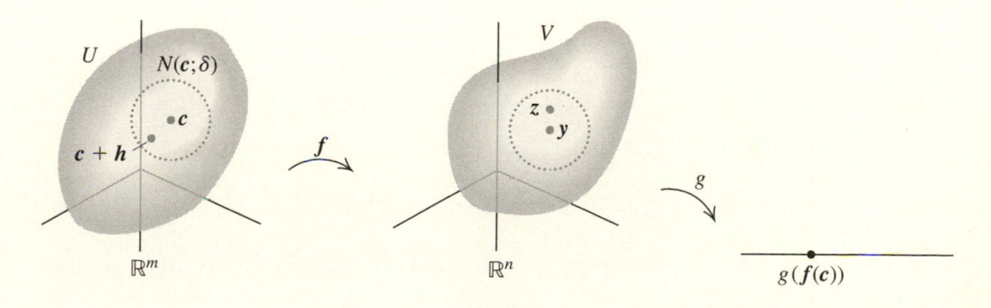

Figure 8.9

LEMMA Given any $\epsilon > 0$, there exists a neighborhood $N(\boldsymbol{0}; \delta_0)$ such that, if $\boldsymbol{h}$ is in $N'(\boldsymbol{0}; \delta_0)$, then

$$\left| \langle \nabla g(\boldsymbol{y}), z - \boldsymbol{y} \rangle - \sum_{k=1}^{n} D_k g(\boldsymbol{y}) \langle \nabla f_k(\boldsymbol{c}), \boldsymbol{h} \rangle \right| < \frac{1}{2} \epsilon ||\boldsymbol{h}||.$$

Proof of the lemma. First recall that $\boldsymbol{y} = \boldsymbol{f}(\boldsymbol{c})$ and that $z = \boldsymbol{f}(\boldsymbol{c} + \boldsymbol{h})$. Let

$$M = \max\{|D_1 g(\boldsymbol{f}(\boldsymbol{c}))|, |D_2 g(\boldsymbol{f}(\boldsymbol{c}))|, \ldots, |D_n g(\boldsymbol{f}(\boldsymbol{c}))|\}.$$

Notice that $M \geq 0$, but it may happen (nontrivially) that $M = 0$.

Since, for $k = 1, 2, \ldots, n$, each f_k is differentiable at $\boldsymbol{c}$, we can choose a neighborhood $N(\boldsymbol{0}; \delta_k)$ such that, whenever $\boldsymbol{h}$ is in $N'(\boldsymbol{0}; \delta_k)$, $\boldsymbol{c} + \boldsymbol{h}$ is in U and,

$$|f_k(\boldsymbol{c} + \boldsymbol{h}) - f_k(\boldsymbol{c}) - \langle \nabla f_k(\boldsymbol{c}), \boldsymbol{h}\rangle| < \frac{\epsilon||\boldsymbol{h}||}{2n(M + 1)}.$$

Let $\delta_0 = \min\{\delta_1, \delta_2, \ldots, \delta_n\}$ and choose any $\boldsymbol{h}$ in $N'(\boldsymbol{0}; \delta_0)$. We observe that

$$\begin{aligned}\langle \nabla g(\boldsymbol{y}), z - \boldsymbol{y}\rangle &= \langle \nabla g(\boldsymbol{f}(\boldsymbol{c})), \boldsymbol{f}(\boldsymbol{c} + \boldsymbol{h}) - \boldsymbol{f}(\boldsymbol{c})\rangle \\ &= \sum_{k=1}^{n} D_k g(\boldsymbol{f}(\boldsymbol{c}))[f_k(\boldsymbol{c} + \boldsymbol{h}) - f_k(\boldsymbol{c})].\end{aligned}$$

Therefore

$$\begin{aligned}&\left|\langle \nabla g(\boldsymbol{y}), z - \boldsymbol{y}\rangle - \sum_{k=1}^{n} D_k g(\boldsymbol{f}(\boldsymbol{c}))\langle \nabla f_k(\boldsymbol{c}), \boldsymbol{h}\rangle\right| \\ &\qquad = \left|\sum_{k=1}^{n} D_k g(\boldsymbol{f}(\boldsymbol{c}))[f_k(\boldsymbol{c} + \boldsymbol{h}) - f_k(\boldsymbol{c}) - \langle \nabla f_k(\boldsymbol{c}), \boldsymbol{h}\rangle]\right| \\ &\qquad \leq \sum_{k=1}^{n} |D_k g(\boldsymbol{f}(\boldsymbol{c}))||f_k(\boldsymbol{c} + \boldsymbol{h}) - f_k(\boldsymbol{c}) - \langle \nabla f_k(\boldsymbol{c}), \boldsymbol{h}\rangle| \\ &\qquad < \frac{nM\epsilon||\boldsymbol{h}||}{2n(M + 1)} < \frac{1}{2}\epsilon||\boldsymbol{h}||,\end{aligned}$$

as claimed in the lemma.

Next we invoke Theorem 8.3.3. Since $\boldsymbol{f}$ is differentiable at $\boldsymbol{c}$, it satisfies a local Lipschitz condition at $\boldsymbol{c}$. We choose a neighborhood $N(\boldsymbol{0}; \delta_0')$ and a constant $K > 0$ such that, for all $\boldsymbol{h}$ in $N(\boldsymbol{0}; \delta_0')$,

$$||\boldsymbol{f}(\boldsymbol{c} + \boldsymbol{h}) - \boldsymbol{f}(\boldsymbol{c})|| \leq K||\boldsymbol{h}||. \tag{8.28}$$

Because g is differentiable at $\boldsymbol{y}$, we can choose a neighborhood $N(\boldsymbol{y})$ such that, if z is in $N'(\boldsymbol{y})$, then

$$|g(z) - g(\boldsymbol{y}) - \langle \nabla g(\boldsymbol{y}), z - \boldsymbol{y}\rangle| < \frac{\epsilon||z - \boldsymbol{y}||}{2K}. \tag{8.29}$$

Now invoke Theorem 8.3.3 again. Since $\boldsymbol{f}$ is continuous at $\boldsymbol{c}$, given the neighborhood $N(\boldsymbol{y})$, we can choose a neighborhood $N(\boldsymbol{0}; \delta_0'')$ such that, if $\boldsymbol{h}$ is in $N(\boldsymbol{0}; \delta_0'')$, then $z = \boldsymbol{f}(\boldsymbol{c} + \boldsymbol{h})$ is in $N(\boldsymbol{y})$.

We assemble all we know. Let $\delta = \min\{\delta_0, \delta_0', \delta_0''\}$ and choose any $\boldsymbol{h}$ in $N'(\boldsymbol{0}; \delta)$. We know that $z = \boldsymbol{f}(\boldsymbol{c} + \boldsymbol{h})$ is in $N(\boldsymbol{y})$. If $z \neq \boldsymbol{y}$, then (8.29) holds; if z should happen to equal y, then most of the quantities in the following computation vanish and the result remains valid. The lemma and (8.28) are also valid, allowing us to write

$$\left|(g \circ f)(c+h) - (g \circ f)(c) - \sum_{k=1}^{n} D_k g(f(c))\langle \nabla f_k(c), h\rangle\right|$$

$$= |g(z) - (g y) - \langle \nabla g(y), z - y\rangle + \langle \nabla g(y), z - y\rangle - \sum_{k=1}^{n} D_k g(f(c))\langle \nabla f_k(c), h\rangle|$$

$$\leq |g(z) - g(y) - \langle \nabla g(y), z - y\rangle| + |\langle \nabla g(y), z - y\rangle - \sum_{k=1}^{n} D_k g(f(c))\langle \nabla f_k(c), h\rangle|$$

$$< \frac{\epsilon||z - y||}{2K} + \frac{1}{2}\epsilon||h|| = \frac{\epsilon||f(c+h) - f(c)||}{2K} + \frac{1}{2}\epsilon||h||$$

$$\leq \frac{\epsilon K||h||}{2K} + \frac{1}{2}\epsilon||h|| = \epsilon||h||.$$

We conclude by Theorem 8.1.1 that $g \circ f$ is differentiable at c and that

$$d(g \circ f)(c; t) = \sum_{k=1}^{n} D_k g(f(c))\langle \nabla f_k(c), t\rangle = \sum_{k=1}^{n} D_k g(f(c)) df_k(c; t)$$

This completes the proof of the theorem. ●

COROLLARY 8.4.2 With the notation of the theorem,

$$\nabla(g \circ f)(c) = \sum_{k=1}^{n} D_k g(f(c)) \nabla f_k(c).$$

Proof. Simply apply Corollary 8.1.2. ●

COROLLARY 8.4.3 The Chain Rule With the notation of the theorem, the m partial derivatives of $g \circ f$ at c are

$$D_j(g \circ f)(c) = \sum_{k=1}^{n} D_k g(f(c)) D_j f_k(c) = \langle (\nabla g)(f(c)), D_j f(c)\rangle,$$

for $j = 1, 2, \ldots, m$.

Proof. The partial derivative of $g \circ f$ with respect to the jth variable x_j, evaluated at c, is

$$D_j(g \circ f)(c) = \langle \nabla(g \circ f)(c), e_j\rangle,$$

where e_j is the jth standard unit vector in $\mathbb{R}^m$. By Corollary 8.4.2, $\nabla(g \circ f)(c) = \sum_{k=1}^{n} D_k g(f(c)) \nabla f_k(c)$. Therefore

$$D_j(g \circ f)(x) = \langle \nabla(g \circ f)(c), e_j\rangle = \sum_{k=1}^{n} D_k g(f(c)) D_j f_k(c)$$

$$= \langle (\nabla g)(f(c)), D_j f(c)\rangle. \quad ●$$

From introductory courses in calculus, you undoubtedly learned to compute partial derivatives using the chain rule. You probably used the traditional notation with

$$y_1 = f_1(x_1, x_2, \ldots, x_m)$$
$$y_2 = f_2(x_1, x_2, \ldots, x_m)$$
$$\vdots \quad \vdots \quad \vdots$$
$$y_n = f_n(x_1, x_2, \ldots, x_m)$$

and $z = g(y_1, y_2, \ldots, y_n)$ given. You then computed in traditional notation the partial derivatives

$$\partial z/\partial x_1 = \partial g/\partial y_1 \partial y_1/\partial x_1 + \partial g/\partial y_2 \partial y_2/\partial x_1 + \cdots + \partial g/\partial y_n \partial y_n/\partial x_1$$
$$\partial z/\partial x_2 = \partial g/\partial y_1 \partial y_1/\partial x_2 + \partial g/\partial y_2 \partial y_2/\partial x_2 + \cdots + \partial g/\partial y_n \partial y_n/\partial x_2$$
$$\vdots \qquad \vdots \qquad \vdots$$
$$\partial z/\partial x_m = \partial g/\partial y_1 \partial y_1/\partial x_m + \partial g/\partial y_2 \partial y_2/\partial x_m + \cdots + \partial g/\partial y_n \partial y_n/\partial x_m.$$

Each of these equations reduces to the single, compact equation of Corollary 8.4.3. Although this traditional notation is suggestive and useful, it does have a serious flaw: There is no indication of the points $\boldsymbol{c}$ or $\boldsymbol{y} = \boldsymbol{f}(\boldsymbol{c})$ where the functions $\partial y_k/\partial x_j$ or $\partial g/\partial y_k$, respectively, are to be evaluated. For performing calculations, this drawback is not significant, but for constructing a proof, such as those you have just studied, this shortcoming of the traditional notation can produce notational nightmares. What's more, it is hard on the eyes. The moral is: Learn both notations; each has its place. But when building a proof, use the modern notation. From this point, we will use whichever notation seems appropriate for the context.

EXAMPLE 4 For $\boldsymbol{x} = (x_1, x_2)$ in $\mathbb{R}^2$, let

$$y_1 = f_1(\boldsymbol{x}) = x_1^2 + x_2^2$$
$$y_2 = f_2(\boldsymbol{x}) = 2x_1x_2.$$

Form the function $\boldsymbol{f} = (f_1, f_2)$. Then

$$\boldsymbol{f}(\boldsymbol{x}) = (x_1^2 + x_2^2, 2x_1x_2)$$

and $\boldsymbol{f}$ maps the entire x_1x_2-plane into the right half-plane in $\mathbb{R}^2$. As you can easily confirm, $\boldsymbol{f}$ is continuously differentiable on all of $\mathbb{R}^2$. Furthermore,

$$D_1 f_1(\boldsymbol{x}) = 2x_1 \qquad D_1 f_2(\boldsymbol{x}) = 2x_2$$
$$D_2 f_1(\boldsymbol{x}) = 2x_2 \qquad D_2 f_2(\boldsymbol{x}) = 2x_1.$$

Suppose that, for $\boldsymbol{y} = (y_1, y_2)$, $g(\boldsymbol{y}) = e^{y_1 - y_2}$. Then, again as you can confirm, g is differentiable on all of $\mathbb{R}^2$. Also, $D_1 g(\boldsymbol{y}) = g(\boldsymbol{y})$ and $D_2 g(\boldsymbol{y}) = -g(\boldsymbol{y})$. When we form $g \circ \boldsymbol{f}$, we obtain the function

$$\begin{aligned}(g \circ \boldsymbol{f})(\boldsymbol{x}) &= \exp[(x_1^2 + x_2^2) - 2x_1 x_2] \\ &= \exp[(x_1 - x_2)^2].\end{aligned}$$

By Theorem 8.4.1, $g \circ \boldsymbol{f}$ is differentiable on all of $\mathbb{R}^2$. We can compute the partial derivatives of $g \circ \boldsymbol{f}$ in two ways: We can use the chain rule (Corollary 8.4.3), or we can compute those derivatives directly. We will do both and compare the results. By the chain rule, we have

$$\begin{aligned}D_1(g \circ \boldsymbol{f})(\boldsymbol{x}) &= D_1 g(\boldsymbol{f}(\boldsymbol{x})) D_1 f_1(\boldsymbol{x}) + D_2 g(\boldsymbol{f}(\boldsymbol{x})) D_1 f_2(\boldsymbol{x}) \\ &= g(\boldsymbol{f}(\boldsymbol{x}))(2x_1) + [-g(\boldsymbol{f}(\boldsymbol{x}))](2x_2) \\ &= 2(x_1 - x_2)\exp[(x_1 - x_2)^2]\end{aligned}$$

and

$$\begin{aligned}D_2(g \circ \boldsymbol{f})(\boldsymbol{x}) &= D_1 g(\boldsymbol{f}(\boldsymbol{x})) D_2 f_1(\boldsymbol{x}) + D_2 g(\boldsymbol{f}(\boldsymbol{x})) D_2 f_2(\boldsymbol{x}) \\ &= g(\boldsymbol{f}(\boldsymbol{x}))(2x_2) + [-g(\boldsymbol{f}(\boldsymbol{x}))](2x_1) \\ &= 2(x_2 - x_1)\exp[(x_1 - x_2)^2].\end{aligned}$$

Computing the derivatives of $(g \circ \boldsymbol{f})(\boldsymbol{x}) = \exp[(x_1 - x_2)^2]$ directly, we have, by the elementary chain rule,

$$\begin{aligned}D_1(g \circ \boldsymbol{f})(\boldsymbol{x}) &= \exp[(x_1 - x_2)^2][2(x_1 - x_2)(1)] \\ &= 2(x_1 - x_2)\exp[(x_1 - x_2)^2]\end{aligned}$$

and

$$\begin{aligned}D_2(g \circ \boldsymbol{f})(\boldsymbol{x}) &= \exp[(x_1 - x_2)^2][2(x_1 - x_2)(-1)] \\ &= 2(x_2 - x_1)\exp[(x_1 - x_2)^2].\end{aligned}$$

Our computations agree, of course. In this example, the choice of functions enabled us to compute the partial derivatives of $g \circ \boldsymbol{f}$ directly. This is not often the case; usually we must rely on the chain rule to find the derivatives of $g \circ \boldsymbol{f}$. ●

EXAMPLE 5 In this example we will use traditional notation. Let $x = f_1(r, \theta) = r\cos\theta$ and $y = f_2(r, \theta) = r\sin\theta$. Then the function $\boldsymbol{f} = (f_1, f_2)$ maps the strip

$$S = \{(r, \theta) : 0 < r < \infty, 0 \leq \theta < 2\pi\}$$

onto the entire xy-plane, omitting the point $\mathbf{0}$. All the points on the line segment $\{(0, \theta) : 0 \leq \theta < 2\pi\}$ are mapped to $\mathbf{0}$. (Refer to Fig. 8.10.) The function $\boldsymbol{f}$ is continuously differentiable on S since the functions

$$\begin{aligned}\partial x/\partial r &= \cos\theta \qquad & \partial x/\partial\theta &= -r\sin\theta \\ \partial y/\partial r &= \sin\theta \qquad & \partial y/\partial\theta &= r\cos\theta\end{aligned}$$

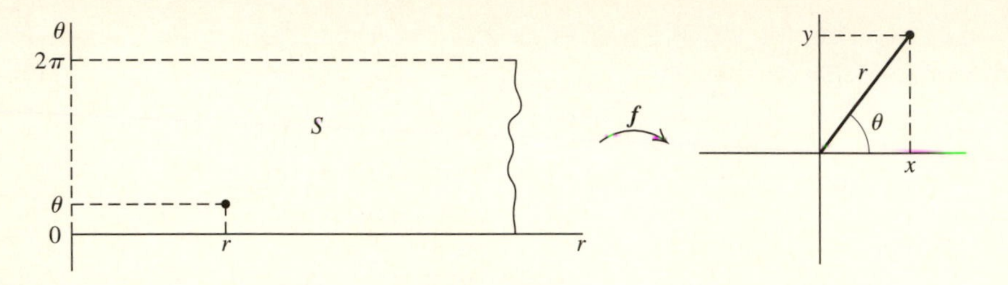

Figure 8.10

are continuous functions of (r, θ) in S. If $z = g(x, y)$ is any differentiable function of (x, y), then

$$z = g(r \cos \theta, r \sin \theta)$$

is a differentiable function of (r, θ) and, whatever may be the function g, the partial derivatives of z with respect to r and θ are given by the chain rule to be

$$\begin{aligned} \partial z/\partial r &= \partial g/\partial x \, \partial x/\partial r + \partial g/\partial y \, \partial y/\partial r \\ &= \partial g/\partial x(\cos \theta) + \partial g/\partial y(\sin \theta) \end{aligned}$$

and

$$\begin{aligned} \partial z/\partial \theta &= \partial g/\partial x \, \partial x/\partial \theta + \partial g/\partial y \, \partial y/\partial \theta \\ &= \partial g/\partial x(-r \sin \theta) + \partial g/\partial y(r \cos \theta). \end{aligned}$$

You will recognize f in this example as the function effecting the "change of coordinates" from Cartesian to polar coordinates. If g measures some quantity expressed in terms of Cartesian coordinates (x, y), then $g \circ f$ measures that same quantity in terms of polar coordinates (r, θ). (See Fig. 8.11.) ●

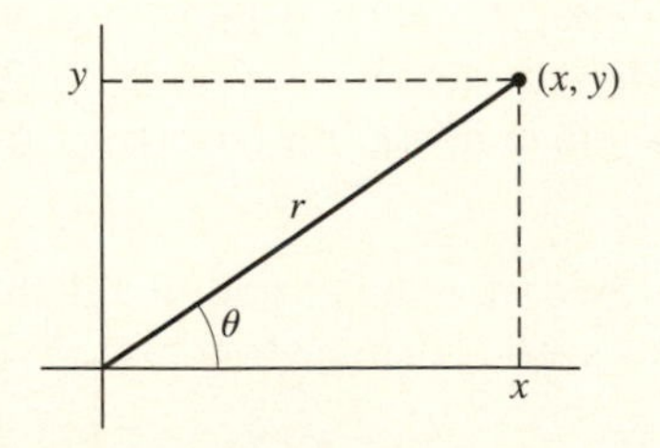

Figure 8.11

EXAMPLE 6 In certain contexts, points in $\mathbb{R}^3$ are usefully represented by spherical coordinates (ρ, θ, ϕ) with $\rho > 0, 0 \leq \theta < 2\pi$, and $0 \leq \phi \leq \pi$. A point in $\mathbb{R}^3$, other than $\mathbf{0}$, described in Cartesian coordinates by the triple (x_1, x_2, x_3) is also uniquely represented in spherical coordinates by a triple (ρ, θ, ϕ) subject to these constraints.

(See Fig. 8.12.) It is useful to think of this *change of coordinates* as a function f mapping the "slab"

$$S = \{(\rho, \theta, \phi) : 0 < \rho < \infty, 0 \le \theta < 2\pi, 0 \le \phi \le \pi\}$$

onto $\{(x_1, x_2, x_3) : x_j \text{ in } \mathbb{R}, (x_1, x_2, x_3) \neq (0, 0, 0)\}$. (See Fig. 8.13.)

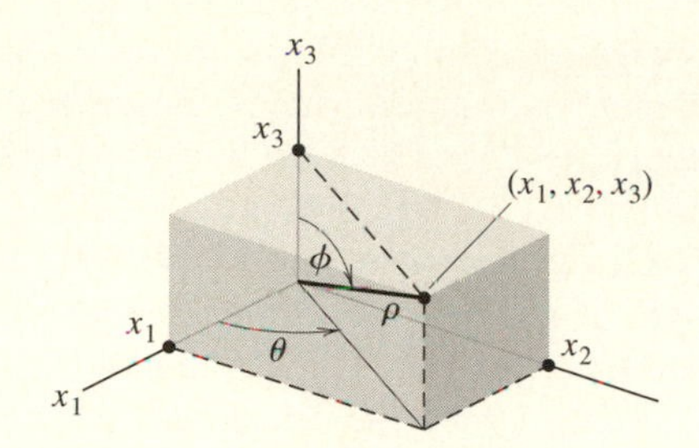

Figure 8.12

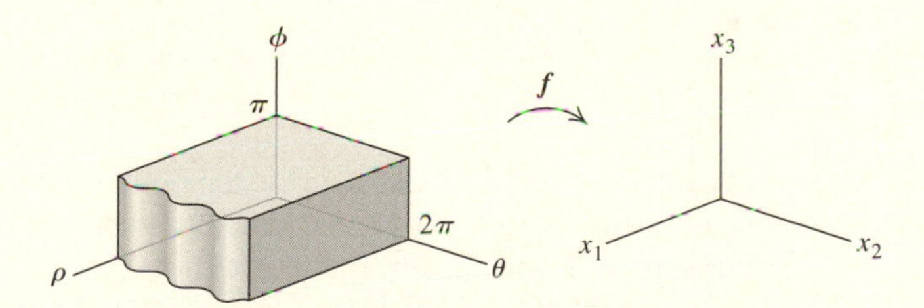

Figure 8.13

To construct f, we define $f = (f_1, f_2, f_3)$ from S to $\mathbb{R}^3$ by

$$x_1 = f_1(\rho, \theta, \phi) = \rho \cos\theta \sin\phi$$

$$x_2 = f_2(\rho, \theta, \phi) = \rho \sin\theta \sin\phi$$

$$x_3 = f_3(\rho, \theta, \phi) = \rho \cos\phi.$$

Of course, these functions result from the trigonometric relationships inherent in Fig. 8.12. The point here is that, if $u = g(x_1, x_2, x_3)$ measures some quantity in terms of the Cartesian coordinates x_1, x_2, and x_3, then $u = (g \circ f)(\rho, \theta, \phi)$ measures that same quantity in terms of the spherical coordinates ρ, θ, and ϕ. The rates of change $\partial u/\partial\rho$, $\partial u/\partial\theta$, and $\partial u/\partial\phi$ are given by the chain rule:

$$\begin{aligned}\partial u/\partial\rho &= \partial g/\partial x_1 \partial x_1/\partial\rho + \partial g/\partial x_2 \partial x_2/\partial\rho + \partial g/\partial x_3 \partial x_3/\partial\rho \\ &= \partial g/\partial x_1(\cos\theta \sin\phi) + \partial g/\partial x_2(\sin\theta \sin\phi) + \partial g/\partial x_3(\cos\phi) \\ \partial u/\partial\theta &= \partial g/\partial x_1 \partial x_1/\partial\theta + \partial g/\partial x_2 \partial x_2/\partial\theta + \partial g/\partial x_3 \partial x_3/\partial\theta \\ &= \partial g/\partial x_1(-\rho \sin\theta \sin\phi) + \partial g/\partial x_2(\rho \cos\theta \sin\phi)\end{aligned}$$

$$\begin{aligned}\partial u/\partial\phi &= \partial g/\partial x_1 \partial x_1/\partial\phi + \partial g/\partial x_2 \partial x_2/\partial\phi + \partial g/\partial x_3 \partial x_3/\partial\phi \\ &= \partial g/\partial x_1(\rho\cos\theta\cos\phi) + \partial g/\partial x_2(\rho\sin\theta\cos\phi) + \partial g/\partial x_3(-\rho\sin\phi). \quad \bullet\end{aligned}$$

EXAMPLE 7 Examples 5 and 6 can be generalized to provide a means for creating new *curvilinear coordinates* in $\mathbb{R}^n$. Suppose that $\boldsymbol{f} = (f_1, f_2, \ldots, f_n)$ is a differentiable function mapping (a subset of) $\mathbb{R}^n$ onto (a subset of) $\mathbb{R}^n$ and that $\boldsymbol{f}$ is one-to-one. Then $\boldsymbol{f}$ induces a new coordinate system in $\mathbb{R}^n$, viewed as the range of $\boldsymbol{f}$. That coordinate system is described by the n hypersurfaces $S_j = \{\boldsymbol{f}(\boldsymbol{x}) : x_j = c\}$ as c varies.

In Example 5, if we hold r constant, say $r = c > 0$, then $\boldsymbol{f}(c, \theta) = (c\cos\theta, c\sin\theta)$ describes a circle of radius c^2 as θ varies over $[0, 2\pi)$. Likewise, if we hold $\theta = \theta_0$ constant and let r vary over $(0, \infty)$, then $f(r, \theta_0) = (r\cos\theta_0, r\sin\theta_0)$ describes a ray emanating from $(0, 0)$, making an angle θ_0 with the positive x-axis. These circles and rays create the polar coordinate system in $\mathbb{R}^2$ with which you are familiar.

Likewise, in Example 6, holding $\rho = c > 0$ fixed yields a sphere of radius c^2. Holding $\theta = \theta_0$ in $[0, 2\pi)$ creates a vertical half-plane having the x_3-axis as an edge and making an angle θ_0 with the x_1x_3-plane. Holding $\phi = \phi_0$ fixed in $(0, \pi)$ describes a cone with vertex $(0, 0, 0)$ that makes an angle $\pi/2 - \phi_0$ with the x_1x_2-plane. These spheres, planes, and cones create the spherical coordinate system in $\mathbb{R}^3$.

If $\boldsymbol{f}$ is a function that defines such a coordinate system in $\mathbb{R}^n$ and if $u = g(\boldsymbol{y})$ measures a quantity relative to the Cartesian system with coordinates $(y_1, y_2, \ldots, y_n)$, then $u = (g \circ \boldsymbol{f})(\boldsymbol{x})$ measures that same quantity relative to the coordinate system induced by $\boldsymbol{f}$. The chain rule applies to enable us to compute the rates of change of u relative to the variables in this new coordinate system, that is $D_j(g \circ \boldsymbol{f})(\boldsymbol{x})$. $\bullet$

8.5 The Mean Value Theorem

The Mean Value Theorem, which has played an essential role in our previous work, lifts easily and naturally to the larger context of $\mathbb{R}^n$. Again, it provides a bridge between the change $f(\boldsymbol{y}) - f(\boldsymbol{x})$ in the value of a differentiable function f and the value of the gradient of f at a point between $\boldsymbol{x}$ and $\boldsymbol{y}$.

Two distinct points in $\mathbb{R}^n$ determine the line segment that joins them. If the line segment includes the two endpoints $\boldsymbol{x}$ and $\boldsymbol{y}$, then it is said to be a *closed line segment* and is described as the set

$$[\boldsymbol{x}, \boldsymbol{y}] = \{(1-t)\boldsymbol{x} + t\boldsymbol{y} : 0 \le t \le 1\}.$$

If the two endpoints are omitted, then the line segment is said to be an *open line segment* and is described as the set

$$(\boldsymbol{x}, \boldsymbol{y}) = \{(1-t)\boldsymbol{x} + t\boldsymbol{y} : 0 < t < 1\}.$$

THEOREM 8.5.1 The Mean Value Theorem Let f be a real-valued, differentiable function defined on an open set U in $\mathbb{R}^n$. Let $\boldsymbol{x}$ and $\boldsymbol{y}$ be any two points in U

such that the closed line segment $[\boldsymbol{x}, \boldsymbol{y}]$ is contained in U. There exists a point z in the open line segment $(\boldsymbol{x}, \boldsymbol{y})$ such that

$$f(\boldsymbol{y}) - f(\boldsymbol{x}) = \langle \nabla \boldsymbol{f}(z), \boldsymbol{y} - \boldsymbol{x} \rangle.$$

Proof. For any t in $[0, 1]$, let $z_t = (1-t)\boldsymbol{x} + t\boldsymbol{y}$. Note that $z_0 = \boldsymbol{x}$ and $z_1 = \boldsymbol{y}$. For intermediate values of t, z_t is in the open line segment $(\boldsymbol{x}, \boldsymbol{y})$ and therefore remains in U. Define $g(t) = f(z_t)$.

Notice that $g(0) = f(\boldsymbol{x})$ and $g(1) = f(\boldsymbol{y})$. We claim that g is differentiable, and therefore also continuous, at every t in $[0, 1]$. In fact, we will show that

$$g'(t) = \langle \nabla \boldsymbol{f}(z_t), \boldsymbol{y} - \boldsymbol{x} \rangle$$

for all t in $[0, 1]$. To prove this, fix any t_0 in $[0, 1]$ and observe that z_{t_0} is a point of U and thus is an interior point of U. Also, f is differentiable at z_{t_0}. Therefore, given $\epsilon > 0$, we can choose a neighborhood $N(z_{t_0}; \delta)$ such that, if z is in $N'(z_{t_0}; \delta)$, then

$$|f(z) - f(z_{t_0}) - \langle \nabla \boldsymbol{f}(z_{t_0}), z - z_{t_0} \rangle| < \epsilon ||z - z_{t_0}||.$$

Let $\delta_1 = \delta / ||\boldsymbol{y} - \boldsymbol{x}||$ and choose any t in $N'(t_0; \delta_1)$. Then

$$\begin{aligned} 0 \neq z_t - z_{t_0} &= [(1-t)\boldsymbol{x} + t\boldsymbol{y}] - [(1-t_0)\boldsymbol{x} + t_0\boldsymbol{y}] \\ &= (t - t_0)(\boldsymbol{y} - \boldsymbol{x}). \end{aligned}$$

It follows first that $0 < ||z_t - z_{t_0}|| < \delta$, so that z_t is in $N'(z_{t_0}; \delta)$. Consequently,

$$\begin{aligned} &|f(z_{t_0}) - f(z_{t_0}) - \langle \nabla \boldsymbol{f}(z_{t_0}), (t - t_0)(\boldsymbol{y} - \boldsymbol{x}) \rangle| \\ &\qquad = |g(t) - g(t_0) - (t - t_0)\langle \nabla \boldsymbol{f}(z_{t_0}), \boldsymbol{y} - \boldsymbol{x} \rangle| \\ &\qquad < \epsilon |t - t_0| \cdot ||\boldsymbol{y} - \boldsymbol{x}||. \end{aligned}$$

Dividing by $|t - t_0|$, we obtain

$$\left| \frac{g(t) - g(t_0)}{t - t_0} - \langle \nabla \boldsymbol{f}(z_{t_0}), \boldsymbol{y} - \boldsymbol{x} \rangle \right| < \epsilon ||\boldsymbol{y} - \boldsymbol{x}||.$$

Because $||\boldsymbol{y} - \boldsymbol{x}||$ is simply a positive constant, we deduce that

$$g'(t_0) = \lim_{t \to t_0} \frac{g(t) - g(t_0)}{t - t_0}$$

exists and equals $\langle \nabla \boldsymbol{f}(z_{t_0}), \boldsymbol{y} - \boldsymbol{x} \rangle$. Thus g is differentiable at every t in $[0, 1]$. By the Mean Value Theorem—that is, Theorem 4.3.3—there exists a c in $(0, 1)$ such that

$$g(1) - g(0) = g'(c).$$

Let $z = z_c$, a point in the open line segment $(\boldsymbol{x}, \boldsymbol{y})$. We conclude that

$$f(\boldsymbol{y}) - f(\boldsymbol{x}) = \langle \nabla \boldsymbol{f}(z), \boldsymbol{y} - \boldsymbol{x} \rangle$$

as promised by the theorem. ●

EXAMPLE 8 For $\boldsymbol{x} = (x_1, x_2)$ in $\mathbb{R}^2$, let

$$f(\boldsymbol{x}) = f(x_1, x_2) = x_1^2 - 3x_1x_2 + 2x_2^2.$$

Then, as you can confirm, f is differentiable on all of $\mathbb{R}^2$ and $D_1 f(\boldsymbol{x}) = 2x_1 - 3x_2$ and $D_2 f(\boldsymbol{x}) = -3x_1 + 4x_2$. Therefore $\nabla \boldsymbol{f}(\boldsymbol{x}) = (2x_1 - 3x_2)\boldsymbol{e}_1 + (-3x_1 + 4x_2)\boldsymbol{e}_2$. The Mean Value Theorem guarantees that there exists a $\boldsymbol{z}$ in the open line segment $(\boldsymbol{x}, \boldsymbol{y})$ joining $\boldsymbol{x}$ to $\boldsymbol{y}$ such that

$$f(\boldsymbol{y}) - f(\boldsymbol{x}) = \langle \nabla \boldsymbol{f}(\boldsymbol{z}), \boldsymbol{y} - \boldsymbol{x} \rangle.$$

Suppose, for example, that $\boldsymbol{x} = (1, 2)$ and $\boldsymbol{y} = (4, 3)$. Then $f(\boldsymbol{y}) - f(\boldsymbol{x}) = -5$ and $\boldsymbol{y} - \boldsymbol{x} = 3\boldsymbol{e}_1 + \boldsymbol{e}_2$. We want to find a $\boldsymbol{z} = (z_1, z_2)$ in the open line segment $(\boldsymbol{x}, \boldsymbol{y})$ such that

$$\begin{aligned} \langle \nabla \boldsymbol{f}(\boldsymbol{z}), \boldsymbol{y} - \boldsymbol{x} \rangle &= (2z_1 - 3z_2)(3) + (-3z_1 + 4z_2)(1) \\ &= 3z_1 - 5z_2 = -5. \end{aligned} \qquad \textbf{(8.30)}$$

Since $\boldsymbol{z}$ is to be on the line segment $(\boldsymbol{x}, \boldsymbol{y})$, we must have $\boldsymbol{z} = (1 - t)\boldsymbol{x} + t\boldsymbol{y}$ for some t in (0, 1) and therefore $z_1 = 1 + 3t$ and $z_2 = 2 + t$.

Substituting these values for z_1 and z_2 in (8.30), we conclude that $t = 1/2$ and thus that $\boldsymbol{z} = (5/2, 5/2)$. As you can easily confirm,

$$\langle \nabla \boldsymbol{f}(\boldsymbol{z}), \boldsymbol{y} - \boldsymbol{x} \rangle = f(\boldsymbol{y}) - f(\boldsymbol{x}),$$

as promised by the Mean Value Theorem. ●

8.6 HIGHER-ORDER PARTIAL DERIVATIVES

Just as higher-order derivatives play a significant role in the analysis of functions of a single variable, so also do higher-order partial derivatives of functions with domain in $\mathbb{R}^n$. Because the notation when treating higher-order derivatives in the general case tends to obscure the content of our discussion, we will focus here initially on real-valued functions of two independent variables defined on an open set U in $\mathbb{R}^2$. Once the issues become clear, we will examine the general case.

We assume that the first-order partial derivatives $D_1 f$ and $D_2 f$ of such a function exist on U. This being the case, then, as functions, $D_1 f$ and $D_2 f$ may have partial derivatives in their own right. These second-order partial derivatives are denoted

$$\begin{aligned} D_{11} f(\boldsymbol{x}) &= D_1[D_1 f](\boldsymbol{x}) = (D_1^2) f(\boldsymbol{x}) \\ D_{21} f(\boldsymbol{x}) &= D_2[D_1 f](\boldsymbol{x}) = (D_2 D_1) f(\boldsymbol{x}) \\ D_{12} f(\boldsymbol{x}) &= D_1[D_2 f](\boldsymbol{x}) = (D_1 D_2) f(\boldsymbol{x}) \\ D_{22} f(\boldsymbol{x}) &= D_2[D_2 f](\boldsymbol{x}) = (D_2^2) f(\boldsymbol{x}), \end{aligned}$$

provided that these derivatives exist. Thus f has four second-order partial derivatives.

EXAMPLE 9 For $\boldsymbol{x} = (x_1, x_2) \neq (0, 0)$ in $\mathbb{R}^2$, let

$$f(\boldsymbol{x}) = f(x_1, x_2) = \ln(3x_1^2 + 5x_2^2).$$

The first-order derivatives of f are

$$D_1 f(\boldsymbol{x}) = \frac{6x_1}{3x_1^2 + 5x_2^2}$$

and

$$D_2 f(\boldsymbol{x}) = \frac{10x_2}{3x_1^2 + 5x_2^2}.$$

The second-order partial derivatives of f are

$$\begin{aligned}
D_{11} f(\boldsymbol{x}) &= D_1 \frac{6x_1}{(3x_1^2 + 5x_2^2)} \\
&= \frac{-6(3x_1^2 - 5x_2^2)}{(3x_1^2 + 5x_2^2)^2}, \\
D_{21} f(\boldsymbol{x}) &= D_2 \frac{6x_1}{(3x_1^2 + 5x_2^2)} \\
&= \frac{-60x_1x_2}{(3x_1^2 + 5x_2^2)^2}, \\
D_{12} f(\boldsymbol{x}) &= D_1 \frac{10x_2}{(3x_1^2 + 5x_2^2)} \\
&= \frac{-60x_1x_2}{(3x_1^2 + 5x_2^2)^2}, \\
D_{22} f(\boldsymbol{x}) &= D_2 \frac{10x_2}{(3x_1^2 + 5x_2^2)} \\
&= \frac{10(3x_1^2 - 5x_2^2)}{(3x_1^2 + 5x_2^2)^2}. \quad \bullet
\end{aligned}$$

You will notice that the mixed partial derivatives $D_{21} f$ and $D_{12} f$ are equal. While this desirable outcome is not accidental, it need not generally occur. In the exercises you will encounter functions for which $D_{21} f \neq D_{12} f$, but we will be more interested below in establishing conditions that suffice to guarantee that the mixed partial derivatives agree.

These second-order partial derivatives may themselves have partial derivatives. If so, they would be $D_{111} f$, $D_{211} f$, $D_{121} f$, $D_{112} f$, $D_{221} f$, $D_{212} f$, $D_{122} f$, and $D_{222} f$. You will notice that there are eight possible third-order partial derivatives of a function f on an open set U in $\mathbb{R}^2$.

More generally, suppose that f is defined on an open set U in $\mathbb{R}^n$ and that all the partial derivatives $D_{j_1} f, D_{j_1 j_2} f, D_{j_1 j_2 j_3} f, \ldots, D_{j_1 j_2 \cdots j_k} f$ exist on U, where the j_i are any of $1, 2, \ldots, n$. Then each of the partial derivatives $D_{j_1 j_2 \cdots j_k} f$ is called a

kth-*order partial derivative of* f. Formally, there are n^k possible kth-order partial derivatives of a function f defined on an open set U in $\mathbb{R}^n$.

DEFINITION 8.6.1 Let U be an open set in $\mathbb{R}^n$ and let f be a real-valued function defined on U. If all the kth-order partial derivatives of f exist and are continuous on U, then f is said to be *k-times continuously differentiable on U*. The collection of all *k-times continuously differentiable functions on U* is denoted $C^{(k)}(U)$. •

A brief comment is in order. If f is in $C^{(k)}(U)$, then all the partial derivatives of f of order lower than k exist on U, have continuous first-order partial derivatives, and therefore by Theorem 8.1.5 are continuously differentiable on U.

The question arises: Are all the possible kth-order partial derivatives of f distinct? We scale down this question to the simplest question first. If f is defined on U in $\mathbb{R}^2$ and if f has second-order partial derivatives, when is it true that $D_{12}f = D_{21}f$? Our next theorem provides sufficient and generally useful conditions.

THEOREM 8.6.1 Let U be an open set in $\mathbb{R}^2$ and let $\boldsymbol{c}$ be a point of U. Suppose that f is a real-valued function on U that is twice continuously differentiable on some neighborhood $N(\boldsymbol{c}; r)$ in U. Then $D_{12}f(\boldsymbol{c}) = D_{21}f(\boldsymbol{c})$.

Proof. Fix $\boldsymbol{x} = (x_1, x_2)$ in $N'(\boldsymbol{c}; r)$. We will assume for concreteness that $c_1 < x_1$ and $c_2 < x_2$. The remaining possible cases are treated similarly. Let

$$R = [c_1, x_1] \times [c_2, x_2],$$

a closed rectangle that is contained entirely in $N(\boldsymbol{c}; r)$. (See Fig. 8.14.) Note that f, $D_1 f$, $D_2 f$, $D_{12}f$, and $D_{21}f$ are all continuous on R. Because $D_2 f$ is continuous on R, we can apply Cauchy's form of the Fundamental Theorem of Calculus to evaluate the following integral. For t_1 in (c_1, x_1), view $D_2 f(t_1, t_2)$ as a continuous function of t_2 in $[c_2, x_2]$. For $c_2 \le t_2 \le x_2$, we have

$$\int_{c_2}^{t_2} D_2 f(t_1, \tau_2)\, d\tau_2 = f(t_1, t_2) - f(t_1, c_2).$$

Let $g(t_1) = f(t_1, c_2)$, a function of t_1 alone. Therefore

$$f(t_1, t_2) = \int_{c_2}^{t_2} D_2 f(t_1, \tau_2) d\tau_2 + g(t_1). \tag{8.31}$$

Because the function $D_{12}f$ exists and is continuous on R, we can differentiate the integral in (8.31) using Leibnitz's Rule (Theorem 6.3.11), to obtain

$$D_1 \left[\int_{c_2}^{t_2} D_2 f(t_1, \tau_2) d\tau_2 \right] = \int_{c_2}^{t_2} D_{12} f(t_1, \tau_2)\, d\tau_2.$$

Therefore, from (8.31),

$$D_1 f(t_1, t_2) = \int_{c_2}^{t_2} D_{12} f(t_1, \tau_2) d\tau_2 + g'(t_1). \tag{8.32}$$

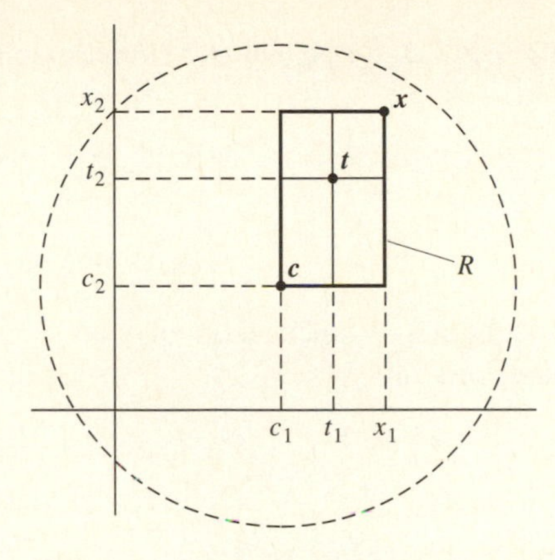

Figure 8.14

Finally, we apply the Fundamental Theorem of Calculus to the integral in (8.32) to deduce that

$$\begin{aligned} D_{21} f(t_1, t_2) &= D_2 \left[\int_{c_2}^{t_2} D_{12} f(t_1, \tau_2) \, d\tau_2 \right] + D_2 g'(t_1) \\ &= D_{12} f(t_1, t_2). \end{aligned}$$

Now let $t = (t_1, t_2)$ tend to $\boldsymbol{c} = (c_1, c_2)$. By the continuity of $D_{12} f$ and $D_{21} f$, we conclude that $D_{12} f(\boldsymbol{c}) = D_{21} f(\boldsymbol{c})$. This proves the theorem. ●

COROLLARY 8.6.2 Let f be a real-valued function defined on an open set U in $\mathbb{R}^2$. Let $\boldsymbol{c}$ be a point of U. Suppose that f is k-times continuously differentiable in a neighborhood of $\boldsymbol{c}$. Fix any $r \leq k$ and let $(j_1' j_2' \cdots j_r')$ be a permutation of $(j_1 j_2 \cdots j_r)$, where each j_p is 1 or 2. Then

$$D_{j_1' j_2' \cdots j_r'} f(\boldsymbol{c}) = D_{j_1 j_2 \cdots j_r} f(\boldsymbol{c}).$$

Remark. By this corollary, if r_1 of the j_p in $(j_1 j_2 \cdots j_r)$ are 1s, then the remaining $r_2 = r - r_1$ of the j_p are 2s; we can write without ambiguity

$$D_{j_1 j_2 \cdots j_r} f(\boldsymbol{c}) = [D_1^{r_1} D_2^{r_2} f](\boldsymbol{c}).$$

While formally a function f with domain in $\mathbb{R}^2$ has 2^k different kth-order partial derivatives, if f is k-times continuously differentiable, then only $k + 1$ of these are distinct. These are $D_1^k D_2^0 f$, $D_1^{k-1} D_2^1 f, \ldots,$ and $D_1^0 D_2^k f$. This simplification will prove useful when we treat Taylor's theorem in Section 8.7.

Proof. The inductive proof consists of recursively applying the theorem for $r = 2, 3, \ldots, k$. The theorem is the case $r = k = 2$. We leave the details for you. ●

COROLLARY 8.6.3 Let f be a real-valued function defined on an open set U in $\mathbb{R}^n$ and let $\boldsymbol{c}$ be a point of U. Suppose that f is twice continuously differentiable on some neighborhood of $\boldsymbol{c}$. Let i and j be any two indices from the set $\{1, 2, \ldots, n\}$. Then $D_{ij} f(\boldsymbol{c}) = D_{ji} f(\boldsymbol{c})$.

Proof. Without loss of generality, $i < j$. We hold all but the ith and jth variables fixed and consider the function

$$g(t_i, t_j) = f(c_1, \ldots, c_{i-1}, t_i, c_{i+1}, \ldots, c_{j-1}, t_j, c_{j+1}, \ldots, c_n).$$

Notice that g can be thought of as a function on an open set in $\mathbb{R}^2$. Furthermore, the relevant partial derivatives of g are

$$\begin{aligned} D_1 g(c_i, c_j) &= D_i f(\boldsymbol{c}) & D_2 g(c_i, c_j) &= D_j f(\boldsymbol{c}) \\ D_{11} g(c_i, c_j) &= D_{ii} f(\boldsymbol{c}) & D_{21} g(c_i, c_j) &= D_{ji} f(\boldsymbol{c}) \\ D_{12} g(c_i, c_j) &= D_{ij} f(\boldsymbol{c}) & D_{22} g(c_i, c_j) &= D_{jj} f(\boldsymbol{c}). \end{aligned}$$

It is straightforward to confirm that each of these first- and second-order partial derivatives of g is continuous on some two-dimensional neighborhood of (c_i, c_j), that is, that g is twice continuously differentiable. By Theorem 8.6.1,

$$D_{12} g(c_i, c_j) = D_{21} g(c_i, c_j).$$

Therefore, $D_{ij} f(\boldsymbol{c}) = D_{ji} f(\boldsymbol{c})$. This proves the corollary. ●

COROLLARY 8.6.4 Let f be a real-valued function defined on an open set U in $\mathbb{R}^n$. Let $\boldsymbol{c}$ be a point of U. Suppose that f is k-times continuously differentiable on some neighborhood of $\boldsymbol{c}$. Fix any $r \le k$ and let $(j_1' j_2' \cdots j_r')$ be a permutation of $(j_1 j_2 \cdots j_r)$, where each j_p is $1, 2, \ldots,$ or n. Then

$$D_{j_1' j_2' \cdots j_r'} f(\boldsymbol{c}) = D_{j_1 j_2 \cdots j_r} f(\boldsymbol{c}).$$

Remark. If, for each j in $\{1, 2, \ldots, n\}$ the symbol j occurs $r_j \ge 0$ times in $(j_1 j_2 \cdots j_r)$, then $r = r_1 + r_2 + \cdots + r_n$ and

$$D_{j_1 j_2 \cdots j_r} f(\boldsymbol{c}) = [D_1^{r_1} D_2^{r_2} \cdots D_n^{r_n} f](\boldsymbol{c}).$$

Proof. Again, the proof is recursive as in the proof of Corollary 8.6.2 and is left for you. ●

EXAMPLE 10 If f is three times continuously differentiable on an open set in $\mathbb{R}^3$, then f has the following (continuous) distinct second- and third-order partial derivatives:

- *Second Order.* $D_1^2 f$, $D_2^2 f$, $D_3^2 f$, $D_1 D_2 f = D_{12} f$, $D_1 D_3 f = D_{13} f$, and $D_2 D_3 f = D_{23} f$ for a total of six.
- *Third Order.* $D_1^3 f$, $D_2^3 f$, $D_3^3 f$, $D_1^2 D_2 f = D_{112} f$, $D_1^2 D_3 f = D_{113} f$, $D_1 D_2^2 f = D_{122} f$, $D_1 D_2 D_3 f = D_{123} f$, $D_1 D_3^2 f = D_{133} f$, $D_2^2 D_3 f = D_{223} f$, and $D_2 D_3^2 f = D_{233} f$ for a total of ten.

You are challenged in the exercises to find the general pattern that will enable you to count the number of distinct jth order partial derivatives of a function in $C^{(k)}(U)$ where U is an open set in $\mathbb{R}^n$. ●

8.7 TAYLOR'S THEOREM

We have seen the power of Taylor's theorem in both theoretical and applied aspects of the analysis of functions of a single variable. The theorem lifts naturally to functions in $C^{(k+1)}(U)$, where U is an open set in $\mathbb{R}^n$. Again, however, for $n > 2$ the notation tends to obscure the elegant content. We will therefore begin with a function f that is $(k+1)$-times continuously differentiable on an open set U in $\mathbb{R}^2$.

Taylor's theorem gives us a polynomial approximation of f in a neighborhood of a point c and provides a formula for the error or remainder which results from using the polynomial to approximate the values of f. We ought to expect that Taylor's theorem in $\mathbb{R}^2$ will show us how to construct a polynomial in *two* variables that will approximate f; we also expect to obtain a formula for the remainder. Initially, to further simplify the notation, we will assume that $\mathbf{0}$ is in U and that f is $(k+1)$-times continuously differentiable in a neighborhood $N(\mathbf{0}; r)$ of $\mathbf{0}$. We will *expand f about **0***. Later we will apply our results to the general case by using a simple shift of variables.

Choose any $\boldsymbol{x} = (x_1, x_2)$ in $N'(\mathbf{0}; r)$. For t in $[0, 1]$, define

$$g(t) = f(t\boldsymbol{x}) = f(tx_1, tx_2).$$

The values of g are simply those of f along the closed line segment $[\mathbf{0}, \boldsymbol{x}]$, which is contained entirely in $N(\mathbf{0}; r)$. Note that $g(0) = f(\mathbf{0})$ and $g(1) = f(\boldsymbol{x})$. We claim that g has $k+1$ derivatives at every point of $[0, 1]$ and that, by Taylor's theorem (Theorem 4.5.2), for every t in $(0, 1]$, there exists a c_t in $(0, t)$ such that

$$g(t) = \sum_{j=0}^{k} \frac{g^{(j)}(0)t^j}{j!} + \frac{g^{(k+1)}(c_t)t^{k+1}}{(k+1)!}. \tag{8.33}$$

Once we have found the derivatives of g, expressed in terms of the partial derivatives of f, substituted them into (8.33), and set $t = 1$, we will have derived Taylor's theorem in $\mathbb{R}^2$. By now this style must be familiar.

LEMMA Let U be an open set in $\mathbb{R}^2$ that contains $\mathbf{0}$. Let f be a real-valued function in $C^{(k+1)}(U)$. Let $N(\mathbf{0}; r)$ be a neighborhood of $\mathbf{0}$ contained entirely in U and let $\boldsymbol{x}$ be any point of $N'(\mathbf{0}; r)$. For t in $[0, 1]$ define $g(t) = f(t\boldsymbol{x})$. Then g has $k+1$ continuous derivatives on $[0, 1]$ given by

$$g^{(j)}(t) = [(x_1 D_1 + x_2 D_2)^j f](t\boldsymbol{x})$$

for $j = 0, 1, 2, \ldots, k+1$.

Remark. As the proof unfolds, the meaning of the notation describing the jth derivative of g will become apparent.

Proof. We let $g(t) = f(t\boldsymbol{x}) = f(tx_1, tx_2)$ as above. The function $\boldsymbol{h}(t) = t\boldsymbol{x}$ is a differentiable function of t on $[0, 1]$: $\boldsymbol{h} = (h_1, h_2)$ where $h_1(t) = tx_1$, $h_2(t) = tx_2$. Recall that, in the notation established in Section 8.4, the derivative of the vector-valued function $\boldsymbol{h}$ is $\boldsymbol{h}'(t) = (h_1'(t), h_2'(t)) = (x_1, x_2)$. Because f is differentiable on $N(\mathbf{0}; r)$, the composite function $(f \circ \boldsymbol{h})(t) = g(t)$ is differentiable on [0, 1]. By the chain rule

$$g'(t) = (f \circ \boldsymbol{h})'(t) = \langle(\nabla \boldsymbol{f})(\boldsymbol{h}(t)), \boldsymbol{h}'(t)\rangle = D_1 f(t\boldsymbol{x})h_1'(t) + D_2 f(t\boldsymbol{x})h_2'(t)$$
$$= [(x_1 D_1 + x_2 D_2) f](t\boldsymbol{x}) = [[(x_1 D_1 + x_2 D_2) f] \circ \boldsymbol{h}](t).$$

This is the formula given in the lemma for $j = 1$. Again, the function $(x_1 D_1 + x_2 D_2) f$ is differentiable on $N(\mathbf{0}; r)$. Therefore the composite function $[(x_1 D_1 + x_2 D_2) f] \circ \boldsymbol{h}$ is also differentiable on [0, 1]. By the chain rule, noting that $D_1 D_2 f = D_{12} f = D_{21} f = D_2 D_1 f$, we have

$$\begin{aligned} g''(t) &= D_1[x_1(D_1 f) + x_2(D_2 f)](t\boldsymbol{x})h_1'(t) \\ &\quad + D_2[x_1(D_1 f) + x_2(D_2 f)](t\boldsymbol{x})h_2'(t) \\ &= [x_1(D_1^2 f) + x_2(D_1 D_2 f)](t\boldsymbol{x})x_1 \\ &\quad + [x_1(D_2 D_1 f) + x_2(D_2^2 f)](t\boldsymbol{x})x_2 \\ &= [x_1^2(D_1^2 f) + 2x_1 x_2(D_1 D_2 f) + x_2^2(D_2^2 f)](t\boldsymbol{x}) \\ &= [(x_1 D_1 + x_2 D_2)^2 f](t\boldsymbol{x}). \end{aligned}$$

In this last step we invoked the binomial theorem. Thus

$$\begin{aligned} g''(t) &= [(x_1 D_1 + x_2 D_2)^2 f](t\boldsymbol{x}) \\ &= [[(x_1 D_1 + x_2 D_2)^2 f] \circ \boldsymbol{h}](t). \end{aligned}$$

This is the formula given in the lemma for $j = 2$. Suppose inductively that, for some $j < k$, we have

$$\begin{aligned} g^{(j)}(t) &= [(x_1 D_1 + x_2 D_2)^j f](t\boldsymbol{x}) \\ &= [[(x_1 D_1 + x_2 D_2)^j f] \circ \boldsymbol{h}](t). \end{aligned} \tag{8.34}$$

The function $\boldsymbol{h}$ is differentiable on [0, 1] and the function $(x_1 D_1 + x_2 D_2)^j f$ is differentiable on $N(\mathbf{0}; r)$. Thus the composite function $[(x_1 D_1 + x_2 D_2)^j f] \circ \boldsymbol{h}$ is differentiable on [0, 1] and the chain rule applies. We use (8.34) to compute

$$\begin{aligned} g^{(j+1)}(t) &= D_1[(x_1 D_1 + x_2 D_2)^j f](t\boldsymbol{x})h_1'(t) \\ &\quad + D_2[(x_1 D_1 + x_2 D_2)^j f](t\boldsymbol{x})h_2'(t). \end{aligned} \tag{8.35}$$

To find $D_1[(x_1 D_1 + x_2 D_2)^j f]$, we use the binomial theorem to write

$$\begin{aligned} D_1(x_1 D_1 + x_2 D_2)^j &= D_1 \sum_{i=0}^{j} C(j, i) x_1^i x_2^{j-i} D_1^i D_2^{j-i} \\ &= \sum_{i=0}^{j} C(j, i) x_1^i x_2^{j-i} D_1^{i+1} D_2^{j-i}. \end{aligned}$$

Similarly,

$$D_2(x_1 D_1 + x_2 D_2)^j = \sum_{i=0}^{j} C(j, i) x_1^i x_2^{j-i} D_1^i D_2^{j+1-i}.$$

Substituting these two expressions in (8.35) and using the fact that $h_1'(t) = x_1$ and $h_2'(t) = x_2$, we obtain

$$g^{(j+1)}(t) = \left[\sum_{i=0}^{j} C(j, i)(x_1 D_1)^{i+1}(x_2 D_2)^{j-i} f\right](t\boldsymbol{x})$$
$$+ \left[\sum_{i=0}^{j} C(j, i)(x_1 D_1)^{i}(x_2 D_2)^{j+1-i} f\right](t\boldsymbol{x}).$$

Changing index in the first sum and rearranging, we next obtain

$$g^{(j+1)}(t)$$
$$= \left[(x_1 D_1)^{j+1} + \sum_{i=1}^{j} [C(j, i-1) + C(j, i)](x_1 D_1)^i (x_2 D_2)^{j+1-i} + (x_2 D_2)^{j+1}\right] f(t\boldsymbol{x})$$
$$= \left[\sum_{i=0}^{j+1} C(j+1, i)(x_1 D_1)^i (x_2 D_2)^{j+1-i}\right] f(t\boldsymbol{x})$$
$$= [(x_1 D_1 + x_2 D_2)^{j+1} f](t\boldsymbol{x})$$

by the binomial theorem.[3] This proves inductively that the derivatives of g are those given by the lemma. ●

The work is behind us. Now we reap the rewards.

THEOREM 8.7.1 Taylor's Theorem in $\mathbb{R}^2$, Form I Let U be an open set in $\mathbb{R}^2$ that contains $\mathbf{0}$. Let f be a real-valued function defined on U that is $(k+1)$-times continuously differentiable on some neighborhood $N(\mathbf{0}; r)$ of $\mathbf{0}$. For any $\boldsymbol{x} = (x_1, x_2)$ in $N'(\mathbf{0}; r)$, there exists a point $\boldsymbol{z_x}$ in the open line segment $(\mathbf{0}, \boldsymbol{x})$ joining $\mathbf{0}$ to $\boldsymbol{x}$ such that

$$f(\boldsymbol{x}) = \sum_{j=0}^{k} \frac{[(x_1 D_1 + x_2 D_2)^j f](\mathbf{0})}{j!} + R_k(\mathbf{0}; \boldsymbol{x})$$

where $R_k(\mathbf{0}; \boldsymbol{x}) = [(x_1 D_1 + x_2 D_2)^{k+1} f](\boldsymbol{z_x})/(k+1)!$.

[3] Recall that the binomial coefficients satisfy, for $i = 1, 2, \ldots, j$,

$$C(j, i-1) + C(j, i) = C(j+1, i).$$

Proof. For t in $[0, 1]$, let $g(t) = f(t\boldsymbol{x})$. We proved in the lemma that g has $k+1$ continuous derivatives on $[0, 1]$. By Theorem 4.5.2 (Taylor's theorem in $\mathbb{R}$), there exists a c in $(0, 1)$ such that

$$g(1) = \sum_{j=0}^{k} \frac{g^{(j)}(0)}{j!} + \frac{g^{(k+1)}(c)}{(k+1)!}. \tag{8.36}$$

By the lemma, for $j = 0, 1, 2, \ldots, k+1$,

$$g^{(j)}(t) = [(x_1 D_1 + x_2 D_2)^j f](t\boldsymbol{x}).$$

Let $\boldsymbol{z}_x = c\boldsymbol{x}$, a point on the open line segment $(\boldsymbol{0}, \boldsymbol{x})$. Since $g(1) = f(\boldsymbol{x})$, substituting in (8.36) yields

$$f(\boldsymbol{x}) = \sum_{j=0}^{k} \frac{[(x_1 D_1 + x_2 D_2)^j f](\boldsymbol{0})}{j!} + R_k(\boldsymbol{0}; \boldsymbol{x}),$$

where

$$R_k(\boldsymbol{0}; \boldsymbol{x}) = \frac{g^{(k+1)}(c)}{(k+1)!} = \frac{[(x_1 D_1 + x_2 D_2)^{k+1} f](\boldsymbol{z}_x)}{(k+1)!}.$$

This completes the proof of the theorem. ●

Now let U be an open set in $\mathbb{R}^2$ and let $\boldsymbol{c}$ be a point in U. Let f be a real-valued function defined on U that is $(k+1)$-times continuously differentiable on a neighborhood $N(\boldsymbol{c}; r)$ of $\boldsymbol{c}$. Change variables by letting $\boldsymbol{y} = \boldsymbol{x} - \boldsymbol{c}$. Define $f_1(\boldsymbol{y}) = f(\boldsymbol{y} + \boldsymbol{c}) = f(\boldsymbol{x})$. Then f_1 is $(k+1)$-times differentiable on $N(\boldsymbol{0}; r)$. By Theorem 8.7.1, for every $\boldsymbol{y}$ in $N'(\boldsymbol{0}; r)$, there exists a $\boldsymbol{z}_y$ in the open segment $(\boldsymbol{0}, \boldsymbol{y})$ such that

$$f_1(\boldsymbol{y}) = \sum_{j=0}^{k} \frac{[(y_1 D_1 + y_2 D_2)^j f_1](\boldsymbol{0})}{j!} + R_k(\boldsymbol{0}; \boldsymbol{y}),$$

where $R_k(\boldsymbol{0}; \boldsymbol{y}) = [(y_1 D_1 + y_2 D_2)^{(k+1)} f_1](\boldsymbol{z}_y)/(k+1)!$. Let $\boldsymbol{z}_x = \boldsymbol{z}_y + \boldsymbol{c}$, a point in the line segment $(\boldsymbol{c}, \boldsymbol{x})$. It follows that

$$f(\boldsymbol{x}) = \sum_{j=0}^{k} \frac{[[(x_1 - c_1) D_1 + (x_2 - c_2) D_2]^j f](\boldsymbol{c})}{j!} + R_k(\boldsymbol{c}; \boldsymbol{x}),$$

where

$$R_k(\boldsymbol{c}; \boldsymbol{x}) = \frac{[[(x_1 - c_1) D_1 + (x_2 - c_2) D_2]^{(k+1)} f](\boldsymbol{z}_x)}{(k+1)!}.$$

We have proved the following theorem.

THEOREM 8.7.2 Taylor's Theorem in $\mathbb{R}^2$, Form II Let U be an open set in $\mathbb{R}^2$ and let $\boldsymbol{c}$ be a point of U. Let f be a real-valued function on U that is $(k+1)$-times continuously differentiable on a neighborhood $N(\boldsymbol{c})$ of $\boldsymbol{c}$. For each $\boldsymbol{x}$ in $N'(\boldsymbol{c})$ there exists a point $z_{\boldsymbol{x}}$ in $(\boldsymbol{c}, \boldsymbol{x})$ such that

$$f(\boldsymbol{x}) = \sum_{j=0}^{k} \frac{[[(x_1 - c_1)D_1 + (x_2 - c_2)D_2]^j f](\boldsymbol{c})}{j!} + R_k(\boldsymbol{c}; \boldsymbol{x}),$$

where

$$R_k(\boldsymbol{c}; \boldsymbol{x}) = \frac{[[(x_1 - c_1)D_1 + (x_2 - c_2)D_2]^{(k+1)} f](z_{\boldsymbol{x}})}{(k+1)!}. \quad \bullet$$

DEFINITION 8.7.1 The polynomial

$$P_k(x_1, x_2) = \sum_{j=0}^{k} \frac{[[(x_1 - c_1)D_1 + (x_2 - c_2)D_2]^j f](\boldsymbol{c})}{j!}$$

is called the kth *Taylor polynomial* of f at $\boldsymbol{c}$. As a polynomial in two variables, P_k has degree less than or equal to k. The quantity $R_k(\boldsymbol{c}; \boldsymbol{x})$ is called the *remainder*. •

EXAMPLE 11 Let U be an open set in $\mathbb{R}^2$, let $\boldsymbol{c} = (c_1, c_2)$ be a point of U, and let f be a function in $C^{(3)}(U)$. The second-degree Taylor polynomial of f at $\boldsymbol{c}$ is

$$\begin{aligned} P_2(x_1, x_2) = {} & f(\boldsymbol{c}) + D_1 f(\boldsymbol{c})(x_1 - c_1) + D_2 f(\boldsymbol{c})(x_2 - c_2) \\ & + \frac{1}{2}\{D_{11} f(\boldsymbol{c})(x_1 - c_1)^2 + 2D_{12} f(\boldsymbol{c})(x_1 - c_1)(x_2 - c_2) \\ & + D_{22} f(\boldsymbol{c})(x_2 - c_2)^2\}, \end{aligned}$$

valid for any point $\boldsymbol{x}$ in a neighborhood $N(\boldsymbol{c}; r)$ contained in U. The remainder $R_2(\boldsymbol{c}; \boldsymbol{x})$ is

$$\begin{aligned} R_2(\boldsymbol{c}; \boldsymbol{x}) = {} & (1/3!)\{D_{111} f(z_{\boldsymbol{x}})(x_1 - c_1)^3 + 3D_{112} f(z_{\boldsymbol{x}})(x_1 - c_1)^2(x_2 - c_2) \\ & + 3D_{122} f(z_{\boldsymbol{x}})(x_1 - c_1)(x_2 - c_2)^2 + D_{222} f(z_{\boldsymbol{x}})(x_2 - c_2)^3\}, \end{aligned}$$

where $z_{\boldsymbol{x}}$ is a point in the open line segment $(\boldsymbol{c}, \boldsymbol{x})$. •

To obtain an estimate for the magnitude of $R_2(\boldsymbol{c}; \boldsymbol{x})$, suppose that all the third-order partial derivatives of f are bounded on $N(\boldsymbol{c}; r)$, with $|D_{j_1 j_2 j_3} f(\boldsymbol{y})| \le M$, for all $\boldsymbol{y}$ in $N(\boldsymbol{c}; r)$. It follows that

$$|R_2(\boldsymbol{c}; \boldsymbol{x})| \le \frac{M(2r)^3}{3!}.$$

Thus, given $\epsilon > 0$, we can guarantee, by choosing r sufficiently small, that $|R_2(\boldsymbol{c}; \boldsymbol{x})| < \epsilon$. Simply choose r less than $\sqrt[3]{3\epsilon/(4M)}$. This observation inspires the following corollary to Taylor's theorem.

COROLLARY 8.7.3 To the hypotheses of Theorem 8.7.2 adjoin the additional assumption that every $(k+1)^{\text{st}}$ partial derivative of f is bounded by M on $N(\boldsymbol{c}; r)$.

i) Then $|R_k(\boldsymbol{c}; \boldsymbol{x})| \le M(2r)^{k+1}/(k+1)!$.

ii) If f has continuous partial derivatives of all orders and if all these partial derivatives are bounded by a single number M, then $\lim_{k\to\infty} R_k(\boldsymbol{c}; \boldsymbol{x}) = 0$.

Proof. Each of the products $(x_1 - c_1)^i (x_2 - c_2)^{k+1-i}$ in the expression for $R_k(\boldsymbol{c}; \boldsymbol{x})$ is bounded by r^{k+1}. Also,[4]

$$\sum_{i=0}^{k+1} C(k+1, i) = 2^{k+1}.$$

Therefore,

$$|R_k(c; \boldsymbol{x})| \le \frac{Mr^{k+1}}{(k+1)!}\left[\sum_{i=0}^{k+1} C(k+1, i)\right] = \frac{M(2r)^{k+1}}{(k+1)!}$$

as claimed in part (i) of the corollary. Since

$$\lim_{k\to\infty} \frac{(2r)^{k+1}}{(k+1)!} = 0,$$

it follows that $\lim_{k\to\infty} R_k(\boldsymbol{c}; \boldsymbol{x}) = 0$ by the Squeeze Play. This completes the proof of the corollary. ●

The proof of Taylor's theorem in $\mathbb{R}^n$ imitates those of Theorems 8.7.1 and 8.7.2, but it uses the multinomial theorem,[5] rather than the binomial theorem, at the key step in the proof of the lemma. The details are left for you to complete.

THEOREM 8.7.4 Taylor's Theorem in $\mathbb{R}^n$ Let U be an open set $\mathbb{R}^n$, let f be a real-valued function defined on U, let $\boldsymbol{c}$ be a point of U, and let $N(\boldsymbol{c}; r)$ be a neighborhood of $\boldsymbol{c}$ on which f is $(k+1)$-times continuously differentiable. For any $\boldsymbol{x}$ in $N'(\boldsymbol{c}; r)$, there exists a point $z_{\boldsymbol{x}}$ on the line segment $(\boldsymbol{c}, \boldsymbol{x})$ such that $f(\boldsymbol{x}) = P_k(\boldsymbol{x}) + R_k(\boldsymbol{c}; \boldsymbol{x})$ where P_k is a polynomial in $x_1, x_2, \ldots, x_n$ given by

$$P_k(x_1, x_2, \ldots, x_n) = \sum_{j=0}^{k} \frac{1}{j!}\left[\left[\sum_{i=1}^{n}(x_i - c_i)D_i\right]^j f\right](\boldsymbol{c})$$

and

$$R_k(\boldsymbol{c}; \boldsymbol{x}) = \frac{1}{(k+1)!}\left[\left[\sum_{i=1}^{n}(x_i - c_i)D_i\right]^{k+1} f\right](z_{\boldsymbol{x}}). \quad ●$$

8.8 Extreme Values of Differentiable Functions

An extremely important problem in both theoretical and applied aspects of analysis is to find points where a function has either a maximum or a minimum value. These

[4] For any k in $\mathbb{N}$, we have $(a+b)^k = \sum_{i=0}^{k} C(k,i)a^i b^{k-i}$. Setting $a = b = 1$ yields $2^k = \sum_{i=0}^{k} C(k,i)$.

[5] $(a_1 + a_2 + \cdots + a_n)^j = \Sigma j!/[r_1!r_2!\ldots r_n!]a^{r_1}a^{r_2}\cdots a^{r_n}$ where the sum is taken over all n-tuples $(r_1, r_2, \ldots, r_n)$ of non-negative integers such that $r_1 + r_2 + \cdots + r_n = j$.

points often represent a significant configuration in whatever situation is being modeled by the mathematics. It is to this problem that we now turn.

DEFINITION 8.8.1 Let f be a real-valued function defined on an open set U in $\mathbb{R}^n$ and let $\boldsymbol{c}$ be a point of U.

i) We say that f has a *local maximum* at $\boldsymbol{c}$ if there exists a neighborhood $N(\boldsymbol{c})$ such that $f(\boldsymbol{x}) < f(\boldsymbol{c})$ for all $\boldsymbol{x}$ in $N'(\boldsymbol{c})$.

ii) We say that f has a *local minimum* at $\boldsymbol{c}$ if there exists a neighborhood $N(\boldsymbol{c})$ such that $f(\boldsymbol{x}) > f(\boldsymbol{c})$ for all $\boldsymbol{x}$ in $N'(\boldsymbol{c})$.

iii) If f has either a local maximum or a local minimum at $\boldsymbol{c}$, then we say that f has a *local extremum* at $\boldsymbol{c}$.

iv) We say that f has a *saddle point* at $\boldsymbol{c}$ if $f - f(\boldsymbol{c})$ takes on both positive and negative values on every deleted neighborhood of $\boldsymbol{c}$. ●

Of course, we immediately want some reasonable means to determine whether a function has a local extremum at a point $\boldsymbol{c}$.The most useful such test is provided by the following theorem.

THEOREM 8.8.1 Let f be a real-valued function defined on an open set U in $\mathbb{R}^n$. Let $\boldsymbol{c}$ be a point of U. Suppose that f is differentiable at $\boldsymbol{c}$. If f has a local extremum at $\boldsymbol{c}$, then $\nabla \boldsymbol{f}(\boldsymbol{c}) = \boldsymbol{0}$. Thus $D_j f(\boldsymbol{c}) = 0$ for $j = 1, 2, \ldots, n$.

Proof. Suppose that f has a local maximum at $\boldsymbol{c}$. Then we have $f(\boldsymbol{x}) < f(\boldsymbol{c})$ for all $\boldsymbol{x}$ in some deleted neighborhood $N'(\boldsymbol{c}; r)$. Let $\boldsymbol{u}$ be any unit vector. For every h in $(0, r)$, $[f(\boldsymbol{c} + h\boldsymbol{u}) - f(\boldsymbol{c})]/h$ is negative. Likewise, for every h in $(-r, 0)$, $[f(\boldsymbol{c} + h\boldsymbol{u}) - f(\boldsymbol{c})]/h$ is positive. Thus

$$D_{\boldsymbol{u}} f(\boldsymbol{c}) = \lim_{h \to 0} \frac{f(\boldsymbol{c} + h\boldsymbol{u}) - f(\boldsymbol{c})}{h},$$

being both nonpositive and nonnegative, must be 0. Since f is differentiable at $\boldsymbol{c}$, we have $D_{\boldsymbol{u}} f(\boldsymbol{c}) = \langle \nabla \boldsymbol{f}(\boldsymbol{c}), \boldsymbol{u} \rangle$. Thus $\langle \nabla \boldsymbol{f}(\boldsymbol{c}), \boldsymbol{u} \rangle = 0$ for every unit vector $\boldsymbol{u}$. We conclude that $\nabla \boldsymbol{f}(\boldsymbol{c}) = \boldsymbol{0}$.

If f has a local minimum at $\boldsymbol{c}$, then an analogous proof, which we leave for you to supply, shows that $\nabla \boldsymbol{f}(\boldsymbol{c}) = \boldsymbol{0}$. This proves the theorem. ●

DEFINITION 8.8.2 Let f be a differentiable, real-valued function on an open set U in $\mathbb{R}^n$. Any point $\boldsymbol{c}$ in U where $\nabla \boldsymbol{f}(\boldsymbol{c}) = \boldsymbol{0}$ is called a *critical point* of f. ●

We hasten to caution: The converse of Theorem 8.8.1 is false; that $\boldsymbol{c}$ is a critical point of f does not imply that f has a local extremum at $\boldsymbol{c}$.

EXAMPLE 12 For $\boldsymbol{x} = (x_1, x_2)$ in $\mathbb{R}^2$, define $f(\boldsymbol{x}) = x_1^2 - x_2^2$. The graph of f is sketched in Fig. 8.15. We have $D_1 f(\boldsymbol{x}) = 2x_1$ and $D_2 f(\boldsymbol{x}) = -2x_2$, both of which vanish at $\boldsymbol{c} = \boldsymbol{0}$. That is, $\nabla \boldsymbol{f}(\boldsymbol{0}) = \boldsymbol{0}$. But f does not have a local extremum at $\boldsymbol{0}$. To see this, let $N(\boldsymbol{0})$ be any neighborhood of $\boldsymbol{0}$. At every $\boldsymbol{x}$ in $N'(\boldsymbol{0})$ of the form $(x_1, 0)$, we have $f(\boldsymbol{x}) > f(\boldsymbol{0}) = 0$; at every $\boldsymbol{x}$ in $N'(\boldsymbol{0})$ of the form $(0, x_2)$, $f(\boldsymbol{x}) < 0$. The conditions for a local extremum are not met. The point $\boldsymbol{0}$ is a saddle point of f. ●

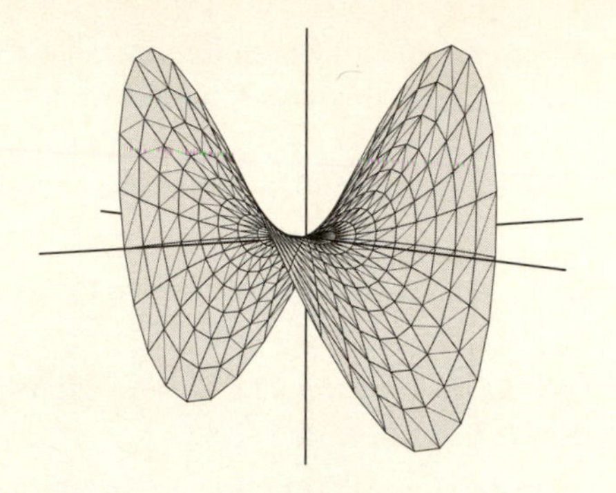

Figure 8.15

EXAMPLE 13 If $f(\boldsymbol{x}) = 2 - x_1^2 - x_2^2$, for $\boldsymbol{x} = (x_1, x_2)$ in $\mathbb{R}^2$, then f has a local maximum value of 2 at $\mathbf{0}$. In this instance, $D_1 f(\boldsymbol{x}) = -2x_1$ and $D_2 f(\boldsymbol{x}) = -2x_2$ both vanish at $\mathbf{0}$; thus $\nabla \boldsymbol{f}(\mathbf{0}) = \mathbf{0}$. If we choose any neighborhood $N(\mathbf{0})$ and any $\boldsymbol{x} = (x_1, x_2)$ in $N'(\mathbf{0})$, then

$$f(\boldsymbol{x}) = 2 - x_1^2 - x_2^2 < 2.$$

Thus f has a local maximum at $\mathbf{0}$. (See Fig. 8.16.) ●

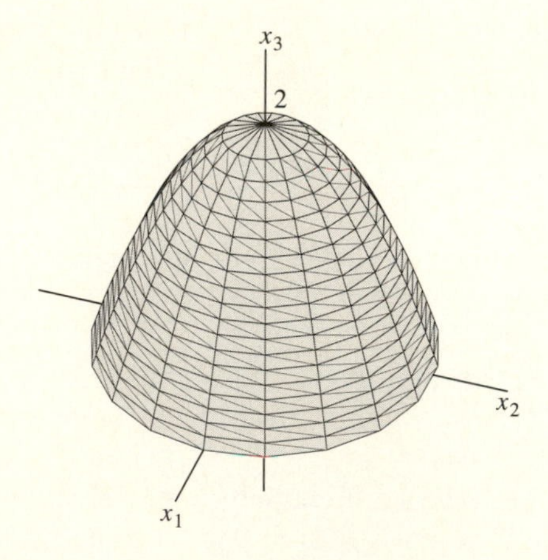

Figure 8.16

Admittedly, Examples 12 and 13 are simple; no algebraic complications interfered with our finding a critical point of f. In practice, however, we may have to resort to numerical techniques to find approximate values for the coordinates of points $\boldsymbol{c}$ where $\nabla \boldsymbol{f}(\boldsymbol{c}) = \mathbf{0}$. Once we have found the critical points of f, there still remains the problem of determining whether we have a local maximum, a local minimum, or a saddle point at each. It is to this problem that we now turn, invoking Taylor's theorem as the tool of choice.

Again, to avoid the complexity of notation, we begin with a function of two variables. Suppose that U is an open set in $\mathbb{R}^2$ and that f is in $C^{(3)}(U)$. Suppose also that $\boldsymbol{c}$ is a critical point of f; that is, $\nabla \boldsymbol{f}(\boldsymbol{c}) = \boldsymbol{0}$. Consequently, the quadratic Taylor polynomial of f at $\boldsymbol{c}$ has no first-degree terms and we can write, for all $\boldsymbol{x}$ in a neighborhood of $\boldsymbol{c}$,

$$\begin{aligned} f(\boldsymbol{x}) &= f(\boldsymbol{c}) \\ &+ \tfrac{1}{2}\{D_{11}f(\boldsymbol{c})(x_1 - c_1)^2 + 2D_{12}f(\boldsymbol{c})(x_1 - c_1)(x_2 - c_2) + D_{22}f(\boldsymbol{c})(x_2 - c_2)^2\} \\ &+ R_2(\boldsymbol{c}; \boldsymbol{x}). \end{aligned} \tag{8.37}$$

Now, all the third-order partial derivatives of f are continuous on U, so we can certainly find a neighborhood of $\boldsymbol{c}$ on which all of them are bounded by, say, M. Therefore by Example 11, given $\epsilon > 0$ we can guarantee that $|R_2(\boldsymbol{c}; \boldsymbol{x})| < \epsilon$ by making r sufficiently small and restricting $\boldsymbol{x}$ to $N'(\boldsymbol{c}; r)$. That being the case, from (8.37) we have

$$f(\boldsymbol{x}) - f(\boldsymbol{c}) = \frac{1}{2}Q(x_1 - c_1, x_2 - c_2) + R_2(\boldsymbol{c}; \boldsymbol{x}),$$

where $Q(t_1, t_2)$ is the *quadratic form*

$$Q(t_1, t_2) = D_{11}f(\boldsymbol{c})t_1^2 + 2D_{12}f(\boldsymbol{c})t_1t_2 + D_{22}f(\boldsymbol{c})t_2^2, \tag{8.38}$$

and where $|R_2(\boldsymbol{c}; \boldsymbol{x})| < \epsilon$.

Since ϵ is arbitrarily small, we deduce that f has a local maximum at $\boldsymbol{c}$ when and only when $f(\boldsymbol{x}) - f(\boldsymbol{c}) < 0$ for all $\boldsymbol{x}$ in $N'(\boldsymbol{c}; r)$, that is, when and only when

$$Q(x_1 - c_1, x_2 - c_2) < 0$$

for all such $\boldsymbol{x}$. Likewise, f has a local minimum at $\boldsymbol{c}$ if and only if $f(\boldsymbol{x}) - f(\boldsymbol{c}) > 0$ for all $\boldsymbol{x}$ in $N'(\boldsymbol{c}; r)$. But this means that

$$Q(x_1 - c_1, x_2 - c_2) > 0$$

for all $\boldsymbol{x}$ in $N'(\boldsymbol{c}; r)$. The only remaining possibility is that f has a saddle point at $\boldsymbol{c}$. In this case,

$$Q(x_1 - c_1, x_2 - c_2)$$

takes on both positive and negative values in any neighborhood of $\boldsymbol{c}$.

The key therefore to determining the behavior of f near $\boldsymbol{c}$ lies in the behavior of Q. For this reason we focus our attention on the quadratic form Q. To simplify notation let $\boldsymbol{y} = \boldsymbol{x} - \boldsymbol{c}$. That $\boldsymbol{x}$ is in $N'(\boldsymbol{c}; r)$ is equivalent to $\boldsymbol{y}$ being in $N'(\boldsymbol{0}; r)$, so we will analyze the behavior of $Q(y_1, y_2)$ near $\boldsymbol{0}$. To simplify notation further, we let

$$\begin{aligned} a_{11} &= D_{11}f(\boldsymbol{c}) \\ a_{22} &= D_{22}f(\boldsymbol{c}) \\ a_{12} &= D_{12}f(\boldsymbol{c}) = D_{21}f(\boldsymbol{c}) = a_{21}. \end{aligned}$$

From (8.38), with these assignments, we have

$$Q(y_1, y_2) = a_{11}y_1^2 + 2a_{12}y_1y_2 + a_{22}y_2^2. \tag{8.39}$$

The complication in describing the behavior of Q results from the presence of the term $2a_{12}y_1y_2$ with mixed product. Were it absent, we could resolve our question immediately. Suppose for a moment that $a_{12} = 0$ so that Q reduces to

$$Q(y_1, y_2) = a_{11}y_1^2 + a_{22}y_2^2.$$

If both a_{11} and a_{22} are positive, then, for all $\boldsymbol{y} \neq \boldsymbol{0}$, it follows that $Q(y_1, y_2) > 0$ and thus f has a local minimum at $\boldsymbol{c}$. On the other hand, if both a_{11} and a_{22} are negative, then $Q(y_1, y_2) < 0$ for all $\boldsymbol{y} \neq \boldsymbol{0}$. In this case f must have a local maximum at $\boldsymbol{c}$. Finally, if $a_{11}a_{22} < 0$, then $Q(y_1, y_2)$ takes on both positive and negative values near $\boldsymbol{0}$ and f has a saddle point at $\boldsymbol{c}$.

These observations motivate us to search for an algebraic procedure to eliminate the mixed term $2a_{12}y_1y_2$ in the formula for Q. To this end, we introduce the 2×2 matrix

$$A = \begin{pmatrix} a_{11} & a_{12} \\ a_{12} & a_{22} \end{pmatrix}.$$

Note that A has the property that $a_{ji} = a_{ij}$ for all i and j; A is said to be a *symmetric matrix*. It defines a linear transformation T_A mapping $\mathbb{R}^2$ into $\mathbb{R}^2$; the effect of T_A on a vector $\boldsymbol{y} = (y_1, y_2)$ is defined by

$$T_A\boldsymbol{y} = A\begin{pmatrix} y_1 \\ y_2 \end{pmatrix} = \begin{pmatrix} a_{11} & a_{12} \\ a_{12} & a_{22} \end{pmatrix}\begin{pmatrix} y_1 \\ y_2 \end{pmatrix} = \begin{pmatrix} a_{11}y_1 + a_{12}y_2 \\ a_{12}y_1 + a_{22}y_2 \end{pmatrix}.$$

That is, $T_A\boldsymbol{y} = (a_{11}y_1 + a_{12}y_2)\boldsymbol{e}_1 + (a_{12}y_1 + a_{22}y_2)\boldsymbol{e}_2$. Now form the inner product of $T_A\boldsymbol{y}$ with $\boldsymbol{y}$:

$$\begin{aligned} \langle T_A\boldsymbol{y}, \boldsymbol{y}\rangle &= (a_{11}y_1 + a_{12}y_2)y_1 + (a_{12}y_1 + a_{22}y_2)y_2 \\ &= a_{11}y_1^2 + 2a_{12}y_1y_2 + a_{22}y_2^2 = Q(y_1, y_2), \end{aligned}$$

by (8.39). Consequently, to study the behavior of Q near $\boldsymbol{0}$ is to study the behavior of the inner product $\langle T_A\boldsymbol{y}, \boldsymbol{y}\rangle$; this reduces our problem to a study of the matrix A itself. This matrix must contain all the information needed to determine whether f has a local maximum, a local minimum, or a saddle point at $\boldsymbol{c}$.

Here we appeal to a major theorem from linear algebra known as the Spectral Theorem or, sometimes, as the Diagonalization Theorem. This is not the place to prove it, its proof being rather complicated, but you certainly want it, ready for use, in your mathematical tool kit.

THEOREM 8.8.2 The Spectral Theorem Let A be an $n \times n$ symmetric matrix of real numbers and let $T = T_A$ denote the linear transformation represented by A relative to the standard orthonormal vectors $\boldsymbol{e}_1, \boldsymbol{e}_2, \ldots, \boldsymbol{e}_n$. There exist n real numbers $\lambda_1, \lambda_2, \ldots, \lambda_n$ and n mutually orthogonal unit vectors $\boldsymbol{u}_1, \boldsymbol{u}_2, \ldots, \boldsymbol{u}_n$ such that, for each $j = 1, 2, \ldots, n$,

$$T\boldsymbol{u}_j = \lambda_j\boldsymbol{u}_j.$$

Relative to the coordinate system induced by the vectors $\boldsymbol{u}_1, \boldsymbol{u}_2, \ldots, \boldsymbol{u}_n$, the matrix that represents T is a diagonal matrix:

$$\Lambda = \begin{pmatrix} \lambda_1 & 0 & 0 & \cdots & 0 \\ 0 & \lambda_2 & 0 & \cdots & 0 \\ 0 & 0 & \lambda_3 & \cdots & 0 \\ \vdots & \vdots & \vdots & \cdots & \vdots \\ 0 & 0 & 0 & \cdots & \lambda_n \end{pmatrix}.$$ ●

Several remarks are in order. The numbers $\lambda_1, \lambda_2, \ldots, \lambda_n$ are called the *eigenvalues* of the transformation T; there are n of them, possibly with repetitions. The corresponding vectors $\boldsymbol{u}_1, \boldsymbol{u}_2, \ldots, \boldsymbol{u}_n$ are called the *eigenvectors* of T; they form an orthonormal set of vectors in $\mathbb{R}^n$. Consequently, if $\boldsymbol{x} = (x_1, x_2, \ldots, x_n) = \sum_{j=1}^{n} x_j \boldsymbol{e}_j$ are the coordinates of a vector $\boldsymbol{x}$ relative to the coordinate system determined by the standard unit vectors $\boldsymbol{e}_1, \boldsymbol{e}_2, \ldots, \boldsymbol{e}_n$, then that same vector has a different set of coordinates $(x_1', x_2', \ldots, x_n')$ relative to the coordinate system induced by the vectors $\boldsymbol{u}_1, \boldsymbol{u}_2, \ldots, \boldsymbol{u}_n$. That is,

$$\boldsymbol{x} = x_1'\boldsymbol{u}_1 + x_2'\boldsymbol{u}_2 + \cdots + x_n'\boldsymbol{u}_n = (x_1', x_2', \ldots, x_n').$$

The vector $\boldsymbol{x}$ remains unchanged, but its *representation* changes, appearing to be a different n-tuple.

The general theory from linear algebra provides us with the means of finding the new coordinates of $\boldsymbol{x}$. First we find the coordinates of the vectors $\boldsymbol{u}_1, \boldsymbol{u}_2, \ldots, \boldsymbol{u}_n$ relative to the standard unit vectors $\boldsymbol{e}_1, \boldsymbol{e}_2, \ldots, \boldsymbol{e}_n$. Suppose that the coordinates we find are those in the following display. (Note the order of the indexing in this array.)

$$\begin{array}{rcllllll} \boldsymbol{u}_1 & = & (u_{11}, & u_{21}, & u_{31}, & \cdots, & u_{n1}) \\ \boldsymbol{u}_2 & = & (u_{12}, & u_{22}, & u_{32}, & \cdots, & u_{n2}) \\ \boldsymbol{u}_3 & = & (u_{13}, & u_{23}, & u_{33}, & \cdots, & u_{n3}) \\ \vdots & & \vdots & \vdots & \vdots & \cdots & \vdots \\ \boldsymbol{u}_n & = & (u_{1n}, & u_{2n}, & u_{3n}, & \cdots, & u_{nn}) \end{array}$$

Then we place the coordinates of $\boldsymbol{u}_j$ down the jth column to form an $n \times n$ matrix U.

$$U = \begin{pmatrix} u_{11} & u_{12} & u_{13} & \cdots & u_{1n} \\ u_{21} & u_{22} & u_{23} & \cdots & u_{2n} \\ u_{31} & u_{32} & u_{33} & \cdots & u_{3n} \\ \vdots & \vdots & \vdots & \cdots & \vdots \\ u_{n1} & u_{n2} & u_{n3} & \cdots & u_{nn} \end{pmatrix}$$

Because the column vectors of U are mutually orthogonal unit vectors—namely, the vectors $\boldsymbol{u}_1, \boldsymbol{u}_2, \ldots, \boldsymbol{u}_n$—it follows that the transposed matrix U^t obtained by interchanging the rows and columns of U is the matrix inverse of U. That is, $UU^t = U^tU = I$, the identity matrix with 1s down the main diagonal and 0s in all entries off the main diagonal.

If $\boldsymbol{x} = (x_1, x_2, \ldots, x_n)$ relative to the standard unit vectors $\boldsymbol{e}_1, \boldsymbol{e}_2, \ldots, \boldsymbol{e}_n$, then $\boldsymbol{x}$ is represented by $(x_1', x_2', \ldots, x_n')$ relative to $\boldsymbol{u}_1, \boldsymbol{u}_2, \ldots, \boldsymbol{u}_n$, where

$$\begin{pmatrix} x_1' \\ x_2' \\ x_3' \\ \vdots \\ x_n' \end{pmatrix} = \begin{pmatrix} u_{11} & u_{21} & u_{31} & \cdots & u_{n1} \\ u_{12} & u_{22} & u_{32} & \cdots & u_{n2} \\ u_{13} & u_{23} & u_{33} & \cdots & u_{n3} \\ \vdots & \vdots & \vdots & \cdots & \vdots \\ u_{1n} & u_{2n} & u_{3n} & \cdots & u_{nn} \end{pmatrix} \begin{pmatrix} x_1 \\ x_2 \\ x_3 \\ \vdots \\ x_n \end{pmatrix} = U^t \boldsymbol{x}.$$

Once we have found the matrix U, we form the matrix product U^tAU. After performing the matrix multiplications, we obtain the diagonal matrix Λ displayed in the Spectral Theorem. The matrix Λ represents the transformation $T = T_A$ relative to the coordinate system induced by $\boldsymbol{u}_1, \boldsymbol{u}_2, \ldots, \boldsymbol{u}_n$.

Before taking up an example, let's see how this theory is pertinent to our problem. We want to determine the behavior of

$$Q(y_1, y_2) = a_{11}y_1^2 + 2a_{12}y_1y_2 + a_{22}y_2^2$$

for $\boldsymbol{y} = (y_1, y_2)$ near $\boldsymbol{0}$. We form the matrix A as above. Recall that $Q(y_1, y_2) = \langle T_A \boldsymbol{y}, \boldsymbol{y} \rangle$, where both A and $\boldsymbol{y}$ are expressed relative to the coordinate system in $\mathbb{R}^2$ determined by $\boldsymbol{e}_1, \boldsymbol{e}_2$. We apply the Spectral Theorem. There exist two real numbers λ_1 and λ_2 and two orthogonal unit vectors $\boldsymbol{u}_1 = (u_{11}, u_{21})$ and $\boldsymbol{u}_2 = (u_{12}, u_{22})$ such that $T_A\boldsymbol{u}_j = \lambda_j\boldsymbol{u}_j$ for $j = 1, 2$. The matrix that represent T_A relative to the coordinate system induced by $\boldsymbol{u}_1$ and $\boldsymbol{u}_2$ is

$$\Lambda = U^tAU = \begin{pmatrix} u_{11} & u_{21} \\ u_{12} & u_{22} \end{pmatrix} \begin{pmatrix} a_{11} & a_{12} \\ a_{21} & a_{22} \end{pmatrix} \begin{pmatrix} u_{11} & u_{12} \\ u_{21} & u_{22} \end{pmatrix} = \begin{pmatrix} \lambda_1 & 0 \\ 0 & \lambda_2 \end{pmatrix}.$$

When all the quantities are expressed relative to the coordinate system of $\boldsymbol{u}_1$ and $\boldsymbol{u}_2$, we obtain

$$T_A \boldsymbol{y} = \begin{pmatrix} \lambda_1 & 0 \\ 0 & \lambda_2 \end{pmatrix} \begin{pmatrix} y_1' \\ y_2' \end{pmatrix} = \begin{pmatrix} \lambda_1 y_1' \\ \lambda_2 y_2' \end{pmatrix}$$

and

$$Q(y_1', y_2') = \langle T_A \boldsymbol{y}, \boldsymbol{y} \rangle = \lambda_1 (y_1')^2 + \lambda_2 (y_2')^2.$$

The theory enables us to transform the appearance of Q and to eliminate the term with mixed product. Our earlier discussion enables us to draw the following conclusions.

i) If both λ_1 and λ_2 are positive, then Q is positive for all $\boldsymbol{y}$ in $N'(\boldsymbol{0}; r)$. Consequently, f has a local minimum at $\boldsymbol{c}$.

ii) If both λ_1 and λ_2 are negative, then Q is negative for all $\boldsymbol{y}$ in $N'(\boldsymbol{0}; r)$. Thus f has a local maximum at $\boldsymbol{c}$.

iii) If one of the eigenvalues is negative and the other is positive, then Q assumes both positive and negative values in any deleted neighborhood $N'(\boldsymbol{0})$ and f has a saddle point at $\boldsymbol{c}$.

The insight provided by these observations convinces us that, in order to solve our problem, we must calculate the values of λ_1 and λ_2 whose existence is guaranteed by the Spectral Theorem. That is, we want to find those λ for which there exists a nonzero vector $\boldsymbol{x}$ with the property that $T_A\boldsymbol{x} = \lambda\boldsymbol{x}$. Equivalently, $(T - \lambda I)\boldsymbol{x} = \mathbf{0}$. According to the theory of linear algebra, $(T - \lambda I)\boldsymbol{x} = \mathbf{0}$ for some nonzero vector $\boldsymbol{x}$ if and only if $\det|T - \lambda I| = 0$. This last statement translates into the equation

$$\begin{aligned} \det|T - I| &= \det\begin{vmatrix} a_{11} - \lambda & a_{12} \\ a_{12} & a_{22} - \lambda \end{vmatrix} \\ &= (a_{11} - \lambda)(a_{22} - \lambda) - a_{12}^2 \\ &= \lambda^2 - (a_{11} + a_{22})\lambda + a_{11}a_{22} - a_{12}^2 = 0. \end{aligned}$$

The quadratic polynomial

$$p(\lambda) = \lambda^2 - (a_{11} + a_{22})\lambda + (a_{11}a_{22} - a_{12}^2)$$

is called the *characteristic polynomial* of the matrix A. Its two roots,

$$\begin{aligned} \lambda &= \tfrac{1}{2}\left[(a_{11} + a_{22}) \pm \sqrt{(a_{11} + a_{22})^2 - 4(a_{11}a_{22} - a_{12}^2)}\right] \\ &= \tfrac{1}{2}\left[(a_{11} + a_{22}) \pm \sqrt{(a_{11} - a_{22})^2 + 4a_{12}^2}\right], \end{aligned}$$

are the eigenvalues of A. Let $\Delta = (a_{11} - a_{22})^2 + 4a_{12}^2$. Note that, since $\Delta \geq 0$, the roots λ_1 and λ_2 are real numbers, as promised by the Spectral Theorem. We let

$$\lambda_1 = \tfrac{1}{2}[(a_{11} + a_{22}) - \sqrt{\Delta}]$$

and

$$\lambda_2 = \tfrac{1}{2}[(a_{11} + a_{22}) + \sqrt{\Delta}].$$

Notice that, with this convention, $\lambda_1 \leq \lambda_2$.

It is guaranteed by the theory of elementary linear algebra that, when, for each $j = 1, 2$, we solve the 2×2 system of linear equations

$$\begin{aligned} (a_{11} - \lambda_j)v_1 + \quad a_{12}v_2 \quad &= 0 \\ a_{12}v_1 \quad + (a_{22} - \lambda_j)v_2 &= 0, \end{aligned}$$

we can always find a nonzero solution vector

$$\boldsymbol{v}_j = (v_{1j}, v_{2j}).$$

We let $\boldsymbol{u}_1 = \boldsymbol{v}_1/||\boldsymbol{v}_1||$ and $\boldsymbol{u}_2 = \boldsymbol{v}_2/||\boldsymbol{v}_2||$. The Spectral Theorem guarantees that $\boldsymbol{u}_1$ and $\boldsymbol{u}_2$ are orthogonal unit vectors such that $T_A\boldsymbol{u}_1 = \lambda_1\boldsymbol{u}_1$ and $T_A\boldsymbol{u}_2 = \lambda_2\boldsymbol{u}_2$. These vectors will provide the new coordinate system relative to which Q will assume the simplified form

$$Q(y_1', y_2') = \lambda_1(y_1')^2 + \lambda_2(y_2')^2,$$

without any mixed product term.

We have indexed the eigenvalues so that $\lambda_1 \leq \lambda_2$. Thus both eigenvalues are positive if and only if λ_1 is positive. Now

$$\lambda_1 = \frac{1}{2}[(a_{11} + a_{22}) - \sqrt{(a_{11} - a_{22})^2 + 4a_{12}^2}]$$

is positive if and only if $a_{11} + a_{22}$ is positive and

$$(a_{11} + a_{22})^2 > (a_{11} - a_{22})^2 + 4a_{12}^2.$$

These conditions are equivalent to

$$a_{11} + a_{22} > 0 \qquad \text{and} \qquad a_{11}a_{22} > a_{12}^2.$$

Thus f has a local minimum at $\boldsymbol{c}$ if and only if

$$D_{11}f(\boldsymbol{c}) + D_{22}f(\boldsymbol{c}) > 0$$

and

$$D_{11}f(\boldsymbol{c})D_{22}f(\boldsymbol{c}) > [D_{12}f(\boldsymbol{c})]^2.$$

Likewise, both eigenvalues are negative if and only if λ_2 is negative. Now, $\lambda_2 = [(a_{11} + a_{22}) + \sqrt{(a_{11} - a_{22})^2 + 4a_{12}^2}]/2$ is negative if and only if $a_{11} + a_{22}$ is negative and

$$(a_{11} + a_{22})^2 > (a_{11} - a_{22})^2 + 4a_{12}^2.$$

Equivalently,

$$a_{11} + a_{22} < 0 \qquad \text{and} \qquad a_{11}a_{22} > a_{12}^2.$$

Thus f has a local maximum at $\boldsymbol{c}$ if and only if

$$D_{11}f(\boldsymbol{c}) + D_{22}f(\boldsymbol{c}) < 0$$

and

$$D_{11}f(\boldsymbol{c})D_{22}f(\boldsymbol{c}) > [D_{12}f(\boldsymbol{c})]^2.$$

The eigenvalues satisfy $\lambda_1 < 0 < \lambda_2$ if and only if $a_{12}^2 > a_{11}a_{22}$. Thus f has a saddle point at $\boldsymbol{c}$ if and only if

$$D_{11}f(\boldsymbol{c})D_{22}f(\boldsymbol{c}) < [D_{12}f(\boldsymbol{c})]^2.$$

If it should happen that either $\lambda_1 = 0$ or $\lambda_2 = 0$, that is, if it should happen that $D_{11}f(\boldsymbol{c})D_{22}f(\boldsymbol{c}) = [D_{12}f(\boldsymbol{c})]^2$, then the nature of the critical point $\boldsymbol{c}$ cannot be determined by this method. We have proved the following theorem.

THEOREM 8.8.3 The Second Derivative Test Let U be an open set in $\mathbb{R}^2$, f be a real-valued function in $C^{(3)}(U)$, and $\boldsymbol{c}$ be a critical point of f in U.

i) The function f has a local minimum at $\boldsymbol{c}$ if and only if $D_{11}f(\boldsymbol{c}) + D_{22}f(\boldsymbol{c}) > 0$ and $D_{11}f(\boldsymbol{c})D_{22}f(\boldsymbol{c}) > [D_{12}f(\boldsymbol{c})]^2$.

ii) The function f has a local maximum at $\boldsymbol{c}$ if and only if $D_{11}f(\boldsymbol{c}) + D_{22}f(\boldsymbol{c}) < 0$ and $D_{11}f(\boldsymbol{c})D_{22}f(\boldsymbol{c}) > [D_{12}f(\boldsymbol{c})]^2$.

iii) The function f has a saddle point at $\boldsymbol{c}$ if and only if $D_{11}f(\boldsymbol{c})D_{22}f(\boldsymbol{c}) < [D_{12}f(\boldsymbol{c})]^2$.

iv) If $D_{11}f(\boldsymbol{c})D_{22}f(\boldsymbol{c}) = [D_{12}f(\boldsymbol{c})]^2$, then the test gives no information about the nature of the critical point $\boldsymbol{c}$. ●

EXAMPLE 14 Suppose that f is in $C^{(3)}(\mathbb{R}^2)$, that $\nabla f(\boldsymbol{c}) = \mathbf{0}$ at some point $\boldsymbol{c}$ in $\mathbb{R}^2$, and that $D_{11}f(\boldsymbol{c}) = 1/2$, $D_{12}f(\boldsymbol{c}) = D_{21}f(\boldsymbol{c}) = -3/2$, and $D_{22}f(\boldsymbol{c}) = 1/2$. The question is: What is the nature of the critical point $\boldsymbol{c}$? Rather than merely applying Theorem 8.8.3 in this example, we work through the argument leading to it. By Taylor's theorem, we first let $y_1 = x_1 - c_1$ and $y_2 = x_2 - c_2$ and write

$$f(\boldsymbol{x}) = f(\boldsymbol{c}) + \tfrac{1}{2}\left\{\tfrac{1}{2}y_1^2 - 3y_1y_2 + \tfrac{1}{2}y_2^2\right\} + R_2(\boldsymbol{c}; \boldsymbol{x}).$$

Our goal is to *diagonalize* the quadratic form

$$Q(y_1, y_2) = \tfrac{1}{2}y_1^2 - 3y_1y_2 + \tfrac{1}{2}y_2^2.$$

To this end, we form the matrix

$$A = \begin{pmatrix} 1/2 & -3/2 \\ -3/2 & 1/2 \end{pmatrix}$$

and find its eigenvalues. The characteristic polynomial of A is

$$\begin{aligned} \det|A - \lambda I| &= \det\begin{vmatrix} 1/2-\lambda & -3/2 \\ -3/2 & 1/2-\lambda \end{vmatrix} = \lambda^2 - \lambda - 2 \\ &= (\lambda + 1)(\lambda - 2). \end{aligned}$$

Therefore, the eigenvalues of A are $\lambda_1 = -1$ and $\lambda_2 = 2$. Next we find the corresponding eigenvectors. First let $\lambda = -1$ and solve the 2×2 system of linear equations $(A + I)\boldsymbol{v}_1 = \mathbf{0}$. That is, solve

$$\begin{aligned} (3/2)v_{11} - (3/2)v_{12} &= 0 \\ (-3/2)v_{11} + (3/2)v_{12} &= 0. \end{aligned}$$

A general, nonzero solution has the form $\boldsymbol{v}_1 = (v, v)$ with $v \neq 0$. A unit vector solution is $\boldsymbol{u}_1 = (1, 1)/\sqrt{2}$.

Next let $\lambda = 2$ and solve the 2×2 system $(A - 2I)\boldsymbol{v} = \mathbf{0}$, that is

$$\begin{aligned} (-3/2)v_{21} - (3/2)v_{22} &= 0 \\ (-3/2)v_{21} - (3/2)v_{22} &= 0. \end{aligned}$$

A general, nonzero solution has the form $\boldsymbol{v}_2 = (v, -v)$, with $v \neq 0$. A unit vector solution is $\boldsymbol{u}_2 = (-1, 1)/\sqrt{2}$.

Notice that $\boldsymbol{u}_1$ and $\boldsymbol{u}_2$ are orthogonal unit vectors and therefore induce a new coordinate system in $\mathbb{R}^2$. As you can compute, $A\boldsymbol{u}_1 = -\boldsymbol{u}_1$ and $A\boldsymbol{u}_2 = 2\boldsymbol{u}_2$.

The existence[6] of λ_1, λ_2 and $\boldsymbol{u}_1$, $\boldsymbol{u}_2$ is guaranteed by the Spectral Theorem; now that we have found them, we use them to diagonalize Q. Let U be the matrix $[\boldsymbol{u}_1\boldsymbol{u}_2]$ with the coordinates of $\boldsymbol{u}_j$ placed down the jth column:

$$U = \begin{pmatrix} 1/\sqrt{2} & -1/\sqrt{2} \\ 1/\sqrt{2} & 1/\sqrt{2} \end{pmatrix} = \frac{1}{\sqrt{2}}\begin{pmatrix} 1 & -1 \\ 1 & 1 \end{pmatrix}.$$

The matrix inverse of U is its transpose

$$U^t = \frac{1}{\sqrt{2}}\begin{pmatrix} 1 & 1 \\ -1 & 1 \end{pmatrix}.$$

As you can compute, $UU^t = U^tU = I$. Actually, the matrix U can be written

$$U = \begin{pmatrix} \cos\pi/4 & -\sin\pi/4 \\ \sin\pi/4 & \cos\pi/4 \end{pmatrix}$$

and thus represents a rotation of the plane through an angle of $\pi/4$. (See Fig. 8.17.) If a vector $\boldsymbol{y}$ has coordinates (y_1, y_2) relative to $\boldsymbol{e}_1$, $\boldsymbol{e}_2$, then $\boldsymbol{y}$ has coordinates (y_1', y_2') relative to $\boldsymbol{u}_1$, $\boldsymbol{u}_2$ where

$$\begin{pmatrix} y_1' \\ y_2' \end{pmatrix} = U^t y = \frac{1}{\sqrt{2}}\begin{pmatrix} 1 & 1 \\ -1 & 1 \end{pmatrix}\begin{pmatrix} y_1 \\ y_2 \end{pmatrix} = \begin{pmatrix} (y_1 + y_2)/\sqrt{2} \\ (-y_1 + y_2)/\sqrt{2} \end{pmatrix}.$$

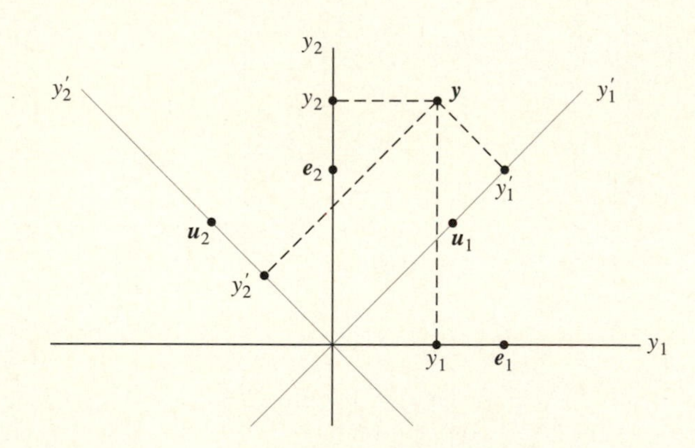

Figure 8.17

By changing to this new coordinate system, the matrix A, which represents T_A, is transformed into

$$U^tAU = \frac{1}{2}\begin{pmatrix} 1 & 1 \\ -1 & 1 \end{pmatrix}\begin{pmatrix} 1/2 & -3/2 \\ -3/2 & 1/2 \end{pmatrix}\begin{pmatrix} 1 & -1 \\ 1 & 1 \end{pmatrix} = \begin{pmatrix} -1 & 0 \\ 0 & 2 \end{pmatrix} = \Lambda.$$

[6]Computing these numbers and these vectors is easy if $n = 2$, but if $n > 2$, finding them can pose significant computational problems. Several efficient and powerful software packages that provide approximate solutions for modest values of n are now available.

Therefore, in these new coordinates,

$$T_A \boldsymbol{y} = \begin{pmatrix} -1 & 0 \\ 0 & 2 \end{pmatrix} \begin{pmatrix} y_1' \\ y_2' \end{pmatrix} = \begin{pmatrix} -y_1' \\ 2y_2' \end{pmatrix}$$

and $Q(y_1', y_2') = \langle T_A \boldsymbol{y}, \boldsymbol{y} \rangle = -(y_1')^2 + 2(y_2')^2$. We deduce that Q has negative values along the new coordinate axis determined by $\boldsymbol{u}_1$ and has positive values along the coordinate axis determined by $\boldsymbol{u}_2$. Therefore, in a deleted neighborhood of $\boldsymbol{c}$, f has values less than $f(\boldsymbol{c})$ along the line $\boldsymbol{c} + t\boldsymbol{u}_1$ and has values greater than $f(\boldsymbol{c})$ along the line $\boldsymbol{c} + t\boldsymbol{u}_2$. We conclude that f must have a saddle point at $\boldsymbol{c}$. ●

Theorem 8.8.3 enables us to reach this same conclusion once we have found $D_{11} f(\boldsymbol{c})$, $D_{12} f(\boldsymbol{c})$, and $D_{22} f(\boldsymbol{c})$. Since, in this example, $D_{11} f(\boldsymbol{c}) D_{22} f(\boldsymbol{c}) = 1/4 < 9/4 = [D_{12} f(\boldsymbol{c})]^2$, we know that f must have a saddle point at $\boldsymbol{c}$.

We have gone into such detail in this example because we intend to generalize the argument to $\mathbb{R}^n$ and, if you have never proved the Spectral Theorem, you may have trouble grasping the argument. This example provides you with the prototype to which you can refer.

Now let f be three times continuously differentiable on an open set U in $\mathbb{R}^n$ and let $\boldsymbol{c}$ be a point of U where $\nabla \boldsymbol{f}(\boldsymbol{c}) = \boldsymbol{0}$. By Taylor's theorem we write

$$f(\boldsymbol{x}) = f(\boldsymbol{c}) + \tfrac{1}{2} Q(y_1, y_2, \ldots, y_n) + R_2(\boldsymbol{c}; \boldsymbol{x}),$$

where, for $j = 1, 2, \ldots, n$, $y_j = x_j - c_j$ and

$$Q(y_1, y_2, \ldots, y_n) = \sum_{i=1}^{n} \sum_{j=1}^{n} D_{ij} f(\boldsymbol{c}) y_i y_j.$$

By the assumed continuous differentiability of f, we have $a_{ij} = D_{ij} f(\boldsymbol{c}) = D_{ji} f(\boldsymbol{c}) = a_{ji}$. Therefore

$$\begin{aligned} Q(y_1, y_2, \ldots, y_n) &= a_{11} y_1^2 + a_{22} y_2^2 + \cdots + a_{nn} y_n^2 \\ &\quad + 2a_{12} y_1 y_2 + 2a_{13} y_1 y_3 + \cdots + 2a_{n-1n} y_{n-1} y_n. \end{aligned}$$

Form the symmetric matrix

$$A = \begin{pmatrix} a_{11} & a_{12} & a_{13} & \cdots & a_{1n} \\ a_{12} & a_{22} & a_{23} & \cdots & a_{2n} \\ a_{13} & a_{23} & a_{33} & \cdots & a_{3n} \\ \vdots & \vdots & \vdots & \cdots & \vdots \\ a_{1n} & a_{2n} & a_{3n} & \cdots & a_{nn} \end{pmatrix}$$

and let T_A denote the linear transformation represented by A relative to $\boldsymbol{e}_1, \boldsymbol{e}_2, \ldots, \boldsymbol{e}_n$. Then

$$Q(y_1, y_2, \ldots, y_n) = \langle T_A \boldsymbol{y}, \boldsymbol{y} \rangle.$$

Apply the Spectral Theorem to the matrix A. There exist n real eigenvalues $\lambda_1, \lambda_2, \ldots, \lambda_n$ and n mutually orthogonal unit vectors $\boldsymbol{u}_1, \boldsymbol{u}_2, \ldots, \boldsymbol{u}_n$, such that

$$T_A \boldsymbol{u}_j = \lambda_j \boldsymbol{u}_j,$$

for $j = 1, 2, \ldots, n$. Form the matrix $U = [\boldsymbol{u}_1\ \boldsymbol{u}_2\ \ldots\ \boldsymbol{u}_n]$. Then, relative to the coordinate system induced by the unit vectors $\boldsymbol{u}_1, \boldsymbol{u}_2, \ldots, \boldsymbol{u}_n$, T_A is represented by the matrix

$$U^t A U = \begin{pmatrix} \lambda_1 & 0 & 0 & \ldots & 0 \\ 0 & \lambda_2 & 0 & \ldots & 0 \\ 0 & 0 & \lambda_3 & \ldots & 0 \\ \vdots & \vdots & \vdots & \ldots & \vdots \\ 0 & 0 & 0 & \ldots & \lambda_n \end{pmatrix} = \Lambda.$$

Express $\boldsymbol{y} = (y_1, y_2, y_3, \ldots, y_n)$ in terms of its coordinates $(y_1', y_2', y_3', \ldots, y_n')$ relative to the new coordinate system. Then

$$Q(y_1', y_2', \ldots, y_n') = \langle T_A \boldsymbol{y}, \boldsymbol{y} \rangle = \sum_{j=1}^{n} \lambda_j (y_j')^2.$$

The preceding discussion proves the following theorem.

THEOREM 8.8.4 Let U be an open set in $\mathbb{R}^n$, let f be in $C^{(3)}(U)$, and let $\boldsymbol{c}$ be a point in U where $\nabla \boldsymbol{f}(\boldsymbol{c}) = \boldsymbol{0}$. Let A be the matrix $[D_{ij} f(\boldsymbol{c})]$. Let $\lambda_1, \lambda_2, \ldots, \lambda_n$ be the eigenvalues of A. Then

i) The function f has a local minimum at $\boldsymbol{c}$ if and only if all the eigenvalues of A are positive.

ii) The function f has a local maximum at $\boldsymbol{c}$ if and only if all the eigenvalues of A are negative.

iii) The function f has a saddle point at $\boldsymbol{c}$ if and only if at least one of the eigenvalues of A is negative and at least one is positive. ●

Remark. The difficulty applying Theorem 8.8.4 lies in finding the eigenvalues of A. Formally, these numbers are the roots of the characteristic polynomial

$$p(\lambda) = \det |A - \lambda I|.$$

That is easily said. If p happens to factor over $\mathbb{Q}$, then we can find the eigenvalues of A. But p is a polynomial of degree n and we know from the algebraic theory of polynomials that, if $\deg p \geq 5$, there is no general method for finding the roots of p; if p does not factor easily, then we may have no algebraic method to find its roots. In practice, we usually must resort to numerical methods in order to compute the eigenvalues of A. There being no general formula for the roots of p (as there is when $n = 2$), we cannot translate the conclusions of Theorem 8.8.4 into statements involving the second-order partial derivatives of f at $\boldsymbol{c}$, as we can in Theorem 8.8.3.

EXERCISES

8.1. For $\boldsymbol{x}$ in $\mathbb{R}^2$, define $f(\boldsymbol{x}) = |x_1 x_2|$.

a) Prove that $D_1 f(\boldsymbol{0}) = D_2 f(\boldsymbol{0}) = 0$.

b) Prove that the graph of f does not have a tangent plane at the point $\boldsymbol{c} = \boldsymbol{0}$. Is f differentiable at $\boldsymbol{c} = \boldsymbol{0}$?

8.2. For $\boldsymbol{x} \neq \mathbf{0}$ in $\mathbb{R}^2$, let $f(\boldsymbol{x}) = x_1x_2/||\boldsymbol{x}||$. Let $f(\mathbf{0}) = 0$. Show that the graph of f does not have a tangent plane at $\boldsymbol{c} = \mathbf{0}$.

8.3. For $\boldsymbol{x}$ in some deleted neighborhood $N'(\mathbf{0})$ in $\mathbb{R}^n$, let $g(\boldsymbol{x})$ be defined and nonzero. Suppose that $\lim_{\boldsymbol{x}\to\mathbf{0}} g(\boldsymbol{x}) = 0$. Let f be a real-valued function defined on $N'(\mathbf{0})$.

- **i)** We say that f vanishes of at least the same order as g at $\boldsymbol{c} = \mathbf{0}$ and write $f(\boldsymbol{x}) = O(g(\boldsymbol{x}))$, read "$f$ is big O of g," if there exists a constant $K > 0$ such that $|f(\boldsymbol{x})/g(\boldsymbol{x})| \leq K$ for all $\boldsymbol{x}$ in $N'(\mathbf{0})$.
- **ii)** We say that f vanishes with exactly the same order as g at $\boldsymbol{c} = \mathbf{0}$ if $|f(\boldsymbol{x})/g(\boldsymbol{x})|$ and $|g(\boldsymbol{x})/f(\boldsymbol{x})|$ are both bounded on some deleted neighborhood $N'(\mathbf{0})$.
- **iii)** We say that f vanishes of a higher order than g at $\boldsymbol{c} = \mathbf{0}$ and write $f(\boldsymbol{x}) = o(g(\boldsymbol{x}))$, read "$f$ is little o of g" if $\lim_{\boldsymbol{x}\to\mathbf{0}} f(\boldsymbol{x})/g(\boldsymbol{x}) = 0$.

Show that:

a) If $f(x_1, x_2) = ax_1^2 + bx_1x_2 + cx_2^2$ and $g(\boldsymbol{x}) = ||\boldsymbol{x}||^2$, then $f(\boldsymbol{x}) = O(g(\boldsymbol{x}))$.

b) Find conditions that ensure that

$$f(\boldsymbol{x}) = ax_1^2 + bx_1x_2 + cx_2^2$$

vanishes with exactly the same order as $g(\boldsymbol{x}) = ||\boldsymbol{x}||^2$.

c) If $f(\boldsymbol{x}) = ||\boldsymbol{x}||$, and $g(\boldsymbol{x}) = 1/\ln(||\boldsymbol{x}||)$, then $f(\boldsymbol{x}) = o(g(\boldsymbol{x}))$.

d) Let f be a real-valued function defined on an open set U in $\mathbb{R}^n$. Let $\boldsymbol{c}$ be a point of U. Show that f is differentiable at $\boldsymbol{c}$ in U if and only if $f(\boldsymbol{c} + \boldsymbol{h}) - f(\boldsymbol{c}) = o(\langle \boldsymbol{v}, \boldsymbol{h} \rangle)$ for some constant vector $\boldsymbol{v}$ in $\mathbb{R}^n$. (The vector $\boldsymbol{h}$ is the variable.)

8.4. For $\boldsymbol{x} = (x_1, x_2) \neq \mathbf{0}$ in $\mathbb{R}^2$, let

$$f(x_1, x_2) = \frac{x_1x_2(x_1^2 - x_2^2)}{x_1^2 + x_2^2}.$$

Let $f(\mathbf{0}) = 0$.

a) Determine whether f is continuous at $\mathbf{0}$.

b) Prove that $D_1 f(\boldsymbol{x})$ and $D_2 f(\boldsymbol{x})$ exist at every point of $\mathbb{R}^2$. Find $D_1 f(\mathbf{0})$ and $D_2 f(\mathbf{0})$.

c) Prove that $D_{21} f(\mathbf{0})$ and $D_{12} f(\mathbf{0})$ exist but are not equal.

8.5. For $\boldsymbol{x} = (x_1, x_2)$ in $\mathbb{R}^2$, define

$$f(\boldsymbol{x}) = \begin{cases} x_1^2 \tan^{-1}(x_2/x_1) - x_2^2 \tan^{-1}(x_1/x_2), & x_1x_2 \neq 0 \\ 0, \quad x_1x_2 = 0. \end{cases}$$

Show that $D_{12} f(\mathbf{0}) = 1$ and $D_{21} f(\mathbf{0}) = -1$.

8.6. Suppose that a real-valued function f is defined on an open set U in $\mathbb{R}^n$ and that all the first-order partial derivatives of f exist at every point in U. Suppose further that there exists a constant $M > 0$ such that $|D_j f(\boldsymbol{x})| \leq M$, for all $\boldsymbol{x}$ in U and each $j = 1, 2, \ldots, n$. Prove that f is continuous on U.

8.7. For $\boldsymbol{x} = (x_1, x_2, \ldots, x_n) \neq 0$ in $\mathbb{R}^n$ and for $j = 1, 2, \ldots, n$, define $f_j(\boldsymbol{x}) = x_j/||\boldsymbol{x}||$.

a) Show that, for each j and for each $\boldsymbol{x} \neq 0$, f_j is differentiable at $\boldsymbol{x}$.

b) Show that $x_1 df_1(\boldsymbol{x}; \boldsymbol{t}) + x_2 df_2(\boldsymbol{x}; \boldsymbol{t}) + \cdots + x_n df_n(\boldsymbol{x}; \boldsymbol{t}) = 0$, for all $\boldsymbol{x} \neq 0$ and all $\boldsymbol{t}$ in $\mathbb{R}^n$.

c) Define $\boldsymbol{f} = (f_1, f_2, \ldots, f_n)$. Explain why $\boldsymbol{f}$ is differentiable at every $\boldsymbol{x} \neq \mathbf{0}$ in $\mathbb{R}^n$. Find the differential of $\boldsymbol{f}$.

8.8. Suppose that f and g are real-valued functions that are differentiable at a point $\boldsymbol{c}$ in $\mathbb{R}^n$ and that $g(\boldsymbol{c}) \neq 0$. Prove that f/g is also differentiable at $\boldsymbol{c}$ and that

$$d(f/g)(\boldsymbol{c};\boldsymbol{t}) = \frac{df(\boldsymbol{c};\boldsymbol{t})g(\boldsymbol{c}) - f(\boldsymbol{c})dg(\boldsymbol{c};\boldsymbol{t})}{g^2(\boldsymbol{c})}.$$

8.9. Let U be an open set in $\mathbb{R}^n$ and let f be a function in $C^{(k)}(U)$. For each $j = 1, 2, \ldots, k$, find the number of *distinct* jth-order partial derivatives of f.

8.10. Find all points where each of the following functions is differentiable:

a) $f(x_1, x_2) = \exp(-x_1^2 - x_2^2)\cos(x_1^2 + x_2^2)$.

b) $f(x_1, x_2) = (x_1 - x_2)^2/(1 - x_1^2 - x_2^2)^{1/2}$.

c) $f(x_1, x_2) = \tan^{-1}\sqrt{4 - x_1^2 - x_2^2}$.

d) $f(x_1, x_2) = (x_1 - x_2)\ln(x_1^2 - 2x_1x_2 + x_2^2)$.

e) $f(x_1, x_2, x_3) = (x_1 + x_2 + x_3)e^{x_1x_2x_3}$.

8.11. In each of the following problems, find the domains of each of the functions f and g. Find all points where f and g are differentiable. Find where $h = (g \circ f)$ is differentiable. Find all first- and second-order derivatives of h. Find the differentials of f, g, and h at all points where the functions are differentiable.

a) For $\boldsymbol{x}$ in $\mathbb{R}^2$, let $y = f(\boldsymbol{x}) = x_1^2 + x_2^2$. Let $g(y) = \sin y$. Let $h = (g \circ f)$.

b) Let $y_1 = f_1(x_1, x_2) = x_1x_2$, $y_2 = f_2(x_1, x_2) = x_1 + x_2$, and $\boldsymbol{f} = (f_1, f_2)$. Let $g(y_1, y_2) = e^{y_1}\sin y_2$ and $h = (g \circ \boldsymbol{f})$.

c) Let $y_1 = f_1(x_1, x_2) = (1 - x_1^2 - x_2^2)^{1/2}$, $y_2 = f_2(x_1, x_2) = \ln(x_1^2 + x_2^2)$. Let $\boldsymbol{f} = (f_1, f_2)$. Let $g(y_1, y_2) = y_1/y_2$ and $h = (g \circ \boldsymbol{f})$.

d) For $\boldsymbol{x}$ in $\mathbb{R}^2$, define $f_1(\boldsymbol{x}) = x_1^2$, $f_2(\boldsymbol{x}) = 2x_1x_2$, and $f_3(\boldsymbol{x}) = x_2^2$. Define $\boldsymbol{f} = (f_1, f_2, f_3)$. For $\boldsymbol{y}$ in $\mathbb{R}^3$ let $g(\boldsymbol{y}) = \exp(y_1 + y_2 + y_3)$. Let $h = (g \circ \boldsymbol{f})$.

e) For $\boldsymbol{x}$ in $\mathbb{R}^2$, let $f_1(\boldsymbol{x}) = x_1^2$, $f_2(\boldsymbol{x}) = x_1x_2$, $f_3(\boldsymbol{x}) = x_2^2$ and $\boldsymbol{f} = (f_1, f_2, f_3)$. Let $g(\boldsymbol{y}) = \tan^{-1}(y_1 + y_2 + y_3)$, for $\boldsymbol{y}$ in $\mathbb{R}^3$. Let $h = (g \circ \boldsymbol{f})$.

f) For $\boldsymbol{x}$ in $\mathbb{R}^3$, let $f_1(\boldsymbol{x}) = x_1x_2$, $f_2(\boldsymbol{x}) = x_1^2 + x_2^2 + x_3^2$, and $\boldsymbol{f} = (f_1, f_2)$. For $\boldsymbol{y}$ in $\mathbb{R}^2$, let $g(\boldsymbol{y}) = y_1\cosh y_2$. Let $h = g \circ \boldsymbol{f}$.

8.12. Let f be a differentiable real-valued function on $\mathbb{R}^3$. Let a_1, a_2, a_3, b_1, b_2, and b_3 be constants. For t in $\mathbb{R}$, define $h(t) = f(a_1 + b_1t, a_2 + b_2t, a_3 + b_3t)$. Show that h is a differentiable function of t and find $h'(t)$ in terms of the partial derivatives of f.

8.13. Let g be a differentiable function on $\mathbb{R}^2$. For $t = (t_1, t_2)$ in $\mathbb{R}^2$, let $x_1 = f_1(\boldsymbol{t}) = t_1 - t_2$ and $x_2 = f_2(\boldsymbol{t}) = t_2 - t_1$. Let $\boldsymbol{f} = (f_1, f_2)$ and $h(\boldsymbol{t}) = (g \circ \boldsymbol{f})(\boldsymbol{t})$. Show that

$$D_1h(\boldsymbol{t}) + D_2h(\boldsymbol{t}) = 0$$

for all $\boldsymbol{t}$ in $\mathbb{R}^2$.

8.14. Let g be differentiable on $\mathbb{R}^2$. For $\boldsymbol{x} = (x_1, x_2, x_3)$ in $\mathbb{R}^3$ with $x_1x_2x_3 \neq 0$, let $y_1 = f_1(\boldsymbol{x}) = (x_2 - x_1)/(x_1x_2)$ and $y_2 = f_2(\boldsymbol{x}) = (x_3 - x_1)/(x_1x_3)$. Let $\boldsymbol{f} = (f_1, f_2)$. Define $h(\boldsymbol{x}) = g(\boldsymbol{f}(\boldsymbol{x}))$. Prove that

$$x_1^2D_1h(\boldsymbol{x}) + x_2^2D_2h(\boldsymbol{x}) + x_3^2D_3h(\boldsymbol{x}) = 0,$$

for all $\boldsymbol{x}$ in R^3 such that $x_1x_2x_3 \neq 0$.

8.15. Let g be a real-valued function in $C^{(2)}(\mathbb{R}^2)$. For $\boldsymbol{x} = (x_1, x_2, x_3)$ in $\mathbb{R}^3$ with $x_1 \neq 0$, let $y_1 = f_1(\boldsymbol{x}) = x_2/x_1$ and $y_2 = f_2(\boldsymbol{x}) = x_3/x_1$. Let $h(\boldsymbol{x}) = x_1^3g(\boldsymbol{f}(\boldsymbol{x}))$.

a) Show that $\langle \nabla \boldsymbol{h}(\boldsymbol{x}), \boldsymbol{x} \rangle = 3h(\boldsymbol{x})$.

b) Show that

$$x_1^2 D_{11}h(\boldsymbol{x}) - x_2^2 D_{22}h(\boldsymbol{x}) - x_3^2 D_{33}h(\boldsymbol{x}) \\ - 2x_2 x_3 D_{23}h(\boldsymbol{x}) + 4x_2 D_2 h(\boldsymbol{x}) + 4x_3 D_3 h(\boldsymbol{x}) = 6h(\boldsymbol{x}).$$

8.16. Suppose that g is a differentiable real-valued function on $\mathbb{R}^2$. For $\boldsymbol{x}$ in $\mathbb{R}^2$, let $y_1 = f_1(\boldsymbol{x}) = x_1^2 - x_2^2$ and $y_2 = f_2(\boldsymbol{x}) = x_2^2 - x_1^2$. Let $\boldsymbol{f} = (f_1, f_2)$ and $h = (g \circ \boldsymbol{f})$. Prove that $x_2 D_1 h(\boldsymbol{x}) + x_1 D_2 h(\boldsymbol{x}) = 0$ for all $\boldsymbol{x}$ in $\mathbb{R}^2$.

8.17. Let f be a differentiable function on $\mathbb{R}^2$. Fix θ. Define

$$y_1 = g_1(\boldsymbol{x}) = x_1 \cos\theta - x_2 \sin\theta$$

$$y_2 = g_2(\boldsymbol{x}) = x_1 \sin\theta + x_2 \cos\theta.$$

Let $\boldsymbol{g} = (g_1, g_2)$ and $h(\boldsymbol{x}) = (f \circ \boldsymbol{g})(\boldsymbol{x})$. Prove that

$$||\nabla \boldsymbol{h}(\boldsymbol{x})|| = ||(\nabla \boldsymbol{f})(g(\boldsymbol{x}))||.$$

(The function $\boldsymbol{g}$ represents a counterclockwise rotation of the plane through an angle θ. Therefore this problem shows that the length of the gradient is unchanged by a rotation.)

8.18. Let f and g be twice differentiable functions on $\mathbb{R}$. Let c be a positive constant. For x_1 and x_2 in $\mathbb{R}$, let $\boldsymbol{x} = (x_1, x_2)$ and define $h(\boldsymbol{x}) = f(x_1 - cx_2) + g(x_1 + cx_2)$. Show that $D_{22}h(\boldsymbol{x}) = c^2 D_{11}h(\boldsymbol{x})$.

8.19. Let f be a real-valued function in $C^{(2)}(\mathbb{R}^2)$. Let c be a positive constant. Define $y_1 = g_1(\boldsymbol{x}) = x_1 + cx_2$ and $y_2 = g_2(\boldsymbol{x}) = x_1 - cx_2$. Let $\boldsymbol{g} = (g_1, g_2)$ and $h = f \circ \boldsymbol{g}$. Find $D_{22}f(\boldsymbol{x}) - c^2 D_{11}f(\boldsymbol{x})$ in terms of the derivatives of h.

8.20. Let f be a differentiable real-valued function on $\mathbb{R}^2$. In Cartesian coordinates, $||\nabla \boldsymbol{f}(\boldsymbol{x})||^2 = D_1 f(\boldsymbol{x})^2 + D_2 f(\boldsymbol{x})^2$. Let $x_1 = g_1(r, \theta) = r\cos\theta$ and $x_2 = g_2(r, \theta) = r\sin\theta$. Let $\boldsymbol{g} = (g_1, g_2)$ and $\boldsymbol{h}(r, \theta) = f(\boldsymbol{g}(r, \theta))$. Show that

$$||(\nabla \boldsymbol{f})(\boldsymbol{g}(r, \theta))||^2 = (\partial h/\partial r)^2 + \left[\frac{1}{r}\partial h/\partial \theta\right]^2.$$

8.21. Let f be a real-valued function in $C^{(2)}(\mathbb{R}^2)$. For $r > 0$ and any θ in $\mathbb{R}$, let $x_1 = g_1(r, \theta) = r\cosh\theta$ and $x_2 = g_2(r, \theta) = r\sinh\theta$. Let $\boldsymbol{g} = (g_1, g_2)$. Define $h(r, \theta) = f(\boldsymbol{g}(r, \theta))$.

a) Find $(\partial h/\partial r)^2 - [\frac{1}{r}\partial h/\partial \theta]^2$ in terms of $D_1 f$ and $D_2 f$.

b) Show that

$$\partial^2 h/\partial r^2 + \frac{1}{r}(\partial h/\partial r) - \frac{1}{r^2}(\partial^2 h/\partial \theta^2) = D_{11}f - D_{22}f.$$

8.22. Let f be a differentiable real-valued function on $\mathbb{R}$.

a) For x and y in $\mathbb{R}$, let $z = yf(x^2 - y^2)$. Show that

$$y(\partial z/\partial x) + x(\partial z/\partial y) = xz/y.$$

b) For x and y in $\mathbb{R}$ with $x \neq 0$, let $z = xy + xf(y/x)$. Show that

$$x(\partial z/\partial x) + y(\partial z/\partial y) = xy + z.$$

c) Suppose that f is twice differentiable on $\mathbb{R}$ and that g is also a twice differentiable real-valued function on $\mathbb{R}$. Let $z = g(x + f(y))$. Show that

$$(\partial z/\partial x)(\partial^2 z/\partial x\,\partial y) = (\partial z/\partial y)(\partial^2 z/\partial x^2).$$

8.23. Suppose that f and g are twice continuously differentiable functions on $\mathbb{R}$. For $\boldsymbol{x} = (x_1, x_2)$ with $x_1 \neq 0$, define $h(x_1, x_2) = x_1^n f(x_2/x_1) + x_1^{-n} g(x_2/x_1)$. Prove that

$$x_1^2 D_{11}h(\boldsymbol{x}) + 2x_1x_2D_{12}h(\boldsymbol{x}) + x_2^2 D_{22}h(\boldsymbol{x}) + x_1 D_1 h(\boldsymbol{x}) + x_2 D_2 h(\boldsymbol{x}) = n^2 h(\boldsymbol{x}).$$

8.24. Let f_1, f_2, and f_3 be differentiable real-valued functions on $\mathbb{R}$. Let

$$F(x_1, x_2, x_3) = \det \begin{vmatrix} f_1(x_1) & f_2(x_1) & f_3(x_1) \\ f_1(x_2) & f_2(x_2) & f_3(x_2) \\ f_1(x_3) & f_2(x_3) & f_3(x_3) \end{vmatrix}.$$

Find $D_1 F$, $D_2 F$, and $D_3 F$ in terms of the derivatives of f_1, f_2, and f_3 and expressed in terms of determinants.

8.25. Let g be a real-valued function in $C^{(2)}(\mathbb{R}^2)$. For $\boldsymbol{x} = (x_1, x_2)$ in $\mathbb{R}^2$ with $x_1x_2 \neq 0$, let $y_1 = f_1(\boldsymbol{x}) = x_2^2 - x_1^2$ and $y_2 = f_2(\boldsymbol{x}) = x_2^2 + x_1^2$. Let $\boldsymbol{f} = (f_1, f_2)$. Define $h = g \circ \boldsymbol{f}$. Prove that $D_{12}h(\boldsymbol{x})/(4x_1x_2) = (D_{22}g)(\boldsymbol{f}(\boldsymbol{x})) - (D_{11}g)(\boldsymbol{f}(\boldsymbol{x}))$ for all $\boldsymbol{x}$ in $\mathbb{R}^2$ with $x_1x_2 \neq 0$.

8.26. Let f be a differentiable real-valued function on an open set V in $\mathbb{R}^3$. Suppose that g is a differentiable real-valued function defined on an open set U in $\mathbb{R}^2$ such that, for every $\boldsymbol{x} = (x_1, x_2)$ in U, the point $(x_1, x_2, g(x_1, x_2))$ is in V. Let $F(x_1, x_2) = f(x_1, x_2, g(x_1, x_2))$. Show that

$$D_1 F(x_1, x_2) = D_1 f(x_1, x_2, g(x_1, x_2)) + D_3 f(x_1, x_2, g(x_1, x_2)) D_1 g(x_1, x_2)$$

and

$$D_2 F(x_1, x_2) = D_2 f(x_1, x_2, g(x_1, x_2)) + D_3 f(x_1, x_2, g(x_1, x_2)) D_2 g(x_1, x_2).$$

8.27. The *Laplacian* of a real-valued function f defined on an open set U in $\mathbb{R}^n$ is $\nabla^2 f(\boldsymbol{x}) = \sum_{j=1}^n D_{jj} f(\boldsymbol{x})$. A real-valued function f in $C^{(2)}(U)$ is said to be *harmonic* on U in $\mathbb{R}^n$ if f satisfies Laplace's equation, $\nabla^2 f(\boldsymbol{x}) = 0$, for all $\boldsymbol{x}$ in U. For each of the following functions f, find $\nabla^2 f$ and determine whether f is harmonic.

a) $f(\boldsymbol{x}) = x_1/(x_1^2 + x_2^2)$.

b) $f(\boldsymbol{x}) = e^{x_1} \sin x_2$.

c) $f(\boldsymbol{x}) = \ln(x_1^2 + x_2^2 + x_3^2)$.

d) $f(\boldsymbol{x}) = (x_1^2 + x_2^2 + \cdots + x_n^2)^{(n-2)/2}$.

8.28. Let g be a twice continuously differentiable function on $\mathbb{R}$. Let c be a positive constant and, for $\boldsymbol{x}$ in $\mathbb{R}^n$, let $r = ||\boldsymbol{x}||$. For t in $\mathbb{R}$, define

$$f(\boldsymbol{x}, t) = f(x_1, x_2, \ldots, x_n, t) = \frac{1}{r} g\left(t - \frac{r}{c}\right).$$

Prove that

$$D_{11} f(\boldsymbol{x}, t) + D_{22} f(\boldsymbol{x}, t) + \cdots + D_{nn} f(\boldsymbol{x}, t) = \frac{1}{c^2} D_{n+1,n+1} f(\boldsymbol{x}, t).$$

8.29. Let f be a function in $C^{(2)}(\mathbb{R}^2)$. Let $\boldsymbol{g} = (g_1, g_2)$ denote the change to polar coordinates:

$$x_1 = g_1(r, \theta) = r\cos\theta$$

$$x_2 = g_2(r, \theta) = r\sin\theta.$$

Let $h = f \circ \boldsymbol{g}$. Show that the Laplacian of f, expressed in terms of polar coordinates, becomes

$$(\nabla^2 f)(\boldsymbol{g}(r,\theta)) = D_{11}h(r,\theta) + \frac{1}{r}D_1 h(r,\theta) + \frac{1}{r}D_{22}h(r,\theta).$$

8.30. Let f be a function in $C^{(2)}(\mathbb{R}^2)$. Fix any angle θ. Define

$$y_1 = g_1(x_1, x_2) = x_1 \cos\theta - x_2 \sin\theta$$
$$y_2 = g_2(x_1, x_2) = x_1 \sin\theta + x_2 \cos\theta.$$

Let $\boldsymbol{g} = (g_1, g_2)$ and $h = f \circ \boldsymbol{g}$. Show that

$$\nabla^2 h(\boldsymbol{x}) = (\nabla^2 f)(\boldsymbol{g}(\boldsymbol{x}))$$

for all $\boldsymbol{x}$ in $\mathbb{R}^2$. (This problem shows that the Laplacian is unchanged by a rotation of the plane through the angle θ.)

8.31. Let f be a function in $C^{(2)}(\mathbb{R}^3)$. Suppose that $f(x_1, x_2, x_3)$ depends only on $r = (x_1^2 + x_2^2 + x_3^2)^{1/2}$; that is, $f(x_1, x_2, x_3) = g(r)$.

a) Find the Laplacian, $\nabla^2 f = D_{11}f + D_{22}f + D_{33}f$, of f in terms of the derivatives of the function g.

b) Prove that, if f satisfies Laplace's equation

$$\nabla^2 f = D_{11}f + D_{22}f + D_{33}f = 0,$$

then $f(x_1, x_2, x_3) = g(r) = a/r + b$, where a and b are constants.

c) Generalize part (b) functions on $\mathbb{R}^n$ for $n \geq 3$.

8.32. Let f be a function in $C^{(2)}(\mathbb{R}^3)$. Let $\boldsymbol{g} = (g_1, g_2, g_3)$ denote the change to spherical coordinates:

$$x_1 = g_1(\rho, \theta, \phi) = \rho \cos\theta \sin\phi$$
$$x_2 = g_2(\rho, \theta, \phi) = \rho \sin\theta \sin\phi$$
$$x_3 = g_3(\rho, \theta, \phi) = \rho \cos\phi.$$

Let $h(\rho, \theta, \phi) = f(\boldsymbol{g}(\rho, \theta, \phi))$. Show that

$$\begin{aligned}(\nabla^2 f)(\boldsymbol{g}(\rho,\theta,\phi)) &= D_{11}h(\rho,\theta,\phi) + \frac{1}{\rho^2 \sin^2\phi} D_{22}h(\rho,\theta,\phi) \\ &\quad + \frac{1}{\rho^2} D_{33}h(\rho,\theta,\phi) + \frac{2}{\rho} D_1 h(\rho,\theta,\phi) \\ &\quad + \frac{\cot\phi}{\rho^2} D_3 h(\rho, 0, \phi),\end{aligned}$$

where $\nabla^2 f(\boldsymbol{x}) = D_{11}f(\boldsymbol{x}) + D_{22}f(\boldsymbol{x}) + D_{33}f(\boldsymbol{x})$.

8.33. Let g be a differentiable real-valued function on $U = \{\boldsymbol{y} \text{ in } \mathbb{R}^2 : y_1 > 0 \text{ and } y_2 > 0\}$. For $\boldsymbol{x} = (x_1, x_2)$ in $\mathbb{R}^2$, let $y_1 = f_1(\boldsymbol{x}) = e^{x_1}$ and $y_2 = f_2(\boldsymbol{x}) = e^{x_2}$. Let $\boldsymbol{f} = (f_1, f_2)$ and $h(\boldsymbol{x}) = (g \circ \boldsymbol{f})(\boldsymbol{x})$. Show that

$$\begin{aligned}&D_{11}h(\boldsymbol{x}) + D_{22}h(\boldsymbol{x}) \\ &\quad = y_1^2 D_{11}g(\boldsymbol{f}(\boldsymbol{x})) + y_2^2 D_{22}g(\boldsymbol{f}(\boldsymbol{x})) + y_1 D_1 g(\boldsymbol{f}(\boldsymbol{x})) + y_2 D_2 g(\boldsymbol{f}(\boldsymbol{x})).\end{aligned}$$

8.34. Let U be an open set in $\mathbb{R}^n$ and let f and g be real-valued functions in $C^{(2)}(U)$. Show that, at any $\boldsymbol{c}$ in U,

$$[\nabla^2(fg)](\boldsymbol{c}) = [(\nabla^2 f)(\boldsymbol{c})]g(\boldsymbol{c}) + 2\langle \nabla f(\boldsymbol{c}), \nabla g(\boldsymbol{c})\rangle + f(\boldsymbol{c})[(\nabla^2 g)(\boldsymbol{c})].$$

8.35. A continuously differentiable function $\boldsymbol{f} = (f_1, f_2)$ mapping from $\mathbb{R}^2$ to $\mathbb{R}^2$ is said to *satisfy the Cauchy–Riemann equations* if $D_1 f_1 = D_2 f_2$ and $D_2 f_1 = -D_1 f_2$. Show that each of the following functions $\boldsymbol{f} = (f_1, f_2)$ satisfies the Cauchy–Riemann equations.

a) $f_1(x_1, x_2) = x_1^2 - x_2^2$ and $f_2(x_1, x_2) = 2x_1x_2$

b) $f_1(x_1, x_2) = e^{x_1}\cos x_2$, $f_2(x_1, x_2) = e^{x_1}\sin x_2$

c) $f_1(\boldsymbol{x}) = x_1/(x_1^2 + x_2^2)$ and $f_2(\boldsymbol{x}) = -x_2/(x_1^2 + x_2^2)$

d) $f_1(\boldsymbol{x}) = \sin x_1 \cosh x_2$ and $f_2(\boldsymbol{x}) = \cos x_1 \sinh x_2$

e) $f_1(\boldsymbol{x}) = \ln (x_1^2 + x_2^2)^{1/2}$ and $f_2(\boldsymbol{x}) = \tan^{-1}(x_2/x_1)$

8.36. Let g be a real-valued function in $C^{(2)}(\mathbb{R}^2)$. For $\boldsymbol{x}$ in $\mathbb{R}^2$, let $y_1 = f_1(x) = e^{x_1}\cos x_2$ and $y_2 = f_2(\boldsymbol{x}) = e^{x_1}\sin x_2$. Let $\boldsymbol{f} = (f_1, f_2)$ and $h = g \circ \boldsymbol{f}$. Show that

$$\begin{aligned}\nabla^2 h(\boldsymbol{x}) &= D_{11}h(\boldsymbol{x}) + D_{22}h(\boldsymbol{x}) = e^{2x_1}[(D_{11}g)(\boldsymbol{f}(\boldsymbol{x})) + (D_{22}g)(\boldsymbol{f}(\boldsymbol{x}))] \\ &= e^{2x_1}(\nabla^2 g)(\boldsymbol{f}(\boldsymbol{x})).\end{aligned}$$

8.37. Suppose that g is a function in $C^{(2)}(\mathbb{R}^2)$ and that $\boldsymbol{f} = (f_1, f_2)$ satisfies the Cauchy–Riemann equations on an open set U in $\mathbb{R}^2$. (See Exercise 8.35.) Let $h = g \circ \boldsymbol{f}$. Prove that

$$\begin{aligned}\nabla^2 h(\boldsymbol{x}) &= D_{11}h(\boldsymbol{x}) + D_{22}h(\boldsymbol{x}) \\ &= [D_{11}g(\boldsymbol{f}(\boldsymbol{x})) + D_{22}g(\boldsymbol{f}(\boldsymbol{x}))]||(\nabla g)(\boldsymbol{f}(\boldsymbol{x}))||^2 \\ &= (\nabla^2 g)(\boldsymbol{f}(\boldsymbol{x}))||(\nabla g)(\boldsymbol{f}(\boldsymbol{x}))||^2.\end{aligned}$$

Thus prove that g is harmonic if and only if $h = g \circ \boldsymbol{f}$ is also harmonic and $\nabla \boldsymbol{g} \neq 0$. (Recall that a harmonic function g is one for which $\nabla^2 \boldsymbol{g} = 0$.)

8.38. Suppose that $\boldsymbol{f} = (f_1, f_2)$ is a continuously differentiable function defined on an open set U in $\mathbb{R}^2$ that satisfies the Cauchy–Riemann equations. (See Exercise 8.35.) Let $\boldsymbol{c}$ be any point of U and let $\boldsymbol{u}$ and $\boldsymbol{v}$ be arbitrary orthogonal unit vectors. Prove that $D_{\boldsymbol{u}} f_1(\boldsymbol{c}) = \pm D_{\boldsymbol{v}} f_2(\boldsymbol{c})$.

8.39. Suppose that the real-valued function f is defined on an open set U in $\mathbb{R}^2$ and is differentiable at a point $\boldsymbol{c} \neq \boldsymbol{0}$ of U. Shift to polar coordinates and let

$$h(r, \theta) = f(r\cos\theta, r\sin\theta).$$

Let $\boldsymbol{u} = \boldsymbol{c}/||\boldsymbol{c}||$ be the unit vector in the direction of the ray emanating from $\boldsymbol{0}$ and passing through $\boldsymbol{c}$. Let $\boldsymbol{v}$ be a unit vector orthogonal to $\boldsymbol{u}$. Prove that $D_{\boldsymbol{u}} f(\boldsymbol{c}) = D_1 h(r_c, \theta_c)$ and that $D_{\boldsymbol{v}} f(\boldsymbol{c}) = \pm(1/r)D_2 h(r_c, \theta_c)$, where (r_c, θ_c) are polar coordinates of the point $\boldsymbol{c}$.

8.40. Let $\boldsymbol{f} = (f_1, f_2)$ be a differentiable function on an open set U in $\mathbb{R}^2$ that satisfies the Cauchy–Riemann equations. Let $x_1 = g_1(r, \theta) = r\cos\theta$, $x_2 = g_2(r, \theta) = r\sin\theta$ and $\boldsymbol{g} = (g_1, g_2)$. Let

$$\begin{aligned}\boldsymbol{h}(r, \theta) = \boldsymbol{f}(\boldsymbol{g}(r, \theta)) &= (f_1(\boldsymbol{g}(r, \theta)), f_2(\boldsymbol{g}(r, \theta))) \\ &= (h_1(r, \theta), h_2(r, \theta)).\end{aligned}$$

Show that

$$D_1 h_1(r, \theta) = \frac{1}{r} D_2 h_2(r, \theta)$$

and that

$$\frac{1}{r}D_2h_1(r,\theta) = -D_1h_2(r,\theta).$$

(See Exercises 8.38 and 8.39.)

8.41. Let f be a function in $C^{(2)}(\mathbb{R}^2)$. Let

$$y_1 = g_1(\boldsymbol{x}) = \frac{1}{3}(x_1 + x_2)$$
$$y_2 = g_2(\boldsymbol{x}) = \frac{1}{3}(x_1 - 2x_2).$$

Let $\boldsymbol{g} = (g_1, g_2)$ and $h = f \circ \boldsymbol{g}$. Show that

$$\nabla^2 h(\boldsymbol{x}) = 5(D_{11}f)(\boldsymbol{g}(\boldsymbol{x})) + 2(D_{12}f)(\boldsymbol{g}(\boldsymbol{x})) + 2(D_{22}f)(\boldsymbol{g}(\boldsymbol{x})).$$

8.42. Let f be a function in $C^{(2)}(\mathbb{R}^2)$. Let

$$y_1 = g_1(x_1, x_2) = a_{11}x_1 + a_{12}x_2$$
$$y_2 = g_2(x_1, x_2) = a_{21}x_1 + a_{22}x_2.$$

Let $\boldsymbol{g} = (g_1, g_2)$ and $h = f \circ \boldsymbol{g}$. Show that, for constants a, b, and c,

$$aD_{11}h(\boldsymbol{x}) + 2bD_{12}h(\boldsymbol{x}) + cD_{22}h(\boldsymbol{x})$$
$$= a_1(D_{11}f)(\boldsymbol{g}(\boldsymbol{x})) + 2b_1(D_{12}f)(\boldsymbol{g}(\boldsymbol{x})) + c_1(D_{22}f)(\boldsymbol{g}(\boldsymbol{x}))$$

where $a_1c_1 - b_1^2 = (ac - b^2)(\det[a_{ij}])^2$.

8.43. Let f be a real-valued function in $C^{(2)}(\mathbb{R}^n)$. Define a change of variables $\boldsymbol{g} = (g_1, g_2, \ldots, g_n)$ by

$$\begin{array}{lllllll} y_1 = g_1(\boldsymbol{x}) = c_{11}x_1 + c_{12}x_2 + \cdots + c_{1n}x_n \\ y_2 = g_2(\boldsymbol{x}) = c_{21}x_1 + c_{22}x_2 + \cdots + c_{2n}x_n \\ \vdots \quad \vdots \qquad \vdots \qquad \vdots \qquad \cdots \quad \vdots \\ y_n = g_n(\boldsymbol{x}) = c_{n1}x_1 + c_{n2}x_2 + \cdots + c_{nn}x_n, \end{array}$$

where the c_{jk} are constants. Let $h = f \circ \boldsymbol{g}$. Let $\{a_{jk} : j, k = 1, 2, \ldots, n\}$ be any n^2 constants. Show that, if

$$\sum_{j=1}^{n}\sum_{k=1}^{n} a_{jk}D_{jk}h(\boldsymbol{x}) = \sum_{p=1}^{n}\sum_{q=1}^{n} b_{pq}(D_{pq}f)(\boldsymbol{g}(\boldsymbol{x})),$$

then $b_{pq} = \sum_{p=1}^{n}\sum_{q=1}^{n} c_{pj}a_{jk}c_{qk}$. That is, $B = CAC^t$ where C is the matrix of coefficients appearing in the function $\boldsymbol{g}$ and A is the arbitrarily preassigned matrix $[a_{jk}]$.

8.44. For points $\boldsymbol{x} = (x_1, x_2)$ in $\mathbb{R}^2$ such that $x_1 + x_2 \neq 0$, let $f(\boldsymbol{x}) = x_1x_2/(x_1 + x_2)$.

a) If $\boldsymbol{x} = (1, 1)$ and $\boldsymbol{y} = (2, 4)$, find all points z in the line segment $(\boldsymbol{x}, \boldsymbol{y})$ such that $\langle \nabla \boldsymbol{f}(z), \boldsymbol{y} - \boldsymbol{x}\rangle = f(\boldsymbol{y}) - f(\boldsymbol{x})$.

b) If $\boldsymbol{x} = (1, 1)$ and $\boldsymbol{y} = (2, 2)$, find all points z in the line segment $(\boldsymbol{x}, \boldsymbol{y})$ such that $\langle \nabla \boldsymbol{f}(z), \boldsymbol{y} - \boldsymbol{x}\rangle = f(\boldsymbol{y}) - f(\boldsymbol{x})$.

c) If $\boldsymbol{x} = (1, 1)$ and $\boldsymbol{y} = (-2, 1)$, does the Mean Value Theorem apply to guarantee the existence of a point z in the line segment $(\boldsymbol{x}, \boldsymbol{y})$ such that $\langle \nabla \boldsymbol{f}(z), \boldsymbol{y} - \boldsymbol{x}\rangle = f(\boldsymbol{y}) - f(\boldsymbol{x})$? Explain. Does such a point exist?

8.45. A real-valued function f defined on an open set U in $\mathbb{R}^n$ is said to be *homogeneous of degree p* if $f(t\boldsymbol{x}) = t^p f(\boldsymbol{x})$ for all $\boldsymbol{x}$ in U and all t in $\mathbb{R}$ for which $t\boldsymbol{x}$ is also in U. The function f is said to be *positively homogeneous of degree* p if $f(t\boldsymbol{x}) = t^p f(\boldsymbol{x})$ for all $\boldsymbol{x}$ in U and all $t > 0$ such that $t\boldsymbol{x}$ is also in U.

Show that each of the following functions is homogeneous or positively homogeneous and find the degree of homogeneity.

a) For $\boldsymbol{x} = (x_1, x_2)$ in $\mathbb{R}^2$ and any natural number p,

$$f(\boldsymbol{x}) = a_1 x_1^p + a_2 x_1^{p-1} x_2 + a_3 x_1^{p-2} x_2^2 + \cdots + a_{p+1} x_2^p,$$

where the coefficients $a_1, a_2, \ldots, a_{p+1}$ are constants.

b) More generally, for $\boldsymbol{x} = (x_1, x_2, \ldots, x_n)$ in $\mathbb{R}^n$, let $a_1, a_2, \ldots, a_q$ be arbitrary constants, let p be any natural number, and, for each $k = 1, 2, \ldots, q$, let p_{1k}, $p_{2k}, \ldots, p_{nk}$ be nonnegative integers such that $p_{1k} + p_{2k} + \cdots + p_{nk} = p$. Let

$$f(\boldsymbol{x}) = \sum_{k=1}^{q} a_k x_1^{p_{1k}} x_2^{p_{2k}} x_3^{p_{3k}} \cdots x_n^{p_{nk}}.$$

c) For $\boldsymbol{x} = (x_1, x_2)$ in $\mathbb{R}^2$ with $x_2 \neq 0$, let $f(\boldsymbol{x}) = \tan(x_1/x_2)$.

d) For $\boldsymbol{x} \neq \boldsymbol{0}$ in $\mathbb{R}^2$, let $f(\boldsymbol{x}) = x_1/(x_1^2 + x_2^2)$.

e) For $\boldsymbol{x} \neq \boldsymbol{0}$ in $\mathbb{R}^2$, let $f(\boldsymbol{x}) = x_1 x_2/(x_1^2 + x_2^2)$.

f) For $\boldsymbol{x} = (x_1, x_2)$ in $\mathbb{R}^2$ with $x_1 x_2 > 0$, let $f(\boldsymbol{x}) = x_1^2 x_2 \ln(x_2/x_1)$.

g) For $\boldsymbol{x} = (x_1, x_2)$ with $x_2 \neq 0$, let $f(\boldsymbol{x}) = x_1^{1/3} + x_1 x_2^{-2/3}$.

h) For $\boldsymbol{x} = (x_1, x_2) \neq \boldsymbol{0}$ in $\mathbb{R}^2$, let $f(\boldsymbol{x}) = (x_2^2 - x_2^2)/(x_1^2 + x_2^2)$.

i) For $\boldsymbol{x}$ in $\mathbb{R}^n$, let $f(\boldsymbol{x}) = ||\boldsymbol{x}||$.

8.46. *Euler's Theorem.* Suppose that f is a real-valued, differentiable function on an open set U in $\mathbb{R}^n$. Prove that, if f is homogeneous of degree p, then, for every $\boldsymbol{x} = (x_1, x_2, \ldots, x_n)$ in U,

$$x_1 D_1 f(\boldsymbol{x}) + x_2 D_2 f(\boldsymbol{x}) + \cdots + x_n D_n f(\boldsymbol{x}) = pf(\boldsymbol{x}). \tag{8.40}$$

8.47. Suppose that f is a real-valued differentiable function defined on an open set U in $\mathbb{R}^n$. Suppose also that

$$x_1 D_1 f(\boldsymbol{x}) + x_2 D_2 f(\boldsymbol{x}) + \cdots + x_n D_n f(\boldsymbol{x}) = pf(\boldsymbol{x}), \tag{8.41}$$

for some constant p in $\mathbb{R}$ and all $\boldsymbol{x}$ in U. Fix $\boldsymbol{x}$ in U. Show that there exists an interval $I = (t_1, t_2)$ in $\mathbb{R}$, with $0 < t_1 < 1 < t_2$, such that, for t in I, $f(t\boldsymbol{x}) = t^p f(\boldsymbol{x})$. (*Hint:* For t in I, define $g(t) = f(t\boldsymbol{x})$. Use (8.41) to show that $tg'(t) = pg(t)$. Deduce that $t^{-p} g(t)$ is constant.)

8.48. Let f be a real-valued function that is differentiable on an open set U in $\mathbb{R}^n$. Suppose that f is homogeneous of degree p. Prove that, for each $j = 1, 2, \ldots, n$, $D_j f$ is homogeneous of degree $p - 1$ on U.

8.49. Suppose that f, $D_1 f$, and $D_2 f$ are differentiable on $\mathbb{R}^2$. Suppose also that f is positively homogeneous of degree p. Prove that, for all $\boldsymbol{x}$ in $\mathbb{R}^2$,

$$x_1^2 D_{22} f(\boldsymbol{x}) + 2x_1 x_2 D_{12} f(\boldsymbol{x}) + x_2^2 D_{22} f(\boldsymbol{x}) = p(p-1) f(\boldsymbol{x}).$$

8.50. Suppose that f is in $C^{(k)}(U)$ and is homogeneous of degree p. Prove that any kth-order partial derivative of f is homogeneous of degree $p - k$.

8.51. Suppose that U is an open set in $\mathbb{R}^2$ and that f is a real-valued function in $C^{(2)}(U)$. Suppose also that f is positively homogeneous of degree p.

a) Fix m in $\mathbb{R}$. For $\boldsymbol{x}$ in U, let $g(\boldsymbol{x}) = r^m f(\boldsymbol{x})$, where $r = (x_1^2 + x_2^2)^{1/2}$. Show that, for all $\boldsymbol{x}$ in U,

$$\nabla^2 g(\boldsymbol{x}) = r^m \nabla^2 f(\boldsymbol{x}) + m(2p + m)r^{m-2} f(\boldsymbol{x}).$$

b) If, in addition, f is harmonic on U, prove that $r^{-2p}f$ is also harmonic on U.

c) Confirm the assertion in part (b) when $f(\boldsymbol{x}) = x_1^2 - x_2^2$, $f(\boldsymbol{x}) = 2x_1x_2$, and $f(\boldsymbol{x}) = 3x_1^2x_2 - x_2^3$.

d) Generalize the result in part (b) to harmonic functions in $C^{(2)}(U)$ that are positively homogeneous of degree p on U, where U is an open set in $\mathbb{R}^n$.

8.52. Suppose that U is an open set in $\mathbb{R}^n$ and that f is a real-valued function in $C^{(2)}(U)$. Suppose also that f is homogeneous of degree $r = 1$ on U. Prove that, for every $\boldsymbol{x} = (x_1, x_2, \ldots, x_n)$ in U,

$$\begin{aligned} &x_1^2 D_{11} f(\boldsymbol{x}) + x_2^2 D_{22} f(\boldsymbol{x}) + \cdots + x_n^2 D_{nn} f(\boldsymbol{x}) \\ &\qquad + 2x_1x_2 D_{12} f(\boldsymbol{x}) + 2x_1x_3 D_{13} f(\boldsymbol{x}) + \cdots + 2x_{n-1}x_n D_{n-1,n} f(\boldsymbol{x}) = 0. \end{aligned}$$

8.53. For $\boldsymbol{x} = (x_1, x_2)$ in $\mathbb{R}^2$ and $\boldsymbol{c} = \boldsymbol{0}$, find the second-degree Taylor polynomial of f about $\boldsymbol{c}$ for each of the following functions. Also, find the remainder $R_2(\boldsymbol{0}; \boldsymbol{x})$.

a) $f(\boldsymbol{x}) = \sin x_1 \sin x_2$.

b) $f(\boldsymbol{x}) = \cos x_1 \cos x_2$.

c) $f(\boldsymbol{x}) = 1/(1 - x_1 - x_2)$.

d) $f(\boldsymbol{x}) = \ln[(x_1 + 1)(x_2 + 1)]$, $x_1 > -1$, $x_2 > -1$.

e) $f(\boldsymbol{x}) = \exp(2x_1x_2 - x_2^2)$.

8.54. For $\boldsymbol{x}$ in $\mathbb{R}^2$ and $\boldsymbol{c} = \boldsymbol{0}$, find the third-degree Taylor polynomial $P_3(\boldsymbol{x})$ of each of the following functions. Also find the remainder $R_3(\boldsymbol{0}; \boldsymbol{x})$.

a) $f(\boldsymbol{x}) = 2x_1^3 - 3x_1^2x_2 + x_1x_2^2 - x_2^3 + 2x_1^2 - x_1x_2 + x_2 + 4$.

b) $f(\boldsymbol{x}) = e^{x_1} \cos x_2$.

c) $f(\boldsymbol{x}) = \cos(x_1^2 + x_2^2)$.

d) $f(\boldsymbol{x}) = \exp(x_1 + x_2)$.

8.55. Work through the details of the proof of Taylor's theorem in $\mathbb{R}^3$ with $\boldsymbol{c} = \boldsymbol{0}$.

8.56. Suppose that f is defined on an open set U in $\mathbb{R}^n$. Suppose that, for some index j, $D_j f(\boldsymbol{c}) = 0$ and $D_{jj} f(\boldsymbol{c}) < 0$. Can f have a local minimum value at $\boldsymbol{c}$? If $D_j f(\boldsymbol{c}) = 0$ and $D_{jj} f(\boldsymbol{c}) > 0$, can f have a local maximum value at $\boldsymbol{c}$? Explain.

8.57. Let a and b be constant real numbers. Let

$$f(x_1, x_2) = x_1x_2 + a/x_1 + b/x_2.$$

a) Suppose that a and b are positive. Show that f has a minimum at its only critical point. Find the minimum value of f.

b) Suppose that a and b are both negative. Determine whether f has a local minimum or maximum at each of its critical points. If so, find the extreme values of f and identify them as to type.

c) Suppose that a and b are of opposite sign. Determine whether f has a local minimum or maximum at each of its critical points. If so, find the extreme values of f and identify them as to type.

8.58. For $\boldsymbol{x}$ in $U = \{\boldsymbol{x} \text{ in } \mathbb{R}^2 : x_1 > 0,\ x_2 > 0\}$, define

$$f(\boldsymbol{x}) = \frac{x_1x_2 - 8x_1 - 4x_2}{x_1^2x_2^2}.$$

Find the maximum value of f on U.

8.59. For $\boldsymbol{x}$ in $U = \{\boldsymbol{x} \text{ in } \mathbb{R}^2 : x_1 > 0, x_2 > 0\}$, define

$$f(\boldsymbol{x}) = 144x_1^3x_2^2(1 - x_1 - x_2).$$

Find the maximum value of f on U.

8.60. Let $S = \{\boldsymbol{x} \text{ in } \mathbb{R}^2 : x_1^2 + x_2^2 \leq 1\}$. For each of the following functions, find all the critical points of f in the interior of S. Find the maximum and the minimum values of f on S.

a) For $\boldsymbol{x}$ in S, let $f(\boldsymbol{x}) = 2x_1x_2 + (1 - x_1^2 - x_2^2)^{1/2}$.

b) For $\boldsymbol{x}$ in S, let $f(\boldsymbol{x}) = x_1x_2 - (1 - x_1^2 - x_2^2)^{1/2}$.

c) For $\boldsymbol{x}$ in S, let $f(\boldsymbol{x}) = 2x_1x_2 - (1 - x_1^2 - x_2^2)^{3/2}$.

8.61. Fix $c > 0$. Find the largest value of

$$f(\boldsymbol{x}) = x_1x_2(c - x_1 - x_2),$$

for $\boldsymbol{x}$ in the closed triangular region with vertices $(0, 0)$, $(c, 0)$, and $(0, c)$.

8.62. Find all the critical points of each of the following functions. Determine whether each critical point gives a local maximum, local minimum, or saddle point of f.

a) $f(\boldsymbol{x}) = x_1^2x_2^2 - 5x_1^2 - 8x_1x_2 - 5x_2^2$.

b) $f(\boldsymbol{x}) = x_1x_2(12 - 3x_1 - 4x_2)$.

c) $f(\boldsymbol{x}) = (1 - x_1)(1 - x_2)(x_1 + x_2 - 1)$.

d) $f(\boldsymbol{x}) = x_1^2x_2(24 - x_1 - x_2)^3$.

e) $f(\boldsymbol{x}) = (x_1x_2)^{-1} - 4(x_1^2x_2)^{-1} - 8(x_1x_2^2)^{-1}$.

f) $f(\boldsymbol{x}) = 12x_1 \sin x_2 - 2x_1^2 \sin x_2 + x_1^2 \sin x_2 \cos x_2$.

8.63. a) Fix a and b such that $0 < a < b$. Find all the critical points of

$$f(\boldsymbol{x}) = (ax_1^2 + bx_2^2)\exp(-x_1^2 - x_2^2).$$

b) Fix constants a, b, and c such that $0 < a < b < c$. For $\boldsymbol{x}$ in $\mathbb{R}^3$, let

$$f(\boldsymbol{x}) = (ax_1^2 + bx_2^2 + cx_3^2)\exp(-x_1^2 - x_2^2 - x_3^2).$$

Find all the critical points of f. Show that there are two points where f has a local maximum, one point where f has a local minimum, and four critical points where f has neither a local minimum nor a local maximum value.

9

Vector-Valued Functions

Building on the foundation laid in Chapter 8, we can now analyze the behavior of differentiable functions of the form $\boldsymbol{f} = (f_1, f_2, \ldots, f_n)$ defined on an open set in $\mathbb{R}^n$ and taking values in $\mathbb{R}^n$. We will be especially interested in three general problems. The first is to find conditions that ensure that such a function $\boldsymbol{f}$ has a *local inverse* in the neighborhood of a point. The second concerns the feasibility of solving an equation of the form $F(\boldsymbol{x}; y) = F(x_1, x_2, \ldots, x_n; y) = a$ for y in terms of the remaining variables. That is, when does the equation $F(\boldsymbol{x}; y) = a$ *implicitly* define a functional relationship between y and the variables $x_1, x_2, \ldots, x_n$? The third general question addresses the issue of finding extreme values of a function subject to constraining side conditions.

Strong connections exist between our results here and those from linear algebra; at crucial points we will rely on certain facts from linear algebra with which we assume you are familiar. In particular, we will use the fact that a system of n linear equations in n variables, written in matrix notation as $A\boldsymbol{x} = \boldsymbol{b}$, has a unique solution if and only if the matrix A of coefficients of the system has a nonzero determinant. If $\det[A] \neq 0$, then the unique solution vector is found either by Gaussian elimination or by Cramer's rule, which expresses the coordinates of the solution vector as quotients of determinants.

9.1 THE JACOBIAN

Consider for a moment a function $\boldsymbol{f} = (f_1, f_2, \ldots, f_n)$ mapping $\mathbb{R}^n$ to $\mathbb{R}^n$ of the following special form. Assume that each component function f_i is linear; there exist real numbers a_{ij} such that, for all $\boldsymbol{x}$ in $\mathbb{R}^n$,

$$\begin{aligned}
f_1(\boldsymbol{x}) &= a_{11}x_1 + a_{12}x_2 + \cdots + a_{1n}x_n \\
f_2(\boldsymbol{x}) &= a_{21}x_1 + a_{22}x_2 + \cdots + a_{2n}x_n \\
f_3(\boldsymbol{x}) &= a_{31}x_1 + a_{32}x_2 + \cdots + a_{3n}x_n \\
\vdots \quad &\quad \;\; \vdots \qquad\quad \vdots \qquad \cdots \quad\; \vdots \\
f_n(\boldsymbol{x}) &= a_{n1}x_1 + a_{n2}x_2 + \cdots + a_{nn}x_n.
\end{aligned}$$

Each component function f_i is differentiable on all of $\mathbb{R}^n$ and $\nabla f_i(\boldsymbol{x})$ is the constant vector $(a_{i1}, a_{i2}, \ldots, a_{in})$. Notice that the effect of $\boldsymbol{f}$ acting on $\boldsymbol{x}$, given by

$$\boldsymbol{f}(\boldsymbol{x}) = (f_1(\boldsymbol{x}), f_2(\boldsymbol{x}), \ldots, f_n(\boldsymbol{x})),$$

is identical to the effect of the matrix $A = [a_{ij}]$ acting on $\boldsymbol{x} = (x_1, x_2, \ldots, x_n)$:

$$A\boldsymbol{x} = \begin{pmatrix} a_{11} & a_{12} & a_{13} & \cdots & a_{1n} \\ a_{21} & a_{22} & a_{23} & \cdots & a_{2n} \\ a_{31} & a_{32} & a_{33} & \cdots & a_{3n} \\ \vdots & \vdots & \vdots & \cdots & \vdots \\ a_{n1} & a_{n2} & a_{n3} & \cdots & a_{nn} \end{pmatrix} \begin{pmatrix} x_1 \\ x_2 \\ x_3 \\ \vdots \\ x_n \end{pmatrix} = \begin{pmatrix} \sum_{j=1}^n a_{1j}x_j \\ \sum_{j=1}^n a_{2j}x_j \\ \sum_{j=1}^n a_{3j}x_j \\ \vdots \\ \sum_{j=1}^n a_{nj}x_j \end{pmatrix} = \begin{pmatrix} f_1(\boldsymbol{x}) \\ f_2(\boldsymbol{x}) \\ f_3(\boldsymbol{x}) \\ \vdots \\ f_n(\boldsymbol{x}) \end{pmatrix}.$$

Thus functions of this simple type fall under the rubric of linear algebra. We say that f is *represented* by the matrix A. Notice that A is obtained by placing the gradient of f_i in the ith row. From the theory of linear algebra we know that $\boldsymbol{f}$ maps $\mathbb{R}^n$ one-to-one onto $\mathbb{R}^n$ and thus has an inverse if and only if $\det[A] \neq 0$.

If $\boldsymbol{g} = (g_1, g_2, \ldots, g_n)$ is another linear function, with $g_i(\boldsymbol{x}) = \sum_{j=1}^n b_{ij}x_j$ for each $i = 1, 2, \ldots, n$ and all $\boldsymbol{x}$ in $\mathbb{R}^n$, then $\boldsymbol{g}$ is represented by the matrix $B = [b_{ij}]$. The composite function $\boldsymbol{f} \circ \boldsymbol{g}$, obtained by applying first $\boldsymbol{g}$ to the vector $\boldsymbol{x}$, then $\boldsymbol{f}$ to $\boldsymbol{g}(\boldsymbol{x})$, is also a linear function mapping $\mathbb{R}^n$ into $\mathbb{R}^n$. In fact, $\boldsymbol{f} \circ \boldsymbol{g}(\boldsymbol{x}) = (f_1(\boldsymbol{g}(\boldsymbol{x})), f_2(\boldsymbol{g}(\boldsymbol{x})), \ldots, f_n(\boldsymbol{g}(\boldsymbol{x})))$, and, for each $i = 1, 2, \ldots, n$ and for each $\boldsymbol{x}$ in $\mathbb{R}^n$,

$$\begin{aligned}
f_i(\boldsymbol{g}(\boldsymbol{x})) &= \sum_{k=1}^n a_{ik} g_k(\boldsymbol{x}) = \sum_{k=1}^n a_{ik} \left[\sum_{j=1}^n b_{kj} x_j \right] \\
&= \sum_{k=1}^n \left[\sum_{j=1}^n a_{ik} b_{kj} \right] x_j = \sum_{j=1}^n c_{ij} x_j,
\end{aligned}$$

where $c_{ij} = \sum_{k=1}^{n} a_{ik}b_{kj}$. Therefore the composite function $\boldsymbol{f} \circ \boldsymbol{g}$ is represented by the matrix C whose entry in the (i, j) position is $c_{ij} = \sum_{k=1}^{n} a_{ik}b_{kj}$. But this is the entry in the (i, j) position of the matrix product

$$AB = [a_{ij}][b_{ij}] = \left[\sum_{k=1}^{n} a_{ik}b_{kj}\right].$$

Thus, the composition of two linear functions $\boldsymbol{f}$ and $\boldsymbol{g}$ on $\mathbb{R}^n$ is represented by the product AB of the matrices A and B, which represent $\boldsymbol{f}$ and $\boldsymbol{g}$, respectively. The ith row of the matrix AB consists of the coordinates of $\nabla(f_i \circ \boldsymbol{g})$; that is, the ith row of AB is $(D_1(f_i \circ \boldsymbol{g}), D_2(f_i \circ \boldsymbol{g}), \ldots, D_n(f_i \circ \boldsymbol{g}))$. To confirm this assertion, note that, by Theorem 8.4.1, $f_i \circ \boldsymbol{g}$ is differentiable on $\mathbb{R}^n$ for each $i = 1, 2, \ldots, n$ and that, by the chain rule,

$$D_j(f_i \circ \boldsymbol{g})(\boldsymbol{x}) = \sum_{k=1}^{n} D_k f_i(\boldsymbol{g}(\boldsymbol{x}))D_j g_k(\boldsymbol{x}) = \sum_{k=1}^{n} a_{ik}b_{kj} = c_{ij}.$$

If $\boldsymbol{g}$ maps $\mathbb{R}^n$ one-to-one onto $\mathbb{R}^n$, then $\boldsymbol{f} = \boldsymbol{g}^{-1}$ exists. We leave for you to prove that $\boldsymbol{f} = \boldsymbol{g}^{-1}$ is also a linear function and that the matrix A representing $\boldsymbol{f} = \boldsymbol{g}^{-1}$ is $A = B^{-1}$.

Finally recall from your study of linear algebra that $\det[AB] = \det[A]\det[B]$. Since $\boldsymbol{f} \circ \boldsymbol{g}$ maps $\mathbb{R}^n$ one-to-one onto $\mathbb{R}^n$ if and only if $\det[AB] \neq 0$, we deduce that $\boldsymbol{f} \circ \boldsymbol{g}$ maps $\mathbb{R}^n$ one-to-one onto $\mathbb{R}^n$ if and only if both $\boldsymbol{f}$ and $\boldsymbol{g}$ separately map $\mathbb{R}^n$ one-to-one onto $\mathbb{R}^n$. If $\boldsymbol{g}$ is one-to-one (so that $\det[B] \neq 0$) and if $f = \boldsymbol{g}^{-1}$, then $(\boldsymbol{f} \circ \boldsymbol{g})(\boldsymbol{x}) = \boldsymbol{x}$ for all $\boldsymbol{x}$ in $\mathbb{R}^n$. This equation corresponds to the matrix equation $AB = I$, where I denotes the $n \times n$ identity matrix. Furthermore,

$$1 = \det[I] = \det[AB] = \det[A]\det[B].$$

Therefore $\det[A] = 1/\det[B]$.

We review these aspects of linear functions from $\mathbb{R}^n$ to $\mathbb{R}^n$ to set the stage for the following question: To what extent can we generalize these properties to functions $\boldsymbol{f} = (f_1, f_2, \ldots, f_n)$ from $\mathbb{R}^n$ to $\mathbb{R}^n$ which may no longer be linear? Clearly, the following definition captures an essential idea from the foregoing discussion.

DEFINITION 9.1.1 Let U be an open set in $\mathbb{R}^n$ and let $\boldsymbol{f} = (f_1, f_2, \ldots, f_n)$ be a differentiable function mapping U into $\mathbb{R}^n$. For every $\boldsymbol{x}$ in U, form the matrix $[D_j f_i(\boldsymbol{x})]$, the entry in the (i, j) position of which is the partial derivative $D_j f_i(\boldsymbol{x})$. The *Jacobian* of $\boldsymbol{f}$ is

$$J_f(\boldsymbol{x}) = \det[D_j f_i(\boldsymbol{x})]. \quad \bullet$$

The Jacobian of $\boldsymbol{f}$ is obtained by placing the components of the gradient of f_i in the ith row of the matrix. As we will see, the Jacobian of $\boldsymbol{f}$ plays a role similar to that of the derivative; to convey this similarity, $J_f(\boldsymbol{x})$ is sometimes denoted $\partial(f_1, f_2, \ldots, f_n)/\partial(x_1, x_2, \ldots, x_n)$. This traditional notation fails to identify the point $\boldsymbol{x}$ where the Jacobian is evaluated; thus we will use it sparingly.

If $\boldsymbol{f}$ is a linear function, such as those we have already discussed, then $\boldsymbol{f}$ is represented by an $n \times n$ matrix A and $J_f(\boldsymbol{x})$ is simply the constant $\det[A]$. But, in general, $J_f(\boldsymbol{x})$ is a real-valued function that varies with $\boldsymbol{x}$; if $\boldsymbol{f}$ is continuously differentiable, then $J_f(\boldsymbol{x})$ is a continuous function of $\boldsymbol{x}$.

EXAMPLE 1 For $r > 0$ and θ in $\mathbb{R}$, define

$$x_1 = f_1(r, \theta) = r\cos\theta$$
$$x_2 = f_2(r, \theta) = r\sin\theta.$$

Let $\boldsymbol{f} = (f_1, f_2)$ denote the resulting function mapping the open right half-plane U in the $r\theta$-plane onto V, the set of all points in the x_1x_2-plane excluding the point $\mathbf{0} = (0, 0)$. The Jacobian of the mapping $\boldsymbol{f}$ is

$$J_f(r, \theta) = \det\begin{vmatrix}\cos\theta & -r\sin\theta \\ \sin\theta & r\cos\theta\end{vmatrix} = r.$$

Note that $\boldsymbol{f}$ is not globally one-to-one, given that $\boldsymbol{f}(r, \theta + 2k\pi) = \boldsymbol{f}(r, \theta)$ for any integer k. If we restrict f to any strip $S = \{(r, \theta) : r > 0, \theta_0 \le \theta < \theta_0 + 2\pi\}$, however, then $\boldsymbol{f}$ is one-to-one on S. If $0 < \theta_2 - \theta_1 < 2\pi$, then f maps the open rectangular region $\{(r, \theta) : r_1 < r < r_2, \theta_1 < \theta < \theta_2\}$ one-to-one onto the polar region drawn in Fig. 9.1. ●

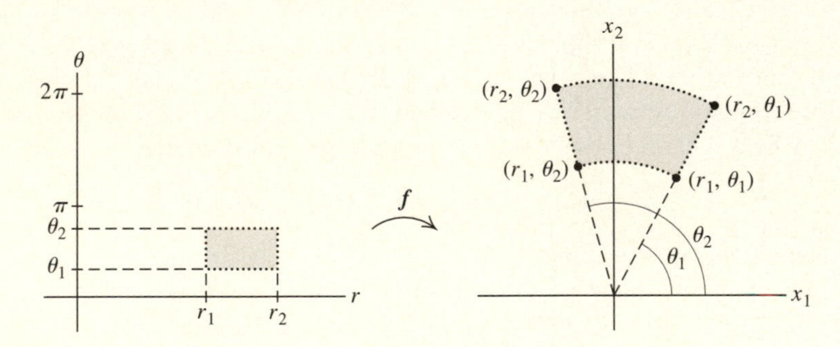

Figure 9.1

EXAMPLE 2 For $\rho > 0, 0 \le \theta < 2\pi$, and $0 \le \phi \le \pi$, define

$$x_1 = f_1(\rho, \theta, \phi) = \rho\cos\theta\sin\phi$$
$$x_2 = f_2(\rho, \theta, \phi) = \rho\sin\theta\sin\phi$$
$$x_3 = f_3(\rho, \theta, \phi) = \rho\cos\phi.$$

Let $\boldsymbol{f} = (f_1, f_2, f_3)$. The function $\boldsymbol{f}$ effects the change to spherical coordinates and is a differentiable function mapping $S = \{(\rho, \theta, \phi) : p > 0, 0 \le \theta < 2\pi, 0 \le \phi \le \pi\}$ one-to-one onto $\{\boldsymbol{x}$ in $\mathbb{R}^3 : \boldsymbol{x} \ne \mathbf{0}\}$; we leave the proof of this claim as an exercise. The Jacobian of $\boldsymbol{f}$ is

$$J_f(\rho, \theta, \phi) = \det\begin{vmatrix}\cos\theta\sin\phi & -\rho\sin\theta\sin\phi & \rho\cos\theta\cos\phi \\ \sin\theta\sin\phi & \rho\cos\theta\sin\phi & \rho\sin\theta\cos\phi \\ \cos\phi & 0 & -\rho\sin\phi\end{vmatrix}.$$

By expanding the determinant along the third row and using the trigonometric identity $\sin^2\alpha + \cos^2\alpha = 1$, we obtain

$$J_f(\rho, \theta, \phi) = -\rho^2 \sin\phi.$$

Notice that J_f does not vanish except at points on the boundary of S where $\rho = 0$ or $\phi = 0$ or π. ●

Our first theorem identifies the relationship between the Jacobian of a composite function $\boldsymbol{f} \circ \boldsymbol{g}$ and the Jacobians of $\boldsymbol{f}$ and $\boldsymbol{g}$.

THEOREM 9.1.1 Let U be an open set in $\mathbb{R}^n$ and let $\boldsymbol{g} = (g_1, g_2, \ldots, g_n)$ be a differentiable function mapping U into $\mathbb{R}^n$. Let V be an open set in $\mathbb{R}^n$ that contains $\boldsymbol{g}(U)$ and let $\boldsymbol{f} = (f_1, f_2, \ldots, f_n)$ be a differentiable function mapping V into $\mathbb{R}^n$. Then $\boldsymbol{f} \circ \boldsymbol{g}$ is a differentiable function mapping U into $\mathbb{R}^n$ and the Jacobian of $\boldsymbol{f} \circ \boldsymbol{g}$ is

$$J_{\boldsymbol{f}\circ\boldsymbol{g}}(\boldsymbol{x}) = J_{\boldsymbol{f}}(\boldsymbol{g}(\boldsymbol{x}))\, J_{\boldsymbol{g}}(\boldsymbol{x}).$$

Remark. The displayed equation of this theorem is strongly reminiscent of the chain rule and suggests that $J_f(\boldsymbol{x})$ can be viewed as a generalization to vector-valued functions of the derivative.

Proof. That $\boldsymbol{f}$ is differentiable at $\boldsymbol{y} = \boldsymbol{g}(\boldsymbol{x})$ in V implies, by Theorem 8.3.2, that each of its component functions f_i is differentiable at $\boldsymbol{y}$. Now, the differentiability of $\boldsymbol{f} \circ \boldsymbol{g} = (f_1 \circ \boldsymbol{g}, f_2 \circ \boldsymbol{g}, \ldots, f_n \circ \boldsymbol{g})$ at $\boldsymbol{x}$ is equivalent to the differentiability of $f_i \circ \boldsymbol{g}$ at $\boldsymbol{x}$ for each $i = 1, \ldots, n$. Since $\boldsymbol{g}$ is assumed to be differentiable at each $\boldsymbol{x}$ in U, Theorem 8.4.1 guarantees the differentiability of each of the component functions of $\boldsymbol{f} \circ \boldsymbol{g}$. Thus, we deduce that $\boldsymbol{f} \circ \boldsymbol{g}$ is also differentiable on U.

By the chain rule, the Jacobian of $\boldsymbol{f} \circ \boldsymbol{g}$ is the determinant

$$\begin{aligned} J_{\boldsymbol{f}\circ\boldsymbol{g}}(\boldsymbol{x}) &= \det\,[D_j(f_i \circ \boldsymbol{g})(\boldsymbol{x})] \\ &= \det\left[\sum_{k=1}^{n} D_k f_i(\boldsymbol{g}(\boldsymbol{x}))\, D_j g_k(\boldsymbol{x})\right]. \end{aligned} \tag{9.1}$$

The Jacobian of $\boldsymbol{g}$, evaluated at $\boldsymbol{x}$, is

$$J_{\boldsymbol{g}}(\boldsymbol{x}) = \det\,[D_j g_i(\boldsymbol{x})],$$

and the Jacobian of $\boldsymbol{f}$, evaluated at $\boldsymbol{g}(\boldsymbol{x})$, is

$$J_{\boldsymbol{f}}(\boldsymbol{g}(\boldsymbol{x})) = \det\,[D_j f_i(\boldsymbol{g}(\boldsymbol{x}))].$$

Notice that the matrix product of $A = [D_j f_i(\boldsymbol{g}(\boldsymbol{x}))]$ and $B = [D_j g_i(\boldsymbol{x})]$ is $AB = [\sum_{k=1}^{n} D_k f_i(\boldsymbol{g}(\boldsymbol{x})) D_j g_k(\boldsymbol{x})]$. Since $\det[AB] = \det[A]\det[B]$, we deduce from (9.1) that

$$J_{\boldsymbol{f}\circ\boldsymbol{g}}(\boldsymbol{x}) = \det\,[AB] = \det\,[A]\det\,[B] = J_{\boldsymbol{f}}(\boldsymbol{g}(\boldsymbol{x}))\, J_{\boldsymbol{g}}(\boldsymbol{x}),$$

thus proving the theorem. ●

EXAMPLE 3 Define a function $\boldsymbol{g} = (g_1, g_2)$ mapping $\mathbb{R}^2$ into $\mathbb{R}^2$ as follows:

$$y_1 = g_1(x_1, x_2) = x_1^2 - x_2^2$$
$$y_2 = g_2(x_1, x_2) = 2x_1x_2.$$

Then

$$J_{\boldsymbol{g}}(\boldsymbol{x}) = \det \begin{vmatrix} 2x_1 & -2x_2 \\ 2x_2 & 2x_1 \end{vmatrix} = 4x_1^2 + 4x_2^2 = 4||\boldsymbol{x}||^2.$$

Define a second function $\boldsymbol{f} = (f_1, f_2)$ by

$$u_1 = f_1(y_1, y_2) = y_1/(y_1^2 + y_2^2)$$
$$u_2 = f_2(y_1, y_2) = y_2/(y_1^2 + y_2^2)$$

for any $\boldsymbol{y} = (y_1, y_2) \neq \boldsymbol{0}$. The Jacobian of $\boldsymbol{f}$ is

$$J_{\boldsymbol{f}}(\boldsymbol{y}) = \det \begin{vmatrix} (-y_1^2 + y_2^2)/||\boldsymbol{y}||^4 & -2y_1y_2/||\boldsymbol{y}||^4 \\ -2y_1y_2/||\boldsymbol{y}||^4 & (y_1^2 - y_2^2)/||\boldsymbol{y}||^4 \end{vmatrix} = -1/||\boldsymbol{y}||^4.$$

Therefore the Jacobian of $\boldsymbol{f} \circ \boldsymbol{g}$, at any $\boldsymbol{x} \neq \boldsymbol{0}$, is

$$\begin{aligned} J_{\boldsymbol{f} \circ \boldsymbol{g}}(\boldsymbol{x}) &= J_{\boldsymbol{f}}(\boldsymbol{g}(\boldsymbol{x}))J_{\boldsymbol{g}}(\boldsymbol{x}) \\ &= -\frac{1}{||\boldsymbol{g}(\boldsymbol{x})||^4} \cdot 4||\boldsymbol{x}||^2 = -\frac{4}{||\boldsymbol{x}||^6}, \end{aligned}$$

since $||\boldsymbol{g}(\boldsymbol{x})|| = ||\boldsymbol{x}||^2$. On the other hand, denote $\boldsymbol{f} \circ \boldsymbol{g}$ by $\boldsymbol{h} = (h_1, h_2)$. You can find h_1 and h_2 by direct substitution. Upon doing so you will obtain, for any $\boldsymbol{x} \neq \boldsymbol{0}$,

$$h_1(x_1, x_2) = \frac{x_1^2 - x_2^2}{(x_1^2 + x_2^2)^2}$$
$$h_2(x_1, x_2) = \frac{2x_1x_2}{(x_1^2 + x_2^2)^2}.$$

From these equations you can compute the Jacobian of $\boldsymbol{h} = \boldsymbol{f} \circ \boldsymbol{g}$ directly in order to confirm our earlier computations above. ●

COROLLARY 9.1.2 Suppose that U is an open set in $\mathbb{R}^n$ and that $\boldsymbol{f}$ is a differentiable function mapping U one-to-one and onto an open set V in $\mathbb{R}^n$. Suppose that $\boldsymbol{f}^{-1}$ is differentiable on V. Then $J_{\boldsymbol{f}^{-1}}(\boldsymbol{f}(\boldsymbol{x})) = 1/J_{\boldsymbol{f}}(\boldsymbol{x})$. ●

9.2 THE INVERSE FUNCTION THEOREM

The main theorem of this section establishes sufficient conditions for the existence of a *local inverse* of a continuously differentiable function $\boldsymbol{f}$ from $\mathbb{R}^n$ to $\mathbb{R}^n$. While it need not be true that $\boldsymbol{f}$ is globally one-to-one, we will show, by restricting to an appropriately chosen open set U_1 containing a point $\boldsymbol{c}$ where $J_{\boldsymbol{f}}(\boldsymbol{c}) \neq 0$ and to a neighborhood $N(\boldsymbol{f}(\boldsymbol{c}))$, that $\boldsymbol{f}$ is one-to-one on U_1, that $\boldsymbol{f}^{-1}$ exists on $N(\boldsymbol{f}(\boldsymbol{c}))$, and that f^{-1} is also continuously differentiable on $N(f(c))$. A first step toward this result is the following theorem.

THEOREM 9.2.1 Let U be an open set in $\mathbb{R}^n$ and let f be a continuously differentiable function mapping U into $\mathbb{R}^n$. Suppose that c is a point of U where $J_f(c) \neq 0$. Then there exists a neighborhood $N(c)$ such that f, restricted to $N(c)$, is one-to-one.

Proof. Since $J_f(c) \neq 0$ and since all the functions $D_j f_i$, for $i, j = 1, 2, \ldots, n$, are assumed to be continuous, there exists a neighborhood $N(c)$ such that, whenever $z^{(1)}, z^{(2)}, z^{(3)}, \ldots, z^{(n)}$ are any n points of $N(c)$, we have

$$\det \begin{vmatrix} D_1 f_1(z^{(1)}) & D_2 f_1(z^{(1)}) & \cdots & D_n f_1(z^{(1)}) \\ D_1 f_2(z^{(2)}) & D_2 f_2(z^{(2)}) & \cdots & D_n f_2(z^{(2)}) \\ D_1 f_3(z^{(3)}) & D_2 f_3(z^{(3)}) & \cdots & D_n f_3(z^{(3)}) \\ \vdots & \vdots & \cdots & \vdots \\ D_1 f_n(z^{(n)}) & D_2 f_n(z^{(n)}) & \cdots & D_n f_n(z^{(n)}) \end{vmatrix} \neq 0. \qquad \mathbf{(9.2)}$$

Choose any two distinct points x and y in $N(c)$. Showing that f is one-to-one on $N(c)$ amounts to showing that $f(x) \neq f(y)$. We assume to the contrary that $f(x) = f(y)$ and obtain a contradiction. Now, $N(c)$ is convex and therefore contains the line segment $[x, y]$. For each $i = 1, 2, \ldots, n$, we apply the Mean Value Theorem to each component function f_i of f to find a point $z^{(i)}$ in the open line segment (x, y) such that $\langle \nabla f_i(z^{(i)}), y - x \rangle = f_i(y) - f_i(x) = 0$. That is, for $i = 1, 2, \ldots, n$,

$$D_1 f_i(z^{(i)})(y_1 - x_1) + \cdots + D_n f_i(z^{(i)})(y_n - x_n) = 0.$$

This is a homogeneous system of n linear equations in the n quantities $y_1 - x_1, y_2 - x_2, \ldots, y_n - x_n$. Since all the points $z^{(1)}, z^{(2)}, \ldots, z^{(n)}$ are in $N(c)$, the matrix $[D_j f_i(z^{(i)})]$ of coefficients of this system has a nonzero determinant by (9.2); therefore, the system of linear equations has the unique solution $y_j - x_j = 0$ for $j = 1, 2, \ldots, n$. Thus $x = y$, contradicting our initial choice of x and y. This proves that f, restricted to $N(c)$, is one-to-one. ●

A second technical issue that will arise in the proof of our main theorem (Theorem 9.2.3) is this. Suppose that f is continuously differentiable and maps an open set U in $\mathbb{R}^n$ to $\mathbb{R}^n$. Suppose that c is a point of U. While c is automatically an interior point of U, we cannot generally assert that $f(c)$ is an interior point of $f(U)$; for all we know, it may happen that $f(c)$ is a boundary point of $f(U)$. We want sufficient conditions that guarantee that $f(c)$ is an interior point. These are provided by the following theorem.

THEOREM 9.2.2 Let U be an open set in $\mathbb{R}^n$, let f be a continuously differentiable function mapping U into $\mathbb{R}^n$, and let c be a point of U. Suppose that $N(c; r)$ is a neighborhood of c whose closure is contained in U, that $J_f(x) \neq 0$ for all x in $N(c; r)$, and that f is one-to-one on $\overline{N(c; r)}$. Then $f(c)$ is an interior point of $f(N(c; r))$.

Proof. For $\boldsymbol{x}$ in $\overline{N(\boldsymbol{c}; r)}$, define $g(\boldsymbol{x}) = ||\boldsymbol{f}(\boldsymbol{x}) - \boldsymbol{f}(\boldsymbol{c})||$. Note that g is a nonnegative, real-valued function on $\overline{N(\boldsymbol{c}; r)}$. Since $\boldsymbol{f}$ is continuous on all of U, g is also continuous on $\overline{N(\boldsymbol{c}; r)}$. Furthermore, $g(\boldsymbol{x}) = 0$ if and only if $\boldsymbol{f}(\boldsymbol{x}) = \boldsymbol{f}(\boldsymbol{c})$. Since $\boldsymbol{f}$ is one-to-one on $\overline{N(\boldsymbol{c}; r)}$, we deduce that $g(\boldsymbol{x}) = 0$ when and only when $\boldsymbol{x} = \boldsymbol{c}$. Therefore $g(\boldsymbol{x}) > 0$ for all $\boldsymbol{x}$ in the compact set $S = \{\boldsymbol{x} : ||\boldsymbol{x} - \boldsymbol{c}|| = r\}$, the hypersphere which forms the boundary of $N(\boldsymbol{c}; r)$. Since g is continuous on S, we can apply Theorem 3.2.4 to deduce that g assumes a minimum value $m > 0$ at some point of S. (See Fig. 9.2.)

We claim that the neighborhood $N(\boldsymbol{f}(\boldsymbol{c}); m/2)$ is contained entirely within $\boldsymbol{f}(N(\boldsymbol{c}; r))$. To prove this, we fix any point $\boldsymbol{y} = (y_1, y_2, \ldots, y_n)$ in $N(\boldsymbol{f}(\boldsymbol{c}); m/2)$ and show that $\boldsymbol{y}$ is in $\boldsymbol{f}(N(\boldsymbol{c}; r))$; that is, we show that there is an $\boldsymbol{x}_0$ in $N(\boldsymbol{c}; r)$ such that $\boldsymbol{f}(\boldsymbol{x}_0) = \boldsymbol{y}$. To this end, define a function h on $\overline{N(\boldsymbol{c}; r)}$ as follows:

$$h(\boldsymbol{x}) = ||\boldsymbol{f}(\boldsymbol{x}) - \boldsymbol{y}|| = \left[\sum_{i=1}^{n}(f_i(\boldsymbol{x}) - y_i)^2\right]^{1/2}.$$

Observe that h is continuous on the compact set $\overline{N(\boldsymbol{c}; r)}$ and therefore, by Theorem 3.2.4, that h has a minimum value at some point $\boldsymbol{x}_0$ in $\overline{N(\boldsymbol{c}; r)}$. We claim that $\boldsymbol{x}_0$ must actually lie in $N(\boldsymbol{c}; r)$ and that $\boldsymbol{f}(\boldsymbol{x}_0) = \boldsymbol{y}$. (See Fig. 9.3.)

To prove these claims, first note that, since $\boldsymbol{y}$ is in $N(\boldsymbol{f}(\boldsymbol{c}); m/2)$, we have $h(\boldsymbol{c}) = ||f(\boldsymbol{c}) - \boldsymbol{y}|| < m/2$. Thus the minimum value of h is certainly less than $m/2$.

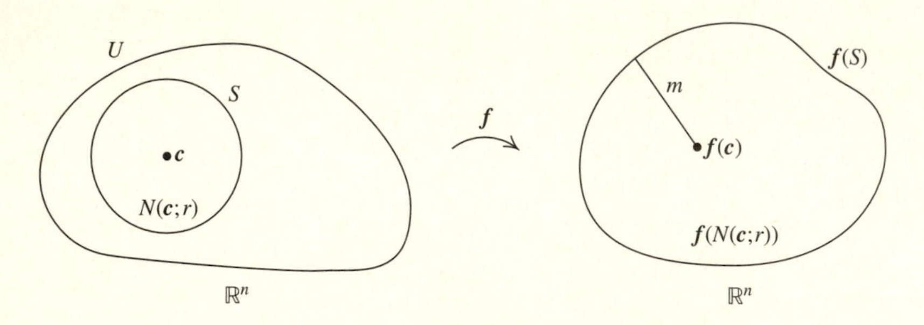

Figure 9.2

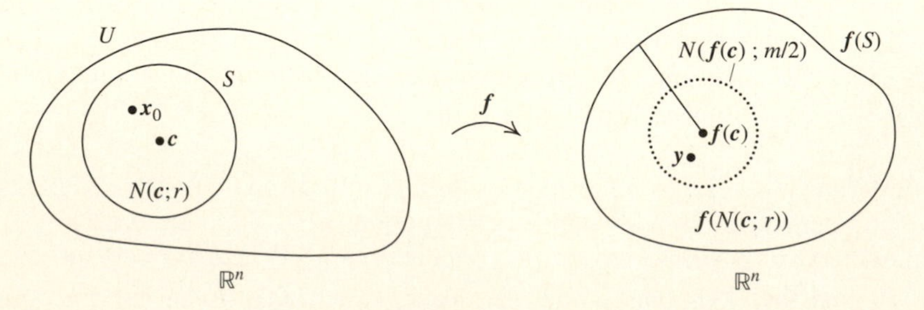

Figure 9.3

But this observation precludes the possibility that $\boldsymbol{x}_0$ is in $S = \{\boldsymbol{x} : ||\boldsymbol{x} - \boldsymbol{c}|| = r\}$. For, if $\boldsymbol{x}_0$ were in S, then

$$h(\boldsymbol{x}_0) = ||\boldsymbol{f}(\boldsymbol{x}_0) - \boldsymbol{y}|| \geq ||\boldsymbol{f}(\boldsymbol{x}_0) - \boldsymbol{f}(\boldsymbol{c})|| - ||\boldsymbol{f}(\boldsymbol{c}) - \boldsymbol{y}||$$
$$> \min_{x \in S} g(\boldsymbol{x}) - m/2 = m/2,$$

whereas the minimum value of h is less than $m/2$. Therefore h has its minimum value at some point $\boldsymbol{x}_0$ in $N(\boldsymbol{c}; r)$. It remains only to show that $\boldsymbol{f}(\boldsymbol{x}_0) = \boldsymbol{y}$.

Since h has a minimum value at $\boldsymbol{x}_0$, so also does h^2. By Theorem 8.8.1, the gradient of h^2 and therefore each of the first order partial derivatives of h^2 must vanish at $\boldsymbol{x}_0$. From the definition of h we have, for $j = 1, 2, \ldots, n$,

$$D_j(h^2)(\boldsymbol{x}_0) = 2\sum_{i=1}^{n} D_j f_i(\boldsymbol{x}_0)[f_i(\boldsymbol{x}_0) - y_i] = 0.$$

Therefore, for $j = 1, 2, \ldots, n$,

$$\sum_{i=1}^{n} D_j f_i(\boldsymbol{x}_0)[f_i(\boldsymbol{x}_0) - y_i] = 0. \tag{9.3}$$

These equations form a system of n linear equations in the n quantities $f_1(\boldsymbol{x}_0) - y_1, f_2(\boldsymbol{x}_0) - y_2, \ldots, f_n(\boldsymbol{x}_0) - y_n$. The matrix of coefficients of this system is the transpose of the matrix $[D_j f_i(\boldsymbol{x}_0)]$. Since $\det[A^t] = \det[A]$ and since J_f does not vanish on $N(\boldsymbol{c}; r)$ (in particular, $J_f(\boldsymbol{x}_0) \neq 0$), we deduce that the system (9.3) has the unique solution $f_i(\boldsymbol{x}_0) - y_i = 0$ for $i = 1, 2, \ldots, n$. That is, $\boldsymbol{f}(\boldsymbol{x}_0) = \boldsymbol{y}$. We conclude that $N(\boldsymbol{f}(\boldsymbol{c}); m/2) \subseteq \boldsymbol{f}(N(\boldsymbol{c}; r))$ and thus that $\boldsymbol{f}(\boldsymbol{c})$ is an interior point of $\boldsymbol{f}(N(\boldsymbol{c}; r))$ as claimed in the theorem. ●

All the pieces are in place; we are ready for the central theorem of this section. Our proof is modeled on that found in Apostol [2].

THEOREM 9.2.3 The Inverse Function Theorem Let U be an open set in $\mathbb{R}^n$ and let $\boldsymbol{f} = (f_1, f_2, \ldots, f_n)$ be a continuously differentiable function mapping U into $\mathbb{R}^n$. Suppose that $\boldsymbol{c}$ is a point of U where $J_f(\boldsymbol{c}) \neq 0$. Then there exist an open set U_1 contained in U that contains $\boldsymbol{c}$, an open set V_1 in $\boldsymbol{f}(U)$ that contains $\boldsymbol{f}(\boldsymbol{c})$, and a function $\boldsymbol{g}$ with domain V_1 such that

i) The function $\boldsymbol{f}$ maps U_1 one-to-one onto V_1.

ii) The function $\boldsymbol{g}$ maps V_1 one-to-one onto U_1.

iii) For all $\boldsymbol{x}$ in U_1, $\boldsymbol{g} \circ \boldsymbol{f}(\boldsymbol{x}) = \boldsymbol{x}$.

iv) The function $\boldsymbol{g}$ is continuously differentiable on V_1.

Remark. The function $\boldsymbol{g}$, whose existence is guaranteed by the theorem, is of course the inverse of $\boldsymbol{f}$.

Proof. Our first step is to use the facts that $J_f(\boldsymbol{c}) \neq 0$ and that $D_j f_i$ is continuous for every $i, j = 1, 2, \ldots, n$ to choose a neighborhood $N(\boldsymbol{c}; r_0)$ contained in U such that, for every choice of $z^{(1)}, z^{(2)}, \ldots, z^{(n)}$ in $N(\boldsymbol{c}; r_0)$,

$$\det \begin{vmatrix} D_1 f_1(z^{(1)}) & D_2 f_1(z^{(1)}) & \cdots & D_n f_1(z^{(1)}) \\ D_1 f_2(z^{(2)}) & D_2 f_2(z^{(2)}) & \cdots & D_n f_2(z^{(2)}) \\ D_1 f_3(z^{(3)}) & D_2 f_3(z^{(3)}) & \cdots & D_n f_3(z^{(3)}) \\ \vdots & \vdots & \cdots & \vdots \\ D_1 f_n(z^{(n)}) & D_2 f_n(z^{(n)}) & \cdots & D_n f_n(z^{(n)}) \end{vmatrix} \neq 0. \tag{9.4}$$

In particular, if $z^{(1)} = z^{(2)} = \cdots = z^{(n)} = \boldsymbol{x}$ is in $N(\boldsymbol{c}; r_0)$, then the determinant in (9.4) becomes $J_f(\boldsymbol{x}) \neq 0$.

Next, invoke Theorem 9.2.1 to choose a neighborhood $N(\boldsymbol{c}; r_1)$ contained in $N(\boldsymbol{c}; r_0)$ such that $\boldsymbol{f}$ is one-to-one on $N(\boldsymbol{c}; r_1)$. Choose any positive $r < r_1$. Consequently $\overline{N(\boldsymbol{c}; r)}$ is contained in $N(\boldsymbol{c}; r_1)$. By Theorem 9.2.2 we deduce that $\boldsymbol{f}(\boldsymbol{c})$ is an interior point of $\boldsymbol{f}(N(\boldsymbol{c}; r))$. Choose a neighborhood $N(\boldsymbol{f}(\boldsymbol{c}))$ that is contained entirely in $\boldsymbol{f}(N(\boldsymbol{c}; r))$. Let $V_1 = N(\boldsymbol{f}(\boldsymbol{c}))$ and $U_1 = \boldsymbol{f}^{-1}(V_1) \cap N(\boldsymbol{c}; r)$. (See Fig. 9.4.) With these definitions, V_1 is open, $\boldsymbol{f}^{-1}(V_1)$ is open (since $\boldsymbol{f}$ is continuous), and therefore U_1 is open.

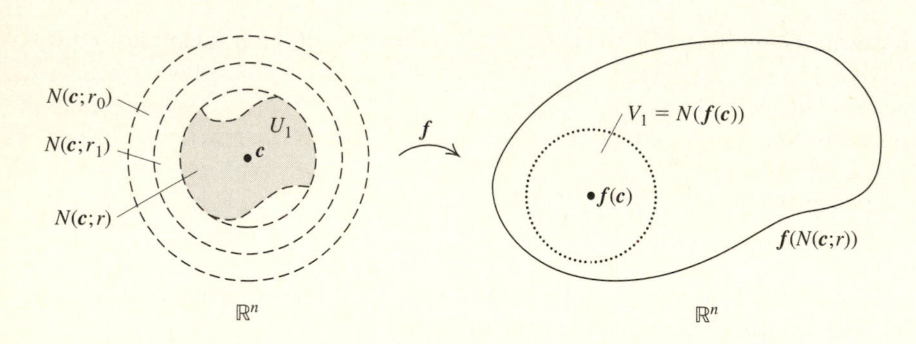

Figure 9.4

There are several minor, yet essential, details to which we must attend. First, $\boldsymbol{f}$ is one-to-one on $N(\boldsymbol{c}; r_1)$ and U_1 is contained in $N(\boldsymbol{c}; r_1)$. Therefore $\boldsymbol{f}$ is one-to-one on U_1. Thus the function $\boldsymbol{g} = \boldsymbol{f}^{-1}$ exists on $\boldsymbol{f}(U_1)$ and $\boldsymbol{g} \circ \boldsymbol{f}(\boldsymbol{x}) = \boldsymbol{x}$ for all $\boldsymbol{x}$ in U_1. Clearly, $\boldsymbol{f}$ maps U_1 into V_1 since U_1 is a subset of $\boldsymbol{f}^{-1}(V_1)$. We claim that $\boldsymbol{f}$ actually maps U_1 *onto* V_1. After all, we chose V_1 to be a subset of $\boldsymbol{f}(N(\boldsymbol{c}; r))$ and every $\boldsymbol{y}$ in V_1 is therefore the image under $\boldsymbol{f}$ of a point $\boldsymbol{x}$ in $N(\boldsymbol{c}; r)$. This $\boldsymbol{x}$ is in $U_1 = \boldsymbol{f}^{-1}(V_1) \cap N(\boldsymbol{c}; r)$. This proves part (i) of the theorem.

For part (ii) note first that, technically, the function $\boldsymbol{g}$ is defined on all of $\boldsymbol{f}(N(\boldsymbol{c}; r_1))$, but we restrict $\boldsymbol{g}$ to V_1. For any $\boldsymbol{y}$ in V_1, $\boldsymbol{g}(\boldsymbol{y})$ is defined to be that unique $\boldsymbol{x}$ in U_1 such that $\boldsymbol{f}(\boldsymbol{x}) = \boldsymbol{y}$. Thus $\boldsymbol{g}$ maps V_1 into U_1. Moreover $\boldsymbol{g}$ is one-to-one because $\boldsymbol{g}$ is the inverse of the function $\boldsymbol{f}$. All that remains is to show that $\boldsymbol{g}$ maps V_1 onto U_1. For any $\boldsymbol{x}$ in $U_1 = \boldsymbol{f}^{-1}(V_1) \cap N(\boldsymbol{c}; r)$, there exists a $\boldsymbol{y}$ in V_1 such that $\boldsymbol{f}(\boldsymbol{x}) = \boldsymbol{y}$. Equivalently, $\boldsymbol{g}(\boldsymbol{y}) = \boldsymbol{x}$. This proves part (ii).

Part (iii) of the theorem is automatic since $\boldsymbol{g} = \boldsymbol{f}^{-1}$. The heart of the proof lies in proving part (iv): that $\boldsymbol{g}$ is continuously differentiable on V_1. Let $\boldsymbol{g} = (g_1, g_2, \ldots, g_n)$. Our task is to prove that, for each $j, k = 1, 2, \ldots, n$ and for each $\boldsymbol{y}$ in V_1, $D_k g_j(\boldsymbol{y})$ exists and that $D_k g_j$ is a continuous function of $\boldsymbol{y}$. We will achieve both proofs by exhibiting formulas for each of the n^2 derivatives $D_k g_j$; these formulas will involve the n^2 continuous derivatives of $\boldsymbol{f}$ in a quotient having the nonzero Jacobian of $\boldsymbol{f}$ as the denominator.

First observe that, since $\boldsymbol{f}$ is continuous and one-to-one on the compact set $\overline{N(\boldsymbol{c}; r)}$, Theorem 3.6.8 implies that $\boldsymbol{g} = \boldsymbol{f}^{-1}$ is also continuous on $\boldsymbol{f}(\overline{N(\boldsymbol{c}; r)})$ and therefore on V_1. Now fix any $\boldsymbol{y}$ in V_1 and choose any j and k in $\{1, 2, \ldots, n\}$. We want to show that $D_k g_j(\boldsymbol{y}) = \lim_{h \to 0}[g_j(\boldsymbol{y} + h\boldsymbol{e}_k) - g_j(\boldsymbol{y})]/h$ exists. For h sufficiently small, $\boldsymbol{y} + h\boldsymbol{e}_k$ is also in V_1 and we let $\boldsymbol{x} = \boldsymbol{g}(\boldsymbol{y})$ and $\boldsymbol{z} = \boldsymbol{g}(\boldsymbol{y} + h\boldsymbol{e}_k)$. Note that $\boldsymbol{x}$ and $\boldsymbol{z}$ are two distinct points of U_1 and therefore are in $N(\boldsymbol{c}; r)$. Since $N(\boldsymbol{c}; r)$ is convex, the closed line segment $[\boldsymbol{x}, \boldsymbol{z}]$ is also contained in $N(\boldsymbol{c}; r)$. The point $\boldsymbol{z}$ depends on h (and k), of course. It is important to observe that

$$\lim_{h \to 0} \boldsymbol{z} = \lim_{h \to 0} \boldsymbol{g}(\boldsymbol{y} + h\boldsymbol{e}_k) = \boldsymbol{g}(\boldsymbol{y}) = \boldsymbol{x},$$

because $\boldsymbol{g}$ is continuous on V_1. Note also that

$$\begin{aligned} \boldsymbol{f}(\boldsymbol{z}) - \boldsymbol{f}(\boldsymbol{x}) &= (\boldsymbol{y} + h\boldsymbol{e}_k) - \boldsymbol{y} = h\boldsymbol{e}_k \\ &= (0, 0, \ldots, 0, h, 0, \ldots, 0), \end{aligned}$$

where h occurs as the kth coordinate of the vector. On the other hand, expressed in terms of coordinates,

$$\boldsymbol{f}(\boldsymbol{z}) - \boldsymbol{f}(\boldsymbol{x}) = (f_1(\boldsymbol{z}) - f_1(\boldsymbol{x}), f_2(\boldsymbol{z}) - f_2(\boldsymbol{x}), \ldots, f_n(\boldsymbol{z}) - f_n(\boldsymbol{x})).$$

Therefore, $f_i(\boldsymbol{z}) - f_i(\boldsymbol{x}) = h\delta_{ik}$, where $\delta_{ik} = 1$ if $i = k$ and $\delta_{ik} = 0$ if $i \neq k$. Now, for each $i = 1, 2, \ldots, n$, apply the Mean Value Theorem to f_i on the line segment $[\boldsymbol{x}, \boldsymbol{z}]$ to find a point $\boldsymbol{z}^{(i)}$ in $(\boldsymbol{x}, \boldsymbol{z})$ such that

$$\begin{aligned} \langle \nabla \boldsymbol{f}_i(\boldsymbol{z}^{(i)}), \boldsymbol{z} - \boldsymbol{x} \rangle &= \langle \nabla \boldsymbol{f}_i(\boldsymbol{z}^{(i)}), \boldsymbol{g}(\boldsymbol{y} + h\boldsymbol{e}_k) - \boldsymbol{g}(\boldsymbol{y}) \rangle \\ &= f_i(\boldsymbol{z}) - f_i(\boldsymbol{x}) = h\delta_{ik}. \end{aligned}$$

Expanding the inner product and dividing by h yields, for each $i = 1, 2, \ldots, n$,

$$\sum_{j=1}^{n} D_j f_i(\boldsymbol{z}^{(i)}) \frac{g_j(\boldsymbol{y} + h\boldsymbol{e}_k) - g_j(\boldsymbol{y})}{h} = \delta_{ik}.$$

These n equations form a nonhomogeneous system of linear equations in the n quantities $[g_1(\boldsymbol{y} + h\boldsymbol{e}_k) - g_1(\boldsymbol{y})]/h$, $[g_2(\boldsymbol{y} + h\boldsymbol{e}_k) - g_2(\boldsymbol{y})]/h, \ldots, [g_n(\boldsymbol{y} + h\boldsymbol{e}_k) - g_n(\boldsymbol{y})]/h$. The matrix of coefficients of the system is $[D_j f_i(\boldsymbol{z}^{(i)})]$. The determinant Δ of this matrix is nonzero by (9.4) since $\boldsymbol{z}^{(1)}, \boldsymbol{z}^{(2)}, \ldots, \boldsymbol{z}^{(n)}$ all belong to $N(\boldsymbol{c}; r)$, a subset of $N(\boldsymbol{c}; r_0)$. By Cramer's rule, we can solve for each of the quantities $[g_j(\boldsymbol{y} + h\boldsymbol{e}_k) - g_j(\boldsymbol{y})]/h$ as a quotient Δ_{jk}/Δ, where Δ_{jk} is the determinant obtained from Δ by replacing the jth column of Δ by the vector $(\delta_{1k}, \delta_{2k}, \ldots, \delta_{nk})$. That is, we have

$$g_j(\boldsymbol{y} + h\boldsymbol{e}_k) - g_j(\boldsymbol{y})/h = \Delta_{jk}/\Delta.$$

To find $D_k g_j(\boldsymbol{y})$, we will pass to the limit as h tends to 0. Recall that $\lim_{h \to 0} \boldsymbol{z} = \boldsymbol{x}$. Since, for each i, $\boldsymbol{z}^{(i)}$ is in the line segment $(\boldsymbol{x}, \boldsymbol{z})$, $\lim_{h \to 0} \boldsymbol{z}^{(i)} = \boldsymbol{x}$ also. When expanded, Δ is a sum of products of the various $D_j f_i(\boldsymbol{z}^{(i)})$ and each of the functions $D_j f_i$ is continuous on all of U; thus $\lim_{h \to 0} D_j f_i(\boldsymbol{z}^{(i)}) = D_j f_i(\boldsymbol{x})$. Consequently $\lim_{h \to 0} \Delta = J_f(\boldsymbol{x}) \neq 0$. Likewise, for each j and k, $\lim_{h \to 0} \Delta_{jk} = [J_f(\boldsymbol{x})]_{jk}$, where $[J_f(\boldsymbol{x})]_{jk}$ denotes the determinant obtained from $J_f(\boldsymbol{x})$ by replacing the jth column of $J_f(\boldsymbol{x})$ by $(\delta_{1k}, \delta_{2k}, \ldots, \delta_{nk})$. We conclude that

$$\begin{aligned} D_k g_j(\boldsymbol{y}) &= \lim_{h \to 0} \frac{g_j(\boldsymbol{y} + h\boldsymbol{e}_k) - g_j(\boldsymbol{y})}{h} \\ &= \lim_{h \to 0} \frac{\Delta_{jk}}{\Delta} = \frac{[J_f(\boldsymbol{x})]_{jk}}{J_f(\boldsymbol{x})}. \end{aligned}$$

Since $\boldsymbol{x} = \boldsymbol{g}(\boldsymbol{y})$, we obtain upon substituting

$$D_k g_j(\boldsymbol{y}) = \frac{[J_f(\boldsymbol{g}(\boldsymbol{y}))]_{jk}}{J_f(\boldsymbol{g}(\boldsymbol{y}))},$$

for each $j, k = 1, 2, \ldots, n$. These are the promised formulas for the partial derivatives $D_k g_j(\boldsymbol{y})$. Since all the functions involved in this quotient are continuous and since $J_f(\boldsymbol{g}(\boldsymbol{y})) \neq 0$, we deduce that $D_k g_j$ is a continuous function of $\boldsymbol{y}$ for every $j, k = 1, 2, \ldots, n$. Therefore, $\boldsymbol{g}$ is continuously differentiable at every point of V_1. This completes the proof of the theorem. ●

EXAMPLE 4 For $\boldsymbol{x} = (x_1, x_2)$ in $\mathbb{R}^2$ define

$$\begin{aligned} y_1 &= f_1(x_1, x_2) = x_1^2 - x_2^2 \\ y_2 &= f_2(x_1, x_2) = 2x_1 x_2. \end{aligned}$$

Let $\boldsymbol{f} = (f_1, f_2)$. The Jacobian of $\boldsymbol{f}$ is $J_f(\boldsymbol{x}) = 4||\boldsymbol{x}||^2$. Since J_f vanishes only at $\boldsymbol{x} = \boldsymbol{0}$, we know that, given any $\boldsymbol{c} \neq \boldsymbol{0}$, there exist an open set U_1 containing $\boldsymbol{c}$ and an open set V_1 containing $\boldsymbol{f}(\boldsymbol{c})$ such that $\boldsymbol{f}$ maps U_1 one-to-one onto V_1. Moreover, there exists a continuously differentiable inverse $\boldsymbol{g} = \boldsymbol{f}^{-1}$ that maps V_1 one-to-one onto U_1.

For insight into the action of $\boldsymbol{f}$ on $\mathbb{R}^2$, we shift to polar coordinates. For any $\boldsymbol{x} = (x_1, x_2) \neq \boldsymbol{0}$, write $x_1 = r\cos\theta$ and $x_2 = r\sin\theta$, where $r = (x_1^2 + x_2^2)^{1/2} = ||\boldsymbol{x}||$ and θ is in $[0, 2\pi)$. Then

$$y_1 = f_1(x_1, x_2) = r^2(\cos^2\theta - \sin^2\theta) = r^2\cos 2\theta$$

and

$$y_2 = f_2(x_1, x_2) = r^2(2\cos\theta\sin\theta) = r^2\sin 2\theta.$$

The point $\boldsymbol{y} = (y_1, y_2) = \boldsymbol{f}(\boldsymbol{x})$ has norm $||\boldsymbol{y}|| = r^2 = ||\boldsymbol{x}||^2$ and polar angle 2θ. Therefore the effect of $\boldsymbol{f}$ on any $\boldsymbol{x} \neq \boldsymbol{0}$ can be described geometrically in the following way: The function $\boldsymbol{f}$ doubles the polar angle and squares the distance of $\boldsymbol{x}$ from $\boldsymbol{0}$. (See Fig. 9.5.) Thus we see that $\boldsymbol{f}$ maps

$$S = \{(x_1, x_2) : x_1 > 0, x_2 \geq 0\} \cup \{(x_1, x_2) : x_1 \leq 0, x_2 > 0\}$$

one-to-one onto $\mathbb{R}^2 \sim \{\boldsymbol{0}\} = \{(y_1, y_2) : (y_1, y_2) \neq (0, 0)\}$.

Observe that $\boldsymbol{f}(-\boldsymbol{x}) = \boldsymbol{f}(\boldsymbol{x})$, so that, globally, $\boldsymbol{f}$ maps $\mathbb{R}^2 \sim \{\boldsymbol{0}\}$ *two*-to-one onto $\mathbb{R}^2 \sim \{\boldsymbol{0}\}$. That is, $\boldsymbol{f}$ maps

$$-S = \{(-x_1, -x_2) : (x_1, x_2) \text{ in } S\}$$

one-to-one onto $\mathbb{R}^2 \sim \{\boldsymbol{0}\}$ also. However, if we restrict to a sufficiently small neighborhood of a point $\boldsymbol{c} \neq \boldsymbol{0}$, we can be assured that $\boldsymbol{f}$ is one-to-one and thus that $\boldsymbol{g} = \boldsymbol{f}^{-1}$ exists. Given the algebraic simplicity of this example, we can actually solve for $\boldsymbol{g} = (g_1, g_2)$. From

$$y_1 = x_1^2 - x_2^2 \qquad \text{and} \qquad y_2 = 2x_1x_2,$$

we have

$$y_1^2 = x_1^4 - 2x_1^2x_2^2 + x_2^4 \qquad \text{and} \qquad y_2^2 = 4x_1^2x_2^2.$$

Adding, we obtain $y_1^2 + y_2^2 = (x_1^2 + x_2^2)^2$, so $x_1^2 + x_2^2 = (y_1^2 + y_2^2)^{1/2}$. Combine this with $x_1^2 - x_2^2 = y_1$ to obtain

$$x_1^2 = \frac{1}{2}[y_1 + (y_1^2 + y_2^2)^{1/2}]$$

and

$$x_2^2 = \frac{1}{2}[-y_1 + (y_1^2 + y_2^2)^{1/2}].$$

Thus,

$$x_1 = g_1(y_1, y_2) = \pm\left[\frac{1}{2}[y_1 + (y_1^2 + y_2^2)^{1/2}]\right]^{1/2}$$

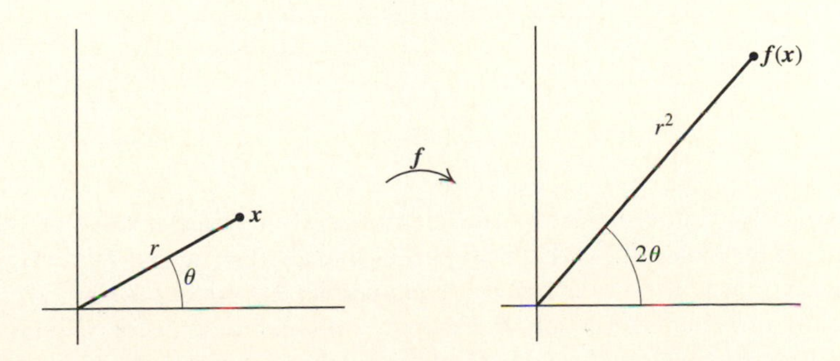

Figure 9.5

and

$$x_2 = g_2(y_1, y_2) = \pm \left[\frac{1}{2}[-y_1 + (y_1^2 + y_2^2)^{1/2}] \right]^{1/2},$$

where we select among the possible choices of positive or negative signs depending on the quadrant in which the original point $\boldsymbol{c}$ lies. ●

In general, although Theorem 9.2.3 ensures the existence of an inverse function, we cannot usually solve for it. However, as we will see in the next section, often only its existence is needed.

9.3 THE IMPLICIT FUNCTION THEOREM

Here we consider a general problem of considerable importance. In its simplest form, we assume that F is a continuously differentiable function of two variables, x and y, defined on an open set U in $\mathbb{R}^2$. We also assume that (c, y_0) is a point of U and that $F(c, y_0) = a$. Without loss of generality, we may assume that $a = 0$; otherwise, we consider the function $F - a$. We form the set

$$C_0 = \{(x, y) \text{ in } U : F(x, y) = 0\},$$

a set in U which is often called the *level curve* of F corresponding to the value $a = 0$. The point (c, y_0) certainly is in the set C_0. (See Fig. 9.6.) The question we ask is this:

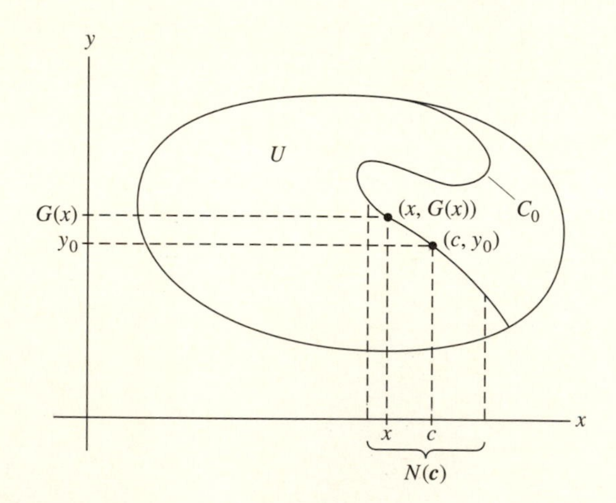

Figure 9.6

Is it possible, by restricting x to a sufficiently small neighborhood $N(c)$, to find a function G defined on $N(c)$ with the properties that $G(c) = y_0$ and $F(x, G(x)) = 0$? If so, we say that $y = G(x)$ is *implicitly defined* by the equation $F(x, y) = 0$. If it were possible to solve the equation $F(x, y) = 0$ algebraically for $y = G(x)$, then we could find the function G *explicitly*. Usually, however, the algebraic problem is intractable. Nevertheless, given an implicit relation $F(x, y) = 0$, we want to know when such a function G exists, even when we cannot actually compute it. We will prove, as a special case of Theorem 9.3.1, that if the partial derivative of F with respect to the variable y does not vanish at (c, y_0)—that is, if $D_y F(c, y_0) \neq 0$—then such an explicit function G must exist on some neighborhood $N(c)$ having the desired properties. Also, G will be continuously differentiable on $N(c)$.

More generally, suppose that F is a continuously differentiable, real-valued function defined on an open set U in $\mathbb{R}^{n+1}$. To focus attention on the variable y, we will write the coordinates of a point in U as $(x_1, x_2, \ldots, x_n, y)$. We will also abbreviate this notation by writing

$$(\boldsymbol{x}; y) = (x_1, x_2, \ldots, x_n, y),$$

where $\boldsymbol{x} = (x_1, x_2, \ldots, x_n)$. For $j = 1, 2, \ldots, n$, the partial derivatives of F with respect to each of the variables $x_1, x_2, \ldots, x_n$ will, as usual, be denoted $D_j F(\boldsymbol{x}; y)$; the partial derivative of F with respect to the variable y will be denoted $D_y F(\boldsymbol{x}; y)$. Again we suppose that $(\boldsymbol{c}; y_0)$ is a point of U where $F(\boldsymbol{c}; y_0) = 0$. Again we ask if it is possible to find a real-valued function G defined on a neighborhood $N(\boldsymbol{c})$ such that (i) $G(\boldsymbol{c}) = y_0$ and (ii) $F(\boldsymbol{x}; G(\boldsymbol{x})) = 0$ for all $\boldsymbol{x}$ in $N(\boldsymbol{c})$. We have little hope of actually computing G; we only want to know if the relation $F(\boldsymbol{x}; y) = 0$ implicitly defines a function $y = G(\boldsymbol{x})$ on some neighborhood of $\boldsymbol{c}$. Again the answer is in the affirmative provided that the partial derivative $D_y F(\boldsymbol{c}; y_0) \neq 0$. The proof is provided in the following theorem.

THEOREM 9.3.1 The Implicit Function Theorem Let U be an open set in $\mathbb{R}^{n+1}$ and let $F(\boldsymbol{x}; y)$ be a continuously differentiable, real-valued function defined on U. Let $(\boldsymbol{c}; y_0)$ be a point of U where $F(\boldsymbol{c}; y_0) = 0$ and $D_y F(\boldsymbol{c}; y_0) \neq 0$. Then there exist an open set V_0 containing $\boldsymbol{c}$ in $\mathbb{R}^n$ and a continuously differentiable, real-valued function G defined on V_0 such that $G(\boldsymbol{c}) = y_0$ and $F(\boldsymbol{x}; G(\boldsymbol{x})) = 0$ for all $\boldsymbol{x}$ in V_0.

Proof. For $(\boldsymbol{x}; y) = (x_1, x_2, \ldots, x_n; y)$ in U and for each $i = 1, 2, \ldots, n$, define $f_i(\boldsymbol{x}; y) = x_i$. Note that f_i is a continuously differentiable function on U and that

$$D_j f_i(\boldsymbol{x}; y) = \delta_{ij} = \begin{cases} 1, & i = j \\ 0, & i \neq j. \end{cases}$$

Also, $D_y f_i(\boldsymbol{x}; y) = 0$. Let $\boldsymbol{f} = (f_1, f_2, \ldots, f_n, F)$. Then $\boldsymbol{f}$ is a continuously differentiable function defined on U that maps U into $\mathbb{R}^{n+1}$. The value of $\boldsymbol{f}$ at a point $(\boldsymbol{x}; y)$ in U is

$$\begin{aligned} \boldsymbol{f}(\boldsymbol{x}; y) &= (f_1(\boldsymbol{x}; y), f_2(\boldsymbol{x}; y), \ldots, f_n(\boldsymbol{x}; y), F(\boldsymbol{x}; y)) \\ &= (x_1, x_2, \ldots, x_n, F(\boldsymbol{x}; y)) = (\boldsymbol{x}; F(\boldsymbol{x}; y)). \end{aligned}$$

We abbreviate the formula for $\boldsymbol{f}$ by writing $\boldsymbol{f} = (\boldsymbol{I}; F)$, where $\boldsymbol{I}(\boldsymbol{x}; y) = \boldsymbol{x}$ for all $(\boldsymbol{x}; y)$ in U. The Jacobian of $\boldsymbol{f}$ at any point $(\boldsymbol{x}; y)$ in U is the $(n+1) \times (n+1)$ determinant

$$\det \begin{vmatrix} 1 & 0 & \cdots & 0 & 0 \\ 0 & 1 & \cdots & 0 & 0 \\ \vdots & \vdots & \cdots & \vdots & \vdots \\ 0 & 0 & \cdots & 1 & 0 \\ D_1 F(\boldsymbol{x}; y) & D_2 F(\boldsymbol{x}; y) & \cdots & D_n F(\boldsymbol{x}; y) & D_y F(\boldsymbol{x}; y) \end{vmatrix}.$$

Therefore, $J_{\boldsymbol{f}}(\boldsymbol{x}; y) = D_y F(\boldsymbol{x}; y)$. In particular, by the hypothesis of the theorem, $J_{\boldsymbol{f}}(\boldsymbol{c}; y_0) \neq 0$. The Inverse Function Theorem (Theorem 9.2.3) guarantees the existence of an open set U_1 in $\mathbb{R}^{n+1}$ containing $(\boldsymbol{c}; y_0)$, an open set V_1 in $\mathbb{R}^{n+1}$ containing $\boldsymbol{f}(\boldsymbol{c}; y_0) = (\boldsymbol{c}; 0)$, and a continuously differentiable function $\boldsymbol{g}$ such that (i) $\boldsymbol{f}$ maps U_1 one-to-one onto V_1; (ii) $\boldsymbol{g}$ maps V_1 one-to-one onto U_1; (iii) $\boldsymbol{g} \circ \boldsymbol{f}(\boldsymbol{x}; y) = (\boldsymbol{x}; y)$ for all $(\boldsymbol{x}; y)$ in U_1. (See Fig. 9.7.)

We want to identify the action of $\boldsymbol{g}$ explicitly. To this end, write $\boldsymbol{g} = (g_1, g_2, \ldots, g_n, h) = (\tilde{\boldsymbol{g}}; h)$, where $\tilde{\boldsymbol{g}} = (g_1, g_2, \ldots, g_n)$. For any $(\boldsymbol{z}; u)$ in V_1, we have

$$\begin{aligned} \boldsymbol{g}(\boldsymbol{z}; u) &= (g_1(\boldsymbol{z}; u), \ldots, g_n(\boldsymbol{z}; u), h(\boldsymbol{z}; u)) \\ &= (\tilde{\boldsymbol{g}}(\boldsymbol{z}; u); h(\boldsymbol{z}; u)). \end{aligned}$$

On the other hand, since $\boldsymbol{f}$ maps U_1 one-to-one onto V_1, there exists a unique point $(\boldsymbol{x}; y)$ in U_1 that is mapped by $\boldsymbol{f}$ to the given point $(\boldsymbol{z}; u)$. That is,

$$\boldsymbol{f}(\boldsymbol{x}; y) = (\boldsymbol{I}(\boldsymbol{x}, y); F(\boldsymbol{x}; y)) = (\boldsymbol{x}; F(\boldsymbol{x}; y)) = (\boldsymbol{z}; u).$$

Therefore, comparing coordinates, we deduce that $\boldsymbol{x} = \boldsymbol{z}$ and $F(\boldsymbol{x}; y) = u$. Furthermore, since $\boldsymbol{g} = \boldsymbol{f}^{-1}$ on V_1 and $\boldsymbol{z} = \boldsymbol{x}$,

$$(\boldsymbol{x}; y) = \boldsymbol{g} \circ \boldsymbol{f}(\boldsymbol{x}; y) = \boldsymbol{g}(\boldsymbol{x}; u) = (\tilde{\boldsymbol{g}}(\boldsymbol{x}; u); h(\boldsymbol{x}; u)).$$

Again comparing coordinates, we deduce that, for any $(\boldsymbol{x}; u)$ in V_1, $\tilde{\boldsymbol{g}}(\boldsymbol{x}; u) = \boldsymbol{x}$ and $h(\boldsymbol{x}; u) = y$, where $(\boldsymbol{x}; y)$ is that unique point in U_1 such that $\boldsymbol{f}(\boldsymbol{x}; y) = (\boldsymbol{x}; u)$.

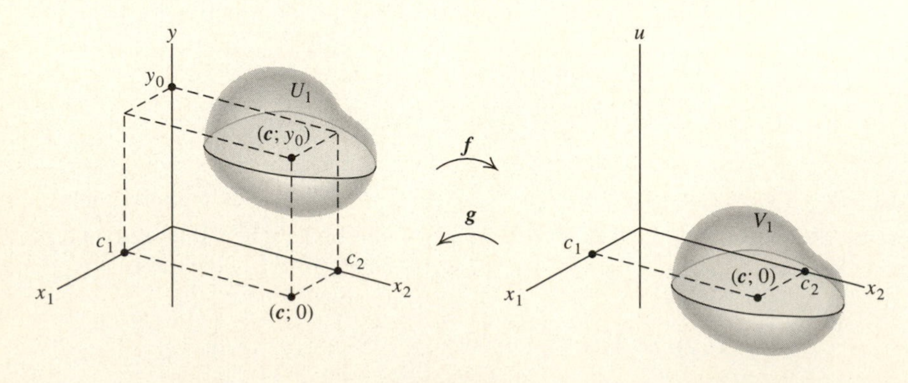

Figure 9.7

Now we can define the open set V_0 and the function G specified in the theorem. Let

$$V_0 = \{\boldsymbol{x} \text{ in } \mathbb{R}^n : (\boldsymbol{x}; 0) \text{ is in } V_1\}.$$

Define G on V_0 as follows: For any $\boldsymbol{x}$ in V_0,

$$G(\boldsymbol{x}) = h(\boldsymbol{x}; 0) = y,$$

where y is that unique real number such that $(\boldsymbol{x}; y)$ is in U_1 and $\boldsymbol{f}(\boldsymbol{x}; y) = (\boldsymbol{x}; 0)$.

We claim first that V_0 is open in $\mathbb{R}^n$. To prove this, we must show that any $\boldsymbol{x}$ in V_0 is an interior point of V_0. Now, $\boldsymbol{x}$ is in V_0 if and only if $(\boldsymbol{x}; 0)$ is in V_1. But V_1 is open in $\mathbb{R}^{n+1}$. Therefore $(\boldsymbol{x}; 0)$ is an interior point of V_1, and there is an $(n+1)$-dimensional neighborhood $N((\boldsymbol{x}; 0); r)$ of $(\boldsymbol{x}; 0)$ that is contained entirely in V_1. The set

$$\{\boldsymbol{z} \text{ in } \mathbb{R}^n : ||\boldsymbol{z} - \boldsymbol{x}|| < r\}$$

is the n-dimensional neighborhood $N(\boldsymbol{x}; r)$ of $\boldsymbol{x}$, which, we claim, is contained entirely in V_0. For if $\boldsymbol{z}$ is in $N(\boldsymbol{x}; r)$, then $||(\boldsymbol{z}; 0) - (\boldsymbol{x}; 0)|| = ||\boldsymbol{z} - \boldsymbol{x}|| < r$. Therefore $(\boldsymbol{z}; 0)$ is in $N((\boldsymbol{x}; 0); r)$, and thus $(\boldsymbol{z}; 0)$ is in V_1. That is, $\boldsymbol{z}$ is in V_0. This proves that $N(\boldsymbol{x}; r)$ is contained in V_0, that $\boldsymbol{x}$ is an interior point of V_0, and that V_0 is open.

Finally, we examine the properties of the function G. Recall that $\boldsymbol{g} = \boldsymbol{f}^{-1}$ is continuously differentiable. Therefore h, being one of the component functions of $\boldsymbol{g}$, is continuously differentiable on V_1. Consequently, G, which is simply the restriction of h to points of the form $(\boldsymbol{x}; 0)$ in V_1, is also continuously differentiable at any point of V_0. Moreover, $G(\boldsymbol{c}) = h(\boldsymbol{c}; 0) = y_0$ because $(\boldsymbol{c}; y_0)$ is that unique point in U_1 such that $\boldsymbol{f}(\boldsymbol{c}; y_0) = (\boldsymbol{c}; 0)$. Finally, for any $\boldsymbol{x}$ in V_0, we have $(\boldsymbol{x}; 0)$ in V_1 and

$$\boldsymbol{g}(\boldsymbol{x}; 0) = (\boldsymbol{x}; y) = (\boldsymbol{x}; G(\boldsymbol{x})),$$

where $(\boldsymbol{x}; y) = (\boldsymbol{x}; G(\boldsymbol{x}))$ is that unique point in U_1 such that

$$\boldsymbol{f}(\boldsymbol{x}; y) = \boldsymbol{f}(\boldsymbol{x}; G(\boldsymbol{x})) = (\boldsymbol{x}; 0).$$

But

$$\boldsymbol{f}(\boldsymbol{x}; G(\boldsymbol{x})) = (\boldsymbol{I}(\boldsymbol{x}; G(\boldsymbol{x})); F(\boldsymbol{x}; G(\boldsymbol{x}))) = (\boldsymbol{x}; F(\boldsymbol{x}; G(\boldsymbol{x}))).$$

Comparing the last coordinates of these two equations, we conclude that $F(\boldsymbol{x}; G(\boldsymbol{x})) = 0$ for all $\boldsymbol{x}$ in V_0. This completes the proof of the theorem. ●

To apply the Implicit Function Theorem to a function F at a point $(\boldsymbol{c}; y_0)$ where $F(\boldsymbol{c}; y_0) = a \neq 0$ and $D_y F(\boldsymbol{c}; y_0) \neq 0$, simply apply it to the function $F_a = F - a$.

COROLLARY 9.3.2 In the notation of the theorem, the partial derivatives of the function G are given by

$$D_j G(\boldsymbol{x}) = -\frac{D_j F(\boldsymbol{x}; G(\boldsymbol{x}))}{D_y F(\boldsymbol{x}; G(\boldsymbol{x}))},$$

for all $\boldsymbol{x}$ in V_0 and all $j = 1, 2, \ldots, n$.

Proof. Since $D_y F(\boldsymbol{c}; y_0) \neq 0$ and since $D_y F$ is continuous, we may assume without loss of generality that $D_y F(\boldsymbol{x}; y) \neq 0$ for all $(\boldsymbol{x}; y)$ in U_1. (Before launching into the proof of the theorem, restrict first to a neighborhood of $(\boldsymbol{c}; y_0)$ where this partial derivative does not vanish. The set U_1 is contained in this neighborhood.) From the theorem, G is a continuously differentiable function on V_0, $y_0 = G(\boldsymbol{c})$, and $(\boldsymbol{x}; G(\boldsymbol{x}))$ is in U_1 for all $\boldsymbol{x}$ in V_0. Thus $D_y F(\boldsymbol{x}; G(\boldsymbol{x})) \neq 0$ for all $\boldsymbol{x}$ in V_0.

Finally, for all $\boldsymbol{x}$ in V_0, define $H(\boldsymbol{x}) = F(\boldsymbol{x}; G(\boldsymbol{x}))$. From the theorem we have

$$H(\boldsymbol{x}) = F(x_1, x_2, \ldots, x_n, G(x_1, x_2, \ldots, x_n)) = 0.$$

Therefore, for each $j = 1, 2, \ldots, n$ and for all $\boldsymbol{x}$ in V_0, $D_j H(\boldsymbol{x}) = 0$. By the chain rule, for $j = 1, 2, \ldots, n$,

$$\begin{aligned} D_j H(\boldsymbol{x}) &= \sum_{k=1}^{n} D_k F(\boldsymbol{x}; G(\boldsymbol{x})) D_j x_k + D_y F(\boldsymbol{x}; G(\boldsymbol{x})) D_j G(\boldsymbol{x}) \\ &= D_j F(\boldsymbol{x}; G(\boldsymbol{x})) + D_y F(\boldsymbol{x}; G(\boldsymbol{x})) D_j G(\boldsymbol{x}) = 0, \end{aligned}$$

since

$$D_j x_k = \delta_{jk} = \begin{cases} 1, & k = j \\ 0, & k \neq j. \end{cases}$$

Solving for $D_j G(\boldsymbol{x})$, we have

$$D_j G(\boldsymbol{x}) = -\frac{D_j F(\boldsymbol{x}; G(\boldsymbol{x}))}{D_y F(\boldsymbol{x}; G(\boldsymbol{x}))},$$

for all $\boldsymbol{x}$ in V_0. This proves the corollary. ●

The Implicit Function Theorem enables us to develop useful geometric insights into the behavior of continuously differentiable functions. To begin, suppose that F is a continuously differentiable, real-valued function on an open set U in $\mathbb{R}^2$. To establish a pattern for our notation, we will denote points of U by $z = (x, y)$. For any a in the range of F, the equation $F(z) = F(x, y) = a$ defines a set $\mathcal{C}_a$ in U. To be specific, $\mathcal{C}_a = \{z \text{ in } U : F(z) = a\}$. The set $\mathcal{C}_a$ is commonly referred to as a *level curve* of F, although, without further restrictions, $\mathcal{C}_a$ need not be a *curve* at all.[1] To see the reason for this terminology, represent the function F by its graph

$$G(F) = \{(x, y, u) \text{ in } \mathbb{R}^3 : z = (x, y) \text{ in } U, u = F(z)\}.$$

Cut $G(F)$ with the horizontal plane $u = a$ in $\mathbb{R}^3$. By doing so, we obtain a subset $\{(z; a) \text{ in } \mathbb{R}^3 : z \text{ in } U, F(z) = a\}$ of the graph of F. Next, project this subset vertically onto the xy-plane. Thus, we obtain the set $\mathcal{C}_a$, defined implicitly by $F(z) = a$. For this reason, $\mathcal{C}_a$ is said to be a "level curve" of F. (See Fig. 9.8.)

Suppose that we can choose a point $z_0 = (x_0, y_0)$ in $\mathcal{C}_a$ where $\nabla \boldsymbol{F}(\boldsymbol{z_0}) \neq \boldsymbol{0}$. Without loss of generality, we assume that $D_2 F(x_0, y_0) \neq 0$. (See Fig. 9.9.) By the

---

[1]The set $\mathcal{C}_a$ may, for example, consist of a single point if $F(z) = a$ at only one point of U. At the other extreme, $\mathcal{C}_a$ may have a nonempty interior—in other words, $\mathcal{C}_a$ may "fill space"—if $F(z) = a$ on some open set contained in U.

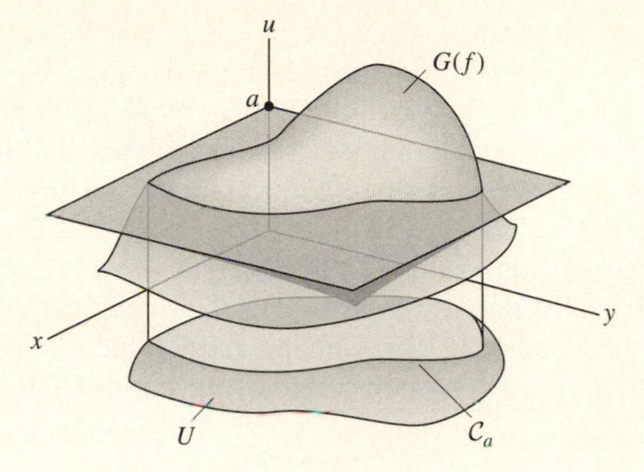

Figure 9.8

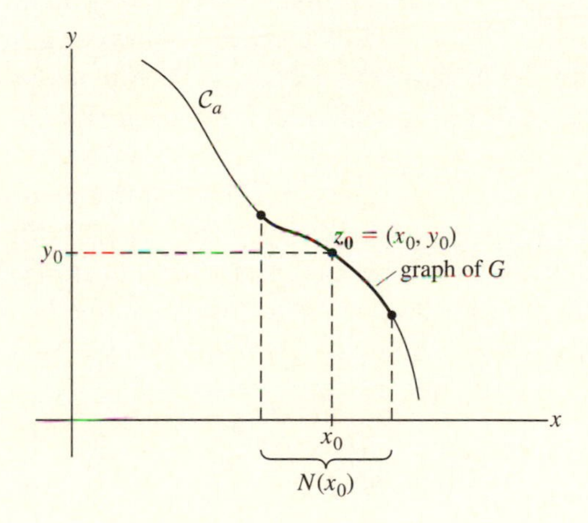

Figure 9.9

Implicit Function Theorem, there exists a neighborhood $N(x_0)$ of the point x_0 and a continuously differentiable function G defined on $N(x_0)$ such that $G(x_0) = y_0$ and $F(x, G(x)) = a$ for all x in $N(x_0)$. That is, for x near x_0, the graph of $y = G(x)$, which is a curve in the plane, coincides with a portion of the set $\mathcal{C}_a$ containing the point (x_0, y_0). Since G is differentiable, the curve $\mathcal{C}_a$ has a tangent line at (x_0, y_0) given by

$$y - y_0 = G'(x_0)(x - x_0). \tag{9.5}$$

By Corollary 9.3.2,

$$G'(x_0) = \frac{-D_1 F(x_0, y_0)}{D_2 F(x_0, y_0)} = \frac{-D_1 F(z_0)}{D_2 F(z_0)}.$$

Substituting in (9.5) and rearranging, we obtain

$$D_1 F(z_0)(x - x_0) + D_2 F(z_0)(y - y_0) = 0$$

for the equation of the line tangent to $\mathcal{C}_a$ at $z_0 = (x_0, y_0)$. We know that the vector $D_1 F(z_0)\boldsymbol{e}_1 + D_2 F(z_0)\boldsymbol{e}_2$ is perpendicular to this line. That is, $\nabla F(z_0)$ is perpendicular to the line tangent to $\mathcal{C}_a$ at z_0. Recall also, by Corollary 8.1.4, that $\nabla F(z_0)$ determines the direction of maximum increase in the values of $|F|$. Therefore the values of F increase (and decrease) near z_0 most rapidly in a direction perpendicular to the curve $\mathcal{C}_a$. (See Fig. 9.10.)

We began our discussion with $n = 2$ because only in this case can we visualize all the graphs involved. But our results generalize to every dimension n. For example, suppose that F is a continuously differentiable, real-valued function defined on an open set U in $\mathbb{R}^3$. Our own physical limitations prevent us from drawing the graph of F because it is the set

$$\{(x_1, x_2, x_3, F(x_1, x_2, x_3)) : (x_1, x_2, x_3) \text{ in } U\}$$

in $\mathbb{R}^4$ and thus is beyond our perceptive capabilities. For any a in the range of F, the equation $F(\boldsymbol{x}) = a$ implicitly defines a set $\mathcal{S}_a = \{\boldsymbol{x} \text{ in } U : F(\boldsymbol{x}) = a\}$ in U, commonly called a *level surface* of F. The set $\mathcal{S}_a$ can be visualized; it is a set in $\mathbb{R}^3$. Suppose that z_0 is a point of $\mathcal{S}_a$ where $\nabla F(z_0) \neq \boldsymbol{0}$. Without loss of generality, we assume that $D_3 F(z_0) \neq 0$. To keep the roles[2] of the variables clear, we denote any point (x_1, x_2, x_3) in U by $z = (\boldsymbol{x}; y)$, where $\boldsymbol{x} = (x_1, x_2)$ and $y = x_3$. In particular, $z_0 = (\boldsymbol{x}_0; \boldsymbol{y}_0)$ where $\boldsymbol{x}_0 = (x_1^{(0)}, x_2^{(0)})$. By the Implicit Function Theorem, there is a neighborhood $N(\boldsymbol{x}_0)$ of the point $\boldsymbol{x}_0$ in $\mathbb{R}^2$ and a continuously differentiable function G defined on $N(\boldsymbol{x}_0)$ such that $G(\boldsymbol{x}_0) = y_0$ and $F(\boldsymbol{x}; G(\boldsymbol{x})) = a$ for all $\boldsymbol{x}$ in $N(\boldsymbol{x}_0)$. That is, for $\boldsymbol{x}$ sufficiently near $\boldsymbol{x}_0$, the surface representing the function $y = G(\boldsymbol{x})$ coincides with part of the set $\mathcal{S}_a$ containing the point $z_0 = (\boldsymbol{x}_0; y_0)$. Because G is

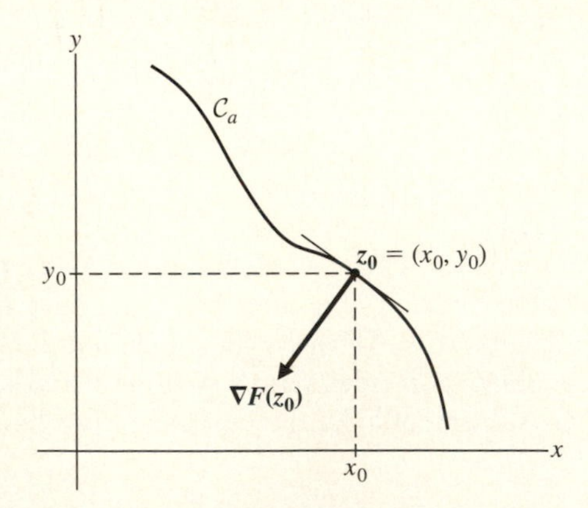

Figure 9.10

[2]This convention corresponds to our assumption that the nonvanishing partial derivative is $D_3 F$ with respect to the *third* independent variable. If our goal is to solve for a different variable—say x_2—in terms of x_1 and x_3, then we will need to know that $D_2 F(z_0) \neq 0$. In this case, we define $y = x_2$, $\boldsymbol{x} = (x_1, x_3)$, and we write $z = (\boldsymbol{x}; y)$.

differentiable at $\boldsymbol{x}_0$, we know by Theorem 8.1.6 that the surface $\mathcal{S}_a$ has a tangent plane P at the point z_0 and that the equation representing the plane P is

$$D_1G(\boldsymbol{x}_0)(x_1 - x_1^{(0)}) + D_2G(\boldsymbol{x}_0)(x_2 - x_2^{(0)}) - (y - G(\boldsymbol{x}_0)) = 0.$$

By Corollary 9.3.2, we have $D_1G(\boldsymbol{x}_0) = -D_1F(z_0)/D_3F(z_0)$ and $D_2G(\boldsymbol{x}_0) = -D_2F(z_0)/D_3F(z_0)$. Substituting these values in (9.5), noting that $G(\boldsymbol{x}_0) = y_0$, and rearranging yields

$$D_1F(z_0)(x_1 - x_1^{(0)}) + D_2F(z_0)(x_2 - x_2^{(0)}) + D_3F(z_0)(y - y_0) = 0$$

as the equation representing the plane tangent to the surface $\mathcal{S}_a$ at the point z_0. (See Fig. 9.11.) From our discussion at the end of Section 8.1, we know that the vector $\boldsymbol{v} = \sum_{j=1}^{3} D_jF(z_0)\boldsymbol{e}_j$ is perpendicular to the tangent plane P. That is, $\nabla\boldsymbol{F}(z_0)$ is perpendicular to P and hence to the surface $\mathcal{S}_a$. Again, the direction of most rapid change in the values of $|F|$ is in the direction determined by $\nabla\boldsymbol{F}(z_0)$, that is, in the direction perpendicular to the surface $\mathcal{S}_a$.

The foregoing discussion generalizes with only slight modifications to apply to a continuously differentiable function F defined on an open set U in $\mathbb{R}^n$. Suppose that F has a nonzero gradient at a point z_0 on a *level hypersurface* $\mathcal{S}_a = \{z = (z_1, z_2, \ldots, z_{n-1}, z_n)$ in $U : F(z) = a\}$. Without loss of generality, $D_nF(z_0) \neq 0$ and we write $z = (\boldsymbol{x}; y)$ with $\boldsymbol{x}$ in $\mathbb{R}^{n-1}$ and y in $\mathbb{R}$. The point $\boldsymbol{z_0}$ is written as $(\boldsymbol{x}_0; y_0)$ where $\boldsymbol{x}_0 = (x_1^{(0)}, x_2^{(0)}, \ldots, x_{n-1}^{(0)})$. Apply the Inverse Function Theorem: There exists a neighborhood $N(\boldsymbol{x_0})$ in $\mathbb{R}^{n-1}$ and a continuously differentiable function G defined on $N(\boldsymbol{x}_0)$ such that $G(\boldsymbol{x}_0) = y_0$ and $F(\boldsymbol{x}; G(\boldsymbol{x})) = a$, for all $\boldsymbol{x}$ in $N(\boldsymbol{x}_0)$. By Theorem 8.1.7, the hyperplane tangent to the hypersurface $\mathcal{S}_a$ at $z_0 = (\boldsymbol{x}_0; y_0)$ exists and is represented by the equation

$$\sum_{j=1}^{n-1} D_jG(\boldsymbol{x}_0)(x_j - x_j^{(0)}) - (y - y_0) = 0.$$

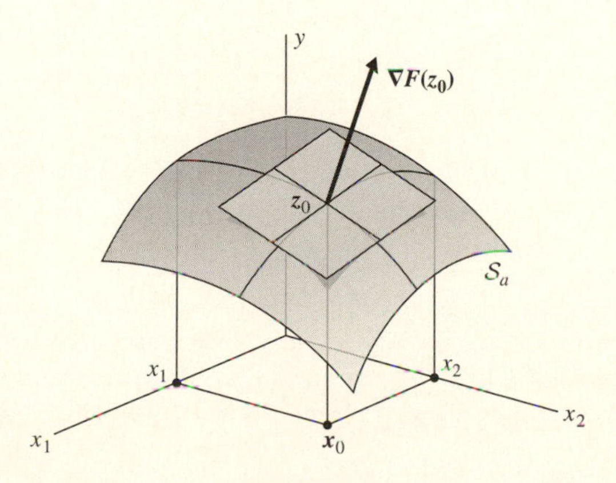

Figure 9.11

By Corollary 9.3.2, $D_j G(\boldsymbol{x}_0) = -D_j F(z_0)/D_n F(z_0)$ for each $j = 1, 2, \ldots, n-1$. Therefore, the equation of the tangent hyperplane becomes

$$\sum_{j=1}^{n-1} D_j F(z_0)(x_j - x_j^{(0)}) + D_n F(z_0)(y - y_0) = 0.$$

Thus the vector $\nabla F(\boldsymbol{z_0})$ is perpendicular to this hyperplane; again, the gradient determines the direction of most rapid change of the values of $|F|$. We summarize the foregoing discussion in the following theorem.

THEOREM 9.3.3 Let F be a continuously differentiable function on an open set U in $\mathbb{R}^n$. Let $\boldsymbol{z_0}$ be any point in U where $\nabla F(\boldsymbol{z_0}) \neq \boldsymbol{0}$. Let $a = F(\boldsymbol{z_0})$. Then the level hypersurface $\mathcal{S}_a$ of F passing through the point $\boldsymbol{z_0}$ has a tangent hyperplane at $\boldsymbol{z_0}$ and $\nabla F(\boldsymbol{z_0})$ is perpendicular to $\mathcal{S}_a$. ●

The Implicit Function Theorem (Theorem 9.3.1) provides conditions under which we can solve for one variable as a function of the others. The natural question arises: Under what conditions can we solve for two (or more) variables in terms of the remaining variables? That is, if $\boldsymbol{F} = (F_1, F_2)$ is a continuously differentiable function defined on an open set U in $\mathbb{R}^{n+2}$ with values in $\mathbb{R}^2$ and if we restrict to the set of points in U where

$$\boldsymbol{F}(x_1, x_2, \ldots, x_n, y_1, y_2) = \boldsymbol{0}, \tag{9.6}$$

do there exist two real-valued, continuously differentiable functions G_1 and G_2 such that $y_1 = G_1(x_1, x_2, \ldots, x_n)$ and $y_2 = G_2(x_1, x_2, \ldots, x_n)$ are defined implicitly by (9.6)? The answer is provided by the following theorem. To streamline notation, we will write $(\boldsymbol{x}; \boldsymbol{y})$ to denote the point $(x_1, x_2, \ldots, x_n, y_1, y_2)$ where $\boldsymbol{x} = (x_1, x_2, \ldots, x_n)$ is a point of $\mathbb{R}^n$ and $\boldsymbol{y} = (y_1, y_2)$ is a point in $\mathbb{R}^2$.

THEOREM 9.3.4 Let U be an open set in $\mathbb{R}^{n+2}$ and let $\boldsymbol{F} = (F_1, F_2)$ be a continuously differentiable function defined on U with values in $\mathbb{R}^2$. Suppose that $(\boldsymbol{c}; \boldsymbol{y}_0)$ is a point of U such that $F(\boldsymbol{c}; \boldsymbol{y}_0) = \boldsymbol{0}$. Suppose also that the determinant

$$\det \begin{vmatrix} D_{y_1} F_1(\boldsymbol{c}; \boldsymbol{y}_0) & D_{y_2} F_1(\boldsymbol{c}; \boldsymbol{y}_0) \\ D_{y_1} F_2(\boldsymbol{c}; \boldsymbol{y}_0) & D_{y_2} F_2(\boldsymbol{c}; \boldsymbol{y}_0) \end{vmatrix} \neq 0.$$

Then there exists an open set V_0 containing $\boldsymbol{c}$ in $\mathbb{R}^n$ and a continuously differentiable function $\boldsymbol{G} = (G_1, G_2)$ defined on V_0 with values in $\mathbb{R}^2$ such that $\boldsymbol{G}(\boldsymbol{c}) = \boldsymbol{y}_0$ and $\boldsymbol{F}(\boldsymbol{x}; \boldsymbol{G}(\boldsymbol{x})) = \boldsymbol{0}$ for all $\boldsymbol{x}$ in V_0.

Proof. Since many of the details of the proof imitate the corresponding steps in the proof of Theorem 9.3.1, we will merely sketch the argument. We leave some details of the proof as an exercise. As in the proof of Theorem 9.3.1, we augment the function $\boldsymbol{F}$ by defining $f_i(\boldsymbol{x}; \boldsymbol{y}) = x_i$ for $i = 1, 2, \ldots, n$ and by forming the function

$$\boldsymbol{f} = (f_1, f_2, \ldots, f_n, F_1, F_2).$$

Again, we write $f = (I; F)$. The Jacobian of f is the determinant

$$\begin{vmatrix} 1 & 0 & \cdots & 0 & 0 & 0 \\ 0 & 1 & \cdots & 0 & 0 & 0 \\ \vdots & \vdots & \cdots & \vdots & \vdots & \vdots \\ 0 & 0 & \cdots & 1 & 0 & 0 \\ D_1 F_1(x; y) & D_2 F_1(x; y) & \cdots & D_n F_1(x; y) & D_{y_1} F_1(x; y) & D_{y_2} F_1(x; y) \\ D_1 F_2(x; y) & D_2 F_2(x; y) & \cdots & D_n F_2(x; y) & D_{y_1} F_2(x; y) & D_{y_2} F_2(x; y) \end{vmatrix}$$

$$= \det \begin{vmatrix} D_{y_1} F_1(x; y) & D_{y_2} F_1(x; y) \\ D_{y_1} F_2(x; y) & D_{y_2} F_2(x; y) \end{vmatrix}.$$

In particular, $J_f(c; y_0) \neq 0$ by the hypothesis. Thus we can invoke the Inverse Function Theorem. There exist open sets U_1 in $\mathbb{R}^{n+2}$ containing $(c; y_0)$ and V_1 in $\mathbb{R}^{n+2}$ containing $f(c; y_0) = (c; 0)$, and a continuously differentiable function g such that (i) f maps U_1 one-to-one onto V_1; (ii) g maps V_1 one-to-one onto U_1; and (iii) $g \circ f(x; y) = (x; y)$ for all points $(x; y)$ in U_1. Again, we write

$$g = (g_1, g_2, \ldots, g_n, h_1, h_2) = (\tilde{g}; h),$$

where $\tilde{g} = (g_1, g_2, \ldots, g_n)$ and $h = (h_1, h_2)$. By an argument similar to that used in the proof of Theorem 9.3.1, we have for any $(x; u)$ in V_1, $\tilde{g}(x; u) = x$ and $h(x; u) = y = (y_1, y_2)$, where $(x; y)$ is that unique point in U_1 such that $f(x; y) = (x; u)$. Again, we define $V_0 = \{x \text{ in } \mathbb{R}^n : (x; 0) \text{ is in } V_1\}$. For x in V_0, define $G(x) = h(x; 0)$. Now $h(x; 0)$ is that unique vector $y = (y_1, y_2)$ in $\mathbb{R}^2$ such that $f(x; y) = (x; 0)$. Therefore $G = (G_1, G_2)$, where $y_1 = G_1(x)$ and $y_2 = G_2(x)$.

As you can prove by imitating the argument in the proof of Theorem 9.3.1, V_0 is open, G is continuously differentiable on V_0, $G(c) = y_0$, and $F(x; G(x)) = 0$ for all x in V_0. This completes the proof of the theorem. ●

Once you have worked through the details of the proof of Theorem 9.3.4, you will see clearly how to treat the general case. The function $F = (F_1, F_2, \ldots, F_k)$ is a continuously differentiable function mapping an open set U in $\mathbb{R}^{n+k}$ into $\mathbb{R}^k$. The point $(x_1, x_2, \ldots, x_n, y_1, y_2, \ldots, y_k)$ is a point of U denoted $(x; y)$. If $(c; y_0)$ is a point of U such that $F(c; y_0) = 0$ and $\det [D_{y_j} F_i(c; y_0)] \neq 0$, then there exist an open set V_0 in $\mathbb{R}^n$ containing c and a continuously differentiable function $G = (G_1, G_2, \ldots, G_k)$ mapping V_0 into $\mathbb{R}^k$ such that $G(c) = y_0$ and $F(x; G(x)) = 0$ for all x in V_0. Thus, for $j = 1, 2, \ldots, k$, we can solve for $y_j = G_j(x)$ in terms of $x_1, x_2, \ldots, x_n$ by restricting x to V_0.

9.4 Constrained Optimization

The problem addressed here is described as follows. Let f and g be two continuously differentiable, real-valued functions defined on an open set U in $\mathbb{R}^n$. Suppose that, for whatever reason, we restrict our attention to the set of points z in U where $g(z) = c$. For example, if g is a function that measures the temperature at points z in space, then $\{z \text{ in } U : g(z) = c\}$ is called an *isotherm*; if g measures pressure, then this same set is called an *isobar*. For simplicity and without loss of generality, we can take $c = 0$; otherwise replace g by $g - c$. Let $Z = \{z \text{ in } U : g(z) = 0\}$. The problem we pose is this. Devise a procedure to find the points in U where f has a local maximum value, not on all of U, but subject to the constraining side condition $g(z) = 0$. That is, we want to find any point z_0 in Z such that, for a neighborhood $N(z_0)$, we have $f(z) \leq f(z_0)$ for z in $Z \cap N(z_0)$. Such points will be called *constrained local extreme points*. We will frame our discussion in terms of local maxima with the understanding that our results, when applied to $-f$, will treat the case when f has a constrained minimum.

To facilitate our search for points where f has a constrained maximum, we want to identify conditions that must hold at such points. These conditions will then suggest a procedure for finding these points. The method is credited to Lagrange and bears the imprint of his personal mathematical style.

A common misconception of beginners is to assume that, since f is to have a local maximum at z_0 subject to the constraint $g = 0$, it must follow that $\nabla \boldsymbol{f}(\boldsymbol{z_0})$ vanishes as in Theorem 8.8.1. Let's settle this question at the outset. The hypotheses of Theorem 8.8.1 stipulate that f is to have a local extremum at a point in an *open set*. The set $Z \cap N(\boldsymbol{z_0})$ generally is *not* an open set. There may well be points in $N(z_0)$ where the value of f exceeds $f(z_0)$, but these points need not be in Z. Thus we have no reason to expect the gradient of f to vanish at a constrained extremum. We mention this warning explicitly because the strategy for the proof of our main theorem (Theorem 9.4.1 below) is to construct an auxiliary function that will have a local extremum at a point in an open set to which Theorem 8.8.1 does apply. From the vanishing of the gradient of the auxiliary function will follow the solution to our problem. But the auxiliary function is not f.

Before launching into the technical details, let us look at the geometrical situation for two continuously differentiable functions f and g defined on U in $\mathbb{R}^2$. Assume that for all points in question, the gradients $\nabla \boldsymbol{f}$ and $\nabla \boldsymbol{g}$ do not vanish. Then the equation $g(z) = 0$ implicitly describes a smooth curve Z in U. Superimpose on the curve Z a number of level curves $\mathcal{C}_a$ of f for various values of a. In particular, suppose that z_0 is a point of Z where f has a local maximum subject to the constraint $g = 0$. Let $a_0 = f(z_0)$ and sketch the graph of $\mathcal{C}_{a_0}$. (See Fig. 9.12.) We claim that the curves $\mathcal{C}_{a_0}$ and Z must be tangent at the point z_0. By this we mean, of course, that $\mathcal{C}_{a_0}$ and Z have the same tangent line at z_0. If we were to choose a to be slightly larger than a_0, then the resulting level curve $\mathcal{C}_a$ of f would fail to intersect Z near z_0. After all, the point z_0 is to yield a local constrained *maximum* value of a_0 for f. On the other hand, if we were to take a value of a_1 slightly less than a_0, then $\mathcal{C}_{a_1}$ meets Z in a point z_1 near z_0 and we can move along the curve Z from z_1 toward z_0, increasing the value of f while satisfying the constraint $g = 0$. That is, $\mathcal{C}_{a_1}$ must "cross" the curve Z at z_1. This discussion does not constitute a proof, but we hope it helps you see that $\mathcal{C}_{a_0}$ and Z must be tangent at z_0. If you will grant this claim for the moment, then we have the answer to our initial problem. The gradient $\nabla \boldsymbol{f}(\boldsymbol{z_0})$ is perpendicular to the curve $\mathcal{C}_{a_0}$ at z_0. Likewise, the gradient $\nabla \boldsymbol{g}(\boldsymbol{z_0})$ is perpendicular to the curve Z at z_0. Since these two curves have the same tangent line at z_0, the vectors $\nabla \boldsymbol{f}(\boldsymbol{z_0})$ and $\nabla \boldsymbol{g}(\boldsymbol{z_0})$ must be parallel. That is, there

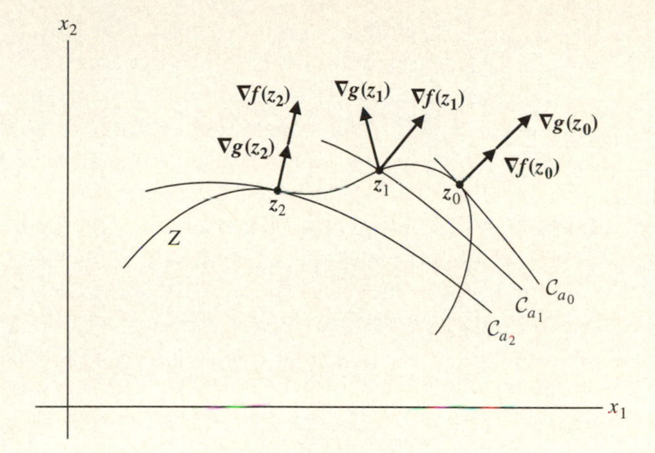

Figure 9.12

must exist a real number λ such that $\nabla \boldsymbol{f}(\boldsymbol{z_0}) = \lambda \nabla \boldsymbol{g}(\boldsymbol{z_0})$. The number λ is called a *Lagrange multiplier*.

Notice that, at a point of intersection of Z and a level curve C_a where f does not have a local constrained extremum, the gradients of f and g may or may not be parallel. (See the point z_2 in Fig. 9.12.) That $\nabla \boldsymbol{f}(\boldsymbol{z_0}) = \lambda \nabla \boldsymbol{g}(\boldsymbol{z_0})$ for some λ is a *necessary*, but insufficient, condition for z_0 to be a point in Z where f has a local constrained extremum.

With this preliminary discussion, we are prepared to prove the following theorem.

THEOREM 9.4.1 Lagrange's Method Let f and g be continuously differentiable, real-valued functions defined on an open set U in $\mathbb{R}^n$, with $n \geq 2$. Let $Z = \{z \text{ in } U : g(z) = 0\}$. Suppose that a point z_0 in Z and a neighborhood $N(\boldsymbol{z_0}) \subseteq U$ exist such that $f(z) \leq f(z_0)$ for all z in $Z \cap N(z_0)$ and such that $\nabla \boldsymbol{g}(z_0) \neq 0$. Then there exists a real number λ such that $\nabla \boldsymbol{f}(\boldsymbol{z_0}) = \lambda \nabla \boldsymbol{g}(\boldsymbol{z_0})$.

Proof. Since $\nabla \boldsymbol{g}(\boldsymbol{z_0}) \neq 0$, at least one of the first-order partial derivatives of g does not vanish at z_0. Without loss of generality, we assume that $D_n g(z_0) \neq 0$. Define

$$\lambda = \frac{D_n f(z_0)}{D_n g(z_0)}.$$

We need to show that, for $j = 1, 2, \ldots, n-1$,

$$D_j f(z_0) = \lambda D_j g(z_0).$$

Once we have done so, it will follow immediately, that

$$\nabla \boldsymbol{f}(z_0) = \sum_{j=1}^{n} D_j f(z_0) \boldsymbol{e}_j = \sum_{j=1}^{n} \lambda D_j g(z_0) \boldsymbol{e}_j = \lambda \nabla \boldsymbol{g}(z_0).$$

For $z = (z_1, z_2, \ldots, z_n)$ in U, write $z = (\boldsymbol{x}; y)$ where $\boldsymbol{x} = (z_1, z_2, \ldots, z_{n-1})$ and $y = z_n$. Write $z_0 = (\boldsymbol{x}_0; y_0)$. Now, g is continuously differentiable on U, $g(\boldsymbol{x}_0; y_0) = 0$, and $D_n g(\boldsymbol{x}_0; y_0) \neq 0$. Therefore, by the Implicit Function Theorem, there exist an open set V_0 containing $\boldsymbol{x}_0$ in $\mathbb{R}^{n-1}$ and a continuously differentiable function G defined on V_0 such that $G(\boldsymbol{x}_0) = y_0$ and $g(\boldsymbol{x}; G(\boldsymbol{x})) = 0$ for all $\boldsymbol{x}$ in V_0. This last condition implies that, for every $\boldsymbol{x}$ in V_0, the point $(\boldsymbol{x}; G(\boldsymbol{x}))$ is in Z.

For the neighborhood $N(z_0)$ given in the statement of the theorem, there exists, by the continuity of G, a neighborhood $N(\boldsymbol{x}_0; r)$ contained in V_0 such that, for all $\boldsymbol{x}$ in $N(\boldsymbol{x}_0; r)$, the point $(\boldsymbol{x}; G(\boldsymbol{x}))$ is in $N(z_0)$. Therefore, if we restrict $\boldsymbol{x}$ to $N(\boldsymbol{x}_0; r)$, then $(\boldsymbol{x}; G(\boldsymbol{x}))$ is in $Z \cap N(z_0)$. Notice also that $(\boldsymbol{x}_0; G(\boldsymbol{x}_0)) = (\boldsymbol{x}_0; y_0) = z_0$.

Now we use the function G to create two new auxiliary functions as follows. For $\boldsymbol{x}$ in the open set $N(\boldsymbol{x}_0; r)$ define

$$F(\boldsymbol{x}) = f(\boldsymbol{x}; G(\boldsymbol{x}))$$

and

$$H(\boldsymbol{x}) = g(\boldsymbol{x}; G(\boldsymbol{x})).$$

Clearly both F and H are continuously differentiable on $N(\boldsymbol{x}_0; r)$. We examine each of these functions in turn.

For each $\boldsymbol{x}$ in $N(\boldsymbol{x}_0; r)$, the point $z = (\boldsymbol{x}; G(\boldsymbol{x}))$ is in $Z \cap N(z_0)$. By the hypothesis of the theorem, it follows that

$$F(\boldsymbol{x}) = f((\boldsymbol{x}; G(\boldsymbol{x})) = f(z) \leq f(z_0) = f(\boldsymbol{x}_0; G(\boldsymbol{x}_0)) = F(\boldsymbol{x}_0).$$

Therefore, we deduce that F has an (unconstrained) local maximum at the point $\boldsymbol{x}_0$ in the open set $N(\boldsymbol{x}_0; r)$. By Theorem 8.8.1, $\nabla \boldsymbol{F}(\boldsymbol{x}_0) = \boldsymbol{0}$. For each $j = 1, 2, \ldots, n - 1$, the partial derivative $D_j F(\boldsymbol{x}_0) = 0$. Since $F(\boldsymbol{x}) = f(\boldsymbol{x}; G(\boldsymbol{x}))$, by the chain rule we have, for $j = 1, 2, \ldots, n - 1$,

$$D_j F(\boldsymbol{x}_0) = \sum_{k=1}^{n-1} D_k f(\boldsymbol{x}_0; G(\boldsymbol{x}_0)) D_j x_k + D_n f(\boldsymbol{x}_0; G(\boldsymbol{x}_0)) D_j G(\boldsymbol{x}_0) = 0.$$

But $D_j x_k = \delta_{jk}$, which is 1 when $j = k$ and is 0 otherwise. Also recall that $(\boldsymbol{x}_0; G(\boldsymbol{x}_0)) = z_0$. Therefore,

$$D_j F(\boldsymbol{x}_0) = D_j f(z_0) + D_n f(z_0) D_j G(\boldsymbol{x}_0) = 0. \qquad \textbf{(9.7)}$$

for $j = 1, 2, \ldots, n - 1$.

Now turn your attention to the function H already defined. For all $\boldsymbol{x}$ in $N(\boldsymbol{x}_0; r)$, the point $(\boldsymbol{x}; G(\boldsymbol{x}))$ is in Z. Therefore, $g(\boldsymbol{x}; G(\boldsymbol{x})) = 0$. That is, $H(\boldsymbol{x}) = 0$ for all $\boldsymbol{x}$ in $N(\boldsymbol{x}_0; r)$. Thus, for each $j = 1, 2, \ldots, n - 1$, $D_j H(\boldsymbol{x}) = 0$ for every $\boldsymbol{x}$ in $N(\boldsymbol{x}_0; r)$. In particular, $D_j H(\boldsymbol{x}_0) = 0$ for every j. Since $H(\boldsymbol{x}) = g(\boldsymbol{x}; G(\boldsymbol{x}))$, an application of the chain rule yields, exactly as for the function F,

$$D_j H(\boldsymbol{x}_0) = D_j g(z_0) + D_n g(z_0) D_j G(\boldsymbol{x}_0) = 0. \qquad \textbf{(9.8)}$$

Finally, multiply Eq. (9.8) by $-\lambda$ as defined at the beginning of the proof and add the result to Eq. (9.7). In this way we obtain, for each $j = 1, 2, \ldots, n-1$,

$$\begin{aligned} 0 &= D_j f(z_0) + D_n f(z_0) D_j G(\boldsymbol{x}_0) - \lambda[D_j g(z_0) + D_n g(z_0) D_j G(\boldsymbol{x}_0)] \\ &= D_j f(z_0) - \lambda D_j g(z_0) + [D_n f(z_0) - \lambda D_n g(z_0)] D_j G(\boldsymbol{x}_0) \\ &= D_j f(z_0) - \lambda D_j g(z_0), \end{aligned}$$

since, by the definition of λ, $D_n f(z_0) - \lambda D_n g(z_0) = 0$. Consequently, for each $j = 1, 2, \ldots, n-1$,

$$D_j f(z_0) = \lambda D_j g(z_0).$$

It follows that $\nabla \boldsymbol{f}(z_0) = \lambda \nabla \boldsymbol{g}(z_0)$, as claimed in the theorem. This completes the proof. ●

The theorem provides us with a procedure for finding points where f has a local maximum or minimum subject to the constraint $g = 0$. At any such point z, the following $n + 1$ equations must hold: For some real number λ and for $j = 1, 2, \ldots, n$,

$$D_j f(z) = \lambda D_j g(z)$$

and

$$g(z) = 0.$$

Faced with these equations, we solve, if possible, for z. The point(s) that solves a particular problem will be among the values of z found by this method. Solutions may provide a maximum constrained value or a minimum constrained value for f or neither. Distinguishing among the possible cases usually depends on the context and is often based on geometric or physical grounds.

EXAMPLE 5 About 50 years ago, the leading Soviet mathematician, Kolmogorov, introduced the concept of the *information content* of a code. Suppose that a system of encoding information utilizes n distinct symbols (such as an alphabet), the jth symbol being used with probability x_j. That is, in a coded message using this system, the jth symbol is used with relative frequency x_j. The information content of the code is a numerical measure of the amount of information that can be contained within an encoded message. On separate grounds that need not concern us here, Kolmogorov proposed the function

$$H(\boldsymbol{x}) = H(x_1, x_2, \ldots, x_n) = -\sum_{j=1}^{n} x_j \ln x_j$$

as the appropriate function to measure information content. Since the numbers x_j are probabilities, they satisfy the conditions $0 \le x_j \le 1$ for $j = 1, 2, \ldots, n$ and $\sum_{j=1}^{n} x_j = 1$. (Note that, since $\lim_{x \to 0^+} x \ln x = 0$, we can allow any of the probabilities x_j to be 0; such probabilities contribute nothing to the value of H.) Kolmogorov raised this question: If we wish to devise a code, with what relative frequencies should each of the symbols be used in order to maximize the information content of the code? The answer is provided easily by Lagrange's method.

We want to maximize $H(\boldsymbol{x}) = -\sum_{j=1}^{n} x_j \ln x_j$ subject to the constraint $g(\boldsymbol{x}) = \sum_{j=1}^{n} x_j - 1 = 0$. Therefore, we set

$$\nabla \boldsymbol{H}(\boldsymbol{x}) = \lambda \nabla \boldsymbol{g}(\boldsymbol{x}),$$

where λ is an unknown Lagrange multiplier. That is, for $j = 1, 2, \ldots, n$, we set $D_j H(\boldsymbol{x}) = \lambda D_j g(\boldsymbol{x})$ and, together with the equation $g(\boldsymbol{x}) = 0$, solve for $\boldsymbol{x}$.

Now, $D_j H(\boldsymbol{x}) = -(1 + \ln x_j)$ and $D_j g(\boldsymbol{x}) = 1$. Therefore, for each $j = 1, 2, \ldots, n$, we have $-(1 + \ln x_j) = \lambda$. It follows that

$$\ln x_1 = \ln x_2 = \cdots = \ln x_n.$$

Therefore, $x_1 = x_2 = \cdots = x_n$. Since $\sum_{j=1}^{n} x_j = 1$, we deduce that $x_j = 1/n$ for each j. Thus, the code with n symbols that yields the greatest information content will use each of its symbols equally often. ●

EXAMPLE 6 Find the largest value of $f(\boldsymbol{x}) = (x_1 x_2 \cdots x_n)^2$ where $\boldsymbol{x}$ is restricted to lie on the hypersphere $\sum_{j=1}^{n} x_j^2 = r^2$. First note that, if any of the x_j is 0, then $f(\boldsymbol{x}) = 0$; this yields a minimum value for f and does not solve our problem. Therefore, none of the coordinates of a solution $\boldsymbol{x}$ is 0. We use Lagrange's method, setting $\nabla \boldsymbol{f}(\boldsymbol{x}) = \lambda \nabla \boldsymbol{g}(\boldsymbol{x})$. Since $D_j f(\boldsymbol{x}) = 2x_j \prod_{i=1, i \neq j}^{n} x_i^2$ and $D_j g(\boldsymbol{x}) = 2x_j$, we obtain the n equations,

$$2x_j \prod_{\substack{i=1 \\ i \neq j}}^{n} x_i^2 = 2\lambda x_j,$$

for $j = 1, 2, \ldots, n$. Therefore, for each $j = 1, 2, \ldots, n$,

$$\lambda = \prod_{\substack{i=1 \\ i \neq j}}^{n} x_i^2.$$

It follows that, for any two distinct indices j and k,

$$\prod_{\substack{i=1 \\ i \neq j}}^{n} x_i^2 = \prod_{\substack{i=1 \\ i \neq k}}^{n} x_i^2.$$

Upon canceling the common factors, we obtain $x_k^2 = x_j^2$. That is, $x_1^2 = x_2^2 = \cdots = x_n^2$. Substituting this common value into $g(\boldsymbol{x}) = \sum_{j=1}^{n} x_j^2 - r^2 = 0$ yields

$$x_1^2 = x_2^2 = \cdots = x_n^2 = r^2/n.$$

Thus, the maximum value of f on the hypersphere $\sum_{j=1}^{n} x_j^2 = r^2$ is $(r^2/n)^n$ and is assumed at any of the 2^n points

$$(\pm r/\sqrt{n}, \pm r/\sqrt{n}, \ldots, \pm r/\sqrt{n}). \quad ●$$

EXAMPLE 7 For $\boldsymbol{x}$ in $\mathbb{R}^3$, consider the quadratic form

$$Q(\boldsymbol{x}) = Q(x_1, x_2, x_3) = a_{11}x_1^2 + a_{22}x_2^2 + a_{33}x_3^2$$
$$+2a_{12}x_1x_2 + 2a_{13}x_1x_3 + 2a_{23}x_2x_3,$$

where the coefficients a_{ij} are real numbers and a_{11}, a_{22}, and a_{33} are positive. The set of all $\boldsymbol{x}$ in $\mathbb{R}^3$ that satisfy the equation $Q(\boldsymbol{x}) = 1$ forms an ellipsoid $\mathcal{S}$ in $\mathbb{R}^3$ that is centered at $\mathbf{0}$. In this example, using Lagrange's method, we develop a procedure to find the lengths of the three semiaxes of this ellipsoid. First we discuss the theory; then we will provide a numerical example below.

The major semiaxis is determined by a point $\boldsymbol{x}$ on $\mathcal{S}$ that is farthest from $\mathbf{0}$. The minor semiaxis is determined by a point $\boldsymbol{x}$ on $\mathcal{S}$ that is closest to $\mathbf{0}$. The third semiaxis has an intermediate length between these two extremes and is measured along the line perpendicular to the plane determined by the major and minor semiaxes from $\mathbf{0}$ to a point $\boldsymbol{x}$ on the ellipsoid; this point $\boldsymbol{x}$ generally is a constrained saddle point. If we use Lagrange's method to optimize the square of the distance from $\mathbf{0}$ to a point $\boldsymbol{x}$ on $\mathcal{S}$, we can expect to find three solutions $\boldsymbol{x}_1$, $\boldsymbol{x}_2$, and $\boldsymbol{x}_3$ such that $0 < ||\boldsymbol{x}_3|| \leq ||\boldsymbol{x}_2|| \leq ||\boldsymbol{x}_1||$. The lengths of these three vectors will be the lengths of the semiaxes of $\mathcal{S}$. Thus, we want to optimize $f(\boldsymbol{x}) = \sum_{j=1}^{3} x_j^2$ subject to the constraint $g(\boldsymbol{x}) = Q(\boldsymbol{x}) - 1 = 0$. By Theorem 9.4.1, there exists a real number λ such that $\nabla \boldsymbol{f}(\boldsymbol{x}) = \lambda \nabla \boldsymbol{g}(\boldsymbol{x})$. That is,

$$2x_1 = \lambda(2a_{11}x_1 + 2a_{12}x_2 + 2a_{13}x_3)$$
$$2x_2 = \lambda(2a_{22}x_2 + 2a_{12}x_1 + 2a_{23}x_3)$$
$$2x_3 = \lambda(2a_{33}x_3 + 2a_{13}x_1 + 2a_{23}x_2).$$

By rearranging, we obtain the following equivalent, homogeneous system of linear equations:

$$\begin{aligned} (a_{11} - 1/\lambda)x_1 + {} & a_{12}x_2 + a_{13}x_3 = 0 \\ a_{12}x_1 + {} & (a_{22} - 1/\lambda)x_2 + a_{23}x_3 = 0 \\ a_{13}x_1 + {} & a_{23}x_2 + (a_{33} - 1/\lambda)x_3 = 0. \end{aligned} \tag{9.9}$$

Let $\mu = 1/\lambda$. We know from elementary linear algebra that this system has a nonzero solution $\boldsymbol{x} = (x_1, x_2, x_3)$ if and only if the determinant of coefficients is 0. That is,

$$p(\mu) = \det \begin{vmatrix} (a_{11} - \mu) & a_{12} & a_{13} \\ a_{12} & (a_{22} - \mu) & a_{23} \\ a_{13} & a_{23} & (a_{33} - \mu) \end{vmatrix} = 0. \tag{9.10}$$

Now, $p(\mu)$ is a cubic polynomial in μ; it is the characteristic polynomial of the symmetric matrix

$$A = \begin{pmatrix} a_{11} & a_{12} & a_{13} \\ a_{12} & a_{22} & a_{23} \\ a_{13} & a_{23} & a_{33} \end{pmatrix}.$$

As we discussed in Section 8.8, the matrix A induces a linear transformation T_A mapping $\mathbb{R}^3$ to $\mathbb{R}^3$ such that, for any $\boldsymbol{x}$ in $\mathbb{R}^3$, $\langle T_A\boldsymbol{x}, \boldsymbol{x}\rangle = Q(\boldsymbol{x})$.

The three roots μ_1, μ_2, and μ_3 of the polynomial p are the eigenvalues of A and are real numbers. Because a_{11}, a_{22}, and a_{33} are positive, it can be shown that all the μ_j are positive. We index them so that $0 < \mu_1 \leq \mu_2 \leq \mu_3$. For each eigenvalue μ_j, we find a corresponding eigenvector $\boldsymbol{u}_j$ such that $||\boldsymbol{u}_j|| = 1$ and $T_A\boldsymbol{u} = \mu_j\boldsymbol{u}_j$. Generally, the vector $\boldsymbol{u}_j$ does not satisfy the constraint equation $Q(\boldsymbol{x}) = 1$ but some constant multiple $\boldsymbol{x}_j = c_j\boldsymbol{u}_j$ does. To find the constant c_j, substitute $\boldsymbol{x}_j = c_j\boldsymbol{u}_j$ into $Q(\boldsymbol{x}) = 1$ and solve for c_j. We have

$$\begin{aligned} 1 = Q(\boldsymbol{x}_j) &= \langle T_A\boldsymbol{x}_j, \boldsymbol{x}_j\rangle = c_j^2\langle T_A\boldsymbol{u}_j, \boldsymbol{u}_j\rangle \\ &= c_j^2\mu_j\langle \boldsymbol{u}_j, \boldsymbol{u}_j\rangle = c_j^2\mu_j||\boldsymbol{u}_j||^2 = c_j^2\mu_j. \end{aligned}$$

Therefore, we can take $c_j = 1/\mu_j^{1/2} = \lambda_j^{1/2}$. The vector $\boldsymbol{x}_j = \lambda_j^{1/2}\boldsymbol{u}_j$ lies in the surface S and satisfies the system of linear equations (9.9) with $\mu_j = 1/\lambda_j$. Finally, since

$$||\boldsymbol{x}_j|| = ||c_j\boldsymbol{u}_j|| = |c_j| \cdot ||\boldsymbol{u}_j|| = |c_j| = \lambda_j^{1/2},$$

we deduce that the lengths of the three semiaxes of the ellipsoid are $\lambda_1^{1/2}$, $\lambda_2^{1/2}$, and $\lambda_3^{1/2}$.

For a numerical example, let

$$Q(\boldsymbol{x}) = 3x_1^2 + 2x_2^2 + 3x_3^2 - 2x_1x_2 - 2x_2x_3.$$

Upon applying Lagrange's method, we obtain the system of linear equations (9.9):

$$\begin{aligned} (3 - 1/\lambda)x_1 - \quad x_2 \qquad\qquad\qquad &= 0 \\ -x_1 \quad + (2 - 1/\lambda)x_2 - \quad x_3 &= 0 \\ -x_2 \quad + (3 - 1/\lambda)x_3 &= 0. \end{aligned}$$

Upon expanding the determinant of coefficients, with $\mu = 1/\lambda$, we find the characteristic polynomial (9.10) to be

$$\begin{aligned} p(\mu) &= -\mu^3 + 8\mu^2 - 19\mu + 12 \\ &= -(\mu - 1)(\mu - 3)(\mu - 4). \end{aligned}$$

Therefore, the eigenvalues are $\mu_1 = 1$, $\mu_2 = 3$, and $\mu_3 = 4$. To find the corresponding eigenvectors, let $\mu_1 = 1$ and solve the system of linear equations

$$\begin{aligned} 2x_1 - x_2 \qquad &= 0 \\ -x_1 + x_2 - x_3 &= 0 \\ -x_2 + 2x_3 &= 0 \end{aligned}$$

for the unit vector $\boldsymbol{u}_1 = (1, 2, 1)/\sqrt{6}$. Let $\mu_2 = 3$ and solve the system of linear equations

$$\begin{aligned} - x_2 \qquad &= 0 \\ -x_1 - x_2 - x_3 &= 0 \\ - x_2 \qquad &= 0 \end{aligned}$$

for the unit vector $\boldsymbol{u}_2 = (1, 0, -1)/\sqrt{2}$. Finally, let $\mu_3 = 4$ and solve the system

$$\begin{aligned} -x_1 - x_2 \qquad &= 0 \\ -x_1 - 2x_2 - x_3 &= 0 \\ - x_2 - x_3 &= 0 \end{aligned}$$

for the unit vector $\boldsymbol{u}_3 = (1, -1, 1)/\sqrt{3}$.

Notice that $\lambda_1^{1/2} = 1$, $\lambda_2^{1/2} = 1/\sqrt{3}$, and $\lambda_3^{1/2} = 1/2$. It follows that the vectors $\boldsymbol{x}_1 = \boldsymbol{u}_1$, $\boldsymbol{x}_2 = \boldsymbol{u}_2/\sqrt{3}$, $\boldsymbol{x}_3 = \boldsymbol{u}_3/2$ give the directions of the three axes of the ellipsoid. Their lengths—1, $1/\sqrt{3}$, and 1/2—give the lengths of the semiaxes. The major semiaxis has length 1; the minor semiaxis has length 1/2; the intermediate semiaxis has length $1/\sqrt{3}$. ●

Heretofore, we have allowed only one constraint $g(\boldsymbol{x}) = 0$, but we can, in fact, admit k constraints, provided that $k < n$. For example, let f be a real-valued, continuously differentiable function on an open set U in $\mathbb{R}^3$. Suppose that we seek the points z where f has a maximum value subject to the two constraints $g_1(z) = 0$ and $g_2(z) = 0$. Again, g_1 and g_2 are assumed to be continuously differentiable on U. Since z is required to belong to the surfaces defined implicitly by $g_1(z) = 0$ and $g_2(z) = 0$, it follows that z is restricted to the curve Z in U, which is the intersection of these two surfaces. If z_0 is a point of Z where f has a local constrained maximum and if $a_0 = f(z_0)$, then Z must be tangent to the level surface S_{a_0} of f at the point z_0. If you will grant this claim for a moment, then our geometric intuition will lead us to a theorem. In this process we will ignore possible special cases requiring explanations that divert us from our goal. The surface $g_1(z) = 0$ has a tangent plane P_1 at z_0; likewise, the surface $g_2(z) = 0$ has a tangent plane P_2 at z_0. The intersection of P_1 and P_2 is the straight line L tangent to the curve Z at z_0. Moreover, since L lies in both P_1 and P_2, the vectors $\nabla \boldsymbol{g}_1(z_0)$ and $\nabla \boldsymbol{g}_2(z_0)$ are perpendicular to L. Assuming that $\nabla \boldsymbol{g}_1(z_0)$ and $\nabla \boldsymbol{g}_2(z_0)$ are not parallel, we deduce that these two vectors determine a plane P perpendicular to L; every vector perpendicular to the line L is a linear combination, $\lambda_1 \nabla \boldsymbol{g}_1(z_0) + \lambda_2 \nabla \boldsymbol{g}_2(z_0)$, of these two vectors.

Now, given that the surface S_{a_0} and the curve Z are tangent at z_0, the line L lies in the plane P_0 tangent to the surface S_{a_0} at z_0. Finally, since the gradient of f is perpendicular to P_0, we deduce that $\nabla \boldsymbol{f}(z_0)$ is perpendicular to L. Therefore $\nabla \boldsymbol{f}(z_0)$ is a linear combination of $\nabla \boldsymbol{g}_1(z_0)$ and $\nabla \boldsymbol{g}_2(z_0)$. That is, there exist real numbers λ_1 and λ_2 such that $\nabla \boldsymbol{f}(z_0) = \lambda_1 \nabla \boldsymbol{g}_1(z_0) + \lambda_2 \nabla \boldsymbol{g}_2(z_0)$. This conclusion leads us to the following theorem. The hypothesis that $\Delta \neq 0$ below guarantees that $\nabla \boldsymbol{g}_1(z_0)$ and $\nabla \boldsymbol{g}_2(z_0)$ are not parallel.

THEOREM 9.4.2 Let f, g_1, and g_2 be real-valued continuously differentiable functions defined on an open set U in $\mathbb{R}^n$ with $n \geq 3$. Let $Z = \{z \text{ in } \mathbb{R}^n : g_1(z) = g_2(z) = 0\}$. Suppose that a point z_0 in Z and a neighborhood $N(z_0) \subseteq U$ exist such that $f(z) \leq f(z_0)$ for all z in $Z \cap N(z_0)$. Suppose also that the determinant

$$\Delta = \det \begin{vmatrix} D_{n-1}g_1(z_0) & D_{n-1}g_2(z_0) \\ D_n g_1(z_0) & D_n g_2(z_0) \end{vmatrix} \neq 0.$$

Then there exist two real numbers λ_1 and λ_2 such that

$$\nabla f(\mathbf{z_0}) = \lambda_1 \nabla \mathbf{g}_1(\mathbf{z_0}) + \lambda_2 \nabla \mathbf{g}_2(\mathbf{z_0}).$$

Remark. The conclusion of the theorem is equivalent to the n equations $D_j f(z_0) = \lambda_1 D_j g_1(z_0) + \lambda_2 D_j g_2(z_0)$, for $j = 1, 2, \ldots, n$.

Proof. First form the system of linear equations

$$\begin{aligned} D_{n-1}g_1(z_0)\lambda_1 + D_{n-1}g_2(z_0)\lambda_2 &= D_{n-1}f(z_0) \\ D_n g_1(z_0)\lambda_1 + D_n g_2(z_0)\lambda_2 &= D_n f(z_0) \end{aligned}$$

in the unknown real numbers λ_1 and λ_2. By the hypothesis of the theorem, the determinant of coefficients of this system is not 0. Thus there exists a unique solution for λ_1 and λ_2. It remains only to show that, for each $j = 1, 2, \ldots, n-2$,

$$D_j g_1(z_0)\lambda_1 + D_j g_2(z_0)\lambda_2 = D_j f(z_0).$$

Once we have done so, we can conclude that

$$\nabla f(z_0) = \lambda_1 \nabla \mathbf{g}_1(z_0) + \lambda_2 \nabla \mathbf{g}_2(z_0).$$

This will prove the theorem.

For a point $z = (z_1, z_2, \ldots, z_{n-2}, z_{n-1}, z_n)$ in $\mathbb{R}^n$, we use the notation $z = (\boldsymbol{x}; \boldsymbol{y})$ where $\boldsymbol{x} = (z_1, z_2, \ldots, z_{n-2})$ in $\mathbb{R}^{n-2}$ and $\boldsymbol{y} = (z_{n-1}, z_n)$ in $\mathbb{R}^2$. Write $z_0 = (\boldsymbol{x}_0; \boldsymbol{y}_0)$. Let $\mathbf{g} = (g_1, g_2)$ denote the continuously differentiable function mapping U into $\mathbb{R}^2$ given in the theorem. Since $\mathbf{g}(z_0) = \mathbf{0}$ and since the determinant $\Delta \neq 0$, we can apply Theorem 9.3.4. (The matrix giving Δ is the transpose of the matrix displayed in the hypotheses of Theorem 9.3.4. Since $\Delta \neq 0$, the determinant of the transpose does not vanish either.) Thus there exists an open set V_0 containing $\boldsymbol{x}_0$ in $\mathbb{R}^{n-2}$ and a continuously differentiable function $\boldsymbol{G} = (G_1, G_2)$ defined on V_0 and with values in $\mathbb{R}^2$ such that $\boldsymbol{G}(\boldsymbol{x}_0) = \boldsymbol{y}_0$ and $\mathbf{g}(\boldsymbol{x}; \boldsymbol{G}(\boldsymbol{x})) = \mathbf{0}$ for all $\boldsymbol{x}$ in V_0. When displayed fully, the equation

$$\mathbf{g}(\boldsymbol{x}; \boldsymbol{G}(\boldsymbol{x})) = \mathbf{0}$$

becomes

$$(g_1(\boldsymbol{x}; G_1(\boldsymbol{x}), G_2(\boldsymbol{x})), g_2(\boldsymbol{x}; G_1(\boldsymbol{x}), G_2(\boldsymbol{x}))) = (0, 0).$$

That is,

$$g_1(\boldsymbol{x}; G_1(\boldsymbol{x}), G_2(\boldsymbol{x})) = 0$$

$$g_2(\boldsymbol{x}; G_1(\boldsymbol{x}), G_2(\boldsymbol{x})) = 0$$

for all $\boldsymbol{x}$ in V_0. Thus, for all $\boldsymbol{x}$ in V_0, $(\boldsymbol{x}; \boldsymbol{G}(\boldsymbol{x}))$ is in Z. Furthermore, $(\boldsymbol{x}_0; \boldsymbol{G}(\boldsymbol{x}_0)) = (\boldsymbol{x}_0; \boldsymbol{y}_0) = z_0$. For the neighborhood $N(z_0)$ given in the statement of the theorem, there exists, by the continuity of $\boldsymbol{G}$, a neighborhood $N(\boldsymbol{x}_0; r) \subseteq V_0$ such that $(\boldsymbol{x}; \boldsymbol{G}(\boldsymbol{x}))$ is in $N(z_0)$ for all $\boldsymbol{x}$ in $N(\boldsymbol{x}_0; r)$. Thus, for $\boldsymbol{x}$ in $N(\boldsymbol{x}_0; r)$, it follows that $(\boldsymbol{x}; \boldsymbol{G}(\boldsymbol{x}))$ is in $Z \cap N(z_0)$.

We now define three auxiliary functions using all the information at hand. For $\boldsymbol{x}$ in $N(\boldsymbol{x}_0; r)$ define

$$F(\boldsymbol{x}) = f(\boldsymbol{x}; \boldsymbol{G}(\boldsymbol{x})) = f(\boldsymbol{x}; G_1(\boldsymbol{x}), G_2(\boldsymbol{x}))$$

$$H_1(\boldsymbol{x}) = g_1(\boldsymbol{x}; \boldsymbol{G}(\boldsymbol{x})) = g_1(\boldsymbol{x}; G_1(\boldsymbol{x}), G_2(\boldsymbol{x}))$$
$$H_2(\boldsymbol{x}) = g_2(\boldsymbol{x}; \boldsymbol{G}(\boldsymbol{x})) = g_2(\boldsymbol{x}; G_1(\boldsymbol{x}), G_2(\boldsymbol{x})).$$

For all $\boldsymbol{x}$ in $N(\boldsymbol{x}_0; r)$, $F(\boldsymbol{x}) = f(\boldsymbol{x}; \boldsymbol{y}) = f(z) \le f(z_0) = F(\boldsymbol{x}_0)$. Therefore $\nabla \boldsymbol{F}(\boldsymbol{x}_0) = \boldsymbol{0}$. That is, for $j = 1, 2, \ldots, n-2$, we have $D_j F(\boldsymbol{x}_0) = 0$. By the chain rule,

$$0 = D_j F(\boldsymbol{x}_0) = D_j f(z_0) + D_{n-1} f(z_0) D_j G_1(\boldsymbol{x}_0) + D_n f(z_0) D_j G_2(\boldsymbol{x}_0), \quad \textbf{(9.11)}$$

for $j = 1, 2, \ldots, n-2$.

Also, $H_1(\boldsymbol{x}) = H_2(\boldsymbol{x}) = 0$ for all $\boldsymbol{x}$ in $N(\boldsymbol{x}_0; r)$. Therefore, $D_j H_1(\boldsymbol{x}_0) = D_j H_2(\boldsymbol{x}_0) = 0$ for all $j = 1, 2, \ldots, n-2$. Again by the chain rule,

$$0 = D_j H_1(\boldsymbol{x}_0) = D_j g_1(z_0) + D_{n-1} g_1(z_0) D_j G_1(\boldsymbol{x}_0) + D_n g_1(z_0) D_j G_2(\boldsymbol{x}_0) \quad \textbf{(9.12)}$$

and

$$0 = D_j H_2(\boldsymbol{x}_0) = D_j g_2(z_0) + D_{n-1} g_2(z_0) D_j G_1(\boldsymbol{x}_0) + D_n g_2(z_0) D_j G_2(\boldsymbol{x}_0). \quad \textbf{(9.13)}$$

Multiply (9.12) by $-\lambda_1$ and multiply (9.13) by $-\lambda_2$ and add the results to (9.11). Upon rearranging the resulting equation, we obtain, for $j = 1, 2, \ldots, n-2$,

$$\begin{aligned} 0 &= [D_j f(z_0) - \lambda_1 D_j g_1(z_0) - \lambda_2 D_j g_2(z_0)] \\ &\quad + [D_{n-1} f(z_0) - \lambda_1 D_{n-1} g_1(z_0) - \lambda_2 D_{n-1} g_2(z_0)] D_j G_1(\boldsymbol{x}_0) \\ &\quad + [D_n f(z_0) - \lambda_1 D_n g_1(z_0) - \lambda_2 D_n g_2(z_0)] D_j G_2(\boldsymbol{x}_0) \\ &= D_j f(z_0) - \lambda_1 D_j g_1(z_0) - \lambda_2 D_j g_2(z_0), \end{aligned}$$

by our initial definition of λ_1 and λ_2. Therefore, $D_j f(z_0) = \lambda_1 D_j g_1(z_0) + \lambda_2 D_j g_2(z_0)$, for all $j = 1, 2, \ldots, n$. Equivalently,

$$\nabla \boldsymbol{f}(z_0) = \lambda_1 \nabla \boldsymbol{g}_1(z_0) + \lambda_2 \nabla \boldsymbol{g}_2(z_0),$$

as claimed in the theorem. ●

By imitating the foregoing proof with only slight modification, you can prove the complete generalization of the Lagrange method given in the following theorem.

THEOREM 9.4.3 Let $f, g_1, g_2, \ldots, g_k$ be real-valued, continuously differentiable functions defined on an open set U in $\mathbb{R}^n$ with $k < n$. Let $Z = \{z \text{ in } U : g_j(z) = 0, j = 1, \ldots, k\}$. Suppose that a point z_0 in Z and a neighborhood $N(z_0) \subseteq U$ exist such that $f(z) \le f(z_0)$ for all z in $Z \cap N(z_0)$. Suppose also that

$$\det \begin{vmatrix} D_{n-k+1} g_1(z_0) & D_{n-k+1} g_2(z_0) & \cdots & D_{n-k+1} g_k(z_0) \\ D_{n-k+2} g_1(z_0) & D_{n-k+2} g_2(z_0) & \cdots & D_{n-k+2} g_k(z_0) \\ \vdots & \vdots & \cdots & \vdots \\ D_n g_1(z_0) & D_n g_2(z_0) & \cdots & D_n g_k(z_0) \end{vmatrix} \ne 0.$$

Then there exist real numbers $\lambda_1, \lambda_2, \ldots, \lambda_k$ such that

$$\nabla f(z_0) = \sum_{i=1}^{k} \lambda_i \nabla g_i(z_0). \quad \bullet$$

Remark. The conclusion of the theorem is equivalent to the validity of the n equations $D_j f(z_0) = \sum_{i=1}^{n} \lambda_i D_j g_i(z_0)$, for $j = 1, 2, \ldots, n$.

EXERCISES

9.1. Let $\boldsymbol{f} = (f_1, f_2, \ldots, f_n)$ be a linear function mapping $\mathbb{R}^n$ one-to-one onto $\mathbb{R}^n$. Prove that $\boldsymbol{g} = \boldsymbol{f}^{-1}$ is also a linear function mapping $\mathbb{R}^n$ one-to-one onto $\mathbb{R}^n$.

9.2. The change of variables from Cartesian to spherical coordinates in $\mathbb{R}^3$ is achieved by $\boldsymbol{f} = (f_1, f_2, f_3)$, where

$$x_1 = f_1(\rho, \theta, \phi) = \rho \cos\theta \sin\phi$$
$$x_2 = f_2(\rho, \theta, \phi) = \rho \sin\theta \sin\phi$$
$$x_3 = f_3(\rho, \theta, \phi) = \rho \cos\theta.$$

Prove that $\boldsymbol{f}$ maps the set

$$S = \{(\rho, \theta, \phi) : \rho > 0, 0 \le \theta < 2\pi, 0 \le \phi < \pi\}$$

one-to-one onto $\mathbb{R}^3 \sim \{\mathbf{0}\}$.

9.3. Each of the following equations implicitly defines a function $x_2 = G(x_1)$ in a neighborhood of the given point. In each case find G', expressed in terms of x_1 and x_2.

a) $x_1^2 + x_1 x_2 + x_2^2 = 27$; $(3, 3)$

b) $x_1 \cos(x_1 x_2) = 0$; $(1, \pi/2)$

c) $x_1 x_2 + \ln(x_1 x_2) = 1$; $(1, 1)$

9.4. For $\boldsymbol{x} = (x_1, x_2)$ in $\mathbb{R}^2$, let $F(\boldsymbol{x}) = x_2^3 - x_1^2$. Show that, although the equation $F(\boldsymbol{x}) = 0$ implicitly defines a function $x_2 = G(x_1)$ for all x_1 in $\mathbb{R}$, the function G fails to be differentiable at $x_1 = 0$. Does this fact violate the Implicit Function Theorem? Explain.

9.5. The *folium of Descartes* is described implicitly by the equation

$$F(\boldsymbol{x}) = F(x_1, x_2) = x_1^3 + x_2^3 - 3c x_1 x_2 = 0, \qquad \textbf{(9.14)}$$

where c is a positive constant.

a) Find all points on the curve $F(\boldsymbol{x}) = 0$ in a neighborhood of which you can solve Eq. (9.14) for x_2 as a function $G(x_1)$. (Do not try to find the function G explicitly.)

b) Find the derivative G' at the points you found in (a), expressed in terms of the partial derivatives of F.

c) Find those values of x_1 where $G'(x_1) = 0$.

d) Show that the graph of $F(\boldsymbol{x}) = 0$ is asymptotic to the line $x_1 + x_2 = -c$.

e) Sketch the graph of $F(\boldsymbol{x}) = 0$.

9.6. Let c be a positive constant. Let $F_1 = (-c, 0)$ and $F_2 = (c, 0)$ denote two points in the plane. Let $\boldsymbol{x}$ be any point in the plane and let r_1 and r_2 denote the distances from $\boldsymbol{x}$ to F_1 and F_2, respectively. The locus of all points $\boldsymbol{x}$ such that $r_1 r_2 = c^2$ is called the *lemniscate* determined by c.

a) Show that the equation

$$F(\mathbf{x}) = F(x_1, x_2) = (x_1^2 + x_2^2)^2 - 2c^2(x_1^2 - x_2^2) = 0 \tag{9.15}$$

represents the lemniscate determined by c.

b) Find all points on the graph of $F(\mathbf{x}) = 0$ in a neighborhood of which Eq. (9.15) can be solved for x_2 as a function G of x_1.

c) Find G' in terms of x_1 and x_2. Find those points where G has an extreme value.

d) Use the change of variables $x_1 = r\cos\theta$, $x_2 = r\sin\theta$ to show that, expressed in terms of polar coordinates, Eq. (9.15) becomes $r^2 = 2c^2\cos(2\theta)$.

e) Sketch the graph of $F(\mathbf{x}) = 0$.

9.7. Let F be a continuously differentiable, real-valued function defined on an open set U in $\mathbb{R}^3$ and let

$$\mathcal{S} = \{\mathbf{x} \text{ in } U : F(\mathbf{x}) = 0\}.$$

For the purposes of this problem, we define $\mathbf{c} = (c_1, c_2, c_3)$ in S to be a *regular point* for F if the equation $F(\mathbf{x}) = 0$ implicitly defines a function of at least one of the forms $x_1 = G_1(x_2, x_3)$, $x_2 = G_2(x_1, x_3)$, or $x_3 = G_3(x_1, x_2)$ on some neighborhood of $\mathbf{c}$.

a) Prove that if $\mathbf{c}$ is in $\mathcal{S}$ and if

$$[D_1F(\mathbf{c})]^2 + [D_2F(\mathbf{c})]^2 + [D_3F(\mathbf{c})]^2 > 0,$$

then $\mathbf{c}$ is a regular point for F.

b) Let $F(\mathbf{x}) = x_1^2 + x_2^2 - x_3^2$. Is $\mathbf{c} = \mathbf{0}$ a regular point for F?

c) Determine whether $\mathbf{0}$ is a regular point for the function

$$F(\mathbf{x}) = x_1 + x_2 + x_3 - \sin(x_1x_2x_3).$$

d) Let a_1, a_2, and a_3 be positive constants and let

$$F(\mathbf{x}) = (x_1^2 + x_2^2 + x_3^2)^2 - a_1x_1^2 - a_2x_2^2 - a_3x_3^2.$$

Show that all points in $\mathcal{S}$ except $\mathbf{0}$ are regular points for F.

9.8. Let $S = \mathbb{R}^2 \sim \{\mathbf{0}\}$. For $\mathbf{x} = (x_1, x_2)$ in S, define $\mathbf{f} = (f_1, f_2)$ as follows:

$$y_1 = f_1(\mathbf{x}) = x_1/(x_1^2 + x_2^2)$$
$$y_2 = f_2(\mathbf{x}) = x_2/(x_1^2 + x_2^2).$$

a) Prove that f is continuously differentiable and maps S one-to-one onto S. Find J_f.

b) Find the inverse map $\mathbf{f}^{-1} = \mathbf{g} = (g_1, g_2)$.

c) Find the partial derivatives D_1g_1, D_2g_1, D_1g_2, and D_2g_2.

9.9. Let $S = \{(x_1, x_2, x_3) : x_1 + x_2 + x_3 \neq -1\}$. For each $j = 1, 2, 3$ and each $\mathbf{x} = (x_1, x_2, x_3)$ in S, let

$$f_j(\mathbf{x}) = x_j/(1 + x_1 + x_2 + x_3).$$

Let $\mathbf{f} = (f_1, f_2, f_3)$.

a) Prove that $\mathbf{f}$ is continuously differentiable on S. Find the Jacobian J_f.

b) Show that $\mathbf{f}$ maps S one-to-one into $\mathbb{R}^3$.

c) Find $\mathbf{f}^{-1} = \mathbf{g} = (g_1, g_2, g_3)$.

d) Compute the first-order partial derivatives of $\mathbf{g}$.

9.10. Let $\boldsymbol{f} = (f_1, f_2)$ be a continuously differentiable function mapping an open set U in $\mathbb{R}^2$ one-to-one into $\mathbb{R}^2$. Assume also that $J_{\boldsymbol{f}}$ does not vanish on U. Denote $\boldsymbol{f}^{-1}$ by $\boldsymbol{g} = (g_1, g_2)$. For $\boldsymbol{x}$ in U, let $\boldsymbol{y} = \boldsymbol{f}(\boldsymbol{x})$. Use the equation $\boldsymbol{y} = \boldsymbol{f}(\boldsymbol{g}(\boldsymbol{y}))$ and the chain rule to show that

$$D_1 g_1(\boldsymbol{y}) = D_2 f_2(\boldsymbol{x})/J_{\boldsymbol{f}}(\boldsymbol{x}), \qquad D_1 g_2(\boldsymbol{y}) = -D_1 f_2(\boldsymbol{x})/J_{\boldsymbol{f}}(\boldsymbol{x}),$$

$$D_2 g_1(\boldsymbol{y}) = -D_2 f_1(\boldsymbol{x})/J_{\boldsymbol{f}}(\boldsymbol{x}), \qquad D_2 g_2(\boldsymbol{y}) = D_1 f_1(\boldsymbol{x})/J_{\boldsymbol{f}}(\boldsymbol{x}).$$

9.11. (See Exercise 9.10.) Let $\boldsymbol{f} = (f_1, f_2)$ be a twice continuously differentiable function mapping an open set U in $\mathbb{R}^2$ one-to-one into $\mathbb{R}^2$. Assume also that $J_{\boldsymbol{f}}$ does not vanish on U. Denote $\boldsymbol{f}^{-1}$ by $\boldsymbol{g} = (g_1, g_2)$. Find the second order partial derivatives of g_1 and g_2. (*Hint*: To compute, for example, $D_{11}g_1(\boldsymbol{y})$ and $D_{11}g_2(\boldsymbol{y})$, begin with the equations

$$D_1 f_1(\boldsymbol{x}) D_1 g_1(\boldsymbol{y}) + D_2 f_1(\boldsymbol{x}) D_1 g_2(\boldsymbol{y}) = 1$$

$$D_1 f_2(\boldsymbol{x}) D_1 g_1(\boldsymbol{y}) + D_2 f_2(\boldsymbol{x}) D_1 g_2(\boldsymbol{y}) = 0$$

from Exercise 9.10. Differentiate each of these equations with respect to y_1 and solve the resulting system of linear equations for $D_{11}g_1(\boldsymbol{y})$ and $D_{11}g_2(\boldsymbol{y})$. Then substitute the results from Exercise 9.10.)

9.12. Let $f_1, f_2, \ldots, f_n$ be continuously differentiable, real-valued functions on $\mathbb{R}$. Let $\boldsymbol{g} = (g_1, g_2, \ldots, g_n)$ be a continuously differentiable function mapping $\mathbb{R}^n$ into $\mathbb{R}^n$. For any point $\boldsymbol{x} = (x_1, x_2, \ldots, x_n)$ in $\mathbb{R}^n$ and for each $j = 1, 2, \ldots, n$, define $h_j(\boldsymbol{x}) = g_j(f_1(x_1), f_2(x_2), \ldots, f_n(x_n))$. Let $\boldsymbol{h} = (h_1, h_2, \ldots, h_n)$. Prove that

$$J_{\boldsymbol{h}}(\boldsymbol{x}) = J_{\boldsymbol{g}}(f_1(x_1), f_2(x_2), \ldots, f_n(x_n)) \prod_{j=1}^{n} f_j'(x_j).$$

9.13. Complete the proof of Theorem 9.3.4.

9.14. Let $\boldsymbol{F} = (F_1, F_2, \ldots, F_k)$ be a continuously differentiable function mapping an open set U in $\mathbb{R}^{n+k}$ into $\mathbb{R}^k$. Denote a point $(x_1, x_2, \ldots, x_n; y_1, y_2, \ldots, y_k)$ in U by $(\boldsymbol{x}; \boldsymbol{y})$. Suppose that $(\boldsymbol{c}; \boldsymbol{y}_0)$ is a point of U such that $\boldsymbol{F}(\boldsymbol{c}; \boldsymbol{y}_0) = \boldsymbol{0}$ and $\det[D_{y_j} F_i(\boldsymbol{c}; \boldsymbol{y}_0)] \neq 0$. Prove that there exists an open set V_0 in $\mathbb{R}^n$ containing $\boldsymbol{c}$ and a continuously differentiable function $\boldsymbol{G} = (G_1, G_2, \ldots, G_k)$ mapping V_0 into $\mathbb{R}^k$ such that $\boldsymbol{G}(\boldsymbol{c}) = \boldsymbol{y}_0$ and $\boldsymbol{F}(\boldsymbol{x}; \boldsymbol{G}(\boldsymbol{x})) = \boldsymbol{0}$ for all $\boldsymbol{x}$ in V_0.

9.15. Let $\boldsymbol{f} = (f_1, f_2)$ be a continuously differentiable function defined on an open set U in $\mathbb{R}^2$ such that ∇f_1 and ∇f_2 do not vanish at any point of U.

a) Suppose that $J_{\boldsymbol{f}}(\boldsymbol{x}) = 0$ for all $\boldsymbol{x}$ in U. Prove that a curve $\mathcal{C}$ in U is a level curve of f_1 if and only if it is also a level curve of f_2.

b) Suppose that f_1 and f_2 have the same level curves in U. Prove that $J_{\boldsymbol{f}}(\boldsymbol{x}) = 0$ for all $\boldsymbol{x}$ in U.

9.16. *Definition:* Let f_1 and f_2 be two continuously differentiable, real-valued functions defined on an open set U in $\mathbb{R}^2$. Then f_1 and f_2 are said to be *functionally dependent* if there exists a differentiable function F defined on the range of $\boldsymbol{f} = (f_1, f_2)$ such that ∇F does not vanish and such that $F(f_1(\boldsymbol{x}), f_2(\boldsymbol{x})) = 0$ for all $\boldsymbol{x}$ in U.

Let $\boldsymbol{f} = (f_1, f_2)$ be a continuously differentiable function defined on an open set U in $\mathbb{R}^2$. (See Exercise 9.15.)

a) Prove that, if f_1 and f_2 are functionally dependent on an open set U, then $J_{\boldsymbol{f}}(\boldsymbol{x}) = 0$ for all $\boldsymbol{x}$ in U.

b) Suppose that $J_f(\boldsymbol{x}) = 0$ for all $\boldsymbol{x}$ in U. Prove that, for each $\boldsymbol{c}$ in U, there exists a neighborhood $N(\boldsymbol{c})$ such that f_1 and f_2 are functionally dependent on $N(\boldsymbol{c})$.

9.17. Let a_1, a_2, and a_3 be positive constants. Let

$$S = \{(x_1, x_2, x_3) : x_j > 0, j = 1, 2, 3\}.$$

For $\boldsymbol{x} = (x_1, x_2, x_3)$ in S, find the maximum value of $f(\boldsymbol{x}) = \sum_{j=1}^{3} x_j$ subject to the constraint $\sum_{j=1}^{3} a_j/x_j = 1$.

9.18. A box with sides parallel to the coordinate axes is to be inscribed within the ellipsoid

$$\frac{x_1^2}{a_1^2} + \frac{x_2^2}{a_2^2} + \frac{x_3^2}{a_3^2} = 1.$$

Find the dimensions of the box with maximum volume.

9.19. Let g be a continuously differentiable function defined on an open set U in $\mathbb{R}^n$. Let $\mathcal{S}$ be the hypersurface in U defined implicitly by $g(\boldsymbol{x}) = 0$. Assume that $\nabla \boldsymbol{g}(\boldsymbol{x}) \neq \boldsymbol{0}$ for all $\boldsymbol{x}$ in $\mathcal{S}$. Let $\boldsymbol{a} = (a_1, a_2, \ldots, a_n)$ be a point in $\mathbb{R}^n$ that is not in $\mathcal{S}$. Use Lagrange's method to prove that the distance from $\boldsymbol{a}$ to $\mathcal{S}$ is measured along a line perpendicular to $\mathcal{S}$.

9.20. Let two planes P_1 and P_2 in $\mathbb{R}^3$ be represented by the equations $\sum_{j=1}^{3} a_j x_j = c_1$ and $\sum_{j=1}^{3} b_j x_j = c_2$, respectively. We assume that P_1 and P_2 are not parallel, that $\sum_{j=1}^{3} a_j^2 \neq 0$ and $\sum_{j=1}^{3} b_j^2 \neq 0$, and that $c_1 \neq 0$ and $c_2 \neq 0$. Thus P_1 and P_2 intersect in a line L in $\mathbb{R}^3$ that does not pass through $\boldsymbol{0}$. Find the point on L that is closest to $\boldsymbol{0}$.

9.21. Let $a_1, a_2, \ldots, a_n$ be n positive numbers. Use our result in Example 6 to show that the geometric average of the numbers $a_1, a_2, \ldots, a_n$ is dominated by the arithmetic average of those numbers. That is,

$$(a_1 a_2 \cdots a_n)^{1/n} \leq \frac{1}{n} \sum_{k=1}^{n} a_k.$$

(*Hint:* For $k = 1, 2, \ldots, n$, let $x_k = r\sqrt{a_k/(\sum_{j=1}^{n} a_j)}$.)

9.22. Find the extreme values of the function $f(\boldsymbol{x}) = x_1 x_2 x_3$ on the set $S = \{(x_1, x_2, x_3) : x_j > 0, j = 1, 2, 3\}$ subject to the constraint $1/x_1 + 1/x_2 + 1/x_3 = 1$.

9.23. Find the shortest distance between the curves represented by the equations $x_1^2 + x_2^2 = 1$ and $x_1^2 x_2 = 2$.

9.24. Let a_1, a_2, and a_3 be positive constants and let b_1, b_2, and b_3 be nonzero constants. The plane $\sum_{j=1}^{3} b_j x_j = 0$ that passes through the point $\boldsymbol{0}$ intersects the ellipsoid $\sum_{j=1}^{3} x_j^2/a_j^2 = 1$ in an ellipse E. Find the lengths of the semiaxes of E.

9.25. A quadric surface $\mathcal{S}$ in $\mathbb{R}^3$ is represented by an equation of the form

$$\begin{aligned} Q(\boldsymbol{x}) = a_{11}x_1^2 + a_{22}x_2^2 + a_{33}x_3^2 \\ + 2a_{12}x_1x_2 + 2a_{13}x_1x_3 + 2a_{23}x_2x_3 = 1. \end{aligned}$$

A plane P that passes through $\boldsymbol{0}$ is represented by an equation of the form $\sum_{j=1}^{3} b_j x_j = 0$. The surface $\mathcal{S}$ and the plane P intersect in a curve C in $\mathbb{R}^3$. Find the smallest and the greatest distances from **0** to C. *Hint:* Show that these extreme distances are $1/\lambda_1$ and $1/\lambda_2$, where λ_1 and λ_2 are solutions of the equation

$$\det \begin{vmatrix} b_1 & b_2 & b_3 & 0 \\ a_{11} - \lambda & a_{12} & a_{13} & b_1 \\ a_{12} & a_{22} - \lambda & a_{23} & b_2 \\ a_{13} & a_{23} & a_{33} - \lambda & b_3 \end{vmatrix} = 0.$$

9.26. Prove Theorem 9.4.3.

10

Multiple Integrals

When constructing the integral of a function f of several variables over its domain, we will find that many of the steps imitate those in the construction of the Riemann integral in Chapter 6. Complicating the construction, however, are some significant differences that mainly concern the domain of the integrand. Since all the complications arise already for a function with domain in $\mathbb{R}^2$ and since our resolution of them can be generalized to enable us to develop the n-fold integral of functions with domain in $\mathbb{R}^n$, we will focus our attention on double integrals of functions of two variables. Later we will merely sketch the treatment of triple integrals.

An interval $[a, b]$ in $\mathbb{R}$ presents us with no complications when we integrate a function f in $R[a, b]$. The counterpart in $\mathbb{R}^2$ of a closed interval is a closed rectangle $R = [a_1, b_1] \times [a_2, b_2]$. Below, when we construct the double integral of a function f over a rectangle R, we will imitate the program from Chapter 6 with relative ease; the complications arise when the domain over which integration is to be performed is no longer rectagular.

10.1 The Double Integral

Let f be a bounded, real-valued function defined on a rectangle $R = [a_1, b_1] \times [a_2, b_2]$. To construct the (double) integral of f over R, we begin by establishing the notation for partitions of sets in $\mathbb{R}^2$.

Let π_1 be a partition in $\Pi[a_1, b_1]$ and let π_2 be a partition in $\Pi[a_2, b_2]$. That is,

$$\pi_1 : a_1 = x_0 < x_1 < \cdots < x_p = b_1,$$

$$\pi_2 : a_2 = y_0 < y_1 < \cdots < y_q = b_2.$$

The lines $x = x_i$ and $y = y_j$, for $i = 0, 1, \ldots, p$ and $j = 0, 1, 2, \ldots, q$, form a grid that partitions R into rectangles as in Fig. 10.1. We will denote this partition of

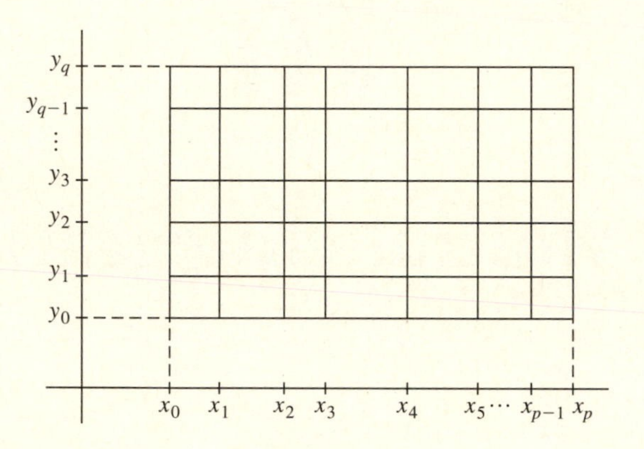

Figure 10.1

R by $\pi = \pi_1 \times \pi_2$. We will refer to the set

$$[x_{i-1}, x_i] \times [y_{j-1}, y_j]$$

as the (i, j) *cell* of this partition and will denote it by C_{ij}. The partition π subdivides R into pq cells. The area of C_{ij} is $\Delta A_{ij} = \Delta x_i \Delta y_j$. The diameter of C_{ij} is $d(C_{ij}) = [(x_i - x_{i-1})^2 + (y_j - y_{j-1})^2]^{1/2}$. The *gauge* of the partition π is

$$||\pi|| = \max\{d(C_{ij}) : i = 1, 2, \ldots, p,\ j = 1, 2, \ldots, q\}.$$

A partition $\pi' = \pi_1' \times \pi_2'$ is a *refinement* of $\pi = \pi_1 \times \pi_2$ if π_1' is a refinement of π_1 and π_2' is a refinement of π_2. Equivalently, each cell C_{ij}' of π' is contained in some cell C_{rs} of π. Clearly, if π' is a refinement of π, then $||\pi'|| \le ||\pi||$.

If $\pi' = \pi_1' \times \pi_2'$ and $\pi'' = \pi_1'' \times \pi_2''$ are two partitions of R, then the *least common refinement* of π' and π'', denoted $\pi = \pi' \vee \pi''$, is the partition $(\pi_1' \vee \pi_1'') \times (\pi_2' \vee \pi_2'')$.

Now let f be a bounded, real-valued function defined on the rectangle R. Let π be any partition of R with cells C_{ij}. For each i and j, choose a point (s_i, t_j) in C_{ij} and form the sum

$$\sum_{i=1}^{p} \sum_{j=1}^{q} f(s_i, t_j) \Delta A_{ij}.$$

DEFINITION 10.1.1 A sum of the form $\sum_{i=1}^{p} \sum_{j=1}^{q} f(s_i, t_j) \Delta A_{ij}$ is a *Riemann sum* for the function f. ●

DEFINITION 10.1.2 Let f be a bounded, real-valued function defined on a rectangle R in $\mathbb{R}^2$. The function f is said to be *(Riemann) integrable* on R if there exists a real number I with the following property:

For any $\epsilon > 0$, there exists a partition π_0 of R such that, for every refinement π of π_0 and for any choice of (s_i, t_j) in the cells of π, we have

$$\left| \sum_{i=1}^{p} \sum_{j=1}^{q} f(s_i, t_j) \Delta A_{ij} - I \right| < \epsilon.$$

The number I is called the (*Riemann*) *integral* of f over R and is denoted $I = \iint_R f(x, y)\, dA$. ●

It is sometimes suggestive to denote the integral $\iint_R f(x, y)\, dA$ by $\iint_R f(x, y)\, dx\, dy$. Basic properties of the integral are listed in the following theorem; its proof is straightforward and is left for you to provide.

THEOREM 10.1.1

i) Let f_1 and f_2 be two integrable functions on a rectangle R. For any real numbers c_1 and c_2,

$$\iint_R [c_1 f_1(x, y) + c_2 f_2(x, y)]\, dA$$
$$= c_1 \iint_R f_1(x, y)\, dA + c_2 \iint_R f_2(x, y)\, dA.$$

ii) Let f be integrable on a rectangle R. Let

$$m = \inf \{f(x, y) : (x, y) \text{ in } R\}$$

and

$$M = \sup \{f(x, y) : (x, y) \text{ in } R\}.$$

Then

$$mA \leq \iint_R f(x, y)\, dA \leq MA,$$

where $A = (b_1 - a_1)(b_2 - a_2)$ denotes the area of R.

iii) Let f and g be two integrable functions on R such that $f(x, y) \leq g(x, y)$ for all (x, y) in R. Then

$$\iint_R f(x, y)\, dA \leq \iint_R g(x, y)\, dA. \quad ●$$

10.1.1 The Riemann Condition

DEFINITION 10.1.3 Let f be any bounded, real-valued function defined on the rectangle R. Let π be any partition of R with cells C_{ij}, for $i = 1, 2, \ldots, p$ and $j = 1, 2, \ldots, q$. Let

$$m_{ij} = \inf \{f(s, t) : (s, t) \text{ in } C_{ij}\}$$

and

$$M_{ij} = \sup \{f(s, t) : (s, t) \text{ in } C_{ij}\}.$$

The sum $L(f, \pi) = \sum_{i=1}^{p} \sum_{j=1}^{q} m_{ij} \Delta A_{ij}$ is called the *lower Riemann sum* for f over R. The sum $U(f, \pi) = \sum_{i=1}^{p} \sum_{j=1}^{q} M_{ij} \Delta A_{ij}$ is called the *upper Riemann sum* for f over R. ●

Evidently, for any partition π of R and for any Riemann sum $S(f, \pi)$ of a bounded, real-valued function f on R, we have $L(f, \pi) \leq S(f, \pi) \leq U(f, \pi)$.

The proof of the following theorem closely imitates that of Theorem 6.2.2 and is left for you to provide.

THEOREM 10.1.2 For any two partitions π' and π'' of R, we have $L(f, \pi') \leq U(f, \pi'')$. ●

Evidently, the set of numbers $\{L(f, \pi) : \pi \text{ a partition of } R\}$ is bounded above. Likewise, $\{U(f, \pi) : \pi \text{ a partition of } R\}$ is bounded below. Therefore, the following concepts are well-defined.

DEFINITION 10.1.4 The *lower Riemann integral* of f over R is defined to be

$$L(f) = \sup \{L(f, \pi) : \pi \text{ a partition of } R\}.$$

The *upper Riemann integral* of f over R is defined to be

$$U(f) = \inf \{U(f, \pi) : \pi \text{ a partition of } R\}. \quad \bullet$$

Clearly, for any bounded, real-valued function f on R, we have $L(f) \leq U(f)$. As in our earlier discussion of integrals, Riemann's condition plays an essential role here as well.

DEFINITION 10.1.5 Riemann's Condition Let f be a bounded, real-valued function defined on a rectangle R. *Riemann's condition* is said to hold for f on R if, for any $\epsilon > 0$, there exists a partition π_0 of R such that, for every refinement π of π_0,

$$0 \leq U(f, \pi) - L(f, \pi) < \epsilon. \quad \bullet$$

The analog of Theorem 6.2.4 provides us with the main tool for determining whether a bounded function is integrable.

THEOREM 10.1.3 Let f be a bounded, real-valued function defined on a rectangle R in $\mathbb{R}^2$. The following are equivalent.

i) The function f is integrable on R.

ii) Riemann's condition holds for f on R.

iii) $L(f) = U(f)$. ●

The proof of Theorem 10.1.3 is virtually identical with that of Theorem 6.2.4 and is left for you as an exercise.

As you might expect, every function that is continuous on the rectangle R is integrable on R.

THEOREM 10.1.4 Let f be a continuous, real-valued function on the rectangle R. Then f is integrable on R.

Proof. Since f is continuous on R and since R is compact, we know that f is uniformly continuous on R. Given $\epsilon > 0$, choose a $\delta > 0$ such that, if (s, t) and (s', t') are two points in R with $[(s - s')^2 + (t - t')^2]^{1/2} < \delta$, then

$$|f(s, t) - f(s', t')| < \frac{\epsilon}{A},$$

where A is the area of the rectangle R. Choose any partition π_0 of R with $||\pi_0|| < \delta$. Choose any refinement π of π_0. Since f is continuous on each of the compact cells C_{ij} of the partition π, f assumes both minimum and maximum values on C_{ij}. That is, there exist points (s_i, t_j) and (s_i', t_j') in C_{ij} such that $f(s_i, t_j) = m_{ij}$ and $f(s_i', t_j') = M_{ij}$. Since (s_i, t_j) and (s_i', t_j') are in C_{ij} and since $d(C_{ij}) < \delta$, it follows that

$$M_{ij} - m_{ij} = |f(s_i', t_j') - f(s_i, t_j)| < \frac{\epsilon}{A}.$$

Therefore,

$$\begin{aligned} U(f, \pi) - L(f, \pi) &= \sum_{i=1}^{p} \sum_{j=1}^{q} (M_{ij} - m_{ij}) \Delta A_{ij} \\ &< \frac{\epsilon}{A} \sum_{i=1}^{p} \sum_{j=1}^{q} \Delta A_{ij} = \epsilon. \end{aligned}$$

That is, Riemann's condition holds for f on R. Thus, by Theorem 10.1.3, f is integrable on R. ●

THEOREM 10.1.5 The Mean Value Theorem for Integrals Let f be a bounded, real-valued function on a rectangle R in $\mathbb{R}^2$. Let A denote the area of R.

i) If f is integrable on R and if $m \leq f(x, y) \leq M$ for all (x, y) in R, then there exists a number K in $[m, M]$ such that

$$\iint_R f(x, y)\, dA = KA.$$

ii) If f is continuous on R, then there exists a point (x_0, y_0) in R such that

$$\iint_R f(x, y)\, dA = f(x_0, y_0) A.$$

Proof. For part (i) simply apply part (ii) of Theorem 10.1.1. For part (ii) apply the Intermediate Value Theorem (Theorem 3.2.6). ●

We next turn our attention to the set over which a double integration is performed. First, notice that the cells of a partition are not disjoint; two adjacent cells share boundary points. But two adjacent cells do not overlap in the sense of the following definition.

DEFINITION 10.1.6 Two closed sets S_1 and S_2 in $\mathbb{R}^2$ are said to be *nonoverlapping* if $(S_1 \cap S_2)^0 = \emptyset$. ●

Evidently, if two closed sets S_1 and S_2 are disjoint, then they are nonoverlapping. Nonoverlapping sets, if they intersect, do so in boundary points.

To treat functions that may not be continuous on all of a rectangle R and also to extend our definition of the integral to sets more general than rectangles, we need one additional concept, that of the *outer content* of a set.

DEFINITION 10.1.7 Let S be a nonempty, bounded set in $\mathbb{R}^2$. Let $\mathcal{R} = \{R_1, R_2, \ldots, R_m\}$ be any finite collection of closed, bounded rectangles, $R_k = [a_1^{(k)}, b_1^{(k)}] \times [a_2^{(k)}, b_2^{(k)}]$, such that $S \subseteq (\cup_{k=1}^{m} R_k)^0$. The collection $\mathcal{R}$ is called a *rectangular cover* of S. Let A_k denote the area of the rectangle R_k in $\mathcal{R}$ and let $|\mathcal{R}| = \sum_{k=1}^{m} A_k$. The *outer content* of S is

$$\bar{c}(S) = \inf \{|\mathcal{R}| : \mathcal{R} \text{ a rectangular cover of } S\}. \quad ●$$

Clearly, for any nonempty, bounded set S, its outer content $\bar{c}(S)$ exists and is nonnegative. It is also clear that we could, if we wished, restrict to nonoverlapping rectangles when computing the outer content of a set. (See Fig. 10.2.)

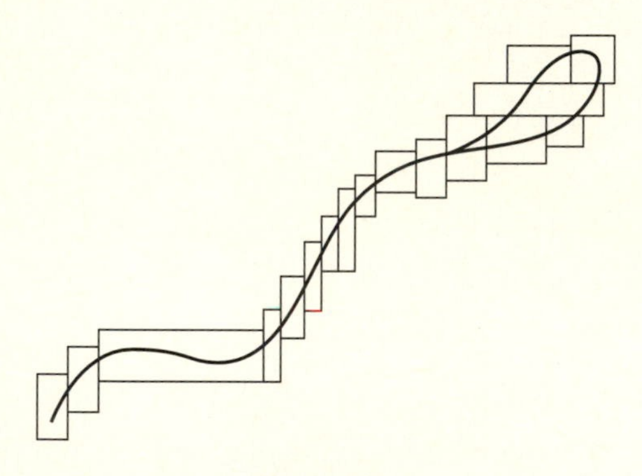

Figure 10.2

The elementary properties of outer content are contained in the next theorem; the proofs are straightforward and are left for you to provide.

THEOREM 10.1.6

i) If S is a finite set, then $\bar{c}(S) = 0$.

ii) If S_1 and S_2 are two nonempty, bounded sets in $\mathbb{R}^2$ such that $S_1 \subseteq S_2$, then $\bar{c}(S_1) \le \bar{c}(S_2)$.

iii) If S_1 and S_2 are two nonempty, bounded sets in $\mathbb{R}^2$, then $\bar{c}(S_1 \cup S_2) \le \bar{c}(S_1) + \bar{c}(S_2)$. ●

Our first significant application of the concept of outer content is provided by the following theorem. The idea behind the theorem is this: If we can enclose the points of discontinuity of the integrand f in a set of arbitrarily small outer content—and thus exclude them from consideration—then we can show that f is integrable.

THEOREM 10.1.7 Let f be a bounded, real-valued function defined on a rectangle R. Let D denote the set of points in R where f is discontinuous. If $\overline{c}(D) = 0$, then f is integrable on R.

Proof. Fix $\epsilon > 0$. Since $\overline{c}(D) = 0$, we can choose a cover $\mathcal{R} = \{R_1, R_2, \ldots, R_m\}$ of closed, bounded, nonoverlapping rectangles such that $D \subseteq (\cup_{k=1}^{m} R_k)^0$ and

$$|\mathcal{R}| < \frac{\epsilon}{4||f||_\infty}.$$

Choose a partition π' of R such that each of the rectangles $R_1, R_2, \ldots, R_m$ is a union of some of the cells of π'.

Let $S = R \cap [(\cup_{k=1}^{m} R_k)^0]^c$. Since $(\cup_{k=1}^{m} R_k)^0$ is open, its complement is closed and S is therefore closed. Since S is also bounded, we deduce that S is compact. Furthermore, all the points of discontinuity of f are contained in $(\cup_{k=1}^{m} R_k)^0$ so that f is continuous on S. Therefore, f is uniformly continuous on S.

Given the above ϵ, choose a $\delta > 0$ such that, if (x, y) and (x', y') are any two points in S with

$$[(x - x')^2 + (y - y')^2]^{1/2} < \delta,$$

we have

$$|f(x, y) - f(x', y')| < \frac{\epsilon}{2A},$$

where A is the area of R.

Choose a partition π'' of R such that $||\pi''|| < \delta$. Let $\pi_0 = \pi' \vee \pi''$ denote the least common refinement of π' and π''. Finally, let π be any refinement of π_0. Since π is also a refinement of π', each of the rectangles R_k is a union of some of the cells of π. We separate the cells C_{ij} of π into two disjoint collections, namely, those contained in $\cup_{k=1}^{m} R_k$ and those contained in S. Let I_1 denote the set of all (i, j) such that $C_{ij} \subseteq \cup_{k=1}^{m} R_k$; let I_2 denote the set of all (i, j) such that $C_{ij} \subseteq S$. We are now in a position to verify that Riemann's condition holds for f on S. Write

$$\begin{aligned} U(f, \pi) - L(f, \pi) &= \sum_{(i,j) \text{ in } I_1} (M_{ij} - m_{ij})\Delta A_{ij} + \sum_{(i,j) \text{ in } I_2} (M_{ij} - m_{ij})\Delta A_{ij} \\ &< 2||f||_\infty |\mathcal{R}| + \frac{\epsilon}{2A} \sum_{(i,j) \text{ in } I_2} \Delta A_{ij} < \frac{\epsilon}{2} + \frac{\epsilon}{2} = \epsilon. \end{aligned}$$

This confirms that Riemann's condition holds for f on R. Therefore, by Theorem 10.1.3, f is integrable on R. ●

To extend the definition of the integral to sets in the plane more general than rectangles, we will use the following fact: Every smooth, rectifiable[1] curve $\mathcal{C}$ in the plane described by parametric equations $x = \phi_1(t)$, $y = \phi_2(t)$, where the parameter t varies in $[0, 1]$, has outer content 0.

THEOREM 10.1.8 Let $\phi = (\phi_1, \phi_2)$ be a continuously differentiable function defined on $[0, 1]$ with values in $\mathbb{R}^2$, the range of which is a smooth rectifiable curve $\mathcal{C}$ in $\mathbb{R}^2$. Then $\overline{c}(\mathcal{C}) = 0$.

Proof. Since ϕ is continuously differentiable, the functions ϕ_1' and ϕ_2' exist and are continuous on $[0, 1]$. Thus there is a constant M such that $||\phi_1'||_\infty \leq M$ and $||\phi_2'||_\infty \leq M$. Given $\epsilon > 0$, choose k in $\mathbb{N}$ such that $4M^2/k < \epsilon$. Partition the interval $[0, 1]$ at the equally spaced points $t_j = j/k$, for $j = 0, 1, 2, \ldots, k$. Let (x_j, y_j) denote the point on the curve $\mathcal{C}$ corresponding to the point t_j in $[0, 1]$. That is, $x_j = \phi_1(t_j)$ and $y_j = \phi_2(t_j)$. At each point (x_j, y_j), center the square

$$R_j = \left[x_j - \frac{M}{k}, x_j + \frac{M}{k}\right] \times \left[y_j - \frac{M}{k}, y_j + \frac{M}{k}\right].$$

For t in $(t_j, t_{j+1}]$, the Mean Value Theorem guarantees the existence of points $c_j^{(1)}$ and $c_j^{(2)}$ in (t_j, t) such that

$$|\phi_1(t) - \phi_1(t_j)| = |\phi_1'(c_j^{(1)})(t - t_j)| \leq \frac{M}{k}$$

and

$$|\phi_2(t) - \phi_2(t_j)| = |\phi_2'(c_j^{(2)})(t - t_j)| \leq \frac{M}{k}.$$

Therefore, $|\phi_1(t) - x_j| \leq M/k$ and $|\phi_2(t) - y_j| \leq M/k$. That is, for any t in $[t_j, t_{j+1}]$, the point $\phi(t) = (\phi_1(t), \phi_2(t))$ is in R_j. Therefore, the curve $\mathcal{C}$ is covered by the closed squares $R_0, R_1, \ldots, R_{k-1}$. The area of each of the squares R_k is

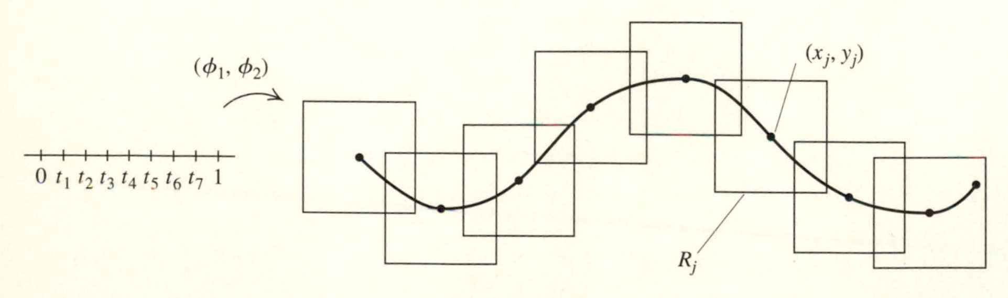

Figure 10.3

[1] Recall that a curve described by the parametric equations $x = \phi_1(t)$, $y = \phi_2(t)$, for t in $[0, 1]$ is *rectifiable* if its arc length, given by $\int_0^1 [\phi_1'(t)^2 + \phi_2'(t)^2]^{1/2}\, dt$, is finite.

$4M^2/k^2$. There are k of them used to cover $\mathcal{C}$. Consequently, $\overline{c}(\mathcal{C}) \leq 4M^2/k < \epsilon$. Since ϵ is arbitrary, we deduce that $\overline{c}(\mathcal{C}) = 0$. (See Fig. 10.3.) ●

In the following definition, we identify the class of sets over which we will perform integration. This class of sets is not the most general possible, but it does contain the domains of greatest interest and utility.

DEFINITION 10.1.8

i) A *region* S in $\mathbb{R}^2$ is an open set U together with some or all of the boundary of U.

ii) A *Riemann domain* in $\mathbb{R}^2$ is a bounded region whose boundary consists of a finite number of smooth rectifiable curves. ●

A typical Riemann domain S is sketched in Fig. 10.4. Note that, since the boundary of S consists of a finite number of smooth rectifiable curves, the outer content of $\text{bd}(S)$ is 0.

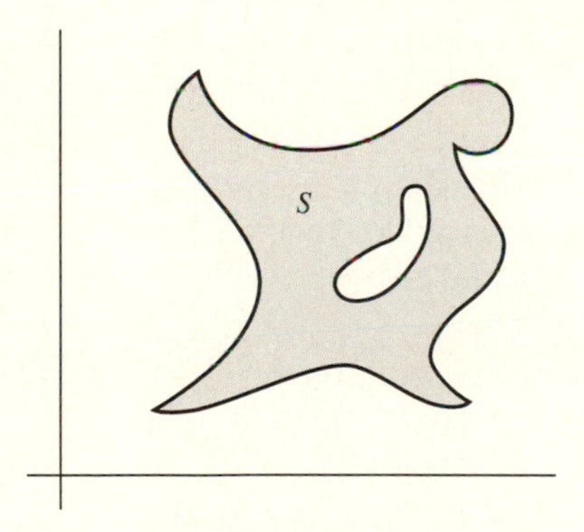

Figure 10.4

Now suppose that S is a Riemann domain in $\mathbb{R}^2$ and that f is a bounded, real-valued function with domain S. Suppose also that the set D of discontinuities of f in S has outer content 0. Since S is bounded, we can enclose S in a bounded, closed rectangle $R = [a_1, b_1] \times [a_2, b_2]$. We extend f to be defined on all of the rectangle R by setting $f(x, y) = 0$ for (x, y) in $R \cap S^c$. By doing so, we may introduce new discontinuities of f, but they will occur only at points of $\text{bd}(S)$. Since $\overline{c}(D) = \overline{c}(\text{bd}(S)) = 0$ and since $\overline{c}(D \cup \text{bd}(S)) \leq \overline{c}(D) + \overline{c}(\text{bd}(S))$, we deduce that the set of discontinuities of the extended function f on R has outer content 0. Therefore, the extended function f is integrable on R. We define

$$\iint_S f(x, y)\, dA = \iint_R f(x, y)\, dA.$$

Clearly, this definition is independent of the particular closed rectangle R chosen to enclose S.

While you undoubtedly have an intuitive sense of the meaning of the word *area*, you must recognize that, technically, the concept has not been defined. Indeed, given the variety of sets in the plane, it is not evident (and it is not true) that every set S in the plane actually has an area. Our construction has relied solely on the area of a rectangle, a concept that is well defined. With the construction completed, we are in a position to define the area of any Riemann domain.

DEFINITION 10.1.9 Let S be a Riemann domain in $\mathbb{R}^2$. The *area* of S is defined to be

$$A(S) = \iint_S dA \quad \bullet$$

We initially constructed the integral $\iint_R f(x, y)\, dA$ by partitioning the rectangle R into nonoverlapping rectangular cells. This approach was necessary because the area of a rectangle is well defined, whereas, at that point of our construction, we had no ready definition of area for more general sets. Now, however, for any Riemann domain S, we can let a *partition* of S be any finite collection of nonoverlapping subsets $\pi = \{S_1, S_2, \ldots, S_p\}$ of S such that (i) each S_j is a Riemann domain and, hence, has a well-defined area, which we will denote ΔA_j, and (ii) $S = \cup_{j=1}^{p} S_j$. The *gauge* $||\pi||$ of π is, again, the smallest diameter of the sets $S_1, S_2, \ldots, S_p$. A *refinement* π' of π is a partition $\{T_1, T_2, \ldots, T_q\}$ of S such that, for each $i = 1, 2, \ldots, q$, there is a j such that $T_i \subseteq S_j$.

Again, for a bounded, real-valued function f defined on S and for any partition $\pi = \{S_1, S_2, \ldots, S_p\}$, we choose a point (s_j, t_j) in each of the partition cells S_j and form the *Riemann sum* $S(f, \pi) = \sum_{j=1}^{p} f(s_j, t_j)\Delta A_j$. The function f is integrable on S if, for any $\epsilon > 0$, there is a partition π_0 of this general type such that, for every refinement π of π_0 and for every choice of the points (s_j, t_j) in the partition sets of π, $|\sum_{j=1}^{p} f(s_j, t_j)\Delta A_j - \iint_S f(x, y)\, dA| < \epsilon$.

We forego the proof of this assertion, leaving its demonstration for you to complete. For all classical applications of integration the construction we have completed for functions that are continuous except on a set of outer content 0 suffices. There remains the problem of evaluating double integrals; it is to this issue that we now turn.

10.2 Evaluation of Double Integrals

As with integrals of a single variable, computation by brute force using the procedure by which the integral is defined is laborious and generally not very fruitful. For single integrals Cauchy's form of the Fundamental Theorem of Calculus proved to be an invaluable tool in evaluating such integrals. Here we address the problem of computing double integrals. We begin with the simplest case, the evaluation of the double integral of an integrable function f defined on a rectangle R in $\mathbb{R}^2$.

THEOREM 10.2.1 Let f be a bounded, real-valued function that is integrable on a rectangle $R = [a_1, b_1] \times [a_2, b_2]$. Suppose that, for each y in $[a_2, b_2]$, the function $f(x, y)$ is an integrable function of x on $[a_1, b_1]$. Then the function $F(y) = \int_{a_1}^{b_1} f(x, y)\, dx$ is an integrable function of y on $[a_2, b_2]$ and

$$\iint_R f(x, y)\, dA = \int_{a_2}^{b_2} \left[\int_{a_1}^{b_1} f(x, y)\, dx \right] dy.$$

Proof. Before launching into the technical details, let's identify exactly what we know and what we need to prove. First, we know that there exists a real number I such that, for a sufficiently refined partition π of R and for arbitrary choices of the points (s_i, t_j) in the cells of π, the difference $|S(f, \pi) - I|$ is arbitrarily small.

Second, we know that, for each y in $[a_2, b_2]$, there is a partition $\pi_1^{(0)}(y)$ of $[a_1, b_1]$ such that, for any refinement π_1 of $\pi_1^{(0)}(y)$ and for any choice of s_i in the intervals of π_1, $|\sum_i f(s_i, y)\Delta x_i - \int_{a_1}^{b_1} f(x, y)\, dx|$ is also arbitrarily small. Note that, in general, the partition $\pi_1^{(0)}(y)$ depends on the choice of y.

We have to prove, first, that $F(y) = \int_{a_1}^{b_1} f(x, y)\, dx$ is an integrable function of y over the interval $[a_2, b_2]$ and, second, that $\int_{a_2}^{b_2} F(y)\, dy = I$. We will achieve both proofs simultaneously by showing that, for a sufficiently refined partition π_2 of $[a_2, b_2]$ and for arbitrary choices of t_j in the intervals of π_2, the difference $|\sum_j F(t_j)\Delta y_j - I|$ can be made arbitrary small.

Now fix $\epsilon > 0$. Choose a partition $\pi_0 = \pi_1^{(0)} \times \pi_2^{(0)}$ of R such that, for any refinement π of π_0 and for any choice of points (s_i, t_j) in the cells of π, we have

$$\left| \sum_i \sum_j f(s_i, t_j)\Delta A_{ij} - I \right| = |S(f, \pi) - I| < \tfrac{1}{2}\epsilon.$$

Set the partition $\pi_1^{(0)}$ aside for later use and focus on the partition $\pi_2^{(0)}$ of $[a_2, b_2]$. Choose and fix any refinement π_2 of $\pi_2^{(0)}$. We write $\pi_2 : a_2 = y_0 < y_1 < y_2 < \cdots < y_q = b_2$. For $j = 1, 2, \ldots, q$, choose an arbitrary t_j in $[y_{j-1}, y_j]$. Below we will choose a partition π_1 of $[a_1, b_1]$ and form the partition $\pi = \pi_1 \times \pi_2$ of R. Then we will choose points in the cells C_{ij} of the partition π. As we do so, we will use just these q values t_j for the second coordinates of the points (s_i, t_j).

Having chosen $t_1, t_2, \ldots, t_q$, we know by the hypothesis of the theorem that, for each $j = 1, 2, \ldots, q$, the function $f(x, t_j)$ is an integrable function of x over $[a_1, b_1]$. Thus, for each j, there exists a partition $\pi_1^{(j)}$ of $[a_1, b_1]$ such that, for every refinement π_1 of $\pi_1^{(j)}$ and for every choice of points s_i in the intervals of π_1, we have

$$\left| \sum_i f(s_i, t_j)\Delta x_i - \int_{a_1}^{b_1} f(x, t_j)\, dx \right| < \frac{\epsilon}{2(b_2 - a_2)}. \qquad \textbf{(10.1)}$$

Now we recall the partition $\pi_1^{(0)}$ above and we let π_1 denote $\pi_1^{(0)} \vee \pi_1^{(1)} \vee \pi_1^{(2)} \vee \cdots \vee \pi_1^{(q)}$, the least common refinement of the partitions $\pi_1^{(0)}, \pi_1^{(1)}, \pi_1^{(2)}, \ldots, \pi_1^{(q)}$. Then, for every choice of s_i in the intervals of π_1, the inequality (10.1) holds for all $j = 1, 2, \ldots, q$ simultaneously.

Finally, let $\pi = \pi_1 \times \pi_2$. Since π_1 is a refinement of $\pi_1^{(0)}$ and since π_2 is a refinement of $\pi_2^{(0)}$, we know that π is a refinement of $\pi_0 = \pi_1^{(0)} \times \pi_2^{(0)}$. For our arbitrary choices of $t_1, t_2, \ldots, t_q$ above, we have

$$\left|\sum_{j=1}^{q} F(t_j)\Delta y_j - I\right|$$

$$\leq \left|\sum_{j=1}^{q} F(t_j)\Delta y_j - \sum_{j=1}^{q}\left[\sum_{i=1}^{p} f(s_i, t_j)\Delta x_i\right]\Delta y_j\right| + |S(f, \pi) - I|$$

$$\leq \sum_{j=1}^{q}\left|\int_{a_1}^{b_1} f(x, t_j)\, dx - \sum_{i=1}^{p} f(s_i, t_j)\Delta x_i\right|\Delta y_j + |S(f, \pi) - I|$$

$$< \frac{\epsilon}{2(b_2 - a_2)}\sum_{j=1}^{q}\Delta y_j + \frac{\epsilon}{2} = \frac{\epsilon}{2} + \frac{\epsilon}{2} = \epsilon.$$

To summarize, we have proved that, for any $\epsilon > 0$, there is a partition $\pi_2^{(0)}$ of $[a_2, b_2]$ such that, for any partition π_2 which refines $\pi_2^{(0)}$ and for any choice of t_j in the partition intervals of π_2,

$$\left|\sum_{j=1}^{q} F(t_j)\Delta y_j - I\right| < \epsilon.$$

Therefore, the function $F(y) = \int_{a_1}^{b_1} f(x, y)\, dx$ is integrable over $[a_2, b_2]$ and

$$\int_{a_2}^{b_2} F(y)\, dy = \int_{a_2}^{b_2}\left[\int_{a_1}^{b_1} f(x, y)\, dx\right] dy = I = \iint_R f(x, y)\, dA.$$

This proves the theorem. ●

Of course, there is the symmetric theorem. If f is integrable on the rectangle R and if, for each x in $[a_1, b_1]$ the function $f(x, y)$ is an integrable function of y over $[a_2, b_2]$, then $G(x) = \int_{a_2}^{b_2} f(x, y)\, dy$ is an integrable function of x over $[a_1, b_1]$ and

$$\int_{a_1}^{b_1} G(x)\, dx = \int_{a_1}^{b_1}\left[\int_{a_2}^{b_2} f(x, y)\, dy\right] dx = \iint_R f(x, y)\, dA.$$

Consequently, if f is integrable over $[a_2, b_2]$ for each x in $[a_1, b_1]$ and if f is integrable over $[a_1, b_1]$ for each y in $[a_2, b_2]$, then

$$\iint_R f(x, y)\, dA = \int_{a_2}^{b_2}\left[\int_{a_1}^{b_1} f(x, y)\, dx\right] dy = \int_{a_1}^{b_1}\left[\int_{a_2}^{b_2} f(x, y)\, dy\right] dx. \tag{10.2}$$

In particular, if f is continuous on R, then the above conditions are met and (10.2) is automatically valid.

The process of writing a double integral as an integral of an integral is called *iteration* and the integrals

$$\int_{a_2}^{b_2}\left[\int_{a_1}^{b_1} f(x,y)\,dx\right]dy \qquad \text{and} \qquad \int_{a_1}^{b_1}\left[\int_{a_2}^{b_2} f(x,y)\,dy\right]dx$$

are called *iterated integrals*. If we can legitimately write a double integral as an iterated integral, then we have the possibility of using Cauchy's form of the Fundamental Theorem of Calculus to evaluate first the inner integral, then the outer one.

EXAMPLE 1 Let $f(x, y) = xe^{xy}$ on $R = [0, 2] \times [0, 1]$. Since f is continuous on R, we can iterate the double integral of f over R in either order. In this instance it is most convenient to write

$$\iint_R xe^{xy}\,dA = \int_0^2\left[\int_0^1 xe^{xy}\,dy\right]dx = \int_0^2\left[e^{xy}\Big|_0^1\right]dx = \int_0^2 (e^x - 1)\,dx$$

$$= (e^x - x)\Big|_0^2 = (e^2 - 2) - (1 - 0) = e^2 - 3,$$

where we have used Cauchy's Fundamental Theorem of Calculus twice. ●

By a straightforward argument we can extend this iteration procedure to Riemann domains. We begin with a Riemann domain S whose boundary consists of the two horizontal lines $y = a_2$ and $y = b_2$ and two curves $x = \phi_1(y)$, $x = \phi_2(y)$ for y in $[a_2, b_2]$ as sketched in Fig. 10.5. Assume that ϕ_1 and ϕ_2 are continuously differentiable on $[a_2, b_2]$ and that $\phi_1(y) \le \phi_2(y)$ for all y in $[a_2, b_2]$. We let

$$a_1 = \min\{\phi_1(y) : y \text{ in } [a_2, b_2]\}$$

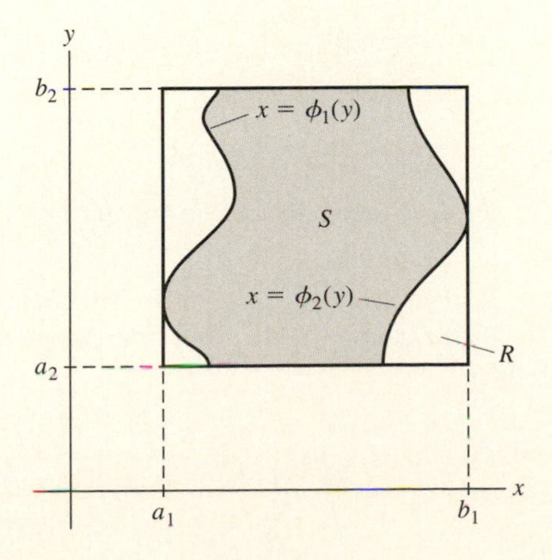

Figure 10.5

and

$$b_1 = \max\{\phi_2(y) : y \text{ in } [a_2, b_2]\}.$$

Thus S is enclosed in the rectangle $R = [a_1, b_1] \times [a_2, b_2]$.

Let f be any bounded real-valued function that is integrable on S. Extend f to have value 0 on $R \cap S^c$. Suppose that for each y in $[a_2, b_2]$, $f(x, y)$ is integrable on $[a_1, b_1]$. Then, by Theorem 10.2.1,

$$\iint_S f(x, y)\, dA = \iint_R f(x, y)\, dA = \int_{a_2}^{b_2} \left[\int_{a_1}^{b_1} f(x, y)\, dx \right] dy.$$

Since f vanishes off S, we have, for each y in $[a_2, b_2]$,

$$\int_{a_1}^{b_1} f(x, y)\, dx = \int_{\phi_1(y)}^{\phi_2(y)} f(x, y)\, dx.$$

Therefore,

$$\iint_S f(x, y)\, dA = \int_{a_2}^{b_2} \left[\int_{\phi_1(y)}^{\phi_2(y)} f(x, y)\, dx \right] dy.$$

Likewise, if the boundary of a Riemann domain S is described by the straight lines $x = a_1$, $x = b_1$ and the two curves $y = \psi_1(x)$, $y = \psi_2(x)$, where ψ_1 and ψ_2 are continuously differentiable and $\psi_1(x) \le \psi_2(x)$ for all x in $[a_1, b_1]$, then we can iterate a double integral over S by integrating first with respect to y, then with respect to x. (See Fig. 10.6.) Let

$$a_2 = \min\{\psi_1(x) : x \text{ in } [a_1, b_1]\}$$

and

$$b_2 = \max\{\psi_2(x) : x \text{ in } [a_1, b_1]\}.$$

Again, S is enclosed in $R = [a_1, b_1] \times [a_2, b_2]$. If f is integrable on S, we extend f to vanish on $R \cap S^c$. If, for each x in $[a_1, b_1]$, the extended function $f(x, y)$ is integrable on $[a_2, b_2]$, then

$$\iint_S f(x, y)\, dA = \int_{a_1}^{b_1} \left[\int_{\psi_1(x)}^{\psi_2(x)} f(x, y)\, dy \right] dx.$$

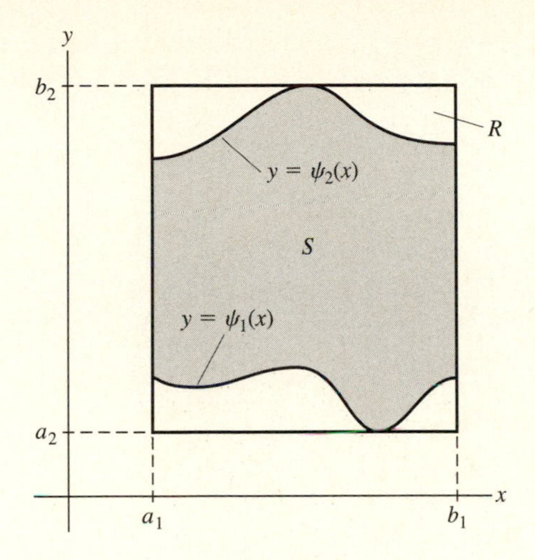

Figure 10.6

Now suppose that a Riemann domain S can be partitioned into nonoverlapping Riemann domains $S_1, S_2, \ldots, S_k$ where each of the S_j is of one of the two simple types described above. (See Fig. 10.7.) If f is integrable on S and also meets the

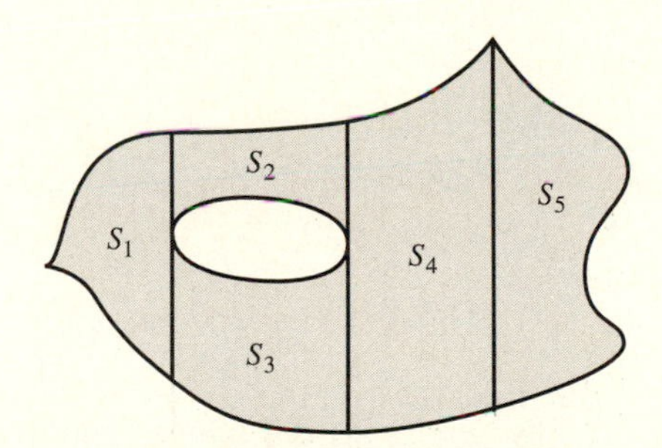

Figure 10.7

additional integrability requirements on each domain S_j, then

$$\iint_S f(x, y)\,dA = \iint_{S_1} f(x, y)\,dA + \iint_{S_2} f(x, y)\,dA + \cdots + \iint_{S_k} f(x, y)\,dA.$$

Since each integral $\iint_{S_j} f(x, y)\,dA$ can be iterated in one or the other of the forms

$$\int_{a_2}^{b_2} \left[\int_{\phi_1(y)}^{\phi_2(y)} f(x, y)\,dx \right] dy \quad \text{or} \quad \int_{a_1}^{b_1} \left[\int_{\psi_1(x)}^{\psi_2(x)} f(x, y)\,dy \right] dx,$$

we can finally write $\iint_S f(x, y)\,dA$ as the sum of iterated integrals. Evaluating each of the iterated integrals and adding yields the value of the orginal integral $\iint_S f(x, y)\,dA$.

EXAMPLE 2 Let r_1 and r_2 be constants with $0 < r_1 < r_2$. Let $S = \{(x, y) : r_1^2 \le x^2 + y^2 \le r_2^2, y > 0\}$ as sketched in Fig. 10.8. Then S is a Riemann domain. Let

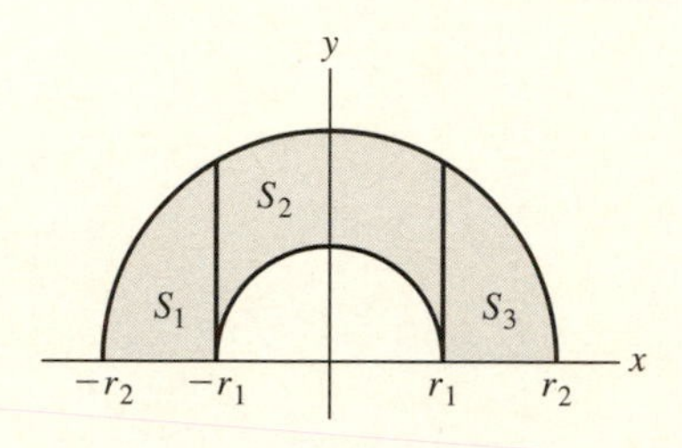

Figure 10.8

$f(x, y) = xy$ on S. Partition S into nonoverlapping domains S_1, S_2, and S_3 as drawn. We write

$$\iint_S xy\,dA = \iint_{S_1} xy\,dA + \iint_{S_2} xy\,dA + \iint_{S_3} xy\,dA$$

and evaluate each of these integrals separately:

$$\iint_{S_1} xy\,dA = \int_{-r_2}^{-r_1}\left[\int_0^{\sqrt{r_2^2-x^2}} xy\,dy\right]dx = \int_{-r_2}^{-r_1} \tfrac{1}{2}x(r_2^2 - x^2)\,dx = \tfrac{-(r_2^2-r_1^2)^2}{8},$$

$$\iint_{S_2} xy\,dA = \int_{-r_1}^{r_1}\left[\int_{\sqrt{r_1^2-x^2}}^{\sqrt{r_2^2-x^2}} xy\,dy\right]dx = \int_{-r_1}^{r_1} \tfrac{1}{2}x(r_2^2 - r_1^2)\,dx = 0,$$

$$\iint_{S_3} xy\,dA = \int_{r_1}^{r_2}\left[\int_0^{\sqrt{r_2^2-x^2}} xy\,dy\right]dx = \int_{r_1}^{r_2} \tfrac{1}{2}x(r_2^2 - x^2)\,dx = \tfrac{(r_2^2-r_1^2)^2}{8}.$$

The value of $\iint_S xy\,dA$ is the sum of these three numbers. Therefore $\iint_S xy\,dA = 0$. ●

10.3 Transformations: Change of Variables

Double integrals are often more difficult to compute than are single integrals. The added difficulty may result from the possibly more complicated form of the integrand, which now can involve more variables. What may be equally important, however, are the complexities introduced, when iterating the integral, by the functions describing the boundary of the set over which the integration is performed. In other words, the domain S itself is a possible source of difficulty. Here we address the problem of changing the domain of integration by means of a mapping from $\mathbb{R}^2$ into $\mathbb{R}^2$.

We begin with transformations of a particularly simple type. We will identify four distinct, albeit similar, types of elementary transformations. In all cases let U be a connected open set in $\mathbb{R}^2$. Suppose that $\phi = (\phi_1, \phi_2)$ is a continuously differentiable function that maps U one-to-one into $\mathbb{R}^2$ defined by the following equations. We will denote the variables in the domain of ϕ by x and y, those in the range of ϕ by u and v. We assume first that

$$u = \phi_1(x, y),$$
$$v = \phi_2(x, y) = y.$$

A transformation of this simple type, which leaves one of the variables unchanged, will be called an *elementary transformation*. The Jacobian of ϕ at a point (x, y) is

$$\det \begin{vmatrix} D_1\phi_1(x, y) & D_2\phi_1(x, y) \\ 0 & 1 \end{vmatrix} = D_1\phi_1(x, y).$$

We assume that $D_1\phi_1$ is never 0 in all of U. Since $D_1\phi_1$ is continuous and since U is connected, it follows that $D_1\phi_1$ is either positive throughout U or negative throughout U. We will be particularly interested in the effect of the transformation ϕ on a closed rectangle $R = [a_1, b_1] \times [a_2, b_2]$ contained in U. In Fig. 10.9 we

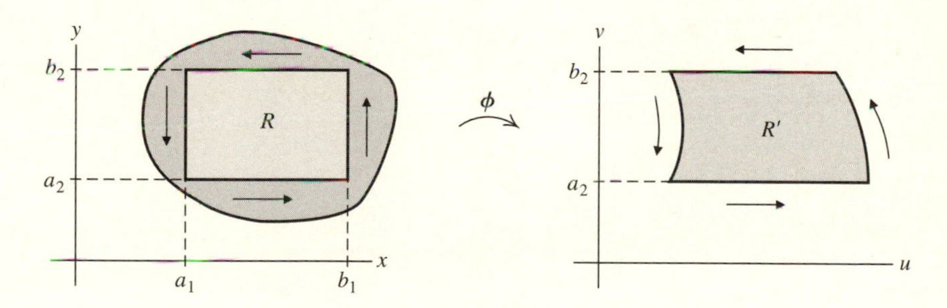

Figure 10.9

sketch a typical situation when $D_1\phi_1$ is positive and $R' = \phi(R)$ denotes the image of R under ϕ. Notice that, if y is any point in $[a_2, b_2]$ and if $a_1 \leq x_1 < x_2 \leq b_1$, then

$$\phi_1(a_1, y) \leq \phi_1(x_1, y) < \phi_1(x_2, y) \leq \phi_1(b_1, y).$$

Thus traversing the boundary of R in a counterclockwise direction results in a similar counterclockwise movement of the image point about the boundary of R'. We say that *orientation is preserved*. In Fig. 10.9 we indicate this fact by the arrows along the boundaries of R and R'.

The portion of the boundary of R described by $x = a_1$ is transformed into the curve $u = \phi_1(a_1, y)$, $a_2 \leq y \leq b_2$. Likewise, the portion of the boundary of R given by $x = b_1$ is transformed into the curve $u = \phi_1(b_1, y)$, $a_2 \leq y \leq b_2$. Thus the area $A(R')$ of R' is given by the integral

$$A(R') = \int_{a_2}^{b_2} [\phi_1(b_1, y) - \phi_1(a_1, y)]\, dy.$$

Since the function $g(y) = \phi_1(b_1, y) - \phi_1(a_1, y)$ is a continuous function of y on $[a_2, b_2]$, we know, by the Mean Value Theorem for Integrals [Theorem 6.3.2], that there is a c_2 in $[a_2, b_2]$ such that

$$A(R') = [\phi_1(b_1, c_2) - \phi_1(a_1, c_2)](b_2 - a_2).$$

By the Mean Value Theorem (Theorem 4.3.3) applied to the function $\phi_1(x, c_2)$ on $[a_1, b_1]$, there is a point c_1 in (a_1, b_1) such that

$$\begin{aligned} A(R') &= D_1\phi_1(c_1, c_2)(b_1 - a_1)(b_2 - a_2) \\ &= J_\phi(c_1, c_2)A(R), \end{aligned}$$

where $A(R)$ denotes the area of R. Thus, the value of the Jacobian is a *magnification factor* for the mapping ϕ. If $0 < J_\phi(c_1, c_2) < 1$, then the area is contracted by the mapping ϕ; if $J_\phi(c_1, c_2) > 1$, then ϕ expands the area.

On the other hand, if, for the elementary transformation ϕ above, we have $D_1\phi_1 < 0$ on U, then, for any y in $[a_2, b_2]$ and $a_1 \le x_1 < x_2 \le b_1$, it follows that

$$\phi_1(a_1, y) \ge \phi_1(x_1, y) > \phi_1(x_2, y) \ge \phi_1(b_1, y).$$

Therefore, the *orientation of R is reversed* by ϕ as it maps R onto R'. (See Fig. 10.10.) In this case, the area $A(R')$ of R' is

$$\begin{aligned} A(R') &= \int_{a_2}^{b_2} [\phi_1(a_1, y) - \phi_1(b_1, y)]\, dy = -J_\phi(c_1, c_2)A(R) \\ &= |J_\phi(c_1, c_2)|A(R), \end{aligned}$$

where (c_1, c_2) is a point of R. Again, $|J_\phi|$ is a *magnification factor* for the mapping ϕ.

A second type of elementary transformation, analogous to the type we have just discussed, is a continuously differentiable mapping $\phi = (\phi_1, \phi_2)$ where

$$\begin{aligned} u &= \phi_1(x, y) = x, \\ v &= \phi_2(x, y). \end{aligned}$$

In this case, $J_\phi(x, y) = D_2\phi_2(x, y)$. Again we assume that $D_2\phi_2$ does not vanish at any point of U, so $D_2\phi_2$ is either positive or negative throughout U. If $D_2\phi_2 > 0$

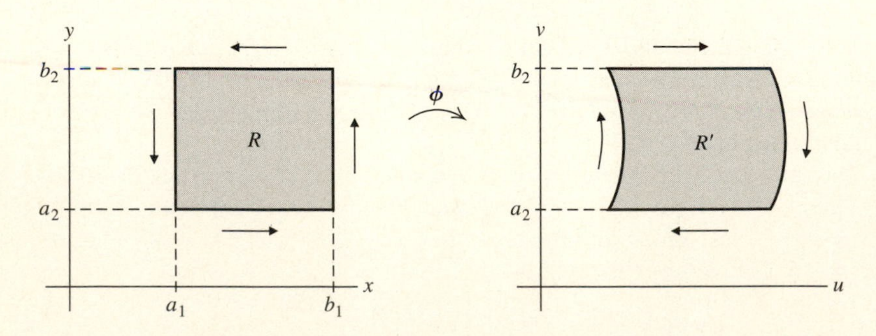

Figure 10.10

on U, then orientation is preserved; if $D_2\phi_2 < 0$, then orientation is reversed. (See Fig. 10.11, where $D_2\phi_2 > 0$, and Fig. 10.12, where $D_2\phi_2 < 0$.)

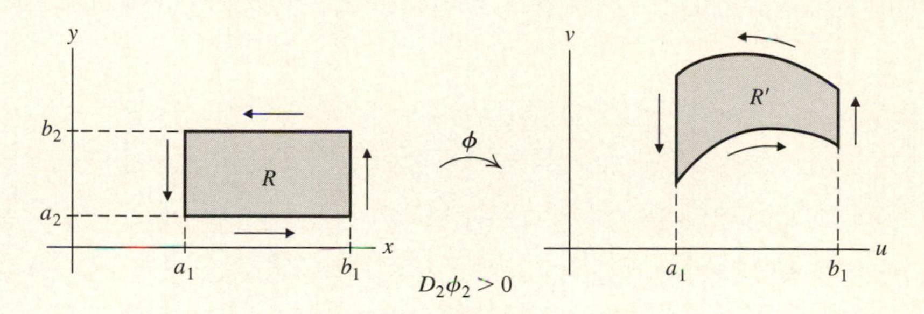

Figure 10.11

Again, whether $D_2\phi_2$ is positive or negative on U, the area of R' is

$$\begin{aligned} A(R') &= \left| \int_{a_1}^{b_1} [\phi_2(x, b_2) - \phi_2(x, a_2)]\, dx \right| \\ &= |D_2\phi_2(c_1, c_2)|(b_1 - a_1)(b_2 - a_2) \\ &= |J_\phi(c_1, c_2)|A(R), \end{aligned}$$

where (c_1, c_2) is a point of R.

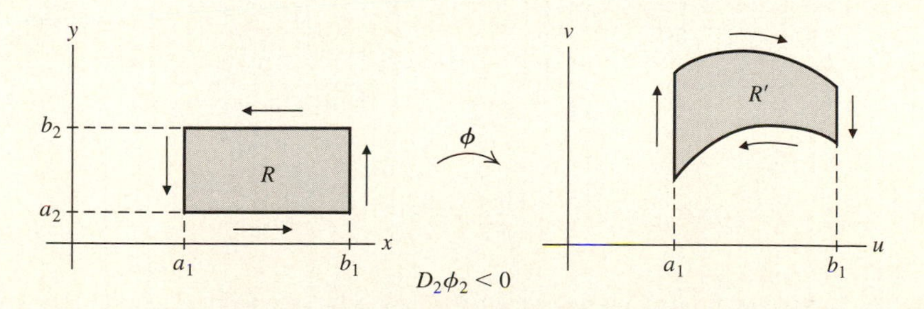

Figure 10.12

The next two types of elementary transformations differ from the previous two only in that the variables are interchanged, introducing a "twist" in the mapping. First, assume that $\phi = (\phi_1, \phi_2)$ is defined by

$$\begin{aligned} u &= \phi_1(x, y) = y, \\ v &= \phi_2(x, y), \end{aligned}$$

where ϕ_2 is continuously differentiable on the connected open set U and $D_1\phi_2$ does not vanish on U. Again, either $D_1\phi_2$ is positive throughout U or it is negative there. The Jacobian of ϕ is $-D_1\phi_2$ and therefore does not vanish on U. If $D_1\phi_2$ is positive—that is, if J_ϕ is negative—on U, then orientation is reversed, as indicated in Fig. 10.13. If $D_1\phi_2 < 0$ on U, then orientation is preserved, as sketched in Fig. 10.14. In both cases, the area $A(R')$ is $|J_\phi(c_1, c_2)|A(R)$ where (c_1, c_2) is a point in R.

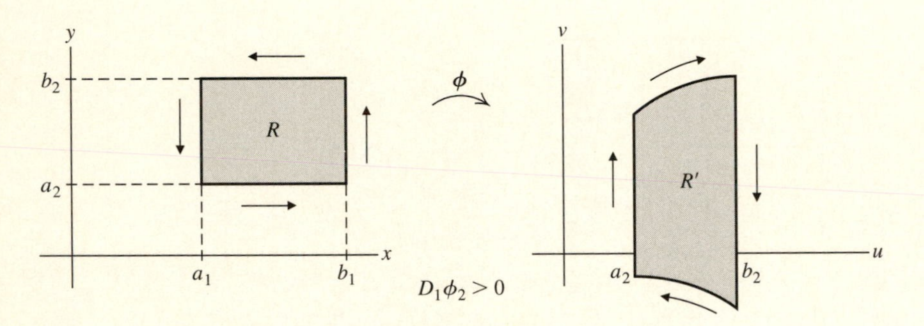

Figure 10.13

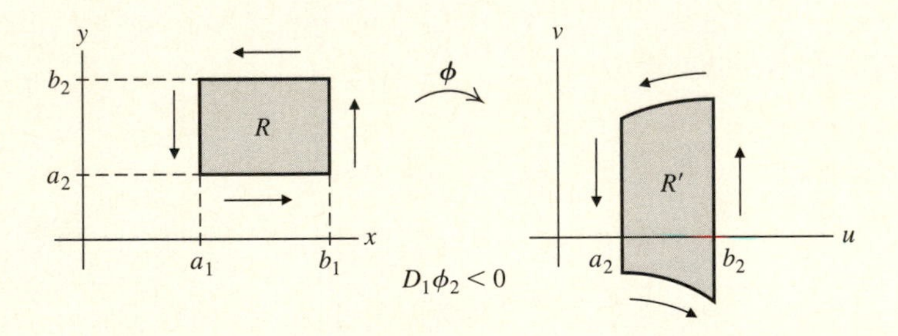

Figure 10.14

The fourth and final elementary transformation has the form $\phi = (\phi_1, \phi_2)$ where

$$u = \phi_1(x, y),$$
$$v = \phi_2(x, y) = x.$$

Again, ϕ_1 is assumed to be continuously differentiable on U and $D_2\phi_1$ is assumed not to vanish on U. Also, $J_\phi = -D_2\phi_1$. And once again, orientation is preserved if $J_\phi > 0$ and is reversed if $J_\phi < 0$. Finally, $A(R') = |J_\phi(c_1, c_2)|A(R)$, where (c_1, c_2) is a point of R. (See Fig. 10.15.)

We can compose any one of these elementary transformations with any of the remaining three types and obtain a (nonelementary) transformation as a result. Rather than belabor the point, we will examine only those compositions that will occur in the proof of Theorem 10.3.1.

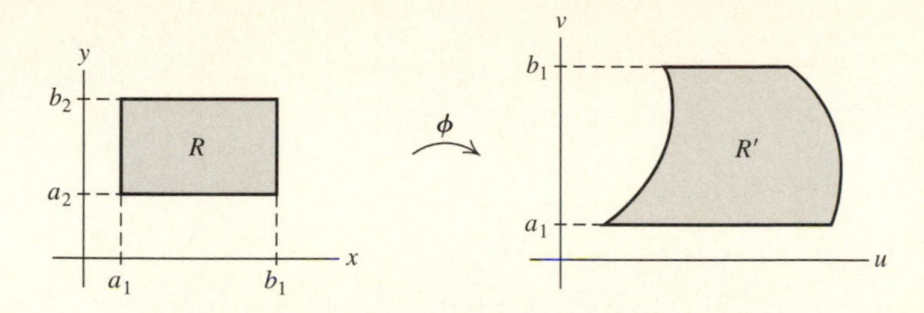

Figure 10.15

First, let $\phi = (\phi_1, \phi_2)$ be a one-to-one transformation defined on a connected open set U in $\mathbb{R}^2$, where

$$u = \phi_1(x, y),$$
$$v = \phi_2(x, y) = y,$$

for (x, y) in U. Suppose also that ϕ_1 is continuously differentiable on U and that $D_1\phi_1$ does not vanish on U. That is, ϕ is an elementary transformation of the first type we discussed earlier. Let U' be a connected open set that contains $\phi(U)$ and let $\psi = (\psi_1, \psi_2)$ be a one-to-one transformation defined by

$$w = \psi_1(u, v) = u,$$
$$z = \psi_2(u, v)$$

for (u, v) in U'. Again, we assume that ψ_2 is continuously differentiable on U' and that $D_2\psi_2$ does not vanish on U'. That is, ψ is an elementary transformation of the second type discussed earlier. Form the transformation $\boldsymbol{h} = \psi \circ \phi$. Then, $\boldsymbol{h}$ maps U one-to-one onto $\psi(\phi(U))$. We write $\boldsymbol{h} = (h_1, h_2)$ where

$$w = h_1(x, y) = \psi_1(\phi_1(x, y), y) = \phi_1(x, y),$$
$$z = h_2(x, y) = \psi_2(\phi_1(x, y), y), \tag{10.3}$$

for all (x, y) in U. (See Fig. 10.16 on page 460.) The Jacobian of $\boldsymbol{h}$ is $J_{\boldsymbol{h}}(x, y) = J_{\psi}(\phi(x, y))J_{\phi}(x, y) = [D_2\psi_2(\phi(x, y))]D_1\phi_1(x, y)$, by Theorem 9.1.1. Thus $J_{\boldsymbol{h}}$ is never 0 on U and hence does not change sign on U.

The second type of composition that will arise is described as follows. For (x, y) in a connected open set U, let

$$u = \phi_1(x, y) = x,$$
$$v = \phi_2(x, y),$$

where ϕ_2 is continuously differentiable on U and $D_2\phi_2$ does not vanish on U. Thus $\phi = (\phi_1, \phi_2)$ is an elementary transformation of the second type discussed earlier. Let U' be a connected open set containing $\phi(U)$ and let $\psi = (\psi_1, \psi_2)$ be an elementary transformation of the third type discussed earlier. That is, for (u, v) in U', let

$$w = \psi_1(u, v) = v,$$

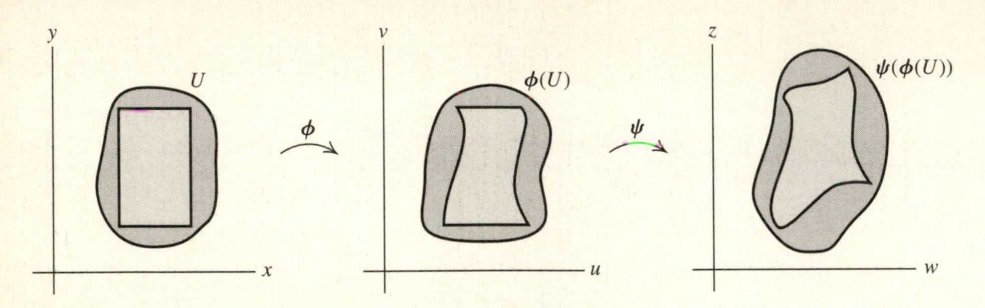

Figure 10.16

$$z = \psi_2(u, v),$$

where ψ_2 is continuously differentiable on U' and $D_1\psi_2$ does not vanish on U'. By forming the composition $\psi \circ \phi$ on U, we obtain a transformation $\boldsymbol{h} = (h_1, h_2)$ where

$$\begin{aligned} w &= h_1(x, y) = \psi_1(x, \phi_2(x, y)) = \phi_2(x, y), \\ z &= h_2(x, y) = \psi_2(x, \phi_2(x, y)). \end{aligned} \tag{10.4}$$

The Jacobian of $\boldsymbol{h}$ is

$$J_{\boldsymbol{h}}(x, y) = J_{\psi}(\phi(x, y))J_{\phi}(x, y) = [-D_1\psi_2(\phi(x, y))]D_2\phi_2(x, y)$$

and does not vanish on U.

Now let $\boldsymbol{h} = (h_1, h_2)$ be any continuously differentiable transformation defined on a connected open set U in $\mathbb{R}^2$ such that $J_{\boldsymbol{h}}$ does not vanish on U. We claim that, in a neighborhood of any point (x_0, y_0) in U, $\boldsymbol{h}$ can be decomposed into the composition of two elementary transformations ϕ and ψ of one of the two forms (10.3) or (10.4) identified above.

THEOREM 10.3.1 Let $\boldsymbol{h} = (h_1, h_2)$ be a continuously differentiable mapping defined on a connected open set U in $\mathbb{R}^2$. Suppose that $J_{\boldsymbol{h}}$ does not vanish on U. Let (x_0, y_0) be a point of U. There exists an open rectangle $R = (x_1, x_2) \times (y_1, y_2)$ that contains (x_0, y_0) in its interior and elementary transformations ϕ defined on R and ψ defined on $\phi(R)$ such that $\boldsymbol{h}(x, y) = (\psi \circ \phi)(x, y)$ for all (x, y) in R.

Proof. Note that, since

$$J_{\boldsymbol{h}} = (D_1h_1)(D_2h_2) - (D_2h_1)(D_1h_2) \neq 0,$$

either $D_1h_1(x_0, y_0) \neq 0$ or $D_2h_1(x_0, y_0) \neq 0$. Assume first that $D_1h_1(x_0, y_0) \neq 0$. The transformation $\boldsymbol{h}$ consists of the component functions $w = h_1(x, y)$ and $z = h_2(x, y)$. Define $H(x, y, w) = h_1(x, y) - w$ on $U \times \mathbb{R}$. The level surface $H = 0$ passes through the point (x_0, y_0, w_0), where $w_0 = h_1(x_0, y_0)$. Note that $D_1H(x_0, y_0, w_0) = D_1h_1(x_0, y_0) \neq 0$. Therefore, by the Implicit Function Theorem (Theorem 9.3.1), there exist open intervals (x_1, x_2), (y_1, y_2), and (w_1, w_2)

containing x_0, y_0, and w_0, respectively, and a continuously differentiable function g defined on $(y_1, y_2) \times (w_1, w_2)$ such that

$$x_0 = g(y_0, w_0)$$

and

$$H(g(y, w), y, w) = 0$$

for all (y, w) in $(y_1, y_2) \times (w_1, w_2)$. That is,

$$w = h_1(g(y, w), y),$$

for all (y, w) in $(y_1, y_2) \times (w_1, w_2)$. Now introduce intermediate variables u and v and define two elementary transformations $\phi = (\phi_1, \phi_2)$ and $\psi = (\psi_1, \psi_2)$ as follows:

$$\begin{aligned} u &= \phi_1(x, y) = h_1(x, y), \\ v &= \phi_2(x, y) = y, \end{aligned}$$

and

$$\begin{aligned} w &= \psi_1(u, v) = u, \\ z &= \psi_2(u, v) = h_2(g(v, u), v). \end{aligned}$$

Then ϕ and ψ are elementary transformations of the first and second types, respectively, such as we have discussed. We need only verify that $\psi \circ \phi = \mathbf{h}$ on $R = (x_1, x_2) \times (y_1, y_2)$. First, for any (x, y) in R, we have

$$w = \psi_1(u, v) = u = h_1(x, y).$$

Next, given (x, y) in R, it follows that $w = h_1(x, y)$ is in (w_1, w_2); equivalently, $x = g(y, w)$. Therefore, since $v = \phi_2(x, y) = y$ and $u = \psi_1(u, v) = w$, we have

$$\begin{aligned} z &= \psi_2(u, v) = h_2(g(v, u), v) \\ &= h_2(g(y, w), y) = h_2(x, y). \end{aligned}$$

We conclude that, in case $D_1h_1(x_0, y_0) \neq 0$, we have $\psi \circ \phi = \mathbf{h}$ on R for the elementary transformations ϕ and ψ defined earlier.

Suppose, on the other hand, that $D_1h_1(x_0, y_0) = 0$ and therefore that $D_2h_1(x_0, y_0) \neq 0$. Again we let $H(x, y, w) = h_1(x, y) - w$. Again, the numbers x_0, y_0, and $w_0 = h_1(x_0, y_0)$ form a point (x_0, y_0, w_0) on the level surface $H = 0$. Since $D_2H(x_0, y_0, w_0) = D_2h_1(x_0, y_0) \neq 0$, the Implicit Function Theorem guarantees the existence of open intervals (x_1, x_2), (y_1, y_2), and (w_1, w_2) and a continuously differentiable function g defined on $(x_1, x_2) \times (w_1, w_2)$ such that (x_0, w_0) is in $(x_1, x_2) \times (w_1, w_2)$ with $y_0 = g(x_0, w_0)$ and such that $H(x, g(x, w), w) = 0$ for all (x, w) in $(x_1, x_2) \times (w_1, w_2)$. That is, $w = h_1(x, g(x, w))$.

Now, for (x, y) in $R = (x_1, x_2) \times (y_1, y_2)$, define a mapping $\phi = (\phi_1, \phi_2)$ as follows:

$$\begin{aligned} u &= \phi_1(x, y) = x, \\ v &= \phi_2(x, y) = h_1(x, y). \end{aligned}$$

Define $\psi = (\psi_1, \psi_2)$ on an open set containing $\phi(R)$ by

$$w = \psi_1(u, v) = v,$$
$$z = \psi_2(u, v) = h_2(u, g(u, v)).$$

Then ϕ and ψ are elementary transformations of the second and third types, respectively, such as we have discussed earlier. We need only confirm that $\psi \circ \phi = \boldsymbol{h}$ on $R = (x_1, x_2) \times (y_1, y_2)$. First, for any (x, y) in R,

$$w = \psi_1(u, v) = h_1(x, y).$$

Next, for (x, y) in R, we have $w = h_1(x, y)$ in (w_1, w_2). Equivalently, $y = g(x, w)$. Also, $u = x$ and $v = w$. Thus

$$\begin{aligned} z &= \psi_2(u, v) = h_2(u, g(u, v)) \\ &= h_2(x, g(x, w)) = h_2(x, y). \end{aligned}$$

This proves that, when $D_2h_1(x_0, y_0) \neq 0$, the transformation $\boldsymbol{h}$ can be written $\boldsymbol{h} = \psi \circ \phi$ on R and completes the proof of the theorem. ●

You will notice that Theorem 10.3.1 is a statement that is local in nature. To apply the theorem, we suppose that $\boldsymbol{h} = (h_1, h_2)$ is a continuously differentiable function defined on a connected open set U in $\mathbb{R}^2$ having a Jacobian that does not vanish on U. We let $R = [a_1, b_1] \times [a_2, b_2]$ be a closed, bounded rectangle contained in U. For each point (x, y) in R, we know by the theorem that there is an open rectangle $R_{(x,y)}$ containing (x, y) and two elementary transformations ϕ defined on $R_{(x,y)}$ and ψ defined on $\phi(R_{(x,y)})$ such that $\boldsymbol{h} = \psi \circ \phi$ on $R_{(x,y)}$. Of course, the *form* of this composition may vary from point to point, but this fact is immaterial to our discussion. To the point, the collection $\{R_{(x,y)} : (x, y)$ in $R\}$ is an open cover of the compact set R. Hence, there exists a finite subcover $\{R_{(x_1,y_1)}, R_{(x_2,y_2)}, \ldots, R_{(x_k,y_k)}\}$ of R. Consequently, we can partition R into the union of finitely many nonoverlapping closed rectangles on each of which $\boldsymbol{h}$ can be written as the composition of elementary transformations.

We take a moment to examine the effect of an elementary transformation ϕ on a partition π of a Riemann domain S. We first enclose S in a rectangle $R = [a_1, b_1] \times [a_2, b_2]$ and we will restrict our discussion to an elementary transformation of the form $\phi = (\phi_1, \phi_2)$ defined on an open set U containing R where, for (x, y) in U,

$$u = \phi_1(x, y),$$
$$v = \phi_2(x, y) = y.$$

We assume, of course, that ϕ is one-to-one, continuously differentiable, and that $D_1\phi_1$ does not vanish. Let $S' = \phi(S)$ and $R' = \phi(R)$. (See Fig. 10.17.)

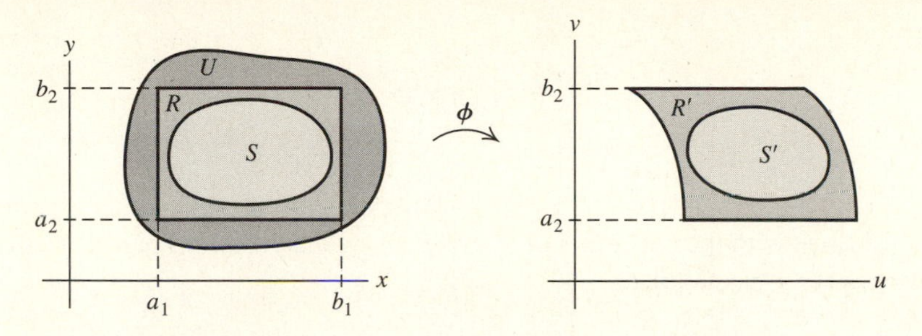

Figure 10.17

Let π be a (rectangular) partition of R and let π' denote the partition of R' induced by ϕ. Specifically, the lines $x = x_i$ and $y = y_j$, which form the boundaries of the cells C_{ij} of the partition π, are transformed into the curves $u = \phi_1(x_i, v)$ and the lines $v_j = y_j$. These curves and lines form the boundaries of the cells C'_{ij} of the (nonrectangular) partition π' as sketched in Fig. 10.18. Recall from our discussion above that the area $\Delta A'_{ij}$ of C'_{ij} is $|D_1\phi_1(c_i, c_j)|\Delta x_i \Delta y_j$, where (c_i, c_j) is a point of C_{ij}.

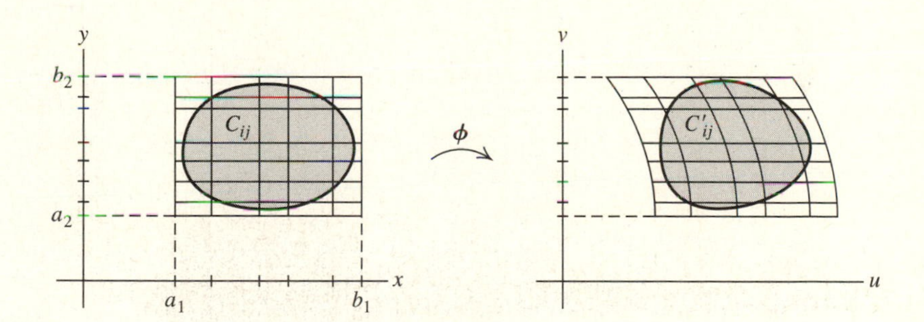

Figure 10.18

It is straightforward to confirm that, if π is a partition of R that refines π_0, then $\pi' = \phi(\pi)$ is a partition of R' that refines $\pi'_0 = \phi(\pi_0)$. Moreover, by the uniform continuity of ϕ_1 on the compact set R, given $\epsilon > 0$, there is a $\delta > 0$ such that, if $[(x - x')^2 + (y - y')^2]^{1/2} < \delta$, then

$$[[\phi_1(x, y) - \phi_1(x', y')]^2 + (y - y')^2]^{1/2} < \epsilon.$$

Thus, if π is a partition of R with $||\pi|| < \delta$, then the gauge of the partition $\pi' = \phi(\pi)$ is less than ϵ.

Finally, notice that, since ϕ maps R one-to-one onto R', the inverse mapping ϕ^{-1} exists. In fact, ϕ^{-1} is also an elementary transformation of the same form as ϕ. In this instance, $\phi^{-1} = (\psi_1, \psi_2)$, where

$$x = \psi_1(u, v),$$
$$y = \psi_2(u, v) = v$$

and

$$(x, y) = \phi^{-1} \circ \phi(x, y) = \phi^{-1}(\phi_1(x, y), y),$$
$$= (\psi_1(\phi_1(x, y), y), y),$$

for all (x, y) in R. Clearly, then, the partitions of R are placed in one-to-one correspondence with the partitions of R' by ϕ, provided we admit partitions that are no longer merely rectangular.

THEOREM 10.3.2 Let f be an integrable function defined on a Riemann domain S'. Let ϕ be a continuously differentiable one-to-one elementary transformation defined on an open set U such that $\phi(U)$ contains S' and such that $J_\phi \neq 0$ on U. Let S be a Riemann domain contained in U that is mapped one-to-one onto S' by ϕ. Then $(f \circ \phi)|J_\phi|$ is integrable on S and

$$\iint_S (f \circ \phi)(x, y)|J_\phi(x, y)|\, dx\, dy = \iint_{S'} f(u, v)\, du\, dv.$$

Proof. We will assume that $\phi = (\phi_1, \phi_2)$ has the form

$$u = \phi_1(x, y),$$
$$v = \phi_2(x, y) = y,$$

and that $D_1\phi_1 \neq 0$ on U, with the understanding that a similar argument will apply to elementary transformations of the other three forms. We assume that we can choose a compact rectangle $R = [a_1, b_2] \times [a_2, b_2]$ such that $S \subseteq R \subseteq U$. Let $R' = \phi(R)$; thus $S' \subseteq R'$. For (u, v) in $R' \cap (S')^c$, we define $f(u, v) = 0$. Let

$$I = \iint_{R'} f(u, v)\, du\, dv = \iint_{S'} f(u, v)\, du\, dv.$$

Fix $\epsilon > 0$. Since $J_\phi = D_1\phi_1$ is assumed to be continuous on U, we know that $|D_1\phi_1|$ is uniformly continuous on the compact set R. Thus there exists a $\delta > 0$ such that, whenever

$$[(x - x')^2 + (y - y')^2]^{1/2} < \delta,$$

it follows that

$$\big||D_1\phi_1(x, y)| - |D_1\phi_1(x', y')|\big| < \frac{\epsilon}{||f||_\infty A(R)}.$$

Choose a sufficiently refined partition π_0' of R' of the form $\pi_0' = \phi(\pi_0)$, where π_0 is a rectangular partition of R, having the following property: If π' is a refinement of π_0' and if the points (u_i, v_j) are chosen arbitrarily in the cells of π', then

$$\left| \sum_i \sum_j f(u_i, v_j)\Delta A'_{ij} - I \right| < \epsilon.$$

Let $\pi_0 = \phi^{-1}(\pi_0')$ denote the corresponding (rectangular) partition of R. Choose any rectangular partition π of R that refines π_0 and has gauge less than δ. Let

$\pi' = \phi(\pi)$. Then π' is a refinement of π_0'. Choose arbitrary points (s_i, t_j) in the cells C_{ij} of π and let $(u_i, v_j) = \phi(s_i, t_j)$. Then

$$\sum_i \sum_j f(u_i, v_j)\Delta A'_{ij} = \sum_i \sum_j (f \circ \phi)(s_i, t_j)|D_1\phi_1(c_i, c_j)|\Delta x_i \Delta y_j,$$

where, for each i and j, (c_i, c_j) is a point of C_{ij}. Finally, inserting and removing

$$\sum_i \sum_j (f \circ \phi)(s_i, t_j)|D_1\phi_1(c_i, c_j)|\Delta x_i \Delta y_j,$$

we calculate

$$\begin{aligned}
&\left|\sum_i \sum_j (f \circ \phi)(s_i, t_j)|D_1\phi_1(s_i, t_j)|\Delta x_i \Delta y_j - I\right| \\
&\quad \le \sum_i \sum_j \left|(f \circ \phi)(s_i, t_j)\right| \left| |D_1\phi_1(s_i, t_j)| - |D_1\phi_1(c_i, c_j)| \right| \Delta x_i \Delta y_j \\
&\quad + \left|\sum_i \sum_j (f \circ \phi)(s_i, t_j)|D_1\phi_1(c_i, c_j)|\Delta x_i \Delta y_j - I\right| \\
&\quad < ||f||_\infty \left[\frac{\epsilon}{(||f||_\infty A(R))}\right] \sum_i \sum_j \Delta x_i \Delta y_j + \epsilon = 2\epsilon,
\end{aligned}$$

since (s_i, t_j) and (c_i, c_j) are in C_{ij} and since $||\pi|| < \delta$. Therefore, $(f \circ \phi)|J_\phi|$ is integrable and

$$\iint_R (f \circ \phi)(x, y)|J_\phi(x, y)|\, dx\, dy = \iint_{R'} f(u, v)\, du\, dv.$$

Since f vanishes off S' and therefore $f \circ \phi$ vanishes off S, this equation is equivalent to

$$\iint_S (f \circ \phi)(x, y)|J_\phi(x, y)|\, dx\, dy = \iint_{S'} f(u, v)\, du\, dv.$$

This proves the theorem. ●

Theorem 10.3.2 leads immediately to the following generalization.

THEOREM 10.3.3 Let f be integrable on a Riemann domain S'' in $\mathbb{R}^2$. Let $\boldsymbol{h}$ be a one-to-one continuously differentiable mapping defined on an open set U in $\mathbb{R}^2$ such that $\boldsymbol{h}(U)$ contains S'' and such that $J_{\boldsymbol{h}} \neq 0$ on U. Let S be a Riemann domain contained in U that is mapped one-to-one onto S'' by $\boldsymbol{h}$. Then $(f \circ \boldsymbol{h})|J_{\boldsymbol{h}}|$ is integrable on S and

$$\iint_{S'} (f \circ \boldsymbol{h})(x, y)|J_{\boldsymbol{h}}(x, y)|\, dx\, dy = \iint_{S''} f(w, z)\, dw\, dz.$$

Proof. First suppose that $\boldsymbol{h}$ is the composition $\psi \circ \phi$ of two one-to-one elementary transformations ϕ on U and ψ on $\phi(U)$. Recall that, for (x, y) in S, $J_{\boldsymbol{h}}(x, y) =$

$J_{\psi}(\phi(x, y))J_{\phi}(x, y)$. Let $S' = \phi(S)$. By Theorem 10.3.2, $(f \circ \psi)|J_{\psi}|$ is integrable on S' and

$$\iint_{S'} (f \circ \psi)(u, v)|J_{\psi}(u, v)|\, du\, dv = \iint_{S''} f(w, z)\, dw\, dz.$$

Apply Theorem 10.3.2 again to deduce that the function

$$[(f \circ \psi) \circ \phi]|J_{\psi} \circ \phi||J_{\phi}| = (f \circ \boldsymbol{h})|J_{\boldsymbol{h}}|$$

is integrable on S and that

$$\iint_{S} (f \circ \boldsymbol{h})(x, y)|J_{\boldsymbol{h}}(x, y)|\, dx\, dy = \iint_{S'} (f \circ \psi)(u, v)|J_{\psi}(u, v)|\, du\, dv$$
$$= \iint_{S''} f(w, z)\, dw\, dz.$$

To complete the proof of the theorem, we assume that we can enclose S in a rectangle R contained in U and partition R into finitely many nonoverlapping closed rectangles $R_1, R_2, \ldots, R_k$ on each of which $\boldsymbol{h}$ can be written as a composition $\psi_j \circ \phi_j$ of one-to-one elementary transformations. Let $R'' = \boldsymbol{h}(R)$ and, for $j = 1, 2, \ldots, k$, let $R_j'' = \boldsymbol{h}(R_j)$. Define f to have value 0 on $R'' \cap (S'')^c$. Then

$$\iint_{R} (f \circ \boldsymbol{h})(x, y)|J_{\boldsymbol{h}}(x, y)|\, dx\, dy = \sum_{j=1}^{k} \iint_{R_j} (f \circ \boldsymbol{h})(x, y)|J_{\boldsymbol{h}}(x, y)|\, dx\, dy$$
$$= \sum_{j=1}^{k} \iint_{R_j''} f(w, z)\, dw\, dz = \iint_{R''} f(w, z)\, dw\, dz.$$

Since $f = 0$ on $R'' \cap (S'')^c$ and, therefore $f \circ \boldsymbol{h} = 0$ on $R \cap S^c$, we have $\iint_S (f \circ \boldsymbol{h})(x, y)|J_{\boldsymbol{h}}(x, y)|\, dx\, dy = \iint_{S''} f(w, z)\, dw\, dz$. This completes the proof of the theorem. ●

EXAMPLE 3 Let $\boldsymbol{h} = (h_1, h_2)$ be the transformation that gives rise to the change of variables from Cartesian to polar coordinates. That is, let

$$x = h_1(r, \theta) = r \cos\theta,$$
$$y = h_2(r, \theta) = r \sin\theta.$$

We know that $\boldsymbol{h}$ is continuously differentiable and that

$$J_{\boldsymbol{h}}(r, \theta) = \det \begin{vmatrix} \cos\theta & -r\sin\theta \\ \sin\theta & r\cos\theta \end{vmatrix} = r.$$

Thus, $J_{\boldsymbol{h}}$ does not vanish on any open set in the $r\theta$-plane that does not contain any points on the θ-axis. Moreover, $\boldsymbol{h}$ is one-to-one on any open set contained in any strip

$$\{(r, \theta) : r > 0, \theta_1 \le \theta < \theta_2\},$$

where $\theta_2 - \theta_1 \le 2\pi$. Thus, given a point (r_0, θ_0) we can find an open rectangle containing (r_0, θ_0) on which we can write $\boldsymbol{h}$ as a composition $\psi \circ \phi$ of two elementary transformations. In this illustration, we assume that $r_0 > 0$ and that $0 < \theta_0 < \pi/2$.

First note that $D_1 h_1(r_0, \theta_0) = \cos\theta_0 \neq 0$. Therefore, in the equation $H(r, \theta, x) = r\cos\theta - x = 0$ we solve for $r = g(\theta, x) = x/\cos\theta$, valid for $r > 0$, $-\pi/2 < \theta < \pi/2$ and $x > 0$. We define the first elementary transformation $\phi = (\phi_1, \phi_2)$ by

$$u = \phi_1(r, \theta) = r\cos\theta,$$
$$v = \phi_2(r, \theta) = \theta.$$

The second elementary transformation $\psi = (\psi_1, \psi_2)$ is defined by

$$x = \psi_1(u, v) = u,$$
$$y = \psi_2(u, v) = h_2(g(v, u), v) = g(v, u)\sin v.$$

Therefore, when we compute $\psi \circ \phi$ we obtain $x = u = r\cos\theta$ and

$$y = \frac{u}{\cos v}\sin v = \frac{r\cos\theta}{\cos\theta}\sin\theta = r\sin\theta.$$

In Fig. 10.19 we sketch these successive mappings and their effects on a rectangle $R = [r_1, r_2] \times [\theta_1, \theta_2]$ containing the point (r_0, θ_0). ●

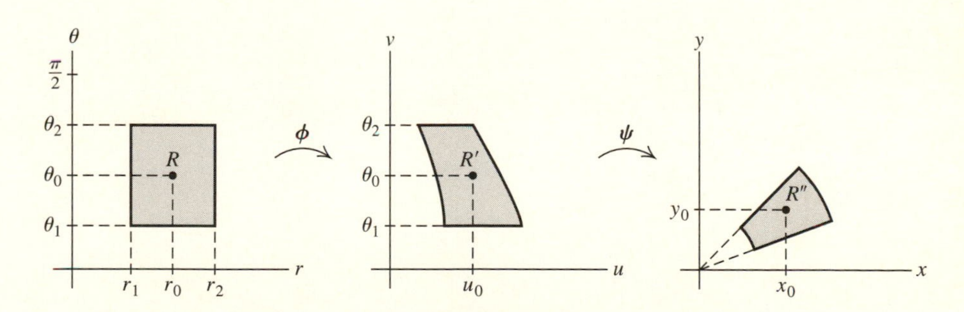

Figure 10.19

EXAMPLE 4 Often the decision to use a particular transformation to evaluate an integral is suggested, not so much by the integrand, but by the region over which the integration is to be performed. For example, suppose we wish to integrate the function $f(x, y) = xy\exp(x^2 + y^2)$ over the region

$$S = \{(x, y) : x \ge 0, y \ge 0, 1 \le x^2 + y^2 \le 4\}.$$

S is sketched in Fig. 10.20. The integral

$$\iint_S xy\, e^{x^2+y^2}\, dx\, dy$$

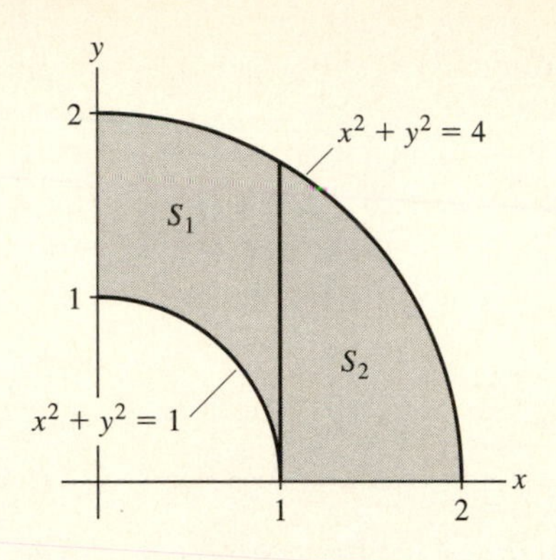

Figure 10.20

can be computed directly by integrating over S_1 and S_2 separately, iterating each, and using methods from Chapter 6. However, the work required can be significantly reduced by observing that, under the transformation $\boldsymbol{h}$ of Example 3 which introduces polar coordinates, S is the image of the rectangle $R = [1, 2] \times [0, \pi/2]$ in the $r\theta$-plane. (See Fig. 10.21.) That is, we use

$$x = h_1(r, \theta) = r\cos\theta,$$
$$y = h_2(r, \theta) = r\sin\theta,$$

to change variables in the integral of f. By Theorem 10.3.3, noting that $J_{\boldsymbol{h}} = r$, we have

$$\iint_S f(x, y)\, dx\, dy = \iint_R (f \circ \boldsymbol{h})(r, \theta)\, r\, dr\, d\theta.$$

This second integral, in terms of the variables r and θ, is easy to iterate and to

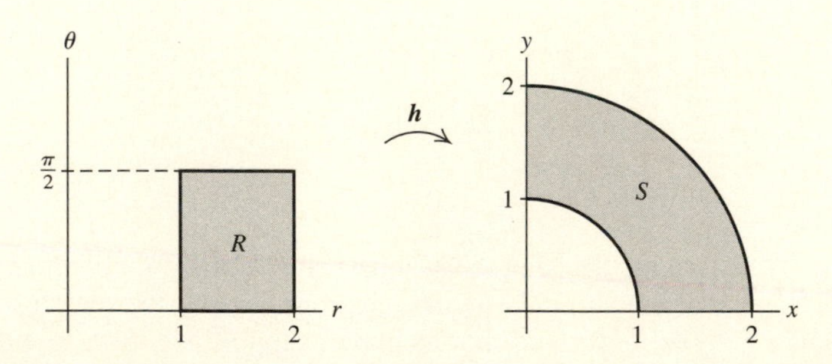

Figure 10.21

compute. We have

$$\iint_S xy\, e^{x^2+y^2}\, dx\, dy = \iint_R (r\cos\theta)(r\sin\theta)\, e^{r^2}\, r dr d\theta$$

$$= \iint_R \sin\theta \cos\theta \, r^3 e^{r^2} \, dr \, d\theta$$

$$= \int_0^{\pi/2} \left[\int_1^2 \sin\theta \cos\theta \, r^3 e^{r^2} \, dr \right] d\theta$$

$$= \left[\int_0^{\pi/2} \sin\theta \cos\theta \, d\theta \right] \left[\int_1^2 r^3 e^{r^2} \, dr \right]$$

$$= \left[\frac{\sin^2\theta}{2} \Big|_0^{\pi/2} \right] \left[\frac{(r^2-1)\, e^{r^2}}{2} \Big|_1^2 \right]$$

$$= \frac{3e^4}{4}.$$

It is worth mentioning that, when changing to polar coordinates, the point $\mathbf{0}$ can often be included in the domain S' of integration, this despite the fact that the Jacobian vanishes there. For example, suppose that

$$S' = \{(x, y) : x^2 + y^2 \leq \rho^2, y \geq 0\}$$

and that f is a bounded, integrable function on S'. Choose any positive ϵ less than ρ. Let $D_\epsilon = \overline{N(0;\epsilon)} \cap S'$ and let $S'_\epsilon = \{(x, y) : \epsilon^2 \leq x^2 + y^2 \leq \rho^2, y \geq 0\}$. Then D_ϵ and S'_ϵ are nonoverlapping Riemann domains and $S' = D_\epsilon \cup S'_\epsilon$. (See Fig. 10.22.) Thus,

$$I = \iint_{S'} f(x, y)\, dx\, dy = \iint_{S'_\epsilon} f(x, y)\, dx\, dy + \iint_{D_\epsilon} f(x, y)\, dx\, dy.$$

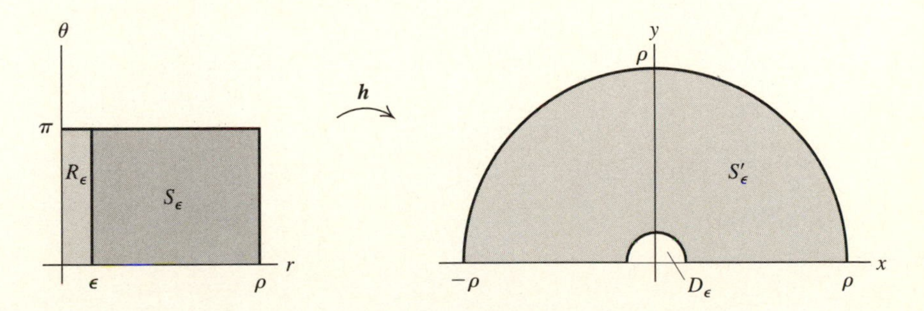

Figure 10.22

The transformation $\boldsymbol{h}$ defined by $x = r\cos\theta$, $y = r\sin\theta$ maps the rectangle $R = \{(r, \theta) : 0 \leq r \leq \rho, 0 \leq \theta \leq \pi\}$ onto S', but fails to be one-to-one along the line segment $r = 0$ where the Jacobian of $\boldsymbol{h}$ vanishes. The thin rectangular strip $R_\epsilon = \{(r, \theta) : 0 \leq r \leq \epsilon, 0 \leq \theta \leq \pi\}$ is mapped onto D_ϵ. The rectangle $S_\epsilon = \{(r, \theta) : \epsilon \leq r \leq \rho, 0 \leq \theta \leq \pi\}$ is mapped one-to-one onto S'_ϵ by $\boldsymbol{h}$.

We can apply Theorem 10.3.3, using $\boldsymbol{h}$, to transform the integral $\iint_{S'_\epsilon} f(x, y)\, dx\, dy$, obtaining $\iint_{S_\epsilon} f(r\cos\theta, r\sin\theta) r\, dr\, d\theta$. This latter integral can be iterated and evaluated, using whatever techniques are suggested by the function f. Thus

$$I = \iint_{S_\epsilon} f(r\cos\theta, r\sin\theta) r\, dr\, d\theta + \iint_{D_\epsilon} f(x, y)\, dx\, dy.$$

Moreover,

$$\left| \iint_{D_\epsilon} f(x, y)\, dx\, dy \right| \le ||f||_\infty A(D_\epsilon) = \frac{||f||_\infty \pi \epsilon^2}{2}.$$

It follows that

$$\left| I - \iint_{S_\epsilon} f(r\cos\theta, r\sin\theta)\, rd\, rd\, \theta \right| \le \frac{||f||_\infty \pi \epsilon^2}{2}$$

and thus that $I = \lim_{\epsilon \to 0^+} \iint_{S_\epsilon} f(r\cos\theta, r\sin\theta) r\, dr\, d\theta$. By this procedure, we excise the set where J_h vanishes and thereby circumvent the technical difficulties that would interfere with the proof of Theorem 10.3.3. A similar maneuver can be applied to some other transformations whose Jacobians vanish. ●

EXAMPLE 5 As an extremely important example of the foregoing technique, we let $f(x, y) = \exp[-(x^2 + y^2)]$ on the domain $S' = \{(x, y) : x \ge 0, y \ge 0, x^2 + y^2 \le \rho^2\}$. Then,

$$I = \iint_{S'} e^{-(x^2+y^2)}\, dx\, dy = \lim_{\epsilon \to 0} \iint_{S_\epsilon} e^{-r^2} r\, dr\, d\theta,$$

where $S_\epsilon = \{(r, \theta) : \epsilon \le r \le \rho, 0 \le 0 \le \pi/2\}$. By iterating the latter integral, we obtain

$$\iint_{S_\epsilon} e^{-r^2} r\, dr\, d\theta = \int_0^{\pi/2} \left[\int_\epsilon^\rho e^{-r^2} r\, dr \right] d\theta = \frac{\pi}{2} \int_\epsilon^\rho e^{-r^2} r\, dr.$$

Letting $u = r^2$, we obtain

$$\int_\epsilon^\rho e^{-r^2} r\, dr = \tfrac{1}{2}[e^{-\epsilon^2} - e^{-\rho^2}].$$

Thus

$$I = \lim_{\epsilon \to 0^+} \frac{\pi}{4}[e^{-\epsilon^2} - e^{-\rho^2}] = \frac{\pi}{4}[1 - e^{-\rho^2}]. \quad ●$$

EXAMPLE 6 Suppose that S is the Riemann domain in the first quadrant bounded by the curves $u^2 - v^2 = 1$, $u^2 - v^2 = 4$, $uv = 1$, and $uv = 2$ as sketched in Fig. 10.23. Rather than partition S into three nonoverlapping simple domains

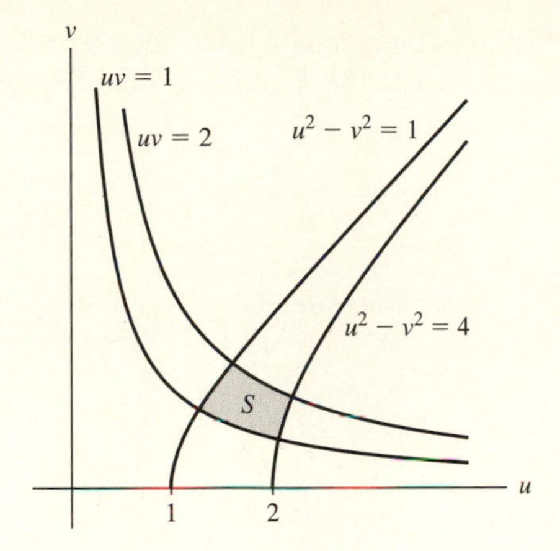

Figure 10.23

and iterate an integral over each of these, we seek a single transformation that will simplify the domain of integration. The boundary curves of S themselves suggest a solution. We define $\boldsymbol{g} = (g_1, g_2)$ by

$$x = g_1(u, v) = u^2 - v^2,$$
$$y = g_2(u, v) = 2uv.$$

Clearly, $\boldsymbol{g}$ is a continuously differentiable transformation. For constant $c \neq 0$, $\boldsymbol{g}$ maps the hyperbola $u^2 - v^2 = c$ into the vertical line $x = c$ and the hyperbola $uv = c$ into the horizontal line $y = 2c$. Consequently $\boldsymbol{g}$ maps S into the rectangle $R = [1, 4] \times [2, 4]$ in the xy-plane. As you can confirm, $\boldsymbol{g}$ maps S one-to-one onto R. The Jacobian of $\boldsymbol{g}$,

$$J_{\boldsymbol{g}}(u, v) = \det \begin{vmatrix} 2u & -2v \\ 2v & 2u \end{vmatrix} = 4(u^2 + v^2),$$

vanishes nowhere in the uv-plane except at the point $\mathbf{0}$. Note that $\boldsymbol{g}$ is not the transformation we want for the change of variables in an integral. We want $\boldsymbol{g}^{-1} = \boldsymbol{h} = (h_1, h_2)$ where

$$u = h_1(x, y) = \left(\frac{x + \sqrt{x^2 + y^2}}{2} \right)^{1/2},$$

$$v = h_2(x, y) = \left(\frac{-x + \sqrt{x^2 + y^2}}{2} \right)^{1/2}.$$

The transformation $\boldsymbol{h}$ is also continuously differentiable and maps the rectangle R one-to-one onto the original Riemann domain S. (See Fig. 10.24.) Since $\boldsymbol{h} = \boldsymbol{g}^{-1}$, the Jacobian of $\boldsymbol{h}$ is

$$J_{\boldsymbol{h}}(x, y) = \frac{1}{J_{\boldsymbol{g}}(\boldsymbol{h}(x, y))}$$

$$= \frac{1}{4[h_1(x, y)^2 + h_2(x, y)^2]}$$

$$= \frac{1}{4\sqrt{x^2 + y^2}}.$$

Thus, $J_{\boldsymbol{h}}$ does not vanish on the rectangle R.

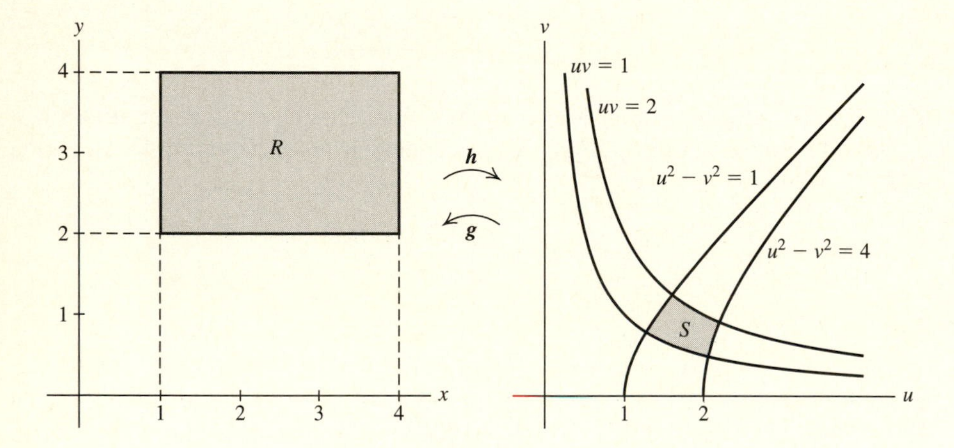

Figure 10.24

Suppose, for example, that we want to compute the integral of $f(u, v) = uv(u^2 + v^2)$ over the domain S. Using the transformation $\boldsymbol{h}$, note first that $f \circ \boldsymbol{h}(x, y) = y\sqrt{x^2 + y^2}/2$. Therefore,

$$\iint_S uv(u^2 + v^2)\, du\, dv = \frac{1}{8} \iint_R y\sqrt{x^2 + y^2} \cdot \frac{1}{\sqrt{x^2 + y^2}}\, dx\, dy$$

$$= \frac{1}{8} \iint_R y\, dx\, dy = \frac{1}{8} \int_1^4 \left[\int_2^4 y\, dy \right] dx$$

$$= \frac{1}{16} \int_1^4 \left[y^2 \Big|_2^4 \right] dx = \frac{1}{16}[16 - 4][4 - 1]$$

$$= \frac{9}{4}. \quad \bullet$$

10.4 MULTIPLE INTEGRALS IN $\mathbb{R}^3$

Our discussion for double integrals in $\mathbb{R}^2$ generalizes in a straightforward way to integrals of functions defined on higher-dimensional Riemann domains. Only the notation and therefore the details of each proof become somewhat more cumbersome. Having treated double integrals in detail, we will merely sketch the theory of triple integrals, leaving a more thorough treatment of multiple integration for those texts in which it plays a more important role. In passing, we mention the usual caveat: The proper arena in which to study integration is Lebesgue's theory where the appropriate tools are available to deal with the technical details.

Let R be a three-dimensional hyper-rectangle in $\mathbb{R}^3$. That is, $R = [a_1, b_1] \times [a_2, b_2] \times [a_3, b_3]$. For $j = 1, 2, 3$, let π_j be a partition of $[a_j, b_j]$. That is,

$$\begin{aligned}
\pi_1 : &\qquad a_1 = x_0 < x_1 < x_2 < \cdots < x_p = b_1, \\
\pi_2 : &\qquad a_2 = y_0 < y_1 < y_2 < \cdots < y_q = b_2, \\
\pi_3 : &\qquad a_3 = z_0 < z_1 < z_2 < \cdots < z_r = b_3.
\end{aligned}$$

As the indices i, j, and k vary, the planes $x = x_i$, $y = y_j$, and $z = z_k$ form the boundaries of three-dimensional hyper-rectangles C_{ijk} that overlap only in their shared boundaries. The union of all the hyper-rectangles C_{ijk} is R. The partition π, denoted $\pi = \pi_1 \times \pi_2 \times \pi_3$, is the collection of these sets C_{ijk}, called the *cells* of π. The *volume* of the cell C_{ijk} is

$$\Delta V_{ijk} = \Delta x_i \Delta y_j \Delta z_k = (x_i - x_{i-1})(y_j - y_{j-1})(z_k - z_{k-1}).$$

The *gauge* $||\pi||$ of π is the smallest diameter of the cells C_{ijk}. A *refinement* of π is a partition $\pi' = \pi_1' \times \pi_2' \times \pi_3'$, where π_j' is a refinement of π_j for each index j.

Let f be a bounded, real-valued function defined on R and let $\pi = \pi_1 \times \pi_2 \times \pi_3$ be any partition of R. For each triple (i, j, k) of the indices, choose a point (s_i, t_j, u_k) in C_{ijk}.

DEFINITION 10.4.1 A *Riemann sum* for f is a sum of the form

$$S(f, \pi) = \sum_{i=1}^{p} \sum_{j=1}^{q} \sum_{k=1}^{r} f(s_i, t_j, u_k) \Delta V_{ijk}. \quad \bullet$$

DEFINITION 10.4.2 We say that f is (Riemann) integrable on R if there exists a number I with the following property: For every $\epsilon > 0$ there exists a partition π_0 of R such that, for every refinement π of π_0 and for every choice of points (s_i, t_j, u_k) in the cells C_{ijk} of π, we have

$$|S(f, \pi) - I| < \epsilon.$$

The number I is said to be the (*Riemann*) *integral* of f over R, and we write

$$I = \iiint_R f(x, y, z)\, dV. \quad \bullet$$

We refer to an integral of the form defined by Definition 10.4.2 as a *triple integral*. Such integrals have the expected properties listed in our next two theorems, the proofs of which are left for you.

THEOREM 10.4.1 Let f, f_1, and f_2 be integrable functions on a hyper-rectangle R in $\mathbb{R}^3$ and let c_1 and c_2 be real numbers.

i) We have

$$\iiint_R [c_1 f_1(x, y, z) + c_2 f_2(x, y, z)]\, dV$$
$$= c_1 \iiint_R f_1(x, y, z)\, dV + c_2 \iiint_R f_2(x, y, z)\, dV.$$

ii) If $f_1 \leq f_2$ on R, then

$$\iiint_R f_1(x, y, z)\, dV \leq \iiint_R f_2(x, y, z)\, dV.$$

iii) If

$$m = \inf\{f(x, y, z) : (x, y, z) \text{ in } R\}$$

and

$$M = \sup\{f(x, y, z) : (x, y, z) \text{ in } R\},$$

then

$$mV(R) \leq \iiint_R f(x, y, z)\, dV \leq MV(R),$$

where $V(R) = (b_1 - a_1)(b_2 - a_2)(b_3 - a_3)$ denotes the volume of R. ●

The discussion leading to Riemann's condition generalizes easily. Again, for any partition π of R, we let m_{ijk} and M_{ijk} denote the infimum and the supremum, respectively, of the bounded function f on the cell C_{ijk} of π. We form

$$L(f, \pi) = \sum_{i=1}^{p} \sum_{j=1}^{q} \sum_{k=1}^{r} m_{ijk} \Delta V_{ijk}$$

and

$$U(f, \pi) = \sum_{i=1}^{p} \sum_{j=1}^{q} \sum_{k=1}^{r} M_{ijk} \Delta V_{ijk},$$

the *lower and upper Riemann sums*, respectively, for f determined by the partition π. A somewhat tedious but straightforward proof, modeled on that of Theorem 10.1.2, will confirm that, for any two partitions π and π' of R, we have $L(f, \pi) \leq U(f, \pi')$. Therefore, the numbers

$$L(f) = \sup\{L(f, \pi) : \pi \text{ a partition of } R\}$$

and

$$U(f) = \inf\{U(f, \pi) : \pi \text{ a partition of } R\}$$

exist and satisfy $L(f) \leq U(f)$ for every f. We omit the proofs of these assertions.

DEFINITION 10.4.3 Riemann's Condition Let f be a bounded, real-valued function on a hyper-rectangle R in $\mathbb{R}^3$. *Riemann's condition* is said to hold for f on R if, for any $\epsilon > 0$, there exists a partition π_0 of R such that for any refinement π of π_0, we have $U(f, \pi) - L(f, \pi) < \epsilon$. ●

THEOREM 10.4.2 Let f be a bounded, real-valued function on a rectangle R in $\mathbb{R}^3$. The following statements are equivalent:

i) The function f is integrable on R.

ii) Riemann's condition holds for f on R.

iii) $L(f) = U(f)$. ●

Again, we forego the proof, being content to refer you to that of Theorem 6.2.4. As in the case of double integrals, Theorem 10.4.2 leads easily to the conclusion that any function that is continuous on R is integrable there; the proof is identical to that of Theorem 10.1.4.

THEOREM 10.4.3 Assume that f is a continuous real-valued function on the hyper-rectangle $R = [a_1, b_1] \times [a_2, b_2] \times [a_3, b_3]$ in $\mathbb{R}^3$. Then f is integrable on R. ●

As we now know to expect, integration is an averaging operation. That is the content of the following theorem, the proof of which mimics that of Theorem 10.1.5.

THEOREM 10.4.4 The Mean Value Theorem for Integrals Let f be a bounded, integrable function on a hyper-rectangle R in $\mathbb{R}^3$.

i) There exists a K between

$$m = \inf\{f(x, y, z) : (x, y, z) \text{ in } R\}$$

and

$$M = \sup\{f(x, y, z) : (x, y, z) \text{ in } R\}$$

such that

$$\iiint_R f(x, y, z)\, dV = K\, V(R),$$

where $V(R)$ denotes the volume of R.

ii) If f is continuous on R, then there exists a point (x_0, y_0, z_0) in R such that

$$\iiint_R f(x, y, z)\, dV = f(x_0, y_0, z_0) V(R). \quad \bullet$$

The value of f at the point (x_0, y_0, z_0) is the *average value* of f on R.

To evaluate a triple integral, we rely on the technique of iteration. If the region over which integration is performed is a hyper-rectangle, the procedure is particularly straightforward.

THEOREM 10.4.5 Let f be a bounded, integrable function defined on a hyper-rectangle $R = [a_1, b_1] \times [a_2, b_2] \times [a_3, b_3]$ in $\mathbb{R}^3$. Suppose that, for each (x, y) in $[a_1, b_1] \times [a_2, b_2]$, the function $f(x, y, z)$ is an integrable function of z on $[a_3, b_3]$. Define

$$g(x, y) = \int_{a_3}^{b_3} f(x, y, z)\, dz.$$

Assume, in turn, that, for each x in $[a_1, b_1]$, $g(x, y)$ is an integrable function of y on $[a_2, b_2]$. Define

$$h(x) = \int_{a_2}^{b_2} g(x, y)\, dy.$$

Finally, assume that h is an integrable function on $[a_1, b_1]$. Then

$$\begin{aligned}\int_{a_1}^{b_1} h(x)\, dx &= \int_{a_1}^{b_1} \left[\int_{a_2}^{b_2} g(x, y)\, dy\right] dx \\ &= \int_{a_1}^{b_1} \left[\int_{a_2}^{b_2} \left[\int_{a_3}^{b_3} f(x, y, z)\, dz\right] dy\right] dx \\ &= \iiint_R f(x, y, z)\, dV. \quad \bullet\end{aligned}$$

The proof of this assertion is modeled on that of Theorem 10.2.1 and is omitted. There are, of course, the analogous theorems, 3! in number, in which the iterations occur in a permuted order. In particular, if f is continuous on the hyper-rectangle R, then we have the following corollary.

COROLLARY 10.4.6 If f is a continuous function on the hyper-rectangle $R = [a_1, b_1] \times [a_2, b_2] \times [a_3, b_3]$, then, for any of the 3! permutations of the subscripts (i, j, k),

$$\int_{a_k}^{b_k} \left[\int_{a_j}^{b_j} \left[\int_{a_i}^{b_i} f(x_1, x_2, x_3)\, dx_i\right] dx_j\right] dx_k = \iiint_R f(x_1, x_2, x_3)\, dV. \quad \bullet$$

EXAMPLE 7 The function $f(x, y, z) = \cos(x + y + z)$ is continuous on the hyper-rectangle

$$R = \left[0, \frac{\pi}{2}\right] \times \left[0, \frac{\pi}{2}\right] \times \left[0, \frac{\pi}{2}\right].$$

Therefore, we can iterate the integral of f in whichever order may be most convenient to obtain:

$$\begin{aligned}\iiint_R f(x, y, z)\, dV &= \int_0^{\pi/2} \left[\int_0^{\pi/2} \left[\int_0^{\pi/2} \cos(x + y + z) dz\right] dy\right] dx \\ &= \int_0^{\pi/2} \left[\int_0^{\pi/2} \left[\sin(x + y + z)\Big|_{z=0}^{\pi/2}\right] dy\right] dx\end{aligned}$$

$$= \int_0^{\pi/2} \left[\int_0^{\pi/2} \left[\sin\left(x + y + \frac{\pi}{2}\right) - \sin(x+y) \right] dy \right] dx$$

$$= \int_0^{\pi/2} \left[-\cos\left(x + y + \frac{\pi}{2}\right) + \cos(x+y) \Big|_{y=0}^{\pi/2} \right] dx$$

$$= \int_0^{\pi/2} \left[-\cos(x+\pi) + 2\cos\left(x + \frac{\pi}{2}\right) - \cos x \right] dx$$

$$= -\sin(x+\pi) + 2\sin\left(x + \frac{\pi}{2}\right) - \sin x \Big|_{x=0}^{\pi/2}$$

$$= -2. \quad \bullet$$

10.4.1 Riemann Domains

As in Section 10.1, we need the following ideas in order to extend the definition of the integral to sets in $\mathbb{R}^3$ more general than three-dimensional hyper-rectangles.

DEFINITION 10.4.4

a) Two sets S_1 and S_2 in $\mathbb{R}^3$ are said to be *non-overlapping* if $(S_1 \cap S_2)^0 = \emptyset$.

b) A finite collection $\mathcal{R} = \{R_1, R_2, \ldots, R_m\}$ of three-dimensional hyper-rectangles of the form

$$R_k = [a_1^{(k)}, b_1^{(k)}] \times [a_2^{(k)}, b_2^{(k)}] \times [a_3^{(k)}, b_3^{(k)}]$$

is a *rectangular cover* of a set S in $\mathbb{R}^3$ if $S \subseteq (\cup_{k=1}^m R_k)^0$.

c) Let V_k denote the volume of R_k and let $|\mathcal{R}| = \sum_{k=1}^m V_k$. The outer content of S is

$$\overline{c}(S) = \inf\{|\mathcal{R}| : \mathcal{R} \text{ a rectangular cover of } S\}. \quad \bullet$$

The elementary properties of outer content are described in the next theorem; its proof is similar to that of Theorem 10.1.6 and is left for you.

THEOREM 10.4.7

i) If S is a finite set in $\mathbb{R}^3$, then $\overline{c}(S) = 0$.

ii) If S_1 and S_2 are two nonempty, bounded sets in $\mathbb{R}^3$ such that $S_1 \subseteq S_2$, then $\overline{c}(S_1) \le \overline{c}(S_2)$.

iii) If S_1 and S_2 are two nonempty, bounded sets in $\mathbb{R}^3$, then $\overline{c}(S_1 \cup S_2) \le \overline{c}(S_1) + \overline{c}(S_2)$. $\bullet$

The proof of Theorem 10.1.7 lifts verbatim to prove the next theorem.

THEOREM 10.4.8 Let f be a bounded, real-valued function defined on a hyper-rectangle R in $\mathbb{R}^3$. Let D denote the set of points of R where f is discontinuous. If $\overline{c}(D) = 0$, then f is integrable on R. $\bullet$

To identify the class of sets in $\mathbb{R}^3$ over which integration can be performed, we discuss briefly the parametric representation of surfaces and the definition of the area of a surface. A more detailed discussion of surface area lies outside the primary focus of this text.

Let U be a connected Riemann domain in $\mathbb{R}^2$. For (u, v) in U, let

$$
\begin{aligned}
x &= \phi_1(u, v)\\
y &= \phi_2(u, v)\\
z &= \phi_3(u, v),
\end{aligned}
$$

where ϕ_1, ϕ_2, and ϕ_3 are continuously differentiable functions defined on U such that the three determinants

$$J_{12}(u, v) = \det\begin{vmatrix} D_1\phi_1(u, v) & D_2\phi_1(u, v)\\ D_1\phi_2(u, v) & D_2\phi_2(u, v)\end{vmatrix},$$

$$J_{23}(u, v) = \det\begin{vmatrix} D_1\phi_2(u, v) & D_2\phi_2(u, v)\\ D_1\phi_3(u, v) & D_2\phi_3(u, v)\end{vmatrix},$$

and

$$J_{31}(u, v) = \det\begin{vmatrix} D_1\phi_3(u, v) & D_2\phi_3(u, v)\\ D_1\phi_1(u, v) & D_2\phi_1(u, v)\end{vmatrix}$$

do not all vanish simultaneously at any point of U. Note that J_{12}, J_{23}, and J_{31} are the Jacobians of the functions (ϕ_1, ϕ_2), (ϕ_2, ϕ_3), and (ϕ_3, ϕ_1), respectively. Let

$$\mathcal{S} = \{(\phi_1(u, v), \phi_2(u, v), \phi_3(u, v)) : (u, v) \text{ in } U\}.$$

Then $\mathcal{S}$ is a surface in $\mathbb{R}^3$ that is represented parametrically by the function $\phi = (\phi_1, \phi_2, \phi_3)$. (See Fig. 10.25.)

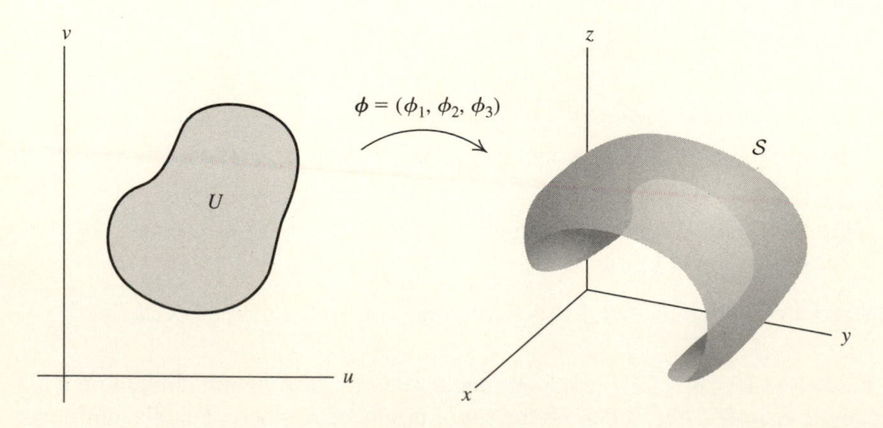

Figure 10.25

DEFINITION 10.4.5 The area A of the surface represented parametrically by $\phi = (\phi_1, \phi_2, \phi_3)$ is defined to be

$$A = \iint_U [J_{12}(u, v)^2 + J_{23}(u, v)^2 + J_{31}(u, v)^2]^{1/2}\, du\, dv. \quad \bullet$$

In passing, if the surface can be described by a continuously differentiable function $z = g(x, y)$ with (x, y) in a Riemann domain U, then we can parametrize the surface by letting $\phi_1(x, y) = x$, $\phi_2(x, y) = y$ and $\phi_3(x, y) = g(x, y)$. With these definitions, we have

$$\begin{aligned} J_{12}(x, y) &= 1, \\ J_{23}(x, y) &= -D_1 g(x, y), \\ J_{31}(x, y) &= -D_2 g(x, y). \end{aligned}$$

The requirement that the determinants J_{12}, J_{23}, and J_{31} not all vanish at any point (x, y) in U is met if $\nabla \mathbf{g} \neq \mathbf{0}$ in U. Moreover, the area of the surface is given by the integral

$$A = \iint_U [1 + D_1 g(x, y)^2 + D_2 g(x, y)^2]^{1/2}\, dx\, dy.$$

EXAMPLE 8 To find the surface area of a sphere with radius R, we can let $z = g(x, y) = [R^2 - x^2 - y^2]^{1/2}$ for (x, y) in the set $U = \{(x, y) : x^2 + y^2 \leq R^2\}$ and double the resulting surface area. It follows that

$$1 + D_1 g(x, y)^2 + D_2 g(x, y)^2 = \frac{R^2}{R^2 - x^2 - y^2}.$$

Thus the surface area of the sphere is given by

$$A = 2R \iint_U \frac{1}{(R^2 - x^2 - y^2)^{1/2}}\, dx\, dy.$$

Shifting to polar coordinates and iterating this integral yields

$$A = 2R \int_0^{2\pi} \left[\int_0^R (R^2 - r^2)^{-1/2} r\, dr \right] d\theta = -4\pi R (R^2 - r^2)^{1/2} \Big|_0^R = 4\pi R^2.$$

We can also compute this surface area by representing the sphere parametrically in spherical coordinates. Let

$$\begin{aligned} x &= \psi_1(\theta, \phi) = R \cos\theta \sin\phi, \\ y &= \psi_2(\theta, \phi) = R \sin\theta \sin\phi, \\ z &= \psi_3(\theta, \phi) = R \cos\phi, \end{aligned}$$

for (θ, ϕ) in the set $U_1 = \{(\theta, \phi) : 0 \leq \theta \leq 2\pi, 0 \leq \phi \leq \pi\}$. With this parametrization we have

$$J_{12}(\theta, \phi) = -R^2 \sin\phi \cos\phi,$$
$$J_{23}(\theta, \phi) = -R^2 \cos\theta \sin^2\phi,$$
$$J_{31}(\theta, \phi) = -R^2 \sin\theta \sin^2\phi.$$

Therefore,

$$\begin{aligned} &J_{12}(\theta, \phi)^2 + J_{23}(\theta, \phi)^2 + J_{31}(\theta, \phi)^2 \\ &= R^4(\sin^2\phi\cos^2\phi + \cos^2\theta\sin^4\phi + \sin^2\theta\sin^4\phi) \\ &= R^4 \sin^2\phi. \end{aligned}$$

Although J_{12}, J_{23}, and J_{31} all vanish when $\phi = 0$ or π, we can circumvent this difficulty as in Example 5 by excising such points. Thus the surface area of the sphere is given by

$$A = \lim_{\epsilon \to 0^+} R^2 \int_0^{2\pi} \left[\int_\epsilon^{\pi-\epsilon} \sin\phi \, d\phi \right] d\theta = \lim_{\epsilon \to 0^+} 2\pi R^2 (-\cos\phi) \Big|_\epsilon^{\pi-\epsilon} = 4\pi R^2,$$

in agreement with our previous computation. ●

EXAMPLE 9 A torus is a doughnut-shaped solid in $\mathbb{R}^3$. The surface of a torus can be described parametrically as follows. [See part (d) of Exercise 3.103.] Fix constants r and R with $0 < r < R$. For (θ, ϕ) in $[0, 2\pi) \times [0, 2\pi)$, define

$$x = \psi_1(\theta, \phi) = (R + r\cos\phi)\cos\theta,$$
$$y = \psi_2(\theta, \phi) = (R + r\cos\phi)\sin\theta,$$
$$z = \psi_3(\theta, \phi) = r\sin\phi.$$

(See Fig. 10.26.) As you can compute,

$$J_{12}(\theta, \phi) = r(R + r\cos\phi)\sin\phi,$$
$$J_{23}(\theta, \phi) = r(R + r\cos\phi)\cos\theta\cos\phi,$$

and

$$J_{31}(\theta, \phi) = r(R + r\cos\phi)\sin\theta\cos\phi.$$

Therefore,

$$J_{12}(\theta, \phi)^2 + J_{23}(\theta, \phi)^2 + J_{31}(\theta, \phi)^2 = r^2(R + r\cos\phi)^2.$$

Consequently, the surface area of the torus is given by

$$\begin{aligned} A &= \iint_U r(R + r\cos\phi)\, d\theta\, d\phi = \int_0^{2\pi} \left[\int_0^{2\pi} r(R + r\cos\phi)\, d\theta \right] d\phi \\ &= r \left[\theta \Big|_0^{2\pi} \right] \left[R\phi + r\sin\phi \Big|_0^{2\pi} \right] = 4\pi^2 Rr. \end{aligned}$$ ●

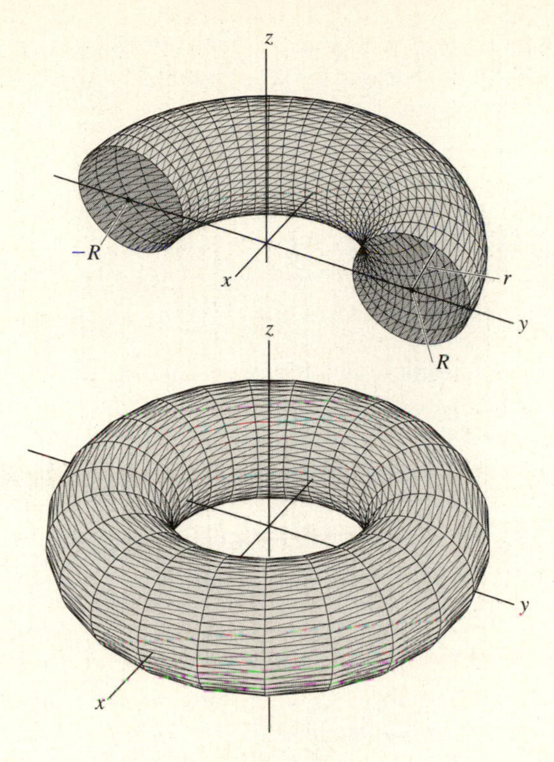

Figure 10.26

Among the exercises you will find surfaces whose areas you can compute. Here we return to the role surfaces play in forming the boundaries of Riemann domains in $\mathbb{R}^3$ and in the iteration of triple integrals. Our interest centers on those surfaces having finite surface area. The essential ideas for the proof of the following theorem are contained in that of Theorem 10.1.8. Here, the surface is to be enclosed in the union of small 3-cubes, the sum of whose volumes is arbitrarily small. We forego the proof.

THEOREM 10.4.9 Let $\mathcal{S}$ be a surface in $\mathbb{R}^3$, represented parametrically by $\phi = (\phi_1, \phi_2, \phi_3)$. Assume that the surface area of $\mathcal{S}$ is finite. Then $\overline{c}(\mathcal{S}) = 0$. ●

DEFINITION 10.4.6 Let S be a bounded open set in $\mathbb{R}^3$. Suppose that the boundary of S consists of a finite number of surfaces each of which can be represented parametrically as above and each with a finite surface area. Then a set consisting of S together with all or part of its boundary is called a *Riemann domain* in $\mathbb{R}^3$. ●

The role of Riemann domains here is analogous to that of such domains in our treatment of double integrals in Section 10.1.

THEOREM 10.4.10 Let f be a bounded, real-valued function defined on a Riemann domain S in $\mathbb{R}^3$. Let D denote the set of points in S where f is discontinuous. If $\overline{c}(D) = 0$, then f is integrable on S.

Proof. We enclose S in a bounded hyper-rectangle R and extend f to have value 0 on $R \cap S^c$. The set of points where the extended function is discontinuous consists of D and all or part of the boundary surfaces of S. This set has zero outer content, so f is integrable on R. It follows that $\iiint_S f(x, y, z)\, dV = \iiint_R f(x, y, z)\, dV$ exists. ●

To iterate a triple integral over a Riemann domain, we first consider domains of a particularly simple type. Let x_i, x_j, x_k be any permutation of the coordinates x_1, x_2, x_3. Assume that S is a Riemann domain such that the perpendicular projection of S onto the $x_i x_j$-plane results in a simple Riemann domain T in $\mathbb{R}^2$. (Refer to Figs. 10.27 and 10.28 where $x_i = x_1, x_j = x_2$, and $x_k = x_3$.) Assume that, as the point (x_i, x_j) varies over T, the set S is bounded between two surfaces $\mathcal{S}_1$ and $\mathcal{S}_2$ represented by $x_k = g_1(x_i, x_j))$ and $x_k = g_2(x_i, x_j)$, respectively, with $g_1 \le g_2$. Assume further that the Riemann domain T is bounded by the straight lines $x_i = a_i$ and $x_i = b_i$ with $a_i < b_i$ and by the two curves $\mathcal{C}_1$ and $\mathcal{C}_2$ represented by the functions $x_j = h_1(x_i)$ and $x_j = h_2(x_i)$, respectively; it is understood that $h_1(x_i) \le h_2(x_i)$ for x_i in $[a_i, b_i]$. A Riemann domain meeting these requirements is called a *simple* Riemann domain in $\mathbb{R}^3$. Assume that, for each (x_i, x_j) in T, the function $f(x_1, x_2, x_3)$ is an integrable function of x_k along the line segment with endpoints $g_1(x_i, x_j)$ and $g_2(x_i, x_j)$. Define

$$F(x_i, x_j) = \int_{g_1(x_i,x_j)}^{g_2(x_i,x_j)} f(x_1, x_2, x_3)\, dx_k.$$

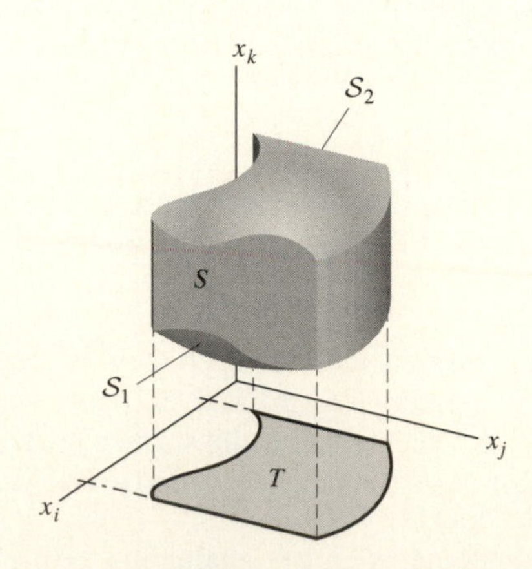

Figure 10.27

Assume further that the double integral of F over T can be iterated to yield

$$\iint_T F(x_i, x_j)\,dA = \int_{a_i}^{b_i} \left[\int_{h_1(x_i)}^{h_2(x_i)} F(x_i, x_j)\,dx_j \right] dx_i.$$

With this notation and these assumptions we have the following theorem.

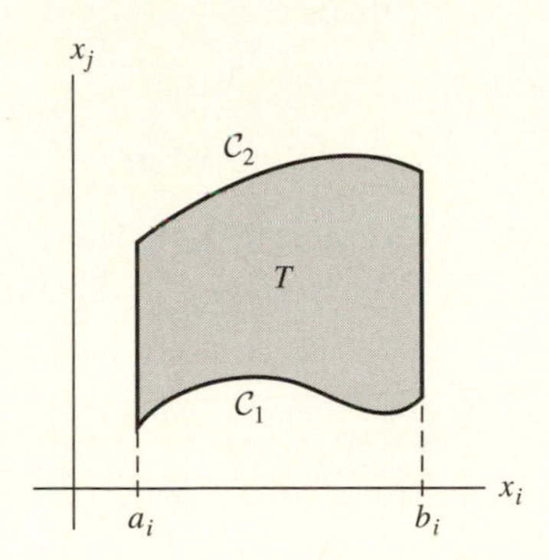

Figure 10.28

THEOREM 10.4.11 Let S be a simple Riemann domain bounded by surfaces S_1 and S_2 represented by $x_k = g_1(x_i, x_j)$ and $x_k = g_2(x_i, x_j)$, respectively, with $g_1 \le g_2$. Let T denote the perpendicular projection of S onto the $x_i x_j$-plane. Assume that T is bounded by curves C_1 and C_2 represented by $x_j = h_1(x_i)$ and $x_j = h_2(x_i)$ respectively, with $h_1 \le h_2$ for x_i in $[a_i, b_i]$. Assume that f is a real-valued function defined on S that satisfies the foregoing integrability assumptions. Then

$$\iiint_S f(x_1, x_2, x_3)\,dV = \iint_T F(x_i, x_j)\,dA = \int_{a_i}^{b_i} \left[\int_{h_1(x_i)}^{h_2(x_i)} F(x_i, x_j)\,dx_j \right] dx_i$$

$$= \int_{a_i}^{b_i} \left[\int_{h_1(x_i)}^{h_2(x_i)} \left[\int_{g_1(x_i,x_j)}^{g_2(x_i,x_j)} f(x_1, x_2, x_3)\,dx_k \right] dx_j \right] dx_i. \quad \bullet$$

Notice that, in order to integrate f over a simple Riemann domain S in $\mathbb{R}^3$, we first integrate from one boundary surface of S to another, then project S onto the simple Riemann domain T in the plane determined by the remaining two variables. We then integrate from one to another boundary curve of T. Finally, we project T onto the remaining axis and integrate from one to another boundary point.

COROLLARY 10.4.12 Let S be a simple Riemann domain. Let f be continuous on S. Then f is integrable on S and, in the notation of Theorem 10.4.10,

$$\iiint_S f(x_1, x_2, x_3)\,dV = \int_{a_i}^{b_i} \left[\int_{h_1(x_i)}^{h_2(x_i)} \left[\int_{g_1(x_i,x_j)}^{g_2(x_i,x_j)} f(x_1, x_2, x_3)\,dx_k \right] dx_j \right] dx_i. \quad \bullet$$

EXAMPLE 10 Fix r, R_1, and R_2 with $0 < r < R_1 < R_2$. Let S be the closed, simple Riemann domain consisting of all points (x, y, z) sketched in Fig. 10.29, where $x^2 + y^2 \leq r^2$ and

$$0 < \sqrt{R_1^2 - x^2 - y^2} \leq z \leq \sqrt{R_2^2 - x^2 - y^2}.$$

The set S is bounded above by a portion of the sphere

$$x^2 + y^2 + z^2 = R_2^2$$

and is bounded below by a portion of the sphere

$$x^2 + y^2 + z^2 = R_1^2.$$

The perpendicular projection of S onto the xy-plane consists of the points in the closed circular disk $T = \{(x, y) : x^2 + y^2 \leq r^2\}$. Then, for any continuous, real-valued function f on S,

$$\iiint_S f(x, y, z)\, dV = \int_{-r}^{r} \left[\int_{\sqrt{r^2-x^2}}^{\sqrt{r^2-x^2}} \left[\int_{\sqrt{R_1^2-x^2-y^2}}^{\sqrt{R_2^2-x^2-y^2}} f(x, y, z)\, dz \right] dy \right] dx.$$

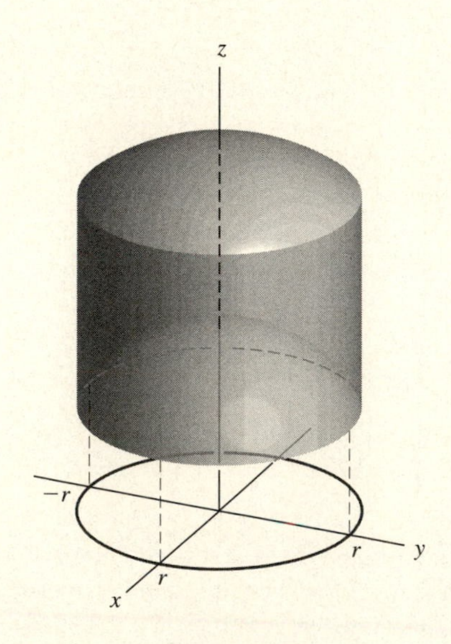

Figure 10.29

It is clear from even this simple example that difficulties in evaluating a triple integral arise not only from the complexity of the integrand f but also from the form of the surfaces that bound S and of the curves that bound its projection T. Often the forms of the boundary surfaces of S and of the boundary curves of T suggest a change of variables to simplify the process of iterating a triple integral. We will turn to that question in a moment. ●

If a Riemann domain S in $\mathbb{R}^3$ is the non-overlapping union of a finite number of simple Riemann domains $S_1, S_2, \ldots, S_k$ and if f is an integrable function on S, then we can write

$$\iiint_S f(x, y, z)\, dV = \sum_{j=1}^{k} \iiint_{S_j} f(x, y, z)\, dV.$$

To evaluate the original integral, apply Theorem 10.4.11 to each integral over each simple Riemann domain S_j.

10.4.2 Change of Variables in Triple Integrals

As was suggested in the foregoing example, when evaluating a triple integral over a Riemann domain we sometimes encounter difficulties caused by the equations that represent the boundary of the domain. These difficulties can sometimes be circumvented by changing variables. There are also theoretical grounds for wanting to transform triple integrals. We will merely sketch the procedure, since the essential ideas for a proof were treated in detail in Section 10.3 in connection with transforming double integrals.

A change of variables is effected by a continuously differentiable function $\boldsymbol{h} = (h_1, h_2, h_3)$ mapping an open set U in $\mathbb{R}^3$ one-to-one into $\mathbb{R}^3$. Suppose also that $J_{\boldsymbol{h}}$ does not vanish on U. If f is a real-valued function defined and integrable on a Riemann domain S' and if $\boldsymbol{h}$ maps a Riemann domain S in U onto S', then $(f \circ \boldsymbol{h})|J_{\boldsymbol{h}}|$ is integrable on S and

$$\iiint_S f \circ \boldsymbol{h}(x, y, z)|J_{\boldsymbol{h}}(x, y, z)|\, dx\, dy\, dz = \iiint_{S'} f(u, v, w)\, du\, dv\, dw.$$

This relationship is, of course, the counterpart for triple integrals of that provided by Theorem 10.3.3 for double integrals. A proof can be constructed, as in the long proof of Theorem 10.3.3, by decomposing $\boldsymbol{h}$ into a composition of three simple transformations, each of which leaves two variables fixed or simply interchanged and functionally transforms the third variable. We omit the proof.

EXAMPLE 11 Fix R_1 and R_2 with $0 < R_1 < R_2$. Let S' be the hemispherical shell

$$S' = \{(x, y, z) : R_1^2 \le x^2 + y^2 + z^2 \le R_2^2, z > 0\}.$$

Then S' is a Riemann domain. Since the upper and lower boundary surfaces of S' are portions of spheres, we can simplify the iteration of an integral over this domain by shifting to spherical coordinates. Let

$$x = h_1(\rho, \theta, \phi) = \rho \cos\theta \sin\phi$$

$$y = h_2(\rho, \theta, \phi) = \rho \sin\theta \sin\phi$$

$$z = h_3(\rho, \theta, \phi) = \rho \cos\phi.$$

Let $\boldsymbol{h} = (h_1, h_2, h_3)$. Then $\boldsymbol{h}$ maps the three-dimensional hyper-rectangle

$$S = \{(\rho, \theta, \phi) : R_1 \le \rho \le R_2, 0 \le \theta < 2\pi, 0 \le \phi \le \frac{\pi}{2}\}$$

in $\theta\phi$-space one-to-one onto S'. As we found in Example 2 of Chapter 9, the Jacobian of $\boldsymbol{h}$ is $-\rho^2 \sin\phi$. Thus, for any f which is integrable on S', we have

$$\iiint_{S'} f(x, y, z)\, dx\, dy\, dz$$

$$= \iiint_S (f \circ \boldsymbol{h})(\rho, \theta, \phi)\rho^2 \sin\phi\, d\rho\, d\theta\, d\phi$$

$$= \int_{\phi=0}^{\pi/2} \left[\int_{\theta=0}^{2\pi} \left[\int_{\rho=R_1}^{R_2} (f \circ \boldsymbol{h})(\rho, \theta, \phi)\rho^2 \sin\phi\, d\rho \right] d\theta \right] d\phi. \quad \bullet$$

EXERCISES

10.1. Integrate $\iint_S x^2y^2\, dA$ over $S = \{(x, y) : x^2 + y^2 \leq r^2\}$.

10.2. Integrate $\iint_R (x^2 + y^2)^{-1/2}\, dA$ over the square $R = \{(x, y) : |x| \leq 1, |y| \leq 1\}$.

10.3. Integrate $\iint_S [x^3 + y^3 - 3xy(x^2 + y^2)]/(x^2 + y^2)^{3/2}\, dA$ over $S = \{(x, y) : x^2 + y^2 \leq 1\}$.

10.4. Integrate $\iiint_S (x^2 + y^2 + z^2)xyz\, dV$ over $S = \{(x, y, z) : x^2 + y^2 + z^2 \leq r^2\}$.

10.5. Integrate $\iiint_S (x + y + z)x^2y^2z^2\, dV$ over the region $S = \{(x, y, z) : x \geq 0,\ y \geq 0,\ z \geq 0,\ x + y + z \leq 1\}$.

10.6. Integrate $\iiint_S z\, dV$ over the region S defined by the inequalities $x^2 + y^2 \leq z^2$ and $x^2 + y^2 + z^2 \leq 1$.

10.7. Integrate $\iiint_S [x^2 + y^2 + (z - 2)^2]^{-1}\, dV$ where S is the ball $S = \{(x, y, z) : x^2 + y^2 + z^2 \leq 1\}$.

10.8. Fix $x > 0$. Let T denote the triangle in $\mathbb{R}^2$ described by $T = \{(u, t) : 0 \leq u \leq x, 0 \leq t \leq u\}$. Let f be any continuous, real-valued function on T.

a) Iterate the integral $\iint_T f(u, t)\, dA$ in two ways.

b) Suppose that f is a function of t alone and is continuous on $[0, x]$. Show that

$$\iint_T f(t)\, dA = \int_0^x f(t)(x - t)\, dt. \quad \textbf{(10.5)}$$

(This last integral is an example of a convolution integral. Convolution will be discussed in detail in Chapter 13.) Compute the integral (10.5) when

i) $f(t) = e^{-t}$

ii) $f(t) = 1/(1 + t^2)$

iii) $f(t) = \cos ct, c > 0$

iv) $f(t) = t \ln(1 + t)$

c) Suppose also that g is a continuously differentiable function on $[0, x]$ such that $g(0) = 0$. Show that

$$\iint_T f(t)g'(u - t)\, dA = \int_0^x f(t)g(x - t)\, dt. \quad \textbf{(10.6)}$$

For the following pairs of functions, compute the integrals in (10.6) by interating the double integral in each order.

i) $f(t) = e^{-t}, g(u) = u^2/2$

ii) $f(t) = e^{-t}$, $g(u) = \sin u$

iii) $f(t) = e^{-t}$, $g(u) = ue^u$

iv) $f(t) = t$, $g(u) = \ln(1 + u)$

10.9. Prove Theorem 10.1.1.

10.10. Prove Theorem 10.1.2 when $\pi' = \pi_1' \times \pi_2'$ and when $\pi'' = \pi_1'' \times \pi_2''$ is obtained from π' by inserting one additional partition point in an interval of π_1' and one additional partition point in an interval of π_2'.

10.11. Prove Theorem 10.1.3.

10.12. For (x, y) in $S = [0, 1] \times [0, 1]$ define

$$f(x, y) = \begin{cases} 1, & \text{if } x \text{ is rational,} \\ 2y, & \text{if } x \text{ is irrational.} \end{cases}$$

a) Prove that f is not Riemann integrable on S.

b) However, show that $\int_0^1 \left[\int_0^1 f(x, y)\, dy \right] dx = 1$.

10.13. **a)** Let f be a continuous, real-valued function on an open set U in $\mathbb{R}^2$. Fix $\boldsymbol{x}_0 = (x_1, x_2)$ in U. Let S_r denote the closed neighborhood $\overline{N(\boldsymbol{x}_0; r)}$, for r so small that $S_r \subset U$. Let $A(S_r)$ denote the area of S_r. Prove that

$$\lim_{r \to 0^+} \frac{1}{A(S_r)} \iint_S f(y_1, y_2)\, dA = f(\boldsymbol{x}_0).$$

b) Analogously, let f be a continuous, real-valued function on an open set U in $\mathbb{R}^3$, fix $\boldsymbol{x}_0$ in U, let $S_r = \overline{N(\boldsymbol{x}_0; r)}$, and let $V(S_r)$ denote the volume of S_r. Prove that

$$\lim_{r \to 0^+} \frac{1}{V(S_r)} \iiint_S f(y_1, y_2, y_3)\, dV = f(\boldsymbol{x}_0).$$

10.14. Define a change of variables by $\boldsymbol{h} = (h_1, h_2)$ where

$$u = h_1(x, y) = x - y,$$
$$v = h_2(x, y) = x + y.$$

a) Find the images under $\boldsymbol{h}$ of the lines $x =$ constant, $y =$ constant, and $x + y =$ constant.

b) Find the image under $\boldsymbol{h}$ of the closed triangular region with vertices $(0, 0)$, $(1, 0)$, and $(0, 1)$. Use elementary geometry to compute the areas of T and $\boldsymbol{h}(T)$.

c) Find the Jacobian $J_{\boldsymbol{h}}$.

d) Use the mapping $\boldsymbol{h}$ to compute the area of $\boldsymbol{h}(T)$ as an integral over T. Compare your answer with that in part (b).

e) Use the change of variables $\boldsymbol{h}$ to evaluate the integral

$$\iint_T \exp\left(\frac{x - y}{x + y}\right) dA.$$

10.15. For (x, y) in $\mathbb{R}^2$, let

$$u = h_1(x, y) = x,$$
$$v = h_2(x, y) = 2y(1 + 2x),$$

and define $\boldsymbol{h} = (h_1, h_2)$.

a) Find the image under $\boldsymbol{h}$ of horizontal lines $y =$ constant. Also find the image under $\boldsymbol{h}$ of vertical lines $x =$ constant. (Distinguish between $x \neq -1/2$ and $x = -1/2$.)

b) Find the Jacobian $J_{\boldsymbol{h}}$.

c) Let $S = [0, 3] \times [1, 3]$. Identify $S' = \boldsymbol{h}(S)$ and use elementary geometry to compute the areas of S and S'. Confirm that orientation of the region S is preserved by $\boldsymbol{h}$.

d) Compute the area of S' as an integral over S. Compare your answer with that in part (b).

e) Separately compute $\iint_S ye^{-x}\,dA$ and the transformed integral over $\boldsymbol{h}(S)$ and compare your answers.

10.16. Let $S = \{(x, y) : x \geq 0, y \geq 0, x + y \leq 2$. For (x, y) in S define

$$u = h_1(x, y) = x + y,$$
$$v = h_2(x, y) = x^2 - y$$

and let $\boldsymbol{h} = (h_1, h_2)$.

a) Find the Jacobian $J_{\boldsymbol{h}}$.

b) Sketch the region $S' = \boldsymbol{h}(S)$ and confirm that the orientation of S is reversed by $\boldsymbol{h}$.

c) Find the inverse map $\boldsymbol{g} = \boldsymbol{h}^{-1}$ and its Jacobian $J_{\boldsymbol{g}}$.

d) Evaluate $\iint_{S'} (1 + 4u + 4v)^{-1/2}\,dA$.

10.17. Let $S = [0, 3] \times [0, 2]$. For (x, y) in S define

$$u = h_1(x, y) = x,$$
$$v = h_2(x, y) = y(1 + x^2)$$

and let $\boldsymbol{h} = (h_1, h_2)$.

a) Find the Jacobian $J_{\boldsymbol{h}}$.

b) Sketch $S' = \boldsymbol{h}(S)$ and compare the orientation of S' with that of S. Is orientation preserved or reversed?

c) Compute the area of S' both as an integral over S' and as an integral over S.

d) Find the inverse mapping $\boldsymbol{g} = \boldsymbol{h}^{-1}$ and its Jacobian $J_{\boldsymbol{g}}$.

e) Compute the integral $\iint_{S'} u\,dA$ directly. Also transform this integral into one over the region S using the transformation $\boldsymbol{h}$ and evaluate the resulting integral. Compare your two answers.

10.18. Use an appropriate change of variables to evaluate the integral $\iint_T (1 + x^2 + y^2)^{-2}\,dA$, where T is the triangle with vertices $(0, 0)$, $(2, 0)$, and $(0, \sqrt{3})$.

10.19. Cylindrical coordinates (r, θ, z) in $\mathbb{R}^3$ are defined by the function $\boldsymbol{h} = (h_1, h_2, h_3)$ where

$$x = h_1(r, \theta, z) = r\cos\theta,$$
$$y = h_2(r, \theta, z) = r\sin\theta,$$
$$z = h_3(r, \theta, z) = z$$

and (r, θ, z) is restricted to the set

$$\{(r, \theta, z) : r > 0, 0 \leq \theta < 2\pi, z \text{ in } \mathbb{R}\}.$$

a) Find the Jacobian of $\boldsymbol{h}$.

b) Fix $c > 0$. Compute $\iiint_S xy\,dV$ where

$$S = \{(x, y, z) : x^2 + y^2 \leq c^2, 0 \leq z \leq 1\}$$

(i) using Cartesian coordinates; (ii) using cylindrical coordinates.

10.20. Find the volume of the solid in $\mathbb{R}^3$ bounded by the cylinder $x^2 + (y - a)^2 = a^2$, the plane $z = 0$, and the paraboloid $z = (x^2 + y^2)/(4a)$. (*Hint:* Change to polar coordinates.)

10.21. Elliptical coordinates in $\mathbb{R}^3$ are described as follows. Fix positive constants a, b, and c. For (ρ, θ, ϕ) in the set $S = \{(\rho, \theta, \phi) : \rho > 0, 0 \leq \theta < 2\pi, 0 \leq \theta \leq \pi\}$, define

$$x = h_1(\rho, \theta, \phi) = a\rho\cos\theta\sin\phi,$$

$$y = h_2(\rho, \theta, \phi) = b\rho\sin\theta\sin\phi,$$

$$z = h_3(\rho, \theta, \phi) = c\rho\cos\phi.$$

Let $\boldsymbol{h} = (h_1, h_2, h_3)$.

a) Show that $x^2/a^2 + y^2/b^2 + z^2/c^2 = \rho^2$.

b) Hence show that, for any $(x, y, z) \neq \boldsymbol{0}$ in $\mathbb{R}^3$, there exists exactly one point (ρ, θ, ϕ) in S such that $\boldsymbol{h}(\rho, \theta, \phi) = (x, y, z)$.

c) Find the Jacobian $J_{\boldsymbol{h}}$.

d) Use elliptical coordinates to find the volume of the ellipse $x^2/a^2 + y^2/b^2 + z^2/c^2 = 1$.

e) Evaluate $\iiint_S xyz\,dV$ where S is the set

$$\left\{(x, y, z) : \frac{x^2}{a^2} + \frac{y^2}{b^2} + \frac{z^2}{c^2} \leq 1\right\}.$$

10.22. Fix $a > 0$.

a) For (x, y) in the half-plane $R = \{(x, y) : x \geq a, y \text{ in } \mathbb{R}\}$ define

$$u = h_1(x, y) = (x^2 + y^2 - a^2)^{1/2},$$

$$v = h_2(x, y) = \frac{y}{x}.$$

Show that $\boldsymbol{h} = (h_1, h_2)$ maps R one-to-one onto the unbounded triangular region $T = \{(u, v) : u \geq 0, -u/a \leq v \leq u/a\}$.

b) Find the Jacobian $J_{\boldsymbol{h}}$. Show that $J_{\boldsymbol{h}}(x, y) = 0$ if and only if $(x, y) = (a, 0)$.

c) For $b \geq a$, let $S(b) = \{(x, y) : x \geq a, x^2 + y^2 \leq a^2 + b^2\}$. Show that $\boldsymbol{h}$ maps $S(b)$ onto the triangular region

$$T(b) = \left\{(u, v) : 0 \leq u \leq b, \frac{-u}{a} \leq v \leq \frac{u}{a}\right\}.$$

d) Use the map $\boldsymbol{h}$ to show that

$$\iint_{S(b)} e^{-(x^2+y^2)}\,dA = -e^{-(a^2+b^2)}\tan^{-1}\left(\frac{b}{a}\right) + ae^{-a^2}\int_o^b \frac{e^{-u^2}}{u^2 + a^2}\,du.$$

e) Hence, show that

$$\lim_{b\to\infty} \iint_{S(b)} e^{-(x^2+y^2)}\, dA = \lim_{b\to\infty} ae^{-a^2} \int_0^b \frac{e^{-u^2}}{u^2+a^2}\, du.$$

10.23. Fix constants a, b, and c, not all zero. Let $r = (a^2+b^2+c^2)^{1/2}$. Let $a_{11} = a/r$, $a_{12} = b/r$, and $a_{13} = c/r$.

a) Show that $\boldsymbol{a}_1 = (a_{11}, a_{12}, a_{13})$ is a unit vector.

b) Choose any two orthogonal unit vectors

$$\boldsymbol{a}_2 = (a_{21}, a_{22}, a_{23}),$$
$$\boldsymbol{a}_3 = (a_{31}, a_{32}, a_{33})$$

each of which is orthogonal to $\boldsymbol{a}_1$. Define

$$u = h_1(x, y, z) = a_{11}x + a_{12}y + a_{13}z$$
$$v = h_2(x, y, z) = a_{21}x + a_{22}y + a_{23}z$$
$$w = h_3(x, y, z) = a_{31}x + a_{32}y + a_{33}z.$$

Show that $\boldsymbol{h} = (h_1, h_2, h_3)$ is an orthogonal transformation of $\mathbb{R}^3$. That is, $\boldsymbol{h}$ is a linear transformation that preserves the lengths of vectors and preserves (or reverses) the angle between any two vectors.

c) Show that $|J_{\boldsymbol{h}}| = |\det[a_{ij}]| = 1$.

d) Show that $\boldsymbol{h}$ maps $S = \{(x, y, z) : x^2 + y^2 + z^2 \leq 1\}$ one-to-one onto S.

e) Hence show that, for any integrable function f on S,

$$\iiint_S f(u, u, w)\, du\, dv\, dw = \iiint_S f(\boldsymbol{h}(x, y, z))\, dx\, dy\, dz.$$

f) Evaluate $\iiint_S \cos(ax + by + cz)\, dV$. (Transform the integral using $\boldsymbol{h}$, then iterate first with respect to v, then w.)

10.24. Find the surface area of that part of the cone $x^2 + y^2 = z^2$ inside the cylinder $x^2 + y^2 = 2ax$.

10.25. Find the surface area of that part of the surface $z = xy$ inside the cylinder $x^2 + y^2 = a^2$.

10.26. Fix constants a, b and c such that $0 < a \leq b \leq c$. Find the area of that portion of the spherical surface $x^2 + y^2 + z^2 = c^2$ that is inside the paraboloid

$$\frac{x^2}{a} + \frac{y^2}{b} = 2(z + c).$$

10.27. For (r, θ) in $[0, 1] \times [0, 2\pi]$, let

$$x = \phi_1(r, \theta) = r\cos\theta,$$
$$y = \phi_2(r, \theta) = r\sin\theta,$$
$$z = \phi_3(r, \theta) = \theta.$$

A surface is described parametrically by $\phi = (\phi_1, \phi_2, \phi_3)$. Sketch the surface and find its surface area.

10.28. Define a surface parametrically by

$$x = \phi_1(u, v) = u - v$$

$$y = \phi_2(u, v) = u + v$$

$$z = \phi_3(u, v) = uv,$$

for (u, v) in $[0, 1] \times [0, 2]$. Find the area of the surface.

10.29. Fix positive constants a and b. For (u, v) in $[0, 2\pi] \times [0, \pi/2]$, define

$$x = \phi_1(u, v) = a \cos u \sin v,$$

$$y = \phi_2(u, v) = a \sin u \sin v,$$

$$z = \phi_3(u, v) = b \cos v.$$

A surface is described parametrically by $\phi = (\phi_1, \phi_2, \phi_3)$. Find its surface area. Treat the cases $a \geq b$ and $b > a$ separately.

10.30. The four-dimensional unit ball S centered at the origin is defined in Cartesian coordinates to be the set

$$S = \{(x_1, x_2, x_3, x_4) : x_1^2 + x_2^2 + x_3^2 + x_4^2 \leq 1\}.$$

Set up a four-fold iterated integral that gives the hypervolume of this ball.

11

Infinite Series

For centuries prior to Newton's work, mathematicians had used primitive arguments involving infinite processes to draw conclusions in the areas of arithmetic and geometry. But it was Newton who first successfully described a given function as an "infinite sum" of more simple functions and thereby initiated a central theme in analysis. Here we begin the study of Newton's theme using the powerful analytic tools developed since Newton's era: the study of the convergence of infinite series. In Chapter 12, treating series of functions, and in Chapter 14, devoted to Fourier series, we will build on the beautiful and powerful results achieved here.

11.1 PRELIMINARIES

It is commonplace to think of an infinite series, $\sum_{j=1}^{\infty} a_j$, as "the sum of the infinitely many numbers a_j." Although this interpretation may appear beneficial, it obscures the subtlety inherent in the subject and could, perhaps, interfere with your ability to formulate successful strategies for proofs. After all, it is humanly impossible to "add infinitely many numbers"; let's agree on that. Therefore it is pointless to think in those terms. Historically, mathematicians enjoyed a significant advance when that misconception was jettisoned and the more careful description of an infinite series was installed in its place.

DEFINITION 11.1.1 Let $\{a_j\}$ be a sequence of real numbers.

i) For each k in $\mathbb{N}$, define $s_k = \sum_{j=1}^{k} a_j$. The sequence $\{s_k\}$ is called the *sequence of kth partial sums.*

ii) An infinite series, denoted $\sum_{j=1}^{\infty} a_j$, is the sequence $\{s_k\}$ of partial sums of the numbers $a_1, a_2, \ldots, a_k, \ldots$. ●

Remark. The notation $\sum_{j=1}^{\infty} a_j$ for an infinite series is simply a convenient way to suggest $\lim_{k\to\infty} \sum_{j=1}^{k} a_j$; it is not meant to indicate that you are to add the infinitely many numbers a_j. For this reason we say that the series is the same as the sequence $\{s_k\}$ of kth partial sums. We have inherited this notation from the earliest years of analysis, before the careful definition of limit was achieved.

DEFINITION 11.1.2 Given an infinite series $\sum_{j=1}^{\infty} a_j$, let $\{s_k\}$ denote its sequence of kth partial sums.

i) If the sequence $\{s_k\}$ converges (to S), then the infinite series $\sum_{j=1}^{\infty} a_j$ is said to *converge* (to S) and we write

$$\sum_{j=1}^{\infty} a_j = S.$$

S is called the *sum* of the series.

ii) If the sequence $\{s_k\}$ diverges, then the infinite series $\sum_{j=1}^{\infty} a_j$ is said to *diverge*. ●

Referring to Definition 1.3.2, we see that the convergence of the infinite series $\sum_{j=1}^{\infty} a_j$ to S is equivalent to the following formulation: Given any $\epsilon > 0$, there exists an index k_0 such that, for $k \geq k_0$, it follows that $|s_k - S| < \epsilon$. That is,

$$\left|\sum_{j=1}^{k} a_j - S\right| < \epsilon.$$

Equivalently, $|\sum_{j=k+1}^{\infty} a_j| < \epsilon$, for all $k \geq k_0$. The quantity $\sum_{j=k+1}^{\infty} a_j$ is called the *tail of the series*. We can recast the definition of convergence of the series $\sum_{j=1}^{\infty} a_j$ in the following equivalent form. It is this version that provides the intuitive form of convergence which you may find most useful.

THEOREM 11.1.1 A series $\sum_{j=1}^{\infty} a_j$ converges if and only if the tail of the series is eventually arbitrarily small. That is, for each $\epsilon > 0$, there exists a k_0 such that, for $k \geq k_0$,

$$\left|\sum_{j=k+1}^{\infty} a_j\right| < \epsilon. \quad ●$$

Our early discussion of Cauchy sequences serves us well at this juncture, especially when we are dealing with an infinite series whose sum is unknown.

THEOREM 11.1.2 An infinite series $\sum_{j=1}^{\infty} a_j$ converges if and only if the sequence $\{s_k\}$ of kth partial sums is Cauchy.

Proof. $\mathbb{R}$ is complete and a sequence in $\mathbb{R}$ is convergent if and only if it is Cauchy. ●

COROLLARY 11.1.3 If the infinite series $\sum_{j=1}^{\infty} a_j$ converges, then $\lim_{j\to\infty} a_j = 0$.

Proof. Since the series $\sum_{j=1}^{\infty} a_j$ converges, the sequence $\{s_k\}$ of kth partial sums is Cauchy. Given $\epsilon > 0$, there exists a k_0 such that, for all k and m greater than k_0, $|s_k - s_m| < \epsilon$. Fix $k \geq k_0$ and let $m = k + 1$. Observe that

$$|a_{k+1}| = |s_{k+1} - s_k| < \epsilon.$$

Thus $\lim_{j\to\infty} a_j = 0$ as claimed. ●

When confronted with a series whose kth term does not tend to 0, you can quickly conclude that the series cannot converge. However, you are forewarned that the converse of Corollary 11.1.3 is false. There exist divergent series whose kth terms tend to 0. Refer to Example 22 of Section 1.3. The sequence of kth partial sums of the series $\sum_{j=1}^{\infty} 1/j$ is not convergent. We conclude that the series $\sum_{j=1}^{\infty} 1/j$, called the *harmonic series*, diverges although $\lim_{j\to\infty} 1/j = 0$.

EXAMPLE 1 The one series that we have used frequently in the earlier chapters of the text is the *geometric series*. If r is any number in $(-1, 1)$, then the geometric series $\sum_{j=0}^{\infty} r^j$ converges; in fact, its sum is $1/(1-r)$. For any other value of r the series diverges. Clearly, we have assumed you are already familiar with this example; we have used it repeatedly. Here let's prove these assertions. The kth partial sum of the series is $s_k = \sum_{j=0}^{k} r^j$. Multiplying s_k by r and subtracting the result from s_k yields

$$(1-r)s_k = \sum_{j=0}^{k} r^j - \sum_{j=1}^{k+1} r^j = 1 - r^{k+1}.$$

Therefore, provided $r \neq 1$, $s_k = (1 - r^{k+1})/(1-r)$ and

$$\lim_{k\to\infty} s_k = \frac{1}{1-r} - \lim_{k\to\infty} \frac{r^{k+1}}{1-r}.$$

This last limit exists if and only if $|r| < 1$, in which case

$$\lim_{k\to\infty} s_k = \frac{1}{1-r},$$

and the geometric series converges to $1/(1-r)$. If $|r| \geq 1$, then the kth term of the series does not converge to 0 and, for these values of r, the geometric series diverges. ●

EXAMPLE 2 Consider $\sum_{j=1}^{\infty} 1/j(j+1)$. Since

$$\frac{1}{j(j+1)} = \frac{1}{j} - \frac{1}{j+1},$$

it follows that the kth partial sum of the series is

$$s_k = \sum_{j=1}^{k} \frac{1}{j(j+1)} = \sum_{j=1}^{k} \left(\frac{1}{j} - \frac{1}{j+1}\right) = 1 - \frac{1}{k+1}.$$

Since $\lim_{k\to\infty} s_k = \lim_{k\to\infty}[1 - 1/(k+1)] = 1$, we deduce that $\sum_{j=1}^{\infty} 1/j(j+1) = 1$. ●

EXAMPLE 3 As we saw in Example 20 in Section 1.3, the sequence of kth partial sums of the series $\sum_{j=0}^{\infty} 1/j!$ is a bounded, monotone increasing sequence that converges to e. Therefore the infinite series $\sum_{j=0}^{\infty} 1/j! = e$. More generally, from Example 10 in Section 4.5, we know that, for any fixed x in $\mathbb{R}$, the sequence of kth partial sums of the infinite series $\sum_{j=0}^{\infty} x^j/j!$ is the sequence $\{p_k(x)\}$ of kth Taylor polynomials of the exponential function evaluated at the number x. From that example, we know that $\{p_k(x)\}$ converges to the number e^x. Therefore, $\sum_{j=0}^{\infty} x^j/j! = e^x$. ●

EXAMPLE 4 Analogously, fix any x in $\mathbb{R}$ and consider the infinite series $\sum_{j=0}^{\infty}(-1)^j x^{2j+1}/(2j+1)!$. The sequence of kth partial sums of this series is the sequence whose kth term is $\sum_{j=0}^{k}(-1)^j x^{2j+1}/(2j+1)!$. But this is the sequence of Taylor polynomials of the sine function evaluated at the number x. From Example 11 in Section 4.5, we know that this sequence converges to $\sin x$. Therefore, we deduce that $\sum_{j=0}^{\infty}(-1)^j x^{2j+1}/(2j+1)! = \sin x$. ●

It is important for you to recognize that the convergence or divergence of a series $\sum_{j=1}^{\infty} a_j$ depends only on the behavior of the tail of the series and not on the first finite number of terms. Omitting a finite number of the terms will affect the value of the sum of the series, provided it converges, but will have no effect whatsoever on its convergence or divergence. Of course, the two distinct problems identified with regard to sequences—convergence or divergence and, if convergent, to what limit—persist here for series as well. Initially we will deal primarily with the first of these questions where the omission of a finite number of terms is immaterial; the identification of the sum of a series is often much more delicate and often requires more subtle tools to be developed later.

We further divide the first question into two. First we will consider only series of positive terms and examine methods for determining whether the series converges. Later we will admit series whose terms are of arbitrary sign. As we will discover, there is a delicate distinction here.

11.2 CONVERGENCE TESTS (POSITIVE SERIES)

Suppose, then, that $\sum_{j=1}^{\infty} a_j$ is a series of positive terms. We want to establish tests that can be easily applied and that enable us to determine whether $\sum_{j=1}^{\infty} a_j$ converges. Our first and simplest test is called the comparison test; it calls for us to compare the corresponding terms of two series.

Suppose that $\sum_{j=1}^{\infty} a_j$ and $\sum_{j=1}^{\infty} b_j$ are two series of positive terms such that, for j in $\mathbb{N}$,

$$0 < a_j \leq b_j. \tag{11.1}$$

We first claim that, if $\sum_{j=1}^{\infty} b_j$ converges, then $\sum_{j=1}^{\infty} a_j$ does also. To prove this claim, let $t_k = \sum_{j=1}^{k} b_j$ denote the kth partial sum of the series $\sum_{j=1}^{\infty} b_j$. Assume that

$\lim_{k\to\infty} t_k = T$. Because the numbers b_j are positive, the sequence $\{t_k\}$ is monotone increasing and, of course, converges to T. Hence

$$\sum_{j=1}^{\infty} b_j = \lim_{k\to\infty} t_k = \sup\{t_k : k \text{ in } \mathbb{N}\} = T.$$

Because the numbers a_j are also positive, the sequence $\{s_k\}$ of kth partial sums of the series $\sum_{j=1}^{\infty} a_j$ is also monotone increasing. Furthermore, by (11.1), $s_k \le t_k \le T$, for all k in $\mathbb{N}$. Thus, $\{s_k\}$ is bounded above. By Theorem 1.3.7, $\{s_k\}$ is convergent. Consequently, the series $\sum_{j=1}^{\infty} a_j$ converges.

Suppose, on the other hand, that (11.1) holds and that $\sum_{j=1}^{\infty} a_j$ diverges. We claim that $\sum_{j=1}^{\infty} b_j$ must also diverge. To prove this claim, notice that the increasing sequence $\{s_k\}$ of kth partial sums of $\sum_{j=1}^{\infty} a_j$ cannot be bounded above and, in fact, must diverge to ∞. The increasing sequence $\{t_k\}$ of kth partial sums of $\sum_{j=1}^{\infty} b_j$ is restricted by the inequality $0 < s_k \le t_k$ for all k in $\mathbb{N}$. Thus $\{t_k\}$ must also diverge as does $\sum_{j=1}^{\infty} b_j$.

In accord with our remarks about the tail of a series, it is worth noting that we need only require that there be an index j_0 such that $0 < a_j \le b_j$, for $j \ge j_0$, to apply our result. We have proved the following theorem.

THEOREM 11.2.1 The Comparison Test Let $\sum_{j=1}^{\infty} a_j$ and $\sum_{j=1}^{\infty} b_j$ be two series of positive terms such that, for some j_0,

$$0 < a_j \le b_j, \text{ for all } j \ge j_0.$$

i) If $\sum_{j=1}^{\infty} b_j$ converges, then $\sum_{j=1}^{\infty} a_j$ also converges.

ii) If $\sum_{j=1}^{\infty} a_j$ diverges, then $\sum_{j=1}^{\infty} b_j$ also diverges. ●

To use the comparison test when confronted with a series, you must first intuitively sense whether it diverges or converges, then select from your known stock of series one whose terms satisfy the inequalities of the theorem. You then compare the given series against the one you chose. As you can see, your intuition must be well tuned and you need a good supply of series whose behavior is known from which to choose a test series. In practice, if the expression giving the jth term of a series is complicated, satisfying the inequalities in Theorem 11.2.1 can be tricky. With just a little effort we can generalize that theorem to a form that is often easier to apply.

Essentially, a series $\sum_{j=1}^{\infty} a_j$ of *positive terms* converges or diverges because of the behavior of the sequence $\{a_j\}$. We know that the convergence of $\sum_{j=1}^{\infty} a_j$ entails $\lim_{j\to\infty} a_j = 0$, but this condition alone will not suffice. If the sequence $\{a_j\}$ converges to 0 *sufficiently rapidly,* then the series $\sum_{j=1}^{\infty} a_j$ converges; otherwise it diverges. Suppose that $\sum_{j=1}^{\infty} b_j$ is another series of positive terms such that

$$\lim_{j\to\infty} \frac{a_j}{b_j} = L > 0.$$

Intuitively, a_j is eventually approximately equal to Lb_j, so that, with regard to the *rate of convergence* to 0, the sequences $\{a_j\}$ and $\{b_j\}$ behave essentially the same. We ought reasonably to expect, then, that if the preceding limit exists, then the series $\sum_{j=1}^{\infty} a_j$ and $\sum_{j=1}^{\infty} b_j$ converge or diverge together. That is the content of the following theorem.

THEOREM 11.2.2 The Limit Comparison Test Suppose that $\sum_{j=1}^{\infty} a_j$ and $\sum_{j=1}^{\infty} b_j$ are two series of positive terms such that $\lim_{j\to\infty} a_j/b_j = L$ exists and is positive. Then $\sum_{j=1}^{\infty} a_j$ converges if and only if $\sum_{j=1}^{\infty} b_j$ converges.

Proof. Choose any ϵ in $(0, L)$ and choose j_0 such that, for $j \geq j_0$, $L - \epsilon < a_j/b_j < L + \epsilon$. Since $b_j > 0$, we have $(L - \epsilon)b_j < a_j < (L + \epsilon)b_j$, for $j \geq j_0$.

Suppose that $\sum_{j=1}^{\infty} a_j$ converges. By Theorem 11.2.1, the series $\sum_{j=1}^{\infty}(L - \epsilon)b_j = (L - \epsilon)\sum_{j=1}^{\infty} b_j$ converges, as does $\sum_{j=1}^{\infty} b_j$.

On the other hand, if $\sum_{j=1}^{\infty} b_j$ converges, then $\sum_{j=1}^{\infty}(L + \epsilon)b_j$ does also. Hence, again by Theorem 11.2.1, $\sum_{j=1}^{\infty} a_j$ converges. ●

Notice that if $\lim_{j\to\infty} a_j/b_j = 0$, then $\{a_j\}$ converges to 0 more rapidly than does $\{b_j\}$. Therefore, we can conclude only that the convergence of $\sum_{j=1}^{\infty} b_j$ implies that of $\sum_{j=1}^{\infty} a_j$.

There still remains the problem of amassing a supply of known series with known behavior against which to compare or limit compare a new series in order to resolve the question of convergence. The geometric series will frequently serve well in this role, but we will need greater variety from which to select a test series. Our next theorem, the integral test, provides some of that.

THEOREM 11.2.3 The Integral Test Suppose that $\sum_{j=1}^{\infty} a_j$ is a series of positive terms and that $\{a_j\}$ is a decreasing sequence converging to 0. Suppose that f is a continuous, monotone decreasing function on $[1, \infty)$ such that $f(j) = a_j$, for $j = 1, 2, 3, \ldots$. The series $\sum_{j=1}^{\infty} a_j$ converges if and only if $\lim_{k\to\infty} \int_1^k f(x)\,dx$ exists.

Proof. For each $j \geq 1$, let

$$b_j = \int_j^{j+1} f(x)\,dx.$$

For $j \geq 2$, let

$$c_j = \int_{j-1}^{j} f(x)\,dx.$$

Then, referring to Fig. 11.1, we have $b_1 \leq a_1$ and

$$b_j \leq a_j \leq c_j,$$

for all $j \geq 2$. If $\sum_{j=1}^{\infty} a_j$ converges, then, by the comparison test, $\sum_{j=1}^{\infty} b_j$ converges, say to B. It follows that

$$B = \sum_{j=1}^{\infty} b_j = \lim_{k\to\infty} \sum_{j=1}^{k} b_j = \lim_{k\to\infty} \sum_{j=1}^{k} \int_j^{j+1} f(x)\,dx = \lim_{k\to\infty} \int_1^{k+1} f(x)\,dx.$$

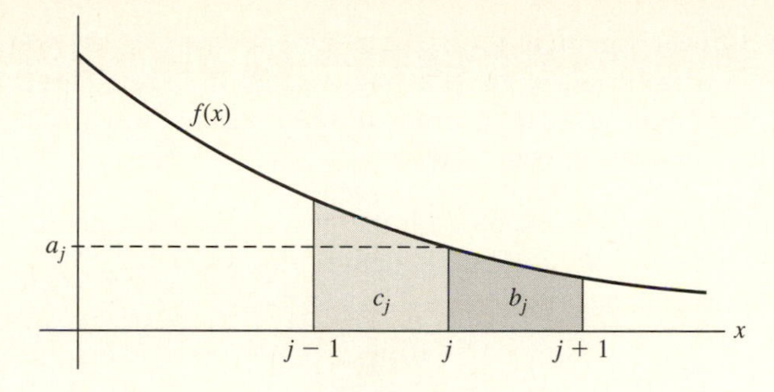

Figure 11.1

Therefore, $\lim_{k\to\infty} \int_1^k f(x)\,dx$ exists.

Conversely, suppose that $\lim_{k\to\infty} \int_1^k f(x)\,dx$ exists. It follows that

$$\sum_{j=2}^{\infty} c_j = \lim_{k\to\infty} \sum_{j=2}^{k} c_j = \lim_{k\to\infty} \sum_{j=2}^{k} \int_{j-1}^{j} f(x)\,dx = \lim_{k\to\infty} \int_1^k f(x)\,dx$$

exists. Therefore, $\sum_{j=2}^{\infty} c_j$ converges. By the comparison test, $\sum_{j=1}^{\infty} a_j$ also converges. ●

Again in accord with our comments regarding convergence correlating with the behavior of the tail of a series, we need only require that the hypotheses of the theorem be met on some interval $[k_0, \infty)$.

EXAMPLE 5 Using the integral test, we can give an easy, alternative proof that the harmonic series diverges. Simply let $f(x) = 1/x$ for x in $[1, \infty)$. Because

$$\lim_{k\to\infty} \int_1^k \frac{1}{x}\,dx = \lim_{k\to\infty} \ln k = \infty,$$

it follows that $\sum_{j=1}^{\infty} 1/j$ diverges. ●

EXAMPLE 6 The series $\sum_{j=2}^{\infty} 1/[j \ln j]$ diverges by the integral test. Again, simply let $f(x) = 1/[x \ln x]$ for x in $[2, \infty)$. As you can compute,

$$\lim_{k\to\infty} \int_2^k \frac{1}{x \ln x}\,dx = \lim_{k\to\infty} [\ln(\ln k) - \ln(\ln 2)] = \infty.$$

Therefore the series diverges. ●

EXAMPLE 7 For any $p > 1$, the series $\sum_{j=1}^{\infty} 1/j^p$ converges. (Such series are called p-series.) Let $f(x) = 1/x^p$ for x in $[1, \infty)$. Then we have

$$\begin{aligned}\lim_{k\to\infty} \int_1^k \frac{1}{x^p}\,dx &= \lim_{k\to\infty} \left.\frac{x^{1-p}}{1-p}\right|_1^k \\ &= \frac{1}{p-1} + \frac{1}{1-p} \lim_{k\to\infty} \frac{1}{k^{p-1}} = \frac{1}{p-1},\end{aligned}$$

since $p - 1 > 0$. Therefore, for $p > 1$, the p-series converges. As you can confirm for yourself, if $p \leq 1$, then the corresponding p-series diverges. This example alone provides you with a large and useful family of series against which to compare some new series to test for convergence. ●

The comparison test combined with the geometric series gives rise to one of the most useful tests in your tool kit. To identify this test, let $\sum_{j=1}^{\infty} a_j$ be a series of positive terms. Assume that

$$\lim_{j\to\infty} \frac{a_{j+1}}{a_j} = L$$

exists. Thus, for any $\epsilon > 0$, there exists a j_0 such that, for $j \geq j_0$,

$$L - \epsilon < \frac{a_{j+1}}{a_j} < L + \epsilon.$$

Equivalently,

$$(L - \epsilon)a_j < a_{j+1} < (L + \epsilon)a_j,$$

for all $j \geq j_0$. It follows, as you can prove easily by induction, that, for j in $\mathbb{N}$,

$$(L - \epsilon)^j a_{j_0} < a_{j_0+j} < (L + \epsilon)^j a_{j_0}. \tag{11.2}$$

Now suppose that L is in $[0, 1)$. Choose ϵ in $(0, 1 - L)$. Then $r = L + \epsilon$ is in $(0, 1)$. Thus the geometric series $\sum_{j=1}^{\infty} r^j$ converges as does $a_{j_0} \sum_{j=1}^{\infty} r^j$. We deduce, by the comparison test and by the right-hand inequality in (11.2), that the original series $\sum_{j=1}^{\infty} a_j$ also converges.

Suppose, however, that $L > 1$. Choose ϵ in $(0, L - 1)$. Then $r = L - \epsilon > 1$ and the geometric series $\sum_{j=1}^{\infty} r^j$ diverges. By the comparison test and the left-hand inequality in (11.2), the original series $\sum_{j=1}^{\infty} a_j$ also diverges. If $L = 1$, then we can draw no conclusions using arguments of this sort. This line of reasoning proves the following theorem.

THEOREM 11.2.4 d'Alembert's Ratio Test Let $\sum_{j=1}^{\infty} a_j$ be a series of positive terms such that

$$\lim_{j\to\infty} \frac{a_{j+1}}{a_j} = L$$

exists.

i) If $L < 1$, then $\sum_{j=1}^{\infty} a_j$ converges.

ii) If $L > 1$, then $\sum_{j=1}^{\infty} a_j$ diverges.

iii) If $L = 1$, no information is gained from the test. ●

The ratio test is so generally useful and so easy to apply that you might consider trying it first when confronted with a series of positive terms.

EXAMPLE 8 Consider the series $\sum_{j=1}^{\infty} 2^j j/[3^j(j+1)]$. Compute

$$\lim_{j\to\infty} \frac{a_{j+1}}{a_j} = \lim_{j\to\infty} \frac{2^{j+1}(j+1)/[3^{j+1}(j+2)]}{2^j j/[3^j(j+1)]}$$

$$= \lim_{j\to\infty} \frac{2(j+1)^2}{3j(j+2)} = \frac{2}{3} < 1.$$

By the ratio test the series converges. ●

EXAMPLE 9 We know, by the integral test, that $\sum_{j=1}^{\infty} 1/j^2$ converges and that $\sum_{j=1}^{\infty} 1/j$ diverges. For each series, $\lim_{j\to\infty} a_{j+1}/a_j = 1$, confirming that the ratio test provides no information in these cases. ●

A somewhat more powerful test, although one more awkward to apply, was found by Cauchy. It is similar to the ratio test in that it depends on the rate at which the jth term of the series becomes small.

THEOREM 11.2.5 Cauchy's Root Test Let $\sum_{j=1}^{\infty} a_j$ be a series of positive terms such that $\lim_{j\to\infty}(a_j)^{1/j} = L$ exists.

i) If $L < 1$, then $\sum_{j=1}^{\infty} a_j$ converges.

ii) If $L > 1$, then $\sum_{j=1}^{\infty} a_j$ diverges.

iii) If $L = 1$, no information is gained from the test.

Proof. Clearly, $L \geq 0$. If $0 \leq L < 1$, choose ϵ in $(0, 1-L)$ and choose j_0 such that, for $j \geq j_0$,

$$(a_j)^{1/j} < L + \epsilon = r < 1.$$

It follows that $a_j < r^j$, for $j \geq j_0$. Since $r < 1$, we deduce that $\sum_{j=1}^{\infty} a_j$ converges by comparing with the tail of the convergent series $\sum_{j=1}^{\infty} r^j$.

If $L > 1$, choose ϵ in $(0, L-1)$ and choose j_0 such that, for $j \geq j_0$,

$$(a_j)^{1/j} > L - \epsilon = r > 1.$$

It follows that $a_j > r^j$, for $j \geq j_0$. Since $r > 1$, we deduce that $\sum_{j=1}^{\infty} a_j$ diverges by comparing with the tail of the divergent series $\sum_{j=1}^{\infty} r^j$.

The series $\sum_{j=1}^{\infty} 1/j$ diverges and $\sum_{j=1}^{\infty} 1/j^2$ converges. In each case $\lim_{j\to\infty}(a_j)^{1/j} = 1$. This confirms that, if $L = 1$, then the test is inconclusive. ●

EXAMPLE 10 Either the ratio test or the root test can be applied to show that $\sum_{j=1}^{\infty} j/2^j$ converges. For example, $\lim_{j\to\infty}(a_j)^{1/j} = \lim_{j\to\infty} j^{1/j}/2 = 1/2$. (See Exercise 1.35.) Therefore, by the root test, the series converges. ●

Both the ratio and the root tests can be strengthened. Whereas the limits that occur in the hypotheses of Theorems 11.2.4 and 11.2.5 may not exist, the limit superior and limit inferior always do, provided we agree to admit ∞ as a possible value. The proofs of the following two theorems are already contained in those of Theorems 11.2.4 and 11.2.5 We leave the details for you to complete.

THEOREM 11.2.6 The Ratio Test, Form II Let $\sum_{j=1}^{\infty} a_j$ be a series of positive terms.

i) If $\limsup a_{j+1}/a_j < 1$, then $\sum_{j=1}^{\infty} a_j$ converges.

ii) If $\liminf a_{j+1}/a_j > 1$, then $\sum_{j=1}^{\infty} a_j$ diverges.

iii) If $\liminf a_{j+1}/a_j \leq 1 \leq \limsup a_{j+1}/a_j$, then the test provides no information. ●

THEOREM 11.2.7 The Root Test, Form II Let $\sum_{j=1}^{\infty} a_j$ be a series of positive terms.

i) If $\limsup (a_j)^{1/j} < 1$, then $\sum_{j=1}^{\infty} a_j$ converges.

ii) If $\limsup (a_j)^{1/j} > 1$, then $\sum_{j=1}^{\infty} a_j$ diverges.

iii) If $\limsup (a_j)^{1/j} = 1$, then the test provides no information. ●

Of course, if $\lim_{j\to\infty} a_{j+1}/a_j$ or $\lim_{j\to\infty}(a_j)^{1/j}$ already exists, then these more general theorems simply reduce to the usual ratio and root tests and it becomes largely a matter of taste and ease which test you prefer to apply. The ratio test is often easier to use; the root test is the more powerful. That is, if the ratio test already resolves the question of convergence of a series of positive terms, then the root test will also. However, there exist series for which the ratio test is inconclusive and the root test resolves the question of convergence. This observation follows from Exercise 1.69 in which you proved that, if $\{a_j\}$ is a sequence of positive numbers, then

$$\liminf \frac{a_{j+1}}{a_j} \leq \liminf (a_j)^{1/j}$$

and

$$\limsup (a_j)^{1/j} \leq \limsup \frac{a_{j+1}}{a_j}.$$

11.3 Absolute Convergence

When we relax our previous restriction and consider series whose terms may be both positive and negative, we encounter a subtle point that deserves your close attention. If the terms of the series are all positive, then the series can converge only by virtue of the rate of convergence of the jth term to 0. If that rate is sufficiently rapid, the series converges; otherwise it diverges. However, if the terms of the series are mixed, some positive, some negative, there is a second possibility. It can happen that the negative terms cancel against the positive terms to such an extent that the series converges, even though the jth term, while converging to 0, does so at a slow rate. If a series converges primarily because the jth term tends to 0 at a sufficiently rapid rate, we say that the series *converges absolutely.* If the series converges only because the positive and negative terms offset each other, we say the series *converges conditionally*.

Of course, such an intuitive description does not constitute an adequate definition, but it does capture the essential difference between the two possible cases. The formal definition follows.

DEFINITION 11.3.1 Let $\sum_{j=1}^{\infty} a_j$ be any series of real numbers. If $\sum_{j=1}^{\infty} |a_j|$ converges, then $\sum_{j=1}^{\infty} a_j$ is said to *converge absolutely*. If the series $\sum_{j=1}^{\infty} a_j$ converges but $\sum_{j=1}^{\infty} |a_j|$ diverges, then the series is said to *converge conditionally*. ●

We shall examine conditional convergence and see examples in Section 11.4. Here we want to explore absolute convergence in some detail. Eventually, we shall come to see that absolute convergence is a strong form of convergence, whereas conditional convergence is significantly weaker and more delicate.

First, review all the tests for convergence presented in Section 11.2. Automatically, they are all tests for absolute convergence. In the hypotheses of each simply replace $\sum_{j=1}^{\infty} a_j$ with $\sum_{j=1}^{\infty} |a_j|$. Thus, to test for absolute convergence, we can use any of those tests. Clearly, any convergent series of positive terms already converges absolutely.

EXAMPLE 11 For $p > 1$, the series $\sum_{j=1}^{\infty} (-1)^j / j^p$ converges absolutely by the integral test. ●

EXAMPLE 12 The series $\sum_{j=0}^{\infty} (-1)^j / j!$ converges absolutely to $1/e$. We already know, from Example 3 in Section 11.1, that the series converges to $1/e$. Since $\sum_{j=0}^{\infty} 1/j!$ converges (to e), that convergence is absolute. The convergence results from the fact that $1/j!$ tends to 0 rapidly rather than from the offsetting of positive and negative terms. ●

A beginner often assumes that the following theorem is trivial, not worthy of attention. That assumption is false.

THEOREM 11.3.1 If a series $\sum_{j=1}^{\infty} a_j$ converges absolutely, then it converges.

Proof. The sequence $\{t_k\}$ of kth partial sums of the series $\sum_{j=1}^{\infty} |a_j|$ is Cauchy. Therefore, given $\epsilon > 0$, there exists a k_0 such that, for $k > m \geq k_0$, we have $t_k - t_m = \sum_{j=m+1}^{k} |a_j| < \epsilon$. Let $\{s_k\}$ denote the sequence of kth partial sums of the series $\sum_{j=1}^{\infty} a_j$. Then, for $k > m \geq k_0$,

$$|s_k - s_m| = \left| \sum_{j=m+1}^{k} a_j \right| \leq \sum_{j=m+1}^{k} |a_j| < \epsilon.$$

We deduce that $\{s_k\}$ is also Cauchy and conclude that $\sum_{j=1}^{\infty} a_j$ must converge. This proves the theorem. ●

Suppose that $\sum_{j=1}^{\infty} a_j$ is an absolutely convergent series. Define $a_j^+ = \max\{a_j, 0\}$ and $a_j^- = \max\{-a_j, 0\}$ and form the two series $\sum_{j=1}^{\infty} a_j^+$ and $\sum_{j=1}^{\infty} a_j^-$. Notice that the nonzero terms of $\sum_{j=1}^{\infty} a_j^+$ are simply the positive terms of the original series; the nonzero terms of $\sum_{j=1}^{\infty} a_j^-$ are the absolute values of the negative terms of that series. From $\sum_{j=1}^{\infty} a_j$ we have extracted two series of nonnegative terms, both of which, we claim, converge (and hence converge absolutely). To verify our claim, fix any $\epsilon > 0$ and choose k_0 such that, for $k \geq k_0$,

$$\sum_{j=k+1}^{\infty} |a_j| < \epsilon.$$

Then, for $k \geq k_0$,

$$0 \leq \sum_{j=k+1}^{\infty} a_j^+ = \sum_{j=k+1}^{\infty} |a_j^+| \leq \sum_{j=k+1}^{\infty} |a_j| < \epsilon,$$

and

$$0 \leq \sum_{j=k+1}^{\infty} a_j^- = \sum_{j=k+1}^{\infty} |a_j^-| \leq \sum_{j=k+1}^{\infty} |a_j| < \epsilon.$$

Thus, $\sum_{j=1}^{\infty} a_j^+$ and $\sum_{j=1}^{\infty} a_j^-$ converge.

Let S^+ and S^-, respectively, denote the sums of these series. We claim that $\sum_{j=1}^{\infty} a_j = S^+ - S^-$ and $\sum_{j=1}^{\infty} |a_j| = S^+ + S^-$. For proof, fix $\epsilon > 0$. Choose an index k_1 such that, for $k \geq k_1$,

$$\left| \sum_{j=1}^{k} a_j^+ - S^+ \right| < \epsilon.$$

Likewise, choose an index k_2 such that, for $k \geq k_2$,

$$\left| \sum_{j=1}^{k} a_j^- - S^- \right| < \epsilon.$$

Let $k_0 = \max\{k_1, k_2\}$ and choose $k \geq k_0$. We have

$$\begin{aligned}\left| \sum_{j=1}^{k} a_j - (S^+ - S^-) \right| &= \left| \left(\sum_{j=1}^{k} a_j^+ - S^+ \right) - \left(\sum_{j=1}^{k} a_j^- - S^- \right) \right| \\ &\leq \left| \sum_{j=1}^{k} a_j^+ - S^+ \right| + \left| \sum_{j=1}^{k} a_j^- - S^- \right| < 2\epsilon.\end{aligned}$$

This proves that $\sum_{j=1}^{\infty} a_j = S^+ - S^-$.

Likewise, for $k \geq k_0$, we have

$$\begin{aligned}\left| \sum_{j=1}^{k} |a_j| - (S^+ + S^-) \right| &= \left| \left(\sum_{j=1}^{k} a_j^+ - S^+ \right) + \left(\sum_{j=1}^{k} a_j^- - S^- \right) \right| \\ &\leq \left| \sum_{j=1}^{k} a_j^+ - S^+ \right| + \left| \sum_{j=1}^{k} a_j^- - S^- \right| < 2\epsilon.\end{aligned}$$

This proves that $\sum_{j=1}^{\infty} |a_j| = S^+ + S^-$. Our discussion proves the next theorem.

THEOREM 11.3.2 If $\sum_{j=1}^{\infty} a_j$ converges absolutely, then the two series $\sum_{j=1}^{\infty} a_j^+$ and $\sum_{j=1}^{\infty} a_j^-$ also converge absolutely. Further, $\sum_{j=1}^{\infty} a_j = \sum_{j=1}^{\infty} a_j^+ - \sum_{j=1}^{\infty} a_j^-$ and $\sum_{j=1}^{\infty} |a_j| = \sum_{j=1}^{\infty} a_j^+ + \sum_{j=1}^{\infty} a_j^-$. ●

In stark contrast, as we will prove in the next section, if $\sum_{j=1}^{\infty} a_j$ converges conditionally, then both $\sum_{j=1}^{\infty} a_j^+$ and $\sum_{j=1}^{\infty} a_j^-$ diverge. Weierstrass used this fact to prove a remarkable theorem: If $\sum_{j=1}^{\infty} a_j$ is a conditionally convergent series and if L is *any real number or either* $\pm\infty$, then, by appropriately rearranging the order of the terms of the series $\sum_{j=1}^{\infty} a_j$, we can obtain a series (with exactly the same terms as the original series but in different order) that converges to L or diverges to either $\pm\infty$. On the other hand, if $\sum_{j=1}^{\infty} a_j$ converges absolutely to S, then any rearrangement of the series $\sum_{j=1}^{\infty} a_j$ also converges to S. Absolute convergence is robust; conditional convergence is delicate.

11.4 Conditional Convergence

The foregoing discussion warns us that conditional convergence is fragile. We ought to expect then that tests for conditional convergence of a series $\sum_j a_j$ might be delicate also. Here we present three of the major tests for such convergence.

THEOREM 11.4.1 The Alternating Series Test If $\{a_j\}$ is a monotone decreasing sequence of positive numbers such that $\lim_{j\to\infty} a_j = 0$, then $\sum_{j=1}^{\infty}(-1)^{j+1}a_j$ converges.

Proof. Since $\{a_j\}$ is a sequence of positive numbers that converges monotonically to 0, the partial sums of the series $\sum_{j=1}^{\infty}(-1)^{j+1}a_j$ with odd index form a monotone decreasing sequence of positive numbers. To confirm this assertion, write

$$s_{2k+1} = (a_1 - a_2) + (a_3 - a_4) + \cdots + (a_{2k-1} - a_{2k}) + a_{2k+1}.$$

Since each of the bracketed terms above is nonnegative and since a_{2k+1} is positive, we recognize that $s_{2k+1} > 0$. Notice that $s_{2k+1} - s_{2k-1} = -a_{2k} + a_{2k+1} \le 0$, since the sequence $\{a_j\}$ is monotone decreasing. Therefore, $\{s_{2k+1}\}$ is monotone decreasing and, by Theorem 1.3.7, converges. Denote the limit of this sequence by S.

The partial sums of $\sum_{j=1}^{\infty}(-1)^{j+1}a_j$ with even index form a monotone increasing sequence of nonnegative numbers. To confirm this claim, write

$$s_{2k} = (a_1 - a_2) + (a_3 - a_4) + \cdots + (a_{2k-1} - a_{2k}).$$

Clearly, s_{2k}, being the sum of nonnegative numbers, is itself nonnegative. Moreover, $s_{2k} - s_{2k-2} = a_{2k-1} - a_{2k}$ is also nonnegative. Thus $\{s_{2k}\}$ is monotone increasing. Furthermore,

$$\lim_{k\to\infty}(s_{2k+1} - s_{2k}) = \lim_{k\to\infty} a_{2k+1} = 0.$$

We deduce that $\lim_{k\to\infty} s_{2k} = \lim_{k\to\infty} s_{2k+1} = S$. This proves that $\sum_{j=1}^{\infty}(-1)^{j+1}a_j$ converges and proves the theorem. ●

EXAMPLE 13 By the alternating series test, for any $p > 0$, the series $\sum_{j=1}^{\infty}(-1)^{j+1}/j^p$ converges. If $p > 1$, the convergence is absolute; if $0 < p \leq 1$, it is conditional. From Example 14 in Section 6.4 we know that $\sum_{j=1}^{\infty}(-1)^{j+1}/j$ converges (slowly) to $\ln 2$. For values of p other than 1 we still lack the tools to determine the value of the sum. ●

EXAMPLE 14 Gregory's series, $\sum_{j=0}^{\infty}(-1)^j/(2j+1)$ converges by the alternating series test. The series does not converge absolutely; therefore convergence is conditional. We will see that the series converges (slowly) to $\tan^{-1} 1 = \pi/4$. ●

The remaining two tests of this section treat series of the form $\sum_{j=1}^{\infty} a_j b_j$ and thus can be useful when the expression giving the general term of a series can be so factored. Their proofs hinge on the following fact.

THEOREM 11.4.2 Abel's Partial Summation Formula Given a series of the form $\sum_{j=1}^{\infty} a_j b_j$, let $s_k = \sum_{j=1}^{k} a_j$. Then

$$\sum_{j=1}^{k} a_j b_j = s_k b_{k+1} - \sum_{j=1}^{k} s_j (b_{j+1} - b_j).$$

Proof. The proof consists of a straightforward computation. Let $s_0 = 0$ and write

$$\sum_{j=1}^{k} a_j b_j = \sum_{j=1}^{k} (s_j - s_{j-1}) b_j = \sum_{j=1}^{k} s_j b_j - \sum_{j=1}^{k} s_j b_{j+1} + s_k b_{k+1}$$

$$= s_k b_{k+1} - \sum_{j=1}^{k} s_j (b_{j+1} - b_j). \quad \bullet$$

Abel used his summation formula to derive a convergence test which now bears his name.

THEOREM 11.4.3 Abel's Test Suppose that $\sum_{j=1}^{\infty} a_j$ is a convergent series and that $\{b_j\}$ is a monotone, convergent sequence. Then the series $\sum_{j=1}^{\infty} a_j b_j$ converges.

Proof. The proof consists of methodically confirming that the quantities that arise in Abel's partial summation formula converge as k tends to infinity. We assume, without loss of generality, that $\{b_j\}$ is monotone increasing and converges to B. Let s_k denote the kth partial sum of $\sum_{j=1}^{\infty} a_j$. By hypothesis, $\lim_{k\to\infty} s_k = S$ exists. Therefore, we know that

$$\lim_{k\to\infty} s_k b_{k+1} = SB. \tag{11.3}$$

Next, since $\{s_k\}$ converges, it is bounded. Choose an M such that $|s_k| \leq M$ for all k in $\mathbb{N}$. Thus, for all j in $\mathbb{N}$,

$$|s_j (b_{j+1} - b_j)| \leq M(b_{j+1} - b_j). \tag{11.4}$$

The kth partial sum t_k of the series $\sum_{j=1}^{\infty} M(b_{j+1} - b_j)$ is

$$t_k = \sum_{j=1}^{k} M(b_{j+1} - b_j) = M(b_{k+1} - b_1).$$

Because $\lim_{k\to\infty} t_k = M(B - b_1)$, we deduce that the series $\sum_{j=1}^{\infty} M(b_{j+1} - b_j)$ converges. By (11.4) and the comparison test it follows that $\sum_{j=1}^{\infty} s_j(b_{j+1} - b_j)$ converges absolutely and hence converges. Thus

$$\lim_{k\to\infty} \sum_{j=1}^{k} s_j(b_{j+1} - b_j) \tag{11.5}$$

exists. Combining (11.3), (11.5), and Abel's partial summation formula confirms that

$$\lim_{k\to\infty} \sum_{j=1}^{k} a_j b_j = \lim_{k\to\infty} s_k b_{k+1} - \lim_{k\to\infty} \sum_{j=1}^{k} s_j(b_{j+1} - b_j)$$

exists. Consequently, $\sum_{j=1}^{\infty} a_j b_j$ converges. ●

EXAMPLE 15 Fix any $c > 0$ and any $d > \max\{c, 1\}$. Consider the series $\sum_{j=1}^{\infty} 1/(c^j - d^j)$. Using Abel's test, we can easily prove that this series converges. Write

$$\frac{1}{c^j - d^j} = \frac{1}{((c/d)^j - 1)d^j}.$$

Let $a_j = (1/d)^j$. Since $d > 1$, the geometric series $\sum_{j=1}^{\infty}(1/d)^j$ converges. Let $b_j = 1/[(c/d)^j - 1]$. Because $0 < c < d$, it follows that $0 < c/d < 1$ and therefore, first, $\lim_{j\to\infty} b_j = -1$ and, second, $b_j < b_{j+1}$. Thus $\{b_j\}$ converges monotonically. By Abel's test, the original series converges. ●

Our final test is similar in spirit to Abel's and is often used in much the same way. For many series of the form $\sum_{j=1}^{\infty} a_j b_j$ one or the other test will resolve the question of convergence.

THEOREM 11.4.4 Dirichlet's Test Let $\sum_{j=1}^{\infty} a_j b_j$ be a series of real numbers. If the sequence $\{s_k\}$ of kth partial sums of the series $\sum_{j=1}^{\infty} a_j$ is bounded and if $\{b_j\}$ is a sequence of positive numbers that converges monotonically to 0, then the series $\sum_{j=1}^{\infty} a_j b_j$ converges.

Proof. We begin with Abel's partial summation formula

$$\sum_{j=1}^{k} a_j b_j = s_k b_{k+1} - \sum_{j=1}^{k} s_j(b_{j+1} - b_j),$$

and examine each part in turn. First, there is an M such that $|s_k| \le M$ for all k in $\mathbb{N}$. Since $\{b_j\}$ converges to 0,

$$0 \le \lim_{k\to\infty} |s_k b_{k+1}| \le M \lim_{k\to\infty} b_{k+1} = 0.$$

By the squeeze play, $\lim_{k\to\infty} s_k b_{k+1} = 0$.

Since $\{b_j\}$ is monotone decreasing and $\{s_k\}$ is bounded above by M,

$$|s_j(b_{j+1} - b_j)| \leq M(b_j - b_{j+1}). \tag{11.6}$$

By telescoping, the kth partial sum of $\sum_{j=1}^{\infty} M(b_j - b_{j+1})$ is $M(b_1 - b_{k+1})$. Because $\{b_j\}$ converges to 0, we deduce that

$$\sum_{j=1}^{\infty} M(b_j - b_{j+1}) = Mb_1.$$

Consequently, by (11.6), the series $\sum_{j=1}^{\infty} s_j(b_{j+1} - b_j)$ converges absolutely. It follows that

$$\lim_{k\to\infty} \sum_{j=1}^{k} s_j(b_{j+1} - b_j)$$

exists. From Abel's formula we deduce that

$$\lim_{k\to\infty} \sum_{j=1}^{k} a_j b_j = \lim_{k\to\infty} s_k b_{k+1} - \lim_{k\to\infty} \sum_{j=1}^{k} s_j(b_{j+1} - b_j)$$

exists. Therefore, $\sum_{j=1}^{\infty} a_j b_j$ converges. This proves that Dirichlet's test is valid. ●

To present examples that arose in Dirichlet's original work (in the study of Fourier series), we take a slight detour to examine the connection between the complex exponential function e^{ix} and the sine and cosine functions and to derive relevant trigonometric identities. The detour is of interest in its own right and will be useful later in Chapter 14. Recall that DeMoivre's theorem states that, for any x in $\mathbb{R}$,

$$e^{ix} = \cos x + i \sin x.$$

Thus, for any x in $\mathbb{R}$ that is not an even multiple of π, we have, by the formula for the sum of a finite geometric series,

$$\sum_{j=1}^{k} \cos(jx) + i \sum_{j=1}^{k} \sin(jx) = \sum_{j=1}^{k} e^{jxi} = e^{xi} \frac{1 - e^{kxi}}{1 - e^{xi}} \tag{11.7}$$

To extract the trigonometric identities hidden in (11.7), we examine the numerator and denominator in the following lemma.

LEMMA

i) $1 - e^{xi} = -2ie^{xi/2} \sin(x/2)$.

ii) $1 - e^{kxi} = -2ie^{kxi/2} \sin(kx/2)$.

Proof. We prove part (ii) since (i) can be derived from (ii) by setting $k = 1$. Factor out $e^{kxi/2}$ from $1 - e^{kxi}$, use DeMoivre's theorem, and simplify to obtain

$$\begin{aligned} 1 - e^{kxi} &= e^{kxi/2}[e^{-kxi/2} - e^{kxi/2}] \\ &= e^{kxi/2}\left[\cos\left(\frac{kx}{2}\right) - i\sin\left(\frac{kx}{2}\right) - \cos\left(\frac{kx}{2}\right) - i\sin\left(\frac{kx}{2}\right)\right] \\ &= -2ie^{kxi/2}\sin\left(\frac{kx}{2}\right). \end{aligned}$$

This proves the lemma. ●

THEOREM 11.4.5 For any real number x that is not an even multiple of π, we have

i) $$\sum_{j=1}^{k} \cos(jx) = \frac{\cos[(k+1)x/2]\sin(kx/2)}{\sin(x/2)}.$$

ii) $$\sum_{j=1}^{k} \sin(jx) = \frac{\sin[(k+1)x/2]\sin(kx/2)}{\sin(x/2)}.$$

Proof. Substitute the identities of the lemma in the identity (11.7) and simplify to obtain

$$\sum_{j=1}^{k} \cos(jx) + i\sum_{j=1}^{k} \sin(jx) = \frac{e^{(k+1)xi/2}\sin(kx/2)}{\sin(x/2)}$$

$$= \frac{[\cos[(k+1)x/2] + i\sin[(k+1)x/2]]\sin(kx/2)}{\sin(x/2)}.$$

Equate the real and imaginary parts of this identity to obtain the two identities of the theorem. ●

EXAMPLE 16 Let x be any real number that is not an even multiple of π. We use Dirichlet's test to show that the two series $\sum_{j=1}^{\infty}(\cos jx)/j$ and $\sum_{j=1}^{\infty}(\sin jx)/j$ converge.

To treat $\sum_{j=1}^{\infty}\cos(jx)/j$, let $a_j = \cos(jx)$ and let $b_j = 1/j$. Then the sequence $\{s_k\}$ of kth partial sums of the series $\sum_{j=1}^{\infty} a_j$ is bounded, as the following computation confirms:

$$\begin{aligned} |s_k| = \left|\sum_{j=1}^{k}\cos(jx)\right| &= \left|\frac{\cos[(k+1)x/2]\sin(kx/2)}{\sin(x/2)}\right| \\ &\le \frac{1}{|\sin(x/2)|}, \end{aligned}$$

for all k in $\mathbb{N}$. Further, $\{b_j\}$ is a sequence of positive numbers that converges monotonically to 0. By Dirichlet's test, the series $\sum_{j=1}^{\infty}\cos(jx)/j$ converges. A completely analogous argument proves that $\sum_{j=1}^{\infty}\sin(jx)/j$ also converges. ●

Notice that Abel's and Dirichlet's tests need not be restricted to series with both positive and negative terms, but if the absolute convergence tests fail or if the form of the general term of the series suggests their use, then they warrant your attention. They give you tools to decide convergence, not absolute convergence, which must be decided on separate grounds.

Now suppose that $\sum_{j=1}^{\infty} a_j$ is a conditionally convergent series. Let $a_j^+ = \max\{a_j, 0\}$ and $a_j^- = \max\{-a_j, 0\}$. Form the series $\sum_{j=1}^{\infty} a_j^+$ and $\sum_{j=1}^{\infty} a_j^-$. If either of these series, say, $\sum_{j=1}^{\infty} a_j^+$, were to converge, then, since it consists only of nonnegative terms, it must converge absolutely. Furthermore, the series $\sum_{j=1}^{\infty} a_j^-$ must also converge because

$$\lim_{k\to\infty} \sum_{j=1}^{k} a_j^- = \lim_{k\to\infty} \sum_{j=1}^{k} a_j^+ - \lim_{k\to\infty} \sum_{j=1}^{k} a_j = \sum_{j=1}^{\infty} a_j^+ - \sum_{j=1}^{\infty} a_j$$

and because both $\sum_{j=1}^{\infty} a_j^+$ and $\sum_{j=1}^{\infty} a_j$ converge. But $\sum_{j=1}^{\infty} a_j^-$ is a series of nonnegative terms and, therefore, if it converges at all, it must converge absolutely. Since

$$\sum_{j=1}^{\infty} a_j = \sum_{j=1}^{\infty} a_j^+ - \sum_{j=1}^{\infty} a_j^-$$

we deduce that $\sum_{j=1}^{\infty} a_j$ must also converge absolutely. But we assumed that $\sum_{j=1}^{\infty} a_j$ converges conditionally, which precludes absolute convergence. We are forced to conclude that neither $\sum_{j=1}^{\infty} a_j^+$ nor $\sum_{j=1}^{\infty} a_j^-$ converges. Since the sequences of partial sums of these series are both monotone increasing and diverge, they must be unbounded. We finally conclude that

$$\sum_{j=1}^{\infty} a_j^+ = \sum_{j=1}^{\infty} a_j^- = \infty.$$

This discussion proves the following theorem.

THEOREM 11.4.6 If $\sum_{j=1}^{\infty} a_j$ converges conditionally, then the two series $\sum_{j=1}^{\infty} a_j^+$ and $\sum_{j=1}^{\infty} a_j^-$ diverge to infinity. ●

With this theorem in hand we can take up Weierstrass's startling result, that a rearrangement of a conditionally convergent series exists to cause it to converge to any prescribed real number or to diverge. A general rearrangement of a series is easy to describe. Let ϕ be any one-to-one function mapping $\mathbb{N}$ onto $\mathbb{N}$. The rearrangement of $\sum_{j=1}^{\infty} a_j$ induced by ϕ is the series $\sum_{j=1}^{\infty} a_{\phi(j)} = a_{\phi(1)} + a_{\phi(2)} + a_{\phi(3)} + \cdots$. To prove Weierstrass's theorem, we need merely construct a function that rearranges the terms of the series appropriately to achieve our objective. Here we will take up the case when L in $(0, \infty)$ is prescribed and will indicate how to construct the rearrangement of $\sum_{j=1}^{\infty} a_j$ so that $\sum_{j=1}^{\infty} a_{\phi(j)}$ converges to L. We leave the remaining cases for you.

The construction depends entirely on the fact that the original series $\sum_{j=1}^{\infty} a_j$ converges, so that its kth term converges to 0, and on the fact that both series $\sum_{j=1}^{\infty} a_j^+$ and $\sum_{j=1}^{\infty} a_j^-$ diverge to ∞. Let t_k denote the kth partial sum of $\sum_{j=1}^{\infty} a_j^+$; let u_k

denote the kth partial sum of $\sum_{j=1}^{\infty} a_j^-$. The sequences $\{t_k\}$ and $\{u_k\}$ are monotone increasing and unbounded. Choose the smallest k_1 such that $t_{k_1} > L$. Choose the smallest m_1 such that $t_{k_1} - u_{m_1} < L$. This is certainly possible because each of the sequences of partial sums diverges monotonically to ∞. Next choose the smallest $k_2 > k_1$ such that

$$t_{k_2} - u_{m_1} > L.$$

This is certainly possible because the sequence $\{t_k\}$ diverges to ∞. Choose the smallest $m_2 > m_1$ such that

$$t_{k_2} - u_{m_1} < L.$$

Again, this can be done because $\{u_k\}$ diverges to ∞. Notice that the well-ordering property of $\mathbb{N}$ is at play here, so that the indices we are choosing are uniquely and well defined. Suppose now that we have chosen two sets of indices $\{k_1, k_2, \ldots, k_n\}$ and $\{m_1, m_2, \ldots, m_n\}$ such that

i) $k_1 < k_2 < \cdots < k_n$ and $m_1 < m_2 < \cdots < m_n$.

ii) For each $i = 1, 2, \ldots, n$, each index is the smallest one having the properties

$$t_{k_i} - u_{m_{i-1}} > L \qquad \text{and} \qquad t_{k_i} - u_{m_i} < L.$$

This procedure is more elaborate in appearance than in practice. All that we have done is first choose a certain number of the positive terms of $\sum_{j=1}^{\infty} a_j$; next we choose a certain number of the negative terms of $\sum_{j=1}^{\infty} a_j$. As we proceed we alternately choose a number of positive, then negative terms. As we methodically choose these terms of the original series we choose terms from smaller index to larger so no term is omitted. We make our choices in such a way that the sums of the terms chosen fall just above and just below the number L.

At the inductive step, choose k_{n+1} to be the smallest index greater than k_n such that $t_{k_{n+1}} - u_{m_n} > L$ and choose m_{n+1} to be the smallest index greater than m_n such that $t_{k_{n+1}} - u_{m_{n+1}} < L$. We can make these choices because each of the sequences $\{t_k\}$ and $\{u_k\}$ diverges to ∞. Inductively this rearranges the series $\sum_{j=1}^{\infty} a_j$, (although, admittedly, writing down the function ϕ which induces this rearrangement would prove cumbersome and not particularly enlightening).

Our claim, of course, is that this rearranged series converges to L. Here the fact that $\lim_{j\to\infty} a_j = 0$ comes into play. Given $\epsilon > 0$, choose a j_0 such that, for $j > j_0$, we have $|a_j| < \epsilon$. Proceed to the nth stage of the construction such that k_n and m_n are both greater than j_0. Thus, the terms we are adding or subtracting from the rearranged series are, at this and all subsequent stages, less than ϵ in magnitude. We have $t_{k_n} - u_{m_n} < L$ and, by the minimality of m_n, it must be that a_{m_n} is negative. (Otherwise a smaller index could have been chosen at the nth stage). Therefore, again since m_n is minimal,

$$t_{k_n} - u_{m_n} < L,$$

but

$$t_{k_n} - (u_{m_n} - a_{m_n}^-) = t_{k_n} - u_{m_n} - a_{m_n} > L.$$

Since $m_n > j_0$, we know that $-\epsilon < a_{m_n} < 0$. It follows that

$$L - \epsilon < t_{k_n} - u_{m_n} < L.$$

Likewise, $t_{k_{n+1}} - u_{m_n} > L$. By the minimality of k_{n+1}, we know that $a_{k_{n+1}}$ must be positive and that

$$(t_{k_{n+1}} - a^+_{k_{n+1}}) - u_{m_n} = (t_{k_{n+1}} - a_{k_{n+1}}) - u_{m_n} < L.$$

Again, since $k_{n+1} > j_0$, it follows that $0 < a_{k_{n+1}} < \epsilon$. Therefore,

$$L < t_{k_{n+1}} - u_{m_n} < L + \epsilon.$$

Since all the intermediate partial sums, indexed between $k_n + m_n$ and $k_{n+1} + m_n$, must fall between these two extremes, we conclude that they all lie in $N(L; \epsilon)$. Furthermore, since n is arbitrary subject to the requirement that k_n and m_n exceed j_0, all partial sums of the rearranged series, with sufficiently large index, lie in the neighborhood $N(L; \epsilon)$. We conclude that the sequence of partial sums of the rearranged series must converge to L and therefore the rearranged series does also. Our reasoning proves part (i) of the following theorem.

THEOREM 11.4.7 Weierstrass Let $\sum_{j=1}^{\infty} a_j$ be a conditionally convergent series.

i) Let L be any real number. Then there exists a rearrangement $\sum_{j=1}^{\infty} a_{\phi(j)}$ of $\sum_{j=1}^{\infty} a_j$ that converges to L.

ii) If L_1 and L_2 are any two real numbers or $\pm\infty$ with $L_1 < L_2$, then there exists a rearrangement $\sum_{j=1}^{\infty} a_{\phi(j)}$ of $\sum_{j=1}^{\infty} a_j$ such that $\liminf t_k = L_1$ and $\limsup t_k = L_2$, where $t_k = \sum_{j=1}^{\infty} a_{\phi(j)}$. ●

We leave the proof of part (ii) for you; simple modifications of the detailed argument we have presented will handle each case. As you can see, all that is required for this proof to work is that the kth term converges to 0 and that the series of positive terms and the series of negative terms diverge. By rearranging such a series, whether it initially converges or not, we can cause some rearrangement to converge to any number whatsoever. Some other rearrangement can be made to oscillate between two extremes or to diverge to $\pm\infty$. Conditionally convergent series converge with great delicacy; when working with them, say numerically, you must sum them in the order in which the terms are presented.

EXAMPLE 17 We know that the series $\sum_{j=1}^{\infty} (-1)^{j+1}/j$ converges conditionally to $\ln 2$ and that the series $\sum_{j=1}^{\infty} 1/j$ diverges. In this example, we show how to rearrange this series to converge to $(3/2)\ln 2$. We first show that the sequence whose terms are

$$c_k = \sum_{j=1}^{k} \frac{1}{j} - \ln k$$

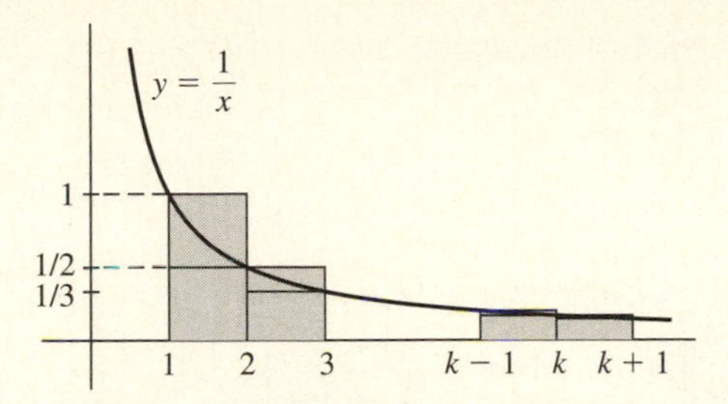

Figure 11.2

converges. From Fig. 11.2 we obtain two useful inequalities. The first is

$$1 + \int_2^{k+1} \frac{1}{t}\,dt < \sum_{j=1}^{k} \frac{1}{j},$$

obtained by summing the upper rectangular approximations to the integral. From this inequality it follows that

$$\begin{aligned} c_k = \sum_{j=1}^{k} \frac{1}{j} - \ln k &> 1 + \int_2^{k+1} \frac{1}{t}\,dt - \ln k \\ &= 1 - \ln 2 + \ln\left(\frac{(k+1)}{k}\right) > 1 - \ln 2 = .30685\ldots. \end{aligned}$$

Therefore the sequence $\{c_k\}$ is bounded below.

Referring again to Fig. 11.2, we obtain the second inequality,

$$\frac{1}{k+1} < \int_k^{k+1} \frac{1}{t}\,dt = \ln\left(\frac{k+1}{k}\right).$$

From this last we obtain

$$c_{k+1} - c_k = \frac{1}{k+1} - \ln\left(\frac{k+1}{k}\right) < 0,$$

and deduce that $\{c_k\}$ is monotone decreasing. Since it is bounded below, $\{c_k\}$ converges. Its limit, known as *Euler's constant* and denoted γ, has an approximate value of .5772 . . . .

Now rearrange the series $\sum_{j=1}^{\infty}(-1)^{j+1}/j$ in such a way that the following series is obtained:

$$1 + \tfrac{1}{3} - \tfrac{1}{2} + \tfrac{1}{5} + \tfrac{1}{7} - \tfrac{1}{4} + \tfrac{1}{9} + \tfrac{1}{11} - \tfrac{1}{6} + \tfrac{1}{13} + \tfrac{1}{15} - \tfrac{1}{8} + \cdots.$$

(After including the next two positive terms, include the next negative one.) We claim that this series converges, not to ln 2, but to (3/2) ln 2. To prove this, let t_k

denote the kth partial sum of the rearranged series and compute first the $3k$th partial sum:

$$t_{3k} = \left[1 + \frac{1}{3} - \frac{1}{2}\right] + \left[\frac{1}{5} + \frac{1}{7} - \frac{1}{4}\right] + \cdots + \left[\frac{1}{4k-3} + \frac{1}{4k-1} - \frac{1}{2k}\right].$$

Add the positive and the negative terms separately to obtain

$$\begin{aligned} t_{3k} &= \left[1 + \frac{1}{3} + \frac{1}{5} + \frac{1}{7} + \cdots + \frac{1}{4k-3} + \frac{1}{4k-1}\right] - \left[\frac{1}{2} + \frac{1}{4} + \frac{1}{6} + \cdots + \frac{1}{2k}\right] \\ &= \sum_{j=1}^{2k} \frac{1}{2j-1} - \frac{1}{2}\sum_{j=1}^{k} \frac{1}{j}. \end{aligned}$$

In the first summation insert the missing terms with even denominators, then remove them. Also use the fact that $c_m = \sum_{j=1}^{m} 1/j - \ln m$ for any m in $\mathbb{N}$.

$$\begin{aligned} t_{3k} &= \sum_{j=1}^{4k} \frac{1}{j} - \sum_{j=1}^{2k} \frac{1}{2j} - \frac{1}{2}\sum_{j=1}^{k} \frac{1}{j} \\ &= \sum_{j=1}^{4k} \frac{1}{j} - \frac{1}{2}\sum_{j=1}^{2k} \frac{1}{j} - \frac{1}{2}\sum_{j=1}^{k} \frac{1}{j} \\ &= [c_{4k} + \ln 4k] - \tfrac{1}{2}[c_{2k} + \ln 2k] - \tfrac{1}{2}[c_k + \ln k] \\ &= \left[c_{4k} - \tfrac{1}{2}c_{2k} - \tfrac{1}{2}c_k\right] + \tfrac{3}{2}\ln 2. \end{aligned}$$

Since $\lim_{m\to\infty} c_m = \gamma$, we deduce that

$$\lim_{k\to\infty} t_{3k} = \left[\gamma - \tfrac{1}{2}\gamma - \tfrac{1}{2}\gamma\right] + \tfrac{3}{2}\ln 2 = \tfrac{3}{2}\ln 2.$$

From the definition of the rearrangement, we see that

$$t_{3k+1} = t_{3k} + \frac{1}{4k+1}$$

and therefore that

$$\lim_{k\to\infty} t_{3k+1} = \lim_{k\to\infty}\left[t_{3k} + \frac{1}{4k+1}\right] = \tfrac{3}{2}\ln 2$$

also. Likewise, $t_{3k+2} = t_{3k+1} + 1/(4k+3)$ and therefore

$$\lim_{k\to\infty} t_{3k+2} = \tfrac{3}{2}\ln 2.$$

This proves that the sequence $\{t_k\}$, and hence the rearranged series, converges to $(3/2)\ln 2$. As you can check with your calculator, this rearranged series is obtained in precise accord with the general method used to prove Weierstrass's theorem; you can apply that general method to any conditionally convergent series to obtain any finite number of rearranged terms whose sum is near any prescribed limit. The degree of accuracy you can achieve will depend only on your persistence and the capabilities of your machine. ●

In contrast to the property of conditionally convergent series revealed by Weierstrass's theorem, we have the following fact about absolutely convergent series.

THEOREM 11.4.8 If $\sum_{j=1}^{\infty} a_j$ converges absolutely to S, then any rearrangement $\sum_{j=1}^{\infty} a_{\phi(j)}$ of $\sum_{j=1}^{\infty} a_j$ also converges to S.

Proof. We consider first the case when the terms of the series are nonnegative. Therefore, the sequence $\{s_k\}$ of kth partial sums of $\sum_{j=1}^{\infty} a_j$ is monotone increasing and converges to S. For each k, let $t_k = \sum_{j=1}^{k} a_{\phi(j)}$ be the kth partial sum of the rearranged series and let

$$m_k = \max\{\phi(1), \phi(2), \ldots, \phi(k)\}.$$

Note that $m_k \geq k$. Since the a_j are nonnegative, it follows that $t_k \leq s_{m_k} \leq S$. Thus, the sequence $\{t_k\}$ is also monotone increasing and is bounded above by S. By Theorem 1.3.7 the sequence $\{t_k\}$ also converges, say to T. Clearly $T \leq S$. In summary, if the terms of the series are nonnegative, then an arbitrary rearrangement $\sum_{j=1}^{\infty} a_{\phi(j)}$ of $\sum_{j=1}^{\infty} a_j$ converges to some number $T \leq S$.

Now, $\sum_{j=1}^{\infty} a_j$ is itself a rearrangement of $\sum_{j=1}^{\infty} a_{\phi(j)}$, effected by the inverse function ϕ^{-1}. Applying the preceding argument, reversing the roles of $\sum_{j=1}^{\infty} a_j$ and $\sum_{j=1}^{\infty} a_{\phi(j)}$, shows that $S \leq T$. Thus $S = T$.

Next let $\sum_{j=1}^{\infty} a_j$ be any absolutely convergent series with sum S. We know, then, that the two series $\sum_{j=1}^{\infty} a_j^+$ and $\sum_{j=1}^{\infty} a_j^-$, each consisting only of nonnegative terms, also converge to S^+ and S^-, respectively. We also know that $S = S^+ - S^-$. Any rearrangement of $\sum_{j=1}^{\infty} a_j$ is also a rearrangement of both $\sum_{j=1}^{\infty} a_j^+$ and $\sum_{j=1}^{\infty} a_j^-$. Since, by the first part of this proof, the convergence of these series of nonnegative terms is unaffected by this rearrangement, so also is that of $\sum_{j=1}^{\infty} a_j$. This proves the theorem. ●

11.5 The Cauchy Product

It is easy to confirm that a linear combination of convergent series is also convergent. It follows that the collection of all convergent series forms a vector space. But when we consider the possibility of defining the product of two convergent series, we find that the situation is significantly more complicated. Given the subtle point regarding the effect of rearrangements on conditionally convergent series, we must take care in any such definition. It seems reasonable that the product of a series $\sum_j a_j$ and a series $\sum_j b_j$ ought to be a series whose terms consist of all possible products $a_i b_j$, but since the *arrangement* of those terms may affect the value of the product series—and even its convergence—we must be more specific. The generally accepted and preferred definition of the product of two series is the *Cauchy product* identified in the following definition. Note that here we begin the indexing of the series with $j = 0$ in order to simplify the notation. This convention also anticipates the application of this product to power series in Chapter 12.

DEFINITION 11.5.1 The Cauchy Product The *Cauchy product* of two series $\sum_{j=0}^{\infty} a_j$ and $\sum_{j=0}^{\infty} b_j$ is defined to be

$$\left(\sum_{j=0}^{\infty} a_j\right)\left(\sum_{j=0}^{\infty} b_j\right) = \sum_{k=0}^{\infty} c_k,$$

where $c_k = \sum_{j=0}^{k} a_j b_{k-j}$. •

The definition is motivated by the observation that, if we *formally multiply* $\sum_{j=0}^{\infty} a_j x^j$ and $\sum_{j=0}^{\infty} b_j x^j$, then collect the coefficients of like powers of x, we obtain

$$\left(\sum_{j=0}^{\infty} a_j x^j\right)\left(\sum_{j=0}^{\infty} b_j x^j\right) = a_0b_0 + (a_0b_1 + a_1b_0)x + (a_0b_2 + a_1b_1 + a_2b_0)x^2 + \cdots$$

$$= \sum_{k=0}^{\infty}\left[\sum_{j=0}^{k} a_j b_{k-j}\right] x^k = \sum_{k=0}^{\infty} c_k x^k.$$

The coefficient of x^k is given by Cauchy's product. Setting $x = 1$ formally yields the foregoing formula; however, such formalism constitutes no proof.

It is reasonable to ask whether the Cauchy product of two convergent series also converges. We also ask, if the product does converge, what is its sum? We will show subsequently that, if $\sum_{j=0}^{\infty} a_j$ and $\sum_{j=0}^{\infty} b_j$ converge absolutely to S and T, respectively, then their Cauchy product also converges absolutely with sum ST. (Actually, we need only require one of the factor series to converge absolutely in order to guarantee that the Cauchy product converges, although the convergence may be conditional.) Accepting this assertion for the time being, we can say, then, that the collection of all absolutely convergent infinite series forms a commutative ring with identity where multiplication is the Cauchy product. The identity is the series $\sum_{j=0}^{\infty} a_j$ where $a_0 = 1$ and $a_j = 0$ for $j \geq 1$.

For conditionally convergent series we can make no such claim; if both the factor series converge conditionally, then the Cauchy product may diverge. Cauchy himself was aware of this shortcoming when he proposed the product; he offered the following example.

EXAMPLE 18 The Cauchy product of the conditionally convergent series $\sum_{j=0}^{\infty}(-1)^j/\sqrt{j+1}$ with itself—in other words, its square—diverges. That product is the series $\sum_{k=0}^{\infty} c_k$ with

$$c_k = \sum_{j=0}^{k} a_j a_{k-j} = \sum_{j=0}^{k}\left[\frac{(-1)^j}{\sqrt{j+1}}\right]\left[\frac{(-1)^{k-j}}{\sqrt{(k-j)+1}}\right]$$

$$= (-1)^k \sum_{j=0}^{k} \frac{1}{\sqrt{(j+1)(k-j+1)}}.$$

For $j = 0, 1, 2, \ldots, k$, we obtain the inequality

$$(j+1)(k-j+1) = jk - j^2 + k + 1 \le jk + k + 1$$
$$< (k+1)^2.$$

It follows that, for these values of j,

$$\frac{1}{\sqrt{(j+1)(k-j+1)}} > \frac{1}{k+1}.$$

We deduce that

$$|c_k| = \sum_{j=0}^{k} \frac{1}{\sqrt{(j+1)(k-j+1)}} > \frac{k+1}{k+1} = 1.$$

Thus, the kth term of the product series does not tend to 0 and $\sum_{k=0}^{\infty} c_k$ must diverge. ●

Let's examine the Cauchy product of $\sum_{j=0}^{\infty} a_j$ and $\sum_{j=0}^{\infty} b_j$ more closely. The pth term c_p of that product is the sum of all terms $a_i b_j$ where $0 \le i \le p, 0 \le j \le p$, and $i + j = p$. The kth partial sum of the Cauchy product is the sum of all such products $a_i b_j$ where $0 \le i + j \le k$. Hence it is the sum over the set of all points (i, j) in the triangular set in Fig. 11.3. Eventually we will partition such a set into carefully chosen subsets and sum over all (i, j) in each separately. We will also sum over all (i, j) in rectangular sets as in Fig. 11.4 on page 518. Such rectangular sums are particularly simple to compute; we have

$$\sum_{i=i_1}^{i_2} \sum_{j=j_1}^{j_2} a_i b_j = \left(\sum_{i=i_1}^{i_2} a_i\right)\left(\sum_{j=j_1}^{j_2} b_j\right).$$

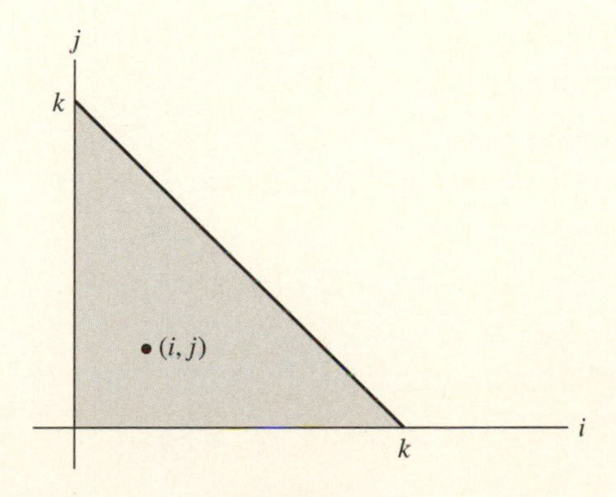

Figure 11.3

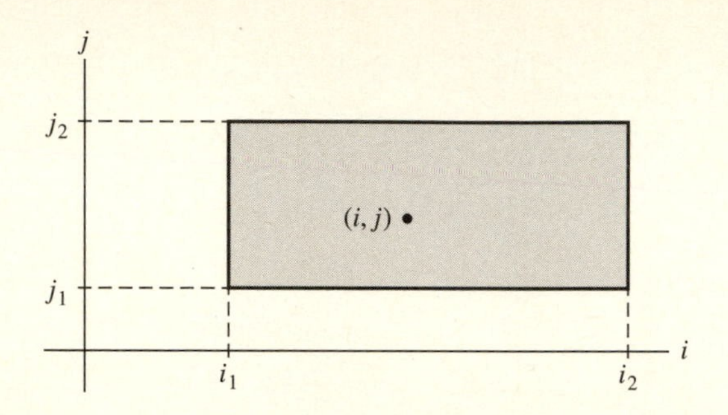

Figure 11.4

Suppose now that $\sum_{j=0}^{\infty} a_j$ and $\sum_{j=0}^{\infty} b_j$ converge absolutely. We want to show that the Cauchy product of these two series also converges and that the convergence is absolute. We know four series to be convergent; denote their partial sums and sums as follows:

$$s_k = \sum_{j=0}^{k} a_j; \qquad \lim_{k\to\infty} s_k = S,$$

$$t_k = \sum_{j=0}^{k} b_j; \qquad \lim_{k\to\infty} t_k = T,$$

$$u_k = \sum_{j=0}^{k} |a_j|; \qquad \lim_{k\to\infty} u_k = U,$$

$$v_k = \sum_{j=0}^{k} |b_j|; \qquad \lim_{k\to\infty} v_k = V.$$

Note that $\{u_k\}$ and $\{v_k\}$ are monotone increasing sequences and that $\{s_k t_k\}$ converges to ST. Let

$$w_k = \sum_{j=0}^{k} c_j = \sum_{j=0}^{k} \left(\sum_{i=0}^{j} a_i b_{j-i} \right).$$

We want to show that $\{w_k\}$ converges to ST.

To this end, fix $\epsilon > 0$ and k_0 such that, for any k, k_1, and k_2 all greater than k_0, we have

$$|s_k t_k - ST| < \epsilon,$$

$$|u_{k_1} - u_{k_2}| < \epsilon,$$

and

$$|v_{k_1} - v_{k_2}| < \epsilon.$$

Fix any $k \geq k_0$. Choose any $m > 2k$ and compute the mth partial sum, w_m, of the Cauchy product. Do so by summing over the six sets indicated in Fig. 11.5. The first sum is

$$S_1 = \sum_{i=0}^{k} \sum_{j=0}^{k} a_i b_j = \left(\sum_{i=0}^{k} a_i \right) \left(\sum_{j=0}^{k} b_j \right) = s_k t_k.$$

To treat the second sum, write

$$|S_2| = \left| \sum_{i=k+1}^{m-k} \sum_{j=0}^{k} a_i b_j \right| = \left| \left(\sum_{i=k+1}^{m-k} a_i \right) \left(\sum_{j=0}^{k} b_j \right) \right| \leq \left(\sum_{i=k+1}^{m-k} |a_i| \right) \left(\sum_{j=0}^{k} |b_j| \right)$$
$$= |u_{m-k} - u_k| v_k < \epsilon V.$$

Similarly,

$$|S_3| = \left| \sum_{i=0}^{k} \sum_{j=k+1}^{m-k} a_i b_j \right| = \left| \left(\sum_{i=0}^{k} a_i \right) \left(\sum_{j=k+1}^{m-k} b_j \right) \right| \leq u_k |v_{m-k} - v_k| < U\epsilon.$$

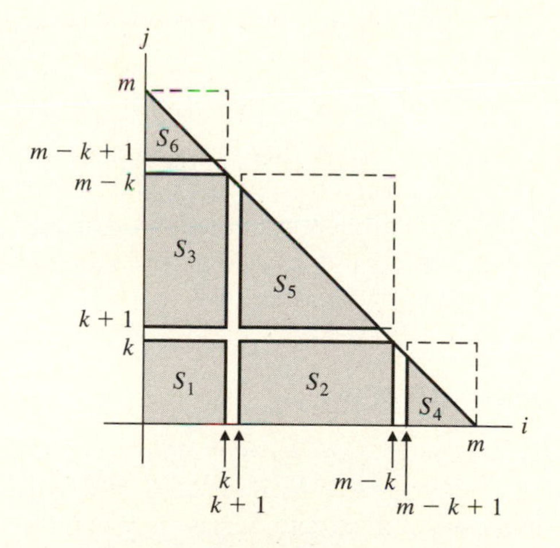

Figure 11.5

The remaining three sums are taken over triangular sets. Enlarge each set to be rectangular as indicated in Fig. 11.5. We have

$$|S_4| \leq \sum_{i=m-k+1}^{m} \sum_{j=0}^{m-i} |a_i|\,|b_j| \leq \sum_{i=m-k+1}^{m} \sum_{j=0}^{k} |a_i|\,|b_j| = \left(\sum_{i=m-k+1}^{m} |a_i| \right) \left(\sum_{j=0}^{k} |b_j| \right)$$
$$= |u_m - u_{m-k}| v_k < \epsilon V.$$

Likewise,

$$|S_5| \leq |u_{m-k} - u_k|\,|v_{m-k} - v_k| < \epsilon^2,$$

and

$$|S_6| \le u_k |v_m - v_{m-k}| < U\epsilon.$$

Thus, for any $m > 2k > k \ge k_0$,

$$|w_m - ST| = \left| \sum_{r=1}^{6} S_r - ST \right| \le |S_1 - ST| + \sum_{r=2}^{6} |S_r|$$
$$< \epsilon + \epsilon V + U\epsilon + \epsilon V + \epsilon^2 + U\epsilon$$
$$= (1 + 2U + 2V + \epsilon)\epsilon.$$

This proves that $\lim_{m\to\infty} u_m = ST$. Hence the Cauchy product of $\sum_{j=0}^{\infty} a_j$ and $\sum_{j=0}^{\infty} b_j$ converges to ST.

It is easy to confirm that $\sum_{k=0}^{\infty} c_k$ actually converges absolutely. The partial sums of $\sum_{k=0}^{\infty} |c_k|$ form an increasing sequence that is bounded above by the partial sums of the Cauchy product of $\sum_{j=0}^{\infty} |a_j|$ and $\sum_{j=0}^{\infty} |b_j|$. As we have just seen, this latter Cauchy product converges monotonically to UV. Thus, we have proved the following theorem.

THEOREM 11.5.1 The Cauchy product of two absolutely convergent series $\sum_{j=0}^{\infty} a_j$ and $\sum_{j=0}^{\infty} b_j$ converges absolutely to $(\sum_{j=0}^{\infty} a_j)(\sum_{j=0}^{\infty} b_j)$. ●

It is straightforward to confirm that the Cauchy product of two series gives a commutative multiplication and that this multiplication distributes over addition (provided that all the products involved converge). The confirmation that the Cauchy product gives an associative multiplication is rather involved and will be omitted. If we restrict ourselves to absolutely convergent series, then the collection of all such series forms a commutative ring with identity.

11.6 CESÀRO SUMMABILITY

The convergence of a series $\sum_{j=1}^{\infty} a_j$ to S is defined to mean that the sequence $\{s_k\}$ of kth partial sums converges to S; if that sequence diverges, then the series diverges. However, even some divergent series may exhibit regular behavior that one might hope to describe. Clearly, any such description requires a generalization of the concept of limit. There now exist several interesting and general extensions of the limit concept, many of them based on the work of Stefan Banach (1892–1945). Historically, most of these generalizations emerged from one classic example, that of the *arithmetic average*. These averages, like all averages, tend to "smooth" the behavior of the terms of a sequence; although $\{s_k\}$ may diverge, it may happen that the sequence of arithmetic averages of the terms of the sequence $\{s_k\}$ converges. (See Exercise 1.70 in Section 1.3.) To be specific, for each k in $\mathbb{N}$, let σ_k denote the arithmetic average of the first k terms of the sequence $\{s_j\}$, that is,

$$\sigma_k = \frac{s_1 + s_2 + \cdots + s_k}{k} = \frac{1}{k} \sum_{j=1}^{k} s_j.$$

DEFINITION 11.6.1 If the sequence $\{\sigma_k\}$ converges to σ, then we say that the series $\sum_{j=1}^{\infty} a_j$ is *Cesàro summable* and converges $(C, 1)$ to σ. We write

$$\sum_{j=1}^{\infty} a_j = \sigma\ (C, 1). \quad \bullet$$

EXAMPLE 19 Consider the series $\sum_{j=1}^{\infty}(-1)^{j-1}$. Obviously, this series diverges; its jth term fails to converge to 0. Its sequence of partial sums oscillates between 0 and 1. To be specific, for k in $\mathbb{N}$, $s_k = [1 - (-1)^k]/2$. We compute the arithmetic average of these partial sums:

$$\sigma_k = \frac{1}{k}\sum_{j=1}^{k} s_j = \frac{1}{k}\sum_{j=1}^{k}\left[\frac{1}{2} - \frac{(-1)^j}{2}\right]$$

$$= \frac{1}{2} - \frac{1}{2k}\sum_{j=1}^{k}(-1)^j = \frac{1}{2} + \frac{1}{4k}[1 + (-1)^{k+1}].$$

Consequently, $\lim_{k\to\infty}\sigma_k = 1/2$. Thus $\sum_{j=1}^{\infty}(-1)^j = 1/2\,(C, 1)$. $\bullet$

It might occur to you that, if $\{\sigma_k\}$ diverges, we can take this process one step further and compute the arithmetic averages of $\{\sigma_k\}$. If this second sequence of averages converges—that is, if $\{\sigma_k\}$ converges $(C, 1)$—then the original series is said to converge $(C, 2)$. And with this insight, you can discern that there is no end to the process; there are infinitely many degrees of Cesàro summability. We shall concern ourselves only with the first degree, $(C, 1)$ convergence, except in this second example.

EXAMPLE 20 Consider $\sum_{j=1}^{\infty}(-1)^{j-1}j$. Clearly this series diverges. As you can easily compute, the partial sums of the series are, for odd indices, $s_{2k-1} = k$ and for even indices, $s_{2k} = -k$. As you can compute, the arithmetic averages of the sequence $\{s_k\}$ are, for even indices, $\sigma_{2k} = 0$, and, for odd indices, $\sigma_{2k-1} = k/(2k-1)$. It follows that $\liminf \sigma_k = 0$ and $\limsup \sigma_k = 1/2$. Hence $\lim_{k\to\infty}\sigma_k$ does not exist. Thus $\sum_{j=1}^{\infty}(-1)^{j-1}j$ does not converge $(C, 1)$. However, if we continue the process and compute the sequence of arithmetic averages of $\{\sigma_k\}$, then we will see that $\sum_{j=1}^{\infty}(-1)^{j-1}j$ converges $(C, 2)$ to $1/4$.

To confirm this assertion, compute first the arithmetic average of the first $2k$ terms of the sequence $\{\sigma_j\}$:

$$\frac{1}{2k}\sum_{j=1}^{2k}\sigma_j = \frac{1}{2k}\left[1 + 0 + \frac{2}{3} + 0 + \frac{3}{5} + \cdots + \frac{k}{2k-1} + 0\right]$$

$$= \frac{1}{2k}\sum_{j=1}^{k}\frac{j}{2j-1} = \frac{1}{2k}\sum_{j=1}^{k}\frac{1}{2}\left(1 + \frac{1}{2j-1}\right)$$

$$= \frac{1}{4} + \frac{1}{4k}\sum_{j=1}^{k}\frac{1}{2j-1}.$$

Inserting and removing the terms with the even denominator in the last summation, we obtain

$$\frac{1}{2k}\sum_{j=1}^{2k}\sigma_j = \frac{1}{4} + \frac{1}{4k}\left[\sum_{j=1}^{2k}\frac{1}{j} - \frac{1}{2}\sum_{j=1}^{k}\frac{1}{j}\right]$$

$$= \frac{1}{4} + \frac{1}{4k}\left[c_{2k} + \ln(2k) - \frac{1}{2}(c_k + \ln k)\right],$$

where $c_k = \sum_{j=1}^{k} 1/j - \ln k$ is the sequence from Example 17 in Section 11.4. Recall that $\{c_k\}$ converges to Euler's constant γ. A simple rearrangement yields

$$\frac{1}{2k}\sum_{j=1}^{2k}\sigma_j = \frac{1}{4} + \frac{2c_{2k} - c_k}{8k} + \frac{\ln 4k}{8k}.$$

It follows that

$$\lim_{k\to\infty}\frac{1}{2k}\sum_{j=1}^{2k}\sigma_j = \frac{1}{4} + \lim_{k\to\infty}\frac{2c_{2k} - c_k}{8k} + \lim_{k\to\infty}\frac{\ln 4k}{8k} = \frac{1}{4}.$$

A completely analogous computation shows that the arithmetic average of the first $(2k-1)$ terms of the sequence $\{\sigma_j\}$ is

$$\frac{1}{2k-1}\sum_{j=1}^{2k-1}\sigma_j = \frac{k}{4k-2} + \frac{2c_{2k} - c_k}{8k-4} + \frac{\ln 4k}{8k-4}.$$

and thus $\lim_{k\to\infty}(1/(2k-1))\sum_{j=1}^{2k-1}\sigma_j = 1/4$. This proves that $\sum_{j=1}^{\infty}(-1)^{j-1}j$ converges $(C, 2)$ to $1/4$. ●

We conclude this section with the demonstration that $(C, 1)$ summability is actually a generalization of convergence.

THEOREM 11.6.1 If $\sum_{j=1}^{\infty} a_j$ converges to S, then $\sum_{j=1}^{\infty} a_j$ is Cesàro summable to S.

Proof. Let $\{s_k\}$ denote the sequence of kth partial sums of the series; $\{s_k\}$, of course, converges to S. Let $\{\sigma_k\}$ denote the sequence of arithmetic averages of $\{s_k\}$. We must show that $\{\sigma_k\}$ converges to S also. For k in $\mathbb{N}$, let $u_k = \sigma_k - S$. Showing that $\lim_{k\to\infty}\sigma_k = S$ is equivalent to showing that $\lim_{k\to\infty}u_k = 0$.

For k in $\mathbb{N}$, let $t_k = s_k - S$. Clearly, $\{t_k\}$ converges to 0 and thus is bounded; there exists a constant M such that $|t_k| \le M$ for all k. Moreover, given any $\epsilon > 0$, there exists a k_0 such that, for $k \ge k_0$, $|t_k| < \epsilon$.

Now fix any $k > k_0$. Notice that, by the definition of σ_k,

$$u_k = \frac{1}{k}\sum_{j=1}^{k}s_j - S = \frac{1}{k}\sum_{j=1}^{k}s_j - \frac{1}{k}kS = \frac{1}{k}\sum_{j=1}^{k}(s_j - S)$$

$$= \frac{1}{k}\sum_{j=1}^{k}t_j = \frac{1}{k}\sum_{j=1}^{k_0}t_j + \frac{1}{k}\sum_{j=k_0+1}^{k}t_j.$$

Therefore,

$$|u_k| \le \frac{1}{k}\sum_{j=1}^{k_0}|t_j| + \frac{1}{k}\sum_{j=k_0+1}^{k}|t_j| < \frac{k_0M}{k} + \frac{(k-k_0)\epsilon}{k} < \frac{k_0M}{k} + \epsilon.$$

It follows that $0 \le \limsup |u_k| \le \epsilon$ for any $\epsilon > 0$. We conclude that $\lim_{k\to\infty} u_k = 0$ and hence that $\sum_{j=1}^{\infty} a_j$ converges $(C, 1)$ to S. ●

We will apply Cesàro summability in the area of Fourier series (Chapter 14). There we will learn that, for a large class of functions, the Fourier series representation of a function f is Cesàro summable to f; at any point c where f has a jump discontinuity, the Fourier series of f is Cesàro summable to the average of $f(c^-)$ and $f(c^+)$.

EXERCISES

11.1. Let $\sum_{j=1}^{\infty} a_j$ be an infinite series with the indicated general term a_j, for appropriate values of j. Test each series for convergence or divergence. If the series converges, determine whether it converges absolutely or conditionally. Justify your conclusions.

a) $a_j = (-1)^j/[j\ln(j+1)]$ **b)** $a_j = \ln(1+1/j)/j$

c) $a_j = [(1+j)/(1+2j)]^j$ **d)** $a_j = (-1)^j(j+1)^2/(j+2)!$

e) $a_j = (-1)^{j+1}(j+1)^j/j^{2j}$ **f)** $a_j = \ln(1+1/j)$

g) $a_j = 2^j j!/(2j)!$ **h)** $a_j = (-1)^j[\ln j]/\sqrt{j}$

i) $a_j = (-1)^j(j+1)^j/j^{j+1}$ **j)** $a_j = (-1)^j j^j/(j+1)^{j+1}$

11.2. Let $\sum_{j=1}^{\infty} a_j$ be an infinite series with the indicated general term a_j, for appropriate values of j. Test each series for convergence or divergence. If the series converges, determine whether it converges absolutely or conditionally. Justify your conclusions.

a) $a_j = (-1)^j(j+2)^2/(2j+1)^3$ **b)** $a_j = (-1)^j(2j)!/[4^j j!(j+1)!]$

c) $a_j = (-2)^j(j!)^2/(2j+2)!$ **d)** $a_j = (-1)^{j+1}[j/(j+1)]^j$

e) $a_j = (-1)^{j+1}(1-3^{1/j})$ **f)** $a_j = (1+1/j)^j/e^j$

g) $a_j = [\sqrt{j+1} - \sqrt{j}]$ **h)** $a_j = j!/j^j$

i) $a_j = j^j/j!$ **j)** $a_j = (-1)^j[\sum_{r=1}^{j} 1/r]/j$

11.3. Use the limit comparison test to prove that the series $\sum_{j=1}^{\infty} 1/[j(j+1)(j+2)]$ converges. Partially fractionate the jth term and use telescoping to find the sum of the series.

11.4. Suppose that $\sum_{j=1}^{\infty} a_j$ and $\sum_{j=1}^{\infty} b_j$ are two series of positive terms such that $\sum_{j=1}^{\infty} b_j$ converges and $\lim_{j\to\infty} a_j/b_j = 0$. Prove that $\sum_{j=1}^{\infty} a_j$ also converges.

11.5. Suppose that $\sum_{j=1}^{\infty} a_j$ and $\sum_{j=1}^{\infty} b_j$ are two series of positive terms. Suppose also that there exists a j_0 such that, for $j \ge j_0$, $a_{j+1}/a_j \le b_{j+1}/b_j$. Prove the following assertions:

a) If $\sum_{j+1}^{\infty} b_j$ converges, then $\sum_{j=1}^{\infty} a_j$ also converges.

b) If $\sum_{j+1}^{\infty} a_j$ diverges, then $\sum_{j=1}^{\infty} b_j$ also diverges.

11.6. Let $\sum_{j=1}^{\infty} a_j$ be a series of positive terms such that, for all j in $\mathbb{N}$,

$$a_{j+1}/a_j \leq 1 - 2/j + 1/j^2.$$

Prove that $\sum_{j=1}^{\infty} a_j$ must converge.

11.7. Suppose that $\sum_{j=1}^{\infty} a_j$ is a series of positive terms. Suppose also that there exists a constant c and an index j_0 such that for all $j \geq j_0$, $a_{j+1}/a_j \geq 1 - c/j$. Prove that $\sum_{j=1}^{\infty} a_j$ must diverge.

11.8. Prove that $\sum_{j=1}^{\infty}(1/j) \ln [(j+1)^2/j(j+2)]$ converges. *Hint:* Show that the sequence $\{s_k\}$ of kth partial sums is monotone and that $s_k < \ln [2(k+1)/(k+2)] < \ln 2$.

11.9. Suppose that $\{a_j\}$ is a sequence of positive numbers that converges monotonically to 0. Suppose also that $\{b_j\}$ is a sequence for which $b_j > 1$ for all j. Assume that there exists an $M > 0$ such that $\prod_{j=1}^{k} b_j \leq M$ for all k in $\mathbb{N}$. Prove that $\sum_{j=1}^{\infty} a_j \ln b_j$ converges.

11.10. For which values of p in $\mathbb{R}$ does $\sum_{j=2}^{\infty} 1/(\ln j)^p$ converge?

11.11. Prove that, if $\sum_{j=1}^{\infty} a_j$ is absolutely convergent, then the series $\sum_{j=1}^{\infty} a_j^2$ also converges. If the series $\sum_{j=1}^{\infty} a_j$ is conditionally convergent, can you draw the same conclusion? Prove your answer.

11.12. Suppose that $\sum_{j=1}^{\infty} a_j^2$ converges. Does it follow that $\sum_{j=1}^{\infty} |a_j|$ also converges? If so, prove it; otherwise, provide a counterexample.

11.13. Let $\sum_{j=1}^{\infty} a_j$ be a series of positive numbers. Prove that the series converges if and only if its sequence $\{s_k\}$ of kth partial sums is bounded. In turn, prove that $\{s_k\}$ is bounded if and only if any one subsequence $\{s_{k_j}\}$ is bounded.

11.14. Let a and b be positive constants. Prove that the series $\sum_{j=1} 1/(aj+b)^p$ converges if $p > 1$ and diverges if $p \leq 1$.

11.15. Suppose that $\sum_{j=1}^{\infty} a_j/j^p$ converges. Prove that, for all $q > p$, $\sum_{j=1}^{\infty} a_j/j^q$ also converges.

11.16. Fix any real number x that is not an even multiple of π and let $\{a_j\}$ be any sequence of positive numbers that converges monotonically to 0. Prove that $\sum_{j=1}^{\infty} a_j \sin(jx)$ and $\sum_{j=1}^{\infty} a_j \cos(jx)$ converge.

11.17. Fix x in $\mathbb{R}$. Prove that $\sum_{j=1}^{\infty} j^{-1/2} \sin(2j-1)x$ converges. *Hint:* Use an appropriate trigonometric identity.

11.18. Let a and b be constants such that $0 < a < b < 1$. Show that

$$1 + a + b + a^2 + b^2 + a^3 + b^3 + a^4 + b^4 \cdots$$

converges.

11.19. For constants a and b such that $0 < a < b$, consider the series

$$1 + a + ab + a^2b + a^2b^2 + a^3b^2 + a^3b^3 + a^4b^3 + \cdots.$$

a) Show that, if $b < 1$, then the ratio test proves that the series converges. Show, however, if $a < 1 < b$, then the ratio test fails to resolve the question.

b) Suppose that $a < 1 < b$. Apply the root test to show that the series converges if $ab < 1$ and diverges if $ab > 1$. Explain separately why the series diverges if $ab = 1$. (This shows that the root test truly is stronger than the ratio test.)

11.20. Consider the series

$$\frac{1}{1^2} + \frac{1}{2^3} + \frac{1}{3^2} + \frac{1}{4^3} + \cdots + \frac{1}{(2j-1)^2} + \frac{1}{(2j)^3} + \cdots.$$

a) Show that neither the ratio test nor the root test resolves the question of convergence of this series.

b) Prove that the series $\sum_{j=1}^{\infty}[1/(2j-1)^2 + 1/(2j)^3]$ converges by using the limit comparison test.

c) Note that the kth partial sum t_k of the series in part (b) is the $2k$th partial sum s_{2k} of the given series. Use this fact to prove that the given series converges. (See Exercise 11.13.)

11.21. Fix a in $\mathbb{R}$ and $b > 0$. Prove that $\sum_{k=1}^{\infty} \prod_{j=1}^{k} (a+j)/(b+j)$ converges if $b > a+1$ and diverges if $b \leq a+1$.

11.22. This exercise guides you through a program to derive Raabe's test for absolute convergence.

THEOREM Raabe's Test Suppose that $\sum_{j=1}^{\infty} a_j$ is a series such that

$$L = \lim_{j\to\infty} j\left(1 - \left|\frac{a_{j+1}}{a_j}\right|\right)$$

exists.

i) If $L > 1$, then $\sum_{j=1}^{\infty} a_j$ converges absolutely.

ii) If $L < 1$, then $\sum_{j=1}^{\infty} a_j$ does not converge absolutely.

iii) If $L = 1$, the test gives no information.

a) In Exercise 11.5 take $b_j = 1/j^p$, for some positive constant p. Show that $b_{j+1}/b_j = (1+1/j)^{-p}$.

b) Suppose that $\sum_{j=1}^{\infty} a_j$ is a series of real numbers such that, for some $p > 1$ and some index j_0,

$$\left|\frac{a_{j+1}}{a_j}\right| \leq \left(1 + \frac{1}{j}\right)^{-p}, \qquad \text{for all } j \geq j_0.$$

Show that $\sum_{j=1}^{\infty} a_j$ converges absolutely.

c) Likewise, suppose that $\sum_{j=1}^{\infty} a_j$ is a series of real numbers such that, for some $p < 1$ and some index j_0,

$$\left(1 + \frac{1}{j}\right)^{-p} \leq \left|\frac{a_{j+1}}{a_j}\right|, \qquad \text{for all } j \geq j_0.$$

Show that $\sum_{j=1}^{\infty} a_j$ does not converge absolutely.

d) For $p > 0$, apply Taylor's theorem[1] to $f(x) = (1+x)^{-p}$ to find the first-degree Taylor polynomial for f at $x_0 = 0$ and find the Lagrange form of the remainder. Show that when $x = 1/j$ there is a constant d_j such that

$$\left(1+\frac{1}{j}\right)^{-p} = 1 - \frac{p}{j} + \frac{d_j}{j^2}.$$

Find d_j explicitly. Show that the sequence $\{d_j\}$ is bounded.

e) Suppose that $\sum_{j=1}^{\infty} a_j$ is a series of real numbers such that $\lim_{j\to\infty} j[1 - |a_{j+1}/a_j|]$ exists.

i) Suppose that there exists a $p > 0$ and a j_0 such that

$$\left|\frac{a_{j+1}}{a_j}\right| \le 1 - \frac{p}{j} + \frac{d_j}{j^2},$$

for $j \ge j_0$. Show that

$$p = \lim_{j\to\infty}\left(p - \frac{d_j}{j}\right) \le \lim_{j\to\infty} j\left(1 - \left|\frac{a_{j+1}}{a_j}\right|\right).$$

ii) Likewise, show that, if there is a $p > 0$ and a j_0 such that

$$\left|\frac{a_{j+1}}{a_j}\right| \ge 1 - \frac{p}{j} + \frac{d_j}{j^2},$$

for $j \ge j_0$, then

$$p \ge \lim_{j\to\infty} j\left(1 - \left|\frac{a_{j+1}}{a_j}\right|\right).$$

f) Now reverse your steps. Suppose that $\sum_{j=1}^{\infty} a_j$ is a series of real numbers such that $L = \lim_{j\to\infty} j(1 - |a_{j+1}/a_j|)$ exists.

i) Suppose that $L > 1$. Show that you can choose a $p > 1$ and an index j_0 such that, for $j \ge j_0$,

$$\left|\frac{a_{j+1}}{a_j}\right| \le \left(1+\frac{1}{j}\right)^{-p}.$$

Hence, conclude that $\sum_{j=1}^{\infty} a_j$ must converge absolutely.

ii) Suppose that $L < 1$. Show that you can choose a $p < 1$ and an index j_0 such that, for $j \ge j_0$,

$$\left(1+\frac{1}{j}\right)^{-p} \le \left|\frac{a_{j+1}}{a_j}\right|.$$

Hence, show that $\sum_{j=1}^{\infty} a_j$ does not converge absolutely.

This completes the proof of Raabe's test.

---

[1] Show that Lagrange's form of the Taylor remainder can be written in the form $R_k(x_0; x) = f^{(k+1)}(x_0 + \theta h)h^{k+1}/(k+1)!$ where θ is some number in $(0, 1)$. Apply this form of the remainder to $(1+x)^{-p}$ with $x_0 = 0$, $h = x$, and $k = 1$. Then set $x = 1/j$.

11.23. Apply Raabe's test to each of the following series and determine whether the test decides absolute convergence.

a) $1/3 + 1/2 + 1/3^2 + 1/2^2 + 1/3^3 + 1/2^3 + \cdots$

b) For a fixed x in $\mathbb{R}$,

$$\left(\tfrac{1}{2}\right)x + \left(\tfrac{1}{2}\right)\left(\tfrac{4}{3}\right)x^2 + \left(\tfrac{1}{2}\right)^2\left(\tfrac{4}{3}\right)x^3 + \left(\tfrac{1}{2}\right)^2\left(\tfrac{4}{3}\right)^2 x^4 + \left(\tfrac{1}{2}\right)^3\left(\tfrac{4}{3}\right)^2 x^5 + \left(\tfrac{1}{2}\right)^3\left(\tfrac{4}{3}\right)^3 x^6 + \cdots$$

c) For a fixed x in $\mathbb{R}$ and $p > 0$, $\sum_{j=1}^{\infty} j^p x^j$.

d) For $p > 0$, $\sum_{j=1}^{\infty}[2^j j!/(5 \cdot 7 \cdot 9 \cdots (2j+3))]^p$.

11.24. This exercise outlines a proof of Gauss's test for absolute convergence. (See Exercises 11.5 and 11.22.)

THEOREM Gauss's Test Suppose that $\sum_{j=1}^{\infty} a_j$ is a series of real numbers and j_0 is an index such that, for $j \geq j_0$,

$$\left|\frac{a_{j+1}}{a_j}\right| = 1 - \frac{p}{j} + \frac{d_j}{j^2}$$

where p is a constant and $\{d_k\}$ is a bounded sequence.

i) If $p > 1$, then $\sum_{j=1}^{\infty} a_j$ converges absolutely.

ii) If $p \leq 1$, then $\sum_{j=1}^{\infty} a_j$ does not converge absolutely.

a) For $p \neq 1$, imitate the proof of Raabe's test.

b) For $p = 1$, let $b_j = 1/[(j-1)\ln(j-1)]$ instead of $b_j = 1/j^p$ and complete the following steps to prove Gauss's test.

i) Use Taylor's theorem with $x_0 = 0$ and $k = 1$ to show that $\ln(1 - 1/j) = -1/j - c_j/j$, where the sequence $\{c_j\}$ is bounded.

ii) Show that there is a C_j such that

$$\frac{b_{j+1}}{b_j} = 1 - \frac{1}{j} - \frac{1}{[j \ln j]} - \frac{C_j}{j^2}.$$

Find C_j explicitly and show that $\{C_j\}$ is bounded.

iii) Hence show that $|a_{j+1}/a_j| \geq b_{j+1}/b_j$ for j sufficiently large. Show how to conclude that $\sum_{j=1}^{\infty} a_j$ does not converge absolutely.

This completes the proof of Gauss's test.

11.25. Apply Gauss's test to the series in Exercise 11.23.

11.26. Prove that, for each of the series $\sum_{j=1}^{\infty}[(2j)!/(2^j j!)^2]^2$ and $\sum_{j=2}^{\infty} 1/[j(\ln j)^2]$, Raabe's test (Exercise 11.22) gives a value of $L = 1$. Use other methods to show that the first series diverges and the second converges, thus showing that Raabe's test truly gives no information when $L = 1$.

11.27. For x in $\mathbb{R}$, consider the series

$$\sum_{j=1}^{\infty} \frac{(-1)^j (2j)! x^j}{[2^{2j}(j!)^2]^2}.$$

a) Prove that this series converges absolutely for x in $(-1, 1)$.

b) Prove that this series converges conditionally at $x = 1$.

c) Prove that the series does not converge absolutely at $x = -1$ and therefore diverges there.

11.28. Suppose that $\sum_{j=1}^{\infty} a_j$ converges to S and suppose that we regroup the terms of this series by the insertion of parentheses around successive groups of any finite number of its terms. For example,

$$(a_1 + a_2) + (a_3) + (a_4 + a_5 + a_6 + a_7) + (a_8 + a_9 + a_{10}) + \cdots$$

is such a regrouping. Let b_k denote the sum of those terms a_j in the kth group. Prove that $\sum_{k=1}^{\infty} b_k$ also converges to S. (Inserting parentheses does not affect the convergence.)

11.29. We know that the series $\sum_{j=1}^{\infty}(-1)^{j+1}/j$ converges to $\ln 2$. Prove that the rearrangement

$$1 + \sum_{j=1}^{\infty}\left(\frac{1}{2j+1} - \frac{1}{2j}\right) = 1 + \frac{1}{3} - \frac{1}{2} + \frac{1}{5} - \frac{1}{4} \cdots$$

also converges to $\ln 2$.

11.30. Find the first 20 terms of a rearrangement of $\sum_{j=1}^{\infty}(-1)^{j+1}/j$ that converges to 0.

11.31. There is a rearrangement of the series $\sum_{j=1}^{\infty}(-1)^{j+1}/j$ that converges to 2. Let s_k denote the kth partial sum of the rearranged series obtained by imitating the proof of Weierstrass's theorem. Find k_0 such that $|s_k - 2| < .05$ for all $k \geq k_0$. List the first k_0 terms (in the correct order) of the rearranged series obtained by Weierstrass's method.

11.32. Rearrange the series $\sum_{j=1}^{\infty}(-1)^{j+1}/j$ so that the next $2k-1$ positive terms are followed by the next $2k$ negative terms. That is, the rearranged series is

$$1 - \tfrac{1}{2} - \tfrac{1}{4} + \tfrac{1}{3} + \tfrac{1}{5} + \tfrac{1}{7} - \tfrac{1}{6} - \tfrac{1}{8} - \tfrac{1}{10} - \tfrac{1}{12} \cdots.$$

Determine whether this series converges.

11.33. Show that the series

$$1 - \tfrac{1}{2} - \tfrac{1}{3} + \tfrac{1}{4} + \tfrac{1}{5} - \tfrac{1}{6} - \tfrac{1}{7} + \tfrac{1}{8} + \tfrac{1}{9} - \cdots$$

converges. (After the first term, two negative terms are followed by two positive terms. Think Euler's constant.)

11.34. Show that the series

$$1 + \tfrac{1}{2} - \tfrac{1}{3} + \tfrac{1}{4} + \tfrac{1}{5} - \tfrac{1}{6} + \tfrac{1}{7} + \tfrac{1}{8} - \tfrac{1}{9} \cdots$$

diverges to ∞. (Two positive terms are followed by one negative term. Think Euler's constant.)

11.35. Determine whether

$$1 - \tfrac{1}{2} - \tfrac{1}{3} + \tfrac{1}{4} + \tfrac{1}{5} + \tfrac{1}{6} - \tfrac{1}{7} - \tfrac{1}{8} - \tfrac{1}{9} - \tfrac{1}{10} + \cdots$$

converges or diverges. The alternating groups of positive and negative terms vary in length: 1 term, 2 terms, 3 terms $\cdots$.

11.36. For k in $\mathbb{N}$, let $d_k = \sum_{j=1}^{k} 1/j^{1/2} - 2(k^{1/2} - 1)$. Show that $\lim_{k\to\infty} d_k$ exists by showing that $\{d_k\}$ is monotone and bounded.

11.37. The series $\sum_{j=1}^{\infty}(-1)^{j+1}/j^{1/2}$ converges conditionally. Consider the rearrangement

$$1 + \frac{1}{\sqrt{3}} - \frac{1}{\sqrt{2}} + \frac{1}{\sqrt{5}} + \frac{1}{\sqrt{7}} - \frac{1}{\sqrt{4}} + \frac{1}{\sqrt{9}} + \frac{1}{\sqrt{11}} - \frac{1}{\sqrt{6}} + \frac{1}{\sqrt{13}} + \frac{1}{\sqrt{15}} - \frac{1}{\sqrt{8}} \cdots$$

with the next two positive terms separated by the next negative term. Show that the rearranged series diverges to ∞. [*Hint:* Let s_{2k} be the $2k$th partial sum of the original series and t_{3k} be the $3k$th partial sum of the rearranged series. Show that $t_{3k} - s_{2k} > k/(4k-1)^{1/2}$. Since $\{s_{2k}\}$ converges, $\{t_{3k}\}$ must be unbounded.]

11.38. Show that the series $\sum_{k=2}^{\infty}(-1)^k \sum_{j=2^{k-1}}^{2^k-1} 1/[j \ln j]$ converges.

11.39. Fix x and y in $\mathbb{R}$. Form the Cauchy product of the series $\sum_{j=0}^{\infty} x^j/j!$ and $\sum_{j=0}^{\infty} y^j/j!$ and show that this product is $\sum_{k=0}^{\infty}(x+y)^k/k!$. Identify the property of the exponential function revealed by this identity.

11.40. Fix x in $\mathbb{R}$. Prove that the Cauchy product of $\sum_{j=0}^{\infty}(-1)^j x^{2j+1}/(2j+1)!$ and $\sum_{j=0}^{\infty}(-1)^j x^{2j}/(2j)!$ is

$$\frac{1}{2}\sum_{k=0}^{\infty}\frac{(-1)^k(2x)^{2k+1}}{(2k+1)!}.$$

Identify the trigonometric identity revealed by this result.

11.41. The series $\sum_{j=1}^{\infty} 1/j$ diverges. Is it $(C, 1)$ summable? If so, find σ such that $\sum_{j=1}^{\infty} 1/j = \sigma\ (C, 1)$; otherwise, prove that it is not $(C, 1)$ summable.

11.42. Let $\sum_{j=1}^{\infty} a_j$ be a convergent series. For each k in $\mathbb{N}$ and for each $j = 1, 2, 3, \ldots, k$, let $w_j = 2j/[k(k+1)]$.

a) Prove that $0 < w_j < 1$ and $\sum_{j=1}^{k} w_j = 1$. The numbers w_j are called *weights*.

b) For each k in $\mathbb{N}$, let $b_k = \sum_{j=1}^{k} w_j a_j$. ($b_k$ is called a *weighted average* of $a_1, a_2, \ldots, a_k$.) Prove that $\sum_{k=1}^{\infty} b_k$ also converges.

11.43. (Cauchy's condensation test) Let $\{a_j\}$ be a monotone decreasing sequence of positive numbers. Prove that $\sum_{j=1}^{\infty} a_j$ converges if and only if $\sum_{j=1}^{\infty} 2^j a_{2^j}$ converges. (*Hint:* Group the terms of $\sum_{j=1}^{\infty} a_j$ into blocks of the appropriate length and use Exercise 11.13.)

11.44. Let $\{a_j\}$ be a monotone decreasing sequence of numbers. Prove that, if $\sum_{j=1}^{\infty} a_j$ converges, then $\lim_{j\to\infty} j a_j = 0$. Is the converse true? If so, prove it; otherwise provide a counterexample.

11.45. A sequence $\{a_j\}$ is said to be of *bounded variation* if $\sum_{j=1}^{\infty} |a_{j+1} - a_j|$ converges.

a) Prove that every sequence that is of bounded variation must converge. Also prove that the converse is false.

b) Which of the following sequences are of bounded variation?

i) $\{(-1)^j/j\}$ **ii)** $\{(-1)^j/j^{1/2}\}$

iii) $\{(-1)^j/j^2\}$ **iv)** $\{(-1)^j/\ln(j+1)\}$

11.46. Let ℓ_1 denote the collection of all sequences $\boldsymbol{x} = \{x_j\}$ of real numbers such that $\sum_{j=1}^{\infty} x_j$ is absolutely convergent.

a) Confirm in detail that ℓ_1 is a vector space.

b) For $\boldsymbol{x}$ in ℓ_1, define $||\boldsymbol{x}||_1 = \sum_{j=1}^{\infty} |x_j|$. Prove that $||\cdot||_1$ is a norm on ℓ_1.

c) For $\boldsymbol{x} = \{x_j\}$ and $\boldsymbol{y} = \{y_j\}$ in ℓ_1, define

$$d_1(\boldsymbol{x}, \boldsymbol{y}) = ||\boldsymbol{x} - \boldsymbol{y}||_1 = \sum_{j=1}^{\infty} |x_j - y_j|.$$

Prove that d_1 is a metric on ℓ_1.

d) Let $\{x_k\}$ be a sequence (of sequences) in ℓ_1. Write $x_k = \{x_j^{(k)};\ j \text{ in } \mathbb{N}\}$. A sequence $\{x_k\}$ of elements in ℓ_1 is said to be a *Cauchy sequence* if, for every $\epsilon > 0$, there exists a k_0 such that, for k and m greater than k_0, we have

$$d_1(x_k, x_m) = ||x_k - x_m||_1 = \sum_{j=1}^{\infty} |x_j^{(k)} - x_j^{(m)}| < \epsilon.$$

Let $\{x_k\}$ be a Cauchy sequence in ℓ_1. Prove that there is an element $x = \{x_j\}$ in ℓ_1 such that $\lim_{k\to\infty} d_1(x_k, x) = 0$. Thus ℓ_1 is Cauchy complete with respect to the metric d_1. *Hint:*

i) Show that, for each j in $\mathbb{N}$, the sequence

$$\{x_j^{(k)} : k \text{ in } \mathbb{N}\}$$

is a Cauchy sequence of real numbers.

ii) Let $x_j = \lim_{k\to\infty} x_j^{(k)}$ and let $x = \{x_j\}$. Prove that $\sum_{j=1}^{\infty} x_j$ is absolutely convergent and that $\lim_{k\to\infty} d_1(x_k, x) = 0$.

11.47. Let ℓ_2 denote the collection of all sequences $x = \{x_j\}$ such that $\sum_{j=1}^{\infty} x_j^2$ is (absolutely) convergent. (See Exercises 11.11 and 11.12.)

a) Prove that ℓ_2 is a vector space. [The main, but not the sole, issue is to prove that, if x and y are in ℓ_2, then $x + y = \{x_j + y_j\}$ is also in ℓ_2. For this you must prove that $\sum_{j=1}^{\infty}(x_j + y_j)^2$ converges. Use the Cauchy–Schwarz inequality, Theorem 2.1.2.]

b) Let $x = \{x_j\}$ and $y = \{y_j\}$ be sequences in ℓ_2. Define

$$\langle x, y\rangle = \sum_{j=1}^{\infty} x_j y_j.$$

Prove that $\langle\cdot, \cdot\rangle$ is an inner product on ℓ_2.

c) For x in ℓ_2, define

$$||x||_2 = \langle x, x\rangle^{1/2} = \left[\sum_{j=1}^{\infty} x_j^2\right]^{1/2}.$$

Prove the Cauchy–Schwarz inequality in ℓ_2: For x and y in ℓ_2,

$$|\langle x, y\rangle| \le ||x||_2 ||y||_2 = \left[\sum_{j=1}^{\infty} x_j^2\right]^{1/2} \left[\sum_{j=1}^{\infty} y_j^2\right]^{1/2}.$$

(See Theorem 2.1.2.)

d) Prove that $||\cdot||_2$ is a norm on ℓ_2.

e) For x and y in ℓ_2, define

$$d_2(x, y) = ||x - y||_2 = \left[\sum_{j=1}^{\infty}(x_j - y_j)^2\right]^{1/2}.$$

Prove that d_2 is a metric on ℓ_2.

f) Let $\{\boldsymbol{x}_k\}$ be a sequence of elements in ℓ_2. We write $\boldsymbol{x}_k = \{x_j^{(k)} : j \text{ in } \mathbb{N}\}$. The sequence $\{\boldsymbol{x}_k\}$ is said to be a *Cauchy sequence* if, for all $\epsilon > 0$, there exists a k_0 such that

$$d_2(\boldsymbol{x}_k, \boldsymbol{x}_m) = ||\boldsymbol{x}_k - \boldsymbol{x}_m||_2 = \left[\sum_{j=1}^{\infty}(x_j^{(k)} - x_j^{(m)})^2\right]^{1/2} < \epsilon,$$

for k and m greater than k_0.

Let $\{\boldsymbol{x}_k\}$ be any Cauchy sequence in ℓ_2. Prove that there exists a sequence $\boldsymbol{x} = \{x_j\}$ in ℓ_2 such that

$$\lim_{k\to\infty} d_2(\boldsymbol{x}_k, \boldsymbol{x}) = 0.$$

Thus, ℓ_2 is Cauchy complete with respect to the metric d_2.

12

Series of Functions

In this chapter we introduce the first and perhaps most useful solution to one of the general questions addressed by classical analysis. The problem is this. Suppose that F is a given real-valued function defined on some domain S. Is it possible to find a sequence $\{f_j\}$ of functions each defined on S, each of some particularly simple type, with the property that, for each $\boldsymbol{x}$ in S, the series $\sum_j f_j(\boldsymbol{x})$ converges to $F(\boldsymbol{x})$? If this is possible, then we say that F is *represented* by the infinite series $\sum_j f_j$.

Turning this question inside out, we can begin with a sequence $\{f_j\}$ of simple functions defined on S and, if the series $\sum_j f_j(\boldsymbol{x})$ converges for each $\boldsymbol{x}$ in S, we can create a function $F(\boldsymbol{x}) = \sum_j f_j(\boldsymbol{x})$ on S. Clearly, this new function F is represented by the given series.

If such a representation is possible, we raise the obvious secondary questions. Does the continuity, differentiability, or integrability of the component functions f_j carry over to the function F being represented? As we shall see, the answer to these questions hinges on the distinction between pointwise and uniform convergence.

12.1 Preliminaries

We begin by establishing the notation to be used throughout this chapter to distinguish between pointwise and uniform convergence of a series $\sum_j f_j$ of functions. Refer to Definition 3.5.4 and the subsequent discussion.

DEFINITION 12.1.1 Let $\sum_{j=1}^{\infty} f_j$ be a series of real-valued functions each of which is defined on a set S in $\mathbb{R}^n$. For each $\boldsymbol{x}$ in S, let $F_k(\boldsymbol{x}) = \sum_{j=1}^{k} f_j(\boldsymbol{x})$. Then

$\{F_k\}$ is a sequence of functions defined on S and is called the *sequence of partial sums* of the series.

i) If $\{F_k\}$ converges pointwise to a function F on S, then we say that the series $\sum_{j=1}^{\infty} f_j$ *converges pointwise* to F on S. We write $\sum_{j=1}^{\infty} f_j = F$ [pointwise].

ii) If $\{F_k\}$ converges uniformly to a function F on S, then we say that the series $\sum_{j=1}^{\infty} f_j$ *converges uniformly* to F on S. We write $\sum_{j=1}^{\infty} f_j = F$ [uniformly]. ●

According to Definition 3.5.4, we have $\sum_{j=1}^{\infty} f_j = F$ [pointwise] if and only if, for each $\epsilon > 0$ and each $\boldsymbol{x}$ in S, there exists a $k_0 = k_0(\epsilon, \boldsymbol{x})$ such that, for all $k \geq k_0$,

$$|F_k(\boldsymbol{x}) - F(\boldsymbol{x})| = \left|\sum_{j=1}^{k} f_j(\boldsymbol{x}) - F(\boldsymbol{x})\right| = \left|\sum_{j=k+1}^{\infty} f_j(\boldsymbol{x})\right| < \epsilon.$$

We emphasize that the index k_0 generally depends not only on ϵ, but on $\boldsymbol{x}$ as well. Furthermore, at any particular point $\boldsymbol{x}$, the convergence of $\sum_{j=1}^{\infty} f_j(\boldsymbol{x})$ to $F(\boldsymbol{x})$ might be absolute or it might be conditional. All the convergence tests of Chapter 11 are available to enable you to determine pointwise convergence and, if pointwise convergence occurs, to determine whether convergence is absolute or conditional.

The uniform convergence of $\sum_{j=1}^{\infty} f_j$ to F means that, for each $\epsilon > 0$, there exists a k_0 such that, for $k \geq k_0$,

$$\begin{aligned} \|F_k - F\|_\infty &= \left\|\sum_{j=1}^{k} f_j - F\right\|_\infty \\ &= \sup\left\{\left|\sum_{j=1}^{k} f_j(\boldsymbol{x}) - F(\boldsymbol{x})\right| : \boldsymbol{x} \text{ in } S\right\} < \epsilon. \end{aligned}$$

That is, provided $k \geq k_0$, we have $|F_k(\boldsymbol{x}) - F(\boldsymbol{x})| < \epsilon$ for all $\boldsymbol{x}$ in S simultaneously. The index k_0 depends only on ϵ and is independent of $\boldsymbol{x}$ in S. Tests to determine whether a given series $\sum_{j=1}^{\infty} f_j$ of functions converges uniformly will be derived in Section 12.3.

Recall from Section 5 of Chapter 3 that, for any nonempty subset S of $\mathbb{R}^n$, we denote by $B(S)$ the vector space of all bounded, real-valued functions defined on S. With the uniform norm, $B(S)$ is Cauchy complete. That is, every uniformly Cauchy sequence $\{f_k\}$ in $B(S)$ is uniformly convergent to a function f in $B(S)$. It is straightforward to translate this fact into the following form, of use when, for example, the limit function is not known.

THEOREM 12.1.1 Cauchy's Criterion for Uniform Convergence Let $\sum_{j=1}^{\infty} f_j$ be a series of bounded, real-valued functions on S and let $\{F_k\}$ denote the sequence of partial sums of $\sum_{j=1}^{\infty} f_j$. The series $\sum_{j=1}^{\infty} f_j$ converges uniformly if and only if $\{F_k\}$ is uniformly Cauchy. ●

It is worth mentioning that there are several distinct types of convergence of a series $\sum_{j=1}^{\infty} f_j$ beyond those we discuss here; their discovery late in the nineteenth century has contributed to the remarkable elaboration of mathematical analysis during the twentieth century.

12.2 Uniform Convergence

We already know from our discussion in Section 3.5 that, if a sequence of functions, each of which is continuous on some set S, converges pointwise, then the limit function need not be continuous on S. The same is true for series. For example let $S = [0, 1]$. For x in S and for j in $\mathbb{N}$, let $f_j(x) = x^j - x^{j-1}$. The kth partial sum of the series $\sum_{j=1}^{\infty} f_j$ telescopes and we obtain $F_k(x) = \sum_{j=1}^{k} f_j(x) = x^k - 1$. The sequence $\{F_k\}$ converges pointwise to the function F that has value -1 on $[0, 1)$ and value 0 at $x = 1$. Although the functions f_j are continuous on $[0, 1]$, the limit function F is discontinuous at $x = 1$. Pointwise convergence of a series of continuous functions does not guarantee the continuity of the sum.

However, if each of the functions f_j is continuous on S and if the series $\sum_{j=1}^{\infty} f_j$ converges uniformly to F on S, then F must also be continuous on S. The proof consists merely of applying Theorem 3.5.3. Since each f_j is continuous on S, the partial sums F_k are also continuous on S and, given that $\{F_k\}$ converges uniformly to F, it follows that F is also continuous on S. This brief argument proves the following theorem.

THEOREM 12.2.1 If each function f_j is continuous on S and if $\sum_{j=1}^{\infty} f_j = F$ [uniformly], then F is continuous on S. ●

It is worth identifying explicitly the general form of the argument used in the foregoing discussion. Let $\{g_j\}$ be any sequence of functions on some set S; let $g_0 = 0$. If we define $f_j = g_j - g_{j-1}$, then, by telescoping, the kth partial sum of the series $\sum_{j=1}^{\infty} f_j$ is $F_k = \sum_{j=1}^{k} f_j = \sum_{j=1}^{k}(g_j - g_{j-1}) = g_k$. Thus we can reproduce the original sequence $\{g_j\}$ as a sequence of partial sums of a series. Convergence, of whatever type, of the sequence $\{g_j\}$ is equivalent to the convergence, of the same type, of the series $\sum_{j=1}^{\infty}(g_j - g_{j-1})$. By means of this technique, you can arrange to use convergence tests for series to determine the convergence of a sequence.

Keeping in mind this connection between the convergence of a sequence of functions and the convergence of a related series, we look more closely at the question of the uniform convergence of a sequence. As our next particularly satisfying theorem shows, subject to fairly common though stringent conditions, pointwise convergence must actually be uniform.

THEOREM 12.2.2 Dini's Theorem Suppose that $\{f_k\}$ is a monotone sequence of continuous functions on a compact set S in $\mathbb{R}^n$ that converges pointwise to a continuous function f. Then the convergence must be uniform.

Proof. Fix $\epsilon > 0$. For each $\boldsymbol{x}$ in S, choose $k_0(\boldsymbol{x})$ such that, for $k \geq k_0(\boldsymbol{x})$, we have $|f_k(\boldsymbol{x}) - f(\boldsymbol{x})| < \epsilon$. Since both f and $f_{k_0(\boldsymbol{x})}$ are assumed to be continuous at each point in S, in particular at $\boldsymbol{x}$, there exists a $\delta = \delta(\boldsymbol{x})$ such that, for $\boldsymbol{y}$ in $S \cap N(\boldsymbol{x}; \delta(\boldsymbol{x}))$, we have both

$$|f(\boldsymbol{y}) - f(\boldsymbol{x})| < \epsilon$$

and

$$|f_{k_0(\boldsymbol{x})}(\boldsymbol{y}) - f_{k_0(\boldsymbol{x})}(\boldsymbol{x})| < \epsilon.$$

Notice that $\delta(\boldsymbol{x})$ depends implicitly on $k_0(\boldsymbol{x})$. For any $\boldsymbol{y}$ in $S \cap N(\boldsymbol{x}; \delta(\boldsymbol{x}))$, we have

$$\begin{aligned}|f_{k_0(\boldsymbol{x})}(\boldsymbol{y}) - f(\boldsymbol{y})| &\leq |f_{k_0(\boldsymbol{x})}(\boldsymbol{y}) - f_{k_0(\boldsymbol{x})}(\boldsymbol{x})| + |f_{k_0(\boldsymbol{x})}(\boldsymbol{x}) - f(\boldsymbol{x})| + |f(\boldsymbol{x}) - f(\boldsymbol{y})| \\ &< 3\epsilon.\end{aligned}$$

Since the [pointwise] convergence of $\{f_k\}$ is monotone, it follows that

$$|f_k(\boldsymbol{y}) - f(\boldsymbol{y})| < 3\epsilon, \tag{12.1}$$

for all $\boldsymbol{y}$ in $S \cap N(\boldsymbol{x}; \delta(\boldsymbol{x}))$ and all $k \geq k_0(\boldsymbol{x})$.

The collection of neighborhoods $\{N(\boldsymbol{x}; \delta(\boldsymbol{x})) : \boldsymbol{x} \text{ in } S\}$ is an open cover of the compact set S and, consequently, there exists a finite subcover

$$\{N(\boldsymbol{x}_1; \delta(\boldsymbol{x}_1)), N(\boldsymbol{x}_2; \delta(\boldsymbol{x}_2)), \ldots, N(\boldsymbol{x}_p; \delta(\boldsymbol{x}_p))\}.$$

Corresponding to each $\boldsymbol{x}_j$ is an index $k_0(\boldsymbol{x}_j)$. Let

$$k_0 = \max\{k_0(\boldsymbol{x}_1), k_0(\boldsymbol{x}_2), \ldots, k_0(\boldsymbol{x}_p)\}.$$

Choose any $\boldsymbol{y}$ in S. Then $\boldsymbol{y}$ is in at least one of the neighborhoods $N(\boldsymbol{x}_j; \delta(\boldsymbol{x}_j))$ in the finite subcover and, for $k \geq k_0 \geq k_0(\boldsymbol{x}_j)$, we have by (12.1),

$$|f_k(\boldsymbol{y}) - f(\boldsymbol{y})| < 3\epsilon.$$

Since the index k_0 is chosen independently of $\boldsymbol{y}$ in S, we conclude that $\lim_{k\to\infty} f_k = f$ [uniformly] on S. ●

COROLLARY 12.2.3 Suppose that $\sum_{i=1}^{\infty} f_j$ is a series of continuous functions on a compact set S in $\mathbb{R}^n$, that the sequence $\{F_k\}$ of partial sums is monotone, and that $\{F_k\}$ converges pointwise to a continuous function F on S. Then

$$\sum_{j=1}^{\infty} f_j = F \quad \text{[uniformly]}. \quad \bullet$$

12.2.1 Convergence and Differentiation

Suppose that S is an interval in $\mathbb{R}$ and that $\{f_k\}$ is a sequence of differentiable real-valued functions defined on S. The mere pointwise convergence of $\{f_k\}$ to a function f is surely insufficient to guarantee the differentiability of f. After all, pointwise convergence fails even to ensure the continuity of f, let alone its differentiability. But suppose it were to happen that $\{f_k\}$ converges to a differentiable function f. Does it necessarily follow that the sequence $\{f_k'\}$ converges to f'? For an affirmative answer, we surely will need some restriction on the sequence $\{f_k'\}$ of derivatives. Consider the following examples.

EXAMPLE 1 For k in $\mathbb{N}$ and x in $S = [-1, 1]$, define

$$f_k(x) = \frac{kx}{1 + k^2x^2}.$$

It is straightforward to confirm the $\lim_{k\to\infty} f_k(x) = 0$ for all x in S; thus the sequence $\{f_k\}$ converges pointwise to the differentiable function $f(x) = 0$ for all x in S. We leave it for you to prove that this convergence is nonuniform on S. Each f_k is differentiable with

$$f_k'(x) = \frac{k(1 - k^2x^2)}{(1 + k^2x^2)^2}$$

on $[-1, 1]$. Clearly, for each $x \neq 0$ in S, $\lim_{k\to\infty} f_k'(x) = 0$. However, $\lim_{k\to\infty} f_k'(0) = \infty$. Thus the convergence of $\{f_k'\}$ cannot be uniform and $\lim_{k\to\infty} f_k'$ differs from the derivative of $\lim_{k\to\infty} f_k$. ●

EXAMPLE 2 For k in $\mathbb{N}$ and x in $S = [0, 1]$, define

$$f_k(x) = xk^2e^{-kx}.$$

For each x in S, $\lim_{k\to\infty} f_k(x) = 0$, so $\{f_k\}$ converges pointwise to 0. To show that this convergence is nonuniform, note that each f_k assumes only nonnegative values and has a maximum value of k/e at $x = 1/k$. (See Fig. 12.1.) For each k in $\mathbb{N}$,

$$f_k'(x) = (1 - kx)k^2e^{-kx}.$$

Note that, for each k, $f_k'(0) = k^2$ so the sequence $\{f_k'(0)\}$ diverges to ∞. For every other x in S, $\lim_{k\to\infty} f_k'(x) = 0$. Thus the convergence of $\{f_k'\}$ cannot be uniform. Again note that $\{f_k'\}$ does not converge to the derivative of $\lim_{k\to\infty} f_k$. ●

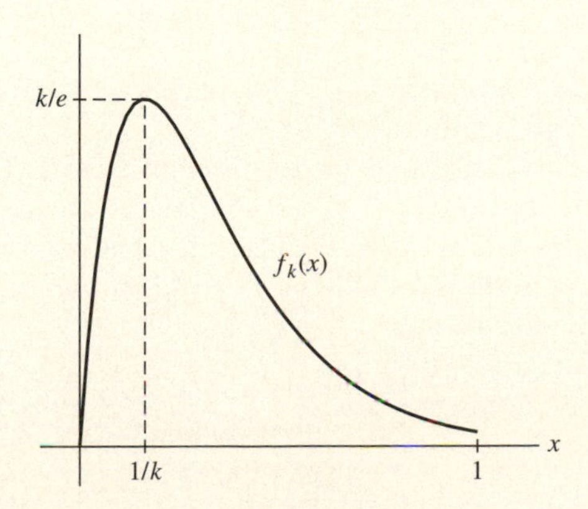

Figure 12.1

EXAMPLE 3 From Example 10 in Section 4.5, we know that the sequence of polynomials

$$p_k(x) = \sum_{j=0}^{k} \frac{x^j}{j!}$$

converges uniformly to $F(x) = e^x$ on any compact interval $[a, b]$. The derivative of p_k is given by

$$p_k'(x) = \sum_{j=0}^{k} \frac{jx^{j-1}}{j!} = \sum_{j=1}^{k} \frac{x^{j-1}}{(j-1)!} = \sum_{j=0}^{k-1} \frac{x_j}{j!} = p_{k-1}(x).$$

It follows that

$$\lim_{k\to\infty} p_k'(x) = \lim_{k\to\infty} p_{k-1}(x) = e^x = F'(x) \quad \text{[uniformly]}$$

on $[a, b]$. Thus, in this example, $\{p_k\}$ converges to F and $\{p_k'\}$ converges to F'. ●

THEOREM 12.2.4 Let $\{f_k\}$ be a sequence of differentiable functions on (a, b) satisfying the following conditions.

i) For some c_0 in (a, b), $\{f_k(c_0)\}$ converges.

ii) For each k, f_k' is bounded on (a, b).

iii) The sequence $\{f_k'\}$ converges uniformly to some function g on (a, b).

Then $\{f_k\}$ converges uniformly to some differentiable function f on (a, b) and $f' = g$.

Proof. First note that, because $\{f_k'\}$ converges uniformly to g on (a, b), the sequence $\{f_k'\}$ is uniformly Cauchy. Thus, given $\epsilon > 0$, there exists a k_0 such that, for k and m greater than k_0, $||f_k' - f_m'||_\infty < \epsilon$. Fix such a k_0.

Next, fix any c in (a, b). We define a new sequence of continuous functions on (a, b) that will depend on the choice of c. For each k in $\mathbb{N}$, let

$$g_k(x) = \begin{cases} \dfrac{f_k(x) - f_k(c)}{x - c}, & \text{for } x \neq c, \\ f_k'(c), & \text{for } x = c. \end{cases}$$

Notice that, for each k, the function g_k is bounded and continuous on (a, b). We claim that $\{g_k\}$ converges uniformly on (a, b).

To prove this, choose any x in (a, b) different from c. Choose any k and m greater than k_0. The function $f_k - f_m$ is continuous on the closed interval and differentiable on the open interval with endpoints c and x. Therefore we can apply the Mean Value Theorem to the function $f_k - f_m$ on this interval. There exists a point d in the interval such that

$$\begin{aligned} f_k'(d) - f_m'(d) &= \frac{[f_k(x) - f_m(x)] - [f_k(c) - f_m(c)]}{x - c} \\ &= \frac{f_k(x) - f_k(c)}{(x - c)} - \frac{f_m(x) - f_m(c)}{x - c} \\ &= g_k(x) - g_m(x). \end{aligned}$$

Notice that $g_k(c) - g_m(c) = f_k'(c) - f_m'(c)$. Therefore, for any k and m greater than k_0 and for any x in (a, b),

$$|g_k(x) - g_m(x)| \leq ||f_k' - f_m'||_\infty < \epsilon,$$

we deduce that $\{g_k\}$ is uniformly Cauchy and therefore uniformly convergent. For each choice of c in (a, b), we construct the sequence $\{g_k\}$ of functions based on c; each such sequence converges uniformly to some continuous function h_c on (a, b). (The subscript is intended to remind you that h_c depends on c.)

In particular, choose c to be that c_0 where, by hypothesis, $\{f_k(c_0)\}$ converges. Construct the corresponding sequence $\{g_k\}$ of functions. From the definition of g_k, we know that

$$f_k(x) = f_k(c_0) + g_k(x)(x - c_0),$$

valid for all x in (a, b) including $x = c_0$. Thus, for any two indices k and m,

$$f_k(x) - f_m(x) = f_k(c_0) - f_m(c_0) + [g_k(x) - g_m(x)](x - c_0).$$

Now, $\{f_k(c_0)\}$ converges and, as we saw above, $\{g_k\}$ converges uniformly. Therefore, for any $\epsilon > 0$, there exists a k_1 such that, for k and m greater than k_1,

$$|f_k(c_0) - f_m(c_0)| < \epsilon \qquad \text{and} \qquad ||g_k - g_m||_\infty < \epsilon.$$

Consequently,

$$\begin{aligned}|f_k(x) - f_m(x)| &\leq |f_k(c_0) - f_m(c_0)| + ||g_k - g_m||_\infty(b - a)\\ &< \epsilon + \epsilon(b - a) = (1 + b - a)\epsilon,\end{aligned}$$

for all x in (a, b). This proves that $\{f_k\}$ is uniformly Cauchy and hence uniformly convergent on (a, b). Let f denote the uniform limit of $\{f_k\}$. We have to prove that f is differentiable on (a, b) and that $f' = g$, the uniform limit of the sequence $\{f_k'\}$.

To compute the derivative $f'(c)$ at an arbitrary point c in (a, b), return to the sequence $\{g_k\}$ constructed above based on the arbitrary point c. The sequence $\{g_k\}$ converges uniformly to the continuous function h_c on (a, b). For any $x \neq c$, we have

$$h_c(x) = \lim_{k\to\infty} g_k(x) = \lim_{k\to\infty} \frac{f_k(x) - f_k(c)}{x - c} = \frac{f(x) - f(c)}{x - c}.$$

Therefore, by the continuity of h_c,

$$\begin{aligned}f'(c) &= \lim_{x\to c} \frac{f(x) - f(c)}{x - c} = \lim_{x\to c} h_c(x)\\ &= h_c(c) = \lim_{k\to\infty} g_k(c) = \lim_{k\to\infty} f_k'(c) = g(c).\end{aligned}$$

We deduce that f is differentiable at c and that $f'(c) = g(c)$. Since c is an arbitrary point of (a, b), we conclude that $f' = g$ on (a, b). This proves the theorem. ●

COROLLARY 12.2.5 Suppose that $\sum_{j=1}^{\infty} f_j$ is a series of functions, each of which is differentiable on (a, b), satisfying the following conditions.

i) For some c_0 in (a, b), $\sum_{j=1}^{\infty} f_j(c_0)$ converges.

ii) For each k, the function $F_k' = \sum_{j=1}^{k} f_j'$ is bounded on (a, b).

iii) The derived series $\sum_{j=1}^{\infty} f_j'$ converges uniformly to some function g.

Then $\sum_{j=1}^{\infty} f_j$ converges uniformly to some differentiable function F on (a, b) and $F' = g$ on (a, b).

Proof. Simply apply Theorem 12.2.4 to the sequences of partial sums of $\sum_{j=1}^{\infty} f_j$ and $\sum_{j=1}^{\infty} f_j'$. ●

12.2.2 Convergence and Integration

By Theorem 6.5.1, we know that, if $\{f_k\}$ is a sequence of integrable functions on $[a, b]$ and if $\lim_{k\to\infty} f_k = f$ [uniformly] then f is also integrable on $[a, b]$. Further, for each x in $[a, b]$, we know that

$$\lim_{k\to\infty} \int_a^x f_k(t)\, dt = \int_a^x f(t)\, dt \quad \text{[uniformly]}.$$

In particular,

$$\lim_{k\to\infty} \int_a^b f_k(x)\, dx = \int_a^b f(x)\, dx.$$

For series Theorem 6.5.1 translates into the following form.

THEOREM 12.2.6 Suppose that $\sum_{j=1}^{\infty} f_j$ is a series of functions each of which is integrable on $[a, b]$. Suppose also that the series $\sum_{j=1}^{\infty} f_j = F$ [uniformly] on $[a, b]$. Then F is integrable on $[a, b]$. For x in $[a, b]$, define

$$G_k(x) = \sum_{j=1}^{k} \int_a^x f_j(t)\, dt,$$

and

$$G(x) = \int_a^x F(t)\, dt.$$

Then $\lim_{k\to\infty} G_k = G$ [uniformly] on $[a, b]$. In particular,

$$\int_a^b F(x)\, dx = \sum_{j=1}^{\infty} \int_a^b f_j(x)\, dx.$$

Proof. Apply Theorem 6.5.1 to the sequence of kth partial sums, $\{F_k\}$, of the series. ●

12.3 Tests for Uniform Convergence

In light of our results in Section 12.2, we clearly need to establish general principles and methods for determining whether a series of functions converges uniformly. The simplest and perhaps most useful test is due to Weierstrass.

THEOREM 12.3.1 Weierstrass's *M*-Test Suppose that $\sum_{j=1}^{\infty} f_j$ is a series of bounded, real-valued functions on a set S in $\mathbb{R}^n$. Assume that there is a convergent series $\sum_{j=1}^{\infty} M_j$ of positive numbers such that, for all j in $\mathbb{N}$, $||f_j||_\infty \leq M_j$. Then the series $\sum_{j=1}^{\infty} f_j$ converges uniformly and absolutely on S.

Note. In accord with our earlier comments regarding convergence being determined by the tail of the series, we need only require that $||f_j||_\infty \leq M_j$ for j greater than some index j_0.

Proof. The proof is an easy application of Cauchy's criterion. The series $\sum_{j=1}^{\infty} M_j$ converges; therefore its sequence of partial sums is Cauchy. Given $\epsilon > 0$, choose k_0 such that, for k and m greater than k_0, with $k > m$,

$$\left| \sum_{j=1}^{k} M_j - \sum_{j=1}^{m} M_j \right| = \sum_{j=m+1}^{k} M_j < \epsilon.$$

It follows, as the following computation confirms, that the sequence $\{F_k\}$ of partial sums of the series $\sum_{j=1}^{\infty} f_j$ is uniformly Cauchy. For k and m greater than k_0, with $k > m$,

$$||F_k - F_m||_\infty = \left|\left| \sum_{j=m+1}^{k} f_j \right|\right|_\infty \leq \sum_{j=m+1}^{k} ||f_j||_\infty \leq \sum_{j=m+1}^{k} M_j < \epsilon.$$

We conclude that $\sum_{j=1}^{\infty} f_j$ converges uniformly on S.

Finally, for any $\boldsymbol{x}$ in S, the series $\sum_{j=1}^{\infty} f_j(\boldsymbol{x})$ converges absolutely by the comparison test: $|f_j(\boldsymbol{x})| \leq M_j$ for $j \geq j_0$ and $\sum_{j=1}^{\infty} M_j$ converges. This proves the theorem. ●

EXAMPLE 4 Fix any r in $(0, 1)$. The geometric series $\sum_{j=0}^{\infty} x^j$ converges uniformly and absolutely on the interval $[-r, r]$. To confirm this assertion, let $M_j = r^j$. Then $\sum_{j=0}^{\infty} M_j = \sum_{j=0}^{\infty} r^j$ is a convergent series of positive numbers. For x in $[-r, r]$ and all j in $\mathbb{N}$, $|f_j(x)| = |x|^j \leq r^j = M_j$. Therefore, by Weierstrass's M-test, the series $\sum_{j=0}^{\infty} x^j$ converges uniformly and absolutely. ●

EXAMPLE 5 The series $\sum_{j=1}^{\infty} \sin(jx)/j^2$ converges uniformly and absolutely on all of $\mathbb{R}$ by Weierstrass's M-test. Merely take $M_j = 1/j^2$ and note that, for all j in $\mathbb{N}$ and all x in $\mathbb{R}$,

$$|f_j(x)| = \left| \frac{\sin jx}{j^2} \right| \leq \frac{1}{j^2} = M_j.$$

Since $\sum_{j=1}^{\infty} 1/j^2$ is a convergent series of positive numbers, we conclude that $\sum_{j=1}^{\infty} \sin(jx)/j^2$ converges uniformly and absolutely on $\mathbb{R}$. ●

Abel's test (Theorem 11.4.3) for convergence of a series $\sum_{j=1}^{\infty} a_j b_j$ of numbers generalizes to apply to a series $\sum_{j=1} f_j g_j$ of functions defined on a set S in $\mathbb{R}^n$.

THEOREM 12.3.2 Abel's Test for Uniform Convergence Let S be a subset of $\mathbb{R}^n$. Suppose that $\{f_j\}$ and $\{g_j\}$ are sequences of bounded, real-valued functions on S such that

i) The series $\sum_{j=1}^{\infty} f_j$ is uniformly convergent on S.

ii) The sequence $\{g_j\}$ is monotone and uniformly bounded on S.

Then $\sum_{j=1}^{\infty} f_j g_j$ converges uniformly on S.

Proof. Let F_k denote the kth partial sum of the series $\sum_{j=1}^{\infty} f_j$. We assume that $\{F_k\}$ converges uniformly to F. Thus, given any $\epsilon > 0$, there exists a k_0 such that, whenever $k \geq k_0$, $||F - F_k||_\infty < \epsilon$. Fix k_0.

The hypotheses on $\{g_j\}$ ensure that there is some constant M such that $||g_j||_\infty \leq M$ for all j in $\mathbb{N}$. For each $\boldsymbol{x}$ in S, $\{g_j(\boldsymbol{x})\}$ is a bounded, monotone sequence of real numbers. By Theorem 1.3.7, we deduce that $\{g_j\}$ converges pointwise. (Note: $\{g_j\}$ need not converge uniformly.) With no loss of generality, we assume that $\{g_j\}$ is monotone decreasing; for each $\boldsymbol{x}$ in S and each j in $\mathbb{N}$, $g_j(\boldsymbol{x}) \geq g_{j+1}(\boldsymbol{x})$.

To prove that $\sum_{j=1}^{\infty} f_j g_j$ converges uniformly, we will show that the sequence $\{G_k\}$ of partial sums of $\sum_{j=1}^{\infty} f_j g_j$ is uniformly Cauchy. To this end, choose any k and m greater than k_0, with $k > m$. Note that, for each $j = 2, 3, 4, \ldots$,

$$f_j g_j = (F_j - F)(g_j - g_{j+1}) + (F - F_{j-1})g_j - (F - F_j)g_{j+1}.$$

Summing on j from $m + 1$ to k yields

$$G_k - G_m = \sum_{j=m+1}^{k} f_j g_j = \sum_{j=m+1}^{k} (F_j - F)(g_j - g_{j+1}) + \sum_{j=m+1}^{k} [(F - F_{j-1})g_j - (F - F_j)g_{j+1}].$$

The last summation above telescopes and we obtain

$$G_k - G_m = \sum_{j=m+1}^{k} (F_j - F)(g_j - g_{j+1}) + (F - F_m)g_{m+1} - (F - F_k)g_{k+1}.$$

It follows that, for any $\boldsymbol{x}$ in S,

$$\begin{aligned}|G_k(\boldsymbol{x}) - G_m(\boldsymbol{x})| &\leq \sum_{j=m+1}^{k} |F_j(\boldsymbol{x}) - F(\boldsymbol{x})|\,|g_j(\boldsymbol{x}) - g_{j+1}(\boldsymbol{x})| \\ &\quad + |F(\boldsymbol{x}) - F_m(\boldsymbol{x})|\,|g_{m+1}(\boldsymbol{x})| + |F(\boldsymbol{x}) - F_k(\boldsymbol{x})|\,|g_{k+1}(\boldsymbol{x})| \\ &\leq \sum_{j=m+1}^{k} ||F_j - F||_\infty [g_j(\boldsymbol{x}) - g_{j+1}(\boldsymbol{x})] \\ &\quad + ||F - F_m||_\infty ||g_{m+1}||_\infty + ||F - F_k||_\infty ||g_{k+1}||_\infty.\end{aligned}$$

Since k and m are greater than k_0, we have $||F_j - F||_\infty < \epsilon$ for all $j = m$, $m + 1, m + 2, \ldots, k$. Therefore,

$$|G_k(\boldsymbol{x}) - G_m(\boldsymbol{x})| < \epsilon \sum_{j=m+1}^{k} [g_j(\boldsymbol{x}) - g_{j+1}(\boldsymbol{x})] + \epsilon\, ||g_{m+1}||_\infty + \epsilon\, ||g_{k+1}||_\infty.$$

Again, the summation telescopes and we obtain

$$|G_k(\boldsymbol{x}) - G_m(\boldsymbol{x})| < \epsilon\,[g_{m+1}(\boldsymbol{x}) - g_{k+1}(\boldsymbol{x})] + \epsilon\,[||g_{m+1}||_\infty + ||g_{k+1}||_\infty] < 4M\epsilon.$$

Since $\boldsymbol{x}$ is arbitrary in S and since k_0 is independent of $\boldsymbol{x}$, we conclude that $\{G_k\}$ is uniformly Cauchy and therefore uniformly convergent. This proves the theorem and establishes Abel's test for uniform convergence. ●

EXAMPLE 6 Fix r in $(0, 1)$. The series $\sum_{j=1}^{\infty} x^j/[j^2(1 - x^{2j})]$ converges uniformly for x in $[-r, r]$ by Abel's test. Let $f_j(x) = x^j/j^2$ and $g_j(x) = 1/(1 - x^{2j})$. The series $\sum_{j=1}^{\infty} f_j = \sum_{j=1}^{\infty} x^j/j^2$ converges uniformly on $[-r, r]$ by Weierstrass's M-test: For all j in $\mathbb{N}$, $||f_j||_\infty \leq 1/j^2$ and $\sum_{j=1}^{\infty} 1/j^2$ converges. The sequence $\{g_j\}$ is monotone decreasing and bounded below. To prove this, note that $0 < (1 - x^{2j}) \leq (1 - x^{2j+2}) \leq 1$ for all x in $[-r, r]$. Thus, for such x,

$$1 \leq \frac{1}{1 - x^{2j+2}} \leq \frac{1}{1 - x^{2j}}.$$

Consequently, Abel's test applies and our series converges uniformly on $[-r, r]$. You might notice that Weierstrass's M-test also applies directly to show that this series converges uniformly and absolutely on $[-r, r]$ because

$$\left|\frac{x^j}{j^2(1 - x^{2j})}\right| \leq \frac{1}{j^2(1 - r^2)} = M_j$$

for all x in $[-r, r]$ and because $\sum_{j=1}^{\infty} 1/[j^2(1 - r^2)]$ converges. ●

Dirichlet's test (Theorem 11.4.4) also generalizes to provide us with a test for uniform convergence.

THEOREM 12.3.3 Dirichlet's Test for Uniform Convergence Let S be a subset of $\mathbb{R}^n$. Suppose that $\{f_j\}$ and $\{g_j\}$ are sequences of bounded, real-valued functions defined on S such that the functions g_j are positive and

i) The sequence $\{F_k\}$ of partial sums of $\sum_{j=1}^{\infty} f_j$ is uniformly bounded on S.

ii) The sequence $\{g_j\}$ converges monotonically and uniformly to 0 on S.

Then the series $\sum_{j=1}^{\infty} f_j g_j$ converges uniformly on S.

Proof. Let G_k denote the kth partial sum of $\sum_{j=1}^{\infty} f_j g_j$. We will show that $\{G_k\}$ is uniformly Cauchy. Our proof relies on Abel's formula, Theorem 11.4.2. In the present context, that formula takes the form

$$G_k = \sum_{j=1}^{k} f_j g_j = F_k g_{k+1} + \sum_{j=1}^{k} F_j(g_j - g_{j+1}).$$

Let M be a uniform bound of the sequence $\{F_k\}$. That is, $||F_k||_\infty \leq M$ for all k in $\mathbb{N}$. Since $\{g_j\}$ converges uniformly to 0, given $\epsilon > 0$, there is a k_0 such that, whenever $k \geq k_0$, $||g_k||_\infty < \epsilon$. Fix such a k_0. Choose any k and m greater than k_0 with $k > m$ and use Abel's formula to compute

$$G_k - G_m = \left[F_k g_{k+1} + \sum_{j=1}^{k} F_j(g_j - g_{j+1}) \right] - \left[F_m g_{m+1} + \sum_{j=1}^{m} F_j(g_j - g_{j+1}) \right]$$

$$= F_k g_{k+1} - F_m g_{m+1} + \sum_{j=m+1}^{k} F_j(g_j - g_{j+1}).$$

For any $\boldsymbol{x}$ in S,

$$|G_k(\boldsymbol{x}) - G_m(\boldsymbol{x})| \leq ||F_k||_\infty g_{k+1}(\boldsymbol{x}) + ||F_m||_\infty g_{m+1}(\boldsymbol{x}) + \sum_{j=m+1}^{k} ||F_j||_\infty [g_j(\boldsymbol{x}) - g_{j+1}(\boldsymbol{x})]$$

$$\leq M[g_{k+1}(\boldsymbol{x}) + g_{m+1}(\boldsymbol{x})] + M \sum_{j=m+1}^{k} [g_j(\boldsymbol{x}) - g_{j+1}(\boldsymbol{x})].$$

The last summation telescopes to yield

$$|G_k(\boldsymbol{x}) - G_m(\boldsymbol{x})| \leq M[g_{k+1}(\boldsymbol{x}) + g_{m+1}(\boldsymbol{x}) + g_{m+1}(\boldsymbol{x}) - g_{k+1}(\boldsymbol{x})] = 2M g_{m+1}(\boldsymbol{x}) < 2M\epsilon.$$

Since $\boldsymbol{x}$ is arbitrary and k_0 was chosen independently of $\boldsymbol{x}$, $\{G_k\}$ is uniformly Cauchy. Therefore, $\sum_{j=1}^{\infty} f_j g_j$ is uniformly convergent. This proves Dirichlet's test. ●

EXAMPLE 7 Fix a and b such that $0 < a < b < 2\pi$. Consider the series $\sum_{j=1}^{\infty} \sin(jx)/j$ for x in $[a, b]$. Let $f_j(x) = \sin(jx)$ and $g_j(x) = 1/j$. The conditions required by Dirichlet's test are met. First, by Theorem 11.4.5, for any x in $[a, b]$, we have

$$|F_k(x)| = \left| \sum_{j=1}^{k} \sin(jx) \right| = \left| \frac{\sin[(k+1)x/2] \sin(kx/2)}{\sin(x/2)} \right| \leq \frac{1}{\sin(x/2)} \leq 1/m,$$

where $m = \min\{\sin(a/2), \sin(b/2)\}$. Thus, $\{F_k\}$ is uniformly bounded by $M = 1/m$ on $[a, b]$. Second, $\{g_j\}$ is a positive sequence that converges monotonically and uniformly to 0. Dirichlet's test implies that $\sum_{j=1}^{\infty} \sin(jx)/j$ converges uniformly on $[a, b]$. Notice that neither Weierstrass's M-test nor Abel's test applies in any obvious way to help us resolve the question of uniform convergence of the series in this example. ●

12.4 Power Series

One of the most useful theories in all of analysis, and one with the most far-reaching ramifications, is that which deals with power series. A large class of functions f, those called *analytic functions*, can be decomposed into a *power series representation*. With this representation in hand, we can calculate the values of the function as accurately as desired. Moreover, by specifying the coefficients of a power series initially, we can construct new functions whose existence you may never have suspected and that behave as we may specify. As we shall see, subject to simple convergence constraints, we can differentiate and integrate power series term-by-term to obtain the derivative and the integral, respectively, of the function being represented. Likewise, subject to convergence constraints, power series converge absolutely and thus can be rearranged and multiplied at will. You can hardly ask for more of a mathematical tool; power series will serve you well.

DEFINITION 12.4.1 A *(real) power series about the point* x_0 in $\mathbb{R}$ is a series of the form $\sum_{j=0}^{\infty} a_j(x - x_0)^j$, where the coefficients a_j are real numbers. ●

In a more general setting the coefficients a_j, the point x_0, and the variable x can all be taken to be complex numbers; virtually all our results carry over and lead to a rich and useful theory of complex analysis. Here we shall be content with considering only real series. Furthermore, to simplify notation, we will usually let $x_0 = 0$; a simple translation will lift our results to the general case.

Let $\sum_{j=0}^{\infty} a_j x^j$ be an arbitrary power series about the point $x_0 = 0$. Apply the root test to determine those values of x where the series converges. Consider

$$\limsup |a_j x^j|^{1/j} = |x| \limsup |a_j|^{1/j}.$$

If this limit superior is less than 1, then Cauchy's root test implies that the series converges absolutely; if this limit superior is greater than 1, then the series diverges.

DEFINITION 12.4.2 Given a power series $\sum_{j=0}^{\infty} a_j x^j$, let

$$\lambda = \limsup |a_j|^{1/j}.$$

If λ is finite and positive, define $R = 1/\lambda$; if $\lambda = 0$, define $R = \infty$; if $\lambda = \infty$, define $R = 0$. The R thus defined is called the *radius of convergence* of $\sum_{j=0}^{\infty} a_j x^j$. The set of points where the series converges is called the *interval of convergence* of the series. ●

Remark. If only finitely many of the a_j are zero, then we can also use the ratio test to find R. Simply compute

$$\limsup \left| \frac{a_{j+1}}{a_j} \right|,$$

considering only those quotients with $a_j \neq 0$. Now let $\lambda = \limsup |a_{j+1}/a_j|$, and define R as above. More generally, for any power series, you can omit all those terms with zero coefficient and apply the ratio test; form the ratio of successive nonzero terms. If the ratio of successive nonzero terms has a limit superior that you can identify, you may be able to use the ratio test to find the radius of convergence R. (See Example 9.)

EXAMPLE 8 The power series $\sum_{j=1}^{\infty} x^j/(j2^j)$ has radius of convergence $R = 2$. We confirm this assertion by computing

$$\lambda = \limsup \frac{1}{(j2^j)^{1/j}} = \frac{1}{2} \limsup \left(\frac{1}{j}\right)^{1/j} = \frac{1}{2}.$$

Therefore $R = 1/\lambda = 2$. In this example, we could also use the ratio test with the same result:

$$\lambda = \limsup \frac{1/[(j+1)2^{j+1}]}{1/(j2^j)} = \frac{1}{2} \limsup \frac{j}{(j+1)} = \frac{1}{2}.$$

Consequently, $R = 1/\lambda = 2$. ●

EXAMPLE 9 The power series $\sum_{j=1}^{\infty} x^{2j}/(j9^j)$ has radius of convergence $R = 3$. Notice that, for this series, a_0 and all coefficients with odd index are 0. To find the radius of convergence using the ratio test, we take the ratio of successive even indexed terms:

$$\left|\frac{a_{2j+2}x^{2j+2}}{a_{2j}x^{2j}}\right| = |x|^2 \frac{1/[(j+1)9^{j+1}]}{1/(j9^j)} = \frac{|x|^2}{9}\left(\frac{j}{j+1}\right).$$

The limit superior of this ratio is $|x|^2/9$ and is less than 1 if and only if $|x|^2 < 9$. Equivalently, $|x| < 3$. Therefore, by the ratio test, the series converges for $|x| < 3$ and diverges for $|x| > 3$. Thus the radius of convergence is $R = 3$. Were you to opt for the use of the root test, you would need to observe again that the odd indexed terms are 0 and that

$$\begin{aligned}
\limsup |a_j|^{1/j} &= \limsup |a_{2j}|^{1/2j} = \limsup \left(\frac{1}{j9^j}\right)^{1/2j} \\
&= \frac{1}{3} \limsup \left(\frac{1}{j}\right)^{1/2j} \\
&= \frac{1}{3}\left[\limsup \left(\frac{1}{j}\right)^{1/j}\right]^{1/2} = \frac{1}{3}.
\end{aligned}$$

Thus, $\lambda = 1/3$ and, again, $R = 3$. ●

As the next theorem reveals, the relatively simple computation required to find the radius of convergence of a power series yields an enormous amount of information.

THEOREM 12.4.1 Let $\sum_{j=0}^{\infty} a_j x^j$ be a power series with radius of convergence R.

i) If $0 < R < \infty$ then $\sum_{j=0}^{\infty} a_j x^j$ converges absolutely for all x in $(-R, R)$. For $|x| > R$, the series diverges. For $x = \pm R$, we have no information.

ii) If $R = \infty$, then $\sum_{j=0}^{\infty} a_j x^j$ converges absolutely for all real numbers x.

iii) If $R = 0$, then $\sum_{j=0}^{\infty} a_j x^j$ converges (trivially) for $x = 0$ and diverges for all other values of x.

Proof. If $0 < R < \infty$ and if $|x| < R$, then

$$\limsup |a_j x^j|^{1/j} = |x| \limsup |a_j|^{1/j} < R\left(\frac{1}{R}\right) = 1.$$

Therefore, by the root test, the series $\sum_{j=0}^{\infty} a_j x^j$ converges absolutely. If $|x| > R$, then the series diverges by that same test. If $|x| = R$, the root test provides no information. The possible convergence of the power series when $x = \pm R$ must be examined separately for each specific series. Thus, if $0 < R < \infty$, then the series converges absolutely on the open interval $(-R, R)$, diverges for $|x| > R$, and may or may not converge at either endpoint $\pm R$.

If $R = \infty$, then $\limsup |a_j|^{1/j} = 0$. Thus, for every x in $\mathbb{R}$, $\limsup |a_j x^j|^{1/j} = |x| \limsup |a_j|^{1/j} = 0 < 1$. Therefore, by the root test, the series converges absolutely on the entire real line.

If $R = 0$, then $\limsup |a_j|^{1/j} = \infty$. Consequently, $\limsup |a_j x^j|^{1/j} = |x| \limsup |a_j|^{1/j} < 1$ if and only if $|x| = 0$; the series converges (trivially) only at $x = 0$. ●

An extremely important feature of power series, described in our next theorem, is that such series always converge uniformly on any compact interval contained strictly within the interval of convergence.

THEOREM 12.4.2 Let $\sum_{j=0}^{\infty} a_j x^j$ be a power series with radius of convergence $R \neq 0$. If $0 < r < R$, then $\sum_{j=0}^{\infty} a_j x^j$ converges uniformly and absolutely to a continuous function on $[-r, r]$.

Proof. If $0 < r < R$, then $\sum_{j=0}^{\infty} |a_j| r^j$ is a convergent series of positive terms and, for all x in $[-r, r]$ and all $j \geq 0$, $|a_j x^j| \leq |a_j| r^j$. Therefore, by Weierstrass's M-test, $\sum_{j=0}^{\infty} a_j x^j$ converges uniformly and absolutely for x in the compact set $[-r, r]$. Since the terms $a_j x^j$ of the series are continuous, so also is the limit function. ●

COROLLARY 12.4.3 The power series $\sum_{j=0}^{\infty} a_j x^j$ converges uniformly and absolutely to a continuous function on any compact interval $[a, b]$ contained in $(-R, R)$.

Proof. The proof consists merely of choosing any r between $\max\{|a|, |b|\}$ and R and applying Theorem 12.4.2 to the interval $[-r, r]$. (See Fig. 12.2.) The series $\sum_{j=0}^{\infty} a_j x^j$ converges uniformly and absolutely to a continuous function on $[-r, r]$; the same is true on the subset $[a, b]$. ●

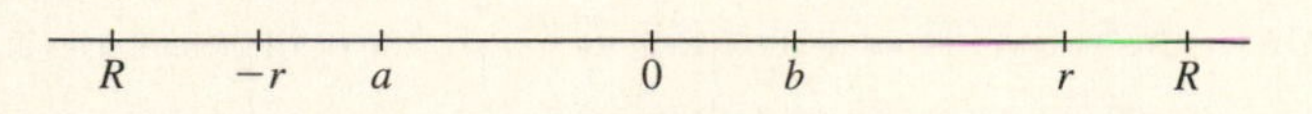

Figure 12.2

EXAMPLE 8 (Revisited) The series $\sum_{j=1} x^j/[j2^j]$ has radius of convergence $R = 2$. Therefore, this series converges absolutely on the interval $(-2, 2)$. It converges uniformly to some continuous function f on any compact $[a, b] \subset (-2, 2)$. Later, we will learn some techniques for identifying the function f to which the series converges. (In this example, it happens that $f(x) = \ln(1 - x/2)$.) If $x = R = 2$, the series reduces to the harmonic series and therefore diverges. However, at the other endpoint $x = -2$, the series becomes $\sum_{j=1}^{\infty}(-1)^j/j$, which converges conditionally by the alternating series test. The interval of convergence is $I = [-2, 2)$. ●

Similarly, the series in Example 9 converges absolutely on $(-3, 3)$, converges conditionally at $x = -3$, diverges at $x = 3$, and converges uniformly to some continuous function on any compact subinterval of $(-3, 3)$.

EXAMPLE 10 Applying the ratio test to successive nonzero terms, we find that the radius of convergence of $\sum_{j=0}^{\infty} x^{2j}/(2j)!$ is $R = \infty$. To confirm this, write

$$\limsup \left| \frac{x^{2j+2}/(2j+2)!}{x^{2j}/(2j)!} \right| = |x|^2 \limsup \frac{1}{(2j+2)(2j+1)} = 0.$$

Therefore $R = \infty$. Thus we deduce that the series converges absolutely for every x in $\mathbb{R}$. It converges uniformly to some continuous function f on every compact subset of $\mathbb{R}$. As we shall see, this particular series happens to converge to the hyperbolic cosine function $f(x) = \cosh x$. ●

Let $\sum_{j=0}^{\infty} a_j x^j$ be a power series with a positive, finite radius of convergence R. We know that the series converges absolutely on $(-R, R)$; it may converge absolutely or conditionally at either or both of the endpoints. We also know that the series converges uniformly on any interval $[a, b]$ contained within $(-R, R)$. Suppose, for the sake of argument, that the series converges (either absolutely or conditionally) at $x = R$. The question we raise is this: Can an interval $[a, b]$, where uniform convergence is known to occur, be extended to include the endpoint R? Of course, the same question can be asked if it should happen that the series converges at $x = -R$. Abel has provided us with the following particularly satisfying result.

THEOREM 12.4.4 Abel's Theorem If $R > 0$ is the radius of convergence of the power series $\sum_{j=0}^{\infty} a_j x^j$ and if the power series converges at $x = R$ (or at $x = -R$), then it converges uniformly on $[0, R]$ (or on $[-R, 0]$).

Proof. To simplify notation we assume that $R = 1$. We lose no generality with this assumption because the series $\sum_{j=0}^{\infty} b_j x^j$ where $b_j = a_j/R^j$, does have radius of convergence 1. If $\sum_{j=0}^{\infty} a_j x^j$ converges at either R or $-R$, then $\sum_{j=0}^{\infty} b_j x^j$ converges correspondingly at 1 or -1. We could simply begin with the series $\sum_{j=0}^{\infty} b_j x^j$.

We assume, then, that $R = 1$ and that $\sum_{j=0}^{\infty} a_j$ converges. Let $\{F_k(x)\}$ denote the sequence of partial sums of the power series $\sum_{j=0}^{\infty} a_j x^j$ and let $\{s_k\}$ denote the sequence of partial sums of the series $\sum_{j=0}^{\infty} a_j$. The sequence $\{s_k\}$ converges and thus is Cauchy. Given $\epsilon > 0$, choose k_0 such that, for k and m greater than k_0 with $k > m$, we have

$$|s_k - s_m| = \left| \sum_{j=m+1}^{k} a_j \right| < \epsilon.$$

Notice that, for any x in $[0, 1]$ and any j in $\mathbb{N}$, $x^j \geq x^{j+1}$. By making the appropriate rearrangements and by telescoping extensively, you can show that, for $k > m > k_0$,

$$\begin{aligned} |F_k(x) - F_m(x)| &= |a_{m+1}x^{m+1} + a_{m+2}x^{m+2} + \cdots + a_k x^k| \\ &= \left| \sum_{j=m+1}^{k} [(x^j - x^{j+1})(s_j - s_m)] + x^{k+1}(s_k - s_m) \right| \\ &< \epsilon \left[\sum_{j=m+1}^{k} (x^j - x^{j+1}) + x^{k+1} \right] = \epsilon x^{m+1} < \epsilon. \end{aligned}$$

Since k_0 is independent of x in $[0, 1]$, we conclude that $||F_k - F_m||_\infty < \epsilon$ for k and m greater than k_0. Hence, $\{F_k\}$ is uniformly Cauchy and thus uniformly convergent on $[0, 1]$. This proves that $\sum_{j=0}^{\infty} a_j x^j$ converges uniformly on $[0, 1]$. We leave the case when the series converges at the negative endpoint for you to prove. ●

Remark. Clearly, if $\sum_{j=0}^{\infty} a_j x^j$ converges at $x = R$ and diverges at $x = -R$, then $\sum_{j=0}^{\infty} a_j x^j$ converges uniformly to some continuous function on any compact interval $[a, R] \subset (-R, R]$. Likewise, if $\sum_{j=0}^{\infty} a_j x^j$ converges at $x = -R$ and diverges at $x = R$, then $\sum_{j=0}^{\infty} a_j x^j$ converges uniformly on $[-R, b] \subset [-R, R)$. If $\sum_{j=0}^{\infty} a_j x^j$ converges at both R and $-R$, then the convergence is uniform on $[-R, R]$. The *interval of convergence* I is either $(-R, R)$, $(-R, R]$, $[-R, R)$, or $[-R, R]$; convergence is uniform on any compact subset of the interval of convergence.

12.4.1 Differentiation of Power Series

Let $\sum_{j=0}^{\infty} a_j x^j$ be any power series with radius of convergence $R \neq 0$. By differentiating each term of this series we obtain its derived series $\sum_{j=1}^{\infty} j a_j x^{j-1}$. Note that, for $j \geq 0$, the coefficient of x^j in this last series is $(j + 1)a_{j+1}$; by shifting the index, we rewrite the derived series as $\sum_{j=0}^{\infty}(j + 1)a_{j+1}x^j$. Since the derived series is also a power series, it has its own radius of convergence R'. To compute that radius use the root test and write

$$\lambda' = \limsup |(j + 1)a_{j+1}|^{1/j} = \limsup (j + 1)^{1/j} |a_{j+1}|^{1/j}.$$

We leave for you to prove as an exercise that

$$\lim_{j \to \infty} (j + 1)^{1/j} = 1$$

and that

$$\begin{aligned} \limsup |a_{j+1}|^{1/j} &= \limsup \left[|a_{j+1}|^{1/(j+1)} \right]^{(j+1)/j} \\ &= \limsup |a_j|^{1/j}. \end{aligned}$$

Applying these two results, we deduce that

$$\lambda' = \limsup |(j+1)a_{j+1}|^{1/j} = \limsup |a_j|^{1/j} = \lambda.$$

It follows that the original series $\sum_{j=0}^{\infty} a_j x^j$ and its derived series $\sum_{j=0}^{\infty} (j+1)a_{j+1}x^j$ have exactly the same radius of convergence; that is, $R' = R$. This argument establishes the starting point for the following theorem. Not only do a power series and its derived series have the same radius of convergence, but, strictly within the interval of convergence, the derived series always converges to the derivative of the function represented by the original power series.

THEOREM 12.4.5 Let $\sum_{j=0} a_j x^j$ be a power series with radius of convergence $R \neq 0$. Let $F(x) = \sum_{j=0}^{\infty} a_j x^j$ for x in the interval of convergence. Then F is differentiable on $(-R, R)$ and $F'(x) = \sum_{j=0}^{\infty}(j+1)a_{j+1}x^j$. The derived series also has radius of convergence R and converges uniformly on any compact subset of $(-R, R)$.

Proof. We already know that $\sum_{j=0}^{\infty}(j+1)a_{j+1}x^j$ has radius of convergence R. We also know that the derived series, as a power series in its own right, converges uniformly to some continuous function on any compact $[a, b] \subset (-R, R)$. Given any c_0 in $(-R, R)$, enclose c_0 in an interval $[a, b]$ with $-R < a \leq c_0 \leq b < R$. Apply Corollary 12.2.5. Since $\sum_{j=0}^{\infty} a_j c_0^j$ converges to $F(c_0)$ and since $\sum_{j=0}^{\infty}(j+1)a_{j+1}x^j$ converges uniformly on $[a, b]$, we conclude that F is differentiable and that $F'(x) = \sum_{j=0}^{\infty}(j+1)a_{j+1}x^j$ [uniformly] on $[a, b]$. ●

Remark. In the linkage between $F(x) = \sum_{j=0}^{\infty} a_j x^j$ and its derivative $F'(x) = \sum_{j=0}^{\infty}(j+1)a_{j+1}x^j$, special attention needs to be paid to the endpoints $\pm R$ of the interval of convergence. While $\sum_{j=0}^{\infty} a_j x^j$ may converge at either R or $-R$, it can happen that its derived series fails to do so. By differentiating you might lose convergence at either or both endpoints. However, if the derived series converges at R (or $-R$), then the original series converges at R (or $-R$) also. You never gain additional points of convergence by differentiating a power series; you may lose endpoints from the interval of convergence. We will prove these claims below.

EXAMPLE 11 The radius of convergence of the power series

$$\sum_{j=0}^{\infty} \frac{(-1)^j x^{2j+1}}{(2j+1)!}$$

is most easily found by applying the ratio test to the successive nonzero terms:

$$\limsup \left| \frac{a_{2j+3}x^{2j+3}}{a_{2j+1}x^{2j+1}} \right| = |x|^2 \limsup \frac{1}{(2j+3)(2j+2)} = 0.$$

Therefore, $R = \infty$ and the series converges absolutely for all x in $\mathbb{R}$ and uniformly on any compact interval $[a, b]$. The partial sums of this series are, as we know from Example 11 of Section 4.5, the Taylor polynomials of the function $\sin x$

about $x_0 = 0$; from that example we know that this sequence of Taylor polynomials converges uniformly to $\sin x$ on any compact set in $\mathbb{R}$. Therefore,

$$\sum_{j=0}^{\infty} \frac{(-1)^j x^{2j+1}}{(2j+1)!} = \sin x.$$

The derived series is

$$\sum_{j=0}^{\infty} \frac{(-1)^j (2j+1) x^{2j}}{(2j+1)!} = \sum_{j=0}^{\infty} \frac{(-1)^j x^{2j}}{(2j)!}.$$

You can easily confirm, by applying the ratio test to successive nonzero terms, that this latter series also converges for all x in $\mathbb{R}$. From Example 12 of Section 4.5 we know that, for every x in $\mathbb{R}$, the derived series converges to $\cos x$, the derivative of the sine function. That convergence is uniform on any compact subset of $\mathbb{R}$. ●

EXAMPLE 12 By the ratio test we find that the radius of convergence of $\sum_{j=1}^{\infty}(-1)^j x^j/\sqrt{j}$ is $R = 1$. Therefore the series converges absolutely on $(-1, 1)$. Furthermore, the series converges conditionally at $x = 1$ by the alternating series test; by the integral test, it diverges at $x = -1$. Abel's theorem tells us that the series converges uniformly on any compact subset of $I = (-1, 1]$ to some continuous function F. The derived series is $\sum_{j=1}^{\infty}(-1)^j j x^{j-1}/\sqrt{j} = \sum_{j=0}^{\infty}(-1)^{j+1}\sqrt{j+1}\,x^j$. Again, you can use the ratio test to show that the radius of convergence of the derived series is $R = 1$. Note that the derived series diverges at both ± 1; in both cases the jth term fails to converge to 0. Therefore, the derived series converges only on $I = (-1, 1)$. At each point in this interval it converges absolutely to F'; the convergence is uniform on any compact subset of $I = (-1, 1)$. ●

Observe that what is true for $F(x) = \sum_{j=0}^{\infty} a_j x^j$ is equally true for its derived series $F'(x) = \sum_{j=0}^{\infty}(j+1)a_{j+1}x^j$; differentiating this last series term-by-term yields a series that has the same radius of convergence R, that converges absolutely on $(-R, R)$, and that also converges uniformly on any compact interval $[a, b]$ contained in $(-R, R)$. Thus, the series $\sum_{j=1}^{\infty}(j+1)j a_{j+1}x^{j-1} = \sum_{j=0}^{\infty}(j+2)(j+1)a_{j+2}x^j$ converges uniformly to $F''(x)$ on $[a, b]$. The function F defined by our original power series is twice differentiable at all points in $(-R, R)$.

Inductively, F has derivatives of all orders at every point in $(-R, R)$. The kth derivative of F is given by the power series

$$F^{(k)}(x) = \sum_{j=k}^{\infty}\left[\frac{j!}{(j-k)!}\right] a_j x^{j-k} = \sum_{j=0}^{\infty}\left[\frac{(j+k)!}{j!}\right] a_{j+k} x^j.$$

For each k in $\mathbb{N}$, each of these series has the same radius of convergence as the original power series. Our discussion proves the following theorem.

THEOREM 12.4.6 Let $\sum_{j=0}^{\infty} a_j x^j$ be a power series with radius of convergence $R \neq 0$. For x in $(-R, R)$, let $F(x) = \sum_{j=0}^{\infty} a_j x^j$. Then the function F has derivatives of all orders on the interval $(-R, R)$. The kth derivative of F is given by

$$F^{(k)}(x) = \sum_{j=0}^{\infty} \frac{(j+k)!}{j!} a_{j+k} x^j. \quad \bullet$$

Finally, we confirm that if the derived series converges at an endpoint of the interval of convergence, then the original power series must also converge there.

THEOREM 12.4.7 Suppose that $\sum_{j=0}^{\infty} a_j x^j$ has radius of convergence R with $0 < R < \infty$. If the series $\sum_{j=0}^{\infty} (j+1) a_{j+1} x^j$ converges at R (or at $-R$), then $\sum_{j=0}^{\infty} a_j x^j$ also converges at R (or at $-R$).

Proof. We assume without loss of generality that $R = 1$ and that $\sum_{j=0}^{\infty} (j+1) a_{j+1} = \sum_{j=1}^{\infty} j a_j$ converges. To show that $\sum_{j=0}^{\infty} a_j$ also converges, let $b_j = j a_j$ and $c_j = 1/j$ for j in $\mathbb{N}$. Then, by hypothesis, the series $\sum_{j=1}^{\infty} b_j$ converges. The sequence $\{c_j\}$ is monotone and convergent. By Abel's test (Theorem 11.4.3), the series $\sum_{j=1}^{\infty} b_j c_j = \sum_{j=1}^{\infty} a_j$ also converges. We leave the case of the negative endpoint for you to prove. $\bullet$

Thus, by Theorem 12.4.7, if the derived series converges at an endpoint of $(-R, R)$, then the original series must also converge at that endpoint. However, Example 12 demonstrates that differentiating a power series term-by-term may cause the loss of convergence at an endpoint.

12.4.2 Integration of Power Series

Suppose that $\sum_{j=0}^{\infty} a_j x^j$ has radius of convergence $R \neq 0$. Let I denote the interval of convergence of the series, consisting of all points in $[-R, R]$ where the series converges. Let $F(x)$ denote the function to which the series converges on I. Finally, let $[a, b]$ be any compact subinterval of I. Each term of the series $\sum_{j=0}^{\infty} a_j x^j$ is integrable on $[a, b]$ and, since the power series converges uniformly to F on $[a, b]$, we know by Theorem 12.2.6 that F is also integrable on $[a, b]$ and that

$$\int_a^b F(x)\,dx = \sum_{j=0}^{\infty} a_j \int_a^b x^j\,dx.$$

Moreover, for x in $[a, b]$, the series of indefinite integrals $\sum_{j=0}^{\infty} a_j \int_a^x t^j\,dt$ converges uniformly to $G(x) = \int_a^x F(t)\,dt$ on $[a, b]$. In particular, if we assume that 0 is in $[a, b]$ and if we integrate from 0 to x, then

$$G_0(x) = \int_0^x F(t)\,dt = \int_0^x \left[\sum_{j=0}^{\infty} a_j t^j\right] dt = \sum_{j=0}^{\infty} a_j \int_0^x t^j\,dt$$

$$= \sum_{j=0}^{\infty} \frac{a_j t^{j+1}}{j+1}\bigg|_0^x = \sum_{j=0}^{\infty} \frac{a_j x^{j+1}}{j+1} = \sum_{j=1}^{\infty} \frac{a_{j-1} x^j}{j} \quad \text{[uniformly]} \qquad \textbf{(12.2)}$$

on $[a, b]$. The indefinite integral of a function F that is represented by a power series is itself represented by a power series uniformly convergent on any compact interval contained in I. Theorem 12.4.5, applied to $\sum_{j=1}^{\infty} a_{j-1}x^j/j$ and its derived series $\sum_{j=0}^{\infty} a_j x^j$, proves that the integrated series and the original series have the same radius of convergence. This argument proves the following theorem.

THEOREM 12.4.8 Let $\sum_{j=0}^{\infty} a_j x^j$ be a power series with radius of convergence $R \neq 0$ and interval of convergence I. For x in I, let $F(x) = \sum_{j=0}^{\infty} a_j x^j$. The series $\sum_{j=1}^{\infty} a_{j-1}x^j/j$ also has radius of convergence R and

$$\int_0^x F(t)\,dt = \sum_{j=1}^{\infty} \frac{a_{j-1}x^j}{j} \qquad \text{[uniformly]}$$

on any compact interval $[a, b]$ that contains 0 and is contained in I. ●

If the original series $\sum_{j=0}^{\infty} a_j x^j$ converges at either R or $-R$, then the integrated series must also converge at that endpoint. However, it may happen that the integrated series converges at either or both endpoints while $\sum_{j=0}^{\infty} a_j x^j$ fails to do so. When integrating a power series, you never lose points of convergence and you may gain one or two. For proof apply Theorem 12.4.7 to $\sum_{j=0}^{\infty} a_j x^j$ and $\sum_{j=1}^{\infty} a_{j-1}x^j/j$.

THEOREM 12.4.9 If $\sum_{j=0}^{\infty} a_j x^j$ has radius of convergence R with $0 < R < \infty$ and if $\sum_{j=0}^{\infty} a_j x^j$ converges at R (or at $-R$), then the integrated series $\sum_{j=1}^{\infty} a_{j-1}x^j/j$ also converges at R (or at $-R$). ●

EXAMPLE 13 The geometric series $\sum_{j=0}^{\infty} x^j$ has radius of convergence $R = 1$, as you can easily confirm using the ratio test and thus converges absolutely on $(-1, 1)$. The series diverges at both endpoints ± 1. By Example 1 of Section 11.1 we know that $\sum_{j=0}^{\infty} x^j = 1/(1-x)$ for x in $(-1, 1)$. Replacing x by $-x$ in this series, we deduce that $\sum_{j=0}^{\infty}(-1)^j x^j$ converges to $F(x) = 1/(1+x)$ on $(-1, 1)$. The convergence is uniform on any compact subset of $(-1, 1)$. [Notice, in passing, that while $\sum_{j=0}^{\infty}(-1)^j x^j$ diverges at $x = 1$, it is summable $(C, 1)$ to $F(1) = 1/2$ at $x = 1$. Refer to Example 19 of Section 11.6.] Integrating this series term-by-term from 0 to x in $(-1, 1)$, we obtain

$$\sum_{j=0}^{\infty}(-1)^j \int_0^x t^j\,dt = \sum_{j=0}^{\infty} \frac{(-1)^j x^{j+1}}{j+1} = \sum_{j=1}^{\infty} \frac{(-1)^{j-1}x^j}{j}$$

$$= \int_0^x \frac{1}{(1+t)}\,dt = \ln(1+x).$$

Therefore, $\ln(1+x) = \sum_{j=1}^{\infty}(-1)^{j-1}x^j/j$ for x in $(-1, 1)$. The series converges conditionally at $x = 1$ by the alternating series test and diverges at $x = -1$ by the integral test. The interval of convergence of the integrated series is $(-1, 1]$ and, by Abel's theorem, that series converges uniformly to $\ln(1+x)$ on any compact subset of $(-1, 1]$. Notice that, by integrating, we gained a point of convergence at $x = 1$; at that point $\sum_{j=1}^{\infty}(-1)^{j-1}/j = \ln 2$. ●

EXAMPLE 14 Extend Example 13 by replacing x with x^2 to obtain the series $\sum_{j=0}^{\infty}(-1)^j x^{2j}$. Clearly, for x in $(-1, 1)$, this series converges to $F(x) = 1/(1+x^2)$; it diverges for $|x| \geq 1$. The integrated series, $\sum_{j=0}^{\infty}(-1)^j x^{2j+1}/(2j+1)$, converges to $G(x) = \int_0^x 1/(1+t^2)\,dt = \tan^{-1} x$ on $(-1, 1)$. By the alternating series test the integrated series converges conditionally at $x = \pm 1$. Abel's theorem tells us that

$$\sum_{j=0}^{\infty} \frac{(-1)^j x^{2j+1}}{2j+1} = \tan^{-1} x \quad \text{[uniformly]}$$

on $[-1, 1]$. In particular, letting $x = 1$, we obtain Gregory's series (Example 14 of Section 11.4):

$$\sum_{j=0}^{\infty} \frac{(-1)^j}{2j+1} = \frac{\pi}{4}. \quad \bullet$$

EXAMPLE 15 By the ratio test, the series $\sum_{j=0}^{\infty} x^j/j!$ has radius of convergence $R = \infty$. We know that, for each x in $\mathbb{R}$, this series converges absolutely to e^x; the convergence is uniform on any compact subset of $\mathbb{R}$. The argument presented in Example 3 in Section 12.2 proves that the derived series is the same as the given series, reflecting the fact that $de^x/dx = e^x$. Likewise,

$$\int_0^x e^t\,dt = \int_0^x \sum_{j=0}^{\infty} \frac{t^j}{j!}\,dt = \sum_{j=0}^{\infty} \frac{1}{j!} \int_0^x t^j\,dt$$
$$= \sum_{j=0}^{\infty} \frac{x^{j+1}}{(j+1)!} = \sum_{j=1}^{\infty} \frac{x^j}{j!} = e^x - 1.$$

Thus this property of the exponential function is reflected in its power series representation. $\bullet$

By using various substitutions we can obtain the power series representations for a variety of functions related to e^x. For example, by first replacing x by $-x$, then by $x^2/2$, we obtain the two power series

$$e^{-x} = \sum_{j=0}^{\infty} \frac{(-1)^j x^j}{j!}$$

and

$$e^{-x^2/2} = \sum_{j=0}^{\infty} \frac{(-1)^j x^{2j}}{2^j j!}.$$

This latter series together with Theorem 12.4.8 enables us to derive a power series representation of the indefinite integral of the function $\exp(-t^2/2)$. We have

$$\int_0^x e^{-t^2/2} = \sum_{j=0}^{\infty} \frac{(-1)^j}{2^j j!} \int_0^x t^{2j}\,dt = \sum_{j=0}^{\infty} \frac{(-1)^j x^{2j+1}}{2^j(2j+1)j!} \quad \text{[uniformly]} \qquad \textbf{(12.3)}$$

on any compact subset of $\mathbb{R}$. As we noted in Chapter 6, this integral is not elementary and cannot be evaluated by using the Fundamental Theorem of Calculus. However, since the series (12.3) converges to the integral, given any x in $\mathbb{R}$ and any $\epsilon > 0$, there is a k_0 such that, for $k \geq k_0$, the kth partial sum $F_k(x)$ of the series differs from the integral by less than ϵ. For example, if $x = 1$, the successive kth partial sums have the following values:

$$\begin{array}{ll} F_0(1) = 1.0000000 & F_4(1) = .8556465 \\ F_1(1) = .8333333 & F_5(1) = .8556228 \\ F_2(1) = .8583333 & F_6(1) = .8556245 \\ F_3(1) = .8553571 & F_7(1) = .8556244. \end{array}$$

Evidently, merely by taking $k = 6$, we obtain a value for the integral that is accurate to six decimal places. However, this demonstration is no proof; we have not proved that the *tail of the series* is bounded by 5×10^{-7}. In an applied setting this demonstration is often taken to be sufficient. After all, in a pragmatic mood, we know that the series converges uniformly and we know that the jth term tends to 0 rapidly; therefore, terms beyond those already calculated contribute little to the value of $F_6(1)$. The series (12.3) converges to the integral rapidly, especially for relatively small values of x.

All our results to this point apply with only minor notational modifications to a power series $\sum_{j=0}^{\infty} a_j(x - x_0)^j$ about a point x_0. The root test applied to $\sum_{j=0}^{\infty} a_j(x - x_0)^j$ yields convergence when

$$|x - x_0| \limsup |a_j|^{1/j} < 1.$$

Again let $\lambda = \limsup |a_j|^{1/j}$ and define the radius of convergence R as in Definition 12.4.2. Then $\sum_{j=0}^{\infty} a_j(x - x_0)^j$ converges absolutely if $|x - x_0| < R$, in other words, if $x_0 - R < x < x_0 + R$. The series diverges when $|x - x_0| > R$. Convergence at the endpoints $x_0 \pm R$ must be examined separately for each series. The *interval of convergence* I of the series is the set of all x in $\mathbb{R}$ where the series converges; I is of one of the forms $(x_0 - R,\ x_0 + R)$, $(x_0 - R,\ x_0 + R]$, $[x_0 - R,\ x_0 + R)$, or $[x_0 - R,\ x_0 + R]$.

THEOREM 12.4.10 Let $\sum_{j=0}^{\infty} a_j(x - x_0)^j$ be a power series about x_0 with radius of convergence $R \neq 0$ and interval of convergence I.

i) The series $\sum_{j=0}^{\infty} a_j(x - x_0)^j$ converges uniformly to a continuous function F on any compact subset of I.

ii) The function F has derivatives of all orders on the interval $(x_0 - R,\ x_0 + R)$ given by

$$F^{(k)}(x) = \sum_{j=0}^{\infty} \frac{(j+k)!}{j!} a_{j+k}(x - x_0)^j.$$

This series converges uniformly to $F^{(k)}$ on any compact subset of the interior I^0 of I. If the kth derived series converges at an endpoint of I, then all the series representing the lower-order derivatives of F converge there also.

iii) $\int_{x_0}^x F(t)\,dt = \sum_{j=0}^\infty a_j \int_{x_0}^x (t-x_0)^j\,dt = \sum_{j=1}^\infty \frac{a_{j-1}(x-x_0)^j}{j}$ [uniformly] on any compact interval in I. If the original power series $\sum_{j=0}^\infty a_j(x-x_0)^j$ converges at an endpoint of I, then the integrated series converges there also.

Proof. The proof consists merely in letting $u = x - x_0$. For x in I, u is in the interval I_0 with endpoints $-R$ and R. The convergence of the original series corresponds to that of $\sum_{j=0}^\infty a_j u^j$. Now apply Corollary 12.4.3 and Theorems 12.4.4 through 12.4.8 to the series $\sum_{j=0}^\infty a_j u^j$. ●

12.4.3 The Algebra of Power Series

Let $\sum_{j=0}^\infty a_j x^j$ have radius of convergence R_1 and let $\sum_{j=0}^\infty b_j x^j$ have radius of convergence R_2. Let $R = \min\{R_1, R_2\}$. We assume that $R > 0$ (otherwise our discussion below reduces to triviality). For any two real numbers c_1 and c_2, the linear combination

$$c_1 \sum_{j=0}^\infty a_j x^j + c_2 \sum_{j=0}^\infty b_j x^j = \sum_{j=0}^\infty [c_1 a_j + c_2 b_j] x^j$$

converges on $(-R, R)$. Moreover, because power series converge absolutely within the interval of convergence, the Cauchy product of $\sum_{j=0}^\infty a_j x^j$ and $\sum_{j=0}^\infty b_j x^j$ converges absolutely on $(-R, R)$ by Theorem 11.5.1. That Cauchy product is, of course,

$$\left[\sum_{j=0}^\infty a_j x^j\right]\left[\sum_{j=0}^\infty b_j x^j\right] = \sum_{k=0}^\infty \left[\sum_{j=0}^k a_j b_{k-j}\right] x^k = \sum_{k=0}^\infty c_k x^k.$$

If $F(x) = \sum_{j=0}^\infty a_j x^j$ and $G(x) = \sum_{j=0}^\infty b_j x^j$, then the Cauchy product $\sum_{k=0}^\infty c_k x^k$ converges to $F(x)G(x)$ on $(-R, R)$. We leave this for you to prove as an exercise. The collection of all power series that converge on $(-R, R)$ is a commutative ring with identity and is denoted $\mathcal{A}$, the ring of all (real) analytic functions on $(-R, R)$. Note that $\mathcal{A}$ contains the ring of all polynomials.

EXAMPLE 16 A power series representation of the function $(e^x + e^{-x})/2$ is obtained by manipulating the power series for e^x and e^{-x}:

$$\frac{e^x + e^{-x}}{2} = \frac{1}{2}\sum_{j=0}^\infty \frac{x^j}{j!} + \frac{1}{2}\sum_{j=0}^\infty \frac{(-1)^j x^j}{j!} = \frac{1}{2}\sum_{j=0}^\infty \frac{[1 + (-1)^j]x^j}{j!}$$

$$= 1 + \frac{x^2}{2!} + \frac{x^4}{4!} + \cdots = \sum_{j=0}^\infty \frac{x^{2j}}{(2j)!}.$$

Likewise,

$$\frac{e^x - e^{-x}}{2} = \sum_{j=0}^\infty \frac{x^{2j+1}}{(2j+1)!}.$$

These two functions are of such importance that we have given them names: the hyperbolic cosine and the hyperbolic sine, respectively:

$$\cosh x = \sum_{j=0}^{\infty} \frac{x^{2j}}{(2j)!},$$

$$\sinh x = \sum_{j=0}^{\infty} \frac{x^{2j+1}}{(2j+1)!}.$$

If we define these two functions simply by their power series, without reference to the derivation of those series, we can deduce all their properties by manipulating the corresponding series. This will demonstrate how to use the algebra of power series to derive information about the functions being represented.

First, by applying the ratio test to each series (omitting zero terms), we find that each has radius of convergence $R = \infty$. Each series converges on the entire real line. Thus $\cosh x$ and $\sinh x$ are continuous on all of $\mathbb{R}$ and have derivatives of all orders at every x. The derivative of $\cosh x$ is obtained by differentiating its series term-by-term:

$$\frac{d(\cosh x)}{dx} = \sum_{j=0}^{\infty} \frac{(2j)x^{2j-1}}{(2j)!} = \sum_{j=1}^{\infty} \frac{x^{2j-1}}{(2j-1)!}$$

$$= \sum_{j=0}^{\infty} \frac{x^{2j+1}}{(2j+1)!} = \sinh x.$$

Likewise, $d(\sinh x)/dx = \cosh x$. Each of these functions is the derivative of the other. Clearly, $d^2(\sinh x)/dx^2 = \sinh x$ and $d^2(\cosh x)/dx^2 = \cosh x$.

Referring again to the series representations of these two functions, it is evident that $\cosh(-x) = \cosh x$ and that $\sinh(-x) = -\sinh x$. Therefore $\cosh x$ is an even function and $\sinh x$ is odd. Again, from the series, note that $\cosh x > 1$ for all x in $\mathbb{R}$. Also, $\cosh 0 = 1$. Note that $\sinh x > 0$ for positive x and is negative when $x < 0$. Also, $\sinh 0 = 0$. This information enables us to sketch the graphs of $\cosh x$ and $\sinh x$ without ever summing a series. Since $\sinh x$ is negative for $x < 0$, the graph of $\cosh x$ is decreasing there; also $y = \cosh x$ increases on $(0, \infty)$. Since $d^2(\cosh x)/dx^2 = \cosh x$ is positive, the graph of $\cosh x$ is concave up on all of $\mathbb{R}$. Similar reasoning yields analogous information about $\sinh x$. The graphs of these two functions are sketched in Figs. 12.3 and 12.4. ●

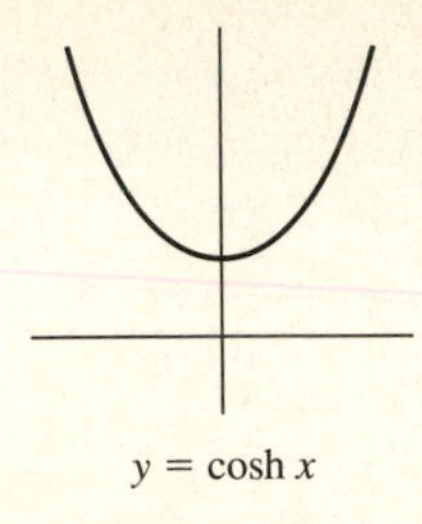

Figure 12.3

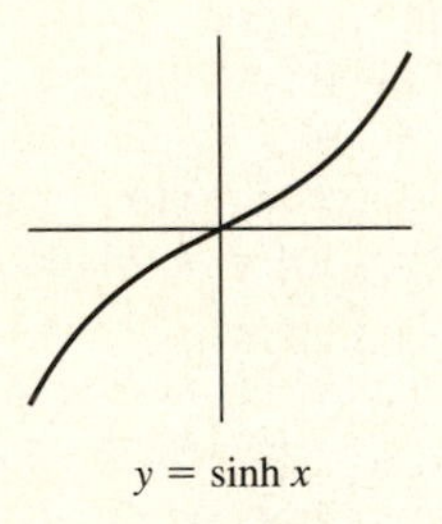

Figure 12.4

The fundamental hypertrigonometric identity, from which all others are derived, is

$$\cosh^2 x - \sinh^2 x = 1.$$

In fact, it is this identity that reveals that these functions are related to a hyperbola, $u^2 - v^2 = 1$. To establish this identity, we first prove three combinatorial facts.

LEMMA 1 For all k in $\mathbb{N}$,

i)
$$\sum_{j=0}^{k} \frac{1}{(2j)!(2k-2j)!} = \frac{1}{(2k)!} \sum_{j=0}^{k} C(2k, 2j).$$

ii)
$$\sum_{j=1}^{k} \frac{1}{(2j-1)!(2k-2j+1)!} = \frac{1}{(2k)!} \sum_{j=1}^{k} C(2k, 2j-1).$$

Proof. The proof consists merely in adjusting each summand and using the definition of the binomial coefficients. ●

LEMMA 2 For all k in $\mathbb{N}$,

$$\sum_{j=0}^{k} C(2k, 2j) = \sum_{j=1}^{k} C(2k, 2j-1) = 2^{2k-1}.$$

Proof. Expand $(1-1)^{2k}$ by the binomial theorem and separate positive and negative terms to obtain

$$0 = (1-1)^{2k} = \sum_{j=0}^{2k} C(2k, j)1^{2k-j}(-1)^j$$
$$= \sum_{j=0}^{k} C(2k, 2j) - \sum_{j=1}^{k} C(2k, 2j-1). \qquad \textbf{(12.4)}$$

Likewise, expand $(1+1)^{2k}$ and separate even and odd indexed terms to obtain

$$2^{2k} = (1+1)^{2k} = \sum_{j=0}^{k} C(2k, 2j) + \sum_{j=1}^{k} C(2k, 2j-1). \qquad \textbf{(12.5)}$$

Solving the two linear equations (12.4) and (12.5) for each of the summations proves the lemma. ●

To establish the identity $\cosh^2 x - \sinh^2 x = 1$ we first use the Cauchy product to compute $\cosh^2 x$ and $\sinh^2 x$; to each product we apply Lemma 1 and Lemma 2. In passing, we derive the power series representations of $\cosh^2 x$ and $\sinh^2 x$:

$$\cosh^2 x = \left[\sum_{j=0}^{\infty} \frac{x^{2j}}{(2j)!}\right]\left[\sum_{j=0}^{\infty} \frac{x^{2j}}{(2j)!}\right] = \sum_{k=0}^{\infty}\left[\sum_{j=0}^{k} \frac{1}{(2j)!(2k-2j)!}\right] x^{2k}$$
$$= \sum_{k=0}^{\infty} \frac{1}{(2k)!}\left[\sum_{j=0}^{k} C(2k, 2j)\right] x^{2k} = 1 + \sum_{k=1}^{\infty} \frac{2^{2k-1}x^{2k}}{(2k)!},$$
$$\sinh^2 x = \left[\sum_{j=0}^{\infty} \frac{x^{2j+1}}{(2j+1)!}\right]\left[\sum_{j=0}^{\infty} \frac{x^{2j+1}}{(2j+1)!}\right] = \sum_{k=1}^{\infty}\left[\sum_{j=1}^{k} \frac{1}{(2j-1)!(2k-2j+1)!}\right] x^{2k}$$
$$= \sum_{k=1}^{\infty} \frac{1}{(2k)!}\left[\sum_{j=1}^{k} C(2k, 2j-1)\right] x^{2k} = \sum_{k=1}^{\infty} \frac{2^{2k-1}x^{2k}}{(2k)!}.$$

Therefore, when we form $\cosh^2 x - \sinh^2 x$, we obtain the desired identity $\cosh^2 x - \sinh^2 x = 1$.

The remaining hyperbolic functions are defined by forming the ratios analogous to those that define the corresponding ordinary trigonometric functions. Namely,

$$\tanh x = \frac{\sinh x}{\cosh x}, \quad \coth x = \frac{\cosh x}{\sinh x}, \quad \operatorname{sech} x = \frac{1}{\cosh x}, \quad \operatorname{csch} x = \frac{1}{\sinh x}.$$

Of course, $\coth x$ and $\operatorname{csch} x$ are defined only where $\sinh x \neq 0$, that is, for $x \neq 0$. The identity we have established above leads to the two corresponding identities for the remaining hyperbolic functions:

$$1 - \tanh^2 x = \operatorname{sech}^2 x \qquad \text{and} \qquad \coth^2 x - 1 = \operatorname{csch}^2 x.$$

EXAMPLE 17 The algebraic and analytic properties of the ring of all convergent power series on $(-R, R)$ enable us to develop a general method for solving an important problem in the theory of differential equations. To be specific, let

$$L(y) = c_2(x)y'' + c_1(x)y' + c_0(x)y$$

denote a linear, second-order differential operator. The coefficient functions $c_0(x), c_1(x)$, and $c_2(x)$ are (real) analytic functions on some interval $(-R, R)$; that is, c_0, c_1, and c_2 can all be represented by convergent (real) power series on $(-R, R)$. To solve the equation $L(y) = 0$ is to find a function $y = f(x)$, defined on $(-R, R)$, such that

$$L(f)(x) = c_2(x)f''(x) + c_1(x)f'(x) + c_0(x)f(x) = 0 \tag{12.6}$$

for all x in $(-R, R)$. Provided that $c_2(x)$ does not vanish on $(-R, R)$, it can be proved that there always exist two linearly independent[1] solutions f_1 and f_2 of $L(y) = 0$ and that these solutions are (real) analytic on $(-R, R)$. Further, every solution, f, of $L(y) = 0$ is a linear combination, $b_1 f_1 + b_2 f_2$, of these two functions; therefore f is also (real) analytic. We will not include the proof of this theorem here, but will content ourselves with indicating how its implementation can be effected.

Suppose that f is an unknown solution of $L(y) = 0$. Let $f(x) = \sum_{j=0}^{\infty} a_j x^j$ be its representation as a power series. The unknown coefficients $a_0, a_1, a_2, \ldots$ are to be determined. Then

$$f'(x) = \sum_{j=0}^{\infty} (j+1)a_{j+1}x^j$$

and

$$f''(x) = \sum_{j=0}^{\infty} (j+2)(j+1)a_{j+2}x^j.$$

We also represent $c_0(x), c_1(x)$, and $c_2(x)$ by their power series. We substitute these six series into (12.6). After forming the Cauchy products and adding the three resulting series we finally arrive at one power series equation of the form

$$L(f)(x) = \sum_{j=0}^{\infty} d_j x^j = 0$$

for all x in $(-R, R)$. Next we invoke a reasonable fact to be proved below; namely, a power series $\sum_{j=0}^{\infty} d_j x^j = 0$ for all x in $(-R, R)$ if and only if $d_j = 0$ for all j. From these last equations we recursively solve for the coefficients $a_0, a_1, a_2, a_3, \ldots$ of the power series representation of f. In general, the work involved is daunting; in practice, for "reasonable" coefficients c_0, c_1, and c_2, the theory applies quite

[1]Two functions f_1 and f_2 are said to be *linearly independent* if a linear combination $b_1 f_1(x) + b_2 f_2(x) = 0$ for all x in $(-R, R)$ when and only when $b_1 = b_2 = 0$.

smoothly. Several classic differential equations of profound significance in mathematical physics can be solved elegantly by this method. For example, the Chebychev equation is

$$L(y) = (1 - x^2)y'' - xy' + \alpha^2 y = 0, \tag{12.7}$$

where α is some constant. The coefficients are polynomials and hence are analytic on all of $\mathbb{R}$; also $c_2(x) = 1 - x^2$ does not vanish on $(-1, 1)$. The general theory ensures that there exist two linearly independent, analytic solutions f_1 and f_2 defined on $(-1, 1)$ and that every solution f is a linear combination of these two. To find f_1 and f_2, suppose that $f(x) = \sum_{j=0}^{\infty} a_j x^j$ is a solution; our task is to find the coefficients a_j. First, $f'(x) = \sum_{j=1}^{\infty} j a_j x^{j-1}$ and therefore $xf'(x) = \sum_{j=1}^{\infty} j a_j x^j$. Next, compute

$$f''(x) = \sum_{j=2}^{\infty} j(j-1)a_j x^{j-2} = \sum_{j=0}^{\infty} (j+2)(j+1)a_{j+2} x^j.$$

Using each of these versions for f'', form the product

$$(1 - x^2)f''(x) = \sum_{j=0}^{\infty} (j+2)(j+1)a_{j+2} x^j - \sum_{j=2}^{\infty} j(j-1)a_j x^j.$$

Notice that we have arranged for all the series involved to be expressed in terms of like powers of x. Therefore substituting in Chebychev's equation (12.7) and collecting, for each j, the coefficient d_j of x^j, we have

$$\begin{aligned} L(f)(x) &= \sum_{j=0}^{\infty} (j+2)(j+1)a_{j+2} x^j \\ &\quad - \sum_{j=2}^{\infty} j(j-1)a_j x^j - \sum_{j=1}^{\infty} j a_j x^j + \alpha^2 \sum_{j=0}^{\infty} a_j x^j \\ &= [2a_2 + \alpha^2 a_0] + [3!a_3 + (\alpha^2 - 1)a_1]x \\ &\quad + \sum_{j=2}^{\infty} [(j+2)(j+1)a_{j+2} + (\alpha^2 - j^2)a_j]x^j = 0. \end{aligned}$$

Since each coefficient must vanish, we deduce that

$$a_2 = -\frac{\alpha^2 a_0}{2},$$

$$a_3 = \frac{-(\alpha^2 - 1)a_1}{3!},$$

and, for $j \geq 2$,

$$a_{j+2} = \frac{-(\alpha^2 - j^2)a_j}{(j+2)(j+1)}. \tag{12.8}$$

From these equations we solve for the coefficients a_j. Note that there are no constraints on the coefficients a_0 and a_1; therefore, these two coefficients are arbitrary.

First, a_0 is arbitrary and $a_2 = -\alpha^2 a_0/2$. From the recursion relation (12.8) we compute the even-indexed coefficients:

$$a_4 = \frac{-(\alpha^2 - 2^2)a_2}{4 \cdot 3} = \frac{(-1)^2(\alpha^2 - 2^2)\alpha^2 a_0}{4!}$$

$$a_6 = \frac{-(\alpha^2 - 4^2)a_4}{6 \cdot 5} = \frac{(-1)^3(\alpha^2 - 4^2)(\alpha^2 - 2^2)\alpha^2 a_0}{6!}$$

$$a_8 = \frac{-(\alpha^2 - 6^2)a_6}{8 \cdot 7} = \frac{(-1)^4(\alpha^2 - 6^2)(\alpha^2 - 4^2)(\alpha^2 - 2^2)\alpha^2 a_0}{8!}.$$

The pattern is clear; recursively, for $k \geq 1$,

$$a_{2k} = \frac{(-1)^k}{(2k)!} \prod_{j=0}^{k-1} [\alpha^2 - (2j)^2] a_0.$$

We next find the coefficients with odd index. Again, a_1 is arbitrary and $a_3 = -(\alpha^2 - 1)a_1/3!$. From the recursion relation (12.8) we find that

$$a_5 = \frac{-(\alpha^2 - 3^2)a_3}{5 \cdot 4} = \frac{(-1)^2(\alpha^2 - 3^2)(\alpha^2 - 1)a_1}{5!}$$

$$a_7 = \frac{-(\alpha^2 - 5^2)a_5}{7 \cdot 6} = \frac{(-1)^3(\alpha^2 - 5^2)(\alpha^2 - 3^2)(\alpha^2 - 1)a_1}{7!}.$$

Recursively, for $k \geq 1$, we obtain

$$a_{2k+1} = \frac{(-1)^k}{(2k+1)!} \prod_{j=1}^{k} [\alpha^2 - (2j-1)^2] a_1.$$

Since power series converge absolutely and can be rearranged without effect, we can collect together the even-indexed terms of $\sum_{j=0}^{\infty} a_j x^j$ and define a function

$$f_1(x) = 1 + \sum_{k=1}^{\infty} \frac{(-1)^k}{(2k)!} \prod_{j=0}^{k-1} [\alpha^2 - (2j)^2] x^{2k}.$$

In effect, f_1 results from assigning values to the arbitrary constants a_0 and a_1: $a_0 = 1$ and $a_1 = 0$. Note that f_1 being a special case of our general solution f, is a solution of $L(y) = 0$.

Likewise, setting $a_0 = 0$ and $a_1 = 1$ results in a series solution of (12.7) consisting of the odd-indexed terms:

$$f_2(x) = x + \sum_{k=1}^{\infty} \frac{(-1)^k}{(2k+1)!} \prod_{j=1}^{k} [\alpha^2 - (2j-1)^2] x^{2k+1}.$$

Provided that α is not an integer, we can apply the ratio test to each of these series. Upon doing so, we find that each has a radius of convergence $R = 1$; each series converges on $(-1, 1)$ to the solutions f_1 and f_2 respectively. Every solution of (12.7) is of the form $f(x) = b_1 f_1 + b_2 f_2$, where b_1 and b_2 are constants.

A special feature of these solutions is revealed if we suppose that α is some nonnegative integer. In this case, one of the functions f_1 or f_2 reduces to a polynomial solution of Chebychev's equation. If α is even, say, $\alpha = 2k_0$, then f_1 reduces to a polynomial of degree $2k_0$. We need only observe that, for $k > k_0$, the coefficient a_{2k} contains $\alpha^2 - (2k_0)^2 = 0$ as a factor and therefore vanishes. Likewise, if α is odd, say, $\alpha = 2k_0 + 1$, then, for $k > k_0$, a_{2k+1} has $\alpha^2 - (2k_0 + 1)^2 = 0$ as a factor; thus f_2 reduces to a polynomial of degree $2k_0 + 1$. Consequently, as α takes on the values $0, 1, 2, \ldots$, we obtain a sequence $\{T_k\}$ of polynomials where T_k is a polynomial of degree k that solves Chebychev's differential equation $(1 - x^2)y'' - xy' + k^2y = 0$ on $(-1, 1)$. We use the formulas for f_1 and f_2 to compute the first several such polynomials:

$$\alpha = 0: \quad f_1(x) = T_0(x) = 1$$

$$\alpha = 1: \quad f_2(x) = T_1(x) = x$$

$$\alpha = 2: \quad f_1(x) = T_2(x) = 1 - 2x^2$$

$$\alpha = 3: \quad f_2(x) = T_3(x) = x - \frac{4}{3}x^3$$

$$\alpha = 4: \quad f_1(x) = T_4(x) = 1 - 8x^2 + 8x^4$$

$$\alpha = 5: \quad f_2(x) = T_5(x) = x - 4x^3 + \frac{16}{5}x^5$$

$$\alpha = 6: \quad f_1(x) = T_6(x) = 1 - 18x^2 + 48x^4 - 32x^6.$$

You may notice that each of these polynomials is a scalar multiple of the corresponding Chebychev polynomial of the same degree defined in Exercise 3.16 to be $P_k(x) = \cos[k \cos^{-1} x]$. This is no accident; as you can show, the polynomial $P_k(x) = \cos[k \cos^{-1} x]$ is a solution of Chebychev's differential equation when $\alpha = k$ and therefore is a linear combination of the two solutions f_1 and f_2 found above. Since only one of these solutions is a polynomial, according to the parity of k, P_k must be a scalar multiple of T_k. ●

12.5 The Taylor Series Representation of Functions

In this section we begin, not with an arbitrary power series but with some continuous function f defined on an interval $[a, b]$. We choose an x_0 in $[a, b]$ and we ask the following question: Is it possible to find a power series of the form $\sum_{j=0}^{\infty} a_j (x - x_0)^j$ that converges to $f(x)$ for all x in $[a, b]$? We are asking, then, whether f can be *represented by a power series about the point* x_0.

We are encouraged in our search for such a power series by Weierstrass's approximation theorem and its converse: A function f is continuous on $[a, b]$ if and only if there exists a sequence of polynomials that converges uniformly to f on $[a, b]$. Further, the partial sums of a power series form a sequence of polynomials that converges uniformly on any compact subset of its interval of convergence to whatever function may be the limit of the series.

However, a cautionary observation warrants explicit mention: The coefficients of the polynomials F_k obtained by taking the partial sums of a power series recur as coefficients of all subsequent polynomials in the sequence; F_{k+1} differs from F_k only in the $(k+1)$st term. By contrast, when in accordance with Weierstrass's approximation theorem we construct a sequence $\{P_k\}$ of polynomials that converges uniformly to a continuous function f, we may find that some or all of the coefficients of P_{k+1} differ from those of P_k. That a sequence of polynomials is the sequence of partial sums of a power series clearly constitutes an extremely special case.

For f to be represented by a power series surely requires restrictions on f beyond mere continuity. After all, as we now know, every power series converges to a function that has derivatives of all orders at every point within the interval of convergence. Were it possible to represent f by a power series, then it must certainly be true, at the very least, that f has derivatives of all orders at each x in $[a, b]$; the function f must be extremely "smooth." Our question really comes to this: Is it sufficient that f has derivatives of all orders at every point of $[a, b]$ to guarantee that f can be represented by a power series about some point x_0 in $[a, b]$?

All our previous work with Taylor polynomials and with the various forms of the remainder has primed us for the complete resolution of the issue addressed in this section. We work backward to discover that answer. Suppose that f is a function that has derivatives of all orders on $[a, b]$, that x_0 is a point in $[a, b]$, and that there is a power series in $(x - x_0)$ that converges uniformly to f on $[a, b]$:

$$f(x) = \sum_{j=0}^{\infty} a_j (x - x_0)^j \quad \text{[uniformly]} \tag{12.9}$$

on $[a, b]$. The power series has a positive radius of convergence R and $[a, b]$ is contained in the interval of convergence. To avoid possible complications at the endpoints, we assume that $[a, b] \subset (x_0 - R, x_0 + R)$. We will show recursively that the coefficients a_j are uniquely determined. By setting $x = x_0$ in (12.9), we see first that $a_0 = f(x_0)$.

Differentiate both sides of (12.9). Theorem 12.4.10 ensures that the resulting series converges uniformly to f' on $[a, b]$ and we obtain

$$f'(x) = \sum_{j=0}^{\infty} (j+1) a_{j+1} (x - x_0)^j \quad \text{[uniformly]}$$

on $[a, b]$. Setting $x = x_0$ in this series gives $f'(x_0) = a_1$.

In general, for each k in $\mathbb{N}$, we compute the kth derivative of f and of the power series representation (12.9) of f. Again by Theorem 12.4.10, we obtain

$$f^{(k)}(x) = \sum_{j=0}^{\infty} \frac{(j+k)!}{j!} a_{j+k} (x - x_0)^j \quad \text{[uniformly]}.$$

When we set $x = x_0$ in this series, all the terms except that corresponding to $j = 0$ vanish. We obtain $f^{(k)}(x_0) = k! a_k$. That is, $a_k = f^{(k)}(x_0)/k!$ for all $k \geq 0$. We have proved the following theorem.

THEOREM 12.5.1 If f can be represented by a power series of the form $\sum_{j=0}^{\infty} a_j(x-x_0)^j$, then $a_j = f^{(j)}(x_0)/j!$ for all j. ●

Remark. The theorem implies that the power series expansion of f about a point x_0 is unique; if such a representation is possible at all, then the coefficients are completely determined by the values of f and of the derivatives of f at x_0.

COROLLARY 12.5.2 If $\sum_{j=0}^{\infty} a_j(x-x_0)^j$ converges to 0 for every x in an interval $[a, b]$, then $a_j = 0$ for all j.

Proof. The function f that is identically 0 has the unique and trivial power series representation, every coefficient of which is 0. The series given in the corollary also represents the function $f = 0$. By Theorem 12.5.1, it follows that $a_j = 0$ for all j. ●

DEFINITION 12.5.1 Let f be a function that has derivatives of all orders on an interval I and let x_0 be a point of I. The corresponding power series

$$\sum_{j=0}^{\infty} \frac{f^{(j)}(x_0)}{j!}(x-x_0)^j$$

is called the *Taylor series of f about the point x_0*. ●

Theorem 12.5.1 tells us that, if f can be represented by a power series about x_0, then that series must be the Taylor series of f about x_0. Notice that the kth partial sum of the Taylor series of f is simply the kth Taylor polynomial $p_k(x) = \sum_{j=0}^{k} f^{(j)}(x_0)(x-x_0)^j/j!$ of f. We already know that the sequence $\{p_k\}$ converges uniformly to f on $[a, b]$ if and only if the remainder $R_k(x_0; x)$ converges uniformly to 0 on $[a, b]$. Further, we have already seen three different forms for that remainder. Theorem 4.5.2 (Taylor's theorem) gives us *Lagrange's form*,

$$R_k(x_0; x) = \frac{f^{(k+1)}(c)(x-x_0)^{k+1}}{(k+1)!}$$

for some c between x_0 and x. Theorem 6.4.3 gives us the *integral form* of the remainder,

$$R_k(x_0; x) = \frac{1}{k!}\int_{x_0}^{x} (x-t)^k f^{(k+1)}(t)\,dt.$$

Finally, in Exercise 6.88 you derived *Cauchy's form* of the remainder,

$$R_k(x_0; x) = \frac{f^{(k+1)}(c)(x-c)^k(x-x_0)}{k!}$$

for some c between x_0 and x. (In Exercise 6.91 you also met the *generalized Cauchy form*, but the three versions of the remainder displayed here will suffice for our purposes.)

For completeness, we include a derivation of Cauchy's form of the remainder. Begin with the integral form. For any x in $[a, b]$ and for all t between x_0 and x, let

$g(t) = (x - t)^k f^{(k+1)}(t)/k!$ and let $h(t) = 1$. By Theorem 6.3.1, there exists a c between x_0 and x such that

$$\int_{x_0}^{x} g(t)h(t)\,dt = g(c)\int_{x_0}^{x} h(t)\,dt.$$

That is,

$$R_k(x_0; x) = \frac{1}{k!}\int_{x_0}^{x} (x - t)^k f^{(k+1)}(t)\,dt$$

$$= g(c)\int_{x_0}^{x} dt = \frac{f^{(k+1)}(c)(x - c)^k(x - x_0)}{k!}.$$

This gives Cauchy's form of the remainder. A variant of Cauchy's remainder is

$$R_k(x_0; x) = \frac{f^{(k+1)}(x_0 + \theta h)h^{k+1}(1 - \theta)^k}{k!},$$

where $h = x - x_0$, θ is in $(0, 1)$, and $c = x_0 + \theta h$. We leave it for you to derive this form as an exercise.

Any function that has derivatives of all orders at x_0 has a formal Taylor series at x_0. Whether that series converges is another matter. If the series does converge, the question remains: Does the series actually converge to f? Our next theorem is the centerpiece of this section; it answers the paramount question raised here. It is the culmination of our work with Taylor's polynomials and the remainder beginning with Section 5 of Chapter 4.

A function f with derivatives of all orders can be represented by a power series about a point x_0 in only one way if at all. That representation can only be its Taylor series. The Taylor series of f, being a power series, must converge absolutely in the interior of its interval of convergence I; further, the convergence is uniform on any compact subset of I. Thus the sequence of partial sums of the Taylor series of f, namely the sequence $\{p_k\}$ of Taylor polynomials of f, must converge in the interior of I. Finally, $\{p_k\}$ converges to f if and only if $\lim_{k\to\infty} R_k(x_0; x) = 0$. Thus to resolve the question raised above, look first at the remainder. If $\{R_k(x_0; x)\}$ converges to 0, then the Taylor series of f converges to $f(x)$; otherwise, even if that series converges, it cannot converge to f.

THEOREM 12.5.3 Let f be a function that has derivatives of all orders on an interval $[a, b]$ and let x_0 be a point in $[a, b]$. Then f can be represented by a power series on $[a, b]$, that is, $f(x) = \sum_{j=0}^{\infty} f^{(j)}(x_0)(x - x_0)^j/j!$ [uniformly] on $[a, b]$, if and only if the sequence $\{p_k\}$ of Taylor polynomials of f converges uniformly to f on $[a, b]$. Equivalently, $\lim_{k\to\infty} R_k(x_0; x) = 0$ [uniformly] on $[a, b]$. ●

In our earlier treatment of Taylor polynomials we have already done the work necessary to derive the Taylor series expansions for several of the elementary functions about $x_0 = 0$ and to show that these series actually converge to the function in question. Abel's theorem (12.4.7) applies at either endpoint of the interval of convergence. For example,

$$e^x = \sum_{j=0}^{\infty} \frac{x^j}{j!} \text{[pointwise] on } \mathbb{R}, \text{[uniformly] on any } [a, b].$$

$$\sin x = \sum_{j=0}^{\infty} \frac{(-1)^j x^{2j+1}}{(2j+1)!} \begin{cases} \text{[pointwise] on } \mathbb{R}. \\ \text{[uniformly] on any } [a, b]. \end{cases}$$

$$\cos x = \sum_{j=0}^{\infty} \frac{(-1)^j x^{2j}}{(2j)!} \begin{cases} \text{[pointwise] on } \mathbb{R}. \\ \text{[uniformly] on any } [a, b]. \end{cases}$$

$$\ln(1+x) = \sum_{j=1}^{\infty} \frac{(-1)^{j-1} x^j}{j} \text{[uniformly] on } [a, b] \subset (-1, 1].$$

$$\tan^{-1} x = \sum_{j=0}^{\infty} \frac{(-1)^j x^{2j+1}}{(2j+1)} \text{[uniformly] on } [-1, 1].$$

These are the standard examples and serve to show you how to proceed when faced with some other function. To illustrate the general procedure and many of its complexities, we will treat Newton's generalized binomial theorem in detail below. First it is worthwhile to consider an example of a Taylor series that cannot converge to the function f that generates it.

EXAMPLE 18 For $x \neq 0$, let $f(x) = \exp(-1/x^2)$; let $f(0) = 0$. You proved in Exercise 4.72 that f has derivatives of all orders on all of $\mathbb{R}$, including at $x_0 = 0$. In fact, by extensive use of l'Hôpital's rule, you showed that $f^{(j)}(0) = 0$ for all j. Consequently, the Taylor series of this particular function is the power series with $a_j = 0$ for all j. But this trivial power series is the Taylor series for the zero function. Therefore, the Taylor series of f cannot converge to f at any point except at $x = 0$. ●

EXAMPLE 19 Finally we take up the example, Newton's generalized binomial theorem, which launched the theory of infinite series. In 1665 the 22-year-old Newton had finished his B.A. degree at Trinity College, Cambridge, and had left London to avoid an outbreak of the plague. During several peaceful months at home he discovered four of what he considered his finest results. One of these, the generalized binomial theorem, was the first use of power series methods in analysis. (Newton's other three favorite results were the invention of the calculus, the law of gravitation, and his theory of color.) At the time, the ordinary binomial theorem had been known for at least 500 years: If n is a positive integer, then $(1+x)^n = \sum_{j=0}^{n} C(n, j)x^j$ for all real x. Suppose, instead, that α is any nonzero number that is not a positive integer. Newton asked if it were possible to expand $f(x) = (1+x)^\alpha$, not as a finite sum, but as an infinite series. If so, for what values of x is the expansion valid? Our theory, finely honed and polished over the intervening 300 years, provides complete answers.

First, compute the successive derivatives $f^{(j)}(x)$ and evaluate these derivatives at $x_0 = 0$. We obtain, for j in $\mathbb{N}$,

$$f^{(j)}(x) = \alpha(\alpha - 1)(\alpha - 2)\cdots(\alpha - j + 1)(1 + x)^{\alpha - j}.$$

Thus,

$$f^{(j)}(0) = \alpha(\alpha - 1)(\alpha - 2)\cdots(\alpha - j + 1)$$

for j in $\mathbb{N}$. Consequently, the Taylor series for f about the point $x_0 = 0$, called the *binomial series*, is

$$1 + \sum_{j=1}^{\infty} \frac{\alpha(\alpha - 1)(\alpha - 2)\cdots(\alpha - j + 1)}{j!} x^j.$$

By the ratio test,

$$\lambda = \limsup \left| \frac{a_{j+1}}{a_j} \right| = \limsup \left| \frac{(\alpha - j)}{(j + 1)} \right| = 1.$$

Therefore, $R = 1$ and the binomial series converges for $|x| < 1$. That's the easy part; next we must check for convergence at the endpoints $x = \pm 1$.

First consider $x = -1$. We claim, in this case, that, whatever the value of α, eventually all the terms of the series are of the same sign. If α is negative and $x = -1$, then the jth term is

$$(-1)^{2j} |\alpha(\alpha - 1)(\alpha - 2)\cdots(\alpha - j + 1)| / j!$$

and is therefore positive for every j. If α is positive, let j_0 denote the integer part of α. For $j \geq j_0 + 2$, the jth term is

$$\frac{(-1)^{2j - j_0 - 1} \alpha(\alpha - 1)\cdots(\alpha - j_0)|\alpha - j_0 - 1|\cdots|\alpha - j + 1|}{j!}.$$

Every factor except $(-1)^{2j - j_0 - 1} = (-1)^{j_0 + 1}$ is positive and $(-1)^{j_0 + 1}$ is positive if and only if j_0 is odd. We deduce that, if $x = -1$, all the terms for $j \geq j_0 + 2$ have the same sign. Therefore, at $x = -1$, either the series converges absolutely or it diverges; conditional convergence is impossible. We emphasize this point because we want to appeal to Gauss's convergence test. That test, which you proved in Exercise 11.24, is a modification of the ratio test and is used to treat the case when the ratio yields a limit superior of 1. ●

THEOREM 12.5.4 Gauss's Convergence Test Let $\sum_{j=1}^{\infty} a_j$ be a series for which $|a_{j+1}/a_j|$ can be written in the form

$$\left| \frac{a_{j+1}}{a_j} \right| = 1 - \frac{c}{j} + \frac{d_j}{j^2},$$

where c is a constant and $\{d_j\}$ is a bounded sequence.

i) If $c > 1$, then $\sum_{j=1}^{\infty} a_j$ converges absolutely.

ii) If $c \le 1$, then $\sum_{j=1}^{\infty} a_j$ either converges conditionally or diverges; it does not converge absolutely. ●

For the binomial series when $x = -1$, we compute the ratio of successive terms and obtain (as you can check), for $j > \alpha$,

$$\left|\frac{a_{j+1}}{a_j}\right| = \frac{j-\alpha}{j+1} = 1 - \frac{\alpha+1}{j} + \frac{d_j}{j^2},$$

where $d_j = j(\alpha+1)/(j+1)$. The sequence $\{d_j\}$ is bounded by $|\alpha+1|$ and $c = \alpha+1$ is constant. Hence by Gauss's test, if $\alpha+1 > 1$—that is, if $\alpha > 0$—then the binomial series converges absolutely at $x = -1$. Also, if $\alpha < 0$, then the binomial series does not converge absolutely at $x = -1$; hence by the preceding discussion, it must diverge when $\alpha < 0$.

Next examine convergence at the endpoint $x = 1$. We claim that, whatever the value of α, the terms of the binomial series eventually alternate in sign when $x = 1$. If α is negative, the jth term is

$$\frac{(-1)^j |\alpha(\alpha-1)\cdots(\alpha-j+1)|}{j!}.$$

These terms alternate in sign. If α is positive and if j_0 denotes the integer part of α, then for $j \ge j_0 + 2$, the jth term is

$$\frac{(-1)^{j-j_0-1}\alpha(\alpha-1)\cdots(\alpha-j_0)|\alpha-j_0-1|\cdots|\alpha-j+1|}{j!}$$

so the terms of the series alternate in sign for $j \ge j_0 + 2$.

To determine convergence when $x = 1$, we consider three cases: $\alpha < -1$, $\alpha = -1$, and $-1 < \alpha$. We examine each of these in turn.

$\boldsymbol{\alpha < -1}$. For all $j \ge 0$, we have $j+1 < j-\alpha = |\alpha - j|$. It follows that the magnitude of the $(j+1)$st term exceeds that of the jth term; that is, for all j,

$$\frac{|\alpha(\alpha-1)(\alpha-2)\cdots(\alpha-j+1)(\alpha-j)|}{(j+1)!} > \frac{|\alpha(\alpha-1)(\alpha-2)\cdots(\alpha-j+1)|}{j!}.$$

Consequently, when $\alpha < -1$ the jth term of the series cannot tend to 0. Thus, the binomial series diverges at $x = 1$ when $\alpha < -1$.

$\boldsymbol{\alpha = -1}$. The binomial series for $f(x) = (1+x)^{-1}$ is $\sum_{j=0}^{\infty}(-1)^j x^j$, which diverges at $x = 1$.

$\boldsymbol{-1 < \alpha}$. An argument analogous to the preceding one shows that, if $-1 < \alpha$ and $x = 1$, then the $(j+1)$st term of the binomial series is strictly smaller in magnitude than the jth term. We showed that the binomial series is eventually an alternating series when $x = 1$. Finally, the jth term tends to 0. By the alternating series test, the binomial series converges at $x = 1$ when $\alpha > -1$. Note that, at $x = 1$, our previous argument (using Gauss's test at $x = -1$) can be applied verbatim to show that the binomial series converges absolutely for $\alpha > 0$ and does not converge absolutely, and therefore converges conditionally, for $-1 < \alpha < 0$.

In summary, we have the following twelve cases describing the convergence of the binomial series:

	$x = -1$	$-1 < x < 1$	$x = 1$
$\alpha < -1$	diverges (Gauss)	converges absolutely (ratio)	diverges
$\alpha = -1$	diverges	converges absolutely (ratio)	diverges
$-1 < \alpha < 0$	diverges (Gauss)	converges absolutely (ratio)	converges conditionally
$\alpha > 0$	converges absolutely (Gauss)	converges absolutely (ratio)	converges absolutely (Gauss)

However, to this point we have not addressed in the slightest the sum to which the series actually converges. This is a separate question requiring us to look at the Taylor remainder $R_k(0; x)$. If we can show that, for any particular values of x, the remainder $R_k(0; x)$ tends to 0, then we can conclude that, for those x, the binomial series converges to $(1 + x)^\alpha$. In view of all the cases in the foregoing table, it should not surprise you that no single argument can suffice to treat all cases simultaneously.

Here we present a sampling of the techniques to resolve the question. Suppose, to begin, that $-1 < x < 1$. To show that, for any α, the binomial series converges to $(1 + x)^\alpha$, we use the variant of Cauchy's form of the remainder discussed above. There is a θ in $(0, 1)$ such that

$$\begin{aligned} R_k(0; x) &= \frac{f^{(k+1)}(\theta x)x^{k+1}(1-\theta)^k}{k!} \\ &= \frac{\alpha(\alpha-1)\cdots(\alpha-k)(1+\theta x)^{\alpha-k-1}x^{k+1}(1-\theta)^k}{k!} \\ &= \left[\frac{\alpha(\alpha-1)\cdots(\alpha-k)}{k!}\right]\left[\frac{1-\theta}{1+\theta x}\right]^k (1+\theta x)^{\alpha-1}x^{k+1}. \end{aligned}$$

Treat the factors of $R_k(0; x)$ methodically. First, it is easy to see that $0 < (1-\theta)/(1+\theta x) < 1$ because $0 < \theta < 1$ and $-1 < x < 1$. Therefore

$$\left|\frac{1-\theta}{1+\theta x}\right|^k < 1.$$

Next consider the quantity $(1 + \theta x)^{\alpha-1}$. If $\alpha \geq 1$, then

$$(1+\theta x)^{\alpha-1} \leq (1+|x|)^{\alpha-1}.$$

If $\alpha < 1$, then

$$(1 + \theta x)^{\alpha-1} \leq (1 - |x|)^{\alpha-1}.$$

Consequently, although θ depends on k, $(1 + \theta x)^{\alpha-1}$ is bounded above by a quantity $g(x, \alpha)$ that is independent of k:

$$g(x, \alpha) = \begin{cases} (1 + |x|)^{\alpha-1}, & \text{if } \alpha \geq 1, \\ (1 - |x|)^{\alpha-1}, & \text{if } \alpha < 1. \end{cases}$$

It follows that

$$|R_k(0; x)| \leq \left| \frac{\alpha(\alpha - 1) \cdots (\alpha - k)}{k!} \right| g(x, \alpha)|x|^{k+1}.$$

Since $g(x, \alpha)$ is independent of k, to show that $\lim_{k\to\infty} R_k(0; x) = 0$ for x in $(-1, 1)$, it suffices to show that

$$\lim_{k\to\infty} \left| \frac{\alpha(\alpha - 1) \cdots (\alpha - k)}{k!} \right| x^{k+1} = 0$$

for $-1 < x < 1$. Apply the ratio test to the series whose kth term is

$$b_k = \left| \frac{\alpha(\alpha - 1) \cdots (\alpha - k)}{k!} \right| x^{k+1}.$$

We have

$$\limsup \left| \frac{b_{k+1}}{b_k} \right| = \limsup \left| \frac{\alpha - k - 1}{k + 1} \right| |x| = |x|.$$

Thus, for $|x| < 1$, the series $\sum_{j=1}^{\infty} b_k$ converges. It follows that the kth term of this series tends to 0; that is,

$$\lim_{k\to\infty} \left| \frac{\alpha(\alpha - 1) \cdots (\alpha - k)}{k!} \right| x^{k+1} = 0.$$

Consequently, by the Squeeze Play, for $-1 < x < 1$,

$$\lim_{k\to\infty} R_k(0; x) = 0.$$

This proves that, for any α and for x in $(-1, 1)$, the binomial series converges to $(1 + x)^{\alpha}$.

At the endpoint $x = 1$ and for $\alpha > -1$, we use the integral form of the remainder. For $k > \alpha - 1$, we have

$$\begin{aligned} R_k(0; 1) &= \frac{1}{k!} \int_0^1 f^{(k+1)}(t)(1 - t)^k \, dt \\ &= \frac{\alpha(\alpha - 1) \cdots (\alpha - k)}{k!} \int_0^1 (1 + t)^{\alpha-k-1}(1 - t)^k \, dt. \end{aligned}$$

Now, $\alpha - k - 1 < 0$ and $0 \leq t \leq 1$. Therefore,

$$0 < (1 + t)^{\alpha-k-1} \leq 1.$$

It follows that

$$\int_0^1 (1+t)^{\alpha-k-1}(1-t)^k\,dt \le \int_0^1 (1-t)^k\,dt = \left.\frac{-(1-t)^{k+1}}{k+1}\right|_0^1 = \frac{1}{k+1}.$$

Consequently,

$$|R_k(0;\,1)| \le \frac{|\alpha(\alpha-1)\cdots(\alpha-k)|}{(k+1)!}.$$

This upper bound for $|R_k(0;\,1)|$ is the magnitude of the $(k+1)$st term of the binomial series evaluated at $x = 1$. Since that series converges for all $\alpha > -1$, its $(k+1)$st term tends to 0. Thus, by the Squeeze Play,

$$\lim_{k\to\infty} R_k(0;\,1) = 0.$$

This proves that, if $\alpha > -1$, then

$$1 + \sum_{j=1}^{\infty} \frac{\alpha(\alpha-1)\cdots(\alpha-j+1)}{j!} = 2^{\alpha}.$$

There remains only the case $x = -1$ and $\alpha > 0$. We leave it for you to show that, under these conditions,

$$1 + \sum_{j=1}^{\infty} \frac{(-1)^j\alpha(\alpha-1)\cdots(\alpha-j+1)}{j!} = 0.$$

This completes the proof that, when the binomial series converges, its sum is $(1+x)^{\alpha}$.

12.6 SOLUTIONS OF FIRST-ORDER DIFFERENTIAL EQUATIONS

Here we present an application of our general theory, which is of profound importance in the study of differential equations. In 1890 Emile Picard published his general *method of successive approximations* to prove the existence of solutions to a large class of first-order differential equations. To set the stage, let F be a continuous function on the closed rectangle $R = \{(x, y)\colon |x - x_0| \le a,\ |y - y_0| \le b\}$ in $\mathbb{R}^2$. We want to show that, under mild restrictions on the function F, we can find a differentiable function f defined on some interval I containing x_0 such that

i) $f(x_0) = y_0$.

ii) $f'(x) = F(x, f(x))$ for all x in I. **(12.10)**

Such a function f is said to *solve the differential equation* $y' = F(x, y)$ with *initial value* $f(x_0) = y_0$ or to *solve the initial value problem* (12.10).

The function f we seek is a differentiable function on I whose graph passes through the point (x_0, y_0) and whose derivative is simply $F(x, f(x))$. It is to be understood that, for (ii) to hold, it must be true that, for all x in I, the point $(x, f(x))$ is in R, the domain of F; otherwise, the expression $F(x, f(x))$ makes no sense. As we construct the desired function f, we will need to take care to address this requirement.

First we show that solving the initial value problem (12.10) is equivalent to solving the *integral equation*

$$f(x) = y_0 + \int_{x_0}^{x} F(t, f(t))\, dt. \tag{12.11}$$

The equation (12.11) is to hold for all x in some interval I containing x_0 and, of course, $(x, f(x))$ is to be in R.

THEOREM 12.6.1 The differentiable function f solves the initial value problem (12.10) if and only if it solves the integral equation (12.11).

Proof. Suppose first that f solves the integral equation

$$f(x) = y_0 + \int_{x_0}^{x} F(t, f(t))\, dt$$

for all x in I. Since f is assumed to be differentiable on I, it is also continuous there. Also, F is assumed to be continuous. Therefore, $g(t) = F(t, f(t))$ is a continuous function of t. By the Fundamental Theorem of Calculus,

$$G(x) = \int_{x_0}^{x} F(t, f(t))\, dt$$

is differentiable on I and, furthermore, $G'(x) = F(x, f(x))$. Since $f = y_0 + G$, it follows that f is differentiable on I and $f'(x) = F(x, f(x))$. It is immediate that $f(x_0) = y_0$. This proves that the solution f of the integral equation (12.11) also solves the initial value problem (12.10).

Suppose next that f solves the initial value problem on some interval I, so that $f'(x) = F(x, f(x))$ for x in I and $f(x_0) = y_0$. It is evident that f', being the composition of continuous functions, is continuous. Therefore, f' is integrable on I. On the one hand,

$$\int_{x_0}^{x} f'(t)\, dt = \int_{x_0}^{x} F(t, f(t))\, dt.$$

On the other hand, by Corollary 6.3.9,

$$\int_{x_0}^{x} f'(t)\, dt = f(x) - f(x_0) = f(x) - y_0.$$

Therefore, $f(x) = y_0 + \int_{x_0}^{x} F(t, f(t))\, dt$. That is, f solves the integral equation (12.11). ●

The point of the theorem, of course, is that, rather than solve the general differential equation (12.10), we can solve the integral equation (12.11) to obtain the desired function f.

To solve the integral equation, first notice that, since F is continuous on the compact set R, it is bounded there. Let $r = \min\{a, b/||F||_\infty\}$ and take I to be the interval $I = \{x \text{ in } \mathbb{R} : |x - x_0| \le r\}$ on which we will construct a solution to (12.11). For that construction, we will recursively define a sequence $\{f_k\}$ of functions defined on I that will converge uniformly to the desired function f.

For x in I, let $f_0(x) = y_0$. While, generally, f_0 does not solve (12.11), it does meet the initial value condition. Also note that, for x in I, the point $(x, f_0(x))$ is in R, the domain of F. Therefore, we can define

$$f_1(x) = y_0 + \int_{x_0}^{x} F(t, f_0(t))\, dt,$$

for x in I.

We are interested, of course, in the properties of f_1. First, $f_1(x_0) = y_0$. Next, by the Fundamental Theorem of Calculus, f_1 is differentiable on the interval I with $f_1'(x) = F(x, f_0(x))$. Finally, for x in I,

$$|f_1(x) - y_0| \leq \int_{x_0}^{x} |F(t, f_0(t))|\, dt \leq ||F||_\infty |x - x_0| \leq b.$$

This proves that, for every x in I, the point $(x, f_1(x))$ is in R, the domain of F. Thus, for x in I, we can define

$$f_2(x) = y_0 + \int_{x_0}^{x} F(t, f_1(t))\, dt.$$

Again, $f_2(x_0) = y_0$, $f_2'(x) = F(x, f_1(x))$, and $|f_2(x) - y_0| \leq ||F||_\infty |x - x_0| \leq b$, for x in I. This guarantees that, for x in I, the point $(x, f_2(x))$ is in R.

Suppose that we have recursively defined functions $f_0, f_1, f_2, \ldots, f_k$ on I by $f_j(x) = y_0 + \int_{x_0}^{x} F(t, f_{j-1}(t))\, dt$, for $j = 1, 2, \ldots, k$. For x in I, define

$$f_{k+1}(x) = y_0 + \int_{x_0}^{x} F(t, f_k(t))\, dt.$$

Then, again, $f_{k+1}(x_0) = y_0$, $f_{k+1}'(x) = F(x, f_k(x))$, for x in I, and

$$|f_{k+1}(x) - y_0| \leq \int_{x_0}^{x} |F(t, f_k(t))|\, dt \leq ||F||_\infty |x - x_0| \leq b.$$

Therefore, for each x in I, the point $(x, f_{k+1}(x))$ is in R, the domain of F, and the recursive procedure can continue.

By this process, we have created a sequence $\{f_k\}$ of functions, each defined on I, having the following properties: For each k in $\mathbb{N}$,

i) $f_k(x_0) = y_0$.

ii) $f_k'(x) = F(x, f_{k-1}(x))$.

iii) $|f_k(x) - y_0| \leq ||F||_\infty |x - x_0| \leq b$.

This last condition, which you will recognize as a Lipschitz condition at x_0, can be described geometrically by saying that the graph of each f_k lies entirely within the subset S of R bounded by the lines

$$y - y_0 = ||F||_\infty (x - x_0)$$

and

$$y - y_0 = -||F||_\infty (x - x_0).$$

The region S is sketched in Figs. 12.5 and 12.6 for the two cases $r = a$ and $r = b/\|F\|_\infty$.

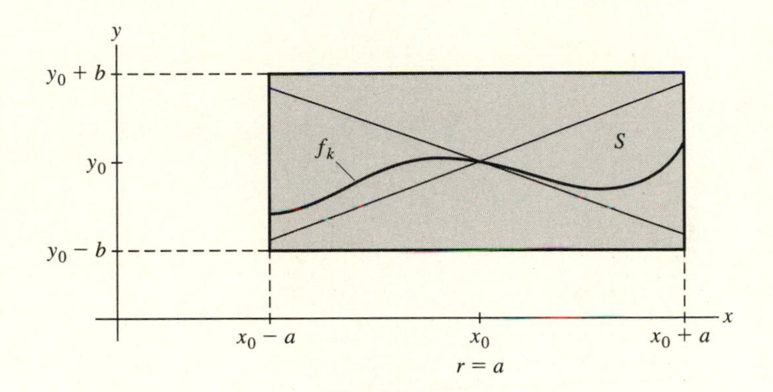

Figure 12.5

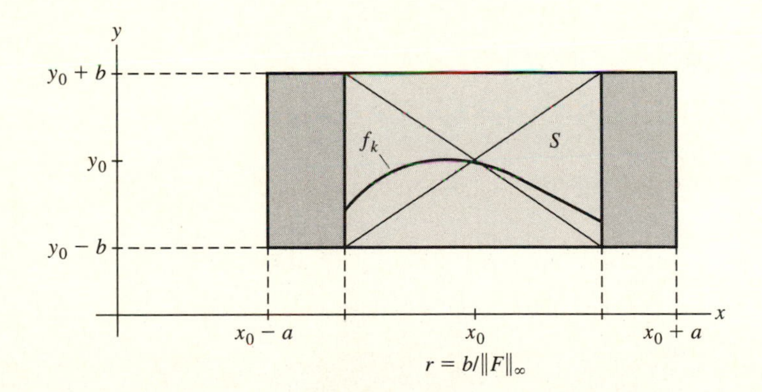

Figure 12.6

With no additional restriction on F, we cannot prove, and it is not generally true, that $\{f_k\}$ necessarily converges to a solution f. The mild condition we do impose is a variant of the Lipschitz conditions with which you are already familiar. Throughout the sequel we will assume that F satisfies the following condition:

There exists a positive constant M such that, for any two points (x, y_1) and (x, y_2) in R,
$$|F(x, y_1) - F(x, y_2)| \leq M|y_1 - y_2|. \tag{12.12}$$

Placing this one additional requirement on F will enable us to show that the sequence $\{f_k\}$ of successive approximations converges to a function f that solves the integral equation (12.11). By Theorem 12.6.1, the sequence $\{f_k\}$ will converge to a solution of the differential equation $y' = F(x, y)$ and will satisfy the initial value condition $f(x_0) = y_0$. The proof will hinge on the following lemma.

LEMMA For k in $\mathbb{N}$ and for x in I,

$$|f_k(x) - f_{k-1}(x)| \leq \frac{||F||_\infty M^{k-1}|x - x_0|^k}{k!}.$$

Proof. We prove the lemma by induction. We already know that

$$|f_1(x) - f_0(x)| \leq ||F||_\infty |x - x_0|.$$

Thus, the lemma holds for $k = 1$. Suppose that, for some k in $\mathbb{N}$ and for all x in I,

$$|f_k(x) - f_{k-1}(x)| \leq \frac{||F||_\infty M^{k-1}|x - x_0|^k}{k!}.$$

Using the definition of the functions f_k and the Lipschitz condition (12.12), consider, for any x in I,

$$\begin{aligned} |f_{k+1}(x) - f_k(x)| &= \left| \left[y_0 + \int_{x_0}^{x} F(t, f_k(t))\, dt \right] - \left[y_0 + \int_{x_0}^{x} F(t, f_{k-1}(t))\, dt \right] \right| \\ &\leq \int_{x_0}^{x} |F(t, f_k(t)) - F(t, f_{k-1}(t))|\, dt \\ &\leq M \int_{x_0}^{x} |f_k(t) - f_{k-1}(t)|\, dt. \end{aligned}$$

By our inductive assumption,

$$|f_k(t) - f_{k-1}(t)| \leq \frac{||F||_\infty M^{k-1}|t - x_0|^k}{k!}$$

for all t in the interval of integration. Therefore,

$$\begin{aligned} |f_{k+1}(x) - f_k(x)| &\leq \frac{||F||_\infty M^k}{k!} \int_{x_0}^{x} |t - x_0|^k\, dt \\ &= \frac{||F||_\infty M^k |x - x_0|^{k+1}}{(k+1)!}. \end{aligned}$$

This proves the lemma. ●

To prove that $\{f_k\}$ converges uniformly to a solution f, use telescoping to write

$$f_k = f_0 + \sum_{j=1}^{k} [f_j - f_{j-1}].$$

Notice that f_k is the kth partial sum of the series

$$f_0 + \sum_{j=1}^{\infty} [f_j - f_{j-1}].$$

Therefore, $\{f_k\}$ converges uniformly if and only if the above series converges uniformly. By the lemma, for x in I and j in $\mathbb{N}$,

$$|f_j(x) - f_{j-1}(x)| \leq \frac{||F||_\infty M^{j-1}|x - x_0|^j}{j!} \leq \frac{||F||_\infty M^{j-1} r^j}{j!}.$$

Note that the series

$$y_0 + \sum_{j=1}^{\infty} \frac{||F||_\infty M^{j-1} r^j}{j!} = y_0 + \frac{||F||_\infty}{M} \sum_{j=1}^{\infty} \frac{(Mr)^j}{j!}$$

converges to $y_0 + ||F||_\infty [e^{Mr} - 1]/M$. Thus, by Weierstrass's M-test, the series $f_0 + \sum_{j=1}^{\infty} [f_j - f_{j-1}]$, and hence the sequence $\{f_k\}$ converges uniformly on I. Let f denote the uniform limit of $\{f_k\}$.

Finally, from the properties of the functions f_k we deduce those of f. First, since each f_k is continuous and since the convergence is uniform, f is also continuous on I. By the Lipschitz condition (12.12)

$$|F(x, f_k(x)) - F(x, f(x))| \leq M|f_k(x) - f(x)| \leq M||f_k - f||_\infty.$$

Therefore, by the Squeeze Play,

$$\lim_{k\to\infty} F(x, f_k(x)) = F(x, f(x)) \quad \text{[uniformly]}$$

on I. Consequently, by Theorem 6.5.1,

$$\lim_{k\to\infty} \int_{x_0}^{x} F(t, f_k(t))\, dt = \int_{x_0}^{x} F(t, f(t))\, dt \quad \text{[uniformly]}$$

on I. We conclude that, for any x in I,

$$\begin{aligned} f(x) &= \lim_{k\to\infty} f_{k+1}(x) = y_0 + \lim_{k\to\infty} \int_{x_0}^{x} F(t, f_k(t))\, dt \\ &= y_0 + \int_{x_0}^{x} F(t, f(t))\, dt. \end{aligned}$$

We have proved the following theorem.

THEOREM 12.6.2 Picard If F satisfies the Lipschitz condition (12.12), then the sequence $\{f_k\}$ constructed above converges uniformly to a solution f of the integral equation

$$f(x) = y_0 + \int_{x_0}^{x} F(t, f(t))\, dt$$

on the interval $I = [x_0 - r, x_0 + r]$. The function f also solves the initial value problem $y' = F(x, y),\ y(x_0) = y_0$. ●

The construction leading to the proof of Theorem 12.6.2 also provides us with the means to estimate the uniform error $||f - f_k||_\infty$ on I. For x in I, write

$$
\begin{aligned}
|f(x) - f_k(x)| &= \left| \sum_{j=k+1}^{\infty} [f_j(x) - f_{j-1}(x)] \right| \leq \sum_{j=k+1}^{\infty} |f_j(x) - f_{j-1}(x)| \\
&\leq \sum_{j=k+1}^{\infty} \frac{||F||_\infty M^{j-1} r^j}{j!} = \frac{||F||_\infty}{M} \sum_{j=k+1}^{\infty} \frac{(Mr)^j}{j!} \qquad (12.13)
\end{aligned}
$$

Let $A = ||F||_\infty / M$. From the last sum in (12.13) factor out $B_k = (Mr)^{k+1}/(k+1)!$ and adjust the index to obtain

$$
\begin{aligned}
|f(x) - f_k(x)| &\leq AB_k \sum_{j=0}^{\infty} \frac{(k+1)!}{(k+j+1)!} (Mr)^j \\
&\leq AB_k \sum_{j=0}^{\infty} \frac{(Mr)^j}{j!} = AB_k e^{Mr},
\end{aligned}
$$

since $(k+1)!/(k+j+1)! \leq 1/j!$. Therefore,

$$
||f - f_k||_\infty \leq AB_k e^{Mr} = \left(\frac{||F||_\infty e^{Mr}}{M} \right) \left(\frac{(Mr)^{k+1}}{(k+1)!} \right).
$$

This argument proves the following theorem.

THEOREM 12.6.3 The kth approximation f_k to the solution f of the initial value problem satisfies

$$
||f - f_k||_\infty \leq \left(\frac{||F||_\infty e^{Mr}}{M} \right) \left(\frac{(Mr)^{k+1}}{(k+1)!} \right). \quad \bullet
$$

EXAMPLE 20 Let $F(x, y) = x - 2xy$ on $R = \{(x, y) : |x| \leq 1, \ |y| \leq 1\}$. Let $x_0 = y_0 = 0$. See Fig. 12.7. We want to solve the initial value problem

$$
y' = x - 2xy, \qquad y(0) = 0.
$$

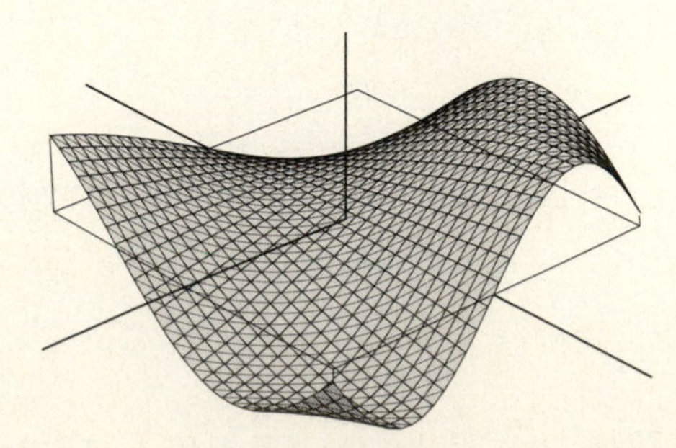

Figure 12.7

Notice that $||F||_\infty = 3$. Therefore $r = 1/3$ and we restrict x to $I = [-1/3, 1/3]$. Also notice that, for two points (x, y_1) and (x, y_2) in R,

$$|F(x, y_1) - F(x, y_2)| = 2|x|\,|y_1 - y_2| \leq 2|y_1 - y_2|.$$

Therefore, we take the Lipschitz constant to be $M = 2$. We set $f_0(x) = 0$ and obtain recursively

$$f_1(x) = \int_0^x t\,dt = \frac{x^2}{2}$$

$$f_2(x) = \int_0^x \left[t - 2t\left(\frac{t^2}{2}\right)\right] dt = \frac{x^2}{2} - \frac{x^4}{4}$$

$$f_3(x) = \int_0^x \left[t - 2t\left(\frac{t^2}{2} - \frac{t^4}{4}\right)\right] dt = \int_0^x \left[t - t^3 + \frac{t^5}{2}\right] dt$$

$$= \frac{x^2}{2} - \frac{x^4}{4} + \frac{x^6}{2 \cdot 6} = -\frac{1}{2}\sum_{j=1}^{3} \frac{(-1)^j x^{2j}}{j!}.$$

These computations suggest a pattern; we prove by induction that the general form of f_k is given by

$$f_k(x) = -\frac{1}{2}\sum_{j=1}^{k} \frac{(-1)^j x^{2j}}{j!}. \tag{12.14}$$

Assuming (12.14), compute

$$f_{k+1}(x) = \int_0^x \left[t - 2t\left(-\frac{1}{2}\sum_{j=1}^{k} \frac{(-1)^j t^{2j}}{j!}\right)\right] dt$$

$$= \int_0^x \left[t + \sum_{j=1}^{k} \frac{(-1)^j t^{2j+1}}{j!}\right] dt = -\frac{1}{2}\sum_{j=1}^{k+1} \frac{(-1)^j x^{2j}}{j!}.$$

This proves by induction that the general form of f_k is given by (12.14). You will recognize $f_k(x)$ as the kth partial sum of the power series expansion for the function $f(x) = (1/2)[1 - \exp(-x^2)]$. Therefore, the function f solves the initial value problem $y' = x - 2xy,\ y(0) = 0$ on I. In fact, the solution is valid on all of $\mathbb{R}$. ●

EXERCISES

12.1. For k in $\mathbb{N}$ and x in $\mathbb{R}$, let $f_k(x) = \tan^{-1} kx$. Find the pointwise limit f of $\{f_k\}$. Prove that the convergence cannot be uniform on all of $\mathbb{R}$. Prove, however, that the convergence is uniform on any closed set S that does not contain 0.

12.2. For k in $\mathbb{N}$ and x in $\mathbb{R}$, let $f_k(x) = (2x/\pi)\tan^{-1} kx$. Find the pointwise limit f of $\{f_k\}$. Show that f is differentiable at all $x \neq 0$. Show that the sequence $\{f_k'\}$ converges at each x in $\mathbb{R}$ but that convergence cannot be uniform on any interval containing 0.

12.3. Define a sequence of functions on [0, 1] as follows. For k in $\mathbb{N}$, let

$$f_k(x) = \begin{cases} 4k^{3/2}x, & 0 \le x \le 1/(2k), \\ -4k^{3/2}x + 4k^{1/2}, & 1/(2k) < x \le 1/k, \\ 0, & 1/k < x \le 1. \end{cases}$$

a) Find the pointwise limit f of $\{f_k\}$ and determine whether that convergence is uniform.

b) For each k in $\mathbb{N}$ compute $\int_0^1 f_k(x)\,dx$ and determine whether

$$\lim_{k\to\infty} \int_0^1 f_k(x)\,dx = \int_0^1 f(x)\,dx.$$

c) Show that for each k, f_k' exists on [0, 1] except at two points a_k and b_k. Define $g_k(x) = f_k'(x)$ for $x \ne a_k$ or b_k; define

$$g_k(a_k) = \frac{f_k'(a_k^+) + f_k'(a_k^-)}{2}$$

and

$$g_k(b_k) = \frac{f_k'(b_k^+) + f_k'(b_k^-)}{2}.$$

For each x in [0, 1] find $\lim_{k\to\infty} g_k(x)$. Determine whether

$$\lim_{k\to\infty} [\lim_{x\to 0^+} g_k(x)] = \lim_{x\to 0^+} [\lim_{k\to\infty} g_k(x)].$$

d) Determine whether $\lim_{k\to\infty} \int_0^1 g_k(x)\,dx = \int_0^1 \lim_{k\to\infty} g_k(x)\,dx$.

12.4. For k in $\mathbb{N}$, let $a_k = 1/2 + 1/4 + \cdots + 1/2^k$. For x in $\mathbb{R}$, define $f_k(x) = x^{a_k}$. Find the pointwise limit f of $\{f_k\}$. Show that, for x in [0, 1], $|f_k(x) - f(x)| \le 1 - a_k$. What can you deduce about the uniform convergence of $\{f_k\}$ to f on [0, 1]?

12.5. For k in $\mathbb{N}$ and x in $[0, \infty)$, define $f_k(x) = kxe^{-kx}$.

a) Show that $\{f_k\}$ converges pointwise on $[0, \infty)$ and find its limit function f. Show that the convergence is not uniform on $[0, \infty)$. Show that, for any $\delta > 0$, the convergence is uniform on $[\delta, \infty)$.

b) For each k in $\mathbb{N}$ and each $b > 0$, compute $\int_0^b f_k(x)\,dx$. Is it true that $\lim_{k\to\infty} \int_0^b f_k(x)\,dx = \int_0^b f(x)\,dx$? Reconcile your answer with Theorem 6.5.1. Is it true that

$$\lim_{b\to\infty} \lim_{k\to\infty} \int_0^b f_k(x)\,dx = \lim_{k\to\infty} \lim_{b\to\infty} \int_0^b f_k(x)\,dx?$$

12.6. For k in $\mathbb{N}$ and x in $[0, \infty)$, define

$$f_k(x) = \frac{xe^{-x/k}}{k}.$$

a) Find the pointwise limit f of $\{f_k\}$. Show that the convergence is not uniform on $[0, \infty)$. However, given any $b > 0$, show that $\{f_k\}$ converges uniformly to f on $[0, b]$.

b) For each $b > 0$ and k in $\mathbb{N}$, compute $\int_0^b f_k(x)\,dx$.

i) Compute $\lim_{k\to\infty} \int_0^b f_k(x)\,dx$.

ii) Compute $\lim_{b\to\infty} \int_0^b f_k(x)\,dx$.

Hence, determine whether

$$\lim_{b\to\infty}\lim_{k\to\infty}\int_0^b f_k(x)\,dx = \lim_{k\to\infty}\lim_{b\to\infty}\int_0^b f_k(x)\,dx.$$

12.7. For k in $\mathbb{N}$ and x in $[0, \infty)$, define

$$f_k(x) = kx\,e^{-k^2x}.$$

a) Show that $\{f_k\}$ converges pointwise on $[0, \infty)$ and find the pointwise limit function f. Determine whether the convergence is uniform on $[0, \infty)$.

b) For each k in $\mathbb{N}$ and each x in $[0, \infty)$, compute $f_k'(x)$ and find the pointwise limit of $\{f_k'\}$. Does $\{f_k'\}$ converge uniformly on $[0, \infty)$? If so, prove it; otherwise, find an additional, nontrivial restriction on a subset S of $[0, \infty)$ that will guarantee uniform convergence of $\{f_k'\}$ on S.

12.8. For k in $\mathbb{N}$ and x in $\mathbb{R}$, define

$$f_k(x) = \frac{e^{-k^2x^2}}{k}.$$

a) Find the pointwise limit f of $\{f_k\}$. Show that $\lim_{k\to\infty} f_k = f$ [uniformly] on $\mathbb{R}$.

b) Compute f_k' and show that $\lim_{k\to\infty} f_k' = f'$ [pointwise] on $\mathbb{R}$. Show that $\{f_k'\}$ does not converge uniformly to f' on any interval containing 0.

12.9. For k in $\mathbb{N}$ and x in $\mathbb{R}$, define $f_k(x) = kx\exp(-kx^2)$.

a) Find the pointwise limit f of $\{f_k\}$. Determine whether $\{f_k\}$ converges uniformly to f on $\mathbb{R}$. If so, prove it; otherwise, find an additional, non-trivial restriction on a subset S of $\mathbb{R}$ that guarantees that $\{f_k\}$ converges uniformly to f on S.

b) For each k in $\mathbb{N}$, compute $f_k'(x)$ and determine the pointwise limit of $\{f_k'\}$. Prove that $\{f_k'\}$ converges uniformly to f' on any closed subset of $\mathbb{R}$ that does not contain the point 0.

c) For each k in $\mathbb{N}$ and $b > 0$, compute $\int_0^b f_k(x)\,dx$. Find $\lim_{k\to\infty}\int_0^b f_k(x)\,dx$ and $\lim_{b\to\infty}\int_0^b f_k(x)\,dx$. Is it true that

i) $\lim_{k\to\infty}\int_0^b f_k(x)\,dx = \int_0^b f(x)\,dx$?

ii) $\lim_{b\to\infty}\lim_{k\to\infty}\int_0^b f_k(x)\,dx = \lim_{k\to\infty}\lim_{b\to\infty}\int_0^b f_k(x)\,dx$?

Justify your answers.

12.10. For each k in $\mathbb{N}$ and x in $[-1, 1]$, define

$$f_k(x) = \frac{2kx}{1+k^2x^2}.$$

a) Find the pointwise limit f of $\{f_k\}$. Determine whether $\{f_k\}$ converges uniformly to f on $[-1, 1]$.

b) For each k in $\mathbb{N}$ and x in $[-1, 1]$, compute $\int_0^x f_k(t)\,dt$. Is it true that $\lim_{k\to\infty}\int_0^x f_k(t)\,dt = \int_0^x f(t)\,dt$?

c) **i)** If your answer in part (b) is yes, is it true that

$$\lim_{k\to\infty}\int_0^x f_k(t)\,dt = \int_0^x f(t)\,dt \text{ [uniformly]}$$

on $[-1, 1]$?

ii) If your answer in part (b) is no, find an additional nontrivial restriction on the interval of integration that guarantees that $\lim_{k\to\infty} \int_0^x f_k(t)\,dt = \int_0^x f(t)\,dt$.

12.11. For x in $\mathbb{R}$, consider the series $\sum_{j=0}^{\infty} x^2/(1+x^2)^j$.

a) Find the function F on $\mathbb{R}$ to which the series converges pointwise. Sketch the graph of $y = F(x)$. At which points is F continuous? Is the convergence uniform on $\mathbb{R}$?

b) Fix any $\delta > 0$. Prove that for all x such that $|x| \geq \delta$,

$$|F(x) - F_k(x)| \leq \frac{1}{(1+\delta^2)^{k-1}},$$

where $F_k(x)$ denotes the kth partial sum of the series. Hence prove that the series converges uniformly to F on $\{x \text{ in } \mathbb{R}\colon |x| \geq \delta\}$.

12.12. Prove that the series $\sum_{j=1}^{\infty} (-1)^{j+1}/(j + x^2)$ converges uniformly and conditionally on $\mathbb{R}$.

12.13. For x in $\mathbb{R}$, consider the series $\sum_{j=1}^{\infty} (-1)^{j+1}/(j^2 + x^2)$.

a) For which values of x does the series converge?

b) For which values of x is the convergence absolute? For which values of x is the convergence conditional?

c) Determine whether the series converges uniformly on $\mathbb{R}$. If so, prove it; otherwise, find a nontrivial restriction on a subset S of $\mathbb{R}$ that guarantees uniform convergence on S.

12.14. For $x > 0$, consider the series $\sum_{j=1}^{\infty} (\ln x)^j$.

a) For which x does the series converge?

b) For which x is the convergence absolute? For which x is the convergence conditional?

c) Find the most general description of a set on which the series converges uniformly.

12.15. For x in $\mathbb{R}$, consider the series $\sum_{j=0}^{\infty} (-1)^j e^{-jx}$.

a) For which x in $\mathbb{R}$ does the series converge?

b) For which x is the convergence absolute? For which x is the convergence conditional?

c) Find the most general description of a set on which the series converges uniformly.

12.16. For x in $\mathbb{R}$, consider the series

$$\sum_{j=1}^{\infty} \frac{\cos(2j-1)x}{2j(2j-1)}$$

a) Find all x where the series converges.

b) For which x is the convergence absolute? For which x is the convergence conditional?

c) Where is the convergence uniform?

12.17. Prove that the series $\sum_{j=0}^{\infty} 2x/(j^2 - x^2)$ converges uniformly on any interval $[a, b]$ that contains no integers.

12.18. For x in $\mathbb{R}$, consider the series $\sum_{j=1}^{\infty} x^j/[j^2(1+x^j)]$.

a) Prove that the series converges for all $x \neq -1$.

b) Prove that the convergence is uniform on $[0, \infty)$.

c) Prove that, for any r in $(0, 1)$, the convergence is uniform on $[-r, r]$.

d) Prove that, for any $b < -1$, the convergence is uniform on $(-\infty, b]$.

e) Explain how to conclude that the convergence is uniform on any closed subset of $\mathbb{R}$ that does not contain the point -1.

12.19. For x in $\mathbb{R}$, consider the series $\sum_{j=0}^{\infty} x^{2j} \exp(jx^2)$.

a) Find those x for which the series converges. (*Hint:* Find an approximate value for the solution u_0 of the equation $e^{-u} = u$; that is, find a fixed point for the function $f(u) = e^{-u}$. Explain how this value applies to this problem.)

b) For which values of x is the convergence absolute? For which values of x is the convergence conditional?

c) Describe the most general interval on which the convergence is uniform.

d) Find $\lim_{x \to 0} \sum_{j=0}^{\infty} x^{2j} \exp(jx^2)$.

12.20. a) Fix $p > 1$. Prove that $\sum_{j=1}^{\infty} 1/j^x$ converges uniformly for x in $[p, \infty)$.

b) More generally, prove that, if $a_j > 0$ and if $\sum_{j=1}^{\infty} a_j/j^p$ converges, then $\sum_{j=1}^{\infty} a_j/j^x$ converges uniformly for x in $[p, \infty)$.

12.21. Define a function f by $f(x) = \sum_{j=1}^{\infty} \cos(jx)/j^2$.

a) Find the domain of f.

b) Find where f is continuous.

c) Find an infinite series that converges to $\int_0^{\pi/2} f(x)\,dx$. Justify all your assertions.

12.22. Define a function f by

$$f(x) = \sum_{j=1}^{\infty} \frac{\sin(2j+1)x}{2j(2j-1)}.$$

a) Find the domain of f.

b) Find where f is continuous.

c) Find an infinite series that converges to $\int_0^{\pi/2} f(x)\,dx$. Justify all your assertions.

12.23. Suppose that $\sum_{j=0}^{\infty} a_j x^j$ has radius of convergence R_1 and that $\sum_{j=0}^{\infty} b_j x^j$ has radius of convergence R_2. Suppose also that for all j greater than some j_0, $|a_j| \le |b_j|$. Prove that $R_1 \ge R_2$.

12.24. a) Prove that if $\{a_j\}$ is a bounded sequence, then the series $\sum_{j=0}^{\infty} a_j x^j$ has radius of convergence $R \ge 1$.

b) Prove that, if $\sum_{j=0}^{\infty} a_j x^j$ has radius of convergence $R > 0$, then $\sum_{j=0}^{\infty} a_j x^{2j}$ has radius of convergence $R^{1/2}$.

12.25. Suppose that $\sum_{j=0}^{\infty} a_j x^j$ converges to $f(x)$ on $(-R_1, R_1)$ and that $\sum_{j=0}^{\infty} b_j x^j$ converges to $g(x)$ on $(-R_2, R_2)$. Let R be $\min\{R_1, R_2\}$. Prove that the Cauchy product of $\sum_{j=0}^{\infty} a_j x^j$ and $\sum_{j=0}^{\infty} b_j x^j$ converges to $f(x)g(x)$ on $(-R, R)$.

12.26. Both $1/(1-x)$ and $\tan^{-1} x$ have power series expansions about $x_0 = 0$ that converge for all x in $(-1, 1)$. By using the Cauchy product of these two series, find the first five terms of the power series expansion of $[\tan^{-1} x]/(1-x)$ about $x_0 = 0$.

12.27. Form the Cauchy product of the appropriate series to find the first five nonzero terms of the power series for $f(x) = [(1+x^2)/(1+x)]^{1/2}$ about $x_0 = 0$. Find all x for which this representation is valid.

12.28. Use the known series for $1/(1-x)$ and $\ln(1-x)$ to prove that the power series representation about $x_0 = 0$ for $[\ln(1-x)]/(1-x)$ is $-\sum_{k=1}^{\infty}[\sum_{j=1}^{k} 1/j]x^k$.

12.29. **a)** Prove that

$$\frac{1}{2}[\ln(1-x)]^2 = \sum_{k=2}^{\infty} \frac{1}{k}\left[\sum_{j=1}^{k-1} \frac{1}{j}\right] x^k$$

for appropriate values of x. Show that the radius of convergence of this series is 1.

b) Show that the series converges conditionally at $x = -1$ and hence find a series that converges to $[\ln 2]^2/2$.

12.30. **a)** Prove that

$$\frac{1}{2}(\tan^{-1} x)^2 = \sum_{k=1}^{\infty} \frac{(-1)^{k+1}}{2k}\left[\sum_{j=1}^{k} \frac{1}{2j-1}\right] x^{2k}$$

for appropriate values of x. Find the radius of convergence of this power series.

b) Determine whether the series in part (a) converges at either endpoint of the interval of convergence. If so, is that convergence absolute or conditional?

c) Use part (a) to find a series whose sum is $\pi^3/72$.

12.31. Use the power series representations developed in the text to establish the hyperbolic identities

$$\cosh^2 x = \frac{1}{2}\cosh(2x) + 1$$

and

$$\sinh^2 x = \frac{1}{2}\cosh(2x) - 1.$$

12.32. **a)** Rearrange the summands, change indices appropriately, and coalesce the sums to show that

$$\sum_{j=0}^{k}(j+1)a_{j+1}b_{k-j} + \sum_{j=0}^{k}(k-j+1)a_j b_{k-j+1} = (k+1)\sum_{j=0}^{k+1} a_j b_{k+1-j}.$$

b) Use the result in part (a) to prove the product rule of differentiation applied to $f(x) = \sum_{j=0}^{\infty} a_j x^j$ and $g(x) = \sum_{j=0}^{\infty} b_j x^j$:

$$\left[\left(\sum_{j=0}^{\infty} a_j x^j\right)\left(\sum_{j=0}^{\infty} b_j x^j\right)\right]'$$

$$= \left(\sum_{j=0}^{\infty} a_j x^j\right)'\left(\sum_{j=0}^{\infty} b_j x^j\right) + \left(\sum_{j=0}^{\infty} a_j x^j\right)\left(\sum_{j=0}^{\infty} b_j x^j\right)'.$$

12.33. **a)** Use the binomial series for $1/(1-t^2)^{1/2}$, the known derivative of $\sin^{-1} x$, and the Fundamental Theorem of Calculus to find the power series representation of $\sin^{-1} x$, valid on $(-1, 1)$.

b) Determine whether the series you found in part (a) converges at the endpoints of its interval of convergence.

c) Use the power series you found in part (a) to express $\pi/2$ and $\pi/6$ as the sums of infinite series.

12.34. Use the result from Exercise 12.33 to show that

$$\frac{\sin^{-1} x}{(1-x^2)^{1/2}} = \sum_{j=0}^{\infty} \frac{[2^j j!]^2 x^{2j+1}}{(2j+1)!}.$$

Find the radius of convergence of this power series and determine whether it converges at either endpoint of its interval of convergence.

12.35. Show that $1/(1-x)^2 = \sum_{j=0}^{\infty}(j+1)x^j$

a) by differentiating the geometric series.

b) by using the binomial series.

c) by forming an appropriate Cauchy product of two series.

d) by computing the Taylor series of $f(x) = 1/(1-x)^2$.

Find all points x where the series converges. Where is the convergence absolute? Where is it conditional?

12.36. Repeat Exercise 12.35 for $f(x) = 1/(1-x)^3$.

12.37. Generalize Exercises 12.35 and 12.36. For each k in $\mathbb{N}$, find the power series representation about $x_0 = 0$ for

$$f(x) = \frac{1}{(1-x)^k}.$$

12.38. Express $\int_0^x e^{-t}/(1+t)\,dt$ as a power series in x and determine where that series converges. Compute an approximate value for $\int_0^1 e^{-t}/(1+t)\,dt$, accurate to four decimal places.

12.39. By integrating $1/(1-t)$ from $-x$ to x, find the power series representation for $(1/2)\ln[(1+x)/(1-x)]$ about $x_0 = 0$. Find all x for which this representation is valid.

12.40. Use the power series expansion of $\sin x$ about $x_0 = 0$ to find a power series for $\int_0^x \sin t/t\,dt$. Find the radius of convergence of the power series you found. Approximate $\int_0^1 \sin t/t\,dt$, accurate to four decimal places.

12.41. Use the power series representation of $\ln(1-x)$ about $x_0 = 0$ to find a power series for $\int_0^x [\ln(1-t)]/t\,dt$. Find all x for which the series converges. Where does the series converge absolutely? conditionally? uniformly? Find an approximate value for $\int_0^1 [\ln(1-t)]/t\,dt$, accurate to four decimal places.

12.42. Use the power series representation for $\cos x$ to find an infinite series which converges to $\int_0^1 [1-\cos x]/x^2\,dx$. Compute an approximate value for this integral, accurate to four decimal places.

12.43. Express $\int_0^x \tan^{-1} t\,dt$ as a power series in x. Find all x where this power series converges and determine where that convergence is conditional or absolute. Hence prove that

$$\frac{\pi}{4} - \ln\sqrt{2} = 1 - \frac{1}{2} - \frac{1}{3} + \frac{1}{4} + \frac{1}{5} - \frac{1}{6} - \frac{1}{7} + \frac{1}{8} \cdots.$$

12.44. For each x in $\mathbb{R}$, represent the function

$$F(x) = \int_0^{\pi/2} (1 - x\sin^2 t)^{1/2}\,dt$$

as a power series in x. Find the radius of convergence of this power series. Hence show that

$$\sum_{j=1}^{\infty}\left[\frac{(2j)!}{4^j(j!)^2}\right]^2\cdot\frac{1}{2j-1}=\frac{\pi-2}{\pi}.$$

12.45. Express $\int_0^{1/2}\exp(-x^2)/\sqrt{1-x^2}\,dx$ as an infinite series.

12.46. For each x in $\mathbb{R}$, define

$$f(t)=\begin{cases}[1-e^{-tx}]/t, & t\neq 0,\\ x, & t=0.\end{cases}$$

a) Prove that f is continuous on $\mathbb{R}$.

b) Define $F(x)=\int_0^1[1-e^{-tx}]/t\,dt$. This integral is nonelementary but can be expressed as a power series in x. Find the power series representation of F about $x_0=0$ and find its radius of convergence.

c) Approximate $F(2)$ accurate to four decimal places.

12.47. Find the radius of convergence of each of the following series.

a) $\sum_{j=0}^{\infty}(2j)!x^j/(j!)^2$

b) $\sum_{j=0}^{\infty}(2j)!x^j/(j!)^3$

c) $\sum_{j=0}^{\infty}(3j)!x^j/(j!)^2$

d) Let p and q be positive integers. Considering the various cases, $p=q,\ p<q,\ p>q$, find the radius of convergence of

$$\sum_{j=0}^{\infty}\frac{(pj)!x^j}{(j!)^q}.$$

12.48. Find the radius of convergence of each series.

a) $\sum_{j=0}^{\infty}(j+3)!x^j/[j!(j+1)!]$

b) $\sum_{j=0}^{\infty}(j+1)^j x^j/j!$

c) $\sum_{j=0}^{\infty}(j+1)!x^j/[j!(j+3)!]$

d) $\sum_{j=0}^{\infty}(2j)!/[2^j j!]^2(x/3)^j$

e) $\sum_{j=0}^{\infty}(j+p)!x^j/[j!(j+q)!]$, p, q fixed in $\mathbb{N}$.

12.49. In each of the following parts a function f is defined by an infinite series. Determine whether f' exists at all the points in the indicated interval. If either endpoint can be included or must be excluded, explain why.

a) $f(x)=\sum_{j=1}^{\infty}(-1)^j x^{2j+1}/(2j),\quad -1\le x\le 1.$

b) $f(x)=\sum_{j=2}^{\infty}x^j/(j\ln j)^2,\quad -1\le x\le 1.$

c) $f(x)=\sum_{j=1}^{\infty}(\sin jx)/j^2,\quad 0<x\le 2\pi.$

d) $f(x)=\sum_{j=1}^{\infty}(-1)^j x^{2j}/[2^j j!]^2,\quad x$ in $\mathbb{R}$.

12.50. The hyperbolic sine function is strictly increasing and maps $\mathbb{R}$ onto $\mathbb{R}$. Therefore its inverse, $\sinh^{-1}x$, exists.

a) Use the identities $\sinh(\sinh^{-1}x)=x$, $\cosh^2 x-\sinh^2 x=1$, and the chain rule to find $d(\sinh^{-1}x)/dx$.

b) Use your result from part (a), the binomial series, and the Fundamental Theorem of Calculus to find the power series representation of $\sinh^{-1} x$ about $x_0 = 0$. Find the radius of convergence of the series you found.

12.51. Express the first derivative of $(e^x - 1)/x$ as a power series in x and find its radius of convergence. Hence show that $\sum_{j=1}^{\infty} j/(j+1)! = 1$.

12.52. Represent the second derivative of $\exp(-x^2)$ as a power series about $x_0 = 0$. Find the radius of convergence of this power series. Show that

$$\sum_{j=1}^{\infty} \frac{(-1)^{j+1}(2j+1)}{2^j j!} = 1.$$

12.53. Define a function f on $\mathbb{R}$ by

$$f(x) = \sum_{j=0}^{\infty} \frac{x^j}{(j+1)^2 j!}.$$

a) Form the power series representation of $x d[xf(x)]/dx$.

b) Identify the power series you found in part (a) as the series representing a well-known, elementary function.

c) Hence express f as an integral of the form

$$f(x) = \frac{1}{x}\int_0^x g(t)\,dt.$$

(The issue is to identify the integrand g.)

12.54. Find the power series representation for $d(x^2e^{-x})/dx$ about $x_0 = 0$. Find the radius of convergence of the series you found. Hence show that $\sum_{j=1}^{\infty}(-2)^{j+1}(j+2)/j! = 4$.

12.55. Suppose that a function f is represented by the power series $\sum_{j=0}^{\infty} a_j x^j$ with positive radius of convergence R.

a) Prove that, if f is an odd function on $(-R, R)$, then the even-indexed coefficients a_{2j} all vanish.

b) Prove that, if f is an even function on $(-R, R)$, then $a_{2j-1} = 0$ for all $j \geq 1$.

12.56. Suppose that f and g are two functions each of which can be represented by a power series on $(-R, R)$ with $R > 0$:

$$f(x) = \sum_{j=0}^{\infty} a_j x^j \qquad \text{and} \qquad g(x) = \sum_{j=0}^{\infty} b_j x^j.$$

Let j_1 be the least index such that $a_j \neq 0$; let j_2 be the least index such that $b_j \neq 0$.

a) Prove that there exists a $\delta > 0$ such that neither f nor g vanishes in $N'(0;\delta)$.

b) Prove that, if $j_1 = j_2$, then $\lim_{x\to 0} f(x)/g(x) = a_{j_1}/b_{j_2}$.

c) Prove that, if $j_1 > j_2$, then $\lim_{x\to 0} f(x)/g(x) = 0$.

d) Prove that, if $j_1 < j_2$, then $\lim_{x\to 0} |f(x)/g(x)| = \infty$.

(*Note:* This proves one form of l'Hôpital's rule for analytic functions f and g and relates that rule to the unique coefficients of their power series representations.)

12.57. Let f be a nonconstant function that can be represented by a power series, $f(x) = \sum_{j=0}^{\infty} a_j x^j$. Let j_0 be the smallest *positive* index such that $a_j \neq 0$.

a) Prove that, if j_0 is odd, then f has neither a relative maximum nor a relative minimum at $x_0 = 0$.

b) Prove that, if j_0 is even, then

i) if $a_{j_0} < 0$, then f has a relative maximum at $x_0 = 0$.

ii) if $a_{j_0} > 0$, then f has a relative minimum at $x_0 = 0$.

12.58. Suppose that α is any constant other than a nonnegative integer and that f is a differentiable function on $(-1, 1)$ for which

$$(1+x)f'(x) = \alpha f(x). \qquad \textbf{(12.15)}$$

a) Let $g(x) = f(x)/(1+x)^\alpha$. Find $g'(x)$ for x in $(-1, 1)$ and show how to conclude that $f(x) = c(1+x)^\alpha$ for some constant c.

b) Assume that a solution of the differential equation (12.15) has the form $f(x) = \sum_{j=0}^{\infty} a_j x^j$ where the coefficients a_j are unknown. Compute the series representation of f', substitute in (12.15), and recursively solve for $a_0, a_1, a_2, \ldots$. Compare the resulting power series for f with the solution found in part (a).

c) Rewrite (12.15) as $y' = \alpha y/(1+x) = F(x, y)$. Restrict F to the rectangle $R = \{(x, y) : |x| \le 1/2,\ |y| \le 1/2\}$.

i) Find $||F||_\infty$ on R.

ii) Find a Lipschitz constant M such that

$$|F(x, y_1) - F(x, y_2)| \le M|y_1 - y_2|$$

for (x, y_1) and (x, y_2) in R.

iii) Use the method presented in Section 12.6 to construct a sequence $\{f_k\}$ of successive approximations to a solution f of the initial value problem $y' = F(x, y),\ y(0) = 1$. Identify the limit f of $\{f_k\}$ and reconcile the solution you obtain by this method with those obtained in parts (a) and (b).

12.59. Hermite's differential equation is

$$y'' - 2xy' + 2\alpha y = 0,$$

where α is a constant.

a) Use the method discussed in Example 17 to solve recursively for the unknown coefficients a_j of a power series solution $f(x) = \sum_{j=0}^{\infty} a_j x^j$.

b) Show that the series solution you found in part (a) separates into a solution f_1 that is an even function and a solution f_2 that is odd. Find the radius of convergence of these two series. Prove that f_1 and f_2 are linearly independent and confirm that they are solutions.

c) Show that, if α is a nonnegative integer, then one of the solutions f_1 or f_2 found in part (b) reduces to a polynomial solution of Hermite's equation. (The polynomial solutions for $\alpha = k = 0, 1, 2, \ldots$ are the *Hermite polynomials* H_k.) Compute H_0, H_1, H_2, H_3, H_4, and H_5.

12.60. Legendre's differential equation is

$$(1-x^2)y'' - 2xy' + \alpha(\alpha+1)y = 0,$$

where α is a constant. Repeat Exercise 12.59 to find two linearly independent power series solutions f_1 and f_2, one even, one odd. Show that, if $\alpha = k$ is a nonnegative integer, then one of the solutions f_1 or f_2 reduces to a polynomial of degree k that solves Legendre's equation. (These polynomial solutions for $\alpha = k = 0, 1, 2, \ldots$, normalized to have value 1 at $x = 1$, are called the *Legendre polynomials*, P_k.) Compute the Legendre polynomials P_0, P_1, P_2, P_3, P_4, and P_5.

In Exercises 12.61 through 12.64 we explore properties of the Legendre polynomials.

12.61. Let $g(x) = (x^2 - 1)^k$ and define $h(x) = g^{(k)}(x)$.

a) Show that $h(x)$ is a polynomial of degree k such that $h(1) = 2^k k!$.

b) Show that $(x^2 - 1)g'(x) - 2kxg(x) = 0$.

c) Differentiate the expression in part (b) $k + 1$ times to show that

$$(1 - x^2)h''(x) - 2xh'(x) + k(k+1)h(x) = 0.$$

Hence show that h is a polynomial solution of Legendre's equation with $\alpha = k$. Show that the kth Legendre polynomial is $P_k(x) = [1/(2^k k!)]h(x)$.

12.62. **a)** Prove by induction that, for every k in $\mathbb{N}$, there exist constants $c_0, c_1, \ldots, c_k$ in $\mathbb{R}$ such that

$$x^k = \sum_{j=0}^{k} c_j P_j(x).$$

b) Use part (a) to prove that every polynomial of degree k is a linear combination $\sum_{j=0}^{k} c_j P_j$ of the Legendre polynomials $P_0, P_1, \ldots, P_k$. (The Legendre polynomials are said to *span* the space of all polynomials.)

12.63. Show that, if $k \neq m$, then

$$\int_{-1}^{1} P_k(x)P_m(x)\,dx = 0.$$

The Legendre polynomials are *orthogonal* over $[-1, 1]$. (*Hint:* Show that

$$[(1 - x^2)P_k']' = -k(k+1)P_k$$

and

$$[(1 - x^2)P_m']' = -m(m+1)P_m. \tag{12.16}$$

Form the difference $P_m[(1 - x^2)P_k']' - P_k[(1 - x^2)P_m']'$, simplify using the above equations (12.16), and integrate from -1 to 1.)

12.64. Prove that, for each k in $\mathbb{N}$, $\int_{-1}^{1} P_k^2(x)\,dx = 2/(2k+1)$. (*Hint:* From Exercise 12.61, $P_k(x) = g^{(k)}(x)/(2^k k!)$, where $g(x) = (x^2 - 1)^k$. Show first that $g^{(j)}(1) = g^{(j)}(-1) = 0$ for all $j = 0, 1, \ldots, k - 1$. Then integrate

$$\int_{-1}^{1} g^{(k)}(x)g^{(k)}(x)\,dx$$

repeatedly by parts to obtain

$$\int_{-1}^{1} g^{(k)}(x)g^{(k)}(x)\,dx = \cdots = (2k)!\int_{-1}^{1} (1 - x^2)^k\,dx.$$

Integrate this last integral by letting $x = \sin\theta$ to obtain

$$\int_{-1}^{1} (1-x^2)^k\,dx = \frac{2(2^k k!)^2}{(2k+1)!}.$$

Fill in the details of these steps and now complete the proof.)

12.65. **a)** Use the method in Example 17 to find two linearly independent functions f_1 and f_2, each of which can be represented by a power series in x, which solve the differential equation $(1-x^2)y'' + 6y = 0$.

b) Find the radius of convergence of each of the series solutions you found in part (a). Confirm that the functions you obtain actually are solutions of the differential equation on the interval of convergence.

c) There is a unique solution f of the differential equation that has the form $f = c_1 f_1 + c_2 f_2$ and that satisfies the initial value conditions $f(0) = 1,\ f'(0) = -1$. Find f as a linear combination of f_1 and f_2.

12.66. Solve the initial value problem

$$y' = F(x, y) = \frac{2 + x - 2y}{4},$$

$y(0) = -2$ on the rectangle $R = \{(x, y) : |x| \le 1,\ |y| \le 1\}$. Identify your solution in closed form.

12.67. Solve the initial value problem

$$y' = F(x, y) = x^2 + y,$$

$y(0) = 1$ on the rectangle $R = \{(x, y) : |x| \le 1,\ |y| \le 1\}$. Identify your solution in closed form. (*Hint:* As you generate the successive approximations f_k, separate the terms into summands that converge to two distinct components of the solution.)

12.68. This exercise guides you through a program to construct a continuous function f on $\mathbb{R}$ that fails to have a derivative at every point. The construction is a variant of that first presented by van der Waerden.

a) For x in $[-2, 2]$, define $f_0(x) = |x|$. Extend f_0 to be periodic with period 4 on all of $\mathbb{R}$: $f_0(x) = f_0(x + 4k)$, for all k in $\mathbb{Z}$.

i) Show that, for all x in $\mathbb{R}$, $f_0(x) = \min\{|x - 4k| : k \text{ in } \mathbb{Z}\}$

ii) Show that, if x and y are the endpoints of an open interval that contains no even integer, then $|f_0(x) - f_0(y)| = |x - y|$.

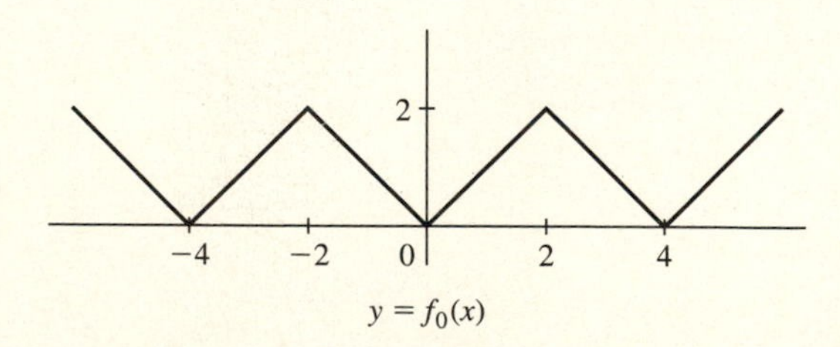

Figure 12.8

b) For each k in $\mathbb{N}$, define $f_k(x) = f_0(4^k x)/4^k$. Sketch the graphs of f_1 and f_2. Show that, for $k \ge 0$, f_k is continuous on $\mathbb{R}$.

c) Define $f = \sum_{k=1}^{\infty} f_k$. Prove that the series converges uniformly on $\mathbb{R}$. Thus f is continuous on $\mathbb{R}$.
We claim that f is not differentiable at any point of $\mathbb{R}$. Fix c in $\mathbb{R}$. To show that f is not differentiable at c, it suffices to construct a sequence $\{x_k\}$ of points distinct from c such that $\lim_{k\to\infty} x_k = c$ and $\lim_{k\to\infty}[f(x_k) - f(c)]/(x_k - c)$ fails to exist. The steps in such a construction are presented in parts (d) through (i).

d) Notice that, for each k in $\mathbb{N}$, the open interval

$$I_k = (4^k c - 1,\ 4^k c + 1)$$

contains at most two integers. Using this property of I_k, we will define a sequence $\{\epsilon_k\}$ where $\epsilon_k = \pm 1$.

i) If $4^k c$ happens to be an integer, then I_k contains only one integer. In this case, let $\epsilon_k = 1$.

ii) If $4^k c$ is not an integer, then I_k contains two adjacent integers, one even, one odd. In this case,

1) if $(4^k c - 1,\ 4^k c)$ contains the even integer, let $\epsilon_k = 1$.

2) if $(4^k c,\ 4^k c + 1)$ contains the even integer, let $\epsilon_k = -1$.

This defines ϵ_k for each k in $\mathbb{N}$.

Show that the open interval with endpoints $4^k c$ and $4^k c + \epsilon_k$ contains no even integer.

e) Define $x_k = c + \epsilon_k / 4^k$. Show that $\lim_{k\to\infty} x_k = c$.

f) Show that, for $j = 1, 2, \ldots, k$, the interval with endpoints $4^j c$ and $4^j c + \epsilon_k / 4^{k-j}$ contains no even integer. (Show that the contrary assumption contradicts the definition of ϵ_k.) Hence prove that, for $j = 1, 2, \ldots, k$,

$$|f_j(x_k) - f_j(c)| = \frac{1}{4^k}.$$

(Use part (a) (ii).)

g) Show that, if $j > k$, then $f_j(x_k) = f_j(c)$. (Use the definition of f_j and the periodicity of f_0.)

h) For k in $\mathbb{N}$, show that $[f(x_k) - f(c)]/(x_k - c)$ reduces to a finite sum of $+1$s and -1s. (*Hint:* Write

$$\begin{aligned}\frac{f(x_k) - f(c)}{x_k - c} &= \frac{4^k}{\epsilon_k}\left[\sum_{j=1}^{\infty} f_j(x_k) - \sum_{j=1}^{\infty} f_j(c)\right] \\ &= \frac{4^k}{\epsilon_k}\sum_{j=1}^{k}\left[f_j(x_k) - f_j(c)\right] + \frac{4^k}{\epsilon_k}\sum_{j=k+1}^{\infty}\left[f_j(x_k) - f_j(c)\right].\end{aligned}$$

The first sum simplifies enormously by part (f); the second sum vanishes by part (g).)

i) Show that $\lim_{k\to\infty}[f(x_k) - f(c)]/(x_k - c)$ cannot exist.
Consequently, f is a function that is continuous on all of $\mathbb{R}$ and is differentiable nowhere.

j) Let g be any function that is differentiable on all of $\mathbb{R}$. Define $h = f + g$. Show that h is also continuous on all of $\mathbb{R}$ and is differentiable nowhere. This proves that there are uncountably many functions that are continuous on $\mathbb{R}$ and differentiable at no point.

13

Improper Integrals

In our treatment of the integral in Chapter 6 we imposed two essential restrictions in order to obtain our results. We required the interval over which we integrated to be bounded and we required the integrand to be bounded on the interval. The first requirement enabled us to partition the interval using a finite partition with an arbitrarily small gauge. The second restriction allowed us to use the bound $||f||_\infty$ in various computations. Here we relax both restrictions and consider the integration theory of possibly unbounded integrands defined on possibly unbounded intervals of integration. We will consider only Riemann integrals with integrator $g(x) = x$, though we hasten to add that many of our results carry over to the more general Riemann–Stieltjes integral. Without intending any moralistic judgment, we say that integrals with unbounded integrand or with unbounded interval of integration are *improper*.

13.1 PRELIMINARIES

When we relax the first restriction—that is, when we allow the interval of integration to be unbounded—we refer to the resulting improper integral as being of the *first kind*. If the interval of integration is bounded but the integrand is unbounded, we say the improper integral is of the *second kind*.

DEFINITION 13.1.1 Let I be an interval of the form $[a, \infty)$, $(-\infty, b]$, or $(-\infty, \infty)$ and let f be a function defined on I. We say that f is *integrable* on I if f is integrable on each compact interval contained in I. The integral of f over I is said to be of the *first kind*. In each case, the integral of f over I is defined as follows:

a) If $I = [a, \infty)$, then $\int_a^\infty f(x)\,dx = \lim_{b\to\infty} \int_a^b f(x)\,dx$.

b) If $I = (-\infty, b]$, then $\int_{-\infty}^b f(x)\,dx = \lim_{a\to-\infty} \int_a^b f(x)\,dx$.

c) If $I = (-\infty, \infty)$, then $\int_{-\infty}^\infty f(x)\,dx = \lim_{\substack{a\to-\infty \\ b\to\infty}} \int_a^b f(x)\,dx$.

If the limit exists and has value L, then the improper integral of f over I is said to *converge* to L. If the limit fails to exist, then the improper integral is said to *diverge*. ●

As we shall see when studying integrals of this type, analogies with the theory of infinite series abound. Also, it will suffice to concentrate on integrals of the form $\int_a^\infty f(x)\,dx$, since the change of variables $u = -x$ transforms the integral $\int_{-\infty}^b f(x)\,dx$ into $\int_{-b}^\infty f(-u)\,du$ and the behavior of the one integral mirrors that of the other. Likewise, integrals of the form $\int_{-\infty}^\infty f(x)\,dx$ can be written as $\int_\infty^0 f(x)\,dx + \int_0^\infty f(x)\,dx$ and the behavior of the former integral is reflected in that of the latter two summands.

When we retain the requirement that the interval of integration I be bounded but relax the restriction that f be bounded on I, we will restrict our attention to the following case. We will assume that $I = (a, b]$, that f has a pole at the endpoint a, and that, for any c in (a, b), f is bounded on $[c, b]$. Arguments analogous to those we will present will treat the case when $I = [a, b)$, when f has a pole at the endpoint b, and when f is bounded on $[a, c]$ for every c in (a, b). We will leave such arguments for you to provide.

DEFINITION 13.1.2 Let $I = (a, b]$ and let f be a real-valued function defined on I. Suppose also that f has a pole at the endpoint a and that, for each c in (a, b), f is bounded on $[c, b]$. Then f is said to be *integrable* on I if f is integrable on $[c, b]$ for each c in $(a, b]$. The *improper integral* of f over I is defined to be

$$\int_a^b f(x)\,dx = \lim_{c\to a^+} \int_c^b f(x)\,dx$$

and is said to be of the *second kind*. If this limit exists and has value L, then the integral is said to *converge* to L. If the limit fails to exist, then the integral is said to *diverge*. ●

If f has a pole at the endpoint b rather than a and if f is bounded and integrable on $[a, c]$ for every c in (a, b), then the integral of f over I is also an improper integral of the second kind defined by

$$\int_a^b f(x)\,dx = \lim_{c\to b^-} \int_a^c f(x)\,dx.$$

Provided f has only finitely many poles in an unbounded interval I, we can, by using the additivity of the integral, decompose an improper integral of mixed type into the sum of integrals each of either the first or the second kind. By applying the theory we will develop to each of the summands, we will be able to treat the general improper integral of an unbounded integrand over an unbounded interval.

As we shall see, the two forms of the Fundamental Theorem of Calculus provide us with powerful tools for treating improper integrals. Although the theory of infinite series often seems somewhat fragmented, involving a variety of tests and specialized tricks to treat different types of series, the powerful theorems for integrals from Chapter 6 enable us to construct a more unified theory for improper integrals, requiring relatively few tests. We begin with improper integrals of the first kind; in Section 13.3 we will treat those of the second kind.

13.2 IMPROPER INTEGRALS OF THE FIRST KIND

Our initial discussion following Definition 13.1.1 shows that, when considering improper integrals of the first kind, it suffices to examine integrals of the form $\int_a^\infty f(x)\,dx$. The integral $\int_a^\infty f(x)\,dx = L$ if and only if, for any $\epsilon > 0$, there is a $b_0 > a$ such that, for all $b \geq b_0$,

$$\left|\int_a^b f(x)\,dx - L\right| < \epsilon.$$

Equivalently, $\int_a^\infty f(x)\,dx$ converges if and only if, for $\epsilon > 0$, there exists a $b_0 > a$ such that, for all $b \geq b_0$,

$$\left|\int_b^\infty f(x)\,dx\right| < \epsilon.$$

The proof of the following theorem is elementary and is left for you.

THEOREM 13.2.1 If f_1 and f_2 are integrable functions on $[a, \infty)$ such that $\int_a^\infty f_1(x)\,dx$ and $\int_a^\infty f_2(x)\,dx$ converge and if c_1 and c_2 are any real numbers, then $\int_a^\infty [c_1 f_1(x) + c_2 f_2(x)]\,dx$ also converges and

$$\int_a^\infty [c_1 f_1(x) + c_2 f_2(x)]\,dx = c_1 \int_a^\infty f_1(x)\,dx + c_2 \int_a^\infty f_2(x)\,dx. \quad \bullet$$

EXAMPLE 1 Fix any $p > 0$. The integral $\int_1^\infty 1/x^p\,dx$ is an improper integral of the first kind. To decide whether this integral converges, consider $\int_1^b 1/x^p\,dx$ for various values of p and, in each case, examine the limiting behavior as b tends to ∞. If $p \neq 1$, then

$$\int_1^b 1/x^p\,dx = \left.\frac{x^{1-p}}{1-p}\right|_1^b = \frac{1-b^{1-p}}{p-1}.$$

Two subcases emerge. If $p > 1$, then $1 - p < 0$ and it follows that $L = \lim_{b\to\infty} (1 - b^{1-p})/(p-1) = 1/(p-1)$ exists. In this case, the integral $\int_1^\infty 1/x^p\,dx$ converges to $1/(p-1)$. If $0 < p < 1$, then $1 - p > 0$ and it follows that $\lim_{b\to\infty} (1 - b^{1-p})/(p-1)$ fails to exist. In this case, the integral diverges.

If $p = 1$, then $\lim_{b\to\infty} \int_1^b 1/x\,dx = \lim_{b\to\infty} \ln b = \infty$. The integral diverges. You have seen the integrals in this example before, of course, in connection with the integral test in Chapter 6. As we will see below, they play an analogous role in the theory of convergence of improper integrals of the first kind. $\bullet$

EXAMPLE 2 For $a > 0$, the improper integral $\int_0^\infty 1/(a^2+x^2)\,dx$ can be rewritten as $(1/a)\int_0^\infty 1/[1+(x/a)^2](1/a)\,dx$. By the change of variables $u = x/a$, we obtain

$$\begin{aligned}\int_0^\infty \frac{1}{a^2+x^2}\,dx &= \frac{1}{a}\lim_{b\to\infty}\int_0^b \frac{1}{1+(x/a)^2}\frac{1}{a}\,dx\\ &= \frac{1}{a}\lim_{b\to\infty}\int_0^{b/a}\frac{1}{1+u^2}\,du\\ &= \frac{1}{a}\lim_{b\to\infty}\tan^{-1}\left(\frac{b}{a}\right) = \frac{\pi}{2a}. \quad \bullet\end{aligned}$$

EXAMPLE 3 For positive a and c, the improper integral $\int_0^\infty e^{-ax}\sin cx\,dx$ can be evaluated by integrating twice by parts. For any $b > 0$, we obtain

$$\begin{aligned}\int_0^b e^{-ax}\sin cx\,dx &= \left.\frac{-e^{-ax}(a\sin cx + c\cos cx)}{a^2+c^2}\right|_0^b\\ &= \frac{c}{a^2+c^2} - \frac{e^{-ab}(a\sin cb + c\cos cb)}{a^2+c^2}.\end{aligned}$$

Thus

$$\int_0^\infty e^{-ax}\sin cx\,dx = \lim_{b\to\infty}\int_0^b e^{-ax}\sin cx\,dx = \frac{c}{a^2+c^2}. \quad \bullet$$

EXAMPLE 4 This example reveals an important distinction between convergent series and convergent integrals of the first kind, one that you will want to keep in mind. Specifically, if $\sum_{j=1}^\infty a_j$ converges, then $\lim_{j\to\infty} a_j = 0$. However, if $\int_a^\infty f(x)\,dx$ converges, we cannot conclude that $\lim_{x\to\infty} f(x) = 0$. The trick is to construct the function f appropriately. Let $\{a_j\}$ and $\{b_j\}$ be two sequences of positive numbers such that $0 < a_j \le 1/2$ and $\{a_j\}$ converges monotonically to 0. For j in $\mathbb{N}$, define $f(j) = b_j$. Next define $f(x) = 0$ for x in the interval $[j + a_j,\ j+1-a_{j+1}]$. Define f to be linear on the intervals $[j,\ j+a_j]$ and $[j+1-a_{j+1},\ j+1]$ in such a way that f is continuous on all of $[1,\infty)$. (See Fig. 13.1.) As you can confirm,

$$\int_1^\infty f(x)\,dx = \frac{a_1b_1}{2} + \sum_{j=2}^\infty a_jb_j.$$

The integral and the series converge or diverge together. If, for example, we take $a_j = 1/(2j^2)$ and $b_j = 1$ for each j in $\mathbb{N}$, then the series, and hence the integral, converges. However, $\lim_{x\to\infty} f(x) \ne 0$; in fact, this limit does not exist. Note, in this case, that f is bounded on $[1,\infty)$.

If we take $a_j = 1/(2j^3)$ and $b_j = j$, again the series $\sum_{j=1}^\infty a_jb_j$ and the integral $\int_1^\infty f(x)\,dx$ converge but, in this case, the integrand is not bounded on $[1,\infty)$.

Finally, if we take $a_j = b_j = 1/(2j^{1/2})$, then both the series and the integral diverge, although $\lim_{x\to\infty} f(x) = 0$. $\bullet$

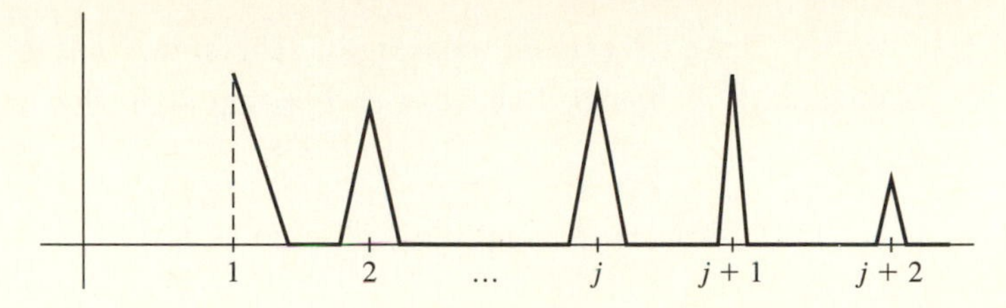

Figure 13.1

13.2.1 Tests for Convergence (Nonnegative Integrands)

As with series, the convergence of improper integrals may occur because the values of the integrand f remain sufficiently controlled as x tends to ∞ or because positive and negative values of f offset each other. In the first case the integral will be said to *converge absolutely*, in the second the convergence will be *conditional*. Here, to establish basic tests for absolute convergence of improper integrals of the first kind, we shall consider only integrands that are nonnegative.

Consider an improper integral of the first kind of the form $\int_a^\infty f(x)\,dx$ where $f \geq 0$. For such integrands it follows that, if $a < b_1 < b_2$, then $\int_a^{b_1} f(x)\,dx \leq \int_a^{b_2} f(x)\,dx$. Note that the set $S = \{\int_a^b f(x)\,dx : b \text{ in } (a, \infty)\}$ is a subset of $[0, \infty)$. If S is bounded above, then $\sup S$ exists; it follows that the integral $\int_a^\infty f(x)\,dx$ converges with value $\sup S$. Conversely, if $\int_a^\infty f(x)\,dx = L$, then S must be bounded above and $L = \sup S$. Analogous statements hold for improper integrals of the form $\int_{-\infty}^b f(x)\,dx$ and $\int_{-\infty}^\infty f(x)\,dx$. Once you provide the details to prove these assertions, you will have proved the following theorem.

THEOREM 13.2.2 Suppose that the integrand f is nonnegative and integrable on $[a, \infty)$. The integral $\int_a^\infty f(x)\,dx$ converges if and only if the set $S = \{\int_a^b f(x)\,dx : b \text{ in } (a, \infty)\}$ is bounded above. In this event, $\int_a^\infty f(x)\,dx = \sup S$. ●

We leave it for you to formulate the corresponding statements for the integrals $\int_{-\infty}^b f(x)\,dx$ and $\int_{-\infty}^\infty f(x)\,dx$.

From Theorem 13.2.2 follows the first test for convergence of improper integrals of the first kind with a nonnegative integrand.

THEOREM 13.2.3 Comparison Test; Integrals of First Kind Let f and g be nonnegative, integrable functions on $[a, \infty)$. Suppose that there exists a $c > a$ such that $f(x) \leq g(x)$ for all x in $[c, \infty)$.

i) If $\int_a^\infty g(x)\,dx$ converges, then $\int_a^\infty f(x)\,dx$ also converges.

ii) If $\int_a^\infty f(x)\,dx$ diverges, then $\int_a^\infty g(x)\,dx$ also diverges.

Proof. Fix $c > a$ such that $f(x) \leq g(x)$ for all x in $[c, \infty)$. For (i), we assume that $\int_a^\infty g(x)\,dx$ converges and must prove that $\int_a^\infty f(x)\,dx$ also converges.

It is easy to see that $\int_c^\infty g(x)\,dx$ converges. Therefore the set $S(g) = \{\int_c^b g(x)\,dx : b \text{ in } (c, \infty)\}$ is bounded above. Let $L = \sup S(g)$. For any $b > c$, write

$$\int_a^b f(x)\,dx = \int_a^c f(x)\,dx + \int_c^b f(x)\,dx.$$

Note that $K = \int_a^c f(x)\,dx$ is just a constant and that, for any $b > c$, $\int_c^b f(x)\,dx \le \int_c^b g(x)\,dx \le L$. Therefore for all $b > c$, $\int_a^b f(x)\,dx \le K + L$. It follows immediately that the set $S(f) = \{\int_a^b f(x)\,dx : b \text{ in } (a, \infty)\}$ is bounded above by $K + L$. Hence $\int_a^\infty f(x)\,dx$ converges.

Suppose, on the other hand, that $\int_a^\infty f(x)\,dx$ diverges. The integral $\int_a^c f(x)\,dx$ is finite and therefore $\int_c^\infty f(x)\,dx$ must diverge; that is, the set $\{\int_c^b f(x)\,dx : b \text{ in } (c, \infty)\}$ is unbounded. For any $b > c$, $f(x) \le g(x)$ for x in $[c, b]$; therefore

$$\int_c^b f(x)\,dx \le \int_c^b g(x)\,dx.$$

Consequently, the set $S(g) = \{\int_c^b g(x)\,dx : b \text{ in } (c, \infty)\}$ is also unbounded. Thus $\int_c^\infty g(x)\,dx$, and hence $\int_a^\infty g(x)\,dx$, diverges. ●

EXAMPLE 5 To show that $\int_0^\infty 1/(1 + x^2)^{1/2}\,dx$ diverges, use the comparison test. The integral is of the first kind and, for $x \ge 0$, we have $0 < f(x) = 1/(1 + x) \le 1/(1 + x^2)^{1/2} = g(x)$. Since $\int_0^\infty 1/(1 + x)\,dx$ diverges, the integral $\int_0^\infty 1/(1 + x^2)^{1/2}\,dx$ diverges also. ●

EXAMPLE 6 To show that $\int_0^\infty \exp(-x^2)\,dx$ converges, compare with $\int_0^\infty e^{-x}\,dx$. Note that $0 < \exp(-x^2) \le e^{-x}$ for all $x > 1$. Furthermore, the integral $\int_0^\infty e^{-x}\,dx$ converges, as the following computation shows.

$$\int_0^\infty e^{-x}\,dx = \lim_{b\to\infty} \int_0^b e^{-x}\,dx = \lim_{b\to\infty} -e^{-x}\Big|_0^b$$
$$= \lim_{b\to\infty} (1 - e^{-b}) = 1.$$

By the comparison test, $\int_0^\infty \exp(-x^2)\,dx$ also converges. ●

As with the corresponding test for series, the comparison test can be strengthened and made more versatile by considering $\lim_{x\to\infty} f(x)/g(x)$ where f and g are two positive integrands. If $\lim_{x\to\infty} f(x)/g(x) = L$ exists and is positive, then f and g share similar essential behavior as x tends to ∞. Consequently, as we might expect, the integrals $\int_a^\infty f(x)\,dx$ and $\int_a^\infty g(x)\,dx$ converge or diverge together.

THEOREM 13.2.4 Limit Comparison Test; Positive Integrands Let f and g be positive functions that are integrable on $[a, \infty)$. If $\lim_{x\to\infty} f(x)/g(x) = L$ exists and is positive, then $\int_a^\infty f(x)\,dx$ converges if and only if $\int_a^\infty g(x)\,dx$ converges.

Proof. Since $\lim_{x\to\infty} f(x)/g(x) = L > 0$, for any ϵ such that $0 < \epsilon < L$, we can choose a $c > a$ such that

$$L - \epsilon < \frac{f(x)}{g(x)} < L + \epsilon$$

for all $x > c$. Thus $(L-\epsilon)g(x) < f(x) < (L+\epsilon)g(x)$ for all such x. If $\int_a^\infty f(x)\,dx$ converges, then, by the comparison test, $\int_a^\infty (L-\epsilon)g(x)\,dx$ also converges. Consequently, $\int_a^\infty g(x)\,dx$ converges. Likewise, if $\int_a^\infty g(x)\,dx$ converges, then so also does $\int_a^\infty (L+\epsilon)g(x)\,dx$. Again by the comparison test, $\int_a^\infty f(x)\,dx$ must also converge. ●

COROLLARY 13.2.5 Let f and g be positive functions that are integrable on $[a, \infty)$.

i) If $\lim_{x\to\infty} f(x)/g(x) = 0$ and if $\int_a^\infty g(x)\,dx$ converges, then $\int_a^\infty f(x)\,dx$ converges.

ii) If $\lim_{x\to\infty} f(x)/g(x) = \infty$ and if $\int_a^\infty g(x)\,dx$ diverges, then $\int_a^\infty f(x)\,dx$ diverges.

Proof. For (i), given $\epsilon > 0$, choose $c > a$ such that, for $x > c$, we have $f(x) \leq \epsilon g(x)$. Part (i) of the corollary follows by the comparison test. We leave the analogous proof of (ii) for you to complete. ●

EXAMPLE 7 Consider the integral $\int_1^\infty 1/(1+x^3)^{1/2}\,dx$. It is improper of the first kind and the integrand is positive and integrable on $[1,\infty)$. Notice that, as x tends to ∞, $f(x) = 1/(1+x^3)^{1/2}$ behaves essentially like $g(x) = 1/x^{3/2}$. This observation suggests that we consider

$$\lim_{x\to\infty} \frac{f(x)}{g(x)} = \lim_{x\to\infty} \frac{x^{3/2}}{(1+x^3)^{1/2}}$$
$$= \lim_{x\to\infty} \left(\frac{x^3}{1+x^3}\right)^{1/2} = 1.$$

Since $g(x)\,dx = \int_1^\infty 1/x^{3/2}\,dx$ converges, we can apply the limit comparison test to deduce that $\int_1^\infty 1/(1+x^3)^{1/2}\,dx$ converges. ●

EXAMPLE 8 Fix any r in $\mathbb{R}$ and any $s > 0$. Then $\int_0^\infty x^r e^{-sx}\,dx$ converges. As we shall see below, this integral is extremely important in various applications of our theory. To prove convergence, let $g(x) = 1/x^2$. Then $f(x) = x^r e^{-sx}$ and $g(x) = 1/x^2$ satisfy the hypotheses of Corollary 13.2.4 on $[1,\infty)$. Furthermore,

$$\lim_{x\to\infty} \frac{f(x)}{g(x)} = \lim_{x\to\infty} \frac{x^{r+2}}{e^{sx}} = 0.$$

Since $\int_1^\infty 1/x^2\,dx$ converges, so also does $\int_1^\infty x^r e^{-sx}\,dx$. Therefore $\int_0^\infty x^r e^{-sx}\,dx$ converges. ●

EXAMPLE 9 The integral $\int_2^\infty 1/(\ln x)^2\,dx$ is of the first kind; the integrand f is positive and integrable on $[2, \infty)$. By Corollary 13.2.5, this integral diverges. To prove this claim, let $g(x) = 1/x$ on $[2, \infty)$ and use l'Hôpital's rule to compute

$$\lim_{x\to\infty} \frac{f(x)}{g(x)} = \lim_{x\to\infty} \frac{x}{(\ln x)^2} = \infty.$$

Since $\int_2^\infty 1/x\,dx$ diverges, so also does $\int_2^\infty 1/(\ln x)^2\,dx$. More generally, for any $p > 0$, $\int_2^\infty 1/(\ln x)^p\,dx$ diverges. To prove this claim, merely limit compare $f(x) = 1/(\ln x)^p$ against $g(x) = 1/x$ as above and use l'Hôpital's rule sufficiently often. (Treat the cases p in $\mathbb{N}$ and p not in $\mathbb{N}$ separately.) •

13.2.2 Conditional Convergence

Of course, not all integrands are positive and, as with series, we draw a distinction between two possible cases.

DEFINITION 13.2.1 Let $\int_a^\infty f(x)\,dx$ be an improper integral of the first kind.

i) If $\int_a^\infty |f(x)|\,dx$ converges, then $\int_a^\infty f(x)\,dx$ is said to *converge absolutely*.

ii) If $\int_a^\infty f(x)\,dx$ converges but $\int_a^\infty |f(x)|\,dx$ diverges, then $\int_a^\infty f(x)\,dx$ is said to *converge conditionally*. •

As with series so here we have the following theorem.

THEOREM 13.2.6 If $\int_a^\infty |f(x)|\,dx$ converges, then $\int_a^\infty f(x)\,dx$ also converges.

Proof. Observe that $0 \le |f(x)| - f(x) \le 2|f(x)|$ for all x in $[a, \infty)$. The assumption that $\int_a^\infty |f(x)|\,dx$ converges and the comparison test imply the convergence of $\int_a^\infty [|f(x)| - f(x)]\,dx$. But then $\int_a^\infty f(x)\,dx$ must converge. •

The comparison test and the limit comparison test, of course, are tests for absolute convergence. Notice that there are no analogs for the ratio or the root tests; nor is there a counterpart for the alternating series test for conditional convergence. The very form of these series tests precludes the possibility of formulating an integral version. Dirichlet's test for convergence of series, however, does have its counterpart for integrals.

THEOREM 13.2.7 Dirichlet's Test Suppose that $\int_a^\infty f(x)g(x)\,dx$ is an improper integral of the first kind. Suppose also that f and g satisfy the following conditions:

i) The function f is continuous on $[a, \infty)$.

ii) The function $F(x) = \int_a^x f(t)\,dt$ is bounded on $[a, \infty)$.

iii) The function g is differentiable on $[a, \infty)$, $g' \le 0$ and $\lim_{x\to\infty} g(x) = 0$.

Then $\int_a^\infty f(x)g(x)\,dx$ converges.

Proof. By the Fundamental Theorem of Calculus $F' = f$ on $\mathbb{R}$. Also, $F(a) = 0$. Thus for any $b > a$, when we integrate by parts, we obtain

$$\int_a^b f(x)g(x)\,dx = F(b)g(b) - \int_a^b F(x)g'(x)\,dx.$$

To show that $\lim_{b\to\infty} \int_a^b f(x)g(x)\,dx$ exists, we note first that

$$0 \le \lim_{b\to\infty} |F(b)g(b)| \le ||F||_\infty \lim_{b\to\infty} g(b) = 0.$$

By the Squeeze Play, $\lim_{b\to\infty} F(b)g(b) = 0$. Therefore,

$$\lim_{b\to\infty} \int_a^b f(x)g(x)\,dx = -\lim_{b\to\infty} \int_a^b F(x)g'(x)\,dx$$

and it suffices to show that $\int_a^\infty F(x)g'(x)\,dx$ converges. Note that

$$\begin{aligned} \int_a^b |F(x)g'(x)|\,dx &\le ||F||_\infty \int_a^b |g'(x)|\,dx \\ &= -||F||_\infty \int_a^b g'(x)\,dx \\ &= ||F||_\infty [g(a) - g(b)]. \end{aligned}$$

Since $\lim_{b\to\infty} ||F||_\infty [g(a) - g(b)] = ||F||_\infty g(a)$, the integral $\int_a^\infty F(x)g'(x)\,dx$ converges absolutely and therefore converges. We conclude that $\lim_{b\to\infty} \int_a^b f(x)g(x)\,dx = \int_a^\infty f(x)g(x)\,dx$ exists. This completes the proof of Dirichlet's test. ●

EXAMPLE 10 The integral $\int_0^\infty (\sin x)/x\,dx$ converges by Dirichlet's test. [The integrand has a removable discontinuity at $x = 0$ because $\lim_{x\to\infty} (\sin x)/x = 1$. It is to be understood that the integrand is to be defined to have value 1 at 0.] To prove that the integral converges, let $f(x) = \sin x$ and $g(x) = 1/x$. With these choices, the hypotheses of Theorem 13.2.7 are satisfied and the integral converges.

The convergence is conditional because, as the following argument reveals, $\int_0^\infty (\sin x)/x\,dx$ does not converge absolutely. For $k = 0, 1, 2, \dots$ and for x in $[k\pi, (k+1)\pi]$, $|(\sin x)/x| \ge |\sin x|/[(k+1)\pi]$ and, therefore,

$$\int_{k\pi}^{(k+1)\pi} \left|\frac{\sin x}{x}\right| dx \ge \frac{1}{(k+1)\pi} \int_{k\pi}^{(k+1)\pi} |\sin x|\,dx = \frac{2}{(k+1)\pi}.$$

It follows that

$$\int_0^\infty \left|\frac{\sin x}{x}\right| dx = \sum_{k=0}^\infty \int_{k\pi}^{(k+1)\pi} \left|\frac{\sin x}{x}\right| dx > \frac{2}{\pi} \sum_{k=0}^\infty \frac{1}{k+1}.$$

The series diverges and, hence, $\int_0^\infty |(\sin x)/x|\,dx$ also diverges. ●

EXAMPLE 11 The integral $\int_0^\infty \sin(x^2)\,dx$ converges conditionally. To prove this, change variables by letting $x(t) = t^{1/2}$. Then $x'(t) = 1/(2t^{1/2})$ and the integral becomes,

$$\int_0^\infty \sin(x^2)\,dx = \frac{1}{2}\int_0^\infty \frac{\sin t}{t^{1/2}}\,dt.$$

Apply Dirichlet's test with $f(t) = \sin t$ and $g(t) = 1/t^{1/2}$. These functions satisfy the hypotheses of that test, and, therefore the transformed integral, and equivalently the original integral, converges. By an argument similar to that presented in Example 10, the integral $\int_0^\infty |\sin t|/t^{1/2}\,dt$ does not converge. We leave the details for you to confirm. ●

Note in passing that although $\int_0^\infty \sin(x^2)\,dx$ converges, the integrand does not tend to 0 at ∞. In fact, $\lim_{x\to\infty} \sin(x^2)$ does not exist.

We close this section with a discussion of the Cauchy principal value of an integral of the form $\int_{-\infty}^\infty f(x)\,dx$. Since this notion will play no significant role in this text, we will keep our discussion brief. Recall that an improper integral of the first kind of the form $\int_{-\infty}^\infty f(x)\,dx$ converges to L provided

$$\lim_{\substack{a\to-\infty \\ b\to\infty}} \int_a^b f(x)\,dx = L.$$

It is to be understood that $a \to -\infty$ and $b \to \infty$ independently in computing this limit. This seemingly minor point, in fact, does matter.

DEFINITION 13.2.2 Let f be an integrable function on $\mathbb{R}$. The *Cauchy principal value* of $\int_{-\infty}^\infty f(x)\,dx$ is $\lim_{a\to\infty} \int_{-a}^a f(x)\,dx$, provided this limit exists. ●

The following theorem is self-evident; its easy proof is left for you.

THEOREM 13.2.8 If $\int_{-\infty}^\infty f(x)\,dx$ converges to L, then the Cauchy principal value of the integral is L. ●

The converse of Theorem 13.2.8, however, is false as the following example demonstrates.

EXAMPLE 12 The Cauchy principal value of $\int_{-\infty}^\infty x\,dx$ is

$$\lim_{a\to\infty} \int_{-a}^a x\,dx = \lim_{a\to\infty} \left.\frac{x^2}{2}\right|_{-a}^a = 0.$$

However, the integral $\int_{-\infty}^\infty x\,dx$ itself does not converge, since neither $\int_{-\infty}^0 x\,dx$ nor $\int_0^\infty x\,dx$ converges. This example alone serves as a warning to you when working with improper integrals of this form; the two unbounded limits of integration must be treated separately. ●

13.3 IMPROPER INTEGRALS OF THE SECOND KIND

In our treatment of improper integrals of the second kind, we will assume that the integrand f has a pole at the left endpoint $x = a$ and will present our results in these terms. All our statements can be modified to treat the case when the integrand has a pole at $x = b$. Recall that, if f has a pole at a, then $\int_a^b f(x)\,dx = L$ if $\lim_{c\to a^+} \int_c^b f(x)\,dx = L$. In other words, given $\epsilon > 0$, there is a c_0 in (a, b) such that, for all c in (a, c_0),

$$\left| \int_c^b f(x)\,dx - L \right| < \epsilon.$$

Equivalently, $|\int_a^c f(x)\,dx| < \epsilon$ for all c in (a, c_0).

EXAMPLE 13 For $p > 0$, the integral $\int_0^1 1/x^p\,dx$ is improper of the second kind. For any c in (0, 1), we integrate from c to 1 to obtain

$$\int_c^1 \frac{1}{x^p}\,dx = \begin{cases} (1 - c^{1-p})/(1-p), & \text{if } p \neq 1, \\ -\ln c, & \text{if } p = 1. \end{cases}$$

Only when $p < 1$ does $\lim_{c\to 0^+} \int_c^1 1/x^p\,dx = 1/(1-p)$ exist. That is, for $0 < p < 1$, the integral $\int_0^1 1/x^p\,dx$ converges. When $p \geq 1$ the integral $\int_0^1 1/x^p\,dx$ diverges because, for such p, $\lim_{c\to 0^+} \int_c^1 1/x^p\,dx$ fails to exist. ●

EXAMPLE 14 The integral $\int_0^1 x/(1-x^2)^{1/2}\,dx$ is improper of the second kind because the integrand has a pole at $b = 1$ and is bounded on any closed subinterval of [0, 1). We change variables, letting $u(x) = 1 - x^2$. Then $u'(x) = -2x$ and, for any c in (0, 1), we obtain

$$\begin{aligned} \int_0^c \frac{x}{(1-x^2)^{1/2}}\,dx &= -\frac{1}{2}\int_1^{1-c^2} \frac{1}{u^{1/2}}\,du \\ &= 1 - (1-c^2)^{1/2}. \end{aligned}$$

Therefore $\int_0^1 x/(1-x^2)^{1/2}\,dx = \lim_{c\to 1^-}[1 - (1-c^2)^{1/2}] = 1$. ●

EXAMPLE 15 Let f be continuous on [0, 1] and consider the integral $\int_0^1 f(x)/(1-x^2)^{1/2}\,dx$. Although this integral is improper of the second kind, it can always be transformed into a proper integral by using the following change of variables. Fix c in (0, 1) and let $x = \sin\theta$ on $[0, c]$. We obtain

$$\begin{aligned} \int_0^1 \frac{f(x)}{(1-x^2)^{1/2}}\,dx &= \lim_{c\to 1^-} \int_0^c \frac{f(x)}{(1-x^2)^{1/2}}\,dx \\ &= \lim_{c\to 1^-} \int_0^{\sin^{-1} c} f(\sin\theta)\,d\theta = \int_0^{\pi/2} f(\sin\theta)\,d\theta. \end{aligned}$$

Since the composition of f with the sine function is continuous, and hence bounded, on $[0, \pi/2]$, this last integral is proper. ●

In general, an improper integral $\int_a^b f(x)\,dx$ of the second kind, with a pole at $x = a$ can be transformed, by means of a change of variables, into an improper integral of the first kind. Simply set $u(x) = 1/(x - a)$ or, equivalently, set $x(u) = a + 1/u$. As $x \to a^+$, $u \to +\infty$, and $u(b) = 1/(b - a)$. Furthermore, $x'(u) = -1/u^2$ so that, upon changing variables, we obtain

$$\begin{aligned}\int_a^b f(x)\,dx &= \int_\infty^{1/(b-a)} f\left(a + \frac{1}{u}\right)\left(-\frac{1}{u^2}\right)du \\ &= \int_{1/(b-a)}^\infty \frac{f(a + 1/u)}{u^2}du.\end{aligned}$$

Let $a_1' = 1/(b - a) > 0$ and fix any $c > a_1'$. As u ranges over $[a_1', c]$, $x = a + 1/u$ ranges over the interval $[a + 1/c, b]$, an interval where f is assumed to be bounded and integrable. Thus $f(a + 1/u)/u^2$ is bounded and integrable on $[a_1', c]$. Thus $\int_{1/(b-a)}^\infty f(a + 1/u)/u^2\,du$ is an improper integral of the first kind. The original integral converges if and only if the transformed integral converges. This observation assures us that the results of Section 13.2 translate into analogous statements about improper integrals of the second kind.

EXAMPLE 16 For $p > 0$, the integral $\int_0^1 (-\ln x)/x^p\,dx$ is improper of the second kind, the integrand having a pole at $a = 0$. By means of the change of variables $u(x) = 1/x$, we obtain

$$\int_0^1 \frac{(-\ln x)}{x^p}dx = \int_1^\infty u^{p-2}\ln u\,du.$$

The latter integral is improper of the first kind. Either integral can be integrated by parts and shown to converge for $0 < p < 1$ to $1/(1 - p)^2$. We leave the details for you. ●

In light of the preceding discussion, we list without proof the counterparts for integrals of the second kind of the theorems we obtained in Section 13.2.

THEOREM 13.3.1 Suppose that the integrand f is nonnegative and integrable on $(a, b]$ and has a pole at $x = a$. The improper integral $\int_a^b f(x)\,dx$ converges if and only if the set $S = \{\int_c^b f(x)\,dx : c \text{ in } (a, b)\}$ is bounded above. In this case, $\int_a^b f(x)\,dx = \sup S$. ●

THEOREM 13.3.2 **Comparison Test; Integrals of Second Kind** Let f and g be nonnegative, integrable functions on $(a, b]$, each having a pole at $x = a$ such that, for some c_0, we have $f(x) \le g(x)$ for all x in $(a, c_0]$.

i) If $\int_a^b g(x)\,dx$ converges, then $\int_a^b f(x)\,dx$ also converges.

ii) If $\int_a^b f(x)\,dx$ diverges, then $\int_a^b g(x)\,dx$ also diverges. ●

We also have a limit comparison test for integrals of the second kind.

THEOREM 13.3.3 Limit Comparison Test; Positive Integrands Let f and g be positive, integrable functions on $(a, b]$, each having a pole at $x = a$. If $\lim_{x\to a^+} f(x)/g(x) = L$ exists and is positive, then $\int_a^b f(x)\,dx$ converges if and only if $\int_a^b g(x)\,dx$ converges. ●

COROLLARY 13.3.4 Let f and g be positive integrable functions on $(a, b]$, each having a pole at $x = a$.

i) If $\lim_{x\to a^+} f(x)/g(x) = 0$ and if $\int_a^b g(x)\,dx$ converges, then $\int_a^b f(x)\,dx$ converges.

ii) If $\lim_{x\to a^+} f(x)/g(x) = \infty$ and if $\int_a^b g(x)\,dx$ diverges, then $\int_a^b f(x)\,dx$ diverges. ●

EXAMPLE 17 To determine whether $\int_0^1 1/(1-x^3)^{1/2}\,dx$ converges, limit compare with $\int_0^1 1/(1-x)^{1/2}$ dx. Since

$$\lim_{x\to 1^-} \frac{1/(1-x^3)^{1/2}}{1/(1-x)^{1/2}} = \lim_{x\to 1^-} \frac{1}{(x^2+x+1)^{1/2}} = \frac{1}{\sqrt{3}},$$

and since $\int_0^1 1/(1-x)^{1/2}\,dx$ converges, we deduce by the limit comparison test that the integral $\int_0^1 1/(1-x^3)^{1/2}\,dx$ also converges. ●

For improper integrals of the second kind there is, of course, the distinction between absolute and conditional convergence.

DEFINITION 13.3.1 Let $\int_a^b f(x)\,dx$ be an improper integral of the second kind.

i) If $\int_a^b |f(x)|\,dx$ converges, then $\int_a^b f(x)\,dx$ is said to *converge absolutely*.

ii) If $\int_a^b f(x)\,dx$ converges but $\int_a^b |f(x)|\,dx$ diverges, then $\int_a^b f(x)\,dx$ is said to *converge conditionally*. ●

The following theorem is, of course, immediate. Its proof is left for you.

THEOREM 13.3.5 If $\int_a^b |f(x)|\,dx$ converges, then $\int_a^b f(x)\,dx$ also converges. ●

EXAMPLE 18 The integral $\int_0^1 x^{-1} \sin(1/x)\,dx$ converges conditionally. To prove this, use the transformation $x = 1/t$ and appeal to Example 10. ●

13.4 Uniform Convergence of Improper Integrals

To begin, we formulate the problem addressed in this section in terms of improper integrals of the first kind. Improper integrals of the second kind are treated analogously; as we proceed we will present the parallel results.

Suppose, then, that S is a subset of $\mathbb{R}$ and that f is a bounded, real-valued function on the unbounded strip

$$[a, \infty) \times S = \{(x_1, x_2) : a \le x_1 < \infty,\ x_2 \text{ in } S\}.$$

Suppose also that, for each x_2 in S, the function $f(t, x_2)$ is integrable on $[a, \infty)$. Define

$$F(x_2) = \int_a^\infty f(t, x_2)\, dt,$$

for x_2 in S. (See Fig. 13.2.)

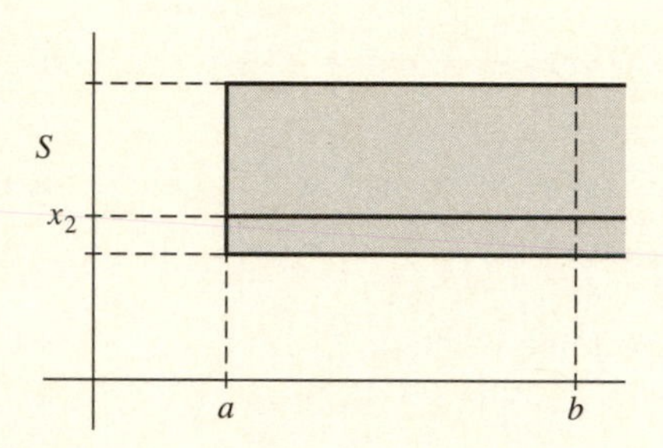

Figure 13.2

Naturally, we want to identify properties of the function F. In particular, we want the counterparts of Theorems 6.3.10, 6.3.11, and 6.3.12. We ask, then, for sufficient conditions that F be continuous on S, that F be differentiable at points of S, and that F be integrable. Notice that each of these issues involves several limits simultaneously and the interchange of two of them. Sometimes this interchange is legitimate, sometimes not; we want sufficient conditions that will justify such interchanges. In all cases when dealing with improper integrals, you must carefully justify such maneuvers. The mere wave of a hand at these subtle points is not only shoddy mathematics; it may lead to false conclusions. As we found for infinite series in Chapter 12, so here the conditions we will develop involve the concept of uniform convergence. We emphasize that our theorems below provide merely sufficient, rather than necessary, conditions.

DEFINITION 13.4.1 Let S be a nonempty subset of $\mathbb{R}$ and let f be a real-valued function defined on $[a, \infty) \times S$ in $\mathbb{R}^2$. Suppose that, for each x_2 in S, $f(t, x_2)$ is integrable on $[a, \infty)$ and that $F(x_2) = \int_a^\infty f(t, x_2)\, dt$ is a convergent improper integral of the first kind. We say that $\int_a^\infty f(t, x_2)\, dt$ *converges uniformly* to F on S if, for every $\epsilon > 0$, there exists a $b_0 > a$ such that, for all $b \geq b_0$ and all x_2 in S,

$$\left| F(x_2) - \int_a^b f(t, x_2)\, dt \right| < \epsilon.$$

We write $\int_a^\infty f(t, x_2)\, dt = F(x_2)$ [uniformly] on S. ●

Equivalently, for $b \geq b_0$, $|\int_b^\infty f(t, x_2)\, dt| < \epsilon$, for all x_2 in S simultaneously. The number b_0 depends on ϵ but is independent of x_2; it is this independence that is crucial to the uniformity of the convergence.

There is, of course, the analogous definition for the uniform convergence of integrals of the second kind. We state the definition for such integrals when the integrand has a pole at the left endpoint a, with the understanding that a completely analogous definition applies when the integrand has a pole at the right endpoint b. (See Fig. 13.3.)

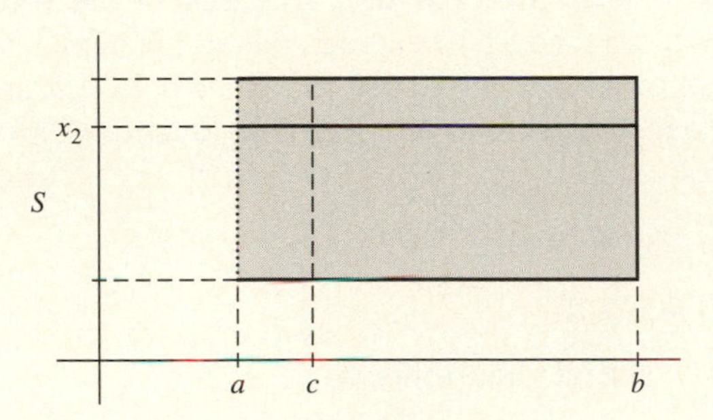

Figure 13.3

DEFINITION 13.4.2 Let S be a nonempty subset of $\mathbb{R}$ and let f be a real-valued function defined on $(a, b] \times S$ in $\mathbb{R}^2$. Suppose that, for each x_2 in S, $f(t, x_2)$ has a pole at $t = a$ and is integrable on $(a, b]$, and that $F(x_2) = \int_a^b f(t, x_2)\,dt$ is a convergent, improper integral of the second kind. We say that $\int_a^b f(t, x_2)\,dt$ *converges uniformly* to F on S if, for each $\epsilon > 0$, there exists a c_0 in (a, b) such that, for all c in (a, c_0) and for all x_2 in S,

$$\left| F(x_2) - \int_c^b f(t, x_2)\,dt \right| < \epsilon.$$

We write $\int_a^b f(t, x_2)\,dt = F(x_2)$ [uniformly] on S. ●

Equivalently, for all c in $(0, c_0)$, $|\int_a^c f(t, x_2)\,dt| < \epsilon$, for all x_2 in S simultaneously.

To implement these definitions, we need straightforward tests to determine uniform convergence. By far the most useful test is the counterpart in the present context of Weierstrass's M-test.

THEOREM 13.4.1 Weierstrass's *M*-test; Integrals of the First Kind Let S be a nonempty subset of $\mathbb{R}$ and let f be a real-valued function defined on $[a, \infty) \times S$ in $\mathbb{R}^2$. Suppose that, for each x_2 in S, $f(t, x_2)$ is integrable on $[a, \infty)$. Suppose also that there exists an integrable function $M(t)$ on $[a, \infty)$ with the following properties:

i) For all (t, x_2) in $[a, \infty) \times S$, $|f(t, x_2)| \le M(t)$.

ii) $\int_a^\infty M(t)\,dt$ converges.

Then $\int_a f(t, x_2)\,dt$ converges uniformly on S.

Proof. Fix $\epsilon > 0$. Since $\int_a^\infty M(t)\,dt$ converges, we can choose a $b_0 > a$ such that, for $b \ge b_0$, $\int_b^\infty M(t)\,dt < \epsilon$. Fix such a b_0. Then, for all $b \ge b_0$,

$$\left| \int_b^\infty f(t, x_2)\,dt \right| \le \int_b^\infty |f(t, x_2)|\,dt \le \int_b^\infty M(t)\,dt < \epsilon,$$

for all x_2 in S simultaneously. Therefore, $\int_a^\infty f(t, x_2)\,dt$ converges uniformly on S. ●

THEOREM 13.4.2 Weierstrass's M-test; Integrals of the Second Kind Let S be a nonempty subset of $\mathbb{R}$ and let f be a real-valued function defined on $(a, b] \times S$ in $\mathbb{R}^2$. Suppose that, for each x_2 in S, $f(t, x_2)$ has a pole at a and is integrable on $(a, b]$. Suppose also that there exists an integrable function $M(t)$ on $(a, b]$ with the following properties:

i) For each (t, x_2) in $(a, b] \times S$, $|f(t, x_2)| \leq M(t)$.

ii) $\int_a^b M(t)\,dt$ converges.

Then $\int_a^b f(t, x_2)\,dt$ converges uniformly on S.

Proof. We leave it for you to provide the appropriate modifications of the proof of Theorem 13.4.1. ●

EXAMPLE 19 The integral $\int_1^\infty t^{x-1}e^{-t}\,dt$ converges uniformly for x in any interval $[c, d] \subset (0, \infty)$. To prove this claim, let $M(t) = t^{d-1}e^{-t}$, for $t \geq 1$. Condition (i) of Theorem 13.4.1 holds for all x in $[c, d]$ and all $t \geq 1$. Also, $\int_1^\infty M(t)\,dt$ converges by Example 8 in Section 13.2. We conclude, by Theorem 13.4.1, that $\int_1^\infty t^{x-1}e^{-t}\,dt$ converges uniformly for x in $[c, d]$. ●

EXAMPLE 20 The integral $\int_0^1 t^{x-1}e^{-t}\,dt$ also converges uniformly for x in any interval $[c, d] \subset (0, \infty)$. In fact, if $1 \leq c$ and if x is restricted to $[c, d]$, then the integrand has no pole at $a = 0$ and the integral is an ordinary integral; there is no problem to be resolved. However, if $0 < c \leq x < 1$, then $\int_0^1 t^{x-1}e^{-t}\,dt$ is an improper integral of the second kind. In this instance let $M(t) = t^{c-1}e^{-t}$. It is straightforward to show that, for all (t, x) in $(0, 1] \times [c, 1]$,

$$M(t) = t^{c-1}e^{-t} \leq t^{c-1}.$$

Also, since $1 - c < 1$, we know by Example 13 in Section 13.3 that $\int_0^1 t^{-(1-c)}\,dt$ converges. It follows, by the comparison test, that $\int_0^1 M(t)\,dt = \int_0^1 t^{c-1}e^{-t}\,dt$ also converges. Therefore, by Theorem 13.4.2, the integral $\int_0^1 t^{x-1}e^{-t}\,dt$ converges uniformly for x in $[c, d]$. ●

The uniform convergence of an improper integral with a continuous integrand ensures the continuity of the function so defined. The details are provided by the next two theorems.

THEOREM 13.4.3 Let S be a nonempty subset of $\mathbb{R}$ and let f be a continuous, real-valued function on $[a, \infty) \times S$ in $\mathbb{R}^2$. Assume that $\int_a^\infty f(t, x_2)\,dt = F(x_2)$ [uniformly] on S. Then F is continuous on S.

Proof. Fix any c_2 in S and any $\epsilon > 0$. We show that F is continuous at c_2. By the uniform convergence of the integral, we can choose $b_0 > a$ such that, for all $b \geq b_0$, $|\int_b^\infty f(t, x_2)\,dt| < \epsilon$ for all x_2 in S. We know by Theorem 6.3.10 that the function

$$G(x_2) = \int_a^b f(t, x_2)\,dt$$

is continuous on S; in particular, G is continuous at c_2. (See Fig. 13.4.) Thus we can choose a $\delta > 0$ such that, if x_2 is in $N(c_2; \delta) \cap S$, then $G(x_2)$ is in $N(G(c_2); \epsilon)$. Therefore

$$\begin{aligned} |F(x_2) - F(c_2)| &= \left| G(x_2) - G(c_2) + \int_b^\infty f(t, x_2)\,dt - \int_b^\infty f(t, c_2)\,dt \right| \\ &\leq |G(x_2) - G(c_2)| + \left| \int_b^\infty f(t, x_2)\,dt \right| + \left| \int_b^\infty f(t, c_2)\,dt \right| < 3\epsilon, \end{aligned}$$

for all x_2 in $N(c_2; \delta) \cap S$. This proves that F is continuous at c_2. Since c_2 is arbitrary in S, it follows that F is continuous on all of S. ●

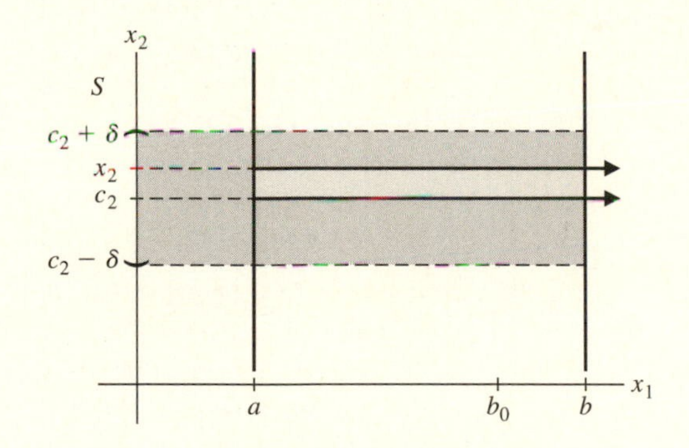

Figure 13.4

The proof of the following analog for improper integrals of the second kind imitates that of Theorem 13.4.3 with few modifications; we leave it for you to complete.

THEOREM 13.4.4 Let S be a nonempty subset of $\mathbb{R}$ and let f be a continuous, real-valued function on $(a, b] \times S$ in $\mathbb{R}^2$. Suppose that for each x_2 in S, $f(t, x_2)$ has a pole at $t = a$ and is integrable on $(a, b]$. Suppose also that the improper integral $\int_a^b f(t, x_2)\,dt = F(x_2)$ [uniformly] on S. Then F is continuous on S. ●

EXAMPLE 21 The integral $\int_0^1 t^{x-1}e^{-t}\,dt$ converges uniformly to some function $F_1(x)$ for x in any $[c, d] \subset (0, \infty)$. Likewise, $\int_1^\infty t^{x-1}e^{-t}\,dt$ converges uniformly to some function $F_2(x)$ on $[c, d]$. Since the integrand $t^{x-1}e^{-t}$ is continuous on the entire strip $(0, \infty) \times [c, d]$, we conclude by Theorems 13.4.3 and 13.4.4 that each of F_1 and F_2 is continuous on any interval $[c, d] \subset (0, \infty)$. Thus,

$$F_1(x) + F_2(x) = \int_0^\infty t^{x-1} e^{-t}\, dt$$

is continuous at every x in $(0, \infty)$. ●

EXAMPLE 22 For any $r > 0$, the integrals $\int_0^\infty e^{-rt} \cos xt\, dt$ and $\int_0^\infty e^{-rt} \sin xt\, dt$ each converge uniformly for x in $\mathbb{R}$. In each case, the integrand is bounded by $M(t) = e^{-rt}$ and the integral $\int_0^\infty e^{-rt}\, dt$ converges; an application of the Weierstrass M-test proves our claim. The functions to which these integrals converge can be found by integrating by parts (twice). Upon completing the details, we obtain

$$\int_0^\infty e^{-rt} \cos xt\, dt = \lim_{b\to\infty} \left. \frac{e^{-rt}(x \sin xt - r \cos xt)}{r^2 + x^2} \right|_0^b = \frac{r}{r^2 + x^2}.$$

Similarly,

$$\int_0^\infty e^{-rt} \sin xt\, dt = \frac{x}{r^2 + x^2}.$$

Each of these functions is continuous on all of $\mathbb{R}$. ●

Under appropriate hypotheses, functions defined by improper integrals are differentiable with the derivative also given by an improper integral. Our next two theorems provide the specifics.

THEOREM 13.4.5 Let f be a real-valued function defined on $[a, \infty) \times [c, d]$ in $\mathbb{R}^2$ such that, for each x_2 in $[c, d]$, the integral $\int_a^\infty f(t, x_2)\, dt$ converges pointwise to a function F on $[c, d]$. Suppose that the partial derivative $D_2 f$ exists and is continuous on $[a, \infty) \times [c, d]$. Suppose further that the improper integral $\int_a^\infty D_2 f(t, x_2)\, dt$ converges uniformly on $[c, d]$. Then F is differentiable on $[c, d]$ and

$$F'(x_2) = \int_a^\infty D_2 f(t, x_2)\, dt.$$

Proof. Fix $\epsilon > 0$. By the uniform convergence of $\int_a^\infty D_2 f(t, x_2)\, dt$, choose $b_0 > a$ such that, for all $b \geq b_0$ and all x_2 in $[c, d]$,

$$\left| \int_b^\infty D_2 f(t, x_2)\, dt \right| < \epsilon.$$

Fix $b \geq b_0$. (See Fig. 13.5.) For x_2 in $[c, d]$, define

$$G(x_2) = \int_a^b f(t, x_2)\, dt.$$

Fix y_2 in $[c, d]$. We want to show that F is differentiable at y_2 and that $F'(y_2) = \int_a^\infty D_2 f(t, y_2)\, dt$.

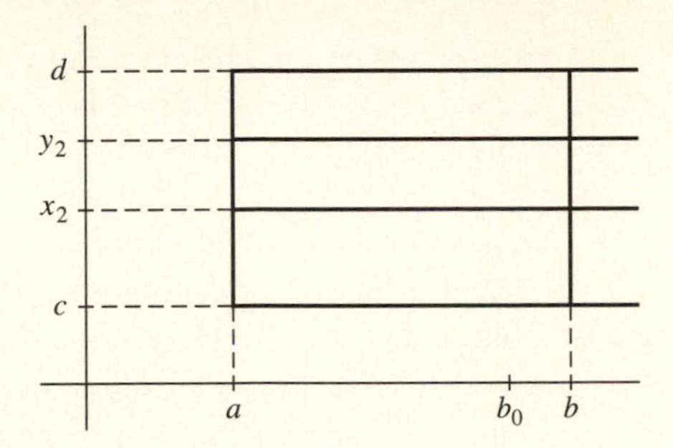

Figure 13.5

By Theorem 6.3.11, the function G is differentiable at the point y_2 and $G'(y_2) = \int_a^b D_2 f(t, y_2)\,dt$. Therefore, given $\epsilon > 0$, there exists a $\delta > 0$ such that, for x_2 in $N'(y_2; \delta) \cap [c, d]$, we have

$$\left| \frac{G(x_2) - G(y_2)}{x_2 - y_2} - \int_a^b D_2 f(t, y_2)\,dt \right| < \epsilon. \tag{13.1}$$

For x_2 in $N'(y_2; \delta) \cap [c, d]$, compute

$$\begin{aligned} \left| \frac{F(x_2) - F(y_2)}{x_2 - y_2} - \int_a^\infty D_2 f(t, y_2)\,dt \right| &\leq \left| \frac{G(x_2) - G(y_2)}{x_2 - y_2} - \int_a^b D_2 f(t, y_2)\,dt \right| \\ &\quad + \left| \frac{1}{x_2 - y_2} \int_b^\infty [f(t, x_2) - f(t, y_2)]\,dt \right| \\ &\quad + \left| \int_b^\infty D_2 f(t, y_2)\,dt \right|. \end{aligned} \tag{13.2}$$

The first summand is less than ϵ by (13.1). The third summand is also less than ϵ by the uniform convergence of the integral of $D_2 f$. To treat the middle summand, apply the Mean Value Theorem to the function $f(t, \cdot)$ on the interval with endpoints x_2 and y_2: choose a z_2 in that interval such that $[f(t, x_2) - f(t, y_2)]/(x_2 - y_2) = D_2 f(t, z_2)$. Thus the middle summand in (13.2) becomes

$$\left| \frac{1}{x_2 - y_2} \int_b^\infty [f(t, x_2) - f(t, y_2)]\,dt \right| = \left| \int_b^\infty D_2 f(t, z_2)\,dt \right| < \epsilon$$

by the uniform convergence of the integral of $D_2 f$. We conclude that, for x_2 in $N'(y_2; \delta) \cap [c, d]$,

$$\left| \frac{F(x_2) - F(y_2)}{x_2 - y_2} - \int_a^\infty D_2 f(t, y_2)\,dt \right| < 3\epsilon.$$

This proves that $F'(y_2)$ exists and equals $\int_a^\infty D_2 f(t, y_2)\,dt$ as promised by the theorem. ●

The modifications that give the corresponding theorem for improper integrals of the second kind follow the usual pattern; the proof of this next theorem is a simple variant of the proof of Theorem 13.4.5 and is left for you.

THEOREM 13.4.6 Let f be a real-valued function on $(a, b] \times [c, d]$ in $\mathbb{R}^2$ such that, for each x_2 in $[c, d]$, the integral $\int_a^b f(t, x_2)\, dt$ is improper of the second kind and converges pointwise to a function F on $[c, d]$. Assume that $D_2 f$ exists and is continuous on $(a, b] \times [c, d]$ and that $\int_a^b D_2 f(t, x_2)\, dt$ is an improper integral of the second kind which converges uniformly on $[c, d]$ to $F(x_2)$. Then F is differentiable at each x_2 in $[c, d]$ and

$$F'(x_2) = \int_a^b D_2 f(t, x_2)\, dt. \quad \bullet$$

EXAMPLE 23 Fix $r > 0$. For x in $\mathbb{R}$, define

$$F(x) = \int_0^\infty \frac{e^{-rt} \sin xt}{t}\, dt.$$

First note that this is an improper integral of the first kind. There is no pole at the endpoint $a = 0$ because $\lim_{t \to 0^+} (e^{-rt} \sin xt)/t = x$. Next, let $[c, d]$ be any interval in $\mathbb{R}$. Let $K = \max\{|c|, |d|\}$ and let $M(t) = Ke^{-rt}$. It follows that

$$\left| \frac{e^{-rt} \sin xt}{t} \right| = |x| e^{-rt} \left| \frac{\sin xt}{xt} \right| \le Ke^{-rt} = M(t)$$

for all $t \ge 0$. Also, $\int_0^\infty M(t)\, dt$ converges. Therefore, by the Weierstrass M-test, the integral $\int_0^\infty (e^{-rt} \sin xt)/t\, dt$ converges uniformly on $[c, d]$ to a function $F(x)$. Since the integrand is continuous on the closed right half-plane $[0, \infty) \times \mathbb{R}$, we know by Theorem 13.4.3 that F is continuous on $[c, d]$. But $[c, d]$ is an arbitrary interval in $\mathbb{R}$. Thus F is continuous on all of $\mathbb{R}$. Note that $F(0) = 0$.

In order to find the function F explicitly, we compute the derivative F', identify it explicitly, and then invoke the Fundamental Theorem of Calculus. To this end, take the partial derivative of $(e^{-rt} \sin xt)/t$ with respect to the variable x. Upon doing so, we obtain

$$\frac{\partial[(e^{-rt} \sin xt)/t]}{\partial x} = e^{-rt} \cos xt.$$

As we saw in Example 22, the integral $\int_0^\infty e^{-rt} \cos xt\, dt$ converges uniformly on $\mathbb{R}$ to $r/(r^2 + x^2)$. We deduce by Theorem 13.4.5 that

$$F'(x) = \int_0^\infty e^{-rt} \cos xt\, dt = \frac{r}{r^2 + x^2},$$

for x in $\mathbb{R}$. Consequently, for any x in $\mathbb{R}$,

$$F(x) = F(x) - F(0) = \int_0^x F'(u)\, du$$

$$= \int_0^x \frac{r}{r^2 + u^2}\, du = \tan^{-1}\left(\frac{x}{r}\right).$$

We have proved that, for $r > 0$ and x in $\mathbb{R}$,

$$\int_0^\infty \frac{e^{-rt} \sin xt}{t}\, dt = \tan^{-1}\left(\frac{x}{r}\right). \quad \bullet$$

EXAMPLE 24 For $x > 0$, define $F(x) = \int_0^1 t^{x-1}e^{-t}\, dt$. Compute the partial derivative of the integrand $t^{x-1}e^{-t}$ with respect to the variable x to obtain

$$\frac{\partial[t^{x-1}e^{-t}]}{\partial x} = t^{x-1}e^{-t} \ln t.$$

We want to show that $\int_0^1 t^{x-1}e^{-t} \ln t\, dt$ converges uniformly for x in any interval $[c, d] \subset (0, \infty)$. Once we have done so, we can conclude by Theorem 13.4.6 that $F'(x) = \int_0^1 t^{x-1}e^{-t} \ln t\, dt$.

First we show that the integral converges. If $x > 1$, the integral $\int_0^1 t^{x-1}e^{-t} \ln t\, dt$ is actually a proper integral since, by l'Hôpital's rule, $\lim_{t\to 0^+} t^r \ln t = 0$ whenever $r > 0$. In this case, the integrand does not have a pole at $a = 0$.

If $0 < x \leq 1$, choose k in $\mathbb{N}$ such that $0 < 1/k < x$ and observe that, for $0 < t \leq 1$, $|\ln t| \leq k/(et^{1/k})$.[1] Thus when $0 < t \leq 1$, we have $0 < e^{-t} < 1$ and

$$|t^{x-1}e^{-t} \ln t| \leq \frac{t^{x-1}k}{et^{1/k}} = \frac{kt^{x-1-1/k}}{e}.$$

Finally, the integral

$$\int_0^1 t^{x-1-1/k}\, dt = \lim_{a\to 0^+} \frac{t^{x-1/k}}{x - 1/k}\bigg|_a^1 = \frac{1}{x - 1/k} - \lim_{a\to 0^+} \frac{a^{x-1/k}}{x - 1/k}$$

is finite, since $x - 1/k > 0$. By the comparison test, the integral $\int_0^1 t^{x-1}e^{-t} \ln t\, dt$ converges.

To show that the integral converges uniformly for x in $[c, d]$, we choose a k in $\mathbb{N}$ such that $0 < 1/k < c$. As in the preceding argument, we note that, for t in $(0, 1]$,

$$|t^{x-1}e^{-t} \ln t| \leq t^{x-1}|\ln t| \leq \frac{t^{c-1}k}{et^{1/k}}$$

$$= \frac{kt^{c-1-1/k}}{e} = M(t).$$

Since $c - 1 - 1/k > -1$, the integral $\int_0^1 M(t)\, dt$ converges. Thus the integral $\int_0^1 t^{x-1}e^{-t} \ln t\, dt$ converges uniformly on $[c, d]$ by Theorem 13.4.2. Applying Theorem 13.4.6 we conclude that

$$F'(x) = \int_0^1 t^{x-1}e^{-t} \ln t\, dt. \quad \bullet$$

[1] You can prove this by showing that the nonnegative function $g(t) = -t^{1/k} \ln t$ on $(0, 1]$ has a maximum value of k/e and that this value is achieved when $t = e^{-k}$.

Finally, we address the question of the integrability of a function F defined by an improper integral. Again, the uniform convergence of that integral provides a sufficient condition whenever the integrand is continuous.

THEOREM 13.4.7 Let f be a continuous, real-valued function on $[a, \infty) \times [c, d]$ in $\mathbb{R}^2$. Assume that $\int_a^\infty f(x_1, x_2)\,dx_1$ is an improper integral of the first kind that converges uniformly for x_2 in $[c, d]$ to a function $F(x_2)$. Then F is integrable on $[c, d]$ and

$$\int_c^d F(x_2)\,dx_2 = \int_a^\infty \left[\int_c^d f(x_1, x_2)\,dx_2\right] dx_1.$$

Remark. This theorem tells us that we can interchange the order of integration in the iterated integrals

$$\int_c^d \left[\int_a^\infty f(x_1, x_2)\,dx_1\right] dx_2 = \int_a^\infty \left[\int_c^d f(x_1, x_2)\,dx_2\right] dx_1.$$

Proof. Since f is continuous on $[a, \infty) \times [c, d]$ and since $F(x_2) = \int_a^\infty f(x_1, x_2)\,dx_1$ [uniformly], we know by Theorem 13.4.3 that F is continuous on $[c, d]$. Therefore F is integrable on $[c, d]$. Since the integral defining F converges uniformly, given $\epsilon > 0$, we can choose a $b_0 > a$ such that, whenever $b \geq b_0$,

$$-\epsilon < F(x_2) - \int_a^b f(x_1, x_2)\,dx_1 < \epsilon, \tag{13.3}$$

for all x_2 in $[c, d]$ simultaneously. By Theorem 6.3.12, the function $G(x_2) = \int_a^b f(x_1, x_2)\,dx_1$ is also integrable on $[c, d]$ so that, integrating (13.3) from c to d gives

$$-\epsilon(d - c) < \int_c^d F(x_2)\,dx_2 - \int_c^d G(x_2)\,dx_2 < \epsilon(d - c), \tag{13.4}$$

whenever $b \geq b_0$. Again by Theorem 6.3.12,

$$\int_c^d G(x_2)\,dx_2 = \int_c^d \left[\int_a^b f(x_1, x_2)\,dx_1\right] dx_2 = \int_a^b \left[\int_c^d f(x_1, x_2)\,dx_2\right] dx_1.$$

Substituting in (13.4) yields

$$\left|\int_c^d F(x_2)\,dx_2 - \int_a^b \left[\int_c^d f(x_1, x_2)\,dx_2\right] dx_1\right| \leq (d - c)\epsilon$$

whenever $b \geq b_0$. But this says that

$$\int_c^d F(x_2)\,dx_2 = \lim_{b\to\infty} \int_a^b \left[\int_c^d f(x_1, x_2)\,dx_2\right] dx_1 = \int_a^\infty \left[\int_c^d f(x_1, x_2)\,dx_2\right] dx_1$$

as claimed in the theorem. •

We have the analogous version of this theorem for integrals of the second kind. Its proof is left for you to complete.

THEOREM 13.4.8 Let f be a continuous, real-valued function on $(a, b] \times [c, d]$. If the improper integral of the second kind $\int_a^b f(x_1, x_2)\, dx_1$ converges uniformly to $F(x_2)$ on $[c, d]$, then F is integrable on $[c, d]$ and

$$\int_c^d F(x_2)\, dx_2 = \int_a^b \left[\int_c^d f(x_1, x_2)\, dx_2\right] dx_1. \quad \bullet$$

EXAMPLE 25 Note that, for $x > 0$, $F(x) = \int_0^\infty e^{-xt}\, dt = 1/x$. If we restrict x to any interval $[c, d] \subset (0, \infty)$, then the integral converges uniformly to $1/x$. To see this, note that, for (t, x) in $[0, \infty) \times [c, d]$, we have $0 < e^{-xt} \leq e^{-ct} = M(t)$. Also, $\int_0^\infty e^{-ct}\, dt$ converges. By Weierstrass's M-test, the original integral $\int_0^\infty e^{-xt}\, dt$ converges uniformly to $F(x) = 1/x$ on $[c, d]$. Therefore, by Theorem 13.4.7,

$$\int_c^d F(x)\, dx = \int_0^\infty \left[\int_c^d e^{-xt}\, dx\right] dt.$$

Now,

$$\int_c^d F(x)\, dx = \int_c^d \frac{1}{x}\, dx = \ln\left(\frac{d}{c}\right),$$

$$\int_c^d e^{-xt}\, dx = \frac{e^{-dt} - e^{-ct}}{t}.$$

Finally, we conclude that

$$\int_0^\infty \frac{e^{-dt} - e^{-ct}}{t}\, dt = \ln\left(\frac{d}{c}\right). \quad \bullet$$

The interchange of two iterated improper integrals is considerably more delicate than the interchanges justified by Theorems 13.4.7 and 13.4.8. Accordingly, we require more stringent hypotheses. The mathematically natural setting for integration theory is that provided by Lebesgue. It includes the machinery to establish a general principle that justifies interchanging iterated integrals. By restricting ourselves to Riemann integrals, we lack that machinery. Consequently our proofs are necessarily more complicated than those possible with Lebesgue's elegant theory. Nevertheless, we want the following theorem.

THEOREM 13.4.9 Let f be continuous on $S = [a, \infty) \times [c, \infty)$. For (x_1, t) in S, define $g(x_1, t) = \int_c^t f(x_1, x_2)\, dx_2$. Assume that

i) $F(x_2) = \int_a^\infty f(x_1, x_2)\, dx_1$ [uniformly] for x_2 in $[c, \infty)$.

ii) $G(t) = \int_a^\infty g(x_1, t)\, dx_1$ [uniformly] for t in $[c, \infty)$.

iii) $H(x_1) = \int_c^\infty f(x_1, x_2)\, dx_2$ [uniformly] for x_1 in $[a, \infty)$.

iv) $\int_a^\infty H(x_1)\, dx_1 = \int_a^\infty \left[\int_c^\infty f(x_1, x_2)\, dx_2\right] dx_1$ converges.

Then

$$\int_a^\infty \left[\int_c^\infty f(x_1, x_2)\, dx_2\right] dx_1 = \int_c^\infty \left[\int_a^\infty f(x_1, x_2)\, dx_1\right] dx_2.$$

Proof. Fix $\epsilon > 0$. By (ii) we can choose $b_1 > a$ such that, for all $b \geq b_1$,

$$\left| G(t) - \int_a^b g(x_1, t)\, dx_1 \right| < \epsilon. \tag{13.5}$$

for all t in $[c, \infty)$. By (iv) we can choose a $b_2 > a$ such that, for all $b \geq b_2$,

$$\left| \int_a^b H(x_1)\, dx_1 - \int_a^\infty H(x_1)\, dx_1 \right| < \epsilon. \tag{13.6}$$

Let $b_0 = \max\{b_1, b_2\}$. For $b \geq b_0$, both (13.5) and (13.6) hold. Fix $b \geq b_0$. Without loss of generality, b is positive.

Next use (iii) and the definition of $g(x_1, t)$ to choose $t_0 > c$ such that, for $t \geq t_0$,

$$\frac{-\epsilon}{b-a} < g(x_1, t) - H(x_1) < \frac{\epsilon}{b-a}$$

for all x_1 in $[a, \infty)$. Integrating from a to b yields

$$\left| \int_a^b g(x_1, t)\, dx_1 - \int_a^b H(x_1)\, dx_1 \right| \leq \epsilon. \tag{13.7}$$

Combining (13.5), (13.6), and (13.7) gives us, for all $t \geq t_0$,

$$\begin{aligned} &\left| G(t) - \int_a^\infty H(x_1)\, dx_1 \right| \\ &\quad \leq \left| G(t) - \int_a^b g(x_1, t)\, dx_1 \right| + \left| \int_a^b g(x_1, t)\, dx_1 - \int_a^b H(x_1)\, dx_1 \right| \\ &\quad + \left| \int_a^b H(x_1)\, dx_1 - \int_a^\infty H(x_1)\, dx_1 \right| < 3\epsilon. \end{aligned}$$

Therefore,

$$\begin{aligned} \lim_{t\to\infty} G(t) &= \lim_{t\to\infty} \int_a^\infty g(x_1, t)\, dx_1 = \int_a^\infty H(x_1)\, dx_1 \\ &= \int_a^\infty \left[\int_c^\infty f(x_1, x_2)\, dx_2 \right] dx_1. \end{aligned} \tag{13.8}$$

But

$$\begin{aligned} \lim_{t\to\infty} \int_a^\infty g(x_1, t)\, dx_1 &= \lim_{t\to\infty} \int_a^\infty \left[\int_c^t f(x_1, x_2)\, dx_2 \right] dx_1 \\ &= \lim_{t\to\infty} \int_c^t \left[\int_a^\infty f(x_1, x_2)\, dx_1 \right] dx_2 \end{aligned}$$

by (i) and Theorem 13.4.7. Combine this result with (13.8) to conclude that

$$\int_a^\infty \left[\int_c^\infty f(x_1, x_2)\, dx_2 \right] dx_1 = \lim_{t\to\infty} \int_c^t \left[\int_a^\infty f(x_1, x_2)\, dx_1 \right] dx_2$$

$$= \int_c^\infty \left[\int_a^\infty f(x_1, x_2)\, dx_1 \right] dx_2.$$

This proves the theorem. ●

13.5 FUNCTIONS DEFINED BY IMPROPER INTEGRALS

To exhibit the usefulness of this theory, we examine several important, classic examples. First we look at integrals of the form $\int_0^\infty [g(t) \sin xt]/t\, dt$, where g is suitably restricted. We want $g(0^+)$ to exist, thus ensuring that the integrand has no pole at $a = 0$. And, of course, we want g to be integrable on $[0, \infty)$. We have already seen one example; in Example 23 we saw that, for $r > 0$ and x in $\mathbb{R}$,

$$\int_0^\infty \frac{e^{-rt} \sin xt}{t}\, dt = \tan^{-1}\left(\frac{x}{r}\right). \tag{13.9}$$

EXAMPLE 26 To evaluate $\int_0^\infty (\sin xt)/t\, dt$ we use our previous result (13.9) and shift our point of view. Now fix x in $\mathbb{R}$ and view r as variable with $r \geq 0$. For such values of r, define

$$G(r) = \int_0^\infty \frac{e^{-rt} \sin xt}{t}\, dt. \tag{13.10}$$

As we have just seen, if $r > 0$, then $G(r) = \tan^{-1}(x/r)$. Clearly, if $r = 0$, then $G(0) = \int_0^\infty (\sin xt)/t\, dt$. It is this last integral that we want to evaluate. Trivially, if $x = 0$, then $\int_0^\infty (\sin xt)/t\, dt = 0$. In the following, we assume then that $x \neq 0$.

We claim that $\int_0^\infty (e^{-rt} \sin xt)/t\, dt$ converges uniformly for $r \geq 0$; that is, given $\epsilon > 0$, we claim that we can choose $b_0 > 0$ with the property that, whenever $b \geq b_0$,

$$\left| \int_b^\infty \frac{e^{-rt} \sin xt}{t}\, dt \right| < \epsilon, \tag{13.11}$$

for all $r \geq 0$. To prove this, we work backwards. First we integrate (13.11) by parts, letting $u(t) = 1/t$ and $v'(t) = e^{-rt} \sin xt$. Then $u'(t) = -1/t^2$ and

$$v(t) = \frac{-e^{-rt}(r \sin xt + x \cos xt)}{r^2 + x^2}.$$

We obtain

$$\begin{aligned}
\left| \int_b^\infty \frac{e^{-rt} \sin xt}{t}\, dt \right| &= \left| \frac{-e^{-rt}(r \sin xt + x \cos xt)}{(r^2 + x^2)t} \right|_{t=b}^{t=\infty} \\
&\quad - \frac{1}{r^2 + x^2} \int_b^\infty \frac{e^{-rt}(r \sin xt + x \cos xt)}{t^2}\, dt \Bigg| \\
&\leq \left| \frac{e^{-rb}(r \sin xb + x \cos xb)}{(r^2 + x^2)b} \right| + \frac{1}{r^2 + x^2} \int_b^\infty \frac{|r \sin xt + x \cos xt|}{t^2}\, dt \\
&\leq \frac{r + |x|}{(r^2 + x^2)b} + \frac{r + |x|}{r^2 + x^2} \int_b^\infty \frac{1}{t^2}\, dt = \frac{2(r + |x|)}{b(r^2 + x^2)}.
\end{aligned}$$

With $b > 0$ and $x \neq 0$ held fixed, you can show that $2(r + |x|)/[b(r^2 + x^2)] \leq 4/(b|x|)$ for all $r \geq 0$. With this information in hand, we now let $\epsilon > 0$ be given and choose a $b_0 > 0$ such that $4/(b_0|x|) < \epsilon$. Then for all $b \geq b_0$,

$$\left| \int_b^{\infty} \frac{e^{-rt} \sin xt}{t} \, dt \right| \leq \frac{4}{(b|x|)} < \epsilon.$$

This proves that $\int_0^{\infty} (e^{-rt} \sin xt)/t \, dt = G(r)$ [uniformly] in r as r varies over the set $S = [0, \infty)$. By Theorem 13.4.3, G is continuous on $[0, \infty)$. In particular, G is continuous at $r = 0$. Thus we conclude that

$$\int_0^{\infty} \frac{\sin xt}{t} \, dt = G(0) = \lim_{r \to 0^+} G(r)$$

$$= \lim_{r \to 0^+} \tan^{-1} \left(\frac{x}{r} \right) = \begin{cases} \pi/2, & \text{if } x > 0, \\ 0, & \text{if } x = 0, \\ -\pi/2, & \text{if } x < 0. \end{cases} \quad \bullet$$

EXAMPLE 27 Fix x and y in $\mathbb{R}$ and consider

$$F(x, y) = \int_0^{\infty} \frac{\cos xt \sin yt}{t} \, dt.$$

Using the trigonometric identity for $\cos u \sin v$, we first obtain

$$F(x, y) = \frac{1}{2} \int_0^{\infty} \frac{\sin (x + y)t}{t} \, dt - \frac{1}{2} \int_0^{\infty} \frac{\sin (x - y)t}{t} \, dt.$$

Applying our result from Example 26 to each of these integrals, we have

$$F(x, y) = \begin{cases} \pi/2, & y > |x|, \\ \pi/4, & y = |x| > 0, \\ 0, & |y| < |x| \text{ or } y = x = 0, \\ -\pi/4, & y = -|x| < 0, \\ -\pi/2, & y < -|x|. \end{cases}$$

The values of F are indicated in various regions of the plane shown in Fig. 13.6. ●

EXAMPLE 28 **The Gamma Function** Among Euler's many remarkable discoveries is the *gamma function*, which, in the intervening centuries, has become extremely important in the theory of probability and statistics and elsewhere in mathematics. For $x > 0$, he defined

$$\Gamma(x) = \int_0^{\infty} t^{x-1} e^{-t} \, dt. \tag{13.12}$$

In Example 21 we saw that the integral defining the gamma function converges uniformly for x in any $[c, d] \subset (0, \infty)$. Therefore Γ is continuous on $(0, \infty)$.

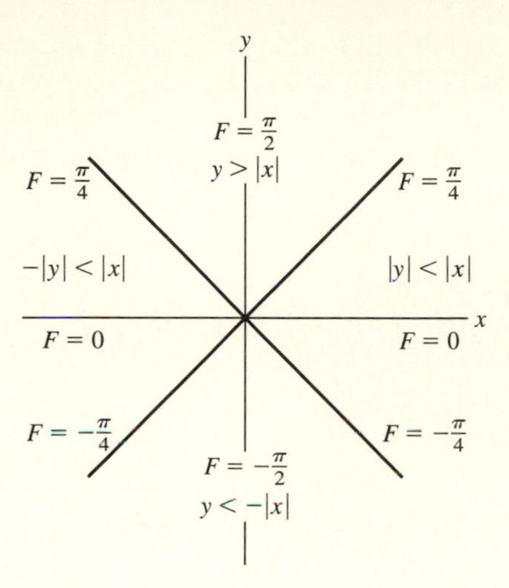

Figure 13.6

One remarkable property of the gamma function is revealed by the following computation. For $x > 0$, we integrate by parts to obtain

$$\Gamma(x+1) = \lim_{b\to\infty} \int_0^b t^x e^{-t}\,dt = \lim_{b\to\infty} \left[-t^x e^{-t} \Big|_0^b + x \int_0^b t^{x-1} e^{-t}\,dt \right]$$

$$= \lim_{b\to\infty} \left[-\frac{b^x}{e^b} + x \int_0^b t^{x-1} e^{-t}\,dt \right] = x\Gamma(x).$$

Thus

$$\Gamma(x+1) = x\Gamma(x)$$

for all $x > 0$. In particular, $\Gamma(1) = \int_0^\infty e^{-t}\,dt = 1$. It follows recursively that

$$\Gamma(2) = 1 \cdot \Gamma(1) = 1,$$

$$\Gamma(3) = 2 \cdot \Gamma(2) = 2,$$

$$\Gamma(4) = 3 \cdot \Gamma(3) = 3 \cdot 2 = 3!,$$

$$\Gamma(5) = 4 \cdot \Gamma(4) = 4 \cdot 3! = 4!.$$

The pattern is clear; as you can prove by induction, $\Gamma(k+1) = k!$ for all k in $\mathbb{N}$. Thus $\Gamma(x)$ is a function that continuously extends the factorial function from the natural numbers to all of the positive real numbers. In passing, mathematicians take $0!$ to have value 1 not merely as a convention but precisely because $0! = \Gamma(1) = 1$.

To study the differentiability of $\Gamma(x)$, recall that $\partial(t^{x-1}e^{-t})/\partial x = t^{x-1}e^{-t}\ln t$. Once we have shown that $\int_0^\infty t^{x-1}e^{-t}\ln t\,dt$ converges uniformly on any interval

$[c, d]$ contained in $(0, \infty)$, we will be able, by applying Theorems 13.4.5 and 13.4.6, to conclude that $\Gamma(x)$ is differentiable at every $x > 0$ and that

$$\Gamma'(x) = \int_0^\infty t^{x-1}e^{-t} \ln t \, dt.$$

We write

$$\int_0^\infty t^{x-1}e^{-t} \ln t \, dt = \int_0^1 t^{x-1}e^{-t} \ln t \, dt + \int_1^\infty t^{x-1}e^{-t} \ln t \, dt$$

and consider each integral in turn.

We already know by Example 24 that $\int_0^1 t^{x-1}e^{-t} \ln t \, dt$ converges uniformly for x in any interval $[c, d] \subset (0, \infty)$ to the derivative (with respect to x) of $\int_0^1 t^{x-1}e^{-t} \, dt$.

The integral $\int_1^\infty t^{x-1}e^{-t} \ln t \, dt$ also converges uniformly on $[c, d]$. To see this, note that, for $1 \le t < \infty$, we have $0 \le \ln t < t$. Hence, for any (t, x) in $[1, \infty) \times [c, d]$,

$$0 \le t^{x-1}e^{-t} \ln t < t^x e^{-t} \le t^d e^{-t} = M(t).$$

Since $\int_1^\infty M(t) \, dt$ converges, Weierstrass's M-test implies that $\int_1^\infty t^{x-1}e^{-t} \ln t \, dt$ converges uniformly on any interval $[c, d]$ in $(0, \infty)$. Combining our results, we deduce that the integral $\int_0^\infty t^{x-1}e^{-t} \ln t \, dt$ converges uniformly on any interval $[c, d] \subset (0, \infty)$. By Theorems 13.4.5 and 13.4.6, we conclude that the gamma function is differentiable at every $x > 0$ and that

$$\Gamma'(x) = \int_0^\infty t^{x-1}e^{-t} \ln t \, dt.$$

By an entirely analogous argument, you can prove that the integral $\int_0^\infty t^{x-1}e^{-t}(\ln t)^2 \, dt = \int_0^\infty \partial[t^{x-1}e^{-t} \ln t]/\partial x \, dt$ converges uniformly on any $[c, d]$ contained in $(0, \infty)$. Thus you can prove that the gamma function is twice differentiable at every $x > 0$ and that

$$\Gamma''(x) = \int_0^\infty t^{x-1}e^{-t}(\ln t)^2 \, dt. \tag{13.13}$$

A proof by induction shows that, for every k in $\mathbb{N}$, the kth order derivative of the gamma function is given by

$$\Gamma^{(k)}(x) = \int_0^\infty t^{x-1}e^{-t}(\ln t)^k \, dt.$$

Notice from (13.13) that $\Gamma''(x) > 0$ for all x in $(0, \infty)$ and therefore that the graph of Γ is concave up. Recall also that $\Gamma(1) = \Gamma(2) = 1$. We conclude that Γ has a minimum value at some point in the interval $(1, 2)$. Finally, to evaluate $\lim_{x \to 0^+} \Gamma(x)$, note that $\Gamma(x) > \int_0^1 t^{x-1}e^{-t} \, dt$, that $e^{-t} \ge e^{-1}$ for $0 \le t \le 1$, and therefore that

$$\Gamma(x) \ge \frac{1}{e} \int_0^1 t^{x-1} \, dt = \frac{1}{e} \lim_{a \to 0^+} \left. \frac{t^x}{x} \right|_{t=a}^{t=1} = \frac{1}{ex}.$$

It follows that $\lim_{x \to 0^+} \Gamma(x) \geq \lim_{x \to 0^+} 1/(ex) = \infty$. We deduce that the gamma function has a pole at 0^+. These considerations enable us to sketch the graph of Γ for $x > 0$. (See Fig. 13.7.)

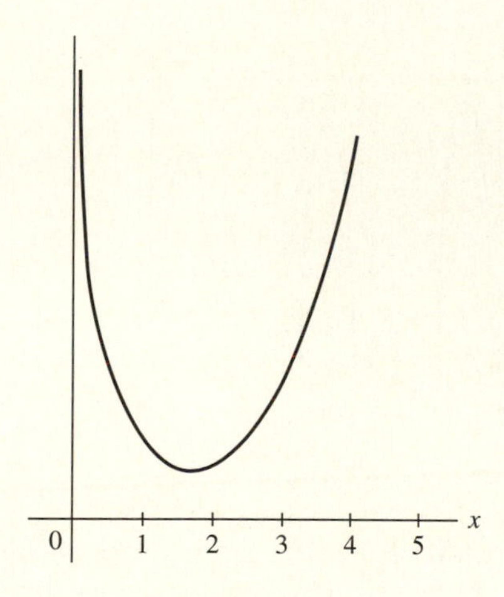

Figure 13.7

Since $\Gamma(x+1) = x\Gamma(x)$ for all $x > 0$, we can extend the domain of Γ recursively to include all negative real numbers that are not integers. To begin, suppose that $-1 < x < 0$. Then $x + 1$ is positive and $\Gamma(x+1)$ is defined. Set

$$\Gamma(x) = \frac{\Gamma(x+1)}{x}.$$

This step extends the domain of Γ to include $(-1, 0)$. Notice that, with this extension, Γ continues to have derivatives of all orders. For x in $(-2, -1)$, the value of $\Gamma(x+1)$ is now defined, and again we set

$$\Gamma(x) = \frac{\Gamma(x+1)}{x}.$$

Thus Γ is defined on $(-2, -1) \cup (-1, 0) \cup (0, \infty)$. Again, Γ so extended has derivatives of all orders. Suppose that we have recursively extended the domain of Γ to include the intervals $(-1, 0)$, $(-2, -1), \ldots,$ and $(-k, -k+1)$. For x in the interval $(-k-1, -k)$, the value of $\Gamma(x+1)$ is now defined and, again, we set $\Gamma(x) = \Gamma(x+1)/x$. By this method we define $\Gamma(x)$ for every x in $\mathbb{R}$ except the nonpositive integers. The graph of the extension of Γ is sketched in Fig. 13.8. ●

EXAMPLE 29 Gauss's Normal Probability Distribution Our first objective in this example is to show that

$$\int_{-\infty}^{\infty} e^{-x^2}\, dx = \sqrt{\pi}.$$

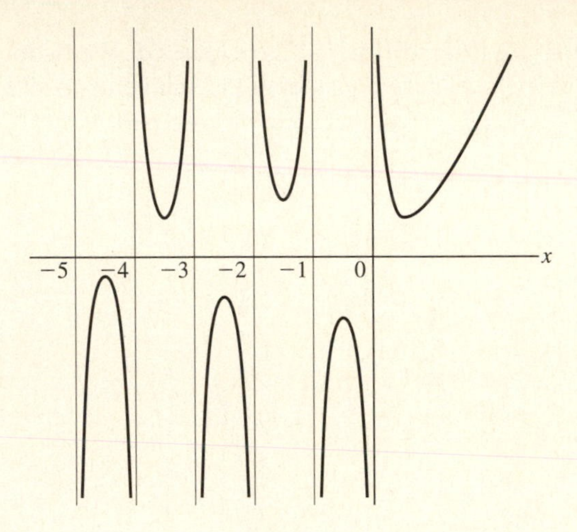

Figure 13.8

This is an improper integral of the first kind, since the integrand is bounded with $0 < \exp(-x^2) \leq 1$ for all x in $\mathbb{R}$. However, since the integrand $\exp(-x^2)$ has no elementary antiderivative, we have no obvious way to use Cauchy's form of the Fundamental Theorem of Calculus. Still, not all is lost.

First we note that $\exp(-x^2)$ is an even function. It follows that $\int_{-\infty}^{\infty} \exp(-x^2)\, dx = 2\int_0^{\infty} \exp(-x^2)\, dx$ and that we need only evaluate $I = \int_0^{\infty} \exp(-x^2)\, dx = \lim_{b\to\infty} \int_0^b \exp(-x^2)\, dx$. For any $b > 0$, denote $\int_0^b \exp(-x^2)\, dx$ by $I(b)$.

The classically elegant solution we present has us compute the double integral of $f(x, y) = \exp(-x^2 - y^2)$ over three nested regions in the first quadrant of $\mathbb{R}^2$. To be specific, choose any $b > 0$ and construct three subsets A_1, A_2, and A_3 in the first quadrant as displayed in Fig. 13.9.

A_1 is the closed quarter-disk with radius b:

$$A_1 = \{(x, y) : x^2 + y^2 \leq b^2;\ x \geq 0, y \geq 0\}.$$

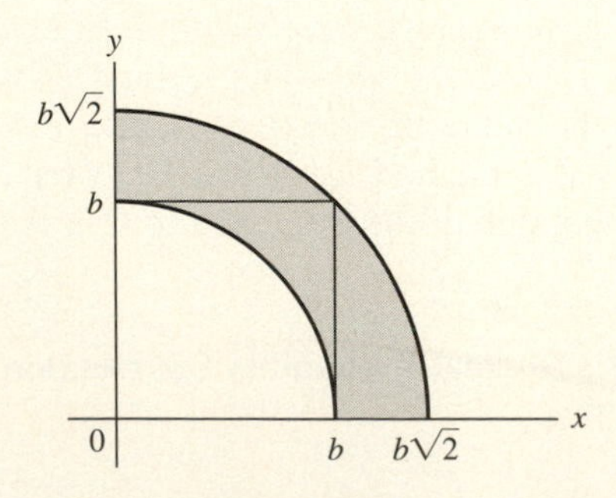

Figure 13.9

A_2 is the closed, solid square with vertices $(0,0)$, $(b,0)$, $(0,b)$, and (b,b):

$$A_2 = \{(x, y) : 0 \le x \le b,\ 0 \le y \le b\}.$$

A_3 is the closed quarter-disk with radius $b\sqrt{2}$:

$$A_3 = \{(x, y) :\ x^2 + y^2 \le 2b^2,\ x \ge 0,\ y \ge 0\}.$$

Since $A_1 \subset A_2 \subset A_3$ and since $\exp(-x^2 - y^2) > 0$, integrating this function over each of A_1, A_2, and A_3 yields

$$\iint_{A_1} e^{-(x^2+y^2)}\, dA \le \iint_{A_2} e^{-(x^2+y^2)}\, dA \le \iint_{A_3} e^{-(x^2+y^2)}\, dA. \qquad \textbf{(13.14)}$$

Since the regions A_1 and A_3 are circular, the first and third integrals are evaluated by shifting to polar coordinates as in Example 5 of Section 10.3. We proved in that example that

$$\iint_{A_1} e^{-(x^2+y^2)}\, dA = \frac{\pi}{4}[1 - e^{-b^2}]$$

and

$$\iint_{A_3} e^{-(x^2+y^2)}\, dA = \frac{\pi}{4}[1 - e^{-2b^2}].$$

The second integral, over a square region, is iterated in Cartesian coordinates:

$$\begin{aligned}\iint_{A_2} e^{-(x^2+y^2)}\, dA &= \int_0^b \int_0^b e^{-x^2} e^{-y^2}\, dx\, dy \\ &= \int_0^b e^{-x^2} \left[\int_0^b e^{-y^2}\, dy\right] dx \\ &= \left[\int_0^b e^{-x^2}\, dx\right]\left[\int_0^b e^{-y^2}\, dy\right] = [I(b)]^2.\end{aligned}$$

Substitute these results in (13.14) to obtain

$$\frac{\pi}{4}(1 - e^{-b^2}) \le [I(b)]^2 \le \frac{\pi}{4}(1 - e^{-2b^2}).$$

Because

$$\lim_{b\to\infty} \frac{\pi}{4}[1 - e^{-b^2}] = \lim_{b\to\infty} \frac{\pi}{4}[1 - e^{-2b^2}] = \frac{\pi}{4}$$

and since

$$\lim_{b\to\infty} I(b)^2 = I^2,$$

the Squeeze Play implies that $I^2 = \pi/4$. Therefore

$$I = \int_0^\infty e^{-x^2}\, dx = \frac{1}{2}\sqrt{\pi}.$$

This remarkable result has consequences in a variety of settings. For example, we can now calculate the values of the gamma function at the points $(2k-1)/2$, for k in $\mathbb{N}$. To begin, $\Gamma(1/2) = \int_0^\infty t^{-1/2}e^{-t}\,dt$. Change variables in this integral by letting $u = u(t) = t^{1/2}$. Then $u'(t) = t^{-1/2}/2$, $u(0) = 0$, and $\lim_{b\to\infty} u(b) = \infty$. When the steps are completed, we obtain

$$\Gamma\left(\frac{1}{2}\right) = \int_0^\infty t^{-1/2}e^{-t}\,dt = 2\int_0^\infty e^{-u^2}\,du = \sqrt{\pi}.$$

From the equation $\Gamma(x+1) = x\Gamma(x)$, we obtain recursively

$$\Gamma\left(\frac{3}{2}\right) = \frac{1}{2}\Gamma\left(\frac{1}{2}\right) = \frac{1}{2}\sqrt{\pi},$$

$$\Gamma\left(\frac{5}{2}\right) = \frac{3}{2}\Gamma\left(\frac{3}{2}\right) = \frac{1\cdot 3}{2^2}\sqrt{\pi} = \frac{4!\sqrt{\pi}}{4^2 2!},$$

$$\Gamma\left(\frac{7}{2}\right) = \frac{5}{2}\Gamma\left(\frac{5}{2}\right) = \frac{1\cdot 3\cdot 5}{2^3}\sqrt{\pi} = \frac{6!\sqrt{\pi}}{4^3 3!},$$

and, in general,

$$\Gamma\left(k+\frac{1}{2}\right) = \frac{(2k)!\sqrt{\pi}}{4^k k!}.$$

The inductive proof is easy and is left for you to complete.

The integral $\int_0^\infty \exp(-x^2/2)\,dx$ can also be evaluated by changing variables; simply let $u = x/\sqrt{2}$. We obtain

$$\int_0^\infty e^{-x^2/2}\,dx = \sqrt{2}\int_0^\infty e^{-u^2}\,du = \sqrt{\pi/2}.$$

As a consequence,

$$\int_{-\infty}^\infty e^{-x^2/2}\,dx = 2\int_0^\infty e^{-x^2/2}\,dx = \sqrt{2\pi}$$

or

$$\frac{1}{\sqrt{2\pi}}\int_{-\infty}^\infty e^{-x^2/2}\,dx = 1.$$

The function F on $\mathbb{R}$ defined by

$$F(x) = \frac{1}{\sqrt{2\pi}}\int_{-\infty}^x e^{-t^2/2}\,dt,$$

has the following properties:

i) F is strictly monotone increasing on $\mathbb{R}$.

ii) $\lim_{x\to-\infty} F(x) = 0^+$; $\lim_{x\to\infty} F(x) = 1^-$; $F(0) = 1/2$.

iii) F is continuous on $\mathbb{R}$.

iv) F is differentiable on $\mathbb{R}$ and

$$F'(x) = \frac{1}{\sqrt{2\pi}}e^{-x^2/2}.$$

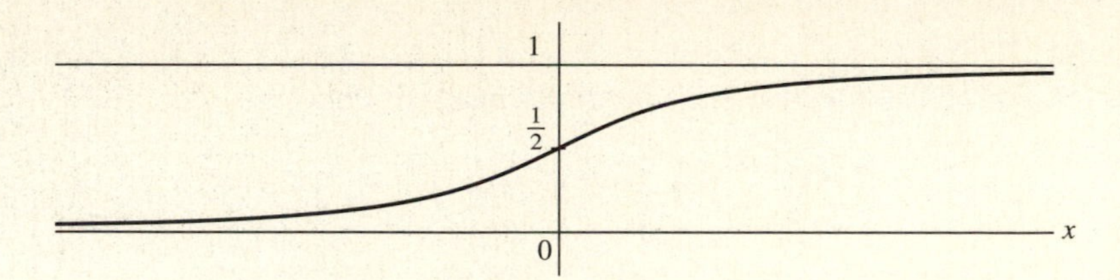

Figure 13.10

We leave it for you to prove that F has each of these properties. The function F is called the *cumulative distribution function* of Gauss's normal probability distribution. The graph of F is sketched in Fig. 13.10. The mean of this probability distribution is defined to be

$$\begin{aligned}\mu &= \frac{1}{\sqrt{2\pi}} \int_{-\infty}^{\infty} x\, e^{-x^2/2}\, dx \\ &= \frac{1}{\sqrt{2\pi}} \left[\int_{-\infty}^{0} x\, e^{-x^2/2}\, dx + \int_{0}^{\infty} x\, e^{-x^2/2}\, dx \right].\end{aligned}$$

We transform the first integral by replacing x with $-x$ to obtain

$$\begin{aligned}\int_{-\infty}^{0} x\, e^{-x^2/2}\, dx &= \int_{\infty}^{0} (-x)\, e^{-x^2/2} (-\, dx) \\ &= -\int_{0}^{\infty} x\, e^{-x^2/2}\, dx.\end{aligned}$$

Therefore, $\mu = 0$. The variance of the distribution is defined to be

$$\begin{aligned}\sigma^2 &= \frac{1}{\sqrt{2\pi}} \int_{-\infty}^{\infty} (x-\mu)^2\, e^{-x^2/2}\, dx = \frac{1}{\sqrt{2\pi}} \int_{-\infty}^{\infty} x^2\, e^{-x^2/2}\, dx \\ &= \frac{2}{\sqrt{2\pi}} \int_{0}^{\infty} x^2\, e^{-x^2/2}\, dx,\end{aligned}$$

since the integrand is even. We transform this last integral by setting $u = x^2/2$. Upon completing the details we obtain

$$\sigma^2 = \frac{2}{\sqrt{\pi}} \int_{0}^{\infty} u^{1/2} e^{-u}\, du = \frac{2}{\sqrt{\pi}} \Gamma\left(\frac{3}{2}\right) = 1. \quad \bullet$$

EXAMPLE 30 The Beta Function For $x > 0$ and $y > 0$, we define another Eulerian function by

$$B(x, y) = \int_{0}^{1} t^{x-1} (1-t)^{y-1}\, dt.$$

Euler called B the *beta function*. When both $x \geq 1$ and $y \geq 1$, this integral is proper. However, if $0 < x < 1$, then the integral is improper of the second kind at the endpoint $a = 0$; if $0 < y < 1$, then the integral is improper of the second kind at the endpoint $b = 1$. To show that these improper integrals converge, write

$$B(x, y) = \int_0^{1/2} t^{x-1}(1-t)^{y-1}\,dt + \int_{1/2}^{1} t^{x-1}(1-t)^{y-1}\,dt$$

and treat each separately. For t in $(0, 1/2]$, x in $(0, 1)$, and for any $y > 0$,

$$0 < t^{x-1}(1-t)^{y-1} \leq 2t^{x-1}.$$

Since $\int_0^{1/2} 2t^{x-1}\,dt$ converges, so also does $\int_0^{1/2} t^{x-1}(1-t)^{y-1}\,dt$. Likewise, for t in $[1/2, 1)$, y in $(0, 1)$, and for any $x > 0$,

$$0 < t^{x-1}(1-t)^{y-1} \leq 2(1-t)^{y-1}.$$

Again, the convergence of $\int_{1/2}^{1} 2(1-t)^{y-1}\,dt$ implies that of $\int_{1/2}^{1} t^{x-1}(1-t)^{y-1}\,dt$. Consequently, the integral defining the beta function converges for all $x, y > 0$. ●

There are several connections between the beta function and other important functions in classic analysis. Here we derive the relationship between the beta function and the gamma function.

THEOREM 13.5.1 For all $x, y > 0$,

$$B(x, y) = \frac{\Gamma(x)\Gamma(y)}{\Gamma(x+y)}.$$

Proof. There are a number of ingeneous proofs of this theorem; the one presented here consists of manipulating the integrals involved in the definitions of B and Γ. First, by changing variables, we transform the integral defining B. Let $u(t) = t/(1-t)$ or $t(u) = u/(1+u)$. If $t = 0$, then $u = 0$ and, as $t \to 1^-$, $u(t) \to +\infty$. Also $t'(u) = 1/(1+u)^2$. This transformation yields

$$\begin{aligned} B(x, y) &= \int_0^1 t^{x-1}(1-t)^{y-1}\,dt \\ &= \int_0^\infty \left(\frac{u}{1+u}\right)^{x-1} \left(\frac{1}{1+u}\right)^{y-1} \left(\frac{1}{1+u}\right)^{2} du \\ &= \int_0^\infty \frac{u^{x-1}}{1+u^{x+y}}\,du. \qquad \textbf{(13.15)} \end{aligned}$$

Set this result aside for the time being and turn your attention to $\Gamma(x) = \int_0^\infty t^{x-1}e^{-t}\,dt$. In this integral change variables by fixing $v > 0$ and letting $u(t) = t/v$ or $t = uv$. Changing variables yields

$$\Gamma(x) = \int_0^\infty t^{x-1}e^{-t}\,dt = \int_0^\infty (uv)^{x-1}e^{-uv}v\,du = \int_0^\infty v^x u^{x-1}e^{-uv}\,du.$$

Multiply by $v^{y-1}e^{-v}$ and integrate with respect to v from 0 to ∞. Doing so yields

$$\Gamma(x)\Gamma(y) = \int_0^\infty \Gamma(x)v^{y-1}e^{-v}\,dv = \int_0^\infty \left[\int_0^\infty v^x u^{x-1}e^{-uv}v^{y-1}e^{-v}\,du\right] dv$$

$$= \int_0^\infty \left[\int_0^\infty u^{x-1} v^{x+y-1} e^{-(u+1)v}\, du \right] dv.$$

Now apply Theorem 13.4.9, which permits the interchange of these two integrals. (We leave for you the task of confirming that the hypotheses of that theorem are met.) Consequently,

$$\Gamma(x)\Gamma(y) = \int_{u=0}^\infty u^{x-1} \left[\int_{v=0}^\infty v^{x+y-1} e^{-(1+u)v}\, dv \right] du.$$

Change variables in the inner integral by letting $w = (1+u)v$ or $v(w) = w/(1+u)$; $w(0) = 0$, $\lim_{v\to\infty} w(v) = \infty$ and $v'(w) = 1/(1+u)$. We obtain

$$\begin{aligned}
\Gamma(x)\Gamma(y) &= \int_{u=0}^\infty u^{x-1} \left[\int_{w=0}^\infty \left(\frac{w}{1+u}\right)^{x+y-1} e^{-w} \left(\frac{1}{1+u}\right) dw \right] du \\
&= \int_{u=0}^\infty \frac{u^{x-1}}{(1+u)^{x+y}} \left[\int_{w=0}^\infty w^{x+y-1} e^{-w}\, dw \right] du \\
&= \int_{u=0}^\infty \frac{u^{x-1}}{(1+u)^{x+y}} \Gamma(x+y)\, du \\
&= \Gamma(x+y) B(x, y),
\end{aligned}$$

by the earlier result (13.15). This proves that

$$B(x, y) = \frac{\Gamma(x)\Gamma(y)}{\Gamma(x+y)},$$

as promised. ●

COROLLARY 13.5.2 Wallis's Product

$$\lim_{k\to\infty} \left(\frac{2 \cdot 4 \cdot 6 \cdot 8 \cdots 2k}{1 \cdot 3 \cdot 5 \cdot 7 \cdots (2k-1)} \right)^2 \cdot \frac{1}{2k} = \frac{\pi}{2}.$$

Equivalently,

$$\lim_{k\to\infty} \frac{4^k [k!]^2}{(2k)!\sqrt{k}} = \sqrt{\pi}.$$

Proof. In the integral defining the beta function, change variables by setting $t = \sin^2\theta$. The integral becomes

$$\begin{aligned}
B(x, y) &= \int_0^{\pi/2} \sin^{2x-2}\theta [1 - \sin^2\theta]^{y-1} 2 \sin\theta \cos\theta\, d\theta \\
&= 2 \int_0^{\pi/2} \sin^{2x-1}\theta \cos^{2y-1}\theta\, d\theta.
\end{aligned}$$

In particular, let $x = k$ in $\mathbb{N}$ and $y = 1/2$. Then

$$\int_0^{\pi/2} \sin^{2k-1}\theta\, d\theta = \frac{B(k, 1/2)}{2} = \frac{\Gamma(k)\Gamma(1/2)}{2\Gamma(k+1/2)}.$$

In Example 29 we found that $\Gamma(k+1/2) = (2k)!\sqrt{\pi}/(4^k k!)$. Therefore

$$\int_0^{\pi/2} \sin^{2k-1}\theta\, d\theta = \frac{4^k k!(k-1)!}{2(2k)!}. \tag{13.16}$$

Likewise, if we set $x = k + 1/2$ and $y = 1/2$, we obtain

$$\begin{aligned}\int_0^{\pi/2} \sin^{2k}\theta\, d\theta &= \frac{B(k+1/2, 1/2)}{2} \\ &= \frac{\Gamma(k+1/2)\Gamma(1/2)}{2\Gamma(k+1)} = \frac{(2k)!\pi}{2\cdot 4^k (k!)^2}.\end{aligned} \tag{13.17}$$

Now, for $0 \le \theta \le \pi/2$, we have $0 \le \sin\theta \le 1$. Consequently, for all k in $\mathbb{N}$,

$$\int_0^{\pi/2} \sin^{2k}\theta\, d\theta \le \int_0^{\pi/2} \sin^{2k-1}\theta\, d\theta \le \int_0^{\pi/2} \sin^{2k-2}\theta\, d\theta.$$

By (13.16) and (13.17), we obtain

$$\frac{(2k)!\pi}{2\cdot 4^k (k!)^2} \le \frac{4^k k!(k-1)!}{2(2k)!} \le \frac{(2k-2)!\pi}{2\cdot 4^{k-1}[(k-1)!]^2}.$$

Multiply these inequalities by $2\cdot 4^k (k!)^2/(2k)!$ and simplify to obtain

$$\pi \le \frac{4^{2k}(k!)^4}{[(2k)!]^2}\cdot\frac{1}{k} \le \frac{2k\pi}{2k-1}.$$

Passing to the limit gives

$$\pi \le \lim_{k\to\infty} \frac{4^{2k}[k!]^4}{[(2k)!]^2}\cdot\frac{1}{k} \le \lim_{k\to\infty} \frac{2k\pi}{(2k-1)} = \pi.$$

Apply the Squeeze Play and take square roots to obtain

$$\lim_{k\to\infty} \frac{4^k [k!]^2}{(2k)!\sqrt{k}} = \sqrt{\pi}.$$

This proves Wallis's product formula. ●

13.6 THE LAPLACE TRANSFORM

The theory of integral transforms, among which the Laplace transform is just one example, has a long and distinguished history. Indeed, the early success of this theory gave strong impetus to profound research in both pure and applied modern analysis. However, at the level of this text we will be able to provide just an introductory glimpse. A thorough and mathematically sound treatment requires the tools and insight afforded by the theory of complex analysis and by the integration theory of Lebesgue, but we can make some progress in this vast and challenging subject.

We let K be an appropriately restricted function of two variables (s, t) that are themselves restricted to a region of the plane (either the entire plane, a half-plane, a quadrant, an unbounded strip...). For various real-valued functions f of t we define

$$F(s) = \int_I K(s, t) f(t)\, dt \tag{13.18}$$

where I is the interval over which the variable t ranges. F is called the *K-transform* of f. It is often convenient to write $T_K f = F$ to express the equation (13.18); T_K is called the *transform* generated by K. The function K is called the *kernel* of the transform. It is easily seen that T_K is a linear transformation on its domain; that is, for f_1 and f_2 in the set of all functions for which the integral (13.18) converges and for all real numbers c_1 and c_2,

$$T_K(c_1 f_1 + c_2 f_2) = c_1 T_K f_1 + c_2 T_K f_2.$$

For a given kernel K, the domain of T_K is completely determined. However, in general, we lack the tools to identify that domain completely. For a given kernel K, the transform T_K will exhibit algebraic and analytic properties that reflect the nature of that kernel. Some kernels yield few useful properties; over time, they have fallen by the wayside. But some other kernels provide a rich harvest of such properties and have been enshrined as part of our mathematical heritage. The Laplace transform, named in honor of Pierre-Simon de Laplace (1749–1827), is one of the earliest integral transforms to be discovered. It provides a remarkably powerful tool for solving problems in mathematical physics and differential equations. Perhaps as important, its usefulness motivated Poisson, Fourier, Abel, Volterra, and many others to search for kernels specifically designed to generate transforms having desirable properties. The kernel for the Laplace transform is taken to be

$$K(s, t) = e^{-st}$$

on the closed first quadrant $[0, \infty) \times [0, \infty)$.

DEFINITION 13.6.1 Let f be a real-valued function on $[0, \infty)$ having the following property: There exists an $s_0 > 0$ such that for $s \geq s_0$, the integral

$$Lf(s) = F(s) = \int_0^{\infty} e^{-st} f(t)\, dt$$

converges. Then L is called the *Laplace transform*, and $Lf = F$ is called the *Laplace transform of f*. The domain of the Laplace transform will be denoted $\mathcal{L}$. •

As a convention and for convenience, we extend f to be defined on all of $\mathbb{R}$ by setting $f(t) = 0$ for t in $(-\infty, 0)$. This stipulation will hold throughout this section. The rapid rate at which the exponential function tends to zero at infinity ensures that the integral defining the Laplace transform converges for a large class of functions. This ensures that the domain of L contains most functions of classical interest. Furthermore, the special properties of the exponential function induce highly desirable algebraic and analytic properties of the transform.

We have already seen a few examples of Laplace transforms (without having identified them as such). To begin, we compute several of the more elementary transforms.

1. Let $f(t) = 1$ for $t \geq 0$. Then, for any $s_0 > 0$,

$$Lf(s) = \int_0^\infty e^{-st}\,dt = \lim_{b\to\infty} \frac{-e^{-st}}{s}\bigg|_0^b = \frac{1}{s}$$

for $s \geq s_0 > 0$.

2. Let $f(t) = t$ for $t \geq 0$. Then, for any $s_0 > 0$,

$$\begin{aligned} Lf(s) = \int_0^\infty te^{-st}\,dt &= \lim_{b\to\infty}\left[\frac{-te^{-st}}{s}\bigg|_0^b\right] + \frac{1}{s}\int_0^\infty e^{-st}\,dt \\ &= \frac{1}{s^2}, \end{aligned}$$

for $s \geq s_0 > 0$. If $f(t) = t^2$ for $t \geq 0$, then

$$\begin{aligned} Lf(s) = \int_0^\infty t^2e^{-st}\,dt &= \lim_{b\to\infty}\left[\frac{-t^2e^{-st}}{s}\bigg|_0^b\right] + \frac{2}{s}\int_0^\infty te^{-st}\,dt \\ &= \frac{2}{s^3}, \end{aligned}$$

for $s \geq s_0 > 0$.

If, for k in $\mathbb{N}$, we set $f(t) = t^k$ for $t \geq 0$, then for any $s_0 > 0$, the Laplace transform of f is

$$Lf(s) = \frac{k!}{s^{k+1}},$$

for $s \geq s_0 > 0$. You can easily prove this assertion by induction.

3. Let c be constant and let $f(t) = e^{ct}$ for $t \geq 0$. For $s \geq s_0 > c$, the Laplace transform of f is

$$Lf(s) = \int_0^\infty e^{(c-s)t}\,dt = \lim_{b\to\infty} \frac{e^{(c-s)t}}{c-s}\bigg|_0^b = \frac{1}{s-c}.$$

If $s \leq c$, the integral defining the Laplace transform fails to converge.

4. Reconsider the result from (3),

$$F(s) = \int_0^\infty e^{(c-s)t}\,dt = \frac{1}{s-c}$$

for $s \geq s_0 > c$. The partial derivative of the integrand with respect to s is $-te^{(c-s)t}$. Also, the integral $\int_0^\infty te^{(c-s)t}\,dt$ converges uniformly for s in any compact interval contained in (c, ∞). Thus by Theorem 13.4.5 we know that

$$F'(s) = -\int_0^\infty te^{(c-s)t}\,dt = \frac{d}{ds}\left(\frac{1}{s-c}\right) = \frac{-1}{(s-c)^2},$$

for $s \geq s_0 > c$. That is,

$$L(te^{ct})(s) = \frac{1}{(s-c)^2}.$$

Using this line of reasoning, you can show recursively that, for all k in $\mathbb{N}$,

$$L(t^k e^{ct})(s) = \frac{k!}{(s-c)^{k+1}},$$

for $s \geq s_0 > c$. As we shall see, the similarity between this result and that in (2) is no mere accident; it is the manifestation of an important property of the Laplace transform.

5. If we let $f(t) = t^{-1/2}$ for $t > 0$, then, for any $s_0 > 0$,

$$Lf(s) = \int_0^\infty e^{-st} t^{-1/2}\, dt.$$

In this integral, change variables by setting $u = st$. Upon completing the details, we have, for $s \geq s_0 > 0$,

$$L(t^{-1/2})(s) = \int_0^\infty e^{-st} t^{-1/2}\, dt = s^{-1/2} \int_0^\infty u^{-1/2} e^{-u}\, du$$

$$= s^{-1/2}\Gamma\left(\frac{1}{2}\right) = \sqrt{\pi} s^{-1/2},$$

for $s \geq s_0$. If $f(t) = t^{1/2}$ for $t > 0$, then the same change of variables yields

$$L(t^{1/2})(s) = \int_0^\infty e^{-st} t^{1/2}\, dt = s^{-3/2} \int_0^\infty u^{1/2} e^{-u}\, du$$

$$= s^{-3/2}\Gamma\left(\frac{3}{2}\right),$$

valid for $s \geq s_0 > 0$. These two results suggest that we consider the Laplace transform of $f(t) = t^x$ for $x > -1$ and $t > 0$. For any $s_0 > 0$ and all $s \geq s_0$, we have

$$L(t^x)(s) = \int_0^\infty e^{-st} t^x\, dt.$$

Again, the change of variables $u = st$ yields

$$L(t^x)(s) = \int_0^\infty e^{-st} t^x\, dt = s^{-(x+1)} \int_0^\infty u^x e^{-u}\, du = \frac{\Gamma(x+1)}{s^{x+1}},$$

for $x > -1$. We leave for you to confirm that, for any $x > -1$ and any c in $\mathbb{R}$,

$$L(t^x e^{ct})(s) = \frac{\Gamma(x+1)}{(s-c)^{x+1}}.$$

6. Let a be a constant. As we saw in Example 22 of Section 13.4, the Laplace transforms of $\sin at$ and $\cos at$ are

$$L(\sin at)(s) = \frac{a}{s^2 + a^2}$$

and

$$L(\cos at)(s) = \frac{s}{s^2 + a^2}$$

valid for all $s \geq s_0 > 0$.

7. Let c be a constant. Using (3), the definitions of $\sinh ct = (e^{ct} - e^{-ct})/2$ and $\cosh ct = (e^{ct} + e^{-ct})/2$, and the linearity of the Laplace transform, we obtain

$$\begin{aligned} L(\sinh ct)(s) &= \frac{1}{2}[L(e^{ct})(s) - L(e^{-ct})(s)] \\ &= \frac{1}{2}\left[\frac{1}{s-c} - \frac{1}{s+c}\right] \\ &= \frac{c}{s^2 - c^2}. \end{aligned}$$

Likewise, $L(\cosh ct)(s) = s/(s^2 - c^2)$. In both cases, we require $s \geq s_0 > |c|$.

8. For constants a and c, the Laplace transform of the function $e^{ct} \sin at$ is

$$L(e^{ct} \sin at)(s) = \int_0^\infty e^{(c-s)t} \sin at\, dt = \frac{a}{(s-c)^2 + a^2},$$

valid for $s \geq s_0 > c$, since it is for such values of s that the integral converges. We leave it for you to compute

$$L(e^{ct} \cos at)(s) = \frac{s-c}{(s-c)^2 + a^2}.$$

9. In Example 23, Section 13.4, we proved that, for $a \neq 0$ and for $s > 0$,

$$\int_0^\infty \frac{e^{-st} \sin at}{t}\, dt = \tan^{-1}\left(\frac{a}{s}\right).$$

That is, if $f(t) = (\sin at)/t$, for $t > 0$, then the Laplace transform of f is $\tan^{-1}(a/s)$, valid for $s \geq s_0 > 0$.

The preceding examples display some of the properties of the Laplace transform. In the following theorem we identify the first general property.

THEOREM 13.6.1 The First Shift Theorem If f is in $\mathcal{L}$, if $F(s) = Lf(s)$, and if c is constant, then

$$L(e^{-ct} f(t))(s) = F(s + c).$$

Proof. Suppose that $F(s) = \int_0^\infty e^{-st} f(t)\,dt$ where the integral converges for $s \geq s_0$. We compute

$$L(e^{-ct} f(t))(s) = \int_0^\infty e^{-st} e^{-ct} f(t)\,dt$$
$$= \int_0^\infty e^{-(c+s)t} f(t)\,dt = F(s+c),$$

valid for $s \geq s_0 - c$. This proves the theorem. ●

THEOREM 13.6.2 The Second Shift Theorem If f is in $\mathcal{L}$ and if $F = Lf$, then, for any constant $c > 0$,

$$L(f(t-c))(s) = e^{-cs} F(s).$$

Proof. Compute $\int_0^\infty e^{-st} f(t-c)\,dt$ using the change of variables $u = t - c$:

$$\int_0^\infty e^{-st} f(t-c)\,dt = \int_{-c}^\infty e^{-s(c+u)} f(u)\,du = e^{-cs} \int_{-c}^\infty e^{-su} f(u)\,du.$$

Recall that we have extended f to have value 0 on $(-\infty, 0)$ so we can adjust the lower limit of integration in this last integral and thus obtain

$$L(f(t-c))(s) = e^{-cs} \int_0^\infty e^{-su} f(u)\,du = e^{-cs} F(s).$$

This proves the theorem. ●

THEOREM 13.6.3 If f is in $\mathcal{L}$ and if $F = Lf$, then, for any constant $c > 0$,

$$L(f(ct))(s) = \frac{1}{c} F\left(\frac{s}{c}\right).$$

Proof. Simply use the change of variables $u = ct$ to evaluate

$$\int_0^\infty e^{-st} f(ct)\,dt = \frac{1}{c} \int_0^\infty e^{-su/c} f(u)\,du = \frac{1}{c} F\left(\frac{s}{c}\right).$$

An equivalent formulation of the result in Theorem 13.6.3 is

$$F(cs) = \frac{1}{c} L\left(f\left(\frac{t}{c}\right)\right)(s),$$

obtained by replacing c by $1/c$ in the theorem. ●

Although we lack the tools to identify all the functions in $\mathcal{L}$, we can formulate conditions that will guarantee the existence of Lf and that are sufficiently general to handle most classical applications. These are described in the following definition.

DEFINITION 13.6.2 Let f be a real-valued function defined on $[0, \infty)$ that has the following properties:

i) For any $b > 0$, f is bounded on $[0, b]$ and is continuous except possibly at finitely many points in $[0, b]$.

ii) There exist constants $a > 0$ and r in $\mathbb{R}$ such that

$$|f(t)| \le ae^{rt},$$

for all t in $[0, \infty)$.

Then f is said to be of *exponential type*. The collection of all *functions of exponential type* is denoted $\mathcal{E}$. ●

The importance to us of the class of functions $\mathcal{E}$ is revealed in the following theorem. It tells us that $\mathcal{E}$ is a commutative ring with identity that is closed under indefinite integration.

THEOREM 13.6.4

i) If f_1 and f_2 are in $\mathcal{E}$ and if c_1 and c_2 are in $\mathbb{R}$, then $c_1 f_1 + c_2 f_2$ and $f_1 f_2$ are in $\mathcal{E}$.

ii) If f is in $\mathcal{E}$ and if we define $g(t) = \int_0^t f(u)\,du$, then g is in $\mathcal{E}$.

Proof. Clearly if each of f_1 and f_2 is bounded on $[0, b]$ and is continuous on $[0, b]$ except at finitely many points, then the same is true of the functions $c_1 f_1 + c_2 f_2$ and $f_1 f_2$. Suppose that for $i = 1, 2$, we have $|f_i(t)| \le a_i e^{r_i t}$, for all $t \ge 0$. Then,

$$|c_1 f_1(t) + c_2 f_2(t)| \le |c_1| a_1 e^{r_1 t} + |c_2| a_2 e^{r_2 t} \le ae^{rt},$$

where $a = |c_1|a_1 + |c_2|a_2$ and $r = \max\{r_1, r_2\}$. Therefore $c_1 f_1 + c_2 f_2$ is in $\mathcal{E}$. Likewise, for all $t \ge 0$,

$$|f_1(t) f_2(t)| \le a_1 a_2 e^{(r_1 + r_2)t} = ae^{rt},$$

where $a = a_1 a_2$ and $r = r_1 + r_2$. Thus $f_1 f_2$ is also in $\mathcal{E}$.

By the Fundamental Theorem of Calculus, the function $g(t) = \int_0^t f(u)\,du$ is continuous on all of $[0, \infty)$. Thus g meets condition (i). For (ii), suppose that $|f(t)| \le ae^{rt}$ for all $t \ge 0$. Without loss of generality, r is positive. Then

$$\begin{aligned} |g(t)| &= \left| \int_0^t f(u)\,du \right| \le \int_0^t |f(u)|du \\ &\le \int_0^t ae^{ru}\,du = \frac{a(e^{rt} - 1)}{r} < \frac{a}{r} e^{rt}. \end{aligned}$$

This proves that g is also of exponential type. ●

THEOREM 13.6.5 Suppose that f is continuous and that f' exists at all but countably many discrete points in $[0, \infty)$. If f' is in $\mathcal{E}$, then f is also in $\mathcal{E}$.

Proof. Since the points where f' fails to exist form a discrete set S, for any $b > 0$ the interval $[0, b]$ can contain only finitely many such points. (Otherwise, the Bolzano–Weierstrass theorem implies that S has a limit point violating the discreteness of the set.) At a point t where f' fails to exist, define $f'(t) = 0$. Since f' is assumed to be in $\mathcal{E}$, we know that, for any $b > 0$, f' is bounded on $[0, b]$ and is continuous at all but finitely many points of $[0, b]$. Also, there exist constants $a > 0$ and r in $\mathbb{R}$ such that $|f'(t)| \le ae^{rt}$ for all t in $[0, \infty)$. Without loss of generality, r is positive. Moreover, by Theorem 13.6.4, we know that the function $g(t) = \int_0^t f'(u)\,du$ is in $\mathcal{E}$.

If f happens to be differentiable at every point of $[0, t]$, then, by Corollary 6.3.9,

$$f(t) = f(0) + \int_0^t f'(u)\,du = f(0) + g(t).$$

The constant $f(0)$ is in $\mathcal{E}$. Thus f is the sum of two functions of exponential type and therefore is also in $\mathcal{E}$.

However, in general, f does not satisfy the hypothesis of Corollary 6.3.9, so we must proceed with more care to confirm that f is in $\mathcal{E}$. We want to show that, in general, $f(t) = f(0) + g(t)$ for all $t \geq 0$. To this end, fix $t > 0$ and any $\epsilon > 0$. Suppose that f' fails to exist at k points in $(0, t)$. Index these points so that

$$0 < x_1 < x_2 < x_3 < \cdots < x_k < t.$$

Since f and the function e^{rx} are uniformly continuous on the compact set $[0, t]$, there exists a single $\delta > 0$ such that, for any two points x and y in $[0, t]$ with $|x - y| < \delta$,

$$|f(x) - f(y)| < \frac{\epsilon}{k+2}$$

and

$$|e^{rx} - e^{ry}| < \frac{\epsilon r}{a(k+2)}.$$

For $j = 1, 2, \ldots, k$, choose u_j and v_j in $N'(x_j; \delta/2)$ with $u_j < x_j < v_j$. Let $u_0 = 0$; choose v_0 in $(0, \delta)$ and u_{k+1} in $(t - \delta, t)$; let $v_{k+1} = t$. (See Fig. 13.11.) By

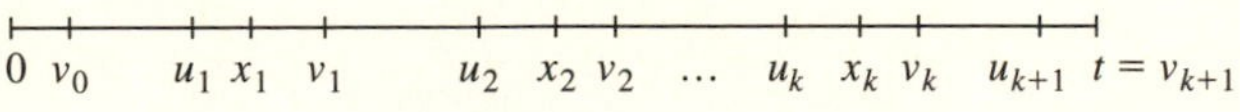

$u_0 = 0 \quad v_0 \quad u_1 \quad x_1 \quad v_1 \quad u_2 \quad x_2 \quad v_2 \quad \ldots \quad u_k \quad x_k \quad v_k \quad u_{k+1} \quad t = v_{k+1}$

Figure 13.11

telescoping and the additivity of the integral, we have

$$g(t) = \int_0^t f'(u)\,du = \sum_{j=0}^{k+1} \int_{u_j}^{v_j} f'(u)\,du + \sum_{j=1}^{k+1} \int_{v_{j-1}}^{u_j} f'(u)\,du$$

and

$$f(t) - f(0) = \sum_{j=0}^{k+1} [f(v_j) - f(u_j)] + \sum_{j=1}^{k+1} [f(u_j) - f(v_{j-1})].$$

Since, for $j = 1, 2, \ldots, k+1$, the function f is differentiable on $[v_{j-1}, u_j]$, Corollary 6.3.9 does apply on these intervals and we have

$$\sum_{j=1}^{k+1} \int_{v_{j-1}}^{u_j} f'(u)\,du = \sum_{j=1}^{k+1} [f(u_j) - f(v_{j-1})].$$

Also, since $|v_j - u_j| < \delta$, we have

$$\sum_{j=0}^{k+1} |f(v_j) - f(u_j)| < \epsilon.$$

Finally, since $|f'(u)| \le ae^{ru}$, we have

$$\sum_{j=0}^{k+1} \int_{u_j}^{v_j} |f'(u)|\, du \le \sum_{j=0}^{k+1} \int_{u_j}^{v_j} ae^{ru}\, du = \frac{a}{r} \sum_{j=0}^{k+1} |e^{rv_j} - e^{ru_j}| < \epsilon.$$

Therefore,

$$\begin{aligned} &|f(t) - [f(0) + g(t)]| \\ &\qquad \le \sum_{j=0}^{k+1} |f(v_j) - f(u_j)| + \sum_{j=0}^{k+1} \int_{u_j}^{v_j} |f'(u)|\, du < 2\epsilon. \end{aligned}$$

Since ϵ is arbitrary, it follows that $f(t) = f(0) + g(t)$. Therefore, f is in $\mathcal{E}$. ●

Our interest in functions of exponential type results from the following theorem.

THEOREM 13.6.6 If f is in $\mathcal{E}$, then the Laplace transform of f exists.

Proof. Suppose that $|f(t)| \le ae^{rt}$ for all $t \ge 0$. Then for all $s > r$,

$$\int_0^\infty e^{-st} |f(t)|\, dt \le \int_0^\infty ae^{(r-s)t}\, dt = \lim_{b\to\infty} \left. \frac{-ae^{(r-s)t}}{s-r} \right|_0^b = \frac{a}{s-r}.$$

That is, $\int_0^\infty e^{-st} f(t)\, dt$ is absolutely convergent and, therefore, convergent for $s > r$. Therefore Lf exists. ●

We must raise the following natural question: Suppose that f and g are two functions in $\mathcal{L}$ and that $Lf(s) = Lg(s)$ for all $s \ge s_0$. Can we conclude that $f = g$? In effect, we are asking whether L is one-to-one and, consequently, whether the inverse Laplace transform L^{-1} exists. Again, we are raising a question whose complete answer requires tools that fall beyond the scope of this text. Here we will be satisfied to state the following theorem without proof.

THEOREM 13.6.7 Uniqueness Theorem Suppose that f and g are in $\mathcal{E}$ and that $Lf(s) = Lg(s)$ for sufficiently large s. Then $f(t) = g(t)$ for all t where f and g are continuous. ●

The next property of the Laplace transform, which makes this transform useful in solving linear differential equations, is identified in the following theorem.

THEOREM 13.6.8 The First Differentiation Theorem If f is continuous on $[0, \infty)$ and if f' is in $\mathcal{E}$, then

$$(Lf')(s) = s(Lf)(s) - f(0).$$

Proof. It is to be understood that, as in Theorem 13.6.5, f' is assumed to exist at all but countably many discrete points in $[0, \infty)$ and, at points where f' fails to exist, we assign value 0 to it. It is also to be understood that $f(0) = \lim_{t\to 0^+} f(t) = f(0^+)$. (Recall that $f(0^-) = 0$ since $f(t) = 0$ for all $t < 0$.) Now, the hypotheses of the present theorem, together with Theorem 13.6.5, imply that f is in $\mathcal{E}$. Therefore, there are constants a and r such that $|f(t)| \le ae^{rt}$, for all $t \ge 0$. Fix $s_0 > r$. For $s \ge s_0$, integrate by parts to obtain

$$\begin{aligned}(Lf')(s) = \int_0^\infty e^{-st} f'(t)\,dt &= \lim_{b\to\infty} e^{-st} f(t)\Big|_0^b + s\int_0^\infty e^{-st} f(t)\,dt \\ &= \lim_{b\to\infty} e^{-sb} f(b) - f(0) + s(Lf)(s).\end{aligned}$$

Since f is of exponential type, we have

$$0 \le \lim_{b\to\infty} e^{-sb}|f(b)| \le \lim_{b\to\infty} ae^{(r-s)b} = 0$$

because $r - s < 0$. By the Squeeze Play, $\lim_{b\to\infty} e^{-sb} f(b) = 0$. We conclude that $(Lf')(s) = s(Lf)(s) - f(0)$, as claimed. ●

COROLLARY 13.6.9 Suppose that f has $k-1$ continuous derivatives on $[0, \infty)$ and that $f^{(k)}$ is in $\mathcal{E}$. Then

$$(Lf^{(k)})(s) = s^k(Lf)(s) - \sum_{j=0}^{k-1} s^{k-j-1} f^{(j)}(0).$$

Proof. The hypotheses imply that, for $j = 0, 1, 2, \ldots, k$, each of the functions $f^{(j)}$ is of exponential type. By the theorem, $Lf'(s) = s(Lf)(s) - f(0)$. Thus

$$\begin{aligned}Lf''(s) &= s(Lf')(s) - f'(0) = s[s(Lf)(s) - f(0)] - f'(0) \\ &= s^2(Lf)(s) - sf(0) - f'(0).\end{aligned}$$

Repeating this maneuver recursively proves the corollary. ●

EXAMPLE 31 To solve the differential equation

$$y'' - y' - 2y = 0 \tag{13.19}$$

with initial conditions $y(0) = 2$, $y'(0) = 3$, assume that $y = f(t)$ is a solution that is in $\mathcal{E}$. Applying the Laplace transform to (13.19) yields

$$Ly'' - Ly' - 2Ly = 0. \tag{13.20}$$

By Theorem 13.6.5 and its corollary with $k = 2$, we have

$$Ly'(s) = s(Ly)(s) - y(0) = s(Ly)(s) - 2$$

and

$$Ly''(s) = s^2(Ly)(s) - sy(0) - y'(0) = s^2(Ly)(s) - 2s - 3.$$

Substitute these quantities into (13.20) to obtain

$$[s^2(Ly)(s) - 2s - 3] - [s(Ly)(s) - 2] - 2Ly(s)$$
$$= (s^2 - s - 2)Ly(s) - 2s - 1 = 0.$$

Solving for $Ly(s)$ and partial fractionating yields

$$Ly(s) = \frac{2s+1}{s^2 - s - 2} = \frac{5}{3}\left(\frac{1}{s-2}\right) + \frac{1}{3}\left(\frac{1}{s+1}\right).$$

Now, from our computed transforms—specifically (3) in that list—the Laplace transform of

$$y_1(t) = \frac{5}{3}e^{2t}$$

is $(5/3)[1/(s-2)]$ and the Laplace transform of

$$y_2(t) = \frac{1}{3}e^{-t}$$

is $(1/3)[1/(s+1)]$. Consequently, by the linearity of L, the Laplace transform of $y_1(t) + y_2(t) = (5/3)e^{2t} + (1/3)e^{-t}$ is

$$\frac{5}{3}\left(\frac{1}{s-2}\right) + \frac{1}{3}\left(\frac{1}{s+1}\right).$$

The Laplace transforms of the (unknown) solution of the equation (13.19) and of the function $y(t) = (5/3)e^{2t} + (1/3)e^{-t}$ agree for $s \geq s_0 > 2$. Therefore, by the Uniqueness Theorem, the solution is $y(t) = (5/3)e^{2t} + (1/3)e^{-t}$. As you can confirm the function we have found satisfies both the differential equation (13.19) and the initial conditions $y(0) = 2$ and $y'(0) = 3$. ●

THEOREM 13.6.10 The First Integration Theorem If f is in $\mathcal{E}$, if $F = Lf$, and if $g(t) = \int_0^t f(u)\,du$, then

$$Lg(s) = \frac{F(s)}{s}.$$

Proof. We will apply Theorem 13.6.8 to the function g. By the Fundamental Theorem of Calculus, g is continuous on $[0, \infty)$ and $g'(t) = f(t)$ at all t where f is continuous. Thus g' is in $\mathcal{E}$. Note also that $g(0) = 0$. Therefore, the formula in the First Differentiation Theorem becomes

$$Lg'(s) = s(Lg)(s) - g(0)$$

or

$$F(s) = Lf(s) = s(Lg)(s).$$

The conclusion of the theorem follows. ●

THEOREM 13.6.11 The Second Differentiation Theorem If f is in $\mathcal{E}$ and if $F = Lf$, then F is differentiable and

$$F'(s) = -L(tf(t))(s)$$

for sufficiently large values of s.

Proof. Since f is in $\mathcal{E}$, there exist constants $a > 0$ and r such that, for all $t \geq 0$, $|f(t)| < ae^{rt}$. Without loss of generality, r is positive. Thus $|tf(t)| \leq ae^{(r+1)t}$ for all $t \geq 0$. Fix $s_0 > r + 1$. We claim that the integral defining $L(tf(t))(s)$ converges uniformly for $s \geq s_0$. To see this, note that for such s and for all $t \geq 0$,

$$|e^{-st}tf(t)| \leq e^{-st}(ae^{(r+1)t}) = ae^{(r+1-s)t} \leq ae^{(r+1-s_0)t} = M(t).$$

Notice also that, since $r + 1 - s_0 < 0$, the integral $\int_0^\infty M(t)\,dt$ converges. Therefore, by Weierstrass's M-test, $\int_0^\infty e^{-st}tf(t)\,dt$ converges uniformly for $s \geq s_0$.

Since the partial derivative of $e^{-st}f(t)$ with respect to the variable s is $-e^{-st}tf(t)$, we deduce, by Theorem 13.4.5, that the function $F(s) = \int_0^\infty e^{-st}f(t)\,dt$ is differentiable for $s \geq s_0$ and that

$$F'(s) = -\int_0^\infty e^{-st}tf(t)\,dt = -L(tf(t))(s)$$

for sufficiently large values of s. ●

THEOREM 13.6.12 The Second Integration Theorem If the function $f(t)/t$ is in $\mathcal{E}$ and if $F = Lf$, then for s sufficiently large, F is integrable on $[s, \infty)$ and

$$\int_s^\infty F(u)\,du = L\left(\frac{f(t)}{t}\right)(s).$$

Proof. We will sketch the proof, leaving for you the task of filling in the details. First write

$$\int_s^\infty F(u)\,du = \int_{u=s}^\infty \left[\int_{t=0}^\infty e^{-ut}f(t)\,dt\right]du.$$

Imitate the maneuvers in the proof of Theorem 13.6.11, using the hypothesis that $f(t)/t$ is of exponential type, to identify an s_0 such that, for $s > s_0$, the hypotheses of Theorem 13.4.9 are satisfied. That theorem, which justifies the interchange of two iterated improper integrals, ensures that we can legitimately write

$$\int_s^\infty F(u)\,du = \int_{t=0}^\infty f(t)\left[\int_{u=s}^\infty e^{-ut}\,du\right]dt, \tag{13.21}$$

for $s \geq s_0$. The inner integral is simply $\int_s^\infty e^{-ut}\,du = e^{-st}/t$. Upon substituting this result in (13.21), we obtain

$$\int_s^\infty F(u)\,du = \int_0^\infty \frac{e^{-st}f(t)}{t}\,dt = L\left(\frac{f(t)}{t}\right)(s),$$

as asserted by the theorem. ●

Our next three theorems treat the limiting behavior of the function $F = Lf$. We are especially interested in discovering what can be said about $\lim_{s\to\infty} F(s)$ and $\lim_{s\to\infty} sF(s)$.

THEOREM 13.6.13 If f is in $\mathcal{E}$ and if $F = Lf$, then there exists an s_0 such that, for $s \geq s_0$, $|sF(s)|$ is bounded. In particular, $\lim_{s\to\infty} F(s) = 0$.

Remark. If f is in $\mathcal{L}$ but is not in $\mathcal{E}$, then it need not be true that $|sF(s)|$ is bounded. However, it remains true that $\lim_{s\to\infty} F(s) = 0$. At this stage, we lack the tools to prove these claims.

Proof. Choose constants a and r such that $|f(t)| \le ae^{rt}$ for all $t \ge 0$. Fix any $s_0 > \max\{r, 0\}$. Then for $s \ge s_0$,

$$|sF(s)| = s\left|\int_0^\infty e^{-st} f(t)\,dt\right| \le s\int_0^\infty ae^{(r-s)t}\,dt$$
$$= \frac{sa}{s-r} = a + \frac{ar}{s-r} \le a + \frac{ar}{s_0 - r}.$$

Therefore, $|sF(s)|$ is bounded. Knowing that $|sF(s)|$ is bounded for $s \ge s_0$, we certainly know, as a consequence, that $\lim_{s\to\infty} F(s) = 0$. This proves the theorem. ●

To illustrate the theorem, let $f(t) = \cos at$. The Laplace transform of f is $F(s) = s/(s^2 + a^2)$. Then $|sF(s)| = s^2/(s^2 + a^2)$ is certainly bounded (by the number 1) and $\lim_{s\to\infty} F(s) = \lim_{s\to\infty} s/(s^2 + a^2) = 0$.

In certain cases we can identify explicitly the limiting value of $sF(s)$ as $s \to \infty$. The details are provided in the following theorem.

THEOREM 13.6.14 If f is continuous on $[0, \infty)$, if f' is in $\mathcal{E}$, and if $F = Lf$, then $\lim_{s\to\infty} sF(s) = f(0)$.

Proof. By Theorem 13.6.8 we have

$$(Lf')(s) = sF(s) - f(0).$$

By Theorem 13.6.13, $\lim_{s\to\infty}(Lf')(s) = 0$. The conclusion of the present theorem follows immediately. ●

To illustrate Theorem 13.6.14, let $f(t) = (\sin at)/t$. Note that

$$f(0) = \lim_{t\to 0^+} \frac{\sin at}{t} = a.$$

The Laplace transform of f is $F(s) = \tan^{-1}(a/s)$. We want to show that $\lim_{s\to\infty} s\tan^{-1}(a/s) = a$. Transform the limit by setting $u = 1/s$ and apply l'Hôpital's rule:

$$\lim_{s\to\infty} s\tan^{-1}\left(\frac{a}{s}\right) = \lim_{u\to 0^+} \frac{\tan^{-1}(au)}{u}$$
$$= \lim_{u\to 0^+} \frac{a}{1 + (au)^2} = a,$$

as guaranteed by the theorem.

THEOREM 13.6.15 If f is in $\mathcal{E}$, if $\lim_{t\to\infty} f(t) = c$, and if $F = Lf$, then $\lim_{s\to 0^+} sF(s) = c$.

Proof. Let $f(t) = c + g(t)$ where $\lim_{t\to\infty} g(t) = 0$. Given $\epsilon > 0$, choose $t_0 > 0$ such that, for $t \geq t_0$, we have $|g(t)| < \epsilon$. The constant function c is in $\mathcal{E}$; hence $g = f - c$ is also in $\mathcal{E}$. Write

$$\begin{aligned} F(s) &= \int_0^\infty e^{-st} f(t)\,dt = \int_0^{t_0} e^{-st} f(t)\,dt + \int_{t_0}^\infty e^{-st} f(t)\,dt \\ &= \int_0^{t_0} e^{-st} f(t)\,dt + c\int_{t_0}^\infty e^{-st}\,dt + \int_{t_0}^\infty e^{-st} g(t)\,dt. \end{aligned} \tag{13.22}$$

Observe that $|\int_0^{t_0} e^{-st} f(t)\,dt| \leq \int_0^{t_0} |f(t)|\,dt = K$ is a nonnegative constant. Also notice that

$$c\int_{t_0}^\infty e^{-st}\,dt = \left.\frac{-ce^{-st}}{s}\right|_{t_0}^\infty = \frac{ce^{-st_0}}{s}. \tag{13.23}$$

Finally, since $|g(t)| < \epsilon$ for all $t \geq t_0$, we have

$$\left|\int_{t_c}^\infty e^{-st} g(t)\,dt\right| < \epsilon\int_{t_0}^\infty e^{-st}\,dt = \left.\frac{-\epsilon e^{-st}}{s}\right|_{t_0}^\infty = \frac{\epsilon e^{-st_0}}{s} < \frac{\epsilon}{s}. \tag{13.24}$$

Combining (13.22), (13.23), and (13.24) and multiplying by s yields

$$|sF(s) - c| < Ks + |ce^{-st_0} - c| + \epsilon.$$

Apply the Squeeze Play to obtain

$$0 \leq \lim_{s\to 0^+} |sF(s) - c| \leq \lim_{s\to 0^+} Ks + \lim_{s\to 0^+} |ce^{-st_0} - c| + \epsilon = \epsilon.$$

Since ϵ is arbitrarily small, we conclude that $\lim_{s\to 0^+} sF(s) = c$ as asserted in the theorem. ●

The converse of Theorem 13.6.15 is false as we can see by considering $f(t) = \sin at$. The Laplace transform of f is $F(s) = a/(s^2 + a^2)$. We have $\lim_{s\to 0^+} sF(s) = \lim_{s\to 0^+} sa/(s^2 + a^2) = 0$, whereas $\lim_{t\to\infty} f(t) = \lim_{t\to\infty} \sin at$ does not exist. On the other hand, if we let $f(t) = e^{-t}\cos at$, then

$$\lim_{t\to\infty} f(t) = \lim_{t\to\infty} e^{-t}\cos at = 0.$$

The Laplace transform of f is

$$F(s) = \frac{s+1}{(s+1)^2 + a^2}$$

and $\lim_{s\to 0^+} sF(s) = 0$ also.

13.6.1 The Convolution Product

Let f and g be two integrable functions on $[0, \infty)$, extended in accord with our current convention to have value 0 at all points of $(-\infty, 0)$.

DEFINITION 13.6.3 The convolution of f and g, denoted $f * g$, is the function whose value at $t \geq 0$ is

$$(f * g)(t) = \int_0^t f(u)g(t-u)\,du. \quad \bullet$$

Several comments are in order. First, consider the similarity between the convolution of f and g and the Cauchy product of two series $\sum_{j=0}^{\infty} a_j$ and $\sum_{j=0}^{\infty} b_j$. Recall that the kth term of the Cauchy product of these two series is $\sum_{j=0}^{k} a_j b_{k-j}$. If we view the index set $\mathbb{N} \cup \{0\}$ as a discrete model for time, then this formula for the kth term is the discrete analog of the convolution integral. Indeed, if we let $f(t) = a_j$ and $g(t) = b_j$ for t in $[j, j+1)$, then

$$\begin{aligned}(f * g)(k+1) &= \int_0^{k+1} f(u)g(k+1-u)\,du \\ &= \sum_{j=0}^{k} \int_j^{j+1} f(u)g(k+1-u)\,du \\ &= \sum_{j=0}^{k} a_j b_{k-j}.\end{aligned}$$

This observation reveals that the convolution integral is a natural extension of the discrete Cauchy product on $\mathbb{N} \cup \{0\}$ to the continuum $[0, \infty)$ with integration replacing summation.

But the convolution integral has more profound significance than mere mimicry. Suppose that we have a linear stimulus-response system and that variable stimulus inputs occur from time 0 to time t. Let $f(u)$ denote the intensity of the stimulus at time u and $f(u)\,du$ the magnitude of the stimulus during a short time interval of duration du. The magnitude of the response to this stimulus is proportional to its magnitude and varies according to the amount of time $t-u$ that has elapsed since the stimulus occurred; usually, although not always, the response diminishes as time passes. Let $g(t-u)$ denote the proportionality parameter; this parameter also may vary in time. Then $f(u)g(t-u)\,du$ is a numerical measure of the system's response at time t resulting from the stimulus that occurred at time u (in the past) in a time period of duration du. Therefore the convolution integral $\int_0^t f(u)g(t-u)\,du$ is a numerical measure of the total response of the system to the stream of stimuli experienced by the system during the time period $[0, t]$.

Finally, it is worth noting that for $u < 0$, $f(u) = 0$ and for $u > t$, we have $t - u < 0$; therefore, $g(t-u) = 0$. These observations allow us to write

$$(f * g)(t) = \int_0^t f(u)g(t-u)\,du = \int_{-\infty}^{\infty} f(u)g(t-u)\,du.$$

This latter form of the convolution integral is often convenient when changing variables; using this form, one need not be concerned about adjusting the transformed integral to have limits of integration 0 and t. While this form of the integral defining $f * g$ appears to be improper, in fact, with our conventions, it is not.

We begin by examining some of the elementary properties of convolution. As we shall see, an appropriately chosen collection of integrable functions on $[0, \infty)$, when endowed with ordinary addition of functions and with convolution as a multiplication, forms a commutative ring. Specifically, we shall restrict our discussion to those functions for which the order of integration of any iterated integrals that arise can be reversed. We will also assume that each function f considered here is defined on $[0, \infty)$ and is bounded and piecewise continuous on every interval $[0, b]$; that is, for every $b > 0$, f is bounded on $[0, b]$ and is continuous on $[0, b]$ except possibly at finitely many points. The functions of exponential type are especially well behaved for our purposes and form a large class of functions of classical importance.

THEOREM 13.6.16 Let f, g, and h be functions that are defined on $[0, \infty)$, vanish on $(-\infty, 0)$, and, for each $b > 0$, are bounded and piecewise continuous on $[0, b]$.

i) The function $f * g$ is also defined on $[0, \infty)$, vanishes on $(-\infty, 0)$, and, for each $b > 0$, is continuous and bounded on $[0, b]$.

ii) Convolution is a commutative operation:

$$f * g = g * f.$$

iii) Convolution is an associative operation:

$$(f * g) * h = f * (g * h).$$

iv) Convolution distributes over addition:

$$f * (g + h) = f * g + f * h.$$

Proof. For (i), note that if $t < 0$, then $(f * g)(t) = \int_0^t f(u)g(t-u)\,du$ certainly has value 0 since $f(u) = 0$ for all u in $[t, 0]$. The continuity of $f * g$ follows immediately from the piecewise continuity of f and g on $[0, t]$ for every $t \geq 0$ and from the Fundamental Theorem of Calculus. Hence, for each $b > 0$, $f * g$ is bounded on $[0, b]$.

The commutativity of convolution follows from a simple change of variables. In the integral $\int_{-\infty}^{\infty} f(u)g(t-u)\,du$, let $w = t - u$. Then

$$\begin{aligned}(f * g)(t) &= \int_{-\infty}^{\infty} f(u)g(t-u)\,du = -\int_{\infty}^{-\infty} \infty f(t-w)g(w)\,dw \\ &= \int_{-\infty}^{\infty} g(w)f(t-w)\,dw = (g * f)(t).\end{aligned}$$

For associativity, begin with

$$\begin{aligned}[(f*g)*h](t) &= \int_{-\infty}^{\infty} (f*g)(u)h(t-u)\,du \\ &= \int_{u=-\infty}^{\infty} \left[\int_{w=-\infty}^{\infty} f(w)g(u-w)h(t-u)\,dw\right] du \\ &= \int_{w=-\infty}^{\infty} f(w) \left[\int_{u=-\infty}^{\infty} g(u-w)h(t-u)\,du\right] dw,\end{aligned}$$

assuming we can interchange the order of integration. Change variables in the inner integral by setting $v = u - w$ to obtain

$$\begin{aligned}[(f*g)*h](t) &= \int_{w=-\infty}^{\infty} f(w) \left[\int_{v=-\infty}^{\infty} g(v)h(t-w-v)\,dv\right] dw \\ &= \int_{w=-\infty}^{\infty} f(w)(g*h)(t-w)\,dw = [f*(g*h)](t).\end{aligned}$$

This proves that, as functions, $(f*g)*h = f*(g*h)$ and therefore that convolution is associative. The distributive law is easily proved. Simply compute

$$\begin{aligned}[f*(g+h)](t) &= \int_{-\infty}^{\infty} f(u)[g(t-u)+h(t-u)]\,du \\ &= \int_{-\infty}^{\infty} f(u)g(t-u)\,du + \int_{-\infty}^{\infty} f(u)h(t-u)\,du \\ &= (f*g)(t) + (f*h)(t).\end{aligned}$$

Therefore $f*(g+h) = f*g + f*h$. This completes the proof of the theorem. ●

Our interest in convolution in this section results from a remarkable fact. Under fairly general hypotheses, the Laplace transform of $f*g$ is the pointwise product $(Lf)(Lg)$. That is, the Laplace transform changes the elaborate multiplication defined by convolution into the simple pointwise multiplication of functions with which you are already familiar. When such a fact emerges in the course of your studies, you know you have found a genuine piece of mathematics.

EXAMPLE 32 For $t \geq 0$, let $f(t) = e^{at}$ and $g(t) = e^{bt}$. If $a = b$, then $(f*g)(t) = \int_0^t e^{au}e^{a(t-u)}\,du = te^{at}$.

If $a \neq b$, then

$$\begin{aligned}(f*g)(t) &= \int_0^t e^{au}e^{b(t-u)}\,du = e^{bt}\int_0^t e^{(a-b)u}\,du \\ &= e^{bt}\frac{e^{(a-b)u}}{a-b}\bigg|_0^b = \frac{e^{at}-e^{bt}}{a-b}.\end{aligned}$$

In each case, we compute $L(f * g)(s)$ from the list of examples at the beginning of this section. If $a = b$, we use the fact that $L(t)(s) = 1/s^2$ and the First Shift Theorem to compute

$$L(f * g)(s) = L(te^{at})(s) = \frac{1}{(s-a)^2} = \left(\frac{1}{s-a}\right)\left(\frac{1}{s-a}\right)$$
$$= L(e^{at})(s) \cdot L(e^{at})(s) = Lf(s) \cdot Lg(s).$$

(Recall that $L(e^{at})(s) = 1/(s-a)$.) If $a \neq b$, then by the linearity of L,

$$L(f * g)(s) = L\left(\frac{e^{at} - e^{bt}}{a-b}\right)(s)$$
$$= \frac{L(e^{at})(s) - L(e^{bt})(s)}{a-b}$$
$$= \frac{1}{a-b}\left[\frac{1}{s-a} - \frac{1}{s-b}\right]$$
$$= \frac{1}{(s-a)(s-b)} = L(e^{at})(s) \cdot L(e^{bt})(s).$$

In each case, this confirms that $L(f * g) = (Lf) \cdot (Lg)$. ●

THEOREM 13.6.17 If f and g are in $\mathcal{E}$, then

i) $f * g$ is also in $\mathcal{E}$.

ii) $L(f * g) = (Lf) \cdot (Lg)$.

Remark. The requirement that f and g are in $\mathcal{E}$ is unnecessarily stringent; in fact, this theorem is valid for a much larger class of functions. However, again, a rigorous proof requires tools beyond the level of this text. We will be satisfied with our version of this theorem using only information available to us.

Proof. Any function in $\mathcal{E}$ is bounded and piecewise continuous on every interval $[0, b]$. Thus $f * g$ is defined and is continuous on $[0, \infty)$. Therefore condition (i) of Definition 13.6.2 holds for $f * g$. Further, we can choose constants a_1, a_2, r_1, and r_2 such that $|f(t)| \leq a_1 e^{r_1 t}$ and $|g(t)| \leq a_2 e^{r_2 t}$, for all $t \geq 0$. Assume, without loss of generality, that $r_1 \neq r_2$. Now compute

$$|(f * g)(t)| = \left|\int_0^t f(u)g(t-u)\,du\right| \leq \int_0^t |f(u)||g(t-u)|\,du$$
$$\leq a_1 a_2 \int_0^t e^{r_1 u} e^{r_2(t-u)}\,du = a_1 a_2 e^{r_2 t} \int_0^t e^{(r_1 - r_2)u}\,du$$
$$= a_1 a_2 e^{r_2 t}\left[\frac{e^{(r_1-r_2)u}}{r_1 - r_2}\Bigg|_0^t\right] = \frac{a_1 a_2 (e^{r_1 t} - e^{r_2 t})}{r_1 - r_2}$$
$$\leq \frac{2a_1 a_2}{|r_1 - r_2|} e^{rt} = ae^{rt},$$

where $a = 2a_1a_2/|r_1 - r_2|$ and $r = \max\{r_1, r_2\}$. Therefore $f * g$ is of exponential type.

To prove (ii), we compute the Laplace transform of $(f * g)$. Assuming that the interchange of the iterated integrals at the last step can be justified below, we obtain

$$L(f * g)(s) = \int_0^\infty e^{-st}(f * g)(t)\,dt = \int_{t=0}^\infty \left[\int_{u=0}^t e^{-st} f(u)g(t-u)\,du\right] dt$$
$$= \int_{t=0}^\infty \left[\int_{u=0}^\infty e^{-st} f(u)g(t-u)\,du\right] dt$$
$$= \int_{u=0}^\infty f(u)\left[\int_{t=0}^\infty e^{-st} g(t-u)\,dt\right] du.$$

Change variables in the inner integral by setting $w = t - u$; after completing the details of this transformation we have

$$L(f * g)(s) = \int_{u=0}^\infty f(u)\left[\int_{w=-u}^\infty e^{-s(u+w)} g(w)\,dw\right] du$$
$$= \int_{u=0}^\infty e^{-su} f(u)\left[\int_{w=0}^\infty e^{-sw} g(w)\,dw\right] du$$
$$= [Lf(s)] \cdot [Lg(s)],$$

as promised in the theorem. (Recall that $f(u) = 0$ for u in $(-\infty, 0)$ and that $g(w) = 0$ for w in $[-u, 0)$.)

All that remains is to justify the interchange of the iterated integrals in the preceding argument. To that end we will appeal to Theorem 13.4.9. In addition, we will assume here that f and g are continuous on $[0, \infty)$ so that the integrand is continuous on $[0, \infty) \times [0, \infty)$. When computing $L(f * g)$, we restrict to $s \ge s_0 > r = \max\{r_1, r_2\}$.

It is immediate that hypothesis (iv) in Theorem 13.4.9 is satisfied. Because $f * g$ is in $\mathcal{E}$, $\int_{t=0}^\infty [\int_{u=0}^\infty e^{-st} f(u)g(t-u)\,du]\,dt$ converges [to $L(f * g)(s)$]. To confirm hypothesis (i) of Theorem 13.4.9 we must prove that

$$\int_0^\infty e^{-st} f(u)g(t-u)\,dt = f(u)\int_0^\infty e^{-st} g(t-u)\,dt \tag{13.25}$$

converges uniformly for u in $[0, \infty)$. In fact, we can show that this integral converges to $e^{-su} f(u)(Lg)(s)$, for $s \ge s_0 > r$. To see this, change variables in the last integral in (13.25) by letting $w = t - u$. We obtain

$$f(u)\int_0^\infty e^{-st} g(t-u)\,dt = f(u)\int_{-u}^\infty e^{-s(u+w)} g(w)\,dw$$
$$= e^{-su} f(u)\int_0^\infty e^{-sw} g(w)\,dw = e^{-su} f(u)(Lg)(s).$$

But to show that the convergence is uniform for u in $[0, \infty)$, we must examine the convergence more carefully. Fix $\epsilon > 0$. Choose b_0 such that $a_1 a_2 e^{-(s-r)b_0}/(s-r) < \epsilon$. For $b \geq b_0$, we compute

$$\left| \int_b^\infty e^{-st} f(u) g(t-u)\, dt \right| \leq a_1 a_2 e^{(r_1 - r_2)u} \int_b^\infty e^{-(s-r_2)t}\, dt$$
$$= \frac{a_1 a_2 e^{(r_1-r_2)u} e^{-(s-r_2)b}}{s - r_2}$$
$$\leq \frac{a_1 a_2 e^{-(s-r)b_0}}{s-r} < \epsilon,$$

for all u in $[0, \infty)$. Thus, the integral (13.25) converges uniformly for u in $[0, \infty)$ and hypotheses (i) of Theorem 13.4.9 is confirmed.

To guarantee that hypothesis (iii) of Theorem 13.4.9 is satisfied, we must show that

$$\int_0^\infty e^{-st} f(u) g(t-u)\, du = e^{-st} \int_0^\infty f(u) g(t-u)\, du$$
$$= e^{-st} (f * g)(t) \qquad \text{[uniformly]} \qquad \textbf{(13.26)}$$

for t in $[0, \infty)$. Notice that we can assume without loss of generality that $r_1 < r_2$; otherwise simply increase r_2 until this inequality is met. Again, given $\epsilon > 0$, choose $b_0 > 0$ such that $a_1 a_2 e^{-(r_2-r_1)b_0}/(r_2 - r_1) < \epsilon$. For $b \geq b_0$, we have, by a computation similar to that we have just seen,

$$\left| \int_b^\infty e^{-st} f(u) g(t-u)\, du \right| \leq \frac{a_1 a_2 e^{-(s-r_1)t} e^{-(r_2-r_1)b_0}}{r_2 - r_1}$$
$$\leq \frac{a_1 a_2 e^{-(r_2-r_1)b_0}}{r_2 - r_1} < \epsilon.$$

This proves that the integral (13.26) converges uniformly for t in $[0, \infty)$ and that hypotheses (iii) of Theorem 13.4.9 is met. For hypothesis (ii) of Theorem 13.4.9, we must show that

$$G(v) = \int_{t=0}^\infty \left[\int_{u=0}^v e^{-st} f(u) g(t-u)\, du \right] dt \qquad \text{[uniformly]} \qquad \textbf{(13.27)}$$

for v in $[0, \infty)$. That is, given $\epsilon > 0$, we must find a $b_0 > 0$ such that, for $b \geq b_0$,

$$\left| \int_{t=b}^\infty \left[\int_{u=0}^v e^{-st} f(u) g(t-u)\, du \right] dt \right| < \epsilon, \qquad \textbf{(13.28)}$$

for all v in $[0, \infty)$. Again, we assume that $r_1 < r_2$ and, when computing the Laplace transform of $f * g$, we restrict s to be greater than $s_0 > r_2$. Choose $b_0 > 0$ such that

$$\frac{a_1 a_2 e^{-(s-r_2)b_0}}{(s-r_2)(r_2-r_1)} < \epsilon.$$

By imitating our reasoning above, you can show that, for $b \geq b_0$, (13.28) is valid and hence that the integral (13.27) converges uniformly for v in $[0, \infty)$. All the hypotheses of Theorem 13.4.9 are met and thus

$$\int_{t=0}^{\infty}\left[\int_{u=0}^{\infty} e^{-st} f(u)g(t-u)\,du\right] dt = \int_{u=0}^{\infty}\left[\int_{t=0}^{\infty} e^{-st} f(u)g(t-u)\,dt\right] du.$$

This completes the proof of Theorem 13.6.17. ●

EXAMPLE 33 To see how this theory can be used, we consider an elementary example. Let

$$y'' + a_1 y' + a_0 y = f(t) \tag{13.29}$$

be a linear differential equation where a_0 and a_1 are constants and f is in $\mathcal{E}$. We want to find a function $y = y(t)$ that satisfies this equation and the initial conditions $y(0) = y'(0) = 0$. We assume that y, y', and y'' are in $\mathcal{E}$. Our first step is to apply the Laplace transform of both sides of (13.29). From Theorem 13.6.8 and its corollary we deduce that $(Ly')(s) = s(Ly)(s)$ and $(Ly'')(s) = s^2(Ly)(s)$. Thus

$$\begin{aligned} L(y'' + a_1 y' + a_0 y) &= s^2(Ly)(s) + a_1 s(Ly)(s) + a_0(Ly)(s) \\ &= (s^2 + a_1 s + a_0)(Ly)(s) = (Lf)(s). \end{aligned}$$

The quadratic polynomial $p(s) = s^2 + a_1 s + a_0$ is called the *characteristic polynomial* of the differential operator $Ty = y'' + a_1 y' + a_0 y$. From this point the details depend on the discriminant $\Delta = a_1^2 - 4a_0$ of the quadratic polynomial $p(s)$.

If $\Delta > 0$, then $p(s)$ has two distinct, real roots r_1 and r_2 and factors into $p(s) = (s - r_1)(s - r_2)$. Therefore, by using a partial fraction decomposition, $1/p(s)$ can be written as

$$\frac{1}{p(s)} = \frac{c_1}{s - r_1} + \frac{c_2}{s - r_2}, \tag{13.30}$$

where $c_1 = 1/(r_1 - r_2)$ and $c_2 = 1/(r_2 - r_1)$. Now, for $j = 1, 2$, the Laplace transform of $c_j e^{r_j t}$ is $c_j/(s - r_j)$. Therefore, by the linearity of L and by (13.30), the function $1/p(s)$ is the Laplace transform of the function

$$g(t) = c_1 e^{r_1 t} + c_2 e^{r_2 t}.$$

That is, $Lg(s) = 1/p(s)$.

If $\Delta = 0$, then $p(s)$ has a double root r; $p(s) = (s - r)^2$. The function $1/p(s) = 1/(s - r)^2$ is the Laplace transform of the function $g(t) = te^{rt}$: $Lg(s) = 1/p(s)$.

If $\Delta < 0$, then p has complex roots $v + iw$ and does not factor over $\mathbb{R}$. By completing the square, we can write $p(s) = (s - v)^2 + w^2$ where v and w are real constants with $w \neq 0$. Therefore

$$\frac{1}{p(s)} = \frac{1}{(s - v)^2 + w^2}.$$

The function $1/[(s - v)^2 + w^2]$ is the Laplace transform of the function $g(t) = (1/w)e^{vt} \sin wt$. In this case again, we have $Lg(s) = 1/p(s)$.

In each case, we have found the unique function g such that $Lg(s) = 1/p(s)$. Since $p(s)(Ly)(s) = (Lf)(s)$, we have

$$\begin{aligned} Ly(s) &= [Lf(s)][1/p(s)] = [Lf(s)] \cdot [Lg(s)] \\ &= L(f * g)(s), \end{aligned}$$

by Theorem 13.6.17. The Uniqueness Theorem guarantees that $y(t) = (f * g)(t)$ is the solution of the differential equation (13.29) that satisfies the initial conditions $y(0) = y'(0) = 0$. Thus, if $\Delta > 0$, then

$$\begin{aligned} y(t) &= c_1 \int_0^t f(u)e^{r_1(t-u)}\, du + c_2 \int_0^t f(u)e^{r_2(t-u)}\, du \\ &= c_1 e^{r_1 t} \int_0^t e^{-r_1 u} f(u)\, du + c_2 e^{r_2 t} \int_0^t e^{-r_2 u} f(u)\, du. \end{aligned}$$

If $\Delta = 0$, then

$$y(t) = \int_0^t f(u)(t-u)e^{r(t-u)}\, du = e^{rt} \int_0^t (t-u)e^{-ru} f(u)\, du.$$

Finally, if $\Delta < 0$, then

$$\begin{aligned} y(t) &= \frac{1}{w} \int_0^t f(u)e^{v(t-u)} \sin w(t-u)\, du \\ &= \frac{e^{vt}}{w} \int_0^t e^{-vu} f(u) \sin w(t-u)\, du. \end{aligned}$$

In each case, the function y solves the initial value problem, whatever may be the function f in $\mathcal{E}$.

Moreover, this technique can be adapted to solve the differential equation (13.29) with general initial conditions $y(0) = b_0$, $y'(0) = b_1$. By Theorem 13.6.8 and its corollary with $y(0) = b_0$ and $y'(0) = b_1$, we have

$$Ly'(s) = s(Ly)(s) - b_0$$

and

$$(Ly'')(s) = s^2(Ly)(s) - b_0 s - b_1.$$

Applying the Laplace transform to (13.29) yields

$$p(s)(Ly)(s) - q(s) = Lf(s)$$

where $q(s) = b_0 s + (b_1 + a_1 b_0)$. Therefore, letting g denote the inverse Laplace transform of $1/p(s)$ as above, we have

$$Ly(s) = [Lf(s)] \cdot \left[\frac{1}{p(s)}\right] + \frac{q(s)}{p(s)} = [Lf(s)] \cdot [Lg(s)] + \frac{q(s)}{p(s)}.$$

Finally, since the degree of q is less than that of p, the rational function $q(s)/p(s)$ can be decomposed uniquely by using partial fractions and its inverse Laplace transform can be found (in the same way that we found g). Let h denote the function such that $Lh(s) = q(s)/p(s)$. It follows that

$$Ly(s) = L(f * g)(s) + Lh(s) = L((f * g) + h)(s).$$

By the Uniqueness Theorem, $y(t) = (f * g)(t) + h(t)$. This function solves the differential equation (13.29) and meets the initial conditions $y(0) = b_0$, $y'(0) = b_1$. The function $f * g$ is the *steady state solution* of (13.29) corresponding to the null initial conditions and the function h represents a *transient* that depends on the initial perturbation given by the nonzero initial conditions. The Laplace transform method automatically produces a solution for (13.29) that is separated into these two components; that is one of its many advantages.

For a concrete example, we find a function $y(t)$ that solves

$$y'' + 2y' + 5y = f(t)$$

such that $y(0) = 2$, $y'(0) = 3$. First, $p(s) = s^2 + 2s + 5$ has complex roots $-1 \pm 2i$. Thus $p(s) = (s+1)^2 + 2^2$ and $q(s) = 2s + 7$. The function $1/p(s) = 1/[(s+1)^2 + 2^2]$ is the Laplace transform of

$$g(t) = \tfrac{1}{2}e^{-t}\sin 2t.$$

Moreover, the rational function

$$\frac{q(s)}{p(s)} = \frac{2s+7}{(s+1)^2+2^2} = \frac{2(s+1)}{(s+1)^2+2^2} + \frac{5}{(s+1)^2+2^2}$$

is the Laplace transform of

$$h(t) = 2e^{-t}\cos 2t + \tfrac{5}{2}e^{-t}\sin 2t.$$

We conclude that the solution of $y'' + 2y' + 2y = f(t)$ that meets the initial conditions $y(0) = 2$, $y'(0) = 3$ is

$$\begin{aligned} y(t) &= 2e^{-t}\cos 2t + \tfrac{5}{2}e^{-t}\sin 2t + \tfrac{1}{2}\textstyle\int_0^t f(u)e^{-(t-u)}\sin 2(t-u)\,du \\ &= 2e^{-t}\cos 2t + \tfrac{5}{2}e^{-t}\sin 2t + \tfrac{1}{2}e^{-t}\textstyle\int_0^t e^u f(u)\sin 2(t-u)\,du. \end{aligned}$$

For a given f in $\mathcal{E}$ we compute this last integral and obtain the solution y in explicit form. ●

It may occur to you that the method we have described here can be generalized to treat linear differential equations of general order k with constant coefficients. We will discuss this generalization in the exercises.

EXAMPLE 34 The Tautochrone Problem Starting from rest, a particle of mass m slides down a smooth, frictionless curve moved solely by the force of gravity. The curve is to have the property that, *regardless of the starting point* $A(u, w)$ on the curve, the total time required for the particle to slide to the bottom point B remains the same. (See Fig. 13.12.) This classic problem challenges us to find a mathematical description of the curve with this property. The word *tautochrone* is derived from two Greek roots meaning "same time."

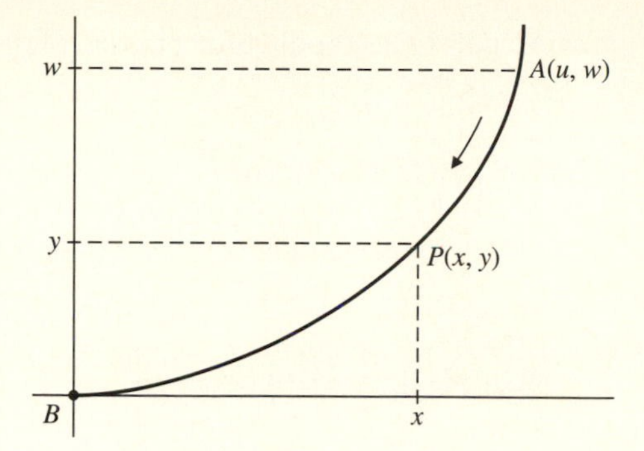

Figure 13.12

We place the point B at (0, 0). Let $P(x, y)$ denote any point on the curve intermediate between $A(u, w)$ and $B(0, 0)$. Let $\sigma(y)$ denote the arc length of the curve, measured from the point B to $P(x, y)$. As the particle slides from A to P, it loses potential energy; the energy lost is $(mg)(w - y)$, where g is a gravitational constant. Since there is no friction, that energy is transformed entirely into kinetic energy, $mv^2/2$, where v is the velocity of the particle at the point P. Therefore our first observation is that

$$\frac{mv^2}{2} = (mg)(w - y).$$

Equivalently,

$$\frac{1}{v} = \frac{1}{\sqrt{2g}}(w - y)^{-1/2}. \tag{13.31}$$

Let T denote the total (constant) time required for the descent from A to B. We express T as an integral and, by repeatedly changing variables and using (13.31), we obtain

$$\begin{aligned} T &= \int_0^T dt = \int_0^{\sigma(w)} dt/d\sigma\, d\sigma = \int_0^{\sigma(w)} \frac{1}{\sigma'(t)}\, d\sigma = \int_0^{\sigma(w)} \frac{1}{v}\, d\sigma \\ &= \int_0^w \frac{1}{v}\sigma'(y)\, dy = \frac{1}{\sqrt{2g}} \int_0^w \sigma'(y)(w - y)^{-1/2}\, dy \\ &= \frac{1}{\sqrt{2g}}(\sigma' * f)(w), \end{aligned}$$

where $f(y) = y^{-1/2}$. Thus our second observation is that

$$(\sigma' * f)(w) = T\sqrt{2g}. \tag{13.32}$$

Now apply the Laplace transform to the equation (13.32). While we may reasonably assume that σ' is in $\mathcal{E}$, $f(y) = y^{-1/2}$ is not. But if we assume (correctly) that L

transforms convolution into a pointwise product for a class of functions that includes f, then we obtain

$$[(L\sigma')(s)] \cdot [Lf(s)] = L(\sigma' * f)(s) = L(T\sqrt{2g}) = \frac{T\sqrt{2g}}{s}.$$

From (5) in our list of elementary Laplace transforms at the beginning of this section, we know that $Lf(s) = \sqrt{\pi}/s^{1/2}$. It follows that

$$[L(\sigma')(s)] \cdot \left(\frac{\sqrt{\pi}}{s^{1/2}}\right) = \frac{T\sqrt{2g}}{s}.$$

Equivalently,

$$\begin{aligned} L(\sigma')(s) &= \left(\frac{T\sqrt{2g}}{\pi}\right)\left(\frac{\sqrt{\pi}}{s^{1/2}}\right) \\ &= \left(\frac{T\sqrt{2g}}{\pi}\right)(Ly^{-1/2})(s). \end{aligned}$$

By the Uniqueness Theorem, $\sigma'(y) = (T\sqrt{2g}/\pi)y^{-1/2}$. Thus we deduce that

$$[\sigma'(y)]^2 = \frac{c^2}{y}, \tag{13.33}$$

where $c = T\sqrt{2g}/\pi$. From your introductory calculus course you will recall that $[\sigma'(y)]^2 = 1 + [dx/dy]^2$. Substituting in (13.33), we find that the differential equation

$$1 + (dx/dy)^2 = \frac{c^2}{y},$$

describes the relationship between x and y for a point $P(x, y)$ on the curve. To solve this nonlinear equation, write it in the equivalent form

$$dx/dy = \left(\frac{c^2}{y} - 1\right)^{1/2}.$$

In this last equation, let $c^2/y = \csc^2\theta$ or $y = c^2\sin^2\theta$. It follows that

$$dx/dy = (\csc^2\theta - 1)^{1/2} = \cot\theta \tag{13.34}$$

and

$$dy/d\theta = 2c^2 \sin\theta\cos\theta.$$

We integrate (13.34) from 0 to θ in order to compute x in terms of the parameter θ.

$$\begin{aligned} x(\theta) &= \int_0^\theta x'(y(t))\, y'(t)\, dt = \int_0^\theta \cot t\,(2c^2 \sin t\cos t)\, dt \\ &= 2c^2 \int_0^\theta \cos^2 t\, dt = \tfrac{1}{2}c^2(2\theta + \sin 2\theta). \end{aligned}$$

Also,

$$y(\theta) = c^2 \sin^2 \theta = \tfrac{1}{2}c^2(1 - \cos 2\theta).$$

Finally, set $\phi = 2\theta$. We recognize the equations

$$x = \tfrac{1}{2}c^2(\phi + \sin \phi),$$
$$y = \tfrac{1}{2}c^2(1 - \cos \phi),$$

as parametric equations for a cycloid with parameter ϕ. It is this curve that solves the tautochrone problem. The *shape* of the curve is completely determined by the mathematics; the *scale* is determined by the constant $c = T\sqrt{2g}/\pi$ and hence depends only on the time T alloted for the descent and the gravitational constant g of the gravitational field in which the particle is sliding. ●

EXERCISES

13.1. Prove Theorem 13.2.1.

13.2. Suppose that f is a nonnegative, integrable function on $[a, \infty)$. Let $S = \{\int_a^b f(x)\,dx : b \text{ in } [a, \infty)\}$. Prove that the integral $\int_a^\infty f(x)\,dx$ converges if and only if S is bounded. If S is bounded, prove that $\int_a^\infty f(x)\,dx = \sup S$. Prove the analogous statements for improper integrals of the first kind of the form $\int_\infty^b f(x)\,dx$ and $\int_{-\infty}^\infty f(x)\,dx$.

13.3. Prove part (ii) of Corollary 13.2.5.

13.4. Prove that, if $\int_{-\infty}^\infty f(x)\,dx = L$, then the Cauchy principal value of $\int_{-\infty}^\infty f(x)\,dx$ is also L.

13.5. Suppose that $\int_a^b f(x)\,dx$ is an improper integral of the second kind because the integrand has a pole at the endpoint a. Prove that, if $\int_a^b |f(x)|\,dx$ converges, then $\int_a^b f(x)\,dx$ also converges.

13.6. Test each of the following integrals for convergence.

a) $\int_0^\infty 1/[(1+x)x^{1/2}]\,dx$ **b)** $\int_0^\infty 1/(1 - \cos x)\,dx$
c) $\int_1^\infty (x+1)/[x(x+2)]\,dx$ **d)** $\int_0^1 x^3 \sin^{-1} x/(1-x^2)^{1/2}\,dx$
e) $\int_0^1 x/(1-x^3)^{1/2}\,dx$ **f)** $\int_0^\infty \tan^{-1} x/(1+x^2)\,dx$
g) $\int_0^\infty x/(1-e^x)\,dx$ **h)** $\int_0^{\pi/2} x^2/(\sin x)^3\,dx$
i) $\int_1^2 x^{1/2}/\ln x\,dx$ **j)** $\int_1^3 1/[(x-1)(3-x)]^{1/2}\,dx$

13.7. For which values of p does $\int_0^\infty x^{p-1}/(1+x)\,dx$ converge?

13.8. For which values of p does $\int_0^{\pi/2} (\sin x)/x^p\,dx$ converge?

13.9. **a)** Fix $a > 1$. For which values of p is $\int_1^a (\ln x)^{-p}\,dx$ improper but convergent?

b) For which values of p is $\int_2^\infty (\ln x)^{-p}\,dx$ convergent?

13.10. Let a and b be two distinct constants. Prove that

$$\lim_{T\to\infty} \frac{1}{T}\int_0^T \sin ax \cos bx\,dx = 0.$$

13.11. Determine whether $\int_0^\infty (\sin x)/(1+x)\,dx$ converges.

13.12. For $y > 0$, evaluate $F(y) = \int_0^1 x^{y-1}(1 + y \ln x)\,dx$.

13.13. Find necessary and sufficient conditions on p and q for $\int_0^{\pi/2} x^p/(\sin x)^q\,dx$ to be a convergent improper integral.

13.14. Prove that $\int_0^1 [\ln(1/x)]^p/x^q\,dx$ is convergent if $p > 0$ and $0 < q < 1$.

13.15. Find all values of p for which $\int_0^\infty x^{p-1}/(1+x)\,dx$ converges.

13.16. Identify conditions on p and q so that the integral $\int_0^\infty 1/[x^p(1+x^q)]\,dx$ converges.

13.17. Find those values of p for which $\int_0^\infty x^{-p}(e^{-x}-1)\,dx$ converges.

13.18. Determine whether each of the following integrals converges. Identify those that converge absolutely.

a) $\int_0^\infty \dfrac{x\cos x}{1+x^2}\,dx$ **b)** $\int_0^\infty \dfrac{x\sin x}{1+x^2}\,dx$

c) $\int_0^\infty \dfrac{\cos x}{(1+x^2)^{1/2}}\,dx$ **d)** $\int_{-\infty}^\infty \dfrac{\cos x}{1-x^2}\,dx$

e) $\int_0^\infty \dfrac{\sin x}{e^{2x}-1}\,dx$ **f)** $\int_0^\infty \dfrac{\sin x}{x(x^2-1)}\,dx$

13.19. **a)** Prove that, for $0 < p < 2$, $\int_0^\infty x^{-p}\sin x\,dx$ converges.

b) Prove that, for $1 < p < 2$, $\int_0^\infty x^{-p}\sin x\,dx$ converges absolutely.

13.20. Prove that $\int_0^\infty x^{-p}(1-\cos x)\,dx$ is absolutely convergent for $1 < p < 3$.

13.21. **a)** Prove that $\int_0^\infty x^{-p}\sin x(1-\cos x)\,dx$ converges for $0 < p < 4$.

b) Prove that $\int_0^\infty x^{-p}\sin x(1-\cos x)\,dx$ converges absolutely for $1 < p < 4$.

13.22. Prove that, if $0 < p < q < 1$, then

$$\int_{-\infty}^{\infty} \frac{e^{-px}-e^{-qx}}{1-e^{-x}}\,dx$$

converges.

13.23. Find the values of p and q for which

$$\int_0^{\infty} \frac{e^{-px}-e^{-qx}}{x(1-e^{-x})}\,dx$$

converges.

13.24. Prove that, for $t > 0$, $(2/\sqrt{\pi})\int_0^\infty \exp(-x^2 t)\,dx = t^{-1/2}$.

13.25. **a)** In Example 11 we showed that $\int_0^\infty \sin(x^2)\,dx$ converges conditionally. Imitate our argument to show that the integral $\int_0^\infty \cos(x^2)\,dx$ also converges conditionally. These two integrals are called *Fresnel integrals* and arise in the study of optics.

b) Write

$$F_1 = \int_{-\infty}^{\infty} \sin(x^2)\,dx = \int_0^\infty \frac{\sin t}{t^{1/2}}\,dt$$

and

$$F_2 = \int_{-\infty}^{\infty} \cos(x^2)\,dx = \int_0^\infty \frac{\cos t}{t^{1/2}}\,dt.$$

Our goal is to evaluate F_1 and F_2. Use the result in Exercise 13.24 to write

$$F_1 = \frac{2}{\sqrt{\pi}}\int_{t=0}^{\infty} \sin t\left[\int_{x=0}^{\infty} \exp(-x^2 t)\,dx\right]dt$$

and

$$F_2 = \frac{2}{\sqrt{\pi}} \int_{t=0}^{\infty} \cos t \left[\int_{x=0}^{\infty} \exp(-x^2 t)\, dx \right] dt.$$

Justify the interchange of the order of integration in each of these integrals. That is, prove that

$$F_1 = \frac{2}{\sqrt{\pi}} \int_{x=0}^{\infty} \left[\int_{t=0}^{\infty} \exp(-x^2 t) \sin t\, dt \right] dx$$

and

$$F_2 = \frac{2}{\sqrt{\pi}} \int_{x=0}^{\infty} \left[\int_{t=0}^{\infty} \exp(-x^2 t) \cos t\, dt \right] dx$$

c) In each case integrate the inner integral by parts to obtain

$$F_1 = \frac{2}{\sqrt{\pi}} \int_0^{\infty} \frac{1}{(1+x^4)}\, dx$$

and

$$F_2 = \frac{2}{\sqrt{\pi}} \int_0^{\infty} \frac{x^2}{(1+x^4)}\, dx.$$

d) Hence show that $F_1 = F_2 = (\pi/2)^{1/2}$. [*Hint:* For F_1, partially fractionate the integrand to obtain

$$\frac{1}{x^4+1} = \frac{ax+b}{x^2+\sqrt{2}x+1} + \frac{cx+d}{x^2-\sqrt{2}x+1}.$$

Find $a,\ b,\ c$, and d. Then proceed to obtain the integral as the sum of logarithms and arctangent functions. For F_2, use the change of variables $u = 1/x$ to show that $F_2 = F_1$.]

13.26. Show that $\int_0^{\infty} 2x \cos(x^4)\, dx$ and $\int_0^{\infty} 2x \sin(x^4)\, dx$ converge although neither integrand is bounded on $[0, \infty)$.

13.27. **a)** Fix any $a > 0$. Prove that, for each $k \geq 0$, the integral $\int_0^{\infty} t^k e^{-xt}\, dt$ converges uniformly on $[a, \infty)$.

b) Beginning with $\int_0^{\infty} e^{-xt}\, dt = 1/x$ for $x \geq a$, recursively differentiate with respect to x in order to find $\int_0^{\infty} t^k e^{-xt}\, dt$ as a function of x for $k \geq 0$ and x in $[a, \infty)$. How do you justify differentiating under the integral sign?

13.28. **a)** Fix $a > 0$. Prove that, for each k in $\mathbb{N}$, the integral $\int_0^{\infty} (x^2+t^2)^{-k}\, dt$ converges uniformly on $[a, \infty)$.

b) Begin with $\int_0^{\infty} (x^2+t^2)^{-1}\, dt = \pi/(2x)$. By recursively differentiating with respect to x, show that, for k in $\mathbb{N}$ and $x \geq a$

$$\int_0^{\infty} (x^2+t^2)^{-k}\, dt = \frac{(2k-2)!\pi}{(2x)^{2k-1}[(k-1)!]^2}.$$

c) Show that

$$\int_0^{\infty} \left(1 + \frac{t^2}{k}\right)^{-k} dt = \frac{(2k-2)!\pi\sqrt{k}}{2^{2k-1}[(k-1)!]^2}.$$

[In part (b), set $x = \sqrt{k}$.]

d) Prove that $\lim_{k\to\infty} \int_0^{\infty} (1 + t^2/k)^{-k}\, dt = \int_0^{\infty} \exp(-t^2)\, dt$.

e) Hence prove that $\lim_{k\to\infty} \dfrac{(2k)!\sqrt{k+1}}{4^k(k!)^2} = 1/\sqrt{\pi}$.

13.29. Show that for $r > 0$ and $x > 0$, $\int_0^\infty t^{x-1}e^{-rt}\,dt = r^{-x}\Gamma(x)$.

13.30. Prove that, for $x > 0$, $\Gamma(x) = 2\int_0^\infty t^{2x-1}\exp(-t^2)\,dt$.

13.31. Evaluate the following integrals.

a) $\int_0^\infty x^2\exp(-x^2)\,dx$ **b)** $\int_0^\infty x^4\exp(-x^2)\,dx$

c) $\int_0^\infty x^{-1/2}e^{-2x}\,dx$ **d)** $\int_0^\infty x^{-3/2}e^{-4x}\,dx$

13.32. **a)** Prove that, for $x > 0$, $\Gamma(x) = \int_0^1 [\ln(1/u)]^{x-1}\,du$.

b) For $r > 0$, $x > 0$, evaluate $\int_0^1 [\ln(1/t)]^{x-1}t^{r-1}\,dt$ in terms of the gamma function. [Change variables in part (a).]

13.33. Prove that, for every k in $\mathbb{N}$, the integral

$$\int_0^\infty t^{x-1}e^{-t}(\ln t)^k\,dt$$

converges uniformly for x in any interval $[c, d] \subset (0, \infty)$.

13.34. Let $T = \{(x, y) : x \geq 0,\ y \geq x\}$. Express the integral $\iint_T e^{-x}(x-y)^{-1/2}\,dA$ in terms of the gamma function.

13.35. Show that, for x in (0, 1),

$$\int_{-\infty}^\infty \frac{e^{xt}}{1+e^t}\,dt = \Gamma(x)\Gamma(1-x).$$

13.36. An improper integral $\int_a^\infty f(x)\,dx$ of the first kind is said to be *Abel summable* to the value I if

$$\lim_{r\to 0^+}\int_a^\infty e^{-rx}f(x)\,dx = I.$$

Find the Abel value of the integrals $\int_0^\infty \sin x\,dx$ and $\int_0^\infty \cos x\,dx$.

13.37. **a)** The Bessel function J_0 of order 0 of the first kind can be defined for x in $\mathbb{R}$ by the integral

$$J_0(x) = \frac{1}{\pi}\int_{-1}^1 \cos xt(1-t^2)^{-1/2}\,dt.$$

J_0 can also be defined by the power series

$$J_0(x) = \sum_{j=0}^\infty \frac{(-1)^j x^{2j}}{2^{2j}(j!)^2}.$$

Starting with each definition, prove that J_0 satisfies the differential equation $x^2y'' + xy' + x^2y = 0$. (Justify differentiating under the integral or the summation when computing J_0' and J_0''.)

b) For k in $\mathbb{N}$, the Bessel function J_k of order k of the first kind can be defined by the integral

$$J_k(x) = \frac{k!(2x)^k}{(2k)!\pi}\int_{-1}^1 \cos xt(1-t^2)^{k-1/2}\,dt.$$

J_k can also be defined by the power series

$$J_k(x) = \left(\frac{x}{2}\right)^k \sum_{j=0}^{\infty} \frac{(-1)^j x^{2j}}{2^{2j} j!(j+k)!}.$$

Starting with each definition, prove that J_k satisfies the differential equation $x^2 y'' + xy' + (x^2 - k^2)y = 0$. (Justify differentiating under the integral or the summation when computing J_k' and J_k''.)

c) Use both the integral and the series definition of the Bessel functions of the first kind to show that $J_1 = -J_0'$ and that $J_{k+1} = J_{k-1} - 2J_k'$ for all k in $\mathbb{N}$.

d) Fix constants a and b. Let $r = (a^2 + b^2)^{1/2}$ and let $D_1 = \{(x, y) : x^2 + y^2 \leq 1\}$. Prove that

$$\iint_{D_1} \cos(ax + by)\, dA = \frac{2J_1(r)}{r}.$$

[Imitate the solution of Exercise 10.23.]

13.38. For x in $\mathbb{R}$, define $F(x) = \int_0^\infty \exp(-t^2 - x^2/t^2)dt$.

a) Prove that the integral defining F converges uniformly for x in $\mathbb{R}$. Hence conclude that F is continuous on $\mathbb{R}$.

b) Restrict now to $x > 0$. Define

$$G(x) = \int_0^\infty \partial[\exp(-t^2 - x^2/t^2)]/\partial x\, dt.$$

After computing the partial derivative of the integrand, use the change of variable $u = x/t$ in order to show that $G(x) = -2F(x)$ for $x > 0$.

c) Assume temporarily that we can invoke Theorem 13.4.5 to deduce that $G(x) = F'(x)$ for $x > 0$. Thus $F'(x) = -2F(x)$. Show that $F(x) = ce^{-2x}$, where c is some constant. Use part (a) and the known value for $F(0)$ to deduce that $c = \sqrt{\pi}/2$. Therefore, $F(x) = (\sqrt{\pi}/2)e^{-2x}$, for $x > 0$.

d) Explain why, for $x < 0$, we have $F(x) = F(-x)$. Conclude that $F(x) = (\sqrt{\pi}/2)e^{-2|x|}$, for all x in $\mathbb{R}$ including $x = 0$.

e) Does $F'(0)$ exist?

f) To justify your use of Theorem 13.4.5, show that the integral defining $G(x)$ converges uniformly on any interval $[a, b] \subset (0, \infty)$. To understand your answer to (e) show that the integral defining $G(x)$ fails to converge uniformly on any interval containing $x = 0$.

g) Prove that, for $p > 0$ and $q \geq 0$,

$$\int_0^\infty e^{-(px^2 + q/x^2)}\, dx = (\pi/4p)^{1/2} e^{-2\sqrt{pq}}.$$

13.39. For x in $\mathbb{R}$ define $F(x) = \int_0^\infty \exp(-t^2) \cos xt\, dt$.

a) Prove that the integral defining F converges uniformly for x in $\mathbb{R}$. Thus, F is continuous on $\mathbb{R}$.

b) Prove that the integral $\int_0^\infty \partial[\exp(-t^2) \cos xt]/\partial x\, dt$ also converges uniformly for x in $\mathbb{R}$.

c) Hence, prove that $F'(x) = -xF(x)/2$. Solve for $F(x)$.

d) Let r be any positive constant. Prove that, for all x in $\mathbb{R}$,

$$\int_0^\infty e^{-rt^2} \cos xt \, dt = \left(\frac{\pi}{4r}\right)^{1/2} e^{-x^2/4r}.$$

13.40. For $k \geq 0$ and $x > 0$, define $F_k(x) = \int_0^\infty t^k \exp(-xt^2)\, dt$.

a) Fix any $a > 0$. Prove that, for each $k \geq 0$, the integral that defines F_k converges uniformly on $[a, \infty)$.

b) Prove that, for each $k \geq 0$ and all $x > 0$,

$$F_k'(x) = -F_{k+2}(x).$$

c) Calculate F_0 directly. Then use induction to prove that

$$F_{2k}(x) = \frac{(2k)!\sqrt{\pi}}{2^{2k+1}k!x^{k+1/2}}.$$

13.41. Prove that

$$\int_0^\infty e^{-t^2} \sin(2xt)\, dt = e^{-x^2} \int_0^x e^{u^2}\, du.$$

13.42. Show that, for $k > 0$,

$$\frac{d^k(1-x^2)^{-1}}{dx^k} = \begin{cases} (-1)^{k/2} \int_0^\infty t^k e^{-t} \cos xt \, dt, & \text{if } k \text{ is even,} \\ (-1)^{(k+1)/2} \int_0^\infty t^k e^{-t} \sin xt \, dt, & \text{if } k \text{ is odd.} \end{cases}$$

13.43. For $x \geq 0$, define $F(x) = \int_0^\infty t^{-2}[1 - \exp(-xt^2)]\, dt$.

a) Show that the integrand has no pole at $t = 0$.

b) Prove that the integral that defines F converges for each $x \geq 0$.

c) Prove that the integral $\int_0^\infty \partial[t^{-2}[1 - \exp(-xt^2)]]/\partial x \, dt$ converges uniformly on any interval $[a, b] \subset (0, \infty)$.

d) Compute $F'(x)$. Hence identify the function F explicitly.

13.44. Begin with $t/(1 + t^2) = \int_0^\infty e^{-tx} \cos x \, dx$, valid for t in $(0, \infty)$.

a) Prove that, for positive constants a and b,

$$\int_0^\infty \frac{e^{-bx} - e^{-ax}}{x} \cos x \, dx = \ln\left(\frac{1 + a^2}{1 + b^2}\right)^{1/2}.$$

b) Assume that, as a function of b, the integral in (a) converges uniformly for b in $[0, \infty)$. Explain carefully why it follows that

$$\int_0^\infty \frac{1 - e^{-ax}}{x} \cos x \, dx = \ln(1 + a^2)^{1/2}.$$

13.45. Prove that $\int_0^\infty (\cos xt)/(1 + t^2)\, dt$ converges uniformly for x in $\mathbb{R}$.

13.46. Fix $a > 0$. Prove that $\int_1^\infty 1/t^{1+x}\, dt$ converges uniformly for x in $[a, \infty)$.

13.47. Let a and b be positive constants with $a < b$.

a) Show that

$$\int_a^b \left[\int_0^\infty (1 + x^2t^2)^{-1}\, dt\right] dx = \int_0^\infty \left[\int_a^b (1 + x^2t^2)^{-1}\, dx\right] dt.$$

b) Evaluate $\int_0^\infty [\tan^{-1}(bt) - \tan^{-1}(at)]/t \, dt$.

13.48. Assume that f is a continuously differentiable function on $\mathbb{R}$ such that the function $g(t, x) = f'(tx)/x$ is continuous for (t, x) in $[a, b] \times \mathbb{R}$ and such that $\int_{-\infty}^{\infty} f'(tx)/x\, dx$ is uniformly convergent for t in $[a, b]$.

a) Show that $\int_a^b [\int_{-\infty}^{\infty} f'(tx)/x\, dx]\, dt = \int_{-\infty}^{\infty} [f(bx) - f(ax)]/x^2\, dx$.

b) Prove that $\int_{-\infty}^{\infty} [\cos bx - \cos ax]/x^2\, dx = \pi(a - b)$.

13.49. Beginning with $t/(1 + t^2) = \int_0^\infty e^{-x} \sin xt\, dx$, prove that

$$\int_0^\infty \frac{e^{-x}(1 - \cos xy)}{x}\, dx = \ln \sqrt{1 + y^2}.$$

13.50. Prove that $\int_0^\infty [\exp(-x^2) \sin(2xy)]/x\, dx = \sqrt{\pi} \int_0^y \exp(-t^2)\, dt$.

13.51. **a)** Show that, for $y > 0$, $\int_0^\infty \exp(-x^2 y^2)\, dx = \sqrt{\pi}/(2y)$.

b) Hence prove that, if a and b are positive constants,

$$\int_0^\infty \frac{e^{-a^2x^2} - e^{-b^2x^2}}{x^2}\, dx = (b - a)\sqrt{\pi}.$$

13.52. Let f be a continuous function on $[0, \infty)$ such that $f'(0^+)$ exists and such that $\int_1^\infty f(t)/t\, dt$ is convergent. Let a and b be positive constants.

a) Show that, for $y > 0$,

$$\int_0^y \frac{f(ax) - f(0)}{x}\, dx = \int_0^{ay} \frac{f(t) - f(0)}{t}\, dt.$$

b) Show that $\lim_{y \to \infty} \int_{ay}^{by} f(t)/t\, dt = 0$.

c) Hence prove that $\int_0^\infty [f(bx) - f(ax)]/x\, dx = f(0) \ln(a/b)$.

d) Show that $f(x) = \cos x$ satisfies the conditions of this exercise and evaluate $\int_0^\infty [\cos bx - \cos ax]/x\, dx$.

13.53. Let f be a continuous function on $[0, \infty)$ that is differentiable on $(0, \infty)$. Assume also that $\lim_{x \to \infty} f(x)$ exists; denote this limit by $f(\infty)$. Let a and b be positive constants with $a < b$.

a) Show that $\int_0^\infty f'(tx)\, dx$ converges uniformly for t in $[a, \infty)$. (Treat the possibility that f' has a pole at $x = 0$; the integral in question may be a sum of integrals of the first and second kinds.)

b) Apply Theorems 13.4.7 and 13.4.8 to the integral $\int_a^b [\int_0^\infty f'(tx)\, dx]\, dt$ in order to show that

$$\int_0^\infty \frac{f(bx) - f(ax)}{x}\, dx = [f(\infty) - f(0)] \ln\left(\frac{b}{a}\right).$$

c) Use the result in part (b) to find $\int_0^\infty [e^{-bx} - e^{-ax}]/x\, dx$ and $\int_0^\infty [\tan^{-1} bx - \tan^{-1} ax]/x\, dx$.

13.54. For (x, y) in $[0, 1] \times [0, \infty)$, define

$$f(x, y) = (2 - xy)xye^{-xy}.$$

a) Show that

$$f(x, y) = \partial(x^2 y e^{-xy})/\partial x$$

and

$$f(x, y) = \partial(xy^2 e^{-xy})/\partial y.$$

b) Show that $\int_0^1 f(x,y)\,dx = \int_0^1 \partial[x^2 y e^{-xy}]/\partial x\,dx = ye^{-y}$, for each y in $[0,\infty)$. Hence, show that

$$\int_0^\infty \left[\int_0^1 f(x,y)\,dx\right] dy = \int_0^\infty ye^{-y}\,dy = 1.$$

c) Show that $\int_0^\infty f(x,y)\,dy = \int_0^\infty \partial[xy^2 e^{-xy}]/\partial y\,dy = 0$ for each x in $[0,1]$. Hence, show that

$$\int_0^1 \left[\int_0^\infty f(x,y)\,dy\right] dx = 0.$$

It follows that $\int_0^\infty [\int_0^1 f(x,y)\,dx]\,dy \neq \int_0^1 [\int_0^\infty f(x,y)\,dy]\,dx$.

13.55. Show that

$$\int_0^1 \left[\int_1^\infty (e^{-xy} - 2e^{-2xy})\,dx\right] dy \neq \int_1^\infty \left[\int_0^1 (e^{-xy} - 2e^{-2xy})\,dy\right] dx.$$

13.56. Fix $b > 0$. Is

$$\int_0^b \left[\int_0^\infty (2y - 2xy^3)\,e^{-xy^2}\,dx\right] dy = \int_0^\infty \left[\int_0^b (2y - 2xy^3)\,e^{-xy^2}\,dy\right] dx?$$

13.57. For x in $\mathbb{R}$, define $F(x) = (1/\sqrt{2\pi}) \int_{-\infty}^x \exp(-t^2/2)\,dt$. Prove each of the following assertions.

a) F is strictly monotone increasing on $\mathbb{R}$.

b) $\lim_{x\to-\infty} F(x) = 0^+$ and $\lim_{x\to\infty} F(x) = 1^-$.

c) F is continuous on $\mathbb{R}$.

d) F is differentiable with $F'(x) = (1/\sqrt{2\pi}) \exp(-x^2/2)$.

13.58. a) Prove that, for $s \geq s_0 > 0$ and any $x > -1$,

$$L(t^x)(s) = \frac{\Gamma(x+1)}{s^{x+1}}.$$

b) Thus show that, for all k in $\mathbb{N}$, $L(t^k)(s) = k!/s^{k+1}$.

c) Let c be a constant. Prove that, for $s \geq s_0 > c$ and any $x > -1$, $L(t^x e^{ct})(s) = \Gamma(x+1)/(s-c)^{x+1}$.

13.59. Use the Laplace transform to find a function $y = f(t)$ such that $y'' + 2y' = e^t$ which satisfies the initial conditions $f(0) = 1,\ f'(0) = 2$.

13.60. Consider the third order linear differential equation

$$y''' + a_2 y'' + a_1 y' + a_0 y = 0. \tag{13.35}$$

with the initial conditions $y(0) = b_0,\ y'(0) = b_1$, and $y''(0) = b_2$. The coefficients $a_0,\ a_1$, and a_2 are real constants as are the numbers $b_0,\ b_1$, and b_2.

a) Apply the Laplace transform to both sides of this equation and invoke Corollary 13.6.9 with $k = 1, 2, 3$. Show that this yields

$$p(s)(Ly)(s) = q(s),$$

where $p(s) = s^3 + a_2 s^2 + a_1 s + a_0$ is the characteristic polynomial of the differential equation and $q(s) = \sum_{j=0}^2 c_j s^j$ is a polynomial of degree no greater than 2. Express the coefficients c_j in terms of the given values for the a_j and b_j.

Subsequent steps in the solution of (13.35) depend upon the partial fraction decomposition of the rational function q/p. The form this takes depends, in turn, on the nature of the roots of p.

b) Suppose that p has three distinct real roots r_1, r_2, and r_3. Find constants d_1, d_2, and d_3, in terms of c_0, c_1, c_2, r_1, r_2, and r_3, such that

$$(Ly)(s) = \frac{q(s)}{p(s)} = \sum_{j=1}^{3} \frac{d_j}{s - r_j}.$$

Use the Uniqueness Theorem to identify the solution of (13.35) that satisfies the given initial conditions.

As a concrete example, use the Laplace transform method to find the function $y(t)$ such that $y''' - 7y' + 6y = 0$ that satisfies the initial conditions $y(0) = 0$, $y'(0) = -1$, and $y''(0) = 3$.

c) Suppose that p has three real roots, r_1, r_1, r_2, one of which is a double root. Find constants d_1, d_2, and d_3, in terms of c_0, c_1, c_2, r_1, and r_2, such that

$$(Ly)(s) = \frac{q(s)}{p(s)} = \frac{d_1}{(s - r_1)} + \frac{d_2}{(s - r_1)^2} + \frac{d_3}{(s - r_2)}.$$

Use the Uniqueness Theorem to identify the solution of (13.35) that satisfies the initial value conditions.

As a concrete example, use the Laplace transform method to find the function $y(t)$ such that $y''' - 3y' - 2y = 0$ that satisfies the initial conditions $y(0) = 1$, $y'(0) = 1$, and $y''(0) = 2$.

d) Discuss the case when p has one real root r and a pair of complex conjugate roots $v \pm iw$, with $w \neq 0$. Identify the form of the partial fraction decomposition of $q(s)/p(s)$. Using the Laplace transform method, find the form of the general solution of (13.35) that satisfies the initial conditions.

As a concrete example, use the Laplace transform method to find the function $y(t)$ such that $y''' + y' + 10y = 0$ that satisfies the initial conditions $y(0) = 1$, $y'(0) = -1$, and $y''(0) = -2$.

13.61. Consider the third order linear differential equation

$$y''' + a_2 y'' + a_1 y' + a_0 y = f(t) \qquad \textbf{(13.36)}$$

where the coefficients are real constants and f is a function of exponential type. Suppose that initial conditions $y(0) = b_0$, $y'(0) = b_1$, and $y''(0) = b_2$ are imposed on the solution y. (Refer to Exercise 13.60.)

a) Show that the solution $y(t)$ satisfies the equation

$$(Ly)(s) = \frac{(Lf)(s)}{p(s)} + \frac{q(s)}{p(s)}$$

where $p(s) = s^3 + a_2 s^2 + a_1 s + a_0$ is the characteristic polynomial of the differential equation and $q(s)$ is the polynomial found in Exercise 13.60 that is determined by the numbers a_0, a_1, a_2, b_0, b_1, and b_2.

b) Let g be the unique function in $\mathcal{E}$ whose Laplace transform is $1/p(s)$: $Lg(s) = 1/p(s)$. Let h be the unique solution of

$$y''' + a_2 y'' + a_1 y' + a_0 y = 0$$

that satisfies the initial conditions $h(0) = b_0$, $h'(0) = b_1$, and $h''(0) = b_2$ as found in Exercise 13.60. Show that $y(t) = (f * g)(t) + h(t)$ solves the equation (13.36) and satisfies the given initial conditions.

c) Find the solution of the equation $y''' - 7y' + 6y = f(t)$ that satisfies $y(0) = 0$, $y'(0) = -1$, and $y''(0) = 3$ when (i) $f(t) = e^t$; (ii) $f(t) = t$; (iii) $f(t) = \sin t$.

d) Find the solution of the equation $y''' - 3y' - 2y = f(t)$ that satisfies $y(0) = 1$, $y'(0) = 1$, and $y''(0) = 2$ when (i) $f(t) = e^t$; (ii) $f(t) = t$; (iii) $f(t) = \sin t$.

e) Find the solution of the equation $y''' + y' + 10y = f(t)$ that satisfies $y(0) = 1$, $y'(0) = -1$, $y''(0) = -2$ when (i) $f(t) = e^t$; (ii) $f(t) = t$; (iii) $f(t) = \sin t$.

13.62. Consider the nth order differential equation

$$y^{(n)} + a_{n-1}y^{(n-1)} + \cdots + a_1 y' + a_0 y = f(t) \tag{13.37}$$

where f is in $\mathcal{E}$. Impose the initial conditions $y(0) = b_0$, $y'(0) = b_1, \ldots, y^{(n-1)}(0) = b_{n-1}$. Describe Laplace's method of finding the solution $y(t)$ in the following steps.

a) Show that $Ly(s) = Lf(s)/p(s) + q(s)/p(s)$ where

$$p(s) = s^n + a_{n-1}s^{n-1} + \cdots + a_1 s + a_0$$

is the characteristic polynomial of the differential equation and $q(s)$ is a polynomial of degree less than n that results from the initial conditions and the repeated application of Corollary 13.6.9.

b) Let g and h be the unique functions in $\mathcal{E}$ such that $Lg = 1/p$ and let $Lh = q/p$. (The specific forms of g and h depend on the factorization of p into irreducible factors.) Show that $y(t) = (f * g)(t) + h(t)$ solves the differential equation (13.37) and satisfies the given initial conditions $y^{(j)}(0) = b_j$ for $j = 0, 1, 2, \ldots, n-1$.

13.63. Let a be a positive constant. Evaluate

$$\iint_{\mathbb{R}^2} (x^2 + y^2 + a^2)^{-3/2}\, dA.$$

(*Hint:* Use polar coordinates.)

13.64. Fix $a > 0$ and define $R = \{(x, y) : x \geq a,\ y \text{ in } \mathbb{R}\}$. Prove that

$$\iint_R e^{-(x^2+y^2)}\, dA = a\, e^{-a^2} \int_0^\infty \frac{e^{-u^2}}{u^2 + a^2}\, du.$$

(See Exercise 10.22.)

13.65. For $a > 0$, let $R(a) = \{(x, y) : 0 \leq x \leq a,\ 0 \leq y \leq a\}$.

a) Show that

$$\iint_{R(a)} y\, e^{-y^2(1+x^2)}\, dA = \frac{1}{2}\tan^{-1} a - \frac{1}{2} a\, e^{-a^2} \int_0^{a^2} \frac{e^{-t^2}}{a^2 + t^2}\, dt.$$

b) Find $\lim_{a\to\infty} \iint_{R(a)} y \exp[-y^2(1 + x^2)]\, dA$.

13.66. Let $D_1 = \{(x, y) : x^2 + y^2 \leq 1\}$ denote the closed unit disk.

a) Note that the integral $\iint_{D_1} \ln(x^2 + y^2)\, dA$ is improper since the integrand has a pole at (0, 0). Prove that

$$\iint_{D_1} \ln(x^2 + y^2)\, dA = \lim_{\epsilon\to 0^+} \iint_{R_\epsilon} \ln(x^2 + y^2)\, dA = -\pi,$$

where $R_\epsilon = \{(x, y) : \epsilon^2 \leq x^2 + y^2 \leq 1\}$.

b) Find those values of p for which $\iint_{D_1} (x^2 + y^2)^{-p/2}\, dA$ converges.

c) Suppose that f is continuous on D_1 except at 0 where f has a pole. Suppose also that there exists an $M > 0$ and a $p < 2$ such that $|f(x, y)| \leq M(x^2 + y^2)^{-p/2}$ for all (x, y) in D_1. Prove that $\iint_{D_1} f(x, y)\, dA = \lim_{\epsilon\to 0^+} \iint_{R_\epsilon} f(x, y)\, dA$ exists.

d) Let $S_1 = \{(x, y, z) : x^2 + y^2 + z^2 \leq 1\}$. Find those values of p for which

$$\iiint_{S_1} (x^2 + y^2 + z^2)^{-p/2}\, dV = \lim_{\epsilon\to 0^+} \iiint_{R_\epsilon} (x^2 + y^2 + z^2)^{-p/2}\, dV$$

converges, where $R_\epsilon = \{(x, y, z) : \epsilon^2 \leq x^2 + y^2 + z^2 \leq 1\}$.

e) Suppose that f is continuous on S_1 except at the point $\mathbf{0}$ where f has a pole. Assume also that there exists an $M > 0$ and a $p < 3$ such that $|f(x, y, z)| \leq M(x^2 + y^2 + z^2)^{-p/2}$, for all (x, y, z) in S_1. Prove that

$$\iiint_{S_1} f(x, y, z)\, dV = \lim_{\epsilon\to 0^+} \iiint_{R_\epsilon} f(x, y, z)\, dV$$

exists. (*Hint:* Use spherical coordinates.)

13.67. Assume that f and g are nonnegative functions on $[0, \infty)$ such that the improper integrals of the first kind, $\int_0^\infty f(x)\, dx$ and $\int_0^\infty g(x)\, dx$, converge. Prove that $\iint_R f(x)g(y)\, dA$ also converges, where $R = [0, \infty) \times [0, \infty)$, and that

$$\iint_R f(x)g(y)\, dA = \left[\int_0^\infty f(x)\, dx\right]\left[\int_0^\infty g(y)\, dy\right].$$

(This result may be false if either f or g assumes both positive and negative values.)

13.68. For (x, y) in the plane $\mathbb{R}^2$, define

$$u = h_1(x, y) = \frac{x}{x^2 + y^2},$$

$$v = h_2(x, y) = \frac{y}{x^2 + y^2}.$$

Let $\boldsymbol{h} = (h_1, h_2)$. The mapping $\boldsymbol{h}$ is called *inversion in the unit circle*; it maps the interior of the unit disk onto its exterior and vice versa along rays emanating from the origin.

a) Find the Jacobian of $\boldsymbol{h}$.

b) Suppose that f is a continuously differentiable function such that $\iint_{\mathbb{R}^2} ||\nabla f||^2\, dA$ converges. Let $g = f \circ \boldsymbol{h}$. Prove that

$$\iint_{\mathbb{R}^2} ||\nabla f(u, v)||^2\, dA = \iint_{\mathbb{R}^2} ||\nabla g(x, y)||^2\, dA.$$

Hint: Let $F(u, v) = ||\nabla f(u, v)||^2$ and find the relationship between $(F \circ \boldsymbol{h})(x, y)$ and $||\nabla g(x, y)||^2$.

14

Fourier Series

The theory of Fourier series has a luminous history spanning nearly two centuries; it has proved its value by its remarkable successes. The theory we present in this chapter provides us with the tools to analyze periodic oscillations and vibrations. Consider for a moment the variety of areas in which oscillatory phenomena occur. The ebb and flow of the tides, the beating of the human heart, the propagation of light in its wave form, the sound of a bird song, complete with all its overtones, are vibratory in nature. The study of electricity and magnetism, and their applications in electroencephalography (EEG), electrocardiography (EKG), and magnetic resonance imaging (MRI) all require analysis of functions whose behavior is essentially periodic. Theoreticians in economics have attempted to apply the theory of Fourier series to gain understanding of the cyclic behavior of economies as predicted by the Keynesian model. The applications in physics, chemistry, and engineering seem endless, from the analysis of the vibrations of airfoils in wind tunnels or the description of heat flow, to a study of vibratory behavior at the atomic level. A single mathematical theory that promises the possibility of addressing such a wide range of phenomena is certainly worth exploring. Fourier series play an essential role in all such studies.

In its essence, the problem comes to this. Let f be a periodic real-valued function on $\mathbb{R}$ with period p; $f(x + kp) = f(x)$ for all x in $\mathbb{R}$ and all k in $\mathbb{Z}$. The value of $f(x)$ recurs at all points that differ from x by an integer multiple of p. The first goal of the theory is to identify a sequence $\phi_1, \phi_2, \phi_3, \ldots$, of simple, periodic functions, each of period p, such that f is *represented* by an infinite series of the form $f(x) = \sum_{j=1}^{\infty} a_j \phi_j(x)$. The possibility of such a representation immediately raises several questions. How can we find the coefficients a_j? Once we have found them and have formed the series $\sum_{j=1}^{\infty} a_j \phi_j(x)$, does the series actually converge to $f(x)$? More carefully, does the series converge pointwise or uniformly or in some other reasonable way to f? Are these coefficients uniquely determined by the function f? If so, then the sequence $\{a_j\}$ of coefficients must correspond exactly to the wave form represented by the function f. Then there are the entirely pragmatic questions. Once we have found such a series representation, what can we do with it? Can we differentiate or integrate the series for f term-by-term and obtain the derivative or the integral of f? Ultimately, how can we apply the theory to be developed in this chapter?

We cannot hope in our limited treatment to address all these questions. On a practical level, a thorough and sound treatment requires the Lebesgue integral and, lacking that tool, we will have to rely on Riemann's integration theory. Actually, what we will accomplish here will suffice for most standard applications, but the subject of Fourier series is vast. Indeed, it has attracted and has served as the wellspring for the research of many of the great mathematical talents of the past two centuries. We will barely scratch its surface.

The subject is named in honor of Joseph Fourier (1768–1830), but earlier several mathematicians, including Euler, Lagrange, Laplace, and D. Bernoulli, had attempted to represent functions by series of sines and cosines in connection with the problem of a vibrating string. Between 1807 and 1822, Fourier conducted extensive researches into the flow of heat in solids and, in the course of his studies, became convinced that *every* real-valued function on $\mathbb{R}$ with period 2π can be represented by a series of the form

$$\frac{a_0}{2} + \sum_{j=1}^{\infty} [a_j \cos jx + b_j \sin jx].$$

As it happened, the problem of identifying the class of functions that can be so represented had been the central issue of an acrimonious, ongoing argument raging between d'Alembert, Euler, Lagrange, Bernoulli, and Laplace, the giants of applied mathematics and analysis of that era. In making his assertion, Fourier stumbled into the midst of that conflict and alienated all these masters except Bernoulli. When Fourier submitted a research paper, including this claim, to the Academy of Sciences of Paris, the referees Lagrange, Laplace, and Legendre rejected it out of hand and criticized Fourier for his lack of rigor. In fact, Fourier's assertion is false. There exist many periodic functions with period 2π that cannot be represented by a series of the form Fourier proposed. In retrospect, we recognize that some of Fourier's work would not meet modern standards. Of course, during Fourier's lifetime, analysis had not yet achieved its eventual high degree of rigor based on the careful work of Cauchy and Weierstrass; many arguments put forward during those early years do not withstand the scrutiny of modern analysts. So it is not surprising that some of Fourier's claims were too grand and that many of his proofs were flawed. He was not alone on this score. Nevertheless, he persisted in his groundbreaking study of trigonometric series in connection with the flow of heat. The core of his intuition was sound and his output was prodigious; for his enormous contributions to the subject of trigonometric series, he is honored today.

14.1 CONVERGENCE IN THE MEAN

In previous chapters we have studied pointwise and uniform convergence of sequences and series of functions. Here we introduce yet another form of convergence, that of *convergence in the mean*. To establish the context in which this form of convergence applies, we need several definitions. You will notice that the concepts introduced here are reminiscent of the ideas introduced in Chapter 2 for $\mathbb{R}^n$.

DEFINITION 14.1.1 For any interval $[a, b]$ and for any two functions f and g in $R[a, b]$, the *inner product* of f and g is defined to be the real number

$$\langle f, g\rangle = \int_a^b f(x)g(x)\,dx. \quad \bullet$$

The elementary properties of the inner product are listed in our first theorem; its proof is easy and is left for you as an exercise.

THEOREM 14.1.1 The inner product is a function from $R[a, b] \times R[a, b]$ into $\mathbb{R}$ with the following properties:

i) The inner product is additive in each of its variables. For all f, f_1, f_2, g, g_1, g_2 in $R[a, b]$, we have

$$\langle f_1 + f_2, g\rangle = \langle f_1, g\rangle + \langle f_2, g\rangle$$

and

$$\langle f, g_1 + g_2\rangle = \langle f, g_1\rangle + \langle f, g_2\rangle.$$

ii) The inner product is homogeneous in each of its variables. For f and g in $R[a, b]$ and for c in $\mathbb{R}$,

$$\langle cf, g\rangle = c\langle f, g\rangle$$

and

$$\langle f, cg\rangle = c\langle f, g\rangle.$$

iii) The inner product is symmetric. For all f and g in $R[a, b]$, $\langle f, g\rangle = \langle g, f\rangle$. $\bullet$

As we saw in Chapter 2 (and hinted at in Exercises 6.45 and 6.46), an inner product on a vector space induces a norm on that space.

DEFINITION 14.1.2 The L_2*-norm* of a function f in $R[a, b]$, denoted $||f||_2$, is

$$||f||_2 = \langle f, f\rangle^{1/2} = \left[\int_a^b f^2(x)\,dx\right]^{1/2}. \quad \bullet$$

It is immediately evident that, for all f in $R[a, b]$, $||f||_2$ is a nonnegative real number. We will confirm below that, restricted to $C([a, b])$, the function $||\cdot||_2$ is positive definite, absolutely homogeneous, and subadditive.

THEOREM 14.1.2 The Cauchy–Schwarz Inequality For all f and g in $R[a, b]$,

$$|\langle f, g\rangle| \leq ||f||_2 ||g||_2.$$

Proof. Let t denote any real number and compute the inner product of $tf + g$ with itself. Use the linearity, homogeneity, and symmetry of the inner product to compute:

$$\begin{aligned} 0 \leq ||tf + g||_2^2 &= \langle tf + g, tf + g\rangle \\ &= t^2\langle f, f\rangle + 2t\langle f, g\rangle + \langle g, g\rangle. \end{aligned}$$

Let $a = \langle f, f\rangle$, $b = \langle f, g\rangle$, and $c = \langle g, g\rangle$. The preceding calculation says that the quadratic polynomial $at^2 + 2bt + c$ is never negative. Therefore the discriminant $\Delta = 4b^2 - 4ac \leq 0$. The Cauchy–Schwarz inequality follows. ●

THEOREM 14.1.3 The function $||\cdot||_2$ is a norm on $C([a, b])$.

Proof. For any f in $R[a, b]$, $||f||_2 = [\int_a^b f^2(x)\,dx]^{1/2} \geq 0$. Furthermore, if f is continuous on $[a, b]$ and if $||f||_2 = 0$, then f must vanish at every x in $[a, b]$. (See Theorem 6.2.9.) Thus $||\cdot||_2$ is positive definite on $C(a, b)$. For f in $R[a, b]$ and c in $\mathbb{R}$,

$$||cf||_2^2 = \langle cf, cf\rangle = c^2||f||_2^2.$$

Taking the square root of both sides gives

$$||cf||_2 = |c| \cdot ||f||_2.$$

Thus $||\cdot||_2$ is absolutely homogeneous. To prove the subadditivity of $||\cdot||_2$, compute

$$\begin{aligned} ||f + g||_2^2 &= \langle f + g, f + g\rangle \\ &= ||f||_2^2 + 2\langle f, g\rangle + ||g||_2^2 \\ &\leq ||f||_2^2 + 2||f||_2||g||_2 + ||g||_2^2 \\ &= (||f||_2 + ||g||_2)^2, \end{aligned}$$

by the Cauchy–Schwarz inequality. Taking square roots yields $||f + g||_2 \leq ||f||_2 + ||g||_2$. This completes the proof of the theorem; $||\cdot||_2$ is a norm on $C([a, b])$. ●

Remark. The foregoing proof shows that all the properties of the norm hold on $R[a, b]$ except positive definiteness.

DEFINITION 14.1.3 The L_2-*metric* on $R[a, b]$ is

$$d_2(f, g) = ||f - g||_2 = \left[\int_a^b [f(x) - g(x)]^2\,dx\right]^{1/2}. \quad ●$$

The significance of $||f - g||_2$ is seen by computing, for each x in $[a, b]$, the non-negative quantity $[f(x) - g(x)]^2$. Next compute the total of all these quantities by integrating. At some values of x, the quantity $[f(x) - g(x)]^2$ might be quite large;

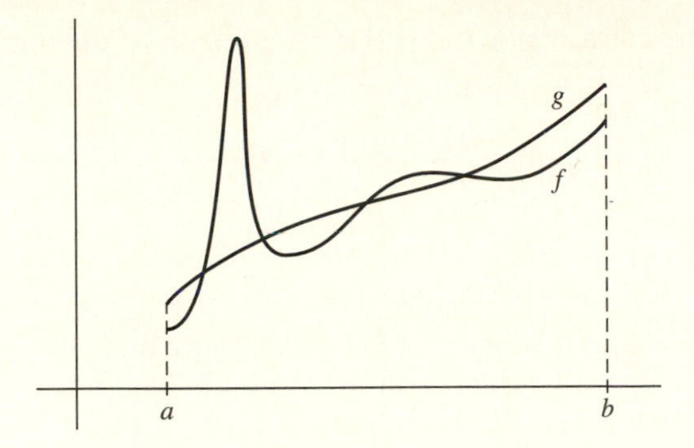

Figure 14.1

however, it contributes little to the total integral if the set of such points x has small length. (Refer to Fig. 14.1.)

DEFINITION 14.1.4 A sequence of functions $\{f_k\}$ in $R[a, b]$ is said to *converge in the mean* to the integrable function f_0 if $\lim_{k\to\infty} ||f_k - f_0||_2 = 0$. ●

Notice that, in the preceding definition, the limit function f_0 is assumed to be (Riemann) integrable. A word of caution is called for: To apply this definition, we need to know, not only that each f_k is integrable, but that the function f_0 is integrable as well. Our warning is based on the following fact. A sequence $\{f_k\}$ in $R[a, b]$ or in $C([a, b])$ that is Cauchy in the mean (that is, for which $||f_k - f_m||_2$ is arbitrarily small when k and m are sufficiently large) may converge in the mean to a function f_0, but it need not be true that f_0 is continuous or even Riemann integrable. In other words, in the language we have established earlier, neither $R[a, b]$ nor $C([a, b])$ is Cauchy complete with respect to the metric d_2. Therefore, we must assume in Definition 14.1.4 that the limit function f_0 is integrable.

It may appear that these facts expose a significant flaw in the concept of mean convergence; they do not. Rather, they reveal the insufficiency of the Riemann integral. By contrast, the Lebesgue integral, which allows the introduction of a vastly larger class of integrable functions than does that of Riemann, provides us with the tools with which we can prove that the class of (Lebesgue) integrable functions on $[a, b]$ is Cauchy complete relative to the norm $||\cdot||_2$.

Later in this chapter we will begin with a Riemann integrable function f and will construct a sequence $\{f_k\}$ that converges in the mean to f. By this approach we bypass the whole issue of completeness. We are not searching for the limit function f; we begin with it. These remarks are intended to warn you that we are in relatively deep water; you must be cautious not to rely too strongly or prematurely on your intuition about the limit in the mean.

EXAMPLE 1 Recall the construction of Cantor's set C in Section 2.2. Let $U_k = \bigcup_{p=1}^{k} \bigcup_{j=1}^{2^{p-1}} I(p, j)$ again denote the union of the intervals removed from $[0, 1]$ by the kth stage of the construction of C. The total length of the intervals comprising U_k is $d_k = 1 - (2/3)^k$. For each k in $\mathbb{N}$, define

$$f_k(x) = \begin{cases} 1, & \text{if } x \text{ is in } U_k, \\ 0, & \text{if } x \text{ is } U_k^c \cap [0, 1]. \end{cases}$$

Since each f_k is a step function on $[0, 1]$, we know that f_k is integrable on $[0, 1]$. Notice that, since f_k assumes only the values 0 and 1, $f_k^2(x) = f_k(x)$ for all x in $[0, 1]$. We claim that $\{f_k\}$ converges in the mean to the function f_0 that has value 1 on all of $[0, 1]$. To verify this claim, merely compute

$$\begin{aligned} ||1 - f_k||_2^2 &= \int_0^1 [1 - f_k(x)]^2\,dx = \int_0^1 [1 - 2f_k(x) + f_k^2(x)]\,dx \\ &= \int_0^1 [1 - f_k(x)]\,dx = 1 - d_k = \left(\frac{2}{3}\right)^k. \end{aligned}$$

Therefore, $\lim_{k\to\infty} ||1 - f_k||_2^2 = \lim_{x\to\infty} (2/3)^k = 0$. This proves that $\{f_k\}$ converges in the mean to 1 as asserted. You will note that, if x is not in C, then for k_0 sufficiently large, x is in U_{k_0} and hence $f_k(x) = 1$ for all $k \geq k_0$. Therefore, $\lim_{k\to\infty} f_k(x) = 1 = f_0(x)$. However, if x is in C, then $f_k(x) = 0$ for every k and $\lim_{k\to\infty} f_k(x) = 0$. We conclude that, although $\{f_k\}$ converges in the mean to 1 on $[0, 1]$ and converges pointwise to 1 on the complement of C in $[0, 1]$, the sequence $\{f_k\}$ fails to converge pointwise to 1 at the uncountably many points in C. ●

EXAMPLE 2 For each k in $\mathbb{N}$, define a function f_k on $[0, 1]$ by

$$f_k(x) = \begin{cases} 1/k, & \text{if } 0 \leq x < 1 - 1/k^3, \\ k, & \text{if } 1 - 1/k^3 \leq x \leq 1. \end{cases}$$

Then

$$\begin{aligned} ||f_k||_2^2 &= \int_0^1 f_k^2(x)\,dx = \left(\frac{1}{k^2}\right)\left(1 - \frac{1}{k^3}\right) + k^2\left(\frac{1}{k^3}\right) \\ &= \frac{1}{k^2} - \frac{1}{k^5} + \frac{1}{k}. \end{aligned}$$

Therefore $\lim_{k\to\infty} ||f_k||_2 = 0$. This proves that $\{f_k\}$ converges in the mean to 0. Notice in this example that, as $k \to \infty$, some of the values of $f_k(x)$ become arbitrarily large. The point is, f_k is large only on a set of small length. ●

As we saw in Example 1, a sequence $\{f_k\}$ can converge in the mean to a function that differs at uncountably many points from the pointwise limit of that same sequence. However, if the convergence is uniform, then $\{f_k\}$ converges in the mean to the same limit function.

THEOREM 14.1.4 If a sequence $\{f_k\}$ of functions in $R[a, b]$ converges uniformly to f_0 on $[a, b]$, then $\{f_k\}$ converges in the mean to f_0 on $[a, b]$.

Proof. Given $\epsilon > 0$, choose a k_0 such that, for all $k \geq k_0$, $||f_0 - f_k||_\infty < \epsilon/(b-a)^{1/2}$. It follows that

$$||f_0 - f_k||_2^2 = \int_a^b [f_0(x) - f_k(x)]^2\, dx$$

$$< \frac{\epsilon^2}{b-a} \int_a^b dx = \epsilon^2.$$

Thus, for $k \geq k_0$, $||f_0 - f_k||_2 < \epsilon$. This proves that $\{f_k\}$ converges to f_0 in the mean. ●

A crucially important property of convergence in the mean is revealed in the following theorem.

THEOREM 14.1.5 Suppose that $\{f_k\}$ is a sequence of functions in $R[a, b]$ that converges in the mean to a function f in $R[a, b]$. Let g be any function in $R[a, b]$. For x in $[a, b]$ and k in $\mathbb{N}$, define $F_k(x) = \int_a^x f_k(t)g(t)\, dt$. Then F_k is in $C([a, b])$ and

$$\lim_{k\to\infty} F_k(x) = F(x) = \int_a^x f(t)g(t)\, dt \quad \text{[uniformly]}$$

on $[a, b]$.

Proof. The Fundamental Theorem of Calculus implies that, for each k in $\mathbb{N}$, F_k is in $C([a, b])$. We have to prove that, for any given $\epsilon > 0$, we can choose a k_0 in $\mathbb{N}$ such that, for all $k \geq k_0$, $||F - F_k||_\infty < \epsilon$. We assume, without loss of generality, that $||g||_2 > 0$.

The key to the proof is to apply the Cauchy–Schwarz inequality to functions in $R[a, x]$ for each x in $[a, b]$. That inequality translates into the following statement. For every x in $[a, b]$,

$$[F(x) - F_k(x)]^2 = \left[\int_a^x [f(t) - f_k(t)]g(t)\, dt\right]^2$$

$$\leq \left[\int_a^x [f(t) - f_k(t)]^2\, dt\right]\left[\int_a^x g^2(t)\, dt\right].$$

It follows that

$$[F(x) - F_k(x)]^2 \leq \left[\int_a^b [f(t) - f_k(t)]^2\, dt\right]\left[\int_a^b g^2(t)\, dt\right]$$

$$= ||f - f_k||_2^2 ||g||_2^2.$$

Now, given $\epsilon > 0$, choose k_0 in $\mathbb{N}$ such that, for $k \geq k_0$,

$$||f - f_k||_2 < \frac{\epsilon}{||g||_2}.$$

Then, for such k, $[F(x) - F_k(x)]^2 < \epsilon^2$ for all x in $[a, b]$. This proves that $||F - F_k||_\infty < \epsilon$ and completes the proof of the theorem. ●

We have the following generalization of the preceding theorem. Its proof, which consists of a straightforward modification of that of Theorem 14.1.5, is left for you as an exercise.

THEOREM 14.1.6 Let $\{f_k\}$ and $\{g_k\}$ be two sequences in $R[a, b]$. Suppose that $\{f_k\}$ converges in the mean to f and that $\{g_k\}$ converges in the mean to g, where both f and g are in $R[a, b]$. For each k in $\mathbb{N}$ and each x in $[a, b]$, define

$$F_k(x) = \int_a^x f_k(t)g_k(t)\,dt.$$

Then F_k is in $C([a, b])$ and

$$\lim_{k\to\infty} F_k(x) = F(x) = \int_a^x f(t)g(t)\,dt \quad \text{[uniformly]}$$

on $[a, b]$. ●

Of course, Definition 14.1.4 can be modified to apply to infinite series simply by considering the sequence of partial sums. Since we will soon be working with series rather than sequences, the following definition will be especially pertinent.

DEFINITION 14.1.5 A series $\sum_{j=1}^{\infty} f_j$ of integrable functions on $[a, b]$ is said to *converge in the mean* to an integrable function F on $[a, b]$ if the sequence $\{F_k\}$ of partial sums of the series converges in the mean to F. That is,

$$\lim_{k\to\infty} ||F - F_k||_2 = \lim_{k\to\infty} ||F - \sum_{j=1}^{k} f_j||_2 = 0. \quad ●$$

Concepts central to all our work in the remainder of this chapter are described in the following definitions.

DEFINITION 14.1.6 Two functions f and g in $R[a, b]$ are said to be *orthogonal* if $\langle f, g\rangle = 0$. ●

DEFINITION 14.1.7 Let $\Phi = \{\phi_1, \phi_2, \phi_3, \ldots\}$ be a countable collection of non-zero functions in $R[a, b]$.

i) Φ is said to be an *orthogonal* collection if, for $j \neq k$, $\langle \phi_j, \phi_k\rangle = 0$. The functions themselves are said to be *mutually orthogonal*.

ii) If each function ϕ in Φ is also normalized so that $||\phi||_2 = 1$, then Φ is said to be an *orthonormal* collection. The functions themselves are said to be *orthonormal*. ●

Any orthogonal collection Φ of functions gives rise to an orthonormal collection. Simply replace every ϕ in Φ with $\phi/||\phi||_2$.

EXAMPLE 3 By far, the most important orthogonal collection in this text is

$$\Phi = \{1, \cos x, \sin x, \cos 2x, \sin 2x, \ldots, \cos kx, \sin kx, \ldots\}$$

on the interval $[-\pi, \pi]$. In Exercises 6.51 and 6.52 you proved that, for any k and m in $\mathbb{N}$,

$$\int_{-\pi}^{\pi} \sin kx \cos mx \, dx = 0.$$

$$\int_{-\pi}^{\pi} \sin kx \sin mx \, dx = \begin{cases} 0, & \text{if } k \neq m, \\ \pi, & \text{if } k = m. \end{cases}$$

$$\int_{-\pi}^{\pi} \cos kx \cos mx \, dx = \begin{cases} 0, & \text{if } k \neq m, \\ \pi, & \text{if } k = m. \end{cases}$$

$$\int_{-\pi}^{\pi} 1 \cos kx \, dx = \int_{-\pi}^{\pi} 1 \sin kx \, dx = 0.$$

This shows that Φ is an orthogonal collection.

The L_2-norm of the function 1 is $\sqrt{2\pi}$; the L_2-norm of each of the functions $\cos kx$ and $\sin kx$ is $\sqrt{\pi}$. Therefore,

$$\left\{ \frac{1}{\sqrt{2\pi}}, \frac{1}{\sqrt{\pi}} \cos kx, \frac{1}{\sqrt{\pi}} \sin kx : k \text{ in } \mathbb{N} \right\}$$

is an orthonormal collection. It was a variant of this orthonormal collection that Fourier and his contemporaries considered when they first attempted to represent a periodic function with period 2π as an infinite series of the form

$$\frac{a_0}{2} + \sum_{j=1}^{\infty} [a_j \cos jx + b_j \sin jx].$$

It is this collection which will form the basis for much of our work in this chapter. ●

EXAMPLE 4 However, the collection of trigonometric functions described in Example 3 is not the only orthogonal collection of interest in mathematics and mathematical physics. In Exercises 12.60–12.64 you constructed the Legendre polynomials $P_0, P_1, P_2, P_3, \ldots$ on $[-1, 1]$ and proved that

$$\langle P_k, P_m \rangle = \int_{-1}^{1} P_k(x) P_m(x) \, dx = \begin{cases} 0, & \text{if } k \neq m, \\ 2/(2k+1), & \text{if } k = m. \end{cases}$$

That is, you proved that $\{P_0, P_1, P_2, P_3, \ldots\}$ is an orthogonal collection in $C([-1, 1])$. The collection

$$\Phi = \left\{ \sqrt{\frac{2k+1}{2}} P_k : k = 0, 1, 2, 3, \ldots \right\}$$

is an orthonormal collection in $C([-1, 1])$. ●

THEOREM 14.1.7 Let $\Phi = \{\phi_1, \phi_2, \phi_3, \ldots\}$ be any orthonormal collection of functions in $R[a, b]$. Suppose that

$$f(x) = \sum_{j=1}^{k} a_j \phi_j(x)$$

is any finite linear combination of the functions in Φ. Then, for each $j = 1, 2, \ldots, k$,

$$a_j = \langle f, \phi_j \rangle = \int_a^b f(x)\phi_j(x)\,dx.$$

Proof. Fix j in $\{1, 2, \ldots, k\}$. Multiply f by ϕ_j and integrate the product from a to b. We obtain

$$\int_a^b f(x)\phi_j(x)\,dx = \sum_{i=1}^{k} a_i \int_a^b \phi_i(x)\phi_j(x)\,dx. \tag{14.1}$$

Because Φ is an orthonormal collection, of the integrals in the sum (14.1) only one does not necessarily vanish and that is $\int_a^b \phi_j(x)\phi_j(x)\,dx = ||\phi_j||_2^2 = 1$. We deduce that

$$\int_a^b f(x)\phi_j(x)\,dx = a_j,$$

as claimed in the theorem. ●

THEOREM 14.1.8 Let $\Phi = \{\phi_1, \phi_2, \phi_3, \ldots\}$ be any orthonormal collection of functions in $R[a, b]$. Suppose that the series $\sum_{j=1}^{\infty} a_j\phi_j$ converges uniformly to a function f on $[a, b]$. Then, for each j in $\mathbb{N}$,

$$a_j = \int_a^b f(x)\phi_j(x)\,dx.$$

Proof. Since $f = \sum_{j=1}^{\infty} a_j\phi_j$ [uniformly] and since each ϕ_j is in $R[a, b]$, Theorem 12.2.6 ensures that f is also in $R[a, b]$. Furthermore, if, for any j in $\mathbb{N}$, we multiply both f and the series by ϕ_j, then the series $\sum_{i=1}^{\infty} a_i\phi_i\phi_j$ also converges uniformly to $f\phi_j$. Again by Theorem 12.2.6, we deduce that

$$\int_a^b f(x)\phi_j(x)\,dx = \sum_{i=1}^{\infty} a_i \int_a^b \phi_i(x)\phi_j(x)\,dx = a_j,$$

proving the theorem. ●

The two preceding theorems motivate the following definition.

DEFINITION 14.1.8 Let $\Phi = \{\phi_j : j \text{ in } \mathbb{N}\}$ be any orthonormal collection of functions in $R[a, b]$ and let f be any function in $R[a, b]$.

i) The *Fourier series* of f with respect to Φ is the infinite series $\sum_{j=1}^{\infty} a_j\phi_j$, where

$$a_j = \langle f, \phi_j \rangle = \int_a^b f(x)\phi_j(x)\,dx.$$

ii) The jth *Fourier coefficient* of f with respect to Φ is the number a_j.

iii) The kth partial sum of the Fourier series of f will be denoted $S_k(f) = \sum_{j=1}^{k} a_j\phi_j$. When there is no possibility of confusion, we will abbreviate $S_k(f)$ to S_k. ●

Naturally, we want to know, for a given f in $R[a, b]$, whether the sequence $\{S_k(f)\}$ converges to f pointwise, uniformly, or in the mean. In general, the question of pointwise convergence of $\{S_k(f)\}$ is subtle; it can happen even for a continuous f that the sequence $\{S_k(x)\}$ fails to converge to $f(x)$ for many values of x. Indeed, it was this very question that led Cantor to study uncountable sets. Although sufficient conditions are known that ensure that $\lim_{k\to\infty} S_k(x) = f(x)$ [pointwise], and we will discuss the most generally useful of these conditions in Section 3, even today necessary conditions are not known. Indeed, it seems likely that necessary conditions do not exist. However the mean convergence of $\{S_k(f)\}$ to f, being a significantly weaker form of convergence, is easier to treat. It is to this question that we first turn.

DEFINITION 14.1.9 Let $\Phi = \{\phi_1, \phi_2, \phi_3, \ldots\}$ be any orthonormal collection of functions in $R[a, b]$ and let f be any function in $R[a, b]$. Let $\{b_j\}$ be any sequence of real numbers. For each k in $\mathbb{N}$, let $g_k = \sum_{j=1}^{k} b_j\phi_j$.

i) The function g_k is called the *kth mean-square approximation* determined by $\{b_j\}$.

ii) The *kth mean-square error* that results from approximating f by g_k is $||f - g_k||_2$. ●

Notice that, among all possible kth mean-square approximations to f, $S_k(f)$ is the special one with $b_j = a_j$ for $j = 1, 2, \ldots, k$. Our first objective is to minimize the kth mean square error. As we shall see, $S_k(f)$ always provides the unique kth mean-square approximation of f for which the kth mean-square error is a minimum. For the remainder of this section we fix an orthonormal collection Φ in $R[a, b]$.

LEMMA 1 If $g_k = \sum_{j=1}^{k} b_j\phi_j$, then $||g_k||_2^2 = \sum_{j=1}^{k} b_j^2$.

Proof. Merely compute

$$||g_k||_2^2 = \langle g_k, g_k\rangle = \left\langle \sum_{i=1}^{k} b_i\phi_i, \sum_{j=1}^{k} b_j\phi_j \right\rangle$$

$$= \sum_{i=1}^{k}\sum_{j=1}^{k} b_i b_j \langle\phi_i, \phi_j\rangle = \sum_{j=1}^{k} b_j^2,$$

since, by the orthonormality of Φ, $\langle\phi_i, \phi_j\rangle = 0$ except when $i = j$, whereupon it has value 1. ●

LEMMA 2 If $g_k = \sum_{j=1}^{k} b_j\phi_j$ and if f is in $R[a, b]$, then

$$\langle f, g_k\rangle = \sum_{j=1}^{k} a_j b_j,$$

where a_j is the jth Fourier coefficient of f with respect to Φ.

Proof. Again merely compute:

$$\langle f, g_k \rangle = \left\langle f, \sum_{j=1}^{k} b_j \phi_j \right\rangle = \sum_{j=1}^{k} b_j \langle f, \phi_j \rangle = \sum_{j=1}^{k} a_j b_j. \quad \bullet$$

LEMMA 3 If $g_k = \sum_{j=1}^{k} b_j \phi_j$ and if f is in $R[a, b]$, then

$$\|f - g_k\|_2^2 = \|f\|_2^2 - \sum_{j=1}^{k} a_j^2 + \sum_{j=1}^{k} (a_j - b_j)^2,$$

where a_j is the jth Fourier coefficient of f with respect to the orthonormal collection Φ.

Remark. If we replace g_k by $S_k(f)$, that is, if we take $b_j = a_j$ for $j = 1, 2, \ldots, k$, then, by this lemma,

$$\|f - S_k(f)\|_2^2 = \|f\|_2^2 - \sum_{j=1}^{k} a_j^2.$$

Proof. Invoking Lemmas 1 and 2, we merely compute

$$\begin{aligned}
\|f - g_k\|_2^2 &= \langle f - g_k, f - g_k \rangle = \langle f, f \rangle - 2\langle f, g_k \rangle + \langle g_k, g_k \rangle \\
&= \|f\|_2^2 - 2\sum_{j=1}^{k} a_j b_j + \sum_{j=1}^{k} b_j^2 \\
&= \|f\|_2^2 - \sum_{j=1}^{k} a_j^2 + \left(\sum_{j=1}^{k} a_j^2 - 2\sum_{j=1}^{k} a_j b_j + \sum_{j=1}^{k} b_j^2 \right) \\
&= \|f\|_2^2 - \sum_{j=1}^{k} a_j^2 + \sum_{j=1}^{k} (a_j - b_j)^2. \quad \bullet
\end{aligned}$$

THEOREM 14.1.9 Of all the kth mean-square approximations of f, the one with the least mean square error is $S_k(f)$.

Proof. By Lemma 3 and the remark following it, for any choice of $\{b_j\}$, we have

$$\begin{aligned}
\|f - g_k\|_2^2 &= \|f\|_2^2 - \sum_{j=1}^{k} a_j^2 + \sum_{j=1}^{k} (a_j - b_j)^2 \\
&= \|f - S_k(f)\|_2^2 + \sum_{j=1}^{k} (a_j - b_j)^2.
\end{aligned}$$

Since $\sum_{j=1}^{k} (a_j - b_j)^2$ is nonnegative, we have

$$\|f - S_k(f)\|_2^2 \leq \|f - g_k\|_2^2. \tag{14.2}$$

Further, we have equality in (14.2) if and only if $b_j = a_j$ for all $j = 1, 2, \ldots, k$, that is, if and only if $g_k = S_k(f)$. ●

THEOREM 14.1.10 Bessel's Inequality Let a_j be the jth Fourier coefficient of f. The series $\sum_{j=1}^{\infty} a_j^2$ converges and

$$\sum_{j=1}^{\infty} a_j^2 \le ||f||_2^2.$$

Proof. In Lemma 3 take $g_k = S_k(f)$ for each k in $\mathbb{N}$. We have

$$0 \le ||f - S_k(f)||_2^2 = ||f||_2^2 - \sum_{j=1}^{k} a_j^2.$$

It follows that, for every k in $\mathbb{N}$,

$$\sum_{j=1}^{k} a_j^2 = ||f||_2^2 - ||f - S_k(f)||_2^2 \le ||f||_2^2.$$

Since the sequence of kth partial sums of the series $\sum_{j=1}^{\infty} a_j^2$ is monotone increasing and is bounded above by $||f||_2^2$, the series converges to a number also bounded above by $||f||_2^2$. Thus $\sum_{j=1}^{\infty} a_j^2 \le ||f||_2^2$. ●

COROLLARY 14.1.11 Let $\{a_j\}$ be the sequence of Fourier coefficients of a function f in $R[a, b]$. Then $\lim_{j\to\infty} a_j = 0$.

Proof. The series $\sum_{j=1}^{\infty} a_j^2$ converges. ●

We are particularly interested in the situation when equality holds in Bessel's inequality.

DEFINITION 14.1.10 Let f be in $R[a, b]$ and let $\{a_j\}$ be the sequence of Fourier coefficients of f (relative to a given orthonormal collection Φ). *Parseval's identity* is said to hold for f if $\sum_{j=1}^{\infty} a_j^2 = ||f||_2^2$. ●

THEOREM 14.1.12 $\lim_{k\to\infty} ||f - S_k(f)||_2 = 0$ if and only if Parseval's identity holds for f.

Proof. Suppose that $\sum_{j=1}^{\infty} a_j^2 = ||f||_2^2$. Given $\epsilon > 0$, we can choose a k_0 such that, for $k \ge k_0$, $0 \le ||f||_2^2 - \sum_{j=1}^{k} a_j^2 < \epsilon$. Since $||f||_2^2 - \sum_{j=1}^{k} a_j^2 = ||f - S_k(f)||_2^2$, we deduce immediately that $\lim_{k\to\infty} ||f - S_k(f)||_2 = 0$.

Conversely, if $\lim_{k\to\infty} ||f - S_k(f)||_2 = 0$, then clearly $||f - S_k(f)||_2^2 = ||f||_2^2 - \sum_{j=1}^{k} a_j^2$ can be made arbitrarily small by taking k to be sufficiently large. It follows that $\sup \{\sum_{j=1}^{k} a_j^2 : k \text{ in } \mathbb{N}\} = ||f||_2^2$ and Parseval's identity holds for f. ●

DEFINITION 14.1.11 An orthonormal collection Φ is said to be *complete* for a collection $\mathcal{F}$ of functions in $R[a, b]$ if Parseval's identity holds for every f in $\mathcal{F}$. ●

THEOREM 14.1.13 Let Φ be an orthonormal collection of functions in $R[a, b]$ that is complete for a collection $\mathcal{F}$ in $R[a, b]$. Then, for every f in $\mathcal{F}$, the Fourier series of f converges in the mean to f.

Proof. Since Φ is complete for $\mathcal{F}$, Parseval's identity holds for every f in $\mathcal{F}$. Equivalently, by Theorem 14.1.12, for each f in $\mathcal{F}$, $\lim_{k\to\infty} ||f - S_k(f)||_2 = 0$. But this is exactly what is meant by saying that the Fourier series of f converges in the mean to f. ●

14.2 TRIGONOMETRIC SERIES

Here we focus on the orthogonal collection provided by the cosine and sine functions on the interval $[-\pi, \pi]$. Given an integrable function f on $[-\pi, \pi]$, we define its Fourier series to be

$$\frac{a_0}{2} + \sum_{j=1}^{\infty} [a_j \cos jx + b_j \sin jx],$$

where the Fourier coefficients a_j and b_j are given by

$$a_j = \frac{1}{\pi} \int_{-\pi}^{\pi} f(x) \cos jx \, dx, \qquad j = 0, 1, 2, \ldots$$

and

$$b_j = \frac{1}{\pi} \int_{-\pi}^{\pi} f(x) \sin jx \, dx, \qquad j = 1, 2, 3, \ldots$$

Not knowing that the Fourier series of f actually converges pointwise to f, we will write

$$f(x) \sim \frac{a_0}{2} + \sum_{j=1}^{\infty} [a_j \cos jx + b_j \cos jx]$$

to indicate the association of f with its Fourier series. It will be convenient to redefine f at the endpoints to have the value $[f(-\pi) + f(\pi)]/2$. Also, if f has a jump discontinuity at a point x in $(-\pi, \pi)$, we redefine f at x to have the value $[f(x^-) + f(x^+)]/2$. Having made these adjustments, we then extend f to all of $\mathbb{R}$ to be periodic with period 2π. Specifically, since for any x in $\mathbb{R}$ there is a unique n_x in $\mathbb{Z}$ such that $x + 2n_x\pi$ is in $(-\pi, \pi]$, we define $f(x) = f(x + 2n_x\pi)$.

Certain observations simplify the computation of the Fourier coefficients whenever f is either even or odd. If, for example, f is even, then the function $f(x) \cos jx$ is even and $f(x) \sin jx$ is odd. In this case,

$$a_j = \frac{2}{\pi} \int_{0}^{\pi} f(x) \cos jx \, dx, \qquad j = 0, 1, 2, \ldots$$

and

$$b_j = \frac{1}{\pi} \int_{-\pi}^{\pi} f(x) \sin jx \, dx = 0, \qquad j = 1, 2, 3, \ldots$$

and the Fourier series of f reduces to a cosine series, involving only cosine terms. Likewise, if f is odd, then

$$a_j = \frac{1}{\pi}\int_{-\pi}^{\pi} f(x)\cos jx\,dx = 0, \qquad j = 0, 1, 2, \ldots$$

and

$$b_j = \frac{2}{\pi}\int_{0}^{\pi} f(x)\sin jx\,dx, \qquad j = 1, 2, 3, \ldots$$

and the Fourier series of f is a sine series. If f is neither even nor odd, then the Fourier series of f will contain both sine and cosine terms.

EXAMPLE 5 Define f on $[-\pi, \pi]$ as follows:

$$f(x) = \begin{cases} 0, & x \text{ in } (-\pi, 0) \\ 1, & x \text{ in } (0, \pi) \\ 1/2, & x = -\pi, 0, \pi. \end{cases}$$

Upon extending f to $\mathbb{R}$ to be periodic with period 2π, we obtain the function whose graph is sketched in Fig. 14.2; this graph describes a *square wave form*. We calculate the Fourier coefficients of f. First,

$$a_0 = \frac{1}{\pi}\int_0^{\pi} dx = 1.$$

For $j = 1, 2, 3, \ldots$

$$a_j = \frac{1}{\pi}\int_0^{\pi} \cos jx\,dx = \left.\frac{\sin jx}{\pi j}\right|_0^{\pi} = 0$$

and

$$b_j = \frac{1}{\pi}\int_0^{\pi} \sin jx\,dx = \left.\frac{-\cos jx}{\pi j}\right|_0^{\pi}$$

$$= \frac{1 - (-1)^j}{j\pi}.$$

Noting that

$$[1 - (-1)^j] = \begin{cases} 2, & \text{if } j \text{ is odd,} \\ 0, & \text{if } j \text{ is even,} \end{cases}$$

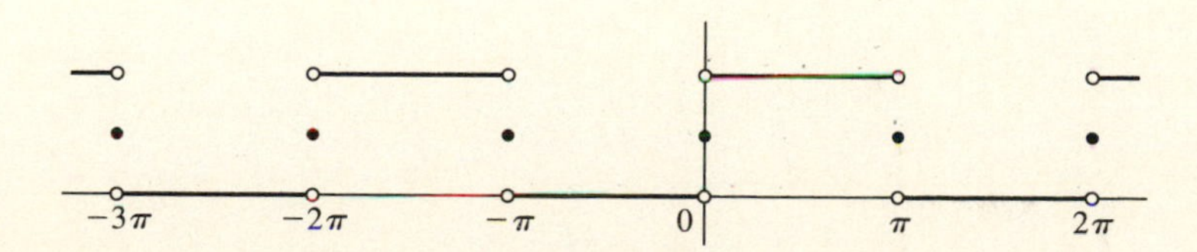

Figure 14.2

we conclude that the Fourier series of f is

$$f(x) \sim \frac{1}{2} + \frac{2}{\pi} \sum_{j=1}^{\infty} \frac{\sin (2j - 1)x}{2j - 1}.$$

Once we have addressed the question of pointwise convergence of Fourier series, we will be able to assign a value to x and obtain a numerical series with sum $f(x)$. For the time being we will assume that the series converges to $f(x)$ at all points where f is continuous, in a neighborhood of which f is of bounded variation. For example, in the above series, set $x = \pi/2$. We obtain

$$1 = \frac{1}{2} + \frac{2}{\pi} \sum_{j=1}^{\infty} \frac{\sin (2j - 1)\pi/2}{2j - 1},$$

or upon rearranging, $\pi/4 = \sum_{j=1}^{\infty} (-1)^{j+1}/(2j - 1)$. This series is familiar; it is the series for $\tan^{-1} x$ evaluated at $x = 1$.

We can modify the function f considered here and obtain the correspondingly modified Fourier series. For example, we can let $g(x) = 2f(x) - 1$; the graph of g is displayed in Fig. 14.3. The Fourier series of g is

$$\frac{4}{\pi} \sum_{j=1}^{\infty} \frac{\sin (2j - 1)x}{2j - 1}. \quad \bullet$$

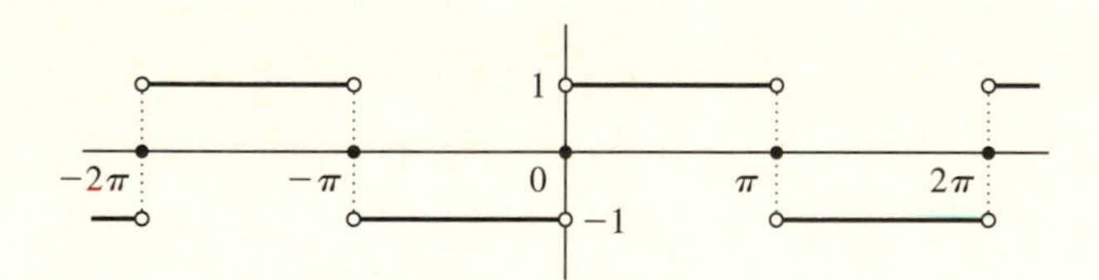

Figure 14.3

EXAMPLE 6 For x in $[-\pi, \pi]$, define

$$f(x) = \begin{cases} x + \pi, & \text{for } x \text{ in } (-\pi, 0), \\ x, & \text{for } x \text{ in } (0, \pi), \\ \pi/2, & \text{for } x = -\pi, 0, \pi. \end{cases}$$

Extend f to be periodic with period 2π. The resulting function has a *saw tooth wave form* as sketched in Fig. 14.4 on page 681. When we compute the Fourier coefficients of f, we obtain

$$\begin{aligned} a_0 &= \frac{1}{\pi} \int_{-\pi}^{0} (x + \pi)\, dx + \frac{1}{\pi} \int_{0}^{\pi} x\, dx \\ &= \frac{1}{\pi} \int_{-\pi}^{\pi} x\, dx + \int_{-\pi}^{0} dx = \pi. \end{aligned}$$

For j in $\mathbb{N}$,

$$a_j = \frac{1}{\pi}\int_{-\pi}^{\pi} x\cos jx\,dx + \int_{-\pi}^{0}\cos jx\,dx = 0$$

and

$$\begin{aligned} b_j &= \frac{1}{\pi}\int_{-\pi}^{\pi} x\sin jx\,dx + \int_{-\pi}^{0}\sin jx\,dx \\ &= \frac{-1+(-1)^j}{j} = \begin{cases} -2/j, & \text{if } j \text{ is even,} \\ 0, & \text{if } j \text{ is odd.} \end{cases} \end{aligned}$$

Thus, the Fourier series of f is

$$\frac{\pi}{2} - \sum_{j=1}^{\infty}\frac{\sin 2jx}{j}. \quad \bullet$$

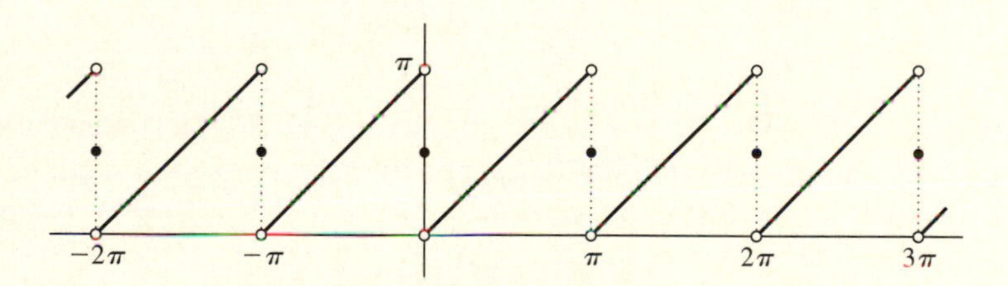

Figure 14.4

EXAMPLE 7 For x in $[-\pi, \pi]$, define $f(x) = x^2$ and extend f to be periodic of period 2π. Notice that f is continuous on all of $\mathbb{R}$. (See Fig. 14.5 on page 682.) Notice also that f is an even function; therefore the Fourier series of f is a cosine series. The Fourier coefficients of f are

$$a_0 = \frac{2\pi^2}{3},$$

$$a_j = \frac{(-1)^j 4}{j^2},$$

and

$$b_j = 0$$

for j in $\mathbb{N}$. Therefore the Fourier series of f is

$$x^2 \sim \frac{\pi^2}{3} + 4\sum_{j=1}^{\infty}\frac{(-1)^j\cos jx}{j^2}.$$

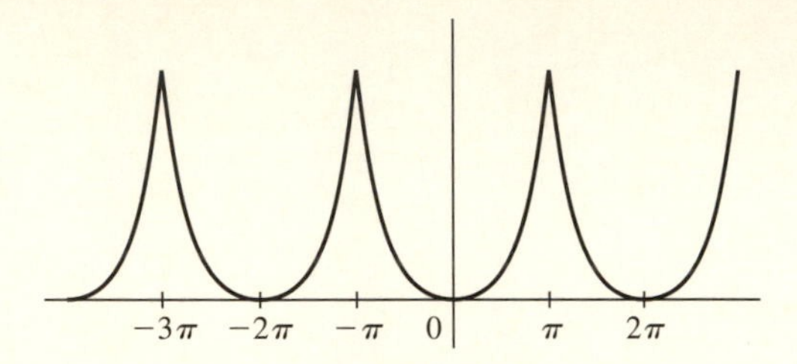

Figure 14.5

Assuming for the time being that this series converges pointwise to f, we set, say, $x = \pi/2$ in this series and deduce that

$$\frac{\pi^2}{4} = \frac{\pi^2}{3} + 4\sum_{j=1}^{\infty} \frac{(-1)^j \cos(j\pi/2)}{j^2}.$$

Upon simplifying and changing index, we obtain

$$\frac{\pi^2}{12} = \sum_{j=1}^{\infty} \frac{(-1)^{j+1}}{j^2}.$$

You will notice that the series converges by the alternating series test (and converges absolutely) but until now we did not know how to identify its sum. Fourier series give us a flexible tool enabling us to identify the sum of a variety of convergent numerical series. ●

14.3 Convergence of Trigonometric Series

Turning our attention to the question of [pointwise] convergence, we examine more closely the kth partial sum, $S_k(x)$, of the Fourier series of f evaluated at a point x. We are especially interested in obtaining a more compact, integral form of this partial sum. Using the definitions of the Fourier coefficients, we obtain

$$\begin{aligned} S_k(x) &= \frac{a_0}{2} + \sum_{j=1}^{k} [a_j \cos jx + b_j \sin jx] \\ &= \frac{1}{2\pi}\int_{-\pi}^{\pi} f(t)\,dt + \sum_{j=1}^{k}\left[\frac{1}{\pi}\int_{-\pi}^{\pi} f(t)[\cos jt \cos jx + \sin jt \sin jx]\right] dt \\ &= \frac{1}{\pi}\int_{-\pi}^{\pi} f(t)\left[\frac{1}{2} + \sum_{j=1}^{k} \cos j(t-x)\right] dt. \end{aligned} \tag{14.3}$$

Let $T_k(z) = 1/2 + \sum_{j=1}^{k} \cos jz$. Notice that T_k is defined on all of $\mathbb{R}$, is even, and is periodic with period 2π. From (14.3) we have

$$S_k(x) = \frac{1}{\pi}\int_{-\pi}^{\pi} f(t) T_k(t-x)\,dt.$$

In this integral, change variables by letting $u = t - x$. The result is

$$S_k(x) = \frac{1}{\pi}\int_{-\pi-x}^{\pi-x} f(x+u)T_k(u)\,du.$$

Since both f and T_k are periodic with period 2π, integrating over any interval of length 2π yields the same value. Thus

$$\begin{aligned} S_k(x) &= \frac{1}{\pi}\int_{-\pi}^{\pi} f(x+u)T_k(u)\,du \\ &= \frac{1}{\pi}\left[\int_{-\pi}^{0} + \int_{0}^{\pi}\right] f(x+u)T_k(u)\,du. \end{aligned}$$

In the integral over $[-\pi, 0]$ let $v = -u$. Because T_k is even, that integral becomes

$$\frac{1}{\pi}\int_{\pi}^{0} f(x-v)T_k(-v)(-dv) = \frac{1}{\pi}\int_{0}^{\pi} f(x-v)T_k(v)\,dv.$$

Since v is a dummy variable,

$$S_k(x) = \frac{2}{\pi}\int_{0}^{\pi} \frac{f(x-u)+f(x+u)}{2} T_k(u)\,du. \tag{14.4}$$

By Theorem 11.4.5, we know that

$$\sum_{j=1}^{k} \cos ju = \frac{\cos((k+1)u/2)\sin(ku/2)}{\sin(u/2)}$$

The trigonometric identity

$$\cos\theta\sin\phi = \frac{\sin(\theta+\phi) - \sin(\theta-\phi)}{2}$$

enables us to write

$$\cos\left(\frac{(k+1)u}{2}\right)\sin\left(\frac{ku}{2}\right) = \frac{\sin[(2k+1)u/2] - \sin(u/2)}{2}.$$

Therefore,

$$\begin{aligned} T_k(u) &= \frac{1}{2} + \frac{\cos[(k+1)u/2]\sin(ku/2)}{\sin(u/2)} \\ &= \frac{1}{2} + \frac{\sin[(2k+1)u/2] - \sin(u/2)}{2\sin(u/2)} \\ &= \frac{\sin[(2k+1)u/2]}{2\sin(u/2)}. \end{aligned}$$

Substitute this result in (14.4) to obtain

$$S_k(x) = \frac{2}{\pi}\int_{0}^{\pi} \frac{f(x-u)+f(x+u)}{2}\,\frac{\sin(2k+1)u/2}{2\sin(u/2)}\,du.$$

Finally, let $u = 2t$ to obtain the integral form of $S_k(x)$ that we want:

$$S_k(x) = \frac{2}{\pi} \int_0^{\pi/2} \frac{f(x-2t) + f(x+2t)}{2} \frac{\sin{(2k+1)t}}{\sin t}\, dt.$$

We have proved the following theorem.

THEOREM 14.3.1 The Fourier series of an integrable function f converges at a point x if and only if

$$\lim_{k\to\infty} \frac{2}{\pi} \int_0^{\pi/2} \frac{f(x-2t) + f(x+2t)}{2} \frac{\sin{(2k+1)t}}{\sin t}\, dt$$

exists. If the limit does exist, its value is the sum of the Fourier series. ●

To use this form of $S_k(x)$, we need an essential fact. The next theorem is the key to all our subsequent work in this section.

THEOREM 14.3.2 The Riemann–Lebesgue Lemma Let f be any integrable function on $[a, b]$. For any values of c_1 and c_2 in $\mathbb{R}$ with $c_1 \neq 0$,

$$\lim_{k\to\infty} \int_a^b f(t) \sin{(c_1 kt + c_2)}\, dt = 0.$$

Proof. Suppose first that f is constant on $[a, b]$. Then the assertion of the theorem certainly holds since

$$\begin{aligned} &\lim_{k\to\infty} \left| \int_a^b \sin{(c_1 kt + c_2)}\, dt \right| \\ &= \lim_{k\to\infty} \frac{|\cos{(c_1 ka + c_2)} - \cos{(c_1 kb + c_2)}|}{|c_1|k} \\ &\leq \lim_{k\to\infty} \frac{2}{|c_1|k} = 0. \end{aligned}$$

Suppose next that f is a step function with possible jumps at $a = x_0 < x_1 < x_2 < x_3 < \cdots < x_p = b$. Then f is constant (except possibly at the endpoints) on $[x_{j-1}, x_j]$ and thus for $j = 1, 2, \ldots, p$,

$$\lim_{k\to\infty} \int_{x_{j-1}}^{x_j} f(t) \sin{(c_1 kt + c_2)}\, dt = 0.$$

Therefore,

$$\lim_{k\to\infty} \int_a^b f(t) \sin{(c_1 kt + c_2)}\, dt = \sum_{j=1}^{p} \lim_{k\to\infty} \int_{x_{j-1}}^{x_j} f(t) \sin{(c_1 kt + c_2)}\, dt = 0.$$

Finally, suppose that f is an arbitrary integrable function on $[a, b]$; Riemann's condition holds for f. Given any $\epsilon > 0$, choose a partition π_0 of $[a, b]$ such that, for every refinement π of π_0,

$$U(f, \pi) - L(f, \pi) < \epsilon.$$

Fix a refinement π of π_0: $\pi = \{a = x_0, x_1, x_2, \ldots, x_p = b\}$. For each $j = 1, 2, \ldots, p$, let

$$M_j = \max\{f(x) : x \text{ in } [x_{j-1}, x_j]\}$$

and

$$m_j = \min\{f(x) : x \text{ in } [x_{j-1}, x_j]\}.$$

Form the step functions $g(x)$ and $h(x)$ such that $g(x) = m_j$ and $h(x) = M_j$ if x is in $[x_{j-1}, x_j)$. (Define $g(b) = m_p$ and $h(b) = M_p$.) Then $g(x) \le f(x) \le h(x)$ for all x in $[a, b]$ and

$$\int_a^b [h(t) - g(t)]\, dt = U(f, \pi) - L(f, \pi) < \epsilon.$$

Since g is a step function, the assertion of the theorem holds for g: there exists a k_0 in $\mathbb{N}$ such that, for $k \ge k_0$,

$$\left| \int_a^b g(t) \sin(c_1 kt + c_2)\, dt \right| < \epsilon.$$

Finally, for any $k \ge k_0$ we compute

$$\begin{aligned} \left| \int_a^b f(t) \sin(c_1 kt + c_2)\, dt \right| &\le \left| \int_a^b [f(t) - g(t)] \sin(c_1 kt + c_2)\, dt \right| \\ &\quad + \left| \int_a^b g(t) \sin(c_1 kt + c_2)\, dt \right| \\ &\le \int_a^b |h(t) - g(t)|\, dt + \epsilon < 2\epsilon. \end{aligned}$$

Therefore, $\lim_{k\to\infty} \int_a^b f(t) \sin(c_1 kt + c_2)\, dt = 0$. This proves the Riemann–Lebesgue lemma. ●

COROLLARY 14.3.3 For any f in $R[-\pi, \pi]$, the sequences $\{a_j\}$ and $\{b_j\}$ of Fourier coefficients of f converge to 0. That is, $\lim_{j\to\infty} \int_{-\pi}^{\pi} f(t) \cos jt\, dt = \lim_{j\to\infty} \int_{-\pi}^{\pi} f(t) \sin jt\, dt = 0$.

Proof. By taking $c_1 = 1$ and $c_2 = 0$ in the theorem, it follows immediately that $\lim_{j\to\infty} b_j = 0$. If we let $c_1 = 1$ and $c_2 = \pi/2$, then

$$\begin{aligned} \lim_{j\to\infty} a_j &= \lim_{j\to\infty} \frac{1}{\pi} \int_{-\pi}^{\pi} f(t) \cos jt\, dt \\ &= \lim_{j\to\infty} \frac{1}{\pi} \int_{-\pi}^{\pi} f(t) \sin\left(jt + \frac{\pi}{2}\right) dt = 0, \end{aligned}$$

again by Theorem 14.3.2. ●

THEOREM 14.3.4 The Fourier series of an integrable function f on $[-\pi, \pi]$ converges at a point x if and only if

$$\lim_{k\to\infty} \frac{2}{\pi} \int_0^{\pi/2} \frac{f(x-2t)+f(x+2t)}{2} \frac{\sin(2k+1)t}{t}\, dt$$

exists. If the limit does exist, its value is the sum of the Fourier series.

Proof. Referring to Theorem 14.3.1, we know that the Fourier series of f converges at x if and only if

$$\lim_{k\to\infty} \frac{2}{\pi} \int_0^{\pi/2} \frac{f(x-2t)+f(x+2t)}{2} \frac{\sin(2k+1)t}{\sin t}\, dt$$

exists. Holding x fixed, let $g(t)$ denote the function $[f(x-2t)+f(x+2t)]/2$ and consider the difference

$$\int_0^{\pi/2} g(t)\frac{\sin(2k+1)t}{\sin t}\, dt - \int_0^{\pi/2} g(t)\frac{\sin(2k+1)t}{t}\, dt$$
$$= \int_0^{\pi/2} g(t)\left(\frac{1}{\sin t} - \frac{1}{t}\right)\sin(2k+1)t\, dt.$$

Define

$$h(t) = \begin{cases} \dfrac{1}{\sin t} - \dfrac{1}{t}, & \text{if } t \neq 0, \\ 0, & \text{if } t = 0. \end{cases}$$

At any point $t \neq 0$, h is certainly continuous. Furthermore, using l'Hôpital's rule, you can show that $\lim_{t\to 0^+} h(t) = 0$. Thus h is continuous at $t = 0$ as well. Therefore, the function gh is integrable on $[0, \pi/2]$. By the Riemann–Lebesgue lemma (Theorem 14.3.2), we deduce that

$$\lim_{k\to\infty} \int_0^{\pi/2} g(t)h(t)\sin(2k+1)t\, dt = 0.$$

Thus,

$$\lim_{k\to\infty} \frac{2}{\pi} \int_0^{\pi/2} \frac{f(x-2t)+f(x+2t)}{2} \frac{\sin(2k+1)t}{\sin t}\, dt$$
$$= \lim_{k\to\infty} \frac{2}{\pi} \int_0^{\pi/2} \frac{f(x-2t)+f(x+2t)}{2} \frac{\sin(2k+1)t}{t}\, dt.$$

This proves the theorem. ●

Theorem 14.3.4, combined with the Riemann–Lebesgue lemma, yields a most startling result, known as *Riemann's localization principle*. Fix any $\delta > 0$ and write the integral in Theorem 14.3.4 as the sum of two integrals:

$$\frac{2}{\pi}\left[\int_0^{\delta} + \int_{\delta}^{\pi/2}\right] \frac{g(t)}{t} \sin(2k+1)t\, dt,$$

where, again, $g(t) = [f(x - 2t) + f(x + 2t)]/2$. For t in $[\delta, \pi/2]$, the function $g(t)/t$ is integrable and therefore by the Riemann–Lebesgue lemma,

$$\lim_{k \to \infty} \frac{2}{\pi} \int_{\delta}^{\pi/2} \frac{g(t)}{t} \sin(2k+1)t \, dt = 0.$$

We have proved the following remarkable fact.

THEOREM 14.3.5 Riemann's Localization Theorem The Fourier series of a function f in $R[a, b]$ converges at a point x if and only if there exists a $\delta > 0$ such that

$$\lim_{k \to \infty} \frac{2}{\pi} \int_{0}^{\delta} \frac{f(x - 2t) + f(x + 2t)}{2} \frac{\sin(2k+1)t}{t} \, dt$$

exists. If the limit does exist, its value is the sum of the Fourier series. ●

The remarkable insight to be gleaned from the localization principle is this: Although the Fourier coefficients of f depend upon the values of f throughout the interval $[-\pi, \pi]$, the convergence of the Fourier series of f at a point x depends only on the behavior of f in a neighborhood $N(x; 2\delta)$ of x. If, for a given x and a given $\delta > 0$, we are able to compute the limit of the integral appearing in Riemann's localization principle, then we can find

$$\lim_{k \to \infty} S_k(x) = \frac{a_0}{2} + \sum_{j=1}^{\infty} [a_j \cos jx + b_j \sin jx]$$

and thus sum the Fourier series of f at x. We will exploit this fact to establish sufficient conditions for the pointwise convergence of a Fourier series. The next step toward that goal is the following theorem.

THEOREM 14.3.6 Jordan Suppose that g is of bounded variation on some interval $[0, \delta]$. Then

$$\lim_{k \to \infty} \frac{2}{\pi} \int_{0}^{\delta} g(t) \frac{\sin kt}{t} \, dt = g(0^+).$$

Proof. We assume without loss of generality that g is monotone increasing on $[0, \delta]$ and we make a number of essential but straightforward observations before launching into the details of the proof. First, we can redefine g to have value $g(0^+)$ at 0 without violating the monotone character of g and without affecting the value of the integral.

Next, we observe that the function $h(t) = (\sin kt)/t$ is continuous on all of $\mathbb{R}$ and that, by Example 26 in Section 13.5, for any positive r,

$$\lim_{k \to \infty} \int_{0}^{r} h(t) \, dt = \lim_{k \to \infty} \int_{0}^{r} \frac{\sin kt}{t} \, dt = \lim_{k \to \infty} \int_{0}^{kr} \frac{\sin u}{u} \, du$$

$$= \int_{0}^{\infty} \frac{\sin u}{u} \, du = \frac{\pi}{2}. \quad \textbf{(14.5)}$$

Consequently, we can choose a positive constant K such that, for any interval $[a, b] \subset [0, \infty)$,

$$\left| \int_a^b \frac{\sin kt}{t} \, dt \right| \le K.$$

Now fix any $\epsilon > 0$. Choose a positive $\delta_1 < \delta$ such that, for all t in $[0, \delta_1]$, $0 \le g(t) - g(0^+) < \epsilon / K$. Write

$$\int_0^\delta g(t)h(t)\, dt = \int_0^{\delta_1} [g(t) - g(0^+)]h(t)\, dt + g(0^+) \int_0^{\delta_1} h(t)\, dt + \int_{\delta_1}^{\delta} g(t)h(t)\, dt. \tag{14.6}$$

We treat each of these integrals in turn. First, let

$$H(x) = \int_0^x h(t)\, dt = \int_0^x \frac{\sin kt}{t}\, dt.$$

Then H is continuous on $[0, \delta_1]$ and $H'(x) = (\sin kx)/x$. By Theorem 7.4.2, there exists a c in $[0, \delta_1]$ such that

$$\begin{aligned}
\int_0^{\delta_1} [g(t) - g(0^+)]h(t)\, dt &= \int_0^{\delta_1} [g(t) - g(0^+)]\, dH(t) \\
&= [g(0) - g(0^+)][H(c) - H(0)] \\
&\quad + [g(\delta_1) - g(0^+)][H(\delta_1) - H(c)] \\
&= [g(\delta_1) - g(0^+)] \int_c^{\delta_1} h(t)\, dt \\
&= [g(\delta_1) - g(0^+)] \int_c^{\delta_1} \frac{\sin kt}{t}\, dt.
\end{aligned}$$

Therefore

$$\begin{aligned}
\left| \int_0^{\delta_1} [g(t) - g(0^+)] \frac{\sin kt}{t}\, dt \right| &= [g(\delta_1) - g(0^+)] \left| \int_c^{\delta_1} \frac{\sin kt}{t}\, dt \right| \\
&< \frac{\epsilon}{K} \cdot K = \epsilon.
\end{aligned}$$

For the second integral in (14.6),

$$\lim_{k \to \infty} g(0^+) \int_0^{\delta_1} \frac{\sin kt}{t}\, dt = g(0^+) \frac{\pi}{2}$$

by (14.5). Finally, apply the Riemann–Lebesgue lemma to the third integral in (14.6):

$$\lim_{k \to \infty} \int_{\delta_1}^{\delta} g(t)h(t)\, dt = \lim_{k \to \infty} \int_{\delta_1}^{\delta} \frac{g(t)}{t} \sin kt\, dt = 0.$$

We conclude that $\lim_{k\to\infty} \int_0^\delta g(t)(\sin kt)/t\, dt = g(0^+)\pi/2$ and hence that

$$\lim_{k\to\infty} \frac{2}{\pi} \int_0^\delta g(t) \frac{\sin kt}{t}\, dt = g(0^+).$$

●

In the early years of analysis, it was a common misconception to believe that a discontinuous function could be "rounded off" and replaced in an argument by a continuous approximation. Moreover, the concept of the total variation of a function had not yet been studied; all functions of interest were implicitly assumed to be of bounded variation. Consequently, it was commonplace to believe, at an intuitive level, that only continuous functions of bounded variation need be considered. In light of this blind spot[1] and the following theorem, it is hardly surprising that Fourier would believe that the trigonometric series of every periodic function with period 2π converges to it. As the French mathematician Jordan proved half a century after Fourier's death, this assertion is true for every continuous function of bounded variation.

THEOREM 14.3.7 Jordan (1881) Assume that f is integrable on $[-\pi, \pi]$ and is extended to be periodic with period 2π. Assume also that f is of bounded variation on an interval $[x - 2\delta, x + 2\delta]$. Then the Fourier series of f converges at the point x to $[f(x^-) + f(x^+)]/2$. If f is continuous at x, then the Fourier series of f converges at x to $f(x)$.

Proof. Let $g(t) = [f(x - 2t) + f(x + 2t)]/2$ for t in the interval $[0, \delta]$. Then g is of bounded variation on $[0, \delta]$ and, by Theorem 14.3.6,

$$\lim_{k\to\infty} \frac{2}{\pi} \int_0^\delta g(t) \frac{\sin (2k+1)t}{t}\, dt = g(0^+).$$

By Riemann's localization theorem, it follows that the Fourier series of f converges at x and that the sum of that series is $g(0^+)$. But $g(0^+) = [f(x^-) + f(x^+)]/2$ and the first claim of the theorem is proved. If f is continuous at x, then $f(x^-) = f(x^+)$ and thus the sum of the Fourier series of f is $f(x)$. ●

14.4 The Cesàro Summability of Fourier Series

While even the continuity of a function f, in the absence of additional hypotheses, does not ensure the convergence of its Fourier series, there remains another line of approach. Recall that, although a series may diverge, it is possible that the sequence of arithmetic averages of the partial sums may converge. The series may be Cesàro summable. As we will see below, Leopold Fejér (1880–1959) took this approach in 1904 with noteworthy success.

---

[1] Many modern students of analysis share this same malady.

Again f is an integrable function on $[-\pi, \pi]$, extended to be periodic with period 2π. For any point x and any n in $\mathbb{N}$, we define

$$\sigma_n(x) = \frac{1}{n} \sum_{k=0}^{n-1} S_k(x).$$

To study the convergence of the sequence $\{\sigma_n(x)\}$, we use the integral form of $S_k(x)$ to derive the corresponding integral form of $\sigma_n(x)$.

$$\begin{aligned} \sigma_n(x) &= \frac{1}{n} \sum_{k=0}^{n-1} \frac{2}{\pi} \int_0^{\pi/2} \frac{f(x-2t) + f(x+2t)}{2} \frac{\sin(2k+1)t}{\sin t} \, dt \\ &= \frac{2}{n\pi} \int_0^{\pi/2} \frac{f(x-2t) + f(x+2t)}{2 \sin t} \sum_{k=0}^{n-1} \sin(2k+1)t \, dt. \end{aligned} \tag{14.7}$$

To simplify the integrand, we find a closed form for the sum $\sum_{k=0}^{n-1} \sin(2k+1)t$. To this end, we begin with

$$\sum_{k=0}^{n-1} e^{(2k+1)ti} = \sum_{k=0}^{n-1} \cos(2k+1)t + i \sum_{k=0}^{n-1} \sin(2k+1)t \tag{14.8}$$

and work with the left-hand side. Provided that t is not an integer multiple of π,

$$\begin{aligned} \sum_{k=0}^{n-1} e^{(2k+1)ti} &= e^{it} \sum_{k=0}^{n-1} (e^{2it})^k = e^{it} \frac{1 - e^{2nti}}{1 - e^{2it}} \\ &= e^{int} \left(\frac{e^{-int} - e^{int}}{e^{-it} - e^{it}} \right) \\ &= \frac{(\cos nt + i \sin nt) \sin nt}{\sin t} \\ &= \frac{\cos nt \sin nt}{\sin t} + i \frac{\sin^2 nt}{\sin t}. \end{aligned} \tag{14.9}$$

By equating the imaginary parts in (14.8) and (14.9) we obtain

$$\sum_{k=0}^{n-1} \sin(2k+1)t = \frac{\sin^2 nt}{\sin t}.$$

We substitute this result in (14.7) and obtain the integral form of $\sigma_n(x)$:

$$\sigma_n(x) = \frac{2}{n\pi} \int_0^{\pi/2} \frac{f(x-2t) + f(x+2t)}{2} \left(\frac{\sin nt}{\sin t} \right)^2 dt.$$

DEFINITION 14.4.1 The function $K_n(t)$ defined by

$$K_n(t) = \frac{2}{n\pi} \left(\frac{\sin nt}{\sin t} \right)^2$$

is called *Fejér's kernel.* ●

Our derivation proves the following theorem.

THEOREM 14.4.1 Let f be a function in $R[-\pi, \pi]$, extended to be periodic with period 2π. Let $\sigma_n(x) = (1/n)\sum_{k=0}^{n-1} S_k(x)$. Then we have

$$\sigma_n(x) = \int_0^{\pi/2} \frac{f(x-2t) + f(x+2t)}{2} K_n(t)\, dt. \quad \bullet$$

LEMMA 4 Fejér's kernel $K_n(t)$ has the property that, for every n in $\mathbb{N}$,

$$\int_0^{\pi/2} K_n(t)\, dt = 1.$$

Proof. Let f be the function identically 1 on $\mathbb{R}$. The Fourier series of f reduces to the single term $a_0/2 = 1$. For each $k = 0, 1, 2, \ldots$, the kth partial sum of this Fourier series is also 1 and, hence, $\sigma_n(x) = 1$ for every n. More, for this choice of f, $[f(x-2t) + f(x+2t)]/2 = 1$ for every x. Consequently, $\sigma_n(x) = \int_0^{\pi/2} K_n(t)\, dt = 1$. $\bullet$

THEOREM 14.4.2 **Fejér 1904** Let f be in $R[-\pi, \pi]$, extended to be periodic with period 2π and let x be any point such that $\lim_{t\to 0^+}[f(x-t) + f(x+t)]/2$ exists.

i) The Fourier series of f, evaluated at x, is $(C, 1)$ summable to this limit. That is,

$$\lim_{n\to\infty} \sigma_n(x) = \lim_{t\to 0^+} \frac{f(x-t) + f(x+t)}{2}$$

whenever the latter limit exists.

ii) If f is continuous at x, then $\lim_{n\to\infty} \sigma_n(x) = f(x)$.

iii) If f is continuous on an interval $[a, b]$, then $\lim_{n\to\infty} \sigma_n = f$ [uniformly] on any interval $[c, d]$ contained in (a, b).

Proof. Let $g(t) = [f(x-2t) + f(x+2t)]/2$ as before and let $S(x) = \lim_{t\to 0^+}[f(x-t) + f(x+t)]/2$ at any x where this limit exists. For each such x, define $h_x(t) = g(t) - S(x)$. Holding x fixed, notice that $\lim_{t\to 0^+} h_x(t) = 0$. Given $\epsilon > 0$, choose $\delta > 0$ such that, whenever $0 < t < \delta$, it follows that $|h_x(t)| < \epsilon$. Without loss of generality, $\delta < \pi/2$ as well. Notice that the function h_x, and hence δ, implicitly depends on x. For part (i) we hold x fixed. Using the preceding lemma, we write

$$\begin{aligned}
\sigma_n(x) - S(x) &= \int_0^{\pi/2} g(t)K_n(t)\, dt - S(x) \\
&= \int_0^{\pi/2} g(t)K_n(t)\, dt - \int_0^{\pi/2} S(x)K_n(t)\, dt \\
&= \int_0^{\pi/2} [g(t) - S(x)]K_n(t)\, dt = \int_0^{\pi/2} h_x(t)K_n(t)\, dt.
\end{aligned}$$

To obtain an estimate for $|\sigma_n(x) - S(x)|$ we write

$$\sigma_n(x) - S(x) = \left[\int_0^{\delta} + \int_{\delta}^{\pi/2}\right] h_x(t) K_n(t)\, dt$$

and, as usual, we treat each of these last two integrals in turn. For the first integral, observe that

$$\left|\int_0^{\delta} h_x(t) K_n(t)\, dt\right| \leq \int_0^{\delta} |h_x(t)| K_n(t)\, dt < \epsilon \int_0^{\delta} K_n(t)\, dt$$
$$\leq \epsilon \int_0^{\pi/2} K_n(t)\, dt = \epsilon.$$

For the second integral, note first that, for t in $[\delta, \pi/2]$, $K_n(t) \leq 2/[n\pi \sin^2 \delta]$. Also let $M = \int_{\delta}^{\pi/2} |h_x(t)|\, dt$, a constant that depends on the fixed choice of x. Then

$$\left|\int_{\delta}^{\pi/2} h_x(t) K_n(t)\, dt\right| \leq \frac{2}{n\pi \sin^2 \delta} \int_{\delta}^{\pi/2} |h_x(t)|\, dt$$
$$\leq \frac{2}{n\pi \sin^2 \delta} \int_{0}^{\pi/2} |h_x(t)|\, dt$$
$$= \frac{2M}{[n\pi \sin^2 \delta]}.$$

Now our way is clear. Choose n_0 such that, for $n \geq n_0$, $0 < 2M/[n\pi \sin^2 \delta] < \epsilon$. We combine our results; for $n \geq n_0$, we have

$$|\sigma_n(x) - S(x)| \leq \left|\int_0^{\delta} h_x(t) K_n(t)\, dt\right| + \left|\int_{\delta}^{\pi/2} h_x(t) K_n(t)\, dt\right| < 2\epsilon.$$

Therefore, $\lim_{n\to\infty} \sigma_n(x) = S(x)$ whenever $S(x)$ exists. This proves (i) of Fejér's theorem. (Notice that M and δ, hence n_0, depend on x.)

Part (ii) follows immediately, since, if f is continuous at x, then $S(x) = [f(x^-) + f(x^+)]/2 = f(x)$.

For (iii), we suppose that f is continuous on $[a, b]$. It follows first that $S(x) = f(x)$ for all x in $[a, b]$. Also, f is uniformly continuous there. Given $\epsilon > 0$, first choose a $\delta_1 > 0$ such that, whenever y and z are in $[a, b]$ and $|y - z| < \delta_1$, then $|f(y) - f(z)| < \epsilon$. Next, given an interval $[c, d]$ contained in (a, b), choose a positive δ that is less than $\min\{\delta_1/2, (c-a)/2, (b-d)/2, \pi/2\}$. Notice that δ is independent of x in $[c, d]$. For this δ and for any x in $[c, d]$, write as in our proof of part (i),

$$|\sigma_n(x) - f(x)| = |\sigma_n(x) - S(x)| \leq \left[\int_0^{\delta} + \int_{\delta}^{\pi/2}\right] |h_x(t)| K_n(t)\, dt.$$

For the first integral, note also that, by our choice of δ, the two points $x \pm 2t$ are in $[a, b] \cap N(x; \delta_1)$ for any x in $[c, d]$ and all t in $[0, \delta]$. Thus, for all t in $[0, \delta]$,

$$|h_x(t)| = \left|\frac{f(x-2t) + f(x+2t)}{2} - f(x)\right|$$

$$\leq \frac{|f(x-2t) - f(x)| + |f(x+2t) - f(x)|}{2} < \epsilon,$$

for all x in $[c, d]$ simultaneously. Consequently,

$$\int_0^{\delta} |h_x(t)| K_n(t)\, dt < \epsilon \int_0^{\pi/2} K_n(t)\, dt = \epsilon,$$

for all x in $[c, d]$ simultaneously.

For the second integral, note that, since f is bounded on $[a, b]$, we have $|h_x(t)| \leq 2||f||_\infty$ for all x and all t in $[0, \pi/2]$. It follows that $\int_0^{\pi/2} |h_x(t)|\, dt \leq \pi ||f||_\infty$. Also, for t in $[\delta, \pi/2]$, we again have $0 < K_n(t) \leq 2/(n\pi \sin^2 \delta)$. We deduce, as in the proof of (i), that

$$\int_{\delta}^{\pi/2} |h_x(t)| K_n(t)\, dt \leq \frac{2||f||_\infty}{n \sin^2 \delta} = \frac{M_1}{n}.$$

Observe that the upper bound M_1/n is independent of x. Now choose n_0, also independent of x, such that, for $n \geq n_0$, $2||f||_\infty/(n \sin^2 \delta) < \epsilon$. Then, for all $n \geq n_0$, we have

$$\int_{\delta}^{\pi/2} |h_x(t)| K_n(t)\, dt \leq \frac{2||f||_\infty}{n \sin^2 \delta} < \epsilon$$

for all x in $[c, d]$ simultaneously. Combining our results proves that, for $n \geq n_0$, $|\sigma_n(x) - f(x)| < 2\epsilon$ for all x in $[c, d]$ simultaneously. That is, $\lim_{n\to\infty} \sigma_n = f$ [uniformly] on $[c, d]$. This proves part (iii) of Fejér's theorem. ●

COROLLARY 14.4.3 If f is continuous on $[-\pi, \pi]$ and is periodic with period 2π, then the sequence $\{\sigma_n\}$ converges in the mean to f.

Proof. Because f is assumed to be continuous on $[-\pi, \pi]$ and periodic with period 2π, f is continuous—in fact, uniformly continuous—on all of $\mathbb{R}$. By Fejér's theorem, the Fourier series of f is uniformly $(C, 1)$ summable to f. That is, $\{\sigma_n\}$ converges uniformly to f on $[-\pi, \pi]$. By Theorem 14.1.4, $\{\sigma_n\}$ converges in the mean to f also. ●

We have followed historically established convention and have used as an orthogonal collection on $[-\pi, \pi]$ the functions $\phi_j(x) = (1/\pi) \cos jx$ and $\psi_j(x) = (1/\pi) \sin jx$. These functions are not actually normalized in the sense of Definition 14.1.7. For our functions, $||\phi_0||_2^2 = 2/\pi$, and, for $j = 1, 2, \ldots ||\phi_j||_2^2 = ||\psi_j||_2^2 = 1/\pi$. Thus, when we compute $||f - S_k(f)||_2^2$, as in the discussion leading to Theorem 14.1.9, we must modify our computations appropriately by including a factor of π. [It will be instructive for you to work through those calculations with the functions $\{\phi_0, \phi_j, \psi_j : j \text{ in } \mathbb{N}\}$.] Thus, for any f in $R[a, b]$,

$$\begin{aligned} ||f - S_k(f)||_2^2 &= \langle f - S_k(f), f - S_k(f) \rangle \\ &= ||f||_2^2 - \pi \left\{ \frac{a_0^2}{2} + \sum_{j=1}^{k} [a_j^2 + b_j^2] \right\}. \end{aligned}$$

It remains true, as in Theorem 14.1.9, that, of all the kth mean-square approximations using these traditional trigonometric functions, $S_k(f)$ is the unique one giving the least mean-square error. That is, if

$$g_k(x) = \frac{c_0}{2} + \sum_{j=1}^{k} [c_j \cos jx + d_j \sin jx],$$

for arbitrary choices of the c_j and the d_j, then

$$||f - S_k(f)||_2 \leq ||f - g_k||_2.$$

Notice, in particular, that the $(k+1)$st arithmetic average, $\sigma_{k+1}(x)$, upon being expanded, is

$$\begin{aligned}\sigma_{k+1}(x) &= \frac{S_0(x) + S_1(x) + \cdots + S_k(x)}{k+1} \\ &= \frac{a_0}{2} + \sum_{j=1}^{k} \frac{k+1-j}{k+1}[a_j \cos jx + b_j \sin jx]\end{aligned}$$

and thus is a linear combination of the form $g_k(x)$ above. We deduce that $||f - S_k(f)||_2 \leq ||f - \sigma_{k+1}||_2$. These observations lead to the following corollary.

COROLLARY 14.4.4 If f is continuous on $[-\pi, \pi]$ and is periodic with period 2π, then $\{S_k(f)\}$ converges in the mean to f.

Proof. By Corollary 14.4.3, the hypotheses imply that $\{\sigma_k\}$ converges in the mean to f. Since

$$||f - S_k(f)||_2 \leq ||f - \sigma_{k+1}||_2,$$

we deduce immediately that $\{S_k(f)\}$ converges in the mean to f also. ●

COROLLARY 14.4.5 Parseval's Identity If f is continuous on $[-\pi, \pi]$ and is periodic with period 2π, then

$$\frac{1}{\pi}||f||_2^2 = \frac{1}{\pi}\int_{-\pi}^{\pi} f^2(x)\,dx = \frac{a_0^2}{2} + \sum_{j=1}^{\infty} [a_j^2 + b_j^2].$$

Proof. Above we computed

$$||f - S_k(f)||_2^2 = ||f||_2^2 - \pi\left\{\frac{a_0^2}{2} + \sum_{j=1}^{k}[a_j^2 + b_j^2]\right\}.$$

Equivalently,

$$\frac{1}{\pi}||f||_2^2 = \frac{a_0^2}{2} + \sum_{j=1}^{k} [a_j^2 + b_j^2] + \frac{1}{\pi}||f - S_k(f)||_2^2.$$

Since $\lim_{k\to\infty} ||f - S_k(f)||_2 = 0$ by Corollary 14.4.4, Parseval's identity follows immediately. ●

COROLLARY 14.4.6 The collection of functions $\phi_0(x) = 1/\pi$, and $\phi_j(x) = (1/\pi)\cos jx$, $\psi_j(x) = (1/\pi)\sin jx$ for j in $\mathbb{N}$, is complete for the collection of all continuous functions with period 2π.

Proof. Parseval's identity holds for every such function. ●

14.4.1 Integration of Fourier Series

A direct application of Theorem 12.2.6 provides us with the following theorem. There is nothing new here; we simply translate the general theorem to the present context.

THEOREM 14.4.7 If the Fourier series

$$\frac{a_0}{2} + \sum_{j=1}^{\infty} [a_j \cos jx + b_j \sin jx]$$

converges uniformly to f on an interval $[a, b]$, then the integrated series

$$\frac{a_0}{2}\int_a^x dt + \sum_{j=1}^{\infty}\left[a_j \int_a^x \cos jt\, dt + b_j \int_a^x \sin jt\, dt\right]$$

converges uniformly to $\int_a^x f(t)\, dt$ for x in $[a, b]$. ●

The remarkable fact about Fourier series is that the hypothesis of Theorem 14.4.7 can be dramatically weakened. Fejér's theorem [14.4.2] provides the key.

THEOREM 14.4.8 If f is continuous on $[-\pi, \pi]$ and is periodic with period 2π, then, for any x in $[-\pi, \pi]$,

$$\int_0^x f(t)\, dt = \frac{a_0 x}{2} + \sum_{j=1}^{\infty}\left[a_j \int_0^x \cos jt\, dt + b_j \int_0^x \sin jt\, dt\right].$$

Furthermore, the series converges uniformly on $[-\pi, \pi]$.

Remark. Notice that the continuity of f, in the absence of any additional conditions, does not imply that the Fourier series of f even converges, let alone converges to f. Nonetheless, the integrated series does converge to the integral of f and that convergence is uniform. We can bypass Theorem 12.2.6 by using the mean convergence of $\{S_k(f)\}$ to f and by invoking Theorem 14.1.5.

Proof. By Corollary 14.4.4, the sequence $\{S_k\}$ converges in the mean to f. Set $g = 1$ in Theorem 14.1.5 to deduce that the sequence with kth term $F_k(x) = \int_0^x S_k(t)\, dt$ converges uniformly to $F(x) = \int_0^x f(t)\, dt$. Now,

$$\int_0^x S_k(t)\, dt = \int_0^x \frac{a_0}{2}\, dt + \int_0^x \sum_{j=1}^{k} [a_j \cos jt + b_j \sin jt]\, dt$$

$$= \frac{a_0 x}{2} + \sum_{j=1}^{k}\left[a_j \int_0^x \cos jt\, dt + b_j \int_0^x \sin jt\, dt\right].$$

Thus the series

$$\frac{a_0 x}{2} + \sum_{j=1}^{\infty} \left[a_j \int_0^x \cos jt\, dt + b_j \int_0^x \sin jt\, dt \right]$$

converges uniformly to $F(x) = \int_0^x f(t)\, dt$. This proves the theorem. ●

We take a moment to examine the function $F(x) = \int_0^x f(t)\, dt$ and the series of integrals which converges to it. Performing the integrations for x in $[-\pi, \pi]$ yields

$$\begin{aligned} F(x) &= \int_0^x f(t)\, dt \\ &= \frac{a_0 x}{2} + \sum_{j=1}^{\infty} \left[\frac{a_j}{j} \sin jx + \frac{b_j}{j} - \frac{b_j}{j} \cos jx \right] \\ &= \frac{a_0 x}{2} + \sum_{j=1}^{\infty} \frac{b_j}{j} - \sum_{j=1}^{\infty} \left[\frac{b_j}{j} \cos jx - \frac{a_j}{j} \sin jx \right]. \end{aligned} \qquad \textbf{(14.10)}$$

Since F is the uniform limit of a series of continuous function, F is continuous on $[-\pi, \pi]$. However, as you can compute,

$$F(-\pi) = \int_0^{-\pi} f(t)\, dt = -\frac{a_0 \pi}{2} + 2 \sum_{j=1}^{\infty} \frac{b_{2j-1}}{2j-1}$$

and

$$F(\pi) = \int_0^{\pi} f(t)\, dt = \frac{a_0 \pi}{2} + 2 \sum_{j=1}^{\infty} \frac{b_{2j-1}}{2j-1}.$$

Thus if $a_0 \neq 0$, the values of F at the endpoints $\pm\pi$ differ. Redefine F at $\pm\pi$ to have value

$$\begin{aligned} \frac{F(-\pi) + F(\pi)}{2} &= \frac{1}{2} \left[\int_0^{-\pi} f(t)\, dt + \int_0^{\pi} f(t)\, dt \right] \\ &= \int_0^{\pi} \frac{f(t) - f(-t)}{2}\, dt = 2 \sum_{j=1}^{\infty} \frac{b_{2j-1}}{2j-1}. \end{aligned}$$

(By redefining F we may have introduced discontinuities at the endpoints.) Now extend F to all of $\mathbb{R}$ to be periodic with period 2π. To find the Fourier series of F, find the Fourier series of the function $a_0 x/2$, substitute in (14.10), and rearrange. As you can easily compute, the Fourier series of

$$g(x) = \begin{cases} a_0 x/2, & x \text{ in } (-\pi, \pi), \\ 0, & x = \pm\pi, \end{cases}$$

is $a_0 \sum_{j=1}^{\infty} [(-1)^{j+1}/j] \sin jx$. Substitute this series into (14.10) and rearrange to obtain

$$\sum_{j=1}^{\infty} \frac{b_j}{j} - \sum_{j=1}^{\infty} \left[\frac{b_j}{j} \cos jx - \frac{a_j - (-1)^j a_0}{j} \sin jx \right]. \qquad \textbf{(14.11)}$$

The series $\sum_{j=1}^{\infty} b_j/j$ converges by Dirichlet's test: $\{b_j\}$ converges to 0, hence is bounded, and $\{1/j\}$ converges monotonically to 0. The series $\sum_{j=1}^{\infty}(b_j/j)\cos jx$ and $\sum_{j=1}^{\infty}(1/j)[a_j - (-1)^j a_0]\cos jx$ converge also. Thus the Fourier series (14.11) converges to F and is the Fourier series of F. We know, by the Fundamental Theorem of Calculus, that F is of bounded variation on $[-\pi, \pi]$ and is continuous on $(-\pi, \pi)$. (Recall that F may be discontinuous at the endpoints since we have redefined F there.) Thus the series converges pointwise to the redefined F; the convergence is uniform on any compact set contained in $(-\pi, \pi)$.

EXAMPLE 8 The function $f(x) = x^2$ on $[-\pi, \pi]$, extended to be periodic with period 2π, meets the hypotheses of Fejér's theorem and its corollaries. We found in Example 7 that its Fourier series is

$$x^2 = \frac{\pi^2}{3} + 4\sum_{j=1}^{\infty} \frac{(-1)^j \cos jx}{j^2}. \tag{14.12}$$

The series converges pointwise to the periodic extension of x^2 by Theorem 14.3.7; we can legitimately use the equality sign rather than $\sim$. Fejér's theorem (Theorem 14.4.2) ensures that the series is uniformly $(C, 1)$ summable to the periodic extension of x^2 on any interval $[a, b]$. Hence, by Corollary 14.4.3, the sequence $\{\sigma_n\}$ of arithmetic averages of the partial sums of the series (14.12) converges in the mean to the periodic extension of x^2. Consequently, by Corollary 14.4.4, the sequence $\{S_k\}$ of partial sums of the series (14.12) converges in the mean to the periodic extension of x^2 also. Moreover, Parseval's identity holds for this series:

$$\begin{aligned}\frac{1}{\pi}\int_{-\pi}^{\pi} x^4\,dx = \frac{2\pi^4}{5} &= \frac{a_0^2}{2} + \sum_{j=1}^{\infty}[a_j^2 + b_j^2] = \frac{[2\pi^2/3]^2}{2} + \sum_{j=1}^{\infty}\frac{16}{j^4} \\ &= \frac{2\pi^4}{9} + 16\sum_{j=1}^{\infty}\frac{1}{j^4}.\end{aligned}$$

Upon simplifying, we deduce that $\sum_{j=1}^{\infty} 1/j^4 = \pi^4/90$.

Finally, we integrate as in Theorem 14.4.8 to obtain $F(x) = x^3/3$ for x in $[-\pi, \pi]$. Note that $F(-\pi) = -F(\pi) = -\pi^3/3$. Therefore, we redefine F to have value 0 at the endpoints. Integrating the Fourier series of $f(x) = x^2$ yields

$$\begin{aligned}\frac{\pi^2 x}{3} + 4\sum_{j=1}^{\infty}\frac{(-1)^j}{j^2}\int_0^x \cos jt\,dt &= \frac{2\pi^2}{3}\sum_{j=1}^{\infty}\frac{(-1)^{j+1}}{j}\sin jx + 4\sum_{j=1}^{\infty}\frac{(-1)^j}{j^3}\sin jx \\ &= \frac{2}{3}\sum_{j=1}^{\infty}\frac{(-1)^j(6 - \pi^2 j^2)}{j^3}\sin jx.\end{aligned}$$

Therefore, the Fourier series of the periodic extension of x^3 is

$$x^3 = 2\sum_{j=1}^{\infty} \frac{(-1)^j(6 - \pi^2 j^2)}{j^3} \sin jx.$$

If we compute the Fourier series of x^3 directly, we note first that, since x^3 is an odd function, its Fourier series is a sine series; only $b_j = (2/\pi)\int_0^\pi x^3 \sin jx\, dx$ fails to vanish. Further, integration by parts yields the function

$$\left(\frac{-x^3}{j} + \frac{6x}{j^3}\right)\cos jx + \left(\frac{3x^2}{j^2} - \frac{6}{j^4}\right)\sin jx$$

as an antiderivative of $x^3 \sin jx$. Therefore, direct computation gives

$$b_j = \frac{2}{\pi}\left(-\frac{x^3}{j} + \frac{6x}{j^3}\right)\cos jx\,\Bigg|_0^\pi = \frac{2(-1)^j(6 - \pi^2 j^2)}{j^3}.$$

This confirms directly that the Fourier series of x^3 is

$$x^3 = 2\sum_{j=1}^{\infty} \frac{(-1)^j(6 - \pi^2 j^2)}{j^3} \sin jx,$$

in agreement with the series we found above. This series converges uniformly to x^3 on any compact interval $[a, b]$ in $(-\pi, \pi)$. Suppose, for example, that we set $x = \pi/2$. Then we obtain

$$\frac{\pi^3}{8} = 2\sum_{j=1}^{\infty} \frac{(-1)^j(6 - \pi^2 j^2)}{j^3} \sin(j\pi/2).$$

Since for i in $\mathbb{N}$,

$$\sin\left(\frac{j\pi}{2}\right) = \begin{cases} 1, & \text{if } j = 4i - 3, \\ 0, & \text{if } j = 4i - 2, \\ -1, & \text{if } j = 4i - 1, \\ 0, & \text{if } j = 4i, \end{cases}$$

we have

$$\begin{aligned} \frac{\pi^3}{16} &= -\frac{(6 - \pi^2 1^2)}{1^3} + \frac{(6 - \pi^2 3^2)}{3^3} - \frac{(6 - \pi^2 5^2)}{5^3} + \cdots \\ &= -6\sum_{j=1}^{\infty} \frac{(-1)^{j+1}}{(2j-1)^3} + \pi^2 \sum_{j=1}^{\infty} \frac{(-1)^{j+1}}{(2j-1)}. \end{aligned}$$

Since $\sum_{j=1}^{\infty}(-1)^{j+1}/(2j-1) = \tan^{-1} 1 = \pi/4$, we obtain upon rearranging

$$\sum_{j=1}^{\infty} \frac{(-1)^{j+1}}{(2j-1)^3} = \frac{\pi^3}{32}.$$

This result demonstrates again how manipulation of Fourier series can result in the identification of the sum of certain series. Additional examples appear in the exercises. ●

14.4.2 Differentiation of Fourier Series

In Chapter 12 we obtained conditions that permit the term-by-term differentiation of a series. In the present context we have the following version.

THEOREM 14.4.9 Let $[a, b]$ be a subset of $[-\pi, \pi]$. Suppose that the Fourier series $a_0/2 + \sum_{j=1}^{\infty}[a_j \cos jx + b_j \sin jx]$ of f converges at a point c_0 in (a, b). Suppose also that each of the kth partial sums of the derived series is bounded on (a, b) and that the derived series $\sum_{j=1}^{\infty}[jb_j \cos jx - ja_j \sin jx]$ converges uniformly to a function g on (a, b). Then f is differentiable on (a, b) and

$$f'(x) = g(x) = \sum_{j=1}^{\infty}[jb_j \cos jx - ja_j \sin jx] \qquad \text{[uniformly]}. \quad \bullet$$

EXAMPLE 9 Straightforward computations, which you can complete as an exercise, show that the Fourier series of $f(x) = |x|$, for x in $[-\pi, \pi]$, is

$$\frac{\pi}{2} - \frac{4}{\pi}\sum_{j=1}^{\infty}\frac{\cos(2j-1)x}{(2j-1)^2}.$$

The periodic extension of f is continuous on $\mathbb{R}$. Since f is of bounded variation on any compact interval, the Fourier series of f converges to f at every point. The derivative f' exists at all points that are not integer multiples of π and is, of course, the function

$$f'(x) = \begin{cases} -1, & -\pi < x < 0, \\ 1, & 0 < x < \pi. \end{cases}$$

The derived series is $(4/\pi)\sum_{j=1}^{\infty}[\sin(2j-1)x]/(2j-1)$. As we saw in Example 5, this is the Fourier series of the function f', redefined at $-\pi$, 0, and π to have value 0. On any interval $[a, b]$ contained in $(-\pi, \pi)$ that does not contain 0, Theorem 12.3.3 ensures the uniform convergence of the derived series. It converges to f' on (a, b). $\bullet$

14.5 Additional Topics

Finally, we discuss a variety of interesting, though somewhat specialized, results involving Fourier series. We also take up the question of other widely used notations that you may encounter in your future studies in this area.

EXAMPLE 10 If f is a polynomial of degree k, then we can factor f as $f(x) = a\prod_{j=1}^{k}(x - r_j)$, where $r_1, r_2, \ldots, r_k$ are the (possibly complex and possibly repeated) roots of f. Now, the function $\sin \pi t$, while not a polynomial, has a zero at every j in $\mathbb{Z}$. Our goal in this example is to obtain the corresponding factorization for the sine function, that is, to show that, for any t in an interval $[a, b] \subset [0, 1)$,

$$\sin \pi t = \pi t \prod_{j=1}^{\infty}\left[1 - \left(\frac{t}{j}\right)^2\right].$$

This form is called the *infinite product for the sine function* and can be derived using our results concerning Fourier series. To obtain this product, suppose that α is any real number that is not an integer. For x in $[-\pi, \pi]$, let $f(x) = \cos \alpha x$. On all of $\mathbb{R}$, f is periodic with period $2\pi/\alpha$ but we are selecting only the values on the interval $[-\pi, \pi]$. Notice that $\cos \alpha(-\pi) = \cos \alpha\pi$, and thus, upon extending f from $[-\pi, \pi]$ to $\mathbb{R}$ to be periodic with period 2π, we obtain an extended function that is continuous on all of $\mathbb{R}$ and is of bounded variation on any compact interval. See Fig. 14.6.

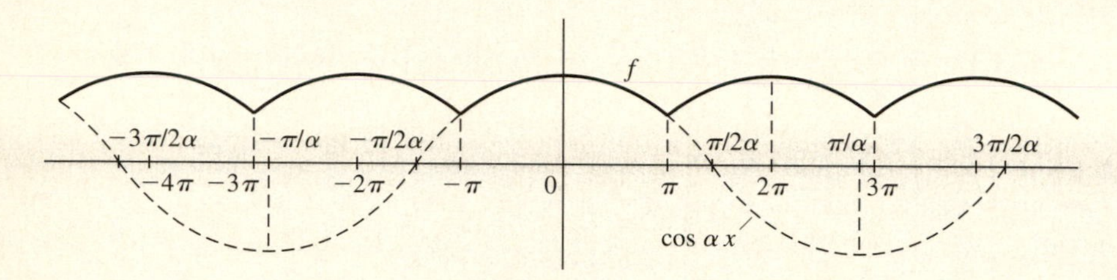

Figure 14.6

Notice also that f is an even function and therefore that the Fourier series of f is a cosine series with

$$\begin{aligned} a_j &= \frac{2}{\pi}\int_0^\pi \cos\alpha x \cos jx\, dx \\ &= \frac{1}{\pi}\int_0^\pi [\cos(\alpha + j)x + \cos(\alpha - j)x]\, dx \\ &= \frac{1}{\pi}\left(\frac{\sin(\alpha + j)\pi}{\alpha + j} + \frac{\sin(\alpha - j)\pi}{\alpha - j}\right) \\ &= \frac{2\alpha}{\pi}\frac{(-1)^j \sin(\alpha\pi)}{\alpha^2 - j^2}, \end{aligned}$$

for $j = 0, 1, 2, \ldots$. Therefore we have

$$\cos\alpha x = \frac{\sin\alpha\pi}{\alpha\pi} + \frac{2\alpha\sin\alpha\pi}{\pi}\sum_{j=1}^{\infty}\frac{(-1)^j}{\alpha^2 + j^2}\cos jx$$

for all x in $\mathbb{R}$. Set $x = \pi$ and divide by $\sin(\alpha\pi)$ to obtain

$$\cot\alpha\pi = \frac{1}{\alpha\pi} + \frac{2\alpha}{\pi}\sum_{j=1}^{\infty}\frac{1}{\alpha^2 - j^2}.$$

Now treat α as a variable, say $\alpha = t$, and restrict t to $(0, b]$ where $b < 1$. We have expanded $\cot\pi t$ as the infinite series

$$\cot\pi t = \frac{1}{\pi t} - \frac{1}{\pi}\sum_{j=1}^{\infty}\frac{2t}{j^2 - t^2},$$

a form that can be viewed as a decomposition of $\cot \pi t$ into partial fractions. Equivalently,

$$\pi\left(\cot \pi t - \frac{1}{\pi t}\right) = -\sum_{j=1}^{\infty} \frac{2t}{j^2 - t^2}. \tag{14.13}$$

For t in $(0, b]$ and j in $\mathbb{N}$, $|2t/(j^2 - t^2)| \leq 2/(j^2 - b^2)$. Since the series $\sum_{j=1}^{\infty} 2/(j^2 - b^2)$ converges, Weierstrass's M-test ensures the uniform convergence of the series (14.13). Thus we can integrate the series term-by-term. Integrating the left-hand side of (14.13) from $a > 0$ to t gives

$$\pi \int_a^t \left(\cot \pi u - \frac{1}{\pi u}\right) du = \ln\left(\frac{\sin \pi t}{\pi t}\right) - \ln\left(\frac{\sin \pi a}{\pi a}\right).$$

Because $\lim_{a\to 0^+} \ln[(\sin \pi a)/(\pi a)] = 0$, we deduce that the improper integral $\pi \int_0^t [\cot(\pi u) - 1/(\pi u)]\, du$ converges to $\ln[(\sin \pi t)/(\pi t)]$. Integrating the right-hand side of (14.13) yields

$$\sum_{j=1}^{\infty} \int_0^t -\frac{2u}{j^2 - u^2}\, du = \sum_{j=1}^{\infty} \ln(j^2 - u^2)\Big|_0^t = \sum_{j=1}^{\infty} \ln\left[1 - \left(\frac{t}{j}\right)^2\right].$$

Thus,

$$\ln\left(\frac{\sin \pi t}{\pi t}\right) = \sum_{j=1}^{\infty} \ln\left[1 - \left(\frac{t}{j}\right)^2\right].$$

By exponentiating, we obtain

$$\frac{\sin \pi t}{\pi t} = \exp\left(\sum_{j=1}^{\infty} \ln\left[1 - \left(\frac{t}{j}\right)^2\right]\right).$$

For each n in $\mathbb{N}$, $\exp\left[\sum_{j=1}^{n} \ln(1 - (t/j)^2)\right] = \prod_{j=1}^{n}[1 - (t/j)^2]$. Thus, by the continuity of the exponential function,

$$\frac{\sin \pi t}{\pi t} = \prod_{j=1}^{\infty}\left[1 - \left(\frac{t}{j}\right)^2\right].$$

This is the infinite product factorization of $\sin(\pi t)$.

As a special case, take $t = 1/2$ to obtain on the left-hand side $2/\pi$ and, on the right-hand side,

$$\prod_{j=1}^{\infty}\left(1 - \frac{1}{4j^2}\right) = \prod_{j=1}^{\infty} \frac{4j^2 - 1}{4j^2}.$$

Therefore,

$$\frac{\pi}{2} = \prod_{j=1}^{\infty} \frac{4j^2}{4j^2 - 1} = \prod_{j=1}^{\infty}\left(\frac{2j}{2j-1}\right)\cdot\left(\frac{2j}{2j+1}\right)$$

$$= \frac{2}{1} \cdot \frac{2}{3} \cdot \frac{4}{3} \cdot \frac{4}{5} \cdot \frac{6}{5} \cdot \frac{6}{7} \cdot \frac{8}{7} \cdot \frac{8}{9} \cdots = \left(\frac{2}{1} \cdot \frac{4}{3} \cdot \frac{6}{5} \cdot \frac{8}{7} \cdots \right)^2.$$

Consequently, we obtain Wallis's product:

$$\frac{\pi}{2} = \prod_{j=1}^{\infty} \left(\frac{2j}{2j-1} \right)^2. \quad \bullet$$

Throughout Sections 14.2, 14.3, and 14.4 we chose to base our computations on the interval $[-\pi, \pi]$ but this is not necessary. Any interval of length 2π would serve as well. Some texts use $[0, 2\pi]$. We made our choice because the effect of f being odd or even on the form of its Fourier series is most readily apparent with this choice.

Nor must the length of the base interval be 2π. After all, there are periodic functions having periods other than 2π; our theory ought to be capable of handling these as well. Here we sketch the simple modifications which must be made to treat a function with period $2T > 0$.

Consider the functions $\cos(j\pi x/T)$ and $\sin(j\pi x/T)$ for j in $\mathbb{N}$. They are all periodic with period $2T$. Furthermore, the integrals

$$\int_{-T}^{T} \cos\left(\frac{j\pi x}{T}\right) \cos\left(\frac{k\pi x}{T}\right) dx,$$

$$\int_{-T}^{T} \sin\left(\frac{j\pi x}{T}\right) \sin\left(\frac{k\pi x}{T}\right) dx.$$

and

$$\int_{-T}^{T} \sin\left(\frac{j\pi x}{T}\right) \cos\left(\frac{k\pi x}{T}\right) dx,$$

upon changing variables with $u = \pi x/T$, are transformed into

$$\frac{T}{\pi} \int_{-\pi}^{\pi} \cos ju \cos ku \, du = \begin{cases} T, & \text{if } j = k, \\ 0, & \text{if } j \neq k, \end{cases}$$

$$\frac{T}{\pi} \int_{-\pi}^{\pi} \sin ju \sin ku \, du = \begin{cases} T, & \text{if } j = k, \\ 0, & \text{if } j \neq k, \end{cases}$$

$$\frac{T}{\pi} \int_{-\pi}^{\pi} \sin ju \cos ku \, du = 0.$$

Now let f be a function in $R[-T, T]$, modified at the endpoints to have value $[f(-T) + f(T)]/2$ and at any point x in $(-T, T)$ where f has a jump discontinuity to have value $[f(x^-) + f(x^+)]/2$. Then extend f to all of $\mathbb{R}$ to be periodic with period $2T$. The Fourier series of f is

$$f(x) \sim \frac{a_0}{2} + \sum_{j=1}^{\infty} \left[a_j \cos\left(\frac{j\pi x}{T}\right) + b_j \sin\left(\frac{j\pi x}{T}\right) \right]$$

where

$$a_j = \frac{1}{T}\int_{-T}^{T} f(x)\cos\left(\frac{j\pi x}{T}\right) dx, \qquad j = 0, 1, 2, \ldots$$

and

$$b_j = \frac{1}{T}\int_{-T}^{T} f(x)\sin\left(\frac{j\pi x}{T}\right) dx, \qquad j = 1, 2, 3, \ldots T.$$

With this starting point, all our theory remains valid; only the computations become rather more tedious.

EXAMPLE 11 For x in $(-1, 1)$, let $f(x) = e^{-x}$. Adjust f to have value $(e + e^{-1})/2$ at ± 1. Now extend f to be periodic with period $2T = 2$ on all of $\mathbb{R}$. The graph of f, thus extended, appears in Fig. 14.7. To find the Fourier series of

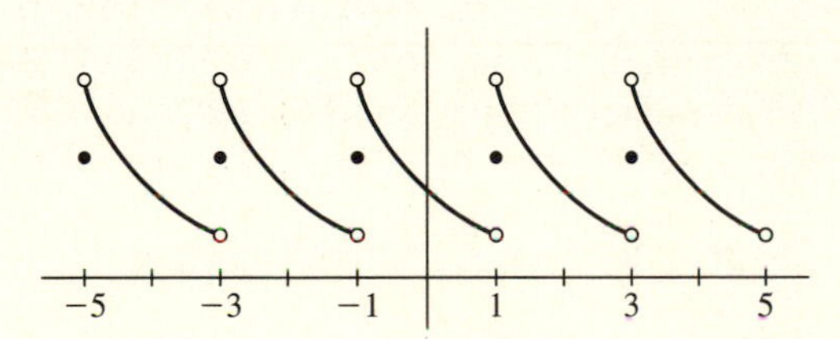

Figure 14.7

f we calculate the integrals indicated above. First,

$$a_0 = \int_{-1}^{1} e^{-x}\, dx = e - e^{-1}.$$

For $j = 1, 2, 3, \ldots$

$$a_j = \int_{-1}^{1} e^{-x}\cos(j\pi x)\, dx = \frac{(-1)^j(e - e^{-1})}{1 + j^2\pi^2}$$

and

$$b_j = \int_{-1}^{1} e^{-x}\sin(j\pi x)\, dx = \frac{(-1)^j(e - e^{-1})j\pi}{1 + j^2\pi^2}.$$

The Fourier series of f is

$$(e - e^{-1})\left\{\frac{1}{2} + \sum_{j=1}^{\infty}\frac{(-1)^j[\cos(j\pi x) + j\pi\sin(j\pi x)]}{1 + j^2\pi^2}\right\}.$$

This series converges at every x that is not an integer to the periodic extension of e^{-x}. At any integer value of x it converges to $(e + e^{-1})/2$.

More generally, given an integrable function f on an interval $[a, b]$, we sketch briefly the adjustments that lead to the Fourier series for f. Let $T = (b - a)/2$ and $\alpha = (a + b)/2$. Let $u = x - \alpha$. As x varies in $[a, b]$, u varies over $[-T, T]$. See

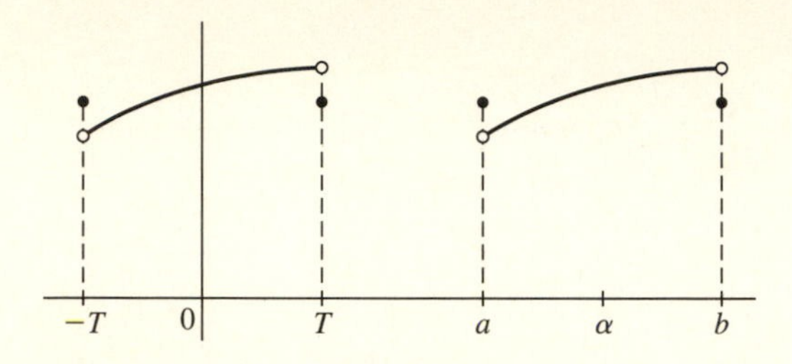

Figure 14.8

Fig. 14.8. Define an integrable function g on $[-T, T]$ by $g(u) = f(u + \alpha) = f(x)$. Redefining g at the endpoints $\pm T$ to have the value $[g(-T) + g(T)]/2$ is equivalent to redefining f at the endpoints a and b to have value $[f(a) + f(b)]/2$. Likewise, if f has a jump discontinuity at x_0, then g has a jump discontinuity at $u_0 = x_0 - \alpha$. Redefining g to have value $[g(u_0^-) + g(u_0^+)]/2$ is equivalent to redefining f at x_0 to have value $[f(x_0^-) + f(x_0^+)]/2$. Having made these adjustments to g, we extend g to $\mathbb{R}$ with period $2T$; also extend f to be periodic with period $2T$ on $\mathbb{R}$. We compute the Fourier series of g:

$$g(u) \sim \frac{a_0}{2} + \sum_{j=1}^{\infty} \left[a_j \cos\left(\frac{j\pi u}{T}\right) + b_j \sin\left(\frac{j\pi u}{T}\right) \right]$$

with

$$a_j = \frac{1}{T} \int_{-T}^{T} g(u) \cos\left(\frac{j\pi u}{T}\right) du$$

and

$$b_j = \frac{1}{T} \int_{-T}^{T} g(u) \sin\left(\frac{j\pi u}{T}\right) du.$$

Equivalently, changing variables by letting $u = x - \alpha$, we find that

$$a_j = \frac{1}{T} \int_{a}^{b} f(x) \cos\left(\frac{j\pi(x - \alpha)}{T}\right) dx$$

and

$$b_j = \frac{1}{T} \int_{a}^{b} f(x) \sin\left(\frac{j\pi(x - \alpha)}{T}\right) dx.$$

The same change of variables yields the Fourier series for f:

$$\frac{a_0}{2} + \sum_{j=1}^{\infty} \left\{ a_j \cos\left(\frac{j\pi(x - \alpha)}{T}\right) + b_j \sin\left(\frac{j\pi(x - \alpha)}{T}\right) \right\}. \tag{14.14}$$

The quantity α/T is called the *phase displacement* or, sometimes, the *lag*. See Fig. 14.9 on page 705. All our results on the convergence of Fourier series translate to apply to the series (14.14). ●

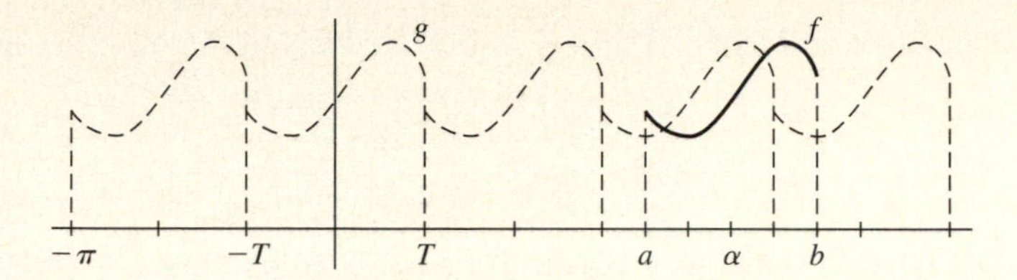

Figure 14.9

14.5.1 Gibb's Phenomenon

We finish this section with a detailed examination of an example of a curious form of behavior exhibited by some Fourier series. Upon superimposing the graph of $S_k(f)$ on that of f, we may find that, in a neighborhood of a point x where f has a jump discontinuity, the graph of $S_k(f)$ overshoots that of f. In some cases, as k tends to ∞, the total variation of $S_k(f)$ in some neighborhood of x strictly exceeds the magnitude $|f(x^+) - f(x^-)|$ of the jump of f at x. This curious behavior is called *Gibb's phenomenon*. It can be characterized mathematically by the existence of a monotone sequence $\{\delta_k\}$ that converges to 0 such that

$$\limsup\,[\sup\{S_k(y) - S_k(z) : \ y, z \text{ in } N'(x; \delta_k)\}] > |f(x^+) - f(x^-)|.$$

EXAMPLE 12 We examine an instance of Gibb's phenomenon by taking up a simple example. We let

$$f(x) = \begin{cases} -1, & -\pi < x < 0, \\ 1, & 0 < x < \pi, \\ 0, & x = -\pi, 0, \pi, \end{cases}$$

and we extend f to be periodic with period 2π. As we know, the Fourier series of f is $(4/\pi)\sum_{j=1}^{\infty}[\sin(2j-1)x]/(2j-1)$. We know that at $x = 0$, the series converges to 0. Furthermore, if $0 < \delta < \pi$, and if x is in $(0, \delta)$, then $\{S_k(x)\}$ converges to $f(x) = 1$; if x is in $(-\delta, 0)$, then $\{S_k(x)\}$ converges to -1. See Fig. 14.10. We are especially interested in a fine analysis of the behavior of the Fourier series of f in a small neighborhood of 0. In all that follows, think of x as positive but small. Both f and its Fourier series are odd functions, so it suffices to analyze the behavior of

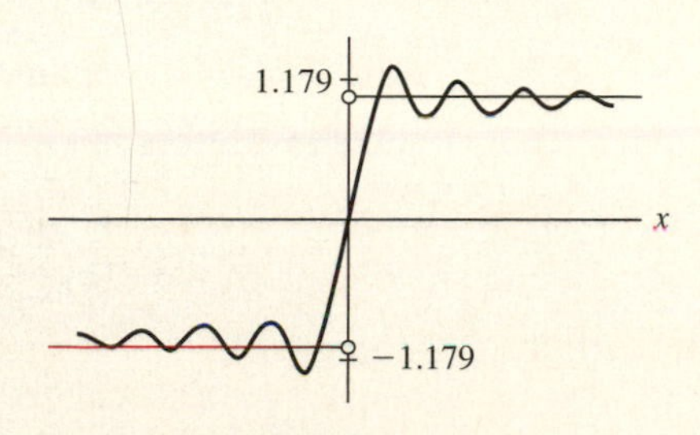

Figure 14.10

$S_k(x)$ for x positive and small. The key to our analysis is the observation that, for this series,

$$S_k(x) = \frac{2}{\pi}\int_0^x \frac{\sin 2kt}{\sin t}\, dt.$$

We leave the derivation of this equation for you as an exercise. It follows, by the Fundamental Theorem of Calculus, that

$$S_k'(x) = \frac{2}{\pi}\frac{\sin 2kx}{\sin x}.$$

Therefore S_k has relative extrema in $(0, \pi)$ at the points where $\sin 2kx = 0$, that is, at $x = j\pi/(2k)$, $j = 1, 2, \ldots, 2k-1$. Furthermore,

$$S_k''(x) = \frac{2}{\pi}\frac{2k\cos 2kx \sin x - \sin 2kx \cos x}{\sin^2 x}.$$

Since $S_k''(x) < 0$ when $x = (2j-1)\pi/(2k)$ and $S_k''(x) > 0$ when $x = 2j\pi/(2k)$, we deduce that S_k has relative maxima M_j at odd multiples $(2j-1)\pi/(2k)$ of $\pi/(2k)$ and has relative minima m_j at even multiples $2j\pi/(2k)$ of $\pi/(2k)$.

We claim that, of all the relative maxima of S_k in $(0, \pi)$, the one that occurs when $x = \pi/(2k)$ is the largest. To prove this, consider any difference

$$\begin{aligned} M_{j+1} - M_j &= S_k\left(\frac{(2j+1)\pi}{2k}\right) - S_k\left(\frac{(2j-1)\pi}{2k}\right) \\ &= \frac{2}{\pi}\int_{(2j-1)\pi/2k}^{(2j+1)\pi/2k} \frac{\sin 2kt}{\sin t}\, dt. \end{aligned}$$

We claim that, for any $j = 1, 2, \ldots, k-1$, this difference is negative. To prove this, we break the integral into two components

$$M_{j+1} - M_j = \frac{2}{\pi}\left[\int_{(2j-1)\pi/2k}^{j\pi/k} + \int_{j\pi/k}^{(2j+1)\pi/2k}\right] \frac{\sin 2kt}{\sin t}\, dt.$$

In order to prove our claim, we need to show that

$$\int_{(2j-1)\pi/2k}^{j\pi/k} \frac{\sin 2kt}{\sin t}\, dt + \int_{j\pi/k}^{(2j+1)\pi/2k} \frac{\sin 2kt}{\sin t}\, dt < 0. \tag{14.15}$$

To this end, change variables in the first integral by letting $u = t + \pi/2k$. That integral becomes

$$\int_{j\pi/k}^{(2j+1)\pi/2k} \frac{\sin 2k(u - \pi/2k)}{\sin (u - \pi/2k)}\, du = -\int_{j\pi/k}^{(2j+1)\pi/2k} \frac{\sin 2ku}{\sin (u - \pi/2k)}\, du.$$

Replacing the dummy variable u by t and substituting in (14.15), we obtain

$$\int_{j\pi/k}^{(2j+1)\pi/2k} \sin 2kt \left[\frac{1}{\sin t} - \frac{1}{\sin (t - \pi/2k)}\right] dt < 0,$$

since, for t in the interval $(j\pi/k, (2j+1)\pi/2k)$,

$$\sin 2kt \left[\frac{1}{\sin t} - \frac{1}{\sin (t - \pi/2k)}\right] < 0.$$

Therefore, $M_{j+1} - M_j < 0$. It follows that

$$M_1 > M_1 + \sum_{j=1}^{p-1}(M_{j+1} - M_j) = M_p,$$

for any $p = 2, 3, \ldots, k$. We deduce that $S_k(x)$ takes its maximum value on $(0, \pi)$ at the point $x = \pi/(2k)$.

Now look more closely at the partial sum S_k evaluated at the point $x = \pi/(2k)$,

$$S_k\left(\frac{\pi}{2k}\right) = \frac{4}{\pi}\sum_{j=1}^{k}\frac{\sin[(2j-1)\pi/(2k)]}{2j-1},$$

and interpret this sum as a Riemann sum approximating the integral of $g(x) = (2/\pi)[\sin t]/t$ over the interval $[0, \pi]$. Use the uniform partition with gauge $\pi/(2k)$ to partition $[0, \pi]$. That is, let $x_j = j\pi/(2k)$, $j = 0, 1, 2, \ldots, 2k$. On a subinterval of the form $[2i\pi/(2k),\ (2i+1)\pi/(2k)]$, choose s_i to be the right endpoint $(2i+1)\pi/(2k)$; on a subinterval of the form $[(2i-1)\pi/(2k),\ 2i\pi/(2k)]$, choose s_i to be the left endpoint $(2i-1)\pi/(2k)$. With these choices for the points s_i, form the Riemann sum

$$\begin{aligned}\sum_{i=1}^{2k} g(s_i)\Delta x_i &= \frac{2}{\pi}\sum_{j=1}^{k}\frac{2\sin[(2j-1)\pi/(2k)]}{(2j-1)\pi/(2k)}\left(\frac{\pi}{2k}\right)\\ &= \frac{4}{\pi}\sum_{j=1}^{k}\frac{\sin[(2j-1)\pi/(2k)]}{2j-1}\\ &= S_k\left(\frac{\pi}{2k}\right).\end{aligned}$$

Since g is continuous on $[0, \pi]$, we have

$$\lim_{k\to\infty}\sum_{i=1}^{2k} g(s_i)\Delta x_i = \int_0^{\pi} g(x)\,dx = \frac{2}{\pi}\int_0^{\pi}\frac{\sin x}{x}\,dx.$$

Thus

$$\lim_{k\to\infty} S_k\left(\frac{\pi}{2k}\right) = \frac{2}{\pi}\int_0^{\pi}\frac{\sin x}{x}\,dx. \qquad \textbf{(14.16)}$$

The integral in (14.16) is nonelementary but can be approximated using numerical methods. Its value is approximately 1.179, accurate to three decimal places.

Finally, for each k in $\mathbb{N}$, let $\delta_k = \pi/k$. Since, for each k, S_k is an odd function, our work above shows that

$$\sup\{S_k(y) - S_k(z) : y, z \text{ in } N'(0; \delta_k)\} = S_k\left(\frac{\pi}{2k}\right) - S_k\left(-\frac{\pi}{2k}\right) = 2S_k\left(\frac{\pi}{2k}\right).$$

Therefore, $\lim_{k\to\infty}[\sup\{S_k(y) - S_k(z) : y, z \text{ in } N'(0; \delta_k)\}]$ exists and equals

$$\lim_{k\to\infty} 2S_k\left(\frac{\pi}{2k}\right) = \frac{4}{\pi}\int_0^{\pi}\frac{\sin x}{x}\,dx \approx 2.358.$$

Since the magnitude of the jump of f at 0 is 2, we have explicitly exhibited an instance of Gibb's phenomenon. ●

EXERCISES

14.1. **a)** Show that, for any x which is not an odd integer multiple of π,

$$A_k(x) = \sum_{j=1}^{k}(-1)^{j+1}\cos jx = \frac{\cos[(k+1)x/2]\cos(kx/2)}{\cos(x/2)}$$

and

$$B_k(x) = \sum_{j=1}^{k}(-1)^{j+1}\sin jx = \frac{\sin[(k+1)x/2]\cos(kx/2)}{\cos(x/2)}.$$

b) Let $[a, b]$ be any interval that does not contain an odd integer multiple of π. Show that there is a constant $M > 0$ such that $|A_k(x)| \le M$ and $|B_k(x)| \le M$ for all x in $[a, b]$ and all k in $\mathbb{N}$. (These and similar uniform bounds are useful when applying Dirichlet's test for uniform convergence below.)

14.2. Let $f(x) = x$ on $(-\pi, \pi)$. Define f appropriately at $\pm\pi$ and extend f to be periodic with period 2π.

a) Sketch the graph of f.

b) Find the Fourier series of f.

c) Discuss the pointwise convergence of the Fourier series of f. Where is the convergence uniform? (See Exercise 14.1.)

14.3. For x in $[-\pi, \pi]$, define

$$f(x) = \begin{cases} 0, & -\pi < x < 0, \\ x, & 0 \le x < \pi, \\ \pi/2, & x = \pm\pi. \end{cases}$$

a) Extend f to be periodic with period 2π. Sketch the graph of f.

b) Find the Fourier series of f.

c) For each x in $\mathbb{R}$, identify the pointwise limit of the Fourier series of f. Where does the series converge uniformly?

14.4. Let $f(x) = |x|$ on $[-\pi, \pi]$; extend f to be periodic with period 2π.

a) Compute the Fourier coefficients of f.

b) Prove that the Fourier series of f converges uniformly to f on $\mathbb{R}$.

c) Integrate the Fourier series of f in order to find the Fourier series of $F(x) = \int_0^x |t|\,dt$.

14.5. For x in $[-\pi, \pi]$, define

$$f(x) = \begin{cases} -x^2/2, & -\pi < x < 0, \\ x^2/2 & 0 \le x < \pi, \\ 0, & x = \pm\pi. \end{cases}$$

Extend f to be periodic with period 2π.

a) Sketch the graph of f.

b) Find the Fourier series of f.

c) Identify the pointwise limit of the Fourier series of f at each point of $\mathbb{R}$. Where is the convergence uniform?

d) Show that the series you obtained in part (b) is the same as that you found in part (c) of Exercise 14.4.

e) Differentiate the series you found in part (b) and show that the resulting series is the Fourier series for $|x|$ that you found in Exercise 14.4.

14.6. For x in $[-\pi, \pi]$, let $f(x) = \cos x/2$. Extend f to be periodic with period 2π.

a) Sketch the graph of f

b) Find the Fourier series of f.

c) Prove that the Fourier series of f converges uniformly to f on $\mathbb{R}$.

14.7. For x in $[-\pi, \pi]$, define

$$f(x) = \begin{cases} -\cos x/2, & -\pi \leq x < 0, \\ \cos x/2, & 0 < x \leq \pi, \\ 0, & x = 0. \end{cases}$$

Extend f to be periodic with period 2π.

a) Sketch the graph of f.

b) Find the Fourier series of f.

c) Show that the Fourier series of f converges pointwise to f at every point of $\mathbb{R}$.

d) Let $[a, b]$ be any compact interval in $[-\pi, \pi]$ that does not contain $x = 0$. Prove that the Fourier series of f converges uniformly to f on $[a, b]$.

14.8. Let $f(x) = x \cos x$ on $(-\pi, \pi)$. Define f appropriately at $\pm\pi$ and extend f to be periodic with period 2π.

a) Sketch the graph of f.

b) Find the Fourier series of f.

c) At each point x in $\mathbb{R}$, identify the pointwise limit of the Fourier series you found in part (b). Where is the convergence uniform?

d) Integrate the series you found in part (b) and thus find the Fourier series of the periodic extension of $x \sin x$.

14.9. For x in $[-\pi, \pi]$, let $f(x) = x(1 + \cos x)$. Extend f to be periodic with period 2π.

a) Sketch the graph of f.

b) Find the Fourier series of f.

c) Combine your answers from Exercises 14.2 and 14.8 in order to find the Fourier series of f. Compare your answer with the series you found in part (b).

d) Identify the pointwise limit of the Fourier series you found in part (b). Where does the series converge uniformly?

e) Integrate the series you found in part (b) and thus find the Fourier series of $\int_0^x t(1 + \cos t)\, dt$.

14.10. For x in $\mathbb{R}$, let $f(x) = |\sin x|$.

a) Sketch the graph of f.

b) Find the Fourier series of f.

c) Identify the pointwise limit of the Fourier series of f at each point of $\mathbb{R}$. Where is the convergence uniform?

d) Compute f' and find the Fourier series of f'.

e) Differentiate the Fourier series of f term-by-term and compare the resulting series with that obtained in (d). How can you justify differentiating term-by-term?

f) Show that, for $0 \le x \le \pi$,

$$\frac{\pi}{4}(\cos x - 1) + \frac{x}{2} = \sum_{j=1}^{\infty} \frac{\sin 2jx}{(2j-1)(2j)(2j+1)}.$$

g) Hence show that

$$\sum_{j=1}^{\infty} \frac{(-1)^{j+1}}{(4j-3)(4j-2)(4j-1)} = \frac{\pi}{8}(\sqrt{2}-1).$$

14.11. Define $f(x) = 0$ for x in $[-\pi, 0)$ and $f(x) = \sin x$ for x in $[0, \pi]$. Extend f to be periodic with period 2π.

a) Sketch the graph of f.

b) Find the Fourier series of f.

c) Identify the pointwise limit of f. Where is the convergence uniform?

d) Integrate the Fourier series for f term-by-term and thus find the Fourier series for $\int_0^x f(t)\,dt$. Discuss the convergence of the integrated series for x in $[-\pi, \pi]$.

14.12. For x in $[-\pi, \pi]$, define

$$f(x) = \begin{cases} -x(\pi + x), & -\pi \le x < 0, \\ x(\pi - x), & 0 \le x \le \pi. \end{cases}$$

a) Extend f to have period 2π and sketch the graph of f.

b) Find the Fourier series of f.

c) Identify the pointwise limit of the series at each x in $\mathbb{R}$. Where is the convergence uniform?

14.13. Use the appropriate Fourier series and, when appropriate, Parseval's identity to sum the following series.

a) $\sum_{j=1}^{\infty} 1/j^2$

b) $\sum_{j=1}^{\infty} 1/(2j-1)^2$

c) $\sum_{j=1}^{\infty} 1/(2j-1)^4$

d) $\sum_{j=1}^{\infty} (-1)^{j+1}/j^4$

e) $\sum_{j=1}^{\infty} 1/j^6$

f) $\sum_{j=1}^{\infty} 1/(2j-1)^6$

14.14. Let α be any positive number that is not an integer. Define $f(x) = \sin \alpha x$ for x in $(-\pi, \pi)$. Let $f(-\pi) = f(\pi) = 0$ and extend f to be periodic with period 2π.

a) Sketch the graph of f when $0 < \alpha < 1$ and when $1 < \alpha < 2$.

b) Show that the Fourier series of f is

$$f(x) \sim \frac{-2 \sin \alpha\pi}{\pi} \sum_{j=1}^{\infty} \frac{(-1)^{j+1} \sin jx}{\alpha^2 - j^2}.$$

c) Identify the pointwise limit of the series at each point of $\mathbb{R}$.

d) Prove that the Fourier series converges uniformly to f on any interval $[a, b]$ that does not contain an odd integer multiple of π.

e) Let $x = \pi/2$ and $t = \alpha/2$. Use the trigonometric identity $\sin 2\pi t = 2 \sin \pi t \cos \pi t$ in order to show that

$$\pi \sec \pi t = -4 \sum_{j=1}^{\infty} \frac{(-1)^{j+1}}{4t^2 - (2j-1)^2}.$$

This is the partial fraction decomposition of the secant function.

14.15. For x in $[-\pi, \pi]$, define

$$g(x) = \begin{cases} -\frac{1}{2} - \frac{x}{2\pi}, & -\pi \leq x < 0, \\ 0, & x = 0 \\ \frac{1}{2} - \frac{x}{2\pi}, & 0 < x \leq \pi. \end{cases}$$

a) Extend g to be periodic with period 2π; sketch the graph of g.

b) Find the Fourier series of g.

c) Prove that the Fourier series of g converges pointwise to g at every x in $\mathbb{R}$. Prove that the convergence is uniform on any interval $[a, b]$ that does not contain an even integer multiple of π.

14.16. Throughout this problem, assume that f is twice continuously differentiable on $\mathbb{R}$ and is periodic with period 2π. Let $f(x) \sim a_0/2 + \sum_{j=1}^{\infty}[a_j \cos jx + b_j \sin jx]$.

a) Explain why, under these conditions, the Fourier series of f converges pointwise to f on $\mathbb{R}$. Our objective in this problem is to show that the convergence must actually be uniform on $\mathbb{R}$.

b) Integrate by parts in order to show that,

$$a_j = \frac{1}{j^2\pi} \int_{-\pi}^{\pi} f''(x) \cos jx \, dx, \quad j = 0, 1, 2, 3, \ldots$$

and

$$b_j = \frac{1}{j^2\pi} \int_{-\pi}^{\pi} f''(x) \sin jx \, dx, \quad j = 1, 2, 3, \ldots.$$

c) Show that $|a_j| \leq 2\|f''\|_\infty / j^2$ and $|b_j| \leq 2\|f''\|_\infty / j^2$ for all j.

d) Hence prove that the Fourier series of f converges uniformly to f on $\mathbb{R}$.

14.17. This problem generalizes the result proved in Exercise 14.16. Assume that f is continuous on $\mathbb{R}$ and periodic with period 2π. Assume also that f' and f'' exist at all but finitely many points in $[-\pi, \pi]$ and that each of f' and f'' is bounded and piecewise continuous, having only finitely many jump discontinuities. Let

$$f(x) \sim \frac{a_0}{2} + \sum_{j=1}^{\infty} [a_j \cos jx + b_j \sin jx].$$

a) Prove that f must be of bounded variation on $[-\pi, \pi]$. Thus prove that the Fourier series of f converges pointwise to f.

b) Imitate the proof in parts (b) and (c) of Exercise 14.16 to show that $|a_j| \leq 2\|f''\|_\infty / j^2$ and $|b_j| \leq 2\|f''\|_\infty / j^2$ for all j.

c) Hence prove that the Fourier series of f converges uniformly to f on $\mathbb{R}$.

14.18. In this problem we generalize the result proved in Exercise 14.17. Assume that f is periodic with period 2π and is continuous on $[-\pi, \pi]$ except at $x = 0$ where f has a jump discontinuity of magnitude $c_0 = f(0^+) - f(0^-)$. Adjust f by defining $f(0) = [f(0^-) + f(0^+)]/2$. Assume also that f' and f'' exist at all but finitely many points in $[-\pi, \pi]$ and that each of f' and f'' is bounded and piecewise continuous, having only finitely many jump discontinuities. Let

$$f(x) \sim \frac{a_0}{2} + \sum_{j=1}^{\infty} [a_j \cos jx + b_j \sin jx].$$

a) Prove that the Fourier series of f converges pointwise to f at all points of $\mathbb{R}$.

b) For x in $[-\pi, \pi]$, define

$$g(x) = \begin{cases} -\frac{1}{2} - \frac{x}{2\pi}, & -\pi \le x < 0, \\ 0, & x = 0, \\ \frac{1}{2} - \frac{x}{2\pi}, & 0 < x \le \pi, \end{cases}$$

as in Exercise 14.15. Recall that the Fourier series of g converges pointwise to g on $\mathbb{R}$ and converges uniformly on any interval $[a, b]$ that does not contain an even integer multiple of π. For x in $\mathbb{R}$, define $h(x) = f(x) - c_0 g(x)$. Prove that h is periodic with period 2π and is continuous on $\mathbb{R}$.

c) Prove that h' and h'' exist at all but finite points in $[-\pi, \pi]$ and that each of h' and h'' is bounded and piecewise continuous, having only finitely many jump discontinuities.

d) Use Exercises 14.15 and 14.17 to prove that the Fourier series of $f = h + c_0 g$ converges uniformly to f on any interval $[a, b]$ that contains no even integer multiple of π.

14.19. This problem is the culmination of the results obtained in Exercises 14.16 through 14.18. Assume that f is periodic with period 2π and that f is continuous except at finitely many points $x_1, x_2, \ldots, x_p$ in $[-\pi, \pi]$. Adjust f at each x_k by defining $f(x_k) = [f(x_k^-) + f(x_k^+)]/2$. For $k = 1, 2, \ldots, p$, let $c_k = f(x_k^+) - f(x_k^-)$. Assume also that f' and f'' exist at all but finitely many points in $[-\pi, \pi]$ and that each of f' and f'' is bounded and piecewise continuous, having only jump discontinuities. Define

$$h(x) = f(x) - \sum_{k=1}^{p} c_k g(x - x_k),$$

where g is the function defined in Exercise 14.15. Prove that h is continuous and periodic with period 2π on $\mathbb{R}$. Also prove that h' and h'' exist at all but finitely many points in $[-\pi, \pi]$ and that each of h' and h'' are bounded and piecewise continuous, having only jump discontinuities. Hence prove that the Fourier series of f converges uniformly to f on any interval $[a, b]$ that contains none of the points

$$\{x_k + 2n\pi : k = 1, 2, \ldots, p, \ n \text{ in } \mathbb{Z}\}.$$

14.20. Confirm that, for the trigonometric orthogonal collection on $[-\pi, \pi]$, Parseval's identity becomes

$$\frac{1}{\pi}\int_{-\pi}^{\pi} f^2(x)\,dx = \frac{a_0^2}{2} + \sum_{j=1}^{\infty}(a_j^2 + b_j^2).$$

for f in $R[-\pi, \pi]$.

14.21. Assume that f is integrable on $[-\pi, \pi]$ and that

$$f(x) \sim \frac{a_0}{2} + \sum_{j=1}^{\infty} [a_j \cos jx + b_j \sin jx].$$

a) Let $\sigma_n(x) = (1/n) \sum_{k=0}^{n-1} S_k(x)$, where S_k denotes the kth partial sum of the Fourier series of f. Prove that

$$\sigma_n(x) = \frac{a_0}{2} + \sum_{k=1}^{n-1} \left(1 - \frac{k}{n}\right) (a_k \cos kx + b_k \sin kx).$$

b) Prove that

$$\begin{aligned} \|f - \sigma_n\|_2^2 &= \int_{-\pi}^{\pi} [f(x) - \sigma_n(x)]^2 \, dx \\ &= \|f\|_2^2 - \pi \left[\frac{a_0^2}{2} + \sum_{k=1}^{n-1} \left(1 - \frac{k^2}{n^2}\right) (a_k^2 + b_k^2) \right]. \end{aligned}$$

c) Prove that if f is continuous and periodic with period 2π, then

$$\lim_{n\to\infty} \pi/n^2 \sum_{k=1}^{n} k^2 (a_k^2 + b_k^2) = 0.$$

14.22. Suppose that $\{f_k\}$ and $\{g_k\}$ are two sequences of integrable functions on $[a, b]$ that converge in the mean to integrable functions f and g, respectively. For k in $\mathbb{N}$ and x in $[a, b]$, define

$$F_k(x) = \int_a^x f_k(t) g_k(t) \, dt$$

and

$$F(x) = \int_a^x f(t) g(t) \, dt.$$

Prove that F_k is in $C([a, b])$ and $\lim_{k\to\infty} F_k = F$ [uniformly] on $[a, b]$.

14.23. For k in $\mathbb{N}$ and x in $[-1, 1]$, define

$$f_k(x) = k^{3/2} x \, e^{-k^2 x^2}.$$

a) Prove that $\lim_{k\to\infty} f_k(x) = 0$ [pointwise] on $[-1, 1]$.

b) Prove that $\{f_k\}$ does not converge to 0 in the mean on $[-1, 1]$.

14.24. For k in $\mathbb{N}$ and x in $[-\pi/2, \pi/2]$, let $f_k(x) = \sin^k x$.

a) Prove that the sequence $\{f_k\}$ converges to 0 in the mean on $[-\pi/2, \pi/2]$.

b) Does $\{f_k\}$ converge pointwise on $[-\pi/2, \pi/2]$? If so, to what limit? If not, explain why not.

14.25. Suppose that $\Phi = \{\phi_j : j \text{ in } \mathbb{N}\}$ is an orthonormal collection of integrable functions on $[a, b]$ that is complete for a class $\mathcal{F}$ of functions in $R[a, b]$. Suppose also that f and g are in $\mathcal{F}$ and that $f \sim \sum_{j=1}^{\infty} a_j \phi_j$ and $g \sim \sum_{j=1} b_j \phi_j$. Prove that $\sum_{j=1}^{\infty} a_j b_j$ converges to $\langle f, g \rangle$.

14.26. Assume that $\{f_k\}$ is a sequence of integrable functions on $[a, b]$ that converges pointwise to f on $[a, b]$ and that converges in the mean to g on $[a, b]$. Assume also that both f and g are continuous. Prove that $f = g$.

14.27. Let $S_k(x)$ denote the kth partial sum of the Fourier series of the function

$$f(x) = \begin{cases} -1, & -\pi \le x < 0, \\ 0, & x = 0, \\ 1, & 0 < x \le \pi. \end{cases}$$

Prove that

$$S_k(x) = \frac{2}{\pi} \int_0^x \frac{\sin 2kt}{\sin t}\, dt.$$

A

Axioms for the Real Numbers $\mathbb{R}$

The arithmetic properties of the set $\mathbb{R}$ of real numbers are described by the following familiar axioms, definitions, and theorems.

AXIOM A.1 Let a, b, and c be any real numbers.

i) The sum $a + b$ of two real numbers is a real number. The binary operation $+$ is called *addition* and has the following properties:

- **a)** *Commutativity.* $a + b = b + a$.
- **b)** *Associativity.* $a + (b + c) = (a + b) + c$.
- **c)** There is a real number, denoted 0 and called the additive identity, such that $a + 0 = a$ for all a in $\mathbb{R}$.
- **d)** Given any a in $\mathbb{R}$, there is a number in $\mathbb{R}$, denoted $-a$ and called the *additive inverse of* a, such that $a + (-a) = 0$.

ii) The product $a \cdot b$, also denoted ab, of two real numbers is a real number. The binary operation $\cdot$ is called *multiplication* and has the following properties:

- **a)** *Commutativity.* $ab = ba$.
- **b)** *Associativity.* $a(bc) = (ab)c$.
- **c)** There is a real number, denoted 1 and called the multiplicative identity, such that $a1 = a$ for all a in $\mathbb{R}$.
- **d)** Given any $a \neq 0$, there exists a real number, denoted a^{-1} or $1/a$ and called the *multiplicative inverse* of a, such that $aa^{-1} = 1$.

iii) *The Distributive Law.* $a(b + c) = ab + ac$. ●

Remarks. (i) The additive identity 0 is unique. (ii) The additive inverse $-a$ of a is unique. (iii) The multiplicative identity 1 is unique. (iv) The multiplicative inverse a^{-1} of $a \neq 0$ is unique. If $a \neq 0$, so that a^{-1} exists, then a is called a *unit*. We leave the proofs of these assertions and of the following theorems for you to provide; we merely list the facts for your reference.

THEOREM A.1 Given any two real numbers a and b, there exists a unique real number x such that $x + b = a$. ●

Remark. The number x is written $x = a - b$, defined by

$$x = a - b = a + (-b).$$

We say that b is *subtracted* from a. The process of forming $a - b$ is called *subtraction*.

THEOREM A.2 Given two real numbers a and b with $b \neq 0$, there exists a unique real number x such that $bx = a$. ●

Remark. The number x is written $x = a/b$ defined by

$$x = \frac{a}{b} = a(b^{-1}) = a\left(\frac{1}{b}\right).$$

The number x is also called the *quotient* of a and b, in that order. We say that a is *divided* by b. The process of forming a/b is called *division*.

THEOREM A.3 Let $a, \ b, \ c$, and d be real numbers.

i) We have $ab = 0$ if and only if either $a = 0$ or $b = 0$ (or both).

ii) If $ab = ac$ and if $a \neq 0$, then $b = c$.

iii) $-(ab) = (-a)b = a(-b) = (-1)ab$.

iv) $(-a)(-b) = ab$.

v) $-(-a) = a$.

vi) If $a \neq 0$, then $1/(1/a) = a$.

vii) If $a \neq 0$ and $b \neq 0$, then $(ab)^{-1} = a^{-1}b^{-1}$.

viii) If $b \neq 0$ and $d \neq 0$, then $(a/b) + (c/d) = (ad + bc)/(ad)$. ●

AXIOM A.2 There exists an order relation on $\mathbb{R}$, written $a \leq b$ and read "a is less than or equal to b," with the following properties for all $a, \ b$, and c in $\mathbb{R}$.

i) *Reflexivity.* $a \leq a$.

ii) *Antisymmetry.* If $a \leq b$ and $b \leq a$, then $a = b$.

iii) *Transitivity.* If $a \leq b$ and $b \leq c$, then $a \leq c$. ●

Remark. The relation $a \leq b$ is also written equivalently as $b \geq a$, read "b is greater than or equal to a."

DEFINITION A.1 If $a \leq b$ and $a \neq b$, then $a < b$ is read "a is less than b." ●

Remark. The statement $b > a$, read "b is greater than a," is equivalent to $a < b$. The inequality $<$ is said to be *strict*.

AXIOM A.3 The Trichotomy Law Given any two real numbers a and b, one and only one of the following three statements is valid: either $a < b$, or $a = b$, or $b < a$. ●

Remark. Because the Trichotomy Law holds in $\mathbb{R}$, the relation $<$ is said to be a *linear order* on $\mathbb{R}$.

DEFINITION A.2 The set $\mathbb{R}^+$ of *positive real numbers* is the set $\mathbb{R}^+ = \{a : a \text{ in } \mathbb{R},\ 0 < a\}$. The set $\mathbb{R}^-$ of *negative real numbers* is the set $\mathbb{R}^- = \{a : a \text{ in } \mathbb{R},\ a < 0\}$. ●

Remark. By the Trichotomy Law we can write $\mathbb{R}$ as the disjoint union $\mathbb{R} = \mathbb{R}^+ \cup \{0\} \cup \mathbb{R}^-$ since, for any a in $\mathbb{R}$, either $0 < a$ or $a = 0$ or $a < 0$, and since only one of these statements can hold.

The following theorem lists basic, useful properties of inequalities. The list is neither exhaustive nor free of redundancy.

THEOREM A.4 Let a, b, c, and d be real numbers.

i) If $a \le b$ and $c \le d$, then $a + c \le b + d$. If either of the inequalities in the hypothesis is strict, then the inequality in the conclusion is also strict.

ii) If $0 < a$ and $0 < b$, then $0 < a + b$. That is, $\mathbb{R}^+$ is closed under addition.

iii) A real number a is positive if and only if $-a$ is negative. That is, $\mathbb{R}^- = -\mathbb{R}^+ = \{-a : a \text{ in } \mathbb{R}^+\}$.

iv) If $0 < a$ and $0 < b$, then $0 < ab$. That is, $\mathbb{R}^+$ is closed under multiplication.

v) If $0 < a$, then $ac \le ad$ if and only if $c \le d$.

vi) If $0 < a \le b$, then $ac \le bd$ if and only if $c \le d$.

vii) If $a < 0$ and $b < 0$, then $ab > 0$.

viii) If $a < 0$, then $ac \le ad$ if and only if $c \ge d$.

ix) If $a \le b < 0$, then $ac \le bd$ if and only if $c \ge d$.

x) If $a < 0$ and $b > 0$, then $ab < 0$.

xi) If $a \ne 0$, then $a^2 > 0$.

xii) $1 > 0$. Therefore, every natural number is in $\mathbb{R}^+$. ●

DEFINITION A.3 The *absolute value* of a real number a is

$$|a| = \begin{cases} a, & \text{if } a > 0, \\ 0, & \text{if } a = 0, \\ -a, & \text{if } a < 0. \end{cases}$$ ●

THEOREM A.5 The absolute value function has the following properties. Let a and b be real numbers.

i) $|a| \geq 0$; $|a| = 0$ if and only if $a = 0$.

ii) $|ab| = |a|\,|b|$.

iii) $|a + b| \leq |a| + |b|$.

iv) $a \leq |a|$.

v) $|a| - |b| \leq \big|\,|a| - |b|\,\big| \leq |a - b|$. ●

THEOREM A.6 Well-Ordering of $\mathbb{N}$ Let S be any nonempty set of natural numbers. There exists a k_0 in S such that $k_0 \leq k$ for all k in S. ●

B

Set Theory

Modern set theory was created almost singlehandedly by Georg Cantor during the years 1872 to 1890. He developed the theory of infinite sets during 1872–1878 to deal with practical problems associated with the exceptional sets where Fourier series fail to converge. In 1890 he gave the proof we presented in Section 1.6, that the unit interval $[0, 1]$ is uncountable. Concurrently he developed the major concepts of topology (limit points, open and closed sets, density ...) and established their main properties in Euclidean spaces. He recognized earlier than most that a rigorous development of analysis required a logically sound theory of sets, free of paradox. In an 1895 paper he advanced the following definition of a set:

> *By a set we shall understand any collection into a whole of definite distinguishable objects of our intuition or thought. The objects will be called members of the collection.*

He also proved that the power set $\mathcal{P}(S)$ of any set S has strictly greater cardinality than the set S itself. He did this by showing that any function f from S to $\mathcal{P}(S)$ cannot map S *onto* $\mathcal{P}(S)$; thus $\mathcal{P}(S)$ must contain more elements than S does. For suppose f is any function mapping S onto $\mathcal{P}(S)$. For each s in S, either s is an element of the set $f(s)$ or it is not. Let C denote the subset of S consisting of those s in S that *do not belong to* $f(s)$. Since C is an element of $\mathcal{P}(S)$ and f is *onto*, there must exist a c in S such that $f(c) = C$. All this is entirely straightforward. Then Cantor sprung his trap. He asked the question, "Does c belong to C or doesn't it?" If c is in C, then by the definition of C, the point c is not in $f(c) = C$. On the other hand, if c is not in $C = f(c)$, then by the definition of C, the point c is in C. The very construction of C produces a paradox that cannot be resolved. Cantor concluded that the function f mapping S onto $\mathcal{P}(S)$ cannot exist. $\mathcal{P}(S)$ is larger than S.

In 1902, motivated perhaps by the spirit of Cantor's proof, Bertrand Russell proposed a set in accord with Cantor's definition that revealed an underlying logical pitfall. Since the foundations of analysis must be free of ambiguity, paradox, and illogic, Russell's paradox required resolution and a more careful formulation of how a "set" is to be defined. The paradox is based on *self-reference*, a frequent source of that eerie feeling inspired by paradox: "Who is my self observing myself?" Russell proposed S to be that set whose elements are those sets that do not contain themselves as members; that is $S = \{A : A \text{ is not in } A\}$. Then S can neither belong to itself nor not belong to itself. If S were to belong to itself, then S contains itself as a member and hence is not an element of S. On the other hand, if S were not an element of S, then, by the defining property of S, the set S does contain itself as an element. The question cannot be resolved.

In 1918 Russell gave a layperson's version of the same paradox: Imagine a small village having only one barber. The barber shaves exactly those people who do not shave themselves. Who then shaves the barber?

During the decades from 1880 to 1910 mathematicians were successfully showing that the theory of sets could be used to provide a rigorous foundation for analysis. Cantor, Dedekind, Borel, Lebesgue, Hilbert, and others had used set theoretic techniques extensively in their research. Consequently, Russell's paradox was viewed as a disaster. One could not tolerate the presence of paradox within the very foundations of the edifice.

Russell's paradox forced the redefinition of the starting point of set theory. To eliminate the possibility of self-reference, we require, first, that a "universal set" U be established and second that all elements of all sets being discussed must be elements of the universal set U. With this convention, a subset of U cannot be defined *prior to* the specification of the elements of U; therefore, no subset of U can be an element of U. Russell's set S is not a set *whose elements are in* U and hence, by our acceptance of this convention, is excluded from set theoretical discussions. No longer does the question of a set belonging or not belonging to itself make any sense. The paradox is ruled out of consideration.

We use conventional notation for set theoretic containment and the operations of union, intersection, and complementation.

DEFINITION B.1 Let U denote the prescribed universal set and let S and T denote arbitrary subsets of U.

i) S is *contained* in T or T *contains* S if and only if every element of S is also an element of T. We write $S \subseteq T$ or $T \supseteq S$.

ii) Two sets S and T are said to be *equal* if and only if $S \subseteq T$ and $T \subseteq S$.

iii) The *union* of S and T is the set

$$S \cup T = \{x \text{ in } U : x \text{ is in } S \text{ or } x \text{ is in } T\}.$$

iv) The *intersection* of S and T, is the set

$$S \cap T = \{x \text{ in } U : x \text{ is in } S \text{ and } x \text{ is in } T\}.$$

v) Two sets S and T are said to be *disjoint* if $S \cap T = \emptyset$, the empty set containing no elements.

vi) The *complement* of S is the set

$$S^c = \{x \text{ in } U : \ x \text{ is not in } S\}. \quad \bullet$$

DEFINITION B.2 Let A be any set of indices. For each α in A, let S_α be a subset of U.

i) The *union* of all the sets S_α is

$$\bigcup_{\alpha \in A} S_\alpha = \{x \text{ in } U\text{: } x \text{ is in } S_\alpha \text{ for some } \alpha \text{ in } A\}.$$

ii) The *intersection* of all the sets S_α is

$$\bigcap_{\alpha \in A} S_\alpha = \{x \text{ in } U\text{: } x \text{ is in } S_\alpha \text{ for every } \alpha \text{ in } A\}. \quad \bullet$$

THEOREM B.1 DeMorgan's Laws The union, intersection, and complement of sets are related as follows:

i)

$$\left(\bigcup_{\alpha \in A} S_\alpha\right)^c = \bigcap_{\alpha \in A} S_\alpha^c.$$

ii)

$$\left(\bigcap_{\alpha \in A} S_\alpha\right)^c = \bigcup_{\alpha \in A} S_\alpha^c. \quad \bullet$$

DEFINITION B.3 Let S and T be any two sets. The *Cartesian product* of S and T, denoted $S \times T$, is the set

$$S \times T = \{(s, t) : s \text{ in } S, t \text{ in } T\}. \quad \bullet$$

Note that, when defining the Cartesian product of S and T, the order of the factors matters. In general, $S \times T$ and $T \times S$ are conceptually different sets. If $S = T$, we can simplify the notation by writing S^2 for $S \times S$. The Cartesian product generalizes; recursively define $S_1 \times S_2 \times \cdots \times S_n$ to be $\{(s_1, s_2, \ldots, s_n) : s_j \text{ in } S_j\}$. If $S_1 = S_2 = \cdots = S_n = S$, then we simplify the notation for the Cartesian product $S_1 \times S_2 \times \cdots \times S_n$ by writing S^n.

C

Functions

DEFINITION C.1

i) A *relation* R on S and T, is a subset of $S \times T$. We write sRt (read "s is related to t") if (s, t) is in R.

ii) The set of all first elements in the ordered pairs that comprise R is called the *domain* of R. The set of all second elements is called the *range* of R. The set T is called the *codomain* of R. ●

DEFINITION C.2

i) A *function* is a relation in which no two ordered pairs have the same first element. If the function f is the set $f = \{(x, y) : x \text{ in } S, y \text{ in } T\}$, we write $y = f(x)$ and say that f is a function *from* S to T. The range of f is often denoted $f(S)$, a subset of T.

ii) Two functions f and g are said to be *equal*, denoted $f = g$, if f and g have the same domain S and the same codomain T and $f(s) = g(s)$ for all s in S.

iii) The *graph* of the function f is

$$G(f) = \{(x, f(x)) : x \text{ in } S\}. \quad \bullet$$

Notice that the graph of f is a subset of $S \times T$.

DEFINITION C.3

i) A function f from S to T is said to be *one-to-one* if, whenever x_1 and x_2 are in S and $f(x_1) = f(x_2)$, it follows that $x_1 = x_2$.

ii) A function f from S to T is said to be *onto* if, for any y in T, there exists at least one x in S such that $f(x) = y$.

iii) A function f from S to T is said to be a *one-to-one correspondence* if f is both one-to-one and onto. ●

That a function is one-to-one means that at most one point in the domain is mapped to any point in the range. That a function is onto means that, for each y in the codomain, at least one point x in the domain is mapped to y.

Beginners often confuse the definition of a function with the property of being one-to-one. We urge you to note well the difference. For f to be a function we require that each x in the domain be mapped to only one y in the range. For the function f to be one-to-one it must be true that only one x in the domain is mapped to a y in the range.

DEFINITION C.4 If f is a function from S to T and if g is a function from T to U, then $h = g \circ f$, called the *composition* of f and g, is the function from S to U defined by

$$h(s) = (g \circ f)(s) = g(f(s)), \qquad \text{for } s \text{ in } S.$$

The domain of h is S; the range of h is $g(f(S))$. ●

DEFINITION C.5

i) The *identity function* on a set S is the function e_S with domain S and codomain S such that $e_S(s) = s$ for all s in S.

ii) Suppose that f is a function from S to T. The *inverse* of f, if it exists, is a function g from $f(S)$ to S such that $g \circ f = e_S$. That is, $(g \circ f)(s) = s$ for all s in S. The inverse of f, if it exists, is denoted f^{-1}. ●

THEOREM C.1 Suppose that f is a function from S to T. The inverse of f exists if and only if f is one-to-one. If f exists, then $f \circ f^{-1} = e_{f(S)}$. If f is also onto T, then the domain of f^{-1} is T and $f \circ f^{-1} = e_T$. ●

DEFINITION C.6 Suppose that S and T are subsets of $\mathbb{R}$ and that f is a function from S to T.

i) f is said to be *increasing* or *monotone increasing* on S if, whenever x_1 and x_2 are in S with $x_1 < x_2$, it follows that $f(x_1) \leq f(x_2)$.

ii) f is said to be *strictly monotone increasing* on S if, whenever x_1 and x_2 are in S with $x_1 < x_2$, it follows that $f(x_1) < f(x_2)$.

iii) f is said to be *decreasing* or *monotone decreasing* on S if, whenever x_1 and x_2 are in S with $x_1 < x_2$, it follows that $f(x_1) \geq f(x_2)$.

iv) f is said to be *strictly monotone decreasing* on S if, whenever x_1 and x_2 are in S with $x_1 < x_2$, it follows that $f(x_1) > f(x_2)$.

v) f is said to be *monotone* if any of (i)–(iv) apply; f is said to be *strictly monotone* if either (ii) or (iv) apply. ●

THEOREM C.2 Suppose that S and T are subsets of $\mathbb{R}$ and that f is a function from S to T. If f is strictly monotone on S, then f is one-to-one and f^{-1} exists. ●

D

Polynomials

We list here the pertinent facts about a special collection of functions, the set P of all polynomials with real coefficients.

DEFINITION D.1 A *[real] polynomial* f is a function of the form $f(x) = \sum_{j=0}^{\infty} a_j x^j$, where the coefficients a_j are real and where only finitely many of the coefficients a_j are not zero. ●

We usually will write $f(x) = a_0 + a_1 x + \cdots + a_k x^k$ where it is to be understood that all coefficients a_j, with $j > k$, are 0. However, to facilitate the statement of the definitions below, it will be convenient to use the more formal definition above.

DEFINITION D.2

i) There is a binary operation, called *addition*, on P. Let $f(x) = \sum_{j=0}^{\infty} a_j x^j$ and $g(x) = \sum_{j=0}^{\infty} b_j x^j$ be two polynomials. The *sum* $f + g$ is the polynomial

$$(f+g)(x) = \sum_{j=0}^{\infty} (a_j + b_j) x^j.$$

ii) There is a binary operation, called *multiplication*, on P. Let $f(x) = \sum_{j=0}^{\infty} a_j x^j$ and $g(x) = \sum_{j=0}^{\infty} b_j x^j$ be two polynomials. The *product* fg is the polynomial

$$(fg)(x) = \sum_{k=0}^{\infty} \left[\sum_{j=0}^{k} a_j b_{k-j} \right] x^k.$$

iii) There is a *scalar multiplication* defined for c in $\mathbb{R}$ and $f(x) = \sum_{j=0}^{\infty} a_j x^j$ in P by

$$(cf)(x) = \sum_{j=0}^{\infty} c a_j x^j. \quad \bullet$$

In practice, the product fg is computed by multiplying out the expression

$$(a_0 + a_1 x + \cdots + a_k x^k)(b_0 + b_1 x + \cdots + b_m x^m),$$

as you learned in your studies of elementary algebra, then collecting together the coefficient of each power of x. The operations of addition and multiplication enjoy the usual algebraic properties with which you are familiar. Compare our next theorem with Axiom A.1.1.

THEOREM D.1 Let f, g, and h be arbitrary polynomials.

i) Addition has the following properties.

- **a)** *Commutativity.* $f + g = g + f$.
- **b)** *Associativity.* $f + (g + h) = (f + g) + h$.
- **c)** There is a unique polynomial, denoted 0 and called the zero polynomial or the additive identity, such that $f + 0 = f$ for all f in P.
- **d)** Given f in P with $f(x) = \sum_{j=0}^{\infty} a_j x^j$, there exists a unique polynomial, denoted $-f$, called the additive inverse of f, and defined by $(-f)(x) = \sum_{j=0}^{\infty} (-a_j) x^j$, such that $f + (-f) = 0$.

ii) Multiplication has the following properties.

- **a)** *Commutativity.* $fg = gf$.
- **b)** *Associativity.* $f(gh) = (fg)h$.
- **c)** There is a unique polynomial, denoted 1 and called the multiplicative identity, such that $f \cdot 1 = f$ for all f in P.

iii) *The Distributive Law.* $f(g + h) = fg + fh$.

iv) Scalar multiplication has the following properties: Let c and d be real numbers and let f and g be in P.

- **a)** $c(f + g) = cf + cg$ and $(c + d)f = cf + df$.
- **b)** $c(fg) = (cf)g = f(cg)$ and $(cd)f = c(df) = d(cf)$.
- **c)** $0 \cdot f = 0$ and $1 \cdot f = f$.
- **d)** $(-c)f = c(-f) = -(cf)$. $\bullet$

Remarks. You will notice that, given an arbitrary, nonzero polynomial there is, in general, no multiplicative inverse. That is, $1/f$ is not generally a polynomial. Thus we cannot generally divide one polynomial by another to obtain a third. Notice also that the real number c can be identified with the polynomial $f_c(x) = c$, consisting only of the constant term c and that scalar multiplication by c corresponds to polynomial multiplication by f_c. Thus scalar multiplication is subsumed in polynomial multiplication and all the properties listed in part (iv) of Theorem D.1.1 can be deduced from parts (i), (ii), and (iii) of that theorem.

DEFINITION D.3 The *degree* of a nonzero polynomial $f(x) = \sum_{j=0}^{\infty} a_j x^j$ is the largest index k such that $a_k \neq 0$. We write $k = \deg f$. If $\deg f = k$ and if $a_k = 1$, then f is said to be a *monic polynomial.* ●

Notice that the zero polynomial has no degree; thus, whenever the degree of a polynomial arises in our discussions below, we must distinguish between all nonzero polynomials and the zero polynomial. Also, if

$$f(x) = \sum_{j=0}^{\infty} a_j x^j = a_0 + a_1 x + a_2 x^2 + \cdots + a_k x^k,$$

with $a_k \neq 0$, then $\deg f = k$ is the highest power of x that occurs with a nonzero coefficient in the polynomial f. The polynomials of degree 0 are the constant polynomials $f_c(x) = c \neq 0$. Polynomials f of degree 0, and only these, have the property that $1/f$ is also a polynomial. Polynomials of degree 1 are said to be *linear*; they are of the form $f(x) = a_0 + a_1 x$, with $a_1 \neq 0$. Polynomials of degree 2 are called *quadratic* and are of the form $f(x) = a_0 + a_1 x + a_2 x^2$ with $a_2 \neq 0$. *Cubic* polynomials have the form $f(x) = a_0 + a_1 x + a_2 x^2 + a_3 x^3$ with $a_3 \neq 0$.

THEOREM D.2 Let f and g be two nonzero polynomials.

i) $\deg(f + g) \leq \max\{\deg f, \ \deg g\}$.

ii) $\deg(fg) = \deg f + \deg g$. ●

THEOREM D.3 Let f, g, and h be in P.

i) If $fg = 0$, then either $f = 0$ or $g = 0$.

ii) If $fg = fh$ and if $f \neq 0$, then $g = h$. ●

Part (i) of Theorem D.1.3 is expressed by saying that P has no *zero-divisors*. Part (ii) is called the *cancellation law*.

THEOREM D.4 The Division Algorithm Given polynomials f and g with $g \neq 0$, there exist unique polynomials q and r such that $f = qg + r$, where either $r = 0$ or $\deg r < \deg g$. ●

The polynomial q whose existence is guaranteed by the theorem is called the *quotient* obtained upon dividing f by g. The polynomial r is called the *remainder*.

DEFINITION D.4 A nonzero polynomial g *divides* a polynomial f if, upon dividing f by g, we obtain the *remainder* $r = 0$. That is, $f = qg$. ●

THEOREM D.5 Let f be any nonzero polynomial, let c be a real number, and let $g(x) = x - c$. The remainder, upon dividing f by g, is $r = f(c)$, the value of f at $x = c$. ●

DEFINITION D.5 A complex number c is a *root* of a polynomial f if $f(c) = 0$. A root of f is also called a *zero* of f. ●

THEOREM D.6 A real number c is a root of the nonzero polynomial f if and only if $g(x) = x - c$ divides f. ●

THEOREM D.7 Let f be a nonzero polynomial with real coefficients.

i) A complex number $c = u + iv$ is a root of f if and only if $\overline{c} = u - iv$ is also a root of f.

ii) The complex number $c = u + iv$ is a root of f if and only if the quadratic polynomial

$$\begin{aligned} g(x) &= [x - (u + iv)][x - (u - iv)] \\ &= x^2 - 2ux + (u^2 + v^2), \end{aligned}$$

with real coefficients, divides f. ●

THEOREM D.8 The Fundamental Theorem of Algebra Every nonzero polynomial f with $\deg f = k$ has exactly k roots in the set of all complex numbers, taking into account possible repetitions among the roots. ●

Consequently, every polynomial f with real coefficients and with degree $k \geq 1$ can be written as a product of the form

$$f(x) = a_k(x - c_1)(x - c_2) \cdots (x - c_p)g_1(x)g_2(x) \cdots g_r(x),$$

where $c_1, c_2, \ldots, c_p$ are the real roots of f (possibly with repetitions) and $g_1, g_2, \ldots, g_r$ are quadratic polynomials of the form

$$g_j(x) = x^2 - 2u_jx + (u_j^2 + v_j^2)$$

corresponding to the complex roots $u_j + iv_j$ and $u_j - iv_j$. Since there are k roots altogether, since there are p real roots, and since the complex roots occur in conjugate pairs, it follows that $p + 2r = k$. In particular, if all the roots of f are real, then

$$f(x) = a_k(x - c_1)(x - c_2) \cdots (x - c_k),$$

where $k = \deg f$.

COROLLARY D.9 If f is a polynomial that is known either to be the zero polynomial or to have degree no more than k, for some k in $\mathbb{N}$, and if f is known to vanish at $k + 1$ distinct points, then f must be the zero polynomial. ●

References and Additional Readings

Abbott, Edwin A. *Flatland. A Romance of Many Dimensions*, 2nd ed. New York: Dover, 1884.

Apostol, Tom M. *Mathematical Analysis: A Modern Approach to Advanced Calculus*, 2nd ed. Reading, MA: Addison-Wesley, 1974.

Bartle, Robert G. and Donald R. Sherbert. *Introduction to Real Analysis*. New York: John Wiley & Sons, 1982.

Byron, Frederick W. Jr, and Robert W. Fuller, *Mathematics of Classical and Quantum Physics*, 2 vols. Reading, MA: Addison-Wesley, 1970.

Churchill, Ruel V., and James W. Brown. *Fourier Series and Boundary Value Problems*, 4th ed. New York: McGraw-Hill, 1987.

Coddington, Earl A. *An Introduction to Ordinary Differential Equations*. Englewood Cliffs, NJ: Prentice-Hall, 1961.

Courant, Richard. *Differential and Integral Calculus*, trans. by E.J McShane, 2 vols, 2nd ed. Interscience, 1937.

Halmos, Paul. *Finite-Dimensional Vector Spaces*, 2nd ed. New York: Van Nostrand Company, 1958.

James, Robert C. *Advanced Calculus*. Belmont, CA: Wadsworth, 1966.

Kaplan, Wilfred. *Advanced Calculus*, 4th ed. Reading, MA: Addison-Wesley, 1991.

Knopp, K. *Theory and Application of Infinite Series*, trans. by R. C. Young. New York: Blackie and Son, 1928.

Rudin, Walter. *Principles of Mathematical Analysis*. New York: McGraw-Hill, 1964.

Widder, David V. *The Laplace Transform*, Princeton, NJ. Princeton University Press, 1946.

Zygmund, A. *Trigonometric Series*, 2 vols. Cambridge: Cambridge University Press, 1959.

Index

Abel
partial summation formula 506
summability of improper integrals 656 [Exer. 13.36]
test for convergence 506
test for uniform convergence 541
theorem on uniform convergence of power series 548
absolute convergence 502*ff*
Alexandroff's one-point compactification 113 [Exer. 2.76]
algebra
of absolutely convergent series 520
of continuous functions 148*ff*
of convergent sequences 45*ff*
of differentiable functions on $\mathbb{R}$ 185, on $\mathbb{R}^n$ 348*ff*
of functions of bounded variation 224*ff*
of functions of exponential type 633
of power series 556
of Riemann integrable functions 246–247
of Riemann–Stieltjes integrable functions 311, 313
alternating series test 505
Archimedes' principle 3
area of a set in $\mathbb{R}^2$ 448
area of a surface in $\mathbb{R}^3$ 479

Bernoulli's inequality: 10 [Exer. 1.16]
Bernštein, S.
polynomial 168
approximation theorem 170
Bessel
functions of the first kind 656 [Exer. 13.37]
inequality 677
Bolzano–Weierstrass
property 16
Theorem 15 (in $\mathbb{R}$), 72 (in $\mathbb{R}^n$)
Bonnet's Mean Value Theorem for Riemann integrals 259
boundary of a set 87
point 79
bounded
set 3 (in $\mathbb{R}$), 69 (in $\mathbb{R}^n$)
variation 221*ff*
variation of sequences 529 [Exer. 11.45]

Cantor
Criterion for completeness in $\mathbb{R}^n$ 94
diagonal technique 56*ff*

function 135 [Exer. 3.23]
Nested Interval Theorem 33 [Exer. 1.50], 93
set 85*ff*, 92 [Exer. 2.47]
cardinality 53*ff*
countable 54
finite 54
uncountable 55
Cauchy 12
completeness 37 (for $\mathbb{R}$), 72 (for $\mathbb{R}^n$), 166 (for $C_\infty(S)$), 534 (for $B(S)$)
condensation test, infinite series 529 [Exer. 11.43]
condition 36 (in $\mathbb{R}$), 72 (in $\mathbb{R}^n$)
criterion, uniform convergence of a series of functions 534
form of remainder (Taylor's theorem) 295 [Exer. 6.88]
form of the Fundamental Theorem of Calculus 260
generalized Mean Value Theorem 193
principal value of an improper integral 602
product of infinite series 516
and absolutely convergent series 520
and conditionally convergent series 516
root test 501, 502
sequence 36*ff* (in $\mathbb{R}$), 72 (in $\mathbb{R}^n$)
uniformly, sequence of functions 159
Cauchy–Riemann equations 396 [Exer. 8.35]
Cauchy–Schwarz inequality
in $\mathbb{R}^n$ 63
for integrals 290 [Exer. 6.45], 668
Cesàro summability
of a series of numbers 520*ff*
of Fourier series 689*ff*
chain rule
functions on $\mathbb{R}$ 189
functions on $\mathbb{R}^n$ 355*ff*, 361
characteristic polynomial
of a matrix 385
of a linear differential operator 648
change of variables
in Riemann integrals 269
in Riemann–Stieltjes integrals 303
in double integrals 465
in triple integrals 485
Chebychev
differential equation 561
polynomials 134 [Exer. 3.16], 329 [Exer. 7.24], 561–563
clopen set 81
closed set 79*ff*
relatively 98
closure of a set 87
cluster point 18 (in $\mathbb{R}$), 70 (in $\mathbb{R}^n$)
compactness 105*ff*, 108
local 113 [Exer. 2.74]
Riesz' formulation of compactness 113 [Exer. 2.75]
comparison test 497
completeness
axiom for $\mathbb{R}$ 5
of $\mathbb{R}^n$ 93*ff*
of a subspace X of $\mathbb{R}^n$ 99*ff*
uniform 166
composition of continuous functions 140*ff*, 176
connectedness 102
constrained optimization 423*ff*, 425, 431, 433
continuity 120, of vector-valued functions 175
uniform 150*ff*, of vector-valued functions 177
contractive sequence 40
convergence
of an infinite series of numbers 494
absolute convergence of a series 502*ff*
conditional convergence of a series 503, 505*ff*
Weierstrass's theorem 512
of a sequence of numbers 18
of a sequence of vectors 70
of a sequence of functions 158*ff*
Dini's theorem 535
in the mean 669*ff*
pointwise 158
uniform 158
—and continuity 164
—and differentiability 538
—and Riemann integrability 279
—and Riemann–Stieltjes integrability 326
of a series of functions 533*ff*

Cauchy criterion for uniform convergence 534
in the mean 672
pointwise 534
uniform 534
—and continuity 535
—and differentiability 539
—and integrability 540
of trigonometric series 682*ff*
(C,1) 691
in the mean 694
pointwise 689
convergence tests for series of numbers
Abel's test 506
alternating series test 505
Cauchy's condensation test 529 [Exer. 11.43]
comparison test 497
Dirichlet's test 507
Gauss' test 527 [Exer. 11.24], 568
integral test 498
limit comparison test 498
Raabe's test 525 [Exer. 11.22]
ratio test 500, generalized 502
root test 501, generalized 502
convergence tests for uniform convergence of series
Abel's test 541
Dirichlet's test 543
Weierstrass's M-test 540
convex set 340
convolution 641*ff*
countable set 54
countable subcover 107
countably infinite set 54
critical point 191 (in $\mathbb{R}$), 379 (in $\mathbb{R}^n$)
curvilinear coordinates 366
cylindrical coordinates 488 [Exer. 10.19]

D'Alembert's ratio test 500, 502
Darboux's Intermediate Value Theorem 196 [Exer. 4.34]
DeMoivre's theorem 508
dense 58
uniformly 167
derivative of a function:
in $\mathbb{R}$ 182
higher order 187
directional in $\mathbb{R}^n$ 332
partial in $\mathbb{R}^n$ 333
higher order 368*ff*
derived set 87
diagonalization technique 56
diameter of a set 89
differentiability
of a function on $\mathbb{R}$ 182*ff*
of a function on $\mathbb{R}^n$ 334
continuous differentiability 342, 370
geometric interpretation 342–347, 418–422
of vector-valued functions on $\mathbb{R}^n$ 352
differential
of functions on $\mathbb{R}$ 184
of functions on $\mathbb{R}^n$ 334
of vector-valued functions on $\mathbb{R}^n$ 352
Dini's theorem on uniform convergence 535
directional derivative 332
Dirichlet
test for convergence 507
test for uniform convergence 543
disconnectedness 102
discontinuity of functions on $\mathbb{R}$
essential 129
jump 129
oscillatory 130
pole 128
removeable 130
discrete set 15
divergence
of a sequence of numbers 18
of a sequence of vectors 70
of an infinite series of numbers 494
double integrals 439*ff*
Duns Scotus 54

eigenvalue 383
eigenvector 383
elliptical coordinates 489 [Exer. 10.21]
exponential type 633
extreme values of a function 190 (in $\mathbb{R}$), 378*ff* (in $\mathbb{R}^n$)
Euler
constant 513
functions
beta 625*ff*
gamma 618*ff*
theorem on homogeneous functions 398 [Exer. 8.46]

Fejér's theorem 691

kernel 690
finite set 54
finite subcover 193, 198
finite intersection property 207 [Exer. 2.75]
folium of Descartes 434 [Exer. 9.5]
Fourier coefficients 674, 678
Fourier series 665*ff*
Cesàro summability 691
integrability 695
mean convergence 694
pointwise convergence 689
Fresnel integrals 654 [Exer. 13.25]
functions of bounded variation 222*ff*,
Fundamental Theorem of Calculus 254
Cauchy's form 260
Fundamental Theorem of Riemann–Stieltjes integrals 322

Galileo 54
gamma function 618*ff*
gauge of a partition 215
Gauss' normalized probability distribution 621*ff*
geometric series 495
Gibb's phenomenon 705*ff*
gradient 334
greatest lower bound: see infimum
Gregory's series 506, 554

Hamilton 62
harmonic function 394 [Exer. 8.27]
harmonic series 495
Heine–Borel Theorem 106 [original version], 108 (in $\mathbb{R}^n$)
Hermite's differential equation 588
Hilbert, David 55
Hobbes 54
homogeneous function on $\mathbb{R}^n$ 398 [Exer. 8.45]
hyper-rectangle 69
hypersphere 66

Implicit Function Theorem 415, 422
improper integrals 593*ff*
convergence 593–594,
absolute 600, 605
conditional 600, 605
of first kind 593, 595*ff*
of second kind 594, 603*ff*
principal value 602
tests for convergence
comparison test 597, 604
Dirichlet's test 600
limit comparison test 598, 605
Weierstrass's M-test for uniform convergence 607, 608
uniform convergence 605*ff*
and continuity 608, 609
and differentiability 610, 612
and integrability 614, 615
infimum 4
infinite product for the sine function 699*ff*
infinite series 494
partial sums 493
information content 427
inner product 62 (in $\mathbb{R}^n$), 667 (in $\mathbb{R}[a, b]$)
integral equations 573
integral test for series 498
integration by parts
Riemann integrals 273
Riemann–Stieltjes integrals 304
interior of a set 87
interior point 79
Intermediate Value Theorem 138, 140
Inverse Function Theorem 409
iterated integrals 448*ff* (in $\mathbb{R}^2$), 476, 483 (in $\mathbb{R}^3$)

Jacobian 402*ff*
Jordan's Theorems 687, 689
jump discontinuity 129

Kolmogorov 427
Kronecker 55

Lagrange
form of remainder [Taylor's theorem] 206
multiplier method 425, 431, 433
Laplace transform 628–653
Laplacian of a function 394 [Exer. 8.27]
least upper bound: see supremum
Leibnitz' rule 265
Legendre's differential equation 588 [Exer. 12.60–64]
lemniscate 434 [Exer. 9.6]
level curve 418, surface 420, hypersurface 421
L'Hôpital's rules 198*ff*
limit comparison test

for improper integrals 598, 605
for series 498
limit of functions
behavior at infinity 142*ff*
of a function 115 (in $\mathbb{R}$), 117 (in $\mathbb{R}^n$)
left-hand limit 121
right-hand limit 121
of vector-valued functions 175
limit of sequences 18 (in $\mathbb{R}$), 70 (in $\mathbb{R}^n$)
limit inferior (lim inf) 28
limit point 12 (in $\mathbb{R}$), 67 (in $\mathbb{R}^n$), 158 (uniform)
limit superior (lim sup) 28
Lindelöf's theorem 107
Lipschitz condition 135 [Exer. 3.21], 188 [Exer. 4.8]
local condition 336
and Picard's method 575
local minimum, maximum 190 (in $\mathbb{R}$), 379 (in $\mathbb{R}^n$)
Locke 54

mean-square approximation 675
mean-square error 675
Mean Value Theorem
for functions on $\mathbb{R}$ 190*ff*
for functions on $\mathbb{R}^n$ 366*ff*
Cauchy's form 193
generalization 204
Mean Value Theorem for Riemann integrals 253 (in $\mathbb{R}$), 443 (in $\mathbb{R}^2$), 475 (in $\mathbb{R}^3$)
Mean Value Theorems, Riemann–Stieltjes integrals 320, 321
method of successive approximations 572*ff*
metric 64
Euclidean in $\mathbb{R}^n$ 65
uniform 158
L_2-metric in $R[a, b]$ 290, 668
monotone functions 215*ff*

n-cube 69
neighborhood 12 (in $\mathbb{R}$), 66 (in $\mathbb{R}^n$)
deleted 12 (in $\mathbb{R}$), 66 (in $\mathbb{R}^n$)
relative 98
uniform 158
Newton
fluxions 11
general binomial theorem 212 [Exer. 4.84], 567–572

Nicholas of Cusa 54
nonoverlapping sets 444 (in $\mathbb{R}^2$), 477 (in $\mathbb{R}^3$)
norm 64
Euclidean in $\mathbb{R}^n$ 63
L_2-norm in $R[a, b]$ 290, 667
uniform 157

open cover 106*ff*
open set 79*ff*
relatively 98
orientation in $\mathbb{R}^2$ 455
orthogonal
vectors 66
collection in $R[a, b]$ 672
functions in $R[a, b]$ 672
orthonormal collection in $R[a, b]$ 672
completeness 677
outer content of a set 444 (in $\mathbb{R}^2$), 477 (in $\mathbb{R}^3$)

parametric representation of surfaces in $\mathbb{R}^3$ 478
Parseval's identity 677, 694
partial derivatives 333*ff*, higher order 368*ff*
partitions in $\mathbb{R}$ 213*ff*
gauge 215
least common refinement 214
refinement 213
partitions 440, 448 (in $\mathbb{R}^2$), 473 (in $\mathbb{R}^3$)
gauge 440, 448 (in $\mathbb{R}^2$), 473 (in $\mathbb{R}^3$)
least common refinement 440
refinement 440 (in $\mathbb{R}^2$), 473 (in $\mathbb{R}^3$)
phase displacement 704
Picard's method of successive approximations 572*ff*
point at infinity 113 [Exer. 2.76]
pointwise convergence
of a sequence of functions 158
of series of functions 534
polar coordinates 364
power series 545*ff*
absolute convergence of 546, 555
and continuity 547, 555
and differentiation 550, 555
and integration 553, 556
interval of convergence 545, 549, 555
radius of convergence 545, 555
representation of functions 563*ff*

solutions of linear differential equations 560–563
uniform convergence of 547, 555

quadratic form 381
quaternions 78 [Exer. 2.18]

radius of convergence 545, 555
ratio test 500, generalized 502
rearrangements of series
of absolutely convergent series 515
of conditionally convergent series 510–514
Weierstrass's theorem 512
rectifiable curve in $\mathbb{R}^2$ 446
region 447 (in $\mathbb{R}^2$)
relatively closed set 98
relatively open set 98
Riemann integral in $\mathbb{R}$ 233*ff*
existence of the integral 239*ff*, 248
integrability 234–235
integrand 235
lower integral 242
Riemann's Condition 242
sum 234, lower sum 239, upper sum 239
upper integral 242
Riemann integral in $\mathbb{R}^2$, $\mathbb{R}^3$
condition 442 (in $\mathbb{R}^2$), 475 (in $\mathbb{R}^3$)
domain 447 (in $\mathbb{R}^2$), 481 (in $\mathbb{R}^3$)
area 448
simple 451*ff* (in $\mathbb{R}^2$), 482 (in $\mathbb{R}^3$)
integrability 440 (in $\mathbb{R}^2$), 473 (in $\mathbb{R}^3$)
iterated 448*ff* (in $\mathbb{R}^2$), 482 (in $\mathbb{R}^3$)
lower integral 442 (in $\mathbb{R}^2$), 474 (in $\mathbb{R}^3$)
sum 440 (in $\mathbb{R}^2$), 473 (in $\mathbb{R}^3$)
lower sum 442 (in $\mathbb{R}^2$), 474 (in $\mathbb{R}^3$)
upper sum 442 (in $\mathbb{R}^2$), 474 (in $\mathbb{R}^3$)
upper integral 442 (in $\mathbb{R}^2$), 474 (in $\mathbb{R}^3$)
transformation of 454*ff*, 465 (in $\mathbb{R}^2$), 485 (in $\mathbb{R}^3$)
Riemann–Lebesgue lemma 684
Riemann's localization principle 687
Riemann–Stieljes integral 299*ff*
change of variables 303
Fundamental Theorem, Riemann–Stieltjes Integration 322
integrability 300
integrand 300
integration by parts 304
integrator 300
lower integral 309
Mean Value Theorem I 320
Mean Value Theorem II 321
reduction to a Riemann integral 315
Riemann's condition for existence 310
sum 299, lower sum 308, upper sum 308
upper integral 309
ring
absolutely convergent series 520
continuous functions 148*ff*
convergent sequences in $\mathbb{R}$ 45*ff*
differentiable functions on $\mathbb{R}$ 185, on $\mathbb{R}^n$ 348*ff*
functions of bounded variation 224*ff*
functions of exponential type 633
power series 556
Riemann integrable functions 246–247
Riemann–Stieltjes integrable functions 311, 313
Rolle's theorem 191
root test 501, generalized 502
Russell, Bertrand 54, 720

saddle point 379
saltus functions 217*ff*
saw-tooth wave form 680
second derivative test for extrema in $\mathbb{R}^2$ 386
sequences of functions 158*ff*
pointwise convergence 158
uniform convergence 158
uniformly Cauchy 159
sequence of numbers
Cauchy 35*ff* (in $\mathbb{R}$), 72 (in $\mathbb{R}^n$)
contractive 40
convergent 18 (in $\mathbb{R}$), 70 (in $\mathbb{R}^n$)
divergent 18 (in $\mathbb{R}$), 70 (in $\mathbb{R}^n$)
monotone 23
series of functions 533*ff*
convergence 534
pointwise 534
uniform 534
partial sum 534
Spectral Theorem 382
spherical coordinates 364
square wave form 679
squeeze play

for sequences of numbers 23
for real-valued functions 124
for real-valued functions at infinity 143
steady state solution 650
subsequence 26 (in $\mathbb{R}$), 74 (in $\mathbb{R}^n$)
supremum 4
surface area 479
symmetric matrix 382, 389

tangent line 183
plane 344
hyperplane 347
tautochrone problem 650
Taylor's polynomial 204 (in $\mathbb{R}$), 377 (in $\mathbb{R}^n$)
Taylor series representation of a function 563*ff*
Taylor's Theorem in $\mathbb{R}$ 203*ff*, 206
Cauchy's form of the remainder 295 [Exer. 6.88], 565
generalized version 295 [Exer. 6.91]
integral form of the remainder 275, 565
Lagrange form of the remainder 206, 565
Taylor's theorem in $\mathbb{R}^n$ 373*ff*
statement of for $\mathbb{R}^2$ (form I) 375, (form II) 376
statement of for $\mathbb{R}^n$ 378
torus, surface area 480
total variation
of a function 223
function V_f 227
negative variation 229 [Exer. 5.18]
positive variation 229 [Exer. 5.18]
transformations 454*ff* (in $\mathbb{R}^2$)
elementary 455*ff* (in $\mathbb{R}^2$)
transient solution 650
trigonometric series 678*ff*

uncountable set 55
uniform continuity 150*ff*
of vector-valued functions 177
uniform convergence
Cauchy's criterion for series 534
and continuity 164, 535, 608, 609
and differentiability 538*ff*, 610, 612
and integration 277*ff*, 540, 614, 615
of a sequence of functions 158
of series of functions 534*ff*
uniform limit point 158
uniform metric 158
uniform neighborhood 158
uniform norm 157
uniformly Cauchy sequence of functions 159
uniformly dense 167

variation function V_f 227

Wallis' product 627, 702
Weierstrass
approximation theorem 168–172, 281–285
M-test:
for integrals 607, 608
for series of functions 540
theorem on conditionally convergent series 512
well-ordering of $\mathbb{N}$ 718

Zeno's Paradox 53